MyLab Math for *Calculus: Early Transcendentals*, 3e

(access code required)

Used by over 3 million students a year, MyLab™ Math is the world's leading online program for teaching and learning mathematics. MyLab Math for *Calculus: Early Transcendentals*, 3e delivers text-specific assessment, tutorials, and multimedia resources that provide engaging and personalized experiences, so learning can happen in any environment and course format.

eBook with Interactive Figures

This groundbreaking eBook includes approximately 700 figures that can be manipulated to provide a deeper geometric understanding of key concepts. Plus, all-new **Enhanced Interactive Figures** incorporate functionality from several standard Interactive Figures, making each one mathematically richer and ideal for in-class demonstrations.

Questions that Deepen Understanding

MyLab Math includes a variety of question types designed to help students succeed in the course. In **Setup & Solve** questions, students show how they set up a problem as well as the solution, better mirroring what is required on tests. **Additional Conceptual Questions** were written by faculty at Cornell University to support deeper, theoretical understanding of the key concepts in calculus.

ALL NEW! Instructional Videos

For each section of the text, there is a newly recorded full-lecture video. Many videos make use of Interactive Figures to enhance understanding of concepts. To make them easier to navigate, each lecture video is segmented into parts (Introduction, Example, or Summary). The videos are assignable within MyLab Math, and a Guide to Video-Based Assignments shows which MyLab Math exercises correspond to each video.

ALGEBRA

Exponents and Radicals

$$x^a x^b = x^{a+b} \qquad \frac{x^a}{x^b} = x^{a-b} \qquad x^{-a} = \frac{1}{x^a} \qquad (x^a)^b = x^{ab} \qquad \left(\frac{x}{y}\right)^a = \frac{x^a}{y^a}$$

$$x^{1/n} = \sqrt[n]{x} \qquad x^{m/n} = \sqrt[n]{x^m} = (\sqrt[n]{x})^m \qquad \sqrt[n]{xy} = \sqrt[n]{x}\sqrt[n]{y} \qquad \sqrt[n]{x/y} = \sqrt[n]{x}/\sqrt[n]{y}$$

Factoring Formulas

$$a^2 - b^2 = (a - b)(a + b) \qquad\qquad a^2 + b^2 \text{ does not factor over real numbers.}$$
$$a^3 - b^3 = (a - b)(a^2 + ab + b^2) \qquad a^3 + b^3 = (a + b)(a^2 - ab + b^2)$$
$$a^n - b^n = (a - b)(a^{n-1} + a^{n-2}b + a^{n-3}b^2 + \cdots + ab^{n-2} + b^{n-1})$$

Binomials

$$(a \pm b)^2 = a^2 \pm 2ab + b^2$$
$$(a \pm b)^3 = a^3 \pm 3a^2b + 3ab^2 \pm b^3$$

Binomial Theorem

$$(a + b)^n = a^n + \binom{n}{1}a^{n-1}b + \binom{n}{2}a^{n-2}b^2 + \cdots + \binom{n}{n-1}ab^{n-1} + b^n,$$

$$\text{where } \binom{n}{k} = \frac{n(n-1)(n-2)\cdots(n-k+1)}{k(k-1)(k-2)\cdots3\cdot2\cdot1} = \frac{n!}{k!(n-k)!}$$

Quadratic Formula

The solutions of $ax^2 + bx + c = 0$ are

$$x = \frac{-b \pm \sqrt{b^2 - 4ac}}{2a}.$$

GEOMETRY

Parallelogram

$$A = bh$$

Triangle

$$A = \frac{1}{2}bh$$

Trapezoid

$$A = \frac{1}{2}(a + b)h$$

Circle

$$A = \pi r^2$$
$$C = 2\pi r$$

Sector

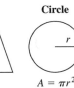

$$A = \frac{1}{2}r^2\theta$$
$$s = r\theta \ (\theta \text{ in radians})$$

Cylinder

$$V = \pi r^2 h$$
$$S = 2\pi rh$$
(lateral surface area)

Cone

$$V = \frac{1}{3}\pi r^2 h$$
$$S = \pi r\ell$$
(lateral surface area)

Sphere

$$V = \frac{4}{3}\pi r^3$$
$$S = 4\pi r^2$$

Equations of Lines and Circles

$$m = \frac{y_2 - y_1}{x_2 - x_1} \qquad \text{slope of line through } (x_1, y_1) \text{ and } (x_2, y_2)$$

$$y - y_1 = m(x - x_1) \qquad \text{point–slope form of line through } (x_1, y_1) \text{ with slope } m$$

$$y = mx + b \qquad \text{slope–intercept form of line with slope } m \text{ and } y\text{-intercept } (0, b)$$

$$(x - h)^2 + (y - k)^2 = r^2 \qquad \text{circle of radius } r \text{ with center } (h, k)$$

TRIGONOMETRY

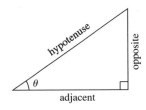

$$\cos\theta = \frac{\text{adj}}{\text{hyp}} \qquad \sin\theta = \frac{\text{opp}}{\text{hyp}} \qquad \tan\theta = \frac{\text{opp}}{\text{adj}}$$

$$\sec\theta = \frac{\text{hyp}}{\text{adj}} \qquad \csc\theta = \frac{\text{hyp}}{\text{opp}} \qquad \cot\theta = \frac{\text{adj}}{\text{opp}}$$

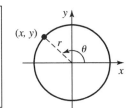

$$\cos\theta = \frac{x}{r} \qquad \sec\theta = \frac{r}{x}$$

$$\sin\theta = \frac{y}{r} \qquad \csc\theta = \frac{r}{y}$$

$$\tan\theta = \frac{y}{x} \qquad \cot\theta = \frac{x}{y}$$

(Continued)

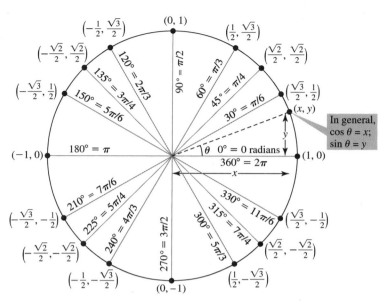

Reciprocal Identities

$$\tan \theta = \frac{\sin \theta}{\cos \theta} \qquad \cot \theta = \frac{\cos \theta}{\sin \theta} \qquad \sec \theta = \frac{1}{\cos \theta} \qquad \csc \theta = \frac{1}{\sin \theta}$$

Pythagorean Identities

$$\sin^2 \theta + \cos^2 \theta = 1 \qquad \tan^2 \theta + 1 = \sec^2 \theta \qquad 1 + \cot^2 \theta = \csc^2 \theta$$

Sign Identities

$$\sin(-\theta) = -\sin \theta \qquad \cos(-\theta) = \cos \theta \qquad \tan(-\theta) = -\tan \theta$$
$$\csc(-\theta) = -\csc \theta \qquad \sec(-\theta) = \sec \theta \qquad \cot(-\theta) = -\cot \theta$$

Double-Angle Identities

$$\sin 2\theta = 2 \sin \theta \cos \theta \qquad \cos 2\theta = \cos^2 \theta - \sin^2 \theta$$
$$= 2\cos^2 \theta - 1$$
$$\tan 2\theta = \frac{2 \tan \theta}{1 - \tan^2 \theta} \qquad = 1 - 2\sin^2 \theta$$

Half-Angle Identities

$$\cos^2 \theta = \frac{1 + \cos 2\theta}{2} \qquad \sin^2 \theta = \frac{1 - \cos 2\theta}{2}$$

Addition Formulas

$$\sin(\alpha + \beta) = \sin \alpha \cos \beta + \cos \alpha \sin \beta \qquad \sin(\alpha - \beta) = \sin \alpha \cos \beta - \cos \alpha \sin \beta$$
$$\cos(\alpha + \beta) = \cos \alpha \cos \beta - \sin \alpha \sin \beta \qquad \cos(\alpha - \beta) = \cos \alpha \cos \beta + \sin \alpha \sin \beta$$
$$\tan(\alpha + \beta) = \frac{\tan \alpha + \tan \beta}{1 - \tan \alpha \tan \beta} \qquad \tan(\alpha - \beta) = \frac{\tan \alpha - \tan \beta}{1 + \tan \alpha \tan \beta}$$

Law of Sines

$$\frac{\sin \alpha}{a} = \frac{\sin \beta}{b} = \frac{\sin \gamma}{c}$$

Law of Cosines

$$a^2 = b^2 + c^2 - 2bc \cos \alpha$$

Graphs of Trigonometric Functions and Their Inverses

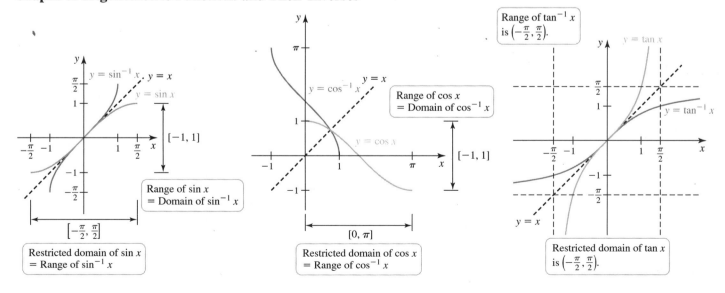

Calculus

MULTIVARIABLE

Third Edition

WILLIAM BRIGGS
University of Colorado, Denver

LYLE COCHRAN
Whitworth University

BERNARD GILLETT
University of Colorado, Boulder

ERIC SCHULZ
Walla Walla Community College

Director, Portfolio Management:	Deirdre Lynch
Executive Editor:	Jeff Weidenaar
Content Producer:	Rachel S. Reeve
Managing Producer:	Scott Disanno
Producer:	Stephanie Green
Manager, Courseware QA:	Mary Durnwald
Manager, Content Development:	Kristina Evans
Product Marketing Manager:	Emily Ockay
Field Marketing Manager:	Evan St. Cyr
Marketing Assistants:	Shannon McCormack, Erin Rush
Senior Author Support/Technology Specialist:	Joe Vetere
Manager, Rights and Permissions:	Gina Cheselka
Manufacturing Buyer:	Carol Melville, LSC Communications
Text and Cover Design, Production Coordination, Composition:	Cenveo® Publisher Services
Illustrations:	Network Graphics, Cenveo® Publisher Services
Cover Image:	Nutexzles/Moment/Getty Images

Library of Congress Cataloging-in-Publication Data
Names: Briggs, William L., author. | Cochran, Lyle, author. | Gillett,
 Bernard, author. | Schulz, Eric P., author. | Briggs, William L. Calculus.
 Early transcendentals.
Title: Calculus: multivariable / William Briggs (University of Colorado,
 Denver), Lyle Cochran (Whitworth University), Bernard Gillett (University
 of Colorado, Boulder), Eric Schulz (Walla Walla Community College).
Description: Third edition. | New York, NY : Pearson, [2019] | Previous
 title: Calculus. Early transcendentals. | Includes index.
Identifiers: LCCN 2017052175 | ISBN 9780134766799 (pbk.) | ISBN 0134766792
 (pbk.)
Subjects: LCSH: Calculus—Textbooks. | Transcendental functions—Textbooks.
Classification: LCC QA303.2 .B7525 2019 | DDC 515—dc23
LC record available at https://lccn.loc.gov/2017052175

1 18

ISBN 13: 978-0-13-476679-9
ISBN 10: 0-13-476679-2

For Julie, Susan, Sally, Sue,
Katie, Jeremy, Elise, Mary, Claire, Katie, Chris, and Annie,
whose support, patience, and encouragement made this book possible.

Contents

Preface

The third edition of *Calculus: Early Transcendentals* supports a three-semester or four-quarter calculus sequence typically taken by students studying mathematics, engineering, the natural sciences, or economics. The third edition has the same goals as the first edition:

• to motivate the essential ideas of calculus with a lively narrative, demonstrating the utility of calculus with applications in diverse fields;

• to introduce new topics through concrete examples, applications, and analogies, appealing to students' intuition and geometric instincts to make calculus natural and believable; and

• once this intuitive foundation is established, to present generalizations and abstractions and to treat theoretical matters in a rigorous way.

The third edition both builds on the success of the previous two editions and addresses the feedback we have received. We have listened to and learned from the instructors who used the text. They have given us wise guidance about how to make the third edition an even more effective learning tool for students and a more powerful resource for instructors. Users of the text continue to tell us that it mirrors the course they teach—and, more important, that students actually read it! Of course, the third edition also benefits from our own experiences using the text, as well as from our experiences teaching mathematics at diverse institutions over the past 30 years.

New to the Third Edition

Exercises

The exercise sets are a major focus of the revision. In response to reviewer and instructor feedback, we've made some significant changes to the exercise sets by rearranging and relabeling exercises, modifying some exercises, and adding many new ones. Of the approximately 10,400 exercises appearing in this edition, 18% are new, and many of the exercises from the second edition were revised for this edition. We analyzed aggregated student usage and performance data from MyLab™ Math for the previous edition of this text. The results of this analysis helped us improve the quality and quantity of exercises that matter the most to instructors and students. We have also simplified the structure of the exercises sets from five parts to the following three:

1. **Getting Started** contains some of the former Review Questions but goes beyond those to include more conceptual exercises, along with new basic skills and short-answer exercises. Our goal in this section is to provide an excellent overall assessment of understanding of the key ideas of a section.

2. **Practice Exercises** consist primarily of exercises from the former Basic Skills, but they also include intermediate-level exercises from the former Further Explorations and Application sections. Unlike previous editions, these exercises are not necessarily organized into groups corresponding to specific examples. For instance, instead of separating out Product Rule exercises from Quotient Rule exercises in Section 3.4, we

have merged these problems into one larger group of exercises. Consequently, specific instructions such as "Use the Product Rule to find the derivative of the following functions" and "Use the Quotient Rule to find the derivative of the given functions" have been replaced with the general instruction "Find the derivative of the following functions." With Product Rule and Quotient Rule exercises mixed together, students must first choose the correct method for evaluating derivatives before solving the problems.

3. **Explorations and Challenges** consist of more challenging problems and those that extend the content of the section.

We no longer have a section of the exercises called "Applications," but (somewhat ironically) in eliminating this section, we feel we are providing better coverage of applications because these exercises have been placed strategically *throughout the exercise sets*. Some are in Getting Started, most are in Practice Exercises, and some are in Explorations and Challenges. The applications nearly always have a boldface heading so that the topic of the application is readily apparent.

Regarding the boldface heads that precede exercises: These heads provide instructors with a quick way to discern the topic of a problem when creating assignments. We heard from users of earlier editions, however, that some of these heads provided too much guidance in how to solve a given problem. In this edition, therefore, we eliminated or reworded run-in heads that provided too much information about the solution method for a problem.

Finally, the **Chapter Review exercises** received a major revamp to provide more exercises (particularly intermediate-level problems) and more opportunities for students to choose a strategy of solution. More than 26% of the Chapter Review exercises are new.

Content Changes

Below are noteworthy changes from the previous edition of the text. Many other detailed changes, not noted here, were made to improve the quality of the narrative and exercises. Bullet points with a 🔑 icon represent major content changes from the previous edition.

Chapter 1 Functions

• Example 2 in Section 1.1 was modified with more emphasis on using algebraic techniques to determine the domain and range of a function. To better illustrate a common feature of limits, we replaced part (c) with a rational function that has a common factor in the numerator and denominator.

• Examples 7 and 8 in Section 1.1 from the second edition (2e) were moved forward in the narrative so that students get an intuitive feel for the composition of two functions using graphs and tables; compositions of functions using algebraic techniques follow.

• Example 10 in Section 1.1, illustrating the importance of secant lines, was made more relevant to students by using real data from a GPS watch during a hike. Corresponding exercises were also added.

• Exercises were added to Section 1.3 to give students practice at finding inverses of functions using the properties of exponential and logarithmic functions.

• New application exercises (investment problems and a biology problem) were added to Section 1.3 to further illustrate the usefulness of logarithmic and exponential functions.

Chapter 2 Limits

• Example 4 in Section 2.2 was revised, emphasizing an algebraic approach to a function with a jump discontinuity, rather than a graphical approach.

• Theorems 2.3 and 2.13 were modified, simplifying the notation to better connect with upcoming material.

• Example 7 in Section 2.3 was added to solidify the notions of left-, right-, and two-sided limits.

• The material explaining the end behavior of exponential and logarithmic functions was reworked, and Example 6 in Section 2.5 was added to show how substitution is used in evaluating limits.

• Exercises were added to Section 2.5 to illustrate the similarities and differences between limits at infinity and infinite limits. We also included some easier exercises in Section 2.5 involving limits at infinity of functions containing square roots.

• Example 5 in Section 2.7 was added to demonstrate an epsilon-delta proof of a limit of a quadratic function.

• We added 17 epsilon-delta exercises to Section 2.7 to provide a greater variety of problems involving limits of quadratic, cubic, trigonometric, and absolute value functions.

Chapter 3 Derivatives

• Chapter 3 now begins with a look back at average and instantaneous velocity, first encountered in Section 2.1, with a corresponding revised example in Section 3.1.

• 🔑 The derivative at a point and the derivative as a function are now treated separately in Sections 3.1 and 3.2.

• After defining the derivative at a point in Section 3.1 with a supporting example, we added a new subsection: Interpreting the Derivative (with two supporting examples).

• Several exercises were added to Section 3.3 that require students to use the Sum and Constant Rules, together with geometry, to evaluate derivatives.

• 🔑 The Power Rule for derivatives in Section 3.4 is stated for all real powers (later proved in Section 3.9). Example 4 in Section 3.4 includes two additional parts to highlight this change, and subsequent examples in upcoming sections rely on the more robust version of the Power Rule. The Power Rule for Rational Exponents in Section 3.8 was deleted because of this change.

- We combined the intermediate-level exercises in Section 3.4 involving the Product Rule and Quotient Rule together under one unified set of directions.

- ⚬⌐ The derivative of e^x still appears early in the chapter, but the derivative of e^{kx} is delayed; it appears only after the Chain Rule is introduced in Section 3.7.

- In Section 3.7, we deleted references to Version 1 and Version 2 of the Chain Rule. Additionally, Chain Rule exercises involving repeated use of the rule were merged with the standard exercises.

- In Section 3.8, we added emphasis on simplifying derivative formulas for implicitly defined functions; see Examples 4 and 5.

- Example 3 in Section 3.11 was replaced; the new version shows how similar triangles are used in solving a related-rates problem.

Chapter 4 Applications of the Derivative

- ⚬⌐ The Mean Value Theorem (MVT) was moved from Section 4.6 to 4.2 so that the proof of Theorem 4.7 is not delayed. We added exercises to Section 4.2 that help students better understand the MVT geometrically, and we included exercises where the MVT is used to prove some well-known identities and inequalities.

- Example 5 in Section 4.5 was added to give guidance on a certain class of optimization problems.

- Example 3b in Section 4.7 was replaced to better drive home the need to simplify after applying l'Hôpital's Rule.

- Most of the intermediate exercises in Section 4.7 are no longer separated out by the type of indeterminate form, and we added some problems in which l'Hôpital's Rule does not apply.

- ⚬⌐ Indefinite integrals of trigonometric functions with argument ax (Table 4.9) were relocated to Section 5.5, where they are derived with the Substitution Rule. A similar change was made to Table 4.10.

- Example 7b in Section 4.9 was added to foreshadow a more complete treatment of the domain of an initial value problem found in Chapter 9.

- We added to Section 4.9 a significant number of intermediate antiderivative exercises that require some preliminary work (e.g., factoring, cancellation, expansion) before the antiderivatives can be determined.

Chapter 5 Integration

- Examples 2 and 3 in Section 5.1 on approximating areas were replaced with a friendlier function where the grid points are more transparent; we return to these approximations in Section 5.3, where an exact result is given (Example 3b).

- Three properties of integrals (bounds on definite integrals) were added in Section 5.2 (Table 5.5); the last of these properties is used in the proof of the Fundamental Theorem (Section 5.3).

- Exercises were added to Sections 5.1 and 5.2 where students are required to evaluate Riemann sums using graphs or tables instead of formulas. These exercises will help students better understand the geometric meaning of Riemann sums.

- We added to Section 5.3 more exercises in which the integrand must be simplified before the integrals can be evaluated.

- A proof of Theorem 5.7 is now offered in Section 5.5.

- Table 5.6 lists the general integration formulas that were relocated from Section 4.9 to Section 5.5; Example 4 in Section 5.5 derives these formulas.

Chapter 6 Applications of Integration
Chapter 7 Logarithmic, Exponential, and Hyperbolic Functions

- ⚬⌐ Chapter 6 from the 2e was split into two chapters in order to match the number of chapters in *Calculus* (Late Transcendentals). The result is a compact Chapter 7.

- Exercises requiring students to evaluate net change using graphs were added to Section 6.1.

- Exercises in Section 6.2 involving area calculations with respect to x and y are now combined under one unified set of directions (so that students must first determine the appropriate variable of integration).

- We increased the number of exercises in Sections 6.3 and 6.4 in which curves are revolved about lines other than the x- and y-axes. We also added introductory exercises that guide students, step by step, through the processes used to find volumes.

- A more gentle introduction to lifting problems (specifically, lifting a chain) was added in Section 6.7 and illustrated in Example 3, accompanied by additional exercises.

- The introduction to exponential growth (Section 7.2) was rewritten to make a clear distinction between the relative growth rate (or percent change) of a quantity and the rate constant k. We revised the narrative so that the equation $y = y_0 e^{kt}$ applies to both growth and decay models. This revision resulted in a small change to the half-life formula.

- The variety of applied exercises in Section 7.2 was increased to further illustrate the utility of calculus in the study of exponential growth and decay.

Chapter 8 Integration Techniques

- Table 8.1 now includes four standard trigonometric integrals that previously appeared in the section Trigonometric Integrals (8.3); these integrals are derived in Examples 1 and 2 in Section 8.1.

- ⚬⌐ A new section (8.6) was added so that students can master integration techniques (that is, choose a strategy) apart from the context given in the previous five sections.

- In Section 8.5 we increased the number and variety of exercises where students must set up the appropriate form of the partial fraction decomposition of a rational function, including more with irreducible quadratic factors.

- A full derivation of Simpson's Rule was added to Section 8.8, accompanied by Example 7, additional figures, and an expanded exercise set.

• 🔑 The Comparison Test for improper integrals was added to Section 8.9, accompanied by Example 7, a two-part example. New exercises in Section 8.9 include some covering doubly infinite improper integrals over infinite intervals.

Chapter 9 Differential Equations

• 🔑 The chapter on differential equations that was available only online in the 2e was converted to a chapter of the text, replacing the single-section coverage found in the 2e.

• More attention was given to the domain of an initial value problem, resulting in the addition and revision of several examples and exercises throughout the chapter.

Chapter 10 Sequences and Infinite Series

• 🔑 The second half of Chapter 10 was reordered: Comparison Tests (Section 10.5), Alternating Series (Section 10.6, which includes the topic of absolute convergence), The Ratio and Root Tests (Section 10.7), and Choosing a Convergence Test (Section 10.8; new section). We split the 2e section that covered the comparison, ratio, and root tests to avoid overwhelming students with too many tests at one time. Section 10.5 focuses entirely on the comparison tests; 39% of the exercises are new. The topic of alternating series now appears before the Ratio and Root Tests so that the latter tests may be stated in their more general form (they now apply to any series rather than only to series with positive terms). The final section (10.8) gives students an opportunity to master convergence tests after encountering each of them separately.

• The terminology associated with sequences (10.2) now includes *bounded above, bounded below,* and *bounded* (rather than only *bounded,* as found in earlier editions).

• Theorem 10.3 (Geometric Sequences) is now developed in the narrative rather than within an example, and an additional example (10.2.3) was added to reinforce the theorem and limit laws from Theorem 10.2.

• Example 5c in Section 10.2 uses mathematical induction to find the limit of a sequence defined recursively; this technique is reinforced in the exercise set.

• Example 3 in Section 10.3 was replaced with telescoping series that are not geometric and that require re-indexing.

• We increased the number and variety of exercises where the student must determine the appropriate series test necessary to determine convergence of a given series.

• We added some easier intermediate-level exercises to Section 10.6, where series are estimated using nth partial sums for a given value of n.

• Properties of Convergent Series (Theorem 10.8) was expanded (two more properties) and moved to Section 10.3 to better balance the material presented in Sections 10.3 and 10.4. Example 4 in Section 10.3 now has two parts to give students more exposure to the theorem.

Chapter 11 Power Series

• Chapter 11 was revised to mesh with the changes made in Chapter 10.

• We included in Section 11.2 more exercises where the student must find the radius and interval of convergence.

• Example 2 in Section 11.3 was added to illustrate how to choose a different center for a series representation of a function when the original series for the function converges to the function on only part of its domain.

• We addressed an issue with the exercises in Section 11.2 of the previous edition by adding more exercises where the intervals of convergence either are closed or contain one, but not both, endpoints.

• We addressed an issue with exercises in the previous edition by adding many exercises that involve power series centered at locations other than 0.

Chapter 12 Parametric and Polar Curves

• 🔑 The arc length of a two-dimensional curve described by parametric equations was added to Section 12.1, supported by two examples and additional exercises. Area and surfaces of revolution associated with parametric curves were also added to the exercises.

• In Example 3 in Section 12.2, we derive more general polar coordinate equations for circles.

• The arc length of a curve described in polar coordinates is given in Section 12.3.

Chapter 13 Vectors and the Geometry of Space

• 🔑 The material from the 2e chapter Vectors and Vector-Valued Functions is now covered in this chapter and the following chapter.

• Example 5c in Section 13.1 was added to illustrate how to express a vector as a product of its magnitude and its direction.

• We increased the number of applied vector exercises in Section 13.1, starting with some easier exercises, resulting in a wider gradation of exercises.

• 🔑 We adopted a more traditional approach to lines and planes; these topics are now covered together in Section 13.5, followed by cylinders and quadric surfaces in Section 13.6. This arrangement gives students early exposure to all the basic three-dimensional objects that they will encounter throughout the remainder of the text.

• 🔑 A discussion of the distance from a point to a line was moved from the exercises into the narrative, supported with Example 3 in Section 13.5. Example 4 finds the point of intersection of two lines. Several related exercises were added to this section.

• In Section 13.6 there is a larger selection of exercises where the student must identify the quadric surface associated with a given equation. Exercises are also included where students design shapes using quadric surfaces.

Chapter 14 Vector-Valued Functions

• More emphasis was placed on the surface(s) on which a space curve lies in Sections 14.1 and 14.3.

- We added exercises in Section 14.1 where students are asked to find the curve of intersection of two surfaces and where students must verify that a curve lies on a given surface.

- Example 3c in Section 14.3 was added to illustrate how a space curve can be mapped onto a sphere.

- Because the arc length of plane curves (described parametrically in Section 12.1 and with polar coordinates in Section 12.3) was moved to an earlier location in the text, Section 14.4 is now a shorter section.

Chapter 15 Functions of Several Variables

- ⚬➤ Equations of planes and quadric surfaces were removed from this chapter and now appear in Chapter 13.

- The notation in Theorem 15.2 was simplified to match changes made to Theorem 2.3.

- Example 7 in Section 15.4 was added to illustrate how the Chain Rule is used to compute second partial derivatives.

- We added more challenging partial derivative exercises to Section 15.3 and more challenging Chain Rule exercises to Section 15.4.

- Example 7 in Section 15.5 was expanded to give students more practice finding equations of curves that lie on surfaces.

- Theorem 15.13 was added in Section 15.5; it's a three-dimensional version of Theorem 15.11.

- Example 7 in Section 15.7 was replaced with a more interesting example; the accompanying figure helps tell the story of maximum/minimum problems and can be used to preview Lagrange multipliers.

- We added to Section 15.7 some basic exercises that help students better understand the second derivative test for functions of two variables.

- ⚬➤ Example 1 in Section 15.8 was modified so that using Lagrange multipliers is the clear path to a solution, rather than eliminating one of the variables and using standard techniques. We also make it clear that care must be taken when using the method of Lagrange multipliers on sets that are not closed and bounded (absolute maximum and minimum values may not exist).

Chapter 16 Multiple Integration

- Example 2 in Section 16.3 was modified because it was too similar to Example 1.

- More care was given to the notation used with polar, cylindrical, and spherical coordinates (see, for example, Theorem 16.3 and the development of integration in different coordinate systems).

- Example 3 in Section 16.4 was modified to make the integration a little more transparent and to show that changing variables to polar coordinates is permissible in more than just the xy-plane.

- More multiple integral exercises were added to Sections 16.1, 16.2, and 16.4, where integration by substitution or integration by parts is needed to evaluate the integrals.

- In Section 16.4 we added more exercises in which the integrals must first be evaluated with respect to x or y instead of z. We also included more exercises that require triple integrals to be expressed in several orderings.

Chapter 17 Vector Calculus

- ⚬➤ Our approach to scalar line integrals was streamlined; Example 1 in Section 17.2 was modified to reflect this fact.

- We added basic exercises in Section 17.2 emphasizing the geometric meaning of line integrals in a vector field. A subset of exercises was added where line integrals are grouped so that the student must determine the type of line integral before evaluating the integral.

- Theorem 17.5 was added to Section 17.3; it addresses the converse of Theorem 17.4. We also promoted the area of a plane region by a line integral to theorem status (Theorem 17.8 in Section 17.4).

- Example 3 in Section 17.7 was replaced to give an example of a surface whose bounding curve is not a plane curve and to provide an example that buttresses the claims made at the end of the section (that is, Two Final Notes on Stokes' Theorem).

- More line integral exercises were added to Section 17.3 where the student must first find the potential function before evaluating the line integral over a conservative vector field using the Fundamental Theorem of Line Integrals.

- We added to Section 17.7 more challenging surface integrals that are evaluated using Stokes' Theorem.

New to MyLab Math

- **Assignable Exercises** To better support students and instructors, we made the following changes to the assignable exercises:

  ◦ Updated the solution processes in Help Me Solve This and View an Example to better match the techniques used in the text.

  ◦ Added more Setup & Solve exercises to better mirror the types of responses that students are expected to provide on tests. We also added a parallel "standard" version of each Setup & Solve exercise, to allow the instructor to determine which version to assign.

  ◦ Added exercises corresponding to new exercises in the text.

- Added exercises where MyLab Math users had identified gaps in coverage in the 2e.

- Added extra practice exercises to each section (clearly labeled EXTRA). These "beyond the text" exercises are perfect for chapter reviews, quizzes, and tests.

- Analyzed aggregated student usage and performance data from MyLab Math for the previous edition of this text. The results of this analysis helped improve the quality and quantity of exercises that matter the most to instructors and students.

- **Instructional Videos** For each section of the text, there is now a new full-lecture video. Many of these videos make use of Interactive Figures to enhance student understanding of concepts. To make it easier for students to navigate to the specific content they need, each lecture video is segmented into shorter clips (labeled Introduction, Example, or Summary). Both the full lectures and the video segments are assignable within MyLab Math. The videos were created by the following team: Matt Hudelson (Washington State University), Deb Carney and Rebecca Swanson (Colorado School of Mines), Greg Wisloski and Dan Radelet (Indiana University of Pennsylvania), and Nick Ormes (University of Denver).

- **Enhanced Interactive Figures** Incorporating functionality from several standard Interactive Figures makes Enhanced Interactive Figures mathematically richer and ideal for in-class demonstrations. Using a single figure, instructors can illustrate concepts that are difficult for students to visualize and can make important connections to key themes of calculus.

- **Enhanced Sample Assignments** These section-level assignments address gaps in pre-calculus skills with a personalized review of prerequisites, help keep skills fresh with spaced practice using key calculus concepts, and provide opportunities to work exercises without learning aids so students can check their understanding. They are assignable and editable.

- **Quick Quizzes** have been added to Learning Catalytics™ (an in-class assessment system) for every section of the text.

- **Maple™, Mathematica®, and Texas Instruments® Manuals and Projects** have all been updated to align with the latest software and hardware.

Noteworthy Features

Figures

Given the power of graphics software and the ease with which many students assimilate visual images, we devoted considerable time and deliberation to the figures in this text. Whenever possible, we let the figures communicate essential ideas using annotations reminiscent of an instructor's voice at the board. Readers will quickly find that the figures facilitate learning in new ways.

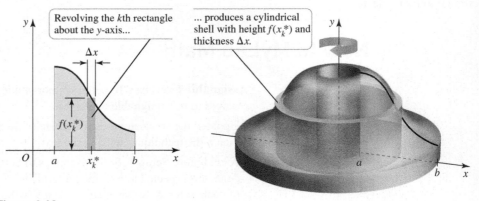

Figure 6.40

Annotated Examples

Worked-out examples feature annotations in blue to guide students through the process of solving the example and to emphasize that each step in a mathematical argument must be rigorously justified. These annotations are designed to echo how instructors "talk through" examples in lecture. They also provide help for students who may struggle with the algebra and trigonometry steps within the solution process.

Quick Checks

The narrative is interspersed with Quick Check questions that encourage students to do the calculus as they are reading about it. These questions resemble the kinds of questions instructors pose in class. Answers to the Quick Check questions are found at the end of the section in which they occur.

Guided Projects

MyLab Math contains 78 Guided Projects that allow students to work in a directed, step-by-step fashion, with various objectives: to carry out extended calculations, to derive physical models, to explore related theoretical topics, or to investigate new applications of calculus. The Guided Projects vividly demonstrate the breadth of calculus and provide a wealth of mathematical excursions that go beyond the typical classroom experience. A list of related Guided Projects is included at the end of each chapter.

Incorporating Technology

We believe that a calculus text should help students strengthen their analytical skills and demonstrate how technology can extend (not replace) those skills. Calculators and graphing utilities are additional tools in the kit, and students must learn when and when not to use them. Our goal is to accommodate the different policies regarding technology adopted by various instructors.

Throughout the text, exercises marked with **T** indicate that the use of technology—ranging from plotting a function with a graphing calculator to carrying out a calculation using a computer algebra system—may be needed. See page xx for information regarding our technology resource manuals covering Maple, Mathematica, and Texas Instruments graphing calculators.

Text Versions

- **eBook with Interactive Figures** The text is supported by a groundbreaking and award-winning electronic book created by Eric Schulz of Walla Walla Community College. This "live book" runs in Wolfram CDF Player (the free version of Mathematica) and contains the complete text of the print book plus interactive versions of approximately 700 figures. Instructors can use these interactive figures in the classroom to illustrate the important ideas of calculus, and students can explore them while they are reading the text. Our experience confirms that the interactive figures help build students' geometric intuition of calculus. The authors have written Interactive Figure Exercises that can be assigned via MyLab Math so that students can engage with the figures outside of class in a directed way. Available only within MyLab Math, the eBook provides instructors with powerful new teaching tools that expand and enrich the learning experience for students.

- **Other eBook Formats** The text is also available in various stand-alone eBook formats. These are listed in the Pearson online catalog: **www.pearson.com**. MyLab Math also contains an HTML eBook that is screen-reader accessible.

- **Other Print Formats** The text is also available in split editions (Single Variable [Chapters 1–12] and Multivariable [Chapters 10–17]) and in unbound (3-hole punched) formats. Again, see the Pearson online catalog for details: **www.pearson.com**.

Acknowledgments

We would like to express our thanks to the following people, who made valuable contributions to this edition as it evolved through its many stages.

Accuracy Checkers

Jennifer Blue

Lisa Collette

Nathan Moyer

Tom Wegleitner

Reviewers

Siham Alfred, *Germanna Community College*

Darry Andrews, *Ohio State University*

Pichmony Anhaouy, *Langara College*

Raul Aparicio, *Blinn College*

Anthony Barcellos, *American River College*

Rajesh K. Barnwal, *Middle Tennessee State University*

Susan Barton, *Brigham Young University, Hawaii*

Aditya Baskota, *Kauai Community College*

Al Batten, *University of Colorado, Colorado Springs*

Jeffrey V. Berg, *Arapahoe Community College*

Martina Bode, *University of Illinois, Chicago*

Kaddour Boukaabar, *California University of Pennsylvania*

Paul W. Britt, *Our Lady of the Lake College*

Steve Brosnan, *Belmont Abbey College*

Brendan Burns Healy, *Tufts University*

MK Choy, *CUNY Hunter College*

Nathan P. Clements, *University of Wyoming*

Gregory M. Coldren, *Frederick Community College*

Brian Crimmel, *U.S. Coast Guard Academy*

Robert D. Crise, Jr, *Crafton Hills College*

Dibyajyoti Deb, *Oregon Institute of Technology*

Elena Dilai, *Monroe Community College*

Johnny I. Duke, *Georgia Highlands College*

Fedor Duzhin, *Nanyang Technological University*

Houssein El Turkey, *University of New Haven*

J. Robson Eby, *Blinn College*

Amy H. Erickson, *Georgia Gwinnett College*

Robert A. Farinelli, *College of Southern Maryland*

Rosemary Farley, *Manhattan College*

Lester A. French, Jr, *University of Maine, Augusta*

Pamela Frost, *Middlesex Community College*

Scott Gentile, *CUNY Hunter College*

Stephen V. Gilmore, *Johnson C. Smith University*

Joseph Hamelman, *Virginia Commonwealth University*

Jayne Ann Harderq, *Oral Roberts University*

Miriam Harris-Botzum, *Lehigh Carbon Community College*

Mako Haruta, *University of Hartford*

Ryan A. Hass, *Oregon State University*

Christy Hediger, *Lehigh Carbon Community College*

Joshua Hertz, *Northeastern University*

Peter Hocking, *Brunswick Community College*

Farhad Jafari, *University of Wyoming*

Yvette Janecek, *Blinn College*

Tom Jerardi, *Howard Community College*

Karen Jones, *Ivy Tech Community College*

Bir Kafle, *Purdue University Northwest*

Mike Kawai, *University of Colorado, Denver*

Mahmoud Khalili, *Oakton Community College*

Lynne E. Kowski, *Raritan Valley Community College*

Tatyana Kravchuk, *Northern Virginia Community College*

Lorraine Krusinski, *Brunswick High School*

Frederic L. Lang, *Howard Community College*

Robert Latta, *University of North Carolina, Charlotte*

Angelica Lyubenko, *University of Colorado, Denver*

Darren E. Mason, *Albion College*

John C. Medcalf, *American River College*

Elaine Merrill, *Brigham Young University, Hawaii*

Robert Miller, *Langara College*

Mishko Mitkovski, *Clemson University*

Carla A. Monticelli, *Camden County College*

Charles Moore, *Washington State University*

Humberto Munoz Barona, *Southern University*

Clark Musselman, *University of Washington, Bothell*

Glenn Newman, *Newbury College*

Daniela Nikolova-Popova, *Florida Atlantic University*

Janis M. Oldham, *North Carolina A&T State University*

Byungdo Park, *CUNY Hunter College*

Denise Race, *Eastfield College*

Hilary Risser, *Montana Tech*

Sylvester Roebuck, Jr, *Olive Harvey College*

John P. Roop, *North Carolina A&T State University*

Paul N. Runnion, *Missouri University of Science and Technology*

Kegan Samuel, *Middlesex Community College*

Steve Scarborough, *Oregon State University*

Vickey Sealy, *West Virginia University*

Ruth Seidel, *Blinn College*

Derek Smith, *Nashville State Community College*

James Talamo, *Ohio State University*

Yan Tang, *Embry-Riddle Aeronautical University*

Kye Taylor, *Tufts University*

Daniel Triolo, *Lake Sumter State College*

Cal Van Niewaal, *Coe College*

Robin Wilson, *California Polytechnic State University, Pomona*

Kurt Withey, *Northampton Community College*

Amy Yielding, *Eastern Oregon University*

Prudence York-Hammons, *Temple College*

Bradley R. Young, *Oakton Community College*

Pablo Zafra, *Kean University*

The following faculty members provided direction on the development of the MyLab Math course for this edition:

Heather Albrecht, *Blinn College*

Terry Barron, *Georgia Gwinnett College*

Jeffrey V. Berg, *Arapahoe Community College*

Joseph Bilello, *University of Colorado, Denver*

Mark Bollman, *Albion College*

Mike Bostick, *Central Wyoming College*

Laurie L. Boudreaux, *Nicholls State University*

Steve Brosnan, *Belmont Abbey College*

Debra S. Carney, *Colorado School of Mines*

Robert D. Crise, Jr, *Crafton Hills College*

Shannon Dingman, *University of Arkansas*

Deland Doscol, *North Georgia Technical College*

J. Robson Eby, *Blinn College*

Stephanie L. Fitch, *Missouri University of Science and Technology*

Dennis Garity, *Oregon State University*

Monica Geist, *Front Range Community College*

Brandie Gilchrist, *Central High School*

Roger M. Goldwyn, *Florida Atlantic University*

Maggie E. Habeeb, *California University of Pennsylvania*

Ryan A. Hass, *Oregon State University*

Danrun Huang, *St. Cloud State University*

Zhongming Huang, *Midland University*

Christa Johnson, *Guilford Technical Community College*

Bir Kafle, *Purdue University Northwest*

Semra Kilic-Bahi, *Colby-Sawyer College*

Kimberly Kinder, *Missouri University of Science and Technology*

Joseph Kotowski, *Oakton Community College*

Lynne E. Kowski, *Raritan Valley Community College*

Paula H. Krone, *Georgia Gwinnett College*

Daniel Lukac, *Miami University*

Jeffrey W. Lyons, *Nova Southeastern University*

James Magee, *Diablo Valley College*

Erum Marfani, *Frederick Community College*

Humberto Munoz Barona, *Southern University*

James A. Mendoza Alvarez, *University of Texas, Arlington*

Glenn Newman, *Newbury College*

Peter Nguyen, *Coker College*

Mike Nicholas, *Colorado School of Mines*

Seungly Oh, *Western New England University*

Denise Race, *Eastfield College*

Paul N. Runnion, *Missouri University of Science and Technology*

Polly Schulle, *Richland College*

Rebecca Swanson, *Colorado School of Mines*

William Sweet, *Blinn College*

M. Velentina Vega-Veglio, *William Paterson University*

Nick Wahle, *Cincinnati State Technical and Community College*

Patrick Wenkanaab, *Morgan State University*

Amanda Wheeler, *Amarillo College*

Diana White, *University of Colorado, Denver*

Gregory A. White, *Linn Benton Community College*

Joseph White, *Olympic College*

Robin T. Wilson, *California Polytechnic State University, Pomona*

Deborah A. Zankofski, *Prince George's Community College*

The following faculty were members of the Engineering Review Panel. This panel made recommendations to improve the text for engineering students.

Al Batten, *University of Colorado, Colorado Springs*

Josh Hertz, *Northeastern University*

Daniel Latta, *University of North Carolina, Charlotte*

Yan Tang, *Embry-Riddle Aeronautical University*

MyLab Math Online Course
for *Calculus: Early Transcendentals, 3e*

(access code required)

MyLab™ Math is available to accompany Pearson's market-leading text offerings. To give students a consistent tone, voice, and teaching method, each text's flavor and approach are tightly integrated throughout the accompanying MyLab Math course, making learning the material as seamless as possible.

PREPAREDNESS

One of the biggest challenges in calculus courses is making sure students are adequately prepared with the prerequisite skills needed to successfully complete their course work. MyLab Math supports students with just-in-time remediation and review of key concepts.

Integrated Review Course

These special MyLab courses contain pre-made, assignable quizzes to assess the prerequisite skills needed for each chapter, plus personalized remediation for any gaps in skills that are identified. Each student, therefore, receives the appropriate level of help—no more, no less.

DEVELOPING DEEPER UNDERSTANDING

MyLab Math provides content and tools that help students build a deeper understanding of course content than would otherwise be possible.

eBook with Interactive Figures

The eBook includes approximately 700 figures that can be manipulated by students to provide a deeper geometric understanding of key concepts and examples as they read and learn new material. Students get unlimited access to the eBook within any MyLab Math course using that edition of the text. The authors have written Interactive Figure Exercises that can be assigned for homework so that students can engage with the figures outside of the classroom.

Exercises with Immediate Feedback

The over 8000 homework and practice exercises for this text regenerate algorithmically to give students unlimited opportunity for practice and mastery. MyLab Math provides helpful feedback when students enter incorrect answers and includes the optional learning aids Help Me Solve This, View an Example, videos, and/or the eBook.

NEW! Enhanced Sample Assignments

These section-level assignments include just-in-time review of prerequisites, help keep skills fresh with spaced practice of key concepts, and provide opportunities to work exercises without learning aids so students can check their understanding. They are assignable and editable within MyLab Math.

Additional Conceptual Questions

Additional Conceptual Questions focus on deeper, theoretical understanding of the key concepts in calculus. These questions were written by faculty at Cornell University under an NSF grant and are also assignable through Learning Catalytics™.

Setup & Solve Exercises

These exercises require students to show how they set up a problem, as well as the solution, thus better mirroring what is required on tests. This new type of exercise was widely praised by users of the second edition, so more were added to the third edition.

ALL NEW! Instructional Videos

For each section of the text, there is now a new full-lecture video. Many of these videos make use of Interactive Figures to enhance student understanding of concepts. To make it easier for students to navigate to the content they need, each lecture video is segmented into shorter clips (labeled Introduction, Example, or Summary). Both the video lectures and the video segments are assignable within MyLab Math. The Guide to Video-Based Assignments makes it easy to assign videos for homework by showing which MyLab Math exercises correspond to each video.

UPDATED! Technology Manuals (downloadable)

- Maple™ Manual and Projects by Kevin Reeves, East Texas Baptist University
- Mathematica® Manual and Projects by Todd Lee, Elon University
- TI-Graphing Calculator Manual by Elaine McDonald-Newman, Sonoma State University

These manuals cover *Maple* 2017, *Mathematica 11,* and the TI-84 Plus and TI-89, respectively. Each manual provides detailed guidance for integrating the software package or graphing calculator throughout the course, including syntax and commands. The projects include instructions and ready-made application files for Maple and Mathematica. The files can be downloaded from within MyLab Math.

Student's Solutions Manuals (softcover and downloadable)

Single Variable Calculus: Early Transcendentals (Chapters 1–12)
    ISBN: 0-13-477048-X | 978-0-13-477048-2
Multivariable Calculus (Chapters 10–17)
    ISBN: 0-13-476682-2 | 978-0-13-476682-9
Written by Mark Woodard (Furman University), the Student's Solutions Manual contains worked-out solutions to all the odd-numbered exercises. This manual is available in print and can be downloaded from within MyLab Math.

SUPPORTING INSTRUCTION

MyLab Math comes from an experienced partner with educational expertise and an eye on the future. It provides resources to help you assess and improve student results at every turn and unparalleled flexibility to create a course tailored to you and your students.

NEW! Enhanced Interactive Figures

Incorporating functionality from several standard Interactive Figures makes Enhanced Interactive Figures mathematically richer and ideal for in-class demonstrations. Using a single enhanced figure, instructors can illustrate concepts that are difficult for students to visualize and can make important connections to key themes of calculus.

Learning Catalytics

Now included in all MyLab Math courses, this student response tool uses students' smartphones, tablets, or laptops to engage them in more interactive tasks and thinking during lecture. Learning Catalytics™ fosters student engagement and peer-to-peer learning with real-time analytics. Access pre-built exercises created specifically for calculus, including Quick Quiz exercises for each section of the text.

PowerPoint Lecture Resources (downloadable)

Slides contain presentation resources such as key concepts, examples, definitions, figures, and tables from this text. They can be downloaded from within MyLab Math or from Pearson's online catalog at **www.pearson.com**.

Comprehensive Gradebook

The gradebook includes enhanced reporting functionality, such as item analysis and a reporting dashboard to enable you to efficiently manage your course. Student performance data are presented at the class, section, and program levels in an accessible, visual manner so you'll have the information you need to keep your students on track.

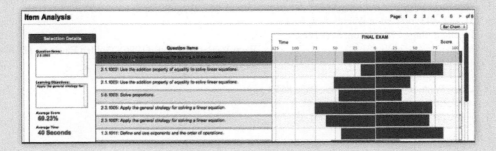

TestGen

TestGen® (**www.pearson.com/testgen**) enables instructors to build, edit, print, and administer tests using a computerized bank of questions developed to cover all the objectives of the text. TestGen is algorithmically based, allowing instructors to create multiple, but equivalent, versions of the same question or test with the click of a button. Instructors can also modify test bank questions and/or add new questions. The software and test bank are available for download from Pearson's online catalog, **www.pearson.com**. The questions are also assignable in MyLab Math.

Instructor's Solutions Manual (downloadable)

Written by Mark Woodard (Furman University), the Instructor's Solutions Manual contains complete solutions to all the exercises in the text. It can be downloaded from within MyLab Math or from Pearson's online catalog, **www.pearson.com**.

Instructor's Resource Guide (downloadable)

This resource includes Guided Projects that require students to make connections between concepts and applications from different parts of the calculus course. They are correlated to specific chapters of the text and can be assigned as individual or group work. The files can be downloaded from within MyLab Math or from Pearson's online catalog, **www.pearson.com**.

Accessibility

Pearson works continuously to ensure that our products are as accessible as possible to all students. We are working toward achieving WCAG 2.0 Level AA and Section 508 standards, as expressed in the Pearson Guidelines for Accessible Educational Web Media, **www.pearson.com/mylab/math/accessibility**.

Credits

Chapter opener art: Andrey Pavlov/123RF

Chapter 10

Page 682, Fleron, J. F. (1999, January). Gabriel's Wedding Cake. *The College Mathematics Journal*, 30(1), 35–38; Selden, A. & Selden, J. (1993, November). Collegiate Mathematics Education Research: What Would That Be Like? *The College Mathematics Journal*, 24(5), 431–445. **Pages 682–683,** Chen, H. & Kennedy, C. (2012, May). Harmonic Series Meets Fibonacci Sequence. *The College Mathematics Journal*, 43(3), 237–243.

Chapter 12

Page 766, Wagon, S. (2010). *Mathematica in Action*. Springer; created by Norton Starr, Amherst College. **Page 767,** Brannen, N. S. (2001, September). The Sun, the Moon, and Convexity. *The College Mathematics Journal*, 32(4), 268–272. **Page 774,** Fray, T. H. (1989). The butterfly curve. *American Mathematical Monthly*, 95(5), 442–443; revived in Wagon, S. & Packel, E. (1994). *Animating Calculus*. Freeman. **Page 778,** Wagon, S. & Packel, E. (1994). *Animating Calculus*. Freeman.

Chapter 13

Page 816, Strang, G. (1991). Calculus. Wellesley-Cambridge Press. **Page 864,** Model Based 3D Tracking of an Articulated Hand, B. Stenger, P. R. S. Mendonça, R. Cipolla, CVPR, Vol. II, p. 310–315, December 2001. CVPR 2001: PROCEEDINGS OF THE 2001 IEEE COMPUTER SOCIETY CONFERENCE ON COMPUTER VISION AND PATTERN RECOGNITION by IEEE Computer Society Reproduced with permission of IEEE COMPUTER SOCIETY PRESS in the format Republish in a book via Copyright Clearance Center.

Chapter 15

Pages 929–930, www.nfl.com. **Page 994,** Karim, R. (2014, December). Optimization of pectin isolation method from pineapple (ananas comosus l.) waste. *Carpathian Journal of Food Science and Technology*, 6(2), 116–122. **Page 996,** Rosenholtz, I. (1985, May). "The Only Critical Point in Town" Test. *Mathematics Magazine*, 58(3), 149–150 and Gillet, P. (1984). *Calculus and Analytical Geometry*, 2nd edition; Rosenholtz, I. (1987, February). Two mountains without a valley. *Mathematics Magazine*, 60(1); Math Horizons, Vol. 11, No. 4, April 2004.

Chapter 16

Page 1072, Glaister, P. (1996). Golden Earrings. *Mathematical Gazette*, 80, 224–225.

10

Sequences and Infinite Series

Chapter Preview This chapter covers topics that lie at the foundation of calculus—indeed, at the foundation of mathematics. The first task is to make a clear distinction between a *sequence* and an *infinite series*. A sequence is an ordered *list* of numbers, $a_1, a_2, \ldots$, while an infinite series is a *sum* of numbers, $a_1 + a_2 + \cdots$. The idea of convergence to a limit is important for both sequences and series, but convergence is analyzed differently in the two cases. To determine limits of sequences, we use the same tools used for limits of functions at infinity. The analysis of infinite series requires new methods, which are introduced in this chapter. The study of infinite series begins with *geometric series*, which have theoretical importance and are used to answer many practical questions (When is your auto loan paid off? How much antibiotic is in your blood if you take three pills per day?). We then present several tests that are used to determine whether a given series converges. In the final section of the chapter, we offer guidelines to help choose an appropriate convergence test for a given series.

10.1 An Overview

To understand sequences and series, you must understand how they differ and how they are related. The purposes of this opening section are to introduce sequences and series in concrete terms, and to illustrate both their differences and their relationships with each other.

Examples of Sequences

Consider the following *list* of numbers:

$$\{1, 4, 7, 10, 13, 16, \ldots\}.$$

Each number in the list is obtained by adding 3 to the previous number in the list. With this rule, we could extend the list indefinitely.

This list is an example of a *sequence*, where each number in the sequence is called a **term** of the sequence. We denote sequences in any of the following forms:

$$\{a_1, a_2, a_3, \ldots, a_n, \ldots\}, \qquad \{a_n\}_{n=1}^{\infty}, \quad \text{or} \quad \{a_n\}.$$

The subscript n that appears in a_n is called an **index**, and it indicates the order of terms in the sequence. The choice of a starting index is arbitrary, but sequences usually begin with $n = 0$ or $n = 1$.

> ➤ In keeping with common practice, the terms *series* and *infinite series* are used interchangeably throughout this chapter.

> ➤ The dots (..., an ellipsis) after the last number of a sequence mean that the list continues indefinitely.

The sequence $\{1, 4, 7, 10, \ldots\}$ can be defined in two ways. First, we have the rule that each term of the sequence is 3 more than the previous term; that is, $a_2 = a_1 + 3$, $a_3 = a_2 + 3$, $a_4 = a_3 + 3$, and so forth. In general, we see that

$$a_1 = 1 \quad \text{and} \quad a_{n+1} = a_n + 3, \qquad \text{for } n = 1, 2, 3, \ldots.$$

This way of defining a sequence is called a *recurrence relation* (or an *implicit formula*). It specifies the initial term of the sequence (in this case, $a_1 = 1$) and gives a general rule for computing the next term of the sequence from previous terms. For example, if you know a_{100}, the recurrence relation can be used to find a_{101}.

Suppose instead you want to find a_{147} directly without computing the first 146 terms of the sequence. The first four terms of the sequence can be written

$$a_1 = 1 + (3 \cdot 0), \qquad a_2 = 1 + (3 \cdot 1), \qquad a_3 = 1 + (3 \cdot 2), \qquad a_4 = 1 + (3 \cdot 3).$$

Observe the pattern: The nth term of the sequence is 1 plus 3 multiplied by $n - 1$, or

$$a_n = 1 + 3(n - 1) = 3n - 2, \qquad \text{for } n = 1, 2, 3, \ldots.$$

QUICK CHECK 1 Find a_{10} for the sequence $\{1, 4, 7, 10, \ldots\}$ using the recurrence relation and then again using the explicit formula for the nth term. ◄

With this *explicit formula*, the nth term of the sequence is determined directly from the value of n. For example, with $n = 147$,

$$a_{147} = 3 \cdot \underbrace{147}_{n} - 2 = 439.$$

➤ When defined by an explicit formula $a_n = f(n)$, it is evident that sequences are functions. The domain is generally a subset of the nonnegative integers, and one real number a_n is assigned to each integer n in the domain.

DEFINITION Sequence

A **sequence** $\{a_n\}$ is an ordered list of numbers of the form

$$\{a_1, a_2, a_3, \ldots, a_n, \ldots\}.$$

A sequence may be generated by a **recurrence relation** of the form $a_{n+1} = f(a_n)$, for $n = 1, 2, 3, \ldots$, where a_1 is given. A sequence may also be defined with an **explicit formula** of the form $a_n = f(n)$, for $n = 1, 2, 3, \ldots$.

EXAMPLE 1   Explicit formulas Use the explicit formula for $\{a_n\}_{n=1}^{\infty}$ to write the first four terms of each sequence. Sketch a graph of the sequence.

a. $a_n = \dfrac{1}{2^n}$ **b.** $a_n = \dfrac{(-1)^n n}{n^2 + 1}$

SOLUTION

a. Substituting $n = 1, 2, 3, 4, \ldots$ into the explicit formula $a_n = \dfrac{1}{2^n}$, we find that the terms of the sequence are

$$\left\{\frac{1}{2}, \frac{1}{2^2}, \frac{1}{2^3}, \frac{1}{2^4}, \ldots\right\} = \left\{\frac{1}{2}, \frac{1}{4}, \frac{1}{8}, \frac{1}{16}, \ldots\right\}.$$

The graph of a sequence is the graph of a function that is defined only on a set of integers. In this case, we plot the coordinate pairs (n, a_n), for $n = 1, 2, 3, \ldots$, resulting in a graph consisting of individual points. The graph of the sequence $a_n = \dfrac{1}{2^n}$ suggests that the terms of this sequence approach 0 as n increases (**Figure 10.1**).

b. Substituting $n = 1, 2, 3, 4, \ldots$ into the explicit formula, the terms of the sequence are

$$\left\{\frac{(-1)^1 (1)}{1^2 + 1}, \frac{(-1)^2 2}{2^2 + 1}, \frac{(-1)^3 3}{3^2 + 1}, \frac{(-1)^4 4}{4^2 + 1}, \ldots\right\} = \left\{-\frac{1}{2}, \frac{2}{5}, -\frac{3}{10}, \frac{4}{17}, \ldots\right\}.$$

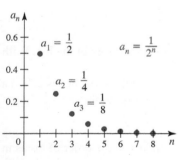

Figure 10.1

➤ The "switch" $(-1)^n$ is used frequently to alternate the signs of the terms of sequences and series.

Figure 10.2

From the graph (**Figure 10.2**), we see that the terms of the sequence alternate in sign and appear to approach 0 as n increases.

Related Exercises 13, 16 ◄

EXAMPLE 2 Recurrence relations Use the recurrence relation for $\{a_n\}_{n=1}^{\infty}$ to write the first four terms of the sequences

$$a_{n+1} = 2a_n + 1, a_1 = 1 \quad \text{and} \quad a_{n+1} = 2a_n + 1, a_1 = -1.$$

SOLUTION Notice that the recurrence relation is the same for the two sequences; only the first term differs. The first four terms of each sequence are as follows.

n	a_n with $a_1 = 1$	a_n with $a_1 = -1$
1	$a_1 = 1$ (given)	$a_1 = -1$ (given)
2	$a_2 = 2a_1 + 1 = 2 \cdot 1 + 1 = 3$	$a_2 = 2a_1 + 1 = 2(-1) + 1 = -1$
3	$a_3 = 2a_2 + 1 = 2 \cdot 3 + 1 = 7$	$a_3 = 2a_2 + 1 = 2(-1) + 1 = -1$
4	$a_4 = 2a_3 + 1 = 2 \cdot 7 + 1 = 15$	$a_4 = 2a_3 + 1 = 2(-1) + 1 = -1$

QUICK CHECK 2 Find an explicit formula for the sequence $\{1, 3, 7, 15, \ldots\}$ (Example 2). ◄

We see that the terms of the first sequence increase without bound, while all terms of the second sequence are -1. Clearly, the initial term of the sequence may determine the behavior of the entire sequence.

Related Exercises 21–22 ◄

EXAMPLE 3 Working with sequences Consider the following sequences.

a. $\{a_n\} = \{-2, 5, 12, 19, \ldots\}$ **b.** $\{b_n\} = \{3, 6, 12, 24, 48, \ldots\}$

(i) Find the next two terms of the sequence.

(ii) Find a recurrence relation that generates the sequence.

(iii) Find an explicit formula for the nth term of the sequence.

SOLUTION

a. (i)   Each term is obtained by adding 7 to its predecessor. The next two terms are $19 + 7 = 26$ and $26 + 7 = 33$.

> In Example 3, we chose the starting index $n = 0$. Other choices are possible.

(ii) Because each term is seven more than its predecessor, a recurrence relation is

$$a_{n+1} = a_n + 7, a_0 = -2, \quad \text{for } n = 0, 1, 2, \ldots.$$

(iii) Notice that $a_0 = -2$, $a_1 = -2 + (1 \cdot 7)$, and $a_2 = -2 + (2 \cdot 7)$, so an explicit formula is

$$a_n = 7n - 2, \quad \text{for } n = 0, 1, 2, \ldots.$$

b. (i)   Each term is obtained by multiplying its predecessor by 2. The next two terms are $48 \cdot 2 = 96$ and $96 \cdot 2 = 192$.

(ii) Because each term is two times its predecessor, a recurrence relation is

$$a_{n+1} = 2a_n, a_0 = 3, \quad \text{for } n = 0, 1, 2, \ldots.$$

(iii) To obtain an explicit formula, note that $a_0 = 3$, $a_1 = 3(2^1)$, and $a_2 = 3(2^2)$. In general,

$$a_n = 3(2^n), \quad \text{for } n = 0, 1, 2, \ldots.$$

Related Exercises 27–28 ◄

Limit of a Sequence

Perhaps the most important question about a sequence is this: If you go farther and farther out in the sequence, $a_{100}, \ldots, a_{10,000}, \ldots, a_{100,000}, \ldots$, how do the terms of the sequence behave? Do they approach a specific number, and if so, what is that number? Or do they grow in magnitude without bound? Or do they wander around with or without a pattern?

The long-term behavior of a sequence is described by its **limit**. The limit of a sequence is defined rigorously in the next section. For now, we work with an informal definition.

> **DEFINITION Limit of a Sequence**
>
> If the terms of a sequence $\{a_n\}$ approach a unique number L as n increases—that is, if a_n can be made arbitrarily close to L by taking n sufficiently large—then we say $\lim\limits_{n \to \infty} a_n = L$ exists, and the sequence **converges** to L. If the terms of the sequence do not approach a single number as n increases, the sequence has no limit, and the sequence **diverges**.

EXAMPLE 4   Limits of sequences Write the first four terms of each sequence. If you believe the sequence converges, make a conjecture about its limit. If the sequence appears to diverge, explain why.

a. $\left\{ \dfrac{(-1)^n}{n^2 + 1} \right\}_{n=1}^{\infty}$ Explicit formula

b. $\left\{ \cos n\pi \right\}_{n=1}^{\infty}$ Explicit formula

c. $\{a_n\}_{n=1}^{\infty}$, where $a_{n+1} = -2a_n,\ a_1 = 1$ Recurrence relation

SOLUTION

a. Beginning with $n = 1$, the first four terms of the sequence are

$$\left\{ \frac{(-1)^1}{1^2 + 1}, \frac{(-1)^2}{2^2 + 1}, \frac{(-1)^3}{3^2 + 1}, \frac{(-1)^4}{4^2 + 1}, \dots \right\} = \left\{ -\frac{1}{2}, \frac{1}{5}, -\frac{1}{10}, \frac{1}{17}, \dots \right\}.$$

The terms decrease in magnitude and approach zero with alternating signs. The limit appears to be 0 (**Figure 10.3**).

b. The first four terms of the sequence are

$$\{ \cos \pi, \cos 2\pi, \cos 3\pi, \cos 4\pi, \dots \} = \{ -1, 1, -1, 1, \dots \}.$$

In this case, the terms of the sequence alternate between -1 and $+1$, and never approach a single value. Therefore, the sequence diverges (**Figure 10.4**).

c. The first four terms of the sequence are

$$\{ 1, -2a_1, -2a_2, -2a_3, \dots \} = \{ 1, -2, 4, -8, \dots \}.$$

Because the magnitudes of the terms increase without bound, the sequence diverges (**Figure 10.5**).

Figure 10.3

Figure 10.4

Figure 10.5

Related Exercises 35–36 ◄

EXAMPLE 5   Limit of a sequence Enumerate and graph the terms of the following sequence, and make a conjecture about its limit.

$$a_n = \frac{4n^3}{n^3 + 1}, \qquad \text{for } n = 1, 2, 3, \ldots. \qquad \text{Explicit formula}$$

SOLUTION The first 14 terms of the sequence $\{a_n\}$ are tabulated in Table 10.1 and graphed in **Figure 10.6**. The terms appear to approach 4.

The sequence values appear to approach 4 as n increases.

$$a_n = \frac{4n^3}{n^3 + 1}$$

Table 10.1

n	a_n	n	a_n
1	2.000	8	3.992
2	3.556	9	3.995
3	3.857	10	3.996
4	3.938	11	3.997
5	3.968	12	3.998
6	3.982	13	3.998
7	3.988	14	3.999

Figure 10.6

Related Exercise 45 ◀

The height of each bounce of the basketball is 0.8 of the height of the previous bounce.

20 ft

16 ft

12.8 ft

10.24 ft

Time

Figure 10.7

EXAMPLE 6   A bouncing ball A basketball tossed straight up in the air reaches a high point and falls to the floor. Each time the ball bounces on the floor it rebounds to 0.8 of its previous height. Let h_n be the high point after the nth bounce, with the initial height being $h_0 = 20$ ft.

a. Find a recurrence relation and an explicit formula for the sequence $\{h_n\}$.

b. What is the high point after the 10th bounce? After the 20th bounce?

c. Speculate on the limit of the sequence $\{h_n\}$.

SOLUTION

a. We first write and graph the heights of the ball for several bounces using the rule that each height is 0.8 of the previous height (**Figure 10.7**). For example, we have

$$\begin{aligned}
h_0 &= 20 \text{ ft}, \\
h_1 &= 0.8 \, h_0 = 16 \text{ ft}, \\
h_2 &= 0.8 \, h_1 = 0.8^2 \, h_0 = 12.80 \text{ ft}, \\
h_3 &= 0.8 \, h_2 = 0.8^3 \, h_0 = 10.24 \text{ ft, and} \\
h_4 &= 0.8 \, h_3 = 0.8^4 \, h_0 \approx 8.19 \text{ ft}.
\end{aligned}$$

Each number in the list is 0.8 of the previous number. Therefore, the recurrence relation for the sequence of heights is

$$h_{n+1} = 0.8 \, h_n, \quad h_0 = 20, \qquad \text{for } n = 0, 1, 2, 3, \ldots.$$

To find an explicit formula for the nth term, note that

$$h_1 = h_0 \cdot 0.8, \qquad h_2 = h_0 \cdot 0.8^2, \qquad h_3 = h_0 \cdot 0.8^3, \qquad \text{and} \qquad h_4 = h_0 \cdot 0.8^4.$$

In general, we have

$$h_n = h_0 \cdot 0.8^n = 20 \cdot 0.8^n, \qquad \text{for } n = 0, 1, 2, 3, \ldots,$$

which is an explicit formula for the terms of the sequence.

b. Using the explicit formula for the sequence, we see that after $n = 10$ bounces, the next height is

$$h_{10} = 20 \cdot 0.8^{10} \approx 2.15 \text{ ft}.$$

$h_n = 20 \cdot 0.8^n$

The sequence values approach 0.

Figure 10.8

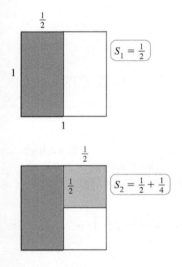

$S_1 = \frac{1}{2}$

$S_2 = \frac{1}{2} + \frac{1}{4}$

$S_3 = \frac{1}{2} + \frac{1}{4} + \frac{1}{8}$

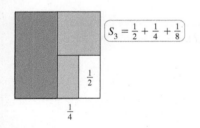

$S_4 = \frac{1}{2} + \frac{1}{4} + \frac{1}{8} + \frac{1}{16}$

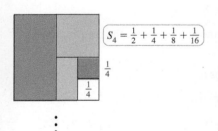

$S_n = \frac{1}{2} + \frac{1}{4} + \cdots + \frac{1}{2^n}$

Figure 10.9

After $n = 20$ bounces, the next height is

$$h_{20} = 20 \cdot 0.8^{20} \approx 0.23 \text{ ft.}$$

c. The terms of the sequence (**Figure 10.8**) appear to decrease and approach 0. A reasonable conjecture is that $\lim\limits_{n \to \infty} h_n = 0$.

Related Exercises 57–60 ◄

Infinite Series and the Sequence of Partial Sums

An infinite series can be viewed as a *sum* of an infinite set of numbers; it has the form

$$a_1 + a_2 + \cdots + a_n + \cdots,$$

where the terms of the series, $a_1, a_2, \ldots$, are real numbers. We first answer the question: How is it possible to sum an infinite set of numbers and produce a finite number? Here is an informative example.

Consider a unit square (sides of length 1) that is subdivided as shown in **Figure 10.9**. We let S_n be the area of the colored region in the nth figure of the progression. The area of the colored region in the first figure is

$$S_1 = \frac{1}{2} \cdot 1 = \frac{1}{2}. \quad \frac{1}{2} = \frac{2^1 - 1}{2^1}$$

The area of the colored region in the second figure is S_1 plus the area of the smaller blue square, which is $\frac{1}{2} \cdot \frac{1}{2} = \frac{1}{4}$. Therefore,

$$S_2 = \frac{1}{2} + \frac{1}{4} = \frac{3}{4}. \quad \frac{3}{4} = \frac{2^2 - 1}{2^2}$$

The area of the colored region in the third figure is S_2 plus the area of the smaller green rectangle, which is $\frac{1}{2} \cdot \frac{1}{4} = \frac{1}{8}$. Therefore,

$$S_3 = \frac{1}{2} + \frac{1}{4} + \frac{1}{8} = \frac{7}{8}. \quad \frac{7}{8} = \frac{2^3 - 1}{2^3}$$

Continuing in this manner, we find that

$$S_n = \frac{1}{2} + \frac{1}{4} + \frac{1}{8} + \cdots + \frac{1}{2^n} = \frac{2^n - 1}{2^n}.$$

If this process is continued indefinitely, the area of the colored region S_n approaches the area of the unit square, which is 1. So it is plausible that

$$\lim_{n \to \infty} S_n = \underbrace{\frac{1}{2} + \frac{1}{4} + \frac{1}{8} + \cdots}_{\text{sum continues indefinitely}} = 1.$$

The explicit formula $S_n = \dfrac{2^n - 1}{2^n}$ can be analyzed to verify our assertion that $\lim\limits_{n \to \infty} S_n = 1$; we turn to that task in Section 10.2.

This example shows that it is possible to sum an infinite set of numbers and obtain a finite number—in this case, the sum is 1. The sequence $\{S_n\}$ generated in this example is extremely important. It is called a *sequence of partial sums*, and its limit is the value of the infinite series $\frac{1}{2} + \frac{1}{4} + \frac{1}{8} + \cdots$. The idea of a sequence of partial sums is illustrated by the decimal expansion of 1.

EXAMPLE 7　Working with series Consider the infinite series

$$0.9 + 0.09 + 0.009 + 0.0009 + \cdots,$$

where each term of the sum is $\frac{1}{10}$ of the previous term.

a. Find the sum of the first one, two, three, and four terms of the series.

b. What value would you assign to the infinite series $0.9 + 0.09 + 0.009 + \cdots$?

SOLUTION

a. Let S_n denote the sum of the first n terms of the given series. Then

$$S_1 = 0.9,$$
$$S_2 = 0.9 + 0.09 = 0.99,$$
$$S_3 = 0.9 + 0.09 + 0.009 = 0.999, \text{ and}$$
$$S_4 = 0.9 + 0.09 + 0.009 + 0.0009 = 0.9999.$$

b. The sums $S_1, S_2, \ldots, S_n$ form a sequence $\{S_n\}$, which is a sequence of partial sums. As more and more terms are included, the values of S_n approach 1. Therefore, a reasonable conjecture for the value of the series is 1:

$$\underbrace{\underbrace{\underbrace{0.9 + 0.09}_{S_2 = 0.99} + 0.009}_{S_3 = 0.999} + 0.0009 + \cdots}_{S_1 = 0.9} = 1.$$

> **QUICK CHECK 3** Reasoning as in Example 7, what is the value of $0.3 + 0.03 + 0.003 + \cdots$? ◄

Related Exercises 61–62 ◄

> Recall the summation notation introduced in Chapter 5: $\sum_{k=1}^{n} a_k$ means $a_1 + a_2 + \cdots + a_n$.

The nth term of the sequence of partial sums is

$$S_n = \underbrace{0.9 + 0.09 + 0.009 + \cdots + 0.0\ldots09}_{n \text{ terms}} = \sum_{k=1}^{n} 9 \cdot 0.1^k.$$

We observed that $\lim_{n \to \infty} S_n = 1$. For this reason, we write

$$\lim_{n \to \infty} S_n = \lim_{n \to \infty} \underbrace{\sum_{k=1}^{n} 9 \cdot 0.1^k}_{S_n} = \underbrace{\sum_{k=1}^{\infty} 9 \cdot 0.1^k}_{\text{new object}} = 1.$$

By letting $n \to \infty$, a new mathematical object $\sum_{k=1}^{\infty} 9 \cdot 0.1^k$ is created. It is an infinite series, and its value is the *limit* of the sequence of partial sums.

> The term *series* is used for historical reasons. When you see *series*, you should think *sum*.

DEFINITION Infinite Series

Given a sequence $\{a_1, a_2, a_3, \ldots\}$, the sum of its terms

$$a_1 + a_2 + a_3 + \cdots = \sum_{k=1}^{\infty} a_k$$

is called an **infinite series**. The **sequence of partial sums** $\{S_n\}$ associated with this series has the terms

$$S_1 = a_1$$
$$S_2 = a_1 + a_2$$
$$S_3 = a_1 + a_2 + a_3$$
$$\vdots$$
$$S_n = a_1 + a_2 + a_3 + \cdots + a_n = \sum_{k=1}^{n} a_k, \quad \text{for } n = 1, 2, 3, \ldots.$$

If the sequence of partial sums $\{S_n\}$ has a limit L, the infinite series **converges** to that limit, and we write

$$\sum_{k=1}^{\infty} a_k = \lim_{n \to \infty} \underbrace{\sum_{k=1}^{n} a_k}_{S_n} = \lim_{n \to \infty} S_n = L.$$

> **QUICK CHECK 4** Do the series $\sum_{k=1}^{\infty} 1$ and $\sum_{k=1}^{\infty} k$ converge or diverge? ◄

If the sequence of partial sums diverges, the infinite series also **diverges**.

EXAMPLE 8   Sequence of partial sums Consider the infinite series

$$\sum_{k=1}^{\infty} \frac{1}{k(k+1)}.$$

a. Find the first four terms of the sequence of partial sums.

b. Find an expression for S_n and make a conjecture about the value of the series.

SOLUTION

a. The sequence of partial sums can be evaluated explicitly:

$$S_1 = \sum_{k=1}^{1} \frac{1}{k(k+1)} = \frac{1}{2},$$

$$S_2 = \sum_{k=1}^{2} \frac{1}{k(k+1)} = \frac{1}{2} + \frac{1}{6} = \frac{2}{3},$$

$$S_3 = \sum_{k=1}^{3} \frac{1}{k(k+1)} = \frac{1}{2} + \frac{1}{6} + \frac{1}{12} = \frac{3}{4}, \text{ and}$$

$$S_4 = \sum_{k=1}^{4} \frac{1}{k(k+1)} = \frac{1}{2} + \frac{1}{6} + \frac{1}{12} + \frac{1}{20} = \frac{4}{5}.$$

b. Based on the pattern in the sequence of partial sums, a reasonable conjecture is that

$$S_n = \frac{n}{n+1}, \text{ for } n = 1, 2, 3, \ldots, \text{ which produces the sequence } \left\{ \frac{1}{2}, \frac{2}{3}, \frac{3}{4}, \frac{4}{5}, \frac{5}{6}, \ldots \right\}.$$

It appears that the values of S_n approach 1 (**Figure 10.10**), so a reasonable conjecture is that $\displaystyle\lim_{n \to \infty} S_n = \lim_{n \to \infty} \frac{n}{n+1} = 1$. Therefore, we claim that

$$\sum_{k=1}^{\infty} \frac{1}{k(k+1)} = \lim_{n \to \infty} S_n = 1.$$

Related Exercise 67 ◄

Figure 10.10

QUICK CHECK 5 Find the first four terms of the sequence of partial sums for the series $\displaystyle\sum_{k=1}^{\infty} (-1)^k k$. Does the series converge or diverge? ◄

Summary

This section features three key ideas to keep in mind.

• A *sequence* $\{a_1, a_2, \ldots, a_n, \ldots\}$ is an ordered *list* of numbers.

• An *infinite series* $\displaystyle\sum_{k=1}^{\infty} a_k = a_1 + a_2 + a_3 + \cdots$ is a *sum* of numbers.

• A *sequence of partial sums* $\{S_1, S_2, S_3, \ldots\}$, where $S_n = a_1 + a_2 + \cdots + a_n$, is used to evaluate the series $\displaystyle\sum_{k=1}^{\infty} a_k$.

For sequences, we ask about the behavior of the individual terms as we go out farther and farther in the list; that is, we ask about $\displaystyle\lim_{n \to \infty} a_n$. For infinite series, we examine the sequence of partial sums related to the series. If the sequence of partial sums $\{S_n\}$ has a limit, then the infinite series $\displaystyle\sum_{k=1}^{\infty} a_k$ converges to that limit. If the sequence of partial sums does not have a limit, the infinite series diverges.

Table 10.2 shows the correspondences between sequences/series and functions, and between summation and integration. For a sequence, the index n plays the role of the independent variable and takes on integer values; the terms of the sequence $\{a_n\}$ correspond to the dependent variable.

With sequences $\{a_n\}$, the idea of accumulation corresponds to summation, whereas with functions, accumulation corresponds to integration. A finite sum is analogous to integrating a function over a finite interval. An infinite series is analogous to integrating a function over an infinite interval.

Table 10.2

	Sequences/Series	Functions
Independent variable	n	x
Dependent variable	a_n	$f(x)$
Domain	Integers e.g., $n = 1, 2, 3, \ldots$	Real numbers e.g., $\{x : x \geq 1\}$
Accumulation	Sums	Integrals
Accumulation over a finite interval	$\displaystyle\sum_{k=1}^{n} a_k$	$\displaystyle\int_{1}^{n} f(x)\, dx$
Accumulation over an infinite interval	$\displaystyle\sum_{k=1}^{\infty} a_k$	$\displaystyle\int_{1}^{\infty} f(x)\, dx$

SECTION 10.1 EXERCISES

Getting Started

1. Define *sequence* and give an example.

2. Suppose the sequence $\{a_n\}$ is defined by the explicit formula $a_n = 1/n$, for $n = 1, 2, 3, \ldots$. Write out the first five terms of the sequence.

3. Suppose the sequence $\{a_n\}$ is defined by the recurrence relation $a_{n+1} = na_n$, for $n = 1, 2, 3, \ldots$, where $a_1 = 1$. Write out the first five terms of the sequence.

4. Find two different explicit formulas for the sequence $\{-1, 1, -1, 1, -1, 1, \ldots\}$.

5. Find two different explicit formulas for the sequence $\{1, -2, 3, -4, -5, \ldots\}$.

6. The first ten terms of the sequence $\{2 \tan^{-1} 10^n\}_{n=1}^{\infty}$ are rounded to 8 digits right of the decimal point (see table). Make a conjecture about the limit of the sequence.

n	a_n
1	2.94225535
2	3.12159332
3	3.13959265
4	3.14139265
5	3.14157265
6	3.14159065
7	3.14159245
8	3.14159263
9	3.14159265
10	3.14159265

7. The first ten terms of the sequence $\left\{\left(1 + \dfrac{1}{10^n}\right)^{10^n}\right\}_{n=1}^{\infty}$ are rounded to 8 digits right of the decimal point (see table). Make a conjecture about the limit of the sequence.

n	a_n
1	2.59374246
2	2.70481383
3	2.71692393
4	2.71814593
5	2.71826824
6	2.71828047
7	2.71828169
8	2.71828179
9	2.71828204
10	2.71828203

8. Does the sequence $\{1, 2, 1, 2, 1, 2, \ldots\}$ converge? Explain.

9. The terms of a sequence of partial sums are defined by
$$S_n = \sum_{k=1}^{n} k^2, \text{ for } n = 1, 2, 3, \ldots. \text{ Evaluate the first four terms of}$$
the sequence.

10. Given the series $\displaystyle\sum_{k=1}^{\infty} k$, evaluate the first four terms of its sequence of partial sums $S_n = \displaystyle\sum_{k=1}^{n} k$.

11. Use sigma notation to write an infinite series whose first four successive partial sums are 10, 20, 30, and 40.

12. Consider the infinite series $\displaystyle\sum_{k=1}^{\infty} \dfrac{1}{k}$. Evaluate the first four terms of the sequence of partial sums.

Practice Exercises

13–20. Explicit formulas *Write the first four terms of the sequence* $\{a_n\}_{n=1}^{\infty}$.

13. $a_n = \dfrac{1}{10^n}$

14. $a_n = 3n + 1$

15. $a_n = \dfrac{(-1)^n}{2^n}$

16. $a_n = 2 + (-1)^n$

17. $a_n = \dfrac{2^{n+1}}{2^n + 1}$

18. $a_n = n + \dfrac{1}{n}$

19. $a_n = 1 + \sin \dfrac{\pi n}{2}$

20. $a_n = n!$ (*Hint:* Recall that $n! = n(n-1)(n-2)\cdots 2\cdot 1$.)

21–26. Recurrence relations *Write the first four terms of the sequence* $\{a_n\}$ *defined by the following recurrence relations.*

21. $a_{n+1} = 2a_n;\ a_1 = 2$

22. $a_{n+1} = \dfrac{a_n}{2};\ a_1 = 32$

23. $a_{n+1} = 3a_n - 12;\ a_1 = 10$

24. $a_{n+1} = a_n^2 - 1;\ a_1 = 1$

25. $a_{n+1} = \dfrac{1}{1 + a_n};\ a_0 = 1$

26. $a_{n+1} = a_n + a_{n-1};\ a_1 = 1, a_0 = 1$

27–34. Working with sequences *Several terms of a sequence* $\{a_n\}_{n=1}^{\infty}$ *are given.*

a. *Find the next two terms of the sequence.*
b. *Find a recurrence relation that generates the sequence (supply the initial value of the index and the first term of the sequence).*
c. *Find an explicit formula for the nth term of the sequence.*

27. $\left\{ 1, \dfrac{1}{2}, \dfrac{1}{4}, \dfrac{1}{8}, \dfrac{1}{16}, \ldots \right\}$

28. $\{2, 5, 8, 11, \ldots\}$

29. $\{1, 2, 4, 8, 16, \ldots\}$

30. $\{64, 32, 16, 8, 4, \ldots\}$

31. $\{1, 3, 9, 27, 81, \ldots\}$

32. $\{1, 4, 9, 16, 25, \ldots\}$

33. $\{-5, 5, -5, 5, \ldots\}$

34. $\{1, 0, 1, 0, 1, 0, 1, \ldots\}$

35–44. Limits of sequences *Write the terms* a_1, a_2, a_3, *and* a_4 *of the following sequences. If the sequence appears to converge, make a conjecture about its limit. If the sequence diverges, explain why.*

35. $a_n = 10^n - 1;\ n = 1, 2, 3, \ldots$

36. $a_{n+1} = \dfrac{10}{a_n};\ a_1 = 1$

37. $a_n = \dfrac{1}{10^n};\ n = 1, 2, 3, \ldots$

38. $a_{n+1} = \dfrac{a_n}{10};\ a_0 = 1$

39. $a_n = 3 + \cos \pi n;\ n = 1, 2, 3, \ldots$

40. $a_n = 1 - 10^{-n};\ n = 1, 2, 3, \ldots$

41. $a_{n+1} = 1 + \dfrac{a_n}{2};\ a_0 = 2$

42. $a_{n+1} = 1 - \dfrac{a_n}{2};\ a_0 = \dfrac{2}{3}$

T 43. $a_{n+1} = \dfrac{a_n}{11} + 50;\ a_0 = 50$

44. $a_{n+1} = 10a_n - 1;\ a_0 = 0$

T 45–48. Explicit formulas for sequences *Consider the formulas for the following sequences* $\{a_n\}_{n=1}^{\infty}$. *Make a table with at least ten terms and determine a plausible limit of the sequence or state that the sequence diverges.*

45. $a_n = \dfrac{5^n}{5^n + 1}$

46. $a_n = 2^n \sin(2^{-n})$

47. $a_n = n^2 + n$

48. $a_n = \dfrac{100n - 1}{10n}$

49–50. Limits from graphs *Consider the following sequences.*

a. *Find the first four terms of the sequence.*
b. *Based on part (a) and the figure, determine a plausible limit of the sequence.*

49. $a_n = 2 + 2^{-n};\ n = 1, 2, 3, \ldots$

50. $a_n = \dfrac{n^2}{n^2 - 1};\ n = 2, 3, 4, \ldots$

T 51–56. Recurrence relations *Consider the following recurrence relations. Make a table with at least ten terms and determine a plausible limit of the sequence or state that the sequence diverges.*

51. $a_{n+1} = \dfrac{1}{2}a_n + 2;\ a_1 = 3$

52. $a_n = \dfrac{1}{4}a_{n-1} - 3;\ a_0 = 1$

53. $a_{n+1} = 4a_n + 1;\ a_0 = 1$

54. $a_{n+1} = \dfrac{a_n}{10} + 3;\ a_0 = 10$

55. $a_{n+1} = \dfrac{1}{2}\sqrt{a_n} + 3;\ a_1 = 8$

56. $a_{n+1} = \sqrt{8a_n + 9};\ a_1 = 10$

57–60. Heights of bouncing balls *A ball is thrown upward to a height of h_0 meters. After each bounce, the ball rebounds to a fraction r of its previous height. Let h_n be the height after the nth bounce. Consider the following values of h_0 and r.*

a. *Find the first four terms of the sequence of heights $\{h_n\}$.*
b. *Find an explicit formula for the nth term of the sequence $\{h_n\}$.*

57. $h_0 = 20$, $r = 0.5$ **58.** $h_0 = 10$, $r = 0.9$

59. $h_0 = 30$, $r = 0.25$ **60.** $h_0 = 20$, $r = 0.75$

61–66. Sequences of partial sums *For the following infinite series, find the first four terms of the sequence of partial sums. Then make a conjecture about the value of the infinite series or state that the series diverges.*

61. $0.3 + 0.03 + 0.003 + \cdots$

62. $0.6 + 0.06 + 0.006 + \cdots$

63. $4 + 0.9 + 0.09 + 0.009 + \cdots$

64. $\displaystyle\sum_{k=1}^{\infty} 10^k$ **65.** $\displaystyle\sum_{k=1}^{\infty} \frac{6}{10^k}$

66. $\displaystyle\sum_{k=1}^{\infty} \cos \pi k$

67–70. Formulas for sequences of partial sums *Consider the following infinite series.*

a. *Find the first four partial sums S_1, S_2, S_3, and S_4 of the series.*
b. *Find a formula for the nth partial sum S_n of the infinite series. Use this formula to find the next four partial sums S_5, S_6, S_7, and S_8 of the infinite series.*
c. *Make a conjecture for the value of the series.*

67. $\displaystyle\sum_{k=1}^{\infty} \frac{2}{(2k-1)(2k+1)}$ **68.** $\displaystyle\sum_{k=1}^{\infty} \frac{1}{2^k}$

69. $\displaystyle\sum_{k=1}^{\infty} 90(0.1)^k$ **T 70.** $\displaystyle\sum_{k=1}^{\infty} \frac{2}{3^{k-1}}$

71. Explain why or why not Determine whether the following statements are true and give an explanation or counterexample.

  a. The sequence of partial sums for the series $1 + 2 + 3 + \cdots$ is $\{1, 3, 6, 10, \ldots\}$.
  b. If a sequence of positive numbers converges, then the sequence is decreasing.
  c. If the terms of the sequence $\{a_n\}$ are positive and increasing, then the sequence of partial sums for the series $\displaystyle\sum_{k=1}^{\infty} a_k$ diverges.

T 72–75. Practical sequences *Consider the following situations that generate a sequence.*

a. *Write out the first five terms of the sequence.*
b. *Find an explicit formula for the terms of the sequence.*
c. *Find a recurrence relation that generates the sequence.*
d. *Using a calculator or a graphing utility, estimate the limit of the sequence or state that it does not exist.*

72. Population growth When a biologist begins a study, a colony of prairie dogs has a population of 250. Regular measurements reveal that each month the prairie dog population increases by 3%. Let p_n be the population (rounded to whole numbers) at the end of the nth month, where the initial population is $p_0 = 250$.

73. Radioactive decay A material transmutes 50% of its mass to another element every 10 years due to radioactive decay. Let M_n be the mass of the radioactive material at the end of the nth decade, where the initial mass of the material is $M_0 = 20$ g.

74. Consumer Price Index The Consumer Price Index (the CPI is a measure of the U.S. cost of living) is given a base value of 100 in the year 1984. Assume the CPI has increased by an average of 3% per year since 1984. Let c_n be the CPI n years after 1984, where $c_0 = 100$.

75. Drug elimination Jack took a 200-mg dose of a pain killer at midnight. Every hour, 5% of the drug is washed out of his bloodstream. Let d_n be the amount of drug in Jack's blood n hours after the drug was taken, where $d_0 = 200$ mg.

Explorations and Challenges

T 76–77. Distance traveled by bouncing balls *A ball is thrown upward to a height of h_0 meters. After each bounce, the ball rebounds to a fraction r of its previous height. Let h_n be the height after the nth bounce and let S_n be the total distance the ball has traveled at the moment of the nth bounce.*

a. *Find the first four terms of the sequence $\{S_n\}$.*
b. *Make a table of 20 terms of the sequence $\{S_n\}$ and determine a plausible value for the limit of $\{S_n\}$.*

76. $h_0 = 20$, $r = 0.5$ **77.** $h_0 = 20$, $r = 0.75$

T 78. A square root finder A well-known method for approximating $\sqrt{c}$ for positive real numbers c consists of the following recurrence relation (based on Newton's method; see Section 4.8). Let $a_0 = c$ and

$$a_{n+1} = \frac{1}{2}\left(a_n + \frac{c}{a_n}\right), \quad \text{for } n = 0, 1, 2, 3, \ldots.$$

  a. Use this recurrence relation to approximate $\sqrt{10}$. How many terms of the sequence are needed to approximate $\sqrt{10}$ with an error less than 0.01? How many terms of the sequence are needed to approximate $\sqrt{10}$ with an error less than 0.0001? (To compute the error, assume a calculator gives the exact value.)
  b. Use this recurrence relation to approximate $\sqrt{c}$, for $c = 2, 3, \ldots, 10$. Make a table showing the number of terms of the sequence needed to approximate $\sqrt{c}$ with an error less than 0.01.

T 79–80. Fixed-point iteration *A method for estimating a solution to the equation $x = f(x)$, known as **fixed-point iteration**, is based on the following recurrence relation. Let $x_0 = c$ and $x_{n+1} = f(x_n)$, for $n = 1, 2, 3, \ldots$ and a real number c. If the sequence $\{x_n\}_{n=0}^{\infty}$ converges to L, then L is a solution to the equation $x = f(x)$ and L is called a fixed point of f. To estimate L with p digits of accuracy to the right of the decimal point, we can compute the terms of the sequence $\{x_n\}_{n=0}^{\infty}$ until two successive values agree to p digits of accuracy. Use fixed-point iteration to find a solution to the following equations with $p = 3$ digits of accuracy using the given value of x_0.*

79. $x = \cos x$; $x_0 = 0.8$ **80.** $x = \dfrac{\sqrt{x^3 + 1}}{20}$; $x_0 = 5$

QUICK CHECK ANSWERS

1. $a_{10} = 28$ **2.** $a_n = 2^n - 1$, $n = 1, 2, 3, \ldots$
3. $0.33333\ldots = \frac{1}{3}$ **4.** Both diverge. **5.** $S_1 = -1$, $S_2 = 1$, $S_3 = -2$, $S_4 = 2$; the series diverges. ◄

10.2 Sequences

The previous section sets the stage for an in-depth investigation of sequences and infinite series. This section is devoted to sequences, and the remainder of the chapter deals with series.

Limit of a Sequence and Limit Laws

A fundamental question about sequences concerns the behavior of the terms as we go out farther and farther in the sequence. For example, in the sequence

$$\{a_n\}_{n=0}^{\infty} = \left\{\frac{1}{n^2+1}\right\}_{n=0}^{\infty} = \left\{1, \frac{1}{2}, \frac{1}{5}, \frac{1}{10}, \ldots\right\},$$

the terms remain positive and decrease to 0. We say that this sequence converges and its limit is 0, written $\lim\limits_{n\to\infty} a_n = 0$. Similarly, the terms of the sequence

$$\{b_n\}_{n=1}^{\infty} = \left\{(-1)^n \frac{n(n+1)}{2}\right\}_{n=1}^{\infty} = \{-1, 3, -6, 10, \ldots\}$$

increase in magnitude and do not approach a unique value as n increases. In this case, we say that the sequence diverges.

Limits of sequences are really no different from limits at infinity of functions except that the variable n assumes only integer values as $n \to \infty$. This idea works as follows.

Given a sequence $\{a_n\}$, we define a function f such that $f(n) = a_n$ for all indices n. For example, if $a_n = n/(n+1)$, then we let $f(x) = x/(x+1)$. By the methods of Section 2.5, we know that $\lim\limits_{x\to\infty} f(x) = 1$; because the terms of the sequence lie on the graph of f, it follows that $\lim\limits_{n\to\infty} a_n = 1$ (**Figure 10.11**). This reasoning is the basis of the following theorem.

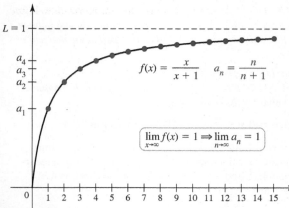

$$f(x) = \frac{x}{x+1} \qquad a_n = \frac{n}{n+1}$$

$$\boxed{\lim_{x\to\infty} f(x) = 1 \Rightarrow \lim_{n\to\infty} a_n = 1}$$

Figure 10.11

> ➤ The converse of Theorem 10.1 is not true. For example, if $a_n = \cos 2\pi n$, then $\lim\limits_{n\to\infty} a_n = 1$, but $\lim\limits_{x\to\infty} \cos 2\pi x$ does not exist.

THEOREM 10.1   Limits of Sequences from Limits of Functions

Suppose f is a function such that $f(n) = a_n$, for positive integers n. If $\lim\limits_{x\to\infty} f(x) = L$, then the limit of the sequence $\{a_n\}$ is also L, where L may be $\pm\infty$.

Because of the correspondence between limits of sequences and limits of functions at infinity, we have the following properties that are analogous to those for functions given in Theorem 2.3.

> ➤ The limit of a sequence $\{a_n\}$ is determined by the terms in the *tail* of the sequence—the terms with large values of n. If the sequences $\{a_n\}$ and $\{b_n\}$ differ in their first 100 terms but have identical terms for $n > 100$, then they have the same limit. For this reason, the initial index of a sequence (for example, $n = 0$ or $n = 1$) is often not specified.

THEOREM 10.2   Limit Laws for Sequences

Assume the sequences $\{a_n\}$ and $\{b_n\}$ have limits A and B, respectively. Then

1. $\lim\limits_{n\to\infty} (a_n \pm b_n) = A \pm B$

2. $\lim\limits_{n\to\infty} ca_n = cA$, where c is a real number

3. $\lim\limits_{n\to\infty} a_n b_n = AB$

4. $\lim\limits_{n\to\infty} \dfrac{a_n}{b_n} = \dfrac{A}{B}$, provided $B \neq 0$.

EXAMPLE 1   Limits of sequences Determine the limits of the following sequences.

a. $a_n = \dfrac{3n^3}{n^3+1}$ **b.** $b_n = \left(\dfrac{n+5}{n}\right)^n$ **c.** $c_n = n^{1/n}$

SOLUTION

a. A function with the property that $f(n) = a_n$ is $f(x) = \dfrac{3x^3}{x^3 + 1}$. Dividing numerator and denominator by x^3 (or appealing to Theorem 2.7), we find that $\lim\limits_{x \to \infty} f(x) = 3$. (Alternatively, we can apply l'Hôpital's Rule and obtain the same result.) We conclude that $\lim\limits_{n \to \infty} a_n = 3$.

b. The limit

$$\lim_{n \to \infty} b_n = \lim_{n \to \infty} \left(\frac{n + 5}{n} \right)^n = \lim_{n \to \infty} \left(1 + \frac{5}{n} \right)^n$$

has the indeterminate form 1^∞. Recall that for this limit (Section 4.7), we first evaluate

$$L = \lim_{n \to \infty} \ln \left(1 + \frac{5}{n} \right)^n = \lim_{n \to \infty} n \ln \left(1 + \frac{5}{n} \right),$$

> It is customary to treat n as a continuous variable and differentiate with respect to n, rather than write the sequence as a function of x, as was done in Example 1a.

and then, if L exists, $\lim\limits_{n \to \infty} b_n = e^L$. Using l'Hôpital's Rule for the indeterminate form $0/0$, we have

$$L = \lim_{n \to \infty} n \ln \left(1 + \frac{5}{n} \right) = \lim_{n \to \infty} \frac{\ln \left(1 + (5/n) \right)}{1/n} \qquad \text{Indeterminate form } 0/0$$

> For a review of l'Hôpital's Rule, see Section 4.7, where we showed that
> $$\lim_{x \to \infty} \left(1 + \frac{a}{x} \right)^x = e^a.$$

$$= \lim_{n \to \infty} \frac{\dfrac{1}{1 + (5/n)} \left(-\dfrac{5}{n^2} \right)}{-1/n^2} \qquad \text{L'Hôpital's Rule}$$

$$= \lim_{n \to \infty} \frac{5}{1 + (5/n)} = 5. \qquad \text{Simplify; } 5/n \to 0 \text{ as } n \to \infty.$$

Because $\lim\limits_{n \to \infty} b_n = e^L = e^5$, we have $\lim\limits_{n \to \infty} \left(\dfrac{5 + n}{n} \right)^n = e^5$.

c. The limit has the indeterminate form ∞^0, so we first evaluate $L = \lim\limits_{n \to \infty} \ln n^{1/n} = \lim\limits_{n \to \infty} \dfrac{\ln n}{n}$; if L exists, then $\lim\limits_{n \to \infty} c_n = e^L$. Using either l'Hôpital's Rule or the relative growth rates in Section 4.7, we find that $L = 0$. Therefore, $\lim\limits_{n \to \infty} c_n = e^0 = 1$.

Related Exercises 14, 38, 39 ◄

Terminology for Sequences

We now introduce some terminology for sequences that is similar to that used for functions. The following terms are used to describe sequences $\{a_n\}$.

> Because an increasing sequence is, by definition, nondecreasing, it is also monotonic. Similarly, a decreasing sequence is monotonic.

DEFINITIONS Terminology for Sequences

- $\{a_n\}$ is **increasing** if $a_{n+1} > a_n$; for example, $\{0, 1, 2, 3, \dots\}$.
- $\{a_n\}$ is **nondecreasing** if $a_{n+1} \geq a_n$; for example, $\{1, 1, 2, 2, 3, 3, \dots\}$.
- $\{a_n\}$ is **decreasing** if $a_{n+1} < a_n$; for example, $\{2, 1, 0, -1, \dots\}$.
- $\{a_n\}$ is **nonincreasing** if $a_{n+1} \leq a_n$; for example, $\{0, -1, -1, -2, -2, \dots\}$.
- $\{a_n\}$ is **monotonic** if it is either nonincreasing or nondecreasing (it moves in one direction).
- $\{a_n\}$ is **bounded above** if there is a number M such that $a_n \leq M$, for all relevant values of n, and $\{a_n\}$ is **bounded below** if there is a number N such that $a_n \geq N$, for all relevant values of n.
- If $\{a_n\}$ is bounded above and bounded below, then we say that $\{a_n\}$ is a **bounded** sequence.

For example, the sequence

$$\{a_n\} = \left\{1 - \frac{1}{n}\right\}_{n=1}^{\infty} = \left\{0, \frac{1}{2}, \frac{2}{3}, \frac{3}{4}, \dots\right\}$$

satisfies $0 \leq a_n \leq 1$, for $n \geq 1$, and its terms are increasing in size. Therefore, the sequence is bounded below, bounded above, and increasing; it is also monotonic (**Figure 10.12**). The sequence

$$\{b_n\} = \left\{1 + \frac{1}{n}\right\}_{n=1}^{\infty} = \left\{2, \frac{3}{2}, \frac{4}{3}, \frac{5}{4}, \dots\right\}$$

satisfies $1 \leq b_n \leq 2$, for $n \geq 1$, and its terms are decreasing in size. Therefore, the sequence is bounded below, bounded above, and decreasing; it is also monotonic (Figure 10.12). Because $\{a_n\}$ and $\{b_n\}$ are bounded below and above, both are bounded sequences.

QUICK CHECK 1 Classify the following sequences as bounded, monotonic, or neither.

a. $\left\{\frac{1}{2}, \frac{3}{4}, \frac{7}{8}, \frac{15}{16}, \dots\right\}$

b. $\left\{1, -\frac{1}{2}, \frac{1}{4}, -\frac{1}{8}, \frac{1}{16}, \dots\right\}$

c. $\{1, -2, 3, -4, 5, \dots\}$

d. $\{1, 1, 1, 1, \dots\}$ ◄

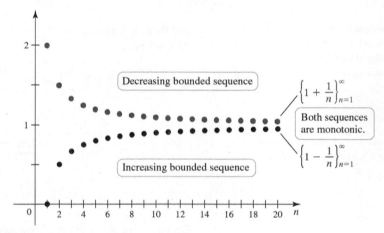

Figure 10.12

EXAMPLE 2   Limits of sequences and graphing Compare and contrast the behavior of $\{a_n\}$ and $\{b_n\}$ as $n \to \infty$.

a. $a_n = \dfrac{n^{3/2}}{n^{3/2} + 1}$ b. $b_n = \dfrac{(-1)^n \, n^{3/2}}{n^{3/2} + 1}$

SOLUTION

a. The terms of $\{a_n\}$ are positive, increasing, and bounded (**Figure 10.13**). Dividing the numerator and denominator of a_n by $n^{3/2}$, we see that

$$\lim_{n \to \infty} a_n = \lim_{n \to \infty} \frac{n^{3/2}}{n^{3/2} + 1} = \lim_{n \to \infty} \frac{1}{1 + \underbrace{\frac{1}{n^{3/2}}}_{\text{approaches } 0 \text{ as } n \to \infty}} = 1.$$

Figure 10.13

b. The terms of the bounded sequence $\{b_n\}$ alternate in sign. Using the result of part (a), it follows that the even terms form an increasing sequence that approaches 1, and the odd terms form a decreasing sequence that approaches -1 (**Figure 10.14**). Therefore, the sequence diverges, illustrating the fact that the presence of $(-1)^n$ may significantly alter the behavior of a sequence.

Related Exercises 55, 60 ◄

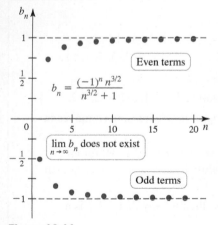

Figure 10.14

Geometric Sequences

Geometric sequences have the property that each term is obtained by multiplying the previous term by a fixed constant, called the **ratio**. They have the form $\{r^n\}$ or $\{ar^n\}$, where the ratio r and $a \neq 0$ are real numbers. The value of r determines the behavior of the sequence.

When a number less than 1 in magnitude is raised to increasing powers, the resulting numbers decrease to zero. For example, the geometric sequence $\{0.75^n\}$, where $r = 0.75$, converges to zero and is monotonic (**Figure 10.15**). If we choose $r = -0.75$, we have $\{(-0.75)^n\} = \{(-1)^n 0.75^n\}$. Observe that the factor $(-1)^n$ oscillates between 1 and -1, while 0.75^n decreases to zero as n increases. Therefore, the sequence oscillates and converges to zero (**Figure 10.16**).

Figure 10.15

Figure 10.16

When a number greater than 1 in magnitude is raised to increasing powers, the resulting numbers increase in magnitude. For example, letting $r = 1.15$ results in the sequence $\{1.15^n\}$, whose terms are positive and increase without bound. In this case, the sequence diverges and is monotonic (**Figure 10.17**). Choosing $r = -1.15$, we have $\{(-1.15)^n\} = \{(-1)^n 1.15^n\}$. As before, the factor $(-1)^n$ oscillates between 1 and -1, while 1.15^n increases without bound as n increases. The terms of the sequence increase in magnitude without bound and alternate in sign. In this case, the sequence oscillates and diverges (**Figure 10.18**).

Figure 10.17

Figure 10.18

QUICK CHECK 2 Describe the behavior of $\{r^n\}$ in the cases $r = -1$ and $r = 1$. ◀

The preceding discussion and Quick Check 2 are summarized in the following theorem.

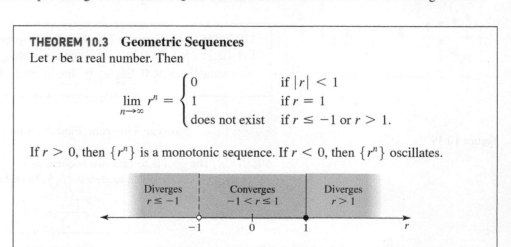

THEOREM 10.3   Geometric Sequences
Let r be a real number. Then

$$\lim_{n \to \infty} r^n = \begin{cases} 0 & \text{if } |r| < 1 \\ 1 & \text{if } r = 1 \\ \text{does not exist} & \text{if } r \le -1 \text{ or } r > 1. \end{cases}$$

If $r > 0$, then $\{r^n\}$ is a monotonic sequence. If $r < 0$, then $\{r^n\}$ oscillates.

EXAMPLE 3   Using Limit Laws Determine the limits of the following sequences.

a. $a_n = 5(0.6)^n - \dfrac{1}{3^n}$ **b.** $b_n = \dfrac{2n^2 + n}{2^n(3n^2 - 4)}$

SOLUTION

a. Using Laws 1 and 2 from Theorem 10.2, we have

$$\lim_{n\to\infty} a_n = \lim_{n\to\infty}\left(5(0.6)^n - \frac{1}{3^n}\right) = 5\lim_{n\to\infty}(0.6)^n - \lim_{n\to\infty}\left(\frac{1}{3}\right)^n.$$

Both $\{(0.6)^n\}$ and $\{(1/3)^n\}$ are geometric sequences with $|r| < 1$, which implies that

$$\lim_{n\to\infty} a_n = 5\underbrace{\lim_{n\to\infty}(0.6)^n}_{0} - \underbrace{\lim_{n\to\infty}\left(\frac{1}{3}\right)^n}_{0} = 5\cdot 0 - 0 = 0. \quad \text{Theorem 10.3}$$

b. We write the limit of the sequence as a product of limits:

$$\lim_{n\to\infty} b_n = \lim_{n\to\infty}\frac{2n^2 + n}{2^n(3n^2 - 4)} = \lim_{n\to\infty}\left(\frac{1}{2}\right)^n \cdot \lim_{n\to\infty}\frac{2n^2 + n}{3n^2 - 4}. \quad \text{Law 3, Theorem 10.2}$$

The first of the new limits involves a geometric sequence with $r = 1/2$, and the second limit yields to Theorem 2.7:

$$\lim_{n\to\infty} b_n = \underbrace{\lim_{n\to\infty}\left(\frac{1}{2}\right)^n}_{0} \cdot \underbrace{\lim_{n\to\infty}\frac{2n^2 + n}{3n^2 - 4}}_{2/3} = 0 \cdot \frac{2}{3} = 0. \quad \text{Evaluate limits.}$$

Related Exercises 26–27 ◀

The previous examples show that a sequence may display any of the following behaviors:

• It may converge to a single value, which is the limit of the sequence.

• Its terms may increase in magnitude without bound (either with one sign or with mixed signs), in which case the sequence diverges.

• Its terms may remain bounded but settle into an oscillating pattern in which the terms approach two or more values; in this case, the sequence diverges.

Not illustrated in the preceding examples is one other type of behavior: The terms of a sequence may remain bounded, but wander chaotically forever without a pattern. In this case, the sequence also diverges (see the Guided Project *Chaos!*)

The Squeeze Theorem

We cite two theorems that are used to evaluate limits and to establish that limits exist. The first theorem is a direct analog of the Squeeze Theorem in Section 2.3.

Squeeze Theorem

$a_n \le b_n \le c_n$ for all n

Figure 10.19

> **THEOREM 10.4   Squeeze Theorem for Sequences**
> Let $\{a_n\}$, $\{b_n\}$, and $\{c_n\}$ be sequences with $a_n \le b_n \le c_n$, for all integers n greater than some index N. If $\lim_{n\to\infty} a_n = \lim_{n\to\infty} c_n = L$, then $\lim_{n\to\infty} b_n = L$ (**Figure 10.19**).

EXAMPLE 4   Squeeze Theorem Find the limit of the sequence $b_n = \dfrac{\cos n}{n^2 + 1}$.

SOLUTION The goal is to find two sequences $\{a_n\}$ and $\{c_n\}$ whose terms lie below and above the terms of the given sequence $\{b_n\}$. Note that $-1 \le \cos n \le 1$, for all n. Therefore,

$$\underbrace{-\frac{1}{n^2 + 1}}_{a_n} \le \underbrace{\frac{\cos n}{n^2 + 1}}_{b_n} \le \underbrace{\frac{1}{n^2 + 1}}_{c_n}.$$

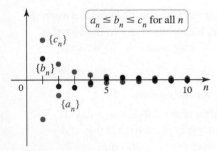

Figure 10.20

Letting $a_n = -\dfrac{1}{n^2 + 1}$ and $c_n = \dfrac{1}{n^2 + 1}$, we have $a_n \le b_n \le c_n$, for $n \ge 1$. Furthermore, $\lim\limits_{n \to \infty} a_n = \lim\limits_{n \to \infty} c_n = 0$. By the Squeeze Theorem, $\lim\limits_{n \to \infty} b_n = 0$ (**Figure 10.20**).

Related Exercises 65, 67 ◀

Bounded Monotonic Sequence Theorem

Suppose you pour a cup of hot coffee and put it on your desk to cool. Assume every minute you measure the temperature of the coffee to create a sequence of temperature readings $\{T_1, T_2, T_3, \ldots\}$. This sequence has two notable properties: First, the terms of the sequence are decreasing (because the coffee is cooling); and second, the sequence is bounded below (because the temperature of the coffee cannot be less than the temperature of the surrounding room). In fact, if the measurements continue indefinitely, the sequence of temperatures converges to the temperature of the room. This example illustrates an important theorem that characterizes convergent sequences in terms of boundedness and monotonicity. The theorem is easy to believe, but its proof is beyond the scope of this text.

THEOREM 10.5 Bounded Monotonic Sequence
A bounded monotonic sequence converges.

> **Some optional terminology:** M is called an *upper bound* of the sequence in Figure 10.21a, and N is a *lower bound* of the sequence in Figure 10.21b. The number M^* is the *least upper bound* of a sequence (or a set) if it is the smallest of all the upper bounds. It is a fundamental property of the real numbers that if a sequence (or a nonempty set) is bounded above, then it has a least upper bound. It can be shown that an increasing sequence that is bounded above converges to its least upper bound. Similarly, a decreasing sequence that is bounded below converges to its greatest lower bound.

Figure 10.21 shows the two cases of this theorem. In the first case, we see a nondecreasing sequence, all of whose terms are less than M. It must converge to a limit less than or equal to M. Similarly, a nonincreasing sequence, all of whose terms are greater than N, must converge to a limit greater than or equal to N.

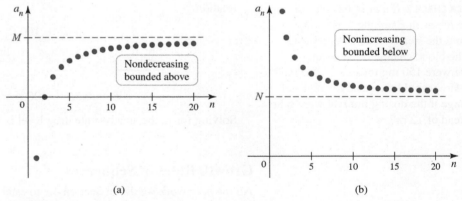

(a)   (b)

Figure 10.21

An Application: Recurrence Relations

> Most drugs decay exponentially in the bloodstream and have a characteristic half-life assuming the drug absorbs quickly into the blood.

EXAMPLE 5 Sequences for drug doses Suppose your doctor prescribes a 100-mg dose of an antibiotic to be taken every 12 hours. Furthermore, the drug is known to have a half-life of 12 hours; that is, every 12 hours half of the drug in your blood is eliminated.

a. Find the sequence that gives the amount of drug in your blood immediately after each dose.

b. Use a graph to propose the limit of this sequence; that is, in the long run, how much drug do you have in your blood?

c. Find the limit of the sequence directly.

SOLUTION

a. Let d_n be the amount of drug in the blood immediately following the nth dose, where $n = 1, 2, 3, \ldots$ and $d_1 = 100$ mg. We want to write a recurrence relation that gives the amount of drug in the blood after the $(n + 1)$st dose (d_{n+1}) in terms of the

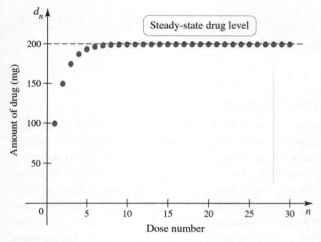

Figure 10.22

> *Mathematical induction* is used to prove statements involving natural numbers. In the first step, we show that the statement holds for $n = 1$. In the second step, we show that if the statement is true for $n = k$, then the statement is also true for $n = k + 1$, where k is any natural number greater than 1. If the proof of the second step is successful, it implies that the statement holds for any natural number.

QUICK CHECK 3 If a drug has the same half-life as in Example 5, (i) how would the steady-state level of drug in the blood change if the regular dose were 150 mg instead of 100 mg? (ii) How would the steady-state level change if the dosing interval were 6 hr instead of 12 hr? ◄

amount of drug after the nth dose (d_n). In the 12 hours between the nth dose and the $(n + 1)$st dose, half of the drug in the blood is eliminated *and* another 100 mg of drug is added. So we have

$$d_{n+1} = 0.5\, d_n + 100, \qquad \text{for } n = 1, 2, 3, \ldots, \text{ with } d_1 = 100,$$

which is the recurrence relation for the sequence $\{d_n\}$.

b. We see from **Figure 10.22** that after about 10 doses (5 days) the amount of antibiotic in the blood is close to 200 mg, and—importantly for your body—it never exceeds 200 mg.

c. Figure 10.22 gives evidence that the terms of the sequence are increasing and bounded above by 200. To prove these facts, note that $d_1 = 100 < 200$. Now suppose $d_k < 200$, for any $k > 1$. The recurrence relation in part (a) implies that $d_{k+1} = 0.5d_k + 100$. Because $d_k < 200$, we have

$$d_{k+1} = 0.5\, \underbrace{d_k}_{<200} + 100 < 0.5(200) + 100 = 200.$$

We conclude by mathematical induction that $d_k < 200$, for all k. To prove that $\{d_k\}$ is increasing (and therefore monotonic), we need to show that $d_{k+1} - d_k > 0$. Note that

$$d_{k+1} - d_k = \underbrace{0.5d_k + 100}_{\text{recurrence relation}} - d_k = 100 - 0.5d_k > 0;$$

the final inequality follows from the fact that $d_k < 200$. Because $\{d_k\}$ is increasing, it is bounded below by its first term, which is $d_1 = 100$, and therefore $\{d_k\}$ is bounded. By the Bounded Monotonic Sequence Theorem, the sequence has a limit; therefore, $\lim\limits_{n\to\infty} d_n = L$ and $\lim\limits_{n\to\infty} d_{n+1} = L$. We now take the limit of both sides of the recurrence relation:

$$d_{n+1} = 0.5\, d_n + 100 \qquad \text{Recurrence relation}$$

$$\underbrace{\lim_{n\to\infty} d_{n+1}}_{L} = 0.5 \underbrace{\lim_{n\to\infty} d_n}_{L} + \lim_{n\to\infty} 100 \qquad \text{Limits of both sides}$$

$$L = 0.5L + 100. \qquad \text{Substitute } L.$$

Solving for L, the steady-state drug level is $L = 200$.

Related Exercises 71, 89 ◄

Growth Rates of Sequences

All the hard work we did in Section 4.7 to establish the relative growth rates of functions is now applied to sequences. Here is the question: Given two nondecreasing sequences of positive terms $\{a_n\}$ and $\{b_n\}$, which sequence grows faster as $n \to \infty$? As with functions, to compare growth rates, we evaluate $\lim\limits_{n\to\infty} a_n/b_n$. If $\lim\limits_{n\to\infty} a_n/b_n = 0$, then $\{b_n\}$ grows faster than $\{a_n\}$. If $\lim\limits_{n\to\infty} a_n/b_n = \infty$, then $\{a_n\}$ grows faster than $\{b_n\}$.

Using the results of Section 4.7, we immediately arrive at the following ranking of growth rates of sequences as $n \to \infty$, with positive real numbers p, q, r, s, and $b > 1$:

$$\{\ln^q n\} \ll \{n^p\} \ll \{n^p \ln^r n\} \ll \{n^{p+s}\} \ll \{b^n\} \ll \{n^n\}.$$

As before, the notation $\{a_n\} \ll \{b_n\}$ means $\{b_n\}$ *grows faster than* $\{a_n\}$ as $n \to \infty$. Another important sequence that should be added to the list is the **factorial sequence** $\{n!\}$, where $n! = n(n-1)(n-2)\cdots 2 \cdot 1$. Where does the factorial sequence $\{n!\}$ appear in the list? The following argument provides some intuition. Notice that

$$n^n = \underbrace{n \cdot n \cdot n \cdots n}_{n \text{ factors}}, \qquad \text{whereas}$$

$$n! = \underbrace{n \cdot (n-1) \cdot (n-2) \cdots 2 \cdot 1}_{n \text{ factors}}.$$

> 0! = 1 (by definition)
> 1! = 1
> 2! = 2·1! = 2
> 3! = 3·2! = 6
> 4! = 4·3! = 24
> 5! = 5·4! = 120
> 6! = 6·5! = 720

The nth term of both sequences involves the product of n factors; however, the factors of $n!$ decrease, while the factors of n^n are the same. Based on this observation, we claim that $\{n^n\}$ grows faster than $\{n!\}$, and we have the ordering $\{n!\} \ll \{n^n\}$. But where does $\{n!\}$ appear in the list relative to $\{b^n\}$? Again, some intuition is gained by noting that

$$b^n = \underbrace{b \cdot b \cdot b \cdots b}_{n \text{ factors}}, \qquad \text{whereas}$$

$$n! = \underbrace{n \cdot (n-1) \cdot (n-2) \cdots 2 \cdot 1}_{n \text{ factors}}.$$

The nth term of both sequences involves a product of n factors; however, the factors of b^n remain constant as n increases, while the factors of $n!$ increase with n. So we claim that $\{n!\}$ grows faster than $\{b^n\}$. This conjecture is supported by computation, although the outcome of the race may not be immediately evident if b is large (Exercise 111).

THEOREM 10.6 Growth Rates of Sequences
The following sequences are ordered according to increasing growth rates as $n \to \infty$; that is, if $\{a_n\}$ appears before $\{b_n\}$ in the list, then $\displaystyle\lim_{n\to\infty} \frac{a_n}{b_n} = 0$ and $\displaystyle\lim_{n\to\infty} \frac{b_n}{a_n} = \infty$:

$$\{\ln^q n\} \ll \{n^p\} \ll \{n^p \ln^r n\} \ll \{n^{p+s}\} \ll \{b^n\} \ll \{n!\} \ll \{n^n\}.$$

The ordering applies for positive real numbers p, q, r, s, and $b > 1$.

QUICK CHECK 4 Which sequence grows faster: $\{\ln n\}$ or $\{n^{1.1}\}$? What is $\displaystyle\lim_{n\to\infty} \frac{n^{1,000,000}}{e^n}$? ◀

It is worth noting that the rankings in Theorem 10.6 do not change if a sequence is multiplied by a positive constant (Exercise 110).

EXAMPLE 6 Convergence and growth rates Compare growth rates of sequences to determine whether the following sequences converge.

a. $\left\{ \dfrac{\ln n^{10}}{0.00001n} \right\}$ b. $\left\{ \dfrac{n^8 \ln n}{n^{8.001}} \right\}$ c. $\left\{ \dfrac{n!}{10^n} \right\}$

SOLUTION

a. Because $\ln n^{10} = 10 \ln n$, the sequence in the numerator is a constant multiple of the sequence $\{\ln n\}$. Similarly, the sequence in the denominator is a constant multiple of the sequence $\{n\}$. By Theorem 10.6, $\{n\}$ grows faster than $\{\ln n\}$ as $n \to \infty$; therefore, the sequence $\left\{ \dfrac{\ln n^{10}}{0.00001n} \right\}$ converges to zero.

b. The sequence in the numerator is $\{n^p \ln^r n\}$ of Theorem 10.6 with $p = 8$ and $r = 1$. The sequence in the denominator is $\{n^{p+s}\}$ of Theorem 10.6 with $p = 8$ and $s = 0.001$. Because $\{n^{p+s}\}$ grows faster than $\{n^p \ln^r n\}$ as $n \to \infty$, we conclude that $\left\{ \dfrac{n^8 \ln n}{n^{8.001}} \right\}$ converges to zero.

c. Using Theorem 10.6, we see that $n!$ grows faster than any exponential function as $n \to \infty$. Therefore, $\displaystyle\lim_{n\to\infty} \frac{n!}{10^n} = \infty$, and the sequence diverges. **Figure 10.23** gives a visual comparison of the growth rates of $\{n!\}$ and $\{10^n\}$. Because these sequences grow so quickly, we plot the logarithm of the terms. The exponential sequence $\{10^n\}$ dominates the factorial sequence $\{n!\}$ until $n = 25$ terms. At that point, the factorial sequence overtakes the exponential sequence.

Related Exercises 76, 78 ◀

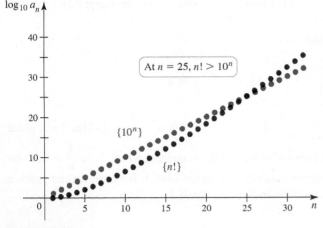

At $n = 25$, $n! > 10^n$

$\{10^n\}$

$\{n!\}$

Figure 10.23

Formal Definition of a Limit of a Sequence

As with limits of functions, there is a formal definition of the limit of a sequence.

DEFINITION Limit of a Sequence

The sequence $\{a_n\}$ converges to L provided the terms of a_n can be made arbitrarily close to L by taking n sufficiently large. More precisely, $\{a_n\}$ has the unique limit L if, given any $\varepsilon > 0$, it is possible to find a positive integer N (depending only on ε) such that

$$|a_n - L| < \varepsilon \qquad \text{whenever } n > N.$$

If the **limit of a sequence** is L, we say the sequence **converges** to L, written

$$\lim_{n \to \infty} a_n = L.$$

A sequence that does not converge is said to **diverge**.

When $n > N$,
$|a_n - L| < \varepsilon$

Figure 10.24

The formal definition of the limit of a convergent sequence is interpreted in much the same way as the limit at infinity of a function. Given a small tolerance $\varepsilon > 0$, how far out in the sequence must you go so that all succeeding terms are within ε of the limit L (Figure 10.24)? Given *any* value of $\varepsilon > 0$ (no matter how small), you must find a value of N such that all terms beyond a_N are within ε of L.

EXAMPLE 7   Limits using the formal definition Consider the claim that

$$\lim_{n \to \infty} a_n = \lim_{n \to \infty} \frac{n}{n-1} = 1.$$

a. Given $\varepsilon = 0.01$, find a value of N that satisfies the conditions of the limit definition.

b. Prove that $\lim_{n \to \infty} a_n = 1$.

SOLUTION

a. We must find an integer N such that $|a_n - 1| < \varepsilon = 0.01$ whenever $n > N$. This condition can be written

$$|a_n - 1| = \left| \frac{n}{n-1} - 1 \right| = \left| \frac{1}{n-1} \right| < 0.01.$$

Noting that $n > 1$, the absolute value can be removed. The condition on n becomes $n - 1 > 1/0.01 = 100$, or $n > 101$. Therefore, we take $N = 101$ or any larger number. This means that $|a_n - 1| < 0.01$ whenever $n > 101$.

b. Given *any* $\varepsilon > 0$, we must find a value of N (depending on ε) that guarantees $|a_n - 1| = \left| \dfrac{n}{n-1} - 1 \right| < \varepsilon$ whenever $n > N$. For $n > 1$, the inequality $\left| \dfrac{n}{n-1} - 1 \right| < \varepsilon$ implies that

$$\left| \frac{n}{n-1} - 1 \right| = \frac{1}{n-1} < \varepsilon.$$

➤ In general, $1/\varepsilon + 1$ is not an integer, so N should be the least integer greater than $1/\varepsilon + 1$ or any larger integer.

Solving for n, we find that $\dfrac{1}{n-1} < \varepsilon$ or $n - 1 > \dfrac{1}{\varepsilon}$ or $n > \dfrac{1}{\varepsilon} + 1$. Therefore, given a tolerance $\varepsilon > 0$, we must look beyond a_N in the sequence, where $N \geq \dfrac{1}{\varepsilon} + 1$, to be sure that the terms of the sequence are within ε of the limit 1. Because we can provide a value of N for any $\varepsilon > 0$, the limit exists and equals 1.

Related Exercise 105 ◀

SECTION 10.2 EXERCISES

Getting Started

1. Give an example of a nonincreasing sequence with a limit.

2. Give an example of a nondecreasing sequence without a limit.

3. Give an example of a bounded sequence that has a limit.

4. Give an example of a bounded sequence without a limit.

5. For what values of r does the sequence $\{r^n\}$ converge? Diverge?

6–9. *Determine whether the following sequences converge or diverge, and state whether they are monotonic or whether they oscillate. Give the limit when the sequence converges.*

6. $\{0.2^n\}$

7. $\{1.00001^n\}$

8. $\{(-2.5)^n\}$

9. $\{(-0.7)^n\}$

10. Find the limit of the sequence $\{a_n\}$ if $1 - \dfrac{1}{n} < a_n < 1 + \dfrac{1}{n}$, for every integer $n \geq 1$.

11. Compare the growth rates of $\{n^{100}\}$ and $\{e^{n/100}\}$ as $n \to \infty$.

12. Use Theorem 10.6 to evaluate $\displaystyle\lim_{n\to\infty} \dfrac{n^{100}}{n^n}$.

Practice Exercises

13–52. Limits of sequences *Find the limit of the following sequences or determine that the sequence diverges.*

13. $\left\{\dfrac{n^3}{n^4 + 1}\right\}$

14. $\left\{\dfrac{n^{12}}{3n^{12} + 4}\right\}$

15. $\left\{\dfrac{3n^3 - 1}{2n^3 + 1}\right\}$

16. $\left\{\dfrac{n^5 + 3n}{10n^3 + n}\right\}$

17. $\left\{\tan^{-1}\left(\dfrac{10n}{10n + 4}\right)\right\}$

18. $\left\{\cot\left(\dfrac{n\pi}{2n + 2}\right)\right\}$

19. $\left\{1 + \cos\dfrac{1}{n}\right\}$

20. $\{\ln(n^3 + 1) - \ln(3n^3 + 10n)\}$

21. $\left\{\ln\sin\dfrac{1}{n} + \ln n\right\}$

22. $\left\{\dfrac{k}{\sqrt{9k^2 + 1}}\right\}$

23. $\left\{\dfrac{\sqrt{4n^4 + 3n}}{8n^2 + 1}\right\}$

24. $\left\{\dfrac{2e^n + 1}{e^n}\right\}$

25. $\left\{\dfrac{\ln n^2}{\ln 3n}\right\}$

26. $\{5(-1.01)^n\}$

27. $\{2^{n+1}3^{-n}\}$

28. $\{100(-0.003)^n\}$

29. $\{(0.5)^n + 3(0.75)^n\}$

30. $\left\{\dfrac{e^n + \pi^n}{e^n}\right\}$

31. $\left\{\dfrac{3^{n+1} + 3}{3^n}\right\}$

32. $\left\{\dfrac{3^n}{3^n + 4^n}\right\}$

33. $\left\{\dfrac{(n + 1)!}{n!}\right\}$

34. $\left\{\dfrac{(2n)!\, n^2}{(2n + 2)!}\right\}$

35. $\{\tan^{-1} n\}$

36. $\left\{\dfrac{e^{n/10}}{2^n}\right\}$

37. $\{\sqrt{n^2 + 1} - n\}$

38. $\{n^{2/n}\}$

39. $\left\{\left(1 + \dfrac{2}{n}\right)^n\right\}$

40. $\left\{\dfrac{\ln n}{n^{1.1}}\right\}$

41. $\{\sqrt[n]{e^{3n+4}}\}$

42. $\left\{\left(\dfrac{n}{n + 5}\right)^n\right\}$

43. $\left\{\sqrt{\left(1 + \dfrac{1}{2n}\right)^n}\right\}$

44. $\left\{\left(1 + \dfrac{4}{n}\right)^{3n}\right\}$

45. $\left\{\dfrac{n}{e^n + 3n}\right\}$

46. $\left\{\dfrac{\ln(1/n)}{n}\right\}$

47. $\left\{\left(\dfrac{1}{n}\right)^{1/n}\right\}$

48. $\left\{\left(1 - \dfrac{4}{n}\right)^n\right\}$

49. $\{b_n\}$, where $b_n = \begin{cases} \dfrac{n}{n + 1} & \text{if } n \leq 5000 \\ ne^{-n} & \text{if } n > 5000 \end{cases}$

50. $\left\{n\left(1 - \cos\dfrac{1}{n}\right)\right\}$

51. $\left\{n\sin\dfrac{6}{n}\right\}$

52. $\left\{\dfrac{n^8 + n^7}{n^7 + n^8\ln n}\right\}$

53. Plot a graph of the sequence $\{a_n\}$, for $a_n = \sin\dfrac{n\pi}{2}$. Then determine the limit of the sequence or explain why the sequence diverges.

54. Plot a graph of the sequence $\{a_n\}$, for $a_n = \dfrac{(-1)^n n}{n + 1}$. Then determine the limit of the sequence or explain why the sequence diverges.

55–70. More sequences *Find the limit of the following sequences or determine that the sequence diverges.*

55. $\left\{\dfrac{(-1)^n}{2^n}\right\}$

56. $\left\{\dfrac{(-1)^n}{n}\right\}$

57. $a_n = (-1)^n \sqrt[n]{n}$

58. $\left\{\dfrac{\cos(n\pi/2)}{\sqrt{n}}\right\}$

59. $\left\{\dfrac{n\sin^3(n\pi/2)}{n + 1}\right\}$

60. $\left\{\dfrac{(-1)^{n+1} n^2}{2n^3 + n}\right\}$

61. $a_n = e^{-n}\cos n$

62. $a_n = \dfrac{e^{-n}}{2\sin(e^{-n})}$

63. $\left\{\dfrac{\tan^{-1} n}{n}\right\}$

64. $\left\{\dfrac{(2n^3 + n)\tan^{-1} n}{n^3 + 4}\right\}$

65. $\left\{\dfrac{\cos n}{n}\right\}$

66. $\left\{\dfrac{\sin 6n}{5n}\right\}$

67. $\left\{\dfrac{\sin n}{2^n}\right\}$

68. $\left\{\displaystyle\int_1^n x^{-2}\, dx\right\}$

69. $\left\{\dfrac{75^{n-1}}{99^n} + \dfrac{5^n \sin n}{8^n}\right\}$

70. $\left\{\cos(0.99^n) + \dfrac{7^n + 9^n}{63^n}\right\}$

T 71. Periodic dosing Many people take aspirin on a regular basis as a preventive measure for heart disease. Suppose a person takes 80 mg of aspirin every 24 hours. Assume aspirin has a half-life of 24 hours; that is, every 24 hours, half of the drug in the blood is eliminated.

  a. Find a recurrence relation for the sequence $\{d_n\}$ that gives the amount of drug in the blood after the nth dose, where $d_1 = 80$.
  b. Use a calculator to estimate this limit. In the long run, how much drug is in the person's blood?
  c. Assuming the sequence has a limit, confirm the result of part (b) by finding the limit of $\{d_n\}$ directly.

T 72. A car loan Marie takes out a $20,000 loan for a new car. The loan has an annual interest rate of 6% or, equivalently, a monthly interest rate of 0.5%. Each month, the bank adds interest to the loan balance (the interest is always 0.5% of the current balance), and then Marie makes a $200 payment to reduce the loan balance. Let B_n be the loan balance immediately after the nth payment, where $B_0 = \$20,000$.

  a. Write the first five terms of the sequence $\{B_n\}$.
  b. Find a recurrence relation that generates the sequence $\{B_n\}$.
  c. Determine how many months are needed to reduce the loan balance to zero.

T 73. A savings plan James begins a savings plan in which he deposits $100 at the beginning of each month into an account that earns 9% interest annually or, equivalently, 0.75% per month. To be clear, on the first day of each month, the bank adds 0.75% of the current balance as interest, and then James deposits $100. Let B_n be the balance in the account after the nth payment, where $B_0 = \$0$.

  a. Write the first five terms of the sequence $\{B_n\}$.
  b. Find a recurrence relation that generates the sequence $\{B_n\}$.
  c. How many months are needed to reach a balance of $5000?

T 74. Diluting a solution A tank is filled with 100 L of a 40% alcohol solution (by volume). You repeatedly perform the following operation: Remove 2 L of the solution from the tank and replace them with 2 L of 10% alcohol solution.

  a. Let C_n be the concentration of the solution in the tank after the nth replacement, where $C_0 = 40\%$. Write the first five terms of the sequence $\{C_n\}$.
  b. After how many replacements does the alcohol concentration reach 15%?
  c. Determine the limiting (steady-state) concentration of the solution that is approached after many replacements.

75–82. Growth rates of sequences *Use Theorem 10.6 to find the limit of the following sequences or state that they diverge.*

75. $\left\{\dfrac{n!}{n^n}\right\}$ **76.** $\left\{\dfrac{3^n}{n!}\right\}$ **77.** $\left\{\dfrac{n^{10}}{\ln^{20} n}\right\}$

78. $\left\{\dfrac{n^{10}}{\ln^{1000} n}\right\}$ **79.** $\left\{\dfrac{n^{1000}}{2^n}\right\}$ **80.** $a_n = \dfrac{4^n + 5n!}{n! + 2^n}$

81. $a_n = \dfrac{6^n + 3^n}{6^n + n^{100}}$ **82.** $a_n = \dfrac{7^n}{n^7 5^n}$

83. Explain why or why not Determine whether the following statements are true and give an explanation or counterexample.

  a. If $\lim\limits_{n\to\infty} a_n = 1$ and $\lim\limits_{n\to\infty} b_n = 3$, then $\lim\limits_{n\to\infty} \dfrac{b_n}{a_n} = 3$.

  b. If $\lim\limits_{n\to\infty} a_n = 0$ and $\lim\limits_{n\to\infty} b_n = \infty$, then $\lim\limits_{n\to\infty} a_n b_n = 0$.

  c. The convergent sequences $\{a_n\}$ and $\{b_n\}$ differ in their first 100 terms, but $a_n = b_n$ for $n > 100$. It follows that $\lim\limits_{n\to\infty} a_n = \lim\limits_{n\to\infty} b_n$.

  d. If $\{a_n\} = \left\{1, \dfrac{1}{2}, \dfrac{1}{3}, \dfrac{1}{4}, \dfrac{1}{5}, \dots\right\}$ and $\{b_n\} = \left\{1, 0, \dfrac{1}{2}, 0, \dfrac{1}{3}, 0, \dfrac{1}{4}, 0, \dots\right\}$, then $\lim\limits_{n\to\infty} a_n = \lim\limits_{n\to\infty} b_n$.

  e. If the sequence $\{a_n\}$ converges, then the sequence $\{(-1)^n a_n\}$ converges.

  f. If the sequence $\{a_n\}$ diverges, then the sequence $\{0.000001 a_n\}$ diverges.

T 84–87. Sequences by recurrence relations *The following sequences, defined by a recurrence relation, are monotonic and bounded, and therefore converge by Theorem 10.5.*

  a. *Examine the first three terms of the sequence to determine whether the sequence is nondecreasing or nonincreasing.*
  b. *Use analytical methods to find the limit of the sequence.*

84. $a_{n+1} = \dfrac{1}{2} a_n + 2;\ a_0 = 1$

85. $a_{n+1} = 2a_n(1 - a_n);\ a_0 = 0.3$

86. $a_{n+1} = \dfrac{1}{2}\left(a_n + \dfrac{2}{a_n}\right);\ a_0 = 2$

T 87. $a_{n+1} = \sqrt{2 + a_n};\ a_0 = 3$

T 88. Fish harvesting A fishery manager knows that her fish population naturally increases at a rate of 1.5% per month, while 80 fish are harvested each month. Let F_n be the fish population after the nth month, where $F_0 = 4000$ fish.

  a. Write out the first five terms of the sequence $\{F_n\}$.
  b. Find a recurrence relation that generates the sequence $\{F_n\}$.
  c. Does the fish population decrease or increase in the long run?
  d. Determine whether the fish population decreases or increases in the long run if the initial population is 5500 fish.
  e. Determine the initial fish population F_0 below which the population decreases.

T 89. Drug Dosing A patient takes 75 mg of a medication every 12 hours; 60% of the medication in the blood is eliminated every 12 hours.

  a. Let d_n equal the amount of medication (in mg) in the bloodstream after n doses, where $d_1 = 75$. Find a recurrence relation for d_n.
  b. Show that $\{d_n\}$ is monotonic and bounded, and therefore converges.
  c. Find the limit of the sequence. What is the physical meaning of this limit?

T 90. Sleep model After many nights of observation, you notice that if you oversleep one night, you tend to undersleep the following night, and vice versa. This pattern of compensation is described by the relationship

$$x_{n+1} = \frac{1}{2}(x_n + x_{n-1}), \quad \text{for } n = 1, 2, 3, \dots,$$

where x_n is the number of hours of sleep you get on the nth night, and $x_0 = 7$ and $x_1 = 6$ are the number of hours of sleep on the first two nights, respectively.

  a. Write out the first six terms of the sequence $\{x_n\}$ and confirm that the terms alternately increase and decrease.

b. Show that the explicit formula

$$x_n = \frac{19}{3} + \frac{2}{3}\left(-\frac{1}{2}\right)^n, \quad \text{for } n \geq 0,$$

generates the terms of the sequence in part (a).

c. Assume the limit of the sequence exists. What is the limit of the sequence?

T 91. Calculator algorithm The CORDIC (COordinate Rotation DIgital Calculation) algorithm is used by most calculators to evaluate trigonometric and logarithmic functions. An important number in the CORDIC algorithm, called the *aggregate constant*, is given by the infinite product $\prod_{n=0}^{\infty} \dfrac{2^n}{\sqrt{1 + 2^{2n}}}$, where $\prod_{n=0}^{N} a_n$ is the *partial product* $a_0 \cdot a_1 \cdots a_N$. This infinite product is the limit of the sequence

$$\left\{ \prod_{n=0}^{0} \frac{2^n}{\sqrt{1 + 2^{2n}}}, \prod_{n=0}^{1} \frac{2^n}{\sqrt{1 + 2^{2n}}}, \prod_{n=0}^{2} \frac{2^n}{\sqrt{1 + 2^{2n}}}, \ldots \right\}.$$

Estimate the value of the aggregate constant. (See the Guided Project *CORDIC algorithms: How your calculator works.*)

92. A sequence of products Find the limit of the sequence

$$\{a_n\}_{n=2}^{\infty} = \left\{ \left(1 - \frac{1}{2}\right)\left(1 - \frac{1}{3}\right) \cdots \left(1 - \frac{1}{n}\right) \right\}_{n=2}^{\infty}.$$

Explorations and Challenges

93. Suppose the sequence $\{a_n\}_{n=0}^{\infty}$ is defined by the recurrence relation $a_{n+1} = \dfrac{1}{3}a_n + 6; a_0 = 3$.

a. Prove that the sequence is increasing and bounded.

b. Explain why $\{a_n\}_{n=0}^{\infty}$ converges and find the limit.

94. Suppose the sequence $\{a_n\}_{n=0}^{\infty}$ is defined by the recurrence relation $a_{n+1} = \sqrt{a_n} + 20; a_0 = 6$.

a. Prove that the sequence is decreasing and bounded.

b. Explain why $\{a_n\}_{n=0}^{\infty}$ converges and find the limit.

T 95. Repeated square roots Consider the sequence defined by $a_{n+1} = \sqrt{2 + a_n}, a_0 = \sqrt{2}$, for $n = 0, 1, 2, 3, \ldots$.

a. Evaluate the first four terms of the sequence $\{a_n\}$. State the exact values first, and then the approximate values.

b. Show that the sequence is increasing and bounded.

c. Assuming the limit exists, use the method of Example 5 to determine the limit exactly.

T 96. Towers of powers For a positive real number p, the tower of exponents $p^{p^{p^{\cdot^{\cdot^{\cdot}}}}}$ continues indefinitely but the expression is ambiguous. The tower could be built from the top as the limit of the sequence $\{p^p, (p^p)^p, ((p^p)^p)^p, \ldots\}$, in which case the sequence is defined recursively as

$$a_{n+1} = a_n^p \text{ (building from the top)}, \qquad (1)$$

where $a_1 = p^p$. The tower could also be built from the bottom as the limit of the sequence $\{p^p, p^{(p^p)}, p^{(p^{(p^p)})}, \ldots\}$, in which case the sequence is defined recursively as

$$a_{n+1} = p^{a_n} \text{ (building from the bottom)}, \qquad (2)$$

where again $a_1 = p^p$.

a. Estimate the value of the tower with $p = 0.5$ by building from the top. That is, use tables to estimate the limit of the sequence defined recursively by (1) with $p = 0.5$. Estimate the maximum value of $p > 0$ for which the sequence has a limit.

b. Estimate the value of the tower with $p = 1.2$ by building from the bottom. That is, use tables to estimate the limit of the sequence defined recursively by (2) with $p = 1.2$. Estimate the maximum value of $p > 1$ for which the sequence has a limit.

T 97. Fibonacci sequence The famous Fibonacci sequence was proposed by Leonardo Pisano, also known as Fibonacci, in about A.D. 1200 as a model for the growth of rabbit populations. It is given by the recurrence relation $f_{n+1} = f_n + f_{n-1}$, for $n = 1, 2, 3, \ldots$, where $f_0 = 1$ and $f_1 = 1$. Each term of the sequence is the sum of its two predecessors.

a. Write out the first ten terms of the sequence.

b. Is the sequence bounded?

c. Estimate or determine $\varphi = \lim_{n \to \infty} \dfrac{f_{n+1}}{f_n}$, the ratio of the successive terms of the sequence. Provide evidence that $\varphi = \dfrac{1 + \sqrt{5}}{2}$, a number known as the *golden mean*.

d. Use induction to verify the remarkable result that

$$f_n = \frac{1}{\sqrt{5}}\left(\varphi^n - (-1)^n \varphi^{-n}\right).$$

98. Arithmetic-geometric mean Pick two positive numbers a_0 and b_0 with $a_0 > b_0$ and write out the first few terms of the two sequences $\{a_n\}$ and $\{b_n\}$:

$$a_{n+1} = \frac{a_n + b_n}{2}, \quad b_{n+1} = \sqrt{a_n b_n}, \quad \text{for } n = 0, 1, 2, \ldots.$$

Recall that the arithmetic mean $A = (p + q)/2$ and the geometric mean $G = \sqrt{pq}$ of two positive numbers p and q satisfy $A \geq G$.

a. Show that $a_n > b_n$ for all n.

b. Show that $\{a_n\}$ is a decreasing sequence and $\{b_n\}$ is an increasing sequence.

c. Conclude that $\{a_n\}$ and $\{b_n\}$ converge.

d. Show that $a_{n+1} - b_{n+1} < \dfrac{a_n - b_n}{2}$ and conclude that $\lim_{n \to \infty} a_n = \lim_{n \to \infty} b_n$. The common value of these limits is called the arithmetic-geometric mean of a_0 and b_0, denoted $\text{AGM}(a_0, b_0)$.

e. Estimate $\text{AGM}(12, 20)$. Estimate Gauss' constant $\dfrac{1}{\text{AGM}(1, \sqrt{2})}$.

T 99. Continued fractions The expression

$$1 + \cfrac{1}{1 + \cfrac{1}{1 + \cfrac{1}{1 + \cfrac{1}{1 + \ddots}}}},$$

where the process continues indefinitely, is called a *continued fraction*.

a. Show that this expression can be built in steps using the recurrence relation $a_0 = 1, a_{n+1} = 1 + \dfrac{1}{a_n}$ for $n = 0, 1, 2, 3, \ldots$. Explain why the value of the expression can be interpreted as $\lim_{n \to \infty} a_n$, provided the limit exists.

b. Evaluate the first five terms of the sequence $\{a_n\}$.

c. Using computation and/or graphing, estimate the limit of the sequence.

d. Assuming the limit exists, use the method of Example 5 to determine the limit exactly. Compare your estimate with $\dfrac{1 + \sqrt{5}}{2}$, a number known as the *golden mean*.

e. Assuming the limit exists, use the same ideas to determine the value of

$$a + \cfrac{b}{a + \cfrac{b}{a + \cfrac{b}{a + \cfrac{b}{a + \ddots}}}},$$

where a and b are positive real numbers.

100. The hailstone sequence Here is a fascinating (unsolved) problem known as the hailstone problem (or the Ulam Conjecture or the Collatz Conjecture). It involves sequences in two different ways. First, choose a positive integer N and call it a_0. This is the *seed* of a sequence. The rest of the sequence is generated as follows: For $n = 0, 1, 2, \ldots$

$$a_{n+1} = \begin{cases} a_n/2 & \text{if } a_n \text{ is even} \\ 3a_n + 1 & \text{if } a_n \text{ is odd.} \end{cases}$$

However, if $a_n = 1$ for any n, then the sequence terminates.

a. Compute the sequence that results from the seeds $N = 2, 3, 4, \ldots, 10$. You should verify that in all these cases, the sequence eventually terminates. The hailstone conjecture (still unproved) states that for all positive integers N, the sequence terminates after a finite number of terms.

b. Now define the hailstone sequence $\{H_k\}$, which is the number of terms needed for the sequence $\{a_n\}$ to terminate starting with a seed of k. Verify that $H_2 = 1$, $H_3 = 7$, and $H_4 = 2$.

c. Plot as many terms of the hailstone sequence as is feasible. How did the sequence get its name? Does the conjecture appear to be true?

101–106. Formal proofs of limits *Use the formal definition of the limit of a sequence to prove the following limits.*

101. $\displaystyle\lim_{n\to\infty} \frac{1}{n} = 0$

102. $\displaystyle\lim_{n\to\infty} \frac{1}{n^2} = 0$

103. $\displaystyle\lim_{n\to\infty} \frac{3n^2}{4n^2 + 1} = \frac{3}{4}$

104. $\displaystyle\lim_{n\to\infty} b^{-n} = 0$, for $b > 1$

105. $\displaystyle\lim_{n\to\infty} \frac{cn}{bn + 1} = \frac{c}{b}$, for real numbers $b > 0$ and $c > 0$

106. $\displaystyle\lim_{n\to\infty} \frac{n}{n^2 + 1} = 0$

▣ 107. Comparing sequences with a parameter For what values of a does the sequence $\{n!\}$ grow faster than the sequence $\{n^{an}\}$? (*Hint:* Stirling's formula is useful: $n! \approx \sqrt{2\pi n}\, n^n e^{-n}$, for large values of n.)

108–109. Reindexing *Express each sequence $\{a_n\}_{n=1}^{\infty}$ as an equivalent sequence of the form $\{b_n\}_{n=3}^{\infty}$.*

108. $\{2n + 1\}_{n=1}^{\infty}$

109. $\{n^2 + 6n - 9\}_{n=1}^{\infty}$

110. Prove that if $\{a_n\} \ll \{b_n\}$ (as used in Theorem 10.6), then $\{ca_n\} \ll \{db_n\}$, where c and d are positive real numbers.

▣ 111. Crossover point The sequence $\{n!\}$ ultimately grows faster than the sequence $\{b^n\}$, for any $b > 1$, as $n \to \infty$. However, b^n is generally greater than $n!$ for small values of n. Use a calculator to determine the smallest value of n such that $n! > b^n$ for each of the cases $b = 2$, $b = e$, and $b = 10$.

QUICK CHECK ANSWERS

1. a. Bounded, monotonic; **b.** Bounded, not monotonic; **c.** Not bounded, not monotonic; **d.** Bounded, monotonic (both nonincreasing and nondecreasing) **2.** If $r = -1$, the sequence is $\{-1, 1, -1, 1, \ldots\}$, the terms alternate in sign, and the sequence diverges. If $r = 1$, the sequence is $\{1, 1, 1, 1, \ldots\}$, the terms are constant, and the sequence converges to 1. **3.** Both changes would increase the steady-state level of drug. **4.** $\{n^{1.1}\}$ grows faster; the limit is 0. ◀

10.3 Infinite Series

> The sequence of partial sums may be visualized nicely as follows:
>
> $$\underbrace{a_1 + a_2 + a_3 + a_4 + \cdots}$$
> $\underbrace{}_{S_1}$
> $\underbrace{}_{S_2}$
> $\underbrace{}_{S_3}$
> $\qquad\qquad\ddots$
>
> When the initial value of the index is not 1, the nth partial sum is still the sum of the first n terms of the series. For example, the nth partial sum of $\displaystyle\sum_{k=3}^{\infty} a_k$ is
>
> $$S_n = \underbrace{a_3 + a_4 + \cdots + a_{n+2}}_{n \text{ terms}}.$$

We begin our discussion of infinite series with *geometric series*. These series arise more frequently than any other infinite series, they are used in many practical problems, and they illustrate all the essential features of infinite series in general. First let's summarize some important ideas from Section 10.1.

Recall that every infinite series $\displaystyle\sum_{k=1}^{\infty} a_k$ has a sequence of partial sums:

$$S_1 = a_1, \qquad S_2 = a_1 + a_2, \qquad S_3 = a_1 + a_2 + a_3,$$

and in general, $\displaystyle S_n = \sum_{k=1}^{n} a_k$, for $n = 1, 2, 3, \ldots$.

If the sequence of partial sums $\{S_n\}$ converges—that is, if $\displaystyle\lim_{n\to\infty} S_n = L$—then the value of the infinite series is also L. If the sequence of partial sums diverges, then the infinite series also diverges.

QUICK CHECK 1 Write the nth partial sum of the series $\sum\limits_{k=0}^{\infty} b_n$. ◄

In summary, to evaluate an infinite series, it is necessary to determine a formula for the sequence of partial sums $\{S_n\}$ and then find its limit. This procedure can be carried out with the series that we discuss in this section: geometric series and telescoping series.

Geometric Sums and Series

➤ Geometric *sequences* have the form $\{r^k\}$ or $\{ar^k\}$. Geometric *sums* and *series* have the form $\sum r^k$ or $\sum ar^k$.

As a preliminary step to geometric series, we study geometric sums, which are *finite sums* in which each term in the sum is a constant multiple of the previous term. A **geometric sum** with n terms has the form

$$S_n = a + ar + ar^2 + \cdots + ar^{n-1} = \sum_{k=0}^{n-1} ar^k,$$

where $a \neq 0$ and r are real numbers; r is called the **ratio** of the sum and a is its first term. For example, the geometric sum with $r = 0.1$, $a = 0.9$, and $n = 4$ is

$$0.9 + 0.09 + 0.009 + 0.0009 = 0.9(1 + 0.1 + 0.01 + 0.001)$$

$$= \sum_{k=0}^{3} 0.9(0.1^k).$$

QUICK CHECK 2 Which of the following sums are not geometric sums?

a. $\sum\limits_{k=0}^{10} \left(\frac{1}{2}\right)^k$ **b.** $\sum\limits_{k=0}^{20} \frac{1}{k}$

c. $\sum\limits_{k=0}^{30} (2k + 1)$ ◄

Our goal is to find a formula for the value of the geometric sum

$$S_n = a + ar + ar^2 + \cdots + ar^{n-1}, \tag{1}$$

for any values of $a \neq 0$, r, and the positive integer n. Doing so requires a clever maneuver. The first step is to multiply both sides of equation (1) by the ratio r:

$$rS_n = r(a + ar + ar^2 + ar^3 + \cdots + ar^{n-1})$$
$$= ar + ar^2 + ar^3 + \cdots + ar^{n-1} + ar^n. \tag{2}$$

➤ The notation $\sum\limits_{k=0}^{\infty} ar^k$ appears to have an undefined first term when $r = 0$. The notation is understood to mean $a + ar + ar^2 + \cdots$ and therefore, the series has a value of a when $r = 0$.

We now subtract equation (2) from equation (1). Notice how most of the terms on the right sides of these equations cancel, leaving

$$S_n - rS_n = a - ar^n.$$

Assuming $r \neq 1$ and solving for S_n results in a general formula for the value of a geometric sum:

$$S_n = a\frac{1 - r^n}{1 - r}. \tag{3}$$

QUICK CHECK 3 Verify that the geometric sum formula gives the correct result for the sums $1 + \frac{1}{2}$ and $\frac{1}{2} + \frac{1}{4} + \frac{1}{8}$. ◄

Having dealt with geometric sums, it is a short step to *geometric series*. We simply note that the geometric sums $S_n = \sum\limits_{k=0}^{n-1} ar^k$ form the sequence of partial sums for the geometric series $\sum\limits_{k=0}^{\infty} ar^k$. The value of the geometric series is the limit of its sequence of partial sums (provided that limit exists). Using equation (3), we have

$$\underbrace{\sum_{k=0}^{\infty} ar^k}_{\text{geometric series}} = \lim_{n\to\infty} \underbrace{\sum_{k=0}^{n-1} ar^k}_{\text{geometric sum } S_n} = \lim_{n\to\infty} a\frac{1 - r^n}{1 - r}.$$

To compute this limit, we must examine the behavior of r^n as $n \to \infty$. Recall from our work with geometric sequences (Section 10.2) that

$$\lim_{n\to\infty} r^n = \begin{cases} 0 & \text{if } |r| < 1 \\ 1 & \text{if } r = 1 \\ \text{does not exist} & \text{if } r \leq -1 \text{ or } r > 1. \end{cases}$$

Case 1: $|r| < 1$ Because $\lim\limits_{n\to\infty} r^n = 0$, we have

$$\lim_{n\to\infty} S_n = \lim_{n\to\infty} a\frac{1 - r^n}{1 - r} = a\frac{1 - \overbrace{\lim_{n\to\infty} r^n}^{0}}{1 - r} = \frac{a}{1 - r}.$$

In the case that $|r| < 1$, the geometric series *converges* to $\dfrac{a}{1 - r}$.

Case 2: $|r| > 1$ In this case, $\lim_{n \to \infty} r^n$ does not exist, so $\lim_{n \to \infty} S_n$ does not exist and the series *diverges*.

Case 3: $|r| = 1$ If $r = 1$, then the geometric series is $\sum_{k=0}^{\infty} a = a + a + a + \cdots$, which

diverges. If $r = -1$, the geometric series is $a \sum_{k=0}^{\infty} (-1)^k = a - a + a - \cdots$, which also

diverges (because the sequence of partial sums oscillates between 0 and a). We summarize these results in Theorem 10.7.

> **QUICK CHECK 4** Evaluate
> $\frac{1}{2} + \frac{1}{4} + \frac{1}{8} + \frac{1}{16} + \cdots$. ◄

THEOREM 10.7   Geometric Series

Let $a \neq 0$ and r be real numbers. If $|r| < 1$, then $\sum_{k=0}^{\infty} ar^k = \dfrac{a}{1-r}$. If $|r| \geq 1$, then the series diverges.

> **QUICK CHECK 5** Explain why $\sum_{k=0}^{\infty} 0.2^k$
> converges and why $\sum_{k=0}^{\infty} 2^k$ diverges. ◄

EXAMPLE 1   Geometric series Evaluate the following geometric series or state that the series diverges.

a. $\displaystyle\sum_{k=0}^{\infty} 1.1^k$ **b.** $\displaystyle\sum_{k=0}^{\infty} e^{-k}$ **c.** $\displaystyle\sum_{k=2}^{\infty} 3(-0.75)^k$

SOLUTION

a. The ratio of this geometric series is $r = 1.1$. Because $|r| \geq 1$, the series diverges.

b. Note that $e^{-k} = \dfrac{1}{e^k} = \left(\dfrac{1}{e}\right)^k$. Therefore, the ratio of the series is $r = \dfrac{1}{e}$, and its first

term is $a = 1$. Because $|r| < 1$, the series converges and its value is

$$\sum_{k=0}^{\infty} e^{-k} = \sum_{k=0}^{\infty} \left(\frac{1}{e}\right)^k = \frac{1}{1 - (1/e)} = \frac{e}{e-1} \approx 1.582.$$

> ► The series in Example 1c is called an *alternating series* because the terms alternate in sign. Such series are discussed in detail in Section 10.6.

c. Writing out the first few terms of the series is helpful:

$$\sum_{k=2}^{\infty} 3(-0.75)^k = \underbrace{3(-0.75)^2}_{a} + \underbrace{3(-0.75)^3}_{ar} + \underbrace{3(-0.75)^4}_{ar^2} + \cdots.$$

We see that the first term of the series is $a = 3(-0.75)^2$ and that the ratio of the series is $r = -0.75$. Because $|r| < 1$, the series converges, and its value is

$$\sum_{k=2}^{\infty} 3(-0.75)^k = \underbrace{\frac{3(-0.75)^2}{1 - (-0.75)}}_{\frac{a}{1-r}} = \frac{27}{28}.$$

Related Exercises 24, 27, 29 ◄

EXAMPLE 2   Decimal expansions as geometric series Write $1.0\overline{35} = 1.0353535\ldots$ as a geometric series and express its value as a fraction.

SOLUTION Notice that the decimal part of this number is a convergent geometric series with $a = 0.035$ and $r = 0.01$:

$$1.0353535\ldots = 1 + \underbrace{0.035 + 0.00035 + 0.0000035 + \cdots}_{\text{geometric series with } a = 0.035 \text{ and } r = 0.01}.$$

Evaluating the series, we have

$$1.0353535\ldots = 1 + \frac{a}{1 - r} = 1 + \frac{0.035}{1 - 0.01} = 1 + \frac{35}{990} = \frac{205}{198}.$$

Related Exercises 49–50 ◄

Telescoping Series

With geometric series, we carried out the entire evaluation process by finding a formula for the sequence of partial sums and evaluating the limit of the sequence. Not many infinite series can be subjected to this sort of analysis. With another class of series, called **telescoping series**, it can also be done. Here is an example.

EXAMPLE 3   Telescoping series Evaluate the following series.

a. $\displaystyle\sum_{k=1}^{\infty}\left(\cos\frac{1}{k} - \cos\frac{1}{k+1}\right)$ **b.** $\displaystyle\sum_{k=3}^{\infty}\frac{1}{(k-2)(k-1)}$

SOLUTION

a. The nth term of the sequence of partial sums is

$$S_n = \sum_{k=1}^{n}\left(\cos\frac{1}{k} - \cos\frac{1}{k+1}\right)$$

$$= \left(\cos 1 - \cos\frac{1}{2}\right) + \left(\cos\frac{1}{2} - \cos\frac{1}{3}\right) + \cdots + \left(\cos\frac{1}{n} - \cos\frac{1}{n+1}\right)$$

$$= \cos 1 + \underbrace{\left(-\cos\frac{1}{2} + \cos\frac{1}{2}\right)}_{0} + \cdots + \underbrace{\left(-\cos\frac{1}{n} + \cos\frac{1}{n}\right)}_{0} - \cos\frac{1}{n+1} \qquad \text{Regroup terms.}$$

$$= \cos 1 - \cos\frac{1}{n+1}. \qquad \text{Simplify.}$$

Observe that the interior terms of the sum cancel (or telescope), leaving a simple expression for S_n. Taking the limit, we find that

$$\sum_{k=1}^{\infty}\left(\cos\frac{1}{k} - \cos\frac{1}{k+1}\right) = \lim_{n\to\infty} S_n$$

$$= \lim_{n\to\infty}\left(\cos 1 - \underbrace{\cos\frac{1}{n+1}}_{\to\,\cos 0\,=\,1}\right) = \cos 1 - 1 \approx -0.46.$$

b. Because the initial value of the index is $k = 3$, the nth partial sum S_n involves n terms that run from $k = 3$ to $k = n + 2$, which is somewhat awkward. The nth partial sums are most easily written when the initial value of the index is $k = 1$. We *reindex* the original sum by using the substitution $j = k - 2$, which implies that $j = 1$ when $k = 3$ and that $k - 1 = j + 1$:

$$\sum_{k=3}^{\infty}\frac{1}{(k-2)(k-1)} = \sum_{j=1}^{\infty}\frac{1}{j(j+1)} \qquad \text{Let } j = k - 2.$$

$$= \sum_{k=1}^{\infty}\frac{1}{k(k+1)}.$$

The variable used as an index does not affect the value of a series, so it is permissible to return to the index k.

Using the method of partial fractions, the sequence of partial sums is

> ► See Section 8.5 for a review of partial fractions.

$$S_n = \sum_{k=1}^{n}\frac{1}{k(k+1)} = \sum_{k=1}^{n}\left(\frac{1}{k} - \frac{1}{k+1}\right).$$

Writing out this sum, we see that

$$S_n = \left(1 - \frac{1}{2}\right) + \left(\frac{1}{2} - \frac{1}{3}\right) + \left(\frac{1}{3} - \frac{1}{4}\right) + \cdots + \left(\frac{1}{n} - \frac{1}{n+1}\right)$$

$$= 1 + \underbrace{\left(-\frac{1}{2} + \frac{1}{2}\right)}_{0} + \underbrace{\left(-\frac{1}{3} + \frac{1}{3}\right)}_{0} + \cdots + \underbrace{\left(-\frac{1}{n} + \frac{1}{n}\right)}_{0} - \frac{1}{n+1}$$

$$= 1 - \frac{1}{n+1}.$$

Again, the sum telescopes and all the interior terms cancel. The result is a simple formula for the nth term of the sequence of partial sums. The value of the series is

$$\sum_{k=1}^{\infty} \frac{1}{k(k+1)} = \lim_{n \to \infty} S_n = \lim_{n \to \infty} \left(1 - \frac{1}{n+1}\right) = 1.$$

Related Exercises 59, 69 ◄

Properties of Convergent Series

We close this section with several properties that are useful in upcoming work. The notation $\sum a_k$, without initial and final values of k, is used to refer to a general infinite series whose terms may be positive or negative (or both).

> ➤ The **leading terms** of an infinite series are those at the beginning with a small index. The **tail** of an infinite series consists of the terms at the "end" of the series with a large and increasing index. The convergence or divergence of an infinite series depends on the tail of the series, while the value of a convergent series is determined primarily by the leading terms.

THEOREM 10.8   Properties of Convergent Series

1. Suppose $\sum a_k$ converges to A and c is a real number. The series $\sum ca_k$ converges, and $\sum ca_k = c \sum a_k = cA$.
2. Suppose $\sum a_k$ diverges. Then $\sum ca_k$ also diverges, for any real number $c \neq 0$.
3. Suppose $\sum a_k$ converges to A and $\sum b_k$ converges to B. The series $\sum (a_k \pm b_k)$ converges, and $\sum (a_k \pm b_k) = \sum a_k \pm \sum b_k = A \pm B$.
4. Suppose $\sum a_k$ diverges and $\sum b_k$ converges. Then $\sum (a_k \pm b_k)$ diverges.
5. If M is a positive integer, then $\displaystyle\sum_{k=1}^{\infty} a_k$ and $\displaystyle\sum_{k=M}^{\infty} a_k$ either both converge or both diverge. In general, *whether* a series converges does not depend on a finite number of terms added to or removed from the series. However, the *value* of a convergent series does change if nonzero terms are added or removed.

Proof: These properties are proved using properties of finite sums and limits of sequences. To prove Property 1, assume $\displaystyle\sum_{k=1}^{\infty} a_k$ converges to A and note that

$$\sum_{k=1}^{\infty} ca_k = \lim_{n \to \infty} \sum_{k=1}^{n} ca_k \qquad \text{Definition of infinite series}$$

$$= \lim_{n \to \infty} c \sum_{k=1}^{n} a_k \qquad \text{Property of finite sums}$$

$$= c \lim_{n \to \infty} \sum_{k=1}^{n} a_k \qquad \text{Property of limits}$$

$$= c \sum_{k=1}^{\infty} a_k \qquad \text{Definition of infinite series}$$

$$= cA. \qquad \text{Value of the series}$$

The proofs of Properties 2 and 3 are similar (Exercises 96–97); we offer Property 4 with no proof.

Property 5 follows by noting that for finite sums with $1 < M < n$,

$$\sum_{k=M}^{n} a_k = \sum_{k=1}^{n} a_k - \sum_{k=1}^{M-1} a_k.$$

Letting $n \to \infty$ in this equation and assuming the value of the infinite series is $\sum_{k=1}^{\infty} a_k = A$, it follows that

$$\sum_{k=M}^{\infty} a_k = \underbrace{\sum_{k=1}^{\infty} a_k}_{A} - \underbrace{\sum_{k=1}^{M-1} a_k}_{\text{finite number}}.$$

Because the right side has a finite value, $\sum_{k=M}^{\infty} a_k$ converges. Similarly, if $\sum_{k=M}^{\infty} a_k$ converges, then $\sum_{k=1}^{\infty} a_k$ converges. By an analogous argument, if one of these series diverges, then the other series diverges.

QUICK CHECK 6 Explain why if $\sum_{k=1}^{\infty} a_k$ converges, then the series $\sum_{k=5}^{\infty} a_k$ (with a different starting index) also converges. Do the two series have the same value? ◄

EXAMPLE 4 Using properties of series Evaluate the following infinite series.

a. $\sum_{k=2}^{\infty} \left(\frac{1}{3^k} - \frac{2}{3^{k+2}} \right)$ **b.** $\sum_{k=1}^{\infty} \left(5\left(\frac{2}{3}\right)^k - \frac{7^{k-1}}{6^k} \right)$

SOLUTION

a. As long as both $\sum_{k=2}^{\infty} \frac{1}{3^k}$ and $\sum_{k=2}^{\infty} \frac{2}{3^{k+1}}$ converge, we can use Properties 1 and 3 of Theorem 10.8 to write

$$\sum_{k=2}^{\infty} \left(\frac{1}{3^k} - \frac{2}{3^{k+1}} \right) = \sum_{k=2}^{\infty} \frac{1}{3^k} - 2\sum_{k=2}^{\infty} \frac{1}{3^{k+1}}.$$

Both of the new series are geometric, and the ratio of each series is $r = \frac{1}{3} < 1$, so each series converges. The first term of $\sum_{k=2}^{\infty} \frac{1}{3^k}$ is $a = \frac{1}{9}$ and the first term of $\sum_{k=2}^{\infty} \frac{1}{3^{k+1}}$ is $a = \frac{1}{27}$, so we have

$$\sum_{k=2}^{\infty} \left(\frac{1}{3^k} - \frac{2}{3^{k+1}} \right) = \sum_{k=2}^{\infty} \frac{1}{3^k} - 2\sum_{k=2}^{\infty} \frac{1}{3^{k+1}} = \frac{\frac{1}{9}}{1 - \frac{1}{3}} - 2\frac{\frac{1}{27}}{1 - \frac{1}{3}} = \frac{1}{9} \cdot \frac{3}{2} - \frac{2}{27} \cdot \frac{3}{2} = \frac{1}{18}.$$

b. We examine the two series $\sum_{k=1}^{\infty} 5\left(\frac{2}{3}\right)^k$ and $\sum_{k=1}^{\infty} \frac{7^{k-1}}{6^k}$ individually. The first series is a geometric series and is evaluated using the methods of Section 10.3. Its first few terms are

$$\sum_{k=1}^{\infty} 5\left(\frac{2}{3}\right)^k = 5\left(\frac{2}{3}\right) + 5\left(\frac{2}{3}\right)^2 + 5\left(\frac{2}{3}\right)^3 + \cdots.$$

The first term of the series is $a = 5\left(\frac{2}{3}\right)$ and the ratio is $r = \frac{2}{3} < 1$; therefore,

$$\sum_{k=1}^{\infty} 5\left(\frac{2}{3}\right)^k = \frac{a}{1-r} = \frac{5\left(\frac{2}{3}\right)}{1 - \frac{2}{3}} = 10.$$

Writing out the first few terms of the second series, we see that it, too, is a geometric series:

$$\sum_{k=1}^{\infty} \frac{7^{k-1}}{6^k} = \frac{1}{6} + \frac{7}{6^2} + \frac{7^2}{6^3} + \cdots.$$

QUICK CHECK 7 For a series with positive terms, explain why the sequence of partial sums $\{S_n\}$ is an increasing sequence. ◄

Because the ratio is $r = \frac{7}{6} > 1$, $\sum_{k=1}^{\infty} \frac{7^{k-1}}{6^k}$ diverges. By Property 4 of Theorem 10.8, the original series also diverges.

Related Exercises 81–82 ◄

SECTION 10.3 EXERCISES

Getting Started

1. What is meant by the *ratio* of a geometric series?

2. What is the difference between a geometric sum and a geometric series?

3. Does a geometric series always have a finite value?

4. Does a geometric sum always have a finite value?

5. Find the first term a and the ratio r of each geometric series.

   a. $\displaystyle\sum_{k=0}^{\infty} \frac{2}{3}\left(\frac{1}{5}\right)^k$ **b.** $\displaystyle\sum_{k=2}^{\infty} \frac{1}{3}\left(-\frac{1}{3}\right)^k$

6. What is the condition for convergence of the geometric series $\displaystyle\sum_{k=0}^{\infty} ar^k$?

7. Find a formula for the *n*th partial sum S_n of $\displaystyle\sum_{k=1}^{\infty} \left(\frac{1}{k+3} - \frac{1}{k+4}\right)$. Use your formula to find the sum of the first 36 terms of the series.

8. Reindex the series $\displaystyle\sum_{k=5}^{\infty} \frac{3}{4k^2 - 63}$ so that it starts at $k = 1$.

Practice Exercises

9–15. Geometric sums *Evaluate each geometric sum.*

9. $\displaystyle\sum_{k=0}^{8} 3^k$

T 10. $\displaystyle\sum_{k=0}^{10} \left(\frac{1}{4}\right)^k$

T 11. $\displaystyle\sum_{k=0}^{20} \left(\frac{2}{5}\right)^{2k}$

12. $\displaystyle\sum_{k=4}^{12} 2^k$

T 13. $\displaystyle\sum_{k=0}^{9} \left(-\frac{3}{4}\right)^k$

14. $\displaystyle\sum_{k=0}^{6} \pi^k$

T 15. $\dfrac{1}{4} + \dfrac{1}{12} + \dfrac{1}{36} + \dfrac{1}{108} + \cdots + \dfrac{1}{2916}$

T 16–17. Periodic savings *Suppose you deposit m dollars at the beginning of every month in a savings account that earns a monthly interest rate of r, which is the annual interest rate divided by 12 (for example, if the annual interest rate is 2.4%, r = 0.024/12 = 0.002). For an initial investment of m dollars, the amount of money in your account at the beginning of the second month is the sum of your second deposit and your initial deposit plus interest, or m + m(1 + r). Continuing in this fashion, it can be shown that the amount of money in your account after n months is $A_n = m + m(1 + r) + \cdots + m(1 + r)^{n-1}$. Use geometric sums to determine the amount of money in your savings account after 5 years (60 months) using the given monthly deposit and interest rate.*

16. Monthly deposits of $100 at an annual interest rate of 1.8%

17. Monthly deposits of $250 at a monthly interest rate of 0.2%

18–20. Evaluating geometric series two ways *Evaluate each geometric series two ways.*

a. Find the nth partial sum S_n of the series and evaluate $\lim_{n\to\infty} S_n$.
b. Evaluate the series using Theorem 10.7.

18. $\displaystyle\sum_{k=0}^{\infty} \left(\frac{2}{5}\right)^k$

19. $\displaystyle\sum_{k=0}^{\infty} \left(-\frac{2}{7}\right)^k$

20. $2 + \dfrac{4}{3} + \dfrac{8}{9} + \dfrac{16}{27} + \cdots$

21–42. Geometric series *Evaluate each geometric series or state that it diverges.*

21. $\displaystyle\sum_{k=0}^{\infty} \left(\frac{1}{4}\right)^k$

22. $\displaystyle\sum_{k=0}^{\infty} \left(\frac{3}{5}\right)^k$

23. $\displaystyle\sum_{k=0}^{\infty} \left(-\frac{9}{10}\right)^k$

24. $\displaystyle\sum_{k=1}^{\infty} \left(-\frac{2}{3}\right)^k$

25. $\displaystyle\sum_{k=0}^{\infty} 0.9^k$

26. $1 + \dfrac{2}{7} + \dfrac{2^2}{7^2} + \dfrac{2^3}{7^3} + \cdots$

27. $1 + 1.01 + 1.01^2 + 1.01^3 + \cdots$

28. $1 + \dfrac{1}{\pi} + \dfrac{1}{\pi^2} + \dfrac{1}{\pi^3} + \cdots$

29. $\displaystyle\sum_{k=1}^{\infty} e^{-2k}$

30. $\displaystyle\sum_{m=2}^{\infty} \frac{5}{2^m}$

31. $\displaystyle\sum_{k=1}^{\infty} 2^{-3k}$

32. $\displaystyle\sum_{k=3}^{\infty} \frac{3 \cdot 4^k}{7^k}$

33. $\displaystyle\sum_{k=4}^{\infty} \frac{1}{5^k}$

34. $\displaystyle\sum_{k=0}^{\infty} \left(\frac{4}{3}\right)^{-k}$

35. $\displaystyle\sum_{k=0}^{\infty} 3(-\pi)^{-k}$

36. $\displaystyle\sum_{k=1}^{\infty} (-e)^{-k}$

37. $1 + \dfrac{e}{\pi} + \dfrac{e^2}{\pi^2} + \dfrac{e^3}{\pi^3} + \cdots$

38. $\dfrac{1}{16} + \dfrac{3}{64} + \dfrac{9}{256} + \dfrac{27}{1024} + \cdots$

39. $\displaystyle\sum_{k=2}^{\infty} (-0.15)^k$

40. $\displaystyle\sum_{k=1}^{\infty} 3\left(-\frac{1}{8}\right)^{3k}$

41. $\displaystyle\sum_{k=1}^{\infty} \frac{4}{12^k}$

42. $\displaystyle\sum_{k=2}^{\infty} 3e^{-k}$

43–44. Periodic doses *Suppose you take a dose of m mg of a particular medication once per day. Assume f equals the fraction of the medication that remains in your blood one day later. Just after taking another dose of medication on the second day, the amount of medication in your blood equals the sum of the second dose and the fraction of the first dose remaining in your blood, which is m + mf. Continuing in this fashion, the amount of medication in your blood just after your nth dose is $A_n = m + mf + \cdots + mf^{n-1}$. For the given values of f and m, calculate A_5, A_{10}, A_{30}, and $\lim_{n\to\infty} A_n$. Interpret the meaning of the limit $\lim_{n\to\infty} A_n$.*

43. $f = 0.25, m = 200$ mg

44. $f = 0.4, m = 150$ mg

45. **Periodic doses** Suppose you take 200 mg of an antibiotic every 6 hr. The half-life of the drug (the time it takes for half of the drug to be eliminated from your blood) is 6 hr. Use infinite series to find the long-term (steady-state) amount of antibiotic in your blood. You may assume the steady-state amount is finite.

46–53. Decimal expansions *Write each repeating decimal first as a geometric series and then as a fraction (a ratio of two integers).*

46. $0.\overline{6} = 0.666\ldots$

47. $0.\overline{3} = 0.333\ldots$

48. $0.\overline{09} = 0.090909\ldots$

49. $0.\overline{037} = 0.037037\ldots$

50. $1.\overline{25} = 1.252525\ldots$ **51.** $0.\overline{456} = 0.456456456\ldots$

52. $5.1\overline{283} = 5.12838383\ldots$ **53.** $0.00\overline{952} = 0.00952952\ldots$

54–69. Telescoping series *For the following telescoping series, find a formula for the nth term of the sequence of partial sums* $\{S_n\}$. *Then evaluate* $\lim\limits_{n\to\infty} S_n$ *to obtain the value of the series or state that the series diverges.*

54. $\displaystyle\sum_{k=1}^{\infty}\left(\frac{1}{k+2} - \frac{1}{k+3}\right)$ **55.** $\displaystyle\sum_{k=1}^{\infty}\left(\frac{1}{k+1} - \frac{1}{k+2}\right)$

56. $\displaystyle\sum_{k=1}^{\infty}\frac{20}{25k^2 + 15k - 4}$ **57.** $\displaystyle\sum_{k=1}^{\infty}\frac{1}{(k+6)(k+7)}$

58. $\displaystyle\sum_{k=-3}^{\infty}\frac{10}{4k^2 + 32k + 63}$ **59.** $\displaystyle\sum_{k=3}^{\infty}\frac{4}{(4k-3)(4k+1)}$

60. $\displaystyle\sum_{k=0}^{\infty}\frac{1}{(3k+1)(3k+4)}$ **61.** $\displaystyle\sum_{k=1}^{\infty}\ln\frac{k+1}{k}$

62. $\displaystyle\sum_{k=3}^{\infty}\frac{2}{(2k-1)(2k+1)}$

63. $\displaystyle\sum_{k=1}^{\infty}\frac{1}{(k+p)(k+p+1)}$, where p is a positive integer

64. $\displaystyle\sum_{k=1}^{\infty}\frac{1}{(ak+1)(ak+a+1)}$, where a is a positive integer

65. $\displaystyle\sum_{k=1}^{\infty}\left(\frac{1}{\sqrt{k+1}} - \frac{1}{\sqrt{k+3}}\right)$

66. $\displaystyle\sum_{k=1}^{\infty}\frac{6}{k^2 + 2k}$ **67.** $\displaystyle\sum_{k=1}^{\infty}\frac{3}{k^2 + 5k + 4}$

68. $\displaystyle\sum_{k=1}^{\infty}(\sqrt{k+1} - \sqrt{k})$ **69.** $\displaystyle\sum_{k=1}^{\infty}(\tan^{-1}(k+1) - \tan^{-1}k)$

70. Evaluating an infinite series two ways Evaluate the series $\displaystyle\sum_{k=1}^{\infty}\left(\frac{1}{2^k} - \frac{1}{2^{k+1}}\right)$ two ways.

a. Use a telescoping series argument.

b. Use a geometric series argument with Theorem 10.8.

71. Evaluating an infinite series two ways Evaluate the series $\displaystyle\sum_{k=1}^{\infty}\left(\frac{4}{3^k} - \frac{4}{3^{k+1}}\right)$ two ways.

a. Use a telescoping series argument.

b. Use a geometric series argument with Theorem 10.8.

72–86. Evaluating series *Evaluate each series or state that it diverges.*

72. $\displaystyle\sum_{k=1}^{\infty}\frac{1}{9k^2 + 39k + 40}$ **73.** $\displaystyle\sum_{k=0}^{\infty}\frac{1}{16k^2 + 8k - 3}$

74. $\displaystyle\sum_{k=0}^{\infty}\left(\sin\left(\frac{(k+1)\pi}{2k+1}\right) - \sin\left(\frac{k\pi}{2k-1}\right)\right)$

75. $\displaystyle\sum_{k=0}^{\infty}\left(\frac{1}{4}\right)^k 5^{3-k}$ **76.** $\displaystyle\sum_{k=2}^{\infty}\left(\frac{3}{8}\right)^{3k}$

77. $\displaystyle\sum_{k=1}^{\infty}\frac{(-2)^k}{3^{k+1}}$

78. $\displaystyle\sum_{k=1}^{\infty}\left(\sin^{-1}\left(\frac{1}{k}\right) - \sin^{-1}\left(\frac{1}{k+1}\right)\right)$

79. $\displaystyle\sum_{k=2}^{\infty}\frac{\ln((k+1)k^{-1})}{(\ln k)\ln(k+1)}$ **80.** $\displaystyle\sum_{k=1}^{\infty}\frac{\pi^k}{e^{k+1}}$

81. $\displaystyle\sum_{k=0}^{\infty}\left(3\left(\frac{2}{5}\right)^k - 2\left(\frac{5}{7}\right)^k\right)$ **82.** $\displaystyle\sum_{k=1}^{\infty}\left(2\left(\frac{3}{5}\right)^k + 3\left(\frac{4}{9}\right)^k\right)$

83. $\displaystyle\sum_{k=1}^{\infty}\left(\frac{1}{3}\left(\frac{5}{6}\right)^k + \frac{3}{5}\left(\frac{7}{9}\right)^k\right)$ **84.** $\displaystyle\sum_{k=0}^{\infty}\left(\frac{1}{2}(0.2)^k + \frac{3}{2}(0.8)^k\right)$

85. $\displaystyle\sum_{k=1}^{\infty}\left(\left(\frac{1}{6}\right)^k + \left(\frac{1}{3}\right)^{k-1}\right)$ **86.** $\displaystyle\sum_{k=0}^{\infty}\frac{2 - 3^k}{6^k}$

87. Explain why or why not Determine whether the following statements are true and give an explanation or counterexample.

a. If $\displaystyle\sum_{k=1}^{\infty}a_k$ converges, then $\displaystyle\sum_{k=10}^{\infty}a_k$ converges.

b. If $\displaystyle\sum_{k=1}^{\infty}a_k$ diverges, then $\displaystyle\sum_{k=10}^{\infty}a_k$ diverges.

c. If $\displaystyle\sum a_k$ converges, then $\displaystyle\sum(a_k + 0.0001)$ converges.

d. If $\displaystyle\sum p^k$ diverges, then $\displaystyle\sum(p + 0.001)^k$ diverges, for a fixed real number p.

e. $\displaystyle\sum_{k=1}^{\infty}\left(\frac{\pi}{e}\right)^{-k}$ is a convergent geometric series.

f. If the series $\displaystyle\sum_{k=1}^{\infty}a^k$ converges and $|a| < |b|$, then the series $\displaystyle\sum_{k=1}^{\infty}b^k$ converges.

g. Viewed as a function of r, the series $1 + r + r^2 + r^3 + \cdots$ takes on all values in the interval $\left(\frac{1}{2}, \infty\right)$.

88–89. Binary numbers *Humans use the ten digits 0 through 9 to form base-10 or decimal numbers, whereas computers calculate and store numbers internally as binary numbers—numbers consisting entirely of 0's and 1's. For this exercise, we consider binary numbers that have the form* $0.b_1b_2b_3\ldots$, *where each of the digits* $b_1, b_2, b_3,\ldots$ *is either 0 or 1. The base-10 representation of the binary number* $0.b_1b_2b_3\ldots$ *is the infinite series* $\frac{b_1}{2^1} + \frac{b_2}{2^2} + \frac{b_3}{2^3} + \cdots$.

88. Verify that the base-10 representation of the binary number $0.010101\ldots$ (which can also be written as $0.\overline{01}$) is $\frac{1}{3}$.

89. Computers can store only a finite number of digits and therefore numbers with nonterminating digits must be rounded or truncated before they can be used and stored by a computer.

a. Find the base-10 representation of the binary number $0.\overline{0011}$.

b. Suppose a computer rounds the binary number $0.\overline{0011}$ to $b = 0.0011001100110011$ before storing it. Find the approximate base-10 representation of b, rounding your answer to 8 digits to the right of the decimal place.

90. For what value of r does $1 + r + r^2 + r^3 + \cdots = 10$?

91. For what value of r does $\displaystyle\sum_{k=3}^{\infty}r^{2k} = 10$?

92. Value of a series

  a. Evaluate the series

$$\sum_{k=1}^{\infty} \frac{3^k}{(3^{k+1} - 1)(3^k - 1)}.$$

  (*Hint:* Find constants b and c so that

$$\frac{3^k}{(3^{k+1} - 1)(3^k - 1)} = \frac{b}{3^k - 1} + \frac{c}{3^{k+1} - 1}.)$$

  b. For what values of a does the series

$$\sum_{k=1}^{\infty} \frac{a^k}{(a^{k+1} - 1)(a^k - 1)}$$

  converge, and in those cases, what is its value?

93–95. Loans *Suppose you borrow P dollars from a bank at a monthly interest rate of r and you make monthly payments of M dollars/month to pay off the loan (for example, if your borrow* $30,000 *at a rate of* 1.5% *per month, then* $P = 30,000$ *and* $r = 0.015$). *Each month, the bank first adds interest to the loan balance and then subtracts your monthly payment to determine the new balance on your loan. So if* A_0 *is your original loan balance and* A_n *equals your loan balance after n payments, then your balance after n payments is given by the recursive formula* $A_n = A_{n-1}(1 + r) - M$, $A_0 = P$.

93. Use the recursive formula and geometric sums to obtain the explicit formula $A_n = \left(A_0 - \dfrac{M}{r}\right)(1 + r)^n + \dfrac{M}{r}$.

T 94. Car loan Suppose you borrow $20,000 for a new car at a monthly interest rate of 0.75%. If you make payments of $600/month, after how many months will the loan balance be zero?

T 95. House loan Suppose you take out a home mortgage for $180,000 at a monthly interest rate of 0.5%. If you make payments of $1000/month, after how many months will the loan balance be zero?

Explorations and Challenges

96. Properties proof Use the ideas in the proof of Property 1 of Theorem 10.8 to prove Property 3 of Theorem 10.8.

97. Property of divergent series Prove Property 2 of Theorem 10.8: If Σa_k diverges, then $\Sigma c a_k$ also diverges, for any real number $c \neq 0$.

98. Zeno's paradox The Greek philosopher Zeno of Elea (who lived about 450 B.C.) invented many paradoxes, the most famous of which tells of a race between the swift warrior Achilles and a tortoise. Zeno argued as follows.

The slower when running will never be overtaken by the quicker; for that which is pursuing must first reach the point from which that which is fleeing started, so that the slower must necessarily always be some distance ahead.

In other words, giving the tortoise a head start ensures that Achilles will never overtake the tortoise because every time Achilles reaches the point where the tortoise was, the tortoise has moved ahead. Resolve this paradox by assuming Achilles gives the tortoise a 1-mi head start and runs 5 mi/hr to the tortoise's 1 mi/hr. How far does Achilles run before he overtakes the tortoise, and how long does it take?

99. Archimedes' quadrature of the parabola The Greeks solved several calculus problems almost 2000 years before the discovery of calculus. One example is Archimedes' calculation of the area of the region R bounded by a segment of a parabola, which he did using the "method of exhaustion." As shown in the figure, the idea was to fill R with an infinite sequence of triangles. Archimedes began with an isosceles triangle inscribed in the parabola, with area A_1, and proceeded in stages, with the number of new triangles doubling at each stage. He was able to show (the key to the solution) that at each stage, the area of a new triangle is $1/8$ of the area of a triangle at the previous stage; for example, $A_2 = 1/8\,A_1$, and so forth. Show, as Archimedes did, that the area of R is $4/3$ times the area of A_1.

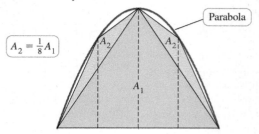

$A_2 = \frac{1}{8}A_1$ Parabola

100. Double glass An insulated window consists of two parallel panes of glass with a small spacing between them. Suppose each pane reflects a fraction p of the incoming light and transmits the remaining light. Considering all reflections of light between the panes, what fraction of the incoming light is ultimately transmitted by the window? Assume the amount of incoming light is 1.

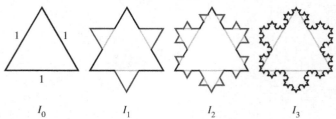

101. Snowflake island fractal The fractal called the *snowflake island* (or *Koch island*) is constructed as follows: Let I_0 be an equilateral triangle with sides of length 1. The figure I_1 is obtained by replacing the middle third of each side of I_0 with a new outward equilateral triangle with sides of length $1/3$ (see figure). The process is repeated, where I_{n+1} is obtained by replacing the middle third of each side of I_n with a new outward equilateral triangle with sides of length $\dfrac{1}{3^{n+1}}$. The limiting figure as $n \to \infty$ is called the snowflake island.

  a. Let L_n be the perimeter of I_n. Show that $\lim_{n\to\infty} L_n = \infty$.

  b. Let A_n be the area of I_n. Find $\lim_{n\to\infty} A_n$. It exists!

I_0 I_1 I_2 I_3

102. Bubbles Imagine a stack of hemispherical soap bubbles with decreasing radii $r_1 = 1, r_2, r_3, \ldots$ (see figure). Let h_n be the distance between the diameters of bubble n and bubble $n + 1$, and let H_n be the total height of the stack with n bubbles.

a. Use the Pythagorean theorem to show that in a stack with n bubbles, $h_1^2 = r_1^2 - r_2^2$, $h_2^2 = r_2^2 - r_3^2$, and so forth. Note that for the last bubble $h_n = r_n$.

b. Use part (a) to show that the height of a stack with n bubbles is

$$H_n = \sqrt{r_1^2 - r_2^2} + \sqrt{r_2^2 - r_3^2} + \cdots + \sqrt{r_{n-1}^2 - r_n^2} + r_n.$$

c. The height of a stack of bubbles depends on how the radii decrease. Suppose $r_1 = 1, r_2 = a, r_3 = a^2, \ldots, r_n = a^{n-1}$, where $0 < a < 1$ is a fixed real number. In terms of a, find the height H_n of a stack with n bubbles.

d. Suppose the stack in part (c) is extended indefinitely ($n \to \infty$). In terms of a, how high would the stack be?

e. Challenge problem: Fix n and determine the sequence of radii $r_1, r_2, r_3, \ldots, r_n$ that maximizes H_n, the height of the stack with n bubbles.

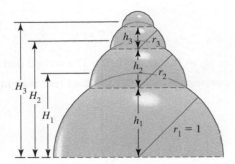

103. Remainder term Consider the geometric series $S = \sum\limits_{k=0}^{\infty} r^k$, which

has the value $\dfrac{1}{1 - r}$ provided $|r| < 1$. Let $S_n = \sum\limits_{k=0}^{n-1} r^k = \dfrac{1 - r^n}{1 - r}$

be the sum of the first n terms. The magnitude of the remainder R_n is the error in approximating S by S_n. Show that

$$R_n = S - S_n = \frac{r^n}{1 - r}.$$

104–107. Comparing remainder terms *Use Exercise 103 to determine how many terms of each series are needed so that the partial sum is within 10^{-6} of the value of the series (that is, to ensure $|R_n| < 10^{-6}$).*

104. a. $\sum\limits_{k=0}^{\infty} 0.6^k$ **b.** $\sum\limits_{k=0}^{\infty} 0.15^k$

105. a. $\sum\limits_{k=0}^{\infty} (-0.8)^k$ **b.** $\sum\limits_{k=0}^{\infty} 0.2^k$

106. a. $\sum\limits_{k=0}^{\infty} 0.72^k$ **b.** $\sum\limits_{k=0}^{\infty} (-0.25)^k$

107. a. $\sum\limits_{k=0}^{\infty} \left(\dfrac{1}{\pi}\right)^k$ **b.** $\sum\limits_{k=0}^{\infty} \left(\dfrac{1}{e}\right)^k$

108. Function defined by a series Suppose a function f is defined by the geometric series $f(x) = \sum\limits_{k=0}^{\infty} x^k$.

a. Evaluate $f(0), f(0.2), f(0.5), f(1)$, and $f(1.5)$, if possible.
b. What is the domain of f?

109. Function defined by a series Suppose a function f is defined by the geometric series $f(x) = \sum\limits_{k=0}^{\infty} (-1)^k x^k$.

a. Evaluate $f(0), f(0.2), f(0.5), f(1)$, and $f(1.5)$, if possible.
b. What is the domain of f?

110. Function defined by a series Suppose a function f is defined by the geometric series $f(x) = \sum\limits_{k=0}^{\infty} x^{2k}$.

a. Evaluate $f(0), f(0.2), f(0.5), f(1)$, and $f(1.5)$, if possible.
b. What is the domain of f?

111. Series in an equation For what values of x does the geometric series

$$f(x) = \sum_{k=0}^{\infty} \left(\frac{1}{1 + x}\right)^k$$

converge? Solve $f(x) = 3$.

QUICK CHECK ANSWERS

1. $S_n = b_0 + b_1 + \cdots + b_{n-1}$ **2.** b and c   **3.** Using the formula, the values are $\frac{3}{2}$ and $\frac{7}{8}$.   **4.** 1   **5.** The first converges because $|r| = 0.2 < 1$; the second diverges because $|r| = 2 > 1$.   **6.** Removing a finite number of terms does not change whether the series converges. It generally changes the value of the series.   **7.** Given the nth term of the sequence of partial sums S_n, the next term is obtained by adding a positive number. So $S_{n+1} > S_n$, which means the sequence is increasing. ◄

10.4 The Divergence and Integral Tests

With geometric series and telescoping series, the sequence of partial sums can be found and its limit can be evaluated (when it exists). Unfortunately, it is difficult or impossible to find an explicit formula for the sequence of partial sums for most infinite series. Therefore, it is difficult to obtain the exact value of most convergent series.

In light of these observations, we now shift our focus and ask a simple *yes* or *no* question: Given an infinite series, does it converge? If the answer is *no*, the series diverges and there are no more questions to ask. If the answer is *yes*, the series converges and it may be possible to estimate its value.

The Divergence Test

One of the simplest and most useful tests determines whether an infinite series *diverges*. Though our focus in this section and the next is on series with positive terms, the Divergence Test applies to series with arbitrary terms.

THEOREM 10.9 Divergence Test

If $\sum a_k$ converges, then $\lim_{k \to \infty} a_k = 0$. Equivalently, if $\lim_{k \to \infty} a_k \neq 0$, then the series diverges.

Important note: Theorem 10.9 cannot be used to conclude that a series converges.

Proof: Let $\{S_n\}$ be the sequence of partial sums for the series $\sum a_k$. Assuming the series converges, it has a finite value, call it S, where

$$S = \lim_{n \to \infty} S_n = \lim_{n \to \infty} S_{n-1}.$$

Note that $S_n - S_{n-1} = a_n$. Therefore,

$$\lim_{n \to \infty} a_n = \lim_{n \to \infty} (S_n - S_{n-1}) = S - S = 0;$$

that is, $\lim_{n \to \infty} a_n = 0$ (which implies $\lim_{k \to \infty} a_k = 0$; **Figure 10.25**). The second part of the test follows immediately because it is the *contrapositive* of the first part (see margin note). ◄

> If the statement *if p, then q* is true, then its contrapositive, *if (not q), then (not p)*, is also true. However its converse, *if q, then p*, is not necessarily true. Try it out on the true statement, *if I live in Paris, then I live in France.*

$$\sum a_k \text{ converges} \Rightarrow \lim_{n \to \infty} S_n = S \Rightarrow \lim_{k \to \infty} a_k = 0$$

Figure 10.25

EXAMPLE 1 Using the Divergence Test Determine whether the following series diverge or state that the Divergence Test is inconclusive.

a. $\displaystyle\sum_{k=0}^{\infty} \frac{k}{k+1}$ **b.** $\displaystyle\sum_{k=1}^{\infty} \frac{1+3^k}{2^k}$ **c.** $\displaystyle\sum_{k=1}^{\infty} \frac{1}{k}$ **d.** $\displaystyle\sum_{k=1}^{\infty} \frac{1}{k^2}$

SOLUTION By the Divergence Test, if $\lim_{k \to \infty} a_k \neq 0$, then the series $\sum a_k$ diverges.

a. $\displaystyle\lim_{k \to \infty} a_k = \lim_{k \to \infty} \frac{k}{k+1} = 1 \neq 0$

The terms of the series do not approach zero, so the series diverges by the Divergence Test.

b. $\lim_{k\to\infty} a_k = \lim_{k\to\infty} \dfrac{1 + 3^k}{2^k}$

$\qquad\qquad = \lim_{k\to\infty} \left(\underbrace{2^{-k}}_{\to\, 0} + \underbrace{\left(\dfrac{3}{2}\right)^k}_{\to\, \infty} \right)$ Simplify.

$\qquad\qquad = \infty$

In this case, $\lim_{k\to\infty} a_k \neq 0$, so the corresponding series $\displaystyle\sum_{k=1}^{\infty} \dfrac{1 + 3^k}{2^k}$ diverges by the Divergence Test.

c. $\lim_{k\to\infty} a_k = \lim_{k\to\infty} \dfrac{1}{k} = 0$

In this case, the terms of the series approach zero, so the Divergence Test is inconclusive. Remember, the Divergence Test cannot be used to prove that a series converges.

QUICK CHECK 1 Apply the Divergence Test to the geometric series $\sum r^k$. For what values of r does the series diverge? ◄

d. $\lim_{k\to\infty} a_k = \lim_{k\to\infty} \dfrac{1}{k^2} = 0$

As in part (c), the terms of the series approach 0, so the Divergence Test is inconclusive.

Related Exercises 9–10 ◄

To summarize: If the terms a_k of a given series do *not* approach zero as $k \to \infty$, then the series diverges. Unfortunately, the test is easy to misuse. It's tempting to conclude that if the terms of the series approach zero, then the series converges. However, look again at the series in Examples 1c and 1d. Although it is true that $\lim_{k\to\infty} a_k = 0$ for both series, we will soon discover that one of them converges while the other diverges. We cannot tell which behavior to expect based only on the observation that $\lim_{k\to\infty} a_k = 0$.

The Harmonic Series

We now look at an example with a surprising result. Consider the infinite series

$$\sum_{k=1}^{\infty} \frac{1}{k} = 1 + \frac{1}{2} + \frac{1}{3} + \frac{1}{4} + \frac{1}{5} + \cdots,$$

a famous series known as the **harmonic series**. Does it converge? As explained in Example 1c, this question cannot be answered by the Divergence Test, despite the fact that $\lim_{k\to\infty} \dfrac{1}{k} = 0$. Suppose instead you try to answer the convergence question by writing out the terms of the sequence of partial sums:

$$S_1 = 1, \qquad\qquad\qquad S_2 = 1 + \frac{1}{2} = \frac{3}{2},$$

$$S_3 = 1 + \frac{1}{2} + \frac{1}{3} = \frac{11}{6}, \qquad S_4 = 1 + \frac{1}{2} + \frac{1}{3} + \frac{1}{4} = \frac{25}{12},$$

and in general,

$$S_n = \sum_{k=1}^{n} \frac{1}{k} = 1 + \frac{1}{2} + \frac{1}{3} + \frac{1}{4} + \cdots + \frac{1}{n}.$$

There is no obvious pattern in this sequence, and in fact, no simple explicit formula for S_n exists, so we analyze the sequence numerically. Have a look at the first 200 terms of the sequence of partial sums shown in **Figure 10.26**. What do you think—does the series converge? The terms of the sequence of partial sums increase, but at a decreasing rate. They could approach a limit or they could increase without bound.

Computing additional terms of the sequence of partial sums does not provide conclusive evidence. Table 10.3 shows that the sum of the first million terms is less than 15; the sum of the first 10^{40} terms—an unimaginably large number of terms—is less than 100. This is a case in which computation alone is not sufficient to determine whether a series converges. We need another way to determine whether the series converges.

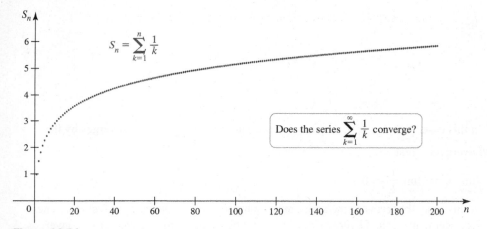

Figure 10.26

Table 10.3

n	S_n	n	S_n
10^3	≈ 7.49	10^{10}	≈ 23.60
10^4	≈ 9.79	10^{20}	≈ 46.63
10^5	≈ 12.09	10^{30}	≈ 69.65
10^6	≈ 14.39	10^{40}	≈ 92.68

Does the series $\sum\limits_{k=1}^{\infty} \dfrac{1}{k}$ converge?

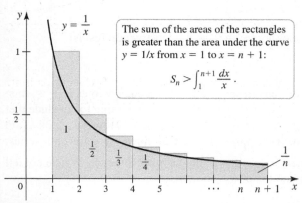

Figure 10.27

The sum of the areas of the rectangles is greater than the area under the curve $y = 1/x$ from $x = 1$ to $x = n + 1$:

$$S_n > \int_1^{n+1} \frac{dx}{x}.$$

▶ Recall that $\int \dfrac{dx}{x} = \ln |x| + C$. In Section 8.9, we showed that $\int_1^{\infty} \dfrac{dx}{x^P}$ diverges for $p \leq 1$. Therefore, $\int_1^{\infty} \dfrac{dx}{x}$ diverges.

Observe that the nth term of the sequence of partial sums,

$$S_n = \sum_{k=1}^{n} \frac{1}{k} = 1 + \frac{1}{2} + \frac{1}{3} + \frac{1}{4} + \cdots + \frac{1}{n},$$

is represented geometrically by a left Riemann sum of the function $y = \dfrac{1}{x}$ on the interval $[1, n + 1]$ (**Figure 10.27**). This fact follows by noticing that the areas of the rectangles, from left to right, are $1, \dfrac{1}{2}, \ldots,$ and $\dfrac{1}{n}$. Comparing the sum of the areas of these n rectangles with the area under the curve, we see that $S_n > \displaystyle\int_1^{n+1} \frac{dx}{x}$. We know that $\displaystyle\int_1^{n+1} \frac{dx}{x} = \ln(n + 1)$ increases without bound as n increases. Because S_n exceeds $\displaystyle\int_1^{n+1} \frac{dx}{x}$, S_n also increases without bound; therefore, $\lim\limits_{n \to \infty} S_n = \infty$ and the harmonic series $\sum\limits_{k=1}^{\infty} \dfrac{1}{k}$ diverges. This argument justifies the following theorem.

THEOREM 10.10   Harmonic Series

The harmonic series $\sum\limits_{k=1}^{\infty} \dfrac{1}{k} = 1 + \dfrac{1}{2} + \dfrac{1}{3} + \dfrac{1}{4} + \dfrac{1}{5} + \cdots$ diverges—even though the terms of the series approach zero.

The ideas used to demonstrate that the harmonic series diverges are now used to prove a new and powerful convergence test. This test and those presented in Section 10.5 apply only to series with positive terms.

The Integral Test

The fact that infinite series are sums and that integrals are limits of sums suggests a connection between series and integrals. The Integral Test exploits this connection.

THEOREM 10.11   Integral Test

Suppose f is a continuous, positive, decreasing function, for $x \geq 1$, and let $a_k = f(k)$, for $k = 1, 2, 3, \ldots$. Then

$$\sum_{k=1}^{\infty} a_k \quad \text{and} \quad \int_1^{\infty} f(x)\, dx$$

either both converge or both diverge. In the case of convergence, the value of the integral is *not* equal to the value of the series.

▶ The Integral Test also applies if the terms of the series a_k are decreasing for $k > N$ for some finite number $N > 1$. The proof can be modified to account for this situation.

Proof: By comparing the shaded regions in **Figure 10.28**, it follows that

$$\sum_{k=2}^{n} a_k < \int_1^n f(x)\, dx < \sum_{k=1}^{n-1} a_k. \tag{1}$$

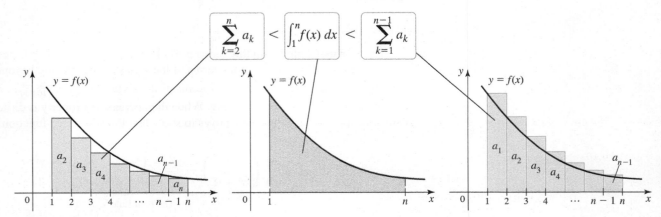

Figure 10.28

The proof must demonstrate two results: If the improper integral $\int_1^\infty f(x)\, dx$ has a finite value, then the infinite series converges, *and* if the infinite series converges, then the improper integral has a finite value. First suppose the improper integral $\int_1^\infty f(x)\, dx$ has a finite value, say I. We have

$$\sum_{k=1}^{n} a_k = a_1 + \sum_{k=2}^{n} a_k \qquad \text{Separate the first term of the series.}$$

$$< a_1 + \int_1^n f(x)\, dx \qquad \text{Left inequality in expression (1)}$$

$$< a_1 + \int_1^\infty f(x)\, dx \qquad f \text{ is positive, so } \int_1^n f(x)\, dx < \int_1^\infty f(x)\, dx.$$

$$= a_1 + I.$$

> In this proof, we rely twice on the Bounded Monotonic Sequence Theorem of Section 10.2: A bounded monotonic sequence converges.

This argument implies that the terms of the sequence of partial sums $S_n = \sum_{k=1}^{n} a_k$ are bounded above by $a_1 + I$. Because $\{S_n\}$ is also increasing (the series consists of positive terms), the sequence of partial sums converges, which means the series $\sum_{k=1}^{\infty} a_k$ converges (to a value less than or equal to $a_1 + I$).

Now suppose the infinite series $\sum_{k=1}^{\infty} a_k$ converges and has a value S. We have

$$\int_1^n f(x)\, dx < \sum_{k=1}^{n-1} a_k \qquad \text{Right inequality in expression (1)}$$

$$< \sum_{k=1}^{\infty} a_k \qquad \text{Terms } a_k \text{ are positive.}$$

$$= S. \qquad \text{Value of infinite series}$$

> An extended version of this proof can be used to show that, in fact,
> $$\int_1^\infty f(x)\, dx < \sum_{k=1}^{\infty} a_k \text{ (strict inequality)}$$
> in all cases.

We see that the sequence $\left\{\int_1^n f(x)\, dx\right\}$ is increasing (because $f(x) > 0$) and bounded above by a fixed number S. Therefore, the improper integral $\int_1^\infty f(x)\, dx = \lim_{n\to\infty} \int_1^n f(x)\, dx$ has a finite value (less than or equal to S).

We have shown that if $\int_1^\infty f(x)\, dx$ is finite, then Σa_k converges, and vice versa. The same inequalities imply that $\int_1^\infty f(x)\, dx$ and Σa_k also diverge together. ◄

The Integral Test is used to determine *whether* a series converges or diverges. For this reason, adding or subtracting a few terms in the series *or* changing the lower limit of integration to another finite point does not change the outcome of the test. Therefore, the test depends on neither the lower index of the series nor the lower limit of the integral.

EXAMPLE 2   Applying the Integral Test Determine whether the following series converge.

a. $\displaystyle\sum_{k=1}^{\infty} \frac{k}{k^2 + 1}$ **b.** $\displaystyle\sum_{k=3}^{\infty} \frac{1}{\sqrt{2k - 5}}$ **c.** $\displaystyle\sum_{k=0}^{\infty} \frac{1}{k^2 + 4}$

SOLUTION

a. The function associated with this series is $f(x) = x/(x^2 + 1)$, which is positive, for $x \geq 1$. We must also show that the terms of the series are decreasing beyond some fixed term of the series. The first few terms of the series are $\left\{ \frac{1}{2}, \frac{2}{5}, \frac{3}{10}, \frac{4}{17}, \ldots \right\}$, and it appears that the terms are decreasing. When the decreasing property is difficult to confirm, one approach is to use derivatives to show that the associated function is decreasing. In this case, we have

$$f'(x) = \underbrace{\frac{d}{dx}\left(\frac{x}{x^2 + 1} \right) = \frac{x^2 + 1 - 2x^2}{(x^2 + 1)^2}}_{\text{Quotient Rule}} = \frac{1 - x^2}{(x^2 + 1)^2}.$$

For $x > 1$, $f'(x) < 0$, which implies that the function and the terms of the series are decreasing. The integral that determines convergence is

$$\int_1^{\infty} \frac{x}{x^2 + 1}\, dx = \lim_{b \to \infty} \int_1^b \frac{x}{x^2 + 1}\, dx \qquad \text{Definition of improper integral}$$

$$= \lim_{b \to \infty} \frac{1}{2} \ln(x^2 + 1)\Big|_1^b \qquad \text{Evaluate integral.}$$

$$= \frac{1}{2} \lim_{b \to \infty} \left(\ln(b^2 + 1) - \ln 2 \right) \qquad \text{Simplify.}$$

$$= \infty. \qquad \lim_{b \to \infty} \ln(b^2 + 1) = \infty$$

Because the integral diverges, the series diverges.

b. The Integral Test may be modified to accommodate initial indices other than $k = 1$. The terms of this series decrease, for $k \geq 3$. In this case, the relevant integral is

$$\int_3^{\infty} \frac{dx}{\sqrt{2x - 5}} = \lim_{b \to \infty} \int_3^b \frac{dx}{\sqrt{2x - 5}} \qquad \text{Definition of improper integral}$$

$$= \lim_{b \to \infty} \sqrt{2x - 5}\,\Big|_3^b \qquad \text{Evaluate integral.}$$

$$= \infty. \qquad \lim_{b \to \infty} \sqrt{2b - 5} = \infty$$

Because the integral diverges, the series also diverges.

c. The terms of the series are positive and decreasing, for $k \geq 0$. The relevant integral is

$$\int_0^{\infty} \frac{dx}{x^2 + 4} = \lim_{b \to \infty} \int_0^b \frac{dx}{x^2 + 4} \qquad \text{Definition of improper integral}$$

$$= \lim_{b \to \infty} \frac{1}{2} \tan^{-1} \frac{x}{2}\Big|_0^b \qquad \text{Evaluate integral.}$$

$$= \frac{1}{2} \underbrace{\lim_{b \to \infty} \tan^{-1} \frac{b}{2}}_{\frac{\pi}{2}} - \tan^{-1} 0 \qquad \text{Simplify.}$$

$$= \frac{\pi}{4}. \qquad \tan^{-1} x \to \frac{\pi}{2}, \text{ as } x \to \infty.$$

Because the integral is finite (equivalently, it converges), the infinite series also converges $\left(\text{but not to } \dfrac{\pi}{4} \right)$.

Related Exercises 17, 20 ◀

The *p*-series

The Integral Test is used to prove Theorem 10.12, which addresses the convergence of an entire family of infinite series known as the *p-series*.

THEOREM 10.12   Convergence of the *p*-series

The *p*-series $\sum_{k=1}^{\infty} \dfrac{1}{k^p}$ converges for $p > 1$ and diverges for $p \le 1$.

QUICK CHECK 2 Which of the following series are *p*-series, and which series converge?

a. $\sum_{k=1}^{\infty} k^{-0.8}$ **b.** $\sum_{k=1}^{\infty} 2^{-k}$ **c.** $\sum_{k=10}^{\infty} k^{-4}$ ◄

Proof: To apply the Integral Test, observe that the terms of the given series are positive and decreasing, for $p > 0$. The function associated with the series is $f(x) = \dfrac{1}{x^p}$. The relevant integral is $\int_{1}^{\infty} \dfrac{dx}{x^p}$. Appealing to Example 2 in Section 8.9, recall that this improper integral converges for $p > 1$ and diverges for $p \le 1$. Therefore, by the Integral Test, the *p*-series $\sum_{k=1}^{\infty} \dfrac{1}{k^p}$ converges for $p > 1$ and diverges for $0 < p \le 1$. For $p \le 0$, the series diverges by the Divergence Test.   ◄

EXAMPLE 3   Using the *p*-series test   Determine whether the following series converge or diverge.

a. $\sum_{k=1}^{\infty} k^{-3}$ **b.** $\sum_{k=1}^{\infty} \dfrac{1}{\sqrt[4]{k^3}}$ **c.** $\sum_{k=4}^{\infty} \dfrac{1}{(k-1)^2}$

SOLUTION

a. Because $\sum_{k=1}^{\infty} k^{-3} = \sum_{k=1}^{\infty} \dfrac{1}{k^3}$ is a *p*-series with $p = 3$, it converges by Theorem 10.12.

b. This series is a *p*-series with $p = \frac{3}{4}$. By Theorem 10.12, it diverges.

c. The series

$$\sum_{k=4}^{\infty} \frac{1}{(k-1)^2} = \sum_{k=3}^{\infty} \frac{1}{k^2} = \frac{1}{3^2} + \frac{1}{4^2} + \frac{1}{5^2} + \cdots$$

is a convergent *p*-series ($p = 2$) without the first two terms. Recall that adding or removing a finite number of terms does not affect the convergence of a series (Property 5, Theorem 10.8). Therefore, the given series converges.

Related Exercises 23–24 ◄

Estimating the Value of Infinite Series

The Integral Test is powerful in its own right, but it comes with an added bonus. It can be used to estimate the value of a convergent series with positive terms. We define the **remainder** to be the error in approximating a convergent series by the sum of its first *n* terms; that is,

$$R_n = \underbrace{\sum_{k=1}^{\infty} a_k}_{\substack{\text{value of} \\ \text{series}}} - \underbrace{\sum_{k=1}^{n} a_k}_{\substack{\text{approximation based} \\ \text{on first } n \text{ terms}}} = a_{n+1} + a_{n+2} + a_{n+3} + \cdots.$$

QUICK CHECK 3 If Σa_k is a convergent series of positive terms, why is $R_n > 0$? ◄

The remainder consists of the *tail* of the series—those terms beyond a_n. For series with positive terms, the remainder is positive.

We now argue much as we did in the proof of the Integral Test. Let f be a continuous, positive, decreasing function such that $f(k) = a_k$, for all relevant k. From **Figure 10.29**, we see that $\int_{n+1}^{\infty} f(x)\, dx < R_n$.

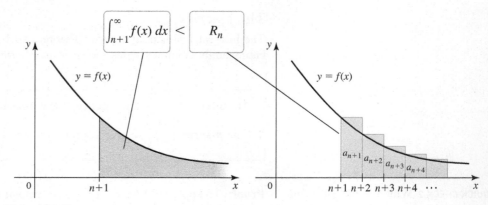

Figure 10.29

Similarly, **Figure 10.30** shows that $R_n < \int_n^\infty f(x)\, dx$.

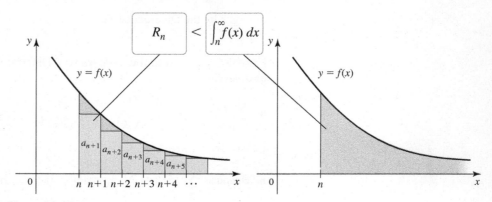

Figure 10.30

Combining these two inequalities, the remainder is squeezed between two integrals:

$$\int_{n+1}^\infty f(x)\, dx < R_n < \int_n^\infty f(x)\, dx. \tag{2}$$

If the integrals can be evaluated, this result provides an estimate of the remainder.

There is, however, another equally useful way to express this result. Notice that the value of the series is

$$S = \sum_{k=1}^\infty a_k = \underbrace{\sum_{k=1}^n a_k}_{S_n} + R_n,$$

which is the sum of the first n terms S_n and the remainder R_n. Adding S_n to each term of (2), we have

$$\underbrace{S_n + \int_{n+1}^\infty f(x)\, dx}_{L_n} < \underbrace{\sum_{k=1}^\infty a_k}_{S_n + R_n = S} < \underbrace{S_n + \int_n^\infty f(x)\, dx}_{U_n}.$$

These inequalities can be abbreviated as $L_n < S < U_n$, where S is the exact value of the series, and L_n and U_n are lower and upper bounds for S, respectively. If the integrals in these bounds can be evaluated, it is straightforward to compute S_n (by summing the first n terms of the series) and to compute both L_n and U_n.

> **THEOREM 10.13 Estimating Series with Positive Terms**
> Let f be a continuous, positive, decreasing function, for $x \geq 1$, and let $a_k = f(k)$,
> for $k = 1, 2, 3, \ldots$. Let $S = \sum_{k=1}^{\infty} a_k$ be a convergent series and let $S_n = \sum_{k=1}^{n} a_k$ be
> the sum of the first n terms of the series. The remainder $R_n = S - S_n$ satisfies
>
> $$R_n < \int_n^{\infty} f(x)\, dx.$$
>
> Furthermore, the exact value of the series is bounded as follows:
>
> $$L_n = S_n + \int_{n+1}^{\infty} f(x)\, dx < \sum_{k=1}^{\infty} a_k < S_n + \int_n^{\infty} f(x)\, dx = U_n.$$

EXAMPLE 4 Approximating a *p*-series

a. How many terms of the convergent *p*-series $\sum_{k=1}^{\infty} \dfrac{1}{k^2}$ must be summed to obtain an approximation that is within 10^{-3} of the exact value of the series?

b. Find an approximation to the series using 50 terms of the series.

SOLUTION The function associated with this series is $f(x) = 1/x^2$.

a. Using the bound on the remainder, we have

$$R_n < \int_n^{\infty} f(x)\, dx = \int_n^{\infty} \frac{dx}{x^2} = \frac{1}{n}.$$

To ensure that $R_n < 10^{-3}$, we must choose n so that $1/n < 10^{-3}$, which implies that $n > 1000$. In other words, we must sum at least 1001 terms of the series to be sure that the remainder is less than 10^{-3}.

b. Using the bounds on the series, we have $L_n < S < U_n$, where S is the exact value of the series, and

$$L_n = S_n + \int_{n+1}^{\infty} \frac{dx}{x^2} = S_n + \frac{1}{n+1} \quad \text{and} \quad U_n = S_n + \int_n^{\infty} \frac{dx}{x^2} = S_n + \frac{1}{n}.$$

> The values of *p*-series with even values of *p* are generally known. For example, with $p = 2$, the series converges to $\pi^2/6$ (a proof is outlined in Exercise 68); with $p = 4$, the series converges to $\pi^4/90$. The values of *p*-series with odd values of *p* are not known.

Therefore, the series is bounded as follows:

$$S_n + \frac{1}{n+1} < S < S_n + \frac{1}{n},$$

where S_n is the sum of the first n terms. Using a calculator to sum the first 50 terms of the series, we find that $S_{50} \approx 1.625133$. The exact value of the series is in the interval

$$S_{50} + \frac{1}{50 + 1} < S < S_{50} + \frac{1}{50},$$

or $1.644741 < S < 1.645133$. Taking the average of these two bounds as our approximation of S, we find that $S \approx 1.644937$. This estimate is better than simply using S_{50}. **Figure 10.31a** shows the lower and upper bounds, L_n and U_n, respectively, for $n = 1, 2, \ldots, 50$. **Figure 10.31b** shows these bounds on an enlarged scale for $n = 50, 51, \ldots, 100$. These figures illustrate how the exact value of the series is squeezed into a narrowing interval as n increases.

(a)

(b)

Figure 10.31

Related Exercises 41, 43 ◀

SECTION 10.4 EXERCISES

Getting Started

1. If we know that $\lim\limits_{k\to\infty} a_k = 1$, then what can we say about $\sum\limits_{k=1}^{\infty} a_k$?

2. Is it true that if the terms of a series of positive terms decrease to zero, then the series converges? Explain using an example.

3. If we know that $\sum\limits_{k=1}^{\infty} a_k = 10,000$, then what can we say about $\lim\limits_{k\to\infty} a_k$?

4. For what values of p does the series $\sum\limits_{k=1}^{\infty} \dfrac{1}{k^p}$ converge? For what values of p does it diverge?

5. For what values of p does the series $\sum\limits_{k=10}^{\infty} \dfrac{1}{k^p}$ converge (initial index is 10)? For what values of p does it diverge?

6. Explain why the sequence of partial sums for a series of positive terms is an increasing sequence.

7. Define the remainder of an infinite series.

8. If a series of positive terms converges, does it follow that the remainder R_n must decrease to zero as $n \to \infty$? Explain.

Practice Exercises

9–16. Divergence Test *Use the Divergence Test to determine whether the following series diverge or state that the test is inconclusive.*

9. $\sum\limits_{k=0}^{\infty} \dfrac{k}{2k+1}$

10. $\sum\limits_{k=1}^{\infty} \dfrac{k}{k^2+1}$

11. $\sum\limits_{k=0}^{\infty} \dfrac{1}{1000+k}$

12. $\sum\limits_{k=1}^{\infty} \dfrac{k^3}{k^3+1}$

13. $\sum\limits_{k=2}^{\infty} \dfrac{k}{\ln k}$

14. $\sum\limits_{k=1}^{\infty} \dfrac{k^2}{2^k}$

15. $\sum\limits_{k=2}^{\infty} \dfrac{\sqrt{k}}{\ln^{10} k}$

16. $\sum\limits_{k=1}^{\infty} \dfrac{k^3}{k!}$

17–22. Integral Test *Use the Integral Test to determine whether the following series converge after showing that the conditions of the Integral Test are satisfied.*

17. $\sum\limits_{k=0}^{\infty} \dfrac{1}{\sqrt{k+8}}$

18. $\sum\limits_{k=1}^{\infty} \dfrac{1}{(2k+4)^2}$

19. $\sum\limits_{k=1}^{\infty} \dfrac{1}{\sqrt[3]{5k+3}}$

20. $\sum\limits_{k=1}^{\infty} \dfrac{e^k}{1+e^{2k}}$

21. $\sum\limits_{k=1}^{\infty} ke^{-2k^2}$

22. $\sum\limits_{k=2}^{\infty} \dfrac{1}{k(\ln k)^2}$

23–38. Divergence, Integral, and *p*-series Tests *Use the Divergence Test, the Integral Test, or the p-series test to determine whether the following series converge.*

23. $\sum\limits_{k=1}^{\infty} k^{-1/5}$

24. $\sum\limits_{k=1}^{\infty} \dfrac{1}{\sqrt{k^3}}$

25. $\sum\limits_{k=1}^{\infty} \dfrac{k^3}{e^{k^4}}$

26. $\sum\limits_{k=1}^{\infty} \dfrac{k^2}{k^3+5}$

27. $\sum\limits_{k=1}^{\infty} k^{1/k}$

28. $\sum\limits_{k=1}^{\infty} \dfrac{\sqrt{k^2+1}}{k}$

29. $\sum\limits_{k=1}^{\infty} \dfrac{1}{k^{10}}$

30. $\sum\limits_{k=2}^{\infty} \dfrac{k^e}{k^\pi}$

31. $\sum\limits_{k=3}^{\infty} \dfrac{1}{(k-2)^4}$

32. $\sum\limits_{k=1}^{\infty} 2k^{-3/2}$

33. $\sum\limits_{k=1}^{\infty} \dfrac{k}{e^k}$

34. $\sum\limits_{k=3}^{\infty} \dfrac{1}{k(\ln k)\ln\ln k}$

35. $\sum\limits_{k=1}^{\infty} \left(\dfrac{k}{k+10}\right)^k$

36. $\sum\limits_{k=1}^{\infty} \dfrac{k}{(k^2+1)^3}$

37. $\sum\limits_{k=1}^{\infty} \dfrac{1}{\sqrt[3]{k}}$

38. $\sum\limits_{k=1}^{\infty} \dfrac{1}{\sqrt[3]{27k^2}}$

T 39–40. Lower and upper bounds of a series *For each convergent series and given value of n, use Theorem 10.13 to complete the following.*

a. *Use S_n to estimate the sum of the series.*
b. *Find an upper bound for the remainder R_n.*
c. *Find lower and upper bounds (L_n and U_n, respectively) for the exact value of the series.*

39. $\displaystyle\sum_{k=1}^{\infty} \frac{1}{k^7}; n = 2$ **40.** $\displaystyle\sum_{k=1}^{\infty} \frac{1}{k^3}; n = 5$

T 41–44. Remainders and estimates *Consider the following convergent series.*

a. *Find an upper bound for the remainder in terms of n.*
b. *Find how many terms are needed to ensure that the remainder is less than 10^{-3}.*
c. *Find lower and upper bounds (L_n and U_n, respectively) on the exact value of the series.*
d. *Find an interval in which the value of the series must lie if you approximate it using ten terms of the series.*

41. $\displaystyle\sum_{k=1}^{\infty} \frac{1}{k^6}$ **42.** $\displaystyle\sum_{k=2}^{\infty} \frac{1}{k(\ln k)^2}$

43. $\displaystyle\sum_{k=1}^{\infty} \frac{1}{3^k}$ **44.** $\displaystyle\sum_{k=1}^{\infty} ke^{-k^2}$

T 45. Estimate the series $\displaystyle\sum_{k=1}^{\infty} \frac{1}{k^7}$ to within 10^{-4} of its exact value.

T 46. Estimate the series $\displaystyle\sum_{k=1}^{\infty} \frac{1}{(3k + 2)^2}$ to within 10^{-3} of its exact value.

47. Explain why or why not Determine whether the following statements are true and give an explanation or counterexample.

a. The sum $\displaystyle\sum_{k=1}^{\infty} \frac{1}{3^k}$ is a p-series.

b. The sum $\displaystyle\sum_{k=3}^{\infty} \frac{1}{\sqrt{k} - 2}$ is a p-series.

c. Suppose f is a continuous, positive, decreasing function, for $x \geq 1$, and $a_k = f(k)$, for $k = 1, 2, 3, \ldots$. If $\displaystyle\sum_{k=1}^{\infty} a_k$ converges to L, then $\displaystyle\int_1^{\infty} f(x)\, dx$ converges to L.

d. Every partial sum S_n of the series $\displaystyle\sum_{k=1}^{\infty} \frac{1}{k^2}$ underestimates the exact value of $\displaystyle\sum_{k=1}^{\infty} \frac{1}{k^2}$.

e. If $\sum k^{-p}$ converges, then $\sum k^{-p+0.001}$ converges.

f. If $\displaystyle\lim_{k \to \infty} a_k = 0$, then $\sum a_k$ converges.

48–63. Choose your test *Determine whether the following series converge or diverge using the properties and tests introduced in Sections 10.3 and 10.4.*

48. $\displaystyle\sum_{k=1}^{\infty} \ln\left(\frac{2k^6}{1 + k^6}\right)$ **49.** $\displaystyle\sum_{k=2}^{\infty} \frac{1}{e^k}$

50. $\displaystyle\sum_{k=1}^{\infty} \sqrt{\frac{k + 1}{k}}$ **51.** $\displaystyle\sum_{k=1}^{\infty} \frac{1}{(3k + 1)(3k + 4)}$

52. $\displaystyle\sum_{k=0}^{\infty} \frac{10}{k^2 + 9}$ **53.** $\displaystyle\sum_{k=1}^{\infty} \frac{k}{\sqrt{k^2 + 1}}$

54. $\displaystyle\sum_{k=1}^{\infty} \frac{2^k + 3^k}{4^k}$ **55.** $\displaystyle\sum_{k=3}^{\infty} \frac{4}{k\sqrt{\ln k}}$

56. $\displaystyle\sum_{k=1}^{\infty} \frac{1}{k^2 + 7k + 12}$ **57.** $\displaystyle\sum_{k=1}^{\infty} \left(\frac{5}{6}\right)^{-k}$

58. $\displaystyle\sum_{k=1}^{\infty} \left(\frac{k + 6}{k}\right)^k$ **59.** $\displaystyle\sum_{k=4}^{\infty} \frac{3}{(k - 3)^4}$

60. $\displaystyle\sum_{k=1}^{\infty} \frac{3^k}{k^2 + 1}$ **61.** $\displaystyle\frac{2}{4^2} + \frac{2}{5^2} + \frac{2}{6^2} + \cdots$

62. $\displaystyle\frac{1}{3} + \frac{1}{5} + \frac{1}{7} + \cdots$ **63.** $\displaystyle\sum_{k=1}^{\infty} \frac{3^{k+2}}{5^k}$

64. Log p-series Consider the series $\displaystyle\sum_{k=2}^{\infty} \frac{1}{k(\ln k)^p}$, where p is a real number.

a. Use the Integral Test to determine the values of p for which this series converges.
b. Does this series converge faster for $p = 2$ or $p = 3$? Explain.

65. Loglog p-series Consider the series $\displaystyle\sum_{k=2}^{\infty} \frac{1}{k(\ln k)(\ln \ln k)^p}$, where p is a real number.

a. For what values of p does this series converge?
b. Which of the following series converges faster? Explain.

$$\sum_{k=2}^{\infty} \frac{1}{k(\ln k)^2} \quad \text{or} \quad \sum_{k=3}^{\infty} \frac{1}{k(\ln k)(\ln \ln k)^2}?$$

Explorations and Challenges

66. Prime numbers The prime numbers are those positive integers that are divisible by only 1 and themselves (for example, 2, 3, 5, 7, 11, 13, ...). A celebrated theorem states that the sequence of prime numbers $\{p_k\}$ satisfies $\displaystyle\lim_{k \to \infty} \frac{p_k}{k \ln k} = 1$. Show that $\displaystyle\sum_{k=2}^{\infty} \frac{1}{k \ln k}$ diverges, which implies that the series $\displaystyle\sum_{k=1}^{\infty} \frac{1}{p_k}$ diverges.

T 67. The zeta function The Riemann zeta function is the subject of extensive research and is associated with several renowned unsolved problems. It is defined by $\displaystyle\zeta(x) = \sum_{k=1}^{\infty} \frac{1}{k^x}$. When x is a real number, the zeta function becomes a p-series. For even positive integers p, the value of $\zeta(p)$ is known exactly. For example,

$$\sum_{k=1}^{\infty} \frac{1}{k^2} = \frac{\pi^2}{6}, \quad \sum_{k=1}^{\infty} \frac{1}{k^4} = \frac{\pi^4}{90}, \quad \text{and} \quad \sum_{k=1}^{\infty} \frac{1}{k^6} = \frac{\pi^6}{945}, \ldots$$

Use the estimation techniques described in the text to approximate $\zeta(3)$ and $\zeta(5)$ (whose values are not known exactly) with a remainder less than 10^{-3}.

68. Showing that $\displaystyle\sum_{k=1}^{\infty} \frac{1}{k^2} = \frac{\pi^2}{6}$ In 1734, Leonhard Euler informally proved that $\displaystyle\sum_{k=1}^{\infty} \frac{1}{k^2} = \frac{\pi^2}{6}$. An elegant proof is outlined here that uses the inequality

$$\cot^2 x < \frac{1}{x^2} < 1 + \cot^2 x \left(\text{provided } 0 < x < \frac{\pi}{2}\right)$$

and the identity

$$\sum_{k=1}^{n} \cot^2 k\theta = \frac{n(2n-1)}{3}, \text{ for } n = 1, 2, 3, \ldots, \text{ where } \theta = \frac{\pi}{2n+1}.$$

a. Show that $\displaystyle\sum_{k=1}^{n} \cot^2 k\theta < \frac{1}{\theta^2} \sum_{k=1}^{n} \frac{1}{k^2} < n + \sum_{k=1}^{n} \cot^2 k\theta$.

b. Use the inequality in part (a) to show that

$$\frac{n(2n-1)\pi^2}{3(2n+1)^2} < \sum_{k=1}^{n} \frac{1}{k^2} < \frac{n(2n+2)\pi^2}{3(2n+1)^2}.$$

c. Use the Squeeze Theorem to conclude that $\displaystyle\sum_{k=1}^{\infty} \frac{1}{k^2} = \frac{\pi^2}{6}$.

(*Source: The College Mathematics Journal, 24, 5, Nov 1993*)

69. Reciprocals of odd squares Assume $\displaystyle\sum_{k=1}^{\infty} \frac{1}{k^2} = \frac{\pi^2}{6}$ (Exercises 67 and 68) and the terms of this series may be rearranged without changing the value of the series. Determine the sum of the reciprocals of the squares of the odd positive integers.

70. Gabriel's wedding cake Consider a wedding cake of infinite height, each layer of which is a right circular cylinder of height 1. The bottom layer of the cake has a radius of 1, the second layer has a radius of $1/2$, the third layer has a radius of $1/3$, and the nth layer has a radius of $1/n$ (see figure).

a. Find the area of the lateral (vertical) sides of the wedding cake.
b. Determine the volume of the cake. (*Hint:* Use the result of Exercise 68.)
c. Comment on your answers to parts (a) and (b).

(*Source: The College Mathematics Journal, 30, 1, Jan 1999*)

$\vdots$

71. A divergence proof Give an argument similar to that given in the text for the harmonic series to show that $\displaystyle\sum_{k=1}^{\infty} \frac{1}{\sqrt{k}}$ diverges.

T 72. Shifted p-series Consider the sequence $\{F_n\}$ defined by

$$F_n = \sum_{k=1}^{\infty} \frac{1}{k(k+n)},$$

for $n = 0, 1, 2, \ldots$. When $n = 0$, the series is a p-series, and we have $F_0 = \dfrac{\pi^2}{6}$ (Exercises 67 and 68).

a. Explain why $\{F_n\}$ is a decreasing sequence.
b. Plot $\{F_n\}$, for $n = 1, 2, \ldots, 20$.
c. Based on your experiments, make a conjecture about $\displaystyle\lim_{n\to\infty} F_n$.

73. A sequence of sums Consider the sequence $\{x_n\}$ defined for $n = 1, 2, 3, \ldots$ by

$$x_n = \sum_{k=n+1}^{2n} \frac{1}{k} = \frac{1}{n+1} + \frac{1}{n+2} + \cdots + \frac{1}{2n}.$$

a. Write out the terms x_1, x_2, x_3.
b. Show that $\dfrac{1}{2} \le x_n < 1$, for $n = 1, 2, 3, \ldots$.
c. Show that x_n is the right Riemann sum for $\displaystyle\int_1^2 \frac{dx}{x}$ using n subintervals.
d. Conclude that $\displaystyle\lim_{n\to\infty} x_n = \ln 2$.

T 74. The harmonic series and Euler's constant

a. Sketch the function $f(x) = 1/x$ on the interval $[1, n+1]$, where n is a positive integer. Use this graph to verify that

$$\ln(n+1) < 1 + \frac{1}{2} + \frac{1}{3} + \cdots + \frac{1}{n} < 1 + \ln n.$$

b. Let S_n be the sum of the first n terms of the harmonic series, so part (a) says $\ln(n+1) < S_n < 1 + \ln n$. Define the new sequence $\{E_n\}$ by

$$E_n = S_n - \ln(n+1), \quad \text{for } n = 1, 2, 3, \ldots.$$

Show that $E_n > 0$ for $n = 1, 2, 3, \ldots$.
c. Using a figure similar to that used in part (a), show that

$$\frac{1}{n+1} > \ln(n+2) - \ln(n+1).$$

d. Use parts (a) and (c) to show that $\{E_n\}$ is an increasing sequence ($E_{n+1} > E_n$).
e. Use part (a) to show that $\{E_n\}$ is bounded above by 1.
f. Conclude from parts (d) and (e) that $\{E_n\}$ has a limit less than or equal to 1. This limit is known as **Euler's constant** and is denoted γ (the Greek lowercase gamma).
g. By computing terms of $\{E_n\}$, estimate the value of γ and compare it to the value $\gamma \approx 0.5772$. (It has been conjectured that γ is irrational.)
h. The preceding arguments show that the sum of the first n terms of the harmonic series satisfy $S_n \approx 0.5772 + \ln(n+1)$. How many terms must be summed for the sum to exceed 10?

75. The harmonic series and the Fibonacci sequence The Fibonacci sequence $\{1, 1, 2, 3, 5, 8, 13, \ldots\}$ is generated by the recurrence relation

$$f_{n+1} = f_n + f_{n-1}, \text{ for } n = 1, 2, 3, \ldots, \text{ where } f_0 = 1, f_1 = 1.$$

a. It can be shown that the sequence of ratios of successive terms of the sequence $\left\{\dfrac{f_{n+1}}{f_n}\right\}$ has a limit φ. Divide both sides of the recurrence relation by f_n, take the limit as $n \to \infty$, and show that $\varphi = \displaystyle\lim_{n\to\infty} \frac{f_{n+1}}{f_n} = \frac{1+\sqrt{5}}{2} \approx 1.618$.

b. Show that $\displaystyle\lim_{n\to\infty} \frac{f_{n-1}}{f_{n+1}} = 1 - \frac{1}{\varphi} \approx 0.382$.

c. Now consider the harmonic series and group terms as follows:

$$\sum_{k=1}^{\infty} \frac{1}{k} = 1 + \frac{1}{2} + \frac{1}{3} + \underbrace{\left(\frac{1}{4} + \frac{1}{5}\right)}_{2 \text{ terms}} + \underbrace{\left(\frac{1}{6} + \frac{1}{7} + \frac{1}{8}\right)}_{3 \text{ terms}}$$

$$+ \underbrace{\left(\frac{1}{9} + \cdots + \frac{1}{13}\right)}_{5 \text{ terms}} + \cdots.$$

With the Fibonacci sequence in mind, show that

$$\sum_{k=1}^{\infty} \frac{1}{k} \geq 1 + \frac{1}{2} + \frac{1}{3} + \frac{2}{5} + \frac{3}{8} + \frac{5}{13} + \cdots = 1 + \sum_{k=1}^{\infty} \frac{f_{k-1}}{f_{k+1}}.$$

d. Use part (b) to conclude that the harmonic series diverges.

(*Source: The College Mathematics Journal*, 43, May 2012)

76. Stacking dominoes Consider a set of identical dominoes that are 2 in long. The dominoes are stacked on top of each other with their long edges aligned so that each domino overhangs the one beneath it *as far as possible* (see figure).

a. If there are n dominoes in the stack, what is the *greatest* distance that the top domino can be made to overhang the bottom domino? (*Hint:* Put the nth domino beneath the previous $n - 1$ dominoes.)

b. If we allow for infinitely many dominoes in the stack, what is the greatest distance that the top domino can be made to overhang the bottom domino?

QUICK CHECK ANSWERS

1. The series diverges for $|r| \geq 1$. **2. a.** Divergent p-series
b. Convergent geometric series **c.** Convergent p-series
3. The remainder is $R_n = a_{n+1} + a_{n+2} + \cdots$, which consists of positive numbers. ◄

10.5 Comparison Tests

The Integral Test from the previous section is easily applied to a series such as $\displaystyle\sum_{k=1}^{\infty} \frac{4}{4 + k^2}$ because $\displaystyle\int_{1}^{\infty} \frac{4}{4 + x^2}\, dx$ is a standard integral. However, using the Integral Test to determine whether $\displaystyle\sum_{k=1}^{\infty} \frac{5k^4 - 2k^2 + 3}{2k^6 - k + 5}$ converges is decidedly more difficult. For this reason, we develop two additional convergence tests, called *comparison tests*, and we use one of them in Example 2a to investigate the convergence of $\displaystyle\sum_{k=1}^{\infty} \frac{5k^4 - 2k^2 + 3}{2k^6 - k + 5}$. As with the Integral Test, both comparison tests apply only to series with positive terms.

The Comparison Test

Comparison tests use known series to test the convergence of unknown series. The first test is the Direct Comparison Test or simply the Comparison Test.

> Whether a series converges depends on the behavior of terms in the tail (large values of the index). So the inequalities $a_k \leq b_k$ and $b_k \leq a_k$ need not hold for all terms of the series. They must hold for all $k > N$ for some positive integer N.

THEOREM 10.14   Comparison Test

Let Σa_k and Σb_k be series with positive terms.

1. If $a_k \leq b_k$ and Σb_k converges, then Σa_k converges.
2. If $b_k \leq a_k$ and Σb_k diverges, then Σa_k diverges.

Proof: Assume Σb_k converges, which means that Σb_k has a finite value B. The sequence of partial sums for Σa_k satisfies

$$S_n = \sum_{k=1}^{n} a_k \leq \sum_{k=1}^{n} b_k \quad a_k \leq b_k$$

$$< \sum_{k=1}^{\infty} b_k \qquad \text{Positive terms are added to a finite sum.}$$

$$= B. \qquad \text{Value of series}$$

Therefore, the sequence of partial sums for Σa_k is increasing and bounded above by B. By the Bounded Monotonic Sequence Theorem (Theorem 10.5), the sequence of partial sums of Σa_k has a limit, which implies that Σa_k converges. The second case of the theorem is proved in a similar way. ◄

The Comparison Test can be illustrated with graphs of sequences of partial sums. Consider the series

$$\sum_{k=1}^{\infty} a_k = \sum_{k=1}^{\infty} \frac{1}{k^2 + 10} \quad \text{and} \quad \sum_{k=1}^{\infty} b_k = \sum_{k=1}^{\infty} \frac{1}{k^2}.$$

Because $\frac{1}{k^2 + 10} < \frac{1}{k^2}$, it follows that $a_k < b_k$, for $k \geq 1$. Furthermore, $\sum b_k$ is a convergent p-series. By the Comparison Test, we conclude that $\sum a_k$ also converges (**Figure 10.32**). The second case of the Comparison Test is illustrated with the series

$$\sum_{k=4}^{\infty} a_k = \sum_{k=4}^{\infty} \frac{1}{\sqrt{k} - 3} \quad \text{and} \quad \sum_{k=4}^{\infty} b_k = \sum_{k=4}^{\infty} \frac{1}{\sqrt{k}}.$$

Now $\frac{1}{\sqrt{k}} < \frac{1}{\sqrt{k} - 3}$, for $k \geq 4$. Therefore, $b_k < a_k$, for $k \geq 4$. Because $\sum b_k$ is a divergent p-series, by the Comparison Test, $\sum a_k$ also diverges. **Figure 10.33** shows that the sequence of partial sums for $\sum a_k$ lies above the sequence of partial sums for $\sum b_k$. Because the sequence of partial sums for $\sum b_k$ diverges, the sequence of partial sums for $\sum a_k$ also diverges.

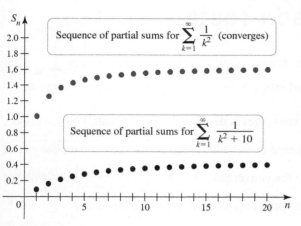

Figure 10.32

Figure 10.33

The key in using the Comparison Test is finding an appropriate comparison series. Plenty of practice will enable you to spot patterns and choose good comparison series.

EXAMPLE 1   Using the Comparison Test Determine whether the following series converge.

a. $\displaystyle\sum_{k=1}^{\infty} \frac{k^3}{2k^4 - 1}$ **b.** $\displaystyle\sum_{k=2}^{\infty} \frac{\ln k}{k^3}$

SOLUTION In using comparison tests, it's helpful to get a feel for how the terms of the given series behave.

> If $\sum a_k$ diverges, then $\sum c a_k$ also diverges for any constant $c \neq 0$ (Exercise 97 of Section 10.3).

a. As we go farther and farther out in this series ($k \to \infty$), the terms behave like

$$\frac{k^3}{2k^4 - 1} \approx \frac{k^3}{2k^4} = \frac{1}{2k}.$$

So a reasonable choice for a comparison series is the divergent series $\sum \frac{1}{2k}$. We must now show that the terms of the given series are *greater* than the terms of the comparison series. It is done by noting that $2k^4 - 1 < 2k^4$. Inverting both sides, we have

$$\frac{1}{2k^4 - 1} > \frac{1}{2k^4}, \quad \text{which implies that} \quad \frac{k^3}{2k^4 - 1} > \frac{k^3}{2k^4} = \frac{1}{2k}.$$

Because $\sum \frac{1}{2k}$ diverges, case (2) of the Comparison Test implies that the given series also diverges.

b. We note that $\ln k < k$, for $k \geq 2$, and then divide by k^3:

$$\frac{\ln k}{k^3} < \frac{k}{k^3} = \frac{1}{k^2}.$$

Therefore, an appropriate comparison series is the convergent p-series $\sum \frac{1}{k^2}$. Because $\sum \frac{1}{k^2}$ converges, the given series converges.

Related Exercises 13, 19 ◄

QUICK CHECK 1 Explain why it is difficult to use the divergent series $\sum 1/k$ as a comparison series to test $\sum 1/(k + 1)$. ◄

The Limit Comparison Test

The Comparison Test should be tried if there is an obvious comparison series and the necessary inequality is easily established. Notice, however, that if the series in Example 1a were $\sum_{k=1}^{\infty} \frac{k^3}{2k^4 + 10}$ instead of $\sum_{k=1}^{\infty} \frac{k^3}{2k^4 - 1}$, then the comparison to the series $\sum \frac{1}{2k}$ would not work. Rather than fiddling with inequalities, it is often easier to use a more refined test called the *Limit Comparison Test*.

QUICK CHECK 2 For case (1) of the Limit Comparison Test, we must have $0 < L < \infty$. Why can either a_k or b_k be chosen as the known comparison series? That is, why can L be the limit of either a_k/b_k or b_k/a_k? ◄

THEOREM 10.15   Limit Comparison Test

Let $\sum a_k$ and $\sum b_k$ be series with positive terms and let

$$\lim_{k \to \infty} \frac{a_k}{b_k} = L.$$

1. If $0 < L < \infty$ (that is, L is a finite positive number), then $\sum a_k$ and $\sum b_k$ either both converge or both diverge.
2. If $L = 0$ and $\sum b_k$ converges, then $\sum a_k$ converges.
3. If $L = \infty$ and $\sum b_k$ diverges, then $\sum a_k$ diverges.

► Recall that $|x| < a$ is equivalent to $-a < x < a$.

Proof (Case 1): Recall the definition of $\lim_{k \to \infty} \frac{a_k}{b_k} = L$: Given any $\varepsilon > 0$, $\left| \frac{a_k}{b_k} - L \right| < \varepsilon$ provided k is sufficiently large. In this case, let's take $\varepsilon = L/2$. It then follows that for sufficiently large k, $\left| \frac{a_k}{b_k} - L \right| < \frac{L}{2}$, or (removing the absolute value) $-\frac{L}{2} < \frac{a_k}{b_k} - L < \frac{L}{2}$. Adding L to all terms in these inequalities, we have

$$\frac{L}{2} < \frac{a_k}{b_k} < \frac{3L}{2}.$$

These inequalities imply that for sufficiently large k,

$$\frac{Lb_k}{2} < a_k < \frac{3Lb_k}{2}.$$

We see that the terms of $\sum a_k$ are sandwiched between multiples of the terms of $\sum b_k$. By the Comparison Test, it follows that the two series converge or diverge together. Cases (2) and (3) ($L = 0$ and $L = \infty$, respectively) are treated in Exercise 65.   ◄

EXAMPLE 2   Using the Limit Comparison Test Determine whether the following series converge.

a. $\sum_{k=1}^{\infty} \frac{5k^4 - 2k^2 + 3}{2k^6 - k + 5}$ **b.** $\sum_{k=1}^{\infty} \frac{\ln k}{k^2}$.

SOLUTION In both cases, we must find a comparison series whose terms behave like the terms of the given series as $k \to \infty$.

a. As $k \to \infty$, a rational function behaves like the ratio of the leading (highest-power) terms. In this case, as $k \to \infty$,

$$\frac{5k^4 - 2k^2 + 3}{2k^6 - k + 5} \approx \frac{5k^4}{2k^6} = \frac{5}{2k^2}.$$

Therefore, a reasonable comparison series is the convergent p-series $\sum\limits_{k=1}^{\infty} \dfrac{1}{k^2}$ (the factor of $5/2$ does not affect whether the given series converges). Having chosen a comparison series, we compute the limit L:

$$L = \lim_{k \to \infty} \frac{(5k^4 - 2k^2 + 3)/(2k^6 - k + 5)}{1/k^2} \qquad \text{Ratio of terms of series}$$

$$= \lim_{k \to \infty} \frac{k^2(5k^4 - 2k^2 + 3)}{2k^6 - k + 5} \qquad \text{Simplify.}$$

$$= \lim_{k \to \infty} \frac{5k^6 - 2k^4 + 3k^2}{2k^6 - k + 5} = \frac{5}{2}. \qquad \text{Simplify and evaluate the limit.}$$

We see that $0 < L < \infty$; therefore, the given series converges.

b. Why is this series interesting? We know that $\sum\limits_{k=1}^{\infty} \dfrac{1}{k^2}$ converges and that $\sum\limits_{k=1}^{\infty} \dfrac{1}{k}$ diverges. The given series $\sum\limits_{k=1}^{\infty} \dfrac{\ln k}{k^2}$ is "between" these two series. This observation suggests that we use either $\sum\limits_{k=1}^{\infty} \dfrac{1}{k^2}$ or $\sum\limits_{k=1}^{\infty} \dfrac{1}{k}$ as a comparison series. In the first case, letting $a_k = \ln k/k^2$ and $b_k = 1/k^2$, we find that

$$L = \lim_{k \to \infty} \frac{a_k}{b_k} = \lim_{k \to \infty} \frac{\ln k/k^2}{1/k^2} = \lim_{k \to \infty} \ln k = \infty.$$

Case (3) of the Limit Comparison Test does not apply here because the comparison series $\sum\limits_{k=1}^{\infty} \dfrac{1}{k^2}$ converges; case (3) is conclusive only when the comparison series *diverges*.

If, instead, we use the comparison series $\sum b_k = \sum \dfrac{1}{k}$, then

$$L = \lim_{k \to \infty} \frac{a_k}{b_k} = \lim_{k \to \infty} \frac{\ln k/k^2}{1/k} = \lim_{k \to \infty} \frac{\ln k}{k} = 0.$$

Case (2) of the Limit Comparison Test does not apply here because the comparison series $\sum\limits_{k=1}^{\infty} \dfrac{1}{k}$ diverges; case (2) is conclusive only when the comparison series *converges*.

With a bit more cunning, the Limit Comparison Test becomes conclusive. A series that lies "between" $\sum\limits_{k=1}^{\infty} \dfrac{1}{k^2}$ and $\sum\limits_{k=1}^{\infty} \dfrac{1}{k}$ is the convergent p-series $\sum\limits_{k=1}^{\infty} \dfrac{1}{k^{3/2}}$; we try it as a comparison series. Letting $a_k = \ln k/k^2$ and $b_k = 1/k^{3/2}$, we find that

$$L = \lim_{k \to \infty} \frac{a_k}{b_k} = \lim_{k \to \infty} \frac{\ln k/k^2}{1/k^{3/2}} = \lim_{k \to \infty} \frac{\ln k}{\sqrt{k}} = 0.$$

(This limit is evaluated using l'Hôpital's Rule or by recalling that $\ln k$ grows more slowly than any positive power of k.) Now case (2) of the Limit Comparison Test applies; the comparison series $\sum \dfrac{1}{k^{3/2}}$ converges, so the given series converges.

Related Exercises 11, 30 ◀

SECTION 10.5 EXERCISES

Getting Started

1. Explain how the Limit Comparison Test works.

2. Explain why, with a series of positive terms, the sequence of partial sums is an increasing sequence.

3. What comparison series would you use with the Comparison Test to determine whether $\sum_{k=1}^{\infty} \dfrac{1}{k^2 + 1}$ converges?

4. Can the Comparison Test, with the comparison series $\sum_{k=1}^{\infty} \dfrac{1}{k^4}$, be used to show that the series $\sum_{k=1}^{\infty} \dfrac{1}{k^4 - 1}$ converges?

5. What comparison series would you use with the Comparison Test to determine whether $\sum_{k=1}^{\infty} \dfrac{2^k}{3^k + 1}$ converges?

6. Determine whether $\sum_{k=2}^{\infty} \dfrac{1}{\sqrt{k} - 1}$ converges using the Comparison Test with the comparison series $\sum_{k=1}^{\infty} \dfrac{1}{\sqrt{k}}$.

7. What comparison series would you use with the Limit Comparison Test to determine whether $\sum_{k=1}^{\infty} \dfrac{k^2 + k + 5}{k^3 + 3k + 1}$ converges?

8. Determine whether $\sum_{k=1}^{\infty} \dfrac{k + 1}{3k^3 + 2}$ converges using the Limit Comparison Test with the comparison series $\sum_{k=1}^{\infty} \dfrac{1}{k^2}$.

Practice Exercises

9–36. Comparison tests *Use the Comparison Test or the Limit Comparison Test to determine whether the following series converge.*

9. $\sum_{k=1}^{\infty} \dfrac{1}{k^2 + 4}$

10. $\sum_{k=1}^{\infty} \dfrac{k^2 + k - 1}{k^4 + 4k^2 - 3}$

11. $\sum_{k=1}^{\infty} \dfrac{k^2 - 1}{k^3 + 4}$

12. $\sum_{k=1}^{\infty} \dfrac{0.0001}{k + 4}$

13. $\sum_{k=1}^{\infty} \dfrac{\sqrt{k}}{k^2 + 3}$

14. $\sum_{k=1}^{\infty} \dfrac{1}{5^k + 3}$

15. $\sum_{k=1}^{\infty} \dfrac{4^k}{5^k - 3}$

16. $\sum_{k=1}^{\infty} \dfrac{\sin^2 k}{k^2}$

17. $\sum_{k=1}^{\infty} \dfrac{1}{k^{3/2} + 1}$

18. $\sum_{k=1}^{\infty} \dfrac{1}{k \cdot 10^k}$

19. $\sum_{k=4}^{\infty} \dfrac{1 + \cos^2 k}{k - 3}$

20. $\sum_{k=1}^{\infty} \dfrac{k^2 + k + 2}{6^k(k^2 + 1)}$

21. $\sum_{k=1}^{\infty} \dfrac{(3k^3 + 4)(7k^2 + 1)}{(2k^3 + 1)(4k^3 - 1)}$

22. $\sum_{k=1}^{\infty} \sqrt{\dfrac{k}{k^3 + 1}}$

23. $\sum_{k=1}^{\infty} \dfrac{\sin(1/k)}{k^2}$

24. $\sum_{k=1}^{\infty} \dfrac{1}{3^k - 2^k}$

25. $\sum_{k=1}^{\infty} \dfrac{1}{2k - \sqrt{k}}$

26. $\sum_{k=1}^{\infty} \dfrac{1}{k\sqrt{k} + 2}$

27. $\sum_{k=1}^{\infty} \dfrac{2 + (-1)^k}{k^2}$

28. $\sum_{k=1}^{\infty} \dfrac{2 + \sin k}{k}$

29. $\sum_{k=1}^{\infty} \dfrac{\sqrt[3]{k^2 + 1}}{\sqrt{k^3 + 2}}$

30. $\sum_{k=3}^{\infty} \dfrac{1}{(k \ln k)^2}$

31. $\sum_{k=1}^{\infty} \dfrac{20}{\sqrt[3]{k} + \sqrt{k}}$

32. $\sum_{k=3}^{\infty} \dfrac{\sqrt{\ln k}}{k}$

33. $\sum_{k=1}^{\infty} \dfrac{1}{2^{\ln k}}$

34. $\sum_{k=2}^{\infty} \dfrac{\ln^2 k}{k^4}$

T 35. $\sum_{k=1}^{\infty} \dfrac{1}{4^{\ln k}}$

36. $\sum_{k=1}^{\infty} \dfrac{5}{\pi^k + e^k - 2}$

37. **Explain why or why not** Determine whether the following statements are true and give an explanation or counterexample.

   a. Suppose $0 < a_k < b_k$. If $\sum a_k$ converges, then $\sum b_k$ converges.

   b. Suppose $0 < a_k < b_k$. If $\sum a_k$ diverges, then $\sum b_k$ diverges.

   c. Suppose $0 < b_k < c_k < a_k$. If $\sum a_k$ converges, then $\sum b_k$ and $\sum c_k$ converge.

   d. When applying the Limit Comparison Test, an appropriate comparison series for $\sum_{k=1}^{\infty} \dfrac{k^2 + 2k + 1}{k^5 + 5k + 7}$ is $\sum_{k=1}^{\infty} \dfrac{1}{k^3}$.

38–39. Examining a series two ways *Determine whether the follow series converge using either the Comparison Test or the Limit Comparison Test. Then use another method to check your answer.*

38. $\sum_{k=1}^{\infty} \dfrac{e^{2k}}{e^{4k} + 1}$

39. $\sum_{k=1}^{\infty} \dfrac{1}{k^2 + 2k + 1}$

40–62. Choose your test *Use the test of your choice to determine whether the following series converge.*

40. $\left(\dfrac{1}{2}\right)^2 + \left(\dfrac{2}{3}\right)^3 + \left(\dfrac{3}{4}\right)^4 + \cdots$

41. $\sum_{k=1}^{\infty} \left(1 + \dfrac{2}{k}\right)^k$

42. $\sum_{k=1}^{\infty} \dfrac{e^{2k} + k}{e^{5k} - k^2}$

43. $\sum_{k=1}^{\infty} \dfrac{k^2 + 2k + 1}{3k^2 + 1}$

44. $\sum_{k=1}^{\infty} \dfrac{2^k}{e^k - 1}$

45. $\sum_{k=3}^{\infty} \dfrac{1}{\ln k}$

46. $\sum_{k=1}^{\infty} \dfrac{1}{5^k - 1}$

47. $\sum_{k=1}^{\infty} \dfrac{1}{\sqrt{k^3 - k + 1}}$

48. $\sum_{k=3}^{\infty} \dfrac{1}{5^k - 3^k}$

49. $\sum_{k=1}^{\infty} \left(\dfrac{1}{k} + 2^{-k}\right)$

50. $\sum_{k=2}^{\infty} \dfrac{5 \ln k}{k}$

51. $\sum_{k=1}^{\infty} \dfrac{k^8}{k^{11} + 3}$

52. $\sum_{k=2}^{\infty} \dfrac{1}{k^2 \ln k}$

53. $\sum_{k=1}^{\infty} \dfrac{1}{k^{1+p}}, \, p > 0$

54. $\sum_{k=1}^{\infty} \dfrac{\tan^{-1} k}{k^2}$

55. $\displaystyle\sum_{k=1}^{\infty} \ln\left(\frac{k+2}{k+1}\right)$

56. $\displaystyle\sum_{k=1}^{\infty} k^{-1/k}$

57. $\displaystyle\sum_{k=2}^{\infty} \frac{1}{k^{\ln k}}$

58. $\displaystyle\sum_{k=1}^{\infty} \sin^2 \frac{1}{k}$

59. $\displaystyle\sum_{k=1}^{\infty} \tan \frac{1}{k}$

60. $\displaystyle\sum_{k=1}^{\infty} \sqrt{\frac{1}{2^{2k}} + \frac{1}{k^2}}$

61. $\dfrac{1}{1\cdot 3} + \dfrac{1}{3\cdot 5} + \dfrac{1}{5\cdot 7} + \cdots$

62. $\dfrac{1}{2^2} + \dfrac{2}{3^2} + \dfrac{3}{4^2} + \cdots$

63. Series of squares Prove that if $\sum a_k$ is a convergent series of positive terms, then the series $\sum a_k^2$ also converges.

64. Two sine series Determine whether the following series converge.

   a. $\displaystyle\sum_{k=1}^{\infty} \sin \frac{1}{k}$
        b. $\displaystyle\sum_{k=1}^{\infty} \frac{1}{k}\sin \frac{1}{k}$

Explorations and Challenges

65. Limit Comparison Test proof Use the proof of case (1) of the Limit Comparison Test (Theorem 10.15) to prove cases (2) and (3).

66. Infinite products An infinite product $P = a_1 a_2 a_3 \ldots$, which is denoted $\displaystyle\prod_{k=1}^{\infty} a_k$, is the limit of the *sequence of partial products* $\{a_1, a_1 a_2, a_1 a_2 a_3, \ldots\}$. Assume $a_k > 0$ for all k and L is a finite constant.

   a. Evaluate $\displaystyle\prod_{k=1}^{\infty}\left(\frac{k}{k+1}\right) = \frac{1}{2}\cdot\frac{2}{3}\cdot\frac{3}{4}\cdot\frac{4}{5}\cdots$.

   b. Show that if $\displaystyle\sum_{k=1}^{\infty} \ln a_k = L$, then $\displaystyle\prod_{k=1}^{\infty} a_k = e^L$.

   c. Use the result of part (b) to evaluate
$$\prod_{k=0}^{\infty} e^{1/2^k} = e\cdot e^{1/2}\cdot e^{1/4}\cdot e^{1/8}\cdots.$$

67. An early limit Working in the early 1600s, the mathematicians Wallis, Pascal, and Fermat wanted to calculate the area of the region under the curve $y = x^p$ between $x = 0$ and $x = 1$, where p is a positive integer. Using arguments that predated the Fundamental Theorem of Calculus, they were able to prove that
$$\lim_{n\to\infty} \frac{1}{n}\sum_{k=0}^{n-1}\left(\frac{k}{n}\right)^p = \frac{1}{p+1}.$$
Use what you know about Riemann sums and integrals to verify this limit.

QUICK CHECK ANSWERS

1. To use the Comparison Test, we would need to show that $1/(k+1) > 1/k$, which is not true.   **2.** If $\displaystyle\lim_{k\to\infty}\frac{a_k}{b_k} = L$ for $0 < L < \infty$, then $\displaystyle\lim_{k\to\infty}\frac{b_k}{a_k} = \frac{1}{L}$, where $0 < 1/L < \infty$. ◄

10.6 Alternating Series

Our previous discussion focused on infinite series with positive terms, which is certainly an important part of the entire subject. But there are many interesting series with terms of mixed sign. For example, the series

$$1 + \frac{1}{2} - \frac{1}{3} - \frac{1}{4} + \frac{1}{5} + \frac{1}{6} - \frac{1}{7} - \frac{1}{8} + \cdots$$

has the pattern that two positive terms are followed by two negative terms, and vice versa. Clearly, infinite series could have endless sign patterns, so we need to restrict our attention.

Fortunately, the simplest sign pattern is also the most important. We consider **alternating series** in which the signs strictly alternate, as in the series

$$\sum_{k=1}^{\infty} \frac{(-1)^{k+1}}{k} = 1 - \frac{1}{2} + \frac{1}{3} - \frac{1}{4} + \frac{1}{5} - \frac{1}{6} + \frac{1}{7} - \frac{1}{8} + \cdots.$$

The factor $(-1)^{k+1}$ (or $(-1)^k$) has the pattern $\{\ldots, 1, -1, 1, -1, \ldots\}$ and provides the alternating signs.

Alternating Harmonic Series

Let's see what is different about alternating series by working with the series $\displaystyle\sum_{k=1}^{\infty}\frac{(-1)^{k+1}}{k}$, which is called the **alternating harmonic series**. Recall that this series *without* the alternating signs, $\displaystyle\sum_{k=1}^{\infty}\frac{1}{k}$, is the *divergent* harmonic series. So an immediate question is whether the presence of alternating signs affects the convergence of a series.

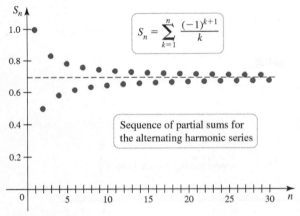

Figure 10.34

$$S_n = \sum_{k=1}^{n} \frac{(-1)^{k+1}}{k}$$

Sequence of partial sums for the alternating harmonic series

We investigate this question by looking at the sequence of partial sums for the series. In this case, the first four terms of the sequence of partial sums are

$$S_1 = 1$$

$$S_2 = 1 - \frac{1}{2} = \frac{1}{2}$$

$$S_3 = 1 - \frac{1}{2} + \frac{1}{3} = \frac{5}{6}$$

$$S_4 = 1 - \frac{1}{2} + \frac{1}{3} - \frac{1}{4} = \frac{7}{12}.$$

Plotting the first 30 terms of the sequence of partial sums results in **Figure 10.34**, which has several noteworthy features.

- The terms of the sequence of partial sums appear to converge to a limit; if they do, it means that, while the harmonic series diverges, the *alternating* harmonic series converges. We will soon learn that taking a divergent series with positive terms and making it an alternating series *may* turn it into a convergent series.

- For series with *positive* terms, the sequence of partial sums is necessarily an increasing sequence. Because the terms of an alternating series alternate in sign, the sequence of partial sums is not increasing (Figure 10.34).

- For the alternating harmonic series, the odd terms of the sequence of partial sums form a decreasing sequence and the even terms form an increasing sequence. As a result, the limit of the sequence of partial sums lies between any two consecutive terms of the sequence.

QUICK CHECK 1 Write out the first few terms of the sequence of partial sums for the alternating series $1 - 2 + 3 - 4 + 5 - 6 + \cdots$. Does this series appear to converge or diverge? ◄

Alternating Series Test

We now consider alternating series in general, which are written $\sum (-1)^{k+1} a_k$, where $a_k > 0$. With the exception of the Divergence Test, none of the convergence tests for series with positive terms applies to alternating series. Fortunately, one test works for most alternating series—and it is easy to use.

▶ Recall that the Divergence Test of Section 10.4 applies to all series: If the terms of *any* series (including an alternating series) do not tend to zero, then the series diverges.

▶ Depending on the sign of the first term of the series, an alternating series may be written with $(-1)^k$ or $(-1)^{k+1}$. Theorem 10.16 also applies to series of the form $\sum (-1)^k a_k$.

> **THEOREM 10.16 Alternating Series Test**
> The alternating series $\sum (-1)^{k+1} a_k$ converges provided
>
> **1.** the terms of the series are nonincreasing in magnitude ($0 < a_{k+1} \le a_k$, for k greater than some index N) and
>
> **2.** $\lim_{k \to \infty} a_k = 0$.

There is potential for confusion here. *For series of positive terms,* $\lim_{k \to \infty} a_k = 0$ ***does not*** *imply convergence. For alternating series with nonincreasing terms,* $\lim_{k \to \infty} a_k = 0$ ***does*** *imply convergence.*

Proof: The proof is short and instructive; it relies on **Figure 10.35**. We consider an alternating series in the form

$$\sum_{k=1}^{\infty} (-1)^{k+1} a_k = a_1 - a_2 + a_3 - a_4 + \cdots.$$

Because the terms of the series are nonincreasing in magnitude, the even terms of the sequence of partial sums $\{S_{2k}\} = \{S_2, S_4, \dots\}$ form a nondecreasing sequence that is bounded above by S_1. By the Bounded Monotonic Sequence Theorem (Section 10.2), this sequence has a limit; call it L. Similarly, the odd terms of the sequence of partial sums $\{S_{2k-1}\} = \{S_1, S_3, \dots\}$ form a nonincreasing sequence that is bounded below by S_2. By

Figure 10.35

the Bounded Monotonic Sequence Theorem, this sequence also has a limit; call it L'. At the moment, we cannot conclude that $L = L'$. However, notice that $S_{2k} = S_{2k-1} - a_{2k}$. By the condition that $\lim_{k \to \infty} a_k = 0$, it follows that

$$\underbrace{\lim_{k \to \infty} S_{2k}}_{L} = \underbrace{\lim_{k \to \infty} S_{2k-1}}_{L'} - \underbrace{\lim_{k \to \infty} a_{2k}}_{0},$$

or $L = L'$. Therefore, the sequence of partial sums converges to a (unique) limit, and the corresponding alternating series converges to that limit.   ◄

Now we can confirm that the alternating harmonic series $\sum_{k=1}^{\infty} \dfrac{(-1)^{k+1}}{k}$ converges. This fact follows immediately from the Alternating Series Test because the terms $a_k = \dfrac{1}{k}$ decrease and $\lim_{k \to \infty} a_k = 0$.

> $\sum_{k=1}^{\infty} \dfrac{1}{k}$
> • Diverges
> • Partial sums increase.

$\sum_{k=1}^{\infty} \dfrac{(-1)^{k+1}}{k}$
• Converges
• Partial sums bound the series above and below.

THEOREM 10.17   Alternating Harmonic Series

The alternating harmonic series $\sum_{k=1}^{\infty} \dfrac{(-1)^{k+1}}{k}$ converges (even though the harmonic series $\sum_{k=1}^{\infty} \dfrac{1}{k}$ diverges).

QUICK CHECK 2 Explain why the value of a convergent alternating series, with terms that are nonincreasing in magnitude, is trapped between successive terms of the sequence of partial sums. ◄

EXAMPLE 1   Alternating Series Test Determine whether the following series converge or diverge.

a. $\sum_{k=1}^{\infty} \dfrac{(-1)^{k+1}}{k^2}$ **b.** $2 - \dfrac{3}{2} + \dfrac{4}{3} - \dfrac{5}{4} + \cdots$ **c.** $\sum_{k=2}^{\infty} \dfrac{(-1)^k \ln k}{k}$

SOLUTION

a. The terms of this series decrease in magnitude, for $k \geq 1$. Furthermore,

$$\lim_{k \to \infty} a_k = \lim_{k \to \infty} \dfrac{1}{k^2} = 0.$$

Therefore, the series converges.

b. Using sigma notation, this series is written as $\sum_{k=1}^{\infty} (-1)^{k+1} \dfrac{k+1}{k}$. Because $\lim_{k \to \infty} \dfrac{k+1}{k} = 1$, the sequence $\left\{ (-1)^{k+1} \dfrac{k+1}{k} \right\}$ oscillates and diverges. Therefore, $\sum_{k=1}^{\infty} (-1)^{k+1} \dfrac{k+1}{k}$ diverges by the Divergence Test.

c. The first step is to show that the terms decrease in magnitude after some fixed term of the series. One way to proceed is to look at the function $f(x) = \dfrac{\ln x}{x}$, which generates the terms of the series. By the Quotient Rule, $f'(x) = \dfrac{1 - \ln x}{x^2}$. The fact that $f'(x) < 0$, for $x > e$, implies that the terms $\dfrac{\ln k}{k}$ decrease, for $k \geq 3$. As long as the terms of the series decrease for all k greater than some fixed integer, the first condition of the test is met. Furthermore, using l'Hôpital's Rule or the fact that $\{\ln k\}$ increases more slowly than $\{k\}$ (Section 10.2), we see that

$$\lim_{k \to \infty} a_k = \lim_{k \to \infty} \frac{\ln k}{k} = 0.$$

The conditions of the Alternating Series Test are met and the series converges.

Related Exercises 14, 15, 18 ◄

Remainders in Alternating Series

Recall that if a series converges to a value S, then the remainder is $R_n = S - S_n$, where S_n is the sum of the first n terms of the series. The magnitude of the remainder is the *absolute error* in approximating S by S_n.

An upper bound on the magnitude of the remainder in an alternating series arises from the following observation: When the terms are nonincreasing in magnitude, the value of the series is always trapped between successive terms of the sequence of partial sums. Therefore, as shown in **Figure 10.36**,

$$|R_n| = |S - S_n| \leq |S_{n+1} - S_n| = a_{n+1}.$$

This argument justifies the following theorem.

Figure 10.36

THEOREM 10.18   Remainder in Alternating Series

Let $\displaystyle\sum_{k=1}^{\infty} (-1)^{k+1} a_k$ be a convergent alternating series with terms that are nonincreasing in magnitude. Let $R_n = S - S_n$ be the remainder in approximating the value of that series by the sum of its first n terms. Then $|R_n| \leq a_{n+1}$. In other words, the magnitude of the remainder is less than or equal to the magnitude of the first neglected term.

EXAMPLE 2   Remainder in an alternating series

a. In turns out that $\ln 2 = 1 - \dfrac{1}{2} + \dfrac{1}{3} - \dfrac{1}{4} + \cdots = \displaystyle\sum_{k=1}^{\infty} \dfrac{(-1)^{k+1}}{k}$. How many terms of the series are required to approximate $\ln 2$ with an error less than 10^{-6}? The exact value of the series is given but is not needed to answer the question.

b. Consider the series $-1 + \dfrac{1}{2!} - \dfrac{1}{3!} + \dfrac{1}{4!} - \cdots = \displaystyle\sum_{k=1}^{\infty} \dfrac{(-1)^k}{k!}$. Find an upper bound for the magnitude of the error in approximating the value of the series (which is $e^{-1} - 1$) with $n = 9$ terms.

SOLUTION Notice that both series meet the conditions of Theorem 10.18.

a. The series is expressed as the sum of the first n terms plus the remainder:

$$\sum_{k=1}^{\infty} \frac{(-1)^{k+1}}{k} = \underbrace{1 - \frac{1}{2} + \frac{1}{3} - \frac{1}{4} + \cdots + \frac{(-1)^{n+1}}{n}}_{S_n \,=\, \text{the sum of the first } n \text{ terms}} + \overbrace{\underbrace{\frac{(-1)^{n+2}}{n+1}}_{\substack{|R_n| = |S - S_n| \text{ is less} \\ \text{than the magnitude} \\ \text{of this term}}} + \cdots}^{R_n}.$$

In magnitude, the remainder is less than or equal to the magnitude of the $(n + 1)$st term:

$$|R_n| = |S - S_n| \le a_{n+1} = \frac{1}{n + 1}.$$

To ensure that the error is less than 10^{-6}, we require that

$$a_{n+1} = \frac{1}{n + 1} < 10^{-6}, \quad \text{or} \quad n + 1 > 10^6.$$

Therefore, it takes 1 million terms of the series to approximate ln 2 with an error less than 10^{-6}.

b. The series may be expressed as the sum of the first nine terms plus the remainder:

$$\sum_{k=1}^{\infty} \frac{(-1)^k}{k!} = \underbrace{-1 + \frac{1}{2!} - \frac{1}{3!} + \cdots - \frac{1}{9!}}_{S_9 \ =\ \text{sum of first 9 terms}} + \overbrace{\underbrace{\frac{1}{10!}}_{\substack{|R_9| = |S - S_9| \\ \text{is less than} \\ \text{this term}}} - \cdots}^{R_9}.$$

The error committed when using the first nine terms to approximate the value of the series satisfies

$$|R_9| = |S - S_9| \le a_{10} = \frac{1}{10!} \approx 2.8 \times 10^{-7}.$$

Therefore, the error is no greater than 2.8×10^{-7}. As a check, the difference between the sum of the first nine terms, $\sum_{k=1}^{9} \frac{(-1)^k}{k!} \approx -0.632120811$, and the exact value, $S = e^{-1} - 1 \approx -0.632120559$, is approximately 2.5×10^{-7}. Therefore, the actual error satisfies the bound given by Theorem 10.18.

Related Exercises 32–33 ◀

QUICK CHECK 3 Compare and comment on the speed of convergence of the two series in Example 2. Why does one series converge more rapidly than the other? ◀

Absolute and Conditional Convergence

In this final segment, we introduce some terminology that is needed in Chapter 11. We now let the notation $\sum a_k$ denote any series—a series of positive terms, an alternating series, or even a more general infinite series.

Look again at the convergent alternating harmonic series $\sum (-1)^{k+1}/k$. The corresponding series of positive terms, $\sum 1/k$, is the divergent harmonic series. In contrast, we saw in Example 1a that the alternating series $\sum (-1)^{k+1}/k^2$ converges, and the corresponding p-series of positive terms $\sum 1/k^2$ also converges. These examples illustrate that removing the alternating signs in a convergent series *may* or *may not* result in a convergent series. The terminology that we now introduce distinguishes between these cases.

DEFINITION Absolute and Conditional Convergence

If $\sum |a_k|$ converges, then $\sum a_k$ **converges absolutely**.

If $\sum |a_k|$ diverges and $\sum a_k$ converges, then $\sum a_k$ **converges conditionally**.

The series $\sum (-1)^{k+1}/k^2$ is an example of an absolutely convergent series because the series of absolute values,

$$\sum_{k=1}^{\infty} \left| \frac{(-1)^{k+1}}{k^2} \right| = \sum_{k=1}^{\infty} \frac{1}{k^2},$$

is a convergent p-series. In this case, removing the alternating signs in the series does *not* affect its convergence.

On the other hand, the convergent alternating harmonic series $\sum (-1)^{k+1}/k$ has the property that the corresponding series of absolute values,

$$\sum_{k=1}^{\infty} \left| \frac{(-1)^{k+1}}{k} \right| = \sum_{k=1}^{\infty} \frac{1}{k},$$

does *not* converge. In this case, removing the alternating signs in the series *does* affect convergence, so this series does not converge absolutely. Instead, we say it *converges conditionally*. A convergent series (such as $\sum (-1)^{k+1}/k$) may not converge absolutely. However, if a series converges absolutely, then it converges.

> **THEOREM 10.19 Absolute Convergence Implies Convergence**
> If $\sum |a_k|$ converges, then $\sum a_k$ converges (absolute convergence implies convergence). Equivalently, if $\sum a_k$ diverges, then $\sum |a_k|$ diverges.

Proof: Because $|a_k| = a_k$ or $|a_k| = -a_k$, it follows that $0 \le a_k + |a_k| \le 2|a_k|$. By assumption, $\sum |a_k|$ converges, which, in turn, implies that $2\sum |a_k|$ converges. Using the Comparison Test and the inequality $0 \le a_k + |a_k| \le 2|a_k|$, it follows that $\sum (a_k + |a_k|)$ converges. Now note that

$$\sum a_k = \sum (a_k + |a_k| - |a_k|) = \underbrace{\sum (a_k + |a_k|)}_{\text{converges}} - \underbrace{\sum |a_k|}_{\text{converges}}.$$

We see that $\sum a_k$ is the sum of two convergent series, so it also converges. The second statement of the theorem is logically equivalent to the first statement. ◄

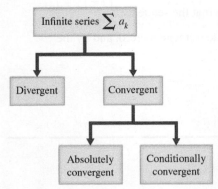

Figure 10.37

Figure 10.37 gives an overview of absolute and conditional convergence. It shows the universe of all infinite series, split first according to whether they converge or diverge. Convergent series are further divided between absolutely and conditionally convergent series. Here are a few more consequences of these definitions.

- The distinction between absolute and conditional convergence is relevant only for series of mixed sign, which includes alternating series. If a series of positive terms converges, it converges absolutely; conditional convergence does not apply.

- To test for absolute convergence, we test the series $\sum |a_k|$, which is a series of positive terms. Therefore, the convergence tests of Sections 10.4 and 10.5 (for positive-term series) are used to determine absolute convergence.

QUICK CHECK 4 Explain why a convergent series of positive terms converges absolutely. ◄

EXAMPLE 3 Absolute and conditional convergence Determine whether the following series diverge, converge absolutely, or converge conditionally.

a. $\displaystyle\sum_{k=1}^{\infty} \frac{(-1)^{k+1}}{\sqrt{k}}$ **b.** $\displaystyle\sum_{k=1}^{\infty} \frac{(-1)^{k+1}}{\sqrt{k^3}}$ **c.** $\displaystyle\sum_{k=1}^{\infty} \frac{\sin k}{k^2}$ **d.** $\displaystyle\sum_{k=1}^{\infty} \frac{(-1)^k k}{k+1}$

SOLUTION

a. We examine the series of absolute values,

$$\sum_{k=1}^{\infty} \left| \frac{(-1)^{k+1}}{\sqrt{k}} \right| = \sum_{k=1}^{\infty} \frac{1}{\sqrt{k}},$$

which is a divergent p-series (with $p = \frac{1}{2} < 1$). Therefore, the given alternating series does not converge absolutely. To determine whether the series converges conditionally, we look at the original series—with alternating signs. The magnitude of the terms of this series decrease with $\lim\limits_{k\to\infty} 1/\sqrt{k} = 0$, so by the Alternating Series Test, the series converges. Because this series converges, but not absolutely, it converges conditionally.

b. To assess absolute convergence, we look at the series of absolute values,

$$\sum_{k=1}^{\infty}\left|\frac{(-1)^{k+1}}{\sqrt{k^3}}\right| = \sum_{k=1}^{\infty}\frac{1}{k^{3/2}},$$

which is a convergent p-series (with $p = \frac{3}{2} > 1$). Therefore, the original alternating series converges absolutely (and by Theorem 10.19 it converges).

c. The terms of this series do not strictly alternate in sign (the first few signs are $+++---$), so the Alternating Series Test does not apply. Because $|\sin k| \le 1$, the terms of the series of absolute values satisfy

$$\left|\frac{\sin k}{k^2}\right| = \frac{|\sin k|}{k^2} \le \frac{1}{k^2}.$$

The series $\sum \dfrac{1}{k^2}$ is a convergent p-series. Therefore, by the Comparison Test, the

series $\sum \left|\dfrac{\sin k}{k^2}\right|$ converges, which implies that the series $\sum \dfrac{\sin k}{k^2}$ converges absolutely (and by Theorem 10.19 it converges).

d. Notice that $\lim\limits_{k\to\infty} k/(k+1) = 1$, which implies that the sequence $\left\{\dfrac{(-1)^k k}{k+1}\right\}$ oscillates and diverges. The terms of the series do not tend to zero, and by the Divergence Test, the series diverges.

Related Exercises 45, 51, 53 ◄

SECTION 10.6 EXERCISES

Getting Started

1. Explain why the sequence of partial sums for an alternating series is not an increasing sequence.

2. Describe how to apply the Alternating Series Test.

3. Why does the value of a converging alternating series with terms that are nonincreasing in magnitude lie between any two consecutive terms of its sequence of partial sums?

4. Suppose an alternating series with terms that are nonincreasing in magnitude converges to a value L. Explain how to estimate the remainder that occurs when the series is terminated after n terms.

5. Explain why the magnitude of the remainder in an alternating series (with terms that are nonincreasing in magnitude) is less than or equal to the magnitude of the first neglected term.

6. Give an example of a convergent alternating series that fails to converge absolutely.

7. Is it possible for a series of positive terms to converge conditionally? Explain.

8. Does the geometric series $\sum\limits_{k=0}^{\infty}(-0.5)^k$ converge absolutely?

9. Is it possible for an alternating series to converge absolutely but not conditionally?

10. Determine the values of p for which the series $\sum\limits_{k=0}^{\infty}\dfrac{(-1)^k}{k^p}$ converges conditionally.

Practice Exercises

11–27. Alternating Series Test *Determine whether the following series converge.*

11. $\sum\limits_{k=0}^{\infty}\dfrac{(-1)^k}{2k+1}$

12. $\sum\limits_{k=1}^{\infty}\dfrac{(-1)^k}{\sqrt{k}}$

13. $\sum\limits_{k=1}^{\infty}\dfrac{(-1)^k k}{3k+2}$

14. $\sum\limits_{k=1}^{\infty}(-1)^k\left(1+\dfrac{1}{k}\right)^k$

15. $\sum\limits_{k=1}^{\infty}\dfrac{(-1)^{k+1}}{k^3}$

16. $\sum\limits_{k=0}^{\infty}\dfrac{(-1)^k}{k^2+10}$

17. $\sum\limits_{k=1}^{\infty}(-1)^{k+1}\dfrac{k^2}{k^3+1}$

18. $\sum\limits_{k=2}^{\infty}(-1)^k\dfrac{\ln k}{k^2}$

19. $\sum\limits_{k=2}^{\infty}(-1)^k\dfrac{k^2-1}{k^2+3}$

20. $\sum\limits_{k=0}^{\infty}\left(-\dfrac{1}{5}\right)^k$

21. $\sum\limits_{k=2}^{\infty}(-1)^k\left(1+\dfrac{1}{k}\right)$

22. $\sum\limits_{k=1}^{\infty}\dfrac{\cos\pi k}{k^2}$

23. $\sum\limits_{k=1}^{\infty}(-1)^k\dfrac{k^{11}+2k^5+1}{4k(k^{10}+1)}$

24. $\sum\limits_{k=2}^{\infty}\dfrac{(-1)^k}{k\ln^2 k}$

25. $\sum\limits_{k=1}^{\infty}(-1)^{k+1}k^{1/k}$

26. $\sum\limits_{k=1}^{\infty}(-1)^k k\sin\dfrac{1}{k}$

27. $\sum\limits_{k=0}^{\infty}\dfrac{(-1)^k}{\sqrt{k^2+4}}$

T 28–32. Estimating errors in partial sums *For each of the following convergent alternating series, evaluate the nth partial sum for the given value of n. Then use Theorem 10.18 to find an upper bound for the error $|S - S_n|$ in using the nth partial sum S_n to estimate the value of the series S.*

28. $\displaystyle\sum_{k=1}^{\infty} \frac{(-1)^{k+1}}{k^3 + 1}; n = 3$ **29.** $\displaystyle\sum_{k=1}^{\infty} \frac{(-1)^k}{k^4}; n = 4$

30. $\displaystyle\sum_{k=0}^{\infty} \frac{(-1)^k}{(3k+1)^4}; n = 2$ **31.** $\displaystyle\sum_{k=0}^{\infty} \frac{(-1)^k}{k^4 + k^2 + 1}; n = 5$

32. $\displaystyle\sum_{k=0}^{\infty} \frac{(-1)^k}{(2k+1)!}; n = 3$

T 33–38. Remainders in alternating series *Determine how many terms of the following convergent series must be summed to be sure that the remainder is less than 10^{-4} in magnitude. Although you do not need it, the exact value of the series is given in each case.*

33. $\ln 2 = \displaystyle\sum_{k=1}^{\infty} \frac{(-1)^{k+1}}{k}$ **34.** $\dfrac{1}{e} = \displaystyle\sum_{k=0}^{\infty} \frac{(-1)^k}{k!}$

35. $\dfrac{\pi}{4} = \displaystyle\sum_{k=0}^{\infty} \frac{(-1)^k}{2k+1}$ **36.** $\dfrac{\pi^2}{12} = \displaystyle\sum_{k=1}^{\infty} \frac{(-1)^{k+1}}{k^2}$

37. $\dfrac{7\pi^4}{720} = \displaystyle\sum_{k=1}^{\infty} \frac{(-1)^{k+1}}{k^4}$ **38.** $\dfrac{\pi^3}{32} = \displaystyle\sum_{k=0}^{\infty} \frac{(-1)^k}{(2k+1)^3}$

T 39–44. Estimating infinite series *Estimate the value of the following convergent series with an absolute error less than 10^{-3}.*

39. $\displaystyle\sum_{k=1}^{\infty} \frac{(-1)^k}{k^5}$ **40.** $\displaystyle\sum_{k=1}^{\infty} \frac{(-1)^k}{(2k+1)^3}$

41. $\displaystyle\sum_{k=1}^{\infty} \frac{(-1)^k k}{k^2 + 1}$ **42.** $\displaystyle\sum_{k=1}^{\infty} \frac{(-1)^k k}{k^4 + 1}$

43. $\displaystyle\sum_{k=1}^{\infty} \frac{(-1)^k}{k^k}$ **44.** $\displaystyle\sum_{k=1}^{\infty} \frac{(-1)^{k+1}}{(2k+1)!}$

45–63. Absolute and conditional convergence *Determine whether the following series converge absolutely, converge conditionally, or diverge.*

45. $\displaystyle\sum_{k=1}^{\infty} \frac{(-1)^k}{k^{2/3}}$ **46.** $\displaystyle\sum_{k=0}^{\infty} \frac{(-1)^k 2k}{\sqrt{k^2 + 9}}$

47. $\displaystyle\sum_{k=1}^{\infty} \frac{(-1)^{k+1}}{(k+1)!}$ **48.** $\displaystyle\sum_{k=1}^{\infty} \left(-\frac{1}{3}\right)^k$

49. $\displaystyle\sum_{k=1}^{\infty} \left(\frac{3}{4}\right)^k$ **50.** $\displaystyle\sum_{k=1}^{\infty} \frac{(-1)^{k+1}}{k^{0.99}}$

51. $\displaystyle\sum_{k=1}^{\infty} \frac{\cos k}{k^3}$ **52.** $\displaystyle\sum_{k=1}^{\infty} \frac{(-1)^k k^2}{\sqrt{k^6 + 1}}$

53. $\displaystyle\sum_{k=1}^{\infty} (-1)^k \tan^{-1} k$ **54.** $\displaystyle\sum_{k=1}^{\infty} (-1)^k e^{-k}$

55. $\displaystyle\sum_{k=1}^{\infty} \frac{(-1)^{k+1}}{2\sqrt{k} - 1}$ **56.** $\displaystyle\sum_{k=2}^{\infty} \frac{(-1)^k}{3^k - 1}$

57. $\displaystyle\sum_{k=1}^{\infty} \frac{(-1)^k k}{2k+1}$ **58.** $\displaystyle\sum_{k=3}^{\infty} \frac{(-1)^k}{\ln k}$

59. $\displaystyle\sum_{k=1}^{\infty} \frac{(-1)^k \tan^{-1} k}{k^3}$ **60.** $\displaystyle\sum_{k=1}^{\infty} (-1)^k \frac{k^2 + 1}{3k^4 + 3}$

61. $\displaystyle\sum_{k=2}^{\infty} (-1)^k \frac{k^2 + 1}{k^3 - 1}$ **62.** $\displaystyle\sum_{k=1}^{\infty} \frac{\sin k}{3^k + 4^k}$

63. $\displaystyle\sum_{k=1}^{\infty} (-1)^{k+1} \frac{k!}{k^k}$ (*Hint:* Show that $\dfrac{k!}{k^k} \leq \dfrac{2}{k^2}$, for $k \geq 3$.)

64. Alternating Series Test Show that the series

$$\frac{1}{3} - \frac{2}{5} + \frac{3}{7} - \frac{4}{9} + \cdots = \sum_{k=1}^{\infty} (-1)^{k+1} \frac{k}{2k+1}$$

diverges. Which condition of the Alternating Series Test is not satisfied?

65. Explain why or why not Determine whether the following statements are true and give an explanation or counterexample.

 a. A series that converges must converge absolutely.
 b. A series that converges absolutely must converge.
 c. A series that converges conditionally must converge.
 d. If $\sum a_k$ diverges, then $\sum |a_k|$ diverges.
 e. If $\sum a_k^2$ converges, then $\sum a_k$ converges.
 f. If $a_k > 0$ and $\sum a_k$ converges, then $\sum a_k^2$ converges.
 g. If $\sum a_k$ converges conditionally, then $\sum |a_k|$ diverges.

Explorations and Challenges

66. Rearranging series It can be proved that if a series converges absolutely, then its terms may be summed in any order without changing the value of the series. However, if a series converges conditionally, then the value of the series depends on the order of summation. For example, the (conditionally convergent) alternating harmonic series has the value

$$1 - \frac{1}{2} + \frac{1}{3} - \frac{1}{4} + \cdots = \ln 2.$$

Show that by rearranging the terms (so the sign pattern is $++-$),

$$1 + \frac{1}{3} - \frac{1}{2} + \frac{1}{5} + \frac{1}{7} - \frac{1}{4} + \cdots = \frac{3}{2} \ln 2.$$

67. Alternating p-series Given that $\displaystyle\sum_{k=1}^{\infty} \frac{1}{k^2} = \frac{\pi^2}{6}$, show that $\displaystyle\sum_{k=1}^{\infty} \frac{(-1)^{k+1}}{k^2} = \frac{\pi^2}{12}$. (Assume the result of Exercise 66.)

68. Alternating p-series Given that $\displaystyle\sum_{k=1}^{\infty} \frac{1}{k^4} = \frac{\pi^4}{90}$, show that $\displaystyle\sum_{k=1}^{\infty} \frac{(-1)^{k+1}}{k^4} = \frac{7\pi^4}{720}$. (Assume the result of Exercise 66.)

69. A fallacy Explain the fallacy in the following argument. Let $x = 1 + \frac{1}{3} + \frac{1}{5} + \frac{1}{7} + \cdots$ and $y = \frac{1}{2} + \frac{1}{4} + \frac{1}{6} + \frac{1}{8} + \cdots$. It follows that $2y = x + y$, which implies that $x = y$. On the other hand,

$$x - y = \underbrace{\left(1 - \frac{1}{2}\right)}_{>0} + \underbrace{\left(\frac{1}{3} - \frac{1}{4}\right)}_{>0} + \underbrace{\left(\frac{1}{5} - \frac{1}{6}\right)}_{>0} + \cdots > 0$$

is a sum of positive terms, so $x > y$. Therefore, we have shown that $x = y$ and $x > y$.

70. Conditions of the Alternating Series Test It can be shown that if the sequence $\{a_{2n}\} = \{a_2, a_4, a_6, \ldots\}$ and the sequence $\{a_{2n-1}\} = \{a_1, a_3, a_5, \ldots\}$ both converge to L, then the sequence $\{a_n\} = \{a_1, a_2, a_3, \ldots\}$ converges to L. It is also the case that if $\{a_{2n}\}$ or $\{a_{2n-1}\}$ diverges, then $\{a_n\}$ diverges. Use these results in this exercise. Consider the alternating series

$$\sum_{k=1}^{\infty}(-1)^{k+1}a_k, \text{ where } a_k = \begin{cases} \dfrac{4}{k+1} & \text{if } k \text{ is odd} \\ \dfrac{2}{k} & \text{if } k \text{ is even.} \end{cases}$$

a. Show that $\{a_n\}$ converges to 0.

b. Show that $S_{2n} = \sum_{k=1}^{n}\dfrac{1}{k}$ and explain why $\lim_{k\to\infty} S_{2n} = \infty$.

c. Explain why the series $\sum_{k=1}^{\infty}(-1)^{k+1}a_k$ diverges even though $\{a_n\}$ converges to 0. Explain why this result does not contradict the Alternating Series Test.

71. A diverging alternating series Consider the alternating series

$$\sum_{k=1}^{\infty}(-1)^{k+1}a_k = \frac{1}{1^2} - \frac{1}{1} + \frac{1}{2^2} - \frac{1}{2} + \frac{1}{3^2} - \frac{1}{3} + \cdots.$$

a. Show that the individual terms of the series converge to 0. (*Hint:* See Exercise 70.)

b. Find a formula for S_{2n}, the sum of the first $2n$ terms of the series.

c. Explain why the alternating series diverges even though individual terms of the series converge to 0. Explain why this result does not contradict the Alternating Series Test.

QUICK CHECK ANSWERS

1. $1, -1, 2, -2, 3, -3, \ldots$; series diverges. **2.** The even terms of the sequence of partial sums approach the value of the series from one side; the odd terms of the sequence of partial sums approach the value of the series from the other side. **3.** The second series with $k!$ in the denominator converges much more quickly than the first series because $k!$ increases much faster than k as $k \to \infty$. **4.** If a series has positive terms, the series of absolute values is the same as the series itself. ◄

10.7 The Ratio and Root Tests

We now consider two additional convergence tests for general series: the Ratio Test and the Root Test. Both tests are used frequently throughout the next chapter. As in previous sections, these tests determine *whether* an infinite series converges, but they do not establish the value of the series.

The Ratio Test

The tests we have developed so far to determine whether a series converges are powerful, but they all have limitations. For example, the Integral Test requires evaluating integrals, and it is difficult to use with a series such as $\sum_{k=1}^{\infty}\dfrac{1}{k!}$ because integrating $1/k!$ is problematic. Further, the Integral Test and the comparison tests work only with series that have positive terms.

The Ratio Test, which we introduce in the next theorem, significantly enlarges the set of infinite series that we can analyze. It can be applied to any series, though its success depends on evaluating the limit of the ratio of successive terms in a series. The conclusions reached by the Ratio Test rely on the idea of absolute convergence and Theorem 10.19 from Section 10.6. We assume $a_k \neq 0$, for all k, so that the ratio a_{k+1}/a_k is defined.

THEOREM 10.20   Ratio Test

Let $\sum a_k$ be an infinite series, and let $r = \lim_{k\to\infty}\left|\dfrac{a_{k+1}}{a_k}\right|$.

1. If $r < 1$, the series converges absolutely, and therefore it converges (by Theorem 10.19).

2. If $r > 1$ (including $r = \infty$), the series diverges.

3. If $r = 1$, the test is inconclusive.

> See Appendix A for a formal proof of Theorem 10.20.

Proof (outline): The idea behind the proof provides insight. Let's assume the limit $r \geq 0$ exists. Then as k gets large and the ratio $|a_{k+1}/a_k|$ approaches r, we have $|a_{k+1}| \approx r|a_k|$. Therefore, as one goes farther and farther out in the series, it behaves like

$$|a_k| + |a_{k+1}| + |a_{k+2}| + \cdots \approx |a_k| + r|a_k| + r^2|a_k| + r^3|a_k| + \cdots$$
$$= |a_k|(1 + r + r^2 + r^3 + \cdots).$$

The tail of the series, which determines whether the series converges, behaves like a geometric series with ratio r. We know that if $r < 1$, the geometric series converges, and if $r > 1$, the series diverges, which is the conclusion of the Ratio Test. ◄

EXAMPLE 1   Using the Ratio Test Use the Ratio Test to determine whether the following series converge.

a. $\displaystyle\sum_{k=1}^{\infty} \frac{10^k}{k!}$ **b.** $\displaystyle\sum_{k=1}^{\infty} \frac{(-1)^k k^k}{k!}$ **c.** $\displaystyle\sum_{k=1}^{\infty} (-1)^{k+1} e^{-k} (k^2 + 4)$

SOLUTION In each case, the limit of the ratio of successive terms is determined.

> Recall that
> $$k! = k \cdot (k-1) \cdots 2 \cdot 1.$$
> Therefore,
> $$(k+1)! = (k+1)\underbrace{k(k-1)\cdots 1}_{k!}$$
> $$= (k+1)k!.$$

a. In this case, we have a series with positive terms. Therefore,

$$r = \lim_{k \to \infty} \left| \frac{a_{k+1}}{a_k} \right| = \lim_{k \to \infty} \frac{10^{k+1}/(k+1)!}{10^k/k!} \qquad \text{Substitute } a_{k+1} \text{ and } a_k.$$

$$= \lim_{k \to \infty} \frac{10^{k+1}}{10^k} \cdot \frac{k!}{(k+1)k!} \qquad \text{Invert and multiply.}$$

$$= \lim_{k \to \infty} \frac{10}{k+1} = 0 \qquad \text{Simplify and evaluate the limit.}$$

Because $r = 0 < 1$, the series converges by the Ratio Test.

b. $r = \lim_{k \to \infty} \left| \frac{a_{k+1}}{a_k} \right| = \lim_{k \to \infty} \left| \frac{(-1)^{k+1}(k+1)^{k+1}/(k+1)!}{(-1)^k k^k/k!} \right| \qquad \text{Substitute } a_{k+1} \text{ and } a_k.$

$$= \lim_{k \to \infty} \left(\frac{k+1}{k} \right)^k \qquad \text{Simplify.}$$

> Recall from Section 4.7 that
> $$\lim_{k \to \infty} \left(1 + \frac{1}{k}\right)^k = e \approx 2.718.$$

$$= \lim_{k \to \infty} \left(1 + \frac{1}{k}\right)^k = e \qquad \text{Simplify and evaluate the limit.}$$

Because $r = e > 1$, the series diverges by the Ratio Test. Alternatively, we could have noted that $\lim_{k \to \infty} k^k/k! = \infty$ (Theorem 10.6) and used the Divergence Test to reach the same conclusion.

c. $r = \lim_{k \to \infty} \left| \frac{a_{k+1}}{a_k} \right| = \lim_{k \to \infty} \frac{e^{-(k+1)}((k+1)^2 + 4)}{e^{-k}(k^2 + 4)} \qquad \text{Substitute } a_{k+1} \text{ and } a_k.$

$$= \lim_{k \to \infty} \frac{e^{-k}e^{-1}(k^2 + 2k + 5)}{e^{-k}(k^2 + 4)} \qquad \text{Simplify.}$$

$$= e^{-1} \lim_{k \to \infty} \underbrace{\frac{k^2 + 2k + 5}{k^2 + 4}}_{1} \qquad \text{Simplify.}$$

$$= e^{-1}$$

QUICK CHECK 1 Evaluate $10!/9!$, $(k+2)!/k!$, and $k!/(k+1)!$ ◄

Because $e^{-1} = \dfrac{1}{e} < 1$, the series converges absolutely by the Ratio Test, which implies that it converges.

Related Exercises 10, 14 ◄

> ▶ In Section 10.8, we offer guidelines to help you to decide which convergence test is best suited for a given series.

The Ratio Test is conclusive for many series. Nevertheless, observe what happens when the Ratio Test is applied to the harmonic series $\sum_{k=1}^{\infty} \frac{1}{k}$:

$$r = \lim_{k \to \infty} \left| \frac{a_{k+1}}{a_k} \right| = \lim_{k \to \infty} \frac{1/(k+1)}{1/k} = \lim_{k \to \infty} \frac{k}{k+1} = 1,$$

which means the test is inconclusive. We know the harmonic series diverges, yet the Ratio Test cannot be used to establish this fact. Like all the convergence tests encountered in this chapter, the Ratio Test works only for certain classes of series.

QUICK CHECK 2 Verify that the Ratio Test is inconclusive for $\sum_{k=1}^{\infty} \frac{1}{k^2}$. What test could be applied to show that $\sum_{k=1}^{\infty} \frac{1}{k^2}$ converges? ◀

The Root Test

The Root Test is similar to the Ratio Test and is well suited to series whose general term involves powers.

THEOREM 10.21 Root Test

Let $\sum a_k$ be an infinite series, and let $\rho = \lim_{k \to \infty} \sqrt[k]{|a_k|}$.

1. If $\rho < 1$, the series converges absolutely, and therefore it converges (by Theorem 10.19).
2. If $\rho > 1$ (including $\rho = \infty$), the series diverges.
3. If $\rho = 1$, the test is inconclusive.

> ▶ See Appendix A for a formal proof of Theorem 10.21.

Proof (outline): Assume the limit ρ exists. If k is large, we have $\rho \approx \sqrt[k]{|a_k|}$ or $|a_k| \approx \rho^k$. For large values of k, the tail of the series, which determines whether a series converges, behaves like

$$|a_k| + |a_{k+1}| + |a_{k+2}| + \cdots \approx \rho^k + \rho^{k+1} + \rho^{k+2} + \cdots.$$

Therefore, the tail of the series is approximately a geometric series with ratio ρ. If $\rho < 1$, the geometric series converges, and if $\rho > 1$, the series diverges, which is the conclusion of the Root Test. ◀

EXAMPLE 2 Using the Root Test Use the Root Test to determine whether the following series converge.

a. $\sum_{k=1}^{\infty} \left(\frac{3 - 4k^2}{7k^2 + 6} \right)^k$ **b.** $\sum_{k=1}^{\infty} \frac{(-2)^k}{k^{10}}$

SOLUTION

a. The required limit is

$$\rho = \lim_{k \to \infty} \sqrt[k]{\left| \left(\frac{3 - 4k^2}{7k^2 + 6} \right)^k \right|} = \lim_{k \to \infty} \frac{4k^2 - 3}{7k^2 + 6} = \frac{4}{7}. \quad |3 - 4k^2| = 4k^2 - 3, \text{ for } k > 1$$

Because $\rho < 1$, the series converges absolutely by the Root Test (and therefore it converges).

b. In this case,

$$\rho = \lim_{k \to \infty} \sqrt[k]{\left| \frac{(-2)^k}{k^{10}} \right|} = \lim_{k \to \infty} \frac{2}{k^{10/k}} = \lim_{k \to \infty} \frac{2}{\left(k^{1/k} \right)^{10}} = 2. \quad \lim_{k \to \infty} k^{1/k} = 1$$

Because $\rho > 1$, the series diverges by the Root Test.

    We could have used the Ratio Test for both series in this example, but the Root Test is easier to apply in each case. In part (b), the Divergence Test leads to the same conclusion.

Related Exercises 17, 32 ◀

SECTION 10.7 EXERCISES

Getting Started

1. Explain how the Ratio Test works.

2. Explain how the Root Test works.

3. Evaluate $1000!/998!$ without a calculator.

4. Evaluate $(100!)^2/(99!)^2$ without a calculator.

5. Simplify $\dfrac{k!}{(k+2)!}$, for any integer $k \geq 0$.

6. Simplify $\dfrac{(2k+1)!}{(2k-1)!}$, for any integer $k \geq 1$.

7. What test is advisable if a series involves a factorial term?

8. Can the value of a series be determined using the Root Test or the Ratio Test?

Practice Exercises

9–30. The Ratio and Root Tests *Use the Ratio Test or the Root Test to determine whether the following series converge absolutely or diverge.*

9. $\displaystyle\sum_{k=1}^{\infty} \frac{(-1)^k}{k!}$

10. $\displaystyle\sum_{k=1}^{\infty} \frac{(-2)^k}{k!}$

11. $\displaystyle\sum_{k=1}^{\infty} (-1)^{k+1} \left(\frac{10k^3 + k}{9k^3 + k + 1} \right)^k$

12. $\displaystyle\sum_{k=1}^{\infty} \left(-\frac{2k}{k+1} \right)^k$

13. $\displaystyle\sum_{k=1}^{\infty} \frac{k^2}{4^k}$

14. $\displaystyle\sum_{k=1}^{\infty} \frac{k^k}{2^k}$

15. $\displaystyle\sum_{k=1}^{\infty} (-1)^{k+1} k e^{-k}$

16. $\displaystyle\sum_{k=1}^{\infty} (-1)^k \frac{k!}{k^k}$

17. $\displaystyle\sum_{k=1}^{\infty} \frac{(-7)^k}{k^2}$

18. $\displaystyle\sum_{k=1}^{\infty} \left(1 + \frac{3}{k} \right)^{k^2}$

19. $\displaystyle\sum_{k=1}^{\infty} \frac{2^k}{k^{99}}$

20. $\displaystyle\sum_{k=1}^{\infty} \frac{(-1)^k k!}{k^6}$

21. $\displaystyle\sum_{k=1}^{\infty} \frac{(k!)^2}{(2k)!}$

22. $2 + \dfrac{4}{16} + \dfrac{8}{81} + \dfrac{16}{256} + \cdots$

23. $\displaystyle\sum_{k=1}^{\infty} \left(\frac{k}{k+1} \right)^{2k^2}$

24. $\displaystyle\sum_{k=1}^{\infty} (-1)^{k+1} \frac{3^{k^2}}{k^k}$

25. $\displaystyle\sum_{k=1}^{\infty} (-1)^{k+1} \left(\frac{3k^2 + 4k}{2k^2 + 1} \right)^k$

26. $\displaystyle\sum_{k=1}^{\infty} \left(\frac{1}{\ln(k+1)} \right)^k$

27. $1 + \left(\dfrac{1}{2} \right)^2 + \left(\dfrac{1}{3} \right)^3 + \left(\dfrac{1}{4} \right)^4 + \cdots$

28. $\displaystyle\sum_{k=1}^{\infty} \frac{k}{e^k}$

29. $\displaystyle\sum_{k=1}^{\infty} \frac{(-1)^{k+1} k^{2k}}{k! k!}$

30. $\displaystyle\sum_{k=1}^{\infty} \left(\frac{k}{1 - 5k} \right)^k$

31. **Explain why or why not** Determine whether the following statements are true and give an explanation or counterexample.

a. $n!n! = (2n)!$, for all positive integers n.

b. $\dfrac{(2n)!}{(2n-1)!} = 2n$.

c. If $\displaystyle\lim_{k\to\infty} \sqrt[k]{|a_k|} = \frac{1}{4}$, then $\displaystyle\sum 10a_k$ converges absolutely.

d. The Ratio Test is always inconclusive when applied to $\displaystyle\sum a_k$, where a_k is a nonzero rational function of k.

32–49. Choose your test *Use the test of your choice to determine whether the following series converge absolutely, converge conditionally, or diverge.*

32. $\displaystyle\sum_{k=1}^{\infty} \left(\frac{4k^3 + 7}{9k^2 - 1} \right)^k$

33. $\displaystyle\sum_{k=1}^{\infty} (-1)^k \frac{2^{k+1}}{9^{k-1}}$

34. $\displaystyle\sum_{k=1}^{\infty} \frac{k!}{(2k+1)!}$

35. $\displaystyle\sum_{k=1}^{\infty} \frac{(2k+1)!}{(k!)^2}$

36. $\displaystyle\sum_{k=1}^{\infty} \left(\frac{k^2}{2k^2 + 1} \right)^k$

37. $\displaystyle\sum_{k=1}^{\infty} \frac{k^{100}}{(k+1)!}$

38. $\displaystyle\sum_{k=1}^{\infty} k^3 \sin(1/k^3)$

39. $\displaystyle\sum_{k=1}^{\infty} \frac{(-1)^k k^3}{\sqrt{k^8 + 1}}$

40. $\displaystyle\sum_{k=1}^{\infty} \frac{1}{(1 + p)^k}, \; p > 0$

41. $\displaystyle\sum_{k=1}^{\infty} (\sqrt[k]{k} - 1)^{2k}$

42. $\displaystyle\sum_{k=1}^{\infty} \frac{(k!)^3}{(3k)!}$

43. $\displaystyle\sum_{k=1}^{\infty} \frac{2^k k!}{k^k}$

44. $\displaystyle\sum_{k=1}^{\infty} \left(1 - \frac{1}{k} \right)^{k^2}$

45. $\displaystyle\sum_{k=1}^{\infty} \frac{(-1)^k}{k^{0.99}}$

46. $\displaystyle\sum_{k=2}^{\infty} 100k^{-k}$

47. $\dfrac{1}{1!} + \dfrac{4}{2!} + \dfrac{9}{3!} + \dfrac{16}{4!} + \cdots$

48. $\displaystyle\sum_{k=1}^{\infty} \frac{(-1)^k}{\tan^{-1} k}$

49. $\displaystyle\sum_{k=1}^{\infty} \frac{(-1)^k}{\sqrt{k^{3/2} + k}}$

Explorations and Challenges

50–57. Convergence parameter *Find the values of the parameter $p > 0$ for which the following series converge.*

50. $\displaystyle\sum_{k=2}^{\infty} \frac{1}{(\ln k)^p}$

51. $\displaystyle\sum_{k=2}^{\infty} \frac{\ln k}{k^p}$

52. $\displaystyle\sum_{k=2}^{\infty} \frac{1}{k(\ln k)(\ln \ln k)^p}$

53. $\displaystyle\sum_{k=2}^{\infty} \left(\frac{\ln k}{k} \right)^p$

54. $\displaystyle\sum_{k=0}^{\infty} \frac{k! p^k}{(k+1)^k}$

(*Hint:* Stirling's formula is useful: $k! \approx \sqrt{2\pi k}\, k^k e^{-k}$, for large k.)

55. $\displaystyle\sum_{k=1}^{\infty} \frac{k p^k}{k + 1}$

56. $\displaystyle\sum_{k=1}^{\infty} \ln \left(\frac{k}{k+1} \right)^p$

57. $\displaystyle\sum_{k=1}^{\infty} \left(1 - \frac{p}{k} \right)^k$

58–63. A glimpse ahead to power series *Use the Ratio Test or the Root Test to determine the values of x for which each series converges.*

58. $\displaystyle\sum_{k=1}^{\infty} \frac{x^k}{k!}$ **59.** $\displaystyle\sum_{k=1}^{\infty} x^k$ **60.** $\displaystyle\sum_{k=1}^{\infty} \frac{x^k}{k}$

61. $\displaystyle\sum_{k=1}^{\infty} \frac{x^k}{k^2}$ **62.** $\displaystyle\sum_{k=1}^{\infty} \frac{x^{2k}}{k^2}$ **63.** $\displaystyle\sum_{k=1}^{\infty} \frac{x^k}{2^k}$

64. Stirling's formula Complete the following steps to find the values of $p > 0$ for which the series $\displaystyle\sum_{k=1}^{\infty} \frac{1 \cdot 3 \cdot 5 \cdots (2k-1)}{p^k k!}$ converges.

    a. Use the Ratio Test to show that $\displaystyle\sum_{k=1}^{\infty} \frac{1 \cdot 3 \cdot 5 \cdots (2k-1)}{p^k k!}$ converges for $p > 2$.

    b. Use Stirling's formula, $k! \approx \sqrt{2\pi k}\, k^k e^{-k}$ for large k, to determine whether the series converges with $p = 2$.

       (*Hint:* $1 \cdot 3 \cdot 5 \cdots (2k-1) = \dfrac{1 \cdot 2 \cdot 3 \cdot 4 \cdot 5 \cdot 6 \cdots (2k-1)2k}{2 \cdot 4 \cdot 6 \cdots 2k}$.)

       (See the Guided Project *Stirling's formula and n!* for more on this topic.)

QUICK CHECK ANSWERS

1. 10; $(k+2)(k+1)$; $1/(k+1)$ **2.** The Integral Test or *p*-series with $p = 2$ ◄

10.8 Choosing a Convergence Test

In previous sections, we presented methods for determining whether an infinite series converges, and in some cases we discovered that it is possible to determine the exact value of a convergent series. In this section, we step back and focus on general strategies for classifying infinite series and deciding which tests are appropriate for a specific series.

Series Strategies

It is helpful if you are already familiar with the convergence tests summarized in Table 10.4. With these tests in mind, we explore various strategies and illustrate them with examples. These strategies can be applied sequentially, more or less in the order in which they are given.

Categorize the Series   The first step in approaching an infinite series is to note some of the distinctive properties that help to classify it. Is it a series with positive terms? An alternating series? Or something more general? Does the general term consist of rational functions, exponential functions, or factorials? Can the series be simplified in any obvious ways before applying convergence tests? Here is a quick suggestive example.

EXAMPLE 1   **Categorize the series**   Discuss the series $\displaystyle\sum_{k=1}^{\infty} \frac{2^k + \cos(\pi k)\sqrt{k}}{3^{k+1}}$.

SOLUTION   A good first step in analyzing this series would be to split it into two series that can be examined separately. We begin by writing

$$\sum_{k=1}^{\infty} \frac{2^k + \cos(\pi k)\sqrt{k}}{3^{k+1}} = \underbrace{\sum_{k=1}^{\infty} \frac{2^k}{3^{k+1}}}_{\text{geometric series}} + \underbrace{\sum_{k=1}^{\infty} \frac{\cos(\pi k)\sqrt{k}}{3^{k+1}}}_{\text{alternating series}}.$$

QUICK CHECK 1 Evaluate the geometric series in Example 1. Does the alternating series in Example 1 converge? ◄

The general term of the first series is the ratio of exponential functions, so this series is a geometric series that can be evaluated explicitly. Because $\cos(\pi k) = (-1)^k$, we see that the second series is an alternating series, whose convergence can also be tested. Notice that for the original series to converge, both the geometric series and the alternating series must converge.

Related Exercise 8 ◄

Apply the Divergence Test   Almost without exception, the Divergence Test is the easiest test to apply because it requires nothing more than taking a limit. Furthermore, it can lead to a quick conclusion, namely that the given series diverges.

Table 10.4 Special Series and Convergence Tests

Series or Test	Form of Series	Condition for Convergence	Condition for Divergence	Comments
Geometric series	$\displaystyle\sum_{k=0}^{\infty} ar^k, a \neq 0$	$\lvert r \rvert < 1$	$\lvert r \rvert \geq 1$	If $\lvert r \rvert < 1$, then $\displaystyle\sum_{k=0}^{\infty} ar^k = \frac{a}{1-r}$.
Divergence Test	$\displaystyle\sum_{k=1}^{\infty} a_k$	Does not apply	$\displaystyle\lim_{k \to \infty} a_k \neq 0$	Cannot be used to prove convergence
Integral Test	$\displaystyle\sum_{k=1}^{\infty} a_k$, where $a_k = f(k)$ and f is continuous, positive, and decreasing	$\displaystyle\int_1^{\infty} f(x)\, dx$ converges.	$\displaystyle\int_1^{\infty} f(x)\, dx$ diverges.	The value of the integral is not the value of the series.
p-series	$\displaystyle\sum_{k=1}^{\infty} \frac{1}{k^p}$	$p > 1$	$p \leq 1$	Useful for comparison tests
Ratio Test	$\displaystyle\sum_{k=1}^{\infty} a_k$	$\displaystyle\lim_{k \to \infty} \left\lvert \frac{a_{k+1}}{a_k} \right\rvert < 1$	$\displaystyle\lim_{k \to \infty} \left\lvert \frac{a_{k+1}}{a_k} \right\rvert > 1$	Inconclusive if $\displaystyle\lim_{k \to \infty} \left\lvert \frac{a_{k+1}}{a_k} \right\rvert = 1$
Root Test	$\displaystyle\sum_{k=1}^{\infty} a_k$	$\displaystyle\lim_{k \to \infty} \sqrt[k]{\lvert a_k \rvert} < 1$	$\displaystyle\lim_{k \to \infty} \sqrt[k]{\lvert a_k \rvert} > 1$	Inconclusive if $\displaystyle\lim_{k \to \infty} \sqrt[k]{\lvert a_k \rvert} = 1$
Comparison Test	$\displaystyle\sum_{k=1}^{\infty} a_k$, where $a_k > 0$	$a_k \leq b_k$ and $\displaystyle\sum_{k=1}^{\infty} b_k$ converges.	$b_k \leq a_k$ and $\displaystyle\sum_{k=1}^{\infty} b_k$ diverges.	$\displaystyle\sum_{k=1}^{\infty} a_k$ is given; you supply $\displaystyle\sum_{k=1}^{\infty} b_k$.
Limit Comparison Test	$\displaystyle\sum_{k=1}^{\infty} a_k$, where $a_k > 0, b_k > 0$	$0 \leq \displaystyle\lim_{k \to \infty} \frac{a_k}{b_k} < \infty$ and $\displaystyle\sum_{k=1}^{\infty} b_k$ converges.	$\displaystyle\lim_{k \to \infty} \frac{a_k}{b_k} > 0$ and $\displaystyle\sum_{k=1}^{\infty} b_k$ diverges.	$\displaystyle\sum_{k=1}^{\infty} a_k$ is given; you supply $\displaystyle\sum_{k=1}^{\infty} b_k$.
Alternating Series Test	$\displaystyle\sum_{k=1}^{\infty} (-1)^k a_k$, where $a_k > 0$	$\displaystyle\lim_{k \to \infty} a_k = 0$ and $0 < a_{k+1} \leq a_k$	$\displaystyle\lim_{k \to \infty} a_k \neq 0$	Remainder R_n satisfies $\lvert R_n \rvert \leq a_{n+1}$
Absolute Convergence	$\displaystyle\sum_{k=1}^{\infty} a_k, a_k$ arbitrary	$\displaystyle\sum_{k=1}^{\infty} \lvert a_k \rvert$ converges.		Applies to arbitrary series

EXAMPLE 2   Divergence Test Does the series $\displaystyle\sum_{k=1}^{\infty} \left(1 - \frac{10}{k}\right)^k$ converge or diverge?

SOLUTION The terms of this series alternate in sign for $k < 10$, and then remain positive and close to zero for $k > 10$. Applying the Divergence Test, we evaluate the limit of the general term using l'Hôpital's Rule and find that

$$\lim_{k \to \infty} \left(1 - \frac{10}{k}\right)^k = e^{-10} \neq 0.$$

QUICK CHECK 2 Show the steps in evaluating the limit in Example 2. ◄

Because the limit of the general term is non-zero, the series diverges (very slowly).

Related Exercise 18 ◄

Identify Special Series We have studied several special families of series that are readily identified and analyzed for convergence. Specifically, you should look for geometric series, p-series, and (less obvious and less common) telescoping series.

EXAMPLE 3 Special series Does the series $\displaystyle\sum_{k=4}^{\infty}\frac{1}{\sqrt[4]{k^2-6k+9}}$ converge or diverge?

SOLUTION The Divergence Test is inconclusive for this series of positive terms. One approach is to rewrite the series in the following steps:

$$\sum_{k=4}^{\infty}\frac{1}{\sqrt[4]{k^2-6k+9}} = \sum_{k=4}^{\infty}\frac{1}{\sqrt[4]{(k-3)^2}} \qquad \text{Factor denominator.}$$

$$= \sum_{j=1}^{\infty}\frac{1}{\sqrt[4]{j^2}} \qquad \text{Shift index; let } j = k-3.$$

$$= \sum_{k=1}^{\infty}\frac{1}{\sqrt{k}}. \qquad \text{Simplify.}$$

We now identify this series as a p-series with $p = \dfrac{1}{2} < 1$; therefore, it diverges.

Related Exercise 24 ◄

Integral, Ratio, and Root Tests You have probably evaluated enough integrals by now to guess whether the general term of a series can be integrated easily. There are also some immediate clues: If a series involves factorials (something involving $k!$) or terms of the form $f(k)^k$, the Integral Test is not likely to help. On the other hand, series with terms of the form $k!$, k^k, or a^k, where a is a constant, often submit to the Ratio Test or the Root Test.

> ➤ When using the Integral Test, make sure the integrand satisfies the conditions given in Theorem 10.11: It must be continuous, positive, and decreasing on $[k, \infty)$ for some finite number k. In Example 4, $f(x) = x^2 e^{-2x}$ satisfies these conditions on $[1, \infty)$.

EXAMPLE 4 More than one test Does the series $\displaystyle\sum_{k=1}^{\infty}k^2 e^{-2k}$ converge or diverge?

SOLUTION The general term of this series of positive terms can be integrated using integration by parts, which reveals that $\displaystyle\int_{1}^{\infty}x^2 e^{-2x}dx = \frac{5e^{-2}}{4} < \infty$. Because the integral converges, the infinite series also converges. However, it may be easier to apply the Ratio Test. We find that

$$\lim_{k\to\infty}\frac{(k+1)^2 e^{-2(k+1)}}{k^2 e^{-2k}} = e^{-2}\underbrace{\lim_{k\to\infty}\left(\frac{k+1}{k}\right)^2}_{1} = e^{-2}\cdot 1 \approx 0.135 < 1.$$

> **QUICK CHECK 3** Verify the limit in the Ratio Test in Example 4. ◄

Because the ratio of successive terms in less than 1, the series converges.

Related Exercise 25 ◄

Rational Terms If a series involves a rational function, or ratios of roots or algebraic functions, then a comparison test is a likely candidate. When the ordinary Comparison Test does not work easily, the more powerful Limit Comparison Test may lead to a conclusion. As shown in the following example, analyzing the end behavior (as $k \to \infty$) of the general term is often the trick needed to find a good comparison series.

EXAMPLE 5 Comparison tests Does the series $\displaystyle\sum_{k=2}^{\infty}\sqrt[3]{\frac{k^2-1}{k^4+4}}$ converge or diverge?

SOLUTION The Divergence Test is inconclusive. Ruling out the Integral, Ratio, and Root Tests as impractical, we are left with a comparison test. The end behavior of the general term is determined by working with the leading terms of the numerator and denominator. We find that for large values of k,

$$\sqrt[3]{\frac{k^2-1}{k^4+4}} \approx \sqrt[3]{\frac{k^2}{k^4}} = \frac{1}{k^{2/3}}.$$

Therefore, a good comparison series is $\displaystyle\sum k^{-2/3}$, which is a divergent p-series. The ordinary Comparison Test is difficult to apply in this case, so we turn to the Limit

Comparison Test. Taking the limit of the ratio of terms of the given series and the comparison series, we find that

$$L = \lim_{k \to \infty} \frac{\sqrt[3]{\dfrac{k^2 - 1}{k^4 + 4}}}{k^{-2/3}} = \lim_{k \to \infty} \sqrt[3]{\frac{k^2 - 1}{k^4 + 4}} \cdot k^{2/3} = \lim_{k \to \infty} \sqrt[3]{\frac{k^4 - k^2}{k^4 + 4}} = \sqrt[3]{\lim_{k \to \infty} \frac{1 - k^{-2}}{1 + 4k^{-4}}} = 1.$$

Because $0 < L < \infty$ and the comparison series diverges, the given series also diverges.

Related Exercise 49 ◄

The guidelines presented in this section are a good start, but in the end, convergence tests are mastered through practice. It's your turn.

SECTION 10.8 EXERCISES

Getting Started

1–10. Choosing convergence tests *Identify a convergence test for each of the following series. If necessary, explain how to simplify or rewrite the series before applying the convergence test. You do not need to carry out the convergence test.*

1. $\displaystyle\sum_{k=1}^{\infty} (-1)^k \left(2 + \frac{1}{k^2}\right)^k$

2. $\displaystyle\sum_{k=3}^{\infty} \frac{2}{k^2 - k - 2}$

3. $\displaystyle\sum_{k=3}^{\infty} \frac{2k^2}{k^2 - k - 2}$

4. $\displaystyle\sum_{k=3}^{\infty} \frac{1}{k \ln^7 k}$

5. $\displaystyle\sum_{k=10}^{\infty} \frac{1}{(k - 9)^5}$

6. $\displaystyle\sum_{k=10}^{\infty} \frac{100^k}{k! \, k^2}$

7. $\displaystyle\sum_{k=1}^{\infty} \frac{k^2}{k^4 + k^3 + 1}$

8. $\displaystyle\sum_{k=1}^{\infty} \frac{(-3)^k}{4^{k+1}}$

9. $\displaystyle\sum_{k=1}^{\infty} \frac{(-1)^{k+1}}{\sqrt{2^k + \ln k}}$

10. $\displaystyle\sum_{k=1}^{\infty} (\tan^{-1} 2k - \tan^{-1}(2k - 2))$

Practice Exercises

11–86. Applying convergence tests *Determine whether the following series converge. Justify your answers.*

11. $\displaystyle\sum_{k=1}^{\infty} \frac{2k^4 + k}{4k^4 - 8k}$

12. $\displaystyle\sum_{k=1}^{\infty} 7^{-2k}$

13. $\displaystyle\sum_{k=3}^{\infty} \frac{5}{2 + \ln k}$

14. $\displaystyle\sum_{k=1}^{\infty} \frac{7k^2 - k - 2}{4k^4 - 3k + 1}$

15. $\displaystyle\sum_{k=1}^{\infty} \frac{(-7)^k}{k!}$

16. $\displaystyle\sum_{k=1}^{\infty} \frac{7^k}{k! + 10}$

17. $\displaystyle\sum_{k=1}^{\infty} \frac{(-k)^3}{3k^3 + 2}$

18. $\displaystyle\sum_{k=1}^{\infty} \left(1 + \frac{a}{k}\right)^k,\ a \text{ is real}$

19. $\displaystyle\sum_{k=0}^{\infty} \frac{3^{k+4}}{5^{k-2}}$

20. $\displaystyle\sum_{j=1}^{\infty} \frac{4j^{10}}{j^{11} + 1}$

21. $\displaystyle\sum_{k=1}^{\infty} \frac{(-1)^k k}{k^3 + 1}$

22. $\displaystyle\sum_{k=1}^{\infty} \left(\frac{e + 1}{\pi}\right)^k$

23. $\displaystyle\sum_{k=1}^{\infty} \frac{k^5}{5^k}$

24. $\displaystyle\sum_{k=1}^{\infty} \frac{4}{(k + 3)^3}$

25. $\displaystyle\sum_{k=1}^{\infty} \frac{1}{\sqrt{k} e^{\sqrt{k}}}$

26. $\displaystyle\sum_{k=1}^{\infty} \frac{5 + \sin k}{\sqrt{k}}$

27. $\displaystyle\sum_{k=1}^{\infty} \frac{3 + \cos 5k}{k^3}$

28. $\displaystyle\sum_{k=3}^{\infty} \frac{(-1)^k \ln k}{k^{1/3}}$

29. $\displaystyle\sum_{k=1}^{\infty} \frac{10^k + 1}{k^{10}}$

30. $\displaystyle\sum_{k=3}^{\infty} \frac{1}{k^3 \ln k}$

31. $\displaystyle\sum_{j=1}^{\infty} \frac{5}{j^2 + 4}$

32. $\displaystyle\sum_{k=1}^{\infty} \frac{k^k}{(k + 2)^k}$

33. $\displaystyle\sum_{k=3}^{\infty} \frac{1}{k^{1/3} \ln k}$

34. $\displaystyle\sum_{k=1}^{\infty} \frac{(-1)^k 5k^2}{\sqrt{3k^5 + 1}}$

35. $\displaystyle\sum_{k=1}^{\infty} \frac{2^k 3^k}{k^k}$

36. $\displaystyle\sum_{k=1}^{\infty} (-1)^{k+1} \frac{2k + 1}{k!}$

37. $\displaystyle\sum_{k=1}^{\infty} (-1)^k \left(\frac{5k}{3k + 7}\right)^k$

38. $\displaystyle\sum_{k=1}^{\infty} \frac{2^k (k - 1)!}{k!}$

39. $\displaystyle\sum_{k=1}^{\infty} \frac{5^k (k!)^2}{(2k)!}$

40. $\displaystyle\sum_{j=1}^{\infty} \frac{\cos((2j + 1)\pi)}{j^2 + 1}$

41. $\displaystyle\sum_{k=1}^{\infty} \frac{2^k}{3^k - 2^k}$

42. $\displaystyle\sum_{k=1}^{\infty} \frac{2 + (-1)^k}{2^k}$

43. $\displaystyle\sum_{k=1}^{\infty} \cos \frac{(3k - 1)\pi}{3}$

44. $\displaystyle\sum_{k=10}^{\infty} \frac{1}{\sqrt{k(k - 9)}}$

45. $\displaystyle\sum_{k=1}^{\infty} \frac{k^4}{e^{k^5}}$

46. $\displaystyle\sum_{k=1}^{\infty} \frac{1}{(k + 1)! - k!}$

47. $\displaystyle\sum_{k=1}^{\infty} \frac{(4k)!}{(k!)^4}$

48. $\dfrac{2}{3} + \dfrac{3}{8} + \dfrac{4}{15} + \dfrac{5}{24} + \dfrac{6}{35} + \cdots$

49. $\displaystyle\sum_{k=1}^{\infty} \frac{\sqrt[5]{k}}{\sqrt[5]{k^7 + 1}}$

50. $\displaystyle\sum_{k=1}^{\infty} e^{-k^3}$

51. $\displaystyle\sum_{k=1}^{\infty} \frac{7^k + 11^k}{11^k}$

52. $\displaystyle\sum_{k=1}^{\infty} \frac{7^k + 11^k}{13^k}$

53. $\displaystyle\sum_{k=1}^{\infty} \sin \frac{1}{k^9}$

54. $\displaystyle\sum_{j=1}^{\infty} j^9 \sin \frac{1}{j^9}$

55. $\displaystyle\sum_{k=1}^{\infty} \cos \frac{1}{k^9}$

56. $\displaystyle\sum_{k=1}^{\infty} \left(\frac{k}{k + 10}\right)^{-k^2}$

57. $\displaystyle\sum_{k=1}^{\infty} 5^{1-2k}$

58. $\displaystyle\sum_{k=1}^{\infty} \frac{(-1)^k}{\sqrt{\ln(k + 2)}}$

59. $\displaystyle\sum_{k=1}^{\infty} \frac{k!}{k^k + 3}$

60. $\displaystyle\sum_{k=1}^{\infty} \left(\frac{1}{k^2} + \frac{1}{k^5}\right)$

61. $\displaystyle\sum_{k=1}^{\infty} \frac{1}{\ln(e^k + 1)}$

62. $\displaystyle\sum_{k=0}^{\infty} \left(\frac{\tan^{-1}k}{\pi}\right)^k$

63. $\displaystyle\sum_{k=1}^{\infty} \left(\frac{k + a}{k}\right)^{k^2}, a > 0$

64. $\displaystyle\frac{1}{1 \cdot 4} + \frac{1}{2 \cdot 7} + \frac{1}{3 \cdot 10} + \frac{1}{4 \cdot 13} + \cdots$

65. $\displaystyle\sum_{k=1}^{\infty} \left(\cos \frac{1}{k} - \cos \frac{1}{k + 1}\right)$

66. $\displaystyle\sum_{k=1}^{\infty} \frac{4^{k^2}}{k!}$

67. $\displaystyle\sum_{j=1}^{\infty} \frac{\cot^{-1} \frac{1}{j}}{2^j}$

68. $\displaystyle\sum_{k=2}^{\infty} \frac{\sqrt{k + 1} - \sqrt{k}}{\sqrt{k^2 - k}}$

69. $\displaystyle\sum_{k=1}^{\infty} \left(1 + \frac{1}{2k}\right)^k$

70. $\displaystyle\sum_{k=0}^{\infty} e^{k/100}$

71. $\displaystyle\sum_{k=1}^{\infty} \frac{\ln^2 k}{k^{3/2}}$

72. $\displaystyle\sum_{k=1}^{\infty} (-1)^k \cos \frac{1}{k}$

73. $\displaystyle\sum_{k=0}^{\infty} k^2 \cdot 1.001^{-k}$

74. $\displaystyle\sum_{k=0}^{\infty} k \cdot 0.999^{-k}$

75. $\displaystyle\sum_{k=1}^{\infty} \left(-\frac{1}{k}\right)^k$

76. $\displaystyle\sum_{k=3}^{\infty} \frac{2}{k(2 - k)}$

77. $\displaystyle\sum_{k=0}^{\infty} \frac{3k}{\sqrt[4]{k^4 + 3}}$

78. $\displaystyle\sum_{k=0}^{\infty} k \cdot 3^{-k^2}$

79. $\displaystyle\sum_{k=1}^{\infty} \tan^{-1} \frac{1}{\sqrt{k}}$

80. $\displaystyle\sum_{k=1}^{\infty} \left(\frac{3k^8 - 2}{3k^9 + 2}\right)^k$

81. $\displaystyle\sum_{k=1}^{\infty} \left(\frac{1}{\sqrt{k + 2}} - \frac{1}{\sqrt{k}}\right)$

82. $\displaystyle\sum_{k=0}^{\infty} \left(\left(\frac{2}{3}\right)^{k+1} - \left(\frac{3}{2}\right)^{-k+1}\right)$

83. $\displaystyle\sum_{j=2}^{\infty} \frac{1}{j \ln^{10} j}$

84. $\displaystyle\sum_{k=1}^{\infty} \tan^{-1} \frac{1}{k^3}$

85. $\displaystyle\frac{1}{2 \cdot 3} + \frac{1}{4 \cdot 5} + \frac{1}{6 \cdot 7} + \frac{1}{8 \cdot 9} + \cdots$

86. $\displaystyle\sum_{k=1}^{\infty} \frac{a^k}{k!}, a \neq 0$

87. Explain why or why not Determine whether the following statements are true and give an explanation or counterexample.

  a. If the Limit Comparison Test can be applied successfully to a given series with a certain comparison series, the Comparison Test also works with the same comparison series.

  b. The series $\displaystyle\sum_{k=3}^{\infty} \frac{1}{k \ln^p k}$ converges for the same values of p as the series $\displaystyle\sum_{k=3}^{\infty} \frac{1}{k^p}$.

  c. Both the Ratio Test and the Root Test can be applied conclusively to a geometric series.

  d. The Alternating Series Test can be used to show that some series diverge.

Explorations and Challenges

88–93. A few more series *Determine whether the following series converge. Justify your answers.*

88. $\displaystyle\sum_{k=1}^{\infty} \left(\sqrt{k^4 + 1} - k^2\right)$

89. $\displaystyle\sum_{k=1}^{\infty} \frac{k}{1 + 2 + \cdots + k}$

90. $\displaystyle\sum_{k=1}^{\infty} \left(\frac{1^3 + 2^3 + 3^3 + \cdots + k^3}{k^5}\right)$

91. $\displaystyle\sum_{k=0}^{\infty} \ln\left(\frac{2k + 1}{2k + 4}\right)$

92. $\displaystyle\sum_{k=1}^{\infty} \frac{1}{10^{\ln k}}$

93. $\displaystyle\sum_{k=1}^{\infty} \frac{1}{2^{\ln k} + 2}$

QUICK CHECK ANSWERS

1. $2/3$; yes ◄

CHAPTER 10   REVIEW EXERCISES

1. Explain why or why not Determine whether the following statements are true and give an explanation or counterexample.

  a. The terms of the sequence $\{a_n\}$ increase in magnitude, so the limit of the sequence does not exist.

  b. The terms of the series $\displaystyle\sum \frac{1}{\sqrt{k}}$ approach zero, so the series converges.

  c. The terms of the sequence of partial sums of the series $\sum a_k$ approach $\frac{5}{2}$, so the infinite series converges to $\frac{5}{2}$.

  d. An alternating series that converges absolutely must converge conditionally.

  e. The sequence $a_n = \dfrac{n^2}{n^2 + 1}$ converges.

  f. The sequence $a_n = \dfrac{(-1)^n n^2}{n^2 + 1}$ converges.

  g. The series $\displaystyle\sum_{k=1}^{\infty} \frac{k^2}{k^2 + 1}$ converges.

  h. The sequence of partial sums associated with the series $\displaystyle\sum_{k=1}^{\infty} \frac{1}{k^2 + 1}$ converges.

2. **Related sequences**

   a. Find the limit of the sequence $\left\{\dfrac{12n^3 + 4n}{6n^3 + 5}\right\}$.

   b. Explain why the sequence $\left\{(-1)^n \dfrac{12n^3 + 4n}{6n^3 + 5}\right\}$ diverges.

3. **Geometric sums** Evaluate the geometric sums $\displaystyle\sum_{k=0}^{9}(0.2)^k$ and $\displaystyle\sum_{k=2}^{9}(0.2)^k$.

4. **A savings plan** Suppose you open a savings account by depositing $100. The account earns interest at an annual rate of 3% per year (0.25% per month). At the end of each month, you earn interest on the current balance, and then you deposit $100. Let B_n be the balance at the beginning of the nth month, where $B_0 = \$100$.

   a. Find a recurrence relation for the sequence $\{B_n\}$.

   b. Find an explicit formula that gives B_n, for $n = 0, 1, 2, 3, \ldots$.

   c. Find B_{30}, the balance at the beginning of the 30th month.

5. **Partial sums** Let S_n be the nth partial sum of $\displaystyle\sum_{k=1}^{\infty}a_k = 8$. Find $\displaystyle\lim_{k\to\infty}a_k$ and $\displaystyle\lim_{n\to\infty}S_n$.

6. Find the value of r for which $\displaystyle\sum_{k=0}^{\infty}3r^k = 6$.

7. Give an example (if possible) of a sequence $\{a_k\}$ that converges, while the series $\displaystyle\sum_{k=1}^{\infty}a_k$ diverges.

8. Give an example (if possible) of a series $\displaystyle\sum_{k=1}^{\infty}a_k$ that converges, while the sequence $\{a_k\}$ diverges.

9. **Sequences versus series**

   a. Find the limit of the sequence $\left\{\left(-\dfrac{4}{5}\right)^k\right\}$.

   b. Evaluate $\displaystyle\sum_{k=0}^{\infty}\left(-\dfrac{4}{5}\right)^k$.

10. **Sequences versus series**

    a. Find the limit of the sequence $\left\{\dfrac{1}{k} - \dfrac{1}{k+1}\right\}$.

    b. Evaluate $\displaystyle\sum_{k=1}^{\infty}\left(\dfrac{1}{k} - \dfrac{1}{k+1}\right)$.

11. a. Does the sequence $\left\{\dfrac{k}{k+1}\right\}$ converge? Why or why not?

    b. Does the series $\displaystyle\sum_{k=1}^{\infty}\dfrac{k}{k+1}$ converge? Why or why not?

12–24. Limits of sequences *Evaluate the limit of the sequence or state that it does not exist.*

12. $a_n = \dfrac{3n^3 + 4n}{6n^3 + 5}$

13. $a_n = (-1)^n \dfrac{3n^3 + 4n}{6n^3 + 5}$

14. $a_n = \dfrac{(2n+2)!}{(2n)!\,n^2}$

15. $a_n = \dfrac{2^n + 5^{n+1}}{5^n}$

16. $a_n = \dfrac{n^2 + 4}{\sqrt{4n^4 + 1}}$

17. $a_n = \dfrac{8^n}{n!}$

18. $a_n = \left(1 + \dfrac{3}{n}\right)^{2n}$

19. $a_n = \dfrac{(3n^2 + 2n + 1)\sin n}{4n^3 + n}$ (*Hint:* Use the Squeeze Theorem.)

20. $a_n = n - \sqrt{n^2 - 1}$

21. $a_n = \left(\dfrac{1}{n}\right)^{1/\ln n}$

22. $a_n = \sin\dfrac{\pi n}{6}$

23. $a_n = \dfrac{(-1)^n}{0.9^n}$

24. $a_n = \dfrac{\pi}{2} - \tan^{-1}n$

25–26. Recursively defined sequences *The following sequences $\{a_n\}_{n=0}^{\infty}$ are defined by a recurrence relation. Assume each sequence is monotonic and bounded.*

   a. *Find the first five terms $a_0, a_1, \ldots, a_4$ of each sequence.*

   b. *Determine the limit of each sequence.*

25. $a_{n+1} = \dfrac{1}{2}a_n + 8;\ a_0 = 80$

T 26. $a_{n+1} = \sqrt{\dfrac{1}{3}a_n + 34};\ a_0 = 81$

27–37. Evaluating series *Evaluate the following infinite series or state that the series diverges.*

27. $\displaystyle\sum_{k=1}^{\infty}3(1.001)^k$

28. $\displaystyle\sum_{k=1}^{\infty}\left(\dfrac{9}{10}\right)^k$

29. $\displaystyle\sum_{k=0}^{\infty}\left(\left(\dfrac{1}{3}\right)^k + \left(\dfrac{4}{3}\right)^k\right)$

30. $\displaystyle\sum_{k=1}^{\infty}\dfrac{3^k - 5^k}{10^k}$

31. $\displaystyle\sum_{k=1}^{\infty}\ln\left(\dfrac{2k+1}{2k-1}\right)$

32. $\displaystyle\sum_{k=0}^{\infty}\left(-\dfrac{1}{5}\right)^k$

33. $\displaystyle\sum_{k=0}^{\infty}(\tan^{-1}(k+2) - \tan^{-1}k)$

34. $\displaystyle\sum_{k=2}^{\infty}\left(\dfrac{1}{\sqrt{k}} - \dfrac{1}{\sqrt{k-1}}\right)$

35. $\displaystyle\sum_{k=1}^{\infty}\dfrac{9}{(3k-2)(3k+1)}$

36. $\displaystyle\sum_{k=1}^{\infty}4^{-3k}$

37. $\displaystyle\sum_{k=1}^{\infty}\dfrac{2^k}{3^{k+2}}$

38. Express $0.29292929\ldots$ as a ratio of two integers.

39. Express $0.314141414\ldots$ as a ratio of two integers.

40. **Finding steady states through sequences** Suppose you take 100 mg of aspirin once per day. Assume the aspirin has a half-life of one day; that is, every day, half of the aspirin in your blood is eliminated. Assume d_n is the amount of aspirin in your blood after the nth dose, where $d_1 = 100$.

    a. Find a recurrence relation for the sequence $\{d_n\}$.

    b. Assuming the sequence $\{d_n\}$ converges, find the long-term (steady-state) amount of aspirin in your blood.

41. **Finding steady states using infinite series** Solve Exercise 40 by expressing the amount of aspirin in your blood as a geometric series and evaluating the series.

42–76. Convergence or divergence *Use a convergence test of your choice to determine whether the following series converge.*

42. $\displaystyle\sum_{k=1}^{\infty} \frac{k}{k^{5/2} + 1}$

43. $\displaystyle\sum_{k=1}^{\infty} k^{-2/3}$

44. $\displaystyle\sum_{j=1}^{\infty} \frac{2j^4 + j^3 + 1}{3j^6 + 4}$

45. $\displaystyle\sum_{k=1}^{\infty} \frac{2k^4 + k^3 + 1}{3k^4 + 4}$

46. $\displaystyle\sum_{k=1}^{\infty} \frac{k^{k/2} 2^k}{\sqrt{k!}}$

47. $\displaystyle\sum_{k=1}^{\infty} \frac{7 + \sin k}{k^2}$

48. $\displaystyle\sum_{k=1}^{\infty} \frac{2k^4 + 1}{\sqrt{9k^8 + 2}}$

49. $\displaystyle\sum_{k=1}^{\infty} \frac{k^4}{\sqrt{9k^{12} + 2}}$

50. $\displaystyle\sum_{k=1}^{\infty} \sqrt{\frac{k!}{k^k}}$

51. $\displaystyle\sum_{k=1}^{\infty} \frac{2^k}{e^k}$

52. $\displaystyle\sum_{k=1}^{\infty} \left(\frac{k}{k + 3}\right)^{2k}$

53. $\displaystyle\sum_{k=1}^{\infty} \left(\frac{k}{2k + 3}\right)^{k}$

54. $\displaystyle\sum_{k=1}^{\infty} \sin \frac{1}{k^4}$

55. $\displaystyle\sum_{k=1}^{\infty} \frac{k!}{e^k k^k}$

56. $\displaystyle\sum_{k=1}^{\infty} \frac{1}{\sqrt{k}\sqrt{k + 1}}$

57. $\displaystyle\sum_{k=1}^{\infty} \frac{5^k}{2^{2k+1}}$

58. $\displaystyle\sum_{k=1}^{\infty} k e^{-k}$

59. $\displaystyle\sum_{j=0}^{\infty} \frac{2 \cdot 4^j}{(2j + 1)!}$

60. $\displaystyle\sum_{k=0}^{\infty} \frac{9^k}{(2k)!}$

61. $\displaystyle\sum_{k=3}^{\infty} \frac{\ln k}{k^{3/2}}$

62. $\displaystyle\sum_{k=1}^{\infty} \frac{k^2}{(2k - 1)!}$

63. $\displaystyle\sum_{k=1}^{\infty} \frac{3}{2 + e^k}$

64. $\displaystyle\sum_{k=1}^{\infty} k \sin \frac{1}{k}$

65. $\displaystyle\sum_{k=1}^{\infty} \frac{\sqrt[k]{k}}{k^3}$

66. $\displaystyle\sum_{k=1}^{\infty} \frac{1}{1 + \ln k}$

67. $\displaystyle\sum_{k=1}^{\infty} k^5 e^{-k}$

68. $\displaystyle\sum_{k=4}^{\infty} \frac{2}{k^2 - 10}$

69. $\displaystyle\sum_{k=1}^{\infty} \left(1 - \cos \frac{1}{k}\right)$

70. $\displaystyle\sum_{k=1}^{\infty} \left(\frac{\pi}{2} - \tan^{-1} k\right)$

71. $\displaystyle\sum_{k=1}^{\infty} \left(1 - \cos \frac{1}{k}\right)^2$

72. $\displaystyle\sum_{k=1}^{\infty} \left(\frac{\pi}{2} - \tan^{-1} k\right)^2$

73. $\displaystyle\sum_{k=1}^{\infty} \frac{\coth k}{k}$

74. $\displaystyle\sum_{k=1}^{\infty} \frac{1}{\sinh k}$

75. $\displaystyle\sum_{k=1}^{\infty} \tanh k$

76. $\displaystyle\sum_{k=0}^{\infty} \operatorname{sech} k$

77–87. Absolute or conditional convergence *Determine whether the following series converge absolutely, converge conditionally, or diverge.*

77. $\displaystyle\sum_{k=1}^{\infty} \frac{(-1)^{k+1}}{k^{3/7}}$

78. $\displaystyle\sum_{k=1}^{\infty} \frac{(-1)^k}{k^{5/2}}$

79. $\displaystyle\sum_{k=2}^{\infty} \frac{(-1)^k}{k^2 - 1}$

80. $\displaystyle\sum_{j=1}^{\infty} \frac{(-1)^j}{\ln(j + 3)}$

81. $\displaystyle\sum_{k=1}^{\infty} \frac{(-1)^{k+1}(k^2 + 4)}{2k^2 + 1}$

82. $\displaystyle\sum_{k=1}^{\infty} \frac{(-1)^k}{\sqrt{k^2 + 1}}$

83. $\displaystyle\sum_{k=1}^{\infty} (-1)^k k e^{-k}$

84. $\displaystyle\sum_{k=0}^{\infty} \frac{(-1)^k}{e^k + e^{-k}}$

85. $\displaystyle\sum_{k=1}^{\infty} \frac{(-1)^{k+1} 10^k}{k!}$

86. $\displaystyle\sum_{k=2}^{\infty} \frac{(-1)^k}{k \ln k}$

87. $\displaystyle\sum_{k=1}^{\infty} \frac{(-2)^{k+1}}{k^2}$

88. Logarithmic *p*-series Show that the series $\displaystyle\sum_{k=2}^{\infty} \frac{1}{k(\ln k)^p}$ converges provided $p > 1$.

T 89–90. Lower and upper bounds of a series *For each convergent series and given value of n, complete the following.*

a. *Use the nth partial sum S_n to estimate the value of the series.*
b. *Find an upper bound for the remainder R_n.*
c. *Find lower and upper bounds (L_n and U_n, respectively) for the exact value of the series.*

89. $\displaystyle\sum_{k=1}^{\infty} \frac{1}{k^5}; n = 5$

90. $\displaystyle\sum_{k=1}^{\infty} \frac{1}{k^2 + 1}; n = 20$

91. Estimate the value of the series $\displaystyle\sum_{k=1}^{\infty} \frac{1}{(2k + 5)^3}$ to within 10^{-4} of its exact value.

92. Estimate the value of the series $\displaystyle\sum_{k=1}^{\infty} \frac{(-1)^{k+1}}{(2k + 5)^3}$ using the fifth partial sum S_5 and then find an upper bound for the error $|S - S_5|$.

T 93. Error in a finite alternating sum How many terms of the series $\displaystyle\sum_{k=1}^{\infty} \frac{(-1)^{k+1}}{k^4}$ must be summed to ensure that the approximation error is less than 10^{-8}?

94. Equations involving series Solve the following equations for x.

a. $\displaystyle\sum_{k=0}^{\infty} e^{kx} = 2$

b. $\displaystyle\sum_{k=0}^{\infty} (3x)^k = 4$

95. Building a tunnel—first scenario A crew of workers is constructing a tunnel through a mountain. Understandably, the rate of construction decreases because rocks and earth must be removed a greater distance as the tunnel gets longer. Suppose each week the crew digs 0.95 of the distance it dug the previous week. In the first week, the crew constructed 100 m of tunnel.

a. How far does the crew dig in 10 weeks? 20 weeks? N weeks?
b. What is the longest tunnel the crew can build at this rate?

96. Building a tunnel—second scenario As in Exercise 95, a crew of workers is constructing a tunnel. The time required to dig 100 m increases by 10% each week, starting with 1 week to dig the first 100 m. Can the crew complete a 1.5-km (1500-m) tunnel in 30 weeks? Explain.

97. Pages of circles On page 1 of a book, there is one circle of radius 1. On page 2, there are two circles of radius $1/2$. On page n, there are 2^{n-1} circles of radius 2^{-n+1}.

a. What is the sum of the areas of the circles on page n of the book?
b. Assuming the book continues indefinitely ($n \to \infty$), what is the sum of the areas of all the circles in the book?

T 98. Max sine sequence Let $a_n = \max\{\sin 1, \sin 2, \ldots, \sin n\}$, for $n = 1, 2, 3, \ldots$, where max $\{\ldots\}$ denotes the maximum element of the set. Does $\{a_n\}$ converge? If so, make a conjecture about the limit.

99. Sierpinski triangle The fractal called the *Sierpinski triangle* is the limit of a sequence of figures. Starting with the equilateral triangle with sides of length 1, an inverted equilateral triangle with sides of length $1/2$ is removed. Then, three inverted equilateral triangles with sides of length $1/4$ are removed from this figure (see figure). The process continues in this way. Let T_n be the total area of the removed triangles after stage n of the process. The area of an equilateral triangle with side length L is $A = \dfrac{\sqrt{3}L^2}{4}$.

a. Find T_1 and T_2, the total area of the removed triangles after stages 1 and 2, respectively.
b. Find T_n, for $n = 1, 2, 3, \ldots$.
c. Find $\lim\limits_{n \to \infty} T_n$.
d. What is the area of the original triangle that remains as $n \to \infty$?

Initial stage   First stage   Second stage

100. Multiplier effect Imagine that the government of a small community decides to give a total of $\$W$, distributed equally, to all its citizens. Suppose each month each citizen saves a fraction p of his or her new wealth and spends the remaining $1 - p$ in the community. Assume no money leaves or enters the community, and all the spent money is redistributed throughout the community.

a. If this cycle of saving and spending continues for many months, how much money is ultimately spent? Specifically, by what factor is the initial investment of $\$W$ increased? Economists refer to this increase in the investment as the *multiplier effect*.
b. Evaluate the limits $p \to 0$ and $p \to 1$, and interpret their meanings. (See the Guided Project *Economic stimulus packages* for more on stimulus packages.)

101. Bouncing ball for time Suppose a rubber ball, when dropped from a given height, returns to a fraction p of that height. In the absence of air resistance, a ball dropped from a height h requires $\sqrt{\dfrac{2h}{g}}$ seconds to fall to the ground, where $g \approx 9.8 \text{ m/s}^2$ is the acceleration due to gravity. The time taken to bounce *up* to a given height equals the time to fall from that height to the ground. How long does it take for a ball dropped from 10 m to come to rest?

Chapter 10 Guided Projects

Applications of the material in this chapter and related topics can be found in the following Guided Projects. For additional information, see the Preface.

- Chaos!
- Financial matters
- Periodic drug dosing
- Economic stimulus packages

- The mathematics of loans
- Archimedes' approximation to π
- Exact values of infinite series
- Conditional convergence in a crystal lattice

11

Power Series

Chapter Preview   Until now, you have worked with infinite series consisting of real numbers. In this chapter, we make a seemingly small, but significant, change by considering infinite series whose terms include powers of a variable. With this change, an infinite series becomes a *power series*. One of the most fundamental ideas in all of calculus is that functions can be represented by power series. As a first step toward this result, we look at approximating functions using polynomials. The transition from polynomials to power series is then straightforward, and we learn how to represent the familiar functions of mathematics in terms of power series called *Taylor series*. The remainder of the chapter is devoted to the properties and many uses of Taylor series.

11.1 Approximating Functions with Polynomials

Power series provide a way to represent familiar functions and to define new functions. For this reason, power series—like sets and functions—are among the most fundamental entities in mathematics.

What Is a Power Series?

A *power series* is an infinite series of the form

$$\sum_{k=0}^{\infty} c_k x^k = \underbrace{c_0 + c_1 x + c_2 x^2 + \cdots + c_n x^n}_{n\text{th-degree polynomial}} + \underbrace{c_{n+1} x^{n+1} + \cdots}_{\text{terms continue}},$$

or, more generally,

$$\sum_{k=0}^{\infty} c_k (x - a)^k = \underbrace{c_0 + c_1(x - a) + \cdots + c_n(x - a)^n}_{n\text{th-degree polynomial}} + \underbrace{c_{n+1}(x - a)^{n+1} + \cdots}_{\text{terms continue}},$$

where the *center* of the series a and the coefficients c_k are constants. This type of series is called a power series because it consists of powers of x or $(x - a)$.

Viewed another way, a power series is built up from polynomials of increasing degree, as shown in the following progression.

$$
\left.
\begin{aligned}
&\text{Degree } 0: \ c_0 \\
&\text{Degree } 1: \ c_0 + c_1 x \\
&\text{Degree } 2: \ c_0 + c_1 x + c_2 x^2 \\
&\quad\ \ \vdots \qquad \vdots \qquad \vdots \\
&\text{Degree } n: \ c_0 + c_1 x + c_2 x^2 + \cdots + c_n x^n = \sum_{k=0}^{n} c_k x^k \\
&\quad\ \ \vdots \qquad \vdots \qquad \vdots
\end{aligned}
\right\}\ \text{Polynomials}
$$

$$
\left.
c_0 + c_1 x + c_2 x^2 + \cdots + c_n x^n + \cdots = \sum_{k=0}^{\infty} c_k x^k
\right\}\ \text{Power series}
$$

According to this perspective, a power series is a "super-polynomial." Therefore, we begin our exploration of power series by using polynomials to approximate functions.

Polynomial Approximation

An important observation motivates our work. To evaluate a polynomial $\left(\text{such as } f(x) = x^8 - 4x^5 + \frac{1}{2}\right)$, all we need is arithmetic—addition, subtraction, multiplication, and division. However, algebraic functions $\left(\text{say, } f(x) = \sqrt[3]{x^4 - 1}\right)$ and the trigonometric, logarithmic, and exponential functions usually cannot be evaluated exactly using arithmetic. Therefore, it makes practical sense to use the simplest of functions, polynomials, to approximate more complicated functions.

Linear and Quadratic Approximation

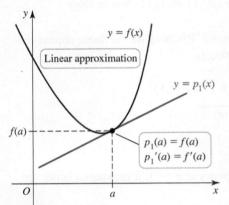

Figure 11.1

In Section 4.6, you learned that if a function f is differentiable at a point a, then it can be approximated near a by its tangent line, which is the linear approximation to f at the point a. The linear approximation at a is given by

$$
y - f(a) = f'(a)(x - a) \quad \text{or} \quad y = f(a) + f'(a)(x - a).
$$

Because the linear approximation is a first-degree polynomial, we name it p_1:

$$
p_1(x) = f(a) + f'(a)(x - a).
$$

This polynomial has some important properties: It matches f in *value* and in *slope* at a. In other words (**Figure 11.1**),

$$
p_1(a) = f(a) \quad \text{and} \quad p_1{}'(a) = f'(a).
$$

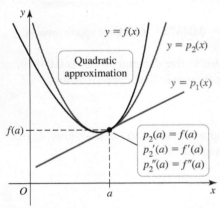

Figure 11.2

▶ Matching concavity (second derivatives) ensures that the graph of p_2 bends in the same direction as the graph of f at a.

Linear approximation works well if f has a fairly constant slope near a. However, if f has a lot of curvature near a, then the tangent line may not provide an accurate approximation. To remedy this situation, we create a quadratic approximating polynomial by adding one new term to the linear polynomial. Denoting this new polynomial p_2, we let

$$
p_2(x) = \underbrace{f(a) + f'(a)(x - a)}_{p_1(x)} + \underbrace{c_2(x - a)^2}_{\text{quadratic term}}.
$$

The new term consists of a coefficient c_2 that must be determined and a quadratic factor $(x - a)^2$.

To determine c_2 and to ensure that p_2 is a good approximation to f near the point a, we require that p_2 agree with f in value, slope, and concavity at a; that is, p_2 must satisfy the matching conditions

$$
p_2(a) = f(a), \quad p_2{}'(a) = f'(a), \quad \text{and} \quad p_2{}''(a) = f''(a),
$$

where we assume f and its first and second derivatives exist at a (**Figure 11.2**).

Substituting $x = a$ into p_2, we see immediately that $p_2(a) = f(a)$, so the first matching condition is met. Differentiating p_2 once, we have

$$
p_2{}'(x) = f'(a) + 2c_2(x - a).
$$

So $p_2'(a) = f'(a)$, and the second matching condition is also met. Because $p_2''(a) = 2c_2$, the third matching condition is

$$p_2''(a) = 2c_2 = f''(a).$$

It follows that $c_2 = \frac{1}{2}f''(a)$; therefore, the quadratic approximating polynomial is

$$p_2(x) = \underbrace{f(a) + f'(a)(x - a)}_{p_1(x)} + \frac{f''(a)}{2}(x - a)^2.$$

EXAMPLE 1 Linear and quadratic approximations for $\ln x$

a. Find the linear approximation to $f(x) = \ln x$ at $x = 1$.

b. Find the quadratic approximation to $f(x) = \ln x$ at $x = 1$.

c. Use these approximations to estimate $\ln 1.05$.

SOLUTION

a. Note that $f(1) = 0$, $f'(x) = 1/x$, and $f'(1) = 1$. Therefore, the linear approximation to $f(x) = \ln x$ at $x = 1$ is

$$p_1(x) = f(1) + f'(1)(x - 1) = 0 + 1(x - 1) = x - 1.$$

As shown in **Figure 11.3**, p_1 matches f in value $(p_1(1) = f(1))$ and in slope $(p_1'(1) = f'(1))$ at $x = 1$.

b. We first compute $f''(x) = -1/x^2$ and $f''(1) = -1$. Building on the linear approximation found in part (a), the quadratic approximation is

$$p_2(x) = \underbrace{x - 1}_{p_1(x)} + \underbrace{\frac{1}{2}f''(1)(x - 1)^2}_{c_2}$$

$$= x - 1 - \frac{1}{2}(x - 1)^2.$$

Because p_2 matches f in value, slope, and concavity at $x = 1$, it provides a better approximation to f near $x = 1$ (Figure 11.3).

c. To approximate $\ln 1.05$, we substitute $x = 1.05$ into each polynomial approximation:

$$p_1(1.05) = 1.05 - 1 = 0.05 \text{ and} \qquad \text{Linear approximation}$$

$$p_2(1.05) = 1.05 - 1 - \frac{1}{2}(1.05 - 1)^2 = 0.04875. \quad \text{Quadratic approximation}$$

The value of $\ln 1.05$ given by a calculator, rounded to five decimal places, is 0.04879, showing the improvement in quadratic approximation over linear approximation.

Related Exercises 10, 13 ◀

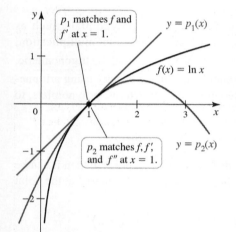

Figure 11.3

We now extend the idea of linear and quadratic approximation to obtain higher-degree polynomials that generally provide better approximations.

Taylor Polynomials

▶ Building on ideas that were already circulating in the early 18th century, Brook Taylor (1685–1731) published Taylor's Theorem in 1715. He is also credited with discovering integration by parts.

Assume f and its first n derivatives exist at a; our goal is to find an nth-degree polynomial that approximates the values of f near a. The first step is to use p_2 to obtain a cubic polynomial p_3 of the form

$$p_3(x) = p_2(x) + c_3(x - a)^3$$

that satisfies the four matching conditions

$$p_3(a) = f(a), \quad p_3'(a) = f'(a), \quad p_3''(a) = f''(a), \quad \text{and} \quad p_3'''(a) = f'''(a).$$

Because p_3 is built "on top of" p_2, the first three matching conditions are met. The last condition, $p_3'''(a) = f'''(a)$, is used to determine c_3. A short calculation shows that $p_3'''(x) = 3 \cdot 2 c_3 = 3! c_3$, so the last matching condition is $p_3'''(a) = 3! c_3 = f'''(a)$.

Solving for c_3, we have $c_3 = \dfrac{f'''(a)}{3!}$. Therefore, the cubic approximating polynomial is

$$p_3(x) = \underbrace{f(a) + f'(a)(x - a) + \frac{f''(a)}{2!}(x - a)^2}_{p_2(x)} + \frac{f'''(a)}{3!}(x - a)^3.$$

> Recall that $2! = 2 \cdot 1$, $3! = 3 \cdot 2 \cdot 1$, $k! = k \cdot (k - 1)!$, and by definition, $0! = 1$.

QUICK CHECK 1 Verify that p_3 satisfies $p_3^{(k)}(a) = f^{(k)}(a)$, for $k = 0, 1, 2,$ and 3. ◄

Continuing in this fashion (Exercise 66), building each new polynomial on the previous polynomial, the nth approximating polynomial for f at a is

$$p_n(x) = \underbrace{f(a) + f'(a)(x - a) + \frac{f''(a)}{2!}(x - a)^2 + \cdots + \frac{f^{(n)}(a)}{n!}(x - a)^n}_{p_{n-1}(x)}.$$

It satisfies the $n + 1$ matching conditions

$$p_n(a) = f(a), \quad p_n'(a) = f'(a), \quad p_n''(a) = f''(a), \ldots, \text{ and } p_n^{(n)}(a) = f^{(n)}(a).$$

These conditions ensure that the graph of p_n conforms as closely as possible to the graph of f near a (**Figure 11.4**).

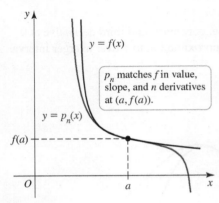

$y = f(x)$

p_n matches f in value, slope, and n derivatives at $(a, f(a))$.

$y = p_n(x)$

$f(a)$

Figure 11.4

> Recall that $f^{(n)}$ denotes the nth derivative of f. By convention, the zeroth derivative $f^{(0)}$ is f itself.

DEFINITION Taylor Polynomials

Let f be a function with f', f'', ..., and $f^{(n)}$ defined at a. The **nth-order Taylor polynomial** for f with its **center** at a, denoted p_n, has the property that it matches f in value, slope, and all derivatives up to the nth derivative at a; that is,

$$p_n(a) = f(a), p_n'(a) = f'(a), \ldots, \text{ and } p_n^{(n)}(a) = f^{(n)}(a).$$

The nth-order Taylor polynomial centered at a is

$$p_n(x) = f(a) + f'(a)(x - a) + \frac{f''(a)}{2!}(x - a)^2 + \cdots + \frac{f^{(n)}(a)}{n!}(x - a)^n.$$

More compactly, $p_n(x) = \displaystyle\sum_{k=0}^{n} c_k(x - a)^k$, where the **coefficients** are

$$c_k = \frac{f^{(k)}(a)}{k!}, \quad \text{for } k = 0, 1, 2, \ldots, n.$$

EXAMPLE 2   Taylor polynomials for $\sin x$ Find the Taylor polynomials $p_1, \ldots, p_7$ centered at $x = 0$ for $f(x) = \sin x$.

SOLUTION We begin by differentiating f repeatedly and evaluating the derivatives at 0; these calculations enable us to compute c_k, for $k = 0, 1, \ldots, 7$. Notice that a pattern emerges:

$$f(x) = \sin x \Rightarrow f(0) = 0$$
$$f'(x) = \cos x \Rightarrow f'(0) = 1$$
$$f''(x) = -\sin x \Rightarrow f''(0) = 0$$
$$f'''(x) = -\cos x \Rightarrow f'''(0) = -1$$
$$f^{(4)}(x) = \sin x \Rightarrow f^{(4)}(0) = 0.$$

The derivatives of $\sin x$ at 0 cycle through the values $\{0, 1, 0, -1\}$. Therefore, $f^{(5)}(0) = 1$, $f^{(6)}(0) = 0$, and $f^{(7)}(0) = -1$.

We now construct the Taylor polynomials that approximate $f(x) = \sin x$ near 0, beginning with the linear polynomial. The polynomial of order $n = 1$ is

$$p_1(x) = f(0) + f'(0)(x - 0) = x,$$

whose graph is the line through the origin with slope 1 (**Figure 11.5**). Notice that f and p_1 agree in value ($f(0) = p_1(0) = 0$) and in slope ($f'(0) = p_1'(0) = 1$) at 0. We see that p_1 provides a good fit to f near 0, but the graphs diverge visibly for $|x| > \pi/4$.

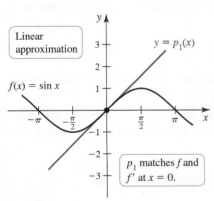

Linear approximation

$y = p_1(x)$

$f(x) = \sin x$

p_1 matches f and f' at $x = 0$.

Figure 11.5

➤ It is worth repeating that the next polynomial in the sequence is obtained by adding one new term to the previous polynomial. For example,

$$p_3(x) = p_2(x) + \frac{f'''(a)}{3!}(x-a)^3.$$

The polynomial of order $n = 2$ is

$$p_2(x) = \underbrace{f(0)}_{0} + \underbrace{f'(0)x}_{1} + \underbrace{\frac{f''(0)}{2!}x^2}_{0} = x,$$

so p_2 is the same as p_1.

The polynomial of order $n = 3$ is

$$p_3(x) = \underbrace{f(0) + f'(0)x + \frac{f''(0)}{2!}x^2}_{p_2(x) = x} + \underbrace{\frac{f'''(0)}{3!}x^3}_{-1/3!} = x - \frac{x^3}{6}.$$

QUICK CHECK 2 Verify the following properties for $f(x) = \sin x$ and $p_3(x) = x - x^3/6$:

$$f(0) = p_3(0),$$
$$f'(0) = p_3'(0),$$
$$f''(0) = p_3''(0), \text{ and }$$
$$f'''(0) = p_3'''(0). \blacktriangleleft$$

We have designed p_3 to agree with f in value, slope, concavity, and third derivative at 0 (**Figure 11.6**). Consequently, p_3 provides a better approximation to f over a larger interval than p_1.

Figure 11.6

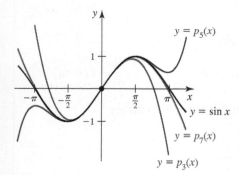

Figure 11.7

QUICK CHECK 3 Why do the Taylor polynomials for $\sin x$ centered at 0 consist only of odd powers of x? ◄

The procedure for finding Taylor polynomials may be extended to polynomials of any order. Because the even derivatives of $f(x) = \sin x$ are zero at $x = 0$, $p_4(x) = p_3(x)$. For the same reason, $p_6(x) = p_5(x)$:

$$p_6(x) = p_5(x) = x - \frac{x^3}{3!} + \frac{x^5}{5!}. \quad c_5 = \frac{f^{(5)}(0)}{5!} = \frac{1}{5!}$$

Finally, the Taylor polynomial of order $n = 7$ is

$$p_7(x) = x - \frac{x^3}{3!} + \frac{x^5}{5!} - \frac{x^7}{7!}. \quad c_7 = \frac{f^{(7)}(0)}{7!} = -\frac{1}{7!}$$

From **Figure 11.7** we see that as the order of the Taylor polynomials increases, more accurate approximations to $f(x) = \sin x$ are obtained over larger intervals centered at 0. For example, p_7 is a good fit to $f(x) = \sin x$ over the interval $[-\pi, \pi]$. Notice that $\sin x$ and its Taylor polynomials (centered at 0) are all odd functions.   *Related Exercises 17, 20* ◄

Approximations with Taylor Polynomials

Taylor polynomials find widespread use in approximating functions, as illustrated in the following examples.

EXAMPLE 3 Taylor polynomials for e^x

a. Find the Taylor polynomials of order $n = 0, 1, 2$, and 3 for $f(x) = e^x$ centered at 0. Graph f and the polynomials.

➤ Recall that if c is an approximation to x, the absolute error in c is $|x - c|$ and the relative error in c is $|x - c|/|x|$. We use *error* to refer to *absolute error*.

b. Use the polynomials in part (a) to approximate $e^{0.1}$ and $e^{-0.25}$. Find the absolute errors, $|f(x) - p_n(x)|$, in the approximations. Use calculator values for the exact values of f.

SOLUTION

a. Recall that the coefficients for the Taylor polynomials centered at 0 are

$$c_k = \frac{f^{(k)}(0)}{k!}, \quad \text{for } k = 0, 1, 2, \dots, n.$$

With $f(x) = e^x$, we have $f^{(k)}(x) = e^x$, $f^{(k)}(0) = 1$, and $c_k = 1/k!$, for $k = 0, 1, 2, 3, \ldots$. The first four polynomials are

$$p_0(x) = f(0) = 1,$$

$$p_1(x) = \underbrace{f(0)}_{p_0(x)\,=\,1} + \underbrace{f'(0)x}_{1} = 1 + x,$$

$$p_2(x) = \underbrace{f(0) + f'(0)x}_{p_1(x)\,=\,1+x} + \underbrace{\frac{f''(0)}{2!}}_{1/2}x^2 = 1 + x + \frac{x^2}{2}, \text{ and}$$

$$p_3(x) = \underbrace{f(0) + f'(0)x + \frac{f''(0)}{2!}x^2}_{p_2(x)\,=\,1+x+x^2/2} + \underbrace{\frac{f^{(3)}(0)}{3!}}_{1/6}x^3 = 1 + x + \frac{x^2}{2} + \frac{x^3}{6}.$$

Taylor polynomials for $f(x) = e^x$ centered at 0. Approximations improve as n increases.

Figure 11.8

Notice that each successive polynomial provides a better fit to $f(x) = e^x$ near 0 (**Figure 11.8**). Continuing the pattern in these polynomials, the nth-order Taylor polynomial for e^x centered at 0 is

$$p_n(x) = 1 + x + \frac{x^2}{2!} + \frac{x^3}{3!} + \cdots + \frac{x^n}{n!} = \sum_{k=0}^{n} \frac{x^k}{k!}.$$

b. We evaluate $p_n(0.1)$ and $p_n(-0.25)$, for $n = 0, 1, 2,$ and 3, and compare these values to the calculator values of $e^{0.1} \approx 1.1051709$ and $e^{-0.25} \approx 0.77880078$. The results are shown in Table 11.1. Observe that the errors in the approximations decrease as n increases. In addition, the errors in approximating $e^{0.1}$ are smaller in magnitude than the errors in approximating $e^{-0.25}$ because $x = 0.1$ is closer to the center of the polynomials than $x = -0.25$. Reasonable approximations based on these calculations are $e^{0.1} \approx 1.105$ and $e^{-0.25} \approx 0.78$.

➤ A rule of thumb in finding estimates based on several approximations: Keep all the digits that are common to the last two approximations after rounding.

QUICK CHECK 4 Write out the next two Taylor polynomials p_4 and p_5 for $f(x) = e^x$ in Example 3. ◄

Table 11.1

n	Approximation $p_n(0.1)$	Absolute Error $\lvert e^{0.1} - p_n(0.1) \rvert$	Approximation $p_n(-0.25)$	Absolute Error $\lvert e^{-0.25} - p_n(-0.25) \rvert$
0	1	1.1×10^{-1}	1	2.2×10^{-1}
1	1.1	5.2×10^{-3}	0.75	2.9×10^{-2}
2	1.105	1.7×10^{-4}	0.78125	2.4×10^{-3}
3	1.105167	4.3×10^{-6}	0.778646	1.5×10^{-4}

Related Exercise 31 ◄

EXAMPLE 4 Approximating a real number using Taylor polynomials Use polynomials of order $n = 0, 1, 2,$ and 3 to approximate $\sqrt{18}$.

SOLUTION Letting $f(x) = \sqrt{x}$, we choose the center $a = 16$ because it is near 18, and f and its derivatives are easy to evaluate at 16. The Taylor polynomials have the form

$$p_n(x) = f(16) + f'(16)(x - 16) + \frac{f''(16)}{2!}(x - 16)^2 + \cdots + \frac{f^{(n)}(16)}{n!}(x - 16)^n.$$

We now evaluate the required derivatives:

$$f(x) = \sqrt{x} \Rightarrow f(16) = 4,$$

$$f'(x) = \frac{1}{2}x^{-1/2} \Rightarrow f'(16) = \frac{1}{8},$$

$$f''(x) = -\frac{1}{4}x^{-3/2} \Rightarrow f''(16) = -\frac{1}{256}, \text{ and}$$

$$f'''(x) = \frac{3}{8}x^{-5/2} \Rightarrow f'''(16) = \frac{3}{8192}.$$

Therefore, the polynomial p_3 (which includes p_0, p_1, and p_2) is

$$p_3(x) = \underbrace{\underbrace{\underbrace{4}_{p_0(x)} + \frac{1}{8}(x - 16)}_{p_1(x)} - \frac{1}{512}(x - 16)^2 + \frac{1}{16{,}384}(x - 16)^3.}_{p_2(x)}$$

The Taylor polynomials (**Figure 11.9**) give better approximations to f as the order of the approximation increases.

Figure 11.9

Letting $x = 18$, we obtain the approximations to $\sqrt{18}$ and the associated absolute errors shown in Table 11.2. (A calculator is used for the value of $\sqrt{18}$.) As expected, the errors decrease as n increases. Based on these calculations, a reasonable approximation is $\sqrt{18} \approx 4.24$.

Table 11.2

n	Approximation $p_n(18)$	Absolute Error $\lvert \sqrt{18} - p_n(18) \rvert$
0	4	2.4×10^{-1}
1	4.25	7.4×10^{-3}
2	4.242188	4.5×10^{-4}
3	4.242676	3.5×10^{-5}

Related Exercises 37–38 ◀

QUICK CHECK 5 At what point would you center the Taylor polynomials for $\sqrt{x}$ and $\sqrt[4]{x}$ to approximate $\sqrt{51}$ and $\sqrt[4]{15}$, respectively? ◀

Remainder in a Taylor Polynomial

Taylor polynomials provide good approximations to functions near a specific point. But how accurate are the approximations? To answer this question we define the *remainder* in a Taylor polynomial. If p_n is the Taylor polynomial for f of order n, then the remainder at the point x is

$$R_n(x) = f(x) - p_n(x).$$

The absolute value of the remainder is the error made in approximating $f(x)$ by $p_n(x)$. Equivalently, we have $f(x) = p_n(x) + R_n(x)$, which says that f consists of two components: the polynomial approximation and the associated remainder.

DEFINITION Remainder in a Taylor Polynomial

Let p_n be the Taylor polynomial of order n for f. The **remainder** in using p_n to approximate f at the point x is

$$R_n(x) = f(x) - p_n(x).$$

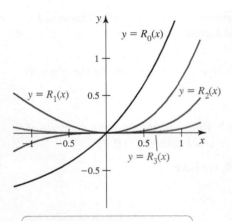

Remainders increase in magnitude as $|x|$ increases. Remainders decrease in magnitude to zero as n increases.

Figure 11.10

➤ The remainder R_n for a Taylor polynomial can be expressed in several different forms. The form stated in Theorem 11.1 is called the *Lagrange form* of the remainder.

The idea of a remainder is illustrated in **Figure 11.10**, where we see the remainders associated with various Taylor polynomials for $f(x) = e^x$ centered at 0 (Example 3). For fixed order n, the remainders tend to increase in magnitude as x moves farther from the center of the polynomials (in this case 0). And for fixed x, remainders decrease in magnitude to zero with increasing n.

The remainder for a Taylor polynomial may be written quite concisely, which enables us to estimate remainders. The following result is known as *Taylor's Theorem* (or the *Remainder Theorem*).

THEOREM 11.1   Taylor's Theorem (Remainder Theorem)

Let f have continuous derivatives up to $f^{(n+1)}$ on an open interval I containing a. For all x in I,

$$f(x) = p_n(x) + R_n(x),$$

where p_n is the nth-order Taylor polynomial for f centered at a and the remainder is

$$R_n(x) = \frac{f^{(n+1)}(c)}{(n+1)!}(x-a)^{n+1},$$

for some point c between x and a.

Discussion: We make two observations about Theorem 11.1 and outline a proof in Exercise 82. First, the case $n = 0$ is the Mean Value Theorem (Section 4.2), which states that

$$\frac{f(x) - f(a)}{x - a} = f'(c),$$

where c is a point between x and a. Rearranging this expression, we have

$$f(x) = \underbrace{f(a)}_{p_0(x)} + \underbrace{f'(c)(x-a)}_{R_0(x)}$$
$$= p_0(x) + R_0(x),$$

which is Taylor's Theorem with $n = 0$. Not surprisingly, the term $f^{(n+1)}(c)$ in Taylor's Theorem comes from a Mean Value Theorem argument.

The second observation makes the remainder easier to remember. If you write the $(n + 1)$st Taylor polynomial p_{n+1}, the highest-degree term is $\dfrac{f^{(n+1)}(a)}{(n+1)!}(x-a)^{n+1}$. Replacing $f^{(n+1)}(a)$ with $f^{(n+1)}(c)$ results in the remainder for p_n.

Estimating the Remainder

The remainder has both practical and theoretical importance. We deal with practical matters now and theoretical matters in Section 11.3. The remainder is used to estimate errors in approximations and to determine the number of terms of a Taylor polynomial needed to achieve a prescribed accuracy.

Because c is generally unknown, the difficulty in estimating the remainder is finding a bound for $|f^{(n+1)}(c)|$. Assuming this can be done, the following theorem gives a standard estimate for the remainder term.

THEOREM 11.2   Estimate of the Remainder

Let n be a fixed positive integer. Suppose there exists a number M such that $|f^{(n+1)}(c)| \le M$, for all c between a and x inclusive. The remainder in the nth-order Taylor polynomial for f centered at a satisfies

$$|R_n(x)| = |f(x) - p_n(x)| \le M \frac{|x-a|^{n+1}}{(n+1)!}.$$

Proof: The proof requires taking the absolute value of the remainder in Theorem 11.1, replacing $|f^{(n+1)}(c)|$ with a larger quantity M, and forming an inequality. ◄

We now give three examples that demonstrate how an upper bound for the remainder is computed and used in different ways.

EXAMPLE 5   Estimating the remainder for cos x   Find a bound for the magnitude of the remainder for the Taylor polynomials of $f(x) = \cos x$ centered at 0.

SOLUTION According to Theorem 11.1 with $a = 0$, we have

$$R_n(x) = \frac{f^{(n+1)}(c)}{(n+1)!} x^{n+1},$$

where c is a point between 0 and x. Notice that $f^{(n+1)}(c) = \pm \sin c$, or $f^{(n+1)}(c) = \pm \cos c$, depending on the value of n. In all cases, $|f^{(n+1)}(c)| \leq 1$. Therefore, we take $M = 1$ in Theorem 11.2, and the absolute value of the remainder can be bounded as

$$|R_n(x)| = \left| \frac{f^{(n+1)}(c)}{(n+1)!} x^{n+1} \right| \leq \frac{|x|^{n+1}}{(n+1)!}.$$

For example, if we approximate cos 0.1 using the Taylor polynomial p_{10}, the remainder satisfies

$$|R_{10}(0.1)| \leq \frac{0.1^{11}}{11!} \approx 2.5 \times 10^{-19}.$$

Related Exercises 41–42 ◄

EXAMPLE 6   Estimating a remainder   Consider again Example 4, in which we approximated $\sqrt{18}$ using the Taylor polynomial

$$p_3(x) = 4 + \frac{1}{8}(x - 16) - \frac{1}{512}(x - 16)^2 + \frac{1}{16{,}384}(x - 16)^3.$$

Find an upper bound on the magnitude of the remainder when using $p_3(x)$ to approximate $\sqrt{18}$.

SOLUTION In Example 4, we computed the error in the approximation knowing the exact value of $\sqrt{18}$ (obtained with a calculator). In the more realistic case in which we do not know the exact value, Theorem 11.2 allows us to estimate remainders (or errors). Applying this theorem with $n = 3$, $a = 16$, and $x = 18$, we find that the remainder in approximating $\sqrt{18}$ by $p_3(18)$ satisfies the bound

$$|R_3(18)| \leq M \frac{(18 - 16)^4}{4!} = \frac{2}{3} M,$$

where M is a number that satisfies $|f^{(4)}(c)| \leq M$, for all c between 16 and 18 inclusive. In this particular problem, we find that $f^{(4)}(c) = -\frac{15}{16} c^{-7/2}$, so M must be chosen (as small as possible) such that $|f^{(4)}(c)| = \frac{15}{16} c^{-7/2} = \frac{15}{16c^{7/2}} \leq M$, for $16 \leq c \leq 18$. You can verify that $\frac{15}{16c^{7/2}}$ is a decreasing function of c on $[16, 18]$ and has a maximum value of approximately 5.7×10^{-5} at $c = 16$ (**Figure 11.11**). Therefore, a bound on the remainder is

$$|R_3(18)| \leq \frac{2}{3} M \approx \frac{2}{3} \cdot 5.7 \times 10^{-5} \approx 3.8 \times 10^{-5}.$$

Notice that the actual error computed in Example 4 (Table 11.2) is 3.5×10^{-5}, which is less than the bound on the remainder—as it should be.

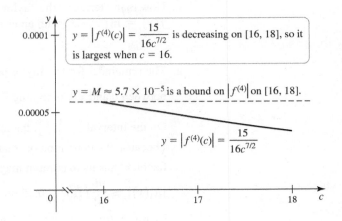

Figure 11.11

Related Exercise 52 ◄

EXAMPLE 7   Estimating the remainder for e^x Find a bound on the remainder in approximating $e^{0.45}$ using the Taylor polynomial of order $n = 6$ for $f(x) = e^x$ centered at 0.

SOLUTION Using Theorem 11.2, a bound on the remainder is given by

$$|R_n(x)| \leq M \frac{|x-a|^{n+1}}{(n+1)!},$$

where M is chosen such that $|f^{(n+1)}(c)| \leq M$, for all c between a and x inclusive. Notice that $f(x) = e^x$ implies that $f^{(k)}(x) = e^x$, for $k = 0, 1, 2, \dots$. In this particular problem, we have $n = 6$, $a = 0$, and $x = 0.45$, so the bound on the remainder takes the form

$$|R_6(0.45)| \leq M \frac{|0.45 - 0|^7}{7!} \approx 7.4 \times 10^{-7} M,$$

where M is chosen such that $|f^{(7)}(c)| = e^c \leq M$, for all c in the interval $[0, 0.45]$. Because e^c is an increasing function of c, its maximum value on the interval $[0, 0.45]$ occurs at $c = 0.45$ and is $e^{0.45}$. However, $e^{0.45}$ cannot be evaluated exactly (it is the number we are approximating), so we must find a number M such that $e^{0.45} \leq M$. Here is one of many ways to obtain a bound: We observe that $e^{0.45} < e^{1/2} < 4^{1/2} = 2$ and take $M = 2$ (**Figure 11.12**). Therefore, a bound on the remainder is

$$|R_6(0.45)| \leq 7.4 \times 10^{-7} M \approx 1.5 \times 10^{-6}.$$

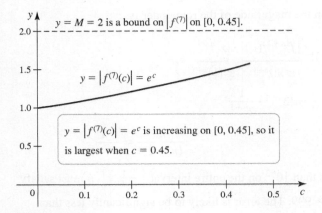

Figure 11.12

Using the Taylor polynomial derived in Example 3 with $n = 6$, the resulting approximation to $e^{0.45}$ is

$$p_6(0.45) = \sum_{k=0}^{6} \frac{0.45^k}{k!} \approx 1.5683114;$$

▶ Recall that if $f(x) = e^x$, then

$$p_n(x) = \sum_{k=0}^{n} \frac{x^k}{k!}.$$

it has an error that does not exceed 1.5×10^{-6}.

Related Exercise 49 ◄

QUICK CHECK 6 In Example 7, find an approximate upper bound for $R_7(0.45)$. ◄

EXAMPLE 8   Working with the remainder In Example 4b of Section 11.2, we show that the nth-order Taylor polynomial for $f(x) = \ln(1-x)$ centered at 0 is

$$p_n(x) = -\sum_{k=1}^{n} \frac{x^k}{k} = -x - \frac{x^2}{2} - \frac{x^3}{3} - \cdots - \frac{x^n}{n}.$$

a. Find a bound on the error in approximating $\ln(1-x)$ by $p_3(x)$ for values of x in the interval $\left[-\frac{1}{2}, \frac{1}{2}\right]$.

b. How many terms of the Taylor polynomial are needed to approximate values of $f(x) = \ln(1-x)$ with an error less than 10^{-3} on the interval $\left[-\frac{1}{2}, \frac{1}{2}\right]$?

SOLUTION

a. The remainder for the Taylor polynomial p_3 is $R_3(x) = \dfrac{f^{(4)}(c)}{4!}x^4$, where c is between 0 and x. Computing four derivatives of f, we find that $f^{(4)}(x) = -\dfrac{6}{(1-x)^4}$.

On the interval $\left[-\frac{1}{2}, \frac{1}{2}\right]$, the maximum magnitude of this derivative occurs at $x = \frac{1}{2}$ $\left(\text{because the denominator is smallest at } x = \frac{1}{2}\right)$ and is $6/\left(\frac{1}{2}\right)^4 = 96$. Similarly, the factor x^4 has its maximum magnitude at $x = \pm\frac{1}{2}$, and it is $\left(\frac{1}{2}\right)^4 = \frac{1}{16}$. Therefore,

$$|R_3(x)| \le \frac{96}{4!}\left(\frac{1}{16}\right) = 0.25 \text{ on the interval } \left[-\frac{1}{2}, \frac{1}{2}\right].$$ The error in approximating $f(x)$ by $p_3(x)$, for $-\frac{1}{2} \le x \le \frac{1}{2}$, does not exceed 0.25.

b. For any positive integer n, the remainder is $R_n(x) = \dfrac{f^{(n+1)}(c)}{(n+1)!}x^{n+1}$. Differentiating f several times reveals that

$$f^{(n+1)}(x) = -\frac{n!}{(1-x)^{n+1}}.$$

On the interval $\left[-\frac{1}{2}, \frac{1}{2}\right]$, the maximum magnitude of this derivative occurs at $x = \frac{1}{2}$ and is $n!/\left(\frac{1}{2}\right)^{n+1}$. Similarly, x^{n+1} has its maximum magnitude at $x = \pm\frac{1}{2}$, and it is $\left(\frac{1}{2}\right)^{n+1}$. Therefore, a bound on the magnitude of the remainder is

$$|R_n(x)| = \frac{1}{(n+1)!} \cdot \underbrace{|f^{(n+1)}(c)|}_{\le\, n!2^{n+1}} \cdot \underbrace{|x|^{n+1}}_{\le\, \left(\frac{1}{2}\right)^{n+1}}$$

$$\le \frac{1}{(n+1)!} \cdot n!2^{n+1} \cdot \frac{1}{2^{n+1}}$$

$$= \frac{1}{n+1}. \qquad\qquad \frac{n!}{(n+1)!} = \frac{1}{n+1}$$

To ensure that the error is less than 10^{-3} on the entire interval $\left[-\frac{1}{2}, \frac{1}{2}\right]$, n must satisfy $|R_n| \le \dfrac{1}{n+1} < 10^{-3}$ or $n > 999$. The error is likely to be significantly less than 10^{-3} if x is near 0.

Related Exercises 57, 62 ◀

SECTION 11.1 EXERCISES

Getting Started

1. Suppose you use a second-order Taylor polynomial centered at 0 to approximate a function f. What matching conditions are satisfied by the polynomial?

2. Does the accuracy of an approximation given by a Taylor polynomial generally increase or decrease with the order of the approximation? Explain.

3. The first three Taylor polynomials for $f(x) = \sqrt{1+x}$ centered at 0 are $p_0 = 1$, $p_1 = 1 + \dfrac{x}{2}$, and $p_2 = 1 + \dfrac{x}{2} - \dfrac{x^2}{8}$. Find three approximations to $\sqrt{1.1}$.

4. Suppose $f(0) = 1$, $f'(0) = 2$, and $f''(0) = -1$. Find the quadratic approximating polynomial for f centered at 0 and use it to approximate $f(0.1)$.

5. Suppose $f(0) = 1$, $f'(0) = 0$, $f''(0) = 2$, and $f^{(3)}(0) = 6$. Find the third-order Taylor polynomial for f centered at 0 and use it to approximate $f(0.2)$.

6. How is the remainder $R_n(x)$ in a Taylor polynomial defined?

7. Suppose $f(2) = 1$, $f'(2) = 1$, $f''(2) = 0$, and $f^{(3)}(2) = 12$. Find the third-order Taylor polynomial for f centered at 2 and use this polynomial to estimate $f(1.9)$.

8. Suppose you want to estimate $\sqrt{26}$ using a fourth-order Taylor polynomial centered at $x = a$ for $f(x) = \sqrt{x}$. Choose an appropriate value for the center a.

Practice Exercises

[T] **9–16. Linear and quadratic approximation**

a. Find the linear approximating polynomial for the following functions centered at the given point a.
b. Find the quadratic approximating polynomial for the following functions centered at a.
c. Use the polynomials obtained in parts (a) and (b) to approximate the given quantity.

9. $f(x) = 8x^{3/2}$, $a = 1$; approximate $8 \cdot 1.1^{3/2}$.

10. $f(x) = \dfrac{1}{x}$, $a = 1$; approximate $\dfrac{1}{1.05}$.

11. $f(x) = e^{-2x}$, $a = 0$; approximate $e^{-0.2}$.

12. $f(x) = \sqrt{x}$, $a = 4$; approximate $\sqrt{3.9}$.

13. $f(x) = (1 + x)^{-1}$, $a = 0$; approximate $\dfrac{1}{1.05}$.

14. $f(x) = \cos x$, $a = \dfrac{\pi}{4}$; approximate $\cos(0.24\pi)$.

15. $f(x) = x^{1/3}$, $a = 8$; approximate $7.5^{1/3}$.

16. $f(x) = \tan^{-1} x$, $a = 0$; approximate $\tan^{-1} 0.1$.

17. Find the Taylor polynomials $p_1, \ldots, p_4$ centered at $a = 0$ for $f(x) = \cos 6x$.

18. Find the Taylor polynomials $p_1, \ldots, p_5$ centered at $a = 0$ for $f(x) = e^{-x}$.

19. Find the Taylor polynomials p_3 and p_4 centered at $a = 0$ for $f(x) = (1 + x)^{-3}$.

20. Find the Taylor polynomials p_4 and p_5 centered at $a = \pi/6$ for $f(x) = \cos x$.

21. Find the Taylor polynomials p_1, p_2, and p_3 centered at $a = 1$ for $f(x) = x^3$.

22. Find the Taylor polynomials p_3 and p_4 centered at $a = 1$ for $f(x) = 8\sqrt{x}$.

23. Find the Taylor polynomial p_3 centered at $a = e$ for $f(x) = \ln x$.

24. Find the Taylor polynomial p_2 centered at $a = 8$ for $f(x) = \sqrt[3]{x}$.

[T] **25–28. Graphing Taylor polynomials**

a. Find the nth-order Taylor polynomials for the following functions centered at the given point a, for $n = 1$ and $n = 2$.
b. Graph the Taylor polynomials and the function.

25. $f(x) = \ln(1 - x)$, $a = 0$ **26.** $f(x) = (1 + x)^{-1/2}$, $a = 0$

27. $f(x) = \sin x$, $a = \dfrac{\pi}{4}$ **28.** $f(x) = \sqrt{x}$, $a = 9$

[T] **29–32. Approximations with Taylor polynomials**

a. Use the given Taylor polynomial p_2 to approximate the given quantity.
b. Compute the absolute error in the approximation, assuming the exact value is given by a calculator.

29. Approximate $\sqrt{1.05}$ using $f(x) = \sqrt{1 + x}$ and $p_2(x) = 1 + x/2 - x^2/8$.

30. Approximate $1/\sqrt{1.08}$ using $f(x) = 1/\sqrt{1 + x}$ and $p_2(x) = 1 - x/2 + 3x^2/8$.

31. Approximate $e^{-0.15}$ using $f(x) = e^{-x}$ and $p_2(x) = 1 - x + x^2/2$.

32. Approximate $\ln 1.06$ using $f(x) = \ln(1 + x)$ and $p_2(x) = x - x^2/2$.

[T] **33–40. Approximations with Taylor polynomials**

a. Approximate the given quantities using Taylor polynomials with $n = 3$.
b. Compute the absolute error in the approximation, assuming the exact value is given by a calculator.

33. $e^{0.12}$ **34.** $\cos(-0.2)$ **35.** $\tan(-0.1)$

36. $\ln 1.05$ **37.** $\sqrt{1.06}$ **38.** $\sqrt[4]{79}$

39. $\sinh 0.5$ **40.** $\tanh 0.5$

41–46. Remainders Find the remainder R_n for the nth-order Taylor polynomial centered at a for the given functions. Express the result for a general value of n.

41. $f(x) = \sin x$, $a = 0$ **42.** $f(x) = \cos 2x$, $a = 0$

43. $f(x) = e^{-x}$, $a = 0$ **44.** $f(x) = \cos x$, $a = \dfrac{\pi}{2}$

45. $f(x) = \sin x$, $a = \dfrac{\pi}{2}$ **46.** $f(x) = \dfrac{1}{1 - x}$, $a = 0$

[T] **47–52. Estimating errors** Use the remainder to find a bound on the error in approximating the following quantities with the nth-order Taylor polynomial centered at 0. Estimates are not unique.

47. $\sin 0.3$, $n = 4$ **48.** $\cos 0.45$, $n = 3$

49. $e^{0.25}$, $n = 4$ **50.** $\tan 0.3$, $n = 2$

51. $e^{-0.5}$, $n = 4$ **52.** $\ln 1.04$, $n = 3$

[T] **53–58. Maximum error** Use the remainder term to find a bound on the error in the following approximations on the given interval. Error bounds are not unique.

53. $\sin x \approx x - \dfrac{x^3}{6}$ on $\left[-\dfrac{\pi}{4}, \dfrac{\pi}{4}\right]$

54. $\cos x \approx 1 - \dfrac{x^2}{2}$ on $\left[-\dfrac{\pi}{4}, \dfrac{\pi}{4}\right]$

55. $e^x \approx 1 + x + \dfrac{x^2}{2}$ on $\left[-\dfrac{1}{2}, \dfrac{1}{2}\right]$

56. $\tan x \approx x$ on $\left[-\dfrac{\pi}{6}, \dfrac{\pi}{6}\right]$

57. $\ln(1 + x) \approx x - \dfrac{x^2}{2}$ on $[-0.2, 0.2]$

58. $\sqrt{1 + x} \approx 1 + \dfrac{x}{2}$ on $[-0.1, 0.1]$

59–64. Number of terms What is the minimum order of the Taylor polynomial required to approximate the following quantities with an absolute error no greater than 10^{-3}? (The answer depends on your choice of a center.)

59. $e^{-0.5}$ **60.** $\sin 0.2$ **61.** $\cos(-0.25)$

62. $\ln 0.85$ **63.** $\sqrt{1.06}$ **64.** $\dfrac{1}{\sqrt{0.85}}$

65. Explain why or why not Determine whether the following statements are true and give an explanation or counterexample.

a. Only even powers of x appear in the Taylor polynomials for $f(x) = e^{-2x}$ centered at 0.

b. Let $f(x) = x^5 - 1$. The Taylor polynomial for f of order 10 centered at 0 is f itself.

c. Only even powers of x appear in the nth-order Taylor polynomial for $f(x) = \sqrt{1 + x^2}$ centered at 0.

d. Suppose f'' is continuous on an interval that contains a, where f has an inflection point at a. Then the second-order Taylor polynomial for f at a is linear.

66. Taylor coefficients for $x = a$ Follow the procedure in the text to show that the nth-order Taylor polynomial that matches f and its derivatives up to order n at a has coefficients

$$c_k = \frac{f^{(k)}(a)}{k!}, \quad \text{for } k = 0, 1, 2, \ldots, n.$$

67. Matching functions with polynomials Match functions a–f with Taylor polynomials A–F (all centered at 0). Give reasons for your choices.

a. $\sqrt{1 + 2x}$

b. $\dfrac{1}{\sqrt{1 + 2x}}$

c. e^{2x}

d. $\dfrac{1}{1 + 2x}$

e. $\dfrac{1}{(1 + 2x)^3}$

f. e^{-2x}

A. $p_2(x) = 1 + 2x + 2x^2$

B. $p_2(x) = 1 - 6x + 24x^2$

C. $p_2(x) = 1 + x - \dfrac{x^2}{2}$

D. $p_2(x) = 1 - 2x + 4x^2$

E. $p_2(x) = 1 - x + \dfrac{3}{2}x^2$

F. $p_2(x) = 1 - 2x + 2x^2$

T 68. Dependence of errors on x Consider $f(x) = \ln(1 - x)$ and its Taylor polynomials given in Example 8.

a. Graph $y = |f(x) - p_2(x)|$ and $y = |f(x) - p_3(x)|$ on the interval $[-1/2, 1/2]$ (two curves).

b. At what points of $[-1/2, 1/2]$ is the error largest? Smallest?

c. Are these results consistent with the theoretical error bounds obtained in Example 8?

T 69–76. Small argument approximations *Consider the following common approximations when x is near zero.*

a. *Estimate $f(0.1)$ and give a bound on the error in the approximation.*

b. *Estimate $f(0.2)$ and give a bound on the error in the approximation.*

69. $f(x) = \sin x \approx x$

70. $f(x) = \tan x \approx x$

71. $f(x) = \cos x \approx 1 - \dfrac{x^2}{2}$

72. $f(x) = \tan^{-1}x \approx x$

73. $f(x) = \sqrt{1 + x} \approx 1 + \dfrac{x}{2}$

74. $f(x) = \ln(1 + x) \approx x - \dfrac{x^2}{2}$

75. $f(x) = e^x \approx 1 + x$

76. $f(x) = \sin^{-1}x \approx x$

Explorations and Challenges

T 77. Errors in approximations Suppose you approximate $f(x) = \sec x$ at the points $x = -0.2, -0.1, 0.0, 0.1, 0.2$ using the Taylor polynomials $p_2(x) = 1 + x^2/2$ and $p_4(x) = 1 + x^2/2 + 5x^4/24$. Assume the exact value of $\sec x$ is given by a calculator.

a. Complete the table showing the absolute errors in the approximations at each point. Show three significant digits.

b. In each error column, how do the errors vary with x? For what values of x are the errors largest and smallest in magnitude?

x	$|\sec x - p_2(x)|$	$|\sec x - p_4(x)|$
−0.2		
−0.1		
0.0		
0.1		
0.2		

T 78–79. Errors in approximations *Carry out the procedure described in Exercise 77 with the following functions and Taylor polynomials.*

78. $f(x) = \cos x$, $p_2(x) = 1 - \dfrac{x^2}{2}$, $p_4(x) = 1 - \dfrac{x^2}{2} + \dfrac{x^4}{24}$

79. $f(x) = e^{-x}$, $p_1(x) = 1 - x$, $p_2(x) = 1 - x + \dfrac{x^2}{2}$

T 80. Best center point Suppose you wish to approximate $\cos(\pi/12)$ using Taylor polynomials. Is the approximation more accurate if you use Taylor polynomials centered at 0 or at $\pi/6$? Use a calculator for numerical experiments and check for consistency with Theorem 11.2. Does the answer depend on the order of the polynomial?

T 81. Best center point Suppose you wish to approximate $e^{0.35}$ using Taylor polynomials. Is the approximation more accurate if you use Taylor polynomials centered at 0 or at $\ln 2$? Use a calculator for numerical experiments and check for consistency with Theorem 11.2. Does the answer depend on the order of the polynomial?

82. Proof of Taylor's Theorem There are several proofs of Taylor's Theorem, which lead to various forms of the remainder. The following proof is instructive because it leads to two different forms of the remainder and it relies on the Fundamental Theorem of Calculus, integration by parts, and the Mean Value Theorem for Integrals. Assume f has at least $n + 1$ continuous derivatives on an interval containing a.

a. Show that the Fundamental Theorem of Calculus can be written in the form

$$f(x) = f(a) + \int_a^x f'(t)\,dt.$$

b. Use integration by parts ($u = f'(t), dv = dt$) to show that

$$f(x) = f(a) + (x - a)f'(a) + \int_a^x (x - t)f''(t)\,dt.$$

c. Show that n integrations by parts give

$$f(x) = f(a) + \frac{f'(a)}{1!}(x - a) + \frac{f''(a)}{2!}(x - a)^2 + \cdots$$
$$+ \frac{f^{(n)}(a)}{n!}(x - a)^n + \underbrace{\int_a^x \frac{f^{(n+1)}(t)}{n!}(x - t)^n\,dt}_{R_n(x)}.$$

d. *Challenge:* The result in part (c) has the form $f(x) = p_n(x) + R_n(x)$, where p_n is the nth-order Taylor polynomial and R_n is a new form of the remainder, known as the integral form of the remainder. Use the Mean Value

Theorem for Integrals (Section 5.4) to show that R_n can be expressed in the form

$$R_n(x) = \frac{f^{(n+1)}(c)}{(n+1)!}(x-a)^{n+1},$$

where c is between a and x.

83. Tangent line is p_1 Let f be differentiable at $x = a$.

  a. Find the equation of the line tangent to the curve $y = f(x)$ at $(a, f(a))$.

  b. Verify that the Taylor polynomial p_1 centered at a describes the tangent line found in part (a).

84. Local extreme points and inflection points Suppose f has continuous first and second derivatives at a.

  a. Show that if f has a local maximum at a, then the Taylor polynomial p_2 centered at a also has a local maximum at a.

  b. Show that if f has a local minimum at a, then the Taylor polynomial p_2 centered at a also has a local minimum at a.

  c. Is it true that if f has an inflection point at a, then the Taylor polynomial p_2 centered at a also has an inflection point at a?

  d. Are the converses in parts (a) and (b) true? If p_2 has a local extreme point at a, does f have the same type of point at a?

▣ 85. Approximating $\sin x$ Let $f(x) = \sin x$, and let p_n and q_n be nth-order Taylor polynomials for f centered at 0 and π, respectively.

  a. Find p_5 and q_5.

  b. Graph f, p_5, and q_5 on the interval $[-\pi, 2\pi]$. On what interval is p_5 a better approximation to f than q_5? On what interval is q_5 a better approximation to f than p_5?

  c. Complete the following table showing the errors in the approximations given by p_5 and q_5 at selected points.

x	$|\sin x - p_5(x)|$	$|\sin x - q_5(x)|$
$\pi/4$		
$\pi/2$		
$3\pi/4$		
$5\pi/4$		
$7\pi/4$		

  d. At which points in the table is p_5 a better approximation to f than q_5? At which points do p_5 and q_5 give equal approximations to f? Explain your observations.

▣ 86. Approximating $\ln x$ Let $f(x) = \ln x$, and let p_n and q_n be nth-order Taylor polynomials for f centered at 1 and e, respectively.

  a. Find p_3 and q_3.

  b. Graph f, p_3, and q_3 on the interval $(0, 4]$.

  c. Complete the following table showing the errors in the approximations given by p_3 and q_3 at selected points.

x	$|\ln x - p_3(x)|$	$|\ln x - q_3(x)|$
0.5		
1.0		
1.5		
2		
2.5		
3		
3.5		

  d. At which points in the table is p_3 a better approximation to f than q_3? Explain your observations.

▣ 87. Approximating square roots Let p_1 and q_1 be the first-order Taylor polynomials for $f(x) = \sqrt{x}$, centered at 36 and 49, respectively.

  a. Find p_1 and q_1.

  b. Complete the following table showing the errors when using p_1 and q_1 to approximate $f(x)$ at $x = 37, 39, 41, 43, 45$, and 47. Use a calculator to obtain an exact value of $f(x)$.

x	$|\sqrt{x} - p_1(x)|$	$|\sqrt{x} - q_1(x)|$
37		
39		
41		
43		
45		
47		

  c. At which points in the table is p_1 a better approximation to f than q_1? Explain this result.

▣ 88. A different kind of approximation When approximating a function f using a Taylor polynomial, we use information about f and its derivative at one point. An alternative approach (called *interpolation*) uses information about f at several different points. Suppose we wish to approximate $f(x) = \sin x$ on the interval $[0, \pi]$.

  a. Write the (quadratic) Taylor polynomial p_2 for f centered at $\pi/2$.

  b. Now consider a quadratic interpolating polynomial $q(x) = ax^2 + bx + c$. The coefficients a, b, and c are chosen such that the following conditions are satisfied:

$$q(0) = f(0),\ q\left(\frac{\pi}{2}\right) = f\left(\frac{\pi}{2}\right),\ \text{and } q(\pi) = f(\pi).$$

    Show that $q(x) = -\dfrac{4}{\pi^2}x^2 + \dfrac{4}{\pi}x.$

  c. Graph f, p_2, and q on $[0, \pi]$.

  d. Find the error in approximating $f(x) = \sin x$ at the points $\dfrac{\pi}{4}, \dfrac{\pi}{2}, \dfrac{3\pi}{4}$, and π using p_2 and q.

  e. Which function, p_2 or q, is a better approximation to f on $[0, \pi]$? Explain.

QUICK CHECK ANSWERS

3. $f(x) = \sin x$ is an odd function, and its even-ordered derivatives are zero at 0, so its Taylor polynomials are also odd functions. **4.** $p_4(x) = p_3(x) + \dfrac{x^4}{4!}$; $p_5(x) = p_4(x) + \dfrac{x^5}{5!}$

5. $x = 49$ and $x = 16$ are good choices. **6.** Because $e^{0.45} < 2$, $|R_7(0.45)| < 2\dfrac{0.45^8}{8!} \approx 8.3 \times 10^{-8}$. ◄

11.2 Properties of Power Series

The preceding section demonstrated that Taylor polynomials provide accurate approximations to many functions and that, in general, the approximations improve as the degree of the polynomials increases. In this section, we take the next step and let the degree of the Taylor polynomials increase without bound to produce a *power series*.

Geometric Series as Power Series

A good way to become familiar with power series is to return to *geometric series*, first encountered in Section 10.3. Recall that for a fixed number r,

$$\sum_{k=0}^{\infty} r^k = 1 + r + r^2 + \cdots = \frac{1}{1 - r}, \quad \text{provided } |r| < 1.$$

It's a small change to replace the real number r with the variable x. In doing so, the geometric series becomes a new representation of a familiar function:

$$\sum_{k=0}^{\infty} x^k = 1 + x + x^2 + \cdots = \frac{1}{1 - x}, \quad \text{provided } |x| < 1.$$

This infinite series is a power series and it is a representation of the function $1/(1 - x)$ that is valid on the interval $|x| < 1$.

In general, power series are used to represent familiar functions such as trigonometric, exponential, and logarithmic functions. They are also used to define new functions. For example, consider the function defined by

$$g(x) = \sum_{k=1}^{\infty} \frac{(-1)^k \, k}{4^k} x^{2k}.$$

The term *function* is used advisedly because it's not yet clear whether g really is a function. If so, is it continuous? Does it have a derivative? Judging by its graph (**Figure 11.13**), g appears to be an ordinary continuous and differentiable function on $(-2, 2)$ (which is identified at the end of the chapter). In fact, power series satisfy the defining property of all functions: For each admissible value of x, a power series has at most one value. For this reason, we refer to a power series as a function, although the domain, properties, and identity of the function may need to be discovered.

Convergence of Power Series

First let's establish some terminology associated with power series.

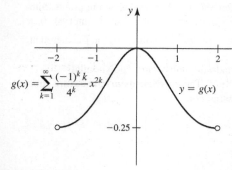

$$g(x) = \sum_{k=1}^{\infty} \frac{(-1)^k \, k}{4^k} x^{2k}$$

Figure 11.13

➤ Figure 11.13 shows an approximation to the graph of g made by summing the first 500 terms of the power series at selected values of x on the interval $(-2, 2)$.

QUICK CHECK 1 By substituting $x = 0$ in the power series for g, evaluate $g(0)$ for the function in Figure 11.13. ◄

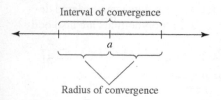

Figure 11.14

DEFINITION Power Series

A **power series** has the general form

$$\sum_{k=0}^{\infty} c_k (x - a)^k,$$

where a and c_k are real numbers, and x is a variable. The c_k's are the **coefficients** of the power series, and a is the **center** of the power series. The set of values of x for which the series converges is its **interval of convergence**. The **radius of convergence** of the power series, denoted R, is the distance from the center of the series to the boundary of the interval of convergence (**Figure 11.14**).

How do we determine the interval of convergence for a given power series? The presence of the term x^k or $(x - a)^k$ in a power series suggests using the Ratio Test or the Root Test to determine the values of x for which the series converges absolutely. By Theorem 10.19, if we determine the interval on which the series converges absolutely, we have a set of values for which the series converges.

Theorem 11.3 spells out the ways in which a power series can converge; its proof is given in Appendix A.

> Theorem 11.3 implies that the interval of convergence is symmetric about the center of the series; the radius of convergence R is determined by analyzing r from the Ratio Test (or ρ from the Root Test). The theorem says nothing about convergence at the endpoints. For example, the intervals of convergence $(2, 6)$, $(2, 6]$, $[2, 6)$, and $[2, 6]$ all have a radius of convergence of $R = 2$.

THEOREM 11.3   Convergence of Power Series

A power series $\displaystyle\sum_{k=0}^{\infty} c_k(x - a)^k$ centered at a converges in one of three ways:

1. The series converges absolutely for all x. It follows, by Theorem 10.19, that the series converges for all x, in which case the interval of convergence is $(-\infty, \infty)$ and the radius of convergence is $R = \infty$.

2. There is a real number $R > 0$ such that the series converges absolutely (and therefore converges) for $|x - a| < R$ and diverges for $|x - a| > R$, in which case the radius of convergence is R.

3. The series converges only at a, in which case the radius of convergence is $R = 0$.

The following examples illustrate how the Ratio and Root Tests are used to determine the radius and interval of convergence, and how the cases of Theorem 11.3 arise.

EXAMPLE 1   Radius and interval of convergence Find the radius and interval of convergence for each power series.

a. $\displaystyle\sum_{k=0}^{\infty} \frac{x^k}{k!}$ **b.** $\displaystyle\sum_{k=0}^{\infty} \frac{(-1)^k(x - 2)^k}{4^k}$ **c.** $\displaystyle\sum_{k=1}^{\infty} k!\, x^k$

SOLUTION

a. The center of the power series is 0 and the terms of the series are $x^k/k!$. Due to the presence of the factor $k!$, we test the series for absolute convergence using the Ratio Test:

$$r = \lim_{k \to \infty} \frac{|x^{k+1}/(k + 1)!|}{|x^k/k!|} \qquad \text{Ratio Test}$$

$$= \lim_{k \to \infty} \frac{|x|^{k+1}}{|x|^k} \cdot \frac{k!}{(k + 1)!} \qquad \text{Invert and multiply.}$$

$$= |x| \lim_{k \to \infty} \frac{1}{k + 1} = 0. \qquad \text{Simplify and take the limit with } x \text{ fixed.}$$

Notice that in taking the limit as $k \to \infty$, x is held fixed. Because $r = 0$ for all real numbers x, the series converges absolutely for all x, which implies that the series converges for all x. Therefore, the interval of convergence is $(-\infty, \infty)$ (**Figure 11.15**) and the radius of convergence is $R = \infty$.

b. We test for absolute convergence using the Root Test:

$$\rho = \lim_{k \to \infty} \sqrt[k]{\left| \frac{(-1)^k(x - 2)^k}{4^k} \right|} = \frac{|x - 2|}{4}.$$

In this case, ρ depends on the value of x. For absolute convergence, x must satisfy

$$\rho = \frac{|x - 2|}{4} < 1,$$

which implies that $|x - 2| < 4$. Using standard techniques for solving inequalities, the solution set is $-4 < x - 2 < 4$, or $-2 < x < 6$. We conclude that the series converges absolutely (and therefore converges) on $(-2, 6)$, which means the radius of convergence is $R = 4$. When $-\infty < x < -2$ or $6 < x < \infty$, we have $\rho > 1$, so the series diverges on these intervals (the terms of the series do not approach 0 as $k \to \infty$ and the Divergence Test applies).

The Root Test does not give information about convergence at the endpoints $x = -2$ and $x = 6$, because at these points, the Root Test results in $\rho = 1$. To test for

Center of
power series

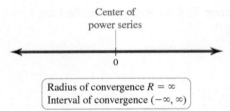

0

Radius of convergence $R = \infty$
Interval of convergence $(-\infty, \infty)$

Figure 11.15

> Either the Ratio Test or the Root Test works for the power series in Example 1b.

convergence at the endpoints, we substitute each endpoint into the series and carry out separate tests. At $x = -2$, the power series becomes

$$\sum_{k=0}^{\infty} \frac{(-1)^k (x-2)^k}{4^k} = \sum_{k=0}^{\infty} \frac{4^k}{4^k} \quad \text{Substitute } x = -2 \text{ and simplify.}$$

$$= \sum_{k=0}^{\infty} 1. \quad \text{Diverges by Divergence Test}$$

> The Ratio and Root Tests determine the radius of convergence conclusively. However, the interval of convergence is not determined until the endpoints are tested.

The series clearly diverges at the left endpoint. At $x = 6$, the power series is

$$\sum_{k=0}^{\infty} \frac{(-1)^k (x-2)^k}{4^k} = \sum_{k=0}^{\infty} (-1)^k \frac{4^k}{4^k} \quad \text{Substitute } x = 6 \text{ and simplify.}$$

$$= \sum_{k=0}^{\infty} (-1)^k. \quad \text{Diverges by Divergence Test}$$

This series also diverges at the right endpoint. Therefore, the interval of convergence is $(-2, 6)$, excluding the endpoints (**Figure 11.16**).

Radius of convergence $R = 4$
Interval of convergence $(-2, 6)$

Figure 11.16

c. In this case, the Ratio Test is preferable:

$$r = \lim_{k \to \infty} \frac{|(k+1)! \, x^{k+1}|}{|k! \, x^k|} \quad \text{Ratio Test}$$

$$= |x| \lim_{k \to \infty} \frac{(k+1)!}{k!} \quad \text{Simplify.}$$

$$= |x| \lim_{k \to \infty} (k+1) \quad \text{Simplify.}$$

$$= \infty. \quad \text{If } x \neq 0$$

Radius of convergence $R = 0$
Interval of convergence $\{0\}$

Figure 11.17

We see that $r > 1$ for all $x \neq 0$, so the series diverges on $(-\infty, 0)$ and $(0, \infty)$. The only way to satisfy $r < 1$ is to take $x = 0$, in which case the power series has a value of 0. The interval of convergence of the power series consists of the single point $x = 0$ (**Figure 11.17**), and the radius of convergence is $R = 0$.

Related Exercises 10–12 ◄

QUICK CHECK 2 What are the radius and interval of convergence of the geometric series $\sum x^k$? ◄

EXAMPLE 2   Radius and interval of convergence Use the Ratio Test to find the radius and interval of convergence of $\displaystyle\sum_{k=1}^{\infty} \frac{(x-2)^k}{\sqrt{k}}$.

SOLUTION

> The power series in Example 2 could also be analyzed using the Root Test.

$$r = \lim_{k \to \infty} \frac{|(x-2)^{k+1}/\sqrt{k+1}|}{|(x-2)^k/\sqrt{k}|} \quad \text{Ratio Test}$$

$$= |x-2| \lim_{k \to \infty} \sqrt{\frac{k}{k+1}} \quad \text{Simplify.}$$

$$= |x-2| \sqrt{\underbrace{\lim_{k \to \infty} \frac{k}{k+1}}_{1}} \quad \text{Limit Law}$$

$$= |x-2| \quad \text{Limit equals 1.}$$

The series converges absolutely (and therefore converges) for all x such that $r < 1$, which implies $|x-2| < 1$, or $1 < x < 3$; we see that the radius of convergence is $R = 1$. On the intervals $-\infty < x < 1$ and $3 < x < \infty$, we have $r > 1$, so the series diverges.

We now test the endpoints. Substituting $x = 1$ gives the series

$$\sum_{k=1}^{\infty} \frac{(x-2)^k}{\sqrt{k}} = \sum_{k=1}^{\infty} \frac{(-1)^k}{\sqrt{k}}.$$

Center of power series

Radius of convergence $R = 1$
Interval of convergence $[1, 3)$

Figure 11.18

This series converges by the Alternating Series Test (the terms of the series decrease in magnitude and approach 0 as $k \to \infty$). Substituting $x = 3$ gives the series

$$\sum_{k=1}^{\infty} \frac{(x-2)^k}{\sqrt{k}} = \sum_{k=1}^{\infty} \frac{1}{\sqrt{k}},$$

which is a divergent p-series. We conclude that the interval of convergence is $1 \le x < 3$ (**Figure 11.18**).

Related Exercise 18 ◄

The technique used in Examples 1 and 2 to determine the radius and interval of convergence for a given power series is summarized below.

SUMMARY Determining the Radius and Interval of Convergence of $\sum c_k(x - a)^k$

1. Use the Ratio Test or the Root Test to find the interval $(a - R, a + R)$ on which the series converges absolutely; the radius of convergence for the series is R.

2. Use the *radius* of convergence to find the *interval* of convergence:
    • If $R = \infty$, the interval of convergence is $(-\infty, \infty)$.
    • If $R = 0$, the interval of convergence is the single point $x = a$.
    • If $0 < R < \infty$, the interval of convergence consists of the interval $(a - R, a + R)$ and possibly one or both of its endpoints. Determining whether the series $\sum c_k(x - a)^k$ converges at the endpoints $x = a - R$ and $x = a + R$ amounts to analyzing the series $\sum c_k(-R)^k$ and $\sum c_k R^k$.

Although the procedure just outlined can be applied to any power series, it is generally used when nothing is known about the series in question. As we learn shortly in Theorems 11.4 and 11.5, new power series can be built from known power series, and in these cases, other methods may be used to find the interval of convergence.

Combining Power Series

A power series defines a function on its interval of convergence. When power series are combined algebraically, new functions are defined. The following theorem, stated without proof, gives three common ways to combine power series.

➤ New power series can also be defined as the product and quotient of power series. The calculation of the coefficients of such series is more challenging (Exercise 77).

➤ Theorem 11.4 also applies to power series centered at points other than $x = 0$. Property 1 applies directly; Properties 2 and 3 apply with slight modifications.

THEOREM 11.4 Combining Power Series
Suppose the power series $\sum c_k x^k$ and $\sum d_k x^k$ converge to $f(x)$ and $g(x)$, respectively, on an interval I.

1. **Sum and difference:** The power series $\sum (c_k \pm d_k)x^k$ converges to $f(x) \pm g(x)$ on I.

2. **Multiplication by a power:** Suppose m is an integer such that $k + m \ge 0$, for all terms of the power series $x^m \sum c_k x^k = \sum c_k x^{k+m}$. This series converges to $x^m f(x)$, for all $x \ne 0$ in I. When $x = 0$, the series converges to $\lim_{x \to 0} x^m f(x)$.

3. **Composition:** If $h(x) = bx^m$, where m is a positive integer and b is a nonzero real number, the power series $\sum c_k (h(x))^k$ converges to the composite function $f(h(x))$, for all x such that $h(x)$ is in I.

EXAMPLE 3 Combining power series Given the geometric series

$$\frac{1}{1 - x} = \sum_{k=0}^{\infty} x^k = 1 + x + x^2 + x^3 + \cdots, \qquad \text{for } |x| < 1,$$

find the power series and interval of convergence for the following functions.

a. $\dfrac{x^5}{1 - x}$ 　　　 b. $\dfrac{1}{1 - 2x}$ 　　　 c. $\dfrac{1}{1 + x^2}$

SOLUTION

a.
$$\frac{x^5}{1-x} = x^5\left(1 + x + x^2 + \cdots\right) \quad \text{Theorem 11.4, Property 2}$$

$$= x^5 + x^6 + x^7 + \cdots$$

$$= \sum_{k=0}^{\infty} x^{k+5}$$

This geometric series has a ratio $r = x$ and converges when $|r| = |x| < 1$. The interval of convergence is $|x| < 1$.

b. We substitute $2x$ for x in the power series for $\dfrac{1}{1-x}$:

$$\frac{1}{1-2x} = 1 + (2x) + (2x)^2 + \cdots \quad \text{Theorem 11.4, Property 3}$$

$$= \sum_{k=0}^{\infty} (2x)^k.$$

This geometric series has a ratio $r = 2x$ and converges provided $|r| = |2x| < 1$ or $|x| < \frac{1}{2}$. The interval of convergence is $|x| < \frac{1}{2}$.

c. We substitute $-x^2$ for x in the power series for $\dfrac{1}{1-x}$:

$$\frac{1}{1+x^2} = 1 + (-x^2) + (-x^2)^2 + \cdots \quad \text{Theorem 11.4, Property 3}$$

$$= 1 - x^2 + x^4 - \cdots$$

$$= \sum_{k=0}^{\infty} (-1)^k x^{2k}.$$

This geometric series has a ratio of $r = -x^2$ and converges provided $|r| = |-x^2| = |x^2| < 1$ or $|x| < 1$.

Related Exercises 41–42 ◄

Differentiating and Integrating Power Series

Some properties of polynomials carry over to power series, but others do not. For example, a polynomial is defined for all values of x, whereas a power series is defined only on its interval of convergence. In general, the properties of polynomials carry over to power series when the power series is restricted to its interval of convergence. The following result illustrates this principle.

> ▶ Theorem 11.5 makes no claim about the convergence of the differentiated or integrated series at the endpoints of the interval of convergence.

> **THEOREM 11.5   Differentiating and Integrating Power Series**
> Suppose the power series $\sum c_k(x-a)^k$ converges for $|x-a| < R$ and defines a function f on that interval.
>
> **1.** Then f is differentiable (which implies continuous) for $|x-a| < R$, and f' is found by differentiating the power series for f term by term; that is,
> $$f'(x) = \sum kc_k(x-a)^{k-1},$$
> for $|x-a| < R$.
>
> **2.** The indefinite integral of f is found by integrating the power series for f term by term; that is,
> $$\int f(x)\,dx = \sum c_k \frac{(x-a)^{k+1}}{k+1} + C,$$
> for $|x-a| < R$, where C is an arbitrary constant.

The proof of this theorem requires advanced ideas and is omitted. However, some discussion is in order before turning to examples. The statements in Theorem 11.5 about term-by-term differentiation and integration say two things. First, the differentiated and integrated power series converge, provided x belongs to the interior of the interval of convergence. But the theorem claims more than convergence. According to the theorem, the differentiated and integrated power series converge to the derivative and indefinite integral of f, respectively, on the interior of the interval of convergence. Let's use this theorem to develop new power series.

EXAMPLE 4 Differentiating and integrating power series Consider the geometric series

$$f(x) = \frac{1}{1-x} = \sum_{k=0}^{\infty} x^k = 1 + x + x^2 + x^3 + \cdots, \quad \text{for } |x| < 1.$$

a. Differentiate this series term by term to find the power series for f' and identify the function it represents.

b. Integrate this series term by term and identify the function it represents.

SOLUTION

a. We know that $f'(x) = (1-x)^{-2}$. Differentiating the series, we find that

$$f'(x) = \frac{d}{dx}(1 + x + x^2 + x^3 + \cdots) \qquad \text{Differentiate the power series for } f.$$

$$= 1 + 2x + 3x^2 + \cdots \qquad \text{Differentiate term by term.}$$

$$= \sum_{k=0}^{\infty} (k+1)\, x^k. \qquad \text{Summation notation}$$

Therefore, on the interval $|x| < 1$,

$$f'(x) = (1-x)^{-2} = \sum_{k=0}^{\infty} (k+1)\, x^k.$$

Theorem 11.5 makes no claim about convergence of the differentiated series to f' at the endpoints. In this case, substituting $x = \pm 1$ into the power series for f' reveals that the series diverges at both endpoints.

b. Integrating f and integrating the power series term by term, we have

$$\int \frac{dx}{1-x} = \int (1 + x + x^2 + x^3 + \cdots)\, dx,$$

which implies that

$$-\ln|1-x| = x + \frac{x^2}{2} + \frac{x^3}{3} + \frac{x^4}{4} + \cdots + C,$$

where C is an arbitrary constant. Notice that the left side is 0 when $x = 0$. The right side is 0 when $x = 0$ provided we choose $C = 0$. Because $|x| < 1$, the absolute value sign on the left side may be removed. Multiplying both sides by -1, we have a series representation for $\ln(1-x)$:

$$\ln(1-x) = -x - \frac{x^2}{2} - \frac{x^3}{3} - \frac{x^4}{4} - \cdots = -\sum_{k=1}^{\infty} \frac{x^k}{k}.$$

It is interesting to test the endpoints of the interval $|x| < 1$. When $x = 1$, the series is (a multiple of) the divergent harmonic series, and when $x = -1$, the series is the convergent alternating harmonic series (Section 10.6). So the interval of convergence is $-1 \le x < 1$. Although we know the series converges at $x = -1$, Theorem 11.5 guarantees convergence to $\ln(1-x)$ only at the interior points. We cannot use Theorem 11.5 to claim that the series converges to $\ln 2$ at $x = -1$. In fact, it does, as shown in Section 11.3.

QUICK CHECK 3 Use the result of Example 4 to write a series representation for $\ln \frac{1}{2} = -\ln 2$. ◀

Related Exercises 51, 55 ◀

EXAMPLE 5 Functions to power series Find power series representations centered at 0 for the following functions and give their intervals of convergence.

a. $\tan^{-1} x$ **b.** $\ln\left(\dfrac{1+x}{1-x}\right)$

SOLUTION In both cases, we work with known power series and use differentiation, integration, and other combinations.

a. The key is to recall that

$$\int \frac{dx}{1+x^2} = \tan^{-1} x + C$$

and that, by Example 3c,

$$\frac{1}{1+x^2} = 1 - x^2 + x^4 - \cdots, \qquad \text{provided } |x| < 1.$$

We now integrate both sides of this last expression:

$$\int \frac{dx}{1+x^2} = \int \left(1 - x^2 + x^4 - \cdots\right) dx,$$

which implies that

$$\tan^{-1} x = x - \frac{x^3}{3} + \frac{x^5}{5} - \cdots + C.$$

Substituting $x = 0$ and noting that $\tan^{-1} 0 = 0$, the two sides of this equation agree provided we choose $C = 0$. Therefore,

$$\tan^{-1} x = x - \frac{x^3}{3} + \frac{x^5}{5} - \cdots = \sum_{k=0}^{\infty} \frac{(-1)^k x^{2k+1}}{2k+1}.$$

▶ Again, Theorem 11.5 does not guarantee that the power series in Example 5a converges to $\tan^{-1} x$ at $x = \pm 1$. In fact, it does.

By Theorem 11.5, this power series converges to $\tan^{-1} x$ for $|x| < 1$. Testing the endpoints separately, we find that it also converges at $x = \pm 1$. Therefore, the interval of convergence is $[-1, 1]$.

b. We have already seen (Example 4) that

$$\ln(1 - x) = -x - \frac{x^2}{2} - \frac{x^3}{3} - \cdots.$$

▶ Nicolaus Mercator (1620–1687) and Sir Isaac Newton (1642–1727) independently derived the power series for $\ln(1 + x)$, which is called the *Mercator series*.

Replacing x with $-x$ (Property 3 of Theorem 11.4), we have

$$\ln(1 - (-x)) = \ln(1 + x) = x - \frac{x^2}{2} + \frac{x^3}{3} - \cdots.$$

Subtracting these two power series gives

$$\ln\left(\frac{1+x}{1-x}\right) = \ln(1+x) - \ln(1-x) \quad \text{Properties of logarithms}$$

$$= \underbrace{\left(x - \frac{x^2}{2} + \frac{x^3}{3} - \cdots\right)}_{\ln(1+x)} - \underbrace{\left(-x - \frac{x^2}{2} - \frac{x^3}{3} - \cdots\right)}_{\ln(1-x)}, \quad \text{for } |x| < 1$$

$$= 2\left(x + \frac{x^3}{3} + \frac{x^5}{5} + \cdots\right) \quad \text{Combine; use Property 1 of Theorem 11.4.}$$

$$= 2\sum_{k=0}^{\infty} \frac{x^{2k+1}}{2k+1}. \quad \text{Summation notation}$$

QUICK CHECK 4 Verify that the power series in Example 5b does not converge at the endpoints $x = \pm 1$. ◀

This power series is the difference of two power series, both of which converge on the interval $|x| < 1$. Therefore, by Theorem 11.4, the new series also converges on $|x| < 1$.

Related Exercises 57, 62 ◀

If you look carefully, every example in this section is ultimately based on the geometric series. Using this single series, we were able to develop power series for many other functions. Imagine what we could do with a few more basic power series. The following section accomplishes precisely that end. There, we discover power series for many of the standard functions of calculus.

SECTION 11.2 EXERCISES

Getting Started

1. Write the first four terms of a power series with coefficients $c_0, c_1, c_2,$ and c_3 centered at 0.

2. Is $\displaystyle\sum_{k=0}^{\infty}(5x - 20)^k$ a power series? If so, find the center a of the power series and state a formula for the coefficients c_k of the power series.

3. What tests are used to determine the radius of convergence of a power series?

4. Is $\displaystyle\sum_{k=0}^{\infty}x^{2k}$ a power series? If so, find the center a of the power series and state a formula for the coefficients c_k of the power series.

5. Do the radius and interval of convergence of a power series change when the series is differentiated or integrated? Explain.

6. Suppose a power series converges if $|x - 3| < 4$ and diverges if $|x - 3| \geq 4$. Determine the radius and interval of convergence.

7. Suppose a power series converges if $|4x - 8| \leq 40$ and diverges if $|4x - 8| > 40$. Determine the radius and interval of convergence.

8. Suppose the power series $\displaystyle\sum_{k=0}^{\infty}c_k(x - a)^k$ has an interval of convergence of $(-3, 7]$. Find the center a and the radius of convergence R.

Practice Exercises

9–36. Radius and interval of convergence *Determine the radius and interval of convergence of the following power series.*

9. $\displaystyle\sum_{k=0}^{\infty}(2x)^k$

10. $\displaystyle\sum_{k=0}^{\infty}\frac{(x - 1)^k}{k!}$

11. $\displaystyle\sum_{k=1}^{\infty}(kx)^k$

12. $\displaystyle\sum_{k=0}^{\infty}k!(x - 10)^k$

13. $\displaystyle\sum_{k=1}^{\infty}\sin^k\left(\frac{1}{k}\right)x^k$

14. $\displaystyle\sum_{k=2}^{\infty}\frac{2^k(x - 3)^k}{k}$

15. $\displaystyle\sum_{k=0}^{\infty}\left(\frac{x}{3}\right)^k$

16. $\displaystyle\sum_{k=0}^{\infty}(-1)^k\frac{x^k}{5^k}$

17. $\displaystyle\sum_{k=1}^{\infty}\frac{x^k}{k^k}$

18. $\displaystyle\sum_{k=1}^{\infty}\frac{(-1)^k x^k}{\sqrt{k}}$

19. $\displaystyle\sum_{k=0}^{\infty}\frac{x^k}{2^k + 1}$

20. $\displaystyle\sum_{k=0}^{\infty}\frac{(2x)^k}{k!}$

21. $-\dfrac{x^2}{1!} + \dfrac{x^4}{2!} - \dfrac{x^6}{3!} + \dfrac{x^8}{4!} - \cdots$

22. $x - \dfrac{x^3}{4} + \dfrac{x^5}{9} - \dfrac{x^7}{16} + \cdots$

23. $\displaystyle\sum_{k=1}^{\infty}\frac{(-1)^{k+1}(x - 1)^k}{k}$

24. $\displaystyle\sum_{k=0}^{\infty}(-1)^k\frac{k(x - 4)^k}{2^k}$

25. $\displaystyle\sum_{k=0}^{\infty}\frac{(4x - 1)^k}{k^2 + 4}$

26. $\displaystyle\sum_{k=1}^{\infty}\frac{(3x + 2)^k}{k}$

27. $\displaystyle\sum_{k=0}^{\infty}\frac{k^{10}(2x - 4)^k}{10^k}$

28. $\displaystyle\sum_{k=2}^{\infty}\frac{(x + 3)^k}{k\ln^2 k}$

29. $\displaystyle\sum_{k=0}^{\infty}\frac{k^2 x^{2k}}{k!}$

30. $\displaystyle\sum_{k=0}^{\infty}k(x - 1)^k$

31. $\displaystyle\sum_{k=1}^{\infty}\frac{x^{2k+1}}{3^{k-1}}$

32. $\displaystyle\sum_{k=0}^{\infty}\left(-\frac{x}{10}\right)^{2k}$

33. $\displaystyle\sum_{k=1}^{\infty}\frac{(x - 1)^k k^k}{(k + 1)^k}$

34. $\displaystyle\sum_{k=0}^{\infty}\frac{(-2)^k(x + 3)^k}{3^{k+1}}$

35. $\displaystyle\sum_{k=0}^{\infty}\frac{k^{20} x^k}{(2k + 1)!}$

36. $\displaystyle\sum_{k=0}^{\infty}(-1)^k\frac{x^{3k}}{27^k}$

37–40. Radius of convergence *Find the radius of convergence for the following power series.*

37. $\displaystyle\sum_{k=1}^{\infty}\frac{k!x^k}{k^k}$

38. $\displaystyle\sum_{k=1}^{\infty}\left(1 + \frac{1}{k}\right)^{k^2}x^k$

39. $\displaystyle\sum_{k=1}^{\infty}\left(\frac{k}{k + 4}\right)^{k^2}x^k$

40. $\displaystyle\sum_{k=1}^{\infty}\left(1 - \cos\frac{1}{2^k}\right)x^k$

41–46. Combining power series *Use the geometric series*

$$f(x) = \frac{1}{1 - x} = \sum_{k=0}^{\infty}x^k, \quad \text{for } |x| < 1,$$

to find the power series representation for the following functions (centered at 0). Give the interval of convergence of the new series.

41. $f(3x) = \dfrac{1}{1 - 3x}$

42. $g(x) = \dfrac{x^3}{1 - x}$

43. $h(x) = \dfrac{2x^3}{1 - x}$

44. $f(x^3) = \dfrac{1}{1 - x^3}$

45. $p(x) = \dfrac{4x^{12}}{1 - x}$

46. $f(-4x) = \dfrac{1}{1 + 4x}$

47–50. Combining power series *Use the power series representation*

$$f(x) = \ln(1 - x) = -\sum_{k=1}^{\infty}\frac{x^k}{k}, \quad \text{for } -1 \leq x < 1,$$

to find the power series for the following functions (centered at 0). Give the interval of convergence of the new series.

47. $f(3x) = \ln(1 - 3x)$

48. $g(x) = x^3\ln(1 - x)$

49. $p(x) = 2x^6\ln(1 - x)$

50. $f(x^3) = \ln(1 - x^3)$

51–56. Differentiating and integrating power series *Find the power series representation for g centered at 0 by differentiating or integrating the power series for f (perhaps more than once). Give the interval of convergence for the resulting series.*

51. $g(x) = \dfrac{2}{(1-2x)^2}$ using $f(x) = \dfrac{1}{1-2x}$

52. $g(x) = \dfrac{1}{(1-x)^3}$ using $f(x) = \dfrac{1}{1-x}$

53. $g(x) = -\dfrac{1}{(1+x)^2}$ using $f(x) = \dfrac{1}{1+x}$

54. $g(x) = \dfrac{x}{(1+x^2)^2}$ using $f(x) = \dfrac{1}{1+x^2}$

55. $g(x) = \ln(1-3x)$ using $f(x) = \dfrac{1}{1-3x}$

56. $g(x) = \ln(1+x^2)$ using $f(x) = \dfrac{x}{1+x^2}$

57–62. Functions to power series *Find power series representations centered at 0 for the following functions using known power series. Give the interval of convergence for the resulting series.*

57. $f(x) = \dfrac{2x}{(1+x^2)^2}$ **58.** $f(x) = \dfrac{1}{1-x^4}$

59. $f(x) = \dfrac{3}{3+x}$ **60.** $f(x) = \ln\sqrt{1-x^2}$

61. $f(x) = \ln\sqrt{4-x^2}$ **62.** $f(x) = \tan^{-1}(4x^2)$

63. Explain why or why not Determine whether the following statements are true and give an explanation or counterexample.

   a. The interval of convergence of the power series $\sum c_k(x-3)^k$ could be $(-2, 8)$.

   b. The series $\displaystyle\sum_{k=0}^{\infty}(-2x)^k$ converges on the interval $-\frac{1}{2} < x < \frac{1}{2}$.

   c. If $f(x) = \displaystyle\sum_{k=0}^{\infty} c_k x^k$ on the interval $|x| < 1$, then $f(x^2) = \sum c_k x^{2k}$ on the interval $|x| < 1$.

   d. If $f(x) = \displaystyle\sum_{k=0}^{\infty} c_k x^k = 0$, for all x on an interval $(-a, a)$, then $c_k = 0$, for all k.

64. Scaling power series If the power series $f(x) = \sum c_k x^k$ has an interval of convergence of $|x| < R$, what is the interval of convergence of the power series for $f(ax)$, where $a \neq 0$ is a real number?

65. Shifting power series If the power series $f(x) = \sum c_k x^k$ has an interval of convergence of $|x| < R$, what is the interval of convergence of the power series for $f(x-a)$, where $a \neq 0$ is a real number?

66. A useful substitution Replace x with $x - 1$ in the series
$$\ln(1+x) = \sum_{k=1}^{\infty}\frac{(-1)^{k+1}x^k}{k}$$
to obtain a power series for $\ln x$ centered at $x = 1$. What is the interval of convergence for the new power series?

67–71. Series to functions *Find the function represented by the following series, and find the interval of convergence of the series. (Not all these series are power series.)*

67. $\displaystyle\sum_{k=0}^{\infty}(\sqrt{x}-2)^k$ **68.** $\displaystyle\sum_{k=1}^{\infty}\frac{x^{2k}}{4^k}$ **69.** $\displaystyle\sum_{k=0}^{\infty}e^{-kx}$

70. $\displaystyle\sum_{k=1}^{\infty}\frac{(x-2)^k}{3^{2k}}$ **71.** $\displaystyle\sum_{k=0}^{\infty}\left(\frac{x^2-1}{3}\right)^k$

72–74. Exponential function *In Section 11.3, we show that the power series for the exponential function centered at 0 is*
$$e^x = \sum_{k=0}^{\infty}\frac{x^k}{k!}, \quad \text{for } -\infty < x < \infty.$$

Use the methods of this section to find the power series centered at 0 for the following functions. Give the interval of convergence for the resulting series.

72. $f(x) = e^{2x}$ **73.** $f(x) = e^{-3x}$ **74.** $f(x) = x^2 e^x$

Explorations and Challenges

75. Powers of x multiplied by a power series Prove that if
$$f(x) = \sum_{k=0}^{\infty}c_k x^k$$
converges with radius of convergence R, then the power series for $x^m f(x)$ also converges with radius of convergence R, for positive integers m.

T 76. Remainders Let
$$f(x) = \sum_{k=0}^{\infty}x^k = \frac{1}{1-x} \quad \text{and} \quad S_n(x) = \sum_{k=0}^{n-1}x^k.$$
The remainder in truncating the power series after n terms is $R_n = f(x) - S_n(x)$, which depends on x.

   a. Show that $R_n(x) = x^n/(1-x)$.

   b. Graph the remainder function on the interval $|x| < 1$, for $n = 1, 2,$ and 3. Discuss and interpret the graph. Where on the interval is $|R_n(x)|$ largest? Smallest?

   c. For fixed n, minimize $|R_n(x)|$ with respect to x. Does the result agree with the observations in part (b)?

   d. Let $N(x)$ be the number of terms required to reduce $|R_n(x)|$ to less than 10^{-6}. Graph the function $N(x)$ on the interval $|x| < 1$. Discuss and interpret the graph.

77. Product of power series Let
$$f(x) = \sum_{k=0}^{\infty}c_k x^k \quad \text{and} \quad g(x) = \sum_{k=0}^{\infty}d_k x^k.$$

   a. Multiply the power series together as if they were polynomials, collecting all terms that are multiples of 1, x, and x^2. Write the first three terms of the product $f(x)g(x)$.

   b. Find a general expression for the coefficient of x^n in the product series, for $n = 0, 1, 2, \ldots$.

78. Inverse sine Given the power series
$$\frac{1}{\sqrt{1-x^2}} = 1 + \frac{1}{2}x^2 + \frac{1\cdot3}{2\cdot4}x^4 + \frac{1\cdot3\cdot5}{2\cdot4\cdot6}x^6 + \cdots,$$
for $-1 < x < 1$, find the power series for $f(x) = \sin^{-1}x$ centered at 0.

QUICK CHECK ANSWERS

1. $g(0) = 0$ **2.** $|x| < 1, R = 1$ **3.** Substituting $x = 1/2, \ln(1/2) = -\ln 2 = -\displaystyle\sum_{k=1}^{\infty}\frac{1}{2^k k}$ ◄

11.3 Taylor Series

In the preceding section, we saw that a power series represents a function on its interval of convergence. This section explores the opposite question: Given a function, what is its power series representation? We have already made significant progress in answering this question because we know how Taylor polynomials are used to approximate functions. We now extend Taylor polynomials to produce power series—called *Taylor series*—that provide series representations for functions.

Taylor Series for a Function

Suppose a function f has derivatives $f^{(k)}(a)$ of *all* orders at the point a. If we write the nth-order Taylor polynomial for f centered at a and allow n to increase indefinitely, a power series is obtained:

$$\underbrace{c_0 + c_1(x - a) + c_2(x - a)^2 + \cdots + c_n(x - a)^n}_{\text{Taylor polynomial of order } n} + \underbrace{\cdots}_{n \to \infty} = \sum_{k=0}^{\infty} c_k(x - a)^k.$$

The coefficients of the Taylor polynomial are given by

$$c_k = \frac{f^{(k)}(a)}{k!}, \quad \text{for } k = 0, 1, 2, \ldots .$$

> Maclaurin series are named after the Scottish mathematician Colin Maclaurin (1698–1746), who described them (with credit to Taylor) in a textbook in 1742.

These coefficients are also the coefficients of the power series, which is called the *Taylor series for f centered at a*. It is the natural extension of the set of Taylor polynomials for f at a. The special case of a Taylor series centered at 0 is called a *Maclaurin series*.

DEFINITION Taylor/Maclaurin Series for a Function

Suppose the function f has derivatives of all orders on an interval centered at the point a. The **Taylor series for f centered at a** is

$$f(a) + f'(a)(x - a) + \frac{f''(a)}{2!}(x - a)^2 + \frac{f^{(3)}(a)}{3!}(x - a)^3 + \cdots$$

$$= \sum_{k=0}^{\infty} \frac{f^{(k)}(a)}{k!}(x - a)^k.$$

A Taylor series centered at 0 is called a **Maclaurin series**.

> There are unusual cases in which the Taylor series for a function converges to a different function (Exercise 92).

For the Taylor series to be useful, we need to know two things:

• the values of x for which the Taylor series converges, and

• the values of x for which the Taylor series for f *equals* f.

> **QUICK CHECK 1** Verify that if the Taylor series for f centered at a is evaluated at $x = a$, then the Taylor series equals $f(a)$. ◄

The second question is subtle and is postponed for a few pages. For now, we find the Taylor series for f centered at a point, but we refrain from saying $f(x)$ equals the power series.

EXAMPLE 1 Maclaurin series and convergence Find the Maclaurin series (which is the Taylor series centered at 0) for the following functions, and then find the interval of convergence.

a. $f(x) = \cos x$ **b.** $f(x) = \dfrac{1}{1 - x}$

SOLUTION The procedure for finding the coefficients of a Taylor series is the same as for Taylor polynomials; most of the work is computing the derivatives of f.

a. The Maclaurin series has the form

$$\sum_{k=0}^{\infty} c_k x^k, \quad \text{where } c_k = \frac{f^{(k)}(0)}{k!}, \quad \text{for } k = 0, 1, 2, \ldots .$$

We evaluate derivatives of $f(x) = \cos x$ at $x = 0$.

$$f(x) = \cos x \;\Rightarrow\; f(0) = 1$$
$$f'(x) = -\sin x \;\Rightarrow\; f'(0) = 0$$
$$f''(x) = -\cos x \;\Rightarrow\; f''(0) = -1$$
$$f'''(x) = \sin x \;\Rightarrow\; f'''(0) = 0$$
$$f^{(4)}(x) = \cos x \;\Rightarrow\; f^{(4)}(0) = 1$$
$$\vdots \qquad\qquad\qquad \vdots$$

▶ In Example 1a, we note that both $\cos x$ and its Maclaurin series are even functions. Be cautious with this observation. A Taylor series for an even function centered at a point different from 0 may be even, odd, or neither. A similar behavior occurs with odd functions.

Because the odd-order derivatives are zero, $c_k = \dfrac{f^{(k)}(0)}{k!} = 0$ when k is odd. Using the even-order derivatives, we have

$$c_0 = f(0) = 1, \qquad c_2 = \frac{f^{(2)}(0)}{2!} = -\frac{1}{2!},$$

$$c_4 = \frac{f^{(4)}(0)}{4!} = \frac{1}{4!}, \quad c_6 = \frac{f^{(6)}(0)}{6!} = -\frac{1}{6!},$$

and in general, $c_{2k} = \dfrac{(-1)^k}{(2k)!}$. Therefore, the Maclaurin series for f is

$$1 - \frac{x^2}{2!} + \frac{x^4}{4!} - \frac{x^6}{6!} + \cdots = \sum_{k=0}^{\infty} \frac{(-1)^k}{(2k)!} x^{2k}.$$

Notice that this series contains all the Taylor polynomials. In this case, it consists only of even powers of x, reflecting the fact that $\cos x$ is an even function.

For what values of x does the series converge? As discussed in Section 11.2, we apply the Ratio Test:

$$r = \lim_{k\to\infty}\left| \frac{(-1)^{k+1}x^{2(k+1)}/(2(k+1))!}{(-1)^k x^{2k}/(2k)!} \right| \qquad r = \lim_{k\to\infty}\left| \frac{a_{k+1}}{a_k} \right|$$

$$= \lim_{k\to\infty}\left| \frac{x^2}{(2k+2)(2k+1)} \right| = 0. \qquad \text{Simplify and take the limit with } x \text{ fixed.}$$

▶ Recall that
$$(2k+2)! = (2k+2)(2k+1)(2k)!.$$
Therefore, $\dfrac{(2k)!}{(2k+2)!} = \dfrac{1}{(2k+2)(2k+1)}$.

In this case, $r < 1$ for all x, so the Maclaurin series converges absolutely for all x, which implies (by Theorem 10.19) that the series converges for all x. We conclude that the interval of convergence is $-\infty < x < \infty$.

b. We proceed in a similar way with $f(x) = 1/(1 - x)$ by evaluating the derivatives of f at 0:

$$f(x) = \frac{1}{1-x} \;\Rightarrow\; f(0) = 1,$$

$$f'(x) = \frac{1}{(1-x)^2} \;\Rightarrow\; f'(0) = 1,$$

$$f''(x) = \frac{2}{(1-x)^3} \;\Rightarrow\; f''(0) = 2!,$$

$$f'''(x) = \frac{3\cdot 2}{(1-x)^4} \;\Rightarrow\; f'''(0) = 3!,$$

$$f^{(4)}(x) = \frac{4\cdot 3\cdot 2}{(1-x)^5} \;\Rightarrow\; f^{(4)}(0) = 4!,$$

and in general, $f^{(k)}(0) = k!$. Therefore, the Maclaurin series coefficients are

$$c_k = \frac{f^{(k)}(0)}{k!} = \frac{k!}{k!} = 1, \text{ for } k = 0, 1, 2, \ldots. \text{ The series for } f \text{ centered at 0 is}$$

$$1 + x + x^2 + x^3 + \cdots = \sum_{k=0}^{\infty} x^k.$$

This power series is familiar! The Maclaurin series for $f(x) = 1/(1 - x)$ is a geometric series. We could apply the Ratio Test, but we have already demonstrated that this series converges for $|x| < 1$.

Related Exercises 12, 18 ◀

QUICK CHECK 2 Based on Example 1b, what is the Taylor series for $f(x) = (1 + x)^{-1}$? ◀

The preceding example has an important lesson. *There is only one power series representation for a given function about a given point; however, there may be several ways to find it.* Another important point is this: The series in Example 1b represents the function $f(x) = 1/(1 - x)$ only on the interval $(-1, 1)$. If we want a power series representation for f on another interval, a new center must be chosen and the coefficients of the power series must be recomputed, as shown in the next example.

EXAMPLE 2 Choosing a different center Find the Taylor series for $f(x) = \dfrac{1}{1 - x}$ centered at 5, and then determine its radius and interval of convergence.

SOLUTION Using the pattern discovered in Example 1b, we find that the kth derivative of f is $f^{(k)}(x) = \dfrac{k!}{(1 - x)^{k+1}}$. Therefore, the Taylor series coefficients are

$$c_k = \frac{f^{(k)}(5)}{k!} = \frac{k!}{k!(1 - 5)^{k+1}} = \frac{(-1)^{k+1}}{4^{k+1}}, \quad \text{for } k = 0, 1, 2, \ldots,$$

and the series for f centered at 5 is

$$\sum_{k=0}^{\infty} \frac{(-1)^{k+1}(x - 5)^k}{4^{k+1}} = -\frac{1}{4} + \frac{x - 5}{4^2} - \frac{(x - 5)^2}{4^3} + \frac{(x - 5)^3}{4^4} \cdots.$$

To determine the interval of convergence for this series, we apply the Ratio Test:

QUICK CHECK 3 Verify that the series $\displaystyle\sum_{k=0}^{\infty} \frac{(-1)^{k+1}(x - 5)^k}{4^{k+1}}$ from Example 2 diverges at $x = 1$ and $x = 9$. ◀

$$r = \lim_{k \to \infty} \left| \frac{(-1)^{k+2}(x - 5)^{k+1}/4^{k+2}}{(-1)^{k+1}(x - 5)^k/4^{k+1}} \right| = \lim_{k \to \infty} \frac{|x - 5|}{4} = \frac{|x - 5|}{4}.$$

The series converges absolutely when $r = \dfrac{|x - 5|}{4} < 1$, or when $1 < x < 9$, so the radius of convergence is 4. The series diverges at $x = 1$ and $x = 9$ (Quick Check 3), so we conclude that the interval of convergence is $(1, 9)$.

The graphs of f and the 7th-order Taylor polynomial p_7 centered at 5 (blue curve) are shown in **Figure 11.19**, along with the graph of p_7 centered at 0 (red curve) found in Example 1b. Notice that in both cases, the interval of convergence extends to the vertical asymptote $(x = 1)$ of $f(x) = \dfrac{1}{1 - x}$. In fact, it can be shown that the Taylor series for f, with any choice of center $a \neq 1$, has an interval of convergence symmetric about a that extends to $x = 1$. Remarkably, Taylor series for f "know" the location of the discontinuity of f.

Related Exercises 29–30 ◀

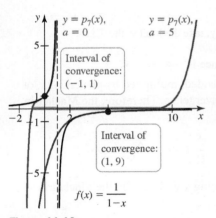

Figure 11.19

EXAMPLE 3 Center other than 0 Find the first four nonzero terms of the Taylor series for $f(x) = \sqrt[3]{x}$ centered at 8.

SOLUTION Notice that f has derivatives of all orders at $x = 8$. The Taylor series centered at 8 has the form

$$\sum_{k=0}^{\infty} c_k(x - 8)^k, \quad \text{where } c_k = \frac{f^{(k)}(8)}{k!}.$$

Next, we evaluate derivatives:

$$f(x) = x^{1/3} \implies f(8) = 2,$$

$$f'(x) = \frac{1}{3}x^{-2/3} \implies f'(8) = \frac{1}{12},$$

$$f''(x) = -\frac{2}{9}x^{-5/3} \implies f''(8) = -\frac{1}{144}, \text{ and}$$

$$f'''(x) = \frac{10}{27}x^{-8/3} \implies f'''(8) = \frac{5}{3456}.$$

We now assemble the power series:

$$2 + \frac{1}{12}(x - 8) + \frac{1}{2!}\left(-\frac{1}{144}\right)(x - 8)^2 + \frac{1}{3!}\left(\frac{5}{3456}\right)(x - 8)^3 + \cdots$$

$$= 2 + \frac{1}{12}(x - 8) - \frac{1}{288}(x - 8)^2 + \frac{5}{20,736}(x - 8)^3 + \cdots.$$

Related Exercise 32 ◄

EXAMPLE 4   Manipulating Maclaurin series Let $f(x) = e^x$.

a. Find the Maclaurin series for f.

b. Find its interval of convergence.

c. Use the Maclaurin series for e^x to find the Maclaurin series for the functions $x^4 e^x$, e^{-2x}, and e^{-x^2}.

SOLUTION

a. The coefficients of the Taylor polynomials for $f(x) = e^x$ centered at 0 are $c_k = 1/k!$ (Example 3, Section 11.1). They are also the coefficients of the Maclaurin series. Therefore, the Maclaurin series for e^x is

$$1 + \frac{x}{1!} + \frac{x^2}{2!} + \cdots + \frac{x^n}{n!} + \cdots = \sum_{k=0}^{\infty} \frac{x^k}{k!}.$$

b. By the Ratio Test,

$$r = \lim_{k \to \infty} \left| \frac{x^{k+1}/(k + 1)!}{x^k/k!} \right| \qquad \text{Substitute } (k + 1)\text{st and } k\text{th terms.}$$

$$= \lim_{k \to \infty} \left| \frac{x}{k + 1} \right| = 0. \qquad \text{Simplify; take the limit with } x \text{ fixed.}$$

Because $r < 1$ for all x, the interval of convergence is $-\infty < x < \infty$.

c. As stated in Theorem 11.4, power series may be added, multiplied by powers of x, or composed with functions on their intervals of convergence. Therefore, the Maclaurin series for $x^4 e^x$ is

$$x^4 \sum_{k=0}^{\infty} \frac{x^k}{k!} = \sum_{k=0}^{\infty} \frac{x^{k+4}}{k!} = x^4 + \frac{x^5}{1!} + \frac{x^6}{2!} + \cdots + \frac{x^{k+4}}{k!} + \cdots.$$

Similarly, e^{-2x} is the composition $f(-2x)$. Replacing x with $-2x$ in the Maclaurin series for f, the series representation for e^{-2x} is

$$\sum_{k=0}^{\infty} \frac{(-2x)^k}{k!} = \sum_{k=0}^{\infty} \frac{(-1)^k (2x)^k}{k!} = 1 - 2x + 2x^2 - \frac{4}{3}x^3 + \cdots.$$

The Maclaurin series for e^{-x^2} is obtained by replacing x with $-x^2$ in the power series for f. The resulting series is

$$\sum_{k=0}^{\infty} \frac{(-x^2)^k}{k!} = \sum_{k=0}^{\infty} \frac{(-1)^k x^{2k}}{k!} = 1 - x^2 + \frac{x^4}{2!} - \frac{x^6}{3!} + \cdots.$$

QUICK CHECK 4 Find the first three terms of the Maclaurin series for $2xe^x$ and e^{-x}. ◄

▶ For nonnegative integers p and k with $0 \leq k \leq p$, the binomial coefficients may also be defined as $\binom{p}{k} = \dfrac{p!}{k!(p-k)!}$, where $0! = 1$. The coefficients form the rows of Pascal's triangle. The coefficients of $(1+x)^5$ form the sixth row of the triangle.

$$
\begin{array}{ccccccccccc}
&&&&&1\\
&&&&1&&1\\
&&&1&&2&&1\\
&&1&&3&&3&&1\\
&1&&4&&6&&4&&1\\
1&&5&&10&&10&&5&&1
\end{array}
$$

QUICK CHECK 5 Evaluate the binomial coefficients $\binom{-3}{2}$ and $\binom{\frac{1}{2}}{3}$. ◄

Because the interval of convergence of $f(x) = e^x$ is $-\infty < x < \infty$, the manipulations used to obtain the series for $x^4 e^x$, e^{-2x}, or e^{-x^2} do not change the interval of convergence. If in doubt about the interval of convergence of a new series, apply the Ratio Test.

Related Exercises 35, 38 ◄

The Binomial Series

We know from algebra that if p is a positive integer, then $(1+x)^p$ is a polynomial of degree p. In fact,

$$(1+x)^p = \binom{p}{0} + \binom{p}{1}x + \binom{p}{2}x^2 + \cdots + \binom{p}{p}x^p,$$

where the binomial coefficients $\binom{p}{k}$ are defined as follows.

> **DEFINITION Binomial Coefficients**
> For real numbers p and integers $k \geq 1$,
> $$\binom{p}{k} = \frac{p(p-1)(p-2)\cdots(p-k+1)}{k!}, \qquad \binom{p}{0} = 1.$$

For example,

$$(1+x)^5 = \underbrace{\binom{5}{0}}_{1} + \underbrace{\binom{5}{1}}_{5}x + \underbrace{\binom{5}{2}}_{10}x^2 + \underbrace{\binom{5}{3}}_{10}x^3 + \underbrace{\binom{5}{4}}_{5}x^4 + \underbrace{\binom{5}{5}}_{1}x^5$$

$$= 1 + 5x + 10x^2 + 10x^3 + 5x^4 + x^5.$$

Our goal is to extend this idea to the functions $f(x) = (1+x)^p$, where $p \neq 0$ is a real number. The result is a Taylor series called the *binomial series*.

> **THEOREM 11.6 Binomial Series**
> For real numbers $p \neq 0$, the Taylor series for $f(x) = (1+x)^p$ centered at 0 is the **binomial series**
> $$\sum_{k=0}^{\infty}\binom{p}{k}x^k = 1 + \sum_{k=1}^{\infty}\frac{p(p-1)(p-2)\cdots(p-k+1)}{k!}x^k$$
> $$= 1 + px + \frac{p(p-1)}{2!}x^2 + \frac{p(p-1)(p-2)}{3!}x^3 + \cdots.$$
>
> The series converges for $|x| < 1$ (and possibly at the endpoints, depending on p). If p is a nonnegative integer, the series terminates and results in a polynomial of degree p.

Proof: We seek a power series centered at 0 of the form

$$\sum_{k=0}^{\infty}c_k x^k, \quad \text{where } c_k = \frac{f^{(k)}(0)}{k!}, \quad \text{for } k = 0, 1, 2, \ldots.$$

▶ To evaluate $\binom{p}{k}$, start with p and successively subtract 1 until k factors are obtained; then take the product of these k factors and divide by $k!$. Recall that $\binom{p}{0} = 1$.

The job is to evaluate the derivatives of f at 0:

$$f(x) = (1+x)^p \Rightarrow f(0) = 1,$$
$$f'(x) = p(1+x)^{p-1} \Rightarrow f'(0) = p,$$
$$f''(x) = p(p-1)(1+x)^{p-2} \Rightarrow f''(0) = p(p-1), \text{ and}$$
$$f'''(x) = p(p-1)(p-2)(1+x)^{p-3} \Rightarrow f'''(0) = p(p-1)(p-2).$$

A pattern emerges: The kth derivative $f^{(k)}(0)$ involves the k factors $p(p-1)(p-2)\cdots(p-k+1)$. In general, we have

$$f^{(k)}(0) = p(p-1)(p-2)\cdots(p-k+1).$$

Therefore,

$$c_k = \frac{f^{(k)}(0)}{k!} = \frac{p(p-1)(p-2)\cdots(p-k+1)}{k!} = \binom{p}{k}, \quad \text{for } k = 0, 1, 2, \ldots.$$

The Taylor series for $f(x) = (1+x)^p$ centered at 0 is

$$\binom{p}{0} + \binom{p}{1}x + \binom{p}{2}x^2 + \binom{p}{3}x^3 + \cdots = \sum_{k=0}^{\infty}\binom{p}{k}x^k.$$

This series has the same general form for all values of p. When p is a nonnegative integer, the series terminates and it is a polynomial of degree p.

The radius of convergence for the binomial series is determined by the Ratio Test. Holding p and x fixed, the relevant limit is

$$r = \lim_{k \to \infty}\left|\frac{x^{k+1}p(p-1)\cdots(p-k+1)(p-k)/(k+1)!}{x^k p(p-1)\cdots(p-k+1)/k!}\right| \qquad \text{Ratio of } (k+1)\text{st to } k\text{th term}$$

$$= |x|\lim_{k \to \infty}\underbrace{\left|\frac{p-k}{k+1}\right|}_{\text{approaches } 1} \qquad\qquad \text{Cancel factors and simplify.}$$

> In Theorem 11.6, it can be shown that the interval of convergence for the binomial series is
> • $(-1, 1)$ if $p \leq -1$,
> • $(-1, 1]$ if $-1 < p < 0$, and
> • $[-1, 1]$ if $p > 0$ and not an integer.

$$= |x|. \qquad\qquad\qquad \text{With } p \text{ fixed,} \quad \lim_{k \to \infty}\left|\frac{p-k}{k+1}\right| = 1.$$

Absolute convergence requires that $r = |x| < 1$. Therefore, the series converges for $|x| < 1$. Depending on the value of p, the interval of convergence may include the endpoints; see margin note. ◄

EXAMPLE 5   Binomial series   Consider the function $f(x) = \sqrt{1+x}$.

a. Find the first four terms of the binomial series for f centered at 0.

b. Approximate $\sqrt{1.15}$ to three decimal places. Assume the series for f converges to f on its interval of convergence, which is $[-1, 1]$.

> A binomial series is a Taylor series. Because the series in Example 5 is centered at 0, it is also a Maclaurin series.

SOLUTION

a. We use the formula for the binomial coefficients with $p = \frac{1}{2}$ to compute the first four coefficients:

$$c_0 = 1, \qquad\qquad c_1 = \binom{\frac{1}{2}}{1} = \frac{\frac{1}{2}}{1!} = \frac{1}{2},$$

$$c_2 = \binom{\frac{1}{2}}{2} = \frac{\frac{1}{2}\left(-\frac{1}{2}\right)}{2!} = -\frac{1}{8}, \text{ and } c_3 = \binom{\frac{1}{2}}{3} = \frac{\frac{1}{2}\left(-\frac{1}{2}\right)\left(-\frac{3}{2}\right)}{3!} = \frac{1}{16}.$$

Table 11.3

n	Approximation $p_n(0.15)$
0	1.0
1	1.075
2	1.0721875
3	1.072398438

The leading terms of the binomial series are

$$1 + \frac{1}{2}x - \frac{1}{8}x^2 + \frac{1}{16}x^3 - \cdots.$$

> The remainder theorem for alternating series (Section 10.6) could be used in Example 5 to estimate the number of terms of the Maclaurin series needed to achieve a desired accuracy.

b. Truncating the binomial series in part (a) produces Taylor polynomials p_n that may be used to approximate $f(0.15) = \sqrt{1.15}$. With $x = 0.15$, we find the polynomial approximations shown in Table 11.3. Four terms of the power series ($n = 3$) give $\sqrt{1.15} \approx 1.072$. Because the approximations with $n = 2$ and $n = 3$ agree to three decimal places, when rounded, the approximation 1.072 is accurate to three decimal places.

QUICK CHECK 6 Use two and three terms of the binomial series in Example 5 to approximate $\sqrt{1.1}$. ◄

Related Exercises 46–47 ◄

EXAMPLE 6   Working with binomial series Consider the functions

$$f(x) = \sqrt[3]{1 + x} \quad \text{and} \quad g(x) = \sqrt[3]{c + x}, \quad \text{where } c > 0 \text{ is a constant.}$$

a. Find the first four terms of the binomial series for f centered at 0.

b. Use part (a) to find the first four terms of the binomial series for g centered at 0.

c. Use part (b) to approximate $\sqrt[3]{23}, \sqrt[3]{24}, \dots, \sqrt[3]{31}$. Assume the series for g converges to g on its interval of convergence.

SOLUTION

a. Because $f(x) = (1 + x)^{1/3}$, we find the binomial coefficients with $p = \frac{1}{3}$:

$$c_0 = \binom{\frac{1}{3}}{0} = 1, \qquad\qquad c_1 = \binom{\frac{1}{3}}{1} = \frac{\frac{1}{3}}{1!} = \frac{1}{3},$$

$$c_2 = \binom{\frac{1}{3}}{2} = \frac{\left(\frac{1}{3}\right)\left(\frac{1}{3} - 1\right)}{2!} = -\frac{1}{9}, \text{ and } c_3 = \binom{\frac{1}{3}}{3} = \frac{\left(\frac{1}{3}\right)\left(\frac{1}{3} - 1\right)\left(\frac{1}{3} - 2\right)}{3!} = \frac{5}{81}.$$

The first four terms of the binomial series are

$$1 + \frac{1}{3}x - \frac{1}{9}x^2 + \frac{5}{81}x^3.$$

b. To avoid deriving a new series for $g(x) = \sqrt[3]{c + x}$, a few steps of algebra allow us to use part (a). Note that

$$g(x) = \sqrt[3]{c + x} = \sqrt[3]{c\left(1 + \frac{x}{c}\right)} = \sqrt[3]{c} \cdot \sqrt[3]{1 + \frac{x}{c}} = \sqrt[3]{c} \cdot f\left(\frac{x}{c}\right).$$

In other words, g can be expressed in terms of f, for which we already have a binomial series. The binomial series for g is obtained by substituting x/c into the binomial series for f and multiplying by $\sqrt[3]{c}$:

$$g(x) = \sqrt[3]{c}\underbrace{\left(1 + \frac{1}{3}\left(\frac{x}{c}\right) - \frac{1}{9}\left(\frac{x}{c}\right)^2 + \frac{5}{81}\left(\frac{x}{c}\right)^3 - \cdots\right)}_{f(x/c)}.$$

It can be shown that the series for f in part (a) converges to $f(x)$ for $|x| \le 1$. Therefore, the series for $f(x/c)$ converges to $f(x/c)$ provided $|x/c| \le 1$, or, equivalently, for $|x| \le c$.

c. The series of part (b) may be truncated after four terms to approximate cube roots. For example, note that $\sqrt[3]{29} = \sqrt[3]{\underbrace{27}_{c} + \underbrace{2}_{x}}$, so we take $c = 27$ and $x = 2$.

The choice $c = 27$ is made because 29 is near 27 and $\sqrt[3]{c} = \sqrt[3]{27} = 3$ is easy to evaluate. Substituting $c = 27$ and $x = 2$, we find that

$$\sqrt[3]{29} \approx \sqrt[3]{27}\left(1 + \frac{1}{3}\left(\frac{2}{27}\right) - \frac{1}{9}\left(\frac{2}{27}\right)^2 + \frac{5}{81}\left(\frac{2}{27}\right)^3\right) \approx 3.0723.$$

The same method is used to approximate the cube roots of 23, 24, ..., 30, 31 (Table 11.4). The absolute error is the difference between the approximation and the value given by a calculator. Notice that the errors increase as we move away from 27.

Related Exercises 52–54 ◀

Table 11.4

	Approximation	Absolute Error
$\sqrt[3]{23}$	2.8439	6.7×10^{-5}
$\sqrt[3]{24}$	2.8845	2.0×10^{-5}
$\sqrt[3]{25}$	2.9240	3.9×10^{-6}
$\sqrt[3]{26}$	2.9625	2.4×10^{-7}
$\sqrt[3]{27}$	3	0
$\sqrt[3]{28}$	3.0366	2.3×10^{-7}
$\sqrt[3]{29}$	3.0723	3.5×10^{-6}
$\sqrt[3]{30}$	3.1072	1.7×10^{-5}
$\sqrt[3]{31}$	3.1414	5.4×10^{-5}

Convergence of Taylor Series

It may seem that the story of Taylor series is over. But there is a technical point that is easily overlooked. Given a function f, we know how to write its Taylor series centered at a point a, and we know how to find its interval of convergence. We still do not know that the series actually converges to f. The remaining task is to determine when the Taylor series for f actually converges to f on its interval of convergence. Fortunately, the necessary tools have already been presented in Taylor's Theorem (Theorem 11.1), which gives the remainder for Taylor polynomials.

Assume f has derivatives of *all* orders on an open interval containing the point a. Taylor's Theorem tells us that

$$f(x) = p_n(x) + R_n(x),$$

where p_n is the nth-order Taylor polynomial for f centered at a, that

$$R_n(x) = \frac{f^{(n+1)}(c)}{(n+1)!}(x-a)^{n+1}$$

is the remainder, and that c is a point between x and a. We see that the remainder, $R_n(x) = f(x) - p_n(x)$, measures the difference between f and the approximating polynomial p_n. When we say the Taylor series converges to f at a point x, we mean the value of the Taylor series at x equals $f(x)$; that is, $\lim_{n\to\infty} p_n(x) = f(x)$. The following theorem makes these ideas precise.

THEOREM 11.7   Convergence of Taylor Series

Let f have derivatives of all orders on an open interval I containing a. The Taylor series for f centered at a converges to f, for all x in I, if and only if $\lim_{n\to\infty} R_n(x) = 0$, for all x in I, where

$$R_n(x) = \frac{f^{(n+1)}(c)}{(n+1)!}(x-a)^{n+1}$$

is the remainder at x, with c between x and a.

Proof: The theorem requires derivatives of *all* orders. Therefore, by Taylor's Theorem (Theorem 11.1), the remainder exists in the given form for all n. Let p_n denote the nth-order Taylor polynomial and note that $\lim_{n\to\infty} p_n(x)$ is the Taylor series for f centered at a, evaluated at a point x in I.

First, assume $\lim_{n\to\infty} R_n(x) = 0$ on the interval I and recall that $p_n(x) = f(x) - R_n(x)$. Taking limits of both sides, we have

$$\underbrace{\lim_{n\to\infty} p_n(x)}_{\text{Taylor series}} = \lim_{n\to\infty} (f(x) - R_n(x)) = \underbrace{\lim_{n\to\infty} f(x)}_{f(x)} - \underbrace{\lim_{n\to\infty} R_n(x)}_{0} = f(x).$$

We conclude that the Taylor series $\lim_{n\to\infty} p_n(x)$ equals $f(x)$, for all x in I.

Conversely, if the Taylor series converges to f, then $f(x) = \lim_{n\to\infty} p_n(x)$ and

$$0 = f(x) - \lim_{n\to\infty} p_n(x) = \lim_{n\to\infty} \underbrace{(f(x) - p_n(x))}_{R_n(x)} = \lim_{n\to\infty} R_n(x).$$

It follows that $\lim_{n\to\infty} R_n(x) = 0$, for all x in I.   ◄

Even with an expression for the remainder, it may be difficult to show that $\lim_{n\to\infty} R_n(x) = 0$. The following examples illustrate cases in which it is possible.

EXAMPLE 7   Remainder in the Maclaurin series for e^x Show that the Maclaurin series for $f(x) = e^x$ converges to $f(x)$, for $-\infty < x < \infty$.

SOLUTION As shown in Example 4, the Maclaurin series for $f(x) = e^x$ is

$$\sum_{k=0}^{\infty} \frac{x^k}{k!} = 1 + x + \frac{x^2}{2!} + \cdots + \frac{x^n}{n!} + \cdots,$$

which converges, for $-\infty < x < \infty$. In Example 7 of Section 11.1 it was shown that the remainder is

$$R_n(x) = \frac{e^c}{(n+1)!}x^{n+1},$$

where c is between 0 and x. Notice that the intermediate point c varies with n, but it is always between 0 and x. Therefore, e^c is between $e^0 = 1$ and e^x; in fact, $e^c \leq e^{|x|}$, for all n. It follows that

$$|R_n(x)| \leq \frac{e^{|x|}}{(n+1)!}|x|^{n+1}.$$

Holding x fixed, we have

$$\lim_{n\to\infty}|R_n(x)| = \lim_{n\to\infty}\frac{e^{|x|}}{(n+1)!}|x|^{n+1} = e^{|x|}\lim_{n\to\infty}\frac{|x|^{n+1}}{(n+1)!} = 0,$$

where we used the fact that $\lim_{n\to\infty} x^n/n! = 0$, for $-\infty < x < \infty$ (Section 10.2). Because $\lim_{n\to\infty}|R_n(x)| = 0$, it follows that for all real numbers x, the Taylor series converges to e^x, or

$$e^x = \sum_{k=0}^{\infty}\frac{x^k}{k!} = 1 + x + \frac{x^2}{2!} + \cdots + \frac{x^n}{n!} + \cdots.$$

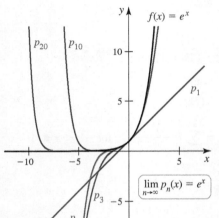

Figure 11.20

The convergence of the Taylor series to e^x is illustrated in **Figure 11.20**, where Taylor polynomials of increasing degree are graphed together with e^x.

Related Exercise 65 ◄

EXAMPLE 8   Maclaurin series convergence for cos x Show that the Maclaurin series for cos x,

$$1 - \frac{x^2}{2!} + \frac{x^4}{4!} - \frac{x^6}{6!} + \cdots = \sum_{k=0}^{\infty}(-1)^k\frac{x^{2k}}{(2k)!},$$

converges to $f(x) = \cos x$, for $-\infty < x < \infty$.

SOLUTION To show that the power series converges to f, we must show that $\lim_{n\to\infty}|R_n(x)| = 0$, for $-\infty < x < \infty$. According to Taylor's Theorem with $a = 0$,

$$R_n(x) = \frac{f^{(n+1)}(c)}{(n+1)!}x^{n+1},$$

where c is between 0 and x. Notice that $f^{(n+1)}(c) = \pm \sin c$ or $f^{(n+1)}(c) = \pm \cos c$. In all cases, $|f^{(n+1)}(c)| \leq 1$. Therefore, the absolute value of the remainder term is bounded as

$$|R_n(x)| = \left|\frac{f^{(n+1)}(c)}{(n+1)!}x^{n+1}\right| \leq \frac{|x|^{n+1}}{(n+1)!}.$$

Figure 11.21

Holding x fixed and using $\lim_{n\to\infty} x^n/n! = 0$, we see that $\lim_{n\to\infty} R_n(x) = 0$ for all x. Therefore, the given power series converges to $f(x) = \cos x$, for all x; that is, $\cos x = \sum_{k=0}^{\infty}\frac{(-1)^k x^{2k}}{(2k)!}$. The convergence of the Taylor series to $\cos x$ is illustrated in **Figure 11.21**.

Related Exercise 63 ◄

The procedure used in Examples 7 and 8 can be carried out for all the Taylor series we have worked with so far (with varying degrees of difficulty). In each case, the Taylor series converges to the function it represents on the interval of convergence. Table 11.5 summarizes commonly used Taylor series centered at 0 and the functions to which they converge.

▶ Table 11.5 asserts, without proof, that in several cases, the Taylor series for f converges to f at the endpoints of the interval of convergence. Proving convergence at the endpoints generally requires advanced techniques. It may also be done using the following theorem:

Suppose the Taylor series for f centered at 0 converges to f on the interval $(-R, R)$. If the series converges at $x = R$, then it converges to $\lim_{x \to R^-} f(x)$. If the series converges at $x = -R$, then it converges to $\lim_{x \to -R^+} f(x)$.

For example, this theorem would allow us to conclude that the series for $\ln(1 + x)$ converges to $\ln 2$ at $x = 1$.

Table 11.5

$$\frac{1}{1 - x} = 1 + x + x^2 + \cdots + x^k + \cdots = \sum_{k=0}^{\infty} x^k, \quad \text{for } |x| < 1$$

$$\frac{1}{1 + x} = 1 - x + x^2 - \cdots + (-1)^k x^k + \cdots = \sum_{k=0}^{\infty} (-1)^k x^k, \quad \text{for } |x| < 1$$

$$e^x = 1 + x + \frac{x^2}{2!} + \cdots + \frac{x^k}{k!} + \cdots = \sum_{k=0}^{\infty} \frac{x^k}{k!}, \quad \text{for } |x| < \infty$$

$$\sin x = x - \frac{x^3}{3!} + \frac{x^5}{5!} - \cdots + \frac{(-1)^k x^{2k+1}}{(2k + 1)!} + \cdots = \sum_{k=0}^{\infty} \frac{(-1)^k x^{2k+1}}{(2k + 1)!}, \quad \text{for } |x| < \infty$$

$$\cos x = 1 - \frac{x^2}{2!} + \frac{x^4}{4!} - \cdots + \frac{(-1)^k x^{2k}}{(2k)!} + \cdots = \sum_{k=0}^{\infty} \frac{(-1)^k x^{2k}}{(2k)!}, \quad \text{for } |x| < \infty$$

$$\ln(1 + x) = x - \frac{x^2}{2} + \frac{x^3}{3} - \cdots + \frac{(-1)^{k+1} x^k}{k} + \cdots = \sum_{k=1}^{\infty} \frac{(-1)^{k+1} x^k}{k}, \quad \text{for } -1 < x \le 1$$

$$-\ln(1 - x) = x + \frac{x^2}{2} + \frac{x^3}{3} + \cdots + \frac{x^k}{k} + \cdots = \sum_{k=1}^{\infty} \frac{x^k}{k}, \quad \text{for } -1 \le x < 1$$

$$\tan^{-1} x = x - \frac{x^3}{3} + \frac{x^5}{5} - \cdots + \frac{(-1)^k x^{2k+1}}{2k + 1} + \cdots = \sum_{k=0}^{\infty} \frac{(-1)^k x^{2k+1}}{2k + 1}, \quad \text{for } |x| \le 1$$

$$\sinh x = x + \frac{x^3}{3!} + \frac{x^5}{5!} + \cdots + \frac{x^{2k+1}}{(2k + 1)!} + \cdots = \sum_{k=0}^{\infty} \frac{x^{2k+1}}{(2k + 1)!}, \quad \text{for } |x| < \infty$$

$$\cosh x = 1 + \frac{x^2}{2!} + \frac{x^4}{4!} + \cdots + \frac{x^{2k}}{(2k)!} + \cdots = \sum_{k=0}^{\infty} \frac{x^{2k}}{(2k)!}, \quad \text{for } |x| < \infty$$

▶ As noted in Theorem 11.6, the binomial series may converge to $(1 + x)^p$ at $x = \pm 1$, depending on the value of p.

$$(1 + x)^p = \sum_{k=0}^{\infty} \binom{p}{k} x^k, \text{ for } |x| < 1 \text{ and } \binom{p}{k} = \frac{p(p - 1)(p - 2) \cdots (p - k + 1)}{k!}, \binom{p}{0} = 1$$

SECTION 11.3 EXERCISES

Getting Started

1. How are the Taylor polynomials for a function f centered at a related to the Taylor series of the function f centered at a?

2. What conditions must be satisfied by a function f to have a Taylor series centered at a?

3. Find a Taylor series for f centered at 2 given that $f^{(k)}(2) = 1$, for all nonnegative integers k.

4. Find a Taylor series for f centered at 0 given that $f^{(k)}(0) = (k + 1)!$, for all nonnegative integers k.

5. Suppose you know the Maclaurin series for f and that it converges to $f(x)$ for $|x| < 1$. How do you find the Maclaurin series for $f(x^2)$ and where does it converge?

6. For what values of p does the Taylor series for $f(x) = (1 + x)^p$ centered at 0 terminate?

7. In terms of the remainder, what does it mean for a Taylor series for a function f to converge to f?

8. Find the Maclaurin series for $\sin(-x)$ using the definition of a Maclaurin series. Check your answer by finding the Maclaurin series for $\sin(-x)$ using Table 11.5.

Practice Exercises

9–26. Taylor series and interval of convergence

a. Use the definition of a Taylor/Maclaurin series to find the first four nonzero terms of the Taylor series for the given function centered at a.
b. Write the power series using summation notation.
c. Determine the interval of convergence of the series.

9. $f(x) = \dfrac{1}{x^2}, a = 1$ 10. $f(x) = \dfrac{1}{x^2}, a = -1$

11. $f(x) = e^{-x}, a = 0$ 12. $f(x) = \cos 2x, a = 0$

13. $f(x) = \dfrac{2}{(1 - x)^3}, a = 0$ 14. $f(x) = x \sin x, a = 0$

15. $f(x) = (1 + x^2)^{-1}, a = 0$ 16. $f(x) = \ln(1 + 4x), a = 0$

17. $f(x) = e^{2x}, a = 0$ 18. $f(x) = (1 + 2x)^{-1}, a = 0$

19. $f(x) = \tan^{-1}\dfrac{x}{2}, a = 0$ 20. $f(x) = \sin 3x, a = 0$

21. $f(x) = 3^x, a = 0$ 22. $f(x) = \log_3(x + 1), a = 0$

23. $f(x) = \cosh 3x, a = 0$ **24.** $f(x) = \sinh 2x, a = 0$

25. $f(x) = \ln(x - 2), a = 3$ **26.** $f(x) = e^x, a = 2$

27–34. Taylor series

a. *Use the definition of a Taylor series to find the first four nonzero terms of the Taylor series for the given function centered at a.*

b. *Write the power series using summation notation.*

27. $f(x) = \sin x, a = \dfrac{\pi}{2}$ **28.** $f(x) = \cos x, a = \pi$

29. $f(x) = \dfrac{1}{x}, a = 1$ **30.** $f(x) = \dfrac{1}{x}, a = 2$

31. $f(x) = \ln x, a = 3$ **32.** $f(x) = e^x, a = \ln 2$

33. $f(x) = 2^x, a = 1$ **34.** $f(x) = x \ln x - x + 1; a = 1$

35–44. Manipulating Taylor series *Use the Taylor series in Table 11.5 to find the first four nonzero terms of the Taylor series for the following functions centered at 0.*

35. $\ln(1 + x^2)$ **36.** $\sin x^2$

37. $\dfrac{1}{1 - 2x}$ **38.** $\ln(1 + 2x)$

39. $\begin{cases} \dfrac{e^x - 1}{x} & \text{if } x \neq 1 \\ 1 & \text{if } x = 1 \end{cases}$ **40.** $\cos x^3$

41. $(1 + x^4)^{-1}$ **42.** $x \tan^{-1} x^2$

43. $\sinh x^2$ **44.** $\cosh(-2x)$

T 45–50. Binomial series

a. *Find the first four nonzero terms of the binomial series centered at 0 for the given function.*

b. *Use the first four terms of the series to approximate the given quantity.*

45. $f(x) = (1 + x)^{-2}$; approximate $\dfrac{1}{1.21} = \dfrac{1}{1.1^2}$.

46. $f(x) = \sqrt{1 + x}$; approximate $\sqrt{1.06}$.

47. $f(x) = \sqrt[3]{1 + x}$; approximate $\sqrt[4]{1.12}$.

48. $f(x) = (1 + x)^{-3}$; approximate $\dfrac{1}{1.331} = \dfrac{1}{1.1^3}$.

49. $f(x) = (1 + x)^{-2/3}$; approximate $1.18^{-2/3}$.

50. $f(x) = (1 + x)^{2/3}$; approximate $1.02^{2/3}$.

51–56. Working with binomial series *Use properties of power series, substitution, and factoring to find the first four nonzero terms of the Maclaurin series for the following functions. Give the interval of convergence for the new series (Theorem 11.4 is useful). Use the Maclaurin series*

$$\sqrt{1 + x} = 1 + \frac{x}{2} - \frac{x^2}{8} + \frac{x^3}{16} - \cdots, \quad \text{for } -1 \leq x \leq 1.$$

51. $\sqrt{1 + x^2}$ **52.** $\sqrt{4 + x}$

53. $\sqrt{9 - 9x}$ **54.** $\sqrt{1 - 4x}$

55. $\sqrt{a^2 + x^2}, a > 0$ **56.** $\sqrt{4 - 16x^2}$

57–62. Working with binomial series *Use properties of power series, substitution, and factoring to find the first four nonzero terms of the Maclaurin series for the following functions. Use the Maclaurin series*

$$(1 + x)^{-2} = 1 - 2x + 3x^2 - 4x^3 + \cdots, \quad \text{for } -1 < x < 1.$$

57. $(1 + 4x)^{-2}$ **58.** $\dfrac{1}{(1 - 4x)^2}$

59. $\dfrac{1}{(4 + x^2)^2}$ **60.** $(x^2 - 4x + 5)^{-2}$

61. $\dfrac{1}{(3 + 4x)^2}$ **62.** $\dfrac{1}{(1 + 4x^2)^2}$

63–66. Remainders *Find the remainder in the Taylor series centered at the point a for the following functions. Then show that $\lim\limits_{n \to \infty} R_n(x) = 0$, for all x in the interval of convergence.*

63. $f(x) = \sin x, a = 0$ **64.** $f(x) = \cos 2x, a = 0$

65. $f(x) = e^{-x}, a = 0$ **66.** $f(x) = \cos x, a = \dfrac{\pi}{2}$

67. Explain why or why not Determine whether the following statements are true and give an explanation or counterexample.

a. The function $f(x) = \sqrt{x}$ has a Taylor series centered at 0.

b. The function $f(x) = \csc x$ has a Taylor series centered at $\pi/2$.

c. If f has a Taylor series that converges only on $(-2, 2)$, then $f(x^2)$ has a Taylor series that also converges only on $(-2, 2)$.

d. If $p(x)$ is the Taylor series for f centered at 0, then $p(x - 1)$ is the Taylor series for f centered at 1.

e. The Taylor series for an even function centered at 0 has only even powers of x.

68–75. Any method

a. *Use any analytical method to find the first four nonzero terms of the Taylor series centered at 0 for the following functions. You do not need to use the definition of the Taylor series coefficients.*

b. *Determine the radius of convergence of the series.*

68. $f(x) = \cos 2x + 2 \sin x$ **69.** $f(x) = \dfrac{e^x + e^{-x}}{2}$

70. $f(x) = \begin{cases} \dfrac{\sin x}{x} & \text{if } x \neq 0 \\ 1 & \text{if } x = 0 \end{cases}$ **71.** $f(x) = (1 + x^2)^{-2/3}$

72. $f(x) = x^2 \cos x^2$ **73.** $f(x) = \sqrt{1 - x^2}$

74. $f(x) = b^x$, for $b > 0, b \neq 1$

75. $f(x) = \dfrac{1}{x^4 + 2x^2 + 1}$

T 76–79. Approximating powers *Compute the coefficients for the Taylor series for the following functions about the given point a, and then use the first four terms of the series to approximate the given number.*

76. $f(x) = \sqrt{x}$ with $a = 36$; approximate $\sqrt{39}$.

77. $f(x) = \sqrt[3]{x}$ with $a = 64$; approximate $\sqrt[3]{60}$.

78. $f(x) = \dfrac{1}{\sqrt{x}}$ with $a = 4$; approximate $\dfrac{1}{\sqrt{3}}$.

79. $f(x) = \sqrt[4]{x}$ with $a = 16$; approximate $\sqrt[4]{13}$.

80. **Geometric/binomial series** Recall that the Taylor series for $f(x) = 1/(1-x)$ centered at 0 is the geometric series $\sum_{k=0}^{\infty} x^k$. Show that this series can also be found as a binomial series.

81. **Integer coefficients** Show that the first five nonzero coefficients of the Taylor series (binomial series) for $f(x) = \sqrt{1+4x}$ centered at 0 are integers. (In fact, *all* the coefficients are integers.)

Explorations and Challenges

82. **Choosing a good center** Suppose you want to approximate $\sqrt{72}$ using four terms of a Taylor series. Compare the accuracy of the approximations obtained using the Taylor series for $\sqrt{x}$ centered at 64 and 81.

83. **Alternative means** By comparing the first four terms, show that the Maclaurin series for $\sin^2 x$ can be found (a) by squaring the Maclaurin series for $\sin x$, (b) by using the identity $\sin^2 x = \dfrac{1 - \cos 2x}{2}$, or (c) by computing the coefficients using the definition.

84. **Alternative means** By comparing the first four terms, show that the Maclaurin series for $\cos^2 x$ can be found (a) by squaring the Maclaurin series for $\cos x$, (b) by using the identity $\cos^2 x = \dfrac{1 + \cos 2x}{2}$, or (c) by computing the coefficients using the definition.

85. **Designer series** Find a power series that has $(2, 6)$ as an interval of convergence.

86. **Composition of series** Use composition of series to find the first three terms of the Maclaurin series for the following functions.
 a. $e^{\sin x}$ b. $e^{\tan x}$ c. $\sqrt{1 + \sin^2 x}$

T 87–88. **Approximations** *Choose a Taylor series and a center point to approximate the following quantities with an error of 10^{-4} or less.*

87. $\cos 40°$ 88. $\sin(0.98\pi)$

89. **Different approximation strategies** Suppose you want to approximate $\sqrt[3]{128}$ to within 10^{-4} of the exact value.
 a. Use a Taylor polynomial for $f(x) = (125 + x)^{1/3}$ centered at 0.
 b. Use a Taylor polynomial for $f(x) = x^{1/3}$ centered at 125.
 c. Compare the two approaches. Are they equivalent?

90. **Nonconvergence to f** Consider the function
$$f(x) = \begin{cases} e^{-1/x^2} & \text{if } x \neq 0 \\ 0 & \text{if } x = 0. \end{cases}$$
 a. Use the definition of the derivative to show that $f'(0) = 0$.
 b. Assume the fact that $f^{(k)}(0) = 0$, for $k = 1, 2, 3, \ldots$. (You can write a proof using the definition of the derivative.) Write the Taylor series for f centered at 0.
 c. Explain why the Taylor series for f does not converge to f for $x \neq 0$.

91. **Version of the Second Derivative Test** Assume f has at least two continuous derivatives on an interval containing a with $f'(a) = 0$. Use Taylor's Theorem to prove the following version of the Second Derivative Test:
 a. If $f''(x) > 0$ on some interval containing a, then f has a local minimum at a.
 b. If $f''(x) < 0$ on some interval containing a, then f has a local maximum at a.

QUICK CHECK ANSWERS

1. When evaluated at $x = a$, all terms of the series are zero except the first term, which is $f(a)$. Therefore, the series equals $f(a)$ at this point. 2. $1 - x + x^2 - x^3 + x^4 - \cdots$
3. The series diverges at both endpoint by the Divergence Test. 4. $2x + 2x^2 + x^3$; $1 - x + x^2/2$ 5. $6, 1/16$
6. $1.05, 1.04875$ ◄

11.4 Working with Taylor Series

We now know the Taylor series for many familiar functions, and we have tools for working with power series. The goal of this final section is to illustrate additional techniques associated with power series. As you will see, power series cover the entire landscape of calculus from limits and derivatives to integrals and approximation. We present five different topics that you can explore selectively.

Limits by Taylor Series

An important use of Taylor series is evaluating limits. Two examples illustrate the essential ideas.

EXAMPLE 1 A limit by Taylor series Evaluate $\displaystyle\lim_{x\to 0} \frac{x^2 + 2\cos x - 2}{3x^4}$.

> L'Hôpital's Rule may be impractical when it must be used more than once on the same limit or when derivatives are difficult to compute.

SOLUTION Because the limit has the indeterminate form $0/0$, l'Hôpital's Rule can be used, which requires four applications of the rule. Alternatively, because the limit involves values of x near 0, we substitute the Maclaurin series for $\cos x$. Recalling that
$$\cos x = 1 - \frac{x^2}{2} + \frac{x^4}{24} - \frac{x^6}{720} + \cdots, \quad \text{Table 11.5}$$

> In using series to evaluate limits, it is often not obvious how many terms of the Taylor series to use. When in doubt, include extra (higher-order) terms. The dots in the calculation stand for powers of x greater than the last power that appears.

QUICK CHECK 1 Use the Taylor series $\sin x = x - x^3/6 + \cdots$ to verify that $\lim_{x \to 0} (\sin x)/x = 1$. ◄

we have

$$\lim_{x \to 0} \frac{x^2 + 2 \cos x - 2}{3x^4} = \lim_{x \to 0} \frac{x^2 + 2\left(1 - \dfrac{x^2}{2} + \dfrac{x^4}{24} - \dfrac{x^6}{720} + \cdots\right) - 2}{3x^4} \quad \text{Substitute for } \cos x.$$

$$= \lim_{x \to 0} \frac{x^2 + \left(2 - x^2 + \dfrac{x^4}{12} - \dfrac{x^6}{360} + \cdots\right) - 2}{3x^4} \quad \text{Simplify.}$$

$$= \lim_{x \to 0} \frac{\dfrac{x^4}{12} - \dfrac{x^6}{360} + \cdots}{3x^4} \quad \text{Simplify.}$$

$$= \lim_{x \to 0} \left(\frac{1}{36} - \frac{x^2}{1080} + \cdots\right) = \frac{1}{36}. \quad \begin{array}{l}\text{Use Theorem} \\ \text{11.4, Property 2;} \\ \text{evaluate limit.}\end{array}$$

Related Exercise 13 ◄

EXAMPLE 2   A limit by Taylor series Evaluate

$$\lim_{x \to \infty} \left(6x^5 \sin \frac{1}{x} - 6x^4 + x^2\right).$$

SOLUTION A Taylor series may be centered at any finite point in the domain of the function, but we don't have the tools needed to expand a function about $x = \infty$. Using a technique introduced earlier, we replace x with $1/t$ and note that as $x \to \infty$, $t \to 0^+$. The new limit becomes

$$\lim_{x \to \infty} \left(6x^5 \sin \frac{1}{x} - 6x^4 + x^2\right) = \lim_{t \to 0^+} \left(\frac{6 \sin t}{t^5} - \frac{6}{t^4} + \frac{1}{t^2}\right) \quad \text{Replace } x \text{ with } 1/t.$$

$$= \lim_{t \to 0^+} \left(\frac{6 \sin t - 6t + t^3}{t^5}\right). \quad \text{Common denominator}$$

This limit has the indeterminate form $0/0$. We now expand $\sin t$ in a Taylor series centered at $t = 0$. Because

$$\sin t = t - \frac{t^3}{6} + \frac{t^5}{120} - \frac{t^7}{5040} + \cdots, \quad \text{Table 11.5}$$

the value of the original limit is

$$\lim_{t \to 0^+} \left(\frac{6 \sin t - 6t + t^3}{t^5}\right)$$

$$= \lim_{t \to 0^+} \left(\frac{6\left(t - \dfrac{t^3}{6} + \dfrac{t^5}{120} - \dfrac{t^7}{5040} + \cdots\right) - 6t + t^3}{t^5}\right) \quad \text{Substitute for } \sin t.$$

$$= \lim_{t \to 0^+} \left(\frac{\dfrac{t^5}{20} - \dfrac{t^7}{840} + \cdots}{t^5}\right) \quad \text{Simplify.}$$

$$= \lim_{t \to 0^+} \left(\frac{1}{20} - \frac{t^2}{840} + \cdots\right) = \frac{1}{20}. \quad \begin{array}{l}\text{Use Theorem 11.4,} \\ \text{Property 2; evaluate limit.}\end{array}$$

Related Exercise 19 ◄

Differentiating Power Series

The following examples illustrate ways in which term-by-term differentiation (Theorem 11.5) may be used.

EXAMPLE 3   Power series for derivatives Differentiate the Maclaurin series for $f(x) = \sin x$ to verify that $\dfrac{d}{dx}(\sin x) = \cos x$.

SOLUTION The Maclaurin series for $f(x) = \sin x$ is

$$\sin x = x - \frac{x^3}{3!} + \frac{x^5}{5!} - \frac{x^7}{7!} + \cdots,$$

and it converges for $-\infty < x < \infty$. By Theorem 11.5, the differentiated series also converges for $-\infty < x < \infty$ and it converges to $f'(x)$. Differentiating, we have

$$\frac{d}{dx}\left(x - \frac{x^3}{3!} + \frac{x^5}{5!} - \frac{x^7}{7!} + \cdots \right) = 1 - \frac{x^2}{2!} + \frac{x^4}{4!} - \frac{x^6}{6!} + \cdots = \cos x.$$

QUICK CHECK 2 Differentiate the power series for $\cos x$ (given in Example 3) and identify the result. ◄

The differentiated series is the Maclaurin series for $\cos x$, confirming that $f'(x) = \cos x$.

Related Exercises 25–26 ◄

EXAMPLE 4   **A differential equation** Find a power series solution of the differential equation $y'(t) = y + 2$, subject to the initial condition $y(0) = 6$. Identify the function represented by the power series.

SOLUTION Because the initial condition is given at $t = 0$, we assume the solution has a Taylor series centered at 0 of the form $y(t) = \sum_{k=0}^{\infty} c_k t^k$, where the coefficients c_k must be determined. Recall that the coefficients of the Taylor series are given by

$$c_k = \frac{y^{(k)}(0)}{k!}, \quad \text{for } k = 0, 1, 2, \dots.$$

If we can determine $y^{(k)}(0)$, for $k = 0, 1, 2, \dots$, the coefficients of the series are also determined.

Substituting the initial condition $t = 0$ and $y = 6$ into the power series

$$y(t) = c_0 + c_1 t + c_2 t^2 + \cdots,$$

we find that

$$6 = c_0 + c_1 (0) + c_2 (0)^2 + \cdots.$$

It follows that $c_0 = 6$. To determine $y'(0)$, we substitute $t = 0$ into the differential equation; the result is $y'(0) = y(0) + 2 = 6 + 2 = 8$. Therefore, $c_1 = y'(0)/1! = 8$.

The remaining derivatives are obtained by successively differentiating the differential equation and substituting $t = 0$. We find that $y''(0) = y'(0) = 8$, $y'''(0) = y''(0) = 8$, and in general, $y^{(k)}(0) = 8$, for $k = 1, 2, 3, 4, \dots$. Therefore,

$$c_k = \frac{y^{(k)}(0)}{k!} = \frac{8}{k!}, \quad \text{for } k = 1, 2, 3, \dots,$$

and the Taylor series for the solution is

$$y(t) = c_0 + c_1 t + c_2 t^2 + \cdots$$

$$= 6 + \frac{8}{1!} t + \frac{8}{2!} t^2 + \frac{8}{3!} t^3 + \cdots.$$

To identify the function represented by this series, we write

$$y(t) = \underbrace{-2 + 8}_{6} + \frac{8}{1!} t + \frac{8}{2!} t^2 + \frac{8}{3!} t^3 + \cdots$$

$$= -2 + 8 \underbrace{\left(1 + t + \frac{t^2}{2!} + \frac{t^3}{3!} + \cdots \right)}_{e^t}.$$

► You should check that $y(t) = -2 + 8e^t$ satisfies $y'(t) = y + 2$ and $y(0) = 6$.

The power series that appears is the Taylor series for e^t. Therefore, the solution is $y = -2 + 8e^t$.

Related Exercise 36 ◄

Integrating Power Series

The following example illustrates the use of power series in approximating integrals that cannot be evaluated by analytical methods.

EXAMPLE 5   Approximating a definite integral   Approximate the value of the integral $\int_0^1 e^{-x^2}\,dx$ with an error no greater than 5×10^{-4}.

SOLUTION   The antiderivative of e^{-x^2} cannot be expressed in terms of familiar functions. The strategy is to write the Maclaurin series for e^{-x^2} and integrate it term by term. Recall that integration of a power series is valid within its interval of convergence (Theorem 11.5). Beginning with the Maclaurin series

$$e^x = 1 + x + \frac{x^2}{2!} + \frac{x^3}{3!} + \cdots + \frac{x^n}{n!} + \cdots,$$

which converges for $-\infty < x < \infty$, we replace x with $-x^2$ to obtain

$$e^{-x^2} = 1 - x^2 + \frac{x^4}{2!} - \frac{x^6}{3!} + \cdots + \frac{(-1)^n x^{2n}}{n!} + \cdots,$$

which also converges for $-\infty < x < \infty$. By the Fundamental Theorem of Calculus,

$$\int_0^1 e^{-x^2}\,dx = \left(x - \frac{x^3}{3} + \frac{x^5}{5 \cdot 2!} - \frac{x^7}{7 \cdot 3!} + \cdots + \frac{(-1)^n x^{2n+1}}{(2n+1)n!} + \cdots \right)\Bigg|_0^1$$

$$= 1 - \frac{1}{3} + \frac{1}{5 \cdot 2!} - \frac{1}{7 \cdot 3!} + \cdots + \frac{(-1)^n}{(2n+1)n!} + \cdots.$$

▶ The integral in Example 5 is important in statistics and probability theory because of its relationship to the *normal distribution*.

Because the definite integral is expressed as an alternating series, the magnitude of the remainder in truncating the series after n terms is less than the magnitude of the first neglected term, which is $\left| \dfrac{(-1)^{n+1}}{(2n+3)(n+1)!} \right|$. By trial and error, we find that the magnitude of this term is less than 5×10^{-4} if $n \geq 5$ (with $n = 5$, we have $\dfrac{1}{13 \cdot 6!} \approx 1.07 \times 10^{-4}$). The sum of the terms of the series up to $n = 5$ gives the approximation

$$\int_0^1 e^{-x^2}\,dx \approx 1 - \frac{1}{3} + \frac{1}{5 \cdot 2!} - \frac{1}{7 \cdot 3!} + \frac{1}{9 \cdot 4!} - \frac{1}{11 \cdot 5!} \approx 0.747.$$

Related Exercises 37–38 ◀

Representing Real Numbers

When values of x are substituted into a convergent power series, the result may be a series representation of a familiar real number. The following example illustrates some techniques.

EXAMPLE 6   Evaluating infinite series

a. Use the Maclaurin series for $f(x) = \tan^{-1} x$ to evaluate

$$1 - \frac{1}{3} + \frac{1}{5} - \cdots = \sum_{k=0}^{\infty} \frac{(-1)^k}{2k+1}.$$

b. Let $f(x) = (e^x - 1)/x$, for $x \neq 0$, and $f(0) = 1$. Use the Maclaurin series for f to evaluate $f'(1)$ and $\displaystyle\sum_{k=1}^{\infty} \frac{k}{(k+1)!}$.

SOLUTION

a. From Table 11.5, we see that for $|x| \leq 1$,

$$\tan^{-1} x = x - \frac{x^3}{3} + \frac{x^5}{5} - \cdots + \frac{(-1)^k x^{2k+1}}{2k+1} + \cdots = \sum_{k=0}^{\infty} \frac{(-1)^k x^{2k+1}}{2k+1}.$$

> The series in Example 6a (known as the *Gregory series*) is one of a multitude of series representations of π. Because this series converges slowly, it does not provide an efficient way to approximate π.

Substituting $x = 1$, we have

$$\tan^{-1} 1 = 1 - \frac{1^3}{3} + \frac{1^5}{5} - \cdots = \sum_{k=0}^{\infty} \frac{(-1)^k}{2k+1}.$$

Because $\tan^{-1} 1 = \pi/4$, the value of the series is $\pi/4$.

b. Using the Maclaurin series for e^x, the series for $f(x) = (e^x - 1)/x$ is

$$f(x) = \frac{e^x - 1}{x} = \frac{1}{x}\left(\underbrace{\left(1 + x + \frac{x^2}{2!} + \frac{x^3}{3!} + \cdots\right)}_{e^x} - 1\right) \quad \text{Substitute series for } e^x.$$

$$= 1 + \frac{x}{2!} + \frac{x^2}{3!} + \frac{x^3}{4!} + \cdots = \sum_{k=1}^{\infty} \frac{x^{k-1}}{k!}, \qquad \text{Theorem 11.4, Property 2}$$

which converges for $-\infty < x < \infty$. By the Quotient Rule,

$$f'(x) = \frac{xe^x - (e^x - 1)}{x^2}.$$

Differentiating the series for f term by term (Theorem 11.5), we find that

$$f'(x) = \frac{d}{dx}\left(1 + \frac{x}{2!} + \frac{x^2}{3!} + \frac{x^3}{4!} + \cdots\right)$$

$$= \frac{1}{2!} + \frac{2x}{3!} + \frac{3x^2}{4!} + \cdots = \sum_{k=1}^{\infty} \frac{kx^{k-1}}{(k+1)!}.$$

QUICK CHECK 3 What value of x would you substitute into the Maclaurin series for $\tan^{-1} x$ to obtain a series representation for $\pi/6$? ◄

We now have two expressions for f'; they are evaluated at $x = 1$ to show that

$$f'(1) = 1 = \sum_{k=1}^{\infty} \frac{k}{(k+1)!}.$$

Related Exercises 50–51 ◄

Representing Functions as Power Series

Power series have a fundamental role in mathematics in defining functions and providing alternative representations of familiar functions. As an overall review, we close this chapter with two examples that use many techniques for working with power series.

EXAMPLE 7   Identify the series Identify the function represented by the power series $\sum_{k=0}^{\infty} \frac{(1-2x)^k}{k!}$ and give its interval of convergence.

SOLUTION The Maclaurin series for the exponential function,

$$e^x = \sum_{k=0}^{\infty} \frac{x^k}{k!},$$

converges for $-\infty < x < \infty$. Replacing x with $1 - 2x$ produces the given series:

$$\sum_{k=0}^{\infty} \frac{(1-2x)^k}{k!} = e^{1-2x}.$$

This replacement is allowed because $1 - 2x$ is within the interval of convergence of the series for e^x; that is, $-\infty < 1 - 2x < \infty$, for all x. Therefore, the given series represents e^{1-2x}, for $-\infty < x < \infty$.

Related Exercises 55–56 ◄

EXAMPLE 8   Mystery series The power series $\sum_{k=1}^{\infty} \frac{(-1)^k k}{4^k} x^{2k}$ appeared in the opening of Section 11.2. Determine the interval of convergence of the power series and find the function it represents on this interval.

SOLUTION Applying the Ratio Test to the series, we determine that it converges when $|x^2/4| < 1$, which implies that $|x| < 2$. A quick check of the endpoints of the original series confirms that it diverges at $x = \pm 2$. Therefore, the interval of convergence is $|x| < 2$.

To find the function represented by the series, we apply several maneuvers until we obtain a geometric series. First note that

$$\sum_{k=1}^{\infty} \frac{(-1)^k k}{4^k} x^{2k} = \sum_{k=1}^{\infty} k\left(-\frac{1}{4}\right)^k x^{2k}.$$

The series on the right is not a geometric series because of the presence of the factor k. The key is to realize that k could appear in this way through differentiation; specifically, something like $\dfrac{d}{dx}(x^{2k}) = 2kx^{2k-1}$. To achieve terms of this form, we write

$$\underbrace{\sum_{k=1}^{\infty} \frac{(-1)^k k}{4^k} x^{2k}}_{\text{original series}} = \sum_{k=1}^{\infty} k\left(-\frac{1}{4}\right)^k x^{2k}$$

$$= \frac{1}{2} \sum_{k=1}^{\infty} 2k\left(-\frac{1}{4}\right)^k x^{2k} \qquad \text{Multiply and divide by 2.}$$

$$= \frac{x}{2} \sum_{k=1}^{\infty} 2k\left(-\frac{1}{4}\right)^k x^{2k-1}. \qquad \text{Remove } x \text{ from the series.}$$

Now we identify the last series as the derivative of another series:

$$\underbrace{\sum_{k=1}^{\infty} \frac{(-1)^k k}{4^k} x^{2k}}_{\text{original series}} = \frac{x}{2} \sum_{k=1}^{\infty} \left(-\frac{1}{4}\right)^k \underbrace{2kx^{2k-1}}_{\frac{d}{dx}(x^{2k})}$$

$$= \frac{x}{2} \sum_{k=1}^{\infty} \left(-\frac{1}{4}\right)^k \frac{d}{dx}(x^{2k}) \qquad \text{Identify a derivative.}$$

$$= \frac{x}{2} \frac{d}{dx}\left(\sum_{k=1}^{\infty} \left(-\frac{x^2}{4}\right)^k\right). \qquad \text{Combine factors; differentiate term by term.}$$

This last series is a geometric series with a ratio $r = -x^2/4$ and first term $-x^2/4$; therefore, its value is $\dfrac{-x^2/4}{1 + (x^2/4)}$, provided $\left|\dfrac{x^2}{4}\right| < 1$, or $|x| < 2$. We now have

$$\underbrace{\sum_{k=1}^{\infty} \frac{(-1)^k k}{4^k} x^{2k}}_{\text{original series}} = \frac{x}{2} \frac{d}{dx}\left(\sum_{k=1}^{\infty} \left(-\frac{x^2}{4}\right)^k\right)$$

$$= \frac{x}{2} \frac{d}{dx}\left(\frac{-x^2/4}{1 + (x^2/4)}\right) \qquad \text{Sum of geometric series.}$$

$$= \frac{x}{2} \frac{d}{dx}\left(\frac{-x^2}{4 + x^2}\right) \qquad \text{Simplify.}$$

$$= -\frac{4x^2}{(4 + x^2)^2}. \qquad \text{Differentiate and simplify.}$$

Therefore, the function represented by the power series on $(-2, 2)$ has been uncovered; it is

$$f(x) = -\frac{4x^2}{(4 + x^2)^2}.$$

Notice that f is defined for $-\infty < x < \infty$ (**Figure 11.22**), but its power series centered at 0 converges to f only on $(-2, 2)$.

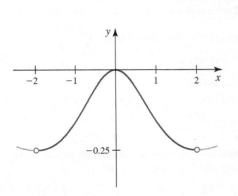

$$\sum_{k=1}^{\infty} \frac{(-1)^k k}{4^k} x^{2k} = -\frac{4x^2}{(4 + x^2)^2} \text{ on } (-2, 2)$$

Figure 11.22

Related Exercises 61, 63 ◄

SECTION 11.4 EXERCISES

Getting Started

1. Explain the strategy presented in this section for evaluating a limit of the form $\lim\limits_{x \to a} \dfrac{f(x)}{g(x)}$, where f and g have Taylor series centered at a.

2. Explain the method presented in this section for approximating $\int_a^b f(x)\, dx$, where f has a Taylor series with an interval of convergence centered at a that includes b.

3. How would you approximate $e^{-0.6}$ using the Taylor series for e^x?

4. Use the Taylor series for $\cos x$ centered at 0 to verify that
$$\lim_{x \to 0} \frac{1 - \cos x}{x} = 0.$$

5. Use the Taylor series for $\sinh x$ and $\cosh x$ to verify that
$$\frac{d}{dx}\sinh x = \cosh x.$$

6. What condition must be met by a function f for it to have a Taylor series centered at a?

Practice Exercises

7–24. Limits *Evaluate the following limits using Taylor series.*

7. $\lim\limits_{x \to 0} \dfrac{e^x - 1}{x}$

8. $\lim\limits_{x \to 0} \dfrac{\tan^{-1}x - x}{x^3}$

9. $\lim\limits_{x \to 0} \dfrac{-x - \ln(1-x)}{x^2}$

10. $\lim\limits_{x \to 0} \dfrac{\sin 2x}{x}$

11. $\lim\limits_{x \to 0} \dfrac{e^x - e^{-x}}{x}$

12. $\lim\limits_{x \to 0} \dfrac{1 + x - e^x}{4x^2}$

13. $\lim\limits_{x \to 0} \dfrac{2\cos 2x - 2 + 4x^2}{2x^4}$

14. $\lim\limits_{x \to \infty} x \sin \dfrac{1}{x}$

15. $\lim\limits_{x \to 0} \dfrac{\ln(1+x) - x + x^2/2}{x^3}$

16. $\lim\limits_{x \to 4} \dfrac{x^2 - 16}{\ln(x-3)}$

17. $\lim\limits_{x \to 0} \dfrac{3\tan^{-1}x - 3x + x^3}{x^5}$

18. $\lim\limits_{x \to 0} \dfrac{\sqrt{1+x} - 1 - (x/2)}{4x^2}$

19. $\lim\limits_{x \to 0} \dfrac{12x - 8x^3 - 6\sin 2x}{x^5}$

20. $\lim\limits_{x \to 1} \dfrac{x - 1}{\ln x}$

21. $\lim\limits_{x \to 2} \dfrac{x - 2}{\ln(x-1)}$

22. $\lim\limits_{x \to \infty} x(e^{1/x} - 1)$

23. $\lim\limits_{x \to 0} \dfrac{e^{-2x} - 4e^{-x/2} + 3}{2x^2}$

24. $\lim\limits_{x \to 0} \dfrac{(1-2x)^{-1/2} - e^x}{8x^2}$

25–32. Power series for derivatives

a. *Differentiate the Taylor series centered at 0 for the following functions.*
b. *Identify the function represented by the differentiated series.*
c. *Give the interval of convergence of the power series for the derivative.*

25. $f(x) = e^x$

26. $f(x) = -\cos x$

27. $f(x) = \ln(1 + x)$

28. $f(x) = \sin x^2$

29. $f(x) = e^{-2x}$

30. $f(x) = (1 - x)^{-1}$

31. $f(x) = \tan^{-1}x$

32. $f(x) = -\ln(1 - x)$

33–36. Differential equations

a. *Find a power series for the solution of the following differential equations, subject to the given initial condition.*
b. *Identify the function represented by the power series.*

33. $y'(t) - y = 0,\ y(0) = 2$

34. $y'(t) + 4y = 8,\ y(0) = 0$

35. $y'(t) - 3y = 10,\ y(0) = 2$

36. $y'(t) = 6y + 9,\ y(0) = 2$

T 37–44. Approximating definite integrals *Use a Taylor series to approximate the following definite integrals. Retain as many terms as needed to ensure the error is less than 10^{-4}.*

37. $\int_0^{0.25} e^{-x^2} dx$

38. $\int_0^{0.2} \sin x^2\, dx$

39. $\int_{-0.35}^{0.35} \cos 2x^2\, dx$

40. $\int_0^{0.2} \sqrt{1 + x^4}\, dx$

41. $\int_0^{0.35} \tan^{-1}x\, dx$

42. $\int_0^{0.4} \ln(1 + x^2)\, dx$

43. $\int_0^{0.5} \dfrac{dx}{\sqrt{1 + x^6}}$

44. $\int_0^{0.2} \dfrac{\ln(1+t)}{t}\, dt$

45–50. Approximating real numbers *Use an appropriate Taylor series to find the first four nonzero terms of an infinite series that is equal to the following numbers.*

45. e^2

46. $\sqrt{e}$

47. $\cos 2$

48. $\sin 1$

49. $\ln \dfrac{3}{2}$

50. $\tan^{-1} \dfrac{1}{2}$

51. **Evaluating an infinite series** Let $f(x) = \dfrac{e^x - 1}{x}$, for $x \neq 0$, and $f(0) = 1$. Use the Taylor series for f centered at 0 to evaluate $f(1)$ and to find the value of $\sum\limits_{k=0}^{\infty} \dfrac{1}{(k+1)!}$.

52. **Evaluating an infinite series** Let $f(x) = \dfrac{e^x - 1}{x}$, for $x \neq 0$, and $f(0) = 1$. Use the Taylor series for f and f' centered at 0 to evaluate $f'(2)$ and to find the value of $\sum\limits_{k=1}^{\infty} \dfrac{k2^{k-1}}{(k+1)!}$.

53. **Evaluating an infinite series** Write the Taylor series for $f(x) = \ln(1 + x)$ centered at 0 and find its interval of convergence. Assume the Taylor series converges to f on the interval of convergence. Evaluate $f(1)$ to find the value of $\sum\limits_{k=1}^{\infty} \dfrac{(-1)^{k+1}}{k}$ (the alternating harmonic series).

54. **Evaluating an infinite series** Write the Maclaurin series for $f(x) = \ln(1 + x)$ and find the interval of convergence. Evaluate $f\left(-\dfrac{1}{2}\right)$ to find the value of $\sum\limits_{k=1}^{\infty} \dfrac{1}{k2^k}$.

55–64. Representing functions by power series *Identify the functions represented by the following power series.*

55. $\sum\limits_{k=0}^{\infty} \dfrac{x^k}{2^k}$

56. $\sum\limits_{k=0}^{\infty} (-1)^k \dfrac{x^k}{3^k}$

57. $\sum\limits_{k=0}^{\infty} (-1)^k \dfrac{x^{2k}}{4^k}$

58. $\sum\limits_{k=0}^{\infty} 2^k x^{2k+1}$

59. $\sum\limits_{k=1}^{\infty} \dfrac{x^k}{k}$

60. $\sum\limits_{k=0}^{\infty} \dfrac{(-1)^k x^{k+1}}{4^k}$

61. $\displaystyle\sum_{k=1}^{\infty} (-1)^k \frac{kx^{k+1}}{3^k}$

62. $\displaystyle\sum_{k=1}^{\infty} \frac{x^{2k}}{k}$

63. $\displaystyle\sum_{k=2}^{\infty} \frac{k(k-1)x^k}{3^k}$

64. $\displaystyle\sum_{k=2}^{\infty} \frac{x^k}{k(k-1)}$

65. Explain why or why not Determine whether the following statements are true and give an explanation or counterexample.

  a. To evaluate $\displaystyle\int_0^2 \frac{dx}{1-x}$, one could expand the integrand in a Taylor series and integrate term by term.

  b. To approximate $\pi/3$, one could substitute $x = \sqrt{3}$ into the Taylor series for $\tan^{-1} x$.

  c. $\displaystyle\sum_{k=0}^{\infty} \frac{(\ln 2)^k}{k!} = 2.$

66–68. Limits with a parameter *Use Taylor series to evaluate the following limits. Express the result in terms of the nonzero real parameter(s).*

66. $\displaystyle\lim_{x\to 0} \frac{e^{ax}-1}{x}$

67. $\displaystyle\lim_{x\to 0} \frac{\sin ax}{\sin bx}$

68. $\displaystyle\lim_{x\to 0} \frac{\sin ax - \tan^{-1} ax}{bx^3}$

Explorations and Challenges

69. A limit by Taylor series Use Taylor series to evaluate

$$\lim_{x\to 0} \left(\frac{\sin x}{x} \right)^{1/x^2}.$$

70. Inverse hyperbolic sine The *inverse of hyperbolic sine* is defined in several ways; among them are

$$\sinh^{-1} x = \ln\left(x + \sqrt{x^2 + 1}\right) = \int_0^x \frac{dt}{\sqrt{1+t^2}}.$$

Find the first four terms of the Taylor series for $\sinh^{-1} x$ using these two definitions (and be sure they agree).

71–74. Derivative trick *Here is an alternative way to evaluate higher derivatives of a function f that may save time. Suppose you can find the Taylor series for f centered at the point a without evaluating derivatives (for example, from a known series). Then $f^{(k)}(a) = k!$ multiplied by the coefficient of $(x - a)^k$. Use this idea to evaluate $f^{(3)}(0)$ and $f^{(4)}(0)$ for the following functions. Use known series and do not evaluate derivatives.*

71. $f(x) = e^{\cos x}$

72. $f(x) = \dfrac{x^2 + 1}{\sqrt[3]{1+x}}$

73. $f(x) = \displaystyle\int_0^x \sin t^2 \, dt$

74. $f(x) = \displaystyle\int_0^x \frac{1}{1+t^4} \, dt$

75. Probability: tossing for a head The expected (average) number of tosses of a fair coin required to obtain the first head is $\displaystyle\sum_{k=1}^{\infty} k\left(\frac{1}{2}\right)^k$. Evaluate this series and determine the expected number of tosses. (*Hint:* Differentiate a geometric series.)

76. Probability: sudden-death playoff Teams A and B go into sudden-death overtime after playing to a tie. The teams alternate possession of the ball, and the first team to score wins. Assume each team has a $1/6$ chance of scoring when it has the ball, and Team A has the ball first.

  a. The probability that Team A ultimately wins is $\displaystyle\sum_{k=0}^{\infty} \frac{1}{6}\left(\frac{5}{6}\right)^{2k}$. Evaluate this series.

  b. The expected number of rounds (possessions by either team) required for the overtime to end is $\dfrac{1}{6}\displaystyle\sum_{k=1}^{\infty} k\left(\frac{5}{6}\right)^{k-1}$. Evaluate this series.

T 77. Elliptic integrals The period of an undamped pendulum is given by

$$T = 4\sqrt{\frac{\ell}{g}} \int_0^{\pi/2} \frac{d\theta}{\sqrt{1 - k^2 \sin^2 \theta}} = 4\sqrt{\frac{\ell}{g}} F(k),$$

where ℓ is the length of the pendulum, $g = 9.8 \text{ m/s}^2$ is the acceleration due to gravity, $k = \sin\dfrac{\theta_0}{2}$, and θ_0 is the initial angular displacement of the pendulum (in radians). The integral in this formula $F(k)$ is called an *elliptic integral*, and it cannot be evaluated analytically. Approximate $F(0.1)$ by expanding the integrand in a Taylor (binomial) series and integrating term by term.

78. Sine integral function The function $\text{Si}(x) = \displaystyle\int_0^x f(t)\,dt$, where

$$f(t) = \begin{cases} \frac{\sin t}{t} & \text{if } t \neq 0 \\ 1 & \text{if } t = 0 \end{cases}, \text{ is called the } \textit{sine integral function}.$$

  a. Expand the integrand in a Taylor series centered at 0.
  b. Integrate the series to find a Taylor series for Si.
  c. Approximate $\text{Si}(0.5)$ and $\text{Si}(1)$. Use enough terms of the series so the error in the approximation does not exceed 10^{-3}.

T 79. Fresnel integrals The theory of optics gives rise to the two *Fresnel integrals*

$$S(x) = \int_0^x \sin t^2 \, dt \quad \text{and} \quad C(x) = \int_0^x \cos t^2 \, dt.$$

  a. Compute $S'(x)$ and $C'(x)$.
  b. Expand $\sin t^2$ and $\cos t^2$ in a Maclaurin series, and then integrate to find the first four nonzero terms of the Maclaurin series for S and C.
  c. Use the polynomials in part (b) to approximate $S(0.05)$ and $C(-0.25)$.
  d. How many terms of the Maclaurin series are required to approximate $S(0.05)$ with an error no greater than 10^{-4}?
  e. How many terms of the Maclaurin series are required to approximate $C(-0.25)$ with an error no greater than 10^{-6}?

T 80. Error function An essential function in statistics and the study of the normal distribution is the *error function*

$$\text{erf}(x) = \frac{2}{\sqrt{\pi}} \int_0^x e^{-t^2} \, dt.$$

  a. Compute the derivative of $\text{erf}(x)$.
  b. Expand e^{-t^2} in a Maclaurin series; then integrate to find the first four nonzero terms of the Maclaurin series for erf.
  c. Use the polynomial in part (b) to approximate $\text{erf}(0.15)$ and $\text{erf}(-0.09)$.
  d. Estimate the error in the approximations of part (c).

T 81. Bessel functions Bessel functions arise in the study of wave propagation in circular geometries (for example, waves on a circular drum head). They are conveniently defined as power series. One of an infinite family of Bessel functions is

$$J_0(x) = \sum_{k=0}^{\infty} \frac{(-1)^k}{2^{2k}(k!)^2} x^{2k}.$$

  a. Write out the first four terms of J_0.
  b. Find the radius and interval of convergence of the power series for J_0.
  c. Differentiate J_0 twice and show (by keeping terms through x^6) that J_0 satisfies the equation $x^2 y''(x) + xy'(x) + x^2 y(x) = 0$.

T 82. Newton's derivation of the sine and arcsine series Newton discovered the binomial series and then used it ingeniously to obtain many more results. Here is a case in point.

a. Referring to the figure, show that $x = \sin s$ or $s = \sin^{-1} x$.

b. The area of a circular sector of radius r subtended by an angle θ is $1/2\, r^2 \theta$. Show that the area of the circular sector APE is $s/2$, which implies that

$$s = 2 \int_0^x \sqrt{1 - t^2}\, dt - x\sqrt{1 - x^2}.$$

c. Use the binomial series for $f(x) = \sqrt{1 - x^2}$ to obtain the first few terms of the Taylor series for $s = \sin^{-1} x$.

d. Newton next inverted the series in part (c) to obtain the Taylor series for $x = \sin s$. He did this by assuming $\sin s = \sum a_k s^k$ and solving $x = \sin(\sin^{-1} x)$ for the coefficients a_k. Find the first few terms of the Taylor series for $\sin s$ using this idea (a computer algebra system might be helpful as well).

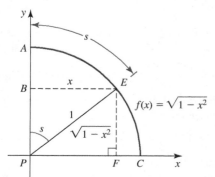

83. L'Hôpital's Rule by Taylor series Suppose f and g have Taylor series about the point a.

a. If $f(a) = g(a) = 0$ and $g'(a) \neq 0$, evaluate $\lim\limits_{x \to a} \dfrac{f(x)}{g(x)}$ by expanding f and g in their Taylor series. Show that the result is consistent with l'Hôpital's Rule.

b. If $f(a) = g(a) = f'(a) = g'(a) = 0$ and $g''(a) \neq 0$, evaluate $\lim\limits_{x \to a} \dfrac{f(x)}{g(x)}$ by expanding f and g in their Taylor series. Show that the result is consistent with two applications of l'Hôpital's Rule.

84. Symmetry

a. Use infinite series to show that $\cos x$ is an even function. That is, show $\cos(-x) = \cos x$.

b. Use infinite series to show that $\sin x$ is an odd function. That is, show $\sin(-x) = -\sin x$.

QUICK CHECK ANSWERS

1. $\dfrac{\sin x}{x} = \dfrac{x - x^3/3! + \cdots}{x} = 1 - \dfrac{x^2}{3!} + \cdots \to 1$ as $x \to 0$

2. The result is the power series for $-\sin x$.   **3.** $x = 1/\sqrt{3}$ (which lies in the interval of convergence) ◄

CHAPTER 11 REVIEW EXERCISES

1. Explain why or why not Determine whether the following statements are true and give an explanation or counterexample.

a. Let p_n be the nth-order Taylor polynomial for f centered at 2. The approximation $p_3(2.1) \approx f(2.1)$ is likely to be more accurate than the approximation $p_2(2.2) \approx f(2.2)$.

b. If the Taylor series for f centered at 3 has a radius of convergence of 6, then the interval of convergence is $[-3, 9]$.

c. The interval of convergence of the power series $\sum c_k x^k$ could be $(-7/3, 7/3)$.

d. The Maclaurin series for $f(x) = (1 + x)^{12}$ has a finite number of nonzero terms.

e. If the power series $\sum c_k(x - 3)^k$ has a radius of convergence of $R = 4$ and converges at the endpoints of its interval of convergence, then its interval of convergence is $[-1, 7]$.

2–9. Taylor polynomials Find the nth-order Taylor polynomial for the following functions centered at the given point a.

2. $f(x) = \sin 2x, \ n = 3, \ a = 0$

3. $f(x) = \cos^3 x, \ n = 2, \ a = 0$

4. $f(x) = \cos^{-1} x, \ n = 2, \ a = \dfrac{1}{2}$

5. $f(x) = e^{1/x - 1}, \ n = 2, \ a = 1$

6. $f(x) = e^{\sin x}, \ n = 2, \ a = 0$

7. $f(x) = \cos(\ln x), \ n = 2, \ a = 1$

8. $f(x) = \sinh(-3x), \ n = 3, \ a = 0$

9. $f(x) = \cosh x, \ n = 3, \ a = \ln 2$

T 10–13. Approximations

a. Find the Taylor polynomials of order $n = 1$ and $n = 2$ for the given functions centered at the given point a.

b. Use the Taylor polynomials to approximate the given expression. Make a table showing the approximations and the absolute error in these approximations using a calculator for the exact function value.

10. $f(x) = \cos x, \ a = 0;\ \cos(-0.08)$

11. $f(x) = e^x, \ a = 0;\ e^{-0.08}$

12. $f(x) = \sqrt{1 + x}, \ a = 0;\ \sqrt{1.08}$

13. $f(x) = \sin x, \ a = \dfrac{\pi}{4};\ \sin\dfrac{\pi}{5}$

14–16. Estimating remainders *Find the remainder term* $R_n(x)$ *for the Taylor series centered at 0 for the following functions. Find an upper bound for the magnitude of the remainder on the given interval for the given value of n. (The bound is not unique.)*

14. $f(x) = e^x$; bound $R_3(x)$, for $|x| < 1$

15. $f(x) = \sin x$; bound $R_3(x)$, for $|x| < \pi$

16. $f(x) = \ln(1 - x)$; bound $R_3(x)$, for $|x| < \dfrac{1}{2}$

17–26. Radius and interval of convergence *Use the Ratio Test or the Root Test to determine the radius of convergence of the following power series. Test the endpoints to determine the interval of convergence, when appropriate.*

17. $\displaystyle\sum_{k=1}^{\infty} \frac{k^2 x^k}{k!}$

18. $\displaystyle\sum_{k=1}^{\infty} \frac{x^{4k}}{k^2}$

19. $\displaystyle\sum_{k=1}^{\infty} \frac{(-1)^k (x + 1)^{2k}}{k!}$

20. $\displaystyle\sum_{k=1}^{\infty} \frac{(x - 1)^k}{k 5^k}$

21. $\displaystyle\sum_{k=0}^{\infty} \left(\frac{x}{9}\right)^{3k}$

22. $\displaystyle\sum_{k=1}^{\infty} \frac{(x + 2)^k}{\sqrt{k}}$

23. $\displaystyle\sum_{k=1}^{\infty} \frac{(x + 2)^k}{2^k \ln k}$

24. $\displaystyle\sum_{k=1}^{\infty} \frac{(2k)!(x - 2)^k}{k!}$

25. $\displaystyle\sum_{k=1}^{\infty} \frac{(-1)^k (2x + 1)^k}{k^2 3^k}$

26. $x + \dfrac{x^3}{3} + \dfrac{x^5}{5} + \dfrac{x^7}{7} + \cdots$

27–28. Radius of convergence *Find the radius of convergence of each series.*

27. $\displaystyle\sum_{k=1}^{\infty} \frac{(3k)! x^k}{(k!)^3}$.

28. $\displaystyle\sum_{k=1}^{\infty} \frac{2^k k! (x - 5)^k}{k^k}$

29–34. Power series from the geometric series *Use the geometric series* $\displaystyle\sum_{k=0}^{\infty} x^k = \frac{1}{1 - x}$, *for* $|x| < 1$, *to determine the Maclaurin series and the interval of convergence for the following functions.*

29. $f(x) = \dfrac{1}{1 - x^2}$

30. $f(x) = \dfrac{1}{1 + x^3}$

31. $f(x) = \dfrac{1}{1 + 5x}$

32. $f(x) = \dfrac{10x}{1 + x}$

33. $f(x) = \dfrac{1}{(1 - 10x)^2}$

34. $f(x) = \ln(1 - 4x)$

35–42. Taylor series *Write out the first three nonzero terms of the Taylor series for the following functions centered at the given point a. Then write the series using summation notation.*

35. $f(x) = e^{3x}$, $a = 0$

36. $f(x) = \dfrac{1}{2x + 1}$, $a = 1$

37. $f(x) = \cos x$, $a = \dfrac{\pi}{2}$

38. $f(x) = \dfrac{x^2}{1 + x}$, $a = 0$

39. $f(x) = \tan^{-1} 4x$, $a = 0$

40. $f(x) = \sin 2x$, $a = -\dfrac{\pi}{2}$

41. $f(x) = \cosh(2x - 2)$, $a = 1$

42. $f(x) = \dfrac{1}{4 + x^2}$, $a = 0$

43–46. Binomial series *Write out the first three terms of the Maclaurin series for the following functions.*

43. $f(x) = (1 + x)^{1/3}$

44. $f(x) = (1 + x)^{-1/2}$

45. $f(x) = \left(1 + \dfrac{x}{2}\right)^{-3}$

46. $f(x) = (1 + 2x)^{-5}$

47–48. Convergence *Write the remainder term* $R_n(x)$ *for the Taylor series for the following functions centered at the given point a. Then show that* $\displaystyle\lim_{n \to \infty} |R_n(x)| = 0$, *for all x in the given interval.*

47. $f(x) = \sinh x + \cosh x$, $a = 0$, $-\infty < x < \infty$

48. $f(x) = \ln(1 + x)$, $a = 0$, $-\dfrac{1}{2} \le x \le \dfrac{1}{2}$

49–54. Limits by power series *Use Taylor series to evaluate the following limits.*

49. $\displaystyle\lim_{x \to 0} \frac{x^2/2 - 1 + \cos x}{x^4}$

50. $\displaystyle\lim_{x \to 0} \frac{2 \sin x - \tan^{-1} x - x}{2x^5}$

51. $\displaystyle\lim_{x \to 4} \frac{\ln(x - 3)}{x^2 - 16}$

52. $\displaystyle\lim_{x \to 0} \frac{\sqrt{1 + 2x} - 1 - x}{x^2}$

53. $\displaystyle\lim_{x \to 0} \frac{\sec x - \cos x - x^2}{x^4}$ *(Hint: The Maclaurin series for $\sec x$ is*

$1 + \dfrac{x^2}{2} + \dfrac{5x^4}{24} + \dfrac{61x^6}{720} + \cdots.)$

54. $\displaystyle\lim_{x \to 0} \frac{(1 + x)^{-2} - \sqrt[3]{1 - 6x}}{2x^2}$

T 55–58. Definite integrals by power series *Use a Taylor series to approximate the following definite integrals. Retain as many terms as necessary to ensure the error is less than* 10^{-3}.

55. $\displaystyle\int_0^{1/2} e^{-x^2} dx$

56. $\displaystyle\int_0^{1/2} \tan^{-1} x \, dx$

57. $\displaystyle\int_0^1 x \cos x \, dx$

58. $\displaystyle\int_0^{1/2} x^2 \tan^{-1} x \, dx$

T 59–62. Approximating real numbers *Use an appropriate Taylor series to find the first four nonzero terms of an infinite series that is equal to the following numbers. There is more than one way to choose the center of the series.*

59. $\sqrt{119}$

60. $\sin 20°$

61. $\tan^{-1}\left(-\dfrac{1}{3}\right)$

62. $\sinh(-1)$

63. A differential equation Find a power series solution of the differential equation $y'(x) - 4y + 12 = 0$, subject to the condition $y(0) = 4$. Identify the solution in terms of known functions.

T 64. Rejected quarters The probability that a random quarter is *not* rejected by a vending machine is given by the integral $11.4 \int_0^{0.14} e^{-102x^2} dx$ (assuming the weights of quarters are normally distributed with a mean of 5.670 g and a standard deviation of 0.07 g). Estimate the value of the integral by using the first two terms of the Maclaurin series for e^{-102x^2}.

T 65. Approximating ln 2 Consider the following three ways to approximate ln 2.

   a. Use the Taylor series for $\ln(1 + x)$ centered at 0 and evaluate it at $x = 1$ (convergence was asserted in Table 11.5). Write the resulting infinite series.

   b. Use the Taylor series for $\ln(1 - x)$ centered at 0 and the identity $\ln 2 = -\ln\dfrac{1}{2}$. Write the resulting infinite series.

   c. Use the property $\ln\dfrac{a}{b} = \ln a - \ln b$ and the series of parts (a) and (b) to find the Taylor series for $f(x) = \ln\left(\dfrac{1 + x}{1 - x}\right)$ centered at 0.

   d. At what value of x should the series in part (c) be evaluated to approximate ln 2? Write the resulting infinite series for ln 2.

   e. Using four terms of the series, which of the three series derived in parts (a)–(d) gives the best approximation to ln 2? Can you explain why?

T 66. Graphing Taylor polynomials Consider the function $f(x) = (1 + x)^{-4}$.

   a. Find the Taylor polynomials p_0, p_1, p_2, and p_3 centered at 0.

   b. Use a graphing utility to plot the Taylor polynomials and f, for $-1 < x < 1$.

   c. For each Taylor polynomial, give the interval on which its graph appears indistinguishable from the graph of f.

Chapter 11 Guided Projects

Applications of the material in this chapter and related topics can be found in the following Guided Projects. For additional information, see the Preface.

- Series approximations to π
- Euler's formula (Taylor series with complex numbers)
- Stirling's formula and $n!$
- Three-sigma quality control
- Fourier series

12

Parametric and Polar Curves

Chapter Preview   Up to this point, we have focused our attention on functions of the form $y = f(x)$ and studied their behavior in the setting of the Cartesian coordinate system. There are, however, alternative ways to generate curves and represent functions. We begin by introducing parametric equations, which are featured prominently in upcoming chapters, to represent curves and trajectories in three-dimensional space. When working with objects that have circular, cylindrical, or spherical shapes, other coordinate systems are often advantageous. In this chapter, we introduce the polar coordinate system for circular geometries. Cylindrical and spherical coordinate systems appear in Chapter 16. After working with parametric equations and polar coordinates, the next step is to investigate calculus in these settings. How do we find slopes of tangent lines, rates of change, areas of regions, and arc length when the curves of interest are described with parametric equations or described in polar coordinates? The chapter ends with the related topic of *conic sections*. Ellipses, parabolas, and hyperbolas (all of which are conic sections) can be represented in both Cartesian and polar coordinates. These important families of curves have many fascinating properties and they appear throughout the remainder of the text.

12.1 Parametric Equations

So far, we have used functions of the form $y = F(x)$ to describe curves in the xy-plane. In this section, we look at another way to define curves, known as *parametric equations*. As you will see, parametric curves enable us to describe both common and exotic curves; they are also indispensable for modeling the trajectories of moving objects.

Basic Ideas

A motor boat travels counterclockwise around a circular course with a radius of 4 miles, completing one lap every 2π hours at a constant speed. Suppose we wish to describe the points on the path of the boat $(x(t), y(t))$ at any time $t \geq 0$, where t is measured in hours. We assume the boat starts on the positive x-axis at the point $(4, 0)$ (**Figure 12.1**). Note that the angle θ corresponding to the position of the boat increases by 2π radians every 2π hours beginning with $\theta = 0$ when $t = 0$; therefore, $\theta = t$, for $t \geq 0$. It follows that the x- and y-coordinates of the boat are

$$x = 4 \cos \theta = 4 \cos t \quad \text{and} \quad y = 4 \sin \theta = 4 \sin t,$$

where $t \geq 0$. You can confirm that when $t = 0$, the boat is at the starting point $(4, 0)$; when $t = 2\pi$, it returns to the starting point.

The equations $x = 4 \cos t$ and $y = 4 \sin t$ are examples of **parametric equations**. They specify x and y in terms of a third variable t called a **parameter**, which often represents time.

Figure 12.1

In general, parametric equations have the form

$$x = f(t), \quad y = g(t),$$

where f and g are given functions and the parameter t typically varies over a specified interval $a \le t \le b$ (**Figure 12.2**). The **parametric curve** described by these equations consists of the points in the plane

$$(x, y) = (f(t), g(t)), \quad \text{for } a \le t \le b.$$

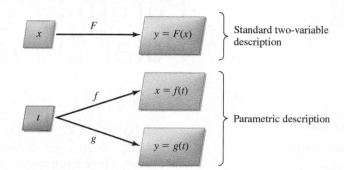

Figure 12.2

> With a parametric description, the parameter t is the independent variable. There are two dependent variables, x and y.

EXAMPLE 1 Parametric parabola Graph and analyze the parametric equations

$$x = f(t) = 2t, \quad y = g(t) = \frac{1}{2}t^2 - 4, \quad \text{for } 0 \le t \le 8.$$

SOLUTION Plotting individual points often helps in visualizing a parametric curve. Table 12.1 shows the values of x and y corresponding to several values of t on the interval $[0, 8]$. By plotting the (x, y) pairs in Table 12.1 and connecting them with a smooth curve, we obtain the graph shown in **Figure 12.3**. As t increases from its initial value of $t = 0$ to its final value of $t = 8$, the curve is generated from the initial point $(0, -4)$ to the final point $(16, 28)$. Notice that the values of the parameter do not appear in the graph. The only signature of the parameter is the direction in which the curve is generated: In this case, it unfolds upward and to the right, as indicated by the arrows on the curve.

Table 12.1

t	x	y	(x, y)
0	0	-4	$(0, -4)$
1	2	$-\frac{7}{2}$	$\left(2, -\frac{7}{2}\right)$
2	4	-2	$(4, -2)$
3	6	$\frac{1}{2}$	$\left(6, \frac{1}{2}\right)$
4	8	4	$(8, 4)$
5	10	$\frac{17}{2}$	$\left(10, \frac{17}{2}\right)$
6	12	14	$(12, 14)$
7	14	$\frac{41}{2}$	$\left(14, \frac{41}{2}\right)$
8	16	28	$(16, 28)$

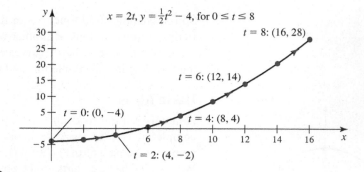

Figure 12.3

Sometimes it is possible to eliminate the parameter from a set of parametric equations and obtain a description of the curve in terms of x and y. In this case, from the x-equation, we have $t = x/2$, which may be substituted into the y-equation to give

$$y = \frac{1}{2}t^2 - 4 = \frac{1}{2}\left(\frac{x}{2}\right)^2 - 4 = \frac{x^2}{8} - 4.$$

QUICK CHECK 1 Identify the graph generated by the parametric equations $x = t^2, y = t$, for $-10 \le t \le 10$. ◄

Expressed in this form, we identify the graph as part of a parabola. Because t lies in the interval $0 \le t \le 8$ and $x = 2t$, it follows that x lies in the interval $0 \le x \le 16$. Therefore, the parametric equations generate the segment of the parabola for $0 \le x \le 16$.

Related Exercise 12 ◄

Given a set of parametric equations, the preceding example shows that as the parameter increases, the corresponding curve unfolds in a particular direction. The following definition captures this fact and is important in upcoming work.

DEFINITION Positive Orientation

The direction in which a parametric curve is generated as the parameter increases is called the **positive orientation** of the curve (and is indicated by arrows on the curve).

EXAMPLE 2   Parametric circle

a. Confirm that the parametric equations

$$x = 4 \cos 2\pi t, \quad y = 4 \sin 2\pi t, \quad \text{for } 0 \le t \le 1$$

describe a circle of radius 4 centered at the origin.

b. Suppose a turtle walks with constant speed in the counterclockwise direction on the circular path from part (a). Starting from the point $(4, 0)$, the turtle completes one lap in 30 minutes. Find a parametric description of the path of the turtle at any time $t \ge 0$, where t is measured in minutes.

SOLUTION

a. For each value of t in Table 12.2, the corresponding ordered pairs (x, y) are recorded. Plotting these points as t increases from $t = 0$ to $t = 1$ results in a graph that appears to be a circle of radius 4; it is generated with positive orientation in the counterclockwise direction, beginning and ending at $(4, 0)$ (**Figure 12.4**). Letting t increase beyond $t = 1$ would simply retrace the same curve.

To identify the curve conclusively, the parameter t is eliminated by observing that

$$x^2 + y^2 = (4 \cos 2\pi t)^2 + (4 \sin 2\pi t)^2$$
$$= 16\underbrace{(\cos^2 2\pi t + \sin^2 2\pi t)}_{1} = 16. \quad \cos^2 \theta + \sin^2 \theta = 1$$

We have confirmed that the graph of the parametric equations is the circle $x^2 + y^2 = 16$.

Table 12.2

t	(x, y)
0	$(4, 0)$
$\frac{1}{8}$	$(2\sqrt{2}, 2\sqrt{2})$
$\frac{1}{4}$	$(0, 4)$
$\frac{3}{8}$	$(-2\sqrt{2}, 2\sqrt{2})$
$\frac{1}{2}$	$(-4, 0)$
$\frac{3}{4}$	$(0, -4)$
1	$(4, 0)$

Figure 12.4

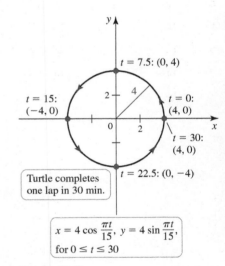

Figure 12.5

b. Duplicating the calculations at the end of part (a) with any nonzero real number b, it can be shown that the parametric equations

$$x = 4 \cos bt, \quad y = 4 \sin bt$$

also describe a circle of radius 4. When $b > 0$, the graph is generated in the counterclockwise direction. The *angular frequency b* must be chosen so that as t varies

> In Example 2b, the constant $|b|$ is called the *angular frequency* because it is the number of radians the object moves per unit time. The turtle travels 2π rad every 30 min, so the angular frequency is $2\pi/30 = \pi/15$ rad/min. Because radians have no units, the angular frequency in this case has units *per minute*, which is written min^{-1}.

from 0 to 30, the product bt varies from 0 to 2π. Specifically, when $t = 30$, we must have $30b = 2\pi$, or $b = \dfrac{\pi}{15}$ rad/min. Therefore, the parametric equations for the turtle's motion are

$$x = 4 \cos \frac{\pi t}{15}, \quad y = 4 \sin \frac{\pi t}{15}, \quad \text{for } 0 \le t \le 30.$$

You should check that as t varies from 0 to 30, the points (x, y) make one complete circuit of a circle of radius 4 (**Figure 12.5**).

Related Exercises 20, 37, 55 ◄

Generalizing Example 2 for nonzero real numbers a and b in the parametric equations $x = a \cos bt$, $y = a \sin bt$, we find that

$$x^2 + y^2 = (a \cos bt)^2 + (a \sin bt)^2$$
$$= a^2 \underbrace{(\cos^2 bt + \sin^2 bt)}_{1} = a^2.$$

> For a nonzero constant b, the functions $\sin bt$ and $\cos bt$ have period $2\pi/|b|$. The equations $x = a \cos bt$, $y = -a \sin bt$ also describe a circle of radius $|a|$, as do the equations $x = \pm a \sin bt$, $y = \pm a \cos bt$, as t varies over an interval of length $2\pi/|b|$.

Therefore, the parametric equations $x = a \cos bt$, $y = a \sin bt$ describe all or part of the circle $x^2 + y^2 = a^2$, centered at the origin with radius $|a|$, for any nonzero value of b. The circle is traversed once as t varies over any interval of length $2\pi/|b|$. If t represents time, the circle is traversed once in $2\pi/|b|$ time units, which means we can vary the speed at which the curve unfolds by varying b. If $b > 0$, the positive orientation is in the counterclockwise direction. If $b < 0$, the curve is generated in the clockwise direction.

More generally, the parametric equations

$$x = x_0 + a \cos bt, \quad y = y_0 + a \sin bt$$

QUICK CHECK 2 Give the center and radius of the circle generated by the equations $x = 3 \sin t$, $y = -3 \cos t$, for $0 \le t \le 2\pi$. Specify the direction of positive orientation. ◄

describe all or part of the circle $(x - x_0)^2 + (y - y_0)^2 = a^2$, centered at (x_0, y_0) with radius $|a|$. If $b > 0$, then the circle is generated in the counterclockwise direction. Example 2 shows that a single curve—for example, a circle of radius 4—may be parameterized in many different ways.

Among the most important of all parametric equations are

$$x = x_0 + at, \, y = y_0 + bt, \quad \text{for } -\infty < t < \infty,$$

where x_0, y_0, a, and b are constants with $a \ne 0$. The curve described by these equations is found by eliminating the parameter. The first step is to solve the x-equation for t, which gives us $t = \dfrac{x - x_0}{a}$. When t is substituted into the y-equation, the result is an equation for y in terms of x:

$$y = y_0 + bt = y_0 + b\left(\frac{x - x_0}{a}\right) \quad \text{or} \quad y - y_0 = \frac{b}{a}(x - x_0).$$

This equation describes a line with slope $\dfrac{b}{a}$ passing through the point (x_0, y_0).

SUMMARY Parametric Equations of a Line

The equations

$$x = x_0 + at, \, y = y_0 + bt, \quad \text{for } -\infty < t < \infty,$$

where x_0, y_0, a, and b are constants with $a \ne 0$, describe a line with slope $\dfrac{b}{a}$ passing through the point (x_0, y_0). If $a = 0$ and $b \ne 0$, the line is vertical.

Notice that the parametric description of a given line is not unique: For example, if k is any nonzero constant, the numbers a and b may be replaced with ka and kb, respectively, and the resulting equations describe the same line (although it may be generated in the opposite direction and at a different speed).

EXAMPLE 3 Parametric equations of lines

a. Consider the parametric equations $x = -2 + 3t$, $y = 4 - 6t$, for $-\infty < t < \infty$, which describe a line. Find the slope-intercept form of the line.

b. Find two pairs of parametric equations for the line with slope $\dfrac{1}{3}$ that passes through the point $(2, 1)$.

c. Find parametric equations for the line segment starting at $P(4, 7)$ and ending at $Q(2, -3)$.

SOLUTION

a. To eliminate the parameter, first solve the x-equation for t to find that $t = \dfrac{x + 2}{3}$.
Replacing t in the y-equation yields

$$y = 4 - 6\left(\frac{x + 2}{3}\right) = 4 - 2x - 4 = -2x.$$

The line $y = -2x$ passes through the origin with slope -2.

b. We use the general parametric equations of a line given in the Summary box. Because the slope of the line is $\dfrac{1}{3}$, we choose $a = 3$ and $b = 1$. Letting $x_0 = 2$ and $y_0 = 1$, parametric equations for the line are $x = 2 + 3t$, $y = 1 + t$, for $-\infty < t < \infty$. The line passes through $(2, 1)$ when $t = 0$ and rises to the right as t increases (**Figure 12.6**). Notice that other choices for a and b also work. For example, with $a = -6$ and $b = -2$, the equations are $x = 2 - 6t$, $y = 1 - 2t$, for $-\infty < t < \infty$. These equations describe the same line, but now, as t increases, the line is generated in the opposite direction (descending to the left).

c. The slope of this line is $\dfrac{7 - (-3)}{4 - 2} = 5$. However, notice that when the line segment is traversed from P to Q, both x and y are decreasing (**Figure 12.7**). To account for the direction in which the line segment is generated, we let $a = -1$ and $b = -5$. Because $P(4, 7)$ is the starting point of the line segment, we choose $x_0 = 4$ and $y_0 = 7$. The resulting equations are $x = 4 - t$, $y = 7 - 5t$. Notice that $t = 0$ corresponds to the starting point $(4, 7)$. Because the equations describe a line *segment*, the interval for t must be restricted. What value of t corresponds to the endpoint of the line segment $Q(2, -3)$? Setting $x = 4 - t = 2$, we find that $t = 2$. As a check, we set $y = 7 - 5t = -3$, which also implies that $t = 2$. (If these two calculations do not give the same value of t, it probably means the slope was not computed correctly.) Therefore, the equations for the line segment are $x = 4 - t$, $y = 7 - 5t$, for $0 \le t \le 2$.

Related Exercises 15, 41–42 ◀

Figure 12.6

➤ The choices $a = -1$ and $b = -5$ result in a slope of $b/a = 5$. These choices also imply that as we move from P to Q, a decrease in x corresponds to a decrease in y.

➤ Lines and line segments may have unexpected parametric representations. For example, the equations $x = \sin t$, $y = 2 \sin t$ represent the line segment $y = 2x$, where $-1 \le x \le 1$ (recall that the range of $x = \sin t$ is $[-1, 1]$).

Figure 12.7

QUICK CHECK 3 Describe the curve generated by $x = 3 + 2t$, $y = -12 - 6t$, for $-\infty < t < \infty$. ◀

EXAMPLE 4 Parametric equations of curves

A common task (particularly in upcoming chapters) is to parameterize curves given either by Cartesian equations or by graphs. Find a parametric representation of the following curves.

a. The segment of the parabola $y = 9 - x^2$, for $-1 \le x \le 3$

b. The complete curve $x = (y - 5)^2 + \sqrt{y}$

c. The piecewise linear path connecting $P(-2, 0)$ to $Q(0, 3)$ to $R(4, 0)$ (in that order), where the parameter varies over the interval $0 \le t \le 2$

SOLUTION

a. The simplest way to represent a curve $y = f(x)$ parametrically is to let $x = t$ and $y = f(t)$, where t is the parameter. We must then find the appropriate interval for the parameter. Using this approach, the curve $y = 9 - x^2$ has the parametric representation

$$x = t, \quad y = 9 - t^2, \quad \text{for} \quad -1 \le t \le 3.$$

Figure 12.8

This representation is not unique. For example, you can verify that the parametric equations

$$x = 1 - t, \quad y = 9 - (1 - t)^2, \quad \text{for} \quad -2 \leq t \leq 2$$

also do the job, although these equations trace the parabola from right to left, while the original equations trace the curve from left to right (**Figure 12.8**).

b. In this case, it is easier to let $y = t$. Then a parametric description of the curve is

$$x = (t - 5)^2 + \sqrt{t}, \quad y = t.$$

Notice that t can take values only in the interval $[0, \infty)$. As $t \to \infty$, we see that $x \to \infty$ and $y \to \infty$ (**Figure 12.9**).

Figure 12.9

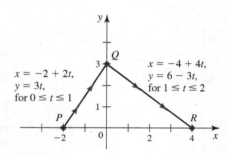

Figure 12.10

c. The path consists of two line segments (**Figure 12.10**) that can be parameterized separately in the form $x = x_0 + at$ and $y = y_0 + bt$. The line segment PQ originates at $P(-2, 0)$ and unfolds in the positive x-direction with slope $\frac{3}{2}$. It can be represented as

$$x = -2 + 2t, \quad y = 3t, \quad \text{for } 0 \leq t \leq 1.$$

Finding the parametric equations for the line segment QR requires some ingenuity. We want the line segment to originate at $Q(0, 3)$ when $t = 1$ and end at $R(4, 0)$ when $t = 2$. Observe that when $t = 1, x = 0$ and when $t = 2, x = 4$. Substituting these pairs of values into the general x-equation $x = x_0 + at$, we obtain the equations

$$x_0 + a = 0 \quad x = 0 \text{ when } t = 1$$
$$x_0 + 2a = 4. \quad x = 4 \text{ when } t = 2$$

Solving for x_0 and a, we find that $x_0 = -4$ and $a = 4$. Applying a similar procedure to the general y-equation $y = y_0 + bt$, the relevant conditions are

$$y_0 + b = 3 \quad y = 3 \text{ when } t = 1$$
$$y_0 + 2b = 0. \quad y = 0 \text{ when } t = 2$$

Solving for y_0 and b, we find that $y_0 = 6$ and $b = -3$. Putting it all together, the equations for the line segment QR are

$$x = -4 + 4t, y = 6 - 3t, \quad \text{for } 1 \leq t \leq 2.$$

QUICK CHECK 4 Find parametric equations for the line segment that goes from $Q(0, 3)$ to $P(-2, 0)$. ◄

You can verify that the points $Q(0, 3)$ and $R(4, 0)$ correspond to $t = 1$ and $t = 2$, respectively. Furthermore, the slope of the line is $\dfrac{b}{a} = -\dfrac{3}{4}$, which is correct.

Related Exercises 43–44, 47 ◄

Many fascinating curves are generated by points on rolling wheels; Examples 5 and 8 investigate two such curves.

EXAMPLE 5   Rolling wheels   The path of a light on the rim of a wheel rolling on a flat surface (**Figure 12.11a**) is a **cycloid**, which has the parametric equations

$$x = a(t - \sin t), \quad y = a(1 - \cos t), \quad \text{for } t \geq 0,$$

where $a > 0$. Use a graphing utility to graph the cycloid with $a = 1$. On what interval does the parameter generate the first arch of the cycloid?

SOLUTION   The graph of the cycloid, for $0 \leq t \leq 3\pi$, is shown in **Figure 12.11b**. The wheel completes one full revolution on the interval $0 \leq t \leq 2\pi$, which gives one arch of the cycloid.

(a)   (b)

Figure 12.11

Related Exercise 58 ◄

Derivatives and Parametric Equations

Parametric equations express a relationship between the variables x and y. Therefore, it makes sense to ask about dy/dx, the rate of change of y with respect to x at a point on a parametric curve. Once we know how to compute dy/dx, it can be used to determine slopes of lines tangent to parametric curves.

Consider the parametric equations $x = f(t), y = g(t)$ on an interval on which both f and g are differentiable. The Chain Rule relates the derivatives $\dfrac{dy}{dt}, \dfrac{dx}{dt}$, and $\dfrac{dy}{dx}$:

$$\frac{dy}{dt} = \frac{dy}{dx}\frac{dx}{dt}.$$

Provided $\dfrac{dx}{dt} \neq 0$, we divide both sides of this equation by $\dfrac{dx}{dt}$ and solve for $\dfrac{dy}{dx}$ to obtain the following result.

Figure 12.12

QUICK CHECK 5   Use Theorem 12.1 to find the slope of the line $x = 4t$, $y = 2t$, for $-\infty < t < \infty$. ◄

THEOREM 12.1   Derivative for Parametric Curves

Let $x = f(t)$ and $y = g(t)$, where f and g are differentiable on an interval $[a, b]$. Then the slope of the line tangent to the curve at the point corresponding to t is

$$\frac{dy}{dx} = \frac{dy/dt}{dx/dt} = \frac{g'(t)}{f'(t)},$$

provided $f'(t) \neq 0$.

Figure 12.12 gives a geometric explanation of Theorem 12.1. The slope of the line tangent to a curve at a point is $\dfrac{dy}{dx} = \lim\limits_{\Delta x \to 0} \dfrac{\Delta y}{\Delta x}$. Using linear approximation (Section 4.6), we have $\Delta x \approx f'(t)\Delta t$ and $\Delta y \approx g'(t)\Delta t$, with these approximations improving as $\Delta t \to 0$. Notice also that $\Delta t \to 0$ as $\Delta x \to 0$. Therefore, the slope of the tangent line is

$$\frac{dy}{dx} = \lim_{\Delta x \to 0} \frac{\Delta y}{\Delta x} = \lim_{\Delta t \to 0} \frac{g'(t)\Delta t}{f'(t)\Delta t} = \frac{g'(t)}{f'(t)}. \quad \text{Cancel } \Delta t \neq 0.$$

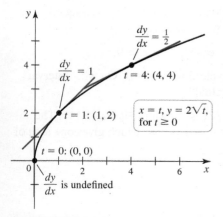

Figure 12.13

➤ In general, the equations $x = a \cos t$, $y = b \sin t$, for $0 \leq t \leq 2\pi$, describe an ellipse. The constants a and b can be seen as horizontal and vertical scalings of the unit circle $x = \cos t$, $y = \sin t$. Ellipses are explored in Exercises 91–94 and in Section 12.4.

Figure 12.14

EXAMPLE 6 Slopes of tangent lines Find $\dfrac{dy}{dx}$ for the following curves. Interpret the result and determine the points (if any) at which the curve has a horizontal or a vertical tangent line.

a. $x = f(t) = t$, $y = g(t) = 2\sqrt{t}$, for $t \geq 0$

b. $x = f(t) = 4 \cos t$, $y = g(t) = 16 \sin t$, for $0 \leq t \leq 2\pi$

SOLUTION

a. We find that $f'(t) = 1$ and $g'(t) = 1/\sqrt{t}$. Therefore,

$$\frac{dy}{dx} = \frac{g'(t)}{f'(t)} = \frac{1/\sqrt{t}}{1} = \frac{1}{\sqrt{t}},$$

provided $t \neq 0$. Notice that $dy/dx \neq 0$ for $t > 0$, so the curve has no horizontal tangent lines. On the other hand, as $t \to 0^+$, we see that $dy/dx \to \infty$. Therefore, the curve has a vertical tangent line at the point $(0, 0)$. To eliminate t from the parametric equations, we substitute $t = x$ into the y-equation to find that $y = 2\sqrt{x}$. Because $y \geq 0$, the curve is the upper half of a parabola (**Figure 12.13**). Slopes of tangent lines at other points on the curve are found by substituting the corresponding values of t. For example, the point $(4, 4)$ corresponds to $t = 4$ and the slope of the tangent line at that point is $1/\sqrt{4} = \frac{1}{2}$.

b. These parametric equations describe an **ellipse** with a major axis of length 32 on the y-axis and a minor axis of length 8 on the x-axis (**Figure 12.14**). In this case, $f'(t) = -4 \sin t$ and $g'(t) = 16 \cos t$. Therefore,

$$\frac{dy}{dx} = \frac{g'(t)}{f'(t)} = \frac{16 \cos t}{-4 \sin t} = -4 \cot t.$$

At $t = 0$ and $t = \pi$, $\cot t$ is undefined. Notice that

$$\lim_{t \to 0^+} \frac{dy}{dx} = \lim_{t \to 0^+} (-4 \cot t) = -\infty \quad \text{and} \quad \lim_{t \to 0^-} \frac{dy}{dx} = \lim_{t \to 0^-} (-4 \cot t) = \infty.$$

Consequently, a vertical tangent line occurs at the point corresponding to $t = 0$, which is $(4, 0)$ (Figure 12.14). A similar argument shows that a vertical tangent line occurs at the point corresponding to $t = \pi$, which is $(-4, 0)$.

At $t = \pi/2$ and $t = 3\pi/2$, $\cot t = 0$ and the curve has horizontal tangent lines at the corresponding points $(0, \pm16)$. Slopes of tangent lines at other points on the curve may be found. For example, the point $(2\sqrt{2}, 8\sqrt{2})$ corresponds to $t = \pi/4$; the slope of the tangent line at that point is $-4 \cot \pi/4 = -4$.

Related Exercises 69–70 ◄

In Section 6.5, we derived the formula for the arc length of curves of the form $y = f(x)$. We conclude this section with a similar derivation for the arc length of curves described parametrically.

Arc Length

Suppose a curve C is given by the parametric equations $x = f(t)$, $y = g(t)$, for $a \leq t \leq b$, where f' and g' are continuous on $[a, b]$. To find the length of the curve between $(f(a), g(a))$ and $(f(b), g(b))$, we first subdivide the interval $[a, b]$ into n subintervals using the grid points

$$a = t_0 < t_1 < t_2 < \cdots < t_n = b.$$

The next step is to connect consecutive points on the curve,

$$(f(t_0), g(t_0)), \ldots, (f(t_k), g(t_k)), \ldots, (f(t_n), g(t_n)),$$

with line segments (**Figure 12.15a**).

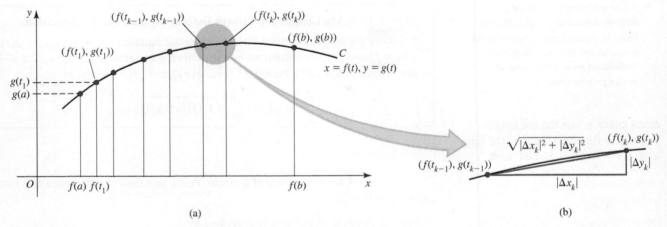

Figure 12.15

The kth line segment is the hypotenuse of a right triangle, whose legs have lengths $|\Delta x_k|$ and $|\Delta y_k|$, where

$$\Delta x_k = f(t_k) - f(t_{k-1}) \quad \text{and} \quad \Delta y_k = g(t_k) - g(t_{k-1}),$$

for $k = 1, 2, \ldots, n$ (**Figure 12.15b**). Therefore, the length of the kth line segment is

$$\sqrt{|\Delta x_k|^2 + |\Delta y_k|^2}.$$

The length of the entire curve L is approximated by the sum of the lengths of the line segments:

$$L \approx \sum_{k=1}^{n} \sqrt{|\Delta x_k|^2 + |\Delta y_k|^2} = \sum_{k=1}^{n} \sqrt{(\Delta x_k)^2 + (\Delta y_k)^2}. \tag{1}$$

The goal is to express this sum as a Riemann sum.

The change in $x = f(t)$ over the kth subinterval is $\Delta x_k = f(t_k) - f(t_{k-1})$. By the Mean Value Theorem, there is a point t_k^* in (t_{k-1}, t_k) such that

$$\frac{\overbrace{f(t_k) - f(t_{k-1})}^{\Delta x_k}}{\underbrace{t_k - t_{k-1}}_{\Delta t_k}} = f'(t_k^*).$$

So the change in x as t changes by $\Delta t_k = t_k - t_{k-1}$ is

$$\Delta x_k = f(t_k) - f(t_{k-1}) = f'(t_k^*)\Delta t_k.$$

Similarly, the change in y over the kth subinterval is

$$\Delta y_k = g(t_k) - g(t_{k-1}) = g'(\hat{t}_k)\Delta t_k,$$

where $\hat{t}_k$ is also a point in (t_{k-1}, t_k). We substitute these expressions for Δx_k and Δy_k into equation (1):

$$L \approx \sum_{k=1}^{n} \sqrt{(\Delta x_k)^2 + (\Delta y_k)^2}$$

$$= \sum_{k=1}^{n} \sqrt{(f'(t_k^*)\Delta t_k)^2 + (g'(\hat{t}_k)\Delta t_k)^2} \quad \text{Substitute for } \Delta x_k \text{ and } \Delta y_k.$$

$$= \sum_{k=1}^{n} \sqrt{f'(t_k^*)^2 + g'(\hat{t}_k)^2}\,\Delta t_k. \quad \text{Factor } \Delta t_k \text{ out of square root.}$$

The intermediate points t_k^* and $\hat{t}_k$ both approach t_k as n increases and as Δt_k approaches zero. Therefore, given the conditions on f' and g', the limit of this sum as $n \to \infty$ and $\Delta t_k \to 0$, for all k, exists and equals a definite integral:

$$L = \lim_{n \to \infty} \sum_{k=1}^{n} \sqrt{f'(t_k^*)^2 + g'(\hat{t}_k)^2}\,\Delta t_k = \int_a^b \sqrt{f'(t)^2 + g'(t)^2}\, dt.$$

We take this integral as the definition of the length of the curve C.

➤ Arc length integrals are usually difficult to evaluate exactly. The few easily evaluated integrals appear in the examples and exercises. Numerical methods are used to approximate challenging integrals.

QUICK CHECK 6 Use the arc length formula to find the length of the line $x = t, y = t$, for $0 \le t \le 1$. ◀

➤ An important fact is that the arc length of a smooth parameterized curve is independent of the choice of parameter (See Section 14.4).

> **DEFINITION** **Arc Length for Curves Defined by Parametric Equations**
>
> Consider the curve described by the parametric equations $x = f(t), y = g(t)$, where f' and g' are continuous, and the curve is traversed once for $a \le t \le b$. The **arc length** of the curve between $(f(a), g(a))$ and $(f(b), g(b))$ is
>
> $$L = \int_a^b \sqrt{f'(t)^2 + g'(t)^2}\, dt.$$

EXAMPLE 7 **Circumference of a circle** Prove that the circumference of a circle of radius $a > 0$ is $2\pi a$.

SOLUTION A circle of radius a is described by

$$x = f(t) = a \cos t, y = g(t) = a \sin t, \quad \text{for } 0 \le t \le 2\pi.$$

Note that $f'(t) = -a \sin t$ and $g'(t) = a \cos t$. The circumference is

$$
\begin{aligned}
L &= \int_0^{2\pi} \sqrt{f'(t)^2 + g'(t)^2}\, dt && \text{Arc length formula} \\
&= \int_0^{2\pi} \sqrt{(-a \sin t)^2 + (a \cos t)^2}\, dt && \text{Substitute for } f' \text{ and } g'. \\
&= a \int_0^{2\pi} \sqrt{\sin^2 t + \cos^2 t}\, dt && \text{Factor } a > 0 \text{ out of square root.} \\
&= a \int_0^{2\pi} 1\, dt && \sin^2 t + \cos^2 t = 1 \\
&= 2\pi a. && \text{Integrate a constant.}
\end{aligned}
$$

Related Exercise 82 ◀

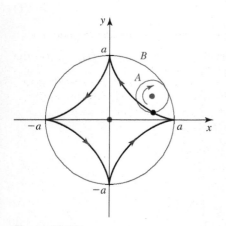

Figure 12.16

EXAMPLE 8 **Astroid arc length** The path of a point on circle A with radius $a/4$ that rolls on the inside of circle B with radius a (**Figure 12.16**) is an **astroid** or a **hypocycloid**. Its parametric equations are

$$x = a \cos^3 t, \quad y = a \sin^3 t, \quad \text{for } 0 \le t \le 2\pi.$$

a. Graph the astroid with $a = 1$ and find its equation in terms of x and y.

b. Find the arc length of the astroid with $a = 1$.

SOLUTION

a. Because both $\cos^3 t$ and $\sin^3 t$ have a period of 2π, the complete curve is generated on the interval $0 \le t \le 2\pi$ (**Figure 12.17**). To eliminate t from the parametric equations, note that $x^{2/3} = \cos^2 t$ and $y^{2/3} = \sin^2 t$. Therefore,

$$x^{2/3} + y^{2/3} = \cos^2 t + \sin^2 t = 1,$$

where the Pythagorean identity has been used. We see that an alternative description of the astroid is $x^{2/3} + y^{2/3} = 1$.

b. The length of the entire curve is four times the length of the curve in the first quadrant. You should verify that the curve in the first quadrant is generated as the parameter varies from $t = 0$ (corresponding to $(1, 0)$) to $t = \pi/2$ (corresponding to $(0, 1)$; **Figure 12.17**). Letting $f(t) = \cos^3 t$ and $g(t) = \sin^3 t$, we have

$$f'(t) = -3 \cos^2 t \sin t \quad \text{and} \quad g'(t) = 3 \sin^2 t \cos t.$$

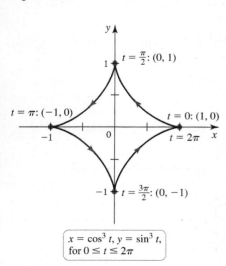

$x = \cos^3 t, y = \sin^3 t,$
for $0 \le t \le 2\pi$

Figure 12.17

The arc length of the full curve is

$$L = 4 \int_0^{\pi/2} \sqrt{f'(t)^2 + g'(t)^2}\, dt \qquad \text{Factor of 4 by symmetry}$$

$$= 4 \int_0^{\pi/2} \sqrt{(-3\cos^2 t \sin t)^2 + (3\sin^2 t \cos t)^2}\, dt \quad \text{Substitute for } f' \text{ and } g'.$$

$$= 4 \int_0^{\pi/2} \sqrt{9\cos^4 t \sin^2 t + 9\cos^2 t \sin^4 t}\, dt \qquad \text{Simplify terms.}$$

$$= 4 \int_0^{\pi/2} 3\sqrt{\cos^2 t \sin^2 t \underbrace{(\cos^2 t + \sin^2 t)}_{1}}\, dt \qquad \text{Factor.}$$

$$= 12 \int_0^{\pi/2} \cos t \sin t\, dt. \qquad \cos t \sin t \geq 0, \text{ for } 0 \leq t \leq \frac{\pi}{2}$$

> The arc length formula from Section 6.5 could be used to compute the length of the astroid in Example 8 because we found a Cartesian equation for the curve in part (a). However, in cases where it is impossible to eliminate the parameter, the arc length formula derived in this section is required.

Letting $u = \sin t$ with $du = \cos t\, dt$, we have

$$L = 12 \int_0^{\pi/2} \cos t \sin t\, dt = 12 \int_0^1 u\, du = 6.$$

The length of the entire hypocycloid is 6 units.

Related Exercise 87 ◄

SECTION 12.1 EXERCISES

Getting Started

1. Explain how a pair of parametric equations generates a curve in the xy-plane.

2. Give two pairs of parametric equations that generate a circle centered at the origin with radius 6.

3. Give parametric equations that describe a circle of radius R, centered at the origin with clockwise orientation, where the parameter varies over the interval $[0, 10]$.

4. Give parametric equations that generate the line with slope -2 passing through $(1, 3)$.

5. Find parametric equations for the complete parabola $x = y^2$. Answers are not unique.

6. Describe the similarities between the graphs of the parametric equations $x = \sin^2 t, y = \sin t$, for $0 \leq t \leq \pi/2$, and $x = \sin^2 t, y = \sin t$, for $\pi/2 \leq t \leq \pi$. Begin by eliminating the parameter t to obtain an equation in x and y.

7. Find the slope of the parametric curve $x = -2t^3 + 1, y = 3t^2$, for $-\infty < t < \infty$, at the point corresponding to $t = 2$.

8. In which direction is the curve $x = -2\sin t, y = 2\cos t$, for $0 \leq t < 2\pi$, generated?

9. Find three different pairs of parametric equations for the line segment that starts at $(0, 0)$ and ends at $(6, 6)$.

10. Use calculus to find the arc length of the line segment $x = 3t + 1$, $y = 4t$, for $0 \leq t \leq 1$. Check your work by finding the distance between the endpoints of the line segment.

Practice Exercises

11–14. Working with parametric equations *Consider the following parametric equations.*

a. *Make a brief table of values of t, x, and y.*
b. *Plot the (x, y) pairs in the table and the complete parametric curve, indicating the positive orientation (the direction of increasing t).*
c. *Eliminate the parameter to obtain an equation in x and y.*
d. *Describe the curve.*

11. $x = 2t, y = 3t - 4; -10 \leq t \leq 10$

12. $x = t^2 + 2, y = 4t; -4 \leq t \leq 4$

13. $x = -t + 6, y = 3t - 3; -5 \leq t \leq 5$

14. $x = t^3 - 1, y = 5t + 1; -3 \leq t \leq 3$

15–30. Working with parametric equations *Consider the following parametric equations.*

a. *Eliminate the parameter to obtain an equation in x and y.*
b. *Describe the curve and indicate the positive orientation.*

15. $x = 3 + t, y = 1 - t; 0 \leq t \leq 1$

16. $x = 4 - 3t, y = -2 + 6t; -1 \leq t \leq 3$

17. $x = \sqrt{t} + 4, y = 3\sqrt{t}; 0 \leq t \leq 16$

18. $x = (t + 1)^2, y = t + 2; -10 \leq t \leq 10$

19. $x = 3\cos t, y = 3\sin t; \pi \leq t \leq 2\pi$

20. $x = 3\cos t, y = 3\sin t; 0 \leq t \leq \pi/2$

21. $x = \cos t, y = \sin^2 t; 0 \leq t \leq \pi$

22. $x = 1 - \sin^2 s,\ y = \cos s;\ \pi \le s \le 2\pi$

23. $x = \cos t,\ y = 1 + \sin t;\ 0 \le t \le 2\pi$

24. $x = \sin t,\ y = 2 \sin t + 1;\ 0 \le t \le \pi/2$

25. $x = r - 1,\ y = r^3;\ -4 \le r \le 4$

26. $x = e^{2t},\ y = e^t + 1;\ 0 \le t \le 25$

27. $x = -7 \cos 2t,\ y = -7 \sin 2t;\ 0 \le t \le \pi$

28. $x = 1 - 3 \sin 4\pi t,\ y = 2 + 3 \cos 4\pi t;\ 0 \le t \le 1/2$

29. $x = 8 + 2t,\ y = 1;\ -\infty < t < \infty$

30. $x = 5,\ y = 3t;\ -\infty < t < \infty$

31–36. Eliminating the parameter *Eliminate the parameter to express the following parametric equations as a single equation in x and y.*

31. $x = 2 \sin 8t,\ y = 2 \cos 8t$ **32.** $x = \sin 8t,\ y = 2 \cos 8t$

33. $x = t,\ y = \sqrt{4 - t^2}$ **34.** $x = \sqrt{t + 1},\ y = \dfrac{1}{t + 1}$

35. $x = \tan t,\ y = \sec^2 t - 1$

36. $x = a \sin^n t,\ y = b \cos^n t$, where a and b are real numbers and n is a positive integer

37–52. Curves to parametric equations *Find parametric equations for the following curves. Include an interval for the parameter values. Answers are not unique.*

37. A circle centered at the origin with radius 4, generated counterclockwise

38. A circle centered at the origin with radius 12, generated clockwise with initial point $(0, 12)$

39. A circle centered at $(2, 3)$ with radius 1, generated counterclockwise

40. A circle centered at $(-2, -3)$ with radius 8, generated clockwise

41. The line segment starting at $P(0, 0)$ and ending at $Q(2, 8)$

42. The line segment starting at $P(-1, -3)$ and ending at $Q(6, -16)$

43. The segment of the parabola $y = 2x^2 - 4$, where $-1 \le x \le 5$

44. The complete curve $x = y^3 - 3y$

45. The vertical line segment starting at $P(2, 3)$ and ending at $Q(2, 9)$

46. The horizontal line segment starting at $P(8, 2)$ and ending at $Q(-2, 2)$

47. The piecewise linear path from $P(-2, 3)$ to $Q(2, -3)$ to $R(3, 5)$, using parameter values $0 \le t \le 2$

48. The path consisting of the line segment from $(-4, 4)$ to $(0, 8)$, followed by the segment of the parabola $y = 8 - 2x^2$ from $(0, 8)$ to $(2, 0)$, using parameter values $0 \le t \le 3$

49. The line that passes through the points $P(1, 1)$ and $Q(3, 5)$, oriented in the direction of increasing x

50. The left half of the parabola $y = x^2 + 1$, originating at $(0, 1)$

51. The upper half of the parabola $x = y^2$, originating at $(0, 0)$

52. The lower half of the circle centered at $(-2, 2)$ with radius 6, oriented in the counterclockwise direction

53–56. Circular motion *Find parametric equations that describe the circular path of the following objects. For Exercises 53–55, assume (x, y) denotes the position of the object relative to the origin at the center of the circle. Use the units of time specified in the problem. There are many ways to describe any circle.*

53. A go-cart moves counterclockwise with constant speed around a circular track of radius 400 m, completing a lap in 1.5 min.

54. The tip of the 15-inch second hand of a clock completes one revolution in 60 seconds.

55. A bicyclist rides counterclockwise with constant speed around a circular velodrome track with a radius of 50 m, completing one lap in 24 seconds.

56. A Ferris wheel has a radius of 20 m and completes a revolution in the clockwise direction at constant speed in 3 min. Assume x and y measure the horizontal and vertical positions of a seat on the Ferris wheel relative to a coordinate system whose origin is at the low point of the wheel. Assume the seat begins moving at the origin.

T 57–62. More parametric curves *Use a graphing utility to graph the following curves. Be sure to choose an interval for the parameter that generates all features of interest.*

57. **Spiral** $x = t \cos t,\ y = t \sin t;\ t \ge 0$

58. **Witch of Agnesi** $x = 2 \cot t,\ y = 1 - \cos 2t$

59. **Folium of Descartes** $x = \dfrac{3t}{1 + t^3},\ y = \dfrac{3t^2}{1 + t^3}$

60. **Involute of a circle** $x = \cos t + t \sin t,\ y = \sin t - t \cos t$

61. **Evolute of an ellipse** $x = \dfrac{a^2 - b^2}{a} \cos^3 t,\ y = \dfrac{a^2 - b^2}{b} \sin^3 t$; $a = 4$ and $b = 3$

62. **Cissoid of Diocles** $x = 2 \sin 2t,\ y = \dfrac{2 \sin^3 t}{\cos t}$

T 63. **Implicit function graph** Explain and carry out a method for graphing the curve $x = 1 + \cos^2 y - \sin^2 y$ using parametric equations and a graphing utility.

64. **Air drop** A plane traveling horizontally at 80 m/s over flat ground at an elevation of 3000 m releases an emergency packet. The trajectory of the packet is given by

$$x = 80t, \quad y = -4.9t^2 + 3000, \quad \text{for } t \ge 0,$$

where the origin is the point on the ground directly beneath the plane at the moment of the release (see figure). Graph the trajectory of the packet and find the coordinates of the point where the packet lands.

65. Air drop—inverse problem A plane traveling horizontally at 100 m/s over flat ground at an elevation of 4000 m must drop an emergency packet on a target on the ground. The trajectory of the packet is given by

$$x = 100t, \quad y = -4.9t^2 + 4000, \quad \text{for } t \geq 0,$$

where the origin is the point on the ground directly beneath the plane at the moment of the release. How many horizontal meters before the target should the packet be released in order to hit the target?

66. Projectile explorations A projectile launched from the ground with an initial speed of 20 m/s and a launch angle θ follows a trajectory approximated by

$$x = (20 \cos \theta)t, \quad y = -4.9t^2 + (20 \sin \theta)t,$$

where x and y are the horizontal and vertical positions of the projectile relative to the launch point $(0, 0)$.

a. Graph the trajectory for various values of θ in the range $0 < \theta < \pi/2$.

b. Based on your observations, what value of θ gives the greatest range (the horizontal distance between the launch and landing points)?

67–72. Derivatives *Consider the following parametric curves.*

a. *Determine dy/dx in terms of t and evaluate it at the given value of t.*

b. *Make a sketch of the curve showing the tangent line at the point corresponding to the given value of t.*

67. $x = 2 + 4t, y = 4 - 8t; t = 2$

68. $x = 3 \sin t, y = 3 \cos t; t = \pi/2$

69. $x = \cos t, y = 8 \sin t; t = \pi/2$

70. $x = 2t, y = t^3; t = -1$

■ 71. $x = t + 1/t, y = t - 1/t; t = 1$

72. $x = \sqrt{t}, y = 2t; t = 4$

73–76. Tangent lines *Find an equation of the line tangent to the curve at the point corresponding to the given value of t.*

73. $x = t^2 - 1, y = t^3 + t; t = 2$

74. $x = \sin t, y = \cos t; t = \pi/4$

75. $x = \cos t + t \sin t, y = \sin t - t \cos t; t = \pi/4$

76. $x = e^t, y = \ln(t + 1); t = 0$

77–80. Slopes of tangent lines *Find all points at which the following curves have the given slope.*

77. $x = 4 \cos t, y = 4 \sin t$; slope $= 1/2$

78. $x = 2 \cos t, y = 8 \sin t$; slope $= -1$

79. $x = t + 1/t, y = t - 1/t$; slope $= 1$

80. $x = 2 + \sqrt{t}, y = 2 - 4t$; slope $= -8$

81–88. Arc length *Find the arc length of the following curves on the given interval.*

81. $x = 3t + 1, y = 4t + 2; 0 \leq t \leq 2$

82. $x = 3 \cos t, y = 3 \sin t + 1; 0 \leq t \leq 2\pi$

83. $x = \cos t - \sin t, y = \cos t + \sin t; 0 \leq t \leq \pi$

84. $x = e^t \sin t, y = e^t \cos t; 0 \leq t \leq 2\pi$

85. $x = t^4, y = \dfrac{t^6}{3}; 0 \leq t \leq 1$

86. $x = 2t \sin t - t^2 \cos t, y = 2t \cos t + t^2 \sin t; 0 \leq t \leq \pi$

87. $x = \cos^3 2t, y = \sin^3 2t; 0 \leq t \leq \pi/4$

88. $x = \sin t, y = t - \cos t; 0 \leq t \leq \pi/2$

89. Explain why or why not Determine whether the following statements are true and give an explanation or counterexample.

a. The equations $x = -\cos t, y = -\sin t$, for $0 \leq t \leq 2\pi$, generate a circle in the clockwise direction.

b. An object following the parametric curve $x = 2 \cos 2\pi t$, $y = 2 \sin 2\pi t$ circles the origin once every 1 time unit.

c. The parametric equations $x = t, y = t^2$, for $t \geq 0$, describe the complete parabola $y = x^2$.

d. The parametric equations $x = \cos t, y = \sin t$, for $-\pi/2 \leq t \leq \pi/2$, describe a semicircle.

e. There are two points on the curve $x = -4 \cos t, y = \sin t$, for $0 \leq t \leq 2\pi$, at which there is a vertical tangent line.

90. Matching curves and equations Match equations a–d with graphs A–D. Explain your reasoning.

a. $x = t^2 - 2, y = t^3 - t$

b. $x = \cos(t + \sin 50t), y = \sin(t + \cos 50t)$

c. $x = t + \cos 2t, y = t - \sin 4t$

d. $x = 2 \cos t + \cos 20t, y = 2 \sin t + \sin 20t$

(A)   (B)

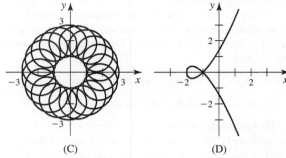

(C)   (D)

91–92. Ellipses *An ellipse (discussed in detail in Section 12.4) is generated by the parametric equations $x = a \cos t, y = b \sin t$. If $0 < a < b$, then the long axis (or **major axis**) lies on the y-axis and the short axis (or **minor axis**) lies on the x-axis. If $0 < b < a$, the axes are reversed. The lengths of the axes in the x- and y-directions are 2a and 2b, respectively. Sketch the graph of the following ellipses. Specify an interval in t over which the entire curve is generated.*

91. $x = 4 \cos t, y = 9 \sin t$

92. $x = 12 \sin 2t, y = 3 \cos 2t$

93–94. Parametric equations of ellipses *Find parametric equations (not unique) of the following ellipses (see Exercises 91–92). Graph the ellipse and find a description in terms of x and y.*

93. An ellipse centered at the origin with major axis of length 6 on the x-axis and minor axis of length 3 on the y-axis, generated counterclockwise

94. An ellipse centered at $(-2, -3)$ with major and minor axes of lengths 30 and 20, parallel to the x- and y-axes, respectively, generated counterclockwise (*Hint:* Shift the parametric equations.)

95. Intersecting lines Consider the following pairs of lines. Determine whether the lines are parallel or intersecting. If the lines intersect, then determine the point of intersection.

a. $x = 1 + s, y = 2s$ and $x = 1 + 2t, y = 3t$
b. $x = 2 + 5s, y = 1 + s$ and $x = 4 + 10t, y = 3 + 2t$
c. $x = 1 + 3s, y = 4 + 2s$ and $x = 4 - 3t, y = 6 + 4t$

96. Multiple descriptions Which of the following parametric equations describe the same curve?

a. $x = 2t^2, y = 4 + t; -4 \le t \le 4$
b. $x = 2t^4, y = 4 + t^2; -2 \le t \le 2$
c. $x = 2t^{2/3}, y = 4 + t^{1/3}; -64 \le t \le 64$

Explorations and Challenges

⊺ 97–100. Beautiful curves *Consider the family of curves*

$$x = \left(2 + \frac{1}{2}\sin at\right)\cos\left(t + \frac{\sin bt}{c}\right),$$

$$y = \left(2 + \frac{1}{2}\sin at\right)\sin\left(t + \frac{\sin bt}{c}\right).$$

Plot a graph of the curve for the given values of a, b, and c with $0 \le t \le 2\pi$.
(*Source: Mathematica in Action*, Stan Wagon, Springer, 2010; created by Norton Starr, Amherst College)

97. $a = b = 5, c = 2$ **98.** $a = 6, b = 12, c = 3$

99. $a = 18, b = 18, c = 7$ **100.** $a = 7, b = 4, c = 1$

⊺ 101–102. Lissajous curves *Consider the following Lissajous curves. Graph the curve, and estimate the coordinates of the points on the curve at which there is (a) a horizontal tangent line or (b) a vertical tangent line. (See the Guided Project Parametric art for more on Lissajous curves.)*

101. $x = \sin 2t, y = 2\sin t; 0 \le t \le 2\pi$

102. $x = \sin 4t, y = \sin 3t; 0 \le t \le 2\pi$

103–105. Area under a curve *Suppose the function $y = h(x)$ is nonnegative and continuous on $[\alpha, \beta]$, which implies that the area bounded by the graph of h and the x-axis on $[\alpha, \beta]$ equals $\int_\alpha^\beta h(x)\,dx$ or $\int_\alpha^\beta y\,dx$. If the graph of $y = h(x)$ on $[\alpha, \beta]$ is traced exactly once by the parametric equations $x = f(t), y = g(t)$, for $a \le t \le b$, then it follows by substitution that the area bounded by h is*

$$\int_\alpha^\beta h(x)\,dx = \int_\alpha^\beta y\,dx = \int_a^b g(t)f'(t)\,dt \text{ if } \alpha = f(a) \text{ and } \beta = f(b)$$

$$\left(\text{or } \int_\alpha^\beta h(x)\,dx = \int_b^a g(t)f'(t)\,dt \text{ if } \alpha = f(b) \text{ and } \beta = f(a)\right).$$

103. Find the area under one arch of the cycloid $x = 3(t - \sin t)$, $y = 3(1 - \cos t)$ (see Example 5).

104. Show that the area of the region bounded by the ellipse $x = 3\cos t, y = 4\sin t$, for $0 \le t \le 2\pi$, equals

$$4\int_{\pi/2}^0 4\sin t(-3\sin t)\,dt.$$ Then evaluate the integral.

105. Find the area of the region bounded by the astroid $x = \cos^3 t, y = \sin^3 t$, for $0 \le t \le 2\pi$ (see Example 8, Figure 12.17).

106–111. Surfaces of revolution *Let C be the curve $x = f(t)$, $y = g(t)$, for $a \le t \le b$, where f' and g' are continuous on $[a, b]$ and C does not intersect itself, except possibly at its endpoints. If g is nonnegative on $[a, b]$, then the area of the surface obtained by revolving C about the x-axis is*

$$S = \int_a^b 2\pi g(t)\sqrt{f'(t)^2 + g'(t)^2}\,dt.$$

Likewise, if f is nonnegative on $[a, b]$, then the area of the surface obtained by revolving C about the y-axis is

$$S = \int_a^b 2\pi f(t)\sqrt{f'(t)^2 + g'(t)^2}\,dt.$$

(These results can be derived in a manner similar to the derivations given in Section 6.6 for surfaces of revolution generated by the curve $y = f(x)$.)

106. Use the parametric equations of a semicircle of radius 1, $x = \cos t$, $y = \sin t$, for $0 \le t \le \pi$, to verify that surface area of a unit sphere is 4π.

107. Consider the curve $x = 3\cos t, y = 3\sin t + 4$, for $0 \le t \le 2\pi$.
a. Describe the curve.
b. If the curve is revolved about the x-axis, describe the shape of the surface of revolution and find the area of the surface.

108. Find the area of the surface obtained by revolving the curve $x = \cos^3 t, y = \sin^3 t$, for $0 \le t \le \pi/2$, about the x-axis.

109. Find the area of the surface obtained by revolving one arch of the cycloid $x = t - \sin t, y = 1 - \cos t$, for $0 \le t \le 2\pi$, about the x-axis.

110. Find the area of the surface obtained by revolving the curve $x = \sqrt{t + 1}, y = t$, for $1 \le t \le 5$, about the y-axis.

⊺ 111. A surface is obtained by revolving the curve $x = e^{3t} + 1$, $y = e^{2t}$, for $0 \le t \le 1$, about the y-axis. Find an integral that gives the area of the surface and approximate the value of the integral.

⊺ 112. Paths of moons A idealized model of the path of a moon (relative to the Sun) moving with constant speed in a circular orbit around a planet, where the planet in turn revolves around the Sun, is given by the parametric equations

$$x(\theta) = a\cos\theta + \cos n\theta, y(\theta) = a\sin\theta + \sin n\theta.$$

The distance from the moon to the planet is taken to be 1, the distance from the planet to the Sun is a, and n is the number of times the moon orbits the planet for every 1 revolution of the planet around the Sun. Plot the graph of the path of a moon for the given constants; then conjecture which values of n produce loops for a fixed value of a.

a. $a = 4, n = 3$ b. $a = 4, n = 4$ c. $a = 4, n = 5$

T 113. Paths of the moons of Earth and Jupiter Use the equations in Exercise 112 to analyze the paths of the following moons in our solar system.

    a. The path of Callisto (one of Jupiter's moons) corresponds to values of $a = 727.5$ and $n = 259.6$. Plot a small portion of the graph to see the detailed behavior of the orbit.

    b. The path of Io (another of Jupiter's moons) corresponds to values of $a = 1846.2$ and $n = 2448.8$. Plot a small portion of the path of Io to see the loops in its orbit.

(Source: The Sun, the Moon, and Convexity, The College Mathematics Journal, 32, Sep 2001)

114. Second derivative Assume a curve is given by the parametric equations $x = f(t)$ and $y = g(t)$, where f and g are twice differentiable. Use the Chain Rule to show that

$$y''(x) = \frac{f'(t)g''(t) - g'(t)f''(t)}{(f'(t))^3}.$$

QUICK CHECK ANSWERS

1. A segment of the parabola $x = y^2$ opening to the right with vertex at the origin **2.** The circle has center $(0, 0)$ and radius 3; it is generated in the counterclockwise direction (positive orientation) starting at $(0, -3)$. **3.** The line $y = -3x - 3$ with slope -3 passing through $(3, -12)$ (when $t = 0$) **4.** One possibility is $x = -2t$, $y = 3 - 3t$, for $0 \le t \le 1$. **5.** $\dfrac{1}{2}$ **6.** $\sqrt{2}$ ◄

12.2 Polar Coordinates

Suppose you work for a company that designs heat shields for space vehicles. The shields are thin plates that are either rectangular or circular in shape. To solve the heat transfer equations for these two shields, you must choose a coordinate system that best fits the geometry of the problem. A Cartesian (rectangular) coordinate system is a natural choice for the rectangular shields (**Figure 12.18a**). However, it does not provide a good fit for the circular shields (**Figure 12.18b**). On the other hand, a **polar coordinate** system, in which the coordinates are constant on circles and rays, is better suited for the circular shields (**Figure 12.18c**).

> Recall that the terms *Cartesian* and *rectangular* both describe the usual *xy*-coordinate system.

(a)

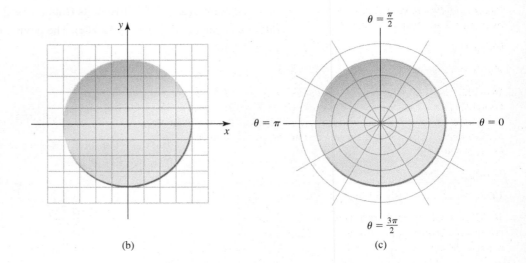

(b)   (c)

Figure 12.18

Defining Polar Coordinates

> Polar points and curves are plotted on a rectangular coordinate system, with standard "*x*" and "*y*" labels attached to the axes. However, plotting polar points and curves is often easier using polar graph paper, which has concentric circles centered at the origin and rays emanating from the origin (Figure 12.18c).

Like Cartesian coordinates, polar coordinates are used to locate points in the plane. When working in polar coordinates, the origin of the coordinate system is also called the **pole**, and the positive *x*-axis is called the **polar axis**. The polar coordinates for a point P have the form (r, θ). **The radial coordinate** r describes the *signed* (or *directed*) distance from the origin to P. The **angular coordinate** θ describes an angle whose initial side is the positive *x*-axis and whose terminal side lies on the ray passing through the origin and P (**Figure 12.19a**). Positive angles are measured counterclockwise from the positive *x*-axis.

    With polar coordinates, points have more than one representation for two reasons. First, angles are determined up to multiples of 2π radians, so the coordinates (r, θ) and

QUICK CHECK 1 Which of the following coordinates represent the same point: $(3, \pi/2)$, $(3, 3\pi/2)$, $(3, 5\pi/2)$, $(-3, -\pi/2)$, and $(-3, 3\pi/2)$? ◄

$(r, \theta \pm 2\pi)$ refer to the same point (**Figure 12.19b**). Second, the radial coordinate may be negative, which is interpreted as follows: The points (r, θ) and $(-r, \theta)$ are reflections of each other through the origin (**Figure 12.19c**). This means that (r, θ), $(-r, \theta + \pi)$, and $(-r, \theta - \pi)$ all refer to the same point. The origin is specified as $(0, \theta)$ in polar coordinates, where θ is any angle.

Figure 12.19

EXAMPLE 1   Points in polar coordinates Graph the following points in polar coordinates: $Q\left(1, \frac{5\pi}{4}\right)$, $R\left(-1, \frac{7\pi}{4}\right)$, and $S\left(2, -\frac{3\pi}{2}\right)$. Give two alternative representations for each point.

SOLUTION The point $Q\left(1, \frac{5\pi}{4}\right)$ is one unit from the origin O on a line OQ that makes an angle of $\frac{5\pi}{4}$ with the positive x-axis (**Figure 12.20a**). Subtracting 2π from the angle, the point Q can be represented as $\left(1, -\frac{3\pi}{4}\right)$. Subtracting π from the angle and negating the radial coordinate implies that Q also has the coordinates $\left(-1, \frac{\pi}{4}\right)$.

To locate the point $R\left(-1, \frac{7\pi}{4}\right)$, it is easiest first to find the point $R^*\left(1, \frac{7\pi}{4}\right)$ in the fourth quadrant. Then $R\left(-1, \frac{7\pi}{4}\right)$ is the reflection of R^* through the origin (**Figure 12.20b**). Other representations of R include $\left(-1, -\frac{\pi}{4}\right)$ and $\left(1, \frac{3\pi}{4}\right)$.

The point $S\left(2, -\frac{3\pi}{2}\right)$ is two units from the origin, found by rotating *clockwise* through an angle of $\frac{3\pi}{2}$ (**Figure 12.20c**). The point S can also be represented as $\left(2, \frac{\pi}{2}\right)$ or $\left(-2, -\frac{\pi}{2}\right)$.

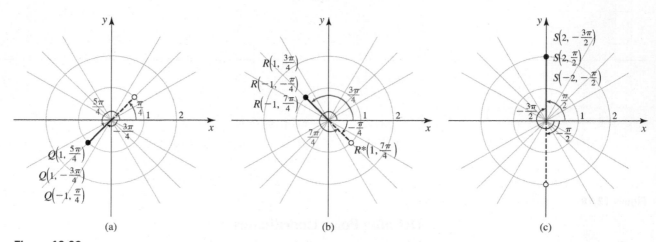

Figure 12.20

Related Exercises 9–13 ◄

Converting Between Cartesian and Polar Coordinates

We often need to convert between Cartesian and polar coordinates. The conversion equations emerge when we look at a right triangle (**Figure 12.21**) in which

$$\cos\theta = \frac{x}{r} \quad \text{and} \quad \sin\theta = \frac{y}{r}.$$

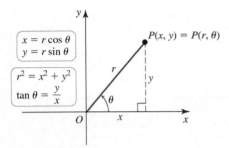

Figure 12.21

QUICK CHECK 2 Draw versions of Figure 12.21 with P in the second, third, and fourth quadrants. Verify that the same conversion formulas hold in all cases. ◀

➤ To determine θ, you may also use the relationships $\cos \theta = x/r$ and $\sin \theta = y/r$. Either method requires checking the signs of x and y to verify that θ is in the correct quadrant.

(a)

(b)

Figure 12.22

QUICK CHECK 3 Give two polar coordinate descriptions of the point with Cartesian coordinates $(1, 0)$. What are the Cartesian coordinates of the point with polar coordinates $\left(2, \frac{\pi}{2}\right)$? ◀

Given a point with polar coordinates (r, θ), we see that its Cartesian coordinates are $x = r \cos \theta$ and $y = r \sin \theta$. Conversely, given a point with Cartesian coordinates (x, y), its radial polar coordinate satisfies $r^2 = x^2 + y^2$. The coordinate θ is determined using the relation $\tan \theta = y/x$, where the quadrant in which θ lies is determined by the signs of x and y. Figure 12.21 illustrates the conversion formulas for a point P in the first quadrant. The same relationships hold if P is in any of the other three quadrants.

PROCEDURE **Converting Coordinates**

A point with polar coordinates (r, θ) has Cartesian coordinates (x, y), where

$$x = r \cos \theta \quad \text{and} \quad y = r \sin \theta.$$

A point with Cartesian coordinates (x, y) has polar coordinates (r, θ), where

$$r^2 = x^2 + y^2 \quad \text{and} \quad \tan \theta = \frac{y}{x}.$$

EXAMPLE 2 **Converting coordinates**

a. Express the point with polar coordinates $P\left(2, \frac{3\pi}{4}\right)$ in Cartesian coordinates.

b. Express the point with Cartesian coordinates $Q(1, -1)$ in polar coordinates.

SOLUTION

a. The point $P\left(2, \frac{3\pi}{4}\right)$ has Cartesian coordinates

$$x = r \cos \theta = 2 \cos \frac{3\pi}{4} = -\sqrt{2} \quad \text{and}$$
$$y = r \sin \theta = 2 \sin \frac{3\pi}{4} = \sqrt{2}.$$

As shown in **Figure 12.22a**, P is in the second quadrant.

b. It's best to locate this point first to be sure that the angle θ is chosen correctly. As shown in **Figure 12.22b**, the point $Q(1, -1)$ is in the fourth quadrant at a distance $r = \sqrt{1^2 + (-1)^2} = \sqrt{2}$ from the origin. The coordinate θ satisfies

$$\tan \theta = \frac{y}{x} = \frac{-1}{1} = -1.$$

The angle in the fourth quadrant with $\tan \theta = -1$ is $\theta = -\frac{\pi}{4}$ or $\frac{7\pi}{4}$. Therefore, two (of infinitely many) polar representations of Q are $\left(\sqrt{2}, -\frac{\pi}{4}\right)$ and $\left(\sqrt{2}, \frac{7\pi}{4}\right)$.

Related Exercises 25, 28, 35 ◀

Basic Curves in Polar Coordinates

A curve in polar coordinates is the set of points that satisfy an equation in r and θ. Some sets of points are easier to describe in polar coordinates than in Cartesian coordinates. Let's begin by examining polar equations of circles, lines, and spirals.

The polar equation $r = 3$ is satisfied by the set of points whose distance from the origin is 3. The angle θ is arbitrary because it is not specified by the equation, so the graph of $r = 3$ is the circle of radius 3 centered at the origin. In general, the equation $r = a$ describes a circle of radius $|a|$ centered at the origin (**Figure 12.23a**).

The equation $\theta = \pi/3$ is satisfied by the points whose angle with respect to the positive x-axis is $\pi/3$. Because r is unspecified, it is arbitrary (and can be positive or negative). Therefore, $\theta = \pi/3$ describes the line through the origin making an angle of $\pi/3$ with the positive x-axis. More generally, $\theta = \theta_0$ describes the line through the origin making an angle of θ_0 with the positive x-axis (**Figure 12.23b**).

▶ If the equation $\theta = \theta_0$ is accompanied by the condition $r \geq 0$, the resulting set of points is a *ray* emanating from the origin.

Figure 12.24

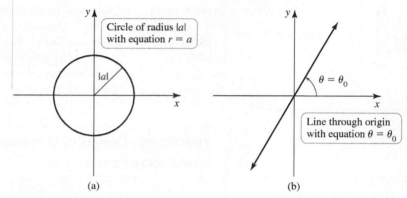

(a)   (b)

Figure 12.23

The simplest polar equation that involves both r and θ is $r = \theta$. Restricting θ to the interval $\theta \geq 0$, we see that as θ increases, r increases. Therefore, as θ increases, the points on the curve move away from the origin as they circle the origin in a counterclockwise direction, generating a spiral (**Figure 12.24**).

QUICK CHECK 4 Describe the polar curves $r = 12$, $r = 6\theta$, and $r \sin \theta = 10$. ◄

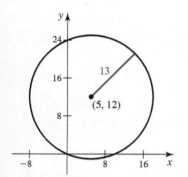

Figure 12.25

EXAMPLE 3   Polar to Cartesian coordinates Convert the polar equation $r = 10 \cos \theta + 24 \sin \theta$ to Cartesian coordinates and describe the corresponding graph.

SOLUTION Multiplying both sides of the equation by r produces the equation $r^2 = 10r \cos \theta + 24r \sin \theta$. Using the conversion relations $r^2 = x^2 + y^2$, $x = r \cos \theta$, and $y = r \sin \theta$, the equation

$$\underbrace{r^2}_{x^2 + y^2} = \underbrace{10r \cos \theta}_{10x} + \underbrace{24r \sin \theta}_{24y}$$

becomes $x^2 + y^2 - 10x - 24y = 0$. Completing the square in both variables gives the equation

$$\underbrace{x^2 - 10x + 25}_{(x-5)^2} - 25 + \underbrace{y^2 - 24y + 144}_{(y-12)^2} - 144 = 0, \text{ or } (x-5)^2 + (y-12)^2 = 169.$$

We recognize $(x - 5)^2 + (y - 12)^2 = 169$ as the equation of a circle of radius 13 centered at $(5, 12)$ (**Figure 12.25**).

Related Exercises 41, 43 ◄

Calculations similar to those in Example 3 lead to the following equations of circles in polar coordinates.

SUMMARY Circles in Polar Coordinates

The equation $r = a$ describes a circle of radius $|a|$ centered at $(0, 0)$.

The equation $r = 2a \cos \theta + 2b \sin \theta$ describes a circle of radius $\sqrt{a^2 + b^2}$ centered at (a, b).

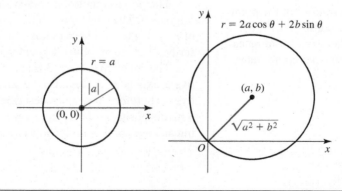

▶ Two important cases deserve mention. The equation $r = 2a \cos \theta$ describes a circle of radius $|a|$ centered at $(a, 0)$. The equation $r = 2b \sin \theta$ describes a circle of radius $|b|$ centered at $(0, b)$.

QUICK CHECK 5 Describe the graph of the equation $r = 2 \cos \theta + 4 \sin \theta$. ◄

Graphing in Polar Coordinates

Equations in polar coordinates often describe curves that are difficult to represent in Cartesian coordinates. Partly for this reason, curve-sketching methods for polar coordinates differ from those used for curves in Cartesian coordinates. Conceptually, the easiest graphing method is to choose several values of θ, calculate the corresponding r-values, and tabulate the coordinates. The points are then plotted and connected with a smooth curve.

> When a curve is described as $r = f(\theta)$, it is natural to tabulate points in θ-r format, just as we list points in x-y format for $y = f(x)$. Despite this fact, the standard form for writing an ordered pair in polar coordinates is (r, θ).

EXAMPLE 4 Plotting a polar curve Graph the polar equation $r = f(\theta) = 1 + \sin \theta$.

SOLUTION The domain of f consists of all real values of θ; however, the complete curve is generated by letting θ vary over any interval of length 2π. Table 12.3 shows several (r, θ) pairs, which are plotted in **Figure 12.26**. The resulting curve, called a **cardioid**, is symmetric about the y-axis.

Table 12.3

θ	$r = 1 + \sin \theta$
0	1
$\pi/6$	$3/2$
$\pi/2$	2
$5\pi/6$	$3/2$
π	1
$7\pi/6$	$1/2$
$3\pi/2$	0
$11\pi/6$	$1/2$
2π	1

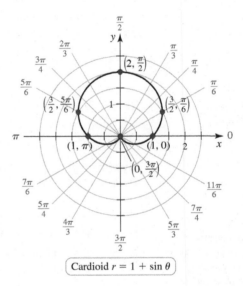

Cardioid $r = 1 + \sin \theta$

Figure 12.26

Related Exercises 53–54 ◀

Cartesian-to-Polar Method Plotting polar curves point by point is time-consuming, and important details may not be revealed. Here is an alternative procedure for graphing polar curves that is usually quicker and more reliable.

PROCEDURE Cartesian-to-Polar Method for Graphing $r = f(\theta)$

1. Graph $r = f(\theta)$ *as if r and θ were Cartesian coordinates* with θ on the horizontal axis and r on the vertical axis. Be sure to choose an interval for θ on which the entire polar curve is produced.

2. Use the Cartesian graph that you created in Step 1 as a guide to sketch the points (r, θ) on the final *polar* curve.

EXAMPLE 5 Plotting polar graphs Use the Cartesian-to-polar method to graph the polar equation $r = 1 + \sin \theta$ (Example 4).

SOLUTION Viewing r and θ as Cartesian coordinates, the graph of $r = 1 + \sin \theta$ on the interval $[0, 2\pi]$ is a standard sine curve with amplitude 1 shifted up 1 unit (**Figure 12.27**). Notice that the graph begins with $r = 1$ at $\theta = 0$, increases to $r = 2$ at $\theta = \pi/2$, decreases to $r = 0$ at $\theta = 3\pi/2$ (which indicates an intersection with the origin on the polar graph), and increases to $r = 1$ at $\theta = 2\pi$. The second row of Figure 12.27 shows the final polar curve (a cardioid) as it is transferred from the Cartesian curve.

r and θ as Cartesian coordinates

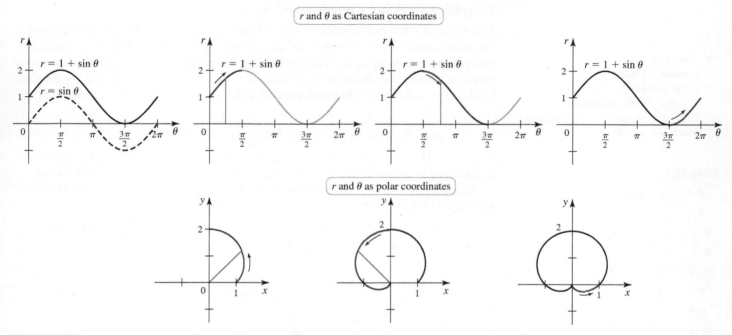

Figure 12.27

Related Exercise 58 ◄

Symmetry   Given a polar equation in r and θ, three types of symmetry are easy to spot (Figure 12.28).

➤ Any two of these three symmetries implies the third. For example, if a graph is symmetric about both the x- and y-axes, then it is symmetric about the origin.

> **SUMMARY Symmetry in Polar Equations**
>
> **Symmetry about the x-axis** occurs if the point (r, θ) is on the graph whenever $(r, -\theta)$ is on the graph.
>
> **Symmetry about the y-axis** occurs if the point (r, θ) is on the graph whenever $(r, \pi - \theta) = (-r, -\theta)$ is on the graph.
>
> **Symmetry about the origin** occurs if the point (r, θ) is on the graph whenever $(-r, \theta) = (r, \theta + \pi)$ is on the graph.

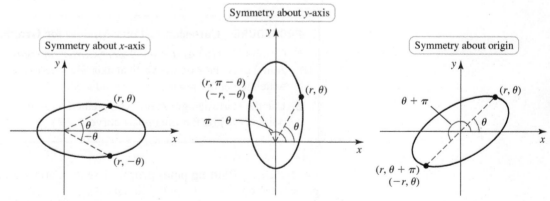

Figure 12.28

QUICK CHECK 6 Identify the symmetry in the graph of (a) $r = 4 + 4 \cos \theta$ and (b) $r = 4 \sin \theta$. ◄

For instance, consider the polar equation $r = 1 + \sin \theta$ in Example 5. If (r, θ) satisfies the equation, then $(r, \pi - \theta)$ also satisfies the equation because $\sin \theta = \sin (\pi - \theta)$. Therefore, the graph is symmetric about the y-axis, as shown in Figure 12.28. Testing for symmetry produces a more accurate graph and often simplifies the task of graphing polar equations.

EXAMPLE 6 Plotting polar graphs Graph the polar equation $r = 3 \sin 2\theta$.

SOLUTION The Cartesian graph of $r = 3 \sin 2\theta$ on the interval $[0, 2\pi]$ has amplitude 3 and period π (Figure 12.29a). The θ-intercepts occur at $\theta = 0, \pi/2, \pi, 3\pi/2$, and 2π, which correspond to the intersections with the origin on the polar graph. Furthermore, the arches of the Cartesian curve between θ-intercepts correspond to loops in the polar curve. The resulting polar curve is a **four-leaf rose** (Figure 12.29b).

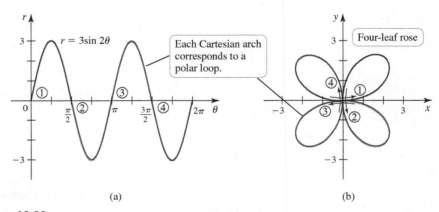

(a)   (b)

Figure 12.29

The graph is symmetric about the x-axis, the y-axis, and the origin. It is instructive to see how these symmetries are justified. To prove symmetry about the y-axis, notice that

$$(r, \theta) \text{ on the graph} \Rightarrow r = 3 \sin 2\theta$$
$$\Rightarrow r = -3 \sin 2(-\theta) \qquad \sin(-\theta) = -\sin \theta$$
$$\Rightarrow -r = 3 \sin 2(-\theta) \qquad \text{Simplify.}$$
$$\Rightarrow (-r, -\theta) \text{ on the graph.}$$

> **Subtle Point**
> The fact that one point has infinitely many representations in polar coordinates presents potential pitfalls. In Example 6, you can show that $(-r, \theta)$ does *not* satisfy the equation $r = 3 \sin 2\theta$ when (r, θ) satisfies the equation. And yet, as shown, the graph is symmetric about the origin because $(r, \theta + \pi)$ satisfies the equation whenever (r, θ) satisfies the equation. Note that $(-r, \theta)$ and $(r, \theta + \pi)$ are the same point.

We see that if (r, θ) is on the graph, then $(-r, -\theta)$ is also on the graph, which implies symmetry about the y-axis. Similarly, to prove symmetry about the origin, notice that

$$(r, \theta) \text{ on the graph} \Rightarrow r = 3 \sin 2\theta$$
$$\Rightarrow r = 3 \sin(2\theta + 2\pi) \qquad \sin(\theta + 2\pi) = \sin \theta$$
$$\Rightarrow r = 3 \sin(2(\theta + \pi)) \qquad \text{Simplify.}$$
$$\Rightarrow (r, \theta + \pi) \text{ on the graph.}$$

We have shown that if (r, θ) is on the graph, then $(r, \theta + \pi)$ is also on the graph, which implies symmetry about the origin. Symmetry about the y-axis and the origin imply symmetry about the x-axis. Had we proved these symmetries in advance, we could have graphed the curve only in the first quadrant—reflections about the x- and y-axes would produce the full curve.

Related Exercises 63–64 ◄

EXAMPLE 7 Plotting polar graphs Graph the polar equation $r^2 = 9 \cos \theta$. Use a graphing utility to check your work.

SOLUTION The graph of this equation has symmetry about the origin (because of the r^2) and about the x-axis (because of $\cos \theta$). These two symmetries imply symmetry about the y-axis.

A preliminary step is required before using the Cartesian-to-polar method for graphing the curve. Solving the given equation for r, we find that $r = \pm 3\sqrt{\cos \theta}$. Notice that $\cos \theta < 0$, for $\pi/2 < \theta < 3\pi/2$, so the curve does not exist on that interval. Therefore, we plot the curve on the intervals $0 \le \theta \le \pi/2$ and $3\pi/2 \le \theta \le 2\pi$ (the interval $\left[-\frac{\pi}{2}, \frac{\pi}{2}\right]$ would also work). Both the positive and negative values of r are included in the Cartesian graph (Figure 12.30a).

Now we are ready to transfer points from the Cartesian graph to the final polar graph (**Figure 12.30b**). The resulting curve is called a **lemniscate**.

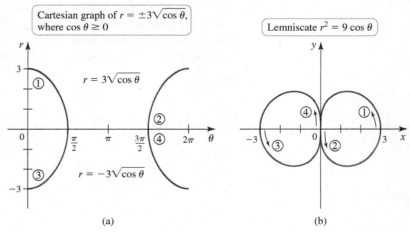

(a)   (b)

Figure 12.30

Related Exercise 61 ◄

EXAMPLE 8   Matching polar and Cartesian graphs The butterfly curve is described by the equation

$$r = e^{\sin \theta} - 2 \cos 4\theta, \quad \text{for } 0 \le \theta \le 2\pi,$$

which is plotted in Cartesian and polar coordinates in **Figure 12.31**. Follow the Cartesian graph through the points $A, B, C, \ldots, N, O$ and mark the corresponding points on the polar curve

SOLUTION Point A in **Figure 12.31a** has the Cartesian coordinates $(\theta = 0, r = -1)$. The corresponding point in the polar plot (**Figure 12.31b**) with polar coordinates $(-1, 0)$ is marked A. Point B in the Cartesian plot is on the θ-axis; therefore, $r = 0$. The corresponding point in the polar plot is the origin. The same argument used to locate B applies to F, H, J, L, and N, all of which appear at the origin in the polar plot. In general, the local and endpoint maxima and minima in the Cartesian graph $(A, C, D, E, G, I, K, M, \text{and } O)$ correspond to the extreme points of the loops of the polar plot and are marked accordingly in Figure 12.31b.

> ➤ See Exercise 105 for a spectacular enhancement of the butterfly curve.

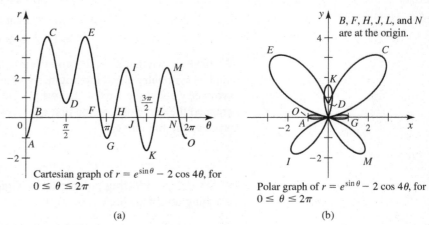

Cartesian graph of $r = e^{\sin \theta} - 2 \cos 4\theta$, for $0 \le \theta \le 2\pi$

(a)

Polar graph of $r = e^{\sin \theta} - 2 \cos 4\theta$, for $0 \le \theta \le 2\pi$

(b)

Figure 12.31

(*Source:* T. H. Fay, *Amer. Math. Monthly* 96, 1989, revived in Wagon and Packel, *Animating Calculus*, Freeman, 1994)

Related Exercises 65–68 ◄

Using Graphing Utilities

When graphing polar curves that eventually close on themselves, it is useful to specify an interval in θ that generates the entire curve. In some cases, this problem is a challenge in itself.

▶ **Using a parametric equation plotter to graph polar curves**
To graph $r = f(\theta)$, treat θ as a parameter and define the parametric equations

$$x = r\cos\theta = \underbrace{f(\theta)}_{r}\cos\theta$$

$$y = r\sin\theta = \underbrace{f(\theta)}_{r}\sin\theta$$

Then graph $(x(\theta), y(\theta))$ as a parametric curve with θ as the parameter.

▶ The prescription given in Example 9 for finding P when working with functions of the form $f(\theta) = \sin\dfrac{p\theta}{q}$ or $f(\theta) = \cos\dfrac{p\theta}{q}$ ensures that the complete curve is generated. Smaller values of P work in some cases.

EXAMPLE 9   Plotting complete curves Consider the closed curve described by $r = \cos(2\theta/5)$. Give an interval in θ that generates the entire curve and then graph the curve.

SOLUTION Recall that $\cos\theta$ has a period of 2π. Therefore, $\cos(2\theta/5)$ completes one cycle when $2\theta/5$ varies from 0 to 2π, or when θ varies from 0 to 5π. Therefore, it is tempting to conclude that the complete curve $r = \cos(2\theta/5)$ is generated as θ varies from 0 to 5π. But you can check that the point corresponding to $\theta = 0$ is *not* the point corresponding to $\theta = 5\pi$, which means the curve does not close on itself over the interval $[0, 5\pi]$ (Figure 12.32a).

To graph the *complete* curve $r = \cos(2\theta/5)$, we must find an interval $[0, P]$, where P is an integer multiple of 5π (so that $f(0) = f(P)$) and an integer multiple of 2π (so that the points $(0, f(0))$ and $(P, f(P))$ are the same). The smallest number satisfying these conditions is 10π. Graphing $r = \cos(2\theta/5)$ over the interval $[0, 10\pi]$ produces the complete curve (Figure 12.32b).

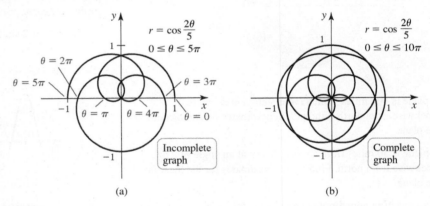

(a)   (b)

Figure 12.32

Related Exercises 72–73 ◀

SECTION 12.2 EXERCISES

Getting Started

1. Plot the points with polar coordinates $(2, \pi/6)$ and $(-3, -\pi/2)$. Give two alternative sets of coordinate pairs for both points.

2. Write the equations that are used to express a point with polar coordinates (r, θ) in Cartesian coordinates.

3. Write the equations that are used to express a point with Cartesian coordinates (x, y) in polar coordinates.

4. What is the polar equation of a circle of radius $\sqrt{a^2 + b^2}$ centered at (a, b)?

5. What is the polar equation of the vertical line $x = 5$?

6. What is the polar equation of the horizontal line $y = 5$?

7. Explain three symmetries in polar graphs and how they are detected in equations.

8. Given three polar coordinate representations for the origin.

Practice Exercises

9–13. *Graph the points with the following polar coordinates. Give two alternative representations of the points in polar coordinates.*

9. $\left(2, \dfrac{\pi}{4}\right)$ 10. $\left(3, \dfrac{2\pi}{3}\right)$ 11. $\left(-1, -\dfrac{\pi}{3}\right)$

12. $\left(2, \dfrac{7\pi}{4}\right)$ 13. $\left(-4, \dfrac{3\pi}{2}\right)$

14. **Points in polar coordinates** Give two sets of polar coordinates for each of the points A–F in the figure.

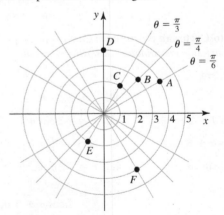

15–22. Sets in polar coordinates *Sketch the following sets of points.*

15. $r \geq 2$ 16. $r = 3$

17. $\theta = \dfrac{2\pi}{3}$ 18. $2 \leq r \leq 8$

19. $\dfrac{\pi}{2} \le \theta \le \dfrac{3\pi}{4}$

20. $1 < r < 2$ and $\dfrac{\pi}{6} \le \theta \le \dfrac{\pi}{3}$

21. $|\theta| \le \dfrac{\pi}{3}$

22. $0 < r < 3$ and $0 \le \theta \le \pi$

23–24. Radar *Airplanes are equipped with transponders that allow air traffic controllers to see their locations on radar screens. Radar gives the distance of the plane from the radar station (located at the origin) and the angular position of the plane, typically measured in degrees clockwise from north.*

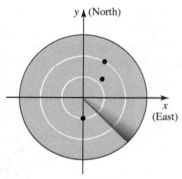

23. A plane is 100 miles from a radar station at an angle of 135° clockwise from north. Find polar coordinates for the location of the plane.

24. A plane is 50 miles from a radar station at an angle of 210° clockwise from north. Find polar coordinates for the location of the plane.

25–30. Converting coordinates *Express the following polar coordinates in Cartesian coordinates.*

25. $\left(3, \dfrac{\pi}{4}\right)$

26. $\left(1, \dfrac{2\pi}{3}\right)$

27. $\left(1, -\dfrac{\pi}{3}\right)$

28. $\left(2, \dfrac{7\pi}{4}\right)$

29. $\left(-4, \dfrac{3\pi}{4}\right)$

30. $(4, 5\pi)$

31–36. Converting coordinates *Express the following Cartesian coordinates in polar coordinates in at least two different ways.*

31. $(2, 2)$

32. $(-1, 0)$

33. $(1, \sqrt{3})$

34. $(0, -9)$

35. $(-4, 4\sqrt{3})$

36. $(-8\sqrt{3}, -8)$

37–48. Polar-to-Cartesian coordinates *Convert the following equations to Cartesian coordinates. Describe the resulting curve.*

37. $r \cos \theta = -4$

38. $r = \cot \theta \csc \theta$

39. $r = 2$

40. $r = 3 \csc \theta$

41. $r = 2 \sin \theta + 2 \cos \theta$

42. $r = -2 \cos \theta$

43. $r = 6 \cos \theta + 8 \sin \theta$

44. $\sin \theta = |\cos \theta|$

45. $r \cos \theta = \sin 2\theta$

46. $r = \sin \theta \sec^2 \theta$

47. $r = 8 \sin \theta$

48. $r = \dfrac{1}{2 \cos \theta + 3 \sin \theta}$

49–52. Cartesian-to-polar coordinates *Convert the following equations to polar coordinates.*

49. $y = x^2$

50. $y = 3$

51. $y = \dfrac{1}{x}$

52. $(x - 1)^2 + y^2 = 1$

53–56. Simple curves *Tabulate and plot enough points to sketch a graph of the following equations.*

53. $r = 8 \cos \theta$

54. $r = 4 + 4 \cos \theta$

55. $r(\sin \theta - 2 \cos \theta) = 0$

56. $r = 1 - \cos \theta$

57–64. Graphing polar curves *Graph the following equations. Use a graphing utility to check your work and produce a final graph.*

57. $r = 2 \sin \theta - 1$

58. $r = 2 - 2 \sin \theta$

59. $r = \sin^2 (\theta/2)$

60. $r^2 = 4 \sin \theta$

61. $r^2 = 16 \cos \theta$

62. $r^2 = 16 \sin 2\theta$

63. $r = \sin 3\theta$

64. $r = 2 \sin 5\theta$

65–68. Matching polar and Cartesian curves *Cartesian and polar graphs of $r = f(\theta)$ are given in the figures. Mark the points on the polar graph that correspond to the lettered points on the Cartesian graph.*

65. $r = 1 - 2 \sin 3\theta$

66. $r = \sin (1 + 3 \cos \theta)$

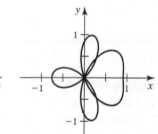

67. $r = \dfrac{1}{4} - \cos 4\theta$

68. $r = \cos\theta + \sin 2\theta$

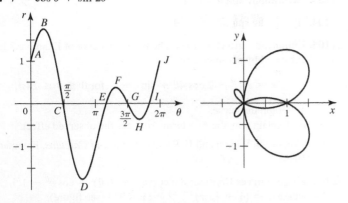

⊤ 69–76. Using a graphing utility *Use a graphing utility to graph the following equations. In each case, give the smallest interval $[0, P]$ that generates the entire curve.*

69. $r = \sin\dfrac{\theta}{4}$ **70.** $r = 2 - 4\cos 5\theta$

71. $r = \cos 3\theta + \cos^2 2\theta$ **72.** $r = 2\sin\dfrac{2\theta}{3}$

73. $r = \cos\dfrac{3\theta}{5}$ **74.** $r = \sin\dfrac{3\theta}{7}$

75. $r = 1 - 3\cos 2\theta$ **76.** $r = 1 - 2\sin 5\theta$

77. Explain why or why not Determine whether the following statements are true and give an explanation or counterexample.

  a. The point with Cartesian coordinates $(-2, 2)$ has polar coordinates $(2\sqrt{2}, 3\pi/4)$, $(2\sqrt{2}, 11\pi/4)$, $(2\sqrt{2}, -5\pi/4)$, and $(-2\sqrt{2}, -\pi/4)$.

  b. The graphs of $r\cos\theta = 4$ and $r\sin\theta = -2$ intersect exactly once.

  c. The graphs of $r = 2$ and $\theta = \pi/4$ intersect exactly once.

  d. The point $(3, \pi/2)$ lies on the graph of $r = 3\cos 2\theta$.

  e. The graphs of $r = 2\sec\theta$ and $r = 3\csc\theta$ are lines.

78. Circles in general Show that the polar equation

$$r^2 - 2r(a\cos\theta + b\sin\theta) = R^2 - a^2 - b^2$$

describes a circle of radius R centered at (a, b).

79. Circles in general Show that the polar equation

$$r^2 - 2rr_0\cos(\theta - \theta_0) = R^2 - r_0{}^2$$

describes a circle of radius R whose center has polar coordinates (r_0, θ_0).

80–83. Equations of circles *Use the results of Exercises 78–79 to describe and graph the following circles.*

80. $r^2 - 6r\cos\theta = 16$

81. $r^2 - 4r\cos(\theta - \pi/3) = 12$

82. $r^2 - 8r\cos(\theta - \pi/2) = 9$

83. $r^2 - 2r(2\cos\theta + 3\sin\theta) = 3$

Explorations and Challenges

84. Equations of circles Find equations of the circles in the figure. Determine whether the combined area of the circles is greater than

or less than the area of the region inside the square but outside the circles.

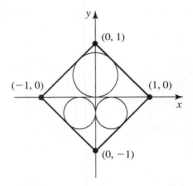

85. Navigating A plane is 150 miles north of a radar station, and 30 minutes later it is 60° east of north at a distance of 100 miles from the radar station. Assume the plane flies on a straight line and maintains constant altitude during this 30-minute period.

  a. Find the distance traveled during this 30-minute period.

  b. Determine the average velocity of the plane (relative to the ground) during this 30-minute period.

86. Lines in polar coordinates

  a. Show that an equation of the line $y = mx + b$ in polar coordinates is $r = \dfrac{b}{\sin\theta - m\cos\theta}$.

  b. Use the figure to find an alternative polar equation of a line, $r\cos(\theta_0 - \theta) = r_0$. Note that $Q(r_0, \theta_0)$ is a fixed point on the line such that OQ is perpendicular to the line and $r_0 \geq 0$; $P(r, \theta)$ is an arbitrary point on the line.

87–90. Equations of lines *Use the result of Exercise 86 to describe and graph the following lines.*

87. $r\cos\left(\dfrac{\pi}{3} - \theta\right) = 3$

88. $r\cos\left(\theta + \dfrac{\pi}{6}\right) = 4$

89. $r(\sin\theta - 4\cos\theta) - 3 = 0$

90. $r(4\sin\theta - 3\cos\theta) = 6$

91. The limaçon family The equations $r = a + b\cos\theta$ and $r = a + b\sin\theta$ describe curves known as *limaçons* (from Latin for *snail*). We have already encountered cardioids, which occur when $|a| = |b|$. The limaçon has an inner loop if $|a| < |b|$. The limaçon has a dent or dimple if $|b| < |a| < 2|b|$. And the limaçon is oval-shaped if $|a| > 2|b|$. Match equations a–f with the limaçons in figures A–F.

a. $r = -1 + \sin\theta$ b. $r = -1 + 2\cos\theta$

c. $r = 2 + \sin\theta$ d. $r = 1 - 2\cos\theta$

e. $r = 1 + 2\sin\theta$ f. $r = 1 + 2/3\sin\theta$

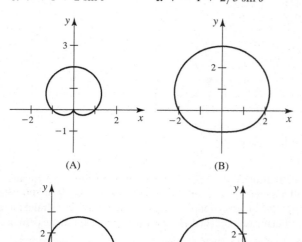

(A)

(B)

(C)

(D)

(E)

(F)

92. Limiting limaçon Consider the family of limaçons
$r = 1 + b\cos\theta$. Describe how the curves change as $b \to \infty$.

93–96. The lemniscate family *Equations of the form $r^2 = a\sin 2\theta$ and $r^2 = a\cos 2\theta$ describe lemniscates (see Example 7). Graph the following lemniscates.*

93. $r^2 = \cos 2\theta$

94. $r^2 = 4\sin 2\theta$

95. $r^2 = -2\sin 2\theta$

96. $r^2 = -8\cos 2\theta$

97–100. The rose family *Equations of the form $r = a\sin m\theta$ or $r = a\cos m\theta$, where a is a real number and m is a positive integer, have graphs known as roses (see Example 6). Graph the following roses.*

97. $r = \sin 2\theta$

98. $r = 4\cos 3\theta$

99. $r = 2\sin 4\theta$

100. $r = 6\sin 5\theta$

101. Number of rose petals Show that the graph of $r = a\sin m\theta$ or $r = a\cos m\theta$ is a rose with m leaves if m is an odd integer and a rose with $2m$ leaves if m is an even integer.

102–104. Spirals *Graph the following spirals. Indicate the direction in which the spiral is generated as θ increases, where $\theta > 0$. Let $a = 1$ and $a = -1$.*

102. Spiral of Archimedes: $r = a\theta$

103. Logarithmic spiral: $r = e^{a\theta}$

104. Hyperbolic spiral: $r = a/\theta$

T 105. Enhanced butterfly curve The butterfly curve of Example 8 is enhanced by adding a term:

$$r = e^{\sin\theta} - 2\cos 4\theta + \sin^5\frac{\theta}{12}, \quad \text{for } 0 \le \theta \le 24\pi.$$

a. Graph the curve.

b. Explain why the new term produces the observed effect.

(*Source:* S. Wagon and E. Packel, *Animating Calculus*, Freeman, 1994)

T 106. Finger curves Consider the curve $r = f(\theta) = \cos a^\theta - 1.5$, where $a = (1 + 12\pi)^{1/(2\pi)} \approx 1.78933$ (see figure).

a. Show that $f(0) = f(2\pi)$ and find the point on the curve that corresponds to $\theta = 0$ and $\theta = 2\pi$.

b. Is the same curve produced over the intervals $[-\pi, \pi]$ and $[0, 2\pi]$?

c. Let $f(\theta) = \cos(a^\theta) - b$, where $a = (1 + 2k\pi)^{1/(2\pi)}$, k is an integer, and b is a real number. Show that $f(0) = f(2\pi)$ and the curve closes on itself.

d. Plot the curve with various values of k. How many fingers can you produce?

T 107. Earth-Mars system A simplified model assumes the orbits of Earth and Mars around the Sun are circular with radii of 2 and 3, respectively, and that Earth completes one orbit in one year while Mars takes two years. When $t = 0$, Earth is at $(2, 0)$ and Mars is at $(3, 0)$; both orbit the Sun (at $(0, 0)$) in the counterclockwise direction. The position of Mars relative to Earth is given by the parametric equations

$$x = (3 - 4\cos\pi t)\cos\pi t + 2, \quad y = (3 - 4\cos\pi t)\sin\pi t.$$

a. Graph the parametric equations, for $0 \le t \le 2$.

b. Letting $r = (3 - 4\cos\pi t)$, explain why the path of Mars relative to Earth is a limaçon (Exercise 91).

108. Channel flow Water flows in a shallow semicircular channel with inner and outer radii of 1 m and 2 m (see figure). At a point $P(r, \theta)$ in the channel, the flow is in the tangential direction (counterclockwise along circles), and it depends only on r, the distance from the center of the semicircles.

a. Express the region formed by the channel as a set in polar coordinates.

b. Express the inflow and outflow regions of the channel as sets in polar coordinates.

c. Suppose the tangential velocity of the water in meters per second is given by $v(r) = 10r$, for $1 \le r \le 2$. Is the velocity greater at $\left(1.5, \frac{\pi}{4}\right)$ or $\left(1.2, \frac{3\pi}{4}\right)$? Explain.

d. Suppose the tangential velocity of the water is given by $v(r) = 20/r$, for $1 \leq r \leq 2$. Is the velocity greater at $(1.8, \pi/6)$ or $(1.3, 2\pi/3)$? Explain.

e. The total amount of water that flows through the channel (across a cross section of the channel $\theta = \theta_0$) is proportional to $\int_1^2 v(r)\, dr$. Is the total flow through the channel greater for the flow in part (c) or part (d)?

Outflow   Inflow

109. Subtle symmetry Without using a graphing utility, determine the symmetries (if any) of the curve $r = 4 - \sin(\theta/2)$.

T 110. Complete curves Consider the polar curve $r = \cos(n\theta/m)$, where n and m are integers.

a. Graph the complete curve when $n = 2$ and $m = 3$.
b. Graph the complete curve when $n = 3$ and $m = 7$.
c. Find a general rule in terms of m and n (where m and n have no common factors) for determining the least positive number P such that the complete curve is generated over the interval $[0, P]$.

111. Cartesian lemniscate Find the equation in Cartesian coordinates of the lemniscate $r^2 = a^2 \cos 2\theta$, where a is a real number.

QUICK CHECK ANSWERS

1. All the points are the same except $(3, 3\pi/2)$.   **3.** Polar coordinates: $(1, 0)$, $(1, 2\pi)$; Cartesian coordinates: $(0, 2)$
4. A circle centered at the origin with radius 12; two spirals; the horizontal line $y = 10$ **5.** A circle of radius $\sqrt{5}$ with center $(1, 2)$ **6.** (a) Symmetric about the x-axis; (b) symmetric about the y-axis ◄

12.3 Calculus in Polar Coordinates

Having learned about the *geometry* of polar coordinates, we now have the tools needed to explore *calculus* in polar coordinates. Familiar topics, such as slopes of tangent lines, areas bounded by curves, and arc length, are now revisited in a different setting.

Slopes of Tangent Lines

Given a function $y = f(x)$, the slope of the line tangent to the graph at a given point is $\dfrac{dy}{dx}$ or $f'(x)$. So it is tempting to conclude that the slope of a curve described by the polar equation $r = f(\theta)$ is $\dfrac{dr}{d\theta} = f'(\theta)$. Unfortunately, it's not that simple.

> ➤ The slope is the change in the vertical coordinate divided by the change in the horizontal coordinate, independent of the coordinate system. In polar coordinates, neither r nor θ corresponds to a vertical or horizontal coordinate.

The key observation is that the slope of a tangent line—in any coordinate system—is the rate of change of the vertical coordinate y with respect to the horizontal coordinate x, which is $\dfrac{dy}{dx}$. We begin by writing the polar equation $r = f(\theta)$ in parametric form with θ as a parameter:

$$x = r \cos \theta = f(\theta) \cos \theta \quad \text{and} \quad y = r \sin \theta = f(\theta) \sin \theta. \tag{1}$$

From Section 12.1, when x and y are defined parametrically as differentiable functions of θ, the derivative is $\dfrac{dy}{dx} = \dfrac{dy/d\theta}{dx/d\theta}$. Using the Product Rule to compute $\dfrac{dy}{d\theta}$ and $\dfrac{dx}{d\theta}$ in equation (1), we have

$$\frac{dy}{dx} = \frac{\overbrace{f'(\theta) \sin \theta + f(\theta) \cos \theta}^{dy/d\theta}}{\underbrace{f'(\theta) \cos \theta - f(\theta) \sin \theta}_{dx/d\theta}}. \tag{2}$$

If the graph passes through the origin for some angle θ_0, then $f(\theta_0) = 0$, and equation (2) simplifies to

$$\frac{dy}{dx} = \frac{\sin \theta_0}{\cos \theta_0} = \tan \theta_0,$$

provided $f'(\theta_0) \neq 0$. Assuming $\cos \theta_0 \neq 0$, $\tan \theta_0$ is the slope of the line $\theta = \theta_0$, which also passes through the origin. In this case, we conclude that if $f(\theta_0) = 0$, then the tangent

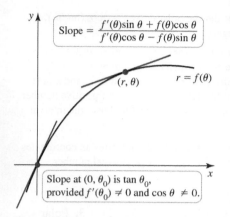

Figure 12.33

QUICK CHECK 1 Verify that if $y = f(\theta) \sin \theta$, then $y'(\theta) = f'(\theta) \sin \theta + f(\theta) \cos \theta$ (which was used earlier to find dy/dx). ◄

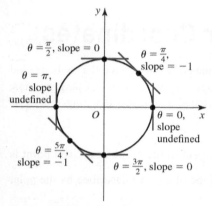

Figure 12.34

line at $(0, \theta_0)$ is simply $\theta = \theta_0$ (**Figure 12.33**). If $f(\theta_0) = 0$, $f'(\theta_0) \neq 0$, and $\cos \theta_0 = 0$, then the graph has a vertical tangent line at the origin.

THEOREM 12.2   Slope of a Tangent Line
Let f be a differentiable function at θ_0. The slope of the line tangent to the curve $r = f(\theta)$ at the point $(f(\theta_0), \theta_0)$ is

$$\frac{dy}{dx} = \frac{f'(\theta_0) \sin \theta_0 + f(\theta_0) \cos \theta_0}{f'(\theta_0) \cos \theta_0 - f(\theta_0) \sin \theta_0},$$

provided the denominator is nonzero at the point. At angles θ_0 for which $f(\theta_0) = 0$, $f'(\theta_0) \neq 0$, and $\cos \theta_0 \neq 0$, the tangent line is $\theta = \theta_0$ with slope $\tan \theta_0$.

EXAMPLE 1   Slopes on a circle Find the slopes of the lines tangent to the circle $r = f(\theta) = 10$.

SOLUTION In this case, $f(\theta)$ is constant (independent of θ). Therefore, $f'(\theta) = 0$, $f(\theta) \neq 0$, and the slope formula becomes

$$\frac{dy}{dx} = \frac{f'(\theta) \sin \theta + f(\theta) \cos \theta}{f'(\theta) \cos \theta - f(\theta) \sin \theta} = -\frac{\cos \theta}{\sin \theta} = -\cot \theta.$$

We can check a few points to see that this result makes sense. With $\theta = 0$ and $\theta = \pi$, the slope $\frac{dy}{dx} = -\cot \theta$ is undefined, which implies the tangent lines are vertical at these points (**Figure 12.34**). With $\theta = \pi/2$ and $\theta = 3\pi/2$, the slope is zero; with $\theta = 3\pi/4$ and $\theta = 7\pi/4$, the slope is 1; and with $\theta = \pi/4$ and $\theta = 5\pi/4$, the slope is -1. At all points $P(r, \theta)$ on the circle, the slope of the line OP from the origin to P is $\tan \theta$, which is the negative reciprocal of $-\cot \theta$. Therefore, OP is perpendicular to the tangent line at all points P on the circle.

Related Exercises 12–13 ◄

EXAMPLE 2   Vertical and horizontal tangent lines Find the points on the interval $-\pi \leq \theta \leq \pi$ at which the cardioid $r = f(\theta) = 1 - \cos \theta$ has a vertical or horizontal tangent line.

SOLUTION Applying Theorem 12.2, we find that

$$\frac{dy}{dx} = \frac{f'(\theta) \sin \theta + f(\theta) \cos \theta}{f'(\theta) \cos \theta - f(\theta) \sin \theta} \qquad f'(\theta) = \sin \theta$$

$$= \frac{\overbrace{\sin \theta \sin \theta}^{\sin^2 \theta = 1 - \cos^2 \theta} + (1 - \cos \theta) \cos \theta}{\underbrace{\sin \theta \cos \theta - (1 - \cos \theta) \sin \theta}_{\sin \theta (2 \cos \theta - 1)}} \qquad \text{Substitute for } f(\theta) \text{ and } f'(\theta).$$

$$= -\frac{(2 \cos^2 \theta - \cos \theta - 1)}{\sin \theta (2 \cos \theta - 1)} \qquad \text{Simplify.}$$

$$= -\frac{(2 \cos \theta + 1)(\cos \theta - 1)}{\sin \theta (2 \cos \theta - 1)}. \qquad \text{Factor the numerator.}$$

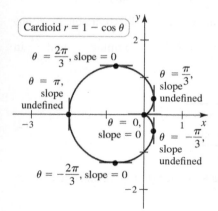

Figure 12.35

The points with a horizontal tangent line satisfy $\frac{dy}{dx} = 0$ and occur where the numerator is zero and the denominator is nonzero. The numerator is zero when $\theta = 0$ and $\pm 2\pi/3$. Because the denominator is *not* zero when $\theta = \pm 2\pi/3$, horizontal tangent lines occur at $\theta = \pm 2\pi/3$ (**Figure 12.35**).

Vertical tangent lines occur where the numerator of $\frac{dy}{dx}$ is nonzero and the denominator is zero. The denominator is zero when $\theta = 0$, $\pm \pi$, and $\pm \pi/3$, and the numerator is not zero when $\theta = \pm\pi$ and $\pm\pi/3$. Therefore, vertical tangent lines occur at $\theta = \pm\pi$ and $\pm\pi/3$.

The point $(0, 0)$ on the curve must be handled carefully because both the numerator and denominator of $\dfrac{dy}{dx}$ equal 0 at $\theta = 0$. Notice that with $f(\theta) = 1 - \cos\theta$, we have $f(0) = f'(0) = 0$. Therefore, $\dfrac{dy}{dx}$ may be computed as a limit using l'Hôpital's Rule.

As $\theta \to 0^+$, we find that

$$\frac{dy}{dx} = \lim_{\theta \to 0^+} \left(-\frac{(2\cos\theta + 1)(\cos\theta - 1)}{\sin\theta\,(2\cos\theta - 1)} \right)$$

$$= \lim_{\theta \to 0^+} \frac{4\cos\theta\sin\theta - \sin\theta}{-2\sin^2\theta + 2\cos^2\theta - \cos\theta} \qquad \text{L'Hôpital's Rule}$$

$$= \frac{0}{1} = 0. \qquad \text{Evaluate the limit.}$$

QUICK CHECK 2 What is the slope of the line tangent to the cardioid in Example 2 at the point corresponding to $\theta = \pi/4$? ◄

A similar calculation using l'Hôpital's Rule shows that as $\theta \to 0^-$, $\dfrac{dy}{dx} \to 0$. Therefore, the curve has a slope of 0 at $(0, 0)$.

Related Exercises 25–26 ◄

Area of Regions Bounded by Polar Curves

The problem of finding the area of a region bounded by polar curves brings us back to the slice-and-sum strategy used extensively in Chapters 5 and 6. The objective is to find the area of the region R bounded by the graph of $r = f(\theta)$ between the two rays $\theta = \alpha$ and $\theta = \beta$ (**Figure 12.36a**). We assume f is continuous and nonnegative on $[\alpha, \beta]$.

The area of R is found by slicing the region in the radial direction, creating wedge-shaped slices. The interval $[\alpha, \beta]$ is partitioned into n subintervals by choosing the grid points

$$\alpha = \theta_0 < \theta_1 < \theta_2 < \cdots < \theta_k < \cdots < \theta_n = \beta.$$

We let $\Delta\theta_k = \theta_k - \theta_{k-1}$, for $k = 1, 2, \ldots, n$, and we let θ_k^* be any point of the interval $[\theta_{k-1}, \theta_k]$. The kth slice is approximated by the sector of a circle swept out by an angle $\Delta\theta_k$ with radius $f(\theta_k^*)$ (**Figure 12.36b**). Therefore, the area of the kth slice is approximately $\frac{1}{2} f(\theta_k^*)^2 \Delta\theta_k$, for $k = 1, 2, \ldots, n$ (**Figure 12.36c**). To find the approximate area of R, we sum the areas of these slices:

$$\text{area} \approx \sum_{k=1}^{n} \frac{1}{2} f(\theta_k^*)^2 \, \Delta\theta_k.$$

Area of circle $= \pi r^2$

Area of $\Delta\theta/(2\pi)$ of a circle

$= \left(\dfrac{\Delta\theta}{2\pi}\right)\pi r^2 = \dfrac{1}{2}r^2\Delta\theta$

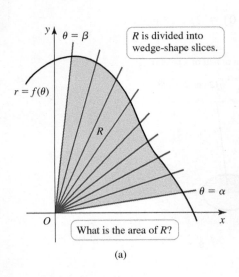

R is divided into wedge-shape slices.

$r = f(\theta)$

$\theta = \beta$

R

$\theta = \alpha$

What is the area of R?

(a)

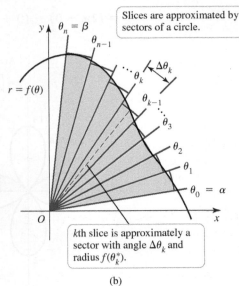

Slices are approximated by sectors of a circle.

$r = f(\theta)$

kth slice is approximately a sector with angle $\Delta\theta_k$ and radius $f(\theta_k^*)$.

(b)

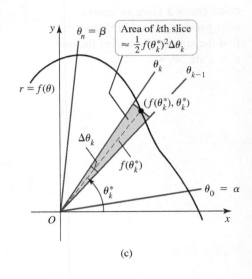

Area of kth slice $\approx \frac{1}{2} f(\theta_k^*)^2 \Delta\theta_k$

$r = f(\theta)$

$(f(\theta_k^*), \theta_k^*)$

$f(\theta_k^*)$

(c)

Figure 12.36

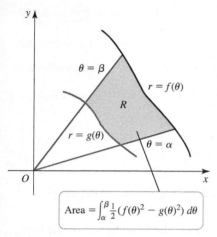

$$\text{Area} = \int_\alpha^\beta \frac{1}{2}(f(\theta)^2 - g(\theta)^2)\, d\theta$$

Figure 12.37

➤ If R is bounded by the graph of $r = f(\theta)$ between $\theta = \alpha$ and $\theta = \beta$, then $g(\theta) = 0$ and the area of R is $\int_\alpha^\beta \frac{1}{2} f(\theta)^2\, d\theta$.

➤ Though we assume $r = f(\theta) \geq 0$ when deriving the formula for the area of a region bounded by a polar curve, the formula is valid when $r < 0$ (see, for example, Exercise 54).

➤ The equation $r = 2 \cos 2\theta$ is unchanged when θ is replaced with $-\theta$ (symmetry about the x-axis) and when θ is replaced with $\pi - \theta$ (symmetry about the y-axis).

QUICK CHECK 3 Give an interval over which you could integrate to find the area of one leaf of the rose $r = 2 \sin 3\theta$. ◄

This approximation is a Riemann sum, and the approximation improves as we take more sectors ($n \to \infty$) and let $\Delta\theta_k \to 0$, for all k. The exact area is given by $\lim\limits_{n \to \infty} \sum\limits_{k=1}^{n} \frac{1}{2} f(\theta_k^*)^2\, \Delta\theta_k$, which we identify as the definite integral $\int_\alpha^\beta \frac{1}{2} f(\theta)^2\, d\theta$.

With a slight modification, a more general result is obtained for the area of a region R bounded by two curves, $r = f(\theta)$ and $r = g(\theta)$, between the rays $\theta = \alpha$ and $\theta = \beta$ (**Figure 12.37**). We assume f and g are continuous and $f(\theta) \geq g(\theta) \geq 0$ on $[\alpha, \beta]$. To find the area of R, we subtract the area of the region bounded by $r = g(\theta)$ from the area of the entire region bounded by $r = f(\theta)$ (all between $\theta = \alpha$ and $\theta = \beta$); that is,

$$\text{area} = \int_\alpha^\beta \frac{1}{2} f(\theta)^2\, d\theta - \int_\alpha^\beta \frac{1}{2} g(\theta)^2\, d\theta = \int_\alpha^\beta \frac{1}{2}(f(\theta)^2 - g(\theta)^2)\, d\theta.$$

DEFINITION Area of Regions in Polar Coordinates

Let R be the region bounded by the graphs of $r = f(\theta)$ and $r = g(\theta)$, between $\theta = \alpha$ and $\theta = \beta$, where f and g are continuous and $f(\theta) \geq g(\theta) \geq 0$ on $[\alpha, \beta]$. The area of R is

$$\int_\alpha^\beta \frac{1}{2}(f(\theta)^2 - g(\theta)^2)\, d\theta.$$

EXAMPLE 3   Area of a polar region Find the area of the four-leaf rose $r = f(\theta) = 2 \cos 2\theta$.

SOLUTION The graph of the rose (**Figure 12.38**) *appears* to be symmetric about the x- and y-axes; in fact, these symmetries can be proved. Appealing to this symmetry, we find the area of one-half of a leaf and then multiply the result by 8 to obtain the area of the full rose. The upper half of the rightmost leaf is generated as θ increases from $\theta = 0$ (when $r = 2$) to $\theta = \pi/4$ (when $r = 0$). Therefore, the area of the entire rose is

$$8 \int_0^{\pi/4} \frac{1}{2} f(\theta)^2\, d\theta = 4 \int_0^{\pi/4} (2 \cos 2\theta)^2\, d\theta \qquad f(\theta) = 2 \cos 2\theta$$

$$= 16 \int_0^{\pi/4} \cos^2 2\theta\, d\theta \qquad \text{Simplify.}$$

$$= 16 \int_0^{\pi/4} \frac{1 + \cos 4\theta}{2}\, d\theta \qquad \text{Half-angle formula}$$

$$= (8\theta + 2 \sin 4\theta) \Big|_0^{\pi/4} \qquad \text{Fundamental Theorem}$$

$$= (2\pi + 0) - (0 + 0) = 2\pi. \quad \text{Simplify.}$$

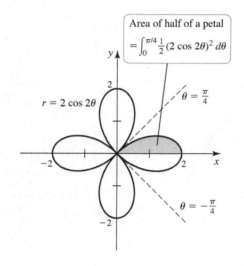

Area of half of a petal $= \int_0^{\pi/4} \frac{1}{2}(2 \cos 2\theta)^2\, d\theta$

$r = 2 \cos 2\theta$

$\theta = \dfrac{\pi}{4}$

$\theta = -\dfrac{\pi}{4}$

Figure 12.38

Related Exercises 38–39 ◄

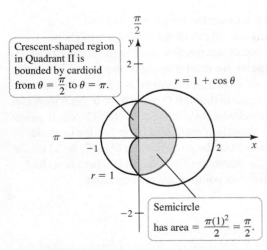

Crescent-shaped region in Quadrant II is bounded by cardioid from $\theta = \dfrac{\pi}{2}$ to $\theta = \pi$.

Semicircle has area $= \dfrac{\pi(1)^2}{2} = \dfrac{\pi}{2}$.

Figure 12.39

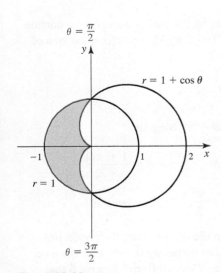

$\theta = \dfrac{\pi}{2}$

$r = 1 + \cos\theta$

$r = 1$

$\theta = \dfrac{3\pi}{2}$

Figure 12.40

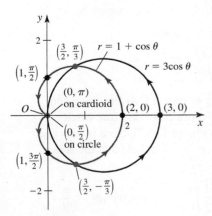

$\left(\dfrac{3}{2}, \dfrac{\pi}{3}\right)$ $r = 1 + \cos\theta$

$\left(1, \dfrac{\pi}{2}\right)$ $r = 3\cos\theta$

$(0, \pi)$ on cardioid

$(2, 0)$ $(3, 0)$

$\left(0, \dfrac{\pi}{2}\right)$ on circle

$\left(1, \dfrac{3\pi}{2}\right)$

$\left(\dfrac{3}{2}, -\dfrac{\pi}{3}\right)$

Figure 12.41

EXAMPLE 4   Areas of polar regions Consider the circle $r = 1$ and the cardioid $r = 1 + \cos\theta$ (Figure 12.39).

a. Find the area of the region inside the circle and inside the cardioid.

b. Find the area of the region inside the circle and outside the cardioid.

SOLUTION

a. The points of intersection of the two curves can be found by solving $1 + \cos\theta = 1$, or $\cos\theta = 0$. The solutions are $\theta = \pm\pi/2$. The region inside the circle and inside the cardioid consists of two subregions (Figure 12.39):

- a semicircle with radius 1 in the first and fourth quadrants bounded by the circle $r = 1$, and

- two crescent-shaped regions in the second and third quadrants bounded by the cardioid $r = 1 + \cos\theta$ and the y-axis.

The area of the semicircle is $\pi/2$. To find the area of the upper crescent-shaped region in the second quadrant, notice that it is bounded by $r = 1 + \cos\theta$, as θ varies from $\pi/2$ to π. Therefore, its area is

$$\int_{\pi/2}^{\pi} \frac{1}{2}(1 + \cos\theta)^2\,d\theta = \int_{\pi/2}^{\pi} \frac{1}{2}(1 + 2\cos\theta + \cos^2\theta)\,d\theta \qquad \text{Expand.}$$

$$= \frac{1}{2}\int_{\pi/2}^{\pi}\left(1 + 2\cos\theta + \frac{1 + \cos 2\theta}{2}\right)d\theta \qquad \text{Half-angle formula}$$

$$= \frac{1}{2}\left(\theta + 2\sin\theta + \frac{\theta}{2} + \frac{\sin 2\theta}{4}\right)\Big|_{\pi/2}^{\pi} \qquad \text{Evaluate integral.}$$

$$= \frac{3\pi}{8} - 1. \qquad \text{Simplify.}$$

The area of the entire region (two crescents and a semicircle) is

$$2\left(\frac{3\pi}{8} - 1\right) + \frac{\pi}{2} = \frac{5\pi}{4} - 2.$$

b. The region inside the circle and outside the cardioid is bounded by the outer curve $r = 1$ and the inner curve $r = 1 + \cos\theta$ on the interval $[\pi/2, 3\pi/2]$ (Figure 12.40). Using the symmetry about the x-axis, the area of the region is

$$2\int_{\pi/2}^{\pi} \frac{1}{2}(1^2 - (1 + \cos\theta)^2)\,d\theta = \int_{\pi/2}^{\pi}(-2\cos\theta - \cos^2\theta)\,d\theta \quad \text{Simplify integrand.}$$

$$= 2 - \frac{\pi}{4}. \qquad \text{Evaluate integral.}$$

Note that the regions in parts (a) and (b) make up the interior of a circle of radius 1; indeed, their areas have a sum of π.

Related Exercise 45 ◀

Part of the challenge in setting up area integrals in polar coordinates is finding the points of intersection of two polar curves. The following example shows some of the subtleties of this process.

EXAMPLE 5   Points of intersection Find the points of intersection of the circle $r = 3\cos\theta$ and the cardioid $r = 1 + \cos\theta$ (Figure 12.41).

SOLUTION The fact that a point has multiple representations in polar coordinates may lead to subtle difficulties in finding intersection points. We first proceed algebraically. Equating the two expressions for r and solving for θ, we have

$$3\cos\theta = 1 + \cos\theta \quad \text{or} \quad \cos\theta = \frac{1}{2},$$

which has roots $\theta = \pm \pi/3$. Therefore, two intersection points are $(3/2, \pi/3)$ and $(3/2, -\pi/3)$ (Figure 12.41). Without examining graphs of the curves, we might be tempted to stop here. Yet the figure shows another intersection point at the origin O that has not been detected. To find this intersection point, we must investigate the way in which the two curves are generated. As θ increases from 0 to 2π, the cardioid is generated counterclockwise, beginning at $(2, 0)$. The cardioid passes through O when $\theta = \pi$. As θ increases from 0 to π, the circle is generated counterclockwise, beginning at $(3, 0)$. The circle passes through O when $\theta = \pi/2$. Therefore, the intersection point O is $(0, \pi)$ on the cardioid (and these coordinates do not satisfy the equation of the circle), while O is $(0, \pi/2)$ on the circle (and these coordinates do not satisfy the equation of the cardioid). There is no foolproof rule for detecting such "hidden" intersection points. Care must be used.

Related Exercise 29 ◄

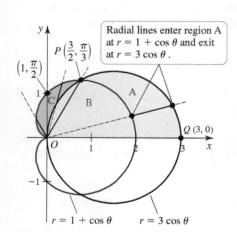

Radial lines enter region A at $r = 1 + \cos\theta$ and exit at $r = 3\cos\theta$.

$P\left(\dfrac{3}{2}, \dfrac{\pi}{3}\right)$ $\left(1, \dfrac{\pi}{2}\right)$ $Q\,(3, 0)$

$r = 1 + \cos\theta$ $r = 3\cos\theta$

Figure 12.42

▶ One way to verify that the inner and outer boundaries of a region have been correctly identified is to draw a ray from the origin through the region—the ray should enter the region at the inner boundary and exit the region at the outer boundary. In Example 6a, this is the case for every ray through region A, for $0 \le \theta \le \pi/3$.

EXAMPLE 6   Computing areas Example 5 discussed the points of intersection of the curves $r = 3 \cos \theta$ (a circle) and $r = 1 + \cos \theta$ (a cardioid). Use those results to compute the areas of the following non-overlapping regions in **Figure 12.42**.

a. region A   **b.** region B   **c.** region C

SOLUTION

a. It is evident that region A is bounded on the inside by the cardioid and on the outside by the circle between the points $Q(\theta = 0)$ and $P(\theta = \pi/3)$. Therefore, the area of region A is

$$\frac{1}{2}\int_0^{\pi/3}\left((3\cos\theta)^2 - (1 + \cos\theta)^2\right)d\theta$$

$$= \frac{1}{2}\int_0^{\pi/3}(8\cos^2\theta - 1 - 2\cos\theta)\,d\theta \qquad \text{Simplify.}$$

$$= \frac{1}{2}\int_0^{\pi/3}(3 + 4\cos 2\theta - 2\cos\theta)\,d\theta \qquad \cos^2\theta = \frac{1 + \cos 2\theta}{2}$$

$$= \frac{1}{2}(3\theta + 2\sin 2\theta - 2\sin\theta)\Big|_0^{\pi/3} = \frac{\pi}{2}. \qquad \text{Evaluate integral.}$$

b. Examining region B, notice that a ray drawn from the origin enters the region immediately. There is no inner boundary, and the outer boundary is $r = 1 + \cos\theta$ on $0 \le \theta \le \pi/3$ and $r = 3\cos\theta$ on $\pi/3 \le \theta \le \pi/2$ (recall from Example 5 that $\theta = \pi/2$ is the angle at which the circle intersects the origin). Therefore, we slice the region into two parts at $\theta = \pi/3$ and write two integrals for its area:

$$\text{area of region B} = \frac{1}{2}\int_0^{\pi/3}(1 + \cos\theta)^2\,d\theta + \frac{1}{2}\int_{\pi/3}^{\pi/2}(3\cos\theta)^2\,d\theta.$$

While these integrals may be evaluated directly, it's easier to notice that

$$\text{area of region B} = \text{area of semicircle } OPQ - \text{area of region A}.$$

Because $r = 3\cos\theta$ is a circle with a radius of $3/2$, we have

$$\text{area of region B} = \frac{1}{2}\cdot\pi\left(\frac{3}{2}\right)^2 - \frac{\pi}{2} = \frac{5\pi}{8}.$$

c. It's easy to *incorrectly* identify the inner boundary of region C as the circle and the outer boundary as the cardioid. While these identifications are true when $\pi/3 \le \theta \le \pi/2$ (notice again the radial lines in Figure 12.42), there is only one boundary curve (the cardioid) when $\pi/2 \le \theta \le \pi$. We conclude that the area of region C is

$$\frac{1}{2}\int_{\pi/3}^{\pi/2}\left((1 + \cos\theta)^2 - (3\cos\theta)^2\right)d\theta + \frac{1}{2}\int_{\pi/2}^{\pi}(1 + \cos\theta)^2\,d\theta = \frac{\pi}{8}.$$

Related Exercise 61 ◄

Arc Length of a Polar Curve

> Recall from Section 12.2 that to convert from polar to Cartesian coordinates we use the relations
>
> $$x = r \cos \theta \quad \text{and} \quad y = r \sin \theta.$$

We now answer the arc length question for polar curves: Given the polar equation $r = f(\theta)$, what is the length of the corresponding curve for $\alpha \le \theta \le \beta$ (assuming the curve does not retrace itself on this interval)? The key idea is to express the polar equation as a set of parametric equations in Cartesian coordinates and then use the arc length formula for parametric equations derived in Section 12.1. Letting θ play the role of a parameter and using $r = f(\theta)$, parametric equations for the polar curve are

$$x = r \cos \theta = f(\theta) \cos \theta \quad \text{and} \quad y = r \sin \theta = f(\theta) \sin \theta,$$

where $\alpha \le \theta \le \beta$. The arc length formula in terms of the parameter θ is

$$L = \int_{\alpha}^{\beta} \sqrt{\left(\frac{dx}{d\theta}\right)^2 + \left(\frac{dy}{d\theta}\right)^2}\, d\theta,$$

where

$$\frac{dx}{d\theta} = f'(\theta) \cos \theta - f(\theta) \sin \theta \quad \text{and} \quad \frac{dy}{d\theta} = f'(\theta) \sin \theta + f(\theta) \cos \theta.$$

When substituted into the arc length formula and simplified, the result is a new arc length integral (Exercise 84).

QUICK CHECK 4 Use the arc length formula to verify that the circumference of the circle $r = f(\theta) = 1$, for $0 \le \theta \le 2\pi$, is 2π. ◄

Arc Length of a Polar Curve

Let f have a continuous derivative on the interval $[\alpha, \beta]$. The **arc length** of the polar curve $r = f(\theta)$ on $[\alpha, \beta]$ is

$$L = \int_{\alpha}^{\beta} \sqrt{f(\theta)^2 + f'(\theta)^2}\, d\theta.$$

EXAMPLE 7   Arc length of polar curves

a. Find the arc length of the spiral $r = f(\theta) = \theta$, for $0 \le \theta \le 2\pi$ (**Figure 12.43**).

b. Find the arc length of the cardioid $r = 1 + \cos \theta$ (**Figure 12.44**).

SOLUTION

a.
$$L = \int_0^{2\pi} \sqrt{\theta^2 + 1}\, d\theta \qquad\qquad\qquad\qquad f(\theta) = \theta \text{ and } f'(\theta) = 1$$

$$= \left(\frac{\theta}{2} \sqrt{\theta^2 + 1} + \frac{1}{2} \ln\left(\theta + \sqrt{\theta^2 + 1} \right) \right)\Big|_0^{2\pi} \qquad \text{Table of integrals or trigonometric substitution}$$

$$= \pi\sqrt{4\pi^2 + 1} + \frac{1}{2} \ln\left(2\pi + \sqrt{4\pi^2 + 1} \right) \qquad \text{Substitute limits of integration.}$$

$$\approx 21.26 \qquad\qquad\qquad\qquad\qquad\qquad\qquad \text{Evaluate.}$$

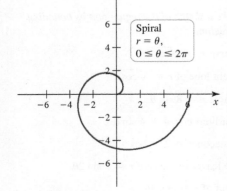

Figure 12.43

b. The cardioid is symmetric about the x-axis, and its upper half is generated for $0 \le \theta \le \pi$. The length of the full curve is twice the length of its upper half:

$$L = 2\int_0^{\pi} \sqrt{(1 + \cos \theta)^2 + (-\sin \theta)^2}\, d\theta \qquad f(\theta) = 1 + \cos \theta;\ f'(\theta) = -\sin \theta$$

$$= 2\int_0^{\pi} \sqrt{2 + 2\cos \theta}\, d\theta \qquad\qquad\qquad\qquad \text{Simplify.}$$

$$= 2\int_0^{\pi} \sqrt{4 \cos^2 (\theta/2)}\, d\theta \qquad\qquad\qquad\qquad 1 + \cos \theta = 2\cos^2 (\theta/2)$$

$$= 4\int_0^{\pi} \cos (\theta/2)\, d\theta \qquad\qquad\qquad\qquad\qquad \cos (\theta/2) \ge 0, \text{ for } 0 \le \theta \le \pi$$

$$= 8 \sin (\theta/2)\Big|_0^{\pi} = 8. \qquad\qquad\qquad\qquad\qquad \text{Integrate and simplify.}$$

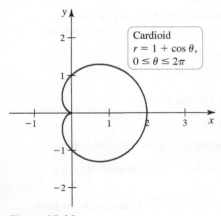

Figure 12.44

Related Exercises 67–68 ◄

SECTION 12.3 EXERCISES

Getting started

1. Express the polar equation $r = f(\theta)$ in parametric form in Cartesian coordinates, where θ is the parameter.

2. Explain why the slope of the line $\theta = \pi/2$ is undefined.

3. Explain why the slope of the line tangent to the polar graph of $r = f(\theta)$ is not $dr/d\theta$.

4. What integral must be evaluated to find the area of the region bounded by the polar graphs of $r = f(\theta)$ and $r = g(\theta)$ on the interval $\alpha \le \theta \le \beta$, where $f(\theta) \ge g(\theta) \ge 0$?

5. What is the slope of the line $\theta = \pi/3$?

6. Find a polar equation of the line with slope -1 that passes through the origin.

7. Find the area of the shaded region.

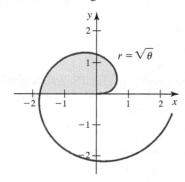

8. Without calculating derivatives, determine the slopes of each of the lines tangent to the curve $r = 8 \cos \theta - 4$ at the origin.

9. Explain why the point with polar coordinates $(0, 0)$ is an intersection point of the curves $r = \cos \theta$ and $r = \sin \theta$ even though $\cos \theta \ne \sin \theta$ when $\theta = 0$.

10. Explain why the point $(-1, 3\pi/2)$ is on the polar graph of $r = 1 + \cos \theta$ even though it does not satisfy the equation $r = 1 + \cos \theta$.

Practice Exercises

11–20. Slopes of tangent lines *Find the slope of the line tangent to the following polar curves at the given points.*

11. $r = 1 - \sin \theta; \left(\dfrac{1}{2}, \dfrac{\pi}{6}\right)$ 12. $r = 4 \cos \theta; \left(2, \dfrac{\pi}{3}\right)$

13. $r = 8 \sin \theta; \left(4, \dfrac{5\pi}{6}\right)$

14. $r = 4 + \sin \theta; (4, 0)$ and $\left(3, \dfrac{3\pi}{2}\right)$

15. $r = 6 + 3 \cos \theta; (3, \pi)$ and $(9, 0)$

16. $r = 2 \sin 3\theta$; at the tips of the leaves

17. $r = 4 \cos 2\theta$; at the tips of the leaves

18. $r = 1 + 2 \sin 2\theta; \left(3, \dfrac{\pi}{4}\right)$ 19. $r^2 = 4 \cos 2\theta; \left(0, \pm\dfrac{\pi}{4}\right)$

20. $r = 2\theta; \left(\dfrac{\pi}{2}, \dfrac{\pi}{4}\right)$

21. **Tangent line at the origin** Find the polar equation of the line tangent to the polar curve $r = \cos \theta + \sin \theta$ at the origin, and then find the slope of this tangent line.

22. **Tangent line at the origin** Find the polar equation of the line tangent to the polar curve $r = 4 \cos \theta$ at the origin. Explain why the slope of this line is undefined.

T 23–24. Multiple tangent lines at a point

a. *Give the smallest interval $[0, P]$ that generates the entire polar curve and use a graphing utility to graph the curve.*

b. *Find a polar equation and the slope of each line tangent to the curve at the origin.*

23. $r = \cos 3\theta + \sin 3\theta$ 24. $r = 1 + 2 \cos 2\theta$

25–28. Horizontal and vertical tangents *Find the points at which the following polar curves have horizontal or vertical tangent lines.*

25. $r = 4 \cos \theta$ 26. $r = 2 + 2 \sin \theta$

T 27. $r = \sin 2\theta$ **T 28.** $r = 3 + 6 \sin \theta$

29–32. Intersection points *Use algebraic methods to find as many intersection points of the following curves as possible. Use graphical methods to identify the remaining intersection points.*

29. $r = 2 \cos \theta$ and $r = 1 + \cos \theta$

30. $r = 1 - \sin \theta$ and $r = 1 + \cos \theta$

31. $r = 1$ and $r = 2 \sin 2\theta$

32. $r = \cos 2\theta$ and $r = \sin 2\theta$

33–40. Areas of regions *Make a sketch of the region and its bounding curves. Find the area of the region.*

33. The region inside the curve $r = \sqrt{\cos \theta}$

34. The region inside the right lobe of $r = \sqrt{\cos 2\theta}$

35. The region inside the circle $r = 8 \sin \theta$

36. The region inside the cardioid $r = 4 + 4 \sin \theta$

37. The region inside the limaçon $r = 2 + \cos \theta$

38. The region inside all the leaves of the rose $r = 3 \sin 2\theta$

39. The region inside one leaf of $r = \cos 3\theta$

40. The region inside the inner loop of $r = \cos \theta - 1/2$

41–44. Intersection points and area

a. *Find all the intersection points of the following curves.*

b. *Find the area of the entire region that lies within both curves.*

41. $r = 3 \sin \theta$ and $r = 3 \cos \theta$

42. $r = 2 + 2 \sin \theta$ and $r = 2 - 2 \sin \theta$

43. $r = 1 + \sin \theta$ and $r = 1 + \cos \theta$

44. $r = 1$ and $r = \sqrt{2} \cos 2\theta$

45–60. Areas of regions *Find the area of the following regions.*

45. The region outside the circle $r = 1/2$ and inside the circle $r = \cos \theta$

46. The region inside the curve $r = \sqrt{\cos \theta}$ and outside the circle $r = 1/\sqrt{2}$

47. The region inside the curve $r = \sqrt{\cos\theta}$ and inside the circle $r = 1/\sqrt{2}$ in the first quadrant

48. The region inside the right lobe of $r = \sqrt{\cos 2\theta}$ and inside the circle $r = 1/\sqrt{2}$ in the first quadrant

49. The region inside one leaf of the rose $r = \cos 5\theta$

50. The region inside the rose $r = 4\cos 2\theta$ and outside the circle $r = 2$

51. The region inside the rose $r = 4\sin 2\theta$ and inside the circle $r = 2$

52. The region inside the lemniscate $r^2 = 2\sin 2\theta$ and outside the circle $r = 1$

53. The region common to the circles $r = 2\sin\theta$ and $r = 1$

54. The region inside the inner loop of the limaçon $r = 2 + 4\cos\theta$

55. The region inside the outer loop but outside the inner loop of the limaçon $r = 3 - 6\sin\theta$

56. The region common to the circle $r = 3\cos\theta$ and the cardioid $r = 1 + \cos\theta$

57. The region inside the lemniscate $r^2 = 6\sin 2\theta$

58. The region inside the limaçon $r = 2 - 4\sin\theta$

59. The region inside the limaçon $r = 4 - 2\cos\theta$

60. The region inside the complete three-leaf rose $r = 2\cos 3\theta$

61. **Two curves, three regions** Determine the intersection points of the polar curves $r = 4 - 2\cos\theta$ and $r = 2 + 2\cos\theta$, and then find areas of regions A, B, and C (see figure).

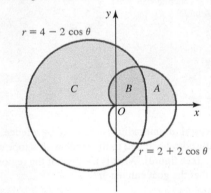

62. Find the area in the first quadrant inside the curve $r = 4 + \cos 2\theta$ but outside the curve $r = 3\cos 2\theta$ (see figure), after first finding the intersection points of the two curves.

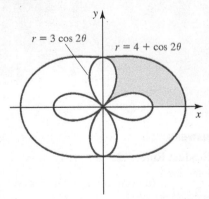

63–74. Arc length of polar curves *Find the length of the following polar curves.*

63. The complete circle $r = a\sin\theta$, where $a > 0$

64. The complete cardioid $r = 2 - 2\sin\theta$

65. The spiral $r = \theta^2$, for $0 \le \theta \le 2\pi$

66. The spiral $r = e^{\theta}$, for $0 \le \theta \le 2\pi n$, where n is a positive integer

67. The complete cardioid $r = 4 + 4\sin\theta$

68. The spiral $r = 4\theta^2$, for $0 \le \theta \le 6$

69. The spiral $r = 2e^{2\theta}$, for $0 \le \theta \le \ln 8$

70. The parabola $r = \dfrac{\sqrt{2}}{1 + \cos\theta}$, for $0 \le \theta \le \dfrac{\pi}{2}$

71. The curve $r = \sin^3\dfrac{\theta}{3}$, for $0 \le \theta \le \dfrac{\pi}{2}$

T 72. The three-leaf rose $r = 2\cos 3\theta$

T 73. The complete limaçon $r = 4 - 2\cos\theta$

T 74. The complete lemniscate $r^2 = 6\sin 2\theta$

75. **Explain why or why not** Determine whether the following statements are true and give an explanation or counterexample.

  a. The area of the region bounded by the polar graph of $r = f(\theta)$ on the interval $[\alpha, \beta]$ is $\int_\alpha^\beta f(\theta)\,d\theta$.

  b. The slope of the line tangent to the polar curve $r = f(\theta)$ at a point (r, θ) is $f'(\theta)$.

  c. There may be more than one line that is tangent to a polar curve at some points on the curve.

Explorations and Challenges

76. **Area calculation** Use polar coordinates to determine the area bounded on the right by the unit circle $x^2 + y^2 = 1$ and bounded on the left by the vertical line $x = \sqrt{2}/2$.

T 77. **Spiral tangent lines** Use a graphing utility to determine the first three points with $\theta \ge 0$ at which the spiral $r = 2\theta$ has a horizontal tangent line. Find the first three points with $\theta \ge 0$ at which the spiral $r = 2\theta$ has a vertical tangent line.

78. **Spiral arc length** Consider the spiral $r = 4\theta$, for $\theta \ge 0$.

  a. Use a trigonometric substitution to find the length of the spiral, for $0 \le \theta \le \sqrt{8}$.

  b. Find $L(\theta)$, the length of the spiral on the interval $[0, \theta]$, for any $\theta \ge 0$.

  c. Show that $L'(\theta) > 0$. Is $L''(\theta)$ positive or negative? Interpret your answer.

79. **Spiral arc length** Find the length of the entire spiral $r = e^{-a\theta}$, for $\theta \ge 0$ and $a > 0$.

80. **Area of roses** Assume m is a positive integer.

  a. *Even number of leaves*: What is the relationship between the total area enclosed by the $4m$-leaf rose $r = \cos(2m\theta)$ and m?

  b. *Odd number of leaves*: What is the relationship between the total area enclosed by the $(2m + 1)$-leaf rose $r = \cos((2m + 1)\theta)$ and m?

81. Regions bounded by a spiral Let R_n be the region bounded by the nth turn and the $(n + 1)$st turn of the spiral $r = e^{-\theta}$ in the first and second quadrants, for $\theta \geq 0$ (see figure).

  a. Find the area A_n of R_n.
  b. Evaluate $\lim\limits_{n \to \infty} A_n$.
  c. Evaluate $\lim\limits_{n \to \infty} \dfrac{A_{n+1}}{A_n}$.

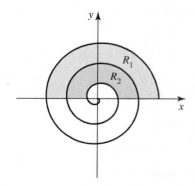

82. Tangents and normals Let a polar curve be described by $r = f(\theta)$, and let ℓ be the line tangent to the curve at the point $P(x, y) = P(r, \theta)$ (see figure).

  a. Explain why $\tan \alpha = dy/dx$.
  b. Explain why $\tan \theta = y/x$.
  c. Let φ be the angle between ℓ and the line through O and P. Prove that $\tan \varphi = f(\theta)/f'(\theta)$.
  d. Prove that the values of θ for which ℓ is parallel to the x-axis satisfy $\tan \theta = -f(\theta)/f'(\theta)$.
  e. Prove that the values of θ for which ℓ is parallel to the y-axis satisfy $\tan \theta = f(\theta)/f'(\theta)$.

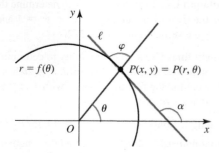

T 83. Isogonal curves Let a curve be described by $r = f(\theta)$, where $f(\theta) > 0$ on its domain. Referring to the figure in Exercise 82, a curve is **isogonal** provided the angle φ is constant for all θ.

  a. Prove that φ is constant for all θ provided $\cot \varphi = \dfrac{f'(\theta)}{f(\theta)}$ is constant, which implies that $\dfrac{d}{d\theta}(\ln f(\theta)) = k$, where k is a constant.
  b. Use part (a) to prove that the family of logarithmic spirals $r = Ce^{k\theta}$ consists of isogonal curves, where C and k are constants.
  c. Graph the curve $r = 2e^{2\theta}$ and confirm the result of part (b).

84. Arc length for polar curves Prove that the length of the curve $r = f(\theta)$, for $\alpha \leq \theta \leq \beta$, is

$$L = \int_\alpha^\beta \sqrt{f(\theta)^2 + f'(\theta)^2}\, d\theta.$$

85–87. Grazing goat problems *Consider the following sequence of problems related to grazing goats tied to a rope. (See the Guided Project* Grazing goat problems.*)*

85. A circular corral of unit radius is enclosed by a fence. A goat inside the corral is tied to the fence with a rope of length $0 \leq a \leq 2$ (see figure). What is the area of the region (inside the corral) that the goat can graze? Check your answer with the special cases $a = 0$ and $a = 2$.

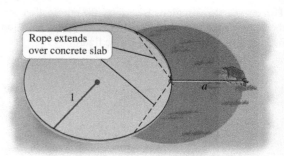

86. A circular concrete slab of unit radius is surrounded by grass. A goat is tied to the edge of the slab with a rope of length $0 \leq a \leq 2$ (see figure). What is the area of the grassy region that the goat can graze? Note that the rope can extend over the concrete slab. Check your answer with the special cases $a = 0$ and $a = 2$.

Rope extends over concrete slab

87. A circular corral of unit radius is enclosed by a fence. A goat is outside the corral and tied to the fence with a rope of length $0 \leq a \leq \pi$ (see figure). What is the area of the region (outside the corral) that the goat can reach?

Rope stretches along fence

QUICK CHECK ANSWERS

1. Apply the Product Rule.   **2.** $\sqrt{2} + 1$
3. $\left[0, \dfrac{\pi}{3}\right]$ or $\left[\dfrac{\pi}{3}, \dfrac{2\pi}{3}\right]$ (among others)   **4.** 2π ◄

12.4 Conic Sections

Conic sections are best visualized as the Greeks did over 2000 years ago by slicing a double cone with a plane (**Figure 12.45**). Three of the seven different sets of points that arise in this way are *ellipses*, *parabolas*, and *hyperbolas*. These curves have practical applications and broad theoretical importance. For example, celestial bodies travel in orbits that are modeled by ellipses and hyperbolas. Mirrors for telescopes are designed using the properties of conic sections. And architectural structures, such as domes and arches, are sometimes based on these curves.

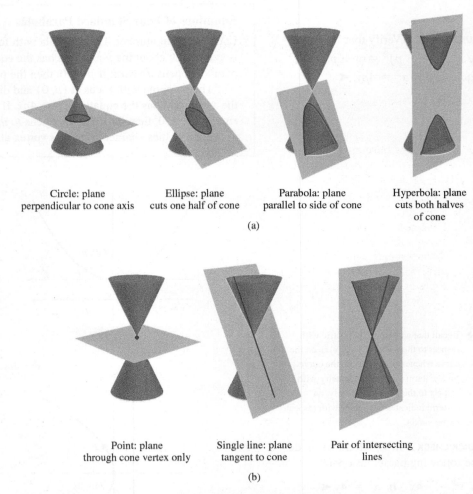

Circle: plane
perpendicular to cone axis

Ellipse: plane
cuts one half of cone

Parabola: plane
parallel to side of cone

Hyperbola: plane
cuts both halves
of cone

(a)

Point: plane
through cone vertex only

Single line: plane
tangent to cone

Pair of intersecting
lines

(b)

Figure 12.45 The standard conic sections (a) are the intersection sets of a double cone and a plane that does not pass through the vertex of the cone. Degenerate conic sections (lines and points) are produced when a plane passes through the vertex of the cone (b).

Parabolas

A **parabola** is the set of points in a plane that are equidistant from a fixed point F (called the **focus**) and a fixed line (called the **directrix**). In the four standard orientations, a parabola may open upward, downward, to the right, or to the left. We derive the equation of the parabola that opens upward.

Suppose the focus F is on the y-axis at $(0, p)$ and the directrix is the horizontal line $y = -p$, where $p > 0$. The parabola is the set of points P that satisfy the defining property $|PF| = |PL|$, where $L(x, -p)$ is the point on the directrix closest to P

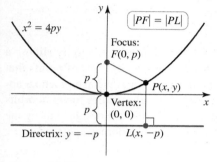

Figure 12.46

QUICK CHECK 1 Verify that
$\sqrt{x^2 + (y - p)^2} = y + p$ is
equivalent to $x^2 = 4py$. ◄

(**Figure 12.46**). Consider an arbitrary point $P(x, y)$ that satisfies this condition. Applying the distance formula, we have

$$\underbrace{\sqrt{x^2 + (y - p)^2}}_{|PF|} = \underbrace{y + p}_{|PL|}.$$

Squaring both sides of this equation and simplifying gives the equation $x^2 = 4py$. This is the equation of a parabola that is symmetric about the y-axis and opens upward. The **vertex** of the parabola is the point closest to the directrix; in this case, it is $(0, 0)$ (which satisfies $|PF| = |PL| = p$).

The equations of the other three standard parabolas are derived in a similar way.

Equations of Four Standard Parabolas

Let p be a real number. The parabola with focus at $(0, p)$ and directrix $y = -p$ is symmetric about the y-axis and has the equation $x^2 = 4py$. If $p > 0$, then the parabola opens *upward*; if $p < 0$, then the parabola opens *downward*.

The parabola with focus at $(p, 0)$ and directrix $x = -p$ is symmetric about the x-axis and has the equation $y^2 = 4px$. If $p > 0$, then the parabola opens *to the right*; if $p < 0$, then the parabola opens *to the left*.

Each of these parabolas has its vertex at the origin (**Figure 12.47**).

➤ Recall that a curve is symmetric with respect to the x-axis if $(x, -y)$ is on the curve whenever (x, y) is on the curve. So a y^2-term indicates symmetry with respect to the x-axis. Similarly, an x^2-term indicates symmetry with respect to the y-axis.

QUICK CHECK 2 In which direction do the following parabolas open?

a. $y^2 = -4x$ **b.** $x^2 = 4y$ ◄

Figure 12.47

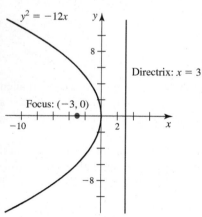

Figure 12.48

EXAMPLE 1   Graphing parabolas Find the focus and directrix of the parabola $y^2 = -12x$. Sketch its graph.

SOLUTION The y^2-term indicates that the parabola is symmetric with respect to the x-axis. Rewriting the equation as $x = -y^2/12$, we see that $x \leq 0$ for all y, implying that the parabola opens to the left. Comparing $y^2 = -12x$ to the standard form $y^2 = 4px$, we see that $p = -3$; therefore, the focus is $(-3, 0)$, and the directrix is $x = 3$ (**Figure 12.48**).

Related Exercises 13–14 ◄

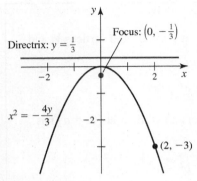

Directrix: $y = \frac{1}{3}$

Focus: $\left(0, -\frac{1}{3}\right)$

$x^2 = -\dfrac{4y}{3}$

$(2, -3)$

Figure 12.49

EXAMPLE 2   Equations of parabolas Find the equation of the parabola with vertex $(0, 0)$ that opens downward and passes through the point $(2, -3)$.

SOLUTION The standard parabola that opens downward has the equation $x^2 = 4py$. The point $(2, -3)$ must satisfy this equation. Substituting $x = 2$ and $y = -3$ into $x^2 = 4py$, we find that $p = -\frac{1}{3}$. Therefore, the focus is at $\left(0, -\frac{1}{3}\right)$, the directrix is $y = \frac{1}{3}$, and the equation of the parabola is $x^2 = -4y/3$, or $y = -3x^2/4$ (**Figure 12.49**).

Related Exercises 35–36 ◄

Reflection Property Parabolas have a property that makes them useful in the design of reflectors and transmitters. A particle approaching a parabola on any line parallel to the axis of the parabola is reflected on a line that passes through the focus (**Figure 12.50**); this property is used to focus incoming light by a parabolic mirror on a telescope. Alternatively, signals emanating from the focus are reflected on lines parallel to the axis, a property used to design radio transmitters and headlights (**Exercise 83**).

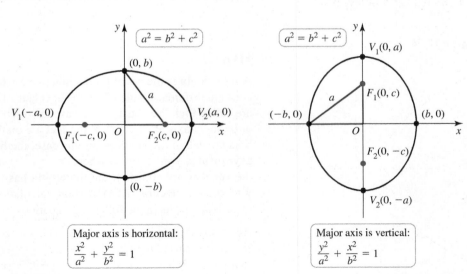

Parabolic reflector

Incoming signal

Receiver

Source

Outgoing light

Parabolic transmitter

Figure 12.50

Ellipses

An **ellipse** is the set of points in a plane whose distances from two fixed points have a constant sum that we denote $2a$ (**Figure 12.51**). Each of the two fixed points is a **focus** (plural **foci**). The equation of an ellipse is simplest if the foci are on the x-axis at $(\pm c, 0)$ or on the y-axis at $(0, \pm c)$. In either case, the **center** of the ellipse is $(0, 0)$. If the foci are on the x-axis, the points $(\pm a, 0)$ lie on the ellipse and are called **vertices**. If the foci are on the y-axis, the vertices are $(0, \pm a)$ (**Figure 12.52**). A short calculation (**Exercise 85**) using the definition of the ellipse results in the following equations for an ellipse.

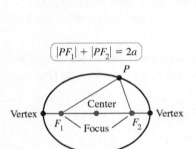

$$|PF_1| + |PF_2| = 2a$$

P

Center

Vertex

Vertex

F_1 Focus F_2

Figure 12.51

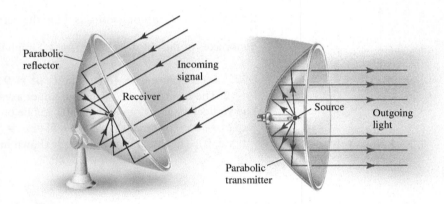

$a^2 = b^2 + c^2$

$(0, b)$

a

$V_1(-a, 0)$

$V_2(a, 0)$

$F_1(-c, 0)$ O $F_2(c, 0)$

$(0, -b)$

Major axis is horizontal:
$$\frac{x^2}{a^2} + \frac{y^2}{b^2} = 1$$

$a^2 = b^2 + c^2$

$V_1(0, a)$

$F_1(0, c)$

a

$(-b, 0)$

$(b, 0)$

O

$F_2(0, -c)$

$V_2(0, -a)$

Major axis is vertical:
$$\frac{y^2}{a^2} + \frac{x^2}{b^2} = 1$$

Figure 12.52

▶ When necessary, we distinguish between the *major-axis vertices* $(\pm a, 0)$ or $(0, \pm a)$, and the *minor-axis vertices* $(\pm b, 0)$ or $(0, \pm b)$. The word *vertices* (without further description) is understood to mean *major-axis vertices*.

QUICK CHECK 3 In the case that the vertices and foci are on the x-axis, show that the length of the minor axis of an ellipse is $2b$. ◄

Equations of Standard Ellipses

An ellipse centered at the origin with foci F_1 and F_2 at $(\pm c, 0)$ and vertices V_1 and V_2 at $(\pm a, 0)$ has the equation

$$\frac{x^2}{a^2} + \frac{y^2}{b^2} = 1, \quad \text{where } a^2 = b^2 + c^2.$$

An ellipse centered at the origin with foci at $(0, \pm c)$ and vertices at $(0, \pm a)$ has the equation

$$\frac{y^2}{a^2} + \frac{x^2}{b^2} = 1, \quad \text{where } a^2 = b^2 + c^2.$$

In both cases, $a > b > 0$ and $a > c > 0$, the length of the long axis (called the **major axis**) is $2a$, and the length of the short axis (called the **minor axis**) is $2b$.

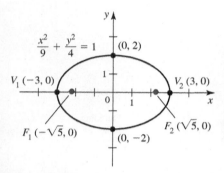

Figure 12.53

EXAMPLE 3 Graphing ellipses Find the vertices, foci, and length of the major and minor axes of the ellipse $\dfrac{x^2}{9} + \dfrac{y^2}{4} = 1$. Graph the ellipse.

SOLUTION Because $9 > 4$, we identify $a^2 = 9$ and $b^2 = 4$. Therefore, $a = 3$ and $b = 2$. The lengths of the major and minor axes are $2a = 6$ and $2b = 4$, respectively. The vertices V_1 and V_2 are at $(\pm 3, 0)$ and lie on the x-axis, as do the foci. The relationship $c^2 = a^2 - b^2$ implies that $c^2 = 5$, or $c = \sqrt{5}$. Therefore, the foci F_1 and F_2 are at $(\pm \sqrt{5}, 0)$. The graph of the ellipse is shown in **Figure 12.53**.

Related Exercises 15–16 ◄

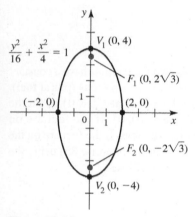

Figure 12.54

EXAMPLE 4 Equation of an ellipse Find the equation of the ellipse centered at the origin with its foci on the y-axis, a major axis of length 8, and a minor axis of length 4. Graph the ellipse.

SOLUTION Because the length of the major axis is 8, the vertices V_1 and V_2 are located at $(0, \pm 4)$, and $a = 4$. Because the length of the minor axis is 4, we have $b = 2$. Therefore, the equation of the ellipse is

$$\frac{y^2}{16} + \frac{x^2}{4} = 1.$$

Using the relation $c^2 = a^2 - b^2$, we find that $c = 2\sqrt{3}$ and the foci F_1 and F_2 are at $(0, \pm 2\sqrt{3})$. The ellipse is shown in **Figure 12.54**.

Related Exercise 39 ◄

Hyperbolas

A **hyperbola** is the set of points in a plane whose distances from two fixed points have a constant difference, either $2a$ or $-2a$ (**Figure 12.55**). As with ellipses, the two fixed points are called **foci**. We consider the case in which the foci are either on the x-axis at $(\pm c, 0)$ or on the y-axis at $(0, \pm c)$. If the foci are on the x-axis, the points $(\pm a, 0)$ on the hyperbola are called the **vertices**. In this case, the hyperbola has no y-intercepts, but it has the **asymptotes** $y = \pm bx/a$, where $b^2 = c^2 - a^2$. Similarly, if the foci are on the y-axis, the vertices are $(0, \pm a)$, the hyperbola has no x-intercepts, and it has the asymptotes $y = \pm ax/b$ (**Figure 12.56**). A short calculation (Exercise 86) using the definition of the hyperbola results in the following equations for standard hyperbolas.

▶ Asymptotes that are not parallel to one of the coordinate axes, as in the case of the standard hyperbolas, are called **oblique**, or **slant**, **asymptotes**.

Figure 12.55

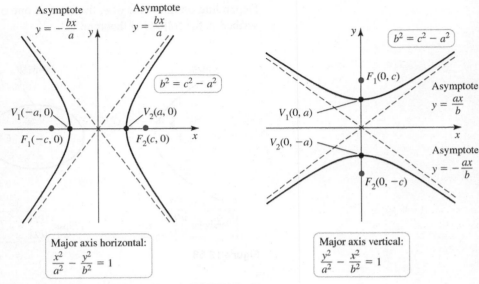

$$\boxed{b^2 = c^2 - a^2}$$

Major axis horizontal:
$$\frac{x^2}{a^2} - \frac{y^2}{b^2} = 1$$

Major axis vertical:
$$\frac{y^2}{a^2} - \frac{x^2}{b^2} = 1$$

Figure 12.56

Equations of Standard Hyperbolas

A hyperbola centered at the origin with foci F_1 and F_2 at $(\pm c, 0)$ and vertices V_1 and V_2 at $(\pm a, 0)$ has the equation

$$\frac{x^2}{a^2} - \frac{y^2}{b^2} = 1, \quad \text{where} \quad b^2 = c^2 - a^2.$$

The hyperbola has **asymptotes** $y = \pm bx/a$.

A hyperbola centered at the origin with foci at $(0, \pm c)$ and vertices at $(0, \pm a)$ has the equation

$$\frac{y^2}{a^2} - \frac{x^2}{b^2} = 1, \quad \text{where} \quad b^2 = c^2 - a^2.$$

The hyperbola has **asymptotes** $y = \pm ax/b$.
    In both cases, $c > a > 0$ and $c > b > 0$.

▶ Notice that the asymptotes for hyperbolas are $y = \pm bx/a$ when the vertices are on the x-axis and $y = \pm ax/b$ when the vertices are on the y-axis (the roles of a and b are reversed).

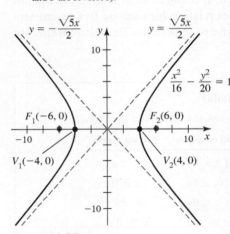

Figure 12.57

QUICK CHECK 4 Identify the vertices and foci of the hyperbola $y^2 - x^2/4 = 1$. ◀

EXAMPLE 5 Graphing hyperbolas Find the equation of the hyperbola centered at the origin with vertices V_1 and V_2 at $(\pm 4, 0)$ and foci F_1 and F_2 at $(\pm 6, 0)$. Graph the hyperbola.

SOLUTION Because the foci are on the x-axis, the vertices are also on the x-axis, and there are no y-intercepts. With $a = 4$ and $c = 6$, we have $b^2 = c^2 - a^2 = 20$, or $b = 2\sqrt{5}$. Therefore, the equation of the hyperbola is

$$\frac{x^2}{16} - \frac{y^2}{20} = 1.$$

The asymptotes are $y = \pm bx/a = \pm\sqrt{5}x/2$ (**Figure 12.57**).

Related Exercise 41 ◀

Eccentricity and Directrix

Parabolas, ellipses, and hyperbolas may also be developed in a single unified way called the *eccentricity-directrix* approach. We let ℓ be a line called the **directrix** and F be a point not on ℓ called a **focus**. The **eccentricity** is a real number $e > 0$. Consider the set C of points P in a plane with the property that the distance $|PF|$ equals e multiplied by the perpendicular distance $|PL|$ from P to ℓ (**Figure 12.58**); that is,

▶ The conic section lies in the plane formed by the directrix and the focus.

$$|PF| = e|PL| \quad \text{or} \quad \frac{|PF|}{|PL|} = e = \text{constant.}$$

Depending on the value of e, the set C is one of the three standard conic sections, as described in the following theorem.

Parabola: $\dfrac{|PF|}{|PL|} = e = 1$ Ellipse: $\dfrac{|PF|}{|PL|} = e, 0 < e < 1$ Hyperbola: $\dfrac{|PF|}{|PL|} = e > 1$

Figure 12.58

> Theorem 12.3 for ellipses and hyperbolas describes how the entire curve is generated using just one focus and one directrix. Nevertheless, all ellipses and hyperbolas have two foci and two directrices.

THEOREM 12.3 Eccentricity-Directrix Theorem

Suppose ℓ is a line, F is a point not on ℓ, and e is a positive real number. Let C be the set of points P in a plane with the property that $\dfrac{|PF|}{|PL|} = e$, where $|PL|$ is the perpendicular distance from P to ℓ.

1. If $e = 1$, C is a **parabola**.
2. If $0 < e < 1$, C is an **ellipse**.
3. If $e > 1$, C is a **hyperbola**.

The proof of this theorem is straightforward; it requires an algebraic calculation that is found in Appendix A. The proof establishes relationships among five parameters a, b, c, d, and e that are characteristic of any ellipse or hyperbola. The relationships are given in the following summary.

SUMMARY Properties of Ellipses and Hyperbolas

An ellipse or a hyperbola centered at the origin has the following properties.

	Foci on x-axis	Foci on y-axis
Major-axis vertices:	$(\pm a, 0)$	$(0, \pm a)$
Minor-axis vertices (for ellipses):	$(0, \pm b)$	$(\pm b, 0)$
Foci:	$(\pm c, 0)$	$(0, \pm c)$
Directrices:	$x = \pm d$	$y = \pm d$

Eccentricity: $0 < e < 1$ for ellipses, $e > 1$ for hyperbolas.

Given any two of the five parameters a, b, c, d, and e, the other three are found using the relations

$$c = ae, \quad d = \frac{a}{e},$$

$$b^2 = a^2 - c^2 \quad \text{(for ellipses)}, \qquad b^2 = c^2 - a^2 \quad \text{(for hyperbolas)}.$$

QUICK CHECK 5 Given an ellipse with $a = 3$ and $e = \frac{1}{2}$, what are the values of b, c, and d? ◄

EXAMPLE 6 Equation of an ellipse Find the equation of the ellipse centered at the origin with foci F_1 and F_2 at $(0, \pm 4)$ and eccentricity $e = \frac{1}{2}$. Give the length of the major and minor axes, the location of the vertices, and the directrices. Graph the ellipse.

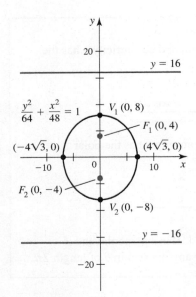

$$\frac{y^2}{64} + \frac{x^2}{48} = 1$$

$V_1(0, 8)$

$F_1(0, 4)$

$(-4\sqrt{3}, 0)$

$(4\sqrt{3}, 0)$

$F_2(0, -4)$

$V_2(0, -8)$

$y = -16$

Figure 12.59

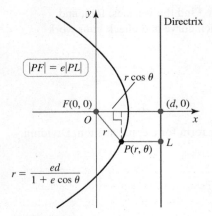

$|PF| = e|PL|$

$r\cos\theta$

Directrix

$F(0, 0)$

$(d, 0)$

$P(r, \theta)$

L

$$r = \frac{ed}{1 + e\cos\theta}$$

Figure 12.60

SOLUTION An ellipse with its major axis along the y-axis has the equation

$$\frac{y^2}{a^2} + \frac{x^2}{b^2} = 1,$$

where a and b must be determined (with $a > b$). Because the foci are at $(0, \pm 4)$, we have $c = 4$. Using $e = \frac{1}{2}$ and the relation $c = ae$, it follows that $a = c/e = 8$. So the length of the major axis is $2a = 16$, and the major-axis vertices V_1 and V_2 are $(0, \pm 8)$. Also, $d = a/e = 16$, so the directrices are $y = \pm 16$. Finally, $b^2 = a^2 - c^2 = 48$, or $b = 4\sqrt{3}$. So the length of the minor axis is $2b = 8\sqrt{3}$, and the minor-axis vertices are $(\pm 4\sqrt{3}, 0)$ (**Figure 12.59**). The equation of the ellipse is

$$\frac{y^2}{64} + \frac{x^2}{48} = 1.$$

Related Exercises 53–54 ◀

Polar Equations of Conic Sections

It turns out that conic sections have a natural representation in polar coordinates, provided we use the eccentricity-directrix approach given in Theorem 12.3. Furthermore, a single polar equation covers parabolas, ellipses, and hyperbolas.

When working in polar equations, the key is to place a focus of the conic section at the origin of the coordinate system. We begin by placing one focus F at the origin and taking a directrix perpendicular to the x-axis through $(d, 0)$, where $d > 0$ (**Figure 12.60**).

We now use the definition $\dfrac{|PF|}{|PL|} = e$, where $P(r, \theta)$ is an arbitrary point on the conic.

As shown in Figure 12.60, $|PF| = r$ and $|PL| = d - r\cos\theta$. The condition $\dfrac{|PF|}{|PL|} = e$

implies that $r = e(d - r\cos\theta)$. Solving for r, we have

$$r = \frac{ed}{1 + e\cos\theta}.$$

A similar derivation (Exercise 84) with the directrix at $x = -d$, where $d > 0$, results in the equation

$$r = \frac{ed}{1 - e\cos\theta}.$$

For horizontal directrices at $y = \pm d$ (**Figure 12.61**), a similar argument (Exercise 84) leads to the equations

$$r = \frac{ed}{1 \pm e\sin\theta}.$$

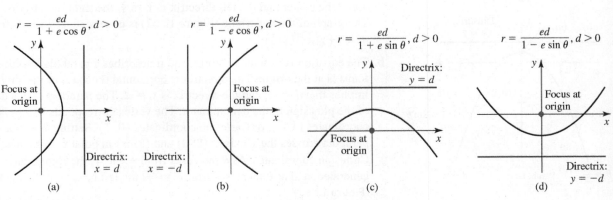

(a) $r = \dfrac{ed}{1 + e\cos\theta}, d > 0$ Focus at origin; Directrix: $x = d$

(b) $r = \dfrac{ed}{1 - e\cos\theta}, d > 0$ Focus at origin; Directrix: $x = -d$

(c) $r = \dfrac{ed}{1 + e\sin\theta}, d > 0$ Directrix: $y = d$; Focus at origin

(d) $r = \dfrac{ed}{1 - e\sin\theta}, d > 0$ Focus at origin; Directrix: $y = -d$

Figure 12.61

> **THEOREM 12.4   Polar Equations of Conic Sections**
> Let $d > 0$. The conic section with a focus at the origin and eccentricity e has the polar equation
>
> $$r = \frac{ed}{1 + e \cos \theta} \quad \text{or} \quad r = \frac{ed}{1 - e \cos \theta}.$$
> $$\underbrace{\phantom{r = \frac{ed}{1 + e \cos \theta}}}_{\text{if one directrix is } x = d} \qquad \underbrace{\phantom{r = \frac{ed}{1 - e \cos \theta}}}_{\text{if one directrix is } x = -d}$$
>
> The conic section with a focus at the origin and eccentricity e has the polar equation
>
> $$r = \frac{ed}{1 + e \sin \theta} \quad \text{or} \quad r = \frac{ed}{1 - e \sin \theta}.$$
> $$\underbrace{\phantom{r = \frac{ed}{1 + e \sin \theta}}}_{\text{if one directrix is } y = d} \qquad \underbrace{\phantom{r = \frac{ed}{1 - e \sin \theta}}}_{\text{if one directrix is } y = -d}$$
>
> If $0 < e < 1$, the conic section is an ellipse; if $e = 1$, it is a parabola; and if $e > 1$, it is a hyperbola. The curves are defined over any interval in θ of length 2π.

QUICK CHECK 6 On which axis do the vertices and foci of the conic section $r = 2/(1 - 2 \sin \theta)$ lie? ◄

EXAMPLE 7   Conic sections in polar coordinates Find the vertices, foci, and directrices of the following conic sections. Graph each curve and check your work with a graphing utility.

a. $r = \dfrac{8}{2 + 3 \cos \theta}$ **b.** $r = \dfrac{2}{1 + \sin \theta}$

SOLUTION

a. The equation must be expressed in standard polar form for a conic section. Dividing numerator and denominator by 2, we have

$$r = \frac{4}{1 + \frac{3}{2} \cos \theta},$$

which allows us to identify $e = \frac{3}{2}$. Therefore, the equation describes a hyperbola (because $e > 1$) with one focus at the origin.

The directrices are vertical (because $\cos \theta$ appears in the equation). Knowing that $ed = 4$, we have $d = 4/e = \frac{8}{3}$, and one directrix is $x = \frac{8}{3}$. Letting $\theta = 0$ and $\theta = \pi$, the polar coordinates of the vertices are $\left(\frac{8}{5}, 0\right)$ and $(-8, \pi)$; equivalently, the vertices are $\left(\frac{8}{5}, 0\right)$ and $(8, 0)$ in Cartesian coordinates (**Figure 12.62**). The center of the hyperbola is halfway between the vertices; therefore, its Cartesian coordinates are $\left(\frac{24}{5}, 0\right)$. The distance between the focus at $(0, 0)$ and the nearest vertex $\left(\frac{8}{5}, 0\right)$ is $\frac{8}{5}$. Therefore, the other focus is $\frac{8}{5}$ units to the right of the vertex $(8, 0)$. So the Cartesian coordinates of the foci are $\left(\frac{48}{5}, 0\right)$ and $(0, 0)$. Because the directrices are symmetric about the center and the left directrix is $x = \frac{8}{3}$, the right directrix is $x = \frac{104}{15} \approx 6.9$. The graph of the hyperbola (Figure 12.62) is generated as θ varies from 0 to 2π (with $\theta \neq \pm \cos^{-1}\left(-\frac{2}{3}\right)$).

b. The equation is in standard form, and it describes a parabola because $e = 1$. The sole focus is at the origin. The directrix is horizontal (because of the $\sin \theta$ term); $ed = 2$ implies that $d = 2$, and the directrix is $y = 2$. The parabola opens downward because of the plus sign in the denominator. The vertex corresponds to $\theta = \frac{\pi}{2}$ and has polar coordinates $\left(1, \frac{\pi}{2}\right)$, or Cartesian coordinates $(0, 1)$. Setting $\theta = 0$ and $\theta = \pi$, the parabola crosses the x-axis at $(2, 0)$ and $(2, \pi)$ in polar coordinates, or $(\pm 2, 0)$ in Cartesian coordinates. As θ increases from $-\frac{\pi}{2}$ to $\frac{\pi}{2}$, the right branch of the parabola is generated, and as θ increases from $\frac{\pi}{2}$ to $\frac{3\pi}{2}$, the left branch of the parabola is generated (**Figure 12.63**).

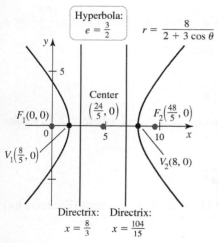

Hyperbola: $e = \frac{3}{2}$ $r = \dfrac{8}{2 + 3 \cos \theta}$

$F_1(0, 0)$

Center $\left(\frac{24}{5}, 0\right)$

$F_2\left(\frac{48}{5}, 0\right)$

$V_1\left(\frac{8}{5}, 0\right)$

$V_2(8, 0)$

Directrix: $x = \frac{8}{3}$ Directrix: $x = \frac{104}{15}$

Figure 12.62

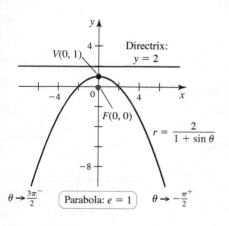

$V(0, 1)$ Directrix: $y = 2$

$F(0, 0)$ $r = \dfrac{2}{1 + \sin \theta}$

$\theta \to \frac{3\pi}{2}^{-}$ Parabola: $e = 1$ $\theta \to -\frac{\pi}{2}^{+}$

Figure 12.63

Related Exercises 57, 60 ◄

Directrix:
$x = 1$
$e = 0.4$
$e = 0.6$
$e = 0.8$
$e = 0.2$

Figure 12.64

EXAMPLE 8 **Conics in polar coordinates** Use a graphing utility to plot the curves

$$r = \frac{e}{1 + e \cos \theta}, \text{ with } e = 0.2, 0.4, 0.6, \text{ and } 0.8. \text{ Comment on the effect of varying the}$$

eccentricity, e.

SOLUTION Because $0 < e < 1$ in each case, all the curves are ellipses. Notice that the equation is in standard form with $d = 1$; therefore, the curves have the same directrix, $x = d = 1$. As the eccentricity increases, the ellipses becomes more elongated. Small values of e correspond to more circular ellipses (**Figure 12.64**).

Related Exercises 65–66 ◄

SECTION 12.4 EXERCISES

Getting Started

1. Give the property that defines all parabolas.

2. Give the property that defines all ellipses.

3. Give the property that defines all hyperbolas.

4. Sketch the three basic conic sections in standard position with vertices and foci on the x-axis.

5. Sketch the three basic conic sections in standard position with vertices and foci on the y-axis.

6. What is the equation of the standard parabola with its vertex at the origin that opens downward?

7. What is the equation of the standard ellipse with vertices at $(\pm a, 0)$ and foci at $(\pm c, 0)$?

8. What is the equation of the standard hyperbola with vertices at $(0, \pm a)$ and foci at $(0, \pm c)$?

9. Given vertices $(\pm a, 0)$ and the eccentricity e, what are the coordinates of the foci of an ellipse and a hyperbola?

10. Give the equation in polar coordinates of a conic section with a focus at the origin, eccentricity e, and a directrix $x = d$, where $d > 0$.

11. What are the equations of the asymptotes of a standard hyperbola with vertices on the x-axis?

12. How does the eccentricity determine the type of conic section?

Practice Exercises

13–30. Graphing conic sections *Determine whether the following equations describe a parabola, an ellipse, or a hyperbola, and then sketch a graph of the curve. For each parabola, specify the location of the focus and the equation of the directrix; for each ellipse, label the coordinates of the vertices and foci, and find the lengths of the major and minor axes; for each hyperbola, label the coordinates of the vertices and foci, and find the equations of the asymptotes.*

13. $x^2 = 12y$

14. $y^2 = 20x$

15. $\dfrac{x^2}{4} + y^2 = 1$

16. $\dfrac{x^2}{9} + \dfrac{y^2}{4} = 1$

17. $x = -\dfrac{y^2}{16}$

18. $4x = -y^2$

19. $\dfrac{x^2}{4} - y^2 = 1$

20. $\dfrac{y^2}{16} - \dfrac{x^2}{9} = 1$

21. $4x^2 - y^2 = 16$

22. $25y^2 - 4x^2 = 100$

23. $8y = -3x^2$

24. $12x = 5y^2$

25. $\dfrac{x^2}{4} + \dfrac{y^2}{16} = 1$

26. $x^2 + \dfrac{y^2}{9} = 1$

27. $\dfrac{x^2}{5} + \dfrac{y^2}{7} = 1$

28. $12x^2 + 5y^2 = 60$

29. $\dfrac{x^2}{3} - \dfrac{y^2}{5} = 1$

30. $10x^2 - 7y^2 = 140$

31–38. Equations of parabolas *Find an equation of the following parabolas. Unless otherwise specified, assume the vertex is at the origin.*

31. A parabola that opens to the right with directrix $x = -4$

32. A parabola that opens downward with directrix $y = 6$

33. A parabola with focus at $(3, 0)$

34. A parabola with focus at $(-4, 0)$

35. A parabola symmetric about the y-axis that passes through the point $(2, -6)$

36. A parabola symmetric about the x-axis that passes through the point $(1, -4)$

37.

38.

39–50. Equations of ellipses and hyperbolas *Find an equation of the following ellipses and hyperbolas, assuming the center is at the origin.*

39. An ellipse whose major axis is on the x-axis with length 8 and whose minor axis has length 6

40. An ellipse with vertices $(\pm 6, 0)$ and foci $(\pm 4, 0)$

41. A hyperbola with vertices $(\pm 4, 0)$ and foci $(\pm 6, 0)$

42. A hyperbola with vertices $(\pm 1, 0)$ that passes through $(5/3, 8)$

43. An ellipse with vertices $(\pm 5, 0)$, passing through the point $(4, 3/5)$

44. An ellipse with vertices $(0, \pm 10)$, passing through the point $(\sqrt{3}/2, 5)$

45. A hyperbola with vertices $(\pm 2, 0)$ and asymptotes $y = \pm 3x/2$

46. A hyperbola with vertices $(0, \pm 4)$ and asymptotes $y = \pm 2x$

47.

48.

49.

50.

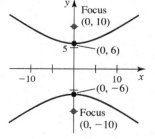

51. Explain why or why not Determine whether the following statements are true and give an explanation or counterexample.

   a. The hyperbola $\dfrac{x^2}{4} - \dfrac{y^2}{9} = 1$ has no y-intercept.

   b. On every ellipse, there are exactly two points at which the curve has slope s, where s is any real number.

   c. Given the directrices and foci of a standard hyperbola, it is possible to find its vertices, eccentricity, and asymptotes.

   d. The point on a parabola closest to the focus is the vertex.

52. Golden Gate Bridge Completed in 1937, San Francisco's Golden Gate Bridge is 2.7 km long and weighs about 890,000 tons. The length of the span between the two central towers is 1280 m; the towers themselves extend 152 m above the roadway. The cables that support the deck of the bridge between the two towers hang in a parabola (see figure). Assuming the origin is midway between the towers on the deck of the bridge, find an equation that describes the cables. How long is a guy wire that hangs vertically from the cables to the roadway 500 m from the center of the bridge?

53–56. Eccentricity-directrix approach *Find an equation of the following curves, assuming the center is at the origin. Sketch a graph labeling the vertices, foci, asymptotes (if they exist), and directrices. Use a graphing utility to check your work.*

53. An ellipse with vertices $(\pm 9, 0)$ and eccentricity $1/3$

54. An ellipse with vertices $(0, \pm 9)$ and eccentricity $1/4$

55. A hyperbola with vertices $(\pm 1, 0)$ and eccentricity 3

56. A hyperbola with vertices $(0, \pm 4)$ and eccentricity 2

57–62. Polar equations for conic sections *Graph the following conic sections, labeling the vertices, foci, directrices, and asymptotes (if they exist). Use a graphing utility to check your work.*

57. $r = \dfrac{4}{1 + \cos\theta}$ **58.** $r = \dfrac{3}{2 + \cos\theta}$

59. $r = \dfrac{1}{2 - \cos\theta}$ **60.** $r = \dfrac{6}{3 + 2\sin\theta}$

61. $r = \dfrac{1}{2 - 2\sin\theta}$ **62.** $r = \dfrac{12}{3 - \cos\theta}$

63–66. Tracing hyperbolas and parabolas *Graph the following equations. Then use arrows and labeled points to indicate how the curve is generated as θ increases from 0 to 2π.*

63. $r = \dfrac{1}{1 + \sin\theta}$ **64.** $r = \dfrac{1}{1 + 2\cos\theta}$

65. $r = \dfrac{3}{1 - \cos\theta}$ **66.** $r = \dfrac{1}{1 - 2\cos\theta}$

T 67. Parabolas with a graphing utility Use a graphing utility to graph the parabolas $r = \dfrac{d}{1 + \cos\theta}$ for $d = 0.25, 0.5, 1, 2, 3,$ and 4 on the same set of axes. Explain how the shapes of the curves vary as d changes.

T 68. Hyperbolas with a graphing utility Use a graphing utility to graph the hyperbolas $r = \dfrac{e}{1 + e\cos\theta}$ for $e = 1.1, 1.3, 1.5, 1.7,$ and 2 on the same set of axes. Explain how the shapes of the curves vary as e changes.

69–72. Tangent lines *Find an equation of the line tangent to the following curves at the given point.*

69. $x^2 = -6y; \; (-6, -6)$ **70.** $y^2 = 8x; \; (8, -8)$

71. $y^2 - \dfrac{x^2}{64} = 1; \; \left(6, -\dfrac{5}{4}\right)$ **72.** $r = \dfrac{1}{1 + \sin\theta}; \; \left(\dfrac{2}{3}, \dfrac{\pi}{6}\right)$

73. Tangent lines for an ellipse Show that an equation of the line tangent to the ellipse $\dfrac{x^2}{a^2} + \dfrac{y^2}{b^2} = 1$ at the point (x_0, y_0) is

$$\frac{xx_0}{a^2} + \frac{yy_0}{b^2} = 1.$$

74. Tangent lines for a hyperbola Find an equation of the line tangent to the hyperbola $\dfrac{x^2}{a^2} - \dfrac{y^2}{b^2} = 1$ at the point (x_0, y_0).

75–76. Graphs to polar equations *Find a polar equation for each conic section. Assume one focus is at the origin.*

75.   **76.**

77. Another construction for a hyperbola Suppose two circles, whose centers are at least $2a$ units apart (see figure), are centered at F_1 and F_2, respectively. The radius of one circle is $2a + r$ and the radius of the other circle is r, where $r \geq 0$. Show that as r increases, the intersection points P_1 and P_2 of the two circles describes one branch of a hyperbola with foci at F_1 and F_2.

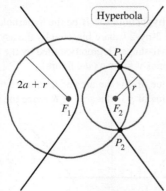

78. The ellipse and the parabola Let R be the region bounded by the upper half of the ellipse $\dfrac{x^2}{2} + y^2 = 1$ and the parabola $y = \dfrac{x^2}{\sqrt{2}}$.

  a. Find the area of R.
  b. Which is greater, the volume of the solid generated when R is revolved about the x-axis or the volume of the solid generated when R is revolved about the y-axis?

79. Volume of an ellipsoid Suppose the ellipse $\dfrac{x^2}{a^2} + \dfrac{y^2}{b^2} = 1$ is revolved about the x-axis. What is the volume of the solid enclosed by the *ellipsoid* that is generated? Is the volume different if the same ellipse is revolved about the y-axis?

80. Area of a sector of a hyperbola Consider the region R bounded by the right branch of the hyperbola $\dfrac{x^2}{a^2} - \dfrac{y^2}{b^2} = 1$ and the vertical line through the right focus.

  a. What is the area of R?
  b. Sketch a graph that shows how the area of R varies with the eccentricity e, for $e > 1$.

81. Volume of a hyperbolic cap Consider the region R bounded by the right branch of the hyperbola $\dfrac{x^2}{a^2} - \dfrac{y^2}{b^2} = 1$ and the vertical line through the right focus.

  a. What is the volume of the solid that is generated when R is revolved about the x-axis?
  b. What is the volume of the solid that is generated when R is revolved about the y-axis?

82. Volume of a paraboloid (Archimedes) The region bounded by the parabola $y = ax^2$ and the horizontal line $y = h$ is revolved about the y-axis to generate a solid bounded by a surface called a *paraboloid* (where $a > 0$ and $h > 0$). Show that the volume of the solid is $3/2$ the volume of the cone with the same base and vertex.

Explorations and Challenges

83. Reflection property of parabolas Consider the parabola $y = \dfrac{x^2}{4p}$ with its focus at $F(0, p)$ (see figure). The goal is to show that the angle of incidence between the ray ℓ and the tangent line L (α in the figure) equals the angle of reflection between the line PF and L (β in the figure). If these two angles are equal, then the reflection property is proved because ℓ is reflected through F.

  a. Let $P(x_0, y_0)$ be a point on the parabola. Show that the slope of the line tangent to the curve at P is $\tan \theta = \dfrac{x_0}{2p}$.

  b. Show that $\tan \varphi = \dfrac{p - y_0}{x_0}$.

  c. Show that $\alpha = \dfrac{\pi}{2} - \theta$; therefore, $\tan \alpha = \cot \theta$.

  d. Note that $\beta = \theta + \varphi$. Use the tangent addition formula $\tan(\theta + \varphi) = \dfrac{\tan \theta + \tan \varphi}{1 - \tan \theta \tan \varphi}$ to show that $\tan \alpha = \tan \beta = \dfrac{2p}{x_0}$.

  e. Conclude that because α and β are acute angles, $\alpha = \beta$.

84. Deriving polar equations for conics Use Figures 12.60 and 12.61 to derive the polar equation of a conic section with a focus at the origin in the following three cases.

  a. Vertical directrix at $x = -d$, where $d > 0$
  b. Horizontal directrix at $y = d$, where $d > 0$
  c. Horizontal directrix at $y = -d$, where $d > 0$

85. Equation of an ellipse Consider an ellipse to be the set of points in a plane whose distances from two fixed points have a constant sum $2a$. Derive the equation of an ellipse. Assume the two fixed points are on the x-axis equidistant from the origin.

86. Equation of a hyperbola Consider a hyperbola to be the set of points in a plane whose distances from two fixed points have a constant difference of $2a$ or $-2a$. Derive the equation of a hyperbola. Assume the two fixed points are on the x-axis equidistant from the origin.

87. Equidistant set Show that the set of points equidistant from a circle and a line not passing through the circle is a parabola. Assume the circle, line, and parabola lie in the same plane.

88. Polar equation of a conic Show that the polar equation of an ellipse or hyperbola with one focus at the origin, major axis of length $2a$ on the x-axis, and eccentricity e is

$$r = \frac{a(1 - e^2)}{1 + e \cos \theta}.$$

89. Shared asymptotes Suppose two hyperbolas with eccentricities e and E have perpendicular major axes and share a set of asymptotes. Show that $e^{-2} + E^{-2} = 1$.

90–94. Focal chords *A **focal chord** of a conic section is a line through a focus joining two points of the curve. The **latus rectum** is the focal chord perpendicular to the major axis of the conic. Prove the following properties.*

90. The lines tangent to the endpoints of any focal chord of a parabola $y^2 = 4px$ intersect on the directrix and are perpendicular.

91. Let L be the latus rectum of the parabola $y^2 = 4px$ for $p > 0$. Let F be the focus of the parabola, P be any point on the parabola to the left of L, and D be the (shortest) distance between P and L. Show that for all P, $D + |FP|$ is a constant. Find the constant.

92. The length of the latus rectum of the parabola $y^2 = 4px$ or $x^2 = 4py$ is $4|p|$.

93. The length of the latus rectum of an ellipse centered at the origin is $\dfrac{2b^2}{a} = 2b\sqrt{1 - e^2}$.

94. The length of the latus rectum of a hyperbola centered at the origin is $\dfrac{2b^2}{a} = 2b\sqrt{e^2 - 1}$.

95. Confocal ellipse and hyperbola Show that an ellipse and a hyperbola that have the same two foci intersect at right angles.

96. Approach to asymptotes Show that the vertical distance between a hyperbola $\dfrac{x^2}{a^2} - \dfrac{y^2}{b^2} = 1$ and its asymptote $y = \dfrac{bx}{a}$ approaches zero as $x \to \infty$, where $0 < b < a$.

97. Sector of a hyperbola Let H be the right branch of the hyperbola $x^2 - y^2 = 1$ and let ℓ be the line $y = m(x - 2)$ that passes through the point $(2, 0)$ with slope m, where $-\infty < m < \infty$. Let R be the region in the first quadrant bounded by H and ℓ (see figure). Let $A(m)$ be the area of R. Note that for some values of m, $A(m)$ is not defined.

a. Find the x-coordinates of the intersection points between H and ℓ as functions of m; call them $u(m)$ and $v(m)$, where $v(m) > u(m) > 1$. For what values of m are there two intersection points?

b. Evaluate $\lim\limits_{m \to 1^+} u(m)$ and $\lim\limits_{m \to 1^+} v(m)$.

c. Evaluate $\lim\limits_{m \to \infty} u(m)$ and $\lim\limits_{m \to \infty} v(m)$.

d. Evaluate and interpret $\lim\limits_{m \to \infty} A(m)$.

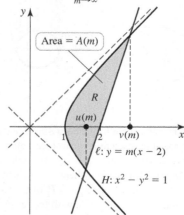

Area = $A(m)$

R

$u(m)$

$v(m)$

$\ell: y = m(x - 2)$

$H: x^2 - y^2 = 1$

T 98. The anvil of a hyperbola Let H be the hyperbola $x^2 - y^2 = 1$ and let S be the 2-by-2 square bisected by the asymptotes of H. Let R be the anvil-shaped region bounded by the hyperbola and the horizontal lines $y = \pm p$ (see figure).

a. For what value of p is the area of R equal to the area of S?

b. For what value of p is the area of R twice the area of S?

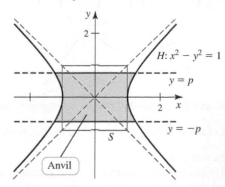

$H: x^2 - y^2 = 1$

$y = p$

$y = -p$

S

Anvil

QUICK CHECK ANSWERS

2. a. Left **b.** Up **3.** The minor-axis vertices are $(0, \pm b)$. The distance between them is $2b$, which is the length of the minor axis. **4.** Vertices: $(0, \pm 1)$; foci: $(0, \pm\sqrt{5})$ **5.** $b = 3\sqrt{3}/2$, $c = 3/2$, $d = 6$ **6.** y-axis ◄

CHAPTER 12   REVIEW EXERCISES

1. Explain why or why not Determine whether the following statements are true and give an explanation or counterexample.

a. A set of parametric equations for a given curve is always unique.

b. The equations $x = e^t$, $y = 2e^t$, for $-\infty < t < \infty$, describe a line passing through the origin with slope 2.

c. The polar coordinates $\left(3, -\dfrac{3\pi}{4}\right)$ and $\left(-3, \dfrac{\pi}{4}\right)$ describe the same point in the plane.

d. The area of the region between the inner and outer loops of the limaçon $r = f(\theta) = 1 - 4\cos\theta$ is $\dfrac{1}{2}\displaystyle\int_0^{2\pi} f(\theta)^2\, d\theta$.

e. The hyperbola $\dfrac{y^2}{2} - \dfrac{x^2}{4} = 1$ has no x-intercept.

f. The equation $x^2 + 4y^2 - 2x = 3$ describes an ellipse.

2. Eliminate the parameter in the parametric equations
$x = 1 + \sin t$, $y = 3 + 2 \sin t$, for $0 \leq t \leq \pi/2$, and describe
the curve, indicating its positive orientation. How does this
curve differ from the curve $x = 1 + \sin t$, $y = 3 + 2 \sin t$, for
$\pi/2 \leq t \leq \pi$?

3–6. Eliminating the parameter *Eliminate the parameter to find a
description of the following curves in terms of x and y. Give a geometric description and the positive orientation of the curve.*

3. $x = 4 \cos t$, $y = 3 \sin t$; $0 \leq t \leq 2\pi$

4. $x = \sin t - 3$, $y = \cos t + 6$; $0 \leq t \leq \pi$

5. $x = t + 1$, $y = \sqrt{t + 1}$; $3 \leq t \leq 8$

6. $x = -\cot^2 t$, $y = 1 + \csc^2 t$; $\dfrac{\pi}{4} \leq t \leq \dfrac{\pi}{2}$

7–8. Parametric curves and tangent lines

a. *Eliminate the parameter to obtain an equation in x and y.*
b. *Find the slope of the curve at the given value of t.*
c. *Make a sketch of the curve showing the positive orientation of the
curve and the tangent line at the point corresponding to the given
value of t.*

7. $x = 8 \cos t + 1$, $y = 8 \sin t + 2$, for $0 \leq t \leq 2\pi$; $t = \pi/3$

8. $x = 4 \sin 2t$, $y = 3 \cos 2t$, for $0 \leq t \leq \pi$; $t = \pi/6$

T 9. Find parametric equations for the curve $x = 5(y - 1)(y - 2) \sin y$
and then plot the curve using a graphing utility.

10–12. Parametric curves

a. *Eliminate the parameter to obtain an equation in x and y.*
b. *Describe the curve, indicating the positive orientation.*
c. *Find the slope of the curve at the specified point.*

10. $x = t^2 + 4$, $y = -t$, for $-2 < t < 0$; $(5, 1)$

11. $x = 3 \cos(-t)$, $y = 3 \sin(-t) - 1$, for $0 \leq t \leq \pi$; $(0, -4)$

12. $x = \ln t$, $y = 8 \ln t^2$, for $1 \leq t \leq e^2$; $(1, 16)$

T 13. Tangent lines Find an equation of the line tangent to the cycloid
$x = t - \sin t$, $y = 1 - \cos t$ at the points corresponding to
$t = \pi/6$ and $t = 2\pi/3$.

14–18. Parametric descriptions *Write parametric equations for the
following curves. Solutions are not unique.*

14. The segment of the curve $x = y^3 + y + 1$ that starts at $(1, 0)$ and
ends at $(11, 2)$.

15. The line segment from $P(-1, 0)$ to $Q(1, 1)$, followed by the line
segment from Q to P (two sets of equations are required)

16. The segment of the curve $f(x) = x^3 + 2x$ from $(0, 0)$ to $(2, 12)$

17. The circle $x^2 + y^2 = 9$, generated clockwise

18. The right side of the ellipse $\dfrac{x^2}{9} + \dfrac{y^2}{4} = 1$, generated
counterclockwise

19–20. Area bounded by parametric curves *Find the area of the
following regions. (Hint: See Exercises 103–105 in Section 12.1.)*

19. The region bounded by the y-axis and the parametric curve
$x = t^2$, $y = t - t^3$, for $0 \leq t \leq 1$

20. The region bounded by the x-axis and the parametric curve
$x = \cos t$, $y = \sin 2t$, for $0 \leq t \leq \pi/2$

T 21. Surface of revolution Find the area of the surface obtained by
revolving the curve $x = \sin t$, $y = \cos 2t$, for $-\pi/4 \leq t \leq \pi/4$,
about the x-axis. (*Hint:* See Exercises 106–111 in Section 12.1.)

22–23. Arc length *Find the length of the following curves.*

22. $x = e^{2t} \sin 3t$, $y = e^{2t} \cos 3t$; $0 \leq t \leq \pi/3$

23. $x = \cos 2t$, $y = 2t - \sin 2t$; $0 \leq t \leq \pi/4$

24–26. Sets in polar coordinates *Sketch the following sets of points.*

24. $\theta = \dfrac{\pi}{6}$

25. $0 \leq r \leq 4$ and $-\dfrac{\pi}{2} \leq \theta \leq -\dfrac{\pi}{3}$

26. $4 \leq r^2 \leq 9$

27–32. Polar curves *Graph the following equations.*

27. $r = 2 - 3 \sin \theta$ **28.** $r = 1 + 2 \cos \theta$

29. $r = 3 \cos 3\theta$ **30.** $r = e^{-\theta/6}$

31. $r = 3 \sin 4\theta$ **32.** $r^2 = 4 \cos \theta$

33. Polar valentine Liz wants to show her love for Jake by passing
him a valentine on her graphing calculator. Sketch each of the
following curves and determine which one Liz should use to get a
heart-shaped curve.

    a. $r = 5 \cos \theta$ **b.** $r = 1 - \sin \theta$ **c.** $r = \cos 3\theta$

34. Jake's response Jake responds to Liz (Exercise 33) with a graph
that shows his love for her is infinite. Sketch each of the following curves. Which one should Jake send to Liz to get an infinity
symbol?

    a. $r = \theta$, for $\theta \geq 0$ **b.** $r = \dfrac{1}{2} + \sin \theta$ **c.** $r^2 = \cos 2\theta$

35. Polar conversion Write the equation $r^2 + r(2 \sin \theta - 6 \cos \theta) = 0$
in Cartesian coordinates and identify the corresponding curve.

36. Polar conversion Consider the equation $r = \dfrac{4}{\sin \theta + \cos \theta}$.

    a. Convert the equation to Cartesian coordinates and identify the
curve it describes.
    b. Graph the curve and indicate the points that correspond to
$\theta = 0, \dfrac{\pi}{2}$, and 2π.
    c. Give an interval in θ on which the entire curve is generated.

37. Parametric to polar equations Find an equation of the following
curve in polar coordinates and describe the curve.

$$x = (1 + \cos t) \cos t, \quad y = (1 + \cos t) \sin t; \quad 0 \leq t \leq 2\pi$$

38. Cartesian conversion Write the equation $x = y^2$ in polar coordinates and state values of θ that produce the entire graph of the
parabola.

39. Cartesian conversion Write the equation $(x - 4)^2 + y^2 = 16$
in polar coordinates and state values of θ that produce the entire
graph of the circle.

T 40–41. Slopes of tangent lines

a. *Find all points where the following curves have vertical and horizontal tangent lines.*

b. *Find the slope of the lines tangent to the curve at the origin (when relevant).*

c. *Sketch the curve and all the tangent lines identified in parts (a) and (b).*

40. $r = 3 - 6 \cos \theta$ **41.** $r = 1 - \sin \theta$

42–43. Intersection points *Find the intersection points of the following curves.*

42. $r = \sqrt{3}$ and $r = 2 \cos 2\theta$ **43.** $r = \sqrt{\cos 3t}$ and $r = \sqrt{\sin 3t}$

44–49. Areas of regions *Find the area of the following regions.*

44. The region inside one leaf of the rose $r = 3 \sin 4\theta$

45. The region inside the inner loop of the limaçon $r = 1 + 2 \sin \theta$

46. The region inside the limaçon $r = 2 + \cos \theta$ and outside the circle $r = 2$

47. The region inside the lemniscate $r^2 = 4 \cos 2\theta$ and outside the circle $r = \sqrt{2}$

48. The region inside $r = \sqrt{2} + \sin \theta$ and inside $r = 3 \sin \theta$

49. The region inside the cardioid $r = 1 + \cos \theta$ and outside the cardioid $r = 1 - \cos \theta$

T 50.   Intersection points Consider the polar equations $r = 1$ and $r = 2 - 4 \cos \theta$.

a. Graph the curves. How many intersection points do you observe?

b. Give approximate polar coordinates of the intersection points.

T 51–52. Arc length of polar curves *Find the approximate length of the following curves.*

51. The limaçon $r = 3 - 6 \cos \theta$

52. The limaçon $r = 3 + 2 \cos \theta$

53–57. Conic sections

a. *Determine whether the following equations describe a parabola, an ellipse, or a hyperbola.*

b. *Use analytical methods to determine the location of the foci, vertices, and directrices.*

c. *Find the eccentricity of the curve.*

d. *Make an accurate graph of the curve.*

53. $x^2 - \dfrac{y^2}{2} = 1$ **54.** $x = 16y^2$

55. $y^2 - 4x^2 = 16$ **56.** $\dfrac{x^2}{4} + \dfrac{y^2}{25} = 1$

57. $4x^2 + 8y^2 = 16$

58–59. Tangent lines *Find an equation of the line tangent to the following curves at the given point. Check your work with a graphing utility.*

58. $\dfrac{x^2}{16} - \dfrac{y^2}{9} = 1; \left(\dfrac{20}{3}, -4 \right)$ **59.** $y^2 = -12x; \left(-\dfrac{4}{3}, -4 \right)$

60.   A polar conic section Consider the equation $r^2 = \sec 2\theta$.

a. Convert the equation to Cartesian coordinates and identify the curve.

b. Find the vertices, foci, directrices, and eccentricity of the curve.

c. Graph the curve. Explain why the polar equation does not have the form given in the text for conic sections in polar coordinates.

61–64. Polar equations for conic sections *Graph the following conic sections, labeling vertices, foci, directrices, and asymptotes (if they exist). Give the eccentricity of the curve. Use a graphing utility to check your work.*

61. $r = \dfrac{2}{1 + \sin \theta}$ **62.** $r = \dfrac{3}{1 - 2 \cos \theta}$

63. $r = \dfrac{4}{2 + \cos \theta}$ **64.** $r = \dfrac{10}{5 + 2 \cos \theta}$

65–68. Eccentricity-directrix approach *Find an equation of the following curves, assuming the center is at the origin. Graph the curve, labeling vertices, foci, asymptotes (if they exist), and directrices.*

65. A hyperbola with vertices $(0, \pm 2)$ and directrices $y = \pm 1$

66. An ellipse with foci $(\pm 4, 0)$ and directrices $x = \pm 8$

67. An ellipse with vertices $(0, \pm 4)$ and directrices $y = \pm 10$

68. A hyperbola with vertices $(\pm 4, 0)$ and directrices $x = \pm 2$

69.   Conic parameters An ellipse has vertices $(0, \pm 6)$ and foci $(0, \pm 4)$. Find the eccentricity, the directrices, and the minor-axis vertices.

70.   Conic parameters A hyperbola has eccentricity $e = 2$ and foci $(0, \pm 2)$. Find the location of the vertices and directrices.

71.   Bisecting an ellipse Let R be the region in the first quadrant bounded by the ellipse $\dfrac{x^2}{a^2} + \dfrac{y^2}{b^2} = 1$. Find the value of m (in terms of a and b) such that the line $y = mx$ divides R into two subregions of equal area.

72.   Parabola-hyperbola tangency Let P be the parabola $y = px^2$ and H be the right half of the hyperbola $x^2 - y^2 = 1$.

a. For what value of p is P tangent to H?

b. At what point does the tangency occur?

c. Generalize your results for the hyperbola $\dfrac{x^2}{a^2} - \dfrac{y^2}{b^2} = 1$.

73.   Another ellipse construction Start with two circles centered at the origin with radii $0 < a < b$ (see figure). Assume the line ℓ through the origin intersects the smaller circle at Q and the larger circle at R. Let $P(x, y)$ have the y-coordinate of Q and the x-coordinate of R. Show that the set of points $P(x, y)$ generated in this way for all lines ℓ through the origin is an ellipse.

T 74. Hyperbolas The graph of the parametric equations $x = a \tan t$, $y = b \sec t$, for $-\pi < t < \pi$ and $|t| \neq \pi/2$, where a and b are nonzero real numbers, is a hyperbola. Graph the hyperbola with $a = b = 1$. Indicate clearly the direction in which the curve is generated as t increases from $t = -\pi$ to $t = \pi$.

T 75. Lamé curves The *Lamé curve* described by $\left|\dfrac{x}{a}\right|^n + \left|\dfrac{y}{b}\right|^n = 1$,

where a, b, and n are positive real numbers, is a generalization of an ellipse.

   a. Express this equation in parametric form (four pairs of equations are needed).

   b. Graph the curve for $a = 4$ and $b = 2$, for various values of n.

   c. Describe how the curves change as n increases.

76. General equations for a circle Prove that the equations

$$x = a \cos t + b \sin t, \quad y = c \cos t + d \sin t$$

where a, b, c, and d are real numbers, describe a circle of radius R provided $a^2 + c^2 = b^2 + d^2 = R^2$ and $ab + cd = 0$.

Chapter 12 Guided Projects

Applications of the material in this chapter and related topics can be found in the following Guided Projects. For additional information, see the Preface.

- The amazing cycloid
- Parametric art
- Polar art
- Grazing goat problems

- Translations and rotations of axes
- Celestial orbits
- Properties of conic sections

13

Vectors and the Geometry of Space

Chapter Preview We now make a significant departure from previous chapters by stepping out of the xy-plane ($\mathbb{R}^2$) into three-dimensional space ($\mathbb{R}^3$). The fundamental concept of a *vector*—a quantity with magnitude and direction—is introduced in two and three dimensions. We then develop the algebra associated with vectors (how to add, subtract, and combine them in various ways), and we define two fundamental operations for vectors: the dot product and the cross product. The chapter concludes with a brief survey of basic objects in three-dimensional geometry, namely lines, planes, and elementary surfaces.

13.1 Vectors in the Plane

Imagine a raft drifting down a river, carried by the current. The speed and direction of the raft at a point may be represented by an arrow (**Figure 13.1**). The length of the arrow represents the speed of the raft at that point; longer arrows correspond to greater speeds. The orientation of the arrow gives the direction in which the raft is headed at that point. The arrows at points A and C in Figure 13.1 have the same length and direction, indicating that the raft has the same speed and heading at these locations. The arrow at B is shorter and points to the left of the rock, indicating that the raft slows down as it nears the rock.

Figure 13.1

Basic Vector Operations

The arrows that describe the raft's motion are examples of *vectors*—quantities that have both *length* (or *magnitude*) and *direction*. Vectors arise naturally in many situations. For example, electric and magnetic fields, the flow of air over an airplane wing, and the velocity and acceleration of elementary particles are described by vectors (**Figure 13.2**). In this section, we examine vectors in the xy-plane; then we extend the concept to three dimensions in Section 13.2.

The vector whose *tail* is at the point P and whose *head* is at the point Q is denoted $\overrightarrow{PQ}$ (**Figure 13.3**). The vector $\overrightarrow{QP}$ has its tail at Q and its head at P. We also label vectors with single boldface characters such as **u** and **v**.

Electric field vectors due to two charges

Velocity vectors of air flowing over an airplane wing

Tracks of elementary particles in a cloud chamber are aligned with the velocity vectors of the particles.

Figure 13.2

Figure 13.3

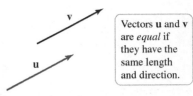

Figure 13.4

Two vectors **u** and **v** are *equal*, written **u** = **v**, if they have equal length and point in the same direction (**Figure 13.4**). An important fact is that equal vectors do not necessarily have the same location. *Any* two vectors with the same length and direction are equal.

Not all quantities are represented by vectors. For example, mass, temperature, and price have magnitude, but no direction. Such quantities are described by real numbers and are called *scalars*.

➤ In this text, *scalar* is another word for *real number*.

➤ The vector **v** is commonly handwritten as $\vec{v}$. The zero vector is handwritten as $\vec{0}$.

Vectors, Equal Vectors, Scalars, Zero Vector

Vectors are quantities that have both length (or magnitude) and direction. Two vectors are **equal** if they have the same magnitude and direction. Quantities having magnitude but no direction are called **scalars**. One exception is the **zero vector**, denoted **0**: It has length 0 and no direction.

Scalar Multiplication

A scalar c and a vector **v** can be combined using scalar-vector multiplication, or simply *scalar multiplication*. The resulting vector, denoted $c\mathbf{v}$, is called a *scalar multiple* of **v**. The length of $c\mathbf{v}$ is $|c|$ multiplied by the length of **v**. The vector $c\mathbf{v}$ has the same direction as **v** if $c > 0$. If $c < 0$, then $c\mathbf{v}$ and **v** point in opposite directions. If $c = 0$, then the product $0\mathbf{v} = \mathbf{0}$ (the zero vector).

For example, the vector $3\mathbf{v}$ is three times as long as **v** and has the same direction as **v**. The vector $-2\mathbf{v}$ is twice as long as **v**, but it points in the opposite direction. The vector $\frac{1}{2}\mathbf{v}$ points in the same direction as **v** and has half the length of **v** (**Figure 13.5**). The vectors **v**, $3\mathbf{v}$, $-2\mathbf{v}$, and $\frac{1}{2}\mathbf{v}$ are examples of *parallel vectors*: Each one is a scalar multiple of the others.

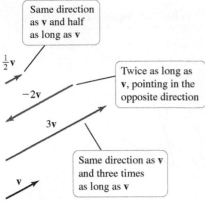

Same direction as **v** and half as long as **v**

$\frac{1}{2}\mathbf{v}$

$-2\mathbf{v}$

Twice as long as **v**, pointing in the opposite direction

$3\mathbf{v}$

v

Same direction as **v** and three times as long as **v**

Figure 13.5

DEFINITION Scalar Multiples and Parallel Vectors

Given a scalar c and a vector **v**, the **scalar multiple** $c\mathbf{v}$ is a vector whose length is $|c|$ multiplied by the length of **v**. If $c > 0$, then $c\mathbf{v}$ has the same direction as **v**. If $c < 0$, then $c\mathbf{v}$ and **v** point in opposite directions. Two vectors are **parallel** if they are scalar multiples of each other.

➤ For convenience, we write $-\mathbf{u}$ for $(-1)\mathbf{u}$, $-c\mathbf{u}$ for $(-c)\mathbf{u}$, and $\mathbf{u}/c$ for $\frac{1}{c}\mathbf{u}$.

QUICK CHECK 1 Describe the length and direction of the vector $-5\mathbf{v}$ relative to $\mathbf{v}$. ◀

(a)

(b)

Figure 13.6

QUICK CHECK 2 Sketch the sum $\mathbf{v}_a + \mathbf{w}$ in Figure 13.7 if the direction of $\mathbf{w}$ is reversed. ◀

Notice that $0\mathbf{v} = \mathbf{0}$ for all vectors $\mathbf{v}$. It follows that *the zero vector is parallel to all vectors*. While it may seem counterintuitive, this result turns out to be a useful convention.

EXAMPLE 1 Parallel vectors Using Figure 13.6a, write the following vectors in terms of $\mathbf{u}$ or $\mathbf{v}$.

a. $\overrightarrow{PQ}$ **b.** $\overrightarrow{QP}$ **c.** $\overrightarrow{QR}$ **d.** $\overrightarrow{RS}$

SOLUTION

a. The vector $\overrightarrow{PQ}$ has the same direction and length as $\mathbf{u}$; therefore, $\overrightarrow{PQ} = \mathbf{u}$. These two vectors are equal even though they have different locations (**Figure 13.6b**).

b. Because $\overrightarrow{QP}$ and $\mathbf{u}$ have equal length but opposite directions, $\overrightarrow{QP} = (-1)\mathbf{u} = -\mathbf{u}$.

c. $\overrightarrow{QR}$ points in the same direction as $\mathbf{v}$ and is twice as long as $\mathbf{v}$, so $\overrightarrow{QR} = 2\mathbf{v}$.

d. $\overrightarrow{RS}$ points in the direction opposite that of $\mathbf{u}$ with three times the length of $\mathbf{u}$. Consequently, $\overrightarrow{RS} = -3\mathbf{u}$.

Related Exercise 15 ◀

Vector Addition and Subtraction

To illustrate the idea of vector addition, consider a plane flying horizontally at a constant speed in a crosswind (**Figure 13.7**). The length of vector $\mathbf{v}_a$ represents the plane's *airspeed*, which is the speed the plane would have in still air; $\mathbf{v}_a$ points in the direction of the nose of the plane. The wind vector $\mathbf{w}$ points in the direction of the crosswind and has a length equal to the speed of the crosswind. The combined effect of the motion of the plane and the wind is the *vector sum* $\mathbf{v}_g = \mathbf{v}_a + \mathbf{w}$, which is the velocity of the plane relative to the ground.

Figure 13.8 illustrates two ways to form the vector sum of two nonzero vectors $\mathbf{u}$ and $\mathbf{v}$ geometrically. The first method, called the **Triangle Rule**, places the tail of $\mathbf{v}$ at the head of $\mathbf{u}$. The sum $\mathbf{u} + \mathbf{v}$ is the vector that extends from the tail of $\mathbf{u}$ to the head of $\mathbf{v}$ (**Figure 13.8b**).

When $\mathbf{u}$ and $\mathbf{v}$ are not parallel, another way to form $\mathbf{u} + \mathbf{v}$ is to use the **Parallelogram Rule**. The *tails* of $\mathbf{u}$ and $\mathbf{v}$ are connected to form adjacent sides of a parallelogram; then the remaining two sides of the parallelogram are sketched. The sum $\mathbf{u} + \mathbf{v}$ is the vector that coincides with the diagonal of the parallelogram, beginning at the tails of $\mathbf{u}$ and $\mathbf{v}$ (**Figure 13.8c**). Both the Triangle Rule and the Parallelogram Rule produce the same vector sum $\mathbf{u} + \mathbf{v}$.

Figure 13.7

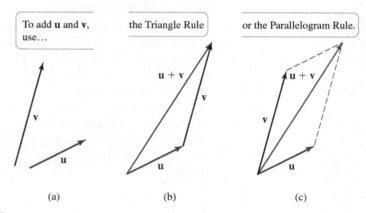

(a)   (b)   (c)

Figure 13.8

QUICK CHECK 3 Use the Triangle Rule to show that the vectors in Figure 13.8 satisfy $\mathbf{u} + \mathbf{v} = \mathbf{v} + \mathbf{u}$. ◀

The difference $\mathbf{u} - \mathbf{v}$ is defined to be the sum $\mathbf{u} + (-\mathbf{v})$. By the Triangle Rule, the tail of $-\mathbf{v}$ is placed at the head of $\mathbf{u}$; then $\mathbf{u} - \mathbf{v}$ extends from the tail of $\mathbf{u}$ to the head of $-\mathbf{v}$ (**Figure 13.9a**). Equivalently, when the tails of $\mathbf{u}$ and $\mathbf{v}$ coincide, $\mathbf{u} - \mathbf{v}$ has its tail at the head of $\mathbf{v}$ and its head at the head of $\mathbf{u}$ (**Figure 13.9b**).

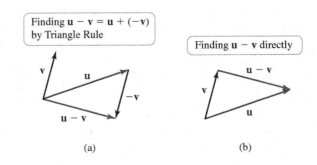

Finding $\mathbf{u} - \mathbf{v} = \mathbf{u} + (-\mathbf{v})$ by Triangle Rule

Finding $\mathbf{u} - \mathbf{v}$ directly

(a)   (b)

Figure 13.9

Figure 13.10

EXAMPLE 2   Vector operations Use **Figure 13.10** to write the following vectors as sums of scalar multiples of **v** and **w**.

a. $\overrightarrow{OP}$ **b.** $\overrightarrow{OQ}$ **c.** $\overrightarrow{QR}$

SOLUTION

a. Using the Triangle Rule, we start at O, move three lengths of **v** in the direction of **v** and then two lengths of **w** in the direction of **w** to reach P. Therefore, $\overrightarrow{OP} = 3\mathbf{v} + 2\mathbf{w}$ (**Figure 13.11a**).

b. The vector $\overrightarrow{OQ}$ coincides with the diagonal of a parallelogram having adjacent sides equal to $3\mathbf{v}$ and $-\mathbf{w}$. By the Parallelogram Rule, $\overrightarrow{OQ} = 3\mathbf{v} - \mathbf{w}$ (**Figure 13.11b**).

c. The vector $\overrightarrow{QR}$ lies on the diagonal of a parallelogram having adjacent sides equal to **v** and $2\mathbf{w}$. Therefore, $\overrightarrow{QR} = \mathbf{v} + 2\mathbf{w}$ (**Figure 13.11c**).

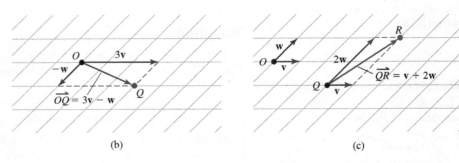

(a)   (b)   (c)

Figure 13.11

Related Exercises 17–18 ◄

Vector Components

So far, vectors have been examined from a geometric point of view. To do calculations with vectors, it is necessary to introduce a coordinate system. We begin by considering a vector **v** whose tail is at the origin in the Cartesian plane and whose head is at the point (v_1, v_2) (**Figure 13.12a**).

> ► Round brackets (a, b) enclose the *coordinates* of a point, while angle brackets $\langle a, b \rangle$ enclose the *components* of a vector. Note that in component form, the zero vector is $\mathbf{0} = \langle 0, 0 \rangle$.

DEFINITION   Position Vectors and Vector Components

A vector **v** with its tail at the origin and head at the point (v_1, v_2) is called a **position vector** (or is said to be in **standard position**) and is written $\langle v_1, v_2 \rangle$. The real numbers v_1 and v_2 are the x- and y-**components** of **v**, respectively. The position vectors $\mathbf{u} = \langle u_1, u_2 \rangle$ and $\mathbf{v} = \langle v_1, v_2 \rangle$ are **equal** if and only if $u_1 = v_1$ and $u_2 = v_2$.

There are infinitely many vectors equal to the position vector **v**, all with the same length and direction (**Figure 13.12b**). It is important to abide by the convention that $\mathbf{v} = \langle v_1, v_2 \rangle$ refers to the position vector **v** *or to any other vector equal to* **v**.

(a)   (b)

Figure 13.12

$$\overrightarrow{PQ} = \langle x_2 - x_1, y_2 - y_1 \rangle$$
$$|\overrightarrow{PQ}| = \sqrt{(x_2 - x_1)^2 + (y_2 - y_1)^2}$$

Figure 13.13

QUICK CHECK 4 Given the points $P(2, 3)$ and $Q(-4, 1)$, find the components of $\overrightarrow{PQ}$. ◄

▶ Just as the absolute value $|p - q|$ gives the distance between the points p and q on the number line, the magnitude $|\overrightarrow{PQ}|$ is the distance between the points P and Q. The magnitude of a vector is also called its **norm**.

Now consider the vector $\overrightarrow{PQ}$ equal to $\mathbf{v} = \langle v_1, v_2 \rangle$, but not in standard position, with its tail at the point $P(x_1, y_1)$ and its head at the point $Q(x_2, y_2)$. The x-component of $\overrightarrow{PQ}$ is the difference in the x-coordinates of Q and P, or $x_2 - x_1$. The y-component of $\overrightarrow{PQ}$ is the difference in the y-coordinates, $y_2 - y_1$ (**Figure 13.13**). Therefore, $\overrightarrow{PQ} = \langle x_2 - x_1, y_2 - y_1 \rangle = \langle v_1, v_2 \rangle = \mathbf{v}$.

As already noted, there are infinitely many vectors equal to a given position vector. All these vectors have the same length and direction; therefore, they are all equal. In other words, two arbitrary vectors are equal if they are equal to the same position vector. For example, the vector $\overrightarrow{PQ}$ from $P(2, 5)$ to $Q(6, 3)$ and the vector $\overrightarrow{AB}$ from $A(7, 12)$ to $B(11, 10)$ are equal because they both equal the position vector $\langle 4, -2 \rangle$.

Magnitude

The magnitude of a vector is simply its length. By the Pythagorean Theorem and Figure 13.13, we have the following definition.

DEFINITION Magnitude of a Vector

Given the points $P(x_1, y_1)$ and $Q(x_2, y_2)$, the **magnitude**, or **length**, of $\overrightarrow{PQ} = \langle x_2 - x_1, y_2 - y_1 \rangle$, denoted $|\overrightarrow{PQ}|$, is the distance between P and Q:

$$|\overrightarrow{PQ}| = \sqrt{(x_2 - x_1)^2 + (y_2 - y_1)^2}.$$

The magnitude of the position vector $\mathbf{v} = \langle v_1, v_2 \rangle$ is $|\mathbf{v}| = \sqrt{v_1^2 + v_2^2}$.

EXAMPLE 3 Calculating components and magnitude Given the points $O(0, 0)$, $P(-3, 4)$, and $Q(6, 5)$, find the components and magnitude of the following vectors.

a. $\overrightarrow{OP}$ **b.** $\overrightarrow{PQ}$

SOLUTION

a. The vector $\overrightarrow{OP}$ is the position vector whose head is located at $P(-3, 4)$. Therefore, $\overrightarrow{OP} = \langle -3, 4 \rangle$ and its magnitude is $|\overrightarrow{OP}| = \sqrt{(-3)^2 + 4^2} = 5$.

b. $\overrightarrow{PQ} = \langle 6 - (-3), 5 - 4 \rangle = \langle 9, 1 \rangle$ and $|\overrightarrow{PQ}| = \sqrt{9^2 + 1^2} = \sqrt{82}$.

Related Exercise 19 ◄

Vector Operations in Terms of Components

We now show how vector addition, vector subtraction, and scalar multiplication are performed using components. Suppose $\mathbf{u} = \langle u_1, u_2 \rangle$ and $\mathbf{v} = \langle v_1, v_2 \rangle$. The vector sum of $\mathbf{u}$ and $\mathbf{v}$ is $\mathbf{u} + \mathbf{v} = \langle u_1 + v_1, u_2 + v_2 \rangle$. This definition of a vector sum is consistent with the Parallelogram Rule given earlier (**Figure 13.14**).

Figure 13.14

For a scalar c and a vector $\mathbf{u}$, the scalar multiple $c\mathbf{u}$ is $c\mathbf{u} = \langle cu_1, cu_2 \rangle$; that is, the scalar c multiplies each component of $\mathbf{u}$. If $c > 0$, $\mathbf{u}$ and $c\mathbf{u}$ have the same direction (**Figure 13.15a**). If $c < 0$, $\mathbf{u}$ and $c\mathbf{u}$ have opposite directions (**Figure 13.15b**). In either case, $|c\mathbf{u}| = |c||\mathbf{u}|$ (Exercise 83).

Notice that $\mathbf{u} - \mathbf{v} = \mathbf{u} + (-\mathbf{v})$, where $-\mathbf{v} = \langle -v_1, -v_2 \rangle$. Therefore, the vector difference of $\mathbf{u}$ and $\mathbf{v}$ is $\mathbf{u} - \mathbf{v} = \langle u_1 - v_1, u_2 - v_2 \rangle$.

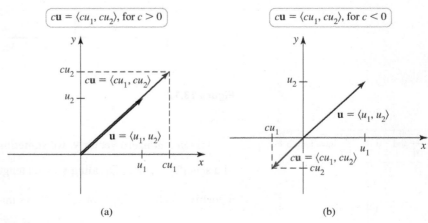

Figure 13.15

DEFINITION Vector Operations in $\mathbb{R}^2$

Suppose c is a scalar, $\mathbf{u} = \langle u_1, u_2 \rangle$, and $\mathbf{v} = \langle v_1, v_2 \rangle$.

$$\mathbf{u} + \mathbf{v} = \langle u_1 + v_1, u_2 + v_2 \rangle \quad \text{Vector addition}$$
$$\mathbf{u} - \mathbf{v} = \langle u_1 - v_1, u_2 - v_2 \rangle \quad \text{Vector subtraction}$$
$$c\mathbf{u} = \langle cu_1, cu_2 \rangle \quad\quad\quad\quad \text{Scalar multiplication}$$

> ▶ Recall that $\mathbb{R}^2$ (pronounced *R-two*) stands for the *xy*-plane or the set of all ordered pairs of real numbers.

EXAMPLE 4 Vector operations Let $\mathbf{u} = \langle -1, 2 \rangle$ and $\mathbf{v} = \langle 2, 3 \rangle$.

a. Evaluate $|\mathbf{u} + \mathbf{v}|$.

b. Simplify $2\mathbf{u} - 3\mathbf{v}$.

c. Find two vectors half as long as $\mathbf{u}$ and parallel to $\mathbf{u}$.

SOLUTION

a. Because $\mathbf{u} + \mathbf{v} = \langle -1, 2 \rangle + \langle 2, 3 \rangle = \langle 1, 5 \rangle$, we have
$$|\mathbf{u} + \mathbf{v}| = \sqrt{1^2 + 5^2} = \sqrt{26}.$$

b. $2\mathbf{u} - 3\mathbf{v} = 2\langle -1, 2 \rangle - 3\langle 2, 3 \rangle = \langle -2, 4 \rangle - \langle 6, 9 \rangle = \langle -8, -5 \rangle$

c. The vectors $\frac{1}{2}\mathbf{u} = \frac{1}{2}\langle -1, 2 \rangle = \langle -\frac{1}{2}, 1 \rangle$ and $-\frac{1}{2}\mathbf{u} = -\frac{1}{2}\langle -1, 2 \rangle = \langle \frac{1}{2}, -1 \rangle$ have half the length of $\mathbf{u}$ and are parallel to $\mathbf{u}$.

Related Exercises 26, 28, 30 ◀

Unit Vectors

A **unit vector** is any vector with length 1. Two useful unit vectors are the **coordinate unit vectors** $\mathbf{i} = \langle 1, 0 \rangle$ and $\mathbf{j} = \langle 0, 1 \rangle$ (**Figure 13.16**). These vectors are directed along the coordinate axes and enable us to express all vectors in an alternative form. For example, by the Triangle Rule (**Figure 13.17a**),

$$\langle 3, 4 \rangle = 3\langle 1, 0 \rangle + 4\langle 0, 1 \rangle = 3\mathbf{i} + 4\mathbf{j}.$$

In general, the vector $\mathbf{v} = \langle v_1, v_2 \rangle$ (**Figure 13.17b**) is also written

$$\mathbf{v} = v_1\langle 1, 0 \rangle + v_2\langle 0, 1 \rangle = v_1\mathbf{i} + v_2\mathbf{j}.$$

Figure 13.16

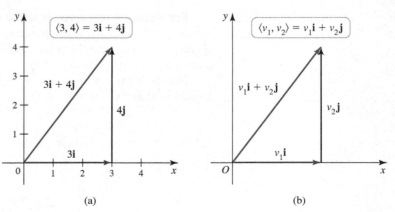

Figure 13.17

▶ Coordinate unit vectors are also called **standard basis vectors**.

$\mathbf{u} = \dfrac{\mathbf{v}}{|\mathbf{v}|}$ and $-\mathbf{u} = -\dfrac{\mathbf{v}}{|\mathbf{v}|}$ have length 1.

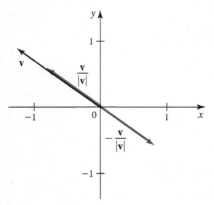

Figure 13.18

QUICK CHECK 5 Find vectors of length 10 parallel to the unit vector $\mathbf{u} = \left\langle \frac{3}{5}, \frac{4}{5} \right\rangle$. ◀

Given a nonzero vector **v**, we sometimes need to construct a new vector parallel to **v** of a specified length. Dividing **v** by its length, we obtain the vector $\mathbf{u} = \dfrac{\mathbf{v}}{|\mathbf{v}|}$. Because **u** is a positive scalar multiple of **v**, it follows that **u** has the same direction as **v**. Furthermore, **u** is a unit vector because $|\mathbf{u}| = \dfrac{|\mathbf{v}|}{|\mathbf{v}|} = 1$. The vector $-\mathbf{u} = -\dfrac{\mathbf{v}}{|\mathbf{v}|}$ is also a unit vector with a direction opposite that of **v** (**Figure 13.18**). Therefore, $\pm \dfrac{\mathbf{v}}{|\mathbf{v}|}$ are unit vectors parallel to **v** that point in opposite directions.

To construct a vector that points in the direction of **v** and has a specified length $c > 0$, we form the vector $\dfrac{c\mathbf{v}}{|\mathbf{v}|}$. It is a positive scalar multiple of **v**, so it points in the direction of **v**, and its length is $\left| \dfrac{c\mathbf{v}}{|\mathbf{v}|} \right| = |c| \dfrac{|\mathbf{v}|}{|\mathbf{v}|} = c$. The vector $-\dfrac{c\mathbf{v}}{|\mathbf{v}|}$ points in the opposite direction and also has length c. With this construction, we can also write **v** as the product of its magnitude and a unit vector in the direction of **v**:

$$\mathbf{v} = \underbrace{|\mathbf{v}|}_{\text{magnitude}} \cdot \underbrace{\frac{\mathbf{v}}{|\mathbf{v}|}}_{\text{direction}}.$$

EXAMPLE 5   Magnitude and unit vectors Consider the points $P(1, -2)$ and $Q(6, 10)$.

a. Find $\overrightarrow{PQ}$ and two unit vectors parallel to $\overrightarrow{PQ}$.

b. Find two vectors of length 2 parallel to $\overrightarrow{PQ}$.

c. Express $\overrightarrow{PQ}$ as the product of its magnitude and a unit vector.

SOLUTION

a. $\overrightarrow{PQ} = \langle 6 - 1, 10 - (-2) \rangle = \langle 5, 12 \rangle$, or $5\mathbf{i} + 12\mathbf{j}$. Because $|\overrightarrow{PQ}| = \sqrt{5^2 + 12^2} = \sqrt{169} = 13$, a unit vector parallel to $\overrightarrow{PQ}$ is

$$\frac{\overrightarrow{PQ}}{|\overrightarrow{PQ}|} = \frac{\langle 5, 12 \rangle}{13} = \left\langle \frac{5}{13}, \frac{12}{13} \right\rangle = \frac{5}{13}\mathbf{i} + \frac{12}{13}\mathbf{j}.$$

The unit vector parallel to $\overrightarrow{PQ}$ with the opposite direction is $\left\langle -\frac{5}{13}, -\frac{12}{13} \right\rangle$.

b. To obtain two vectors of length 2 that are parallel to $\overrightarrow{PQ}$, we multiply the unit vector $\frac{5}{13}\mathbf{i} + \frac{12}{13}\mathbf{j}$ by ± 2:

$$2 \left(\frac{5}{13}\mathbf{i} + \frac{12}{13}\mathbf{j} \right) = \frac{10}{13}\mathbf{i} + \frac{24}{13}\mathbf{j} \quad \text{and} \quad -2 \left(\frac{5}{13}\mathbf{i} + \frac{12}{13}\mathbf{j} \right) = -\frac{10}{13}\mathbf{i} - \frac{24}{13}\mathbf{j}.$$

QUICK CHECK 6 Verify that the vector $\langle \frac{5}{13}, \frac{12}{13} \rangle$ has length 1. ◄

c. The unit vector $\langle \frac{5}{13}, \frac{12}{13} \rangle$ points in the direction of $\vec{PQ}$, so we have

$$\vec{PQ} = |\vec{PQ}|\frac{\vec{PQ}}{|\vec{PQ}|} = 13\left\langle \frac{5}{13}, \frac{12}{13} \right\rangle.$$

Related Exercises 34, 43, 45 ◄

Properties of Vector Operations

► The Parallelogram Rule illustrates the commutative property $\mathbf{u} + \mathbf{v} = \mathbf{v} + \mathbf{u}$.

When we stand back and look at vector operations, ten general properties emerge. For example, the first property says that vector addition is commutative, which means $\mathbf{u} + \mathbf{v} = \mathbf{v} + \mathbf{u}$. This property is proved by letting $\mathbf{u} = \langle u_1, u_2 \rangle$ and $\mathbf{v} = \langle v_1, v_2 \rangle$. By the commutative property of addition for real numbers,

$$\mathbf{u} + \mathbf{v} = \langle u_1 + v_1, u_2 + v_2 \rangle = \langle v_1 + u_1, v_2 + u_2 \rangle = \mathbf{v} + \mathbf{u}.$$

The proofs of other properties are outlined in Exercises 78–81.

> **SUMMARY Properties of Vector Operations**
>
> Suppose $\mathbf{u}$, $\mathbf{v}$, and $\mathbf{w}$ are vectors and a and c are scalars. Then the following properties hold (for vectors in any number of dimensions).
>
> 1. $\mathbf{u} + \mathbf{v} = \mathbf{v} + \mathbf{u}$ Commutative property of addition
> 2. $(\mathbf{u} + \mathbf{v}) + \mathbf{w} = \mathbf{u} + (\mathbf{v} + \mathbf{w})$ Associative property of addition
> 3. $\mathbf{v} + \mathbf{0} = \mathbf{v}$ Additive identity
> 4. $\mathbf{v} + (-\mathbf{v}) = \mathbf{0}$ Additive inverse
> 5. $c(\mathbf{u} + \mathbf{v}) = c\mathbf{u} + c\mathbf{v}$ Distributive property 1
> 6. $(a + c)\mathbf{v} = a\mathbf{v} + c\mathbf{v}$ Distributive property 2
> 7. $0\mathbf{v} = \mathbf{0}$ Multiplication by zero scalar
> 8. $c\mathbf{0} = \mathbf{0}$ Multiplication by zero vector
> 9. $1\mathbf{v} = \mathbf{v}$ Multiplicative identity
> 10. $a(c\mathbf{v}) = (ac)\mathbf{v}$ Associative property of scalar multiplication

These properties allow us to solve vector equations. For example, to solve the equation $\mathbf{u} + \mathbf{v} = \mathbf{w}$ for $\mathbf{u}$, we proceed as follows:

$$(\mathbf{u} + \mathbf{v}) + (-\mathbf{v}) = \mathbf{w} + (-\mathbf{v}) \quad \text{Add } -\mathbf{v} \text{ to both sides.}$$
$$\mathbf{u} + \underbrace{(\mathbf{v} + (-\mathbf{v}))}_{\mathbf{0}} = \mathbf{w} + (-\mathbf{v}) \quad \text{Property 2}$$
$$\mathbf{u} + \mathbf{0} = \mathbf{w} - \mathbf{v} \quad \text{Property 4}$$
$$\mathbf{u} = \mathbf{w} - \mathbf{v}. \quad \text{Property 3}$$

QUICK CHECK 7 Solve $3\mathbf{u} + 4\mathbf{v} = 12\mathbf{w}$ for $\mathbf{u}$. ◄

Applications of Vectors

Vectors have countless practical applications, particularly in the physical sciences and engineering. These applications are explored throughout the remainder of this text. For now, we present two common uses of vectors: to describe velocities and forces.

► *Velocity of the boat relative to the water* means the velocity (direction and speed) the boat has relative to someone traveling with the current.

Velocity Vectors Consider a motorboat crossing a river whose current is everywhere represented by the constant vector $\mathbf{w}$ (**Figure 13.19**); this means that $|\mathbf{w}|$ is the speed of the moving water and $\mathbf{w}$ points in the direction of the moving water. Assume the vector $\mathbf{v}_w$ gives the velocity of the boat relative to the water. The combined effect of $\mathbf{w}$ and $\mathbf{v}_w$ is the sum $\mathbf{v}_g = \mathbf{v}_w + \mathbf{w}$, which is the velocity of the boat that would be observed by someone on the shore (or on the ground).

Figure 13.19

EXAMPLE 6 Speed of a boat in a current Suppose the water in a river moves southwest (45° west of south) at 4 mi/hr and a motorboat travels due east at 15 mi/hr relative to the shore. Determine the speed of the boat and its heading relative to the moving water (Figure 13.19).

SOLUTION To solve this problem, the vectors are placed in a coordinate system (**Figure 13.20**). Because the boat moves east at 15 mi/hr, the velocity relative to the shore is $\mathbf{v}_g = \langle 15, 0 \rangle$. To obtain the components of $\mathbf{w} = \langle w_x, w_y \rangle$, observe that $|\mathbf{w}| = 4$ and that the lengths of the sides of the 45–45–90 triangle in Figure 13.20 are

$$|w_x| = |w_y| = |\mathbf{w}| \cos 45° = \frac{4}{\sqrt{2}} = 2\sqrt{2}.$$

Figure 13.20

Given the orientation of $\mathbf{w}$ (southwest), $\mathbf{w} = \langle -2\sqrt{2}, -2\sqrt{2} \rangle$. Because $\mathbf{v}_g = \mathbf{v}_w + \mathbf{w}$ (Figure 13.20),

$$\mathbf{v}_w = \mathbf{v}_g - \mathbf{w} = \langle 15, 0 \rangle - \langle -2\sqrt{2}, -2\sqrt{2} \rangle$$
$$= \langle 15 + 2\sqrt{2}, 2\sqrt{2} \rangle.$$

The magnitude of $\mathbf{v}_w$ is

$$|\mathbf{v}_w| = \sqrt{(15 + 2\sqrt{2})^2 + (2\sqrt{2})^2} \approx 18.$$

Therefore, the speed of the boat relative to the water is approximately 18 mi/hr.

The heading of the boat is given by the angle θ between $\mathbf{v}_w$ and the positive x-axis. The x-component of $\mathbf{v}_w$ is $15 + 2\sqrt{2}$ and the y-component is $2\sqrt{2}$. Therefore,

$$\theta = \tan^{-1}\left(\frac{2\sqrt{2}}{15 + 2\sqrt{2}}\right) \approx 9°.$$

The heading of the boat is approximately 9° north of east, and its speed relative to the water is approximately 18 mi/hr.

Related Exercises 56–57 ◀

► Recall that the lengths of the legs of a 45–45–90 triangle are equal and are $1/\sqrt{2}$ times the length of the hypotenuse.

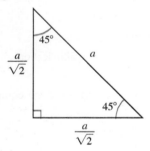

► The magnitude of $\mathbf{F}$ is typically measured in pounds (lb) or newtons (N), where $1\ \text{N} = 1\ \text{kg-m/s}^2$.

► The vector $\langle \cos\theta, \sin\theta \rangle$ is a unit vector. Therefore, any position vector $\mathbf{v}$ may be written $\mathbf{v} = \langle |\mathbf{v}| \cos\theta, |\mathbf{v}| \sin\theta \rangle$, where θ is the angle that $\mathbf{v}$ makes with the positive x-axis.

Force Vectors   Suppose a child pulls on the handle of a wagon at an angle of θ with the horizontal (**Figure 13.21a**). The vector $\mathbf{F}$ represents the force exerted on the wagon; it has a magnitude $|\mathbf{F}|$ and a direction given by θ. We denote the horizontal and vertical components of $\mathbf{F}$ by F_x and F_y, respectively. From **Figure 13.21b**, we see that $F_x = |\mathbf{F}| \cos\theta$, $F_y = |\mathbf{F}| \sin\theta$, and the force vector is $\mathbf{F} = \langle |\mathbf{F}| \cos\theta, |\mathbf{F}| \sin\theta \rangle$.

(a)   (b)

Figure 13.21

Figure 13.22

Figure 13.23

▶ The components of $\mathbf{F}_2$ in Example 8 can also be computed using an angle of 120°. That is, $\mathbf{F}_2 = \langle |\mathbf{F}_2| \cos 120°, |\mathbf{F}_2| \sin 120° \rangle$.

EXAMPLE 7   Finding force vectors A child pulls a wagon (Figure 13.21) with a force of $|\mathbf{F}| = 20$ lb at an angle of $\theta = 30°$ to the horizontal. Find the force vector $\mathbf{F}$.

SOLUTION The force vector (**Figure 13.22**) is

$$\mathbf{F} = \langle |\mathbf{F}| \cos \theta, |\mathbf{F}| \sin \theta \rangle = \langle 20 \cos 30°, 20 \sin 30° \rangle = \langle 10\sqrt{3}, 10 \rangle.$$

Related Exercise 60 ◀

EXAMPLE 8   Balancing forces A 400-lb engine is suspended from two chains that form 60° angles with a horizontal ceiling (**Figure 13.23**). How much weight does each chain support?

SOLUTION Let $\mathbf{F}_1$ and $\mathbf{F}_2$ denote the forces exerted by the chains on the engine, and let $\mathbf{F}_3$ be the downward force due to the weight of the engine (Figure 13.23). Placing the vectors in a standard coordinate system (**Figure 13.24**), we find that $\mathbf{F}_1 = \langle |\mathbf{F}_1| \cos 60°, |\mathbf{F}_1| \sin 60° \rangle$, $\mathbf{F}_2 = \langle -|\mathbf{F}_2| \cos 60°, |\mathbf{F}_2| \sin 60° \rangle$, and $\mathbf{F}_3 = \langle 0, -400 \rangle$.

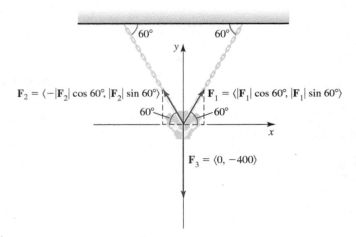

Figure 13.24

Because the engine is in equilibrium (the chains and engine are stationary), the sum of the forces is zero; that is, $\mathbf{F}_1 + \mathbf{F}_2 + \mathbf{F}_3 = \mathbf{0}$ or $\mathbf{F}_1 + \mathbf{F}_2 = -\mathbf{F}_3$. Therefore,

$$\langle |\mathbf{F}_1| \cos 60° - |\mathbf{F}_2| \cos 60°, |\mathbf{F}_1| \sin 60° + |\mathbf{F}_2| \sin 60° \rangle = \langle 0, 400 \rangle.$$

Equating corresponding components, we obtain two equations to be solved for $|\mathbf{F}_1|$ and $|\mathbf{F}_2|$:

$$|\mathbf{F}_1| \cos 60° - |\mathbf{F}_2| \cos 60° = 0 \text{ and}$$
$$|\mathbf{F}_1| \sin 60° + |\mathbf{F}_2| \sin 60° = 400.$$

Factoring the first equation, we find that $(|\mathbf{F}_1| - |\mathbf{F}_2|) \cos 60° = 0$, which implies that $|\mathbf{F}_1| = |\mathbf{F}_2|$. Replacing $|\mathbf{F}_2|$ with $|\mathbf{F}_1|$ in the second equation gives $2|\mathbf{F}_1| \sin 60° = 400$. Noting that $\sin 60° = \sqrt{3}/2$ and solving for $|\mathbf{F}_1|$, we find that $|\mathbf{F}_1| = 400/\sqrt{3} \approx 231$. Each chain must be able to support a weight of approximately 231 lb.

Related Exercise 63 ◀

SECTION 13.1 EXERCISES

Getting Started

1. Interpret the following statement: Points have a location, but no size or direction; nonzero vectors have a size and direction, but no location.

2. What is a position vector?

3. Given a position vector $\mathbf{v}$, why are there infinitely many vectors equal to $\mathbf{v}$?

4. Use the points $P(3, 1)$ and $Q(7, 1)$ to find position vectors equal to $\vec{PQ}$ and $\vec{QP}$.

5. If $\mathbf{u} = \langle u_1, u_2 \rangle$ and $\mathbf{v} = \langle v_1, v_2 \rangle$, how do you find $\mathbf{u} + \mathbf{v}$?

6. Find two unit vectors parallel to $\langle 2, 3 \rangle$.

7. Is $\langle 1, 1 \rangle$ a unit vector? Explain.

8. Evaluate $\langle 3, 1 \rangle + \langle 2, 4 \rangle$ and illustrate the sum geometrically using the Parallelogram Rule.

9. How do you compute the magnitude of $\mathbf{v} = \langle v_1, v_2 \rangle$?

10. Write the vector $\mathbf{v} = \langle v_1, v_2 \rangle$ in terms of the unit vectors $\mathbf{i}$ and $\mathbf{j}$.

11. How do you compute $|\overrightarrow{PQ}|$ from the coordinates of the points P and Q?

12. The velocity of a kayak on a lake is $\mathbf{v} = \langle 2\sqrt{2}, 2\sqrt{2} \rangle$. Find the speed and heading of the kayak. Assume the positive x-axis points east and the positive y-axis points north. Assume the coordinates of $\mathbf{v}$ are in feet per second.

Practice Exercises

13–18. Vector operations *Refer to the figure and carry out the following vector operations.*

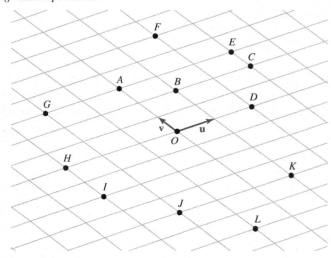

13. Scalar multiples Which of the following vectors equal $\overrightarrow{CE}$? (There may be more than one correct answer.)

    a. $\mathbf{v}$ **b.** $\frac{1}{2}\overrightarrow{HI}$ **c.** $\frac{1}{3}\overrightarrow{OA}$ **d.** $\mathbf{u}$ **e.** $\frac{1}{2}\overrightarrow{IH}$

14. Scalar multiples Which of the following vectors equal $\overrightarrow{BK}$? (There may be more than one correct answer.)

    a. $6\mathbf{v}$ **b.** $-6\mathbf{v}$ **c.** $3\overrightarrow{HI}$ **d.** $3\overrightarrow{IH}$ **e.** $3\overrightarrow{AO}$

15. Scalar multiples Write the following vectors as scalar multiples of $\mathbf{u}$ or $\mathbf{v}$.

    a. $\overrightarrow{OA}$ **b.** $\overrightarrow{OD}$ **c.** $\overrightarrow{OH}$ **d.** $\overrightarrow{AG}$ **e.** $\overrightarrow{CE}$

16. Scalar multiples Write the following vectors as scalar multiples of $\mathbf{u}$ or $\mathbf{v}$.

    a. $\overrightarrow{IH}$ **b.** $\overrightarrow{HI}$ **c.** $\overrightarrow{JK}$ **d.** $\overrightarrow{FD}$ **e.** $\overrightarrow{EA}$

17. Vector addition Write the following vectors as sums of scalar multiples of $\mathbf{u}$ and $\mathbf{v}$.

    a. $\overrightarrow{OE}$ **b.** $\overrightarrow{OB}$ **c.** $\overrightarrow{OF}$ **d.** $\overrightarrow{OG}$ **e.** $\overrightarrow{OC}$
    f. $\overrightarrow{OI}$ **g.** $\overrightarrow{OJ}$ **h.** $\overrightarrow{OK}$ **i.** $\overrightarrow{OL}$

18. Vector addition Write the following vectors as sums of scalar multiples of $\mathbf{u}$ and $\mathbf{v}$.

    a. $\overrightarrow{BF}$ **b.** $\overrightarrow{DE}$ **c.** $\overrightarrow{AF}$ **d.** $\overrightarrow{AD}$ **e.** $\overrightarrow{CD}$
    f. $\overrightarrow{JD}$ **g.** $\overrightarrow{JI}$ **h.** $\overrightarrow{DB}$ **i.** $\overrightarrow{IL}$

19. Components and magnitudes Define the points $O(0, 0)$, $P(3, 2)$, $Q(4, 2)$, and $R(-6, -1)$. For each vector, do the following.

    (i) Sketch the vector in an xy-coordinate system.
    (ii) Compute the magnitude of the vector.

    a. $\overrightarrow{OP}$ **b.** $\overrightarrow{QP}$ **c.** $\overrightarrow{RQ}$

20. Finding vectors from two points Given the points $A(-2, 0)$, $B(6, 16)$, $C(1, 4)$, $D(5, 4)$, $E(\sqrt{2}, \sqrt{2})$, and $F(3\sqrt{2}, -4\sqrt{2})$, find the position vector equal to the following vectors.

    a. $\overrightarrow{AB}$ **b.** $\overrightarrow{AC}$ **c.** $\overrightarrow{EF}$ **d.** $\overrightarrow{CD}$

21–23. Components and equality *Define the points $P(-3, -1)$, $Q(-1, 2)$, $R(1, 2)$, $S(3, 5)$, $T(4, 2)$, and $U(6, 4)$.*

21. Sketch $\overrightarrow{QU}$, $\overrightarrow{PT}$, and $\overrightarrow{RS}$ and their corresponding position vectors.

22. Find the equal vectors among $\overrightarrow{PQ}$, $\overrightarrow{RS}$, and $\overrightarrow{TU}$.

23. Consider the vectors $\overrightarrow{QT}$ and $\overrightarrow{SU}$: Which vector is equal to $\langle 5, 0 \rangle$?

24–27. Vector operations *Let $\mathbf{u} = \langle 4, -2 \rangle$, $\mathbf{v} = \langle -4, 6 \rangle$, and $\mathbf{w} = \langle 0, 8 \rangle$. Express the following vectors in the form $\langle a, b \rangle$.*

24. $\mathbf{u} + \mathbf{v}$ **25.** $\mathbf{w} - \mathbf{u}$

26. $2\mathbf{u} + 3\mathbf{v}$ **27.** $10\mathbf{u} - 3\mathbf{v} + \mathbf{w}$

28–31. Vector operations *Let $\mathbf{u} = \langle 3, -4 \rangle$, $\mathbf{v} = \langle 1, 1 \rangle$, and $\mathbf{w} = \langle -1, 0 \rangle$.*

28. Find $|\mathbf{u} + \mathbf{v} + \mathbf{w}|$.   **29.** Find $|-2\mathbf{v}|$.

30. Find two vectors parallel to $\mathbf{u}$ with four times the magnitude of $\mathbf{u}$.

31. Which has the greater magnitude, $\mathbf{u} - \mathbf{v}$ or $\mathbf{w} - \mathbf{u}$?

32. Find a unit vector in the direction of $\mathbf{v} = \langle -6, 8 \rangle$.

33. Write $\mathbf{v} = \langle -5, 12 \rangle$ as a product of its magnitude and a unit vector in the direction of $\mathbf{v}$.

34. Consider the points $P(2, 7)$ and $Q(6, 4)$. Write $\overrightarrow{PQ}$ as a product of its magnitude and a unit vector in the direction of $\overrightarrow{PQ}$.

35. Find the vector $\mathbf{v}$ of length 6 that has the same direction as the unit vector $\langle 1/2, \sqrt{3}/2 \rangle$.

36. Find the vector $\mathbf{v}$ that has a magnitude of 10 and a direction opposite that of the unit vector $\langle 3/5, -4/5 \rangle$.

37. Find the vector in the direction of $\langle 5, -12 \rangle$ with length 3.

38. Find the vector pointing in the direction opposite that of $\langle 6, -8 \rangle$ with length 20.

39. Find a vector in the same direction as $\langle 3, -2 \rangle$ with length 10.

40. Let $\mathbf{v} = \langle 8, 15 \rangle$.

    a. Find a vector in the direction of $\mathbf{v}$ that is three times as long as $\mathbf{v}$.
    b. Find a vector in the direction of $\mathbf{v}$ that has length 3.

41–46. Unit vectors *Define the points $P(-4, 1)$, $Q(3, -4)$, and $R(2, 6)$.*

41. Express $\overrightarrow{QR}$ in the form $a\mathbf{i} + b\mathbf{j}$.

42. Express $\overrightarrow{PQ}$ in the form $a\mathbf{i} + b\mathbf{j}$.

43. Find two unit vectors parallel to $\overrightarrow{PR}$.

44. Find the unit vector with the same direction as $\overrightarrow{QR}$.

45. Find two vectors parallel to $\overrightarrow{QP}$ with length 4.

46. Find two vectors parallel to $\overrightarrow{RP}$ with length 4.

47. Unit vectors

a. Find two unit vectors parallel to $\mathbf{v} = 6\mathbf{i} - 8\mathbf{j}$.

b. Find b if $\mathbf{v} = \langle 1/3, b \rangle$ is a unit vector.

c. Find all values of a such that $\mathbf{w} = a\mathbf{i} - \dfrac{a}{3}\mathbf{j}$ is a unit vector.

48. Vectors from polar coordinates Suppose O is the origin and P has polar coordinates (r, θ). Show that $\overrightarrow{OP} = \langle r\cos\theta, r\sin\theta \rangle$.

49. Vectors from polar coordinates Find the position vector $\overrightarrow{OP}$ if O is the origin and P has polar coordinates $(8, 5\pi/6)$.

50. Find the velocity $\mathbf{v}$ of an ocean freighter that is traveling northeast (45° east of north) at 40 km/hr.

51. Find the velocity $\mathbf{v}$ of an ocean freighter that is traveling 30° south of east at 30 km/hr.

52. Find a force vector of magnitude 100 that is directed 45° south of east.

⊤ **53–55. Airplanes and crosswinds** *Assume each plane flies horizontally in a crosswind that blows horizontally.*

53. An airplane flies east to west at 320 mi/hr relative to the air in a crosswind that blows at 40 mi/hr toward the southwest (45° south of west).

a. Find the velocity of the plane relative to the air $\mathbf{v}_a$, the velocity of the crosswind $\mathbf{w}$, and the velocity of the plane relative to the ground $\mathbf{v}_g$.

b. Find the ground speed and heading of the plane relative to the ground.

⊤ **54.** A commercial jet flies west to east at 400 mi/hr relative to the air, and it flies at 420 mi/hr at a heading of 5° north of east relative to the ground.

a. Find the velocity of the plane relative to the air $\mathbf{v}_a$, the velocity of the plane relative to the ground $\mathbf{v}_g$, and the crosswind $\mathbf{w}$.

b. Find the speed and heading of the wind.

55. Determine the necessary air speed and heading that a pilot must maintain in order to fly her commercial jet north at a speed of 480 mi/hr relative to the ground in a crosswind that is blowing 60° south of east at 20 mi/hr.

56. A boat in a current The water in a river moves south at 10 mi/hr. A motorboat travels due east at a speed of 20 mi/hr relative to the shore. Determine the speed and direction of the boat relative to the moving water.

57. Another boat in a current The water in a river moves south at 5 km/hr. A motorboat travels due east at a speed of 40 km/hr relative to the water. Determine the speed of the boat relative to the shore.

58. Parachute in the wind In still air, a parachute with a payload falls vertically at a terminal speed of 4 m/s. Find the direction and magnitude of its terminal velocity relative to the ground if it falls in a steady wind blowing horizontally from west to east at 10 m/s.

59. Boat in a wind A sailboat floats in a current that flows due east at 1 m/s. Because of a wind, the boat's actual speed relative to the shore is $\sqrt{3}$ m/s in a direction 30° north of east. Find the speed and direction of the wind.

60. Towing a boat A boat is towed with a force of 150 lb with a rope that makes an angle of 30° to the horizontal. Find the horizontal and vertical components of the force.

61. Pulling a suitcase Suppose you pull a suitcase with a strap that makes a 60° angle with the horizontal. The magnitude of the force you exert on the suitcase is 40 lb.

a. Find the horizontal and vertical components of the force.

b. Is the horizontal component of the force greater if the angle of the strap is 45° instead of 60°?

c. Is the vertical component of the force greater if the angle of the strap is 45° instead of 60°?

62. Which is greater? Which has a greater horizontal component, a 100-N force directed at an angle of 60° above the horizontal or a 60-N force directed at an angle of 30° above the horizontal?

63. Suspended load If a 500-lb load is suspended by two chains (see figure), what is the magnitude of the force each chain must be able to support?

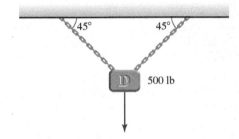

64. Net force Three forces are applied to an object, as shown in the figure. Find the magnitude and direction of the sum of the forces.

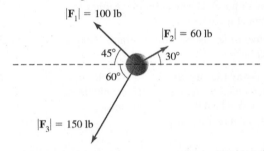

65. Explain why or why not Determine whether the following statements are true and give an explanation or counterexample.

a. José travels from point A to point B in the plane by following vector $\mathbf{u}$, then vector $\mathbf{v}$, and then vector $\mathbf{w}$. If he starts at A and follows $\mathbf{w}$, then $\mathbf{v}$, and then $\mathbf{u}$, he still arrives at B.

b. Maria travels from A to B in the plane by following the vector $\mathbf{u}$. By following $-\mathbf{u}$, she returns from B to A.

c. $|\mathbf{u} + \mathbf{v}| \geq |\mathbf{u}|$, for all vectors $\mathbf{u}$ and $\mathbf{v}$.

d. $|\mathbf{u} + \mathbf{v}| \geq |\mathbf{u}| + |\mathbf{v}|$, for all vectors $\mathbf{u}$ and $\mathbf{v}$.

e. Parallel vectors have the same length.

f. If $\overrightarrow{AB} = \overrightarrow{CD}$, then $A = C$ and $B = D$.

g. If $\mathbf{u}$ and $\mathbf{v}$ are perpendicular, then $|\mathbf{u} + \mathbf{v}| = |\mathbf{u}| + |\mathbf{v}|$.

h. If $\mathbf{u}$ and $\mathbf{v}$ are parallel and have the same direction, then $|\mathbf{u} + \mathbf{v}| = |\mathbf{u}| + |\mathbf{v}|$.

66. Equal vectors For the points $A(3, 4)$, $B(6, 10)$, $C(a + 2, b + 5)$, and $D(b + 4, a - 2)$, find the values of a and b such that $\overrightarrow{AB} = \overrightarrow{CD}$.

67–69. Vector equations *Use the properties of vectors to solve the following equations for the unknown vector* $\mathbf{x} = \langle a, b \rangle$. *Let* $\mathbf{u} = \langle 2, -3 \rangle$ *and* $\mathbf{v} = \langle -4, 1 \rangle$.

67. $10\mathbf{x} = \mathbf{u}$ **68.** $2\mathbf{x} + \mathbf{u} = \mathbf{v}$ **69.** $3\mathbf{x} - 4\mathbf{u} = \mathbf{v}$

70. Solve the pair of equations $2\mathbf{u} + 3\mathbf{v} = \mathbf{i}$, $\mathbf{u} - \mathbf{v} = \mathbf{j}$ for the vectors $\mathbf{u}$ and $\mathbf{v}$. Assume $\mathbf{i} = \langle 1, 0 \rangle$ and $\mathbf{j} = \langle 0, 1 \rangle$.

Explorations and Challenges

71–73. Linear combinations *A sum of scalar multiples of two or more vectors (such as $c_1\mathbf{u} + c_2\mathbf{v} + c_3\mathbf{w}$, where c_i are scalars) is called a **linear combination** of the vectors. Let $\mathbf{i} = \langle 1, 0 \rangle$, $\mathbf{j} = \langle 0, 1 \rangle$, $\mathbf{u} = \langle 1, 1 \rangle$, and $\mathbf{v} = \langle -1, 1 \rangle$.*

71. Express $\langle 4, -8 \rangle$ as a linear combination of $\mathbf{i}$ and $\mathbf{j}$ (that is, find scalars c_1 and c_2 such that $\langle 4, -8 \rangle = c_1\mathbf{i} + c_2\mathbf{j}$).

72. Express $\langle 4, -8 \rangle$ as a linear combination of $\mathbf{u}$ and $\mathbf{v}$.

73. For arbitrary real numbers a and b, express $\langle a, b \rangle$ as a linear combination of $\mathbf{u}$ and $\mathbf{v}$.

74. Ant on a page An ant walks due east at a constant speed of 2 mi/hr on a sheet of paper that rests on a table. Suddenly, the sheet of paper starts moving southeast at $\sqrt{2}$ mi/hr. Describe the motion of the ant relative to the table.

75. Clock vectors Consider the 12 vectors that have their tails at the center of a (circular) clock and their heads at the numbers on the edge of the clock.

    a. What is the sum of these 12 vectors?

    b. If the 12:00 vector is removed, what is the sum of the remaining 11 vectors?

    c. By removing one or more of these 12 clock vectors, explain how to make the sum of the remaining vectors as large as possible in magnitude.

    d. Consider the 11 vectors that originate at the number 12 at the top of the clock and point to the other 11 numbers. What is the sum of these vectors?

(*Source: Calculus*, by Gilbert Strang, Wellesley-Cambridge Press, 1991)

76. Three-way tug-of-war Three people located at A, B, and C pull on ropes tied to a ring. Find the magnitude and direction of the force with which the person at C must pull so that no one moves (the system is at equilibrium).

77–81. *Prove the following vector properties using components. Then make a sketch to illustrate the property geometrically. Suppose $\mathbf{u}$, $\mathbf{v}$, and $\mathbf{w}$ are vectors in the xy-plane and a and c are scalars.*

77. $\mathbf{u} + \mathbf{v} = \mathbf{v} + \mathbf{u}$ Commutative property

78. $(\mathbf{u} + \mathbf{v}) + \mathbf{w} = \mathbf{u} + (\mathbf{v} + \mathbf{w})$ Associative property

79. $a(c\mathbf{v}) = (ac)\mathbf{v}$ Associative property

80. $a(\mathbf{u} + \mathbf{v}) = a\mathbf{u} + a\mathbf{v}$ Distributive property 1

81. $(a + c)\mathbf{v} = a\mathbf{v} + c\mathbf{v}$ Distributive property 2

82. Midpoint of a line segment Use vectors to show that the midpoint of the line segment joining $P(x_1, y_1)$ and $Q(x_2, y_2)$ is the point $\left(\dfrac{x_1 + x_2}{2}, \dfrac{y_1 + y_2}{2} \right)$. (*Hint:* Let O be the origin and let M be the midpoint of PQ. Draw a picture and show that
$$\overrightarrow{OM} = \overrightarrow{OP} + \frac{1}{2}\overrightarrow{PQ} = \overrightarrow{OP} + \frac{1}{2}(\overrightarrow{OQ} - \overrightarrow{OP}).)$$

83. Magnitude of scalar multiple Prove that $|c\mathbf{v}| = |c||\mathbf{v}|$, where c is a scalar and $\mathbf{v}$ is a vector.

84. Equality of vectors Assume $\overrightarrow{PQ}$ equals $\overrightarrow{RS}$. Does it follow that $\overrightarrow{PR}$ is equal to $\overrightarrow{QS}$? Prove your conclusion.

85. Linear independence A pair of nonzero vectors in the plane is *linearly dependent* if one vector is a scalar multiple of the other. Otherwise, the pair is *linearly independent*.

    a. Which pairs of the following vectors are linearly dependent and which are linearly independent: $\mathbf{u} = \langle 2, -3 \rangle$, $\mathbf{v} = \langle -12, 18 \rangle$, and $\mathbf{w} = \langle 4, 6 \rangle$?

    b. Geometrically, what does it mean for a pair of nonzero vectors in the plane to be linearly dependent? Linearly independent?

    c. Prove that if a pair of vectors $\mathbf{u}$ and $\mathbf{v}$ is linearly independent, then given any vector $\mathbf{w}$, there are constants c_1 and c_2 such that $\mathbf{w} = c_1\mathbf{u} + c_2\mathbf{v}$.

86. Perpendicular vectors Show that two nonzero vectors $\mathbf{u} = \langle u_1, u_2 \rangle$ and $\mathbf{v} = \langle v_1, v_2 \rangle$ are perpendicular to each other if $u_1v_1 + u_2v_2 = 0$.

87. Parallel and perpendicular vectors Let $\mathbf{u} = \langle a, 5 \rangle$ and $\mathbf{v} = \langle 2, 6 \rangle$.

    a. Find the value of a such that $\mathbf{u}$ is parallel to $\mathbf{v}$.

    b. Find the value of a such that $\mathbf{u}$ is perpendicular to $\mathbf{v}$.

88. The Triangle Inequality Suppose $\mathbf{u}$ and $\mathbf{v}$ are vectors in the plane.

    a. Use the Triangle Rule for adding vectors to explain why $|\mathbf{u} + \mathbf{v}| \le |\mathbf{u}| + |\mathbf{v}|$. This result is known as the *Triangle Inequality*.

    b. Under what conditions is $|\mathbf{u} + \mathbf{v}| = |\mathbf{u}| + |\mathbf{v}|$?

QUICK CHECK ANSWERS

1. The vector $-5\mathbf{v}$ is five times as long as $\mathbf{v}$ and points in the opposite direction.   **2.** $\mathbf{v}_a + \mathbf{w}$ points in a northeasterly direction.   **3.** Constructing $\mathbf{u} + \mathbf{v}$ and $\mathbf{v} + \mathbf{u}$ using the Triangle Rule produces vectors having the same length and direction.   **4.** $\overrightarrow{PQ} = \langle -6, -2 \rangle$ **5.** $10\mathbf{u} = \langle 6, 8 \rangle$ and $-10\mathbf{u} = \langle -6, -8 \rangle$

6. $\left| \left\langle \dfrac{5}{13}, \dfrac{12}{13} \right\rangle \right| = \sqrt{\dfrac{25 + 144}{169}} = \sqrt{\dfrac{169}{169}} = 1$

7. $\mathbf{u} = -\frac{4}{3}\mathbf{v} + 4\mathbf{w}$ ◄

13.2 Vectors in Three Dimensions

Up to this point, our study of calculus has been limited to functions, curves, and vectors that can be plotted in the two-dimensional xy-plane. However, a two-dimensional coordinate system is insufficient for modeling many physical phenomena. For example, to describe the trajectory of a jet gaining altitude, we need two coordinates, say x and y, to measure east–west and north–south distances. In addition, another coordinate, say z, is needed to measure the altitude of the jet. By adding a third coordinate and creating an ordered triple (x, y, z), the location of the jet can be described. The set of all points described by the triples (x, y, z) is called *three-dimensional space*, *xyz*-space, or $\mathbb{R}^3$. Many of the properties of *xyz*-space are extensions of familiar ideas you have seen in the xy-plane.

The *xyz*-Coordinate System

> The notation $\mathbb{R}^3$ (pronounced *R-three*) stands for the set of all ordered triples of real numbers.

A three-dimensional coordinate system is created by adding a new axis, called the **z-axis**, to the familiar xy-coordinate system. The new z-axis is inserted through the origin perpendicular to the x- and y-axes (**Figure 13.25**). The result is a new coordinate system called the **three-dimensional rectangular coordinate system** or the ***xyz*-coordinate system**.

We use a conventional **right-handed coordinate system**: If the curled fingers of the right hand are rotated from the positive x-axis to the positive y-axis, the thumb points in the direction of the positive z-axis (Figure 13.25).

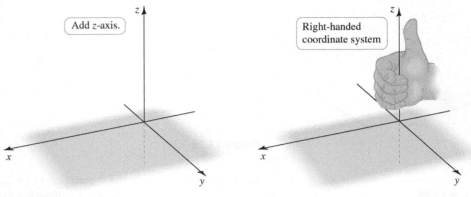

Figure 13.25

The coordinate plane containing the x-axis and y-axis is still called the xy-plane. We now have two new coordinate planes: the ***xz*-plane** containing the x-axis and the z-axis, and the ***yz*-plane** containing the y-axis and the z-axis. Taken together, these three coordinate planes divide *xyz*-space into eight regions called **octants** (**Figure 13.26**).

Figure 13.26

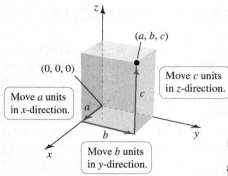

Figure 13.27

The point where all three axes intersect is the **origin**, which has coordinates $(0, 0, 0)$. An ordered triple (a, b, c) refers to the point in xyz-space that is found by starting at the origin, moving a units in the x-direction, b units in the y-direction, and c units in the z-direction. With a negative coordinate, you move in the negative direction along the corresponding coordinate axis. To visualize this point, it's helpful to construct a rectangular box with one vertex at the origin and the opposite vertex at the point (a, b, c) (**Figure 13.27**).

EXAMPLE 1   Plotting points in xyz-space Plot the following points.

a. $(3, 4, 5)$ **b.** $(-2, -3, 5)$

SOLUTION

a. Starting at $(0, 0, 0)$, we move 3 units in the x-direction to the point $(3, 0, 0)$, then 4 units in the y-direction to the point $(3, 4, 0)$, and finally 5 units in the z-direction to reach the point $(3, 4, 5)$ (**Figure 13.28**).

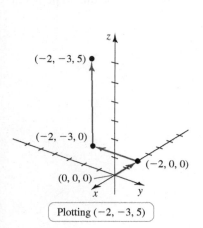

Plotting $(-2, -3, 5)$

Figure 13.29

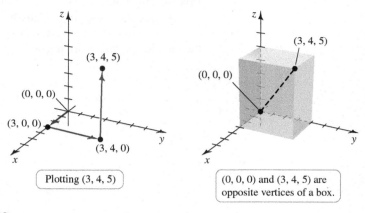

Plotting $(3, 4, 5)$

$(0, 0, 0)$ and $(3, 4, 5)$ are opposite vertices of a box.

Figure 13.28

b. We move -2 units in the x-direction to $(-2, 0, 0)$, -3 units in the y-direction to $(-2, -3, 0)$, and 5 units in the z-direction to reach $(-2, -3, 5)$ (**Figure 13.29**).

Related Exercises 13–14 ◄

QUICK CHECK 1 Suppose the positive x-, y-, and z-axes point east, north, and upward, respectively. Describe the location of the points $(-1, -1, 0)$, $(1, 0, 1)$, and $(-1, -1, -1)$ relative to the origin. ◄

► Planes that are not parallel to the coordinate planes are discussed in Section 13.5.

Equations of Simple Planes

The xy-plane consists of all points in xyz-space that have a z-coordinate of 0. Therefore, the xy-plane is the set $\{(x, y, z): z = 0\}$; it is represented by the equation $z = 0$. Similarly, the xz-plane has the equation $y = 0$, and the yz-plane has the equation $x = 0$.

Planes parallel to one of the coordinate planes are easy to describe. For example, the equation $x = 2$ describes the set of all points whose x-coordinate is 2 and whose y- and z-coordinates are arbitrary; this plane is parallel to and 2 units from the yz-plane. Similarly, the equation $y = a$ describes a plane that is everywhere $|a|$ units from the xz-plane, and $z = a$ is the equation of a horizontal plane $|a|$ units from the xy-plane (**Figure 13.30**).

QUICK CHECK 2 To which coordinate planes are the planes $x = -2$ and $z = 16$ parallel? ◄

Figure 13.30

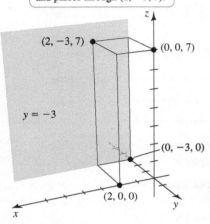

Plane is parallel to the *xz*-plane and passes through $(2, -3, 7)$.

Figure 13.31

EXAMPLE 2 **Parallel planes** Determine the equation of the plane parallel to the *xz*-plane passing through the point $(2, -3, 7)$.

SOLUTION Points on a plane parallel to the *xz*-plane have the same *y*-coordinate. Therefore, the plane passing through the point $(2, -3, 7)$ with a *y*-coordinate of -3 has the equation $y = -3$ (**Figure 13.31**).

Related Exercises 20–22 ◄

Distances in *xyz*-Space

Recall that the distance between two points (x_1, y_1) and (x_2, y_2) in the *xy*-plane is $\sqrt{(x_2 - x_1)^2 + (y_2 - y_1)^2}$. This distance formula is useful in deriving a similar formula for the distance between two points $P(x_1, y_1, z_1)$ and $Q(x_2, y_2, z_2)$ in *xyz*-space.

Figure 13.32 shows the points P and Q, together with the auxiliary point $R(x_2, y_2, z_1)$, which has the same *z*-coordinate as P and the same *x*- and *y*-coordinates as Q. The line segment PR has length $|PR| = \sqrt{(x_2 - x_1)^2 + (y_2 - y_1)^2}$ and is one leg of the right triangle $\triangle PRQ$. The length of the hypotenuse of that triangle is the distance between P and Q:

$$|PQ| = \sqrt{|PR|^2 + |RQ|^2} = \sqrt{\underbrace{(x_2 - x_1)^2 + (y_2 - y_1)^2}_{|PR|^2} + \underbrace{(z_2 - z_1)^2}_{|RQ|^2}}.$$

Figure 13.32

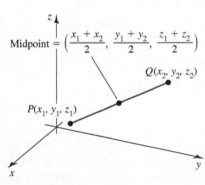

Figure 13.33

Distance Formula in *xyz*-Space

The distance between the points $P(x_1, y_1, z_1)$ and $Q(x_2, y_2, z_2)$ is

$$\sqrt{(x_2 - x_1)^2 + (y_2 - y_1)^2 + (z_2 - z_1)^2}.$$

By using the distance formula, we can derive the formula (Exercise 81) for the **midpoint** of the line segment joining $P(x_1, y_1, z_1)$ and $Q(x_2, y_2, z_2)$, which is found by averaging the *x*-, *y*-, and *z*-coordinates (**Figure 13.33**):

$$\text{Midpoint} = \left(\frac{x_1 + x_2}{2}, \frac{y_1 + y_2}{2}, \frac{z_1 + z_2}{2} \right).$$

Equation of a Sphere

A *sphere* is the set of all points that are a constant distance r from a point (a, b, c); r is the *radius* of the sphere, and (a, b, c) is the *center* of the sphere. A *ball* centered at (a, b, c) with radius r consists of all the points inside and on the sphere centered at (a, b, c) with radius r (**Figure 13.34**). We now use the distance formula to translate these statements.

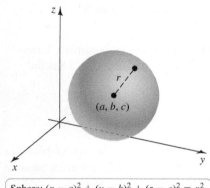

Sphere: $(x - a)^2 + (y - b)^2 + (z - c)^2 = r^2$

Ball: $(x - a)^2 + (y - b)^2 + (z - c)^2 \le r^2$

Figure 13.34

➤ Just as a circle is the boundary of a disk in two dimensions, a *sphere* is the boundary of a *ball* in three dimensions. We have defined a *closed ball*, which includes its boundary. An *open ball* does not contain its boundary.

DEFINITION Spheres and Balls

A **sphere** centered at (a, b, c) with radius r is the set of points satisfying the equation

$$(x - a)^2 + (y - b)^2 + (z - c)^2 = r^2.$$

A **ball** centered at (a, b, c) with radius r is the set of points satisfying the inequality

$$(x - a)^2 + (y - b)^2 + (z - c)^2 \le r^2.$$

EXAMPLE 3   Equation of a sphere Consider the points $P(1, -2, 5)$ and $Q(3, 4, -6)$. Find an equation of the sphere for which the line segment PQ is a diameter.

SOLUTION The center of the sphere is the midpoint of PQ:

$$\left(\frac{1 + 3}{2}, \frac{-2 + 4}{2}, \frac{5 - 6}{2} \right) = \left(2, 1, -\frac{1}{2} \right).$$

The diameter of the sphere is the distance $|PQ|$, which is

$$\sqrt{(3 - 1)^2 + (4 + 2)^2 + (-6 - 5)^2} = \sqrt{161}.$$

Therefore, the sphere's radius is $\frac{1}{2}\sqrt{161}$, its center is $\left(2, 1, -\frac{1}{2} \right)$, and it is described by the equation

$$(x - 2)^2 + (y - 1)^2 + \left(z + \frac{1}{2} \right)^2 = \left(\frac{1}{2}\sqrt{161} \right)^2 = \frac{161}{4}.$$

Related Exercises 27–28 ◀

EXAMPLE 4   Identifying equations Describe the set of points that satisfy the equation $x^2 + y^2 + z^2 - 2x + 6y - 8z = -1$.

QUICK CHECK 3 Describe the solution set of the equation

$(x - 1)^2 + y^2 + (z + 1)^2 + 4 = 0.$ ◀

SOLUTION We simplify the equation by completing the square and factoring:

$$(x^2 - 2x) + (y^2 + 6y) + (z^2 - 8z) = -1 \quad \text{Group terms.}$$
$$(x^2 - 2x + 1) + (y^2 + 6y + 9) + (z^2 - 8z + 16) = 25 \quad \text{Complete the square.}$$
$$(x - 1)^2 + (y + 3)^2 + (z - 4)^2 = 25. \quad \text{Factor.}$$

The equation describes a sphere of radius 5 with center $(1, -3, 4)$.

Related Exercises 31–32 ◀

Vectors in $\mathbb{R}^3$

Vectors in $\mathbb{R}^3$ are straightforward extensions of vectors in the xy-plane; we simply include a third component. The position vector $\mathbf{v} = \langle v_1, v_2, v_3 \rangle$ has its tail at the origin and its head at the point (v_1, v_2, v_3). Vectors having the same length and direction are equal. Therefore, the vector from $P(x_1, y_1, z_1)$ to $Q(x_2, y_2, z_2)$ is denoted $\overrightarrow{PQ}$ and is equal to the position vector $\langle x_2 - x_1, y_2 - y_1, z_2 - z_1 \rangle$. It is also equal to all vectors such as $\overrightarrow{RS}$ (Figure 13.35) that have the same length and direction as $\mathbf{v}$.

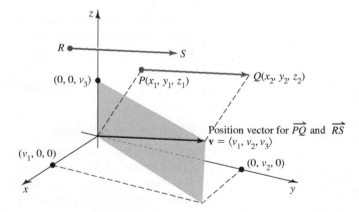

Figure 13.35

QUICK CHECK 4 Which of the following vectors are parallel to each other?

a. $\mathbf{u} = \langle -2, 4, -6 \rangle$

b. $\mathbf{v} = \langle 4, -8, 12 \rangle$

c. $\mathbf{w} = \langle -1, 2, 3 \rangle$ ◄

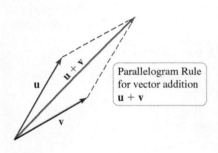

The operations of vector addition and scalar multiplication in $\mathbb{R}^2$ generalize in a natural way to three dimensions. For example, the sum of two vectors is found geometrically using the Triangle Rule or the Parallelogram Rule (Section 13.1). The sum is found analytically by adding the respective components of the two vectors. As with two-dimensional vectors, scalar multiplication corresponds to stretching or compressing a vector, possibly with a reversal of direction. Two nonzero vectors are parallel if one is a scalar multiple of the other (**Figure 13.36**).

> **DEFINITION Vector Operations in $\mathbb{R}^3$**
>
> Let c be a scalar, $\mathbf{u} = \langle u_1, u_2, u_3 \rangle$, and $\mathbf{v} = \langle v_1, v_2, v_3 \rangle$.
>
> $$\mathbf{u} + \mathbf{v} = \langle u_1 + v_1, u_2 + v_2, u_3 + v_3 \rangle \quad \text{Vector addition}$$
>
> $$\mathbf{u} - \mathbf{v} = \langle u_1 - v_1, u_2 - v_2, u_3 - v_3 \rangle \quad \text{Vector subtraction}$$
>
> $$c\mathbf{u} = \langle cu_1, cu_2, cu_3 \rangle \qquad\qquad \text{Scalar multiplication}$$

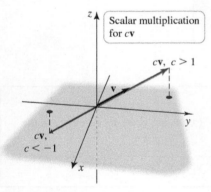

Figure 13.36

EXAMPLE 5 Vectors in $\mathbb{R}^3$ Let $\mathbf{u} = \langle 2, -4, 1 \rangle$ and $\mathbf{v} = \langle 3, 0, -1 \rangle$. Find the components of the following vectors and draw them in $\mathbb{R}^3$.

a. $\dfrac{1}{2}\mathbf{u}$ **b.** $-2\mathbf{v}$ **c.** $\mathbf{u} + 2\mathbf{v}$

SOLUTION

a. Using the definition of scalar multiplication, $\dfrac{1}{2}\mathbf{u} = \dfrac{1}{2}\langle 2, -4, 1 \rangle = \left\langle 1, -2, \dfrac{1}{2} \right\rangle$. The vector $\dfrac{1}{2}\mathbf{u}$ has the same direction as $\mathbf{u}$ with half the length of $\mathbf{u}$ (**Figure 13.37**).

b. Using scalar multiplication, $-2\mathbf{v} = -2\langle 3, 0, -1 \rangle = \langle -6, 0, 2 \rangle$. The vector $-2\mathbf{v}$ has the direction opposite that of $\mathbf{v}$ and twice the length of $\mathbf{v}$ (**Figure 13.38**).

Figure 13.37   Figure 13.38

c. Using vector addition and scalar multiplication,

$$\mathbf{u} + 2\mathbf{v} = \langle 2, -4, 1 \rangle + 2\langle 3, 0, -1 \rangle = \langle 8, -4, -1 \rangle.$$

The vector $\mathbf{u} + 2\mathbf{v}$ is drawn by applying the Parallelogram Rule to $\mathbf{u}$ and $2\mathbf{v}$ (**Figure 13.39**).

Figure 13.39

Related Exercise 39 ◄

Magnitude and Unit Vectors

The magnitude of the vector $\overrightarrow{PQ}$ from $P(x_1, y_1, z_1)$ to $Q(x_2, y_2, z_2)$ is denoted $|\overrightarrow{PQ}|$; it is the distance between P and Q and is given by the distance formula (**Figure 13.40**).

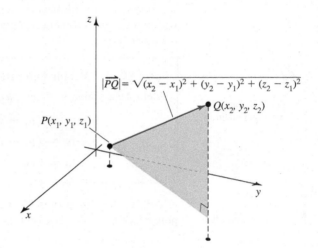

Figure 13.40

DEFINITION Magnitude of a Vector

The **magnitude** (or **length**) of the vector $\overrightarrow{PQ} = \langle x_2 - x_1, y_2 - y_1, z_2 - z_1 \rangle$ is the distance from $P(x_1, y_1, z_1)$ to $Q(x_2, y_2, z_2)$:

$$|\overrightarrow{PQ}| = \sqrt{(x_2 - x_1)^2 + (y_2 - y_1)^2 + (z_2 - z_1)^2}.$$

The coordinate unit vectors introduced in Section 13.1 extend naturally to three dimensions. The three coordinate unit vectors in $\mathbb{R}^3$ (**Figure 13.41**) are

$$\mathbf{i} = \langle 1, 0, 0 \rangle, \quad \mathbf{j} = \langle 0, 1, 0 \rangle, \quad \text{and} \quad \mathbf{k} = \langle 0, 0, 1 \rangle.$$

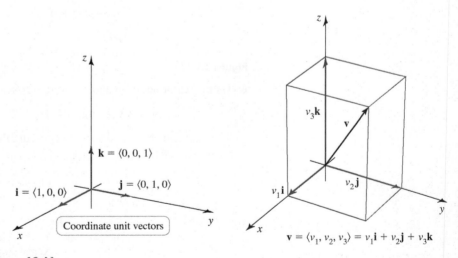

Figure 13.41

These unit vectors give an alternative way of expressing position vectors. If $\mathbf{v} = \langle v_1, v_2, v_3 \rangle$, then we have

$$\mathbf{v} = v_1 \langle 1, 0, 0 \rangle + v_2 \langle 0, 1, 0 \rangle + v_3 \langle 0, 0, 1 \rangle = v_1 \mathbf{i} + v_2 \mathbf{j} + v_3 \mathbf{k}.$$

EXAMPLE 6 **Magnitudes and unit vectors** Consider the points $P(5, 3, 1)$ and $Q(-7, 8, 1)$.

a. Express $\overrightarrow{PQ}$ in terms of the unit vectors $\mathbf{i}$, $\mathbf{j}$, and $\mathbf{k}$.

b. Find the magnitude of $\overrightarrow{PQ}$.

c. Find the position vector of magnitude 10 in the direction of $\overrightarrow{PQ}$.

SOLUTION

a. $\overrightarrow{PQ}$ is equal to the position vector $\langle -7 - 5, 8 - 3, 1 - 1 \rangle = \langle -12, 5, 0 \rangle$. Therefore, $\overrightarrow{PQ} = -12\mathbf{i} + 5\mathbf{j}$.

b. $|\overrightarrow{PQ}| = |-12\mathbf{i} + 5\mathbf{j}| = \sqrt{12^2 + 5^2} = \sqrt{169} = 13$

QUICK CHECK 5 Which vector has the smaller magnitude: $\mathbf{u} = 3\mathbf{i} - \mathbf{j} - \mathbf{k}$ or $\mathbf{v} = 2(\mathbf{i} + \mathbf{j} + \mathbf{k})$? ◄

c. The unit vector in the direction of $\overrightarrow{PQ}$ is $\mathbf{u} = \dfrac{\overrightarrow{PQ}}{|\overrightarrow{PQ}|} = \dfrac{1}{13} \langle -12, 5, 0 \rangle$. Therefore, the vector in the direction of $\mathbf{u}$ with a magnitude of 10 is $10\mathbf{u} = \dfrac{10}{13} \langle -12, 5, 0 \rangle$.

Related Exercises 45, 68 ◄

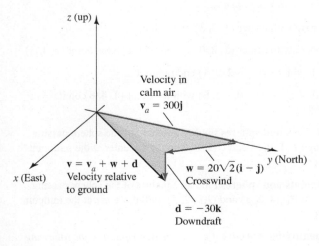

Figure 13.42

EXAMPLE 7 **Flight in crosswinds** A plane is flying horizontally due north in calm air at 300 mi/hr when it encounters a horizontal crosswind blowing southeast at 40 mi/hr and a downdraft blowing vertically downward at 30 mi/hr. What are the resulting speed and direction of the plane relative to the ground?

SOLUTION Let the unit vectors $\mathbf{i}$, $\mathbf{j}$, and $\mathbf{k}$ point east, north, and upward, respectively (**Figure 13.42**). The velocity of the plane relative to the air (300 mi/hr due north) is $\mathbf{v}_a = 300\mathbf{j}$. The crosswind blows 45° south of east, so its component to the east is $40 \cos 45° = 20\sqrt{2}$ (in the $\mathbf{i}$-direction) and its component to the south is $40 \cos 45° = 20\sqrt{2}$ (in the negative $\mathbf{j}$-direction). Therefore, the crosswind may be expressed as $\mathbf{w} = 20\sqrt{2}\mathbf{i} - 20\sqrt{2}\mathbf{j}$. Finally, the downdraft in the negative $\mathbf{k}$-direction is $\mathbf{d} = -30\mathbf{k}$. The velocity of the plane relative to the ground is the sum of $\mathbf{v}_a$, $\mathbf{w}$, and $\mathbf{d}$:

$$\begin{aligned} \mathbf{v} &= \mathbf{v}_a + \mathbf{w} + \mathbf{d} \\ &= 300\mathbf{j} + (20\sqrt{2}\mathbf{i} - 20\sqrt{2}\mathbf{j}) - 30\mathbf{k} \\ &= 20\sqrt{2}\mathbf{i} + (300 - 20\sqrt{2})\mathbf{j} - 30\mathbf{k}. \end{aligned}$$

Figure 13.42 shows the velocity vector of the plane. A quick calculation shows that the speed is $|\mathbf{v}| \approx 275$ mi/hr. The direction of the plane is slightly east of north and downward. In the next section, we present methods for precisely determining the direction of a vector.

Related Exercises 51–52 ◄

SECTION 13.2 EXERCISES

Getting Started

1. Explain how to plot the point $(3, -2, 1)$ in $\mathbb{R}^3$.

2. What is the y-coordinate of all points in the xz-plane?

3. Describe the plane $x = 4$.

4. What position vector is equal to the vector from $(3, 5, -2)$ to $(0, -6, 3)$?

5. Let $\mathbf{u} = \langle 3, 5, -7 \rangle$ and $\mathbf{v} = \langle 6, -5, 1 \rangle$. Evaluate $\mathbf{u} + \mathbf{v}$ and $3\mathbf{u} - \mathbf{v}$.

6. What is the magnitude of a vector joining two points $P(x_1, y_1, z_1)$ and $Q(x_2, y_2, z_2)$?

7. Which point is farther from the origin, $(3, -1, 2)$ or $(0, 0, -4)$?

8. Express the vector from $P(-1, -4, 6)$ to $Q(1, 3, -6)$ as a position vector in terms of $\mathbf{i}$, $\mathbf{j}$, and $\mathbf{k}$.

Practice Exercises

9–12. Points in $\mathbb{R}^3$ *Find the coordinates of the vertices A, B, and C of the following rectangular boxes.*

9.

10.

11.

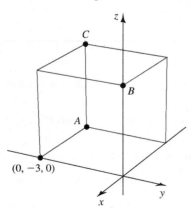

12. Assume all the edges have the same length.

13–14. Plotting points in $\mathbb{R}^3$ *For each point P(x, y, z) given below, let A(x, y, 0), B(x, 0, z), and C(0, y, z) be points in the xy-, xz-, and yz-planes, respectively. Plot and label the points A, B, C, and P in* $\mathbb{R}^3$.

13. a. $P(2, 2, 4)$ **b.** $P(1, 2, 5)$ **c.** $P(-2, 0, 5)$

14. a. $P(-3, 2, 4)$ **b.** $P(4, -2, -3)$ **c.** $P(-2, -4, -3)$

15–20. Sketching planes *Sketch the following planes in the window* $[0, 5] \times [0, 5] \times [0, 5]$.

15. $x = 2$ **16.** $z = 3$ **17.** $y = 2$ **18.** $z = y$

19. The plane that passes through $(2, 0, 0)$, $(0, 3, 0)$, and $(0, 0, 4)$

20. The plane parallel to the xz-plane containing the point $(1, 2, 3)$

21. Planes Sketch the plane parallel to the xy-plane through $(2, 4, 2)$ and find its equation.

22. Planes Sketch the plane parallel to the yz-plane through $(2, 4, 2)$ and find its equation.

23–26. Spheres and balls *Find an equation or inequality that describes the following objects.*

23. A sphere with center $(1, 2, 3)$ and radius 4

24. A sphere with center $(1, 2, 0)$ passing through the point $(3, 4, 5)$

25. A ball with center $(-2, 0, 4)$ and radius 1

26. A ball with center $(0, -2, 6)$ with the point $(1, 4, 8)$ on its boundary

27. Midpoints and spheres Find an equation of the sphere passing through $P(1, 0, 5)$ and $Q(2, 3, 9)$ with its center at the midpoint of PQ.

28. Midpoints and spheres Find an equation of the sphere passing through $P(-4, 2, 3)$ and $Q(0, 2, 7)$ with its center at the midpoint of PQ.

29–38. Identifying sets *Give a geometric description of the following sets of points.*

29. $(x - 1)^2 + y^2 + z^2 - 9 = 0$

30. $(x + 1)^2 + y^2 + z^2 - 2y - 24 = 0$

31. $x^2 + y^2 + z^2 - 2y - 4z - 4 = 0$

32. $x^2 + y^2 + z^2 - 6x + 6y - 8z - 2 = 0$

33. $x^2 + y^2 - 14y + z^2 \geq -13$

34. $x^2 + y^2 - 14y + z^2 \leq -13$

35. $x^2 + y^2 + z^2 - 8x - 14y - 18z \leq 79$

36. $x^2 + y^2 + z^2 - 8x + 14y - 18z \geq 65$

37. $x^2 - 2x + y^2 + 6y + z^2 + 10 = 0$

38. $x^2 - 4x + y^2 + 6y + z^2 + 14 = 0$

39–44. Vector operations *For the given vectors* **u** *and* **v***, evaluate the following expressions.*

a. $3\mathbf{u} + 2\mathbf{v}$ *b.* $4\mathbf{u} - \mathbf{v}$ *c.* $|\mathbf{u} + 3\mathbf{v}|$

39. $\mathbf{u} = \langle 4, -3, 0 \rangle$, $\mathbf{v} = \langle 0, 1, 1 \rangle$

40. $\mathbf{u} = \langle -2, -3, 0 \rangle$, $\mathbf{v} = \langle 1, 2, 1 \rangle$

41. $\mathbf{u} = \langle -2, 1, -2 \rangle$, $\mathbf{v} = \langle 1, 1, 1 \rangle$

42. $\mathbf{u} = \langle -5, 0, 2 \rangle$, $\mathbf{v} = \langle 3, 1, 1 \rangle$

43. $\mathbf{u} = \langle -7, 11, 8 \rangle$, $\mathbf{v} = \langle 3, -5, -1 \rangle$

44. $\mathbf{u} = \langle -4, -8\sqrt{3}, 2\sqrt{2} \rangle$, $\mathbf{v} = \langle 2, 3\sqrt{3}, -\sqrt{2} \rangle$

45–50. Unit vectors and magnitude *Consider the following points P and Q.*

a. *Find $\overrightarrow{PQ}$ and state your answer in two forms: $\langle a, b, c \rangle$ and $a\mathbf{i} + b\mathbf{j} + c\mathbf{k}$.*

b. *Find the magnitude of $\overrightarrow{PQ}$.*

c. *Find two unit vectors parallel to $\overrightarrow{PQ}$.*

45. $P(1, 5, 0), Q(3, 11, 2)$ **46.** $P(5, 11, 12), Q(1, 14, 13)$

47. $P(-3, 1, 0), Q(-3, -4, 1)$ **48.** $P(3, 8, 12), Q(3, 9, 11)$

49. $P(0, 0, 2), Q(-2, 4, 0)$

50. $P(a, b, c), Q(1, 1, -1)$ (a, b, and c are real numbers)

51. Flight in crosswinds A model airplane is flying horizontally due north at 20 mi/hr when it encounters a horizontal crosswind blowing east at 20 mi/hr and a downdraft blowing vertically downward at 10 mi/hr.

    a. Find the position vector that represents the velocity of the plane relative to the ground.

    b. Find the speed of the plane relative to the ground.

52. Another crosswind flight A model airplane is flying horizontally due east at 10 mi/hr when it encounters a horizontal crosswind blowing south at 5 mi/hr and an updraft blowing vertically upward at 5 mi/hr.

    a. Find the position vector that represents the velocity of the plane relative to the ground.

    b. Find the speed of the plane relative to the ground.

53. Crosswinds A small plane is flying horizontally due east in calm air at 250 mi/hr when it encounters a horizontal crosswind blowing southwest at 50 mi/hr and a 30-mi/hr updraft. Find the resulting speed of the plane, and describe with a sketch the approximate direction of the velocity relative to the ground.

54. Combined force An object at the origin is acted on by the forces $\mathbf{F}_1 = 20\mathbf{i} - 10\mathbf{j}$, $\mathbf{F}_2 = 30\mathbf{j} + 10\mathbf{k}$, and $\mathbf{F}_3 = 40\mathbf{j} + 20\mathbf{k}$. Find the magnitude of the combined force, and describe the approximate direction of the force.

55. Submarine course A submarine climbs at an angle of 30° above the horizontal with a heading to the northeast. If its speed is 20 knots, find the components of the velocity in the east, north, and vertical directions.

56. Maintaining equilibrium An object is acted on by the forces $\mathbf{F}_1 = \langle 10, 6, 3 \rangle$ and $\mathbf{F}_2 = \langle 0, 4, 9 \rangle$. Find the force $\mathbf{F}_3$ that must act on the object so that the sum of the forces is zero.

57. Explain why or why not Determine whether the following statements are true and give an explanation or counterexample.

    a. Suppose $\mathbf{u}$ and $\mathbf{v}$ are distinct vectors that both make a 45° angle with $\mathbf{w}$ in $\mathbb{R}^3$. Then $\mathbf{u} + \mathbf{v}$ makes a 45° angle with $\mathbf{w}$.

    b. Suppose $\mathbf{u}$ and $\mathbf{v}$ are distinct vectors that both make a 90° angle with $\mathbf{w}$ in $\mathbb{R}^3$. Then $\mathbf{u} + \mathbf{v}$ can never make a 90° angle with $\mathbf{w}$.

    c. $\mathbf{i} + \mathbf{j} + \mathbf{k} = \mathbf{0}$

    d. The intersection of the planes $x = 1$, $y = 1$, and $z = 1$ is a point.

58–60. Sets of points *Describe with a sketch the sets of points (x, y, z) satisfying the following equations.*

58. $(x + 1)(y - 3) = 0$

59. $x^2 y^2 z^2 > 0$

60. $y - z = 0$

61–64. Sets of points

61. Give a geometric description of the set of points (x, y, z) satisfying the pair of equations $z = 0$ and $x^2 + y^2 = 1$. Sketch a figure of this set of points.

62. Give a geometric description of the set of points (x, y, z) satisfying the pair of equations $z = x^2$ and $y = 0$. Sketch a figure of this set of points.

63. Give a geometric description of the set of points (x, y, z) that lie on the intersection of the sphere $x^2 + y^2 + z^2 = 5$ and the plane $z = 1$.

64. Give a geometric description of the set of points (x, y, z) that lie on the intersection of the sphere $x^2 + y^2 + z^2 = 36$ and the plane $z = 6$.

65. Describing a circle Find a pair of equations describing a circle of radius 3 centered at $(2, 4, 1)$ that lies in a plane parallel to the xz-plane.

66. Describing a line Find a pair of equations describing a line passing through the point $(-2, -5, 1)$ that is parallel to the x-axis.

67. Write the vector $\mathbf{v} = \langle 2, -4, 4 \rangle$ as a product of its magnitude and a unit vector with the same direction as $\mathbf{v}$.

68. Find the vector of length 10 with the same direction as $\mathbf{w} = \langle 2, \sqrt{2}, \sqrt{3} \rangle$.

69. Find a vector of length 5 in the direction opposite that of $\langle 3, -2, \sqrt{3} \rangle$.

70–73. Parallel vectors of varying lengths *Find vectors parallel to $\mathbf{v}$ of the given length.*

70. $\mathbf{v} = \langle 3, -2, 6 \rangle$; length $= 10$

71. $\mathbf{v} = \langle 6, -8, 0 \rangle$; length $= 20$

72. $\mathbf{v} = \overrightarrow{PQ}$ with $P(1, 0, 1)$ and $Q(2, -1, 1)$; length $= 3$

73. $\mathbf{v} = \overrightarrow{PQ}$ with $P(3, 4, 0)$ and $Q(2, 3, 1)$; length $= 3$

74. Collinear points Determine the values of x and y such that the points $(1, 2, 3)$, $(4, 7, 1)$, and $(x, y, 2)$ are collinear (lie on a line).

75. Collinear points Determine whether the points P, Q, and R are collinear (lie on a line) by comparing $\overrightarrow{PQ}$ and $\overrightarrow{PR}$. If the points are collinear, determine which point lies between the other two points.

    a. $P(1, 6, -5), Q(2, 5, -3), R(4, 3, 1)$

    b. $P(1, 5, 7), Q(5, 13, -1), R(0, 3, 9)$

    c. $P(1, 2, 3), Q(2, -3, 6), R(3, -1, 9)$

    d. $P(9, 5, 1), Q(11, 18, 4), R(6, 3, 0)$

76. Lengths of the diagonals of a box What is the longest diagonal of a rectangular 2 ft × 3 ft × 4 ft box?

Explorations and Challenges

77. Three-cable load A 500-lb load hangs from three cables of equal length that are anchored at the points $(-2, 0, 0)$, $(1, \sqrt{3}, 0)$, and $(1, -\sqrt{3}, 0)$. The load is located at $(0, 0, -2\sqrt{3})$. Find the vectors describing the forces on the cables due to the load.

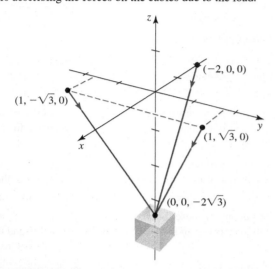

78. Four-cable load A 500-lb load hangs from four cables of equal length that are anchored at the points $(\pm 2, 0, 0)$ and $(0, \pm 2, 0)$. The load is located at $(0, 0, -4)$. Find the vectors describing the forces on the cables due to the load.

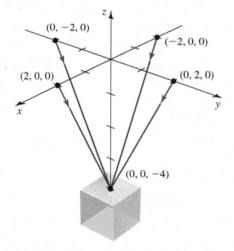

79. Possible parallelograms The points $O(0, 0, 0)$, $P(1, 4, 6)$, and $Q(2, 4, 3)$ lie at three vertices of a parallelogram. Find all possible locations of the fourth vertex.

80. Diagonals of parallelograms Two sides of a parallelogram are formed by the vectors $\mathbf{u}$ and $\mathbf{v}$. Prove that the diagonals of the parallelogram are $\mathbf{u} + \mathbf{v}$ and $\mathbf{u} - \mathbf{v}$.

81. Midpoint formula Prove that the midpoint of the line segment joining $P(x_1, y_1, z_1)$ and $Q(x_2, y_2, z_2)$ is

$$\left(\frac{x_1 + x_2}{2}, \frac{y_1 + y_2}{2}, \frac{z_1 + z_2}{2} \right).$$

82. Equation of a sphere For constants a, b, c, and d, show that the equation

$$x^2 + y^2 + z^2 - 2ax - 2by - 2cz = d$$

describes a sphere centered at (a, b, c) with radius r, where $r^2 = d + a^2 + b^2 + c^2$, provided $d + a^2 + b^2 + c^2 > 0$.

83. Medians of a triangle—without coordinates Assume $\mathbf{u}$, $\mathbf{v}$, and $\mathbf{w}$ are vectors in $\mathbb{R}^3$ that form the sides of a triangle (see figure). Use the following steps to prove that the medians intersect at a point that divides each median in a 2:1 ratio. The proof does not use a coordinate system.

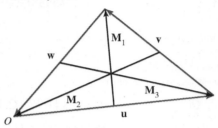

  a. Show that $\mathbf{u} + \mathbf{v} + \mathbf{w} = \mathbf{0}$.
  b. Let $\mathbf{M}_1$ be the median vector from the midpoint of $\mathbf{u}$ to the opposite vertex. Define $\mathbf{M}_2$ and $\mathbf{M}_3$ similarly. Using the geometry of vector addition, show that $\mathbf{M}_1 = \dfrac{\mathbf{u}}{2} + \mathbf{v}$. Find analogous expressions for $\mathbf{M}_2$ and $\mathbf{M}_3$.
  c. Let $\mathbf{a}$, $\mathbf{b}$, and $\mathbf{c}$ be the vectors from O to the points one-third of the way along $\mathbf{M}_1$, $\mathbf{M}_2$, and $\mathbf{M}_3$, respectively. Show that

  $$\mathbf{a} = \mathbf{b} = \mathbf{c} = \frac{\mathbf{u} - \mathbf{w}}{3}.$$

  d. Conclude that the medians intersect at a point that divides each median in a 2:1 ratio.

84. Medians of a triangle—with coordinates In contrast to the proof in Exercise 83, we now use coordinates and position vectors to prove the same result. Without loss of generality, let $P(x_1, y_1, 0)$ and $Q(x_2, y_2, 0)$ be two points in the xy-plane, and let $R(x_3, y_3, z_3)$ be a third point such that P, Q, and R do not lie on a line. Consider $\triangle PQR$.

  a. Let M_1 be the midpoint of the side PQ. Find the coordinates of M_1 and the components of the vector $\overrightarrow{RM_1}$.
  b. Find the vector $\overrightarrow{OZ_1}$ from the origin to the point Z_1 two-thirds of the way along $\overrightarrow{RM_1}$.
  c. Repeat the calculation of part (b) with the midpoint M_2 of RQ and the vector $\overrightarrow{PM_2}$ to obtain the vector $\overrightarrow{OZ_2}$.
  d. Repeat the calculation of part (b) with the midpoint M_3 of PR and the vector $\overrightarrow{QM_3}$ to obtain the vector $\overrightarrow{OZ_3}$.
  e. Conclude that the medians of $\triangle PQR$ intersect at a point. Give the coordinates of the point.
  f. With $P(2, 4, 0)$, $Q(4, 1, 0)$, and $R(6, 3, 4)$, find the point at which the medians of $\triangle PQR$ intersect.

85. The amazing quadrilateral property—without coordinates The points P, Q, R, and S, joined by the vectors $\mathbf{u}$, $\mathbf{v}$, $\mathbf{w}$, and $\mathbf{x}$, are the vertices of a quadrilateral in $\mathbb{R}^3$. *The four points need not lie in a plane* (see figure). Use the following steps to prove that the line segments joining the midpoints of the sides of the quadrilateral form a parallelogram. The proof does not use a coordinate system.

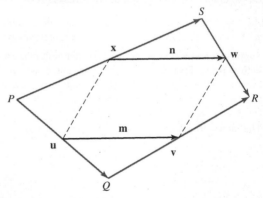

a. Use vector addition to show that $\mathbf{u} + \mathbf{v} = \mathbf{w} + \mathbf{x}$.
b. Let $\mathbf{m}$ be the vector that joins the midpoints of PQ and QR.
Show that $\mathbf{m} = \dfrac{\mathbf{u} + \mathbf{v}}{2}$.
c. Let $\mathbf{n}$ be the vector that joins the midpoints of PS and SR.
Show that $\mathbf{n} = \dfrac{\mathbf{x} + \mathbf{w}}{2}$.
d. Combine parts (a), (b), and (c) to conclude that $\mathbf{m} = \mathbf{n}$.
e. Explain why part (d) implies that the line segments joining the midpoints of the sides of the quadrilateral form a parallelogram.

86. The amazing quadrilateral property—with coordinates Prove the quadrilateral property in Exercise 85, assuming the coordinates of P, Q, R, and S are $P(x_1, y_1, 0)$, $Q(x_2, y_2, 0)$, $R(x_3, y_3, 0)$, and $S(x_4, y_4, z_4)$, where we assume P, Q, and R lie in the xy-plane without loss of generality.

QUICK CHECK ANSWERS

1. Southwest; due east and upward; southwest and downward   **2.** yz-plane; xy-plane   **3.** No solution
4. $\mathbf{u}$ and $\mathbf{v}$ are parallel.   **5.** $|\mathbf{u}| = \sqrt{11}$ and $|\mathbf{v}| = \sqrt{12} = 2\sqrt{3}$; $\mathbf{u}$ has the smaller magnitude. ◀

13.3 Dot Products

> The dot product is also called the *scalar product*, a term we do not use in order to avoid confusion with *scalar multiplication*.

The *dot product* is used to determine the angle between two vectors. It is also a tool for calculating *projections*—the measure of how much of a given vector lies in the direction of another vector.

To see the usefulness of the dot product, consider an example. Recall that the work done by a constant force F in moving an object a distance d is $W = Fd$ (Section 6.7). This rule is valid provided the force acts in the direction of motion (**Figure 13.43a**). Now assume the force is a vector $\mathbf{F}$ applied at an angle θ to the direction of motion; the resulting displacement of the object is a vector $\mathbf{d}$. In this case, the work done by the force is the component of the force in the direction of motion multiplied by the distance moved by the object, which is $W = (|\mathbf{F}| \cos \theta)|\mathbf{d}|$ (**Figure 13.43b**). We call this product of the magnitudes of two vectors and the cosine of the angle between them the dot product.

Horizontal and vertical components of a vector

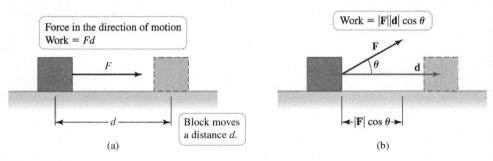

Figure 13.43

Two Forms of the Dot Product

The example of work done by a force leads to our first definition of the dot product. We then give an equivalent formula that is often better suited for computation.

DEFINITION Dot Product

Given two nonzero vectors $\mathbf{u}$ and $\mathbf{v}$ in two or three dimensions, their **dot product** is

$$\mathbf{u} \cdot \mathbf{v} = |\mathbf{u}||\mathbf{v}| \cos \theta,$$

where θ is the angle between $\mathbf{u}$ and $\mathbf{v}$ with $0 \le \theta \le \pi$ (**Figure 13.44**). If $\mathbf{u} = \mathbf{0}$ or $\mathbf{v} = \mathbf{0}$, then $\mathbf{u} \cdot \mathbf{v} = 0$, and θ is undefined.

Figure 13.44

The dot product of two vectors is itself a scalar. Two special cases immediately arise:

• **u** and **v** are parallel ($\theta = 0$ or $\theta = \pi$) if and only if $\mathbf{u} \cdot \mathbf{v} = \pm |\mathbf{u}||\mathbf{v}|$.

• **u** and **v** are perpendicular ($\theta = \pi/2$) if and only if $\mathbf{u} \cdot \mathbf{v} = 0$.

The second case gives rise to the important property of *orthogonality*.

> ➤ In two and three dimensions, the terms *orthogonal* and *perpendicular* are used interchangeably. *Orthogonal* is a more general term that also applies in more than three dimensions.

DEFINITION Orthogonal Vectors

Two vectors **u** and **v** are **orthogonal** if and only if $\mathbf{u} \cdot \mathbf{v} = 0$. The zero vector is orthogonal to all vectors. In two or three dimensions, two nonzero orthogonal vectors are perpendicular to each other.

> **QUICK CHECK 1** Sketch two nonzero vectors **u** and **v** with $\theta = 0$. Sketch two nonzero vectors **u** and **v** with $\theta = \pi$. ◀

EXAMPLE 1   Dot products Compute the dot products of the following vectors.

a. $\mathbf{u} = 2\mathbf{i} - 6\mathbf{j}$ and $\mathbf{v} = 12\mathbf{k}$

b. $\mathbf{u} = \langle \sqrt{3}, 1 \rangle$ and $\mathbf{v} = \langle 0, 1 \rangle$

SOLUTION

a. The vector **u** lies in the *xy*-plane and the vector **v** is perpendicular to the *xy*-plane. Therefore, $\theta = \dfrac{\pi}{2}$, **u** and **v** are orthogonal, and $\mathbf{u} \cdot \mathbf{v} = 0$ (**Figure 13.45a**).

b. As shown in **Figure 13.45b**, **u** and **v** form two sides of a 30–60–90 triangle in the *xy*-plane, with an angle of $\pi/3$ between them. Because $|\mathbf{u}| = 2$, $|\mathbf{v}| = 1$, and $\cos \pi/3 = 1/2$, the dot product is

$$\mathbf{u} \cdot \mathbf{v} = |\mathbf{u}||\mathbf{v}| \cos \theta = 2 \cdot 1 \cdot \frac{1}{2} = 1.$$

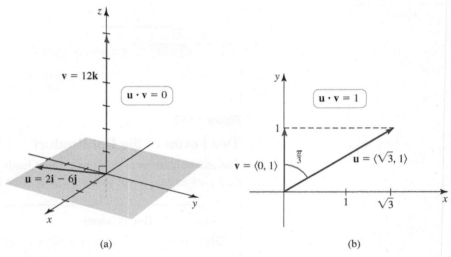

Figure 13.45

Related Exercises 16–17 ◀

Computing a dot product in this manner requires knowing the angle θ between the vectors. Often the angle is not known; in fact, it may be exactly what we seek. For this reason, we present another method for computing the dot product that does not require knowing θ.

> ➤ In $\mathbb{R}^2$ with $\mathbf{u} = \langle u_1, u_2 \rangle$ and $\mathbf{v} = \langle v_1, v_2 \rangle$, $\mathbf{u} \cdot \mathbf{v} = u_1 v_1 + u_2 v_2$.

THEOREM 13.1   Dot Product

Given two vectors $\mathbf{u} = \langle u_1, u_2, u_3 \rangle$ and $\mathbf{v} = \langle v_1, v_2, v_3 \rangle$,

$$\mathbf{u} \cdot \mathbf{v} = u_1 v_1 + u_2 v_2 + u_3 v_3.$$

Figure 13.46

Law of Cosines

$$c^2 = a^2 + b^2 - 2ab \cos \theta$$

Proof: Consider two position vectors $\mathbf{u} = \langle u_1, u_2, u_3 \rangle$ and $\mathbf{v} = \langle v_1, v_2, v_3 \rangle$, and suppose θ is the angle between them. The vector $\mathbf{u} - \mathbf{v}$ forms the third side of a triangle (**Figure 13.46**). By the Law of Cosines,

$$|\mathbf{u} - \mathbf{v}|^2 = |\mathbf{u}|^2 + |\mathbf{v}|^2 - 2\underbrace{|\mathbf{u}||\mathbf{v}| \cos \theta}_{\mathbf{u} \cdot \mathbf{v}}.$$

The definition of the dot product, $\mathbf{u} \cdot \mathbf{v} = |\mathbf{u}||\mathbf{v}| \cos \theta$, allows us to write

$$\mathbf{u} \cdot \mathbf{v} = |\mathbf{u}||\mathbf{v}| \cos \theta = \frac{1}{2}\left(|\mathbf{u}|^2 + |\mathbf{v}|^2 - |\mathbf{u} - \mathbf{v}|^2\right). \qquad (1)$$

Using the definition of magnitude, we find that

$$|\mathbf{u}|^2 = u_1^2 + u_2^2 + u_3^2, \qquad |\mathbf{v}|^2 = v_1^2 + v_2^2 + v_3^2,$$

and

$$|\mathbf{u} - \mathbf{v}|^2 = (u_1 - v_1)^2 + (u_2 - v_2)^2 + (u_3 - v_3)^2.$$

Expanding the terms in $|\mathbf{u} - \mathbf{v}|^2$ and simplifying yields

$$|\mathbf{u}|^2 + |\mathbf{v}|^2 - |\mathbf{u} - \mathbf{v}|^2 = 2(u_1 v_1 + u_2 v_2 + u_3 v_3).$$

Substituting into expression (1) gives a compact expression for the dot product:

$$\mathbf{u} \cdot \mathbf{v} = u_1 v_1 + u_2 v_2 + u_3 v_3. \qquad \blacktriangleleft$$

This new representation of $\mathbf{u} \cdot \mathbf{v}$ has two immediate consequences.

1. Combining it with the definition of dot product gives

$$\mathbf{u} \cdot \mathbf{v} = u_1 v_1 + u_2 v_2 + u_3 v_3 = |\mathbf{u}||\mathbf{v}| \cos \theta.$$

If $\mathbf{u}$ and $\mathbf{v}$ are both nonzero, then

$$\cos \theta = \frac{u_1 v_1 + u_2 v_2 + u_3 v_3}{|\mathbf{u}||\mathbf{v}|} = \frac{\mathbf{u} \cdot \mathbf{v}}{|\mathbf{u}||\mathbf{v}|},$$

QUICK CHECK 2 Use Theorem 13.1 to compute the dot products $\mathbf{i} \cdot \mathbf{j}$, $\mathbf{i} \cdot \mathbf{k}$, and $\mathbf{j} \cdot \mathbf{k}$ for the unit coordinate vectors. What do you conclude about the angles between these vectors? ◄

and we have a way to compute θ.

2. Notice that $\mathbf{u} \cdot \mathbf{u} = u_1^2 + u_2^2 + u_3^2 = |\mathbf{u}|^2$. Therefore, we have a relationship between the dot product and the magnitude of a vector: $|\mathbf{u}| = \sqrt{\mathbf{u} \cdot \mathbf{u}}$ or $|\mathbf{u}|^2 = \mathbf{u} \cdot \mathbf{u}$.

EXAMPLE 2 Dot products and angles Let $\mathbf{u} = \langle \sqrt{3}, 1, 0 \rangle$, $\mathbf{v} = \langle 1, \sqrt{3}, 0 \rangle$, and $\mathbf{w} = \langle 1, \sqrt{3}, 2\sqrt{3} \rangle$.

a. Compute $\mathbf{u} \cdot \mathbf{v}$.

b. Find the angle between $\mathbf{u}$ and $\mathbf{v}$.

c. Find the angle between $\mathbf{u}$ and $\mathbf{w}$.

SOLUTION

a. $\mathbf{u} \cdot \mathbf{v} = \langle \sqrt{3}, 1, 0 \rangle \cdot \langle 1, \sqrt{3}, 0 \rangle = \sqrt{3} + \sqrt{3} + 0 = 2\sqrt{3}$

b. Note that $|\mathbf{u}| = \sqrt{\mathbf{u} \cdot \mathbf{u}} = \sqrt{\langle \sqrt{3}, 1, 0 \rangle \cdot \langle \sqrt{3}, 1, 0 \rangle} = 2$, and similarly, $|\mathbf{v}| = 2$. Therefore,

$$\cos \theta = \frac{\mathbf{u} \cdot \mathbf{v}}{|\mathbf{u}||\mathbf{v}|} = \frac{2\sqrt{3}}{2 \cdot 2} = \frac{\sqrt{3}}{2}.$$

Because $0 \le \theta \le \pi$, it follows that $\theta = \cos^{-1} \frac{\sqrt{3}}{2} = \frac{\pi}{6}$.

c. $\cos \theta = \frac{\mathbf{u} \cdot \mathbf{w}}{|\mathbf{u}||\mathbf{w}|} = \frac{\langle \sqrt{3}, 1, 0 \rangle \cdot \langle 1, \sqrt{3}, 2\sqrt{3} \rangle}{|\langle \sqrt{3}, 1, 0 \rangle||\langle 1, \sqrt{3}, 2\sqrt{3} \rangle|} = \frac{2\sqrt{3}}{2 \cdot 4} = \frac{\sqrt{3}}{4}$

It follows that

$$\theta = \cos^{-1} \frac{\sqrt{3}}{4} \approx 1.12 \text{ rad} \approx 64.3°.$$

Related Exercises 20, 26 ◄

Properties of Dot Products   The properties of the dot product in the following theorem are easily proved using vector components (Exercises 79–81).

> ▶ Theorem 13.1 extends to vectors with any number of components. If $\mathbf{u} = \langle u_1, \ldots, u_n \rangle$ and $\mathbf{v} = \langle v_1, \ldots, v_n \rangle$, then
>
> $$\mathbf{u} \cdot \mathbf{v} = u_1 v_1 + \cdots + u_n v_n.$$
>
> The properties in Theorem 13.2 also apply in two or more dimensions.

THEOREM 13.2   Properties of the Dot Product
Suppose $\mathbf{u}$, $\mathbf{v}$, and $\mathbf{w}$ are vectors and let c be a scalar.

1. $\mathbf{u} \cdot \mathbf{v} = \mathbf{v} \cdot \mathbf{u}$ Commutative property
2. $c(\mathbf{u} \cdot \mathbf{v}) = (c\mathbf{u}) \cdot \mathbf{v} = \mathbf{u} \cdot (c\mathbf{v})$ Associative property
3. $\mathbf{u} \cdot (\mathbf{v} + \mathbf{w}) = \mathbf{u} \cdot \mathbf{v} + \mathbf{u} \cdot \mathbf{w}$ Distributive property

---

Orthogonal Projections

Given vectors $\mathbf{u}$ and $\mathbf{v}$, how closely aligned are they? That is, how much of $\mathbf{u}$ points in the direction of $\mathbf{v}$? This question is answered using *projections*. As shown in **Figure 13.47a**, the projection of the vector $\mathbf{u}$ onto a nonzero vector $\mathbf{v}$, denoted $\text{proj}_\mathbf{v}\mathbf{u}$, is the "shadow" cast by $\mathbf{u}$ onto the line through $\mathbf{v}$. The projection of $\mathbf{u}$ onto $\mathbf{v}$ is itself a vector; it points in the same direction as $\mathbf{v}$ if the angle between $\mathbf{u}$ and $\mathbf{v}$ lies in the interval $0 \leq \theta < \pi/2$ (**Figure 13.47b**); it points in the direction opposite that of $\mathbf{v}$ if the angle between $\mathbf{u}$ and $\mathbf{v}$ lies in the interval $\pi/2 < \theta \leq \pi$ (**Figure 13.47c**). If $\theta = \dfrac{\pi}{2}$, $\mathbf{u}$ and $\mathbf{v}$ are orthogonal, and there is no shadow.

$0 \leq \theta < \frac{\pi}{2}$
$\text{scal}_\mathbf{v}\mathbf{u} = |\mathbf{u}| \cos\theta > 0$

(a)

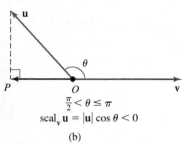

$\frac{\pi}{2} < \theta \leq \pi$
$\text{scal}_\mathbf{v}\mathbf{u} = |\mathbf{u}| \cos\theta < 0$

(b)

Figure 13.48

proj$_\mathbf{v}$ **u**
is the shadow of **u**
on the line through **v**.

(a)   (b)   (c)

Figure 13.47

To find the projection of $\mathbf{u}$ onto $\mathbf{v}$, we proceed as follows: With the tails of $\mathbf{u}$ and $\mathbf{v}$ together, we drop a perpendicular line segment from the head of $\mathbf{u}$ to the point P on the line through $\mathbf{v}$ (**Figure 13.48**). The vector $\overrightarrow{OP}$ is the *orthogonal projection of $\mathbf{u}$ onto $\mathbf{v}$*. An expression for $\text{proj}_\mathbf{v}\mathbf{u}$ is found using two observations.

- If $0 \leq \theta < \pi/2$, then $\text{proj}_\mathbf{v}\mathbf{u}$ has length $|\mathbf{u}| \cos\theta$ and points in the direction of the unit vector $\mathbf{v}/|\mathbf{v}|$ (**Figure 13.48a**). Therefore,

$$\text{proj}_\mathbf{v}\mathbf{u} = \underbrace{|\mathbf{u}| \cos\theta}_{\text{length}} \underbrace{\left(\frac{\mathbf{v}}{|\mathbf{v}|} \right)}_{\text{direction}}.$$

We define the *scalar component of $\mathbf{u}$ in the direction of $\mathbf{v}$* to be $\text{scal}_\mathbf{v}\mathbf{u} = |\mathbf{u}| \cos\theta$. In this case, $\text{scal}_\mathbf{v}\mathbf{u}$ is the length of $\text{proj}_\mathbf{v}\mathbf{u}$.

- If $\pi/2 < \theta \leq \pi$, then $\text{proj}_\mathbf{v}\mathbf{u}$ has length $-|\mathbf{u}| \cos\theta$ (which is positive) and points in the direction of $-\mathbf{v}/|\mathbf{v}|$ (**Figure 13.48b**). Therefore,

$$\text{proj}_\mathbf{v}\mathbf{u} = \underbrace{-|\mathbf{u}| \cos\theta}_{\text{length}} \underbrace{\left(-\frac{\mathbf{v}}{|\mathbf{v}|} \right)}_{\text{direction}} = |\mathbf{u}| \cos\theta \left(\frac{\mathbf{v}}{|\mathbf{v}|} \right).$$

In this case, $\text{scal}_\mathbf{v}\mathbf{u} = |\mathbf{u}| \cos\theta < 0$.

▶ Notice that $\text{scal}_v\mathbf{u}$ may be positive, negative, or zero. However, $|\text{scal}_v\mathbf{u}|$ is the length of $\text{proj}_v\mathbf{u}$. The projection $\text{proj}_v\mathbf{u}$ is defined for all vectors $\mathbf{u}$, but only for nonzero vectors $\mathbf{v}$.

We see that in both cases, the expression for $\text{proj}_v\mathbf{u}$ is the same:

$$\text{proj}_v\mathbf{u} = \underbrace{|\mathbf{u}|\cos\theta}_{\text{scal}_v\mathbf{u}} \left(\frac{\mathbf{v}}{|\mathbf{v}|}\right) = \text{scal}_v\mathbf{u}\left(\frac{\mathbf{v}}{|\mathbf{v}|}\right).$$

Note that if $\theta = \dfrac{\pi}{2}$, then $\text{proj}_v\mathbf{u} = \mathbf{0}$ and $\text{scal}_v\mathbf{u} = 0$.

Using properties of the dot product, $\text{proj}_v\mathbf{u}$ may be written in different ways:

$$\text{proj}_v\mathbf{u} = |\mathbf{u}|\cos\theta\left(\frac{\mathbf{v}}{|\mathbf{v}|}\right)$$

$$= \frac{\mathbf{u}\cdot\mathbf{v}}{|\mathbf{v}|}\left(\frac{\mathbf{v}}{|\mathbf{v}|}\right) \qquad |\mathbf{u}|\cos\theta = \frac{|\mathbf{u}||\mathbf{v}|\cos\theta}{|\mathbf{v}|} = \frac{\mathbf{u}\cdot\mathbf{v}}{|\mathbf{v}|}$$

$$= \underbrace{\left(\frac{\mathbf{u}\cdot\mathbf{v}}{\mathbf{v}\cdot\mathbf{v}}\right)}_{\text{scalar}}\mathbf{v}. \qquad \text{Regroup terms; } |\mathbf{v}|^2 = \mathbf{v}\cdot\mathbf{v}$$

QUICK CHECK 3 Let $\mathbf{u} = 4\mathbf{i} - 3\mathbf{j}$. By inspection (not calculations), find the orthogonal projection of $\mathbf{u}$ onto $\mathbf{i}$ and onto $\mathbf{j}$. Find the scalar component of $\mathbf{u}$ in the direction of $\mathbf{i}$ and in the direction of $\mathbf{j}$. ◀

The first two expressions show that $\text{proj}_v\mathbf{u}$ is a scalar multiple of the unit vector $\dfrac{\mathbf{v}}{|\mathbf{v}|}$, whereas the last expression shows that $\text{proj}_v\mathbf{u}$ is a scalar multiple of $\mathbf{v}$.

DEFINITION (Orthogonal) Projection of u onto v

The **orthogonal projection of u onto v**, denoted $\text{proj}_v\mathbf{u}$, where $\mathbf{v} \neq \mathbf{0}$, is

$$\text{proj}_v\mathbf{u} = |\mathbf{u}|\cos\theta\left(\frac{\mathbf{v}}{|\mathbf{v}|}\right).$$

The orthogonal projection may also be computed with the formulas

$$\text{proj}_v\mathbf{u} = \text{scal}_v\mathbf{u}\left(\frac{\mathbf{v}}{|\mathbf{v}|}\right) = \left(\frac{\mathbf{u}\cdot\mathbf{v}}{\mathbf{v}\cdot\mathbf{v}}\right)\mathbf{v},$$

where the **scalar component of u in the direction of v** is

$$\text{scal}_v\mathbf{u} = |\mathbf{u}|\cos\theta = \frac{\mathbf{u}\cdot\mathbf{v}}{|\mathbf{v}|}.$$

EXAMPLE 3 Orthogonal projections Find $\text{proj}_v\mathbf{u}$ and $\text{scal}_v\mathbf{u}$ for the following vectors and illustrate each result.

a. $\mathbf{u} = \langle 4, 1\rangle$, $\mathbf{v} = \langle 3, 4\rangle$

b. $\mathbf{u} = \langle -4, -3\rangle$, $\mathbf{v} = \langle 1, -1\rangle$

SOLUTION

a. The scalar component of $\mathbf{u}$ in the direction of $\mathbf{v}$ (Figure 13.49) is

$$\text{scal}_v\mathbf{u} = \frac{\mathbf{u}\cdot\mathbf{v}}{|\mathbf{v}|} = \frac{\langle 4, 1\rangle\cdot\langle 3, 4\rangle}{|\langle 3, 4\rangle|} = \frac{16}{5}.$$

Because $\dfrac{\mathbf{v}}{|\mathbf{v}|} = \left\langle\dfrac{3}{5}, \dfrac{4}{5}\right\rangle$, we have

$$\text{proj}_v\mathbf{u} = \text{scal}_v\mathbf{u}\left(\frac{\mathbf{v}}{|\mathbf{v}|}\right) = \frac{16}{5}\left\langle\frac{3}{5}, \frac{4}{5}\right\rangle = \frac{16}{25}\langle 3, 4\rangle.$$

b. Using another formula for $\text{proj}_v\mathbf{u}$, we have

$$\text{proj}_v\mathbf{u} = \left(\frac{\mathbf{u}\cdot\mathbf{v}}{\mathbf{v}\cdot\mathbf{v}}\right)\mathbf{v} = \left(\frac{\langle -4, -3\rangle\cdot\langle 1, -1\rangle}{\langle 1, -1\rangle\cdot\langle 1, -1\rangle}\right)\langle 1, -1\rangle = -\frac{1}{2}\langle 1, -1\rangle.$$

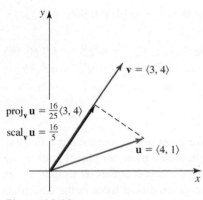

$\text{proj}_v\mathbf{u} = \dfrac{16}{25}\langle 3, 4\rangle$

$\text{scal}_v\mathbf{u} = \dfrac{16}{5}$

Figure 13.49

Figure 13.50

Figure 13.51

▶ If the unit of force is newtons (N) and the distance is measured in meters, then the unit of work is joules (J), where $1 \text{ J} = 1$ N-m. If force is measured in pounds and distance is measured in feet, then work has units of ft-lb.

Only the component of **F** in the direction of **d** contributes to the work.

Figure 13.52

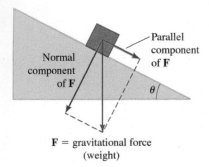

F = gravitational force (weight)

Figure 13.53

The vectors **v** and proj$_v$**u** point in opposite directions because $\pi/2 < \theta \le \pi$ (**Figure 13.50**). This fact is reflected in the scalar component of **u** in the direction of **v**, which is negative:

$$\text{scal}_v\mathbf{u} = \frac{\langle -4, -3 \rangle \cdot \langle 1, -1 \rangle}{|\langle 1, -1 \rangle|} = -\frac{1}{\sqrt{2}}.$$

Related Exercises 35–36 ◀

Applications of Dot Products

Work and Force   In the opening of this section, we observed that if a constant force **F** acts at an angle θ to the direction of motion of an object (**Figure 13.51**), the work done by the force is

$$W = |\mathbf{F}| \cos \theta\, |\mathbf{d}| = \mathbf{F} \cdot \mathbf{d}.$$

Notice that the work is a scalar, and if the force acts in a direction orthogonal to the motion, then $\theta = \pi/2$, $\mathbf{F} \cdot \mathbf{d} = 0$, and no work is done by the force.

DEFINITION Work

Let a constant force **F** be applied to an object, producing a displacement **d**. If the angle between **F** and **d** is θ, then the **work** done by the force is

$$W = |\mathbf{F}||\mathbf{d}| \cos \theta = \mathbf{F} \cdot \mathbf{d}.$$

EXAMPLE 4   Calculating work   A force $\mathbf{F} = \langle 3, 3, 2 \rangle$ (in newtons) moves an object along a line segment from $P(1, 1, 0)$ to $Q(6, 6, 0)$ (in meters). What is the work done by the force? Interpret the result.

SOLUTION   The displacement of the object is $\mathbf{d} = \vec{PQ} = \langle 6 - 1, 6 - 1, 0 - 0 \rangle = \langle 5, 5, 0 \rangle$. Therefore, the work done by the force is

$$W = \mathbf{F} \cdot \mathbf{d} = \langle 3, 3, 2 \rangle \cdot \langle 5, 5, 0 \rangle = 30 \text{ J}.$$

To interpret this result, notice that the angle between the force and the displacement vector satisfies

$$\cos \theta = \frac{\mathbf{F} \cdot \mathbf{d}}{|\mathbf{F}||\mathbf{d}|} = \frac{\langle 3, 3, 2 \rangle \cdot \langle 5, 5, 0 \rangle}{|\langle 3, 3, 2 \rangle||\langle 5, 5, 0 \rangle|} = \frac{30}{\sqrt{22}\sqrt{50}} \approx 0.905.$$

Therefore, $\theta \approx 0.44$ rad $\approx 25°$. The magnitude of the force is $|\mathbf{F}| = \sqrt{22} \approx 4.7$ N, but only the component of that force in the direction of motion, $|\mathbf{F}| \cos \theta \approx \sqrt{22} \cos 0.44 \approx 4.2$ N, contributes to the work (**Figure 13.52**).

Related Exercises 44, 46 ◀

Parallel and Normal Forces   Projections find frequent use in expressing a force in terms of orthogonal components. A common situation arises when an object rests on an inclined plane (**Figure 13.53**). The gravitational force on the object equals its weight, which is directed vertically downward. The projections of the gravitational force in the directions **parallel** to and **normal** (or perpendicular) to the plane are of interest. Specifically, the projection of the force parallel to the plane determines the tendency of the object to slide down the plane, while the projection of the force normal to the plane determines its tendency to "stick" to the plane.

EXAMPLE 5   Components of a force   A 10-lb block rests on a plane that is inclined at 30° above the horizontal. Find the components of the gravitational force parallel to and normal (perpendicular) to the plane.

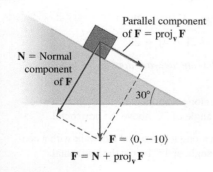

$v = \left\langle \frac{\sqrt{3}}{2}, -\frac{1}{2} \right\rangle$

Unit vector
down the plane

Parallel component
of $\mathbf{F} = \text{proj}_\mathbf{v}\,\mathbf{F}$

$\mathbf{N} = $ Normal
component
of $\mathbf{F}$

$\mathbf{F} = \langle 0, -10 \rangle$

$\mathbf{F} = \mathbf{N} + \text{proj}_\mathbf{v}\,\mathbf{F}$

Figure 13.54

SOLUTION The gravitational force $\mathbf{F}$ acting on the block equals the weight of the block (10 lb); we regard the block as a point mass. Using the coordinate system shown in **Figure 13.54**, the force acts in the negative y-direction; therefore, $\mathbf{F} = \langle 0, -10 \rangle$. The direction *down* the plane is given by the unit vector $\mathbf{v} = \langle \cos(-30°), \sin(-30°) \rangle = \langle \frac{\sqrt{3}}{2}, -\frac{1}{2} \rangle$ (check that $|\mathbf{v}| = 1$). The component of the gravitational force parallel to the plane is

$$\text{proj}_\mathbf{v}\mathbf{F} = \left(\frac{\mathbf{F} \cdot \mathbf{v}}{\underbrace{\mathbf{v} \cdot \mathbf{v}}_{\mathbf{v} \cdot \mathbf{v} = 1}} \right) \mathbf{v} = \left(\underbrace{\langle 0, -10 \rangle}_{\mathbf{F}} \cdot \underbrace{\left\langle \frac{\sqrt{3}}{2}, -\frac{1}{2} \right\rangle}_{\mathbf{v}} \right) \underbrace{\left\langle \frac{\sqrt{3}}{2}, -\frac{1}{2} \right\rangle}_{\mathbf{v}} = 5\left\langle \frac{\sqrt{3}}{2}, -\frac{1}{2} \right\rangle.$$

Let the component of $\mathbf{F}$ normal to the plane be $\mathbf{N}$. Note that $\mathbf{F} = \text{proj}_\mathbf{v}\mathbf{F} + \mathbf{N}$, so

$$\mathbf{N} = \mathbf{F} - \text{proj}_\mathbf{v}\mathbf{F} = \langle 0, -10 \rangle - 5\left\langle \frac{\sqrt{3}}{2}, -\frac{1}{2} \right\rangle = -5\left\langle \frac{\sqrt{3}}{2}, \frac{3}{2} \right\rangle.$$

Figure 13.54 shows how the components of $\mathbf{F}$ parallel to and normal to the plane combine to form the total force $\mathbf{F}$.

Related Exercises 47, 49 ◀

SECTION 13.3 EXERCISES

Getting Started

1. Express the dot product of $\mathbf{u}$ and $\mathbf{v}$ in terms of their magnitudes and the angle between them.

2. Express the dot product of $\mathbf{u}$ and $\mathbf{v}$ in terms of the components of the vectors.

3. Compute $\langle 2, 3, -6 \rangle \cdot \langle 1, -8, 3 \rangle$.

4. Use the definition of the dot product to explain why $\mathbf{v} \cdot \mathbf{v} = |\mathbf{v}|^2$.

5. Explain how to find the angle between two nonzero vectors.

6. Find the angle θ between $\mathbf{u}$ and $\mathbf{v}$ if $\text{scal}_\mathbf{v}\mathbf{u} = -2$ and $|\mathbf{u}| = 4$. Assume $0 \le \theta \le \pi$.

7. Find $\text{proj}_\mathbf{v}\mathbf{u}$ if $\text{scal}_\mathbf{v}\mathbf{u} = -2$ and $\mathbf{v} = \langle 2, -1, -2 \rangle$.

8. Use a dot product to determine whether the vectors $\mathbf{u} = \langle 1, 2, 3 \rangle$ and $\mathbf{v} = \langle 4, 1, -2 \rangle$ are orthogonal.

9. Find $\mathbf{u} \cdot \mathbf{v}$ if $\mathbf{u}$ and $\mathbf{v}$ are unit vectors and the angle between $\mathbf{u}$ and $\mathbf{v}$ is π.

10. Explain how the work done by a force in moving an object is computed using dot products.

11. Suppose $\mathbf{v}$ is a nonzero position vector in the xy-plane. How many position vectors with length 2 in the xy-plane are orthogonal to $\mathbf{v}$?

12. Suppose $\mathbf{v}$ is a nonzero position vector in xyz-space. How many position vectors with length 2 in xyz-space are orthogonal to $\mathbf{v}$?

Practice Exercises

13–16. Dot product from the definition *Consider the following vectors* $\mathbf{u}$ *and* $\mathbf{v}$*. Sketch the vectors, find the angle between the vectors, and compute the dot product using the definition* $\mathbf{u} \cdot \mathbf{v} = |\mathbf{u}||\mathbf{v}|\cos\theta$.

13. $\mathbf{u} = 4\mathbf{i}$ and $\mathbf{v} = 6\mathbf{j}$

14. $\mathbf{u} = \langle -3, 2, 0 \rangle$ and $\mathbf{v} = \langle 0, 0, 6 \rangle$

15. $\mathbf{u} = \langle 10, 0 \rangle$ and $\mathbf{v} = \langle 10, 10 \rangle$

16. $\mathbf{u} = \langle -\sqrt{3}, 1 \rangle$ and $\mathbf{v} = \langle \sqrt{3}, 1 \rangle$

17. **Dot product from the definition** Compute $\mathbf{u} \cdot \mathbf{v}$ if $\mathbf{u}$ and $\mathbf{v}$ are unit vectors and the angle between them is $\pi/3$.

18. **Dot product from the definition** Compute $\mathbf{u} \cdot \mathbf{v}$ if $\mathbf{u}$ is a unit vector, $|\mathbf{v}| = 2$, and the angle between them is $3\pi/4$.

T 19–28. Dot products and angles *Compute the dot product of the vectors* $\mathbf{u}$ *and* $\mathbf{v}$*, and find the angle between the vectors.*

19. $\mathbf{u} = \mathbf{i} + \mathbf{j}$ and $\mathbf{v} = \mathbf{i} - \mathbf{j}$

20. $\mathbf{u} = \langle 10, 0 \rangle$ and $\mathbf{v} = \langle -5, 5 \rangle$

21. $\mathbf{u} = \mathbf{i}$ and $\mathbf{v} = \mathbf{i} + \sqrt{3}\mathbf{j}$

22. $\mathbf{u} = \sqrt{2}\mathbf{i} + \sqrt{2}\mathbf{j}$ and $\mathbf{v} = -\sqrt{2}\mathbf{i} - \sqrt{2}\mathbf{j}$

23. $\mathbf{u} = 4\mathbf{i} + 3\mathbf{j}$ and $\mathbf{v} = 4\mathbf{i} - 6\mathbf{j}$

24. $\mathbf{u} = \langle 3, 4, 0 \rangle$ and $\mathbf{v} = \langle 0, 4, 5 \rangle$

25. $\mathbf{u} = \langle -10, 0, 4 \rangle$ and $\mathbf{v} = \langle 1, 2, 3 \rangle$

26. $\mathbf{u} = \langle 3, -5, 2 \rangle$ and $\mathbf{v} = \langle -9, 5, 1 \rangle$

27. $\mathbf{u} = 2\mathbf{i} - 3\mathbf{k}$ and $\mathbf{v} = \mathbf{i} + 4\mathbf{j} + 2\mathbf{k}$

28. $\mathbf{u} = \mathbf{i} - 4\mathbf{j} - 6\mathbf{k}$ and $\mathbf{v} = 2\mathbf{i} - 4\mathbf{j} + 2\mathbf{k}$

T 29–30. Angles of a triangle *For the given points P, Q, and R, find the approximate measurements of the angles of* $\triangle PQR$.

29. $P(0, -1, 3)$, $Q(2, 2, 1)$, $R(-2, 2, 4)$

30. $P(1, -4)$, $Q(2, 7)$, $R(-2, 2)$

31–34. Sketching orthogonal projections *Find* proj$_v$**u** *and* scal$_v$**u** *by inspection without using formulas.*

31.

32.

33.

34.

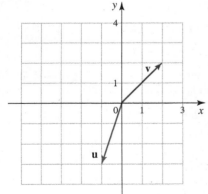

35–40. Calculating orthogonal projections *For the given vectors* **u** *and* **v**, *calculate* proj$_v$**u** *and* scal$_v$**u**.

35. $\mathbf{u} = \langle -1, 4 \rangle$ and $\mathbf{v} = \langle -4, 2 \rangle$

36. $\mathbf{u} = \langle 10, 5 \rangle$ and $\mathbf{v} = \langle 2, 6 \rangle$

37. $\mathbf{u} = \langle -8, 0, 2 \rangle$ and $\mathbf{v} = \langle 1, 3, -3 \rangle$

38. $\mathbf{u} = \langle 5, 0, 15 \rangle$ and $\mathbf{v} = \langle 0, 4, -2 \rangle$

39. $\mathbf{u} = 5\mathbf{i} + \mathbf{j} - 5\mathbf{k}$ and $\mathbf{v} = -\mathbf{i} + \mathbf{j} - 2\mathbf{k}$

40. $\mathbf{u} = \mathbf{i} + 4\mathbf{j} + 7\mathbf{k}$ and $\mathbf{v} = 2\mathbf{i} - 4\mathbf{j} + 2\mathbf{k}$

41–46. Computing work *Calculate the work done in the following situations.*

41. A suitcase is pulled 50 ft along a horizontal sidewalk with a constant force of 30 lb at an angle of 30° above the horizontal.

T 42. A stroller is pushed 20 m along a horizontal sidewalk with a constant force of 10 N at an angle of 15° below the horizontal.

43. A sled is pulled 10 m along horizontal ground with a constant force of 5 N at an angle of 45° above the horizontal.

44. A constant force $\mathbf{F} = \langle 4, 3, 2 \rangle$ (in newtons) moves an object from $(0, 0, 0)$ to $(8, 6, 0)$. (Distance is measured in meters.)

45. A constant force $\mathbf{F} = \langle 40, 30 \rangle$ (in newtons) is used to move a sled horizontally 10 m.

46. A constant force $\mathbf{F} = \langle 2, 4, 1 \rangle$ (in newtons) moves an object from $(0, 0, 0)$ to $(2, 4, 6)$. (Distance is measured in meters.)

47–48. Parallel and normal forces *Find the components of the vertical force* $\mathbf{F} = \langle 0, -10 \rangle$ *in the directions parallel to and normal to the following planes. Show that the total force is the sum of the two component forces.*

47. A plane that makes an angle of $\pi/3$ with the positive x-axis

48. A plane that makes an angle of $\theta = \tan^{-1} \dfrac{4}{5}$ with the positive x-axis

49. **Mass on a plane** A 100-kg object rests on an inclined plane at an angle of 45° to the floor. Find the components of the force parallel to and perpendicular to the plane.

T 50. **Forces on an inclined plane** An object on an inclined plane does not slide, provided the component of the object's weight parallel to the plane $|\mathbf{W}_{par}|$ is less than or equal to the magnitude of the opposing frictional force $|\mathbf{F}_f|$. The magnitude of the frictional force, in turn, is proportional to the component of the object's weight perpendicular to the plane $|\mathbf{W}_{perp}|$ (see figure). The constant of proportionality is the coefficient of static friction $\mu > 0$. Suppose a 100-lb block rests on a plane that is tilted at an angle of $\theta = 20°$ to the horizontal.

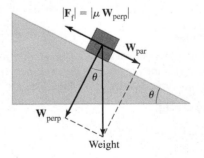

a. Find $|\mathbf{W}_{par}|$ and $|\mathbf{W}_{perp}|$. (*Hint:* It is not necessary to find $\mathbf{W}_{par}$ and $\mathbf{W}_{perp}$ first.)

b. The condition for the block not sliding is $|\mathbf{W}_{par}| \leq \mu |\mathbf{W}_{perp}|$. If $\mu = 0.65$, does the block slide?

c. What is the critical angle above which the block slides?

51. Explain why or why not Determine whether the following statements are true and give an explanation or counterexample.

a. $\text{proj}_{\mathbf{v}}\mathbf{u} = \text{proj}_{\mathbf{u}}\mathbf{v}$.

b. If nonzero vectors $\mathbf{u}$ and $\mathbf{v}$ have the same magnitude, they make equal angles with $\mathbf{u} + \mathbf{v}$.

c. $(\mathbf{u} \cdot \mathbf{i})^2 + (\mathbf{u} \cdot \mathbf{j})^2 + (\mathbf{u} \cdot \mathbf{k})^2 = |\mathbf{u}|^2$.

d. If $\mathbf{u}$ is orthogonal to $\mathbf{v}$ and $\mathbf{v}$ is orthogonal to $\mathbf{w}$, then $\mathbf{u}$ is orthogonal to $\mathbf{w}$.

e. The vectors orthogonal to $\langle 1, 1, 1 \rangle$ lie on the same line.

f. If $\text{proj}_{\mathbf{v}}\mathbf{u} = \mathbf{0}$, then vectors $\mathbf{u}$ and $\mathbf{v}$ (both nonzero) are orthogonal.

52. For what value of a is the vector $\mathbf{v} = \langle 4, -3, 7 \rangle$ orthogonal to $\mathbf{w} = \langle a, 8, 3 \rangle$?

53. For what value of c is the vector $\mathbf{v} = \langle 2, -5, c \rangle$ orthogonal to $\mathbf{w} = \langle 3, 2, 9 \rangle$?

54. Find two vectors that are orthogonal to $\langle 0, 1, 1 \rangle$ and to each other.

55. Let a and b be real numbers. Find all vectors $\langle 1, a, b \rangle$ orthogonal to $\langle 4, -8, 2 \rangle$.

56. Find three mutually orthogonal unit vectors in $\mathbb{R}^3$ besides $\pm\mathbf{i}$, $\pm\mathbf{j}$, and $\pm\mathbf{k}$.

57. Equal angles Consider the set of all unit position vectors $\mathbf{u}$ in $\mathbb{R}^3$ that make a $60°$ angle with the unit vector $\mathbf{k}$ in $\mathbb{R}^3$.

a. Prove that $\text{proj}_{\mathbf{k}}\mathbf{u}$ is the same for all vectors in this set.

b. Is $\text{scal}_{\mathbf{k}}\mathbf{u}$ the same for all vectors in this set?

58–61. Vectors with equal projections *Given a fixed vector $\mathbf{v}$, there is an infinite set of vectors $\mathbf{u}$ with the same value of $\text{proj}_{\mathbf{v}}\mathbf{u}$.*

58. Find another vector that has the same projection onto $\mathbf{v} = \langle 1, 1 \rangle$ as $\mathbf{u} = \langle 1, 2 \rangle$. Draw a picture.

59. Let $\mathbf{v} = \langle 1, 1 \rangle$. Give a description of the position vectors $\mathbf{u}$ such that $\text{proj}_{\mathbf{v}}\mathbf{u} = \text{proj}_{\mathbf{v}}\langle 1, 2 \rangle$.

60. Find another vector that has the same projection onto $\mathbf{v} = \langle 1, 1, 1 \rangle$ as $\mathbf{u} = \langle 1, 2, 3 \rangle$.

61. Let $\mathbf{v} = \langle 0, 0, 1 \rangle$. Give a description of all position vectors $\mathbf{u}$ such that $\text{proj}_{\mathbf{v}}\mathbf{u} = \text{proj}_{\mathbf{v}}\langle 1, 2, 3 \rangle$.

62–65. Decomposing vectors *For the following vectors $\mathbf{u}$ and $\mathbf{v}$, express $\mathbf{u}$ as the sum $\mathbf{u} = \mathbf{p} + \mathbf{n}$, where $\mathbf{p}$ is parallel to $\mathbf{v}$ and $\mathbf{n}$ is orthogonal to $\mathbf{v}$.*

62. $\mathbf{u} = \langle 4, 3 \rangle$, $\mathbf{v} = \langle 1, 1 \rangle$

63. $\mathbf{u} = \langle -2, 2 \rangle$, $\mathbf{v} = \langle 2, 1 \rangle$

64. $\mathbf{u} = \langle 4, 3, 0 \rangle$, $\mathbf{v} = \langle 1, 1, 1 \rangle$

65. $\mathbf{u} = \langle -1, 2, 3 \rangle$, $\mathbf{v} = \langle 2, 1, 1 \rangle$

66–69. An alternative line definition *Given a fixed point $P_0(x_0, y_0)$ and a nonzero vector $\mathbf{n} = \langle a, b \rangle$, the set of points $P(x, y)$ for which $\overrightarrow{P_0P}$ is orthogonal to $\mathbf{n}$ is a line ℓ (see figure). The vector $\mathbf{n}$ is called a normal vector or a vector normal to ℓ.*

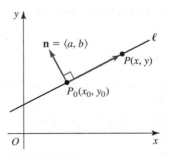

66. Show that the equation of the line passing through $P_0(x_0, y_0)$ with a normal vector $\mathbf{n} = \langle a, b \rangle$ is $a(x - x_0) + b(y - y_0) = 0$. (*Hint:* For a point $P(x, y)$ on ℓ, examine $\mathbf{n} \cdot \overrightarrow{P_0P}$.)

67. Use the result of Exercise 66 to find an equation of the line passing through the point $P_0(2, 6)$ with a normal vector $\mathbf{n} = \langle 3, -7 \rangle$. Write the final answer in the form $ax + by = c$.

68. Use the result of Exercise 66 to find an equation of the line passing through the point $P_0(1, -3)$ with a normal vector $\mathbf{n} = \langle 4, 0 \rangle$.

69. Suppose a line is normal to $\mathbf{n} = \langle 5, 3 \rangle$. What is the slope of the line?

Explorations and Challenges

70–72. Orthogonal unit vectors in $\mathbb{R}^2$ *Consider the vectors $\mathbf{I} = \langle 1/\sqrt{2}, 1/\sqrt{2} \rangle$ and $\mathbf{J} = \langle -1/\sqrt{2}, 1/\sqrt{2} \rangle$.*

70. Show that $\mathbf{I}$ and $\mathbf{J}$ are orthogonal unit vectors.

71. Express $\mathbf{I}$ and $\mathbf{J}$ in terms of the usual unit coordinate vectors $\mathbf{i}$ and $\mathbf{j}$. Then write $\mathbf{i}$ and $\mathbf{j}$ in terms of $\mathbf{I}$ and $\mathbf{J}$.

72. Write the vector $\langle 2, -6 \rangle$ in terms of $\mathbf{I}$ and $\mathbf{J}$.

73. Orthogonal unit vectors in $\mathbb{R}^3$ Consider the vectors $\mathbf{I} = \langle 1/2, 1/2, 1/\sqrt{2} \rangle$, $\mathbf{J} = \langle -1/\sqrt{2}, 1/\sqrt{2}, 0 \rangle$, and $\mathbf{K} = \langle 1/2, 1/2, -1/\sqrt{2} \rangle$.

a. Sketch $\mathbf{I}$, $\mathbf{J}$, and $\mathbf{K}$ and show that they are unit vectors.

b. Show that $\mathbf{I}$, $\mathbf{J}$, and $\mathbf{K}$ are pairwise orthogonal.

c. Express the vector $\langle 1, 0, 0 \rangle$ in terms of $\mathbf{I}$, $\mathbf{J}$, and $\mathbf{K}$.

74. Flow through a circle Suppose water flows in a thin sheet over the xy-plane with a uniform velocity given by the vector $\mathbf{v} = \langle 1, 2 \rangle$; this means that at all points of the plane, the velocity of the water has components 1 m/s in the x-direction and 2 m/s in the y-direction (see figure). Let C be an imaginary unit circle (that does not interfere with the flow).

a. Show that at the point (x, y) on the circle C, the outward-pointing unit vector normal to C is $\mathbf{n} = \langle x, y \rangle$.

b. Show that at the point $(\cos\theta, \sin\theta)$ on the circle C, the outward-pointing unit vector normal to C is also $\mathbf{n} = \langle \cos\theta, \sin\theta \rangle$.

c. Find all points on C at which the velocity is normal to C.

d. Find all points on C at which the velocity is tangential to C.

e. At each point on C, find the component of $\mathbf{v}$ normal to C. Express the answer as a function of (x, y) and as a function of θ.

f. What is the net flow through the circle? Does water accumulate inside the circle?

75. Heat flux Let D be a solid heat-conducting cube formed by the planes $x = 0$, $x = 1$, $y = 0$, $y = 1$, $z = 0$, $z = 1$. The heat flow at every point of D is given by the constant vector $\mathbf{Q} = \langle 0, 2, 1 \rangle$.

a. Through which faces of D does $\mathbf{Q}$ point into D?

b. Through which faces of D does $\mathbf{Q}$ point out of D?

c. On which faces of D is $\mathbf{Q}$ tangential to D (pointing neither in nor out of D)?

d. Find the scalar component of $\mathbf{Q}$ normal to the face $x = 0$.

e. Find the scalar component of $\mathbf{Q}$ normal to the face $z = 1$.

f. Find the scalar component of $\mathbf{Q}$ normal to the face $y = 0$.

76. Hexagonal circle packing The German mathematician Carl Friedrich Gauss proved that the densest way to pack circles with the same radius in the plane is to place the centers of the circles on a hexagonal grid (see figure). Some molecular structures use this packing or its three-dimensional analog. Assume all circles have a radius of 1, and let $\mathbf{r}_{ij}$ be the vector that extends from the center of circle i to the center of circle j, for $i, j = 0, 1, \ldots, 6$.

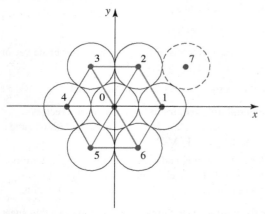

a. Find $\mathbf{r}_{0j}$ for $j = 1, 2, \ldots, 6$.

b. Find $\mathbf{r}_{12}$, $\mathbf{r}_{34}$, and $\mathbf{r}_{61}$.

c. Imagine that circle 7 is added to the arrangement as shown in the figure. Find $\mathbf{r}_{07}$, $\mathbf{r}_{17}$, $\mathbf{r}_{47}$, and $\mathbf{r}_{75}$.

77. Hexagonal sphere packing Imagine three unit spheres (radius equal to 1) with centers at $O(0, 0, 0)$, $P(\sqrt{3}, -1, 0)$, and $Q(\sqrt{3}, 1, 0)$. Now place another unit sphere symmetrically on top of these three spheres with its center at R (see figure).

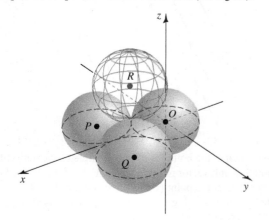

a. Find the coordinates of R. (*Hint:* The distance between the centers of any two spheres is 2.)

b. Let $\mathbf{r}_{IJ}$ be the vector from point I to point J. Find $\mathbf{r}_{OP}$, $\mathbf{r}_{OQ}$, $\mathbf{r}_{PQ}$, $\mathbf{r}_{OR}$, and $\mathbf{r}_{PR}$.

78–81. Properties of dot products *Let* $\mathbf{u} = \langle u_1, u_2, u_3 \rangle$, $\mathbf{v} = \langle v_1, v_2, v_3 \rangle$, *and* $\mathbf{w} = \langle w_1, w_2, w_3 \rangle$. *Prove the following vector properties, where c is a scalar.*

78. $|\mathbf{u} \cdot \mathbf{v}| \le |\mathbf{u}||\mathbf{v}|$

79. $\mathbf{u} \cdot \mathbf{v} = \mathbf{v} \cdot \mathbf{u}$ Commutative property

80. $c(\mathbf{u} \cdot \mathbf{v}) = (c\mathbf{u}) \cdot \mathbf{v} = \mathbf{u} \cdot (c\mathbf{v})$ Associative property

81. $\mathbf{u} \cdot (\mathbf{v} + \mathbf{w}) = \mathbf{u} \cdot \mathbf{v} + \mathbf{u} \cdot \mathbf{w}$ Distributive property

82. Distributive properties

a. Show that $(\mathbf{u} + \mathbf{v}) \cdot (\mathbf{u} + \mathbf{v}) = |\mathbf{u}|^2 + 2\mathbf{u} \cdot \mathbf{v} + |\mathbf{v}|^2$.

b. Show that $(\mathbf{u} + \mathbf{v}) \cdot (\mathbf{u} + \mathbf{v}) = |\mathbf{u}|^2 + |\mathbf{v}|^2$ if $\mathbf{u}$ is orthogonal to $\mathbf{v}$.

c. Show that $(\mathbf{u} + \mathbf{v}) \cdot (\mathbf{u} - \mathbf{v}) = |\mathbf{u}|^2 - |\mathbf{v}|^2$.

83. Direction angles and cosines Let $\mathbf{v} = \langle a, b, c \rangle$ and let α, β, and γ be the angles between $\mathbf{v}$ and the positive x-axis, the positive y-axis, and the positive z-axis, respectively (see figure).

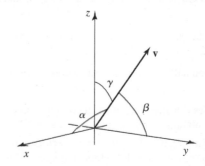

a. Prove that $\cos^2 \alpha + \cos^2 \beta + \cos^2 \gamma = 1$.

b. Find a vector that makes a $45°$ angle with $\mathbf{i}$ and $\mathbf{j}$. What angle does it make with $\mathbf{k}$?

c. Find a vector that makes a $60°$ angle with $\mathbf{i}$ and $\mathbf{j}$. What angle does it make with $\mathbf{k}$?

d. Is there a vector that makes a $30°$ angle with $\mathbf{i}$ and $\mathbf{j}$? Explain.

e. Find a vector $\mathbf{v}$ such that $\alpha = \beta = \gamma$. What is the angle?

84–88. Cauchy-Schwarz Inequality *The definition* $\mathbf{u} \cdot \mathbf{v} = |\mathbf{u}||\mathbf{v}|\cos \theta$ *implies that* $|\mathbf{u} \cdot \mathbf{v}| \le |\mathbf{u}||\mathbf{v}|$ *(because* $|\cos \theta| \le 1$). *This inequality, known as the Cauchy-Schwarz Inequality, holds in any number of dimensions and has many consequences.*

84. What conditions on $\mathbf{u}$ and $\mathbf{v}$ lead to equality in the Cauchy-Schwarz Inequality?

85. Verify that the Cauchy-Schwarz Inequality holds for $\mathbf{u} = \langle 3, -5, 6 \rangle$ and $\mathbf{v} = \langle -8, 3, 1 \rangle$.

86. Geometric-arithmetic mean Use the vectors $\mathbf{u} = \langle \sqrt{a}, \sqrt{b} \rangle$ and $\mathbf{v} = \langle \sqrt{b}, \sqrt{a} \rangle$ to show that $\sqrt{ab} \le \dfrac{a + b}{2}$, where $a \ge 0$ and $b \ge 0$.

87. Triangle Inequality Consider the vectors $\mathbf{u}$, $\mathbf{v}$, and $\mathbf{u} + \mathbf{v}$ (in any number of dimensions). Use the following steps to prove that $|\mathbf{u} + \mathbf{v}| \le |\mathbf{u}| + |\mathbf{v}|$.

a. Show that $|\mathbf{u} + \mathbf{v}|^2 = (\mathbf{u} + \mathbf{v}) \cdot (\mathbf{u} + \mathbf{v}) = |\mathbf{u}|^2 + 2\mathbf{u} \cdot \mathbf{v} + |\mathbf{v}|^2$.

b. Use the Cauchy-Schwarz Inequality to show that
$$|u + v|^2 \le (|u| + |v|)^2.$$

c. Conclude that $|u + v| \le |u| + |v|$.

d. Interpret the Triangle Inequality geometrically in $\mathbb{R}^2$ or $\mathbb{R}^3$.

88. Algebra inequality Show that
$$(u_1 + u_2 + u_3)^2 \le 3(u_1^2 + u_2^2 + u_3^2),$$

for any real numbers u_1, u_2, and u_3. (*Hint:* Use the Cauchy-Schwarz Inequality in three dimensions with $u = \langle u_1, u_2, u_3 \rangle$ and choose v in the right way.)

89. Diagonals of a parallelogram Consider the parallelogram with adjacent sides u and v.

a. Show that the diagonals of the parallelogram are $u + v$ and $u - v$.

b. Prove that the diagonals have the same length if and only if $u \cdot v = 0$.

c. Show that the sum of the squares of the lengths of the diagonals equals the sum of the squares of the lengths of the sides.

QUICK CHECK ANSWERS

1. If $\theta = 0$, then u and v are parallel and point in the same direction. If $\theta = \pi$, then u and v are parallel and point in opposite directions.   **2.** All these dot products are zero, and the unit vectors are mutually orthogonal. The angle between two different unit vectors is $\pi/2$.   **3.** $\text{proj}_i u = 4i$, $\text{proj}_j u = -3j$, $\text{scal}_i u = 4$, $\text{scal}_j u = -3$ ◄

13.4 Cross Products

The dot product combines two vectors to produce a *scalar* result. There is an equally fundamental way to combine two vectors in $\mathbb{R}^3$ and obtain a *vector* result. This operation, known as the *cross product* (or *vector product*), may be motivated by a physical application.

Suppose you want to loosen a bolt with a wrench. As you apply force to the end of the wrench in the plane perpendicular to the bolt, the "twisting power" you generate depends on three variables:

• the magnitude of the force F applied to the wrench;

• the length $|r|$ of the wrench;

• the angle at which the force is applied to the wrench.

The twisting generated by a force acting at a distance from a pivot point is called **torque** (from the Latin *to twist*). The torque is a vector whose magnitude is proportional to $|F|$, $|r|$, and $\sin \theta$, where θ is the angle between F and r (**Figure 13.55**). If the force is applied parallel to the wrench—for example, if you pull the wrench ($\theta = 0$) or push the wrench ($\theta = \pi$)—there is no twisting effect; if the force is applied perpendicular to the wrench ($\theta = \pi/2$), the twisting effect is maximized. The direction of the torque vector is defined to be orthogonal to both F and r. As we will see shortly, the torque is expressed in terms of the cross product of F and r.

Torque

Component of F perpendicular to r has length $|F| \sin \theta$.

Figure 13.55

The Cross Product

The preceding physical example leads to the following definition of the cross product.

DEFINITION Cross Product

Given two nonzero vectors u and v in $\mathbb{R}^3$, the **cross product** $u \times v$ is a vector with magnitude
$$|u \times v| = |u||v| \sin \theta,$$

where $0 \le \theta \le \pi$ is the angle between u and v. The direction of $u \times v$ is given by the **right-hand rule**: When you put the vectors tail to tail and let the fingers of your right hand curl from u to v, the direction of $u \times v$ is the direction of your thumb, orthogonal to both u and v (**Figure 13.56**). When $u \times v = 0$, the direction of $u \times v$ is undefined.

$u \times v$

v

θ

u

Figure 13.56

QUICK CHECK 1 Sketch the vectors $\mathbf{u} = \langle 1, 2, 0 \rangle$ and $\mathbf{v} = \langle -1, 2, 0 \rangle$. Which way does $\mathbf{u} \times \mathbf{v}$ point? Which way does $\mathbf{v} \times \mathbf{u}$ point? ◄

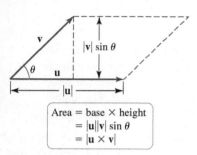

Area = base × height
= $|\mathbf{u}||\mathbf{v}| \sin \theta$
= $|\mathbf{u} \times \mathbf{v}|$

Figure 13.57

$\mathbf{u} \times \mathbf{v}$ is orthogonal to $\mathbf{u}$ and $\mathbf{v}$ with $|\mathbf{u} \times \mathbf{v}| = 2$.

Figure 13.58

The following theorem is an immediate consequence of the definition of the cross product.

THEOREM 13.3   Geometry of the Cross Product
Let $\mathbf{u}$ and $\mathbf{v}$ be two nonzero vectors in $\mathbb{R}^3$.

1. The vectors $\mathbf{u}$ and $\mathbf{v}$ are parallel ($\theta = 0$ or $\theta = \pi$) if and only if $\mathbf{u} \times \mathbf{v} = \mathbf{0}$.

2. If $\mathbf{u}$ and $\mathbf{v}$ are two sides of a parallelogram (**Figure 13.57**), then the area of the parallelogram is

$$|\mathbf{u} \times \mathbf{v}| = |\mathbf{u}||\mathbf{v}| \sin \theta.$$

EXAMPLE 1   A cross product Find the magnitude and direction of $\mathbf{u} \times \mathbf{v}$, where $\mathbf{u} = \langle 1, 1, 0 \rangle$ and $\mathbf{v} = \langle 1, 1, \sqrt{2} \rangle$.

SOLUTION Because $\mathbf{u}$ is one side of a 45–45–90 triangle and $\mathbf{v}$ is the hypotenuse (**Figure 13.58**), we have $\theta = \pi/4$ and $\sin \theta = \frac{1}{\sqrt{2}}$. Also, $|\mathbf{u}| = \sqrt{2}$ and $|\mathbf{v}| = 2$, so the magnitude of $\mathbf{u} \times \mathbf{v}$ is

$$|\mathbf{u} \times \mathbf{v}| = |\mathbf{u}||\mathbf{v}| \sin \theta = \sqrt{2} \cdot 2 \cdot \frac{1}{\sqrt{2}} = 2.$$

The direction of $\mathbf{u} \times \mathbf{v}$ is given by the right-hand rule: $\mathbf{u} \times \mathbf{v}$ is orthogonal to $\mathbf{u}$ and $\mathbf{v}$ (**Figure 13.58**).

Related Exercises 14–15 ◄

Properties of the Cross Product

The cross product has several algebraic properties that simplify calculations. For example, scalars factor out of a cross product; that is, if a and b are scalars, then (Exercise 69)

$$(a\mathbf{u}) \times (b\mathbf{v}) = ab(\mathbf{u} \times \mathbf{v}).$$

The order in which the cross product is performed is important. The magnitudes of $\mathbf{u} \times \mathbf{v}$ and $\mathbf{v} \times \mathbf{u}$ are equal. However, applying the right-hand rule shows that $\mathbf{u} \times \mathbf{v}$ and $\mathbf{v} \times \mathbf{u}$ point in opposite directions. Therefore, $\mathbf{u} \times \mathbf{v} = -(\mathbf{v} \times \mathbf{u})$. There are two distributive properties for the cross product, whose proofs are omitted.

THEOREM 13.4   Properties of the Cross Product
Let $\mathbf{u}$, $\mathbf{v}$, and $\mathbf{w}$ be nonzero vectors in $\mathbb{R}^3$, and let a and b be scalars.

1. $\mathbf{u} \times \mathbf{v} = -(\mathbf{v} \times \mathbf{u})$ Anticommutative property
2. $(a\mathbf{u}) \times (b\mathbf{v}) = ab(\mathbf{u} \times \mathbf{v})$ Associative property
3. $\mathbf{u} \times (\mathbf{v} + \mathbf{w}) = (\mathbf{u} \times \mathbf{v}) + (\mathbf{u} \times \mathbf{w})$ Distributive property
4. $(\mathbf{u} + \mathbf{v}) \times \mathbf{w} = (\mathbf{u} \times \mathbf{w}) + (\mathbf{v} \times \mathbf{w})$ Distributive property

---

QUICK CHECK 2 Explain why the vector $2\mathbf{u} \times 3\mathbf{v}$ points in the same direction as the vector $\mathbf{u} \times \mathbf{v}$. ◄

EXAMPLE 2   Cross products of unit vectors Evaluate all the cross products among the coordinate unit vectors $\mathbf{i}, \mathbf{j}$, and $\mathbf{k}$.

SOLUTION These vectors are mutually orthogonal, which means the angle between any two distinct vectors is $\theta = \pi/2$ and $\sin \theta = 1$. Furthermore, $|\mathbf{i}| = |\mathbf{j}| = |\mathbf{k}| = 1$. Therefore, the cross product of any two distinct vectors has magnitude 1. By the right-hand rule, when the fingers of the right hand curl from $\mathbf{i}$ to $\mathbf{j}$, the thumb points in the direction of the positive z-axis (**Figure 13.59**). The unit vector in the positive z-direction is $\mathbf{k}$, so $\mathbf{i} \times \mathbf{j} = \mathbf{k}$. Similar calculations show that $\mathbf{j} \times \mathbf{k} = \mathbf{i}$ and $\mathbf{k} \times \mathbf{i} = \mathbf{j}$.

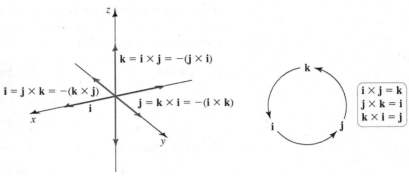

Figure 13.59

By property 1 of Theorem 13.4, $\mathbf{j} \times \mathbf{i} = -(\mathbf{i} \times \mathbf{j}) = -\mathbf{k}$, so $\mathbf{j} \times \mathbf{i}$ and $\mathbf{i} \times \mathbf{j}$ point in opposite directions. Similarly, $\mathbf{k} \times \mathbf{j} = -\mathbf{i}$ and $\mathbf{i} \times \mathbf{k} = -\mathbf{j}$. These relationships are easily remembered with the circle diagram in Figure 13.59. Finally, the angle between any unit vector and itself is $\theta = 0$. Therefore, $\mathbf{i} \times \mathbf{i} = \mathbf{j} \times \mathbf{j} = \mathbf{k} \times \mathbf{k} = \mathbf{0}$.

Related Exercises 17–19 ◀

THEOREM 13.5   Cross Products of Coordinate Unit Vectors

$$\mathbf{i} \times \mathbf{j} = -(\mathbf{j} \times \mathbf{i}) = \mathbf{k} \qquad \mathbf{j} \times \mathbf{k} = -(\mathbf{k} \times \mathbf{j}) = \mathbf{i}$$
$$\mathbf{k} \times \mathbf{i} = -(\mathbf{i} \times \mathbf{k}) = \mathbf{j} \qquad \mathbf{i} \times \mathbf{i} = \mathbf{j} \times \mathbf{j} = \mathbf{k} \times \mathbf{k} = \mathbf{0}$$

What is missing so far is an efficient method for finding the components of the cross product of two vectors in $\mathbb{R}^3$. Let $\mathbf{u} = u_1\mathbf{i} + u_2\mathbf{j} + u_3\mathbf{k}$ and $\mathbf{v} = v_1\mathbf{i} + v_2\mathbf{j} + v_3\mathbf{k}$. Using the distributive properties of the cross product (Theorem 13.4), we have

$$\mathbf{u} \times \mathbf{v} = (u_1\mathbf{i} + u_2\mathbf{j} + u_3\mathbf{k}) \times (v_1\mathbf{i} + v_2\mathbf{j} + v_3\mathbf{k})$$
$$= u_1v_1\underbrace{(\mathbf{i} \times \mathbf{i})}_{0} + u_1v_2\underbrace{(\mathbf{i} \times \mathbf{j})}_{\mathbf{k}} + u_1v_3\underbrace{(\mathbf{i} \times \mathbf{k})}_{-\mathbf{j}}$$
$$+ u_2v_1\underbrace{(\mathbf{j} \times \mathbf{i})}_{-\mathbf{k}} + u_2v_2\underbrace{(\mathbf{j} \times \mathbf{j})}_{0} + u_2v_3\underbrace{(\mathbf{j} \times \mathbf{k})}_{\mathbf{i}}$$
$$+ u_3v_1\underbrace{(\mathbf{k} \times \mathbf{i})}_{\mathbf{j}} + u_3v_2\underbrace{(\mathbf{k} \times \mathbf{j})}_{-\mathbf{i}} + u_3v_3\underbrace{(\mathbf{k} \times \mathbf{k})}_{0}.$$

> The determinant of the matrix A is denoted both $|A|$ and det A. The formula for the determinant of a 3×3 matrix A is
> $$\begin{vmatrix} a_1 & a_2 & a_3 \\ b_1 & b_2 & b_3 \\ c_1 & c_2 & c_3 \end{vmatrix} = a_1\begin{vmatrix} b_2 & b_3 \\ c_2 & c_3 \end{vmatrix} - a_2\begin{vmatrix} b_1 & b_3 \\ c_1 & c_3 \end{vmatrix}$$
> $$+ a_3\begin{vmatrix} b_1 & b_2 \\ c_1 & c_2 \end{vmatrix},$$
> where
> $$\begin{vmatrix} a & b \\ c & d \end{vmatrix} = ad - bc.$$

This formula looks impossible to remember until we see that it fits the pattern used to evaluate 3×3 determinants. Specifically, if we compute the determinant of the matrix

$$\begin{array}{cc} \text{Unit vectors} & \rightarrow \\ \text{Components of } \mathbf{u} & \rightarrow \\ \text{Components of } \mathbf{v} & \rightarrow \end{array} \begin{pmatrix} \mathbf{i} & \mathbf{j} & \mathbf{k} \\ u_1 & u_2 & u_3 \\ v_1 & v_2 & v_3 \end{pmatrix}$$

(expanding about the first row), the following formula for the cross product emerges (see margin note).

THEOREM 13.6   Evaluating the Cross Product
Let $\mathbf{u} = u_1\mathbf{i} + u_2\mathbf{j} + u_3\mathbf{k}$ and $\mathbf{v} = v_1\mathbf{i} + v_2\mathbf{j} + v_3\mathbf{k}$. Then

$$\mathbf{u} \times \mathbf{v} = \begin{vmatrix} \mathbf{i} & \mathbf{j} & \mathbf{k} \\ u_1 & u_2 & u_3 \\ v_1 & v_2 & v_3 \end{vmatrix} = \begin{vmatrix} u_2 & u_3 \\ v_2 & v_3 \end{vmatrix}\mathbf{i} - \begin{vmatrix} u_1 & u_3 \\ v_1 & v_3 \end{vmatrix}\mathbf{j} + \begin{vmatrix} u_1 & u_2 \\ v_1 & v_2 \end{vmatrix}\mathbf{k}.$$

$\overrightarrow{OP} = \langle 2, 3, 4 \rangle$

$P(2, 3, 4)$

$Q(3, 2, 0)$

$\overrightarrow{OQ} = \langle 3, 2, 0 \rangle$

Area of parallelogram
$= |\overrightarrow{OP} \times \overrightarrow{OQ}|$.

Area of triangle
$= \frac{1}{2}|\overrightarrow{OP} \times \overrightarrow{OQ}|$.

Figure 13.60

EXAMPLE 3   Area of a triangle Find the area of the triangle with vertices $O(0, 0, 0)$, $P(2, 3, 4)$, and $Q(3, 2, 0)$ (Figure 13.60).

SOLUTION First consider the parallelogram, two of whose sides are the vectors $\overrightarrow{OP}$ and $\overrightarrow{OQ}$. By Theorem 13.3, the area of this parallelogram is $|\overrightarrow{OP} \times \overrightarrow{OQ}|$. Computing the cross product, we find that

$$\overrightarrow{OP} \times \overrightarrow{OQ} = \begin{vmatrix} \mathbf{i} & \mathbf{j} & \mathbf{k} \\ 2 & 3 & 4 \\ 3 & 2 & 0 \end{vmatrix} = \begin{vmatrix} 3 & 4 \\ 2 & 0 \end{vmatrix}\mathbf{i} - \begin{vmatrix} 2 & 4 \\ 3 & 0 \end{vmatrix}\mathbf{j} + \begin{vmatrix} 2 & 3 \\ 3 & 2 \end{vmatrix}\mathbf{k}$$

$$= -8\mathbf{i} + 12\mathbf{j} - 5\mathbf{k}.$$

Therefore, the area of the parallelogram is

$$|\overrightarrow{OP} \times \overrightarrow{OQ}| = |-8\mathbf{i} + 12\mathbf{j} - 5\mathbf{k}| = \sqrt{233} \approx 15.26.$$

The triangle with vertices O, P, and Q forms half of the parallelogram, so its area is $\sqrt{233}/2 \approx 7.63$.

Related Exercises 34–36 ◄

EXAMPLE 4 Vector orthogonal to two vectors Find a vector orthogonal to the two vectors $\mathbf{u} = -\mathbf{i} + 6\mathbf{k}$ and $\mathbf{v} = 2\mathbf{i} - 5\mathbf{j} - 3\mathbf{k}$.

SOLUTION A vector orthogonal to $\mathbf{u}$ and $\mathbf{v}$ is parallel to $\mathbf{u} \times \mathbf{v}$ (**Figure 13.61**). One such orthogonal vector is

$$\mathbf{u} \times \mathbf{v} = \begin{vmatrix} \mathbf{i} & \mathbf{j} & \mathbf{k} \\ -1 & 0 & 6 \\ 2 & -5 & -3 \end{vmatrix}$$

$$= (0 + 30)\mathbf{i} - (3 - 12)\mathbf{j} + (5 - 0)\mathbf{k}$$

$$= 30\mathbf{i} + 9\mathbf{j} + 5\mathbf{k}.$$

Any scalar multiple of this vector is also orthogonal to $\mathbf{u}$ and $\mathbf{v}$.

Related Exercises 42–44 ◄

Figure 13.61

QUICK CHECK 3 A good check on a cross product calculation is to verify that $\mathbf{u}$ and $\mathbf{v}$ are orthogonal to the computed $\mathbf{u} \times \mathbf{v}$. In Example 4, verify that $\mathbf{u} \cdot (\mathbf{u} \times \mathbf{v}) = 0$ and $\mathbf{v} \cdot (\mathbf{u} \times \mathbf{v}) = 0$. ◄

Applications of the Cross Product

We now investigate two physical applications of the cross product.

Torque Returning to the example of applying a force to a wrench, suppose a force $\mathbf{F}$ is applied to the point P at the head of a vector $\mathbf{r} = \overrightarrow{OP}$ (**Figure 13.62**). The **torque**, or twisting effect, produced by the force about the point O is given by $\boldsymbol{\tau} = \mathbf{r} \times \mathbf{F}$. The torque vector has a magnitude of

$$|\boldsymbol{\tau}| = |\mathbf{r} \times \mathbf{F}| = |\mathbf{r}||\mathbf{F}| \sin \theta,$$

where θ is the angle between $\mathbf{r}$ and $\mathbf{F}$. The direction of the torque is given by the right-hand rule; it is orthogonal to both $\mathbf{r}$ and $\mathbf{F}$. As noted earlier, if $\mathbf{r}$ and $\mathbf{F}$ are parallel, then $\sin \theta = 0$ and the torque is zero. For a given $\mathbf{r}$ and $\mathbf{F}$, the maximum torque occurs when $\mathbf{F}$ is applied in a direction orthogonal to $\mathbf{r}$ ($\theta = \pi/2$).

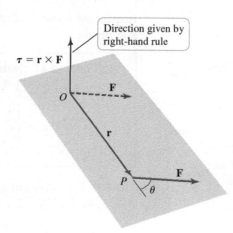

Figure 13.62

EXAMPLE 5   Tightening a bolt A force of 20 N is applied to a wrench attached to a bolt in a direction perpendicular to the bolt (**Figure 13.63**). Which produces more torque: applying the force at an angle of 60° on a wrench that is 0.15 m long or applying the force at an angle of 135° on a wrench that is 0.25 m long? In each case, what is the direction of the torque?

▶ When standard threads are added to the bolt in Figure 13.63, the forces used in Example 5 cause the bolt to move upward into a nut—in the direction of the torque.

(a)   (b)

Figure 13.63

SOLUTION The magnitude of the torque in the first case is

$$|\boldsymbol{\tau}| = |\mathbf{r}||\mathbf{F}| \sin \theta = (0.15 \text{ m})(20 \text{ N}) \sin 60° \approx 2.6 \text{ N-m}.$$

In the second case, the magnitude of the torque is

$$|\boldsymbol{\tau}| = |\mathbf{r}||\mathbf{F}| \sin \theta = (0.25 \text{ m})(20 \text{ N}) \sin 135° \approx 3.5 \text{ N-m}.$$

The second instance gives the greater torque. In both cases, the torque is orthogonal to **r** and **F**, parallel to the shaft of the bolt (Figure 13.63).

Related Exercises 47, 49 ◀

Magnetic Force on a Moving Charge Moving electric charges (either an isolated charge or a current in a wire) experience a force when they pass through a magnetic field. For an isolated charge q, the force is given by $\mathbf{F} = q(\mathbf{v} \times \mathbf{B})$, where **v** is the velocity of the charge and **B** is the magnetic field. The magnitude of the force is

$$|\mathbf{F}| = |q||\mathbf{v} \times \mathbf{B}| = |q||\mathbf{v}||\mathbf{B}| \sin \theta,$$

where θ is the angle between **v** and **B** (**Figure 13.64**). Note that the sign of the charge also determines the direction of the force. If the velocity vector is parallel to the magnetic field, the charge experiences no force. The maximum force occurs when the velocity is orthogonal to the magnetic field.

Path of charged particle

F is orthogonal to **v** and **B**.

Figure 13.64

The force $\mathbf{F} = q\mathbf{v} \times \mathbf{B}$ is orthogonal to **v** and **B** at all points and holds the proton in a circular trajectory.

EXAMPLE 6   Force on a proton A proton with a mass of 1.7×10^{-27} kg and a charge of $q = +1.6 \times 10^{-19}$ coulombs (C) moves along the x-axis with a speed of $|\mathbf{v}| = 9 \times 10^5$ m/s. When it reaches $(0, 0, 0)$, a uniform magnetic field is turned on. The field has a constant strength of 1 tesla (1 T) and is directed along the negative z-axis (**Figure 13.65**).

a. Find the magnitude and direction of the force on the proton at the instant it enters the magnetic field.

b. Assume the proton loses no energy and the force in part (a) acts as a *centripetal force* with magnitude $|\mathbf{F}| = m|\mathbf{v}|^2/R$ that keeps the proton in a circular orbit of radius R. Find the radius of the orbit.

Figure 13.65

▶ The standard unit of magnetic field strength is the tesla (T, named after Nicola Tesla). A typical strong bar magnet has a strength of about 1 T. In terms of other units, 1 T = 1 kg/(C-s), where C is the unit of charge called the *coulomb.* Therefore, the units of force in Example 6a are kg-m/s², or newtons.

SOLUTION

a. Expressed as vectors, we have $\mathbf{v} = 9 \times 10^5\,\mathbf{i}$ and $\mathbf{B} = -\mathbf{k}$. Therefore, the force on the proton in newtons is

$$\mathbf{F} = q(\mathbf{v} \times \mathbf{B}) = 1.6 \times 10^{-19}\big((9 \times 10^5\,\mathbf{i}) \times (-\mathbf{k})\big)$$
$$= 1.44 \times 10^{-13}\mathbf{j}.$$

As shown in Figure 13.65, when the proton enters the magnetic field in the positive *x*-direction, the force acts in the positive *y*-direction, which changes the path of the proton.

b. The magnitude of the force acting on the proton remains 1.44×10^{-13} N at all times (from part (a)). Equating this force to the centripetal force $|\mathbf{F}| = m|\mathbf{v}|^2/R$, we find that

$$R = \frac{m|\mathbf{v}|^2}{|\mathbf{F}|} = \frac{(1.7 \times 10^{-27}\,\text{kg})\,(9 \times 10^5\,\text{m/s})^2}{1.44 \times 10^{-13}\,\text{N}} \approx 0.01\,\text{m}.$$

Assuming no energy loss, the proton moves in a circular orbit of radius 0.01 m.

Related Exercises 55, 67 ◀

SECTION 13.4 EXERCISES

Getting Started

1. What is the magnitude of the cross product of two parallel vectors?

2. If **u** and **v** are orthogonal, what is the magnitude of $\mathbf{u} \times \mathbf{v}$?

3. Suppose **u** and **v** are nonzero vectors. What is the geometric relationship between **u** and **v** under each of the following conditions?

   a. $\mathbf{u} \cdot \mathbf{v} = 0$ **b.** $\mathbf{u} \times \mathbf{v} = \mathbf{0}$

4. Use a geometric argument to explain why $\mathbf{u} \cdot (\mathbf{u} \times \mathbf{v}) = 0$.

5. Compute $|\mathbf{u} \times \mathbf{v}|$ if **u** and **v** are unit vectors and the angle between them is $\pi/4$.

6. Compute $|\mathbf{u} \times \mathbf{v}|$ if $|\mathbf{u}| = 3$ and $|\mathbf{v}| = 4$ and the angle between **u** and **v** is $2\pi/3$.

7. Find $\mathbf{v} \times \mathbf{u}$ if $\mathbf{u} \times \mathbf{v} = 3\mathbf{i} + 2\mathbf{j} - 7\mathbf{k}$.

8. For any vector **v** in $\mathbb{R}^3$, explain why $\mathbf{v} \times \mathbf{v} = \mathbf{0}$.

9. Explain how to use a determinant to compute $\mathbf{u} \times \mathbf{v}$.

10. Explain how to find the torque produced by a force using cross products.

Practice Exercises

11–12. Cross products from the definition *Find the cross product* $\mathbf{u} \times \mathbf{v}$ *in each figure.*

11.

$\mathbf{u} = \langle 3, 0, 0 \rangle$ $\mathbf{v} = \langle 0, 5, 0 \rangle$

12.

$\mathbf{v} = \langle 0, 0, 2 \rangle$ $\mathbf{u} = \langle -4, 0, 0 \rangle$

13–16. Cross products from the definition *Sketch the following vectors* **u** *and* **v**. *Then compute* $|\mathbf{u} \times \mathbf{v}|$ *and show the cross product on your sketch.*

13. $\mathbf{u} = \langle 0, -2, 0 \rangle, \mathbf{v} = \langle 0, 1, 0 \rangle$

14. $\mathbf{u} = \langle 0, 4, 0 \rangle, \mathbf{v} = \langle 0, 0, -8 \rangle$

15. $\mathbf{u} = \langle 3, 3, 0 \rangle, \mathbf{v} = \langle 3, 3, 3\sqrt{2} \rangle$

16. $\mathbf{u} = \langle 0, -2, -2 \rangle, \mathbf{v} = \langle 0, 2, -2 \rangle$

17–22. Coordinate unit vectors *Compute the following cross products. Then make a sketch showing the two vectors and their cross product.*

17. $\mathbf{j} \times \mathbf{k}$ 18. $\mathbf{i} \times \mathbf{k}$ 19. $-\mathbf{j} \times \mathbf{k}$

20. $3\mathbf{j} \times \mathbf{i}$ 21. $-2\mathbf{i} \times 3\mathbf{k}$ 22. $2\mathbf{j} \times (-5)\mathbf{i}$

23–28. Computing cross products *Find the cross products* $\mathbf{u} \times \mathbf{v}$ *and* $\mathbf{v} \times \mathbf{u}$ *for the following vectors* **u** *and* **v**.

23. $\mathbf{u} = \langle 3, 5, 0 \rangle, \mathbf{v} = \langle 0, 3, -6 \rangle$

24. $\mathbf{u} = \langle -4, 1, 1 \rangle, \mathbf{v} = \langle 0, 1, -1 \rangle$

25. $\mathbf{u} = \langle 2, 3, -9 \rangle, \mathbf{v} = \langle -1, 1, -1 \rangle$

26. $\mathbf{u} = \langle 3, -4, 6 \rangle, \mathbf{v} = \langle 1, 2, -1 \rangle$

27. $\mathbf{u} = 3\mathbf{i} - \mathbf{j} - 2\mathbf{k}, \mathbf{v} = \mathbf{i} + 3\mathbf{j} - 2\mathbf{k}$

28. $\mathbf{u} = 2\mathbf{i} - 10\mathbf{j} + 15\mathbf{k}, \mathbf{v} = 0.5\mathbf{i} + \mathbf{j} - 0.6\mathbf{k}$

29–32. Area of a parallelogram *Find the area of the parallelogram that has two adjacent sides* **u** *and* **v**.

29. $\mathbf{u} = 3\mathbf{i} - \mathbf{j}, \mathbf{v} = 3\mathbf{j} + 2\mathbf{k}$

30. $\mathbf{u} = -3\mathbf{i} + 2\mathbf{k}, \mathbf{v} = \mathbf{i} + \mathbf{j} + \mathbf{k}$

31. $\mathbf{u} = 2\mathbf{i} - \mathbf{j} - 2\mathbf{k}, \mathbf{v} = 3\mathbf{i} + 2\mathbf{j} - \mathbf{k}$

32. $\mathbf{u} = 8\mathbf{i} + 2\mathbf{j} - 3\mathbf{k}, \mathbf{v} = 2\mathbf{i} + 4\mathbf{j} - 4\mathbf{k}$

33–38. Areas of triangles *Find the area of the following triangles T.*

33. The vertices of T are $A(0, 0, 0)$, $B(3, 0, 1)$, and $C(1, 1, 0)$.

34. The vertices of T are $O(0, 0, 0)$, $P(1, 2, 3)$, and $Q(6, 5, 4)$.

35. The vertices of T are $A(5, 6, 2)$, $B(7, 16, 4)$, and $C(6, 7, 3)$.

36. The vertices of T are $A(-1, -5, -3)$, $B(-3, -2, -1)$, and $C(0, -5, -1)$.

37. The sides of T are $\mathbf{u} = \langle 3, 3, 3 \rangle$, $\mathbf{v} = \langle 6, 0, 6 \rangle$, and $\mathbf{u} - \mathbf{v}$.

38. The sides of T are $\mathbf{u} = \langle 0, 6, 0 \rangle$, $\mathbf{v} = \langle 4, 4, 4 \rangle$, and $\mathbf{u} - \mathbf{v}$.

39. Collinear points and cross products Explain why the points A, B, and C in $\mathbb{R}^3$ are collinear if and only if $\vec{AB} \times \vec{AC} = \mathbf{0}$.

40–41. Collinear points *Use cross products to determine whether the points A, B, and C are collinear.*

40. $A(3, 2, 1)$, $B(5, 4, 7)$, and $C(9, 8, 19)$

41. $A(-3, -2, 1)$, $B(1, 4, 7)$, and $C(4, 10, 14)$

42–44. Orthogonal vectors *Find a vector orthogonal to the given vectors.*

42. $\langle 1, 2, 3 \rangle$ and $\langle -2, 4, -1 \rangle$

43. $\langle 0, 1, 2 \rangle$ and $\langle -2, 0, 3 \rangle$

44. $\langle 8, 0, 4 \rangle$ and $\langle -8, 2, 1 \rangle$

45–48. Computing torque *Answer the following questions about torque.*

45. Let $\mathbf{r} = \vec{OP} = \mathbf{i} + \mathbf{j} + \mathbf{k}$. A force $\mathbf{F} = \langle 20, 0, 0 \rangle$ is applied at P. Find the torque about O that is produced.

46. Let $\mathbf{r} = \vec{OP} = \mathbf{i} - \mathbf{j} + 2\mathbf{k}$. A force $\mathbf{F} = \langle 10, 10, 0 \rangle$ is applied at P. Find the torque about O that is produced.

47. Let $\mathbf{r} = \vec{OP} = 10\mathbf{i}$. Which is greater (in magnitude): the torque about O when a force $\mathbf{F} = 5\mathbf{i} - 5\mathbf{k}$ is applied at P or the torque about O when a force $\mathbf{F} = 4\mathbf{i} - 3\mathbf{j}$ is applied at P?

48. A pump handle has a pivot at $(0, 0, 0)$ and extends to $P(5, 0, -5)$. A force $\mathbf{F} = \langle 1, 0, -10 \rangle$ is applied at P. Find the magnitude and direction of the torque about the pivot.

49. Tightening a bolt Suppose you apply a force of 20 N to a 0.25-meter-long wrench attached to a bolt in a direction perpendicular to the bolt. Determine the magnitude of the torque when the force is applied at an angle of 45° to the wrench.

50. Opening a laptop A force of 1.5 lb is applied in a direction perpendicular to the screen of a laptop at a distance of 10 in from the hinge of the screen. Find the magnitude of the torque (in ft-lb) that you apply.

51. Bicycle brakes A set of caliper brakes exerts a force on the rim of a bicycle wheel that creates a frictional force **F** of 40 N perpendicular to the radius of the wheel (see figure). Assuming the wheel

has a radius of 33 cm, find the magnitude and direction of the torque about the axle of the wheel.

52. Arm torque A horizontally outstretched arm supports a weight of 20 lb in a hand (see figure). If the distance from the shoulder to the elbow is 1 ft and the distance from the elbow to the hand is 1 ft, find the magnitude and describe the direction of the torque about (a) the shoulder and (b) the elbow. (The units of torque in this case are ft-lb.)

53–56. Force on a moving charge *Answer the following questions about force on a moving charge.*

53. A particle with a unit positive charge ($q = 1$) enters a constant magnetic field $\mathbf{B} = \mathbf{i} + \mathbf{j}$ with a velocity $\mathbf{v} = 20\mathbf{k}$. Find the magnitude and direction of the force on the particle. Make a sketch of the magnetic field, the velocity, and the force.

54. A particle with a unit negative charge ($q = -1$) enters a constant magnetic field $\mathbf{B} = 5\mathbf{k}$ with a velocity $\mathbf{v} = \mathbf{i} + 2\mathbf{j}$. Find the magnitude and direction of the force on the particle. Make a sketch of the magnetic field, the velocity, and the force.

55. An electron ($q = -1.6 \times 10^{-19}$ C) enters a constant 2-T magnetic field at an angle of 45° to the field with a speed of 2×10^5 m/s. Find the magnitude of the force on the electron.

56. A proton ($q = 1.6 \times 10^{-19}$ C) with velocity $2 \times 10^6 \mathbf{j}$ m/s experiences a force in newtons of $\mathbf{F} = 5 \times 10^{-12} \mathbf{k}$ as it passes through the origin. Find the magnitude and direction of the magnetic field at that instant.

57. Explain why or why not Determine whether the following statements are true and give an explanation or counterexample.

    a. The cross product of two nonzero vectors is a nonzero vector.
    b. $|\mathbf{u} \times \mathbf{v}|$ is less than both $|\mathbf{u}|$ and $|\mathbf{v}|$.
    c. If **u** points east and **v** points south, then $\mathbf{u} \times \mathbf{v}$ points west.
    d. If $\mathbf{u} \times \mathbf{v} = \mathbf{0}$ and $\mathbf{u} \cdot \mathbf{v} = 0$, then either $\mathbf{u} = \mathbf{0}$ or $\mathbf{v} = \mathbf{0}$.
    e. Law of Cancellation? If $\mathbf{u} \times \mathbf{v} = \mathbf{u} \times \mathbf{w}$, then $\mathbf{v} = \mathbf{w}$.

58. Finding an unknown Find the value of a such that $\langle a, a, 2 \rangle \times \langle 1, a, 3 \rangle = \langle 2, -4, 2 \rangle$.

59. Vector equation Find all vectors **u** that satisfy the equation

$$\langle 1, 1, 1 \rangle \times \mathbf{u} = \langle -1, -1, 2 \rangle.$$

60. Vector equation Find all vectors **u** that satisfy the equation

$$\langle 1, 1, 1 \rangle \times \mathbf{u} = \langle 0, 0, 1 \rangle.$$

61. Area of a triangle Find the area of the triangle with vertices on the coordinate axes at the points $(a, 0, 0)$, $(0, b, 0)$, and $(0, 0, c)$, in terms of a, b, and c.

Explorations and Challenges

62–66. Scalar triple product *Another operation with vectors is the* **scalar triple product**, *defined to be* $\mathbf{u} \cdot (\mathbf{v} \times \mathbf{w})$, *for nonzero vectors* **u**, **v**, *and* **w** *in* $\mathbb{R}^3$.

62. Express **u**, **v**, and **w** in terms of their components, and show that $\mathbf{u} \cdot (\mathbf{v} \times \mathbf{w})$ equals the determinant

$$\begin{vmatrix} u_1 & u_2 & u_3 \\ v_1 & v_2 & v_3 \\ w_1 & w_2 & w_3 \end{vmatrix}.$$

63. Consider the *parallelepiped* (slanted box) determined by the position vectors **u**, **v**, and **w** (see figure). Show that the volume of the parallelepiped is $|\mathbf{u} \cdot (\mathbf{v} \times \mathbf{w})|$, the absolute value of the scalar triple product.

64. Find the volume of the parallelepiped determined by the position vectors $\mathbf{u} = \langle 3, 1, 0 \rangle$, $\mathbf{v} = \langle 2, 4, 1 \rangle$, and $\mathbf{w} = \langle 1, 1, 5 \rangle$ (see Exercise 63).

65. Explain why the position vectors **u**, **v**, and **w** are coplanar if and only if $|\mathbf{u} \cdot (\mathbf{v} \times \mathbf{w})| = 0$. (*Hint:* See Exercise 63).

66. Prove that $\mathbf{u} \cdot (\mathbf{v} \times \mathbf{w}) = (\mathbf{u} \times \mathbf{v}) \cdot \mathbf{w}$.

67. Electron speed An electron with a mass of 9.1×10^{-31} kg and a charge of -1.6×10^{-19} C travels in a circular path with no loss of energy in a magnetic field of 0.05 T that is orthogonal to the path of the electron (see figure). If the radius of the path is 0.002 m, what is the speed of the electron?

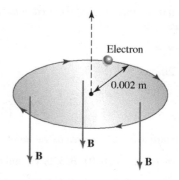

68. Three proofs Prove that $\mathbf{u} \times \mathbf{u} = \mathbf{0}$ in three ways.
   a. Use the definition of the cross product.
   b. Use the determinant formulation of the cross product.
   c. Use the property that $\mathbf{u} \times \mathbf{v} = -(\mathbf{v} \times \mathbf{u})$.

69. Associative property Prove in two ways that for scalars a and b, $(a\mathbf{u}) \times (b\mathbf{v}) = ab(\mathbf{u} \times \mathbf{v})$. Use the definition of the cross product and the determinant formula.

70–72. Possible identities *Determine whether the following statements are true using a proof or counterexample. Assume* **u**, **v**, *and* **w** *are nonzero vectors in* $\mathbb{R}^3$.

70. $\mathbf{u} \times (\mathbf{u} \times \mathbf{v}) = \mathbf{0}$

71. $(\mathbf{u} - \mathbf{v}) \times (\mathbf{u} + \mathbf{v}) = 2\mathbf{u} \times \mathbf{v}$

72. $\mathbf{u} \cdot (\mathbf{v} \times \mathbf{w}) = \mathbf{w} \cdot (\mathbf{u} \times \mathbf{v})$

73–74. Identities *Prove the following identities. Assume* **u**, **v**, **w**, *and* **x** *are nonzero vectors in* $\mathbb{R}^3$.

73. $\mathbf{u} \times (\mathbf{v} \times \mathbf{w}) = (\mathbf{u} \cdot \mathbf{w})\mathbf{v} - (\mathbf{u} \cdot \mathbf{v})\mathbf{w}$ Vector triple product

74. $(\mathbf{u} \times \mathbf{v}) \cdot (\mathbf{w} \times \mathbf{x}) = (\mathbf{u} \cdot \mathbf{w})(\mathbf{v} \cdot \mathbf{x}) - (\mathbf{u} \cdot \mathbf{x})(\mathbf{v} \cdot \mathbf{w})$

75. Cross product equations Suppose **u** and **v** are nonzero vectors in $\mathbb{R}^3$.
   a. Prove that the equation $\mathbf{u} \times \mathbf{z} = \mathbf{v}$ has a nonzero solution **z** if and only if $\mathbf{u} \cdot \mathbf{v} = 0$. (*Hint:* Take the dot product of both sides with **v**.)
   b. Explain this result geometrically.

QUICK CHECK ANSWERS

1. $\mathbf{u} \times \mathbf{v}$ points in the positive z-direction; $\mathbf{v} \times \mathbf{u}$ points in the negative z-direction.   **2.** The vector $2\mathbf{u}$ points in the same direction as **u**, and the vector $3\mathbf{v}$ points in the same direction as **v**. So the right-hand rule gives the same direction for $2\mathbf{u} \times 3\mathbf{v}$ as it does for $\mathbf{u} \times \mathbf{v}$.   **3.** $\mathbf{u} \cdot (\mathbf{u} \times \mathbf{v}) = \langle -1, 0, 6 \rangle \cdot \langle 30, 9, 5 \rangle = -30 + 0 + 30 = 0$. A similar calculation shows that $\mathbf{v} \cdot (\mathbf{u} \times \mathbf{v}) = 0$. ◄

13.5 Lines and Planes in Space

In Chapter 1, we reviewed the catalog of standard functions and their associated graphs. For example, the graph of a linear equation in two variables ($y = mx + b$) is a line, the graph of a quadratic equation ($y = ax^2 + bx + c$) is a parabola, and both of these graphs lie in the xy-plane (two-dimensional space). Our immediate aim is to begin a similar journey through three-dimensional space. What are the basic geometrical objects in three dimensions, and how do we describe them with equations?

Certainly among the most fundamental objects in three dimensions are the line and plane. In this section, we develop equations for both lines and planes and explore their properties and uses.

Lines in Space

Two distinct points in $\mathbb{R}^3$ determine a unique line. Alternatively, one point and a direction also determine a unique line. We use both of these properties to derive two different descriptions of lines in space: one using parametric equations, and one using vector equations.

Let ℓ be the line passing through the point $P_0(x_0, y_0, z_0)$ parallel to the nonzero vector $\mathbf{v} = \langle a, b, c \rangle$, where P_0 and $\mathbf{v}$ are given. The fixed point P_0 is associated with the position vector $\mathbf{r}_0 = \overrightarrow{OP_0} = \langle x_0, y_0, z_0 \rangle$. We let $P(x, y, z)$ be a variable point on ℓ and let $\mathbf{r} = \overrightarrow{OP} = \langle x, y, z \rangle$ be the position vector associated with P (**Figure 13.66**). Because ℓ is parallel to $\mathbf{v}$, the vector $\overrightarrow{P_0P}$ is also parallel to $\mathbf{v}$; therefore, $\overrightarrow{P_0P} = t\mathbf{v}$, where t is a real number. By vector addition, we see that $\overrightarrow{OP} = \overrightarrow{OP_0} + \overrightarrow{P_0P}$, or $\overrightarrow{OP} = \overrightarrow{OP_0} + t\mathbf{v}$. Expressing these vectors in terms of their components, we obtain a *vector equation* for a line:

Component form $\underbrace{\langle x, y, z \rangle}_{\mathbf{r} = \overrightarrow{OP}} = \underbrace{\langle x_0, y_0, z_0 \rangle}_{\mathbf{r}_0 = \overrightarrow{OP_0}} + t\underbrace{\langle a, b, c \rangle}_{\mathbf{v}}$ or

Vector form $\mathbf{r} = \mathbf{r}_0 + t\mathbf{v}.$

Equating components, we obtain *parametric equations* for a line:

$$x = x_0 + at, \qquad y = y_0 + bt, \qquad z = z_0 + ct, \quad \text{for } -\infty < t < \infty.$$

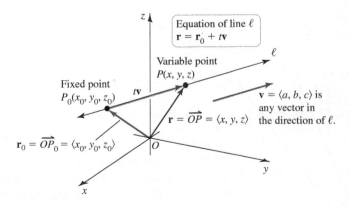

Figure 13.66

QUICK CHECK 1 Describe the line $\mathbf{r} = t\mathbf{k}$, for $-\infty < t < \infty$. Describe the line $\mathbf{r} = t(\mathbf{i} + \mathbf{j} + 0\mathbf{k})$, for $-\infty < t < \infty$. ◄

The parameter t determines the location of points on the line, where $t = 0$ corresponds to P_0. If t increases from 0, we move along the line in the direction of $\mathbf{v}$, and if t decreases from 0, we move along the line in the direction of $-\mathbf{v}$. As t varies over all real numbers ($-\infty < t < \infty$), the entire line ℓ is generated. If, instead of knowing the direction $\mathbf{v}$ of the line, we are given two points $P_0(x_0, y_0, z_0)$ and $P_1(x_1, y_1, z_1)$, then the direction of the line is $\mathbf{v} = \overrightarrow{P_0P_1} = \langle x_1 - x_0, y_1 - y_0, z_1 - z_0 \rangle$.

▶ There are infinitely many equations for the same line. The direction vector is determined only up to a scalar multiple.

Equation of a Line

A **vector equation of the line** passing through the point $P_0(x_0, y_0, z_0)$ in the direction of the vector $\mathbf{v} = \langle a, b, c \rangle$ is $\mathbf{r} = \mathbf{r}_0 + t\mathbf{v}$, or

$$\langle x, y, z \rangle = \langle x_0, y_0, z_0 \rangle + t\langle a, b, c \rangle, \quad \text{for} \quad -\infty < t < \infty.$$

Equivalently, the corresponding **parametric equations of the line** are

$$x = x_0 + at, \quad y = y_0 + bt, \quad z = z_0 + ct, \quad \text{for} \quad -\infty < t < \infty.$$

$\mathbf{r} = \langle 1 + 5t, 2 - 3t, 4 + t \rangle$

$\mathbf{v} = \langle 5, -3, 1 \rangle$

$P_0(1, 2, 4)$

ℓ

x

Projection of line
in xy-plane

$y = -\dfrac{3}{5}x + \dfrac{13}{5}$

Figure 13.67

EXAMPLE 1   Equations of lines   Find a vector equation of the line ℓ that passes through the point $P_0(1, 2, 4)$ in the direction of $\mathbf{v} = \langle 5, -3, 1 \rangle$, and then find the corresponding parametric equations of ℓ.

SOLUTION   We are given $\mathbf{r}_0 = \langle 1, 2, 4 \rangle$. Therefore, an equation of the line is

$$\mathbf{r} = \mathbf{r}_0 + t\mathbf{v} = \langle 1, 2, 4 \rangle + t\langle 5, -3, 1 \rangle = \langle 1 + 5t, 2 - 3t, 4 + t \rangle,$$

for $-\infty < t < \infty$ (**Figure 13.67**). The corresponding parametric equations are

$$x = 1 + 5t, \qquad y = 2 - 3t, \qquad z = 4 + t, \quad \text{for} \quad -\infty < t < \infty.$$

The line is easier to visualize if it is plotted together with its projection in the xy-plane. Setting $z = 0$ (the equation of the xy-plane), parametric equations of the projection line are $x = 1 + 5t$, $y = 2 - 3t$, and $z = 0$. Eliminating t from these equations, an equation of the projection line is $y = -\frac{3}{5}x + \frac{13}{5}$ (**Figure 13.67**).

Related Exercises 11–12 ◀

EXAMPLE 2   Equation of a line and a line segment   Let ℓ be the line that passes through the points $P_0(-3, 5, 8)$ and $P_1(4, 2, -1)$.

a. Find an equation of ℓ.

b. Find parametric equations of the line segment that extends from P_0 to P_1.

SOLUTION

a. The direction of the line is

$$\mathbf{v} = \overrightarrow{P_0P_1} = \langle 4 - (-3), 2 - 5, -1 - 8 \rangle = \langle 7, -3, -9 \rangle.$$

Therefore, with $\mathbf{r}_0 = \langle -3, 5, 8 \rangle$, a vector equation of ℓ is

$$\mathbf{r} = \mathbf{r}_0 + t\mathbf{v}, \quad \text{or}$$
$$\langle x, y, z \rangle = \langle -3, 5, 8 \rangle + t\langle 7, -3, -9 \rangle$$
$$= \langle -3 + 7t, 5 - 3t, 8 - 9t \rangle.$$

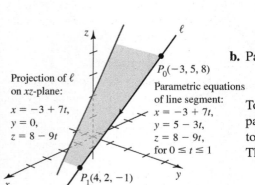

z

ℓ

Projection of ℓ
on xz-plane:

$x = -3 + 7t,$
$y = 0,$
$z = 8 - 9t$

$P_0(-3, 5, 8)$

Parametric equations
of line segment:
$x = -3 + 7t,$
$y = 5 - 3t,$
$z = 8 - 9t,$
for $0 \le t \le 1$

x

$P_1(4, 2, -1)$

y

Figure 13.68

b. Parametric equations for ℓ are

$$x = -3 + 7t, \quad y = 5 - 3t, \quad z = 8 - 9t, \quad \text{for} -\infty < t < \infty.$$

To generate only the line segment from P_0 to P_1, we simply restrict the values of the parameter t. Notice that $t = 0$ corresponds to $P_0(-3, 5, 8)$, and $t = 1$ corresponds to $P_1(4, 2, -1)$. Letting t vary from 0 to 1 generates the line segment from P_0 to P_1. Therefore, parametric equations of the line segment are

$$x = -3 + 7t, \quad y = 5 - 3t, \quad z = 8 - 9t, \quad \text{for } 0 \le t \le 1.$$

The graph of ℓ, which includes the line segment from P_0 to P_1, is shown in **Figure 13.68**, along with the projection of ℓ in the xz-plane. The parametric equations of the projection line are found by setting $y = 0$, which is the equation of the xz-plane.

Related Exercises 16, 29–30 ◀

QUICK CHECK 2   In the equation of the line

$$\langle x, y, z \rangle = \langle x_0, y_0, z_0 \rangle + t\langle x_1 - x_0, y_1 - y_0, z_1 - z_0 \rangle,$$

what value of t corresponds to the point $P_0(x_0, y_0, z_0)$? What value of t corresponds to the point $P_1(x_1, y_1, z_1)$? ◀

Distance from a Point to a Line

Three-dimensional geometry has practical applications in such diverse fields as orbital mechanics, ballistics, computer graphics, and regression analysis. For example, determining the distance from a point to a line is an important calculation in problems such as programming video games. We first derive a formula for this distance and then illustrate how the formula is used to determine whether (virtual) billiard balls collide in a video game.

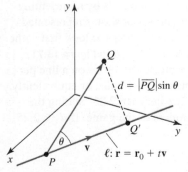

Figure 13.69

Consider a point Q and a line ℓ containing the point P, where ℓ is given by the vector equation $\mathbf{r} = \mathbf{r}_0 + t\mathbf{v}$. Our goal is to find the distance d between Q and ℓ. The geometry of the problem is shown in **Figure 13.69**, where we have placed the tail of $\mathbf{v}$, which is a vector parallel to ℓ, at the point P. We drop a perpendicular from Q to the point Q' on ℓ to form the right triangle PQQ'; the shortest distance d from Q to ℓ is the distance from Q to Q'. From trigonometry, we know that $d = |\overrightarrow{PQ}| \sin\theta$, where θ is the angle between $\mathbf{v}$ and $\overrightarrow{PQ}$. Using the definition of the magnitude of the cross product, we can also write

$$|\mathbf{v} \times \overrightarrow{PQ}| = |\mathbf{v}| \underbrace{|\overrightarrow{PQ}| \sin\theta}_{d} = |\mathbf{v}| d. \quad \text{Cross product definition}$$

Dividing both sides of this equation by $|\mathbf{v}|$ leads to the desired result,

$$d = \frac{|\mathbf{v} \times \overrightarrow{PQ}|}{|\mathbf{v}|}.$$

QUICK CHECK 3 Find the distance between the point $Q(1, 0, 3)$ and the line $\langle x, y, z \rangle = t\langle 2, 1, 2 \rangle$. Note that $P(0, 0, 0)$ lies on the line and $\mathbf{v} = \langle 2, 1, 2 \rangle$ is parallel to the line. ◄

Distance Between a Point and a Line

The distance d between the point Q and the line $\mathbf{r} = \mathbf{r}_0 + t\mathbf{v}$ is

$$d = \frac{|\mathbf{v} \times \overrightarrow{PQ}|}{|\mathbf{v}|},$$

where P is any point on the line and $\mathbf{v}$ is a vector parallel to the line.

The computer program for a billiards video game must keep track of the locations of all the balls on a two-dimensional screen. Although it is possible to write code that tracks each pixel in every ball, it is much simpler to track only the center pixel of each ball. As explained in the next example, the question of whether two balls collide during the game is answered by computing the distance between a point and a line.

EXAMPLE 3   Video game calculation Arria is playing a billiards video game on her iPad. The playing surface is represented in the video game by the rectangle in the first quadrant with opposite corners at the origin and the point $(100, 50)$ (**Figure 13.70**). Suppose the cue ball is located at $P(25, 16)$, and Arria shoots the ball with an angle of 30° above the x-axis, aiming for a target ball located at $Q(75, 46)$. If the cue ball is struck with sufficient force, will it collide with the target ball? Assume the diameter of each ball is 2.25.

Figure 13.70

SOLUTION We assume the video game stores the locations of all the balls by representing each ball with the coordinates of its center, and that the path of a pool shot is represented by the equation of a line. To determine whether two balls collide, it helps to look first at the situation where the cue ball barely touches the target ball. As illustrated in **Figure 13.71**, the balls will *not* collide when the distance between their centers (which lie on a line perpendicular to the line of the shot) is greater than the diameter of the balls, or equivalently, when the distance d between Q and the line of the shot is greater than 2.25. Stating this result in another way leads to a useful test for the programmers of the game: If $d < 2.25$, the balls *will* collide.

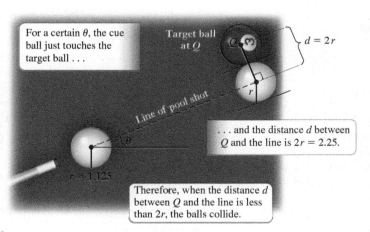

For a certain θ, the cue ball just touches the target ball . . .

Target ball at Q

$d = 2r$

r

Line of pool shot

θ

$r = 1.125$

. . . and the distance d between Q and the line is $2r = 2.25$.

Therefore, when the distance d between Q and the line is less than $2r$, the balls collide.

Figure 13.71

To find d for Arria's attempt, we need a vector parallel to the line of the shot, and we need the vector from the cue ball at $P(25, 16)$ to the target ball at $Q(75, 46)$. Because the cue ball is aimed at an angle $30°$ above the x-axis, a vector parallel to the line of the shot is $\mathbf{v} = \langle \cos 30°, \sin 30° \rangle = \left\langle \dfrac{\sqrt{3}}{2}, \dfrac{1}{2} \right\rangle$. Notice that $\vec{PQ} = \langle 75, 46 \rangle - \langle 25, 16 \rangle = \langle 50, 30 \rangle$. Therefore, the distance between $Q(75, 46)$ and the line of the pool shot is

$$d = \frac{|\mathbf{v} \times \vec{PQ}|}{|\mathbf{v}|} = \frac{\left| \left\langle \dfrac{\sqrt{3}}{2}, \dfrac{1}{2}, 0 \right\rangle \times \langle 50, 30, 0 \rangle \right|}{\left| \left\langle \dfrac{\sqrt{3}}{2}, \dfrac{1}{2}, 0 \right\rangle \right|}.$$

> Example 3 is a two-dimensional problem while the cross product is defined for three-dimensional vectors. Therefore, we must embed the 2D vectors of the example into three dimensions by adding a z-component of 0.

The cross product in the numerator is

$$\begin{vmatrix} \mathbf{i} & \mathbf{j} & \mathbf{k} \\ \sqrt{3}/2 & 1/2 & 0 \\ 50 & 30 & 0 \end{vmatrix} = \left(\frac{\sqrt{3}}{2} 30 - \frac{1}{2} 50 \right) \mathbf{k} \approx 0.98 \mathbf{k}.$$

Because $\mathbf{v}$ is a unit vector, its length is 1, and we find that $d = |0.98\mathbf{k}| = 0.98 < 2.25$; therefore, the balls collide. Additional calculations enable the programmers to determine the directions in which the two balls travel after the collision.

Related Exercise 41 ◄

Determining whether two balls collide in a video game may not seem very important. However, the same mathematical methods used to create realistic video games are employed in flight simulators that train military and civilian pilots. The principles used to design video games are also used to train surgeons and to assist them in performing surgery. Example 4 looks at another crucial calculation used in designing the virtual-world tools that are becoming part of our daily lives: finding points of intersection.

EXAMPLE 4   Points of intersection Determine whether the lines ℓ_1 and ℓ_2 intersect. If they do, find the point of intersection.

a. $\ell_1: x = 2 + 3t, y = 3t, z = 1 - t$ $\ell_2: x = 4 + 2s, y = -3 + 3s, z = -2s$

b. $\ell_1: x = 3 + t, y = 4 - t, z = 5 + 3t$ $\ell_2: x = 2s, y = -1 + 2s, z = 4s$

SOLUTION

a. Let's first check whether the lines are parallel; if they are, there is no point of intersection (unless the lines are identical). Reading the coefficients in the parametric equations for each line, we find that $\mathbf{v}_1 = \langle 3, 3, -1 \rangle$ is parallel to ℓ_1 and that $\mathbf{v}_2 = \langle 2, 3, -2 \rangle$ is parallel to ℓ_2. Because $\mathbf{v}_1$ is not a constant multiple of $\mathbf{v}_2$, the lines are not parallel. In $\mathbb{R}^3$, this fact alone does not guarantee that the lines intersect. Two lines that are neither parallel nor intersecting are said to be **skew**.

> Given two skew lines in $\mathbb{R}^3$, one can always find two parallel planes in which the lines lie: one line in one plane, and the other line in a plane parallel to the first.

Determining whether two lines intersect amounts to solving a system of three linear equations in two variables. We set the x-, y-, and z-components of each line equal to one another, which results in the following system:

$$2 + 3t = 4 + 2s \quad (1) \quad \text{Equate the } x\text{-components.}$$
$$3t = -3 + 3s \quad (2) \quad \text{Equate the } y\text{-components.}$$
$$1 - t = -2s. \quad (3) \quad \text{Equate the } z\text{-components.}$$

When equation (2) is subtracted from equation (1), the result is $2 = 7 - s$, or $s = 5$. Substituting $s = 5$ into equation (1) or (2) yields $t = 4$. However, when these values are substituted into (3), a false statement results, which implies that the system of equations is *inconsistent* and the lines do not intersect. We conclude that ℓ_1 and ℓ_2 are skew.

b. Proceeding as we did in part (a), we note that the lines are not parallel and solve the system

$$3 + t = 2s$$
$$4 - t = -1 + 2s$$
$$5 + 3t = 4s.$$

When the first two equations are added to eliminate the variable t, we find that $s = 2$, which implies that $t = 1$. When these values are substituted into the last equation, a true statement results, which means we have a solution to the system and the lines intersect. To find the point of intersection, we substitute $t = 1$ into the parametric equations for ℓ_1 and arrive at $(4, 3, 8)$. You can verify that when $s = 2$ is substituted into the equations for ℓ_2, the same point of intersection results.

Related Exercises 32–33 ◄

Equations of Planes

Intuitively, a plane is a flat surface with infinite extent in all directions. Three noncollinear points (not all on the same line) determine a unique plane in $\mathbb{R}^3$. A plane in $\mathbb{R}^3$ is also uniquely determined by one point in the plane and any nonzero vector orthogonal (perpendicular) to the plane. Such a vector, called a *normal vector*, specifies the orientation of the plane.

The orientation of a plane is specified by a normal vector **n**. All vectors $\overrightarrow{P_0 P}$ in the plane are orthogonal to **n**.

Figure 13.72

DEFINITION   Plane in $\mathbb{R}^3$

Given a fixed point P_0 and a nonzero **normal vector n**, the set of points P in $\mathbb{R}^3$ for which $\overrightarrow{P_0 P}$ is orthogonal to **n** is called a **plane** (Figure 13.72).

We now derive an equation of the plane passing through the point $P_0(x_0, y_0, z_0)$ with nonzero normal vector $\mathbf{n} = \langle a, b, c \rangle$. Notice that for any point $P(x, y, z)$ in the plane,

▶ Just as the slope determines the orientation of a line in $\mathbb{R}^2$, a normal vector determines the orientation of a plane in $\mathbb{R}^3$.

the vector $\overrightarrow{P_0P} = \langle x - x_0, y - y_0, z - z_0 \rangle$ lies in the plane and is orthogonal to **n**. This orthogonality relationship is written and simplified as follows:

$$\mathbf{n} \cdot \overrightarrow{P_0P} = 0 \qquad \text{Dot product of orthogonal vectors}$$
$$\langle a, b, c \rangle \cdot \langle x - x_0, y - y_0, z - z_0 \rangle = 0 \qquad \text{Substitute vector components.}$$
$$a(x - x_0) + b(y - y_0) + c(z - z_0) = 0 \qquad \text{Expand the dot product.}$$
$$ax + by + cz = d. \qquad d = ax_0 + by_0 + cz_0$$

This important result states that the most general linear equation in three variables, $ax + by + cz = d$, describes a plane in $\mathbb{R}^3$.

▶ A vector $\mathbf{n} = \langle a, b, c \rangle$ is used to describe a *plane* by specifying a direction *orthogonal* to the plane. By contrast, a vector $\mathbf{v} = \langle a, b, c \rangle$ is used to describe a *line* by specifying a direction *parallel* to the line.

General Equation of a Plane in $\mathbb{R}^3$

The plane passing through the point $P_0(x_0, y_0, z_0)$ with a nonzero normal vector $\mathbf{n} = \langle a, b, c \rangle$ is described by the equation

$$a(x - x_0) + b(y - y_0) + c(z - z_0) = 0 \quad \text{or} \quad ax + by + cz = d,$$

where $d = ax_0 + by_0 + cz_0$.

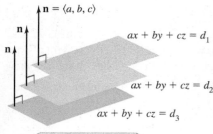

Figure 13.73

The coefficients a, b, and c in the equation of a plane determine the *orientation* of the plane, while the constant term d determines the *location* of the plane. If a, b, and c are held constant and d is varied, a family of parallel planes is generated, all with the same orientation (**Figure 13.73**).

EXAMPLE 5 **Equation of a plane**

a. Find an equation of the plane passing through $P_0(2, -3, 4)$ with a normal vector $\mathbf{n} = \langle -1, 2, 3 \rangle$.

b. Find an equation of the plane passing through $P_0(2, -3, 4)$ that is perpendicular to the line $x = 3 + 2t, y = -4t, z = 1 - 6t$.

SOLUTION

a. Substituting the components of **n** ($a = -1, b = 2$, and $c = 3$) and the coordinates of P_0 ($x_0 = 2, y_0 = -3$, and $z_0 = 4$) into the equation of a plane, we have

$$a(x - x_0) + b(y - y_0) + c(z - z_0) = 0 \qquad \text{General equation of a plane}$$
$$(-1)(x - 2) + 2(y - (-3)) + 3(z - 4) = 0 \qquad \text{Substitute.}$$
$$-x + 2y + 3z = 4. \qquad \text{Simplify.}$$

The plane is shown in **Figure 13.74**.

QUICK CHECK 4 Consider the equation of a plane in the form $\mathbf{n} \cdot \overrightarrow{P_0P} = 0$. Explain why the equation of the plane depends only on the direction, not on the length, of the normal vector **n**. ◀

b. Note that $\mathbf{v} = \langle 2, -4, -6 \rangle$ is parallel to the given line and therefore perpendicular to the plane, so we have a vector normal to the plane. We could carry out calculations similar to those found in part (a) to find the equation of the plane, but here is an easier solution. Observe that **v** is a multiple of the normal vector in part (a) ($\mathbf{v} = -2\mathbf{n}$), and therefore **v** and **n** are parallel, which implies both planes are oriented in the same direction. Because both planes pass through P_0, we conclude that the planes are identical.

Related Exercises 43–44 ◀

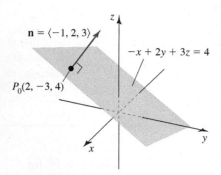

Figure 13.74

EXAMPLE 6 **A plane through three points** Find an equation of the plane that passes through the (noncollinear) points $P(2, -1, 3)$, $Q(1, 4, 0)$, and $R(0, -1, 5)$.

SOLUTION To write an equation of the plane, we must find a normal vector. Because P, Q, and R lie in the plane, the vectors $\overrightarrow{PQ} = \langle -1, 5, -3 \rangle$ and $\overrightarrow{PR} = \langle -2, 0, 2 \rangle$ also lie in the plane. The cross product $\overrightarrow{PQ} \times \overrightarrow{PR}$ is perpendicular to both $\overrightarrow{PQ}$ and $\overrightarrow{PR}$; therefore, a vector normal to the plane is

▶ Three points P, Q, and R determine a plane provided they are not collinear. If P, Q, and R *are* collinear, then the vectors $\overrightarrow{PQ}$ and $\overrightarrow{PR}$ are parallel, which implies that $\overrightarrow{PQ} \times \overrightarrow{PR} = \mathbf{0}$.

$$\mathbf{n} = \overrightarrow{PQ} \times \overrightarrow{PR} = \begin{vmatrix} \mathbf{i} & \mathbf{j} & \mathbf{k} \\ -1 & 5 & -3 \\ -2 & 0 & 2 \end{vmatrix} = 10\mathbf{i} + 8\mathbf{j} + 10\mathbf{k}.$$

QUICK CHECK 5 Verify that in Example 6, the same equation for the plane results if either Q or R is used as the fixed point in the plane. ◄

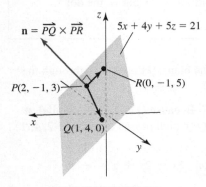

$\overrightarrow{PQ}$ and $\overrightarrow{PR}$ lie in the same plane.
$\overrightarrow{PQ} \times \overrightarrow{PR}$ is orthogonal to the plane.

Figure 13.75

► There is a possibility for confusion here. Working in $\mathbb{R}^3$ with no other restrictions, the equation $-3y - z = 6$ describes a plane that is parallel to the x-axis (because x is unspecified). To make it clear that $-3y - z = 6$ is a line in the yz-plane, the condition $x = 0$ is included.

Two distinct planes are parallel if $\mathbf{n}_1$ and $\mathbf{n}_2$ are parallel.

(a)

Two planes are orthogonal if $\mathbf{n}_1 \cdot \mathbf{n}_2 = 0$.

(b)

Figure 13.77

Any nonzero scalar multiple of $\mathbf{n}$ may be used as the normal vector. Choosing $\mathbf{n} = \langle 5, 4, 5 \rangle$ and $P_0(2, -1, 3)$ as the fixed point in the plane (**Figure 13.75**), an equation of the plane is

$$5(x - 2) + 4(y - (-1)) + 5(z - 3) = 0 \quad \text{or} \quad 5x + 4y + 5z = 21.$$

Using either Q or R as the fixed point in the plane leads to an equivalent equation of the plane.

Related Exercises 49–50 ◄

EXAMPLE 7   Properties of a plane Let Q be the plane described by the equation $2x - 3y - z = 6$.

a. Find a vector normal to Q.

b. Find the points at which Q intersects the coordinate axes and plot Q.

c. Describe the sets of points at which Q intersects the yz-plane, the xz-plane, and the xy-plane.

SOLUTION

a. The coefficients of x, y, and z in the equation of Q are the components of a vector normal to Q. Therefore, a normal vector is $\mathbf{n} = \langle 2, -3, -1 \rangle$ (or any nonzero multiple of $\mathbf{n}$).

b. The point (x, y, z) at which Q intersects the x-axis must have $y = z = 0$. Substituting $y = z = 0$ into the equation of Q gives $x = 3$, so Q intersects the x-axis at $(3, 0, 0)$. Similarly, Q intersects the y-axis at $(0, -2, 0)$, and Q intersects the z-axis at $(0, 0, -6)$. Connecting the three intercepts with straight lines allows us to visualize the plane (**Figure 13.76**).

c. All points in the yz-plane have $x = 0$. Setting $x = 0$ in the equation of Q gives the equation $-3y - z = 6$, which, with the condition $x = 0$, describes a line in the yz-plane. If we set $y = 0$, then Q intersects the xz-plane in the line $2x - z = 6$, where $y = 0$. If $z = 0$, then Q intersects the xy-plane in the line $2x - 3y = 6$, where $z = 0$ (**Figure 13.76**).

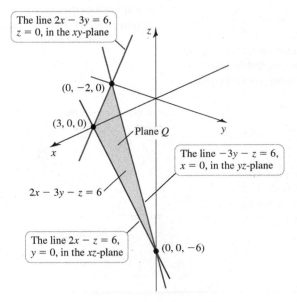

The line $2x - 3y = 6$, $z = 0$, in the xy-plane

$(0, -2, 0)$

$(3, 0, 0)$

Plane Q

The line $-3y - z = 6$, $x = 0$, in the yz-plane

$2x - 3y - z = 6$

The line $2x - z = 6$, $y = 0$, in the xz-plane

$(0, 0, -6)$

Figure 13.76

Related Exercise 61 ◄

Parallel and Orthogonal Planes

The normal vectors of distinct planes tell us about the relative orientation of the planes. Two cases are of particular interest: Two distinct planes may be *parallel* (**Figure 13.77a**) and two intersecting planes may be *orthogonal* (**Figure 13.77b**).

> **DEFINITION Parallel and Orthogonal Planes**
>
> Two distinct planes are **parallel** if their respective normal vectors are parallel (that is, the normal vectors are scalar multiples of each other). Two planes are **orthogonal** if their respective normal vectors are orthogonal (that is, the dot product of the normal vectors is zero).

QUICK CHECK 6 Determine whether the planes $2x - 3y + 6z = 12$ and $6x + 8y + 2z = 1$ are parallel, orthogonal, or neither. ◄

EXAMPLE 8   Parallel planes Find an equation of the plane Q that passes through the point $(-2, 4, 1)$ and is parallel to the plane R: $3x - 2y + z = 4$.

SOLUTION The vector $\mathbf{n} = \langle 3, -2, 1 \rangle$ is normal to R. Because Q and R are parallel, $\mathbf{n}$ is also normal to Q. We conclude that an equation of Q (which passes through $(-2, 4, 1)$ and has a normal vector $\langle 3, -2, 1 \rangle$) is given by

$$3(x + 2) - 2(y - 4) + (z - 1) = 0 \quad \text{or} \quad 3x - 2y + z = -13.$$

Related Exercises 51–52 ◄

EXAMPLE 9   Intersecting planes Find an equation of the line of intersection of the planes Q: $x + 2y + z = 5$ and R: $2x + y - z = 7$.

SOLUTION First note that the vectors normal to the planes, $\mathbf{n}_Q = \langle 1, 2, 1 \rangle$ and $\mathbf{n}_R = \langle 2, 1, -1 \rangle$, are *not* multiples of each other. Therefore, the planes are not parallel and they must intersect in a line; call it ℓ. To find an equation of ℓ, we need two pieces of information: a point on ℓ and a vector pointing in the direction of ℓ. Here is one of several ways to find a point on ℓ. Setting $z = 0$ in the equations of the planes gives equations of the lines in which the planes intersect the xy-plane:

$$x + 2y = 5$$
$$2x + y = 7.$$

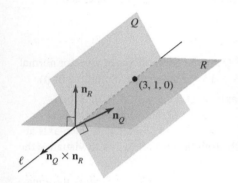

> $\mathbf{n}_Q \times \mathbf{n}_R$ is a vector perpendicular to $\mathbf{n}_Q$ and $\mathbf{n}_R$.
> Line ℓ is perpendicular to $\mathbf{n}_Q$ and $\mathbf{n}_R$.
> Therefore, ℓ and $\mathbf{n}_Q \times \mathbf{n}_R$ are parallel to each other.

Figure 13.78

▶ By setting $z = 0$ and solving the two resulting equations, we find the point that lies on both planes *and* lies in the xy-plane $(z = 0)$.

Solving these equations simultaneously, we find that $x = 3$ and $y = 1$. Combining this result with $z = 0$, we see that $(3, 1, 0)$ is a point on ℓ (**Figure 13.78**).

We next find a vector parallel to ℓ. Because ℓ lies in Q and R, it is orthogonal to the normal vectors $\mathbf{n}_Q$ and $\mathbf{n}_R$. Therefore, the cross product of $\mathbf{n}_Q$ and $\mathbf{n}_R$ is a vector parallel to ℓ (**Figure 13.78**). In this case, the cross product is

$$\mathbf{n}_Q \times \mathbf{n}_R = \begin{vmatrix} \mathbf{i} & \mathbf{j} & \mathbf{k} \\ 1 & 2 & 1 \\ 2 & 1 & -1 \end{vmatrix} = -3\mathbf{i} + 3\mathbf{j} - 3\mathbf{k} = \langle -3, 3, -3 \rangle.$$

An equation of the line ℓ in the direction of the vector $\langle -3, 3, -3 \rangle$ passing through the point $(3, 1, 0)$ is

$$\begin{aligned} \mathbf{r} &= \langle x_0, y_0, z_0 \rangle + t\langle a, b, c \rangle && \text{Equation of a line} \\ &= \langle 3, 1, 0 \rangle + t\langle -3, 3, -3 \rangle && \text{Substitute.} \\ &= \langle 3 - 3t, 1 + 3t, -3t \rangle, && \text{Simplify.} \end{aligned}$$

▶ Any nonzero scalar multiple of $\langle -3, 3, -3 \rangle$ can be used for the direction of ℓ. For example, another equation of ℓ is $\mathbf{r} = \langle 3 + t, 1 - t, t \rangle$.

where $-\infty < t < \infty$. You can check that any point (x, y, z) with $x = 3 - 3t$, $y = 1 + 3t$, and $z = -3t$ satisfies the equations of both planes.

Related Exercise 74 ◄

SECTION 13.5 EXERCISES

Getting Started

1. Find a position vector that is parallel to the line $x = 2 + 4t$, $y = 5 - 8t$, $z = 9t$.

2. Find the parametric equations of the line $\mathbf{r} = \langle 1, 2, 3 \rangle + t\langle 4, 0, -6 \rangle$.

3. Explain how to find a vector in the direction of the line segment from $P_0(x_0, y_0, z_0)$ to $P_1(x_1, y_1, z_1)$.

4. Find the vector equation of the line through the points $P_0(x_0, y_0, z_0)$ and $P_1(x_1, y_1, z_1)$.

5. Determine whether the plane $x + y + z = 9$ and the line $x = t, y = t + 1, z = t + 2$ are parallel, perpendicular, or neither. Be careful.

6. Determine whether the plane $x + y + z = 9$ and the line $x = t, y = -2t + 1, z = t + 2$ are parallel, perpendicular, or neither.

7. Give two pieces of information that, taken together, uniquely determine a plane.

8. Find a vector normal to the plane $-2x - 3y = 12 - 4z$.

9. Where does the plane $-2x - 3y + 4z = 12$ intersect the coordinate axes?

10. Give an equation of the plane with a normal vector $\mathbf{n} = \langle 1, 1, 1 \rangle$ that passes through the point $(1, 0, 0)$.

Practice Exercises

11–26. Equations of lines *Find both the parametric and the vector equations of the following lines.*

11. The line through $(0, 0, 1)$ in the direction of the vector $\mathbf{v} = \langle 4, 7, 0 \rangle$

12. The line through $(-3, 2, -1)$ in the direction of the vector $\mathbf{v} = \langle 1, -2, 0 \rangle$

13. The line through $(0, 0, 1)$ parallel to the y-axis

14. The line through $(-2, 4, 3)$ parallel to the x-axis

15. The line through $(0, 0, 0)$ and $(1, 2, 3)$

16. The line through $(-3, 4, 6)$ and $(5, -1, 0)$

17. The line through $(0, 0, 0)$ that is parallel to the line $\mathbf{r} = \langle 3 - 2t, 5 + 8t, 7 - 4t \rangle$

18. The line through $(1, -3, 4)$ that is parallel to the line $x = 3 + 4t, y = 5 - t, z = 7$

19. The line through $(0, 0, 0)$ that is perpendicular to both $\mathbf{u} = \langle 1, 0, 2 \rangle$ and $\mathbf{v} = \langle 0, 1, 1 \rangle$

20. The line through $(-3, 4, 2)$ that is perpendicular to both $\mathbf{u} = \langle 1, 1, -5 \rangle$ and $\mathbf{v} = \langle 0, 4, 0 \rangle$

21. The line through $(-2, 5, 3)$ that is perpendicular to both $\mathbf{u} = \mathbf{i} + \mathbf{j} - 2\mathbf{k}$ and the x-axis

22. The line through $(0, 2, 1)$ that is perpendicular to both $\mathbf{u} = 4\mathbf{i} + 3\mathbf{j} - 5\mathbf{k}$ and the z-axis

23. The line through $(1, 2, 3)$ that is perpendicular to the lines $x = 3 - 2t, y = 5 + 8t, z = 7 - 4t$ and $x = -2t, y = 5 + t, z = 7 - t$

24. The line through $(1, 0, -1)$ that is perpendicular to the lines $x = 3 + 2t, y = 3t, z = -4t$ and $x = t, y = t, z = -t$

25. The line that is perpendicular to the lines $\mathbf{r} = \langle 4t, 1 + 2t, 3t \rangle$ and $\mathbf{R} = \langle -1 + s, -7 + 2s, -12 + 3s \rangle$, and passes through the point of intersection of the lines $\mathbf{r}$ and $\mathbf{R}$

26. The line that is perpendicular to the lines $\mathbf{r} = \langle -2 + 3t, 2t, 3t \rangle$ and $\mathbf{R} = \langle -6 + s, -8 + 2s, -12 + 3s \rangle$, and passes through the point of intersection of the lines $\mathbf{r}$ and $\mathbf{R}$

27–30. Line segments *Find parametric equations for the line segment joining the first point to the second point.*

27. $(0, 0, 0)$ and $(1, 2, 3)$ **28.** $(1, 0, 1)$ and $(0, -2, 1)$

29. $(2, 4, 8)$ and $(7, 5, 3)$ **30.** $(-1, -8, 4)$ and $(-9, 5, -3)$

31–37. Parallel, intersecting, or skew lines *Determine whether the following pairs of lines are parallel, intersect at a single point, or are skew. If the lines are parallel, determine whether they are the same line (and thus intersect at all points). If the lines intersect at a single point, determine the point of intersection.*

31. $\mathbf{r} = \langle 1, 3, 2 \rangle + t\langle 6, -7, 1 \rangle$; $\mathbf{R} = \langle 10, 6, 14 \rangle + s\langle 3, 1, 4 \rangle$

32. $x = 2t, y = t + 2, z = 3t - 1$ and $x = 5s - 2, y = s + 4,$ $z = 5s + 1$

33. $x = 4, y = 6 - t, z = 1 + t$ and $x = -3 - 7s, y = 1 + 4s,$ $z = 4 - s$

34. $x = 4 + 5t, y = -2t, z = 1 + 3t$ and $x = 10s, y = 6 + 4s,$ $z = 4 + 6s$

35. $x = 1 + 2t, y = 7 - 3t, z = 6 + t$ and $x = -9 + 6t,$ $y = 22 - 9t, z = 1 + 3t$

36. $\mathbf{r} = \langle 3, 1, 0 \rangle + t\langle 4, -6, 4 \rangle$; $\mathbf{R} = \langle 0, 5, 4 \rangle + s\langle -2, 3, -2 \rangle$

37. $\mathbf{r} = \langle 4 + t, -2t, 1 + 3t \rangle$; $\mathbf{R} = \langle 1 - 7s, 6 + 14s, 4 - 21s \rangle$

38. Intersecting lines and colliding particles Consider the lines
$$\mathbf{r} = \langle 2 + 2t, 8 + t, 10 + 3t \rangle \quad \text{and}$$
$$\mathbf{R} = \langle 6 + s, 10 - 2s, 16 - s \rangle.$$

a. Determine whether the lines intersect (have a common point), and if so, find the coordinates of that point.

b. If $\mathbf{r}$ and $\mathbf{R}$ describe the paths of two particles, do the particles collide? Assume $t \geq 0$ and $s \geq 0$ measure time in seconds, and that motion starts at $s = t = 0$.

39–40. Distance from a point to a line *Find the distance between the given point Q and the given line.*

39. $Q(-5, 2, 9)$; $x = 5t + 7, y = 2 - t, z = 12t + 4$

40. $Q(5, 6, 1)$; $x = 1 + 3t, y = 3 - 4t, z = t + 1$

T 41. Billiards shot A cue ball in a billiards video game lies at $P(25, 16)$ (see figure). Refer to Example 3, where we assume the diameter of each ball is 2.25 screen units, and pool balls are represented by the point at their center.

a. The cue ball is aimed at an angle of $58°$ above the negative x-axis toward a target ball at $A(5, 45)$. Do the balls collide?

b. The cue ball is aimed at the point $(50, 25)$ in an attempt to hit a target ball at $B(76, 40)$. Do the balls collide?

c. The cue ball is aimed at an angle θ above the x-axis in the general direction of a target ball at $C(75, 30)$. What range of angles (for $0 \leq \theta \leq \pi/2$) will result in a collision? Express your answer in degrees.

T 42. Bank shot Refer to the figure in Exercise 41. The cue ball lies at $P(25, 16)$ and Jerrod hopes to hit ball $D(59, 17)$; a direct shot isn't an option with other balls blocking the path. Instead, he attempts a bank shot, aiming the cue ball at an angle $45°$ below the x-axis. Will the balls collide? Assume the angles at which the cue ball meets and leaves the bumper are equal and that the diameter of each ball is 2 screen units. (*Hint:* The cue ball will bounce off the bumper when its center hits an "imaginary bumper" one unit above the bumper; see following figure.)

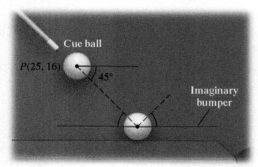

43–58. Equations of planes *Find an equation of the following planes.*

43. The plane passing through the point $P_0(0, 2, -2)$ with a normal vector $\mathbf{n} = \langle 1, 1, -1 \rangle$

44. The plane passing through the point $P_0(2, 3, 0)$ with a normal vector $\mathbf{n} = \langle -1, 2, -3 \rangle$

45. The plane that is parallel to the vectors $\langle 1, 0, 1 \rangle$ and $\langle 0, 2, 1 \rangle$, passing through the point $(1, 2, 3)$

46. The plane that is parallel to the vectors $\langle 1, -3, 1 \rangle$ and $\langle 4, 2, 0 \rangle$, passing through the point $(3, 0, -2)$

47. The plane passing through the origin that is perpendicular to the line $x = t, y = 1 + 4t, z = 7t$

48. The plane passing through the point $(2, -3, 5)$ that is perpendicular to the line $x = 2t, y = 1 + 3t, z = 5 + 4t$

49. The plane passing through the points $(1, 0, 3)$, $(0, 4, 2)$, and $(1, 1, 1)$

50. The plane passing through the points $(2, -1, 4)$, $(1, 1, -1)$, and $(-4, 1, 1)$

51. The plane passing through the point $P_0(1, 0, 4)$ that is parallel to the plane $-x + 2y - 4z = 1$

52. The plane passing through the point $P_0(0, 2, -2)$ that is parallel to the plane $2x + y - z = 1$

53. The plane containing the x-axis and the point $P_0(1, 2, 3)$

54. The plane containing the z-axis and the point $P_0(3, -1, 2)$

55. The plane passing through the origin and containing the line $x = t - 1, y = 2t, z = 3t + 4$

56. The plane passing though the point $P_0(1, -2, 3)$ and containing the line $\mathbf{r} = \langle t, -t, 2t \rangle$

57. The plane passing though the point $P_0(-4, 1, 2)$ and containing the line $\mathbf{r} = \langle 2t - 2, -2t, -4t + 1 \rangle$

58. The plane passing through the origin that contains the line of intersection of the planes $x + y + 2z = 0$ and $x - y = 4$

59. Parallel planes Is the line $x = t + 1, y = 2t + 3, z = 4t + 5$ parallel to the plane $2x - y = -2$? If so, explain why, and then find an equation of the plane containing the line that is parallel to the plane $2x - y = -2$.

60. Do the lines $x = t, y = 2t + 1, z = 3t + 4$ and $x = 2s - 2, y = 2s - 1, z = 3s + 1$ intersect each other at only one point? If so, find a plane that contains both lines.

61–64. Properties of planes *Find the points at which the following planes intersect the coordinate axes, and find equations of the lines where the planes intersect the coordinate planes. Sketch a graph of the plane.*

61. $3x - 2y + z = 6$ **62.** $-4x + 8z = 16$

63. $x + 3y - 5z - 30 = 0$ **64.** $12x - 9y + 4z + 72 = 0$

65–68. Pairs of planes *Determine whether the following pairs of planes are parallel, orthogonal, or neither.*

65. $x + y + 4z = 10$ and $-x - 3y + z = 10$

66. $2x + 2y - 3z = 10$ and $-10x - 10y + 15z = 10$

67. $3x + 2y - 3z = 10$ and $-6x - 10y + z = 10$

68. $3x + 2y + 2z = 10$ and $-6x - 10y + 19z = 10$

69–70. Equations of planes *For the following sets of planes, determine which pairs of planes in the set are parallel, which pairs are orthogonal, and which pairs are identical.*

69. $Q: 3x - 2y + z = 12$; $R: -x + 2y/3 - z/3 = 0$; $S: -x + 2y + 7z = 1$; $T: 3x/2 - y + z/2 = 6$

70. $Q: x + y - z = 0$; $R: y + z = 0$; $S: x - y = 0$; $T: x + y + z = 0$

71–72. Lines normal to planes *Find an equation of the following lines.*

71. The line passing through the point $P_0(2, 1, 3)$ that is normal to the plane $2x - 4y + z = 10$

72. The line passing through the point $P_0(0, -10, -3)$ that is normal to the plane $x + 4z = 2$

73–76. Intersecting planes *Find an equation of the line of intersection of the planes Q and R.*

73. $Q: -x + 2y + z = 1$; $R: x + y + z = 0$

74. $Q: x + 2y - z = 1$; $R: x + y + z = 1$

75. $Q: 2x - y + 3z - 1 = 0$; $R: -x + 3y + z - 4 = 0$

76. $Q: x - y - 2z = 1$; $R: x + y + z = -1$

77–80. Line-plane intersections *Find the point (if it exists) at which the following planes and lines intersect.*

77. $x = 3$ and $\mathbf{r} = \langle t, t, t \rangle$

78. $y = -2$ and $\mathbf{r} = \langle 2t + 1, -t + 4, t - 6 \rangle$

79. $3x + 2y - 4z = -3$ and $x = -2t + 5, y = 3t - 5, z = 4t - 6$

80. $2x - 3y + 3z = 2$ and $x = 3t, y = t, z = -t$

81. Explain why or why not Determine whether the following statements are true and give an explanation or counterexample.

a. The line $\mathbf{r} = \langle 3, -1, 4 \rangle + t \langle 6, -2, 8 \rangle$ passes through the origin.

b. Any two nonparallel lines in $\mathbb{R}^3$ intersect.

c. The plane $x + y + z = 0$ and the line $x = t, y = t, z = t$ are parallel.

d. The vector equations $\mathbf{r} = \langle 1, 2, 3 \rangle + t \langle 1, 1, 1 \rangle$ and $\mathbf{R} = \langle 1, 2, 3 \rangle + t \langle -2, -2, -2 \rangle$ describe the same line.

e. The equations $x + y - z = 1$ and $-x - y + z = 1$ describe the same plane.

f. Any two distinct lines in $\mathbb{R}^3$ determine a unique plane.

g. The vector $\langle -1, -5, 7 \rangle$ is perpendicular to both the line $x = 1 + 5t, y = 3 - t, z = 1$ and the line $x = 7t, y = 3, z = 3 + t$.

82. Distance from a point to a plane Suppose P is a point in the plane $ax + by + cz = d$. The distance from any point Q to the plane equals the length of the orthogonal projection of $\vec{PQ}$ onto a vector $\mathbf{n} = \langle a, b, c \rangle$ normal to the plane. Use this information to show that the distance from Q to the plane is $|\vec{PQ} \cdot \mathbf{n}| / |\mathbf{n}|$.

83. Find the distance from the point $Q(6, -2, 4)$ to the plane $2x - y + 2z = 4$.

84. Find the distance from the point $Q(1, 2, -4)$ to the plane $2x - 5z = 5$.

Explorations and Challenges

85–86. Symmetric equations for a line *If we solve for t in the parametric equations of the line* $x = x_0 + at, y = y_0 + bt, z = z_0 + ct,$ *we obtain the symmetric equations*

$$\frac{x - x_0}{a} = \frac{y - y_0}{b} = \frac{z - z_0}{c},$$

provided a, b, and c do not equal 0.

85. Find symmetric equations of the line $\mathbf{r} = \langle 1, 2, 0 \rangle + t \langle 4, 7, 2 \rangle$.

86. Find parametric and symmetric equations of the line passing through the points $P(1, -2, 3)$ and $Q(2, 3, -1)$.

T 87. Angle between planes The angle between two planes is the smallest angle θ between the normal vectors of the planes, where the directions of the normal vectors are chosen so that $0 \le \theta \le \pi/2$. Find the angle between the planes $5x + 2y - z = 0$ and $-3x + y + 2z = 0$.

88. Intercepts Let a, b, c, and d be constants. Find the points at which the plane $ax + by + cz = d$ intersects the x-, y-, and z-axes.

89. A family of orthogonal planes Find an equation for a family of planes that are orthogonal to the planes $2x + 3y = 4$ and $-x - y + 2z = 8$.

90. Orthogonal plane Find an equation of the plane passing through $(0, -2, 4)$ that is orthogonal to the planes $2x + 5y - 3z = 0$ and $-x + 5y + 2z = 8$.

91. Three intersecting planes Describe the set of all points (if any) at which all three planes $x + 3z = 3$, $y + 4z = 6$, and $x + y + 6z = 9$ intersect.

92. Three intersecting planes Describe the set of all points (if any) at which all three planes $x + 2y + 2z = 3$, $y + 4z = 6$, and $x + 2y + 8z = 9$ intersect.

93. T-shirt profits A clothing company makes a profit of \$10 on its long-sleeved T-shirts and a profit of \$5 on its short-sleeved T-shirts. Assuming there is a \$200 setup cost, the profit on T-shirt sales is $z = 10x + 5y - 200$, where x is the number of long-sleeved T-shirts sold and y is the number of short-sleeved T-shirts sold. Assume x and y are nonnegative.

a. Graph the plane that gives the profit using the window $[0, 40] \times [0, 40] \times [-400, 400]$.

b. If $x = 20$ and $y = 10$, is the profit positive or negative?

c. Describe the values of x and y for which the company breaks even (for which the profit is zero). Mark this set on your graph.

QUICK CHECK ANSWERS

1. The z-axis; the line $y = x$ in the xy-plane   **2.** When $t = 0$, the point on the line is P_0; when $t = 1$, the point on the line is P_1.   **3.** $d = \sqrt{26}/3$ **4.** Because the right side of the equation is 0, the equation can be multiplied by any nonzero constant (changing the length of $\mathbf{n}$) without changing the graph.   **6.** The planes are orthogonal because $\langle 2, -3, 6 \rangle \cdot \langle 6, 8, 2 \rangle = 0$. ◄

13.6 Cylinders and Quadric Surfaces

In Section 13.5, we discovered that lines in three-dimensional space are described by parametric equations (or vector equations) that are linear in the variable. We also saw that planes are described with linear equations in three variables. In this section, we take this progression one step further and investigate the geometry of three-dimensional objects described by quadratic equations in three variables. The result is a collection of *quadric surfaces* that you will encounter frequently throughout the remainder of the text. You saw one such surface in Section 13.2: A sphere with radius a centered at the origin with an equation of $x^2 + y^2 + z^2 = a^2$ is an example of a quadric surface. We also introduce a family of surfaces called *cylinders*, some of which are quadric surfaces.

Cylinders and Traces

In everyday language, we use the word *cylinder* to describe the surface that forms, say, the curved wall of a paint can. In the context of three-dimensional surfaces, the term *cylinder* refers to a surface that is parallel to a line. In this text, we focus on cylinders that are parallel to one of the coordinate axes. Equations for such cylinders are easy to identify: The variable corresponding to the coordinate axis parallel to the cylinder is missing from the equation.

For example, working in $\mathbb{R}^3$, the equation $y = x^2$ does not include z, which means that z is arbitrary and can take on all values. Therefore, $y = x^2$ describes the cylinder consisting of all lines parallel to the z-axis that pass through the parabola $y = x^2$ in the xy-plane (**Figure 13.79a**). In a similar way, the equation $y = z^2$ in $\mathbb{R}^3$ is missing the variable x, so it describes a cylinder parallel to the x-axis. The cylinder consists of lines parallel to the x-axis that pass through the parabola $y = z^2$ in the yz-plane (**Figure 13.79b**).

> ▶ The parabolic cylinder in Figure 13.79a can also be described as the surface swept out by translating the plane curve $y = x^2$ up and down (or parallel to) the z-axis.

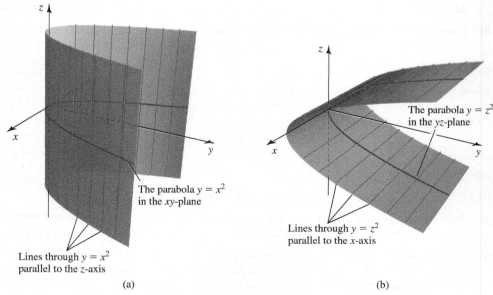

Lines through $y = x^2$
parallel to the z-axis

The parabola $y = x^2$
in the xy-plane

The parabola $y = z^2$
in the yz-plane

Lines through $y = z^2$
parallel to the x-axis

(a)   (b)

Figure 13.79

QUICK CHECK 1 To which coordinate axis in $\mathbb{R}^3$ is the cylinder $z - 2 \ln x = 0$ parallel? To which coordinate axis in $\mathbb{R}^3$ is the cylinder $y = 4z^2 - 1$ parallel? ◀

Graphing surfaces—and cylinders in particular—is facilitated by identifying the *traces* of the surface.

> **DEFINITION Trace**
>
> A **trace** of a surface is the set of points at which the surface intersects a plane that is parallel to one of the coordinate planes. The traces in the coordinate planes are called the **xy-trace**, the **yz-trace**, and the **xz-trace** (**Figure 13.80**).

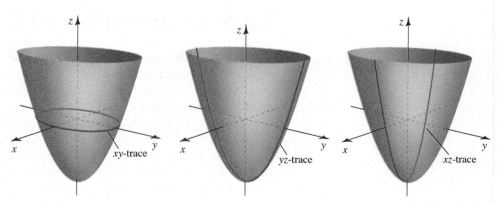

xy-trace yz-trace xz-trace

Figure 13.80

EXAMPLE 1   Graphing cylinders Sketch the graphs of the following cylinders in $\mathbb{R}^3$. Identify the axis to which each cylinder is parallel.

a. $x^2 + 4y^2 = 16$ **b.** $x - \sin z = 0$

SOLUTION

a. As an equation in $\mathbb{R}^3$, the variable z is absent. Therefore, z assumes all real values and the graph is a cylinder consisting of lines parallel to the z-axis passing through the curve $x^2 + 4y^2 = 16$ in the xy-plane. You can sketch the cylinder in the following steps.

1. Rewriting the given equation as $\dfrac{x^2}{4^2} + \dfrac{y^2}{2^2} = 1$, we see that the trace of the cylinder in the xy-plane (the xy-trace) is an ellipse. We begin by drawing this ellipse.

2. Next draw a second trace (a copy of the ellipse in Step 1) in a plane parallel to the xy-plane.

3. Now draw lines parallel to the z-axis through the two traces to fill out the cylinder (**Figure 13.81a**).

    The resulting surface, called an *elliptic cylinder*, runs parallel to the z-axis (**Figure 13.81b**).

b. As an equation in $\mathbb{R}^3$, $x - \sin z = 0$ is missing the variable y. Therefore, y assumes all real values and the graph is a cylinder consisting of lines parallel to the y-axis passing through the curve $x = \sin z$ in the xz-plane. You can sketch the cylinder in the following steps.

1. Graph the curve $x = \sin z$ in the xz-plane, which is the xz-trace of the surface.

2. Draw a second trace (a copy of the curve in Step 1) in a plane parallel to the xz-plane.

3. Draw lines parallel to the y-axis passing through the two traces (**Figure 13.82a**).

    The result is a cylinder, running parallel to the y-axis, consisting of copies of the curve $x = \sin z$ (**Figure 13.82b**).

(a)

(b)

Figure 13.82

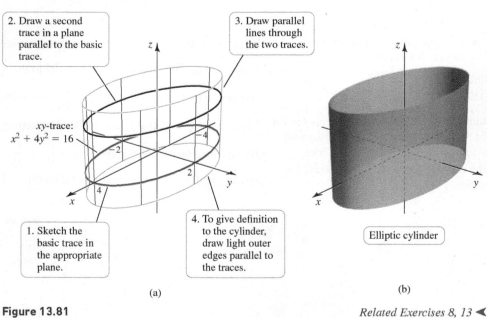

2. Draw a second trace in a plane parallel to the basic trace.

3. Draw parallel lines through the two traces.

xy-trace: $x^2 + 4y^2 = 16$

1. Sketch the basic trace in the appropriate plane.

4. To give definition to the cylinder, draw light outer edges parallel to the traces.

(a)

Elliptic cylinder

(b)

Figure 13.81   *Related Exercises 8, 13* ◄

Quadric Surfaces

Quadric surfaces are described by the general quadratic (second-degree) equation in three variables,

$$Ax^2 + By^2 + Cz^2 + Dxy + Exz + Fyz + Gx + Hy + Iz + J = 0,$$

where the coefficients $A, \ldots, J$ are constants and not all of $A, B, C, D, E,$ and F are zero. We do not attempt a detailed study of this large family of surfaces. However, a few standard surfaces are worth investigating.

Apart from their mathematical interest, quadric surfaces have a variety of practical uses. Paraboloids (defined in Example 3) share the reflective properties of their two-dimensional counterparts (Section 12.4) and are used to design satellite dishes, headlamps, and mirrors in telescopes. Cooling towers for nuclear power plants have the shape of hyperboloids of one sheet. Ellipsoids appear in the design of water tanks and gears.

Making hand sketches of quadric surfaces can be challenging. Here are a few general ideas to keep in mind as you sketch their graphs.

> Working with quadric surfaces requires familiarity with conic sections (Section 12.4).

1. **Intercepts** Determine the points, if any, where the surface intersects the coordinate axes. To find these intercepts, set x, y, and z equal to zero in pairs in the equation of the surface, and solve for the third coordinate.

QUICK CHECK 2 Explain why the elliptic cylinder discussed in Example 1a is a quadric surface. ◄

2. **Traces** As illustrated in the following examples, finding traces of the surface helps visualize the surface. For example, setting $z = 0$ or $z = z_0$ (a constant) gives the traces in planes parallel to the xy-plane.

3. **Completing the figure** Sketch at least two traces in parallel planes (for example, traces with $z = 0$ and $z = \pm 1$). Then draw smooth curves that pass through the traces to fill out the surface.

EXAMPLE 2   An ellipsoid The surface defined by the equation $\dfrac{x^2}{a^2} + \dfrac{y^2}{b^2} + \dfrac{z^2}{c^2} = 1$ is an *ellipsoid*. Graph the ellipsoid with $a = 3$, $b = 4$, and $c = 5$.

SOLUTION Setting x, y, and z equal to zero in pairs gives the intercepts $(\pm 3, 0, 0)$, $(0, \pm 4, 0)$, and $(0, 0, \pm 5)$. Note that points in $\mathbb{R}^3$ with $|x| > 3$ or $|y| > 4$ or $|z| > 5$ do not satisfy the equation of the surface (because the left side of the equation is a sum of nonnegative terms that cannot exceed 1). Therefore, the entire surface is contained in the rectangular box defined by $|x| \le 3$, $|y| \le 4$, and $|z| \le 5$.

The trace in the horizontal plane $z = z_0$ is found by substituting $z = z_0$ into the equation of the ellipsoid, which gives

$$\frac{x^2}{9} + \frac{y^2}{16} + \frac{z_0{}^2}{25} = 1 \quad \text{or} \quad \frac{x^2}{9} + \frac{y^2}{16} = 1 - \frac{z_0{}^2}{25}.$$

> The name *ellipsoid* is used in Example 2 because all traces of this surface, when they exist, are ellipses.

If $|z_0| < 5$, then $1 - \dfrac{z_0{}^2}{25} > 0$, and the equation describes an ellipse in the horizontal plane $z = z_0$. The largest ellipse parallel to the xy-plane occurs with $z_0 = 0$; it is the xy-trace, which is the ellipse $\dfrac{x^2}{9} + \dfrac{y^2}{16} = 1$ with axes of length 6 and 8 (**Figure 13.83a**).

QUICK CHECK 3 Assume $0 < c < b < a$ in the general equation of an ellipsoid. Along which coordinate axis does the ellipsoid have its longest axis? Its shortest axis? ◄

You can check that the yz-trace, found by setting $x = 0$, is the ellipse $\dfrac{y^2}{16} + \dfrac{z^2}{25} = 1$. The xz-trace (set $y = 0$) is the ellipse $\dfrac{x^2}{9} + \dfrac{z^2}{25} = 1$ (**Figure 13.83b**). When we sketch the xy-, xz-, and yz-traces, an outline of the ellipsoid emerges (**Figure 13.83c**).

(a)

(b)

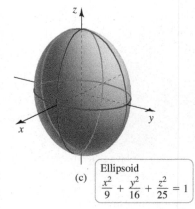

(c)

Figure 13.83

Related Exercise 29 ◄

EXAMPLE 3   An elliptic paraboloid The surface defined by the equation $z = \dfrac{x^2}{a^2} + \dfrac{y^2}{b^2}$ is an *elliptic paraboloid*. Graph the elliptic paraboloid with $a = 4$ and $b = 2$.

SOLUTION Note that the only intercept of the coordinate axes is $(0, 0, 0)$, which is the *vertex* of the paraboloid. The trace in the horizontal plane $z = z_0$, where $z_0 > 0$, satisfies the equation $\dfrac{x^2}{16} + \dfrac{y^2}{4} = z_0$, which describes an ellipse; there is no horizontal trace when $z_0 < 0$ (**Figure 13.84a**). The trace in the vertical plane $x = x_0$ is the parabola $z = \dfrac{x_0^2}{16} + \dfrac{y^2}{4}$ (**Figure 13.84b**); the trace in the vertical plane $y = y_0$ is the parabola $z = \dfrac{x^2}{16} + \dfrac{y_0^2}{4}$ (**Figure 13.84c**).

> The name *elliptic paraboloid* reflects the fact that the traces of this surface are parabolas and ellipses. Two of the three traces in the coordinate planes are parabolas, so this surface is called a paraboloid rather than an ellipsoid.

To graph the surface, we sketch the xz-trace $z = \dfrac{x^2}{16}$ (setting $y = 0$) and the yz-trace $z = \dfrac{y^2}{4}$ (setting $x = 0$). When these traces are combined with an elliptical trace $\dfrac{x^2}{16} + \dfrac{y^2}{4} = z_0$ in a plane $z = z_0$, an outline of the surface appears (**Figure 13.84d**).

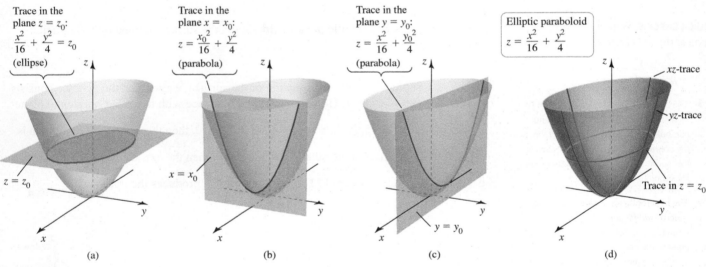

Trace in the plane $z = z_0$:
$\dfrac{x^2}{16} + \dfrac{y^2}{4} = z_0$
(ellipse)

Trace in the plane $x = x_0$:
$z = \dfrac{x_0^2}{16} + \dfrac{y^2}{4}$
(parabola)

Trace in the plane $y = y_0$:
$z = \dfrac{x^2}{16} + \dfrac{y_0^2}{4}$
(parabola)

Elliptic paraboloid
$z = \dfrac{x^2}{16} + \dfrac{y^2}{4}$

$z = z_0$ $x = x_0$ $y = y_0$ xz-trace yz-trace   Trace in $z = z_0$

(a)   (b)   (c)   (d)

Figure 13.84

Related Exercise 32 ◄

QUICK CHECK 4 The elliptic paraboloid $x = \dfrac{y^2}{3} + \dfrac{z^2}{7}$ is a bowl-shaped surface. Along which axis does the bowl open? ◄

> To be completely accurate, this surface should be called an *elliptic hyperboloid of one sheet* because the traces are ellipses and hyperbolas.

EXAMPLE 4   A hyperboloid of one sheet Graph the surface defined by the equation $\dfrac{x^2}{4} + \dfrac{y^2}{9} - z^2 = 1$.

SOLUTION The intercepts of the coordinate axes are $(0, \pm 3, 0)$ and $(\pm 2, 0, 0)$. Setting $z = z_0$, the traces in horizontal planes are ellipses of the form $\dfrac{x^2}{4} + \dfrac{y^2}{9} = 1 + z_0^2$. This equation has solutions for all choices of z_0, so the surface has traces in all horizontal planes. These elliptical traces increase in size as $|z_0|$ increases (**Figure 13.85a**), with the smallest trace being the ellipse $\dfrac{x^2}{4} + \dfrac{y^2}{9} = 1$ in the xy-plane. Setting $y = 0$, the xz-trace is the hyperbola $\dfrac{x^2}{4} - z^2 = 1$; with $x = 0$, the yz-trace is the hyperbola $\dfrac{y^2}{9} - z^2 = 1$ (**Figure 13.85b, c**). In fact, the intersection of the surface with any vertical plane is a hyperbola. The resulting surface is a *hyperboloid of one sheet* (**Figure 13.85d**).

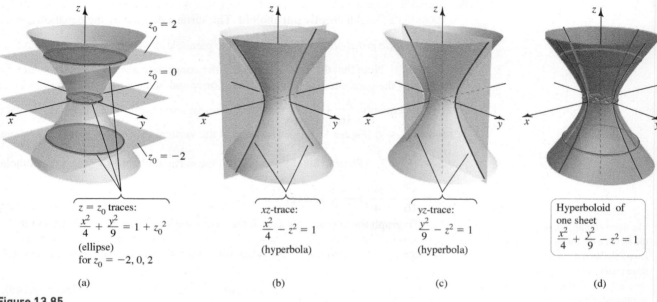

$z = z_0$ traces:
$$\frac{x^2}{4} + \frac{y^2}{9} = 1 + z_0^2$$
(ellipse)
for $z_0 = -2, 0, 2$

(a)

xz-trace:
$$\frac{x^2}{4} - z^2 = 1$$
(hyperbola)

(b)

yz-trace:
$$\frac{y^2}{9} - z^2 = 1$$
(hyperbola)

(c)

Hyperboloid of one sheet
$$\frac{x^2}{4} + \frac{y^2}{9} - z^2 = 1$$

(d)

Figure 13.85

Related Exercise 33 ◄

QUICK CHECK 5 Which coordinate axis is the axis of the hyperboloid
$$\frac{y^2}{a^2} + \frac{z^2}{b^2} - \frac{x^2}{c^2} = 1?$$ ◄

▶ The name *hyperbolic paraboloid* tells us that the traces are hyperbolas and parabolas. Two of the three traces in the coordinate planes are parabolas, so this surface is a paraboloid rather than a hyperboloid.

▶ The hyperbolic paraboloid has a feature called a *saddle point*. For the surface in Example 5, if you walk from the saddle point at the origin in the direction of the *x*-axis, you move uphill. If you walk from the saddle point in the direction of the *y*-axis, you move downhill. Saddle points are examined in detail in Section 15.7.

EXAMPLE 5   A hyperbolic paraboloid Graph the surface defined by the equation
$$z = x^2 - \frac{y^2}{4}.$$

SOLUTION Setting $z = 0$ in the equation of the surface, we see that the *xy*-trace consists of the two lines $y = \pm 2x$. However, slicing the surface with any other horizontal plane $z = z_0$ produces a hyperbola $x^2 - \frac{y^2}{4} = z_0$. If $z_0 > 0$, then the axis of the hyperbola is parallel to the *x*-axis. On the other hand, if $z_0 < 0$, then the axis of the hyperbola is parallel to the *y*-axis (**Figure 13.86a**). Setting $x = x_0$ produces the trace $z = x_0^2 - \frac{y^2}{4}$,

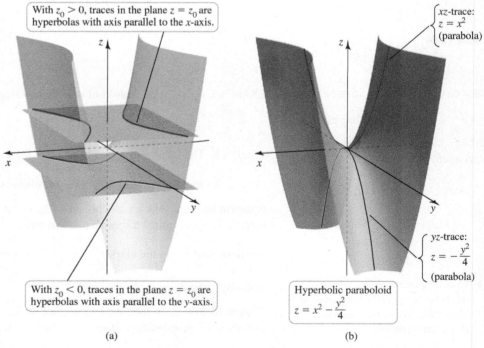

With $z_0 > 0$, traces in the plane $z = z_0$ are hyperbolas with axis parallel to the *x*-axis.

With $z_0 < 0$, traces in the plane $z = z_0$ are hyperbolas with axis parallel to the *y*-axis.

(a)

xz-trace:
$z = x^2$
(parabola)

yz-trace:
$z = -\frac{y^2}{4}$
(parabola)

Hyperbolic paraboloid
$z = x^2 - \frac{y^2}{4}$

(b)

Figure 13.86

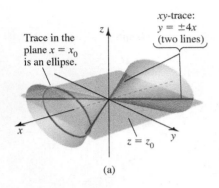

Trace in the plane $x = x_0$ is an ellipse.

xy-trace: $y = \pm 4x$ (two lines)

$z = z_0$

(a)

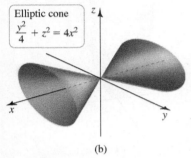

Elliptic cone
$\dfrac{y^2}{4} + z^2 = 4x^2$

(b)

Figure 13.87

▶ The equation $-x^2 - \dfrac{y^2}{4} + \dfrac{z^2}{16} = 1$ describes a hyperboloid of two sheets with its axis on the z-axis. Therefore, the equation in Example 7 describes the same surface shifted 2 units in the positive x-direction.

Hyperboloid of two sheets
$-(x-2)^2 - \dfrac{y^2}{4} + \dfrac{z^2}{16} = 1$

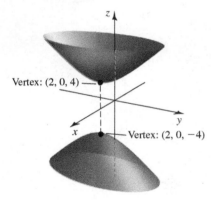

Vertex: $(2, 0, 4)$

Vertex: $(2, 0, -4)$

Figure 13.88

which is the equation of a parabola that opens downward in a plane parallel to the yz-plane. You can check that traces in planes parallel to the xz-plane are parabolas that open upward. The resulting surface is a *hyperbolic paraboloid* (**Figure 13.86b**).

Related Exercise 35 ◀

EXAMPLE 6   Elliptic cones Graph the surface defined by the equation $\dfrac{y^2}{4} + z^2 = 4x^2$.

SOLUTION The only point at which the surface intersects the coordinate axes is $(0, 0, 0)$. Traces in the planes $x = x_0$ are ellipses of the form $\dfrac{y^2}{4} + z^2 = 4x_0{}^2$ that shrink in size as x_0 approaches 0. Setting $y = 0$, the xz-trace satisfies the equation $z^2 = 4x^2$ or $z = \pm 2x$, which are equations of two lines in the xz-plane that intersect at the origin. Setting $z = 0$, the xy-trace satisfies $y^2 = 16x^2$ or $y = \pm 4x$, which describes two lines in the xy-plane that intersect at the origin (**Figure 13.87a**). The complete surface consists of two *cones* opening in opposite directions along the x-axis with a common vertex at the origin (**Figure 13.87b**).

Related Exercise 38 ◀

EXAMPLE 7   A hyperboloid of two sheets Graph the surface defined by the equation
$$-16x^2 - 4y^2 + z^2 + 64x - 80 = 0.$$

SOLUTION We first regroup terms, which yields
$$-16\underbrace{(x^2 - 4x)}_{\text{complete the square}} - 4y^2 + z^2 - 80 = 0,$$

and then complete the square in x:
$$-16\underbrace{(x^2 - 4x + 4}_{(x-2)^2} - 4) - 4y^2 + z^2 - 80 = 0.$$

Collecting terms and dividing by 16 gives the equation
$$-(x-2)^2 - \frac{y^2}{4} + \frac{z^2}{16} = 1.$$

Notice that if $z = 0$, the equation has no solution, so the surface does not intersect the xy-plane. The traces in planes parallel to the xz- and yz-planes are hyperbolas. If $|z_0| \geq 4$, the trace in the plane $z = z_0$ is an ellipse. This equation describes a *hyperboloid of two sheets*, with its axis parallel to the z-axis and shifted 2 units in the positive x-direction (**Figure 13.88**).

Related Exercise 56 ◀

QUICK CHECK 6 In which variable(s) should you complete the square to identify the surface $x = y^2 + 2y + z^2 - 4z + 16$? Name and describe the surface. ◀

Table 13.1 (where a, b, and c are nonzero real numbers) summarizes the standard quadric surfaces. It is important to note that the same surfaces with different orientations are obtained when the roles of the variables are interchanged. For this reason, Table 13.1 summarizes many more surfaces than those listed.

Table 13.1

Name	Standard Equation	Features	Graph
Ellipsoid	$\dfrac{x^2}{a^2} + \dfrac{y^2}{b^2} + \dfrac{z^2}{c^2} = 1$	All traces are ellipses.	
Elliptic paraboloid	$z = \dfrac{x^2}{a^2} + \dfrac{y^2}{b^2}$	Traces with $z = z_0 > 0$ are ellipses. Traces with $x = x_0$ or $y = y_0$ are parabolas.	
Hyperboloid of one sheet	$\dfrac{x^2}{a^2} + \dfrac{y^2}{b^2} - \dfrac{z^2}{c^2} = 1$	Traces with $z = z_0$ are ellipses for all z_0. Traces with $x = x_0$ or $y = y_0$ are hyperbolas.	
Hyperboloid of two sheets	$-\dfrac{x^2}{a^2} - \dfrac{y^2}{b^2} + \dfrac{z^2}{c^2} = 1$	Traces with $z = z_0$ with $|z_0| > |c|$ are ellipses. Traces with $x = x_0$ and $y = y_0$ are hyperbolas.	
Elliptic cone	$\dfrac{x^2}{a^2} + \dfrac{y^2}{b^2} = \dfrac{z^2}{c^2}$	Traces with $z = z_0 \neq 0$ are ellipses. Traces with $x = x_0$ or $y = y_0$ are hyperbolas or intersecting lines.	
Hyperbolic paraboloid	$z = \dfrac{x^2}{a^2} - \dfrac{y^2}{b^2}$	Traces with $z = z_0 \neq 0$ are hyperbolas. Traces with $x = x_0$ or $y = y_0$ are parabolas.	

SECTION 13.6 EXERCISES

Getting Started

1. To which coordinate axes are the following cylinders in $\mathbb{R}^3$ parallel: $x^2 + 2y^2 = 8$, $z^2 + 2y^2 = 8$, and $x^2 + 2z^2 = 8$?

2. Describe the graph of $x = z^2$ in $\mathbb{R}^3$.

3. What is a trace of a surface?

4. What is the name of the surface defined by the equation $y = \dfrac{x^2}{4} + \dfrac{z^2}{8}$?

5. What is the name of the surface defined by the equation $x^2 + \dfrac{y^2}{3} + 2z^2 = 1$?

6. What is the name of the surface defined by the equation $-y^2 - \dfrac{z^2}{2} + x^2 = 1$?

Practice Exercises

7–14. Cylinders in $\mathbb{R}^3$ *Consider the following cylinders in $\mathbb{R}^3$.*

a. Identify the coordinate axis to which the cylinder is parallel.
b. Sketch the cylinder.

7. $z = y^2$

8. $x^2 + 4y^2 = 4$

9. $x^2 + z^2 = 4$

10. $x = z^2 - 4$

11. $y - x^3 = 0$

12. $x - 2z^2 = 0$

13. $z - \ln y = 0$

14. $x - \dfrac{1}{y} = 0$

15–20. Identifying quadric surfaces *Identify the following quadric surfaces by name. Find and describe the xy-, xz-, and yz-traces, when they exist.*

15. $25x^2 + 25y^2 + z^2 = 25$

16. $25x^2 + 25y^2 - z^2 = 25$

17. $25x^2 + 25y^2 - z = 0$

18. $25x^2 - 25y^2 - z = 0$

19. $-25x^2 - 25y^2 + z^2 = 25$

20. $-25x^2 - 25y^2 + z^2 = 0$

21–28. Identifying surfaces *Identify the following surfaces by name.*

21. $y = 4z^2 - x^2$

22. $-y^2 - 9z^2 + \dfrac{x^2}{4} = 1$

23. $y = \dfrac{x^2}{6} + \dfrac{z^2}{16}$

24. $z^2 + 4y^2 - x^2 = 1$

25. $y^2 - z^2 = 2$

26. $x^2 + 4y^2 = 1$

27. $9x^2 + 4z^2 - 36y = 0$

28. $9y^2 + 4z^2 - 36x^2 = 0$

29–51. Quadric surfaces *Consider the following equations of quadric surfaces.*

a. Find the intercepts with the three coordinate axes, when they exist.
b. Find the equations of the xy-, xz-, and yz-traces, when they exist.
c. Identify and sketch a graph of the surface.

29. $x^2 + \dfrac{y^2}{4} + \dfrac{z^2}{9} = 1$

30. $4x^2 + y^2 + \dfrac{z^2}{2} = 1$

31. $x = y^2 + z^2$

32. $z = \dfrac{x^2}{4} + \dfrac{y^2}{9}$

33. $\dfrac{x^2}{25} + \dfrac{y^2}{9} - z^2 = 1$

34. $\dfrac{y^2}{4} + \dfrac{z^2}{9} - \dfrac{x^2}{16} = 1$

35. $z = \dfrac{x^2}{9} - y^2$

36. $y = \dfrac{x^2}{16} - 4z^2$

37. $x^2 + \dfrac{y^2}{4} = z^2$

38. $4y^2 + z^2 = x^2$

39. $\dfrac{x^2}{3} + 3y^2 + \dfrac{z^2}{12} = 3$

40. $\dfrac{x^2}{6} + 24y^2 + \dfrac{z^2}{24} - 6 = 0$

41. $9x - 81y^2 - \dfrac{z^2}{4} = 0$

42. $2y - \dfrac{x^2}{8} - \dfrac{z^2}{18} = 0$

43. $\dfrac{y^2}{16} + 36z^2 - \dfrac{x^2}{4} - 9 = 0$

44. $9z^2 + x^2 - \dfrac{y^2}{3} - 1 = 0$

45. $5x - \dfrac{y^2}{5} + \dfrac{z^2}{20} = 0$

46. $6y + \dfrac{x^2}{6} - \dfrac{z^2}{24} = 0$

47. $\dfrac{z^2}{32} + \dfrac{y^2}{18} = 2x^2$

48. $\dfrac{x^2}{3} + \dfrac{z^2}{12} = 3y^2$

49. $-x^2 + \dfrac{y^2}{4} - \dfrac{z^2}{9} = 1$

50. $-\dfrac{x^2}{6} - 24y^2 + \dfrac{z^2}{24} - 6 = 0$

51. $-\dfrac{x^2}{3} + 3y^2 - \dfrac{z^2}{12} = 1$

52. Describe the relationship between the graphs of the quadric surfaces $x^2 + y^2 - z^2 + 2z = 1$ and $x^2 + y^2 - z^2 = 0$, and state the names of the surfaces.

53. Describe the relationship between the graphs of $x^2 + 4y^2 + 9z^2 = 100$ and $x^2 + 4y^2 + 9z^2 + 54z = 19$, and state the names of the surfaces.

54–58. Identifying surfaces *Identify and briefly describe the surfaces defined by the following equations.*

54. $x^2 + y^2 + 4z^2 + 2x = 0$

55. $9x^2 + y^2 - 4z^2 + 2y = 0$

56. $-x^2 - y^2 + \dfrac{z^2}{9} + 6x - 8y = 26$

57. $\dfrac{x^2}{4} + y^2 - 2x - 10y - z^2 + 41 = 0$

58. $z = -x^2 - y^2$

59. **Explain why or why not** Determine whether the following statements are true and give an explanation or counterexample.

   a. The graph of the equation $y = z^2$ in $\mathbb{R}^3$ is both a cylinder and a quadric surface.

   b. The xy-traces of the ellipsoid $x^2 + 2y^2 + 3z^2 = 16$ and the cylinder $x^2 + 2y^2 = 16$ are identical.

   c. Traces of the surface $y = 3x^2 - z^2$ in planes parallel to the xy-plane are parabolas.

   d. Traces of the surface $y = 3x^2 - z^2$ in planes parallel to the xz-plane are parabolas.

   e. The graph of the ellipsoid $x^2 + 2y^2 + 3(z - 4)^2 = 25$ is obtained by shifting the graph of the ellipsoid $x^2 + 2y^2 + 3z^2 = 25$ down 4 units.

60. Matching graphs with equations Match equations a–f with surfaces A–F.

a. $y - z^2 = 0$

b. $2x + 3y - z = 5$

c. $4x^2 + \dfrac{y^2}{9} + z^2 = 1$

d. $x^2 + \dfrac{y^2}{9} - z^2 = 1$

e. $x^2 + \dfrac{y^2}{9} = z^2$

f. $y = |x|$

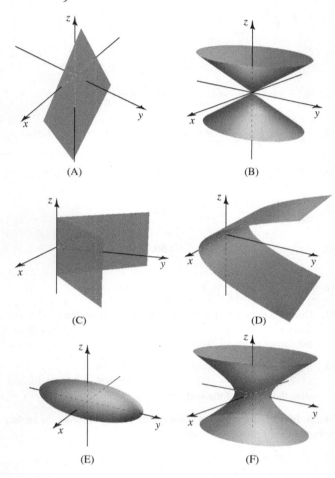

(A)

(B)

(C)

(D)

(E)

(F)

Explorations and Challenges

61. Solids of revolution Which of the quadric surfaces in Table 13.1 can be generated by revolving a curve in one of the coordinate planes about a coordinate axis, assuming $a = b = c \neq 0$?

62. Solids of revolution Consider the ellipse $x^2 + 4y^2 = 1$ in the xy-plane.

a. If this ellipse is revolved about the x-axis, what is the equation of the resulting ellipsoid?

b. If this ellipse is revolved about the y-axis, what is the equation of the resulting ellipsoid?

63. Volume Find the volume of the solid that is bounded between the planes $z = 0$ and $z = 3$ and the cylinders $y = x^2$ and $y = 2 - x^2$.

64. Light cones The idea of a *light cone* appears in the Special Theory of Relativity. The xy-plane (see figure) represents all of three-dimensional space, and the z-axis is the time axis (t-axis). If an event E occurs at the origin, the interior of the future light cone ($t > 0$) represents all events in the future that could be affected by E, assuming no signal travels faster than the speed of light. The interior of the past light cone ($t < 0$) represents all events in the

past that could have affected E, again assuming no signal travels faster than the speed of light.

a. If time is measured in seconds and distance (x and y) is measured in light-seconds (the distance light travels in 1 s), the light cone makes a 45° angle with the xy-plane. Write the equation of the light cone in this case.

b. Suppose distance is measured in meters and time is measured in seconds. Write the equation of the light cone in this case, given that the speed of light is $3 \times 10^8 \, \text{m/s}$.

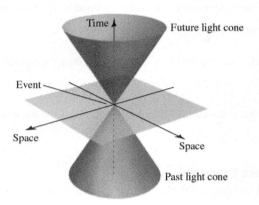

65. Designing an NFL football A *prolate spheroid* is a surface of revolution obtained by rotating an ellipse about its major axis.

a. Explain why one possible equation for a prolate spheroid is
$$\frac{x^2 + z^2}{a^2} + \frac{y^2}{b^2} = 1, \text{ where } b > a > 0.$$

b. According to the National Football League (NFL) rulebook, the shape of an NFL football is required to be a prolate spheroid with a long axis between 11 and 11.25 inches long and a short circumference (the circumference of the xz-trace) between 21 and 21.25 inches. Find an equation for the shape of the football if the long axis is 11.1 inches and the short circumference is 21.1 inches.

66. Hand tracking Researchers are developing hand tracking software that will allow computers to track and recognize detailed hand movements for better human-computer interaction. One three-dimensional hand model under investigation is constructed from a set of truncated quadrics (see figure). For example, the palm of the hand consists of a truncated elliptic cylinder, capped off by the upper half of an ellipsoid. Suppose the palm of the hand is modeled by the truncated cylinder $4x^2/9 + 4y^2 = 1$, for $0 \leq z \leq 3$. Find an equation of the upper half of an ellipsoid, whose bottom corresponds with the top of the cylinder, if the distance from the top of the truncated cylinder to the top of the ellipsoid is $1/2$.

(*Source: Computer Vision and Pattern Recognition*, 2, Dec 2001)

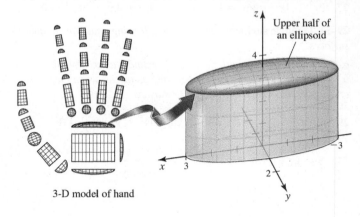

3-D model of hand

67. Designing a snow cone A surface, having the shape of an oblong snow cone, consists of a truncated cone, $\dfrac{x^2}{2} + y^2 = \dfrac{z^2}{8}$, for $0 \leq z \leq 3$, capped off by the upper half of an ellipsoid. Find an equation for the upper half of the ellipsoid so that the bottom edge of the truncated ellipsoid and the top edge of the cone coincide, and the distance from the top of the cone to the top of the ellipsoid is $3/2$.

68. Designing a glass The outer, lateral side of a 6-inch-tall glass has the shape of the truncated hyperboloid of one sheet $\dfrac{x^2}{a^2} + \dfrac{y^2}{b^2} - \dfrac{z^2}{c^2} = 1$, for $0 \leq z \leq 6$. If the base of the glass has a

radius of 1 inch and the top of the glass has a radius of 2 inches, find the values of a^2, b^2, and c^2 that satisfy these conditions. Assume horizontal traces of the glass are circular.

QUICK CHECK ANSWERS

1. y-axis; x-axis   **2.** The equation $x^2 + 4y^2 = 16$ is a special case of the general equation for quadric surfaces; all the coefficients except A, B, and J are zero.   **3.** x-axis; z-axis   **4.** Positive x-axis   **5.** x-axis   **6.** Complete the square in y and z; elliptic paraboloid with its axis parallel to the x-axis ◀

CHAPTER 13   REVIEW EXERCISES

1. Explain why or why not Determine whether the following statements are true and give an explanation or counterexample.

a. Given two vectors $\mathbf{u}$ and $\mathbf{v}$, it is always true that $2\mathbf{u} + \mathbf{v} = \mathbf{v} + 2\mathbf{u}$.

b. The vector in the direction of $\mathbf{u}$ with the length of $\mathbf{v}$ equals the vector in the direction of $\mathbf{v}$ with the length of $\mathbf{u}$.

c. If $\mathbf{u} \neq \mathbf{0}$ and $\mathbf{u} + \mathbf{v} = \mathbf{0}$, then $\mathbf{u}$ and $\mathbf{v}$ are parallel.

d. The lines $x = 3 + t, y = 4 + 2t, z = 2 - t$ and $x = 2t, y = 4t, z = t$ are parallel.

e. The lines $x = 3 + t, y = 4 + 2t, z = 2 - t$ and the plane $x + 2y + 5z = 3$ are parallel.

f. There is always a plane orthogonal to both of two distinct intersecting planes.

2–5. Working with vectors *Let* $\mathbf{u} = \langle 3, -4 \rangle$ *and* $\mathbf{v} = \langle -1, 2 \rangle$. *Evaluate each of the following.*

2. $\mathbf{u} - \mathbf{v}$

3. $-3\mathbf{v}$

4. $\mathbf{u} + 2\mathbf{v}$

5. $2\mathbf{v} - \mathbf{u}$

6–15. Working with vectors *Let* $\mathbf{u} = \langle 2, 4, -5 \rangle$, $\mathbf{v} = \langle -6, 10, 2 \rangle$, *and* $\mathbf{w} = \langle 4, -8, 8 \rangle$.

6. Compute $\mathbf{u} - 3\mathbf{v}$.

7. Compute $|\mathbf{u} + \mathbf{v}|$.

8. Find the unit vector with the same direction as $\mathbf{u}$.

9. Write the vector $\mathbf{w}$ as a product of its magnitude and a unit vector in the direction of $\mathbf{w}$.

10. Find a vector in the direction of $\mathbf{w}$ that is 10 times as long as $\mathbf{w}$.

11. Find a vector in the direction of $\mathbf{w}$ with a length of 10.

12. Compute $\mathbf{u} \cdot \mathbf{v}$

13. Compute $\mathbf{u} \times \mathbf{v}$

14. For what value of a is the vector $\mathbf{v}$ orthogonal to $\mathbf{y} = \langle a, 1, -3 \rangle$?

15. For what value of a is the vector $\mathbf{w}$ parallel to $\mathbf{y} = \langle a, 6, -6 \rangle$?

16. Scalar multiples Find scalars a, b, and c such that
$$\langle 2, 2, 2 \rangle = a\langle 1, 1, 0 \rangle + b\langle 0, 1, 1 \rangle + c\langle 1, 0, 1 \rangle.$$

17. Velocity vectors Assume the positive x-axis points east and the positive y-axis points north.

a. An airliner flies northwest at a constant altitude at 550 mi/hr in calm air. Find a and b such that its velocity may be expressed in the form $\mathbf{v} = a\mathbf{i} + b\mathbf{j}$.

b. An airliner flies northwest at a constant altitude at 550 mi/hr relative to the air in a southerly crosswind $\mathbf{w} = \langle 0, 40 \rangle$. Find the velocity of the airliner relative to the ground.

18. Position vectors Let $\overrightarrow{PQ}$ extend from $P(2, 0, 6)$ to $Q(2, -8, 5)$.

a. Find the position vector equal to $\overrightarrow{PQ}$.

b. Find the midpoint M of the line segment PQ. Then find the magnitude of $\overrightarrow{PM}$.

c. Find a vector of length 8 with direction opposite to that of $\overrightarrow{PQ}$.

19–21. Spheres and balls *Use set notation to describe the following sets.*

19. The sphere of radius 4 centered at $(1, 0, -1)$

20. The points inside the sphere of radius 10 centered at $(2, 4, -3)$

21. The points outside the sphere of radius 2 centered at $(0, 1, 0)$

22–25. Identifying sets *Give a geometric description of the following sets of points.*

22. $x^2 - 6x + y^2 + 8y + z^2 - 2z - 23 = 0$

23. $x^2 - x + y^2 + 4y + z^2 - 6z + 11 \leq 0$

24. $x^2 + y^2 - 10y + z^2 - 6z = -34$

25. $x^2 - 6x + y^2 + z^2 - 20z + 9 > 0$

26. Combined force An object at the origin is acted on by the forces $\mathbf{F}_1 = -10\mathbf{i} + 20\mathbf{k}$, $\mathbf{F}_2 = 40\mathbf{j} + 10\mathbf{k}$, and $\mathbf{F}_3 = -50\mathbf{i} + 20\mathbf{j}$. Find the magnitude of the combined force, and use a sketch to illustrate the direction of the combined force.

27. Falling probe A remote sensing probe falls vertically with a terminal velocity of 60 m/s when it encounters a horizontal crosswind blowing north at 4 m/s and an updraft blowing vertically at 10 m/s. Find the magnitude and direction of the resulting velocity relative to the ground.

28. Crosswinds A small plane is flying north in calm air at 250 mi/hr when it is hit by a horizontal crosswind blowing northeast at 40 mi/hr and a 25 mi/hr downdraft. Find the resulting velocity and speed of the plane.

29. Net force Jack pulls east on a rope attached to a camel with a force of 40 lb. Jill pulls north on a rope attached to the same camel with a force of 30 lb. What is the magnitude and direction of the force on the camel? Assume the vectors lie in a horizontal plane.

30. Canoe in a current A woman in a canoe paddles due west at 4 mi/hr relative to the water in a current that flows northwest at 2 mi/hr. Find the speed and direction of the canoe relative to the shore.

31. Set of points Describe the set of points satisfying both the equations $x^2 + z^2 = 1$ and $y = 2$.

32–33. Angles and projections

a. *Find the angle between* **u** *and* **v**.
b. *Compute* $\text{proj}_v \mathbf{u}$ *and* $\text{scal}_v \mathbf{u}$.
c. *Compute* $\text{proj}_u \mathbf{v}$ *and* $\text{scal}_u \mathbf{v}$.

32. u = −3j + 4k, v = −4i + j + 5k

33. u = −i + 2j + 2k, v = 3i + 6j + 6k

34. Parallelepiped Find the volume of a parallelepiped determined by the position vectors $\mathbf{u} = \langle 2, 4, -5 \rangle$, $\mathbf{v} = \langle -6, 10, 2 \rangle$, and $\mathbf{w} = \langle 4, -8, 8 \rangle$ (see Exercise 63 in Section 13.4).

35–36. Computing work *Calculate the work done in the following situations.*

35. A suitcase is pulled 25 ft along a horizontal sidewalk with a constant force of 20 lb at an angle of 45° above the horizontal.

36. A constant force $\mathbf{F} = \langle 2, 3, 4 \rangle$ (in newtons) moves an object from $(0, 0, 0)$ to $(2, 1, 6)$. (Distance is measured in meters.)

37–38. Inclined plane *A 180-lb man stands on a hillside that makes an angle of 30° with the horizontal, producing a force of* $\mathbf{W} = \langle 0, -180 \rangle$.

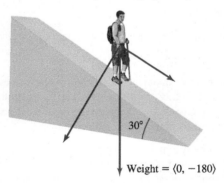

37. Find the component of his weight in the downward direction perpendicular to the hillside and in the downward direction parallel to the hillside.

38. How much work is done when the man moves 10 ft up the hillside?

39. Area of a parallelogram Find the area of the parallelogram with the vertices $(1, 2, 3)$, $(1, 0, 6)$, and $(4, 2, 4)$.

40. Area of a triangle Find the area of the triangle with the vertices $(1, 0, 3)$, $(5, 0, -1)$, and $(0, 2, -2)$.

41. Unit normal vector Find unit vectors normal to the vectors $\langle 2, -6, 9 \rangle$ and $\langle -1, 0, 6 \rangle$.

42. Angle in two ways Find the angle between $\langle 2, 0, -2 \rangle$ and $\langle 2, 2, 0 \rangle$ using (a) the dot product and (b) the cross product.

43. Let $\mathbf{r} = \overrightarrow{OP} = 3\mathbf{i} + 2\mathbf{j} + \mathbf{k}$. A force $\mathbf{F} = \langle 10, 10, 0 \rangle$ is applied at P. Find the torque about O that is produced.

44. Suppose you apply a force of $|\mathbf{F}| = 50$ N near the end of a wrench attached to a bolt (see figure). Determine the magnitude of the torque when the force is applied at an angle of 60° to the wrench. Assume the distance along the wrench from the center of the bolt to the point where the force is applied is $|\mathbf{r}| = 0.25$ m.

45. Knee torque Jan does leg lifts with a 10-kg weight attached to her foot, so the resulting force is $mg \approx 98$ N directed vertically downward (see figure). If the distance from her knee to the weight is 0.4 m and her lower leg makes an angle of θ to the vertical, find the magnitude of the torque about her knee as her leg is lifted (as a function of θ). What are the minimum and maximum magnitudes of the torque? Does the direction of the torque change as her leg is lifted?

46–50. Lines in space *Find an equation of the following lines or line segments.*

46. The line that passes through the points $(2, 6, -1)$ and $(-6, 4, 0)$

47. The line segment that joins the points $(0, -3, 9)$ and $(2, -8, 1)$

48. The line through the point $(0, 1, 1)$ and parallel to the line $\mathbf{R} = \langle 1 + 2t, 3 - 5t, 7 + 6t \rangle$.

49. The line through the point $(0, 1, 1)$ that is orthogonal to both $\langle 0, -1, 3 \rangle$ and $\langle 2, -1, 2 \rangle$.

50. The line through the point $(0, 1, 4)$ and orthogonal to the vector $\langle -2, 1, 7 \rangle$ and the y-axis.

51. Equations of planes Consider the plane passing through the points $(0, 0, 3)$, $(1, 0, -6)$, and $(1, 2, 3)$.

a. Find an equation of the plane.
b. Find the intercepts of the plane with the three coordinate axes.
c. Make a sketch of the plane.

52–53. Intersecting planes *Find an equation of the line of intersection of the planes Q and R.*

52. $Q: 2x + y - z = 0$, $R: -x + y + z = 1$

53. $Q: -3x + y + 2z = 0$, $R: 3x + 3y + 4z - 12 = 0$

54–57. Equations of planes *Find an equation of the following planes.*

54. The plane passing through $(5, 0, 2)$ that is parallel to the plane $2x + y - z = 0$

55. The plane containing the lines $x = 5 + t, y = 3 - 2t, z = 1$ and $x = 4s, y = 5s, z = 3 - 2s$, if possible

56. The plane passing through $(2, -3, 1)$ normal to the line $\langle x, y, z \rangle = \langle 2 + t, 3t, 2 - 3t \rangle$

57. The plane passing through $(-2, 3, 1)$, $(1, 1, 0)$, and $(-1, 0, 1)$

58. Distance from a point to a line Find the distance from the point $(1, 2, 3)$ to the line $x = 2 + t, y = 3, z = 1 - 3t$.

59. Distance from a point to a plane Find the distance from the point $(2, 2, 2)$ to the plane $x + 2y + 2z = 1$.

60–74. Identifying surfaces *Consider the surfaces defined by the following equations.*

a. *Identify and briefly describe the surface.*
b. *Find the xy-, xz-, and yz-traces, when they exist.*
c. *Find the intercepts with the three coordinate axes, when they exist.*
d. *Sketch the surface.*

60. $z - \sqrt{x} = 0$

61. $3z = \dfrac{x^2}{12} - \dfrac{y^2}{48}$

62. $\dfrac{x^2}{100} + 4y^2 + \dfrac{z^2}{16} = 1$

63. $y^2 = 4x^2 + \dfrac{z^2}{25}$

64. $\dfrac{4x^2}{9} + \dfrac{9z^2}{4} = y^2$

65. $4z = \dfrac{x^2}{4} + \dfrac{y^2}{9}$

66. $\dfrac{x^2}{16} + \dfrac{z^2}{36} - \dfrac{y^2}{100} = 1$

67. $y^2 + 4z^2 - 2x^2 = 1$

68. $-\dfrac{x^2}{16} + \dfrac{z^2}{36} - \dfrac{y^2}{25} = 4$

69. $\dfrac{x^2}{4} + \dfrac{y^2}{16} - z^2 = 4$

70. $x = \dfrac{y^2}{64} - \dfrac{z^2}{9}$

71. $\dfrac{x^2}{4} + \dfrac{y^2}{16} + z^2 = 4$

72. $y - e^{-x} = 0$

73. $\dfrac{y^2}{49} + \dfrac{x^2}{9} = \dfrac{z^2}{64}$

74. $y = 4x^2 + \dfrac{z^2}{9}$

75. Matching surfaces Match equations a–d with surfaces A–D.

a. $z = \sqrt{2x^2 + 3y^2 + 1} - 1$
b. $z = -3y^2$
c. $z = 2x^2 - 3y^2 + 1$
d. $z = \sqrt{2x^2 + 3y^2 - 1}$

(A)

(B)

(C)

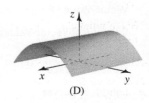
(D)

76. Designing a water bottle The lateral surface of a water bottle consists of a circular cylinder of radius 2 and height 6, topped off by a truncated hyperboloid of one sheet of height 2 (see figure). Assume the top of the truncated hyperboloid has a radius of 1/2. Find two equations that, when graphed together, form the lateral surface of the bottle. Answers may vary.

Chapter 13 Guided Projects

Applications of the material in this chapter and related topics can be found in the following Guided Projects. For additional information, see the Preface.

- Intercepting a UFO
- CORDIC algorithms: How your calculator works

14

Vector-Valued Functions

Chapter Preview In Chapter 13, we used vectors to represent static quantities, such as the constant force applied to the end of a wrench or the constant velocity of a boat in a current. In this chapter, we put vectors in motion by introducing *vector-valued functions*, or simply *vector functions*. Our first task is to investigate the graphs of vector-valued functions and to study them in the setting of calculus. Everything you already know about limits, derivatives, and integrals applies to this new family of functions. Also, with the calculus of vector functions, we can solve a wealth of practical problems involving the motion of objects in space. The chapter closes with an exploration of arc length, curvature, and tangent and normal vectors, all important features of space curves.

14.1 Vector-Valued Functions

Imagine a projectile moving along a path in three-dimensional space; it could be an electron or a comet, a soccer ball or a rocket. If you take a snapshot of the object, its position is described by a static position vector $\mathbf{r} = \langle x, y, z \rangle$. However, if you want to describe the full trajectory of the object as it unfolds in time, you must represent the object's position with a *vector-valued function* such as $\mathbf{r}(t) = \langle x(t), y(t), z(t) \rangle$ whose components change in time (**Figure 14.1**). The goal of this section is to describe continuous motion using vector-valued functions.

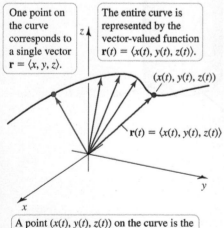

One point on the curve corresponds to a single vector $\mathbf{r} = \langle x, y, z \rangle$.

The entire curve is represented by the vector-valued function $\mathbf{r}(t) = \langle x(t), y(t), z(t) \rangle$.

$(x(t), y(t), z(t))$

$\mathbf{r}(t) = \langle x(t), y(t), z(t) \rangle$

A point $(x(t), y(t), z(t))$ on the curve is the head of the vector $\mathbf{r}(t) = \langle x(t), y(t), z(t) \rangle$.

Figure 14.1

Vector-Valued Functions

A function of the form $\mathbf{r}(t) = \langle x(t), y(t), z(t) \rangle$ may be viewed in two ways.

- It is a set of three parametric equations that describe a curve in space.

- It is also a **vector-valued function**, which means that the three dependent variables $(x, y,$ and $z)$ are the components of $\mathbf{r}$, and each component varies with respect to a single independent variable t (that often represents time).

Here is the connection between these perspectives: As t varies, a point $(x(t), y(t), z(t))$ on a parametric curve is also the head of the position vector $\mathbf{r}(t) = \langle x(t), y(t), z(t) \rangle$. In other words, a vector-valued function is a set of parametric equations written in vector form. It is useful to keep both of these interpretations in mind as you work with vector-valued functions.

Although our focus is on vector functions whose graphs lie in three-dimensional space, vector functions can be given in any number of dimensions. In fact, you became acquainted with the essential ideas behind two-dimensional vector functions in Section 12.1 when you studied parametric equations. For example, recall that the parametric equations

$$x = a \cos t, \quad y = a \sin t, \quad \text{for } 0 \le t \le 2\pi$$

describe a circle of radius a centered at the origin. The corresponding vector function is

$$\mathbf{r}(t) = \langle a \cos t, a \sin t \rangle, \quad \text{for } 0 \le t \le 2\pi.$$

All the plane curves described by parametric equations in Section 12.1 are easily converted to vector functions in the same manner.

Curves in Space

We now explore general vector-valued functions of the form

$$\mathbf{r}(t) = \langle f(t), g(t), h(t) \rangle = f(t)\mathbf{i} + g(t)\mathbf{j} + h(t)\mathbf{k},$$

where f, g, and h are defined on an interval $a \le t \le b$. The **domain** of $\mathbf{r}$ is the largest set of values of t on which all of f, g, and h are defined.

Figure 14.2 illustrates how a parameterized curve is generated by such a function. As the parameter t varies over the interval $a \le t \le b$, each value of t produces a position vector that corresponds to a point on the curve, starting at the initial vector $\mathbf{r}(a)$ and ending at the terminal vector $\mathbf{r}(b)$. The resulting parameterized curve can either have finite length or extend indefinitely. The curve may also cross itself or close and retrace itself. As shown in Example 1, when f, g, and h are linear functions of t, the resulting curve is a line or line segment.

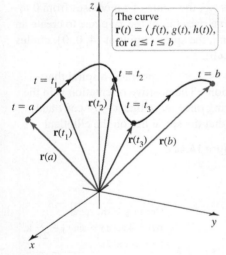

The curve
$\mathbf{r}(t) = \langle f(t), g(t), h(t) \rangle$,
for $a \le t \le b$

Figure 14.2

EXAMPLE 1 Lines as vector-valued functions Find a vector function for the line that passes through the points $P(2, -1, 4)$ and $Q(3, 0, 6)$.

SOLUTION Recall from Section 13.5 that parametric equations of the line parallel to the vector $\mathbf{v} = \langle a, b, c \rangle$ and passing through the point $P_0(x_0, y_0, z_0)$ are

$$x = x_0 + at, \quad y = y_0 + bt, \quad z = z_0 + ct.$$

The vector $\mathbf{v} = \overrightarrow{PQ} = \langle 3 - 2, 0 - (-1), 6 - 4 \rangle = \langle 1, 1, 2 \rangle$ is parallel to the line, and we let $P_0 = P(2, -1, 4)$. Therefore, parametric equations for the line are

$$x = 2 + t, \quad y = -1 + t, \quad z = 4 + 2t,$$

and the corresponding vector function for the line is

$$\mathbf{r}(t) = \langle 2 + t, -1 + t, 4 + 2t \rangle,$$

with a domain of all real numbers. As t increases, the line is generated in the direction of $\overrightarrow{PQ}$. Just as we did with parametric equations, we can restrict the domain to a finite interval to produce a vector function for a line segment (see Quick Check 1).

Related Exercises 9, 13 ◄

QUICK CHECK 1 Restrict the domain of the vector function in Example 1 to produce a line segment that goes from $P(2, -1, 4)$ to $R(5, 1, 8)$. ◄

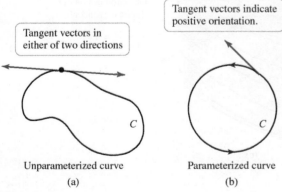

Tangent vectors in either of two directions

Unparameterized curve
(a)

Figure 14.3

Tangent vectors indicate positive orientation.

Parameterized curve
(b)

Orientation of Curves If a smooth curve C is viewed only as a set of points, then at any point of C, it is possible to draw tangent vectors in two directions (**Figure 14.3a**). On the other hand, a parameterized curve described by the function $\mathbf{r}(t)$, where $a \le t \le b$, has a natural direction, or **orientation**. The *positive* orientation is the direction in which the curve is generated as the parameter increases from a to b. For example, the positive orientation of the circle $\mathbf{r}(t) = \langle \cos t, \sin t \rangle$, for $0 \le t \le 2\pi$, is counterclockwise (**Figure 14.3b**), and the positive orientation of the line in Example 1 is in the direction of the vector $\overrightarrow{PQ}$. An important property of all parameterized curves is the relationship between the orientation of a given curve and its tangent vectors (to be defined precisely in Section 14.2): At all points, the tangent vectors point in the direction of the positive orientation of the curve.

EXAMPLE 2   A spiral Graph the curve described by the equation

$$\mathbf{r}(t) = 4\cos t\,\mathbf{i} + \sin t\,\mathbf{j} + \frac{t}{2\pi}\,\mathbf{k},$$

where (a) $0 \le t \le 2\pi$ and (b) $-\infty < t < \infty$.

SOLUTION

a. We begin by setting $z = 0$ to determine the projection of the curve in the xy-plane. The resulting function $\mathbf{r}(t) = 4\cos t\,\mathbf{i} + \sin t\,\mathbf{j}$ implies that $x = 4\cos t$ and $y = \sin t$; these equations describe an ellipse in the xy-plane whose positive direction is counterclockwise (**Figure 14.4a**). Because $z = \frac{t}{2\pi}$, the value of z increases from 0 to 1 as t increases from 0 to 2π. Therefore, the curve rises out of the xy-plane to create an elliptical spiral (or coil). Over the interval $[0, 2\pi]$, the spiral begins at $(4, 0, 0)$, circles the z-axis once, and ends at $(4, 0, 1)$ (**Figure 14.4b**).

b. Letting the parameter vary over the interval $-\infty < t < \infty$ generates a spiral that winds around the z-axis endlessly in both directions. The positive orientation is in the upward direction (increasing z-direction). Noticing once more that $x = 4\cos t$ and $y = \sin t$ are x- and y-components of $\mathbf{r}$, we see that the spiral lies on the elliptical cylinder $\left(\dfrac{x}{4}\right)^2 + y^2 = \cos^2 t + \sin^2 t = 1$ (**Figure 14.4c**).

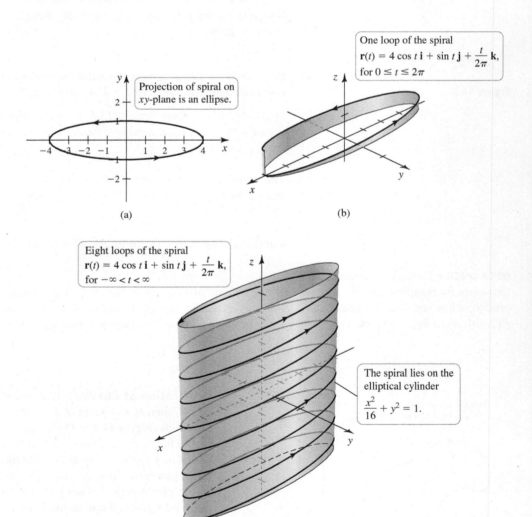

▶ Recall that the functions $\sin at$ and $\cos at$ oscillate a times over the interval $[0, 2\pi]$. Therefore, their period is $2\pi/a$.

Figure 14.4

Related Exercise 24 ◀

EXAMPLE 3 **Roller coaster curve** Graph the curve

$$\mathbf{r}(t) = \cos t\,\mathbf{i} + \sin t\,\mathbf{j} + 0.4 \sin 2t\,\mathbf{k}, \quad \text{for } 0 \le t \le 2\pi.$$

SOLUTION Without the z-component, the resulting function $\mathbf{r}(t) = \cos t\,\mathbf{i} + \sin t\,\mathbf{j}$ describes a circle of radius 1 in the xy-plane. The z-component of the function varies between -0.4 and 0.4 with a period of π units. Therefore, on the interval $[0, 2\pi]$, the z-coordinates of points on the curve oscillate twice between -0.4 and 0.4, while the x- and y-coordinates describe a circle. The result is a curve that circles the z-axis once in the counterclockwise direction with two peaks and two valleys (**Figure 14.5a**).

The space curve in this example is not particularly complicated, but visualizing a given curve is easier when we determine the surface(s) on which it lies. Writing the vector function $\mathbf{r}(t) = \cos t\,\mathbf{i} + \sin t\,\mathbf{j} + 0.4 \sin 2t\,\mathbf{k}$ in parametric form, we have

$$x = \cos t, \quad y = \sin t, \quad z = 0.4 \sin 2t, \quad \text{for } 0 \le t \le 2\pi.$$

Noting that $x^2 + y^2 = \cos^2 t + \sin^2 t = 1$, we conclude that the curve lies on the cylinder $x^2 + y^2 = 1$. In this case, we can also eliminate the parameter by writing

$$
\begin{aligned}
z &= 0.4 \sin 2t \\
&= 0.4(2\,\underbrace{\sin t}_{y}\,\underbrace{\cos t}_{x}) \quad \text{Double angle identity: } \sin 2t = 2 \sin t \cos t \\
&= 0.8xy, \qquad\qquad x = \cos t, y = \sin t
\end{aligned}
$$

which implies that the curve also lies on the hyperbolic paraboloid $z = 0.8xy$ (see margin note). In fact, the roller coaster curve is the curve in which the surfaces $x^2 + y^2 = 1$ and $z = 0.8xy$ intersect, as shown in **Figure 14.5b**.

▶ The graph of $z = 0.8xy$ is a rotation of the quadric surface $z = 0.4(x^2 - y^2)$, which we recognize as a hyperbolic paraboloid. See the Guided Project *Translation and rotation of axes*.

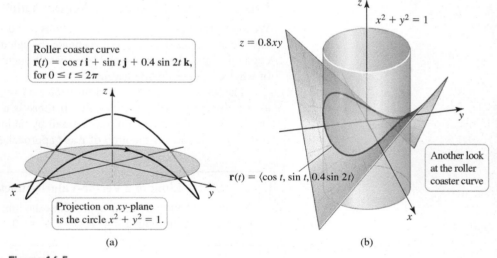

Roller coaster curve
$\mathbf{r}(t) = \cos t\,\mathbf{i} + \sin t\,\mathbf{j} + 0.4 \sin 2t\,\mathbf{k}$, for $0 \le t \le 2\pi$

Projection on xy-plane is the circle $x^2 + y^2 = 1$.

(a)

$x^2 + y^2 = 1$

$z = 0.8xy$

$\mathbf{r}(t) = \langle \cos t, \sin t, 0.4 \sin 2t \rangle$

Another look at the roller coaster curve

(b)

Figure 14.5

Related Exercises 28, 55 ◀

EXAMPLE 4 **Slinky curve** Use a graphing utility to graph the curve

$$\mathbf{r}(t) = (3 + \cos 15t) \cos t\,\mathbf{i} + (3 + \cos 15t) \sin t\,\mathbf{j} + \sin 15t\,\mathbf{k},$$

for $0 \le t \le 2\pi$, and discuss its properties.

SOLUTION The factor $A(t) = 3 + \cos 15t$ that appears in the x- and y-components is a varying amplitude for $\cos t\,\mathbf{i}$ and $\sin t\,\mathbf{j}$. Its effect is seen in the graph of the x-component $A(t) \cos t$ (**Figure 14.6a**). For $0 \le t \le 2\pi$, the curve consists of one period of $3 \cos t$ with 15 small oscillations superimposed on it. As a result, the x-component of $\mathbf{r}$ varies from -4 to 4 with 15 small oscillations along the way. A similar behavior is seen in the y-component of $\mathbf{r}$. Finally, the z-component of $\mathbf{r}$, which is $\sin 15t$, oscillates between -1 and 1 fifteen times over $[0, 2\pi]$. Combining these effects, we discover a coil-shaped curve that circles the z-axis in the counterclockwise direction and closes on itself. **Figures 14.6b** and **14.6c** show two views, one looking along the xy-plane and the other from overhead on the z-axis. It can be shown (Exercise 56) that eliminating the parameter from the parametric equations defining $\mathbf{r}$ leads to a standard equation of a torus in Cartesian coordinates—in this case, $(3 - \sqrt{x^2 + y^2})^2 + z^2 = 1$—and therefore, the curve lies on this torus, as seen in Figure 14.6b.

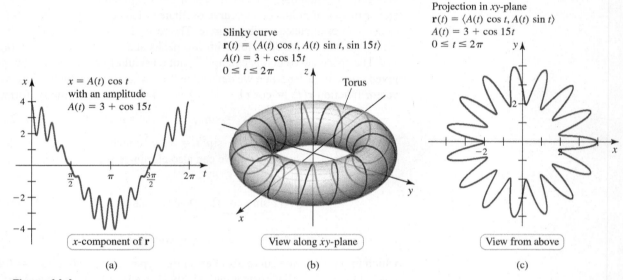

Figure 14.6

Related Exercises 27, 56 ◀

Limits and Continuity for Vector-Valued Functions

We have presented vector valued functions and established their relationship to parametric equations. The next step is to investigate the calculus of vector-valued functions. The concepts of limits, derivatives, and integrals of vector-valued functions are direct extensions of what you have already learned.

The limit of a vector-valued function $\mathbf{r}(t) = f(t)\mathbf{i} + g(t)\mathbf{j} + h(t)\mathbf{k}$ is defined much as it is for scalar-valued functions. If there is a vector $\mathbf{L}$ such that the scalar function $|\mathbf{r}(t) - \mathbf{L}|$ can be made arbitrarily small by taking t sufficiently close to a, then we write $\lim_{t \to a} \mathbf{r}(t) = \mathbf{L}$ and say the limit of $\mathbf{r}$ as t approaches a is $\mathbf{L}$.

DEFINITION Limit of a Vector-Valued Function

A vector-valued function $\mathbf{r}$ approaches the limit $\mathbf{L}$ as t approaches a, written $\lim_{t \to a} \mathbf{r}(t) = \mathbf{L}$, provided $\lim_{t \to a} |\mathbf{r}(t) - \mathbf{L}| = 0$.

Notice that while $\mathbf{r}$ is vector valued, $|\mathbf{r}(t) - \mathbf{L}|$ is a function of the single variable t, to which our familiar limit theorems apply. Therefore, this definition and a short calculation (Exercise 66) lead to a straightforward method for computing limits of the vector-valued function $\mathbf{r} = \langle f, g, h \rangle$. Suppose

$$\lim_{t \to a} f(t) = L_1, \qquad \lim_{t \to a} g(t) = L_2, \qquad \text{and} \qquad \lim_{t \to a} h(t) = L_3.$$

Then

$$\lim_{t \to a} \mathbf{r}(t) = \left\langle \lim_{t \to a} f(t), \lim_{t \to a} g(t), \lim_{t \to a} h(t) \right\rangle = \langle L_1, L_2, L_3 \rangle.$$

In other words, the limit of $\mathbf{r}$ is determined by computing the limits of its components.

The limits laws in Chapter 2 have analogs for vector-valued functions. For example, if $\lim_{t \to a} \mathbf{r}(t)$ and $\lim_{t \to a} \mathbf{s}(t)$ exist and c is a scalar, then

$$\lim_{t \to a} (\mathbf{r}(t) + \mathbf{s}(t)) = \lim_{t \to a} \mathbf{r}(t) + \lim_{t \to a} \mathbf{s}(t) \quad \text{and} \quad \lim_{t \to a} c\mathbf{r}(t) = c\lim_{t \to a} \mathbf{r}(t).$$

The idea of continuity also extends directly to vector-valued functions. A function $\mathbf{r}(t) = f(t)\mathbf{i} + g(t)\mathbf{j} + h(t)\mathbf{k}$ is **continuous at** a provided $\lim_{t \to a} \mathbf{r}(t) = \mathbf{r}(a)$. Specifically,

▶ Continuity is often taken as part of the definition of a parameterized curve.

if the component functions f, g, and h are continuous at a, then $\mathbf{r}$ is also continuous at a, and vice versa. The function $\mathbf{r}$ is **continuous on an interval** I if it is continuous for all t in I.

Continuity has the same intuitive meaning in this setting as it does for scalar-valued functions. If $\mathbf{r}$ is continuous on an interval, the curve it describes has no breaks or gaps, which is an important property when $\mathbf{r}$ describes the trajectory of an object.

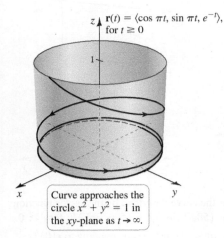

z ▲ $\mathbf{r}(t) = \langle \cos \pi t, \sin \pi t, e^{-t} \rangle$, for $t \geq 0$

1

x y

Curve approaches the circle $x^2 + y^2 = 1$ in the xy-plane as $t \to \infty$.

Figure 14.7

QUICK CHECK 2 Explain why the curve in Example 5 lies on the cylinder $x^2 + y^2 = 1$, as shown in Figure 14.7. ◀

EXAMPLE 5   Limits and continuity Consider the function

$$\mathbf{r}(t) = \cos \pi t\, \mathbf{i} + \sin \pi t\, \mathbf{j} + e^{-t}\, \mathbf{k}, \quad \text{for } t \geq 0.$$

a. Evaluate $\lim\limits_{t \to 2} \mathbf{r}(t)$.

b. Evaluate $\lim\limits_{t \to \infty} \mathbf{r}(t)$.

c. At what points is $\mathbf{r}$ continuous?

SOLUTION

a. We evaluate the limit of each component of $\mathbf{r}$:

$$\lim_{t \to 2} \mathbf{r}(t) = \lim_{t \to 2} (\underbrace{\cos \pi t}_{\to 1}\, \mathbf{i} + \underbrace{\sin \pi t}_{\to 0}\, \mathbf{j} + \underbrace{e^{-t}}_{\to e^{-2}}\mathbf{k}) = \mathbf{i} + e^{-2}\mathbf{k}.$$

b. Note that although $\lim\limits_{t \to \infty} e^{-t} = 0$, $\lim\limits_{t \to \infty} \cos t$ and $\lim\limits_{t \to \infty} \sin t$ do not exist. Therefore, $\lim\limits_{t \to \infty} \mathbf{r}(t)$ does not exist. As shown in **Figure 14.7**, the curve is a coil that approaches the unit circle in the xy-plane.

c. Because the components of $\mathbf{r}$ are continuous for all t, $\mathbf{r}$ is also continuous for all t.

Related Exercise 31 ◀

SECTION 14.1 EXERCISES

Getting Started

1. How many independent variables does the function $\mathbf{r}(t) = \langle f(t), g(t), h(t) \rangle$ have?

2. How many dependent scalar variables does the function $\mathbf{r}(t) = \langle f(t), g(t), h(t) \rangle$ have?

3. Why is $\mathbf{r}(t) = \langle f(t), g(t), h(t) \rangle$ called a vector-valued function?

4. In what plane does the curve $\mathbf{r}(t) = t\mathbf{i} + t^2\mathbf{k}$ lie?

5. How do you evaluate $\lim\limits_{t \to a} \mathbf{r}(t)$, where $\mathbf{r}(t) = \langle f(t), g(t), h(t) \rangle$?

6. How do you determine whether $\mathbf{r}(t) = f(t)\mathbf{i} + g(t)\mathbf{j} + h(t)\mathbf{k}$ is continuous at $t = a$?

7. Find a function $\mathbf{r}(t)$ for the line passing through the points $P(0, 0, 0)$ and $Q(1, 2, 3)$. Express your answer in terms of $\mathbf{i}$, $\mathbf{j}$, and $\mathbf{k}$.

8. Find a function $\mathbf{r}(t)$ whose graph is a circle of radius 1 parallel to the xy-plane and centered at $(0, 0, 10)$.

Practice Exercises

9–14. Lines and line segments *Find a function $\mathbf{r}(t)$ that describes the given line or line segment.*

9. The line through $P(2, 3, 7)$ and $Q(4, 6, 3)$

10. The line through $P(0, -3, 2)$ that is parallel to the line $\mathbf{r}(t) = \langle 4, 6 - t, 1 + t \rangle$

11. The line through $P(3, 4, 5)$ that is orthogonal to the plane $2x - z = 4$

12. The line of intersection of the planes $2x + 3y + 4z = 7$ and $2x + 3y + 5z = 8$

13. The line segment from $P(1, 2, 1)$ to $Q(0, 2, 3)$

14. The line segment from $P(-4, -2, 1)$ to $Q(-2, -2, 3)$

15–26. Graphing curves *Graph the curves described by the following functions, indicating the positive orientation.*

15. $\mathbf{r}(t) = \langle 2 \cos t, 2 \sin t \rangle$, for $0 \leq t \leq 2\pi$

16. $\mathbf{r}(t) = \langle 1 + \cos t, 2 + \sin t \rangle$, for $0 \leq t \leq 2\pi$

17. $\mathbf{r}(t) = \langle t, 2t \rangle$, for $0 \leq t \leq 1$

18. $\mathbf{r}(t) = \langle 3 \cos t, 2 \sin t \rangle$, for $0 \leq t \leq 2\pi$

19. $\mathbf{r}(t) = \langle \cos t, 0, \sin t \rangle$, for $0 \leq t \leq 2\pi$

20. $\mathbf{r}(t) = \langle 0, 4 \cos t, 16 \sin t \rangle$, for $0 \leq t \leq 2\pi$

21. $\mathbf{r}(t) = \cos t\, \mathbf{i} + \mathbf{j} + \sin t\, \mathbf{k}$, for $0 \leq t \leq 2\pi$

22. $\mathbf{r}(t) = 2 \cos t\, \mathbf{i} + 2 \sin t\, \mathbf{j} + 2\, \mathbf{k}$, for $0 \leq t \leq 2\pi$

T 23. $\mathbf{r}(t) = t \cos t\, \mathbf{i} + t \sin t\, \mathbf{j} + t\, \mathbf{k}$, for $0 \leq t \leq 6\pi$

T 24. $\mathbf{r}(t) = 4 \sin t\, \mathbf{i} + 4 \cos t\, \mathbf{j} + e^{-t/10}\, \mathbf{k}$, for $0 \leq t < \infty$

T 25. $\mathbf{r}(t) = e^{-t/20} \sin t\, \mathbf{i} + e^{-t/20} \cos t\, \mathbf{j} + t\, \mathbf{k}$, for $0 \leq t < \infty$

T 26. $\mathbf{r}(t) = e^{-t/10}\mathbf{i} + 3 \cos t\, \mathbf{j} + 3 \sin t\, \mathbf{k}$, for $0 \leq t < \infty$

T **27–30. Exotic curves** *Graph the curves described by the following functions. Use analysis to anticipate the shape of the curve before using a graphing utility.*

27. $\mathbf{r}(t) = \cos 15t\, \mathbf{i} + (4 + \sin 15t) \cos t\, \mathbf{j} + (4 + \sin 15t) \sin t\, \mathbf{k}$, for $0 \leq t \leq 2\pi$

28. $\mathbf{r}(t) = 2 \cos t\, \mathbf{i} + 4 \sin t\, \mathbf{j} + \cos 10t\, \mathbf{k}$, for $0 \leq t \leq 2\pi$

29. $\mathbf{r}(t) = \sin t\,\mathbf{i} + \sin^2 t\,\mathbf{j} + \dfrac{t}{5\pi}\,\mathbf{k}$, for $0 \le t \le 10\pi$

30. $\mathbf{r}(t) = \cos t \sin 3t\,\mathbf{i} + \sin t \sin 3t\,\mathbf{j} + \sqrt{t}\,\mathbf{k}$, for $0 \le t \le 9$

31–36. Limits *Evaluate the following limits.*

31. $\lim\limits_{t\to\pi/2}\left(\cos 2t\,\mathbf{i} - 4\sin t\,\mathbf{j} + \dfrac{2t}{\pi}\,\mathbf{k}\right)$

32. $\lim\limits_{t\to\ln 2}(2e^{t}\mathbf{i} + 6e^{-t}\mathbf{j} - 4e^{-2t}\mathbf{k})$

33. $\lim\limits_{t\to\infty}\left(e^{-t}\mathbf{i} - \dfrac{2t}{t+1}\,\mathbf{j} + \tan^{-1}t\,\mathbf{k}\right)$

34. $\lim\limits_{t\to 2}\left(\dfrac{t}{t^2+1}\,\mathbf{i} - 4e^{-t}\sin \pi t\,\mathbf{j} + \dfrac{1}{\sqrt{4t+1}}\,\mathbf{k}\right)$

35. $\lim\limits_{t\to 0}\left(\dfrac{\sin t}{t}\,\mathbf{i} - \dfrac{e^{t}-t-1}{t}\,\mathbf{j} + \dfrac{\cos t + t^2/2 - 1}{t^2}\,\mathbf{k}\right)$

36. $\lim\limits_{t\to 0}\left(\dfrac{\tan t}{t}\,\mathbf{i} - \dfrac{3t}{\sin t}\,\mathbf{j} + \sqrt{t+1}\,\mathbf{k}\right)$

37. **Explain why or why not** Determine whether the following statements are true and give an explanation or counterexample.

   a. The projection of the curve $\mathbf{r}(t) = \langle t, \cos t, t^2\rangle$ in the xz-plane is a parabola.

   b. The curve $\mathbf{r}(t) = \langle \sin t, \cos t, \sin t\rangle$ lies on a unit sphere.

   c. The curve $\mathbf{r}(t) = \langle e^{-t}, \sin t, -\cos t\rangle$ approaches a circle as $t\to\infty$.

   d. If $\mathbf{r}(t) = e^{-t^2}\langle 1, 1, 1\rangle$, then $\lim\limits_{t\to\infty}\mathbf{r}(t) = \lim\limits_{t\to-\infty}\mathbf{r}(t)$.

38–41. Domains *Find the domain of the following vector-valued functions.*

38. $\mathbf{r}(t) = \dfrac{2}{t-1}\,\mathbf{i} + \dfrac{3}{t+2}\,\mathbf{j}$

39. $\mathbf{r}(t) = \sqrt{t+2}\,\mathbf{i} + \sqrt{2-t}\,\mathbf{j}$

40. $\mathbf{r}(t) = \cos 2t\,\mathbf{i} + e^{\sqrt{t}}\mathbf{j} + \dfrac{12}{t}\,\mathbf{k}$

41. $\mathbf{r}(t) = \sqrt{4-t^2}\,\mathbf{i} + \sqrt{t}\,\mathbf{j} - \dfrac{2}{\sqrt{1+t}}\,\mathbf{k}$

42–44. Curve-plane intersections *Find the points (if they exist) at which the following planes and curves intersect.*

42. $y = 1;\ \mathbf{r}(t) = \langle 10\cos t, 2\sin t, 1\rangle$, for $0 \le t \le 2\pi$

43. $z = 16;\ \mathbf{r}(t) = \langle t, 2t, 4+3t\rangle$, for $-\infty < t < \infty$

44. $y + x = 0;\ \mathbf{r}(t) = \langle \cos t, \sin t, t\rangle$, for $0 \le t \le 4\pi$

45. **Matching functions with graphs** Match functions a–f with the appropriate graphs A–F.

   a. $\mathbf{r}(t) = \langle t, -t, t\rangle$ **b.** $\mathbf{r}(t) = \langle t^2, t, t\rangle$

   c. $\mathbf{r}(t) = \langle 4\cos t, 4\sin t, 2\rangle$ **d.** $\mathbf{r}(t) = \langle 2t, \sin t, \cos t\rangle$

   e. $\mathbf{r}(t) = \langle \sin t, \cos t, \sin 2t\rangle$ **f.** $\mathbf{r}(t) = \langle \sin t, 2t, \cos t\rangle$

(A)

(B)

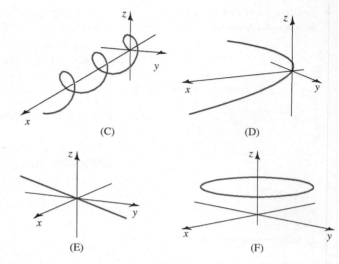

(C)   (D)

(E)   (F)

46. **Upward path** Consider the curve described by the vector function $\mathbf{r}(t) = (50e^{-t}\cos t)\mathbf{i} + (50e^{-t}\sin t)\mathbf{j} + (5 - 5e^{-t})\mathbf{k}$, for $t \ge 0$.

   a. What is the initial point of the path corresponding to $\mathbf{r}(0)$?

   b. What is $\lim\limits_{t\to\infty}\mathbf{r}(t)$?

   c. Eliminate the parameter t to show that the curve $\mathbf{r}(t)$ lies on the surface $z = 5 - r/10$, where $r^2 = x^2 + y^2$.

47–50. Curve of intersection *Find a function $\mathbf{r}(t)$ that describes the curve where the following surfaces intersect. Answers are not unique.*

47. $z = 4;\ z = x^2 + y^2$

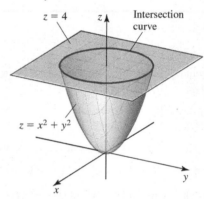

48. $z = 3x^2 + y^2 + 1;\ z = 5 - x^2 - 3y^2$

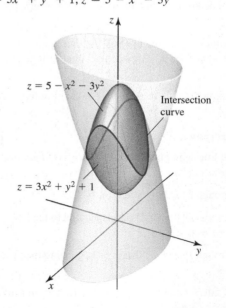

49. $x^2 + y^2 = 25; z = 2x + 2y$

50. $z = y + 1; z = x^2 + 1$

T 51. Golf slice A golfer launches a tee shot down a horizontal fairway; it follows a path given by $\mathbf{r}(t) = \langle at, (75 - 0.1a)t, -5t^2 + 80t \rangle$, where $t \geq 0$ measures time in seconds and $\mathbf{r}$ has units of feet. The y-axis points straight down the fairway and the z-axis points vertically upward. The parameter a is the slice factor that determines how much the shot deviates from a straight path down the fairway.

    a. With no slice ($a = 0$), describe the shot. How far does the ball travel horizontally (the distance between the point where the ball leaves the ground and the point where it first strikes the ground)?

    b. With a slice ($a = 0.2$), how far does the ball travel horizontally?

    c. How far does the ball travel horizontally with $a = 2.5$?

52–56. Curves on surfaces *Verify that the curve* $\mathbf{r}(t)$ *lies on the given surface. Give the name of the surface.*

52. $\mathbf{r}(t) = (t \cos t)\mathbf{i} + (t \sin t)\mathbf{j} + t\,\mathbf{k}; x^2 + y^2 = z^2$

53. $\mathbf{r}(t) = \left(\sqrt{t^2 + 1}\,\cos t\right)\mathbf{i} + \left(\sqrt{t^2 + 1}\,\sin t\right)\mathbf{j} + t\,\mathbf{k};$ $x^2 + y^2 - z^2 = 1$

54. $\mathbf{r}(t) = \langle \sqrt{t}\cos t, \sqrt{t}\sin t, t \rangle; z = x^2 + y^2$

55. $\mathbf{r}(t) = \langle 0, 2\cos t, 3\sin t \rangle; x^2 + \dfrac{y^2}{4} + \dfrac{z^2}{9} = 1$

56. $\mathbf{r}(t) = \langle (3 + \cos 15t)\cos t, (3 + \cos 15t)\sin t, \sin 15t \rangle;$ $\left(3 - \sqrt{x^2 + y^2}\right)^2 + z^2 = 1$ (*Hint:* See Example 4.)

57–58. Closest point on a curve *Find the point P on the curve* $\mathbf{r}(t)$ *that lies closest to* P_0 *and state the distance between* P_0 *and P.*

57. $\mathbf{r}(t) = t^2\mathbf{i} + t\,\mathbf{j} + t\,\mathbf{k}; P_0(1, 1, 15)$

T 58. $\mathbf{r}(t) = \cos t\,\mathbf{i} + \sin t\,\mathbf{j} + t\,\mathbf{k}; P_0(1, 1, 3)$

Explorations and Challenges

59–61. Curves on spheres

T 59. Graph the curve $\mathbf{r}(t) = \left\langle \dfrac{1}{2}\sin 2t, \dfrac{1}{2}(1 - \cos 2t), \cos t \right\rangle$ and prove that it lies on the surface of a sphere centered at the origin.

60. Prove that for integers m and n, the curve

$$\mathbf{r}(t) = \langle a \sin mt \cos nt, b \sin mt \sin nt, c \cos mt \rangle$$

lies on the surface of a sphere provided $a^2 = b^2 = c^2$.

61. Find the period of the function in Exercise 60; that is, in terms of m and n, find the smallest positive real number T such that $\mathbf{r}(t + T) = \mathbf{r}(t)$ for all t.

62–65. Closed plane curves *Consider the curve*

$$\mathbf{r}(t) = \langle a\cos t + b\sin t, c\cos t + d\sin t, e\cos t + f\sin t \rangle,$$

where a, b, c, d, e, and f are real numbers. It can be shown that this curve lies in a plane.

62. Assuming the curve lies in a plane, show that it is a circle centered at the origin with radius R provided $a^2 + c^2 + e^2 = b^2 + d^2 + f^2 = R^2$ and $ab + cd + ef = 0$.

T 63. Graph the following curve and describe it.

$$\mathbf{r}(t) = \left(\frac{1}{\sqrt{2}}\cos t + \frac{1}{\sqrt{3}}\sin t\right)\mathbf{i}$$
$$+ \left(-\frac{1}{\sqrt{2}}\cos t + \frac{1}{\sqrt{3}}\sin t\right)\mathbf{j} + \left(\frac{1}{\sqrt{3}}\sin t\right)\mathbf{k}$$

T 64. Graph the following curve and describe it.

$$\mathbf{r}(t) = (2\cos t + 2\sin t)\mathbf{i} + (-\cos t + 2\sin t)\mathbf{j} + (\cos t - 2\sin t)\mathbf{k}$$

65. Find a general expression for a nonzero vector orthogonal to the plane containing the curve

$$\mathbf{r}(t) = \langle a\cos t + b\sin t, c\cos t + d\sin t, e\cos t + f\sin t \rangle,$$

where $\langle a, c, e \rangle \times \langle b, d, f \rangle \neq \mathbf{0}$.

66. Limits of vector functions Let $\mathbf{r}(t) = \langle f(t), g(t), h(t) \rangle$.

    a. Assume $\lim\limits_{t \to a} \mathbf{r}(t) = \mathbf{L} = \langle L_1, L_2, L_3 \rangle$, which means that $\lim\limits_{t \to a} |\mathbf{r}(t) - \mathbf{L}| = 0$. Prove that

$$\lim_{t \to a} f(t) = L_1, \quad \lim_{t \to a} g(t) = L_2, \quad \text{and} \quad \lim_{t \to a} h(t) = L_3.$$

    b. Assume $\lim\limits_{t \to a} f(t) = L_1, \lim\limits_{t \to a} g(t) = L_2,$ and $\lim\limits_{t \to a} h(t) = L_3$. Prove that $\lim\limits_{t \to a} \mathbf{r}(t) = \mathbf{L} = \langle L_1, L_2, L_3 \rangle$, which means that $\lim\limits_{t \to a} |\mathbf{r}(t) - \mathbf{L}| = 0$.

QUICK CHECK ANSWERS

1. $0 \leq t \leq 2$ **2.** The x- and y-components of the curve are $x = \cos \pi t$ and $y = \sin \pi t$, and $x^2 + y^2 = \cos^2 \pi t + \sin^2 \pi t = 1$. ◄

14.2 Calculus of Vector-Valued Functions

We now turn to the topic of ultimate interest in this chapter: the calculus of vector-valued functions. Everything you learned about differentiating and integrating functions of the form $y = f(x)$ carries over to vector-valued functions $\mathbf{r}(t)$; you simply apply the rules of differentiation and integration to the individual components of $\mathbf{r}$.

The Derivative and Tangent Vector

Consider the function $\mathbf{r}(t) = f(t)\,\mathbf{i} + g(t)\,\mathbf{j} + h(t)\,\mathbf{k}$, where f, g, and h are differentiable functions on an interval $a < t < b$. The first task is to explain the meaning of the *derivative* of a vector-valued function and to show how to compute it. We begin with the definition of the derivative—now from a vector perspective:

$$\mathbf{r}'(t) = \lim_{\Delta t \to 0} \frac{\Delta \mathbf{r}}{\Delta t} = \lim_{\Delta t \to 0} \frac{\mathbf{r}(t + \Delta t) - \mathbf{r}(t)}{\Delta t}.$$

Let's first look at the geometry of this limit. The function $\mathbf{r}(t) = f(t)\,\mathbf{i} + g(t)\,\mathbf{j} + h(t)\,\mathbf{k}$ describes a parameterized curve in space. Let P be a point on that curve associated with the position vector $\mathbf{r}(t)$, and let Q be a nearby point associated with the position vector $\mathbf{r}(t + \Delta t)$, where $\Delta t > 0$ is a small increment in t (**Figure 14.8a**). The difference $\Delta \mathbf{r} = \mathbf{r}(t + \Delta t) - \mathbf{r}(t)$ is the vector $\overrightarrow{PQ}$, where we assume $\Delta \mathbf{r} \neq \mathbf{0}$. Because Δt is a scalar, the direction of $\Delta \mathbf{r}/\Delta t$ is the same as the direction of $\overrightarrow{PQ}$.

As Δt approaches 0, Q approaches P and the vector $\Delta \mathbf{r}/\Delta t$ approaches a limiting vector that we denote $\mathbf{r}'(t)$ (**Figure 14.8b**). This new vector $\mathbf{r}'(t)$ has two important interpretations.

▶ An analogous argument can be given for $\Delta t < 0$, with the same result. Figure 14.8 illustrates the tangent vector $\mathbf{r}'$ for $\Delta t > 0$.

- The vector $\mathbf{r}'(t)$ points in the direction of the curve at P. For this reason, $\mathbf{r}'(t)$ is a *tangent vector* at P (provided it is not the zero vector).

- The vector $\mathbf{r}'(t)$ is the *derivative* of $\mathbf{r}$ with respect to t; it gives the rate of change of the function $\mathbf{r}(t)$ at the point P. In fact, if $\mathbf{r}(t)$ is the position function of a moving object, then $\mathbf{r}'(t)$ is the velocity vector of the object, which always points in the direction of motion, and $|\mathbf{r}'(t)|$ is the speed of the object.

▶ Section 14.3 is devoted to problems of motion in two and three dimensions.

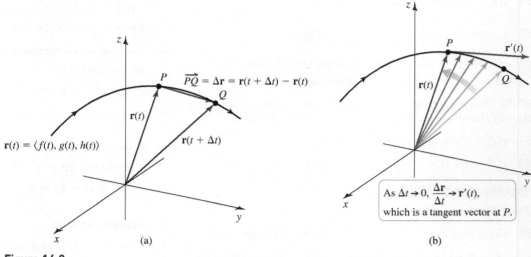

Figure 14.8

We now evaluate the limit that defines $\mathbf{r}'(t)$ by expressing $\mathbf{r}$ in terms of its components and using the properties of limits.

$$
\begin{aligned}
\mathbf{r}'(t) &= \lim_{\Delta t \to 0} \frac{\mathbf{r}(t + \Delta t) - \mathbf{r}(t)}{\Delta t} \\[2mm]
&= \lim_{\Delta t \to 0} \frac{(f(t + \Delta t)\,\mathbf{i} + g(t + \Delta t)\,\mathbf{j} + h(t + \Delta t)\,\mathbf{k}) - (f(t)\,\mathbf{i} + g(t)\,\mathbf{j} + h(t)\,\mathbf{k})}{\Delta t} \\
&\qquad\qquad\qquad\qquad\qquad\qquad\qquad\qquad\qquad\qquad\qquad \text{Substitute components of } \mathbf{r}. \\[2mm]
&= \lim_{\Delta t \to 0} \left(\frac{f(t + \Delta t) - f(t)}{\Delta t}\,\mathbf{i} + \frac{g(t + \Delta t) - g(t)}{\Delta t}\,\mathbf{j} + \frac{h(t + \Delta t) - h(t)}{\Delta t}\,\mathbf{k} \right) \\
&\qquad\qquad\qquad\qquad\qquad\qquad\qquad\qquad\qquad \text{Rearrange terms inside of limit.} \\[2mm]
&= \underbrace{\lim_{\Delta t \to 0} \frac{f(t + \Delta t) - f(t)}{\Delta t}}_{f'(t)}\,\mathbf{i} + \underbrace{\lim_{\Delta t \to 0} \frac{g(t + \Delta t) - g(t)}{\Delta t}}_{g'(t)}\,\mathbf{j} + \underbrace{\lim_{\Delta t \to 0} \frac{h(t + \Delta t) - h(t)}{\Delta t}}_{h'(t)}\,\mathbf{k}
\end{aligned}
$$

Limit of sum equals sum of limits.

Because f, g, and h are differentiable scalar-valued functions of the variable t, the three limits in the last step are identified as the derivatives of f, g, and h, respectively. Therefore, there are no surprises:

$$\mathbf{r}'(t) = f'(t)\,\mathbf{i} + g'(t)\,\mathbf{j} + h'(t)\,\mathbf{k}.$$

In other words, to differentiate the vector-valued function $\mathbf{r}(t)$, we simply differentiate each of its components with respect to t.

DEFINITION Derivative and Tangent Vector

Let $\mathbf{r}(t) = f(t)\,\mathbf{i} + g(t)\,\mathbf{j} + h(t)\,\mathbf{k}$, where f, g, and h are differentiable functions on (a, b). Then $\mathbf{r}$ has a **derivative** (or is **differentiable**) on (a, b) and

$$\mathbf{r}'(t) = f'(t)\,\mathbf{i} + g'(t)\,\mathbf{j} + h'(t)\,\mathbf{k}.$$

Provided $\mathbf{r}'(t) \neq \mathbf{0}$, $\mathbf{r}'(t)$ is a **tangent vector** at the point corresponding to $\mathbf{r}(t)$.

EXAMPLE 1 Derivative of vector functions Compute the derivative of the following functions.

a. $\mathbf{r}(t) = \langle t^3, 3t^2, t^3/6 \rangle$ **b.** $\mathbf{r}(t) = e^{-t}\,\mathbf{i} + 10\sqrt{t}\,\mathbf{j} + 2\cos 3t\,\mathbf{k}$

SOLUTION

a. $\mathbf{r}'(t) = \langle 3t^2, 6t, t^2/2 \rangle$; note that $\mathbf{r}$ is differentiable for all t and $\mathbf{r}'(0) = \mathbf{0}$.

b. $\mathbf{r}'(t) = -e^{-t}\,\mathbf{i} + \dfrac{5}{\sqrt{t}}\,\mathbf{j} - 6\sin 3t\,\mathbf{k}$; the function $\mathbf{r}$ is differentiable for $t > 0$.

Related Exercises 11–12 ◄

QUICK CHECK 1 Let $\mathbf{r}(t) = \langle t, t, t \rangle$. Compute $\mathbf{r}'(t)$ and interpret the result. ◄

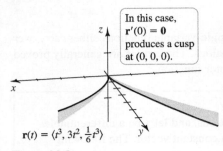

In this case, $\mathbf{r}'(0) = \mathbf{0}$ produces a cusp at $(0, 0, 0)$.

$\mathbf{r}(t) = \langle t^3, 3t^2, \frac{1}{6}t^3 \rangle$

Figure 14.9

► If a curve has a cusp at a point, then $\mathbf{r}'(t) = \mathbf{0}$ at that point. However, the converse is not true; it may happen that $\mathbf{r}'(t) = \mathbf{0}$ at a point that is not a cusp (Exercise 95).

The condition that $\mathbf{r}'(t) \neq \mathbf{0}$ in order for the tangent vector to be defined requires explanation. Consider the function $\mathbf{r}(t) = \langle t^3, 3t^2, t^3/6 \rangle$. As shown in Example 1a, $\mathbf{r}'(0) = \mathbf{0}$; that is, all three components of $\mathbf{r}'(t)$ are zero simultaneously when $t = 0$. We see in **Figure 14.9** that this otherwise smooth curve has a *cusp*, or a sharp point, at the origin. If $\mathbf{r}$ describes the motion of an object, then $\mathbf{r}'(t) = \mathbf{0}$ means that the velocity (and speed) of the object is zero at a point. At such a stationary point, the object *may* change direction abruptly, creating a cusp in its trajectory. For this reason, we say a function $\mathbf{r}(t) = \langle f(t), g(t), h(t) \rangle$ is **smooth** on an interval if f, g, and h are differentiable *and* $\mathbf{r}'(t) \neq \mathbf{0}$ on that interval. Smooth curves have no cusps or corners.

Unit Tangent Vector In situations in which only the direction (but not the length) of the tangent vector is of interest, we work with the *unit tangent vector*. It is the vector with magnitude 1, formed by dividing $\mathbf{r}'(t)$ by its length.

DEFINITION Unit Tangent Vector

Let $\mathbf{r}(t) = f(t)\,\mathbf{i} + g(t)\,\mathbf{j} + h(t)\,\mathbf{k}$ be a smooth parameterized curve, for $a \leq t \leq b$. The **unit tangent vector** for a particular value of t is

$$\mathbf{T}(t) = \frac{\mathbf{r}'(t)}{|\mathbf{r}'(t)|}.$$

QUICK CHECK 2 Suppose $\mathbf{r}'(t)$ has units of m/s. Explain why $\mathbf{T}(t) = \mathbf{r}'(t)/|\mathbf{r}'(t)|$ is dimensionless (has no units) and carries information only about direction. ◄

EXAMPLE 2 Unit tangent vectors Find the unit tangent vectors for the following parameterized curves.

a. $\mathbf{r}(t) = \langle t^2, 4t, 4\ln t \rangle$,   for $t > 0$

b. $\mathbf{r}(t) = \langle 10, 3\cos t, 3\sin t \rangle$,   for $0 \leq t \leq 2\pi$

SOLUTION

a. A tangent vector is $\mathbf{r}'(t) = \langle 2t, 4, 4/t \rangle$, which has a magnitude of

$$|\mathbf{r}'(t)| = \sqrt{(2t)^2 + 4^2 + \left(\frac{4}{t}\right)^2} \quad \text{Definition of magnitude}$$

$$= \sqrt{4t^2 + 16 + \frac{16}{t^2}} \quad \text{Expand.}$$

$$= \sqrt{\left(2t + \frac{4}{t}\right)^2} \quad \text{Factor.}$$

$$= 2t + \frac{4}{t}. \quad \text{Simplify.}$$

Therefore, the unit tangent vector for a particular value of t is

$$\mathbf{T}(t) = \frac{\langle 2t, 4, 4/t \rangle}{2t + 4/t}.$$

As shown in **Figure 14.10**, the unit tangent vectors change direction along the curve but maintain unit length.

b. In this case, $\mathbf{r}'(t) = \langle 0, -3 \sin t, 3 \cos t \rangle$ and

$$|\mathbf{r}'(t)| = \sqrt{0^2 + (-3 \sin t)^2 + (3 \cos t)^2} = \sqrt{9\underbrace{(\sin^2 t + \cos^2 t)}_{1}} = 3.$$

Therefore, the unit tangent vector for a particular value of t is

$$\mathbf{T}(t) = \frac{1}{3}\langle 0, -3 \sin t, 3 \cos t \rangle = \langle 0, -\sin t, \cos t \rangle.$$

The direction of $\mathbf{T}$ changes along the curve, but its length remains 1.

Related Exercises 25, 27 ◄

Unit tangent vectors change direction along the curve, but they always have length 1.

$\mathbf{r}(t) = \langle t^2, 4t, 4 \ln t \rangle$, for $t > 0$

Figure 14.10

Derivative Rules   The rules for derivatives for single-variable functions either carry over directly to vector-valued functions or have close analogs. These rules are generally proved by working on the individual components of the vector function.

> **THEOREM 14.1   Derivative Rules**
> Let $\mathbf{u}$ and $\mathbf{v}$ be differentiable vector-valued functions, and let f be a differentiable scalar-valued function, all at a point t. Let $\mathbf{c}$ be a constant vector. The following rules apply.
>
> 1. $\dfrac{d}{dt}(\mathbf{c}) = \mathbf{0}$ Constant Rule
>
> 2. $\dfrac{d}{dt}(\mathbf{u}(t) + \mathbf{v}(t)) = \mathbf{u}'(t) + \mathbf{v}'(t)$ Sum Rule
>
> 3. $\dfrac{d}{dt}(f(t)\mathbf{u}(t)) = f'(t)\mathbf{u}(t) + f(t)\mathbf{u}'(t)$ Product Rule
>
> 4. $\dfrac{d}{dt}\big(\mathbf{u}(f(t))\big) = \mathbf{u}'(f(t))f'(t)$ Chain Rule
>
> 5. $\dfrac{d}{dt}(\mathbf{u}(t) \cdot \mathbf{v}(t)) = \mathbf{u}'(t) \cdot \mathbf{v}(t) + \mathbf{u}(t) \cdot \mathbf{v}'(t)$ Dot Product Rule
>
> 6. $\dfrac{d}{dt}(\mathbf{u}(t) \times \mathbf{v}(t)) = \mathbf{u}'(t) \times \mathbf{v}(t) + \mathbf{u}(t) \times \mathbf{v}'(t)$ Cross Product Rule

➤ With the exception of the Cross Product Rule, these rules apply to vector-valued functions with any number of components. Notice that we have three new product rules, all of which mimic the original Product Rule. In Rule 4, $\mathbf{u}$ must be differentiable at $f(t)$.

QUICK CHECK 3 Let $\mathbf{u}(t) = \langle t, t, t \rangle$ and $\mathbf{v}(t) = \langle 1, 1, 1 \rangle$. Compute $\dfrac{d}{dt}(\mathbf{u}(t) \cdot \mathbf{v}(t))$ using Derivative Rule 5, and show that it agrees with the result obtained by first computing the dot product and differentiating directly. ◄

The proofs of these rules are assigned in Exercises 92–94 with the exception of the following representative proofs.

Proof of the Chain Rule: Let $\mathbf{u}(t) = \langle u_1(t), u_2(t), u_3(t) \rangle$, which implies that

$$\mathbf{u}(f(t)) = u_1(f(t))\,\mathbf{i} + u_2(f(t))\,\mathbf{j} + u_3(f(t))\,\mathbf{k}.$$

We now apply the ordinary Chain Rule componentwise:

$$\frac{d}{dt}\big(\mathbf{u}(f(t))\big) = \frac{d}{dt}\big(u_1(f(t))\,\mathbf{i} + u_2(f(t))\,\mathbf{j} + u_3(f(t))\,\mathbf{k}\big) \qquad \text{Components of } \mathbf{u}$$

$$= \frac{d}{dt}\big(u_1(f(t))\big)\,\mathbf{i} + \frac{d}{dt}\big(u_2(f(t))\big)\,\mathbf{j} + \frac{d}{dt}\big(u_3(f(t))\big)\,\mathbf{k} \qquad \begin{array}{l}\text{Differentiate each}\\ \text{component.}\end{array}$$

$$= u_1{}'(f(t))\,f'(t)\,\mathbf{i} + u_2{}'(f(t))\,f'(t)\,\mathbf{j} + u_3{}'(f(t))\,f'(t)\,\mathbf{k} \qquad \text{Chain Rule}$$

$$= \big(u_1{}'(f(t))\,\mathbf{i} + u_2{}'(f(t))\,\mathbf{j} + u_3{}'(f(t))\,\mathbf{k}\big)\,f'(t) \qquad \text{Factor } f'(t).$$

$$= \mathbf{u}'(f(t))\,f'(t). \qquad \text{Definition of } \mathbf{u}'$$

◄

Proof of the Dot Product Rule: We use the standard Product Rule on each component. Let $\mathbf{u}(t) = \langle u_1(t), u_2(t), u_3(t)\rangle$ and $\mathbf{v}(t) = \langle v_1(t), v_2(t), v_3(t)\rangle$. Then

$$\frac{d}{dt}(\mathbf{u}\cdot\mathbf{v}) = \frac{d}{dt}(u_1v_1 + u_2v_2 + u_3v_3) \qquad \text{Definition of dot product}$$

$$= u_1{}'\,v_1 + u_1\,v_1{}' + u_2{}'\,v_2 + u_2\,v_2{}' + u_3{}'\,v_3 + u_3\,v_3{}' \qquad \text{Product Rule}$$

$$= \underbrace{u_1{}'\,v_1 + u_2{}'\,v_2 + u_3{}'\,v_3}_{\mathbf{u}'\cdot\mathbf{v}} + \underbrace{u_1\,v_1{}' + u_2\,v_2{}' + u_3\,v_3{}'}_{\mathbf{u}\cdot\mathbf{v}'} \qquad \text{Rearrange.}$$

$$= \mathbf{u}'\cdot\mathbf{v} + \mathbf{u}\cdot\mathbf{v}'. $$

◄

EXAMPLE 3   Derivative rules Compute the following derivatives, where

$$\mathbf{u}(t) = t\,\mathbf{i} + t^2\,\mathbf{j} - t^3\,\mathbf{k} \quad\text{and}\quad \mathbf{v}(t) = \sin t\,\mathbf{i} + 2\cos t\,\mathbf{j} + \cos t\,\mathbf{k}.$$

a. $\dfrac{d}{dt}(\mathbf{v}(t^2))$ **b.** $\dfrac{d}{dt}(t^2\,\mathbf{v}(t))$ **c.** $\dfrac{d}{dt}(\mathbf{u}(t)\cdot\mathbf{v}(t))$

SOLUTION

a. Note that $\mathbf{v}'(t) = \cos t\,\mathbf{i} - 2\sin t\,\mathbf{j} - \sin t\,\mathbf{k}$. Using the Chain Rule, we have

$$\frac{d}{dt}(\mathbf{v}(t^2)) = \mathbf{v}'(t^2)\,\frac{d}{dt}(t^2) = \underbrace{(\cos t^2\,\mathbf{i} - 2\sin t^2\,\mathbf{j} - \sin t^2\,\mathbf{k})}_{\mathbf{v}'(t^2)}(2t).$$

b. $\dfrac{d}{dt}(t^2\,\mathbf{v}(t)) = \dfrac{d}{dt}(t^2)\mathbf{v}(t) + t^2\,\dfrac{d}{dt}(\mathbf{v}(t)) \qquad \text{Product Rule}$

$$= 2t\,\mathbf{v}(t) + t^2\,\mathbf{v}'(t)$$

$$= 2t\underbrace{(\sin t\,\mathbf{i} + 2\cos t\,\mathbf{j} + \cos t\,\mathbf{k})}_{\mathbf{v}(t)} + t^2\underbrace{(\cos t\,\mathbf{i} - 2\sin t\,\mathbf{j} - \sin t\,\mathbf{k})}_{\mathbf{v}'(t)}$$

Differentiate.

$$= (2t\sin t + t^2\cos t)\,\mathbf{i} + (4t\cos t - 2t^2\sin t)\,\mathbf{j} + (2t\cos t - t^2\sin t)\,\mathbf{k}$$

Collect terms.

c. $\dfrac{d}{dt}\big(\mathbf{u}(t)\cdot\mathbf{v}(t)\big) = \mathbf{u}'(t)\cdot\mathbf{v}(t) + \mathbf{u}(t)\cdot\mathbf{v}'(t) \qquad \text{Dot Product Rule}$

$$= (\mathbf{i} + 2t\,\mathbf{j} - 3t^2\,\mathbf{k})\cdot(\sin t\,\mathbf{i} + 2\cos t\,\mathbf{j} + \cos t\,\mathbf{k})$$

$$\quad + (t\,\mathbf{i} + t^2\,\mathbf{j} - t^3\,\mathbf{k})\cdot(\cos t\,\mathbf{i} - 2\sin t\,\mathbf{j} - \sin t\,\mathbf{k}) \qquad \text{Differentiate.}$$

$$= (\sin t + 4t\cos t - 3t^2\cos t) + (t\cos t - 2t^2\sin t + t^3\sin t) \qquad \text{Dot products}$$

$$= (1 - 2t^2 + t^3)\sin t + (5t - 3t^2)\cos t \qquad \text{Simplify.}$$

Note that the result is a scalar. The same result is obtained if you first compute $\mathbf{u}\cdot\mathbf{v}$ and then differentiate.

Related Exercises 33, 36, 37 ◄

Higher-Order Derivatives Higher-order derivatives of vector-valued functions are computed in the expected way: We simply differentiate each component multiple times. Second derivatives feature prominently in the next section, playing the role of acceleration.

EXAMPLE 4 Higher-order derivatives Compute the first, second, and third derivative of $\mathbf{r}(t) = \langle t^2, 8 \ln t, 3e^{-2t} \rangle$.

SOLUTION Differentiating once, we have $\mathbf{r}'(t) = \langle 2t, 8/t, -6e^{-2t} \rangle$. Differentiating again produces $\mathbf{r}''(t) = \langle 2, -8/t^2, 12e^{-2t} \rangle$. Differentiating once more, we have $\mathbf{r}'''(t) = \langle 0, 16/t^3, -24e^{-2t} \rangle$.

Related Exercise 58 ◄

Integrals of Vector-Valued Functions

An **antiderivative** of the vector function $\mathbf{r}$ is a function $\mathbf{R}$ such that $\mathbf{R}' = \mathbf{r}$. If

$$\mathbf{r}(t) = f(t)\,\mathbf{i} + g(t)\,\mathbf{j} + h(t)\,\mathbf{k},$$

then an antiderivative of $\mathbf{r}$ is

$$\mathbf{R}(t) = F(t)\,\mathbf{i} + G(t)\,\mathbf{j} + H(t)\,\mathbf{k},$$

where F, G, and H are antiderivatives of f, g, and h, respectively. This fact follows by differentiating the components of $\mathbf{R}$ and verifying that $\mathbf{R}' = \mathbf{r}$. The collection of all antiderivatives of $\mathbf{r}$ is the *indefinite integral* of $\mathbf{r}$.

DEFINITION Indefinite Integral of a Vector-Valued Function

Let $\mathbf{r}(t) = f(t)\,\mathbf{i} + g(t)\,\mathbf{j} + h(t)\,\mathbf{k}$ be a vector function, and let $\mathbf{R}(t) = F(t)\,\mathbf{i} + G(t)\,\mathbf{j} + H(t)\,\mathbf{k}$, where F, G, and H are antiderivatives of f, g, and h, respectively. The **indefinite integral** of $\mathbf{r}$ is

$$\int \mathbf{r}(t)\,dt = \mathbf{R}(t) + \mathbf{C},$$

where $\mathbf{C}$ is an arbitrary constant vector. Alternatively, in component form,

$$\int \langle f(t), g(t), h(t) \rangle\,dt = \langle F(t), G(t), H(t) \rangle + \langle C_1, C_2, C_3 \rangle.$$

EXAMPLE 5 Indefinite integrals Compute

$$\int \left(\frac{t}{\sqrt{t^2 + 2}}\,\mathbf{i} + e^{-3t}\,\mathbf{j} + (\sin 4t + 1)\,\mathbf{k} \right) dt.$$

> The substitution $u = t^2 + 2$ is used to evaluate the **i**-component of the integral.

SOLUTION We compute the indefinite integral of each component:

$$\int \left(\frac{t}{\sqrt{t^2 + 2}}\,\mathbf{i} + e^{-3t}\,\mathbf{j} + (\sin 4t + 1)\,\mathbf{k} \right) dt$$

$$= (\sqrt{t^2 + 2} + C_1)\,\mathbf{i} + \left(-\frac{1}{3}e^{-3t} + C_2 \right)\mathbf{j} + \left(-\frac{1}{4}\cos 4t + t + C_3 \right)\mathbf{k}$$

$$= \sqrt{t^2 + 2}\,\mathbf{i} - \frac{1}{3}e^{-3t}\,\mathbf{j} + \left(t - \frac{1}{4}\cos 4t \right)\mathbf{k} + \mathbf{C}. \quad \text{Let } \mathbf{C} = C_1\,\mathbf{i} + C_2\,\mathbf{j} + C_3\,\mathbf{k}.$$

QUICK CHECK 4 Let $\mathbf{r}(t) = \langle 1, 2t, 3t^2 \rangle$. Compute $\int \mathbf{r}(t)\,dt$. ◄

The constants C_1, C_2, and C_3 are combined to form one vector constant $\mathbf{C}$ at the end of the calculation.

Related Exercise 63 ◄

EXAMPLE 6 Finding one antiderivative Find $\mathbf{r}(t)$ such that $\mathbf{r}'(t) = \langle 10, \sin t, t \rangle$ and $\mathbf{r}(0) = \mathbf{j}$.

SOLUTION The required function $\mathbf{r}$ is an antiderivative of $\langle 10, \sin t, t \rangle$:

$$\mathbf{r}(t) = \int \langle 10, \sin t, t \rangle\,dt = \left\langle 10t, -\cos t, \frac{t^2}{2} \right\rangle + \mathbf{C},$$

where **C** is an arbitrary constant vector. The condition $\mathbf{r}(0) = \mathbf{j}$ allows us to determine **C**; substituting $t = 0$ implies that $\mathbf{r}(0) = \langle 0, -1, 0 \rangle + \mathbf{C} = \mathbf{j}$, where $\mathbf{j} = \langle 0, 1, 0 \rangle$. Solving for **C**, we have $\mathbf{C} = \langle 0, 1, 0 \rangle - \langle 0, -1, 0 \rangle = \langle 0, 2, 0 \rangle$. Therefore,

$$\mathbf{r}(t) = \left\langle 10t, 2 - \cos t, \frac{t^2}{2} \right\rangle.$$

Related Exercise 65 ◄

Definite integrals are evaluated by applying the Fundamental Theorem of Calculus to each component of a vector-valued function.

DEFINITION Definite Integral of a Vector-Valued Function

Let $\mathbf{r}(t) = f(t)\,\mathbf{i} + g(t)\,\mathbf{j} + h(t)\,\mathbf{k}$, where f, g, and h are integrable on the interval $[a, b]$. The **definite integral** of **r** on $[a, b]$ is

$$\int_a^b \mathbf{r}(t)\, dt = \left(\int_a^b f(t)\, dt \right) \mathbf{i} + \left(\int_a^b g(t)\, dt \right) \mathbf{j} + \left(\int_a^b h(t)\, dt \right) \mathbf{k}.$$

EXAMPLE 7 Definite integrals Evaluate

$$\int_0^\pi \left(\mathbf{i} + 3\cos\frac{t}{2}\,\mathbf{j} - 4t\,\mathbf{k} \right) dt.$$

SOLUTION

$$\int_0^\pi \left(\mathbf{i} + 3\cos\frac{t}{2}\,\mathbf{j} - 4t\,\mathbf{k} \right) dt = t\,\mathbf{i}\Big|_0^\pi + 6\sin\frac{t}{2}\,\mathbf{j}\Big|_0^\pi - 2t^2\,\mathbf{k}\Big|_0^\pi \qquad \text{Evaluate integrals for each component.}$$

$$= \pi\,\mathbf{i} + 6\mathbf{j} - 2\pi^2\,\mathbf{k} \qquad\qquad \text{Simplify.}$$

Related Exercise 75 ◄

With the tools of differentiation and integration in hand, we are prepared to tackle some practical problems, notably the motion of objects in space.

SECTION 14.2 EXERCISES

Getting Started

1. What is the derivative of $\mathbf{r}(t) = \langle f(t), g(t), h(t) \rangle$?

2. Explain the geometric meaning of $\mathbf{r}'(t)$.

3. Given a tangent vector on an oriented curve, how do you find the unit tangent vector?

4. Compute $\mathbf{r}''(t)$ when $\mathbf{r}(t) = \langle t^{10}, 8t, \cos t \rangle$.

5. How do you find the indefinite integral of $\mathbf{r}(t) = \langle f(t), g(t), h(t) \rangle$?

6. How do you evaluate $\int_a^b \mathbf{r}(t)\, dt$?

7. Find **C** if $\mathbf{r}(t) = \langle e^t, 3\cos t, t + 10 \rangle + \mathbf{C}$ and $\mathbf{r}(0) = \langle 0, 0, 0 \rangle$.

8. Find the unit tangent vector at $t = 0$ for the parameterized curve $\mathbf{r}(t)$ if $\mathbf{r}'(t) = \langle e^t + 5, \sin t + 2, \cos t + 2 \rangle$.

Practice Exercises

9–16. Derivatives of vector-valued functions *Differentiate the following functions.*

9. $\mathbf{r}(t) = \langle \cos t, t^2, \sin t \rangle$

10. $\mathbf{r}(t) = 4e^t\mathbf{i} + 5\mathbf{j} + \ln t\,\mathbf{k}$

11. $\mathbf{r}(t) = \left\langle 2t^3, 6\sqrt{t}, \frac{3}{t} \right\rangle$

12. $\mathbf{r}(t) = \langle 4, 3\cos 2t, 2\sin 3t \rangle$

13. $\mathbf{r}(t) = e^t\mathbf{i} + 2e^{-t}\mathbf{j} - 4e^{2t}\mathbf{k}$

14. $\mathbf{r}(t) = \tan t\,\mathbf{i} + \sec t\,\mathbf{j} + \cos^2 t\,\mathbf{k}$

15. $\mathbf{r}(t) = \langle te^{-t}, t\ln t, t\cos t \rangle$

16. $\mathbf{r}(t) = \langle (t + 1)^{-1}, \tan^{-1} t, \ln(t + 1) \rangle$

17–22. Tangent vectors *Find a tangent vector at the given value of t for the following parameterized curves.*

17. $\mathbf{r}(t) = \langle t, 3t^2, t^3 \rangle$, $t = 1$

18. $\mathbf{r}(t) = \langle e^t, e^{3t}, e^{5t} \rangle$, $t = 0$

19. $\mathbf{r}(t) = \langle t, \cos 2t, 2\sin t \rangle$, $t = \dfrac{\pi}{2}$

20. $\mathbf{r}(t) = \left\langle 2\sin t, 3\cos t, \sin\dfrac{t}{2} \right\rangle$, $t = \pi$

21. $\mathbf{r}(t) = 2t^4\mathbf{i} + 6t^{3/2}\mathbf{j} + \dfrac{10}{t}\mathbf{k}$, $t = 1$

22. $\mathbf{r}(t) = 2e^t\mathbf{i} + e^{-2t}\mathbf{j} + 4e^{2t}\mathbf{k}$, $t = \ln 3$

23–28. Unit tangent vectors *Find the unit tangent vector for the following parameterized curves.*

23. $\mathbf{r}(t) = \langle 2t, 2t, t \rangle$, for $0 \le t \le 1$

24. $\mathbf{r}(t) = \langle \cos t, \sin t, 2 \rangle$, for $0 \le t \le 2\pi$

25. $\mathbf{r}(t) = \langle 8, \cos 2t, 2\sin 2t \rangle$, for $0 \le t \le 2\pi$

26. $\mathbf{r}(t) = \langle \sin t, \cos t, \cos t \rangle$, for $0 \le t \le 2\pi$

27. $\mathbf{r}(t) = \left\langle t, 2, \dfrac{2}{t} \right\rangle$, for $t \ge 1$

28. $\mathbf{r}(t) = \langle e^{2t}, 2e^{2t}, 2e^{-3t} \rangle$, for $t \ge 0$

29–32. Unit tangent vectors at a point *Find the unit tangent vector at the given value of t for the following parameterized curves.*

29. $\mathbf{r}(t) = \langle \cos 2t, 4, 3 \sin 2t \rangle$; $t = \dfrac{\pi}{2}$

30. $\mathbf{r}(t) = \langle \sin t, \cos t, e^{-t} \rangle$; $t = 0$

31. $\mathbf{r}(t) = \left\langle 6t, 6, \dfrac{3}{t} \right\rangle$; $t = 1$

32. $\mathbf{r}(t) = \langle \sqrt{7}e^t, 3e^t, 3e^t \rangle$; $t = \ln 2$

33–38. Derivative rules *Let* $\mathbf{u}(t) = 2t^3 \mathbf{i} + (t^2 - 1)\mathbf{j} - 8\mathbf{k}$ *and* $\mathbf{v}(t) = e^t \mathbf{i} + 2e^{-t}\mathbf{j} - e^{2t}\mathbf{k}$. *Compute the derivative of the following functions.*

33. $(t^{12} + 3t)\mathbf{u}(t)$ **34.** $(4t^8 - 6t^3)\mathbf{v}(t)$

35. $\mathbf{u}(t^4 - 2t)$ **36.** $\mathbf{v}(\sqrt{t})$

37. $\mathbf{u}(t) \cdot \mathbf{v}(t)$ **38.** $\mathbf{u}(t) \times \mathbf{v}(t)$

39–42. Derivative rules *Suppose* $\mathbf{u}$ *and* $\mathbf{v}$ *are differentiable functions at* $t = 0$ *with* $\mathbf{u}(0) = \langle 0, 1, 1 \rangle$, $\mathbf{u}'(0) = \langle 0, 7, 1 \rangle$, $\mathbf{v}(0) = \langle 0, 1, 1 \rangle$, *and* $\mathbf{v}'(0) = \langle 1, 1, 2 \rangle$. *Evaluate the following expressions.*

39. $\dfrac{d}{dt}(\mathbf{u} \cdot \mathbf{v})\Big|_{t=0}$ **40.** $\dfrac{d}{dt}(\mathbf{u} \times \mathbf{v})\Big|_{t=0}$

41. $\dfrac{d}{dt}(\cos t\, \mathbf{u}(t))\Big|_{t=0}$ **42.** $\dfrac{d}{dt}(\mathbf{u}(\sin t))\Big|_{t=0}$

43–48. Derivative rules *Let* $\mathbf{u}(t) = \langle 1, t, t^2 \rangle$, $\mathbf{v}(t) = \langle t^2, -2t, 1 \rangle$, *and* $g(t) = 2\sqrt{t}$. *Compute the derivatives of the following functions.*

43. $\mathbf{v}(e^t)$ **44.** $\mathbf{u}(t^3)$ **45.** $\mathbf{v}(g(t))$

46. $g(t)\mathbf{v}(t)$ **47.** $\mathbf{u}(t) \times \mathbf{v}(t)$ **48.** $\mathbf{u}(t) \cdot \mathbf{v}(t)$

49–52. Derivative rules *Compute the following derivatives.*

49. $\dfrac{d}{dt}(t^2(\mathbf{i} + 2\mathbf{i} - 2t\,\mathbf{k}) \cdot (e^t \mathbf{i} + 2e^t\mathbf{j} - 3e^{-t}\mathbf{k}))$

50. $\dfrac{d}{dt}((t^3 \mathbf{i} - 2t\,\mathbf{j} - 2\,\mathbf{k}) \times (t\,\mathbf{i} - t^2\mathbf{j} - t^3\mathbf{k}))$

51. $\dfrac{d}{dt}((3t^2\mathbf{i} + \sqrt{t}\,\mathbf{j} - 2t^{-1}\mathbf{k}) \cdot (\cos t\,\mathbf{i} + \sin 2t\,\mathbf{j} - 3t\,\mathbf{k}))$

52. $\dfrac{d}{dt}((t^3\mathbf{i} + 6\,\mathbf{j} - 2\sqrt{t}\,\mathbf{k}) \times (3t\,\mathbf{i} - 12t^2\mathbf{j} - 6t^{-2}\mathbf{k}))$

53–58. Higher-order derivatives *Compute* $\mathbf{r}''(t)$ *and* $\mathbf{r}'''(t)$ *for the following functions.*

53. $\mathbf{r}(t) = \langle t^2 + 1, t + 1, 1 \rangle$

54. $\mathbf{r}(t) = \langle 3t^{12} - t^2, t^8 + t^3, t^{-4} - 2 \rangle$

55. $\mathbf{r}(t) = \langle \cos 3t, \sin 4t, \cos 6t \rangle$

56. $\mathbf{r}(t) = \langle e^{4t}, 2e^{-4t} + 1, 2e^{-t} \rangle$

57. $\mathbf{r}(t) = \sqrt{t + 4}\,\mathbf{i} + \dfrac{t}{t + 1}\mathbf{j} - e^{-t^2}\mathbf{k}$

58. $\mathbf{r}(t) = \tan t\,\mathbf{i} + \left(t + \dfrac{1}{t} \right)\mathbf{j} - \ln(t + 1)\mathbf{k}$

59–64. Indefinite integrals *Compute the indefinite integral of the following functions.*

59. $\mathbf{r}(t) = \langle t^4 - 3t, 2t - 1, 10 \rangle$

60. $\mathbf{r}(t) = \left\langle 5t^{-4} - t^2, t^6 - 4t^3, \dfrac{2}{t} \right\rangle$

61. $\mathbf{r}(t) = \langle 2 \cos t, 2 \sin 3t, 4 \cos 8t \rangle$

62. $\mathbf{r}(t) = te^t \mathbf{i} + t \sin t^2\mathbf{j} - \dfrac{2t}{\sqrt{t^2 + 4}}\mathbf{k}$

63. $\mathbf{r}(t) = e^{3t}\mathbf{i} + \dfrac{1}{1 + t^2}\mathbf{j} - \dfrac{1}{\sqrt{2t}}\mathbf{k}$

64. $\mathbf{r}(t) = 2^t \mathbf{i} + \dfrac{1}{1 + 2t}\mathbf{j} + \ln t\,\mathbf{k}$

65–70. Finding r from r' *Find the function* $\mathbf{r}$ *that satisfies the given conditions.*

65. $\mathbf{r}'(t) = \langle e^t, \sin t, \sec^2 t \rangle$; $\mathbf{r}(0) = \langle 2, 2, 2 \rangle$

66. $\mathbf{r}'(t) = \langle 0, 2, 2t \rangle$; $\mathbf{r}(1) = \langle 4, 3, -5 \rangle$

67. $\mathbf{r}'(t) = \langle 1, 2t, 3t^2 \rangle$; $\mathbf{r}(1) = \langle 4, 3, -5 \rangle$

68. $\mathbf{r}'(t) = \left\langle \sqrt{t}, \cos \pi t, \dfrac{4}{t} \right\rangle$; $\mathbf{r}(1) = \langle 2, 3, 4 \rangle$

69. $\mathbf{r}'(t) = \langle e^{2t}, 1 - 2e^{-t}, 1 - 2e^t \rangle$; $\mathbf{r}(0) = \langle 1, 1, 1 \rangle$

70. $\mathbf{r}'(t) = \dfrac{t}{t^2 + 1}\mathbf{i} + te^{-t^2}\mathbf{j} - \dfrac{2t}{\sqrt{t^2 + 4}}\mathbf{k}$; $\mathbf{r}(0) = \mathbf{i} + \dfrac{3}{2}\mathbf{j} - 3\mathbf{k}$

71–78. Definite integrals *Evaluate the following definite integrals.*

71. $\displaystyle\int_{-1}^{1} (\mathbf{i} + t\,\mathbf{j} + 3t^2\,\mathbf{k})\,dt$

72. $\displaystyle\int_{1}^{4} (6t^2\mathbf{i} + 8t^3\mathbf{j} + 9t^2\,\mathbf{k})\,dt$

73. $\displaystyle\int_{0}^{\ln 2} (e^t \mathbf{i} + e^t \cos(\pi e^t)\,\mathbf{j})\,dt$

74. $\displaystyle\int_{1/2}^{1} \left(\dfrac{3}{1 + 2t}\mathbf{i} - \pi \csc^2\!\left(\dfrac{\pi}{2}t \right)\mathbf{k} \right)dt$

75. $\displaystyle\int_{-\pi}^{\pi} (\sin t\,\mathbf{i} + \cos t\,\mathbf{j} + 2t\,\mathbf{k})\,dt$

76. $\displaystyle\int_{0}^{\ln 2} (e^{-t}\mathbf{i} + 2e^{2t}\mathbf{j} - 4e^t\,\mathbf{k})\,dt$

77. $\displaystyle\int_{0}^{2} te^t(\mathbf{i} + 2\mathbf{j} - \mathbf{k})\,dt$

78. $\displaystyle\int_{0}^{\pi/4} (\sec^2 t\,\mathbf{i} - 2 \cos t\,\mathbf{j} - \mathbf{k})\,dt$

79. Explain why or why not Determine whether the following statements are true and give an explanation or counterexample.

a. The vectors $\mathbf{r}(t)$ and $\mathbf{r}'(t)$ are parallel for all values of t in the domain.

b. The curve described by the function $\mathbf{r}(t) = \langle t, t^2 - 2t, \cos \pi t \rangle$ is smooth, for $-\infty < t < \infty$.

c. If f, g, and h are odd integrable functions and a is a real number, then

$$\int_{-a}^{a} (f(t)\,\mathbf{i} + g(t)\,\mathbf{j} + h(t)\,\mathbf{k})\,dt = \mathbf{0}.$$

80–83. Tangent lines *Suppose the vector-valued function* $\mathbf{r}(t) = \langle f(t), g(t), h(t) \rangle$ *is smooth on an interval containing the point* t_0. *The line tangent to* $\mathbf{r}(t)$ *at* $t = t_0$ *is the line parallel to the tangent vector* $\mathbf{r}'(t_0)$ *that passes through* $(f(t_0), g(t_0), h(t_0))$. *For each of the following functions, find an equation of the line tangent to the curve at* $t = t_0$. *Choose an orientation for the line that is the same as the direction of* $\mathbf{r}'$.

80. $\mathbf{r}(t) = \langle e^t, e^{2t}, e^{3t} \rangle; t_0 = 0$

81. $\mathbf{r}(t) = \langle 2 + \cos t, 3 + \sin 2t, t \rangle; t_0 = \dfrac{\pi}{2}$

82. $\mathbf{r}(t) = \langle \sqrt{2t + 1}, \sin \pi t, 4 \rangle; t_0 = 4$

83. $\mathbf{r}(t) = \langle 3t - 1, 7t + 2, t^2 \rangle; t_0 = 1$

Explorations and Challenges

84–89. Relationship between r and r′

84. Consider the circle $\mathbf{r}(t) = \langle a \cos t, a \sin t \rangle$, for $0 \le t \le 2\pi$, where a is a positive real number. Compute $\mathbf{r}'$ and show that it is orthogonal to $\mathbf{r}$ for all t.

85. Consider the parabola $\mathbf{r}(t) = \langle at^2 + 1, t \rangle$, for $-\infty < t < \infty$, where a is a positive real number. Find all points on the parabola at which $\mathbf{r}$ and $\mathbf{r}'$ are orthogonal.

86. Consider the curve $\mathbf{r}(t) = \langle \sqrt{t}, 1, t \rangle$, for $t > 0$. Find all points on the curve at which $\mathbf{r}$ and $\mathbf{r}'$ are orthogonal.

87. Consider the helix $\mathbf{r}(t) = \langle \cos t, \sin t, t \rangle$, for $-\infty < t < \infty$. Find all points on the helix at which $\mathbf{r}$ and $\mathbf{r}'$ are orthogonal.

88. Consider the ellipse $\mathbf{r}(t) = \langle 2 \cos t, 8 \sin t, 0 \rangle$, for $0 \le t \le 2\pi$. Find all points on the ellipse at which $\mathbf{r}$ and $\mathbf{r}'$ are orthogonal.

89. Give two families of curves in $\mathbb{R}^3$ for which $\mathbf{r}$ and $\mathbf{r}'$ are parallel for all t in the domain.

90. Motion on a sphere Prove that $\mathbf{r}$ describes a curve that lies on the surface of a sphere centered at the origin ($x^2 + y^2 + z^2 = a^2$ with $a \ge 0$) if and only if $\mathbf{r}$ and $\mathbf{r}'$ are orthogonal at all points of the curve.

91. Vectors r and r′ for lines

a. If $\mathbf{r}(t) = \langle at, bt, ct \rangle$ with $\langle a, b, c \rangle \ne \langle 0, 0, 0 \rangle$, show that the angle between $\mathbf{r}$ and $\mathbf{r}'$ is constant for all $t > 0$.

b. If $\mathbf{r}(t) = \langle x_0 + at, y_0 + bt, z_0 + ct \rangle$, where x_0, y_0, and z_0 are not all zero, show that the angle between $\mathbf{r}$ and $\mathbf{r}'$ varies with t.

c. Explain the results of parts (a) and (b) geometrically.

92. Proof of Sum Rule By expressing $\mathbf{u}$ and $\mathbf{v}$ in terms of their components, prove that

$$\frac{d}{dt}(\mathbf{u}(t) + \mathbf{v}(t)) = \mathbf{u}'(t) + \mathbf{v}'(t).$$

93. Proof of Product Rule By expressing $\mathbf{u}$ in terms of its components, prove that

$$\frac{d}{dt}(f(t)\mathbf{u}(t)) = f'(t)\mathbf{u}(t) + f(t)\mathbf{u}'(t).$$

94. Proof of Cross Product Rule Prove that

$$\frac{d}{dt}(\mathbf{u}(t) \times \mathbf{v}(t)) = \mathbf{u}'(t) \times \mathbf{v}(t) + \mathbf{u}(t) \times \mathbf{v}'(t).$$

There are two ways to proceed: Either express $\mathbf{u}$ and $\mathbf{v}$ in terms of their three components or use the definition of the derivative.

T 95. Cusps and noncusps

a. Graph the curve $\mathbf{r}(t) = \langle t^3, t^3 \rangle$. Show that $\mathbf{r}'(0) = \mathbf{0}$ and the curve does not have a cusp at $t = 0$. Explain.

b. Graph the curve $\mathbf{r}(t) = \langle t^3, t^2 \rangle$. Show that $\mathbf{r}'(0) = \mathbf{0}$ and the curve has a cusp at $t = 0$. Explain.

c. The functions $\mathbf{r}(t) = \langle t, t^2 \rangle$ and $\mathbf{p}(t) = \langle t^2, t^4 \rangle$ both satisfy $y = x^2$. Explain how the curves they parameterize are different.

d. Consider the curve $\mathbf{r}(t) = \langle t^m, t^n \rangle$, where $m > 1$ and $n > 1$ are integers with no common factors. Is it true that the curve has a cusp at $t = 0$ if one (not both) of m and n is even? Explain.

QUICK CHECK ANSWERS

1. $\mathbf{r}(t)$ describes a line, so its tangent vector $\mathbf{r}'(t) = \langle 1, 1, 1 \rangle$ has constant direction and magnitude. **2.** Both $\mathbf{r}'$ and $|\mathbf{r}'|$ have units of m/s. In forming $\mathbf{r}'/|\mathbf{r}'|$, the units cancel and $\mathbf{T}(t)$ is without units. **3.** $\dfrac{d}{dt}(\mathbf{u}(t) \cdot \mathbf{v}(t)) = \langle 1, 1, 1 \rangle \cdot \langle 1, 1, 1 \rangle + \langle t, t, t \rangle \cdot \langle 0, 0, 0 \rangle = 3$.
$\dfrac{d}{dt}(\langle t, t, t \rangle \cdot \langle 1, 1, 1 \rangle) = \dfrac{d}{dt}(3t) = 3$. **4.** $\langle t, t^2, t^3 \rangle + \mathbf{C}$, where $\mathbf{C} = \langle a, b, c \rangle$, and a, b, and c are real numbers ◄

14.3 Motion in Space

It is a remarkable fact that given the forces acting on an object and its initial position and velocity, the motion of the object in three-dimensional space can be modeled for all future times. To be sure, the accuracy of the results depends on how well the various forces on the object are described. For example, it may be more difficult to predict the trajectory of a spinning soccer ball than the path of a space station orbiting Earth. Nevertheless, as shown in this section, by combining Newton's Second Law of Motion with everything we have learned about vectors, it is possible to solve a variety of moving-body problems.

Position, Velocity, Speed, Acceleration

Until now, we have studied objects that move in one dimension (along a line). The next step is to consider the motion of objects in two dimensions (in a plane) and three

Figure 14.11

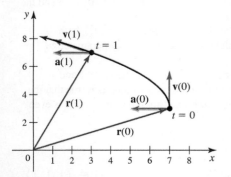

Figure 14.12

▶ In the case of two-dimensional motion, $\mathbf{r}(t) = \langle x(t), y(t) \rangle$, $\mathbf{v}(t) = \mathbf{r}'(t)$, and $\mathbf{a}(t) = \mathbf{r}''(t)$.

QUICK CHECK 1 Given $\mathbf{r}(t) = \langle t, t^2, t^3 \rangle$, find $\mathbf{v}(t)$ and $\mathbf{a}(t)$. ◀

Circular motion: At all times $\mathbf{a}(t) = -\mathbf{r}(t)$ and $\mathbf{v}(t)$ is orthogonal to $\mathbf{r}(t)$ and $\mathbf{a}(t)$.

Figure 14.13

dimensions (in space). We work in a three-dimensional coordinate system and let the vector-valued function $\mathbf{r}(t) = \langle x(t), y(t), z(t) \rangle$ describe the *position* of a moving object at times $t \geq 0$. The curve described by $\mathbf{r}$ is the *path* or *trajectory* of the object (**Figure 14.11**). Just as with one-dimensional motion, the rate of change of the position function with respect to time is the *instantaneous velocity* of the object—a vector with three components corresponding to the velocity in the x-, y-, and z-directions:

$$\mathbf{v}(t) = \mathbf{r}'(t) = \langle x'(t), y'(t), z'(t) \rangle.$$

This expression should look familiar. The velocity vectors of a moving object are simply tangent vectors; that is, at any point, the velocity vector is tangent to the trajectory (Figure 14.11).

As with one-dimensional motion, the *speed* of an object moving in three dimensions is the magnitude of its velocity vector:

$$|\mathbf{v}(t)| = |\langle x'(t), y'(t), z'(t) \rangle| = \sqrt{x'(t)^2 + y'(t)^2 + z'(t)^2}.$$

The speed is a nonnegative scalar-valued function.

Finally, the *acceleration* of a moving object is the rate of change of the velocity:

$$\mathbf{a}(t) = \mathbf{v}'(t) = \mathbf{r}''(t).$$

Although the position vector gives the path of a moving object and the velocity vector is always tangent to the path, the acceleration vector is more difficult to visualize. **Figure 14.12** shows one particular instance of two-dimensional motion. The trajectory is a segment of a parabola and is traced out by the position vectors (shown at $t = 0$ and $t = 1$). As expected, the velocity vectors are tangent to the trajectory. In this case, the acceleration is $\mathbf{a} = \langle -2, 0 \rangle$; it is constant in magnitude and direction for all times. The relationships among $\mathbf{r}$, $\mathbf{v}$, and $\mathbf{a}$ are explored in the coming examples.

DEFINITION   **Position, Velocity, Speed, Acceleration**

Let the **position** of an object moving in three-dimensional space be given by $\mathbf{r}(t) = \langle x(t), y(t), z(t) \rangle$, for $t \geq 0$. The **velocity** of the object is

$$\mathbf{v}(t) = \mathbf{r}'(t) = \langle x'(t), y'(t), z'(t) \rangle.$$

The **speed** of the object is the scalar function

$$|\mathbf{v}(t)| = \sqrt{x'(t)^2 + y'(t)^2 + z'(t)^2}.$$

The **acceleration** of the object is $\mathbf{a}(t) = \mathbf{v}'(t) = \mathbf{r}''(t)$.

EXAMPLE 1   **Velocity and acceleration for circular motion**   Consider the two-dimensional motion given by the position vector

$$\mathbf{r}(t) = \langle x(t), y(t) \rangle = \langle 3 \cos t, 3 \sin t \rangle, \quad \text{for } 0 \leq t \leq 2\pi.$$

a. Sketch the trajectory of the object.

b. Find the velocity and speed of the object.

c. Find the acceleration of the object.

d. Sketch the position, velocity, and acceleration vectors, for $t = 0$, $\pi/2$, π, and $3\pi/2$.

SOLUTION

a. Notice that

$$x(t)^2 + y(t)^2 = 9(\cos^2 t + \sin^2 t) = 9,$$

which is an equation of a circle centered at the origin with radius 3. The object moves on this circle in the counterclockwise direction (**Figure 14.13**).

b. $\mathbf{v}(t) = \langle x'(t), y'(t) \rangle = \langle -3 \sin t, 3 \cos t \rangle$ Velocity vector

$|\mathbf{v}(t)| = \sqrt{x'(t)^2 + y'(t)^2}$ Definition of speed

$= \sqrt{(-3 \sin t)^2 + (3 \cos t)^2}$

$= \sqrt{9\underbrace{(\sin^2 t + \cos^2 t)}_{1}} = 3$

The velocity vector has a constant magnitude and a continuously changing direction.

c. Differentiating the velocity, we find that $\mathbf{a}(t) = \mathbf{v}'(t) = \langle -3 \cos t, -3 \sin t \rangle = -\mathbf{r}(t)$. In this case, the acceleration vector is the negative of the position vector at all times.

d. The relationships among $\mathbf{r}$, $\mathbf{v}$, and $\mathbf{a}$ at four points in time are shown in Figure 14.13. The velocity vector is always tangent to the trajectory and has length 3, while the acceleration vector and position vector each have length 3 and point in opposite directions. At all times, $\mathbf{v}$ is orthogonal to $\mathbf{r}$ and $\mathbf{a}$.

Related Exercise 13 ◄

EXAMPLE 2 Comparing trajectories Consider the trajectories described by the position functions

$$\mathbf{r}(t) = \left\langle t, t^2 - 4, \frac{t^3}{4} - 8 \right\rangle, \quad \text{for } t \geq 0, \text{ and}$$

$$\mathbf{R}(t) = \left\langle t^2, t^4 - 4, \frac{t^6}{4} - 8 \right\rangle, \quad \text{for } t \geq 0,$$

where t is measured in the same time units for both functions.

a. Graph and compare the trajectories using a graphing utility.

b. Find the velocity vectors associated with the position functions.

SOLUTION

a. Plotting the position functions at selected values of t results in the trajectories shown in **Figure 14.14**. Because $\mathbf{r}(0) = \mathbf{R}(0) = \langle 0, -4, -8 \rangle$, both curves have the same initial point. For $t \geq 0$, the two curves consist of the same points, but they are traced out differently. For example, both curves pass through the point $(4, 12, 8)$, but that point corresponds to $\mathbf{r}(4)$ on the first curve and $\mathbf{R}(2)$ on the second curve. In general, $\mathbf{r}(t^2) = \mathbf{R}(t)$, for $t \geq 0$.

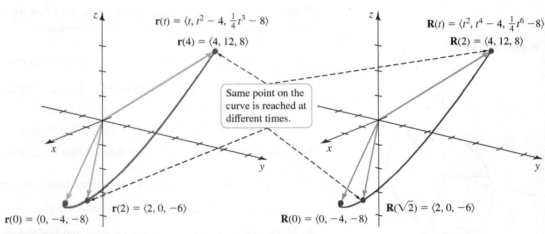

Figure 14.14

b. The velocity vectors are

$$\mathbf{r}'(t) = \left\langle 1, 2t, \frac{3t^2}{4} \right\rangle \quad \text{and} \quad \mathbf{R}'(t) = \left\langle 2t, 4t^3, \frac{3}{2} t^5 \right\rangle.$$

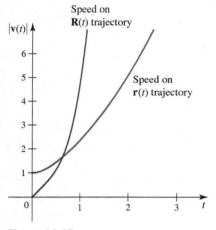

Figure 14.15

QUICK CHECK 2 Find the functions that give the speed of the two objects in Example 2, for $t \geq 0$ (corresponding to the graphs in Figure 14.15). ◄

➤ See Exercise 83 for a discussion of nonuniform straight-line motion.

Figure 14.16

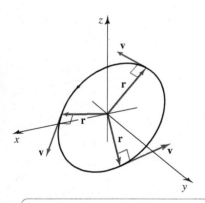

On a trajectory on which $|\mathbf{r}|$ is constant, $\mathbf{v}$ is orthogonal to $\mathbf{r}$ at all points.

Figure 14.17

The difference in the motion on the two curves is revealed by the graphs of the speeds associated with the trajectories (**Figure 14.15**). The object on the first trajectory reaches the point $(4, 12, 8)$ at $t = 4$, where its speed is $|\mathbf{r}'(4)| = |\langle 1, 8, 12 \rangle| \approx 14.5$. The object on the second trajectory reaches the same point $(4, 12, 8)$ at $t = 2$, where its speed is $|\mathbf{R}'(2)| = |\langle 4, 32, 48 \rangle| \approx 57.8$.

Related Exercise 21 ◄

Straight-Line and Circular Motion

Two types of motion in space arise frequently and deserve to be singled out. First consider a trajectory described by the vector function

$$\mathbf{r}(t) = \langle x_0 + at, y_0 + bt, z_0 + ct \rangle, \quad \text{for } t \geq 0,$$

where x_0, y_0, z_0, a, b, and c are constants. This function describes a straight-line trajectory with an initial position $\langle x_0, y_0, z_0 \rangle$ and a direction given by the vector $\langle a, b, c \rangle$ (Section 13.5). The velocity on this trajectory is the constant $\mathbf{v}(t) = \mathbf{r}'(t) = \langle a, b, c \rangle$ in the direction of the trajectory, and the acceleration is $\mathbf{a} = \langle 0, 0, 0 \rangle$. The motion associated with this function is **uniform** (constant velocity) **straight-line motion**.

A different situation is **circular motion** (Example 1). Consider the two-dimensional circular path

$$\mathbf{r}(t) = \langle A \cos t, A \sin t \rangle, \quad \text{for } 0 \leq t \leq 2\pi,$$

where A is a nonzero constant (**Figure 14.16**). The velocity and acceleration vectors are

$$\mathbf{v}(t) = \langle -A \sin t, A \cos t \rangle \quad \text{and}$$
$$\mathbf{a}(t) = \langle -A \cos t, -A \sin t \rangle = -\mathbf{r}(t).$$

Notice that $\mathbf{r}$ and $\mathbf{a}$ are parallel but point in opposite directions. Furthermore, $\mathbf{r} \cdot \mathbf{v} = \mathbf{a} \cdot \mathbf{v} = 0$; therefore, the position and acceleration vectors are both orthogonal to the velocity vectors at any given point (Figure 14.16). Finally, $\mathbf{r}$, $\mathbf{v}$, and $\mathbf{a}$ have constant magnitude A and variable directions. The conclusion that $\mathbf{r} \cdot \mathbf{v} = 0$ applies to any motion for which $|\mathbf{r}|$ is constant; that is, to any motion on a circle or a sphere (**Figure 14.17**).

THEOREM 14.2   Motion with Constant $|\mathbf{r}|$
Let $\mathbf{r}$ describe a path on which $|\mathbf{r}|$ is constant (motion on a circle or sphere centered at the origin). Then $\mathbf{r} \cdot \mathbf{v} = 0$, which means the position vector and the velocity vector are orthogonal at all times for which the functions are defined.

Proof: If $\mathbf{r}$ has constant magnitude, then $|\mathbf{r}(t)|^2 = \mathbf{r}(t) \cdot \mathbf{r}(t) = c$ for some constant c. Differentiating the equation $\mathbf{r}(t) \cdot \mathbf{r}(t) = c$, we have

$$\begin{aligned}
0 &= \frac{d}{dt}(\mathbf{r}(t) \cdot \mathbf{r}(t)) && \text{Differentiate both sides of } |\mathbf{r}(t)|^2 = c. \\
&= \mathbf{r}'(t) \cdot \mathbf{r}(t) + \mathbf{r}(t) \cdot \mathbf{r}'(t) && \text{Derivative of dot product (Theorem 14.1)} \\
&= 2\mathbf{r}'(t) \cdot \mathbf{r}(t) && \text{Simplify.} \\
&= 2\mathbf{v}(t) \cdot \mathbf{r}(t). && \mathbf{r}'(t) = \mathbf{v}(t)
\end{aligned}$$

Because $\mathbf{r}(t) \cdot \mathbf{v}(t) = 0$ for all t, it follows that $\mathbf{r}$ and $\mathbf{v}$ are orthogonal for all t. ◄

EXAMPLE 3   Path on a sphere An object moves on a trajectory described by

$$\mathbf{r}(t) = \langle x(t), y(t), z(t) \rangle = \langle 3 \cos t, 5 \sin t, 4 \cos t \rangle, \quad \text{for } 0 \leq t \leq 2\pi.$$

a. Show that the object moves on a sphere and find the radius of the sphere.

b. Find the velocity and speed of the object.

c. Consider the curve $\mathbf{r}(t) = \langle 5 \cos t, 5 \sin t, 5 \sin 2t \rangle$, which is the roller coaster curve from Example 3 of Section 14.1, with different coefficients. Show that this curve does not lie on a sphere. How could $\mathbf{r}$ be modified so that it describes a curve that lies on a sphere of radius 1, centered at the origin?

▶ For generalizations of this example and explorations of trajectories that lie on spheres and ellipses, see Exercises 79 and 82.

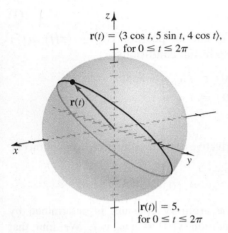

$\mathbf{r}(t) = \langle 3 \cos t, 5 \sin t, 4 \cos t \rangle$, for $0 \le t \le 2\pi$

$\mathbf{r}(t)$

$|\mathbf{r}(t)| = 5$, for $0 \le t \le 2\pi$

Figure 14.18

SOLUTION

a. $\ |\mathbf{r}(t)|^2 = x(t)^2 + y(t)^2 + z(t)^2$ Square of the distance from the origin

$\qquad = (3 \cos t)^2 + (5 \sin t)^2 + (4 \cos t)^2$ Substitute.

$\qquad = 25 \cos^2 t + 25 \sin^2 t$ Simplify.

$\qquad = 25 \underbrace{(\cos^2 t + \sin^2 t)}_{1} = 25$ Factor.

Therefore, $|\mathbf{r}(t)| = 5$, for $0 \le t \le 2\pi$, and the trajectory lies on a sphere of radius 5 centered at the origin (**Figure 14.18**).

b. $\quad \mathbf{v}(t) = \mathbf{r}'(t) = \langle -3 \sin t, 5 \cos t, -4 \sin t \rangle$ Velocity vector

$\qquad |\mathbf{v}(t)| = \sqrt{\mathbf{v}(t) \cdot \mathbf{v}(t)}$ Speed of the object

$\qquad\qquad = \sqrt{9 \sin^2 t + 25 \cos^2 t + 16 \sin^2 t}$ Evaluate the dot product.

$\qquad\qquad = \sqrt{25 \underbrace{(\sin^2 t + \cos^2 t)}_{1}}$ Simplify.

$\qquad\qquad = 5$ Simplify.

The speed of the object is always 5. You should verify that $\mathbf{r}(t) \cdot \mathbf{v}(t) = 0$, for all t, implying that $\mathbf{r}$ and $\mathbf{v}$ are always orthogonal.

c. We first compute the distance from the origin to the curve:

$\qquad |\mathbf{r}(t)| = \sqrt{(5 \cos t)^2 + (5 \sin t)^2 + (5 \sin 2t)^2}$ Distance from origin to curve

$\qquad\qquad = \sqrt{25(\cos^2 t + \sin^2 t + \sin^2 2t)}$ Simplify.

$\qquad\qquad = 5\sqrt{1 + \sin^2 2t}.$ $\cos^2 t + \sin^2 t = 1$

It is clear that $|\mathbf{r}(t)|$ is not constant, and therefore the curve does not lie on a sphere.

One way to modify the curve so that it does lie on a sphere is to divide each output vector $\mathbf{r}(t)$ by its length. In fact, as long as $|\mathbf{r}(t)| \ne 0$ on the interval of interest, we can force any path onto a sphere (centered at the origin) with this modification. The function

$$\mathbf{u}(t) = \frac{\mathbf{r}(t)}{|\mathbf{r}(t)|} = \left\langle \frac{\cos t}{\sqrt{1 + \sin^2 2t}}, \frac{\sin t}{\sqrt{1 + \sin^2 2t}}, \frac{\sin 2t}{\sqrt{1 + \sin^2 2t}} \right\rangle$$

describes a curve on which $|\mathbf{u}(t)|$ is constant because $\dfrac{\mathbf{r}(t)}{|\mathbf{r}(t)|}$ is a unit vector. We conclude that the new curve, which is reminiscent of the seam on a tennis ball (**Figure 14.19**), lies on the unit sphere centered at the origin.

Related Exercises 32–33 ◀

$\mathbf{u}(t) = \mathbf{r}(t)/|\mathbf{r}(t)|$, where $\mathbf{r}(t) = \langle 5 \cos t, 5 \sin t, 5 \sin 2t \rangle$
$0 \le t \le 2\pi$

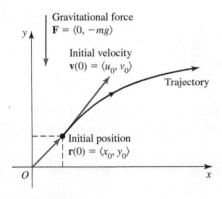

$|\mathbf{u}(t)| = 1$, for $0 \le t \le 2\pi$

Tennis ball curve

Figure 14.19

QUICK CHECK 3 Explain how to modify the curve $\mathbf{r}(t)$ given in Example 3c so that it lies on a sphere of radius 5 centered at the origin. ◀

Two-Dimensional Motion in a Gravitational Field

Newton's Second Law of Motion, which is used to model the motion of most objects, states that

$$\underbrace{\text{mass}}_{m} \cdot \underbrace{\text{acceleration}}_{\mathbf{a}(t) = \mathbf{r}''(t)} = \underbrace{\text{sum of all forces.}}_{\sum \mathbf{F}_k}$$

The governing law says something about the *acceleration* of an object, and in order to describe the motion fully, we must find the velocity and position from the acceleration.

Finding Velocity and Position from Acceleration   We begin with the case of two-dimensional projectile motion in which the only force acting on the object is the gravitational force; for the moment, air resistance and other possible external forces are neglected.

A convenient coordinate system uses a y-axis that points vertically upward and an x-axis that points in the direction of horizontal motion. The gravitational force is in the negative y-direction and is given by $\mathbf{F} = \langle 0, -mg \rangle$, where m is the mass of the object and $g = 9.8 \text{ m/s}^2 = 32 \text{ ft/s}^2$ is the acceleration due to gravity (**Figure 14.20**).

Gravitational force
$\mathbf{F} = \langle 0, -mg \rangle$

Initial velocity
$\mathbf{v}(0) = \langle u_0, v_0 \rangle$

Trajectory

Initial position
$\mathbf{r}(0) = \langle x_0, y_0 \rangle$

Figure 14.20

With these observations, Newton's Second Law takes the form

$$m\mathbf{a}(t) = \mathbf{F} = \langle 0, -mg \rangle.$$

Significantly, the mass of the object cancels, leaving the vector equation

$$\mathbf{a}(t) = \langle 0, -g \rangle. \tag{1}$$

In order to find the velocity $\mathbf{v}(t) = \langle x'(t), y'(t) \rangle$ and the position $\mathbf{r}(t) = \langle x(t), y(t) \rangle$ from this equation, we must be given the following **initial conditions**:

Initial velocity at $t = 0$: $\mathbf{v}(0) = \langle u_0, v_0 \rangle$ and

Initial position at $t = 0$: $\mathbf{r}(0) = \langle x_0, y_0 \rangle.$

We proceed in two steps.

> Recall that an antiderivative of 0 is a constant C and an antiderivative of $-g$ is $-gt + C$.

1. Solve for the velocity The velocity is an antiderivative of the acceleration in equation (1). Integrating the acceleration, we have

$$\mathbf{v}(t) = \int \mathbf{a}(t) \, dt = \int \langle 0, -g \rangle \, dt = \langle 0, -gt \rangle + \mathbf{C},$$

> You have a choice. You may do these calculations in vector notation as we have done here, or you may work with individual components.

where $\mathbf{C}$ is an arbitrary constant vector. The arbitrary constant is determined by substituting $t = 0$ and using the initial condition $\mathbf{v}(0) = \langle u_0, v_0 \rangle$. We find that $\mathbf{v}(0) = \langle 0, 0 \rangle + \mathbf{C} = \langle u_0, v_0 \rangle$, or $\mathbf{C} = \langle u_0, v_0 \rangle$. Therefore, the velocity is

$$\mathbf{v}(t) = \langle 0, -gt \rangle + \langle u_0, v_0 \rangle = \langle u_0, -gt + v_0 \rangle. \tag{2}$$

Notice that the horizontal component of velocity is simply the initial horizontal velocity u_0 for all time. The vertical component of velocity decreases linearly from its initial value of v_0.

2. Solve for the position The position is an antiderivative of the velocity given by equation (2):

$$\mathbf{r}(t) = \int \mathbf{v}(t) \, dt = \int \langle u_0, -gt + v_0 \rangle \, dt = \left\langle u_0 \, t, -\frac{1}{2} gt^2 + v_0 t \right\rangle + \mathbf{C},$$

where $\mathbf{C}$ is an arbitrary constant vector. Substituting $t = 0$, we have $\mathbf{r}(0) = \langle 0, 0 \rangle + \mathbf{C} = \langle x_0, y_0 \rangle$, which implies that $\mathbf{C} = \langle x_0, y_0 \rangle$. Therefore, the position of the object, for $t \geq 0$, is

$$\mathbf{r}(t) = \left\langle u_0 \, t, -\frac{1}{2} gt^2 + v_0 \, t \right\rangle + \langle x_0, y_0 \rangle = \left\langle \underbrace{u_0 \, t + x_0}_{x(t)}, \underbrace{-\frac{1}{2} gt^2 + v_0 \, t + y_0}_{y(t)} \right\rangle.$$

SUMMARY   Two-Dimensional Motion in a Gravitational Field

Consider an object moving in a plane with a horizontal x-axis and a vertical y-axis, subject only to the force of gravity. Given the initial velocity $\mathbf{v}(0) = \langle u_0, v_0 \rangle$ and the initial position $\mathbf{r}(0) = \langle x_0, y_0 \rangle$, the velocity of the object, for $t \geq 0$, is

$$\mathbf{v}(t) = \langle x'(t), y'(t) \rangle = \langle u_0, -gt + v_0 \rangle$$

and the position is

$$\mathbf{r}(t) = \langle x(t), y(t) \rangle = \left\langle u_0 \, t + x_0, -\frac{1}{2} gt^2 + v_0 \, t + y_0 \right\rangle.$$

EXAMPLE 4   Flight of a baseball A baseball is hit from 3 ft above home plate with an initial velocity in ft/s of $\mathbf{v}(0) = \langle u_0, v_0 \rangle = \langle 80, 80 \rangle$. Neglect all forces other than gravity.

a. Find the position and velocity of the ball between the time it is hit and the time it first hits the ground.

b. Show that the trajectory of the ball is a segment of a parabola.

c. Assuming a flat playing field, how far does the ball travel horizontally? Plot the trajectory of the ball.

d. What is the maximum height of the ball?

e. Does the ball clear a 20-ft fence that is 380 ft from home plate (directly under the path of the ball)?

SOLUTION Assume the origin is located at home plate. Because distances are measured in feet, we use $g = 32$ ft/s².

a. Substituting $x_0 = 0$ and $y_0 = 3$ into the equation for **r**, the position of the ball is

$$\mathbf{r}(t) = \langle x(t), y(t) \rangle = \langle 80t, -16t^2 + 80t + 3 \rangle, \quad \text{for } t \geq 0. \qquad (3)$$

We then compute $\mathbf{v}(t) = \mathbf{r}'(t) = \langle 80, -32t + 80 \rangle$.

b. Equation (3) says that the horizontal position is $x = 80t$ and the vertical position is $y = -16t^2 + 80t + 3$. Substituting $t = x/80$ into the equation for y gives

$$y = -16\left(\frac{x}{80}\right)^2 + x + 3 = -\frac{x^2}{400} + x + 3,$$

which is an equation of a parabola.

c. The ball lands on the ground at the value of $t > 0$ at which $y = 0$. Solving $y(t) = -16t^2 + 80t + 3 = 0$, we find that $t \approx -0.04$ and $t \approx 5.04$ s. The first root is not relevant for the problem at hand, so we conclude that the ball lands when $t \approx 5.04$ s. The horizontal distance traveled by the ball is $x(5.04) \approx 403$ ft. The path of the ball in the xy-coordinate system on the time interval $[0, 5.04]$ is shown in **Figure 14.21**.

d. The ball reaches its maximum height at the time its vertical velocity is zero. Solving $y'(t) = -32t + 80 = 0$, we find that $t = 2.5$ s. The height at that time is $y(2.5) = 103$ ft.

e. The ball reaches a horizontal distance of 380 ft (the distance to the fence) when $x(t) = 80t = 380$. Solving for t, we find that $t = 4.75$ s. The height of the ball at that time is $y(4.75) = 22$ ft. So, indeed, the ball clears a 20-ft fence.

Related Exercises 41, 43 ◀

y

Parabolic trajectory of baseball
Time of flight 5.04 s
Range 403 ft
Maximum height 103 ft

120

100

80

60

40

20

Fence

0   100   200   300   400   500   x

Figure 14.21

▶ The equation in Example 4c can be solved using the quadratic formula or a root-finder on a calculator.

QUICK CHECK 4 Write the functions $x(t)$ and $y(t)$ in Example 4 in the case that $x_0 = 0$, $y_0 = 2$, $u_0 = 100$, and $v_0 = 60$. ◀

Range, Time of Flight, Maximum Height

Having solved one specific motion problem, we can make some general observations about two-dimensional projectile motion in a gravitational field. Assume the motion of an object begins at the origin; that is, $x_0 = y_0 = 0$. Also assume the object is launched at an angle of α ($0 \leq \alpha \leq \pi/2$) above the horizontal with an initial speed $|\mathbf{v}_0|$ (**Figure 14.22**). This means that the initial velocity is

$$\langle u_0, v_0 \rangle = \langle |\mathbf{v}_0| \cos \alpha, |\mathbf{v}_0| \sin \alpha \rangle.$$

Substituting these values into the general expressions for the velocity and position, we find that the velocity of the object is

$$\mathbf{v}(t) = \langle u_0, -gt + v_0 \rangle = \langle |\mathbf{v}_0| \cos \alpha, -gt + |\mathbf{v}_0| \sin \alpha \rangle.$$

The position of the object (with $x_0 = y_0 = 0$) is

$$\mathbf{r}(t) = \langle x(t), y(t) \rangle = \langle (|\mathbf{v}_0| \cos \alpha)t, -gt^2/2 + (|\mathbf{v}_0| \sin \alpha)t \rangle.$$

Notice that the motion is determined entirely by the parameters $|\mathbf{v}_0|$ and α. Several general conclusions now follow.

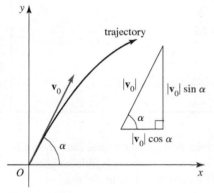

y

trajectory

$\mathbf{v}_0$

$|\mathbf{v}_0|$

$|\mathbf{v}_0| \sin \alpha$

α

$|\mathbf{v}_0| \cos \alpha$

α

O   x

Figure 14.22

▶ The other root of the equation $y(t) = 0$ is $t = 0$, the time the object leaves the ground.

1. Assuming the object is launched from the origin over horizontal ground, it returns to the ground when $y(t) = -gt^2/2 + (|\mathbf{v}_0| \sin \alpha)t = 0$. Solving for t, the **time of flight** is $T = 2|\mathbf{v}_0| (\sin \alpha)/g$.

Figure 14.23

QUICK CHECK 5 Show that the range attained with an angle α equals the range attained with the angle $\pi/2 - \alpha$. ◄

▶ Use caution with the formulas in the summary box: They are applicable only when the initial position of the object is the origin. Exercise 73 addresses the case where the initial position of the object is $\langle 0, y_0 \rangle$.

2. The **range** of the object, which is the horizontal distance it travels, is the x-coordinate of the trajectory when $t = T$:

$$
\begin{aligned}
x(T) &= (|\mathbf{v}_0| \cos \alpha)T \\
&= (|\mathbf{v}_0| \cos \alpha)\frac{2|\mathbf{v}_0| \sin \alpha}{g} \qquad \text{Substitute for } T. \\
&= \frac{2|\mathbf{v}_0|^2 \sin \alpha \cos \alpha}{g} \qquad \text{Simplify.} \\
&= \frac{|\mathbf{v}_0|^2 \sin 2\alpha}{g}. \qquad 2 \sin \alpha \cos \alpha = \sin 2\alpha
\end{aligned}
$$

Note that on the interval $0 \le \alpha \le \pi/2$, $\sin 2\alpha$ has a maximum value of 1 when $\alpha = \pi/4$, so the maximum range is $|\mathbf{v}_0|^2/g$. In other words, in an ideal world, firing an object from the ground at an angle of $\pi/4$ (45°) maximizes its range. Notice that the ranges obtained with the angles α and $\pi/2 - \alpha$ are equal (**Figure 14.23**).

3. The maximum height of the object is reached when the vertical velocity is zero, or when $y'(t) = -gt + |\mathbf{v}_0| \sin \alpha = 0$. Solving for t, the maximum height is reached at $t = |\mathbf{v}_0|(\sin \alpha)/g = T/2$, which is half the time of flight. The object spends equal amounts of time ascending and descending. The maximum height is

$$
y\left(\frac{T}{2}\right) = \frac{(|\mathbf{v}_0| \sin \alpha)^2}{2g}.
$$

4. Finally, by eliminating t from the equations for $x(t)$ and $y(t)$, it can be shown (Exercise 72) that the trajectory of the object is a segment of a parabola.

SUMMARY   Two-Dimensional Motion

Assume an object traveling over horizontal ground, acted on only by the gravitational force, has an initial position $\langle x_0, y_0 \rangle = \langle 0, 0 \rangle$ and initial velocity $\langle u_0, v_0 \rangle = \langle |\mathbf{v}_0| \cos \alpha, |\mathbf{v}_0| \sin \alpha \rangle$. The trajectory, which is a segment of a parabola, has the following properties.

$$
\text{time of flight} = T = \frac{2|\mathbf{v}_0| \sin \alpha}{g}
$$

$$
\text{range} = \frac{|\mathbf{v}_0|^2 \sin 2\alpha}{g}
$$

$$
\text{maximum height} = y\left(\frac{T}{2}\right) = \frac{(|\mathbf{v}_0| \sin \alpha)^2}{2g}
$$

EXAMPLE 5   Flight of a golf ball A golf ball is driven down a horizontal fairway with an initial speed of 55 m/s at an initial angle of 25° (from a tee with negligible height). Neglect all forces except gravity, and assume the ball's trajectory lies in a plane.

a. When the ball first touches the ground, how far has it traveled horizontally and how long has it been in the air?

b. What is the maximum height of the ball?

c. At what angles should the ball be hit to reach a green that is 300 m from the tee?

SOLUTION

a. Using the range formula with $\alpha = 25°$ and $|\mathbf{v}_0| = 55$ m/s, the ball travels

$$
\frac{|\mathbf{v}_0|^2 \sin 2\alpha}{g} = \frac{(55 \text{ m/s})^2 \sin 50°}{9.8 \text{ m/s}^2} \approx 236 \text{ m}.
$$

The time of the flight is

$$T = \frac{2|\mathbf{v_0}| \sin \alpha}{g} = \frac{2(55 \text{ m/s}) \sin 25°}{9.8 \text{ m/s}^2} \approx 4.7 \text{ s}.$$

b. The maximum height of the ball is

$$\frac{(|\mathbf{v_0}| \sin \alpha)^2}{2g} = \frac{((55 \text{ m/s})(\sin 25°))^2}{2(9.8 \text{ m/s}^2)} \approx 27.6 \text{ m}.$$

c. Letting R denote the range and solving the range formula for $\sin 2\alpha$, we find that $\sin 2\alpha = Rg/|\mathbf{v_0}|^2$. For a range of $R = 300$ m and an initial speed of $|\mathbf{v_0}| = 55$ m/s, the required angle satisfies

$$\sin 2\alpha = \frac{Rg}{|\mathbf{v_0}|^2} = \frac{(300 \text{ m})(9.8 \text{ m/s}^2)}{(55 \text{ m/s})^2} \approx 0.972.$$

For the ball to travel a horizontal distance of exactly 300 m, the required angles are $\alpha = \frac{1}{2} \sin^{-1} 0.972 \approx 38.2°$ or $51.8°$.

Related Exercises 42, 45 ◀

Three-Dimensional Motion

To solve three-dimensional motion problems, we adopt a coordinate system in which the x- and y-axes point in two perpendicular horizontal directions (for example, east and north), while the positive z-axis points vertically upward (**Figure 14.24**). Newton's Second Law now has three components and appears in the form

$$m\mathbf{a}(t) = \langle mx''(t), my''(t), mz''(t) \rangle = \mathbf{F}.$$

If only the gravitational force is present (now in the negative z-direction), then the force vector is $\mathbf{F} = \langle 0, 0, -mg \rangle$; the equation of motion is then $\mathbf{a}(t) = \langle 0, 0, -g \rangle$. Other effects, such as crosswinds, spins, or slices, can be modeled by including other force components.

z (up)

Gravitational force
$\mathbf{F} = \langle 0, 0, -mg \rangle$

y (North)

x (East)

Figure 14.24

EXAMPLE 6   Projectile motion A small projectile is fired to the east over horizontal ground with an initial speed of $|\mathbf{v_0}| = 300$ m/s at an angle of $\alpha = 30°$ above the horizontal. A crosswind blows from south to north, producing an acceleration of the projectile of 0.36 m/s^2 to the north.

a. Where does the projectile land? How far does it land from its launch site?

b. In order to correct for the crosswind and make the projectile land due east of the launch site, at what angle from due east must the projectile be fired? Assume the initial speed $|\mathbf{v_0}| = 300$ m/s and the angle of elevation $\alpha = 30°$ are the same as in part (a).

SOLUTION

a. Letting $g = 9.8$ m/s^2, the equations of motion are $\mathbf{a}(t) = \mathbf{v}'(t) = \langle 0, 0.36, -9.8 \rangle$. Proceeding as in the two-dimensional case, the indefinite integral of the acceleration is the velocity function

$$\mathbf{v}(t) = \langle 0, 0.36t, -9.8t \rangle + \mathbf{C},$$

where $\mathbf{C}$ is an arbitrary constant vector. With an initial speed $|\mathbf{v_0}| = 300$ m/s and an angle of elevation of $\alpha = 30°$ (**Figure 14.25a**), the initial velocity is

$$\mathbf{v}(0) = \langle 300 \cos 30°, 0, 300 \sin 30° \rangle = \langle 150\sqrt{3}, 0, 150 \rangle.$$

Substituting $t = 0$ and using the initial condition, we find that $\mathbf{C} = \langle 150\sqrt{3}, 0, 150 \rangle$. Therefore, the velocity function is

$$\mathbf{v}(t) = \langle 150\sqrt{3}, 0.36t, -9.8t + 150 \rangle.$$

Integrating the velocity function produces the position function

$$\mathbf{r}(t) = \langle 150\sqrt{3}t, 0.18t^2, -4.9t^2 + 150t \rangle + \mathbf{C}.$$

Using the initial condition $\mathbf{r}(0) = \langle 0, 0, 0 \rangle$, we find that $\mathbf{C} = \langle 0, 0, 0 \rangle$, and the position function is

$$\mathbf{r}(t) = \langle x(t), y(t), z(t) \rangle = \langle 150\sqrt{3}t, 0.18t^2, -4.9t^2 + 150t \rangle.$$

The projectile lands when $z(t) = -4.9t^2 + 150t = 0$. Solving for t, the positive root, which gives the time of flight, is $T = 150/4.9 \approx 30.6$ s. The x- and y-coordinates at that time are

$$x(T) \approx 7953 \text{ m} \quad \text{and} \quad y(T) \approx 169 \text{ m}.$$

Therefore, the projectile lands approximately 7953 m east and 169 m north of the firing site. Because the projectile started at $(0, 0, 0)$, it traveled a horizontal distance of $\sqrt{7953^2 + 169^2} \approx 7955$ m (Figure 14.25a).

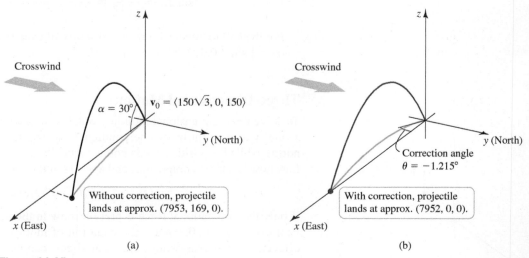

Figure 14.25

b. Keeping the initial speed of the projectile equal to $|\mathbf{v}_0| = 300$ m/s, we decompose the horizontal component of the speed, $150\sqrt{3}$ m/s, into an east component, $u_0 = 150\sqrt{3} \cos \theta$, and a north component, $v_0 = 150\sqrt{3} \sin \theta$, where θ is the angle relative to due east; we must determine the correction angle θ (**Figure 14.25b**). The x- and y-components of the position are

$$x(t) = (150\sqrt{3} \cos \theta)t \quad \text{and} \quad y(t) = 0.18t^2 + (150\sqrt{3} \sin \theta)t.$$

These changes in the initial velocity affect the x- and y-equations, but not the z-equation. Therefore, the time of flight is still $T = 150/4.9 \approx 30.6$ s. The aim is to choose θ so that the projectile lands on the x-axis (due east from the launch site), which means $y(T) = 0$. Solving

$$y(T) = 0.18T^2 + (150\sqrt{3} \sin \theta)T = 0,$$

with $T = 150/4.9$, we find that $\sin \theta \approx -0.0212$; therefore, $\theta \approx -0.0212$ rad $\approx -1.215°$. In other words, the projectile must be fired at a horizontal angle of $1.215°$ to the *south* of east to correct for the northerly crosswind (Figure 14.25b). The landing location of the projectile is $x(T) \approx 7952$ m and $y(T) = 0$.

Related Exercises 52–53 ◀

SECTION 14.3 EXERCISES

Getting Started

1. Given the position function $\mathbf{r}$ of a moving object, explain how to find the velocity, speed, and acceleration of the object.

2. What is the relationship between the position and velocity vectors for motion on a circle?

3. Write Newton's Second Law of Motion in vector form.

4. Write Newton's Second Law of Motion for three-dimensional motion with only the gravitational force (acting in the z-direction).

5. Given the acceleration of an object and its initial velocity, how do you find the velocity of the object, for $t \geq 0$?

6. Given the velocity of an object and its initial position, how do you find the position of the object, for $t \geq 0$?

7. The velocity of a moving object, for $t \geq 0$, is
$\mathbf{r}'(t) = \langle 60, 96 - 32t \rangle$ ft/s.

   a. When is the vertical component of velocity of the object equal to 0?

   b. Find $\mathbf{r}(t)$ if $\mathbf{r}(0) = \langle 0, 3 \rangle$.

8. A baseball is hit 2 feet above home plate, and the position of the ball t seconds later is $\mathbf{r}(t) = \langle 40t, -16t^2 + 31t + 2 \rangle$ ft. Find each of the following values.

   a. The time of flight of the baseball

   b. The range of the baseball

Practice Exercises

9–20. Velocity and acceleration from position *Consider the following position functions.*

a. Find the velocity and speed of the object.

b. Find the acceleration of the object.

9. $\mathbf{r}(t) = \langle 3t^2 + 1, 4t^2 + 3 \rangle$, for $t \geq 0$

10. $\mathbf{r}(t) = \left\langle \dfrac{5}{2}t^2 + 3, 6t^2 + 10 \right\rangle$, for $t \geq 0$

11. $\mathbf{r}(t) = \langle 2 + 2t, 1 - 4t \rangle$, for $t \geq 0$

12. $\mathbf{r}(t) = \langle 1 - t^2, 3 + 2t^3 \rangle$, for $t \geq 0$

13. $\mathbf{r}(t) = \langle 8 \sin t, 8 \cos t \rangle$, for $0 \leq t \leq 2\pi$

14. $\mathbf{r}(t) = \langle 3 \cos t, 4 \sin t \rangle$, for $0 \leq t \leq 2\pi$

15. $\mathbf{r}(t) = \left\langle t^2 + 3, t^2 + 10, \dfrac{1}{2}t^2 \right\rangle$, for $t \geq 0$

16. $\mathbf{r}(t) = \langle 2e^{2t} + 1, e^{2t} - 1, 2e^{2t} - 10 \rangle$, for $t \geq 0$

17. $\mathbf{r}(t) = \langle 3 + t, 2 - 4t, 1 + 6t \rangle$, for $t \geq 0$

18. $\mathbf{r}(t) = \langle 3 \sin t, 5 \cos t, 4 \sin t \rangle$, for $0 \leq t \leq 2\pi$

19. $\mathbf{r}(t) = \langle 1, t^2, e^{-t} \rangle$, for $t \geq 0$

20. $\mathbf{r}(t) = \langle 13 \cos 2t, 12 \sin t\, 2t, 5 \sin 2t \rangle$, for $0 \leq t \leq \pi$

T 21–26. Comparing trajectories *Consider the following position functions* $\mathbf{r}$ *and* $\mathbf{R}$ *for two objects.*

a. Find the interval $[c, d]$ *over which the* $\mathbf{R}$ *trajectory is the same as the* $\mathbf{r}$ *trajectory over* $[a, b]$.

b. Find the velocity for both objects.

c. Graph the speed of the two objects over the intervals $[a, b]$ *and* $[c, d]$, *respectively.*

21. $\mathbf{r}(t) = \langle t, t^2 \rangle$, $[a, b] = [0, 2]$; $\mathbf{R}(t) = \langle 2t, 4t^2 \rangle$ on $[c, d]$

22. $\mathbf{r}(t) = \langle 1 + 3t, 2 + 4t \rangle$, $[a, b] = [0, 6]$; $\mathbf{R}(t) = \langle 1 + 9t, 2 + 12t \rangle$ on $[c, d]$

23. $\mathbf{r}(t) = \langle \cos t, 4 \sin t \rangle$, $[a, b] = [0, 2\pi]$; $\mathbf{R}(t) = \langle \cos 3t, 4 \sin 3t \rangle$ on $[c, d]$

24. $\mathbf{r}(t) = \langle 2 - e^t, 4 - e^{-t} \rangle$, $[a, b] = [0, \ln 10]$; $\mathbf{R}(t) = \langle 2 - t, 4 - 1/t \rangle$ on $[c, d]$

25. $\mathbf{r}(t) = \langle 4 + t^2, 3 - 2t^4, 1 + 3t^6 \rangle$, $[a, b] = [0, 6]$; $\mathbf{R}(t) = \langle 4 + \ln t, 3 - 2 \ln^2 t, 1 + 3 \ln^3 t \rangle$ on $[c, d]$
For graphing, let $c = 1$ and $d = 20$.

26. $\mathbf{r}(t) = \langle 2 \cos 2t, \sqrt{2} \sin 2t, \sqrt{2} \sin 2t \rangle$, $[a, b] = [0, \pi]$; $\mathbf{R}(t) = \langle 2 \cos 4t, \sqrt{2} \sin 4t, \sqrt{2} \sin 4t \rangle$ on $[c, d]$

T 27–28. Carnival rides

27. Consider a carnival ride where Andrea is at point P that moves counterclockwise around a circle centered at C while the arm, represented by the line segment from the origin O to point C, moves counterclockwise about the origin (see figure). Andrea's position (in feet) at time t (in seconds) is
$$\mathbf{r}(t) = \langle 20 \cos t + 10 \cos 5t, 20 \sin t + 10 \sin 5t \rangle.$$

   a. Plot a graph of $\mathbf{r}(t)$, for $0 \leq t \leq 2\pi$.

   b. Find the velocity $\mathbf{v}(t)$.

   c. Show that the speed $|\mathbf{v}(t)| = v(t) = 10\sqrt{29 + 20 \cos 4t}$ and plot the speed, for $0 \leq t \leq 2\pi$. (*Hint:* Use the identity $\sin mx \sin nx + \cos mx \cos nx = \cos((m - n)x)$.)

   d. Determine Andrea's maximum and minimum speeds.

28. Suppose the carnival ride in Exercise 27 is modified so that Andrea's position P (in ft) at time t (in s) is
$$\mathbf{r}(t) = \langle 20 \cos t + 10 \cos 5t, 20 \sin t + 10 \sin 5t, 5 \sin 2t \rangle.$$

   a. Describe how this carnival ride differs from the ride in Exercise 27.

   b. Find the speed function $|\mathbf{v}(t)| = v(t)$ and plot its graph.

   c. Find Andrea's maximum and minimum speeds.

29–32. Trajectories on circles and spheres *Determine whether the following trajectories lie on either a circle in* $\mathbb{R}^2$ *or a sphere in* $\mathbb{R}^3$ *centered at the origin. If so, find the radius of the circle or sphere, and show that the position vector and the velocity vector are everywhere orthogonal.*

29. $\mathbf{r}(t) = \langle 8 \cos 2t, 8 \sin 2t \rangle$, for $0 \leq t \leq \pi$

30. $\mathbf{r}(t) = \langle 4 \sin t, 2 \cos t \rangle$, for $0 \leq t \leq 2\pi$

31. $\mathbf{r}(t) = \langle \sin t + \sqrt{3} \cos t, \sqrt{3} \sin t - \cos t \rangle$, for $0 \leq t \leq 2\pi$

32. $\mathbf{r}(t) = \langle 3 \sin t, 5 \cos t, 4 \sin t \rangle$, for $0 \leq t \leq 2\pi$

33–34. Path on a sphere *Show that the following trajectories lie on a sphere centered at the origin, and find the radius of the sphere.*

33. $\mathbf{r}(t) = \left\langle \dfrac{5 \sin t}{\sqrt{1 + \sin^2 2t}}, \dfrac{5 \cos t}{\sqrt{1 + \sin^2 2t}}, \dfrac{5 \sin 2t}{\sqrt{1 + \sin^2 2t}} \right\rangle$, for $0 \leq t \leq 2\pi$

34. $\mathbf{r}(t) = \left\langle \dfrac{4 \cos t}{\sqrt{4 + t^2}}, \dfrac{2t}{\sqrt{4 + t^2}}, \dfrac{4 \sin t}{\sqrt{4 + t^2}} \right\rangle$, for $0 \leq t \leq 4\pi$

35–40. Solving equations of motion *Given an acceleration vector, initial velocity* $\langle u_0, v_0 \rangle$, *and initial position* $\langle x_0, y_0 \rangle$, *find the velocity and position vectors for* $t \geq 0$.

35. $\mathbf{a}(t) = \langle 0, 1 \rangle$, $\langle u_0, v_0 \rangle = \langle 2, 3 \rangle$, $\langle x_0, y_0 \rangle = \langle 0, 0 \rangle$

36. $\mathbf{a}(t) = \langle 1, 2 \rangle$, $\langle u_0, v_0 \rangle = \langle 1, 1 \rangle$, $\langle x_0, y_0 \rangle = \langle 2, 3 \rangle$

37. $\mathbf{a}(t) = \langle 0, 10 \rangle$, $\langle u_0, v_0 \rangle = \langle 0, 5 \rangle$, $\langle x_0, y_0 \rangle = \langle 1, -1 \rangle$

38. $\mathbf{a}(t) = \langle 1, t \rangle$, $\langle u_0, v_0 \rangle = \langle 2, -1 \rangle$, $\langle x_0, y_0 \rangle = \langle 0, 8 \rangle$

39. $\mathbf{a}(t) = \langle \cos t, 2 \sin t \rangle$, $\langle u_0, v_0 \rangle = \langle 0, 1 \rangle$, $\langle x_0, y_0 \rangle = \langle 1, 0 \rangle$

40. $\mathbf{a}(t) = \langle e^{-t}, 1 \rangle$, $\langle u_0, v_0 \rangle = \langle 1, 0 \rangle$, $\langle x_0, y_0 \rangle = \langle 0, 0 \rangle$

T 41–46. Two-dimensional motion *Consider the motion of the following objects. Assume the x-axis is horizontal, the positive y-axis is vertical, the ground is horizontal, and only the gravitational force acts on the object.*

a. *Find the velocity and position vectors, for $t \geq 0$.*
b. *Graph the trajectory.*
c. *Determine the time of flight and range of the object.*
d. *Determine the maximum height of the object.*

41. A soccer ball has an initial position (in m) of $\langle x_0, y_0 \rangle = \langle 0, 0 \rangle$ when it is kicked with an initial velocity of $\langle u_0, v_0 \rangle = \langle 30, 6 \rangle$ m/s.

42. A golf ball has an initial position (in ft) of $\langle x_0, y_0 \rangle = \langle 0, 0 \rangle$ when it is hit at an angle of 30° with an initial speed of 150 ft/s.

43. A baseball has an initial position (in ft) of $\langle x_0, y_0 \rangle = \langle 0, 6 \rangle$ when it is thrown with an initial velocity of $\langle u_0, v_0 \rangle = \langle 80, 10 \rangle$ ft/s.

44. A baseball is thrown horizontally from a height of 10 ft above the ground with a speed of 132 ft/s.

45. A projectile is launched from a platform 20 ft above the ground at an angle of 60° above the horizontal with a speed of 250 ft/s. Assume the origin is at the base of the platform.

46. A rock is thrown from the edge of a vertical cliff 40 m above the ground at an angle of 45° above the horizontal with a speed of $10\sqrt{2}$ m/s. Assume the origin is at the foot of the cliff.

47–50. Solving equations of motion *Given an acceleration vector, initial velocity $\langle u_0, v_0, w_0 \rangle$, and initial position $\langle x_0, y_0, z_0 \rangle$, find the velocity and position vectors, for $t \geq 0$.*

47. $\mathbf{a}(t) = \langle 0, 0, 10 \rangle$, $\langle u_0, v_0, w_0 \rangle = \langle 1, 5, 0 \rangle$, $\langle x_0, y_0, z_0 \rangle = \langle 0, 5, 0 \rangle$

48. $\mathbf{a}(t) = \langle 1, t, 4t \rangle$, $\langle u_0, v_0, w_0 \rangle = \langle 20, 0, 0 \rangle$, $\langle x_0, y_0, z_0 \rangle = \langle 0, 0, 0 \rangle$

49. $\mathbf{a}(t) = \langle \sin t, \cos t, 1 \rangle$, $\langle u_0, v_0, w_0 \rangle = \langle 0, 2, 0 \rangle$, $\langle x_0, y_0, z_0 \rangle = \langle 0, 0, 0 \rangle$

50. $\mathbf{a}(t) = \langle t, e^{-t}, 1 \rangle$, $\langle u_0, v_0, w_0 \rangle = \langle 0, 0, 1 \rangle$, $\langle x_0, y_0, z_0 \rangle = \langle 4, 0, 0 \rangle$

T 51–56. Three-dimensional motion *Consider the motion of the following objects. Assume the x-axis points east, the y-axis points north, the positive z-axis is vertical and opposite g, the ground is horizontal, and only the gravitational force acts on the object unless otherwise stated.*

a. *Find the velocity and position vectors, for $t \geq 0$.*
b. *Make a sketch of the trajectory.*
c. *Determine the time of flight and range of the object.*
d. *Determine the maximum height of the object.*

51. A bullet is fired from a rifle 1 m above the ground in a northeast direction. The initial velocity of the bullet is $\langle 200, 200, 0 \rangle$ m/s.

52. A golf ball is hit east down a fairway with an initial velocity of $\langle 50, 0, 30 \rangle$ m/s. A crosswind blowing to the south produces an acceleration of the ball of -0.8 m/s².

53. A baseball is hit 3 ft above home plate with an initial velocity of $\langle 60, 80, 80 \rangle$ ft/s. The spin on the baseball produces a horizontal acceleration of the ball of 10 ft/s² in the eastward direction.

54. A baseball is hit 3 ft above home plate with an initial velocity of $\langle 30, 30, 80 \rangle$ ft/s. The spin on the baseball produces a horizontal acceleration of the ball of 5 ft/s² in the northward direction.

55. A small rocket is fired from a launch pad 10 m above the ground with an initial velocity, in m/s, of $\langle 300, 400, 500 \rangle$. A crosswind blowing to the north produces an acceleration of the rocket of 2.5 m/s².

56. A soccer ball is kicked from the point $\langle 0, 0, 0 \rangle$ with an initial velocity of $\langle 0, 80, 80 \rangle$ ft/s. The spin on the ball produces an acceleration of $\langle 1.2, 0, 0 \rangle$ ft/s².

57. **Explain why or why not** Determine whether the following statements are true and give an explanation or counterexample.

a. If the speed of an object is constant, then its velocity components are constant.
b. The functions $\mathbf{r}(t) = \langle \cos t, \sin t \rangle$ and $\mathbf{R}(t) = \langle \sin t^2, \cos t^2 \rangle$ generate the same set of points, for $t \geq 0$.
c. A velocity vector of variable magnitude cannot have a constant direction.
d. If the acceleration of an object is $\mathbf{a}(t) = \mathbf{0}$, for all $t \geq 0$, then the velocity of the object is constant.
e. If you double the initial speed of a projectile, its range also doubles (assume no forces other than gravity act on the projectile).
f. If you double the initial speed of a projectile, its time of flight also doubles (assume no forces other than gravity).
g. A trajectory with $\mathbf{v}(t) = \mathbf{a}(t) \neq \mathbf{0}$, for all t, is possible.

T 58–61. Trajectory properties *Find the time of flight, range, and maximum height of the following two-dimensional trajectories, assuming no forces other than gravity. In each case, the initial position is $\langle 0, 0 \rangle$ and the initial velocity is $\mathbf{v}_0 = \langle u_0, v_0 \rangle$.*

58. $\langle u_0, v_0 \rangle = \langle 10, 20 \rangle$ ft/s

59. Initial speed $|\mathbf{v}_0| = 150$ m/s, launch angle $\alpha = 30°$

60. $\langle u_0, v_0 \rangle = \langle 40, 80 \rangle$ m/s

61. Initial speed $|\mathbf{v}_0| = 400$ ft/s, launch angle $\alpha = 60°$

62. **Motion on the moon** The acceleration due to gravity on the moon is approximately $g/6$ (one-sixth its value on Earth). Compare the time of flight, range, and maximum height of a projectile on the moon with the corresponding values on Earth.

63. **Firing angles** A projectile is fired over horizontal ground from the origin with an initial speed of 60 m/s. What firing angles produce a range of 300 m?

T 64. Firing strategies Suppose you wish to fire a projectile over horizontal ground from the origin and attain a range of 1000 m.

a. Sketch a graph of the initial speed required for all firing angles $0 < \alpha < \pi/2$.
b. What firing angle requires the least initial speed?

65. **Speed on an ellipse** An object moves along an ellipse given by the function $\mathbf{r}(t) = \langle a \cos t, b \sin t \rangle$, for $0 \leq t \leq 2\pi$, where $a > 0$ and $b > 0$.

a. Find the velocity and speed of the object in terms of a and b, for $0 \leq t \leq 2\pi$.

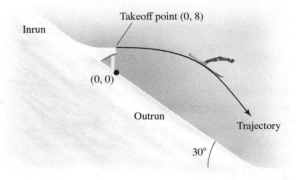

Inrun

Takeoff point $(0, 8)$

$(0, 0)$

Outrun

Trajectory

$30°$

b. With $a = 1$ and $b = 6$, graph the speed function, for $0 \le t \le 2\pi$. Mark the points on the trajectory at which the speed is a minimum and a maximum.

c. Is it true that the object speeds up along the flattest (straightest) parts of the trajectory and slows down where the curves are sharpest?

d. For general a and b, find the ratio of the maximum speed to the minimum speed on the ellipse (in terms of a and b).

Explorations and Challenges

66. Golf shot A golfer stands 390 ft (130 yd) horizontally from the hole and 40 ft below the hole (see figure). Assuming the ball is hit with an initial speed of 150 ft/s, at what angle(s) should it be hit to land in the hole? Assume the path of the ball lies in a plane.

40 ft

390 ft (130 yd)

67. Another golf shot A golfer stands 420 ft (140 yd) horizontally from the hole and 50 ft above the hole (see figure). Assuming the ball is hit with an initial speed of 120 ft/s, at what angle(s) should it be hit to land in the hole? Assume the path of the ball lies in a plane.

50 ft

420 ft (140 yd)

T 68. Initial speed of a golf shot A golfer stands 390 ft horizontally from the hole and 40 ft below the hole (see figure for Exercise 66). If the ball leaves the ground at an initial angle of 45° with the horizontal, with what initial speed should it be hit to land in the hole?

T 69. Initial speed of a golf shot A golfer stands 420 ft horizontally from the hole and 50 ft above the hole (see figure for Exercise 67). If the ball leaves the ground at an initial angle of 30° with the horizontal, with what initial speed should it be hit to land in the hole?

T 70. Ski jump The lip of a ski jump is 8 m above the outrun that is sloped at an angle of 30° to the horizontal (see figure).

a. If the initial velocity of a ski jumper at the lip of the jump is $\langle 40, 0 \rangle$ m/s, what is the length of the jump (distance from the origin to the landing point)? Assume only gravity affects the motion.

b. Assume air resistance produces a constant horizontal acceleration of 0.15 m/s² opposing the motion. What is the length of the jump?

c. Suppose the takeoff ramp is tilted upward at an angle of $\theta°$, so that the skier's initial velocity is $40\langle \cos \theta, \sin \theta \rangle$ m/s. What value of θ maximizes the length of the jump? Express your answer in degrees and neglect air resistance.

71. Designing a baseball pitch A baseball leaves the hand of a pitcher 6 vertical feet above and 60 horizontal feet from home plate. Assume the coordinate axes are oriented as shown in the figure.

6 ft

60 ft

a. Suppose a pitch is thrown with an initial velocity of $\langle 130, 0, -3 \rangle$ ft/s (about 90 mi/hr). In the absence of all forces except gravity, how far above the ground is the ball when it crosses home plate and how long does it take the pitch to arrive?

b. What vertical velocity component should the pitcher use so that the pitch crosses home plate exactly 3 ft above the ground?

c. A simple model to describe the curve of a baseball assumes the spin of the ball produces a constant sideways acceleration (in the y-direction) of c ft/s². Suppose a pitcher throws a curve ball with $c = 8$ ft/s² (one fourth the acceleration of gravity). How far does the ball move in the y-direction by the time it reaches home plate, assuming an initial velocity of $\langle 130, 0, -3 \rangle$ ft/s?

d. In part (c), does the ball curve more in the first half of its trip to the plate or in the second half? How does this fact affect the batter?

e. Suppose the pitcher releases the ball from an initial position of $\langle 0, -3, 6 \rangle$ with initial velocity $\langle 130, 0, -3 \rangle$. What value of the spin parameter c is needed to put the ball over home plate passing through the point $(60, 0, 3)$?

72. Parabolic trajectories Show that the two-dimensional trajectory

$$x(t) = u_0 t + x_0 \text{ and } y(t) = -\frac{gt^2}{2} + v_0 t + y_0, \text{ for } 0 \le t \le T,$$

of an object moving in a gravitational field is a segment of a parabola for some value of $T > 0$. Find T such that $y(T) = 0$.

73. Time of flight, range, height Derive the formulas for time of flight, range, and maximum height in the case that an object is launched from the initial position $\langle 0, y_0 \rangle$ above the horizontal ground with initial velocity $|\mathbf{v}_0| \langle \cos \alpha, \sin \alpha \rangle$.

74. A race Two people travel from $P(4, 0)$ to $Q(-4, 0)$ along the paths given by

$$\mathbf{r}(t) = \left\langle 4 \cos \frac{\pi t}{8}, 4 \sin \frac{\pi t}{8} \right\rangle \quad \text{and}$$

$$\mathbf{R}(t) = \langle 4 - t, (4 - t)^2 - 16 \rangle.$$

a. Graph both paths between P and Q.
b. Graph the speeds of both people between P and Q.
c. Who arrives at Q first?

75. Circular motion Consider an object moving along the circular trajectory $\mathbf{r}(t) = \langle A \cos \omega t, A \sin \omega t \rangle$, where A and ω are constants.

a. Over what time interval $[0, T]$ does the object traverse the circle once?
b. Find the velocity and speed of the object. Is the velocity constant in either direction or magnitude? Is the speed constant?
c. Find the acceleration of the object.
d. How are the position and velocity related? How are the position and acceleration related?
e. Sketch the position, velocity, and acceleration vectors at four different points on the trajectory with $A = \omega = 1$.

76. A linear trajectory An object moves along a straight line from the point $P(1, 2, 4)$ to the point $Q(-6, 8, 10)$.

a. Find a position function $\mathbf{r}$ that describes the motion if it occurs with a constant speed over the time interval $[0, 5]$.
b. Find a position function $\mathbf{r}$ that describes the motion if it occurs with speed e^t.

77. A circular trajectory An object moves clockwise around a circle centered at the origin with radius 5 m beginning at the point $(0, 5)$.

a. Find a position function $\mathbf{r}$ that describes the motion if the object moves with a constant speed, completing 1 lap every 12 s.
b. Find a position function $\mathbf{r}$ that describes the motion if it occurs with speed e^{-t}.

78. A helical trajectory An object moves on the helix $\langle \cos t, \sin t, t \rangle$, for $t \geq 0$.

a. Find a position function $\mathbf{r}$ that describes the motion if it occurs with a constant speed of 10.
b. Find a position function $\mathbf{r}$ that describes the motion if it occurs with speed t.

79. Tilted ellipse Consider the curve $\mathbf{r}(t) = \langle \cos t, \sin t, c \sin t \rangle$, for $0 \leq t \leq 2\pi$, where c is a real number. Assuming the curve lies in a plane, prove that the curve is an ellipse in that plane.

80. Equal area property Consider the ellipse $\mathbf{r}(t) = \langle a \cos t, b \sin t \rangle$, for $0 \leq t \leq 2\pi$, where a and b are real numbers. Let θ be the angle between the position vector and the x-axis.

a. Show that $\tan \theta = \dfrac{b}{a} \tan t$.
b. Find $\theta'(t)$.
c. Recall that the area bounded by the polar curve $r = f(\theta)$ on the interval $[0, \theta]$ is $A(\theta) = \dfrac{1}{2} \displaystyle\int_0^\theta (f(u))^2 \, du$. Letting $f(\theta(t)) = |\mathbf{r}(\theta(t))|$, show that $A'(t) = \dfrac{1}{2} ab$.
d. Conclude that as an object moves around the ellipse, it sweeps out equal areas in equal times.

81. Another property of constant $|\mathbf{r}|$ motion Suppose an object moves on the surface of a sphere with $|\mathbf{r}(t)|$ constant for all t. Show that $\mathbf{r}(t)$ and $\mathbf{a}(t) = \mathbf{r}''(t)$ satisfy $\mathbf{r}(t) \cdot \mathbf{a}(t) = -|\mathbf{v}(t)|^2$.

82. Conditions for a circular/elliptical trajectory in the plane An object moves along a path given by

$$\mathbf{r}(t) = \langle a \cos t + b \sin t, c \cos t + d \sin t \rangle, \text{ for } 0 \leq t \leq 2\pi.$$

a. What conditions on a, b, c, and d guarantee that the path is a circle?
b. What conditions on a, b, c, and d guarantee that the path is an ellipse?

83. Nonuniform straight-line motion Consider the motion of an object given by the position function

$$\mathbf{r}(t) = f(t) \langle a, b, c \rangle + \langle x_0, y_0, z_0 \rangle, \quad \text{for } t \geq 0,$$

where a, b, c, x_0, y_0, and z_0 are constants, and f is a differentiable scalar function, for $t \geq 0$.

a. Explain why $\mathbf{r}$ describes motion along a line.
b. Find the velocity function. In general, is the velocity constant in magnitude or direction along the path?

QUICK CHECK ANSWERS

1. $\mathbf{v}(t) = \langle 1, 2t, 3t^2 \rangle, \mathbf{a}(t) = \langle 0, 2, 6t \rangle$
2. $|\mathbf{r}'(t)| = \sqrt{1 + 4t^2 + 9t^4/16};$
$|\mathbf{R}'(t)| = \sqrt{4t^2 + 16t^6 + 9t^{10}/4}$ 3. The curve $\mathbf{R}(t) = 5\mathbf{r}(t)/|\mathbf{r}(t)|$ lies on a sphere of radius 5.
4. $x(t) = 100t, y(t) = -16t^2 + 60t + 2$
5. $\sin(2(\pi/2 - \alpha)) = \sin(\pi - 2\alpha) = \sin 2\alpha$ ◄

14.4 Length of Curves

We return now to a recurring theme: determining the arc length of a curve. In Section 6.5, we learned how to find the arc length of curves of the form $y = f(x)$, and in Sections 12.1 and 12.3, we discovered formulas for the arc length of a plane curve described parametrically or described in polar coordinates. In this section, we extend these ideas to handle the arc length of a three-dimensional curve described by a vector function. We also discover how to formulate a parametric description of a curve using arc length as a parameter.

Arc Length

Suppose a curve is described by the vector-valued function $\mathbf{r}(t) = \langle f(t), g(t), h(t) \rangle$, for $a \le t \le b$, where f', g', and h' are continuous on $[a, b]$. In Section 12.1, we showed that the arc length L of the two-dimensional curve $\mathbf{r}(t) = \langle f(t), g(t) \rangle$, for $a \le t \le b$ is given by

$$L = \int_a^b \sqrt{f'(t)^2 + g'(t)^2}\, dt.$$

An analogous arc length formula for three-dimensional curves follows using a similar argument. The length of the curve $\mathbf{r}(t) = \langle f(t), g(t), h(t) \rangle$ on the interval $[a, b]$ is

$$L = \int_a^b \sqrt{f'(t)^2 + g'(t)^2 + h'(t)^2}\, dt.$$

Noting that $\mathbf{r}'(t) = \langle f'(t), g'(t), h'(t) \rangle$, we state the following definition.

> ▶ An important fact is that the arc length of a smooth parameterized curve is independent of the choice of parameter (Exercise 54).
>
> ▶ For curves in the xy-plane, we set $h(t) = 0$ in the definition of arc length.

QUICK CHECK 1 What does the arc length formula give for the length of the line $\mathbf{r}(t) = \langle 2t, t, -2t \rangle$, for $0 \le t \le 3$? ◄

DEFINITION Arc Length for Vector Functions

Consider the parameterized curve $\mathbf{r}(t) = \langle f(t), g(t), h(t) \rangle$, where f', g', and h' are continuous, and the curve is traversed once for $a \le t \le b$. The **arc length** of the curve between $(f(a), g(a), h(a))$ and $(f(b), g(b), h(b))$ is

$$L = \int_a^b \sqrt{f'(t)^2 + g'(t)^2 + h'(t)^2}\, dt = \int_a^b |\mathbf{r}'(t)|\, dt.$$

The following application of arc length leads to an integral that is difficult to evaluate exactly.

EXAMPLE 1 Lengths of planetary orbits According to Kepler's first law, the planets revolve about the sun in elliptical orbits. A vector function that describes an ellipse in the xy-plane is

$$\mathbf{r}(t) = \langle a \cos t, b \sin t \rangle, \qquad \text{where } 0 \le t \le 2\pi.$$

If $a > b > 0$, then $2a$ is the length of the major axis and $2b$ is the length of the minor axis (**Figure 14.26**). Verify the lengths of the planetary orbits given in Table 14.1. Distances are given in terms of the astronomical unit (AU), which is the length of the semimajor axis of Earth's orbit, or about 93 million miles.

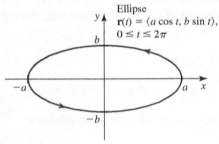

Ellipse
$\mathbf{r}(t) = \langle a \cos t, b \sin t \rangle$,
$0 \le t \le 2\pi$

Figure 14.26

> ▶ The German astronomer and mathematician Johannes Kepler (1571–1630) worked with the meticulously gathered data of Tycho Brahe to formulate three empirical laws obeyed by planets and comets orbiting the sun. The work of Kepler formed the foundation for Newton's laws of gravitation developed 50 years later.
>
> ▶ In September 2006, Pluto joined the ranks of Ceres, Haumea, Makemake, and Eris as one of five dwarf planets in our solar system.

Table 14.1

Planet	Semimajor axis, a (AU)	Semiminor axis, b (AU)	$\alpha = b/a$	Orbit length (AU)
Mercury	0.387	0.379	0.979	2.407
Venus	0.723	0.723	1.000	4.543
Earth	1.000	0.999	0.999	6.280
Mars	1.524	1.517	0.995	9.554
Jupiter	5.203	5.179	0.995	32.616
Saturn	9.539	9.524	0.998	59.888
Uranus	19.182	19.161	0.999	120.458
Neptune	30.058	30.057	1.000	188.857

SOLUTION Using the arc length formula, the length of a general elliptical orbit is

$$
\begin{aligned}
L &= \int_0^{2\pi} \sqrt{x'(t)^2 + y'(t)^2}\, dt \\[2mm]
&= \int_0^{2\pi} \sqrt{(-a \sin t)^2 + (b \cos t)^2}\, dt \quad \text{Substitute for } x'(t) \text{ and } y'(t). \\[2mm]
&= \int_0^{2\pi} \sqrt{a^2 \sin^2 t + b^2 \cos^2 t}\, dt. \quad \text{Simplify.}
\end{aligned}
$$

Factoring a^2 out of the square root and letting $\alpha = b/a$, we have

$$L = \int_0^{2\pi} \sqrt{a^2 \left(\sin^2 t + (b/a)^2 \cos^2 t\right)}\, dt \qquad \text{Factor out } a^2.$$

$$= a \int_0^{2\pi} \sqrt{\sin^2 t + \alpha^2 \cos^2 t}\, dt \qquad\qquad \text{Let } \alpha = b/a.$$

$$= 4a \int_0^{\pi/2} \sqrt{\sin^2 t + \alpha^2 \cos^2 t}\, dt. \qquad\quad \text{Use symmetry; quarter orbit on } [0, \pi/2].$$

> The integral that gives the length of an ellipse is a *complete elliptic integral of the second kind*. Many reference books and software packages provide approximate values of this integral.

Unfortunately, an antiderivative for this integrand cannot be found in terms of elementary functions, so we have two options: This integral is well known and values have been tabulated for various values of α. Alternatively, we may use a calculator to approximate the integral numerically (see Section 8.8). Using numerical integration, the orbit lengths in Table 14.1 are obtained. For example, the length of Mercury's orbit with $a = 0.387$ and $\alpha = 0.979$ is

> Although rounded values for α appear in Table 14.1, the calculations in Example 1 were done in full precision and were rounded to three decimal places only in the final step.

$$L = 4a \int_0^{\pi/2} \sqrt{\sin^2 t + \alpha^2 \cos^2 t}\, dt$$

$$= 1.548 \int_0^{\pi/2} \sqrt{\sin^2 t + 0.959 \cos^2 t}\, dt \qquad \text{Simplify.}$$

$$\approx 2.407. \qquad\qquad\qquad\qquad\qquad \text{Approximate using calculator.}$$

> Recall from Chapter 6 that the distance traveled by an object in one dimension is $\int_a^b |v(t)|\, dt$. The arc length formula generalizes this formula to three dimensions.

The fact that α is close to 1 for all the planets means that their orbits are nearly circular. For this reason, the lengths of the orbits shown in the table are nearly equal to $2\pi a$, which is the length of a circular orbit with radius a.

Related Exercises 27–28 ◄

If the function $\mathbf{r}(t) = \langle x(t), y(t), z(t) \rangle$ is the position function for a moving object, then the arc length formula has a natural interpretation. Recall that $\mathbf{v}(t) = \mathbf{r}'(t)$ is the velocity of the object and $|\mathbf{v}(t)| = |\mathbf{r}'(t)|$ is the speed of the object. Therefore, the arc length formula becomes

$$L = \int_a^b |\mathbf{r}'(t)|\, dt = \int_a^b |\mathbf{v}(t)|\, dt.$$

This formula is an analog of the familiar *distance = speed × elapsed time* formula.

EXAMPLE 2   Flight of an eagle An eagle rises at a rate of 100 vertical ft/min on a helical path given by

$$\mathbf{r}(t) = \langle 250 \cos t, 250 \sin t, 100t \rangle$$

(**Figure 14.27**), where $\mathbf{r}$ is measured in feet and t is measured in minutes. How far does it travel in 10 min?

SOLUTION The speed of the eagle is

$$|\mathbf{v}(t)| = \sqrt{x'(t)^2 + y'(t)^2 + z'(t)^2}$$

$$= \sqrt{(-250 \sin t)^2 + (250 \cos t)^2 + 100^2} \qquad \text{Substitute derivatives.}$$

$$= \sqrt{250^2 (\sin^2 t + \cos^2 t) + 100^2} \qquad\qquad \text{Combine terms.}$$

$$= \sqrt{250^2 + 100^2} \approx 269. \qquad\qquad\qquad \sin^2 t + \cos^2 t = 1$$

The constant speed makes the arc length integral easy to evaluate:

$$L = \int_0^{10} |\mathbf{v}(t)|\, dt \approx \int_0^{10} 269\, dt = 2690.$$

The eagle travels approximately 2690 ft in 10 min.

Related Exercise 25 ◄

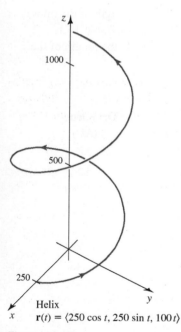

Helix
$\mathbf{r}(t) = \langle 250 \cos t, 250 \sin t, 100t \rangle$

Figure 14.27

> The standard parameterization for a helix winding counterclockwise around the z-axis is $\mathbf{r}(t) = \langle a \cos t, a \sin t, bt \rangle$. A helix has the property that its tangent vector makes a constant angle with the axis around which it winds.

Arc Length as a Parameter

Until now, the parameter t used to describe a curve $\mathbf{r}(t) = \langle f(t), g(t), h(t) \rangle$ has been chosen either for convenience or because it represents time in some specified unit. We now introduce the most natural parameter for describing curves; that parameter is *arc length*. Let's see what it means for a curve to be *parameterized by arc length*.

Consider the following two characterizations of the unit circle centered at the origin:

$$\langle \cos t, \sin t \rangle, \text{ for } 0 \leq t \leq 2\pi \quad \text{and} \quad \langle \cos 2t, \sin 2t \rangle, \text{ for } 0 \leq t \leq \pi.$$

In the first description, as the parameter t increases from $t = 0$ to $t = 2\pi$, the full circle is generated and the arc length s of the curve also increases from $s = 0$ to $s = 2\pi$. In other words, as the parameter t increases, it measures the arc length of the curve that is generated (**Figure 14.28a**).

In the second description, as t varies from $t = 0$ to $t = \pi$, the full circle is generated and the arc length increases from $s = 0$ to $s = 2\pi$. In this case, the length of the interval in t does not equal the length of the curve generated; therefore, the parameter t does not correspond to arc length (**Figure 14.28b**). In general, there are infinitely many ways to parameterize a given curve; however, for a given initial point and orientation, arc length is the parameter for only one of them.

QUICK CHECK 2 Consider the portion of a circle $\mathbf{r}(t) = \langle \cos t, \sin t \rangle$, for $a \leq t \leq b$. Show that the arc length of the curve is $b - a$. ◄

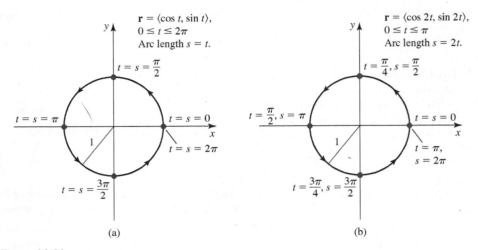

Figure 14.28

The Arc Length Function Suppose a smooth curve is represented by the function $\mathbf{r}(t) = \langle f(t), g(t), h(t) \rangle$, for $t \geq a$, where t is a parameter. Notice that as t increases, the length of the curve also increases. Using the arc length formula, the length of the curve from $\mathbf{r}(a)$ to $\mathbf{r}(t)$ is

▶ Notice that t is the independent variable of the function $s(t)$, so a different symbol u is used for the variable of integration. It is common to use s as the arc length function.

$$s(t) = \int_a^t \sqrt{f'(u)^2 + g'(u)^2 + h'(u)^2} \, du = \int_a^t |\mathbf{v}(u)| \, du.$$

This equation gives the relationship between the arc length of a curve and any parameter t used to describe the curve.

An important consequence of this relationship arises if we differentiate both sides with respect to t using the Fundamental Theorem of Calculus:

$$\frac{ds}{dt} = \frac{d}{dt}\left(\int_a^t |\mathbf{v}(u)| \, du \right) = |\mathbf{v}(t)|.$$

Specifically, if t represents time and $\mathbf{r}$ is the position of an object moving on the curve, then the rate of change of the arc length with respect to time is the speed of the object. Notice that if $\mathbf{r}(t)$ describes a smooth curve, then $|\mathbf{v}(t)| \neq 0$; hence $ds/dt > 0$, and s is an increasing function of t—as t increases, the arc length also increases. If $\mathbf{r}(t)$ is a curve on which $|\mathbf{v}(t)| = 1$, then

$$s(t) = \int_a^t |\mathbf{v}(u)| \, du = \int_a^t 1 \, du = t - a,$$

which means the parameter t corresponds to arc length. These observations are summarized in the following theorem.

> **THEOREM 14.3   Arc Length as a Function of a Parameter**
> Let $\mathbf{r}(t)$ describe a smooth curve, for $t \geq a$. The arc length is given by
> $$s(t) = \int_a^t |\mathbf{v}(u)| \, du,$$
> where $|\mathbf{v}| = |\mathbf{r}'|$. Equivalently, $\dfrac{ds}{dt} = |\mathbf{v}(t)|$. If $|\mathbf{v}(t)| = 1$, for all $t \geq a$, then the parameter t corresponds to arc length.

EXAMPLE 3   Arc length parameterization Consider the helix
$\mathbf{r}(t) = \langle 2 \cos t, 2 \sin t, 4t \rangle$, for $t \geq 0$.

a. Find the arc length function $s(t)$.

b. Find another description of the helix that uses arc length as the parameter.

SOLUTION

a. Note that $\mathbf{r}'(t) = \langle -2 \sin t, 2 \cos t, 4 \rangle$ and

$$
\begin{aligned}
|\mathbf{v}(t)| = |\mathbf{r}'(t)| &= \sqrt{(-2 \sin t)^2 + (2 \cos t)^2 + 4^2} \\
&= \sqrt{4(\sin^2 t + \cos^2 t) + 4^2} && \text{Simplify.} \\
&= \sqrt{4 + 4^2} && \sin^2 t + \cos^2 t = 1 \\
&= \sqrt{20} = 2\sqrt{5}. && \text{Simplify.}
\end{aligned}
$$

Therefore, the relationship between the arc length s and the parameter t is

$$s(t) = \int_a^t |\mathbf{v}(u)| \, du = \int_0^t 2\sqrt{5} \, du = 2\sqrt{5} \, t.$$

An increase of $1/(2\sqrt{5})$ in the parameter t corresponds to an increase of 1 in the arc length. Therefore, the curve is not parameterized by arc length.

b. Substituting $t = s/(2\sqrt{5})$ into the original parametric description of the helix, we find that the description with arc length as a parameter is (using a different function name)

QUICK CHECK 3 Does the line $\mathbf{r}(t) = \langle t, t, t \rangle$ have arc length as a parameter? Explain. ◄

$$\mathbf{r}_1(s) = \left\langle 2 \cos\left(\frac{s}{2\sqrt{5}}\right), 2 \sin\left(\frac{s}{2\sqrt{5}}\right), \frac{2s}{\sqrt{5}} \right\rangle, \quad \text{for } s \geq 0.$$

This description has the property that an increment of Δs in the parameter corresponds to an increment of exactly Δs in the arc length.

Related Exercises 37–39 ◄

As you will see in Section 14.5, using arc length as a parameter—when it can be done—generally leads to simplified calculations.

SECTION 14.4 EXERCISES

Getting Started

1. Find the length of the line given by $\mathbf{r}(t) = \langle t, 2t \rangle$, for $a \leq t \leq b$.

2. Explain how to find the length of the curve
$$\mathbf{r}(t) = \langle f(t), g(t), h(t) \rangle, \text{ for } a \leq t \leq b.$$

3. Express the arc length of a curve in terms of the speed of an object moving along the curve.

4. Suppose an object moves in space with the position function $\mathbf{r}(t) = \langle x(t), y(t), z(t) \rangle$. Write the integral that gives the distance it travels between $t = a$ and $t = b$.

5. An object moves on a trajectory given by
$$\mathbf{r}(t) = \langle 10 \cos 2t, 10 \sin 2t \rangle, \text{ for } 0 \leq t \leq \pi.$$
How far does it travel?

6. Use calculus to find the length of the line segment $\mathbf{r}(t) = \langle t, -8t, 4t \rangle$, for $0 \leq t \leq 2$. Verify your answer without using calculus.

7. Explain what it means for a curve to be parameterized by its arc length.

8. Is the curve $\mathbf{r}(t) = \langle \cos t, \sin t \rangle$ parameterized by its arc length? Explain.

Practice Exercises

9–22. Arc length calculations *Find the length of the following two- and three-dimensional curves.*

9. $\mathbf{r}(t) = \langle 3t^2 - 1, 4t^2 + 5 \rangle$, for $0 \le t \le 1$

10. $\mathbf{r}(t) = \langle 3t - 1, 4t + 5, t \rangle$, for $0 \le t \le 1$

11. $\mathbf{r}(t) = \langle 3\cos t, 3\sin t \rangle$, for $0 \le t \le \pi$

12. $\mathbf{r}(t) = \langle 4\cos 3t, 4\sin 3t \rangle$, for $0 \le t \le 2\pi/3$

13. $\mathbf{r}(t) = \langle \cos t + t\sin t, \sin t - t\cos t \rangle$, for $0 \le t \le \pi/2$

14. $\mathbf{r}(t) = \langle \cos t + \sin t, \cos t - \sin t \rangle$, for $0 \le t \le 2\pi$

15. $\mathbf{r}(t) = \langle 2 + 3t, 1 - 4t, -4 + 3t \rangle$, for $1 \le t \le 6$

16. $\mathbf{r}(t) = \langle 4\cos t, 4\sin t, 3t \rangle$, for $0 \le t \le 6\pi$

17. $\mathbf{r}(t) = \langle t, 8\sin t, 8\cos t \rangle$, for $0 \le t \le 4\pi$

18. $\mathbf{r}(t) = \left\langle \dfrac{t^2}{2}, \dfrac{(2t+1)^{3/2}}{3} \right\rangle$, for $0 \le t \le 2$

19. $\mathbf{r}(t) = \langle e^{2t}, 2e^{2t} + 5, 2e^{2t} - 20 \rangle$, for $0 \le t \le \ln 2$

20. $\mathbf{r}(t) = \langle t^2, t^3 \rangle$, for $0 \le t \le 4$

21. $\mathbf{r}(t) = \langle \cos^3 t, \sin^3 t \rangle$, for $0 \le t \le \pi/2$

22. $\mathbf{r}(t) = \langle 2\cos t, 2\sqrt{3}\cos t, 4\sin t \rangle$, for $0 \le t \le 2\pi$

23–26. Speed and arc length *For the following trajectories, find the speed associated with the trajectory, and then find the length of the trajectory on the given interval.*

23. $\mathbf{r}(t) = \langle 2t^3, -t^3, 5t^3 \rangle$, for $0 \le t \le 4$

24. $\mathbf{r}(t) = \langle 5\cos t^2, 5\sin t^2, 12t^2 \rangle$, for $0 \le t \le 2$

25. $\mathbf{r}(t) = \langle 13\sin 2t, 12\cos 2t, 5\cos 2t \rangle$, for $0 \le t \le \pi$

26. $\mathbf{r}(t) = \langle e^t \sin t, e^t \cos t, e^t \rangle$, for $0 \le t \le \ln 2$

T 27. Speed of Earth Verify that the length of one orbit of Earth is approximately 6.280 AU (see Table 14.1). Then determine the average speed of Earth relative to the sun in miles per hour. (*Hint:* It takes Earth 365.25 days to orbit the sun.)

T 28. Speed of Jupiter Verify that the length of one orbit of Jupiter is approximately 32.616 AU (see Table 14.1). Then determine the average speed of Jupiter relative to the sun in miles per hour. (*Hint:* It takes Jupiter 11.8618 Earth years to orbit the sun.)

T 29–32. Arc length approximations *Use a calculator to approximate the length of the following curves. In each case, simplify the arc length integral as much as possible before finding an approximation.*

29. $\mathbf{r}(t) = \langle 2\cos t, 4\sin t \rangle$, for $0 \le t \le 2\pi$

30. $\mathbf{r}(t) = \langle 2\cos t, 4\sin t, 6\cos t \rangle$, for $0 \le t \le 2\pi$

31. $\mathbf{r}(t) = \langle t, 4t^2, 10 \rangle$, for $-2 \le t \le 2$

32. $\mathbf{r}(t) = \langle e^t, 2e^{-t}, t \rangle$, for $0 \le t \le \ln 3$

33–42. Arc length parametrization *Determine whether the following curves use arc length as a parameter. If not, find a description that uses arc length as a parameter.*

33. $\mathbf{r}(t) = \langle 1, \sin t, \cos t \rangle$, for $t \ge 1$

34. $\mathbf{r}(t) = \left\langle \dfrac{t}{\sqrt{3}}, \dfrac{t}{\sqrt{3}}, \dfrac{t}{\sqrt{3}} \right\rangle$, for $0 \le t \le 10$

35. $\mathbf{r}(t) = \langle t, 2t \rangle$, for $0 \le t \le 3$

36. $\mathbf{r}(t) = \langle t + 1, 2t - 3, 6t \rangle$, for $0 \le t \le 10$

37. $\mathbf{r}(t) = \langle 2\cos t, 2\sin t \rangle$, for $0 \le t \le 2\pi$

38. $\mathbf{r}(t) = \langle 17\cos t, 15\sin t, 8\sin t \rangle$, for $0 \le t \le \pi$

39. $\mathbf{r}(t) = \langle \cos t^2, \sin t^2 \rangle$, for $0 \le t \le \sqrt{\pi}$

40. $\mathbf{r}(t) = \langle t^2, 2t^2, 4t^2 \rangle$, for $1 \le t \le 4$

41. $\mathbf{r}(t) = \langle e^t, e^t, e^t \rangle$, for $t \ge 0$

42. $\mathbf{r}(t) = \left\langle \dfrac{\cos t}{\sqrt{2}}, \dfrac{\cos t}{\sqrt{2}}, \sin t \right\rangle$, for $0 \le t \le 10$

43. Explain why or why not Determine whether the following statements are true and give an explanation or counterexample.

  a. If an object moves on a trajectory with constant speed S over a time interval $a \le t \le b$, then the length of the trajectory is $S(b - a)$.

  b. The curves defined by
$$\mathbf{r}(t) = \langle f(t), g(t) \rangle \quad \text{and} \quad \mathbf{R}(t) = \langle g(t), f(t) \rangle$$
have the same length over the interval $[a, b]$.

  c. The curve $\mathbf{r}(t) = \langle f(t), g(t) \rangle$, for $0 \le a \le t \le b$, and the curve $\mathbf{R}(t) = \langle f(t^2), g(t^2) \rangle$, for $\sqrt{a} \le t \le \sqrt{b}$, have the same length.

  d. The curve $\mathbf{r}(t) = \langle t, t^2, 3t^2 \rangle$, for $1 \le t \le 4$, is parameterized by arc length.

44. Length of a line segment Consider the line segment joining the points $P(x_0, y_0, z_0)$ and $Q(x_1, y_1, z_1)$.

  a. Find a function $\mathbf{r}(t)$ for the segment PQ.

  b. Use the arc length formula to find the length of PQ.

  c. Use geometry (distance formula) to verify the result of part (b).

45. Tilted circles Let the curve C be described by
$$\mathbf{r}(t) = \langle a\cos t, b\sin t, c\sin t \rangle,$$
where a, b, and c are real positive numbers.

  a. Assume C lies in a plane. Show that C is a circle centered at the origin, provided $a^2 = b^2 + c^2$.

  b. Find the arc length of the circle in part (a).

  c. Assuming C lies in a plane, find the conditions for which $\mathbf{r}(t) = \langle a\cos t + b\sin t, c\cos t + d\sin t, e\cos t + f\sin t \rangle$ describes a circle. Then find its arc length.

46. A family of arc length integrals Find the length of the curve $\mathbf{r}(t) = \langle t^m, t^m, t^{3m/2} \rangle$, for $0 \le a \le t \le b$, where m is a real number. Express the result in terms of m, a, and b.

47. A special case Suppose a curve is described by
$$\mathbf{r}(t) = \langle Ah(t), Bh(t) \rangle, \text{ for } a \le t \le b,$$
where A and B are constants and h has a continuous derivative.

  a. Show that the length of the curve is
$$\sqrt{A^2 + B^2} \int_a^b |h'(t)| \, dt.$$

  b. Use part (a) to find the length of the curve $x = 2t^3$, $y = 5t^3$, for $0 \le t \le 4$.

  c. Use part (a) to find the length of the curve $x = 4/t$, $y = 10/t$, for $1 \le t \le 8$.

Explorations and Challenges

T 48. Toroidal magnetic field A circle of radius a that is centered at $(A, 0)$ is revolved about the y-axis to create a torus (assume $a < A$). When current flows through a copper wire that is wrapped around this torus, a magnetic field is created and the strength of this field depends on the amount of copper wire used. If the wire is wrapped evenly around the torus a total of k times, the shape of the wire is modeled by the function

$$\mathbf{r}(t) = \langle (A + a \cos kt) \cos t, (A + a \cos kt) \sin t, a \sin kt \rangle,$$

for $0 \le t \le 2\pi$. Determine the amount of copper required if $A = 4$ in, $a = 1$ in, and $k = 35$.

T 49. Projectile trajectories A projectile (such as a baseball or a cannonball) launched from the origin with an initial horizontal velocity u_0 and an initial vertical velocity v_0 moves in a parabolic trajectory given by

$$\mathbf{r}(t) = \left\langle u_0 t, -\frac{1}{2} g t^2 + v_0 t \right\rangle, \quad \text{for } t \ge 0,$$

where air resistance is neglected and $g = 9.8 \text{ m/s}^2$ is the acceleration due to gravity (see Section 14.3).

a. Let $u_0 = 20$ m/s and $v_0 = 25$ m/s. Assuming the projectile is launched over horizontal ground, at what time does it return to Earth?

b. Find the integral that gives the length of the trajectory from launch to landing.

c. Evaluate the integral in part (b) by first making the change of variables $u = -gt + v_0$. The resulting integral is evaluated either by making a second change of variables or by using a calculator. What is the length of the trajectory?

d. How far does the projectile land from its launch site?

50. Variable speed on a circle Consider a particle that moves in a plane according to the function $\mathbf{r}(t) = \langle \sin t^2, \cos t^2 \rangle$ with an initial position $(0, 1)$ at $t = 0$.

a. Describe the path of the particle, including the time required to return to the initial position.

b. What is the length of the path in part (a)?

c. Describe how the motion of this particle differs from the motion described by the equations $x = \sin t$ and $y = \cos t$.

d. Consider the motion described by $x = \sin t^n$ and $y = \cos t^n$, where n is a positive integer. Describe the path of the particle, including the time required to return to the initial position.

e. What is the length of the path in part (d) for any positive integer n?

f. If you were watching a race on a circular path between two runners, one moving according to $x = \sin t$ and $y = \cos t$ and one according to $x = \sin t^2$ and $y = \cos t^2$, who would win and when would one runner pass the other?

51. Arc length parameterization Prove that the line $\mathbf{r}(t) = \langle x_0 + at, y_0 + bt, z_0 + ct \rangle$ is parameterized by arc length, provided $a^2 + b^2 + c^2 = 1$.

52. Arc length parameterization Prove that the curve $\mathbf{r}(t) = \langle a \cos t, b \sin t, c \sin t \rangle$ is parameterized by arc length, provided $a^2 = b^2 + c^2 = 1$.

53. Lengths of related curves Suppose a curve is given by $\mathbf{r}(t) = \langle f(t), g(t) \rangle$, where f' and g' are continuous, for $a \le t \le b$. Assume the curve is traversed once, for $a \le t \le b$, and the length of the curve between $(f(a), g(a))$ and $(f(b), g(b))$ is L. Prove that for any nonzero constant c, the length of the curve defined by $\mathbf{r}(t) = \langle cf(t), cg(t) \rangle$, for $a \le t \le b$, is $|c| L$.

54. Change of variables Consider the parameterized curves $\mathbf{r}(t) = \langle f(t), g(t), h(t) \rangle$ and $\mathbf{R}(t) = \langle f(u(t)), g(u(t)), h(u(t)) \rangle$, where f, g, h, and u are continuously differentiable functions and u has an inverse on $[a, b]$.

a. Show that the curve generated by $\mathbf{r}$ on the interval $a \le t \le b$ is the same as the curve generated by $\mathbf{R}$ on $u^{-1}(a) \le t \le u^{-1}(b)$ (or $u^{-1}(b) \le t \le u^{-1}(a)$).

b. Show that the lengths of the two curves are equal. (*Hint:* Use the Chain Rule and a change of variables in the arc length integral for the curve generated by $\mathbf{R}$.)

QUICK CHECK ANSWERS

1. 9 **2.** For $a \le t \le b$, the curve C generated is $(b - a)/2\pi$ of a full circle. Because the full circle has a length of 2π, the curve C has a length of $b - a$. **3.** No. If t increases by 1 unit, the length of the curve increases by $\sqrt{3}$ units. ◄

14.5 Curvature and Normal Vectors

We know how to find tangent vectors and lengths of curves in space, but much more can be said about the shape of such curves. In this section, we introduce several new concepts. *Curvature* measures how *fast* a curve turns at a point, the *normal vector* gives the *direction* in which a curve turns, and the *binormal vector* and the *torsion* describe the twisting of a curve.

Curvature

Imagine driving a car along a winding mountain road. There are two ways to change the velocity of the car (that is, to accelerate). You can change the *speed* of the car or you can change the *direction* of the car. A change of speed is relatively easy to describe, so we postpone that discussion and focus on the change of direction. The rate at which the car changes direction is related to the notion of *curvature*.

Unit Tangent Vector   Recall from Section 14.2 that if $\mathbf{r}(t) = \langle x(t), y(t), z(t) \rangle$ is a smooth oriented curve, then the unit tangent vector at a point is the unit vector that points in the direction of the tangent vector $\mathbf{r}'(t)$; that is,

$$\mathbf{T}(t) = \frac{\mathbf{r}'(t)}{|\mathbf{r}'(t)|} = \frac{\mathbf{v}(t)}{|\mathbf{v}(t)|}.$$

Because $\mathbf{T}$ is a unit vector, its length does not change along the curve. The only way $\mathbf{T}$ can change is through a change in direction.

How quickly does $\mathbf{T}$ change (in direction) as we move along the curve? If a small increment in arc length Δs along the curve results in a large change in the direction of $\mathbf{T}$, the curve is turning quickly over that interval and we say it has a large *curvature* (**Figure 14.29a**). If a small increment Δs in arc length results in a small change in the direction of $\mathbf{T}$, the curve is turning slowly over that interval and it has a small curvature (**Figure 14.29b**). The magnitude of the rate at which the direction of $\mathbf{T}$ changes with respect to arc length is the curvature of the curve.

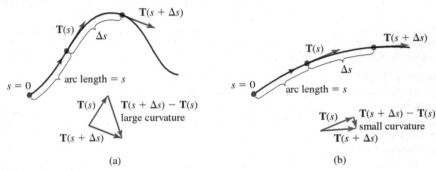

(a)   (b)

Figure 14.29

> Recall that the unit tangent vector at a point depends on the orientation of the curve. The curvature does not depend on the orientation of the curve, but it does depend on the shape of the curve. The Greek letter κ (kappa) is used to denote curvature.

DEFINITION Curvature

Let $\mathbf{r}$ describe a smooth parameterized curve. If s denotes arc length and $\mathbf{T} = \mathbf{r}'/|\mathbf{r}'|$ is the unit tangent vector, the **curvature** is $\kappa(s) = \left| \dfrac{d\mathbf{T}}{ds} \right|$.

Note that κ is a nonnegative scalar-valued function. A large value of κ at a point indicates a tight curve that changes direction quickly. If κ is small, then the curve is relatively flat and its direction changes slowly. The minimum curvature (zero) occurs on a straight line, where the tangent vector never changes direction along the curve.

In order to evaluate $d\mathbf{T}/ds$, a description of the curve in terms of the arc length appears to be needed, but it may be difficult to obtain. A short calculation leads to the first of two practical curvature formulas.

Letting t be an arbitrary parameter, we begin with the Chain Rule and write $\dfrac{d\mathbf{T}}{dt} = \dfrac{d\mathbf{T}}{ds} \cdot \dfrac{ds}{dt}$. After dividing both sides of this equation by $ds/dt = |\mathbf{v}|$, we take absolute values and arrive at

$$\kappa = \left| \frac{d\mathbf{T}}{ds} \right| = \frac{|d\mathbf{T}/dt|}{|ds/dt|} = \frac{1}{|\mathbf{v}|} \left| \frac{d\mathbf{T}}{dt} \right|.$$

This derivation is a proof of the following theorem.

THEOREM 14.4   Curvature Formula

Let $\mathbf{r}(t)$ describe a smooth parameterized curve, where t is any parameter. If $\mathbf{v} = \mathbf{r}'$ is the velocity and $\mathbf{T}$ is the unit tangent vector, then the curvature is

$$\kappa(t) = \frac{1}{|\mathbf{v}|} \left| \frac{d\mathbf{T}}{dt} \right| = \frac{|\mathbf{T}'(t)|}{|\mathbf{r}'(t)|}.$$

EXAMPLE 1   Lines have zero curvature Consider the line

$$\mathbf{r}(t) = \langle x_0 + at, y_0 + bt, z_0 + ct \rangle, \text{ for } -\infty < t < \infty.$$

Show that $\kappa = 0$ at all points on the line.

SOLUTION Note that $\mathbf{r}'(t) = \langle a, b, c \rangle$ and $|\mathbf{r}'(t)| = |\mathbf{v}(t)| = \sqrt{a^2 + b^2 + c^2}$. Therefore,

$$\mathbf{T}(t) = \frac{\mathbf{r}'(t)}{|\mathbf{r}'(t)|} = \frac{\langle a, b, c \rangle}{\sqrt{a^2 + b^2 + c^2}}.$$

Because $\mathbf{T}$ is a constant, $\dfrac{d\mathbf{T}}{dt} = \mathbf{0}$; therefore, $\kappa = 0$ at all points of the line.

Related Exercise 11 ◄

EXAMPLE 2   Circles have constant curvature Consider the circle $\mathbf{r}(t) = \langle R \cos t, R \sin t \rangle$, for $0 \le t \le 2\pi$, where $R > 0$. Show that $\kappa = 1/R$.

SOLUTION We compute $\mathbf{r}'(t) = \langle -R \sin t, R \cos t \rangle$ and

$$\begin{aligned}
|\mathbf{v}(t)| = |\mathbf{r}'(t)| &= \sqrt{(-R \sin t)^2 + (R \cos t)^2} \\
&= \sqrt{R^2 (\sin^2 t + \cos^2 t)} \qquad \text{Simplify.} \\
&= R. \qquad\qquad\qquad\qquad \sin^2 t + \cos^2 t = 1, R > 0
\end{aligned}$$

Therefore,

$$\mathbf{T}(t) = \frac{\mathbf{r}'(t)}{|\mathbf{r}'(t)|} = \frac{\langle -R \sin t, R \cos t \rangle}{R} = \langle -\sin t, \cos t \rangle, \text{ and}$$

$$\frac{d\mathbf{T}}{dt} = \langle -\cos t, -\sin t \rangle.$$

> The curvature of a curve at a point can also be visualized in terms of a **circle of curvature**, which is a circle of radius R that is tangent to the curve at that point. The curvature at the point is $\kappa = 1/R$. See Exercises 70–73.

Combining these observations, the curvature is

$$\kappa = \frac{1}{|\mathbf{v}|}\left|\frac{d\mathbf{T}}{dt}\right| = \frac{1}{R}|\langle -\cos t, -\sin t \rangle| = \frac{1}{R}\underbrace{\sqrt{\cos^2 t + \sin^2 t}}_{1} = \frac{1}{R}.$$

The curvature of a circle is constant; a circle with a small radius has a large curvature, and vice versa.

Related Exercise 12 ◄

QUICK CHECK 1 What is the curvature of the circle $\mathbf{r}(t) = \langle 3 \sin t, 3 \cos t \rangle$? ◄

An Alternative Curvature Formula A second curvature formula, which pertains specifically to trajectories of moving objects, is easier to use in some cases. The calculation is instructive because it relies on many properties of vector functions. In the end, a remarkably simple formula emerges.

Again consider a smooth curve $\mathbf{r}(t) = \langle x(t), y(t), z(t) \rangle$, where $\mathbf{v}(t) = \mathbf{r}'(t)$ and $\mathbf{a}(t) = \mathbf{v}'(t)$ are the velocity and acceleration of an object moving along that curve, respectively. We assume $\mathbf{v}(t) \ne \mathbf{0}$ and $\mathbf{a}(t) \ne \mathbf{0}$. Because $\mathbf{T} = \mathbf{v}/|\mathbf{v}|$, we begin by writing $\mathbf{v} = |\mathbf{v}| \mathbf{T}$ and differentiating both sides with respect to t:

$$\mathbf{a} = \frac{d\mathbf{v}}{dt} = \frac{d}{dt}(|\mathbf{v}(t)| \mathbf{T}(t)) = \frac{d}{dt}(|\mathbf{v}(t)|)\mathbf{T}(t) + |\mathbf{v}(t)|\frac{d\mathbf{T}}{dt}. \quad \text{Product Rule}$$

We now form $\mathbf{v} \times \mathbf{a}$:

> Distributive law for cross products:
> $\mathbf{w} \times (\mathbf{u} + \mathbf{v}) = (\mathbf{w} \times \mathbf{u}) + (\mathbf{w} \times \mathbf{v})$
> $(\mathbf{u} + \mathbf{v}) \times \mathbf{w} = (\mathbf{u} \times \mathbf{w}) + (\mathbf{v} \times \mathbf{w})$

$$\begin{aligned}
\mathbf{v} \times \mathbf{a} &= \underbrace{|\mathbf{v}|\mathbf{T}}_{\mathbf{v}} \times \underbrace{\left(\frac{d}{dt}(|\mathbf{v}|)\mathbf{T} + |\mathbf{v}|\frac{d\mathbf{T}}{dt}\right)}_{\mathbf{a}} \\
&= \underbrace{|\mathbf{v}|\mathbf{T} \times \left(\frac{d}{dt}(|\mathbf{v}|)\right)\mathbf{T}}_{0} + |\mathbf{v}|\mathbf{T} \times |\mathbf{v}|\frac{d\mathbf{T}}{dt} \quad \text{Distributive law for cross products}
\end{aligned}$$

The first term in this expression has the form $a\mathbf{T} \times b\mathbf{T}$, where a and b are scalars. Therefore, $a\mathbf{T}$ and $b\mathbf{T}$ are parallel vectors and $a\mathbf{T} \times b\mathbf{T} = \mathbf{0}$. To simplify the second term, recall that a vector $\mathbf{u}(t)$ of constant length has the property that $\mathbf{u}$ and $d\mathbf{u}/dt$ are orthogonal (Section 14.3). Because **T** is a unit vector, it has constant length, and **T** and $d\mathbf{T}/dt$ are orthogonal. Furthermore, scalar multiples of **T** and $d\mathbf{T}/dt$ are also orthogonal. Therefore, the magnitude of the second term simplifies as follows:

$$\left| |\mathbf{v}|\mathbf{T} \times |\mathbf{v}|\frac{d\mathbf{T}}{dt} \right| = |\mathbf{v}||\mathbf{T}| \left| |\mathbf{v}|\frac{d\mathbf{T}}{dt} \right| \underbrace{\sin\theta}_{1} \quad |\mathbf{u} \times \mathbf{v}| = |\mathbf{u}||\mathbf{v}|\sin\theta$$

$$= |\mathbf{v}|^2 \left| \frac{d\mathbf{T}}{dt} \right| \underbrace{|\mathbf{T}|}_{1} \qquad \text{Simplify, } \theta = \pi/2.$$

$$= |\mathbf{v}|^2 \left| \frac{d\mathbf{T}}{dt} \right|. \qquad |\mathbf{T}| = 1$$

> Recall that the magnitude of the cross product of nonzero vectors is $|\mathbf{u} \times \mathbf{v}| = |\mathbf{u}||\mathbf{v}|\sin\theta$, where θ is the angle between the vectors. If the vectors are orthogonal, $\sin\theta = 1$ and $|\mathbf{u} \times \mathbf{v}| = |\mathbf{u}||\mathbf{v}|$.

The final step is to use Theorem 14.4 and substitute $\left|\dfrac{d\mathbf{T}}{dt}\right| = \kappa|\mathbf{v}|$. Putting these results together, we find that

$$|\mathbf{v} \times \mathbf{a}| = |\mathbf{v}|^2 \left| \frac{d\mathbf{T}}{dt} \right| = |\mathbf{v}|^2 \kappa |\mathbf{v}| = \kappa|\mathbf{v}|^3.$$

> Note that $\mathbf{a}(t) = \mathbf{0}$ corresponds to straight-line motion and $\kappa = 0$. If $\mathbf{v}(t) = \mathbf{0}$, the object is at rest and κ is undefined.

Solving for the curvature gives $\kappa = \dfrac{|\mathbf{v} \times \mathbf{a}|}{|\mathbf{v}|^3}$.

THEOREM 14.5　Alternative Curvature Formula
Let **r** be the position of an object moving on a smooth curve. The **curvature** at a point on the curve is

$$\kappa = \frac{|\mathbf{v} \times \mathbf{a}|}{|\mathbf{v}|^3},$$

where $\mathbf{v} = \mathbf{r}'$ is the velocity and $\mathbf{a} = \mathbf{v}'$ is the acceleration.

QUICK CHECK 2 Use the alternative curvature formula to compute the curvature of the curve $\mathbf{r}(t) = \langle t^2, 10, -10 \rangle$. ◄

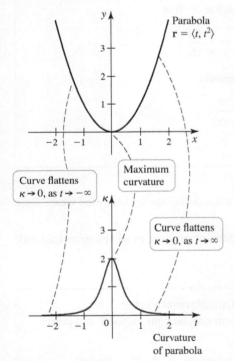

Figure 14.30

EXAMPLE 3　Curvature of a parabola Find the curvature of the parabola $\mathbf{r}(t) = \langle t, at^2 \rangle$, for $-\infty < t < \infty$, where $a > 0$ is a real number.

SOLUTION The alternative formula works well in this case. We find that $\mathbf{v}(t) = \mathbf{r}'(t) = \langle 1, 2at \rangle$ and $\mathbf{a}(t) = \mathbf{v}'(t) = \langle 0, 2a \rangle$. To compute the cross product $\mathbf{v} \times \mathbf{a}$, we append a third component of 0 to each vector:

$$\mathbf{v} \times \mathbf{a} = \begin{vmatrix} \mathbf{i} & \mathbf{j} & \mathbf{k} \\ 1 & 2at & 0 \\ 0 & 2a & 0 \end{vmatrix} = 2a\,\mathbf{k}.$$

Therefore, the curvature is

$$\kappa(t) = \frac{|\mathbf{v} \times \mathbf{a}|}{|\mathbf{v}|^3} = \frac{|2a\,\mathbf{k}|}{|\langle 1, 2at \rangle|^3} = \frac{2a}{(1 + 4a^2 t^2)^{3/2}}.$$

The curvature is a maximum at the vertex of the parabola where $t = 0$ and $\kappa = 2a$. The curvature decreases as one moves along the curve away from the vertex, as shown in Figure 14.30 with $a = 1$.

Related Exercise 23 ◄

EXAMPLE 4　Curvature of a helix Find the curvature of the helix $\mathbf{r}(t) = \langle a \cos t, a \sin t, bt \rangle$, for $-\infty < t < \infty$, where $a > 0$ and $b > 0$ are real numbers.

SOLUTION We use the alternative curvature formula, with

$$\mathbf{v}(t) = \mathbf{r}'(t) = \langle -a \sin t, a \cos t, b \rangle \quad \text{and}$$
$$\mathbf{a}(t) = \mathbf{v}'(t) = \langle -a \cos t, -a \sin t, 0 \rangle.$$

The cross product $\mathbf{v} \times \mathbf{a}$ is

$$\mathbf{v} \times \mathbf{a} = \begin{vmatrix} \mathbf{i} & \mathbf{j} & \mathbf{k} \\ -a \sin t & a \cos t & b \\ -a \cos t & -a \sin t & 0 \end{vmatrix} = ab \sin t \, \mathbf{i} - ab \cos t \, \mathbf{j} + a^2 \, \mathbf{k}.$$

Therefore,

$$\begin{aligned} |\mathbf{v} \times \mathbf{a}| &= |ab \sin t \, \mathbf{i} - ab \cos t \, \mathbf{j} + a^2 \, \mathbf{k}| \\ &= \sqrt{a^2 b^2 \underbrace{(\sin^2 t + \cos^2 t)}_{1} + a^4} \\ &= a\sqrt{a^2 + b^2}. \end{aligned}$$

> In the curvature formula for the helix, if $b = 0$, the helix becomes a circle of radius a with $\kappa = \dfrac{1}{a}$. At the other extreme, holding a fixed and letting $b \to \infty$ stretches and straightens the helix so that $\kappa \to 0$.

By a familiar calculation, $|\mathbf{v}| = |\langle -a \sin t, a \cos t, b \rangle| = \sqrt{a^2 + b^2}$. Therefore,

$$\kappa = \frac{|\mathbf{v} \times \mathbf{a}|}{|\mathbf{v}|^3} = \frac{a\sqrt{a^2 + b^2}}{(\sqrt{a^2 + b^2})^3} = \frac{a}{a^2 + b^2}.$$

A similar calculation shows that all helices of this form have constant curvature.

Related Exercise 22 ◄

Principal Unit Normal Vector

The curvature answers the question of how *fast* a curve turns. The principal unit normal vector determines the *direction* in which a curve turns. Specifically, the magnitude of $d\mathbf{T}/ds$ is the curvature: $\kappa = |d\mathbf{T}/ds|$. What about the direction of $d\mathbf{T}/ds$? If only the direction, but not the magnitude, of a vector is of interest, it is convenient to work with a unit vector that has the same direction as the original vector. We apply this idea to $d\mathbf{T}/ds$. The unit vector that points in the direction of $d\mathbf{T}/ds$ is the *principal unit normal vector*.

> The principal unit normal vector depends on the shape of the curve but not on the orientation of the curve.

DEFINITION Principal Unit Normal Vector

Let $\mathbf{r}$ describe a smooth curve parameterized by arc length. The **principal unit normal vector** at a point P on the curve at which $\kappa \neq 0$ is

$$\mathbf{N}(s) = \frac{d\mathbf{T}/ds}{|d\mathbf{T}/ds|} = \frac{1}{\kappa} \frac{d\mathbf{T}}{ds}.$$

For other parameters, we use the equivalent formula

$$\mathbf{N}(t) = \frac{d\mathbf{T}/dt}{|d\mathbf{T}/dt|},$$

evaluated at the value of t corresponding to P.

The practical formula $\mathbf{N} = \dfrac{d\mathbf{T}/dt}{|d\mathbf{T}/dt|}$ follows from the definition by using the Chain Rule to write $\dfrac{d\mathbf{T}}{ds} = \dfrac{d\mathbf{T}}{dt} \cdot \dfrac{dt}{ds}$ (Exercise 80). Two important properties of the principal unit normal vector follow from the definition.

THEOREM 14.6 Properties of the Principal Unit Normal Vector

Let $\mathbf{r}$ describe a smooth parameterized curve with unit tangent vector $\mathbf{T}$ and principal unit normal vector $\mathbf{N}$.

1. $\mathbf{T}$ and $\mathbf{N}$ are orthogonal at all points of the curve; that is, $\mathbf{T} \cdot \mathbf{N} = 0$ at all points where $\mathbf{N}$ is defined.

2. The principal unit normal vector points to the inside of the curve—in the direction that the curve is turning.

At all points,
$|\mathbf{T}| = |\mathbf{N}| = 1$
and $\mathbf{T} \cdot \mathbf{N} = 0$.

N points to the inside of the curve—in the direction the curve is turning.

Figure 14.31

For small Δs,
$\mathbf{T}(s + \Delta s) - \mathbf{T}(s)$
points to the inside of the curve, as does $d\mathbf{T}/ds$.

$\mathbf{T}(s)$ $\mathbf{T}(s + \Delta s)$
Δs

$\mathbf{T}(s)$
$\mathbf{T}(s + \Delta s) - \mathbf{T}(s)$
$\mathbf{T}(s + \Delta s)$

Figure 14.32

QUICK CHECK 3 Consider the parabola $\mathbf{r}(t) = \langle t, -t^2 \rangle$. Does the principal unit normal vector point in the positive y-direction or the negative y-direction along the curve? ◀

Helix
$\mathbf{r}(t) = \langle a \cos t, a \sin t, bt \rangle$

$\mathbf{T} \cdot \mathbf{N} = 0$ at all points of the curve.
$\mathbf{T}$ points in the direction of the curve.
$\mathbf{N}$ points to the inside of the curve.

Figure 14.33

QUICK CHECK 4 Why is the principal unit normal vector for a straight line undefined? ◀

Proof:

1. As a unit vector, $\mathbf{T}$ has constant length. Therefore, by Theorem 14.2, $\mathbf{T}$ and $d\mathbf{T}/dt$ (or $\mathbf{T}$ and $d\mathbf{T}/ds$) are orthogonal. Because $\mathbf{N}$ is a scalar multiple of $d\mathbf{T}/ds$, $\mathbf{T}$ and $\mathbf{N}$ are orthogonal (**Figure 14.31**).

2. We motivate—but do not prove—this fact by recalling that

$$\frac{d\mathbf{T}}{ds} = \lim_{\Delta s \to 0} \frac{\mathbf{T}(s + \Delta s) - \mathbf{T}(s)}{\Delta s}.$$

Therefore, $d\mathbf{T}/ds$ points in the approximate direction of $\mathbf{T}(s + \Delta s) - \mathbf{T}(s)$ when Δs is small. As shown in **Figure 14.32**, this difference points in the direction in which the curve is turning. Because $\mathbf{N}$ is a positive scalar multiple of $d\mathbf{T}/ds$, it points in the same direction. ◀

EXAMPLE 5 Principal unit normal vector for a helix Find the principal unit normal vector for the helix $\mathbf{r}(t) = \langle a \cos t, a \sin t, bt \rangle$, for $-\infty < t < \infty$, where $a > 0$ and $b > 0$ are real numbers.

SOLUTION Several preliminary calculations are needed. First, we have $\mathbf{v}(t) = \mathbf{r}'(t) = \langle -a \sin t, a \cos t, b \rangle$. Therefore,

$$
\begin{aligned}
|\mathbf{v}(t)| = |\mathbf{r}'(t)| &= \sqrt{(-a \sin t)^2 + (a \cos t)^2 + b^2} \\
&= \sqrt{a^2 (\sin^2 t + \cos^2 t) + b^2} \qquad \text{Simplify.} \\
&= \sqrt{a^2 + b^2}. \qquad\qquad \sin^2 t + \cos^2 t = 1
\end{aligned}
$$

The unit tangent vector is

$$\mathbf{T}(t) = \frac{\mathbf{r}'(t)}{|\mathbf{r}'(t)|} = \frac{\langle -a \sin t, a \cos t, b \rangle}{\sqrt{a^2 + b^2}}.$$

Notice that $\mathbf{T}$ points along the curve in an upward direction (at an angle to the horizontal that satisfies the equation $\tan \theta = b/a$; **Figure 14.33**). We can now calculate the principal unit normal vector. First, we determine that

$$\frac{d\mathbf{T}}{dt} = \frac{d}{dt} \left(\frac{\langle -a \sin t, a \cos t, b \rangle}{\sqrt{a^2 + b^2}} \right) = \frac{\langle -a \cos t, -a \sin t, 0 \rangle}{\sqrt{a^2 + b^2}}$$

and

$$\left| \frac{d\mathbf{T}}{dt} \right| = \frac{a}{\sqrt{a^2 + b^2}}.$$

The principal unit normal vector now follows:

$$\mathbf{N} = \frac{d\mathbf{T}/dt}{|d\mathbf{T}/dt|} = \frac{\dfrac{\langle -a \cos t, -a \sin t, 0 \rangle}{\sqrt{a^2 + b^2}}}{\dfrac{a}{\sqrt{a^2 + b^2}}} = \langle -\cos t, -\sin t, 0 \rangle.$$

Several important checks should be made. First note that $\mathbf{N}$ is a unit vector; that is, $|\mathbf{N}| = 1$. It should also be confirmed that $\mathbf{T} \cdot \mathbf{N} = 0$; that is, the unit tangent vector and the principal unit normal vector are everywhere orthogonal. Finally, $\mathbf{N}$ is parallel to the xy-plane and points inward toward the z-axis, in the direction the curve turns (Figure 14.33). Notice that in the special case $b = 0$, the trajectory is a circle, but the normal vector is still $\mathbf{N} = \langle -\cos t, -\sin t, 0 \rangle$.

Related Exercise 28 ◀

Components of the Acceleration

The vectors $\mathbf{T}$ and $\mathbf{N}$ may be used to gain insight into how moving objects accelerate. Recall the observation made earlier that the two ways to change the velocity of an object (accelerate) are to change its *speed* and to change its *direction* of motion. We show that

> Recall that the speed is $|\mathbf{v}| = ds/dt$, where s is arc length.

changing the speed produces acceleration in the direction of **T** and changing the direction produces acceleration in the direction of **N**.

We begin with the fact that $\mathbf{T} = \dfrac{\mathbf{v}}{|\mathbf{v}|}$ or $\mathbf{v} = \mathbf{T}|\mathbf{v}| = \mathbf{T}\dfrac{ds}{dt}$. Differentiating both sides of $\mathbf{v} = \mathbf{T}\dfrac{ds}{dt}$ with respect to t gives

$$\mathbf{a} = \frac{d\mathbf{v}}{dt} = \frac{d}{dt}\left(\mathbf{T}\frac{ds}{dt}\right)$$

$$= \frac{d\mathbf{T}}{dt}\frac{ds}{dt} + \mathbf{T}\frac{d^2s}{dt^2} \qquad \text{Product Rule}$$

$$= \underbrace{\frac{d\mathbf{T}}{ds}\frac{ds}{dt}}_{\kappa\mathbf{N}}\underbrace{\frac{ds}{dt}}_{|\mathbf{v}|} + \mathbf{T}\frac{d^2s}{dt^2} \qquad \text{Chain Rule: } \frac{d\mathbf{T}}{dt} = \frac{d\mathbf{T}}{ds}\frac{ds}{dt}$$

$$= \kappa\mathbf{N}|\mathbf{v}|^2 + \mathbf{T}\frac{d^2s}{dt^2}. \qquad \text{Substitute.}$$

We now identify the normal and tangential components of the acceleration.

> Note that a_N and a_T are defined even at points where $\kappa = 0$ and **N** is undefined.

THEOREM 14.7 Tangential and Normal Components of the Acceleration
The acceleration vector of an object moving in space along a smooth curve has the following representation in terms of its **tangential component** a_T (in the direction of **T**) and its **normal component** a_N (in the direction of **N**):

$$\mathbf{a} = a_N\mathbf{N} + a_T\mathbf{T},$$

where $a_N = \kappa|\mathbf{v}|^2 = \dfrac{|\mathbf{v}\times\mathbf{a}|}{|\mathbf{v}|}$ and $a_T = \dfrac{d^2s}{dt^2}$.

Tangential component $a_T\mathbf{T}$ Trajectory in $\mathbb{R}^3$ $\mathbf{a} = a_N\mathbf{N} + a_T\mathbf{T}$

Normal component $a_N\mathbf{N}$

Figure 14.34

The tangential component of the acceleration, in the direction of **T**, is the usual acceleration $a_T = d^2s/dt^2$ of an object moving along a straight line (**Figure 14.34**). The normal component, in the direction of **N**, increases with the speed $|\mathbf{v}|$ and with the curvature. Higher speeds on tighter curves produce greater normal accelerations.

EXAMPLE 6 Acceleration on a circular path Find the components of the acceleration on the circular trajectory

$$\mathbf{r}(t) = \langle R\cos\omega t, R\sin\omega t\rangle,$$

where R and ω are positive real numbers.

SOLUTION We find that $\mathbf{r}'(t) = \langle -R\omega\sin\omega t, R\omega\cos\omega t\rangle$, $|\mathbf{v}(t)| = |\mathbf{r}'(t)| = R\omega$, and, by Example 2, $\kappa = 1/R$. Recall that $ds/dt = |\mathbf{v}(t)|$, which is constant; therefore, $d^2s/dt^2 = 0$ and the tangential component of the acceleration is zero. The acceleration is

$$\mathbf{a} = \kappa|\mathbf{v}|^2\mathbf{N} + \underbrace{\frac{d^2s}{dt^2}}_{0}\mathbf{T} = \frac{1}{R}(R\omega)^2\mathbf{N} = R\omega^2\mathbf{N}.$$

On a circular path (traversed at constant speed), the acceleration is entirely in the normal direction, orthogonal to the tangent vectors. The acceleration increases with the radius of the circle R and with the frequency of the motion ω.

Related Exercise 36 ◀

EXAMPLE 7 A bend in the road The driver of a car follows the parabolic trajectory $\mathbf{r}(t) = \langle t, t^2\rangle$, for $-2 \le t \le 2$, through a sharp bend (**Figure 14.35**). Find the tangential and normal components of the acceleration of the car.

Parabolic trajectory $\mathbf{r}(t) = \langle t, t^2\rangle$

$t = -2$ $t = 2$ $t < 0$ $t > 0$ Approaching bend   Leaving bend $t = -1$ $t = 1$ $t = 0$

Figure 14.35

SOLUTION The velocity and acceleration vectors are easily computed:
$\mathbf{v}(t) = \mathbf{r}'(t) = \langle 1, 2t \rangle$ and $\mathbf{a}(t) = \mathbf{r}''(t) = \langle 0, 2 \rangle$. The goal is to express $\mathbf{a} = \langle 0, 2 \rangle$ in terms of $\mathbf{T}$ and $\mathbf{N}$. A short calculation reveals that

$$\mathbf{T} = \frac{\mathbf{v}}{|\mathbf{v}|} = \frac{\langle 1, 2t \rangle}{\sqrt{1 + 4t^2}} \quad \text{and} \quad \mathbf{N} = \frac{d\mathbf{T}/dt}{|d\mathbf{T}/dt|} = \frac{\langle -2t, 1 \rangle}{\sqrt{1 + 4t^2}}.$$

We now have two ways to proceed. One is to compute the normal and tangential components of the acceleration directly using the definitions. More efficient is to note that $\mathbf{T}$ and $\mathbf{N}$ are orthogonal unit vectors, and then to compute the scalar projections of $\mathbf{a} = \langle 0, 2 \rangle$ in the directions of $\mathbf{T}$ and $\mathbf{N}$. We find that

$$a_N = \mathbf{a} \cdot \mathbf{N} = \langle 0, 2 \rangle \cdot \frac{\langle -2t, 1 \rangle}{\sqrt{1 + 4t^2}} = \frac{2}{\sqrt{1 + 4t^2}}$$

and

$$a_T = \mathbf{a} \cdot \mathbf{T} = \langle 0, 2 \rangle \cdot \frac{\langle 1, 2t \rangle}{\sqrt{1 + 4t^2}} = \frac{4t}{\sqrt{1 + 4t^2}}.$$

> Using the fact that $|\mathbf{T}| = |\mathbf{N}| = 1$, we have, from Section 13.3,
>
> $$a_N = \mathrm{scal}_N \mathbf{a} = \frac{\mathbf{a} \cdot \mathbf{N}}{|\mathbf{N}|} = \mathbf{a} \cdot \mathbf{N}$$
>
> and
>
> $$a_T = \mathrm{scal}_T \mathbf{a} = \frac{\mathbf{a} \cdot \mathbf{T}}{|\mathbf{T}|} = \mathbf{a} \cdot \mathbf{T} = \frac{\mathbf{v} \cdot \mathbf{a}}{|\mathbf{v}|}.$$

You should verify that at all times (Exercise 76),

$$\mathbf{a} = a_N \mathbf{N} + a_T \mathbf{T} = \frac{2}{\sqrt{1 + 4t^2}} (\mathbf{N} + 2t\,\mathbf{T}) = \langle 0, 2 \rangle.$$

Let's interpret these results. First notice that the driver negotiates the curve in a sensible way: The speed $|\mathbf{v}| = \sqrt{1 + 4t^2}$ decreases as the car approaches the origin (the tightest part of the curve) and increases as it moves away from the origin (**Figure 14.36**). As the car approaches the origin ($t < 0$), $\mathbf{T}$ points in the direction of the trajectory and $\mathbf{N}$ points to the inside of the curve. However, $a_T = \dfrac{d^2 s}{dt^2} < 0$ when $t < 0$, so $a_T \mathbf{T}$ points in the direction opposite that of $\mathbf{T}$ (corresponding to a deceleration). As the car leaves the origin ($t > 0$), $a_T > 0$ (corresponding to an acceleration) and $a_T \mathbf{T}$ and $\mathbf{T}$ point in the direction of the trajectory. At all times, $\mathbf{N}$ points to the inside of the curve (Figure 14.36; Exercise 78).

Related Exercise 38 ◀

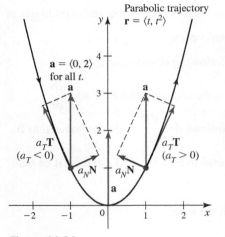

Figure 14.36

QUICK CHECK 5 Verify that $\mathbf{T}$ and $\mathbf{N}$ given in Example 7 satisfy $|\mathbf{T}| = |\mathbf{N}| = 1$ and that $\mathbf{T} \cdot \mathbf{N} = 0$. ◀

The Binormal Vector and Torsion

We have seen that the curvature function and the principal unit normal vector tell us how quickly and in what direction a curve turns. For curves in two dimensions, these quantities give a fairly complete description of motion along the curve. However, in three dimensions, a curve has more "room" in which to change its course, and another descriptive function is often useful. **Figure 14.37** shows a smooth parameterized curve C with its unit tangent vector $\mathbf{T}$ and its principal unit normal vector $\mathbf{N}$ at two different points. These two vectors determine a plane called the *osculating plane* (**Figure 14.37b**). The question we now ask is, How quickly does the curve C move out of the plane determined by $\mathbf{T}$ and $\mathbf{N}$?

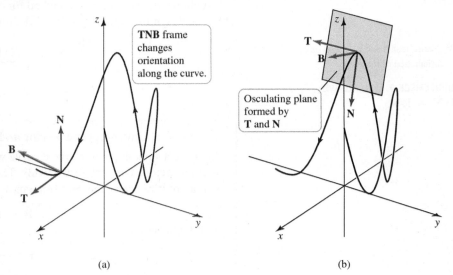

(a)   (b)

Figure 14.37

To answer this question, we begin by defining the *unit binormal vector* $\mathbf{B} = \mathbf{T} \times \mathbf{N}$. By the definition of the cross product, $\mathbf{B}$ is orthogonal to $\mathbf{T}$ and $\mathbf{N}$. Because $\mathbf{T}$ and $\mathbf{N}$ are unit vectors, $\mathbf{B}$ is also a unit vector. Notice that $\mathbf{T}$, $\mathbf{N}$, and $\mathbf{B}$ form a right-handed coordinate system (like the *xyz*-coordinate system) that changes its orientation as we move along the curve. This coordinate system is often called the **TNB frame** (Figure 14.37).

The rate at which the curve C twists out of the plane determined by $\mathbf{T}$ and $\mathbf{N}$ is the rate at which $\mathbf{B}$ changes as we move along C, which is $\dfrac{d\mathbf{B}}{ds}$. A short calculation leads to a practical formula for the twisting of the curve. Differentiating the cross product $\mathbf{T} \times \mathbf{N}$, we find that

$$\frac{d\mathbf{B}}{ds} = \frac{d}{ds}(\mathbf{T} \times \mathbf{N})$$

$$= \underbrace{\frac{d\mathbf{T}}{ds} \times \mathbf{N}}_{\text{parallel vectors}} + \mathbf{T} \times \frac{d\mathbf{N}}{ds} \quad \text{Product Rule for cross products}$$

$$= \mathbf{T} \times \frac{d\mathbf{N}}{ds}. \qquad \frac{d\mathbf{T}}{ds} \text{ and } \mathbf{N} \text{ are parallel; } \frac{d\mathbf{T}}{ds} \times \mathbf{N} = \mathbf{0}.$$

Notice that by definition, $\mathbf{N} = \dfrac{1}{\kappa}\dfrac{d\mathbf{T}}{ds}$, which implies that $\mathbf{N}$ and $\dfrac{d\mathbf{T}}{ds}$ are scalar multiples of each other. Therefore, their cross product is the zero vector.

The properties of $\dfrac{d\mathbf{B}}{ds}$ become clear with the following observations.

- $\dfrac{d\mathbf{B}}{ds}$ is orthogonal to both $\mathbf{T}$ and $\dfrac{d\mathbf{N}}{ds}$, because it is the cross product of $\mathbf{T}$ and $\dfrac{d\mathbf{N}}{ds}$.

- Applying Theorem 14.2 to the unit vector $\mathbf{B}$, it follows that $\dfrac{d\mathbf{B}}{ds}$ is also orthogonal to $\mathbf{B}$.

- By the previous two observations, $\dfrac{d\mathbf{B}}{ds}$ is orthogonal to both $\mathbf{B}$ and $\mathbf{T}$, so it must be parallel to $\mathbf{N}$.

Because $\dfrac{d\mathbf{B}}{ds}$ is parallel to (a scalar multiple of) $\mathbf{N}$, we write

$$\frac{d\mathbf{B}}{ds} = -\tau\mathbf{N},$$

where the scalar τ is the *torsion*. Notice that $\left|\dfrac{d\mathbf{B}}{ds}\right| = |{-\tau\mathbf{N}}| = |{-\tau}|$, so the magnitude of the torsion equals the magnitude of $\dfrac{d\mathbf{B}}{ds}$, which is the rate at which the curve twists out of the **TN**-plane.

A short calculation gives a method for computing the torsion. We take the dot product of both sides of the equation defining the torsion with $\mathbf{N}$:

$$\frac{d\mathbf{B}}{ds} \cdot \mathbf{N} = -\tau\underbrace{\mathbf{N} \cdot \mathbf{N}}_{1}$$

$$\frac{d\mathbf{B}}{ds} \cdot \mathbf{N} = -\tau. \quad \mathbf{N} \text{ is a unit vector.}$$

> The **TNB** frame is also called the Frenet-Serret frame, after two 19th-century French mathematicians, Jean Frenet and Joseph Serret.

QUICK CHECK 6 Explain why $\mathbf{B} = \mathbf{T} \times \mathbf{N}$ is a unit vector. ◄

> Note that $\mathbf{B}$ is a unit vector (of constant length). Therefore, by Theorem 14.2, $\mathbf{B}$ and $\mathbf{B}'(t)$ are orthogonal. Because $\mathbf{B}'(t)$ and $\mathbf{B}'(s)$ are parallel, it follows that $\mathbf{B}$ and $\mathbf{B}'(s)$ are orthogonal.

> The negative sign in the definition of the torsion is conventional. However, τ may be positive or negative (or zero), and in general, it varies along the curve.

> Notice that $\mathbf{B}$ and τ depend on the orientation of the curve.

QUICK CHECK 7 Explain why $\mathbf{N} \cdot \mathbf{N} = 1$. ◄

DEFINITION Unit Binormal Vector and Torsion

Let C be a smooth parameterized curve with unit tangent and principal unit normal vectors $\mathbf{T}$ and $\mathbf{N}$, respectively. Then at each point of the curve at which the curvature is nonzero, the **unit binormal vector** is

$$\mathbf{B} = \mathbf{T} \times \mathbf{N},$$

and the **torsion** is

$$\tau = -\frac{d\mathbf{B}}{ds} \cdot \mathbf{N}.$$

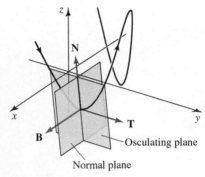

Figure 14.38

▶ The third plane formed by the vectors **T** and **B** is called the *rectifying plane*.

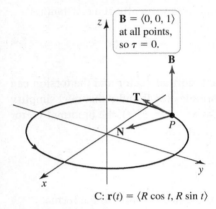

C: $\mathbf{r}(t) = \langle R \cos t, R \sin t \rangle$

Figure 14.39

Figure 14.38 provides some interpretation of the curvature and the torsion. First, we see a smooth curve C passing through a point where the mutually orthogonal vectors **T**, **N**, and **B** are defined. The **osculating plane** is defined by the vectors **T** and **N**. The plane orthogonal to the osculating plane containing **N** is called the **normal plane**. Because **N** and $\dfrac{d\mathbf{B}}{ds}$ are parallel, $\dfrac{d\mathbf{B}}{ds}$ also lies in the normal plane. The torsion, which is equal in magnitude to $\left| \dfrac{d\mathbf{B}}{ds} \right|$, gives the rate at which the curve moves *out of* the osculating plane. In a complementary way, the curvature, which is equal to $\left| \dfrac{d\mathbf{T}}{ds} \right|$, gives the rate at which the curve turns *within* the osculating plane. Two examples will clarify these concepts.

EXAMPLE 8   Unit binormal vector Consider the circle C defined by

$$\mathbf{r}(t) = \langle R \cos t, R \sin t \rangle, \text{ for } 0 \le t \le 2\pi, \text{ with } R > 0.$$

a. Without doing any calculations, find the unit binormal vector **B** and determine the torsion.

b. Use the definition of **B** to calculate **B** and confirm your answer in part (a).

SOLUTION

a. The circle C lies in the xy-plane, so at all points on the circle, **T** and **N** are in the xy-plane. Therefore, at all points of the circle, $\mathbf{B} = \mathbf{T} \times \mathbf{N}$ is the unit vector in the positive z-direction (by the right-hand rule); that is, $\mathbf{B} = \mathbf{k}$. Because **B** changes in neither length nor direction, $\dfrac{d\mathbf{B}}{ds} = \mathbf{0}$ and $\tau = 0$ **(Figure 14.39)**.

b. Building on the calculations of Example 2, we find that

$$\mathbf{T} = \langle -\sin t, \cos t \rangle \quad \text{and} \quad \mathbf{N} = \langle -\cos t, -\sin t \rangle.$$

Therefore, the unit binormal vector is

$$\mathbf{B} = \mathbf{T} \times \mathbf{N} = \begin{vmatrix} \mathbf{i} & \mathbf{j} & \mathbf{k} \\ -\sin t & \cos t & 0 \\ -\cos t & -\sin t & 0 \end{vmatrix} = 0 \cdot \mathbf{i} - 0 \cdot \mathbf{j} + 1 \cdot \mathbf{k} = \mathbf{k}.$$

As in part (a), it follows that the torsion is zero.

Related Exercise 41 ◀

Generalizing Example 8, it can be shown that the binormal vector of any curve that lies in the xy-plane is always parallel to the z-axis; therefore, the torsion of the curve is everywhere zero.

EXAMPLE 9   Torsion of a helix Compute the torsion of the helix $\mathbf{r}(t) = \langle a \cos t, a \sin t, bt \rangle$, for $t \ge 0$, $a > 0$, and $b > 0$.

SOLUTION In Example 5, we found that

$$\mathbf{T} = \frac{\langle -a \sin t, a \cos t, b \rangle}{\sqrt{a^2 + b^2}} \quad \text{and} \quad \mathbf{N} = \langle -\cos t, -\sin t, 0 \rangle.$$

Therefore,

$$\mathbf{B} = \mathbf{T} \times \mathbf{N} = \frac{1}{\sqrt{a^2 + b^2}} \begin{vmatrix} \mathbf{i} & \mathbf{j} & \mathbf{k} \\ -a \sin t & a \cos t & b \\ -\cos t & -\sin t & 0 \end{vmatrix} = \frac{\langle b \sin t, -b \cos t, a \rangle}{\sqrt{a^2 + b^2}}.$$

The next step is to determine $\dfrac{d\mathbf{B}}{ds}$, which we do in the same way we computed $\dfrac{d\mathbf{T}}{ds}$, by writing

$$\frac{d\mathbf{B}}{dt} = \frac{d\mathbf{B}}{ds} \cdot \frac{ds}{dt} \quad \text{or} \quad \frac{d\mathbf{B}}{ds} = \frac{d\mathbf{B}/dt}{ds/dt}.$$

In this case,

$$\frac{ds}{dt} = |\mathbf{r}'(t)| = \sqrt{a^2 \sin^2 t + a^2 \cos^2 t + b^2} = \sqrt{a^2 + b^2}.$$

Computing $\dfrac{d\mathbf{B}}{dt}$, we have

$$\frac{d\mathbf{B}}{ds} = \frac{d\mathbf{B}/dt}{ds/dt} = \frac{\langle b \cos t, b \sin t, 0 \rangle}{a^2 + b^2}.$$

The final step is to compute the torsion:

$$\tau = -\frac{d\mathbf{B}}{ds} \cdot \mathbf{N} = -\frac{\langle b \cos t, b \sin t, 0 \rangle}{a^2 + b^2} \cdot \langle -\cos t, -\sin t, 0 \rangle = \frac{b}{a^2 + b^2}.$$

We see that the torsion is constant over the helix. In Example 4, we found that the curvature of a helix is also constant. This special property of circular helices means that the curve turns about its axis at a constant rate and rises vertically at a constant rate (**Figure 14.40**).

Related Exercise 47 ◀

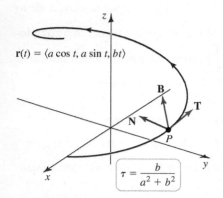

$\mathbf{r}(t) = \langle a \cos t, a \sin t, bt \rangle$

$\tau = \dfrac{b}{a^2 + b^2}$

Figure 14.40

Example 9 suggests that the computation of the binormal vector and the torsion can be involved. We close by stating some alternative formulas for $\mathbf{B}$ and τ that may simplify calculations in some cases. Letting $\mathbf{v} = \mathbf{r}'(t)$ and $\mathbf{a} = \mathbf{v}'(t) = \mathbf{r}''(t)$, the binormal vector can be written compactly as (Exercise 83)

$$\mathbf{B} = \mathbf{T} \times \mathbf{N} = \frac{\mathbf{v} \times \mathbf{a}}{|\mathbf{v} \times \mathbf{a}|}.$$

We also state without proof that the torsion may be expressed in either of the forms

$$\tau = \frac{(\mathbf{v} \times \mathbf{a}) \cdot \mathbf{a}'}{|\mathbf{v} \times \mathbf{a}|^2} \quad \text{or} \quad \tau = \frac{(\mathbf{r}' \times \mathbf{r}'') \cdot \mathbf{r}'''}{|\mathbf{r}' \times \mathbf{r}''|^2}.$$

SUMMARY Formulas for Curves in Space

Position function: $\mathbf{r}(t) = \langle x(t), y(t), z(t) \rangle$

Velocity: $\mathbf{v} = \mathbf{r}'$

Acceleration: $\mathbf{a} = \mathbf{v}'$

Unit tangent vector: $\mathbf{T} = \dfrac{\mathbf{v}}{|\mathbf{v}|}$

Principal unit normal vector: $\mathbf{N} = \dfrac{d\mathbf{T}/dt}{|d\mathbf{T}/dt|}$ (provided $d\mathbf{T}/dt \neq \mathbf{0}$)

Curvature: $\kappa = \left|\dfrac{d\mathbf{T}}{ds}\right| = \dfrac{1}{|\mathbf{v}|}\left|\dfrac{d\mathbf{T}}{dt}\right| = \dfrac{|\mathbf{v} \times \mathbf{a}|}{|\mathbf{v}|^3}$

Components of acceleration: $\mathbf{a} = a_N \mathbf{N} + a_T \mathbf{T}$, where $a_N = \kappa |\mathbf{v}|^2 = \dfrac{|\mathbf{v} \times \mathbf{a}|}{|\mathbf{v}|}$
and $a_T = \dfrac{d^2 s}{dt^2} = \dfrac{\mathbf{v} \cdot \mathbf{a}}{|\mathbf{v}|}$

Unit binormal vector: $\mathbf{B} = \mathbf{T} \times \mathbf{N} = \dfrac{\mathbf{v} \times \mathbf{a}}{|\mathbf{v} \times \mathbf{a}|}$

Torsion: $\tau = -\dfrac{d\mathbf{B}}{ds} \cdot \mathbf{N} = \dfrac{(\mathbf{v} \times \mathbf{a}) \cdot \mathbf{a}'}{|\mathbf{v} \times \mathbf{a}|^2} = \dfrac{(\mathbf{r}' \times \mathbf{r}'') \cdot \mathbf{r}'''}{|\mathbf{r}' \times \mathbf{r}''|^2}$

SECTION 14.5 EXERCISES

Getting Started

1. What is the curvature of a straight line?

2. Explain in words the meaning of *the curvature of a curve*. Is it a scalar function or a vector function?

3. Give a practical formula for computing curvature.

4. Interpret *the principal unit normal vector of a curve*. Is it a scalar function or a vector function?

5. Give a practical formula for computing the principal unit normal vector.

6. Explain how to decompose the acceleration vector of a moving object into its tangential and normal components.

7. Explain how the vectors **T**, **N**, and **B** are related geometrically.

8. How do you compute **B**?

9. Give a geometrical interpretation of torsion.

10. How do you compute torsion?

Practice Exercises

11–20. Curvature *Find the unit tangent vector* **T** *and the curvature* κ *for the following parameterized curves.*

11. $\mathbf{r}(t) = \langle 2t + 1, 4t - 5, 6t + 12 \rangle$

12. $\mathbf{r}(t) = \langle 2 \cos t, -2 \sin t \rangle$

13. $\mathbf{r}(t) = \langle 2t, 4 \sin t, 4 \cos t \rangle$

14. $\mathbf{r}(t) = \langle \cos t^2, \sin t^2 \rangle$

15. $\mathbf{r}(t) = \langle \sqrt{3} \sin t, \sin t, 2 \cos t \rangle$

16. $\mathbf{r}(t) = \langle t, \ln \cos t \rangle$

17. $\mathbf{r}(t) = \langle t, 2t^2 \rangle$

18. $\mathbf{r}(t) = \langle \cos^3 t, \sin^3 t \rangle$

19. $\mathbf{r}(t) = \left\langle \int_0^t \cos \frac{\pi u^2}{2} \, du, \int_0^t \sin \frac{\pi u^2}{2} \, du \right\rangle, t > 0$

20. $\mathbf{r}(t) = \left\langle \int_0^t \cos u^2 \, du, \int_0^t \sin u^2 \, du \right\rangle, t > 0$

21–26. Alternative curvature formula *Use the alternative curvature formula* $\kappa = \dfrac{|\mathbf{v} \times \mathbf{a}|}{|\mathbf{v}|^3}$ *to find the curvature of the following parameterized curves.*

21. $\mathbf{r}(t) = \langle -3 \cos t, 3 \sin t, 0 \rangle$

22. $\mathbf{r}(t) = \langle 4t, 3 \sin t, 3 \cos t \rangle$

23. $\mathbf{r}(t) = \langle 4 + t^2, t, 0 \rangle$

24. $\mathbf{r}(t) = \langle \sqrt{3} \sin t, \sin t, 2 \cos t \rangle$

25. $\mathbf{r}(t) = \langle 4 \cos t, \sin t, 2 \cos t \rangle$

26. $\mathbf{r}(t) = \langle e^t \cos t, e^t \sin t, e^t \rangle$

27–34. Principal unit normal vector *Find the unit tangent vector* **T** *and the principal unit normal vector* **N** *for the following parameterized curves. In each case, verify that* $|\mathbf{T}| = |\mathbf{N}| = 1$ *and* $\mathbf{T} \cdot \mathbf{N} = 0$.

27. $\mathbf{r}(t) = \langle 2 \sin t, 2 \cos t \rangle$ 28. $\mathbf{r}(t) = \langle 4 \sin t, 4 \cos t, 10t \rangle$

29. $\mathbf{r}(t) = \left\langle \dfrac{t^2}{2}, 4 - 3t, 1 \right\rangle$ 30. $\mathbf{r}(t) = \left\langle \dfrac{t^2}{2}, \dfrac{t^3}{3} \right\rangle, t > 0$

31. $\mathbf{r}(t) = \langle \cos t^2, \sin t^2 \rangle$ 32. $\mathbf{r}(t) = \langle \cos^3 t, \sin^3 t \rangle$

33. $\mathbf{r}(t) = \langle t^2, t \rangle$ 34. $\mathbf{r}(t) = \langle t, \ln \cos t \rangle$

35–40. Components of the acceleration *Consider the following trajectories of moving objects. Find the tangential and normal components of the acceleration.*

35. $\mathbf{r}(t) = \langle t, 1 + 4t, 2 - 6t \rangle$ 36. $\mathbf{r}(t) = \langle 10 \cos t, -10 \sin t \rangle$

37. $\mathbf{r}(t) = \langle e^t \cos t, e^t \sin t, e^t \rangle$ 38. $\mathbf{r}(t) = \langle t, t^2 + 1 \rangle$

39. $\mathbf{r}(t) = \langle t^3, t^2 \rangle$ 40. $\mathbf{r}(t) = \langle 20 \cos t, 20 \sin t, 30t \rangle$

41–44. Computing the binormal vector and torsion *In Exercises 27–30, the unit tangent vector* **T** *and the principal unit normal vector* **N** *were computed for the following parameterized curves. Use the definitions to compute their unit binormal vector and torsion.*

41. $\mathbf{r}(t) = \langle 2 \sin t, 2 \cos t \rangle$ 42. $\mathbf{r}(t) = \langle 4 \sin t, 4 \cos t, 10t \rangle$

43. $\mathbf{r}(t) = \left\langle \dfrac{t^2}{2}, 4 - 3t, 1 \right\rangle$ 44. $\mathbf{r}(t) = \left\langle \dfrac{t^2}{2}, \dfrac{t^3}{3} \right\rangle, t > 0$

45–48. Computing the binormal vector and torsion *Use the definitions to compute the unit binormal vector and torsion of the following curves.*

45. $\mathbf{r}(t) = \langle 2 \cos t, 2 \sin t, -t \rangle$

46. $\mathbf{r}(t) = \langle t, \cosh t, -\sinh t \rangle$

47. $\mathbf{r}(t) = \langle 12t, 5 \cos t, 5 \sin t \rangle$

48. $\mathbf{r}(t) = \langle \sin t - t \cos t, \cos t + t \sin t, t \rangle$

49. **Explain why or why not** Determine whether the following statements are true and give an explanation or counterexample.

   a. The position, unit tangent, and principal unit normal vectors (**r**, **T**, and **N**) at a point lie in the same plane.
   b. The vectors **T** and **N** at a point depend on the orientation of a curve.
   c. The curvature at a point depends on the orientation of a curve.
   d. An object with unit speed ($|\mathbf{v}| = 1$) on a circle of radius R has an acceleration of $\mathbf{a} = \mathbf{N}/R$.
   e. If the speedometer of a car reads a constant 60 mi/hr, the car is not accelerating.
   f. A curve in the xy-plane that is concave up at all points has positive torsion.
   g. A curve with large curvature also has large torsion.

50. **Special formula: Curvature for** $y = f(x)$ Assume f is twice differentiable. Prove that the curve $y = f(x)$ has curvature

$$\kappa(x) = \frac{|f''(x)|}{(1 + f'(x)^2)^{3/2}}.$$

   (*Hint:* Use the parametric description $x = t, y = f(t)$.)

51–54. Curvature for $y = f(x)$ *Use the result of Exercise 50 to find the curvature function of the following curves.*

51. $f(x) = x^2$ 52. $f(x) = \sqrt{a^2 - x^2}$

53. $f(x) = \ln x$ 54. $f(x) = \ln \cos x$

55. Special formula: Curvature for plane curves Show that the parametric curve $\mathbf{r}(t) = \langle f(t), g(t) \rangle$, where f and g are twice differentiable, has curvature

$$\kappa(t) = \frac{|f'g'' - f''g'|}{((f')^2 + (g')^2)^{3/2}},$$

where all derivatives are taken with respect to t.

56–59. Curvature for plane curves *Use the result of Exercise 55 to find the curvature function of the following curves.*

56. $\mathbf{r}(t) = \langle a \sin t, a \cos t \rangle$ (circle)

57. $\mathbf{r}(t) = \langle a \sin t, b \cos t \rangle$ (ellipse)

58. $\mathbf{r}(t) = \langle a \cos^3 t, a \sin^3 t \rangle$ (astroid)

59. $\mathbf{r}(t) = \langle t, at^2 \rangle$ (parabola)

When appropriate, consider using the special formulas derived in Exercises 50 and 55 in the remaining exercises.

60–63. Same paths, different velocity *The position functions of objects A and B describe different motion along the same path for $t \ge 0$.*

a. *Sketch the path followed by both A and B.*
b. *Find the velocity and acceleration of A and B and discuss the differences.*
c. *Express the acceleration of A and B in terms of the tangential and normal components and discuss the differences.*

60. $A: \mathbf{r}(t) = \langle 1 + 2t, 2 - 3t, 4t \rangle, B: \mathbf{r}(t) = \langle 1 + 6t, 2 - 9t, 12t \rangle$

61. $A: \mathbf{r}(t) = \langle t, 2t, 3t \rangle, B: \mathbf{r}(t) = \langle t^2, 2t^2, 3t^2 \rangle$

62. $A: \mathbf{r}(t) = \langle \cos t, \sin t \rangle, B: \mathbf{r}(t) = \langle \cos 3t, \sin 3t \rangle$

63. $A: \mathbf{r}(t) = \langle \cos t, \sin t \rangle, B: \mathbf{r}(t) = \langle \cos t^2, \sin t^2 \rangle$

64–67. Graphs of the curvature *Consider the following curves.*

a. *Graph the curve.*
b. *Compute the curvature.*
c. *Graph the curvature as a function of the parameter.*
d. *Identify the points (if any) at which the curve has a maximum or minimum curvature.*
e. *Verify that the graph of the curvature is consistent with the graph of the curve.*

64. $\mathbf{r}(t) = \langle t, t^2 \rangle$, for $-2 \le t \le 2$ (parabola)

65. $\mathbf{r}(t) = \langle t - \sin t, 1 - \cos t \rangle$, for $0 \le t \le 2\pi$ (cycloid)

66. $\mathbf{r}(t) = \langle t, \sin t \rangle$, for $0 \le t \le \pi$ (sine curve)

67. $\mathbf{r}(t) = \left\langle \frac{t^2}{2}, \frac{t^3}{3} \right\rangle$, for $t > 0$

68. Curvature of ln x Find the curvature of $f(x) = \ln x$, for $x > 0$, and find the point at which it is a maximum. What is the value of the maximum curvature?

69. Curvature of e^x Find the curvature of $f(x) = e^x$ and find the point at which it is a maximum. What is the value of the maximum curvature?

70. Circle and radius of curvature Choose a point P on a smooth curve C in the plane. The **circle of curvature** (or **osculating circle**) at P is the circle that (a) is tangent to C at P, (b) has the same curvature as C at P, and (c) lies on the same side of C as the principal unit normal $\mathbf{N}$ (see figure). The **radius of curvature** is

the radius of the circle of curvature. Show that the radius of curvature is $1/\kappa$, where κ is the curvature of C at P.

Circles of curvature

Radius of curvature $= \dfrac{1}{\kappa}$

71–73. Finding the radius of curvature *Find the radius of curvature (see Exercise 70) of the following curves at the given point. Then write an equation of the circle of curvature at the point.*

71. $\mathbf{r}(t) = \langle t, t^2 \rangle$ (parabola) at $t = 0$

72. $y = \ln x$ at $x = 1$

73. $\mathbf{r}(t) = \langle t - \sin t, 1 - \cos t \rangle$ (cycloid) at $t = \pi$

74. Designing a highway curve The function

$$\mathbf{r}(t) = \left\langle \int_0^t \cos \frac{u^2}{2} du, \int_0^t \sin \frac{u^2}{2} du \right\rangle,$$

whose graph is called a **clothoid** or **Euler spiral** (Figure A), has applications in the design of railroad tracks, roller coasters, and highways.

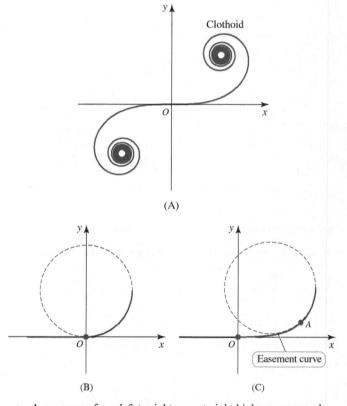

(A)

(B)   (C)

a. A car moves from left to right on a straight highway, approaching a curve at the origin (Figure B). Sudden changes in curvature at the start of the curve may cause the driver to jerk the steering wheel. Suppose the curve starting at the origin is a segment of a circle of radius a. Explain why there is a sudden change in the curvature of the road at the origin. (*Hint:* See Exercise 70.)

b. A better approach is to use a segment of a clothoid as an easement curve, in between the straight highway and a circle, to avoid sudden changes in curvature (Figure C). Assume the easement curve corresponds to the clothoid $\mathbf{r}(t)$, for $0 \le t \le 1.2$. Find the curvature of the easement curve as a function of t, and explain why this curve eliminates the sudden change in curvature at the origin.

c. Find the radius of a circle connected to the easement curve at point A (that corresponds to $t = 1.2$ on the curve $\mathbf{r}(t)$) so that the curvature of the circle matches the curvature of the easement curve at point A.

75. Curvature of the sine curve The function $f(x) = \sin nx$, where n is a positive real number, has a local maximum at $x = \dfrac{\pi}{2n}$. Compute the curvature κ of f at this point. How does κ vary (if at all) as n varies?

76. Parabolic trajectory In Example 7 it was shown that for the parabolic trajectory $\mathbf{r}(t) = \langle t, t^2 \rangle$, $\mathbf{a} = \langle 0, 2 \rangle$ and

$$\mathbf{a} = \frac{2}{\sqrt{1 + 4t^2}} (\mathbf{N} + 2t\,\mathbf{T}).$$ Show that the second expression for $\mathbf{a}$ reduces to the first expression.

T 77. Parabolic trajectory Consider the parabolic trajectory

$$x = (V_0 \cos \alpha)t, \quad y = (V_0 \sin \alpha)t - \frac{1}{2} gt^2,$$

where V_0 is the initial speed, α is the angle of launch, and g is the acceleration due to gravity. Consider all times $[0, T]$ for which $y \ge 0$.

a. Find and graph the speed, for $0 \le t \le T$.

b. Find and graph the curvature, for $0 \le t \le T$.

c. At what times (if any) do the speed and curvature have maximum and minimum values?

Explorations and Challenges

78. Relationship between T, N, and a Show that if an object accelerates in the sense that $\dfrac{d^2s}{dt^2} > 0$ and $\kappa \ne 0$, then the acceleration vector lies between $\mathbf{T}$ and $\mathbf{N}$ in the plane of $\mathbf{T}$ and $\mathbf{N}$. Show that if an object decelerates in the sense that $\dfrac{d^2s}{dt^2} < 0$, then the acceleration vector lies in the plane of $\mathbf{T}$ and $\mathbf{N}$, but not between $\mathbf{T}$ and $\mathbf{N}$.

79. Zero curvature Prove that the curve

$$\mathbf{r}(t) = \langle a + bt^p, c + dt^p, e + ft^p \rangle,$$

where $a, b, c, d, e,$ and f are real numbers and p is a positive integer, has zero curvature. Give an explanation.

80. Practical formula for N Show that the definition of the principal unit normal vector $\mathbf{N} = \dfrac{d\mathbf{T}/ds}{|d\mathbf{T}/ds|}$ implies the practical formula $\mathbf{N} = \dfrac{d\mathbf{T}/dt}{|d\mathbf{T}/dt|}$. Use the Chain Rule and recall that $|\mathbf{v}| = \dfrac{ds}{dt} > 0$.

T 81. Maximum curvature Consider the "superparabolas" $f_n(x) = x^{2n}$, where n is a positive integer.

a. Find the curvature function of f_n, for $n = 1, 2,$ and 3.

b. Plot f_n and their curvature functions, for $n = 1, 2,$ and 3, and check for consistency.

c. At what points does the maximum curvature occur, for $n = 1, 2,$ and 3?

d. Let the maximum curvature for f_n occur at $x = \pm z_n$. Using either analytical methods or a calculator, determine $\lim_{n \to \infty} z_n$. Interpret your result.

82. Alternative derivation of curvature Derive the computational formula for curvature using the following steps.

a. Use the tangential and normal components of the acceleration to show that $\mathbf{v} \times \mathbf{a} = \kappa |\mathbf{v}|^3 \mathbf{B}$. (Note that $\mathbf{T} \times \mathbf{T} = \mathbf{0}$.)

b. Solve the equation in part (a) for κ and conclude that $\kappa = \dfrac{|\mathbf{v} \times \mathbf{a}|}{|\mathbf{v}|^3}$, as shown in the text.

83. Computational formula for B Use the result of part (a) of Exercise 82 and the formula for κ to show that

$$\mathbf{B} = \frac{\mathbf{v} \times \mathbf{a}}{|\mathbf{v} \times \mathbf{a}|}.$$

84. Torsion formula Show that the formula defining torsion, $\tau = -\dfrac{d\mathbf{B}}{ds} \cdot \mathbf{N}$, is equivalent to $\tau = -\dfrac{1}{|\mathbf{v}|} \dfrac{d\mathbf{B}}{dt} \cdot \mathbf{N}$. The second formula is generally easier to use.

85. Descartes' four-circle solution Consider the four mutually tangent circles shown in the figure that have radii $a, b, c,$ and $d,$ and curvatures $A = \dfrac{1}{a}, B = \dfrac{1}{b}, C = \dfrac{1}{c},$ and $D = \dfrac{1}{d}$. Prove Descartes' result (1643) that

$$(A + B + C + D)^2 = 2(A^2 + B^2 + C^2 + D^2).$$

CHAPTER 14 REVIEW EXERCISES

1. **Explain why or why not** Determine whether the following statements are true and give an explanation or counterexample.
   a. If $\mathbf{r}(t) = \langle \cos t, e^t, t \rangle + \mathbf{C}$ and $\mathbf{r}(0) = \langle 0, 0, 0 \rangle$, then $\mathbf{C} = \langle 0, 0, 0 \rangle$.
   b. The curvature of a circle of radius 5 is $\kappa = 1/5$.
   c. The graph of $\mathbf{r}(t) = \langle 3 \cos t, 0, 6 \sin t \rangle$ is an ellipse in the xz-plane.
   d. If $\mathbf{r}'(t) = \mathbf{0}$, then $\mathbf{r}(t) = \langle a, b, c \rangle$, where a, b, and c are real numbers.
   e. The parameterized curve $\mathbf{r}(t) = \langle 5 \cos t, 12 \cos t, 13 \sin t \rangle$ has arc length as a parameter.
   f. The position vector and the principal unit normal are always parallel on a smooth curve.

2. **Sets of points** Describe the set of points satisfying the equations $x^2 + z^2 = 1$ and $y = 2$.

T **3–6. Graphing curves** *Sketch the curves described by the following functions, indicating the orientation of the curve. Use analysis and describe the shape of the curve before using a graphing utility.*

3. $\mathbf{r}(t) = (2t + 1)\mathbf{i} + t\mathbf{j}$

4. $\mathbf{r}(t) = \langle \cos t, 1 + \cos^2 t \rangle$, for $0 \le t \le \pi/2$

5. $\mathbf{r}(t) = 4 \cos t\,\mathbf{i} + \mathbf{j} + 4 \sin t\,\mathbf{k}$, for $0 \le t \le 2\pi$

6. $\mathbf{r}(t) = e^t\mathbf{i} + 2e^t\mathbf{j} + \mathbf{k}$, for $t \ge 0$

7. **Intersection curve** A sphere S and a plane P intersect along the curve $\mathbf{r}(t) = \sin t\,\mathbf{i} + \sqrt{2} \cos t\,\mathbf{j} + \sin t\,\mathbf{k}$, for $0 \le t \le 2\pi$. Find equations for S and P and describe the curve $\mathbf{r}$.

8–13. Vector-valued functions *Find a function $\mathbf{r}(t)$ that describes each of the following curves.*

8. The line segment from $P(2, -3, 0)$ to $Q(1, 4, 9)$

9. The line passing through the point $P(4, -2, 3)$ that is orthogonal to the lines $\mathbf{R}(t) = \langle t, 5t, 2t \rangle$ and $\mathbf{S}(t) = \langle t + 1, -1, 3t - 1 \rangle$

10. A circle of radius 3 centered at $(2, 1, 0)$ that lies in the plane $y = 1$

11. An ellipse in the plane $x = 2$ satisfying the equation $\dfrac{y^2}{9} + \dfrac{z^2}{16} = 1$

12. The projection of the curve onto the xy-plane is the parabola $y = x^2$, and the projection of the curve onto the xz-plane is the line $z = x$.

13. The projection of the curve onto the xy-plane is the unit circle $x^2 + y^2 = 1$, and the projection of the curve onto the yz-plane is the line segment $z = y$, for $-1 \le y \le 1$.

14–15. Intersection curve *Find the curve $\mathbf{r}(t)$ where the following surfaces intersect.*

14. $z = x^2 - 5y^2$; $z = 10x^2 + 4y^2 - 36$

15. $x^2 + 7y^2 + 2z^2 = 9$; $z = y$

16–19. Working with vector-valued functions *For each vector-valued function $\mathbf{r}$, carry out the following steps.*
   a. Evaluate $\lim\limits_{t \to 0} \mathbf{r}(t)$ and $\lim\limits_{t \to \infty} \mathbf{r}(t)$, if each exists.
   b. Find $\mathbf{r}'(t)$ and evaluate $\mathbf{r}'(0)$.

   c. Find $\mathbf{r}''(t)$.
   d. Evaluate $\int \mathbf{r}(t)\,dt$.

16. $\mathbf{r}(t) = \langle t + 1, t^2 - 3 \rangle$

17. $\mathbf{r}(t) = \left\langle \dfrac{1}{2t + 1}, \dfrac{t}{t + 1} \right\rangle$

18. $\mathbf{r}(t) = \langle e^{-2t}, te^{-t}, \tan^{-1} t \rangle$

19. $\mathbf{r}(t) = \langle \sin 2t, 3 \cos 4t, t \rangle$

20–21. Definite integrals *Evaluate the following definite integrals.*

20. $\displaystyle \int_1^3 \left(6t^2\mathbf{i} + 4t\mathbf{j} + \frac{1}{t}\mathbf{k} \right) dt$

21. $\displaystyle \int_{-1}^1 \left(\sin \pi t\,\mathbf{i} + \mathbf{j} + \frac{2}{1 + t^2}\mathbf{k} \right) dt$

22–24. Derivative rules *Suppose $\mathbf{u}$ and $\mathbf{v}$ are differentiable functions at $t = 0$ with $\mathbf{u}(0) = \langle 2, 7, 0 \rangle$, $\mathbf{u}'(0) = \langle 3, 1, 2 \rangle$, $\mathbf{v}(0) = \langle 3, -1, 0 \rangle$, and $\mathbf{v}'(0) = \langle 5, 0, 3 \rangle$. Evaluate the following expressions.*

22. $\dfrac{d}{dt}(\mathbf{u} \cdot \mathbf{v}) \Big|_{t=0}$

23. $\dfrac{d}{dt}(\mathbf{u} \times \mathbf{v}) \Big|_{t=0}$

24. $\dfrac{d}{dt}(\mathbf{u}(e^{5t} - 1)) \Big|_{t=0}$

25–27. Finding r from r′ *Find the function $\mathbf{r}$ that satisfies the given conditions.*

25. $\mathbf{r}'(t) = \langle 1, \sin 2t, \sec^2 t \rangle$; $\mathbf{r}(0) = \langle 2, 2, 2 \rangle$

26. $\mathbf{r}'(t) = \langle e^t, 2e^{2t}, 6e^{3t} \rangle$; $\mathbf{r}(0) = \langle 1, 3, -1 \rangle$

27. $\mathbf{r}'(t) = \left\langle \dfrac{4}{1 + t^2}, 2t + 1, 3t^2 \right\rangle$; $\mathbf{r}(1) = \langle 0, 0, 0 \rangle$

28–29. Unit tangent vectors *Find the unit tangent vector $\mathbf{T}(t)$ for the following parameterized curves. Then determine the unit tangent vector at the given value of t.*

28. $\mathbf{r}(t) = \langle 8, 3 \sin 2t, 3 \cos 2t \rangle$, for $0 \le t \le \pi$; $t = \pi/4$

29. $\mathbf{r}(t) = \langle 2e^t, e^{2t}, t \rangle$, for $0 \le t \le 2\pi$; $t = 0$

30–31. Velocity and acceleration from position *Consider the following position functions.*
   a. Find the velocity and speed of the object.
   b. Find the acceleration of the object.

30. $\mathbf{r}(t) = \left\langle \dfrac{5}{3}t^3 + 1, t^2 + 10t \right\rangle$, for $t \ge 0$

31. $\mathbf{r}(t) = \left\langle e^{4t} + 1, e^{4t}, \dfrac{1}{2}e^{4t} + 1 \right\rangle$, for $t \ge 0$

32–33. Solving equations of motion *Given an acceleration vector, initial velocity $\langle u_0, v_0 \rangle$, and initial position $\langle x_0, y_0 \rangle$, find the velocity and position vectors for $t \ge 0$.*

32. $\mathbf{a}(t) = \langle 1, 4 \rangle$, $\langle u_0, v_0 \rangle = \langle 4, 3 \rangle$, $\langle x_0, y_0 \rangle = \langle 0, 2 \rangle$

33. $\mathbf{a}(t) = \langle \cos t, 2 \sin t \rangle$, $\langle u_0, v_0 \rangle = \langle 2, 1 \rangle$, $\langle x_0, y_0 \rangle = \langle 1, 2 \rangle$

T **34. Orthogonal r and r′** Find all points on the ellipse $\mathbf{r}(t) = \langle 1, 8 \sin t, \cos t \rangle$, for $0 \le t \le 2\pi$, at which $\mathbf{r}(t)$ and $\mathbf{r}'(t)$ are orthogonal. Sketch the curve and the tangent vectors to verify your conclusion.

T 35–36. Modeling motion *Consider the motion of the following objects. Assume the x-axis is horizontal, the positive y-axis is vertical, the ground is horizontal, and only the gravitational force acts on the object.*

a. *Find the velocity and position vectors, for t ≥ 0.*
b. *Determine the time of flight and range of the object.*
c. *Determine the maximum height of the object.*

35. A baseball has an initial position $\langle x_0, y_0 \rangle = \langle 0, 3 \rangle$ ft when it is hit at an angle of 60° with an initial speed of 80 ft/s.

36. A rock is thrown from 2 m above horizontal ground at an angle of 30° above the horizontal with a speed of 6 m/s. Assume the initial position of the rock is $\langle x_0, y_0 \rangle = \langle 0, 2 \rangle$.

T 37. A baseball is hit 2 ft above home plate with an initial velocity of $\langle 40, 20, 40 \rangle$ ft/s. The spin on the baseball produces a horizontal acceleration of the ball of 4 ft/s² in the eastward direction. Assume the positive x-axis points east and the positive y-axis points north.

a. Find the velocity and position vectors, for $t ≥ 0$. Assume the origin is at home plate.
b. When does the ball hit the ground? Round your answer to three digits to the right of the decimal place.
c. How far does it land from home plate? Round your answer to the nearest whole number.

38. Firing angles A projectile is fired over horizontal ground from the ground with an initial speed of 40 m/s. What firing angles produce a range of 100 m?

T 39. Projectile motion A projectile is launched from the origin, which is a point 50 ft from a 30-ft vertical cliff (see figure). It is launched at a speed of $50\sqrt{2}$ ft/s at an angle of 45° to the horizontal. Assume the ground is horizontal on top of the cliff and that only the gravitational force affects the motion of the object.

a. Give the coordinates of the landing spot of the projectile on the top of the cliff.
b. What is the maximum height reached by the projectile?
c. What is the time of flight?
d. Write the integral that gives the length of the trajectory.
e. Approximate the length of the trajectory.
f. What is the range of launch angles needed to clear the edge of the cliff?

40. Baseball motion A toddler on level ground throws a baseball into the air at an angle of 30° with the ground from a height of 2 ft. If the ball lands 10 ft from the child, determine the initial speed of the ball.

T 41. Closest point Find the approximate location of the point on the curve $\mathbf{r}(t) = \langle t, t^2 + 1, 3t \rangle$ that lies closest to the point $P(3, 1, 6)$.

T 42. Basketball shot A basketball is shot at an angle of 45° to the horizontal. The center of the basketball is at the point $A(0, 8)$ at the moment it is released, and it passes through the center of the basketball hoop that is located at the point $B(18, 10)$. Assume the basketball does not hit the front of the hoop (otherwise it might not pass through the basket). The validity of this assumption is investigated in parts (d), (e), and (f).

a. Determine the initial speed of the basketball.
b. Find the initial velocity $\mathbf{v}(0)$ at the moment it is released.
c. Find the position function $\mathbf{r}(t)$ of the center of the basketball t seconds after the ball is released. Assume $\mathbf{r}(0) = \langle 0, 8 \rangle$.
d. Find the distance $s(t)$ between the center of the basketball and the front of the basketball hoop t seconds after the ball is released. Assume the diameter of the basketball hoop is 18 inches.
e. Determine the closest distance (in inches) between the center of the basketball and the front of the basketball hoop.
f. Is the assumption that the basketball does not hit the front of the hoop valid? Use the fact that the diameter of a women's basketball is about 9.23 inches. (*Hint:* The ball hits the front of the hoop if, during its flight, the distance from the center of the ball to the front of the hoop is less than the radius of the basketball.)

43–46. Arc length *Find the arc length of the following curves.*

43. $\mathbf{r}(t) = \left\langle t^2, \dfrac{4\sqrt{2}}{3} t^{3/2}, 2t \right\rangle$, for $1 ≤ t ≤ 3$

44. $\mathbf{r}(t) = \langle 2t^{9/2}, t^3 \rangle$, for $0 ≤ t ≤ 2$

T 45. $\mathbf{r}(t) = \langle \sin t, t + \cos t, 4t \rangle$, for $0 ≤ t ≤ \dfrac{\pi}{2}$

46. $\mathbf{r}(t) = \langle t, \ln \sec t, \ln (\sec t + \tan t) \rangle$, for $0 ≤ t ≤ \dfrac{\pi}{4}$

47. Velocity and trajectory length The acceleration of a wayward firework is given by $\mathbf{a}(t) = \sqrt{2}\,\mathbf{j} + 2t\mathbf{k}$, for $0 ≤ t ≤ 3$. Suppose the initial velocity of the firework is $\mathbf{v}(0) = \mathbf{i}$.

a. Find the velocity of the firework, for $0 ≤ t ≤ 3$.
b. Find the length of the trajectory of the firework over the interval $0 ≤ t ≤ 3$.

48–49. Arc length parameterization *Find a description of the following curves that uses arc length as a parameter.*

48. $\mathbf{r}(t) = (1 + 4t)\mathbf{i} - 3t\mathbf{j}$, for $t ≥ 1$

49. $\mathbf{r}(t) = \left\langle t^2, \dfrac{4\sqrt{2}}{3} t^{3/2}, 2t \right\rangle$, for $t ≥ 0$

50. Tangents and normals for an ellipse Consider the ellipse $\mathbf{r}(t) = \langle 3 \cos t, 4 \sin t \rangle$, for $0 \leq t \leq 2\pi$.

  a. Find the tangent vector $\mathbf{r}'$, the unit tangent vector $\mathbf{T}$, and the principal unit normal vector $\mathbf{N}$ at all points on the curve.
  b. At what points does $|\mathbf{r}'|$ have maximum and minimum values?
  c. At what points does the curvature have maximum and minimum values? Interpret this result in light of part (b).
  d. Find the points (if any) at which $\mathbf{r}$ and $\mathbf{N}$ are parallel.

T **51–54. Properties of space curves** *Do the following calculations.*

  a. *Find the tangent vector and the unit tangent vector.*
  b. *Find the curvature.*
  c. *Find the principal unit normal vector.*
  d. *Verify that $|\mathbf{N}| = 1$ and $\mathbf{T} \cdot \mathbf{N} = 0$.*
  e. *Graph the curve and sketch $\mathbf{T}$ and $\mathbf{N}$ at two points.*

51. $\mathbf{r}(t) = \langle 6 \cos t, 3 \sin t \rangle$, for $0 \leq t \leq 2\pi$

52. $\mathbf{r}(t) = \cos t\,\mathbf{i} + 2 \sin t\,\mathbf{j} + \mathbf{k}$, for $0 \leq t \leq 2\pi$

53. $\mathbf{r}(t) = \cos t\,\mathbf{i} + 2 \cos t\,\mathbf{j} + \sqrt{5} \sin t\,\mathbf{k}$, for $0 \leq t \leq 2\pi$

54. $\mathbf{r}(t) = t\,\mathbf{i} + 2 \cos t\,\mathbf{j} + 2 \sin t\,\mathbf{k}$, for $0 \leq t \leq 2\pi$

55–58. Analyzing motion *Consider the position vector of the following moving objects.*

  a. *Find the normal and tangential components of the acceleration.*
  b. *Graph the trajectory and sketch the normal and tangential components of the acceleration at two points on the trajectory. Show that their sum gives the total acceleration.*

55. $\mathbf{r}(t) = 2 \cos t\,\mathbf{i} + 2 \sin t\,\mathbf{j}$, for $0 \leq t \leq 2\pi$

56. $\mathbf{r}(t) = 3t\,\mathbf{i} + (4 - t)\,\mathbf{j} + t\,\mathbf{k}$, for $t \geq 0$

57. $\mathbf{r}(t) = (t^2 + 1)\,\mathbf{i} + 2t\,\mathbf{j}$, for $t \geq 0$

58. $\mathbf{r}(t) = 2 \cos t\,\mathbf{i} + 2 \sin t\,\mathbf{j} + 10t\,\mathbf{k}$, for $0 \leq t \leq 2\pi$

59. Computing the binormal vector and torsion Compute the unit binormal vector $\mathbf{B}$ and the torsion of the curve $\mathbf{r}(t) = \langle t, t^2, t^3 \rangle$ at $t = 1$.

60–61. Curve analysis *Carry out the following steps for the given curves C.*

  a. *Find $\mathbf{T}(t)$ at all points of C.*
  b. *Find $\mathbf{N}(t)$ and the curvature at all points of C.*
  c. *Sketch the curve and show $\mathbf{T}(t)$ and $\mathbf{N}(t)$ at the points of C corresponding to $t = 0$ and $t = \pi/2$.*
  d. *Are the results of parts (a) and (b) consistent with the graph?*
  e. *Find $\mathbf{B}(t)$ at all points of C.*
  f. *On the graph of part (c), plot $\mathbf{B}(t)$ at the points of C corresponding to $t = 0$ and $t = \pi/2$.*
  g. *Describe three calculations that serve to check the accuracy of your results in parts (a)–(f).*
  h. *Compute the torsion at all points of C. Interpret this result.*

60. $C: \mathbf{r}(t) = \langle 3 \sin t, 4 \sin t, 5 \cos t \rangle$, for $0 \leq t \leq 2\pi$

61. $C: \mathbf{r}(t) = \langle 3 \sin t, 3 \cos t, 4t \rangle$, for $0 \leq t \leq 2\pi$

62. Torsion of a plane curve Suppose $\mathbf{r}(t) = \langle f(t), g(t), h(t) \rangle$, where f, g, and h are the quadratic functions $f(t) = a_1 t^2 + b_1 t + c_1$, $g(t) = a_2 t^2 + b_2 t + c_2$, and $h(t) = a_3 t^2 + b_3 t + c_3$, and where at least one of the leading coefficients a_1, a_2, and a_3 is nonzero. Apart from a set of degenerate cases (for example, $\mathbf{r}(t) = \langle t^2, t^2, t^2 \rangle$, whose graph is a line), it can be shown that the graph of $\mathbf{r}(t)$ is a parabola that lies in a plane (Exercise 63).

  a. Show by direct computation that $\mathbf{v} \times \mathbf{a}$ is constant. Then explain why the unit binormal vector is constant at all points on the curve. What does this result say about the torsion of the curve?
  b. Compute $\mathbf{a}'(t)$ and explain why the torsion is zero at all points on the curve for which the torsion is defined.

63. Families of plane curves Let f and g be continuous on an interval I. Consider the curve

$$C: \mathbf{r}(t) = \langle a_1 f(t) + a_2 g(t) + a_3, \; b_1 f(t) + b_2 g(t) + b_3,$$
$$c_1 f(t) + c_2 g(t) + c_3 \rangle,$$

for t in I, and where a_i, b_i, and c_i, for $i = 1, 2$, and 3, are real numbers.

  a. Show that, in general, C lies in a plane.
  b. Explain why the torsion is zero at all points of C for which the torsion is defined.

64. Length of a DVD groove The capacity of a single-sided, single-layer digital versatile disc (DVD) is approximately 4.7 billion bytes—enough to store a two-hour movie. (Newer double-sided, double-layer DVDs have about four times that capacity, and Blu-ray discs are in the range of 50 gigabytes.) A DVD consists of a single "groove" that spirals outward from the inner edge to the outer edge of the storage region.

  a. First consider the spiral given in polar coordinates by $r = \dfrac{t\theta}{2\pi}$, where $0 \leq \theta \leq 2\pi N$ and successive loops of the spiral are t units apart. Explain why this spiral has N loops and why the entire spiral has a radius of $R = Nt$ units. Sketch three loops of the spiral.
  b. Write an integral for the length L of the spiral with N loops.
  c. The integral in part (b) can be evaluated exactly, but a good approximation can also be made. Assuming N is large, explain why $\theta^2 + 1 \approx \theta^2$. Use this approximation to simplify the integral in part (b) and show that $L \approx t\pi N^2 = \dfrac{\pi R^2}{t}$.
  d. Now consider a DVD with an inner radius of $r = 2.5$ cm and an outer radius of $R = 5.9$ cm. Model the groove by a spiral with a thickness of $t = 1.5$ microns $= 1.5 \times 10^{-6}$ m. Because of the hole in the DVD, the lower limit in the arc length integral is not $\theta = 0$. What are the limits of integration?
  e. Use the approximation in part (c) to find the length of the DVD groove. Express your answer in centimeters and miles.

Chapter 14 Guided Projects

Applications of the material in this chapter and related topics can be found in the following Guided Projects. For additional information, see the Preface.

- Designing a trajectory
- Bezier curves for graphic design
- Kepler's laws

15

Functions of Several Variables

Chapter Preview   The vectors of Chapter 13 and the vector-valued functions of Chapter 14 took us into three-dimensional space for the first time. In this chapter, we step into three-dimensional space along a different path by considering functions with several independent variables and one dependent variable. All the familiar properties of single-variable functions—domains, graphs, limits, continuity, and derivatives—have generalizations for multivariable functions, although you will also see subtle differences and new features. With functions of several independent variables, we work with *partial derivatives*, which, in turn, give rise to directional derivatives and the *gradient*, a fundamental concept in calculus. Partial derivatives allow us to find maximum and minimum values of multivariable functions. We define tangent planes, rather than tangent lines, that enable us to make linear approximations. The chapter ends with a survey of optimization problems in several variables.

15.1 Graphs and Level Curves

In Chapter 14, we discussed vector-valued functions with one independent variable and several dependent variables. We now reverse the situation and consider functions with several independent variables and one dependent variable. Such functions are aptly called *functions of several variables* or *multivariable functions*.

To set the stage, consider the following practical questions that illustrate a few of the many applications of functions of several variables.

- What is the probability that one man selected randomly from a large group of men weighs more than 200 pounds and is over 6 feet tall? (The answer depends on two variables, weight and height.)

- Where on the wing of an airliner flying at a speed of 550 mi/hr is the pressure greatest? (Pressure depends on the x-, y-, and z-coordinates of various points on the wing.)

- A physician knows the optimal blood concentration of an antibiotic needed by a patient. What dose of antibiotic is needed and how often should it be given to reach this optimal level? (The concentration depends (at least) on the amount of the dose, the frequency with which it is administered, and the weight of the patient.)

Although we don't answer these questions immediately, they clearly suggest the scope and importance of the topic. First, we must introduce the idea of a function of several variables.

Functions of Two Variables

The key concepts related to functions of several variables are most easily presented in the case of two independent variables; the extension to three or more variables is then straight-forward. In general, functions of two variables are written *explicitly* in the form

$$z = f(x, y)$$

or *implicitly* in the form

$$F(x, y, z) = 0.$$

Both forms are important, but for now, we consider explicitly defined functions.

The concepts of domain and range carry over directly from functions of a single variable.

DEFINITION Function, Domain, and Range with Two Independent Variables

A **function** $z = f(x, y)$ assigns to each point (x, y) in a set D in $\mathbb{R}^2$ a unique real number z in a subset of $\mathbb{R}$. The set D is the **domain** of f. The **range** of f is the set of real numbers z that are assumed as the points (x, y) vary over the domain (**Figure 15.1**).

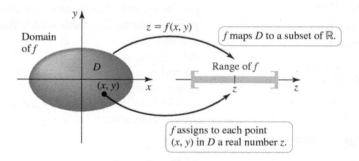

Figure 15.1

As with functions of one variable, a function of several variables may have a domain that is restricted by the context of the problem. For example, if the independent variables correspond to price or length or population, they take only nonnegative values, even though the associated function may be defined for negative values of the variables. If not stated otherwise, D is the set of all points for which the function is defined.

A polynomial in x and y consists of sums and products of polynomials in x and polynomials in y; for example, $f(x, y) = x^2y - 2xy - xy^2$. Such polynomials are defined for all values of x and y, so their domain is $\mathbb{R}^2$. A quotient of two polynomials in x and y, such as $h(x, y) = \dfrac{xy}{x - y}$, is a rational function in x and y. The domain of a rational function excludes points at which the denominator is zero, so the domain of h is $\{(x, y): x \neq y\}$.

EXAMPLE 1   Finding domains Find the domain of the function
$g(x, y) = \sqrt{4 - x^2 - y^2}$.

SOLUTION Because g involves a square root, its domain consists of ordered pairs (x, y) for which $4 - x^2 - y^2 \geq 0$ or $x^2 + y^2 \leq 4$. Therefore, the domain of g is $\{(x, y): x^2 + y^2 \leq 4\}$, which is the set of points on or within the circle of radius 2 centered at the origin in the xy-plane (a *disk* of radius 2) (**Figure 15.2**).

Related Exercises 17–18 ◄

Graphs of Functions of Two Variables

The **graph** of a function f of two variables is the set of points (x, y, z) that satisfy the equation $z = f(x, y)$. More specifically, for each point (x, y) in the domain of f, the point $(x, y, f(x, y))$ lies on the graph of f (**Figure 15.3**). A similar definition applies to relations of the form $F(x, y, z) = 0$.

Domain of
$g(x, y) = \sqrt{4 - x^2 - y^2}$

Figure 15.2

QUICK CHECK 1 Find the domains of $f(x, y) = \sin xy$ and $g(x, y) = \sqrt{(x^2 + 1)y}$. ◄

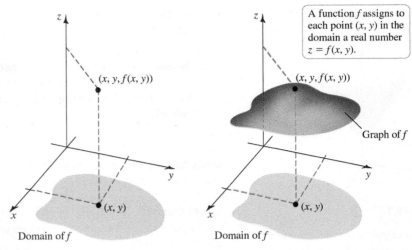

A function f assigns to each point (x, y) in the domain a real number $z = f(x, y)$.

$(x, y, f(x, y))$

$(x, y, f(x, y))$

Graph of f

(x, y)

(x, y)

Domain of f

Domain of f

Figure 15.3

QUICK CHECK 2 Does the graph of a hyperboloid of one sheet represent a function? Does the graph of a cone with its axis parallel to the x-axis represent a function? ◄

An ellipsoid does not pass the vertical line test: not the graph of a function.

This elliptic paraboloid passes the vertical line test: graph of a function.

Figure 15.4

Like functions of one variable, functions of two variables must pass a **vertical line test**. A relation of the form $F(x, y, z) = 0$ is a function provided every line parallel to the z-axis intersects the graph of the relation at most once. For example, an ellipsoid (discussed in Section 13.6) is not the graph of a function because some vertical lines intersect the surface twice. On the other hand, an elliptic paraboloid of the form $z = ax^2 + by^2$ does represent a function (**Figure 15.4**).

EXAMPLE 2   Graphing two-variable functions Find the domain and range of the following functions. Then sketch a graph.

a. $f(x, y) = 2x + 3y - 12$ **b.** $g(x, y) = x^2 + y^2$ **c.** $h(x, y) = \sqrt{1 + x^2 + y^2}$

SOLUTION

a. Letting $z = f(x, y)$, we have the equation $z = 2x + 3y - 12$, or $2x + 3y - z = 12$, which describes a plane with a normal vector $\langle 2, 3, -1 \rangle$ (Section 13.5). The domain consists of all points in $\mathbb{R}^2$, and the range is $\mathbb{R}$. We sketch the surface by noting that the x-intercept is $(6, 0, 0)$ (setting $y = z = 0$); the y-intercept is $(0, 4, 0)$ and the z-intercept is $(0, 0, -12)$ (**Figure 15.5**).

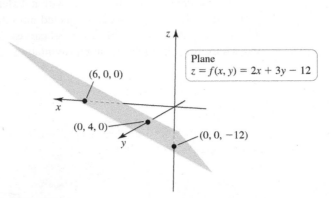

$(6, 0, 0)$

Plane
$z = f(x, y) = 2x + 3y - 12$

$(0, 4, 0)$

$(0, 0, -12)$

Figure 15.5

Paraboloid
$z = f(x, y) = x^2 + y^2$

Figure 15.6

b. Letting $z = g(x, y)$, we have the equation $z = x^2 + y^2$, which describes an elliptic paraboloid that opens upward with vertex $(0, 0, 0)$. The domain is $\mathbb{R}^2$ and the range consists of all nonnegative real numbers (**Figure 15.6**).

Upper sheet of hyperboloid of two sheets
$z = \sqrt{1 + x^2 + y^2}$

Figure 15.7

▶ To anticipate results that appear later in the chapter, notice how the streams in the topographic map—which flow downhill—cross the level curves roughly at right angles.

c. The domain of the function is $\mathbb{R}^2$ because the quantity under the square root is always positive. Note that $1 + x^2 + y^2 \geq 1$, so the range is $\{z: z \geq 1\}$. Squaring both sides of $z = \sqrt{1 + x^2 + y^2}$, we obtain $z^2 = 1 + x^2 + y^2$, or $-x^2 - y^2 + z^2 = 1$. This is the equation of a hyperboloid of two sheets that opens along the z-axis. Because the range is $\{z: z \geq 1\}$, the given function represents only the upper sheet of the hyperboloid (**Figure 15.7**; the lower sheet was introduced when we squared the original equation).

Related Exercises 25, 27, 29 ◀

QUICK CHECK 3 Find a function whose graph is the lower half of the hyperboloid $-x^2 - y^2 + z^2 = 1$. ◀

Level Curves   Functions of two variables are represented by surfaces in $\mathbb{R}^3$. However, such functions can be represented in another illuminating way, which is used to make topographic maps (**Figure 15.8**).

Closely spaced contours: rapid changes in elevation

Widely spaced contours: slow changes in elevation

Figure 15.8

▶ A contour curve is a trace (Section 13.6) in the plane $z = z_0$.

▶ A level curve may not always be a single curve. It might consist of a point ($x^2 + y^2 = 0$) or it might consist of several lines or curves ($xy = 0$).

Consider a surface defined by the function $z = f(x, y)$ (**Figure 15.9**). Now imagine stepping onto the surface and walking along a path on which your elevation has the constant value $z = z_0$. The path you walk on the surface is part of a **contour curve**; the complete contour curve is the intersection of the surface and the horizontal plane $z = z_0$. When the contour curve is projected onto the xy-plane, the result is the curve $f(x, y) = z_0$. This curve in the xy-plane is called a **level curve**.

Imagine repeating this process with a different constant value of z, say, $z = z_1$. The path you walk this time, when projected onto the xy-plane, is part of another level curve $f(x, y) = z_1$. A collection of such level curves, corresponding to different values of z, provides a useful two-dimensional representation of the surface (**Figure 15.10**).

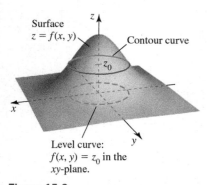

Surface $z = f(x, y)$
Contour curve
z_0

Level curve: $f(x, y) = z_0$ in the xy-plane.

Figure 15.9

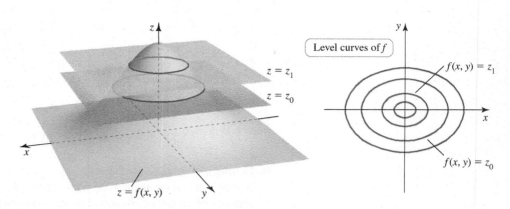

$z = z_1$
$z = z_0$
$z = f(x, y)$

Level curves of f
$f(x, y) = z_1$
$f(x, y) = z_0$

QUICK CHECK 4 Can two level curves of a function intersect? Explain. ◀

Figure 15.10

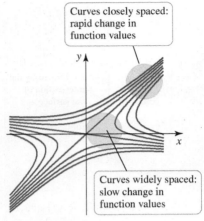

Curves closely spaced: rapid change in function values

Curves widely spaced: slow change in function values

Figure 15.11

QUICK CHECK 5 Describe in words the level curves of the top half of the sphere $x^2 + y^2 + z^2 = 1$. ◀

Assuming two adjacent level curves always correspond to the same change in z, widely spaced level curves indicate gradual changes in z-values, while closely spaced level curves indicate rapid changes in some directions (**Figure 15.11**). Concentric closed level curves generally indicate either a peak or a depression on the surface.

EXAMPLE 3 Level curves Find and sketch the level curves of the following surfaces.

a. $f(x, y) = y - x^2 - 1$ **b.** $f(x, y) = e^{-x^2 - y^2}$

SOLUTION

a. The level curves are described by the equation $y - x^2 - 1 = z_0$, where z_0 is a constant in the range of f. For all values of z_0, these curves are parabolas in the xy-plane, as seen by writing the equation in the form $y = x^2 + z_0 + 1$. For example:

- With $z_0 = 0$, the level curve is the parabola $y = x^2 + 1$; along this curve, the surface has an elevation (z-coordinate) of 0.

- With $z_0 = -1$, the level curve is $y = x^2$; along this curve, the surface has an elevation of -1.

- With $z_0 = 1$, the level curve is $y = x^2 + 2$, along which the surface has an elevation of 1.

As shown in **Figure 15.12a**, the level curves form a family of shifted parabolas. When these level curves are labeled with their z-coordinates, the graph of the surface $z = f(x, y)$ can be visualized (**Figure 15.12b**).

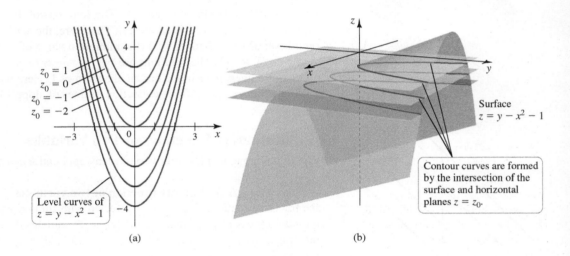

$z_0 = 1$
$z_0 = 0$
$z_0 = -1$
$z_0 = -2$

Level curves of $z = y - x^2 - 1$

Surface $z = y - x^2 - 1$

Contour curves are formed by the intersection of the surface and horizontal planes $z = z_0$.

(a)   (b)

Figure 15.12

b. The level curves satisfy the equation $e^{-x^2 - y^2} = z_0$, where z_0 is a positive constant. Taking the natural logarithm of both sides gives the equation $x^2 + y^2 = -\ln z_0$, which describes circular level curves. These curves can be sketched for all values of z_0 with $0 < z_0 \le 1$ (because the right side of $x^2 + y^2 = -\ln z_0$ must be nonnegative). For example:

- With $z_0 = 1$, the level curve satisfies the equation $x^2 + y^2 = 0$, whose solution is the single point $(0, 0)$; at this point, the surface has an elevation of 1.

- With $z_0 = e^{-1}$, the level curve is $x^2 + y^2 = -\ln e^{-1} = 1$, which is a circle centered at $(0, 0)$ with a radius of 1; along this curve the surface has an elevation of $e^{-1} \approx 0.37$.

In general, the level curves are circles centered at $(0, 0)$; as the radii of the circles increase, the corresponding z-values decrease. **Figure 15.13a** shows the level curves, with larger z-values corresponding to darker shades. From these labeled level curves, we can reconstruct the graph of the surface (**Figure 15.13b**).

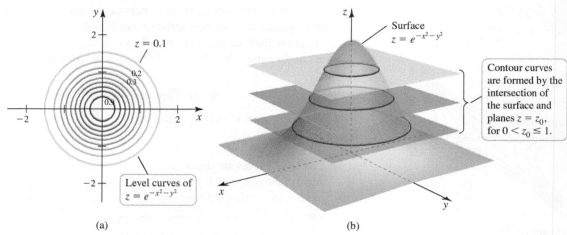

Contour curves
are formed by the
intersection of
the surface and
planes $z = z_0$,
for $0 < z_0 \leq 1$.

(a)   (b)

Figure 15.13

Related Exercises 37, 40 ◄

QUICK CHECK 6 Does the surface in Example 3b have a level curve for $z_0 = 0$? Explain. ◄

EXAMPLE 4   Level curves The graph of the function

$$f(x, y) = 2 + \sin(x - y)$$

is shown in **Figure 15.14a**. Sketch several level curves of the function.

SOLUTION The level curves are $f(x, y) = 2 + \sin(x - y) = z_0$, or $\sin(x - y) = z_0 - 2$. Because $-1 \leq \sin(x - y) \leq 1$, the admissible values of z_0 satisfy $-1 \leq z_0 - 2 \leq 1$, or, equivalently, $1 \leq z_0 \leq 3$. For example, when $z_0 = 2$, the level curves satisfy $\sin(x - y) = 0$. The solutions of this equation are $x - y = k\pi$, or $y = x - k\pi$, where k is an integer. Therefore, the surface has an elevation of 2 on this set of lines. With $z_0 = 1$ (the minimum value of z), the level curves satisfy $\sin(x - y) = -1$. The solutions are $x - y = -\pi/2 + 2k\pi$, where k is an integer; along these lines, the surface has an elevation of 1. Here we have an example in which each level curve is an infinite collection of lines of slope 1 (**Figure 15.14b**).

Related Exercise 43 ◄

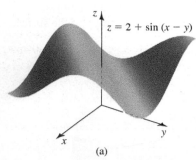

(a)

Applications of Functions of Two Variables

The following examples offer two of many applications of functions of two variables.

EXAMPLE 5   A probability function of two variables Suppose on a particular day, the fraction of students on campus infected with flu is r, where $0 \leq r \leq 1$. If you have n random (possibly repeated) encounters with students during the day, the probability of meeting *at least* one infected person is $p(n, r) = 1 - (1 - r)^n$ (**Figure 15.15a**). Discuss this probability function.

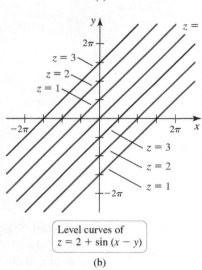

Level curves of
$z = 2 + \sin(x - y)$

(b)

Figure 15.14

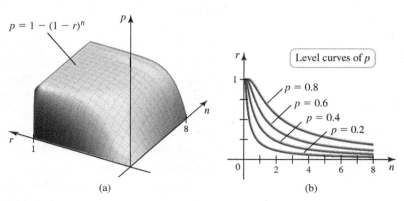

(a)   (b)

Figure 15.15

SOLUTION The independent variable r is restricted to the interval $[0, 1]$ because it is a fraction of the population. The other independent variable n is any nonnegative integer; for the purposes of graphing, we treat n as a real number in the interval $[0, 8]$. With $0 \leq r \leq 1$, note that $0 \leq 1 - r \leq 1$. If n is nonnegative, then $0 \leq (1 - r)^n \leq 1$, and

QUICK CHECK 7 In Example 5, if 50% of the population is infected, what is the probability of meeting at least one infected person in five encounters? ◄

Table 15.1

	n				
	2	**5**	**10**	**15**	**20**
0.05	0.10	0.23	0.40	0.54	0.64
0.1	0.19	0.41	0.65	0.79	0.88
r **0.3**	0.51	0.83	0.97	1	1
0.5	0.75	0.97	1	1	1
0.7	0.91	1	1	1	1

it follows that $0 \le p(n, r) \le 1$. Therefore, the range of the function is $[0, 1]$, which is consistent with the fact that p is a probability.

The level curves (**Figure 15.15b**) show that for a fixed value of n, the probability of at least one encounter increases with r; and for a fixed value of r, the probability increases with n. Therefore, as r increases or as n increases, the probability approaches 1 (at a surprising rate). If 10% of the population is infected ($r = 0.1$) and you have $n = 10$ encounters, then the probability of at least one encounter with an infected person is $p(10, 0.1) \approx 0.651$, which is about 2 in 3.

A numerical view of this function is given in Table 15.1, where we see probabilities tabulated for various values of n and r (rounded to two digits). The numerical values confirm the preceding observations.

Related Exercise 44 ◄

EXAMPLE 6   Electric potential function in two variables The electric field at points in the xy-plane due to two point charges located at $(0, 0)$ and $(1, 0)$ is related to the electric potential function

$$\varphi(x, y) = \frac{2}{\sqrt{x^2 + y^2}} + \frac{2}{\sqrt{(x - 1)^2 + y^2}}.$$

Discuss the electric potential function.

> The electric potential function, often denoted φ (pronounced *fee* or *fie*), is a scalar-valued function from which the electric field can be computed. Potential functions are discussed in detail in Chapter 17.

> A function that grows without bound near a point, as in the case of the electric potential function, is said to have a *singularity* at that point. A singularity is analogous to a vertical asymptote in a function of one variable.

SOLUTION The domain of the function contains all points of $\mathbb{R}^2$ except $(0, 0)$ and $(1, 0)$ where the charges are located. As these points are approached, the potential function becomes arbitrarily large (**Figure 15.16a**). The potential approaches zero as x or y increases in magnitude. These observations imply that the range of the potential function is all positive real numbers. The level curves of φ are closed curves, encircling either a single charge (at small distances) or both charges (at larger distances; **Figure 15.16b**).

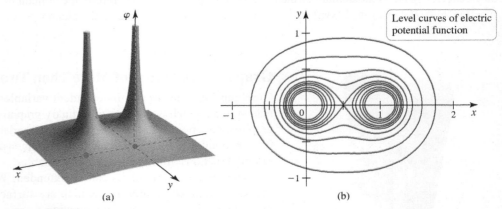

Level curves of electric potential function

(a)   (b)

Figure 15.16

Related Exercise 45 ◄

QUICK CHECK 8 In Example 6, what is the electric potential at the point $\left(\frac{1}{2}, 0\right)$? ◄

Functions of More Than Two Variables

Many properties of functions of two independent variables extend naturally to functions of three or more variables. A function of three variables is defined explicitly in the form $w = f(x, y, z)$ and implicitly in the form $F(x, y, z, w) = 0$. With more than three independent variables, the variables are usually written $x_1, \ldots, x_n$. Table 15.2 shows the progression of functions of several variables.

Table 15.2

Number of Independent Variables	Explicit Form	Implicit Form	Graph Resides In ...
1	$y = f(x)$	$F(x, y) = 0$	$\mathbb{R}^2$ (xy-plane)
2	$z = f(x, y)$	$F(x, y, z) = 0$	$\mathbb{R}^3$ (xyz-space)
3	$w = f(x, y, z)$	$F(x, y, z, w) = 0$	$\mathbb{R}^4$
n	$x_{n+1} = f(x_1, x_2, \ldots, x_n)$	$F(x_1, x_2, \ldots, x_n, x_{n+1}) = 0$	$\mathbb{R}^{n+1}$

Page 926

CHAPTER 15 • FUNCTIONS OF SEVERAL VARIABLES

The concepts of domain and range extend from the one- and two-variable cases in an obvious way.

DEFINITION Function, Domain, and Range with n Independent Variables

The **function** $x_{n+1} = f(x_1, x_2, \ldots, x_n)$ assigns a unique real number x_{n+1} to each point $(x_1, x_2, \ldots, x_n)$ in a set D in $\mathbb{R}^n$. The set D is the **domain** of f. The **range** is the set of real numbers x_{n+1} that are assumed as the points $(x_1, x_2, \ldots, x_n)$ vary over the domain.

EXAMPLE 7 Finding domains Find the domain of the following functions.

a. $g(x, y, z) = \sqrt{16 - x^2 - y^2 - z^2}$ b. $h(x, y, z) = \dfrac{12y^2}{z - y}$

SOLUTION

a. Values of the variables that make the argument of a square root negative must be excluded from the domain. In this case, the quantity under the square root is nonnegative provided

$$16 - x^2 - y^2 - z^2 \geq 0, \quad \text{or} \quad x^2 + y^2 + z^2 \leq 16.$$

Therefore, the domain of g is a closed ball in $\mathbb{R}^3$ of radius 4 centered at the origin.

> Recall that a closed ball of radius r is the set of all points on or within a sphere of radius r.

b. Values of the variables that make a denominator zero must be excluded from the domain. In this case, the denominator vanishes for all points in $\mathbb{R}^3$ that satisfy $z - y = 0$, or $y = z$. Therefore, the domain of h is the set $\{(x, y, z): y \neq z\}$. This set is $\mathbb{R}^3$ excluding the points on the plane $y = z$.

Related Exercises 51–52 ◄

QUICK CHECK 9 What is the domain of the function $w = f(x, y, z) = 1/xyz$? ◄

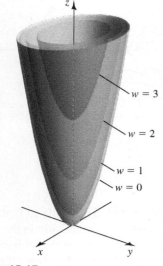

Figure 15.17

Graphs of Functions of More Than Two Variables

Graphing functions of *two* independent variables requires a three-dimensional coordinate system, which is the limit of ordinary graphing methods. Clearly, difficulties arise in graphing functions with three or more independent variables. For example, the graph of the function $w = f(x, y, z)$ resides in four dimensions. Here are two approaches to representing functions of three independent variables.

The idea of level curves can be extended. With the function $w = f(x, y, z)$, level curves become **level surfaces**, which are surfaces in $\mathbb{R}^3$ on which w is constant. For example, the level surfaces of the function

$$w = f(x, y, z) = \sqrt{z - x^2 - 2y^2}$$

satisfy $w = \sqrt{z - x^2 - 2y^2} = C$, where C is a nonnegative constant. This equation is satisfied when $z = x^2 + 2y^2 + C^2$. Therefore, the level surfaces are elliptic paraboloids, stacked one inside another (**Figure 15.17**).

Another approach to displaying functions of three variables is to use colors to represent the fourth dimension. **Figure 15.18a** shows the electrical activity of the heart at one snapshot in time. The three independent variables correspond to locations in the heart. At each point, the value of the electrical activity, which is the dependent variable, is coded by colors.

In **Figure 15.18b**, the dependent variable is the switching speed in an integrated circuit, again represented by colors, as it varies over points of the domain. Software to produce such images, once expensive and inefficient, has become much more accessible.

(a)   (b)

Figure 15.18

SECTION 15.1 EXERCISES

Getting Started

1. A function is defined by $z = x^2y - xy^2$. Identify the independent and dependent variables.

2. What is the domain of $f(x, y) = x^2y - xy^2$?

3. What is the domain of $g(x, y) = \dfrac{1}{xy}$?

4. What is the domain of $h(x, y) = \sqrt{x - y}$?

5. How many axes (or how many dimensions) are needed to graph the function $z = f(x, y)$? Explain.

6. Explain how to graph the level curves of a surface $z = f(x, y)$.

7. Given the function $f(x, y) = \sqrt{10 - x + y}$, evaluate $f(2, 1)$ and $f(-9, -3)$.

8. Given the function $g(x, y, z) = \dfrac{x + y}{z}$, evaluate $g(1, 5, 3)$ and $g(3, 7, 2)$.

9–10. *The function $z = f(x, y)$ gives the elevation z (in hundreds of feet) of a hillside above the point (x, y). Use the level curves of f to answer the following questions (see figure).*

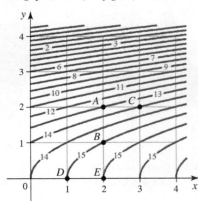

9. Katie and Zeke are standing on the surface above the point $A(2, 2)$.

    a. At what elevation are Katie and Zeke standing?

    b. Katie hikes south to the point on the surface above $B(2, 1)$ and Zeke hikes east to the point on the surface above $C(3, 2)$. Who experienced the greater elevation change and what is the difference in their elevations?

10. Katie and Zeke are standing on the surface above $D(1, 0)$. Katie hikes on the surface above the level curve containing $D(1, 0)$ to $B(2, 1)$ and Zeke walks east along the surface to $E(2, 0)$. What can be said about the elevations of Katie and Zeke during their hikes?

11. Describe in words the level curves of the paraboloid $z = x^2 + y^2$.

12. How many axes (or how many dimensions) are needed to graph the level surfaces of $w = f(x, y, z)$? Explain.

13. The domain of $Q = f(u, v, w, x, y, z)$ lies in $\mathbb{R}^n$ for what value of n? Explain.

14. Give two methods for graphically representing a function with three independent variables.

Practice Exercises

15–24. Domains *Find the domain of the following functions.*

15. $f(x, y) = 2xy - 3x + 4y$ **16.** $f(x, y) = \cos(x^2 - y^2)$

17. $f(x, y) = \sqrt{25 - x^2 - y^2}$ **18.** $f(x, y) = \dfrac{1}{\sqrt{x^2 + y^2 - 25}}$

19. $f(x, y) = \sin\dfrac{x}{y}$ **20.** $f(x, y) = \dfrac{12}{y^2 - x^2}$

21. $g(x, y) = \ln(x^2 - y)$ **22.** $f(x, y) = \sin^{-1}(y - x^2)$

23. $g(x, y) = \sqrt{\dfrac{xy}{x^2 + y^2}}$ **24.** $h(x, y) = \sqrt{x - 2y + 4}$

25–33. Graphs of familiar functions *Use what you learned about surfaces in Sections 13.5 and 13.6 to sketch a graph of the following functions. In each case, identify the surface and state the domain and range of the function.*

25. $f(x, y) = 6 - x - 2y$ **26.** $g(x, y) = 4$

27. $p(x, y) = x^2 - y^2$ **28.** $h(x, y) = 2x^2 + 3y^2$

29. $G(x, y) = -\sqrt{1 + x^2 + y^2}$

30. $F(x, y) = \sqrt{1 - x^2 - y^2}$

31. $P(x, y) = \sqrt{x^2 + y^2 - 1}$

32. $H(x, y) = \sqrt{x^2 + y^2}$ **33.** $g(x, y) = \sqrt{16 - 4x^2}$

34. Matching level curves with surfaces Match surfaces a–f in the figure with level curves A–F.

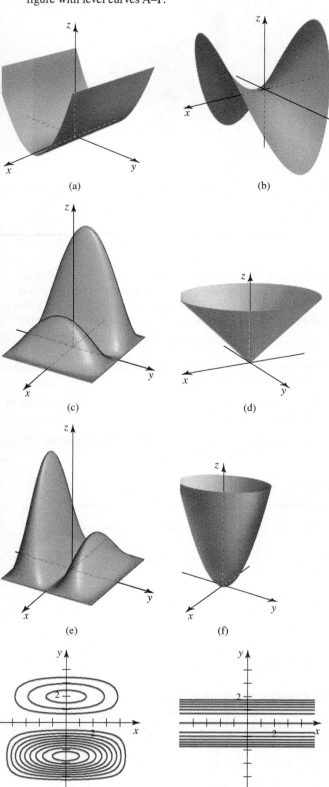

(a)

(b)

(c)

(d)

(e)

(f)

(A)

(B)

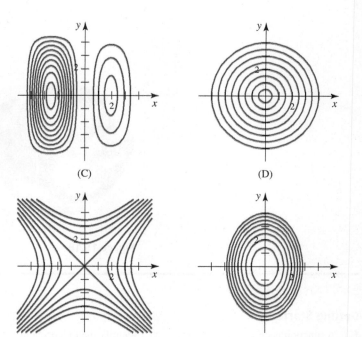

(C)

(D)

(E)

(F)

35. Matching surfaces Match functions a–d with surfaces A–D in the figure.

a. $f(x, y) = \cos xy$

b. $g(x, y) = \ln (x^2 + y^2)$

c. $h(x, y) = \dfrac{1}{x - y}$

d. $p(x, y) = \dfrac{1}{1 + x^2 + y^2}$

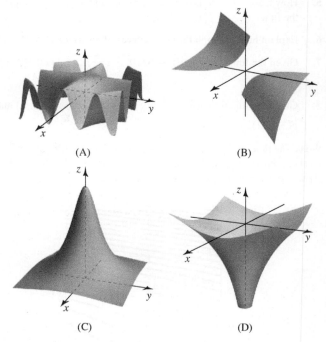

(A)

(B)

(C)

(D)

T 36–43. Level curves *Graph several level curves of the following functions using the given window. Label at least two level curves with their z-values.*

36. $z = x^2 + y^2$; $[-4, 4] \times [-4, 4]$

37. $z = x - y^2$; $[0, 4] \times [-2, 2]$

38. $z = 2x - y$; $[-2, 2] \times [-2, 2]$

39. $z = \sqrt{x^2 + 4y^2}$; $[-8, 8] \times [-8, 8]$

40. $z = e^{-x^2 - 2y^2}$; $[-2, 2] \times [-2, 2]$

41. $z = \sqrt{25 - x^2 - y^2}$; $[-6, 6] \times [-6, 6]$

42. $z = \sqrt{y - x^2 - 1}$; $[-5, 5] \times [-5, 5]$

43. $z = 3\cos(2x + y)$; $[-2, 2] \times [-2, 2]$

44. Earned run average A baseball pitcher's earned run average (ERA) is $A(e, i) = 9e/i$, where e is the number of earned runs given up by the pitcher and i is the number of innings pitched. Good pitchers have low ERAs. Assume $e \geq 0$ and $i > 0$ are real numbers.

   a. The single-season major league record for the lowest ERA was set by Dutch Leonard of the Detroit Tigers in 1914. During that season, Dutch pitched a total of 224 innings and gave up just 24 earned runs. What was his ERA?

   b. Determine the ERA of a relief pitcher who gives up 4 earned runs in one-third of an inning.

   c. Graph the level curve $A(e, i) = 3$ and describe the relationship between e and i in this case.

T 45. Electric potential function The electric potential function for two positive charges, one at $(0, 1)$ with twice the strength of the charge at $(0, -1)$, is given by

$$\varphi(x, y) = \frac{2}{\sqrt{x^2 + (y - 1)^2}} + \frac{1}{\sqrt{x^2 + (y + 1)^2}}.$$

   a. Graph the electric potential using the window $[-5, 5] \times [-5, 5] \times [0, 10]$.

   b. For what values of x and y is the potential φ defined?

   c. Is the electric potential greater at $(3, 2)$ or $(2, 3)$?

   d. Describe how the electric potential varies along the line $y = x$.

T 46. Cobb-Douglas production function The output Q of an economic system subject to two inputs, such as labor L and capital K, is often modeled by the Cobb-Douglas production function $Q(L, K) = cL^a K^b$, where a, b, and c are positive real numbers. When $a + b = 1$, the case is called *constant returns to scale*. Suppose $a = 1/3$, $b = 2/3$, and $c = 40$.

   a. Graph the output function using the window $[0, 20] \times [0, 20] \times [0, 500]$.

   b. If L is held constant at $L = 10$, write the function that gives the dependence of Q on K.

   c. If K is held constant at $K = 15$, write the function that gives the dependence of Q on L.

T 47. Resistors in parallel Two resistors wired in parallel in an electrical circuit give an effective resistance of $R(x, y) = \dfrac{xy}{x + y}$, where x and y are the positive resistances of the individual resistors (typically measured in ohms).

   a. Graph the resistance function using the window $[0, 10] \times [0, 10] \times [0, 5]$.

   b. Estimate the maximum value of R, for $0 < x \leq 10$ and $0 < y \leq 10$.

   c. Explain what it means to say that the resistance function is symmetric in x and y.

T 48. Level curves of a savings account Suppose you make a one-time deposit of P dollars into a savings account that earns interest at an annual rate of $p\%$ compounded continuously. The balance in the account after t years is $B(P, r, t) = Pe^{rt}$, where $r = p/100$ (for example, if the annual interest rate is 4%, then $r = 0.04$). Let the interest rate be fixed at $r = 0.04$.

   a. With a target balance of $2000, find the set of all points (P, t) that satisfy $B = 2000$. This curve gives all deposits P and times t that result in a balance of $2000.

   b. Repeat part (a) with $B = $500, $1000, $1500, and $2500, and draw the resulting level curves of the balance function.

   c. In general, on one level curve, if t increases, does P increase or decrease?

T 49. Level curves of a savings plan Suppose you make monthly deposits of P dollars into an account that earns interest at a *monthly* rate of $p\%$. The balance in the account after t years is

$$B(P, r, t) = P\left(\frac{(1 + r)^{12t} - 1}{r}\right), \text{ where } r = \frac{p}{100} \text{ (for example,}$$

if the annual interest rate is 9%, then $p = \dfrac{9}{12} = 0.75$ and $r = 0.0075$). Let the time of investment be fixed at $t = 20$ years.

   a. With a target balance of $20,000, find the set of all points (P, r) that satisfy $B = 20,000$. This curve gives all deposits P and monthly interest rates r that result in a balance of $20,000 after 20 years.

   b. Repeat part (a) with $B = $5000, $10,000, $15,000, and $25,000, and draw the resulting level curves of the balance function.

50–56. Domains of functions of three or more variables *Find the domain of the following functions. If possible, give a description of the domains (for example, all points outside a sphere of radius 1 centered at the origin).*

50. $f(x, y, z) = 2xyz - 3xz + 4yz$

51. $g(x, y, z) = \dfrac{1}{x - z}$

52. $p(x, y, z) = \sqrt{x^2 + y^2 + z^2 - 9}$

53. $f(x, y, z) = \sqrt{y - z}$

54. $Q(x, y, z) = \dfrac{10}{1 + x^2 + y^2 + 4z^2}$

55. $F(x, y, z) = \sqrt{y - x^2}$

56. $f(w, x, y, z) = \sqrt{1 - w^2 - x^2 - y^2 - z^2}$

57. Explain why or why not Determine whether the following statements are true and give an explanation or counterexample.

   a. The domain of the function $f(x, y) = 1 - |x - y|$ is $\{(x, y): x \geq y\}$.

   b. The domain of the function $Q = g(w, x, y, z)$ is a region in $\mathbb{R}^3$.

   c. All level curves of the plane $z = 2x - 3y$ are lines.

58. Quarterback passer ratings One measurement of the quality of a quarterback in the National Football League is known as the *quarterback passer rating*. The rating formula is

$$R(c, t, i, y) = \frac{50 + 20c + 80t - 100i + 100y}{24}, \text{ where } c\% \text{ of}$$

a quarterback's passes were completed, $t\%$ of his passes were thrown for touchdowns, $i\%$ of his passes were intercepted, and an average of y yards were gained per attempted pass.

a. In the 2016/17 NFL playoffs, Atlanta Falcons quarterback Matt Ryan completed 71.43% of his passes, 9.18% of his passes were thrown for touchdowns, none of his passes were intercepted, and he gained an average of 10.35 yards per passing attempt. What was his passer rating in the 2016 playoffs?

b. In the 2016 regular season, New England Patriots quarterback Tom Brady completed 67.36% of his passes, 6.48% of his passes were thrown for touchdowns, 0.46% of his passes were intercepted, and he gained an average of 8.23 yards per passing attempt. What was his passer rating in the 2016 regular season?

c. If c, t, and y remain fixed, what happens to the quarterback passer rating as i increases? Explain your answer with and without mathematics.

(*Source:* www.nfl.com)

T 59. **Ideal Gas Law** Many gases can be modeled by the Ideal Gas Law, $PV = nRT$, which relates the temperature (T, measured in kelvins (K)), pressure (P, measured in pascals (Pa)), and volume (V, measured in m^3) of a gas. Assume the quantity of gas in question is $n = 1$ mole (mol). The gas constant has a value of $R = 8.3$ m^3 Pa/mol-K.

a. Consider T to be the dependent variable, and plot several level curves (called *isotherms*) of the temperature surface in the region $0 \le P \le 100{,}000$ and $0 \le V \le 0.5$.

b. Consider P to be the dependent variable, and plot several level curves (called *isobars*) of the pressure surface in the region $0 \le T \le 900$ and $0 < V \le 0.5$.

c. Consider V to be the dependent variable, and plot several level curves of the volume surface in the region $0 \le T \le 900$ and $0 < P \le 100{,}000$.

Explorations and Challenges

T 60. **Water waves** A snapshot of a water wave moving toward shore is described by the function $z = 10 \sin(2x - 3y)$, where z is the height of the water surface above (or below) the xy-plane, which is the level of undisturbed water.

a. Graph the height function using the window $[-5, 5] \times [-5, 5] \times [-15, 15]$.

b. For what values of x and y is z defined?

c. What are the maximum and minimum values of the water height?

d. Give a vector in the xy-plane that is orthogonal to the level curves of the crests and troughs of the wave (which is parallel to the direction of wave propagation).

T 61. **Approximate mountains** Suppose the elevation of Earth's surface over a 16-mi by 16-mi region is approximated by the function

$$z = 10e^{-(x^2+y^2)} + 5e^{-((x+5)^2+(y-3)^2)/10} + 4e^{-2((x-4)^2+(y+1)^2)}.$$

a. Graph the height function using the window $[-8, 8] \times [-8, 8] \times [0, 15]$.

b. Approximate the points (x, y) where the peaks in the landscape appear.

c. What are the approximate elevations of the peaks?

T 62–68. Graphing functions

a. Determine the domain and range of the following functions.

b. Graph each function using a graphing utility. Be sure to experiment with the window and orientation to give the best perspective on the surface.

62. $g(x, y) = e^{-xy}$ **63.** $f(x, y) = |xy|$

64. $p(x, y) = 1 - |x - 1| + |y + 1|$

65. $h(x, y) = \dfrac{x + y}{x - y}$

66. $G(x, y) = \ln(2 + \sin(x + y))$

67. $F(x, y) = \tan^2(x - y)$ **68.** $P(x, y) = \cos x \sin 2y$

T 69–72. Peaks and valleys *The following functions have exactly one isolated peak or one isolated depression (one local maximum or minimum). Use a graphing utility to approximate the coordinates of the peak or depression.*

69. $f(x, y) = x^2 y^2 - 8x^2 - y^2 + 6$

70. $g(x, y) = (x^2 - x - 2)(y^2 + 2y)$

71. $h(x, y) = 1 - e^{-(x^2 + y^2 - 2x)}$

72. $p(x, y) = 2 + |x - 1| + |y - 1|$

73. **Level curves of planes** Prove that the level curves of the plane $ax + by + cz = d$ are parallel lines in the xy-plane, provided $a^2 + b^2 \ne 0$ and $c \ne 0$.

74–77. Level surfaces *Find an equation for the family of level surfaces corresponding to f. Describe the level surfaces.*

74. $f(x, y, z) = \dfrac{1}{x^2 + y^2 + z^2}$ **75.** $f(x, y, z) = x^2 + y^2 - z$

76. $f(x, y, z) = x^2 - y^2 - z$ **77.** $f(x, y, z) = \sqrt{x^2 + 2z^2}$

78–81. Challenge domains *Find the domain of the following functions. Specify the domain mathematically, and then describe it in words or with a sketch.*

78. $g(x, y, z) = \dfrac{10}{x^2 - (y + z)x + yz}$

79. $f(x, y) = \sin^{-1}(x - y)^2$

80. $f(x, y, z) = \ln(z - x^2 - y^2 + 2x + 3)$

81. $h(x, y, z) = \sqrt[4]{z^2 - xz + yz - xy}$

82. **Other balls** The closed unit ball in $\mathbb{R}^3$ centered at the origin is the set $\{(x, y, z): x^2 + y^2 + z^2 \le 1\}$. Describe the following alternative unit balls.

a. $\{(x, y, z): |x| + |y| + |z| \le 1\}$

b. $\{(x, y, z): \max\{|x|, |y|, |z|\} \le 1\}$, where $\max\{a, b, c\}$ is the maximum value of a, b, and c

QUICK CHECK ANSWERS

1. $\mathbb{R}^2$; $\{(x, y): y \ge 0\}$ **2.** No; no
3. $z = -\sqrt{1 + x^2 + y^2}$ **4.** No; otherwise the function would have two values at a single point.   **5.** Concentric circles   **6.** No; $z = 0$ is not in the range of the function.
7. 0.97   **8.** 8   **9.** $\{(x, y, z): x \ne 0 \text{ and } y \ne 0 \text{ and } z \ne 0\}$ (which is $\mathbb{R}^3$, excluding the coordinate planes) ◄

15.2 Limits and Continuity

You have now seen examples of functions of several variables, but calculus has not yet entered the picture. In this section, we revisit topics encountered in single-variable calculus and see how they apply to functions of several variables. We begin with the fundamental concepts of limits and continuity.

Limit of a Function of Two Variables

A function f of two variables has a limit L as $P(x, y)$ approaches a fixed point $P_0(a, b)$ if $|f(x, y) - L|$ can be made arbitrarily small for all P in the domain that are sufficiently close to P_0. If such a limit exists, we write

$$\lim_{(x, y) \to (a, b)} f(x, y) = \lim_{P \to P_0} f(x, y) = L.$$

To make this definition more precise, *close to* must be defined carefully.

A point x on the number line is close to another point a provided the distance $|x - a|$ is small (**Figure 15.19a**). In $\mathbb{R}^2$, a point $P(x, y)$ is close to another point $P_0(a, b)$ if the distance between them $|PP_0| = \sqrt{(x - a)^2 + (y - b)^2}$ is small (**Figure 15.19b**). When we say *for all P close to P_0*, it means that $|PP_0|$ is small for points P *on all sides of P_0.*

With this understanding of closeness, we can give a formal definition of a limit with two independent variables. This definition parallels the formal definition of a limit given in Section 2.7 (**Figure 15.20**).

DEFINITION Limit of a Function of Two Variables

The function f has the **limit** L as $P(x, y)$ approaches $P_0(a, b)$, written

$$\lim_{(x, y) \to (a, b)} f(x, y) = \lim_{P \to P_0} f(x, y) = L,$$

if, given any $\varepsilon > 0$, there exists a $\delta > 0$ such that

$$|f(x, y) - L| < \varepsilon$$

whenever (x, y) is in the domain of f and

$$0 < |PP_0| = \sqrt{(x - a)^2 + (y - b)^2} < \delta.$$

The condition $|PP_0| < \delta$ means that the distance between $P(x, y)$ and $P_0(a, b)$ is less than δ as P approaches P_0 from all possible directions (**Figure 15.21**). Therefore, the limit exists only if $f(x, y)$ approaches L as P approaches P_0 *along all possible paths* in the domain of f. As shown in upcoming examples, this interpretation is critical in determining whether a limit exists.

As with functions of one variable, we first establish limits of the simplest functions.

THEOREM 15.1   Limits of Constant and Linear Functions
Let a, b, and c be real numbers.

1. Constant function $f(x, y) = c$: $\displaystyle\lim_{(x, y) \to (a, b)} c = c$

2. Linear function $f(x, y) = x$: $\displaystyle\lim_{(x, y) \to (a, b)} x = a$

3. Linear function $f(x, y) = y$: $\displaystyle\lim_{(x, y) \to (a, b)} y = b$

Proof:

1. Consider the constant function $f(x, y) = c$ and assume $\varepsilon > 0$ is given. To prove that the value of the limit is $L = c$, we must produce a $\delta > 0$ such that $|f(x, y) - L| < \varepsilon$

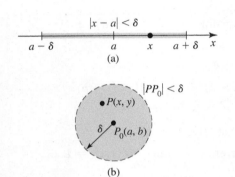

$$|x - a| < \delta$$

(a)

Figure 15.19

> The formal definition extends naturally to any number of variables. With n variables, the limit point is $P_0(a_1, \ldots, a_n)$, the variable point is $P(x_1, \ldots, x_n)$, and $|PP_0| = \sqrt{(x_1 - a_1)^2 + \cdots + (x_n - a_n)^2}$.

$f(x, y)$ is between $L - \varepsilon$ and $L + \varepsilon$ whenever $P(x, y)$ is within δ of P_0.

Figure 15.20

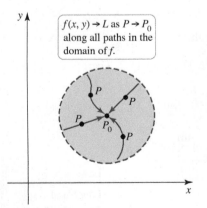

$f(x, y) \to L$ as $P \to P_0$ along all paths in the domain of f.

Figure 15.21

whenever $0 < \sqrt{(x-a)^2 + (y-b)^2} < \delta$. For constant functions, we may use *any* $\delta > 0$. Then, for every (x, y) in the domain of f,

$$|f(x, y) - L| = |f(x, y) - c| = |c - c| = 0 < \varepsilon$$

whenever $0 < \sqrt{(x-a)^2 + (y-b)^2} < \delta$.

2. Assume $\varepsilon > 0$ is given and take $\delta = \varepsilon$. The condition $0 < \sqrt{(x-a)^2 + (y-b)^2} < \delta$ implies that

$$0 < \sqrt{(x-a)^2 + (y-b)^2} < \varepsilon \qquad \delta = \varepsilon$$

$$\sqrt{(x-a)^2} < \varepsilon \qquad (x-a)^2 \le (x-a)^2 + (y-b)^2$$

$$|x - a| < \varepsilon. \qquad \sqrt{x^2} = |x| \text{ for real numbers } x$$

Because $f(x, y) = x$ and $L = a$, we have shown that $|f(x, y) - L| < \varepsilon$ whenever $0 < \sqrt{(x-a)^2 + (y-b)^2} < \delta$. Therefore, $\displaystyle\lim_{(x, y) \to (a, b)} f(x, y) = L$, or

$\displaystyle\lim_{(x, y) \to (a, b)} x = a$. The proof that $\displaystyle\lim_{(x, y) \to (a, b)} y = b$ is similar (Exercise 86). ◀

Using the three basic limits in Theorem 15.1, we can compute limits of more complicated functions. The only tools needed are limit laws analogous to those given in Theorem 2.3. The proofs of these laws are examined in Exercises 88–89.

THEOREM 15.2   Limit Laws for Functions of Two Variables
Let L and M be real numbers and suppose $\displaystyle\lim_{(x, y) \to (a, b)} f(x, y) = L$ and

$\displaystyle\lim_{(x, y) \to (a, b)} g(x, y) = M$. Assume c is a constant, and $n > 0$ is an integer.

1. Sum $\displaystyle\lim_{(x, y) \to (a, b)} (f(x, y) + g(x, y)) = L + M$

2. Difference $\displaystyle\lim_{(x, y) \to (a, b)} (f(x, y) - g(x, y)) = L - M$

3. Constant multiple $\displaystyle\lim_{(x, y) \to (a, b)} cf(x, y) = cL$

4. Product $\displaystyle\lim_{(x, y) \to (a, b)} f(x, y)g(x, y) = LM$

5. Quotient $\displaystyle\lim_{(x, y) \to (a, b)} \frac{f(x, y)}{g(x, y)} = \frac{L}{M}$, provided $M \ne 0$

6. Power $\displaystyle\lim_{(x, y) \to (a, b)} (f(x, y))^n = L^n$

7. Root $\displaystyle\lim_{(x, y) \to (a, b)} (f(x, y))^{1/n} = L^{1/n}$, where we assume $L > 0$ if n is even.

▶ Recall that a polynomial in two variables consists of sums and products of polynomials in x and polynomials in y. A rational function is the quotient of two polynomials.

Combining Theorems 15.1 and 15.2 allows us to find limits of polynomial, rational, and algebraic functions in two variables.

EXAMPLE 1   Limits of two-variable functions Evaluate $\displaystyle\lim_{(x, y) \to (2, 8)} (3x^2y + \sqrt{xy})$.

SOLUTION All the operations in this function appear in Theorem 15.2. Therefore, we can apply the limit laws directly.

$$\lim_{(x, y) \to (2, 8)} (3x^2y + \sqrt{xy}) = \lim_{(x, y) \to (2, 8)} 3x^2y + \lim_{(x, y) \to (2, 8)} \sqrt{xy} \quad \text{Law 1}$$

$$= 3 \lim_{(x, y) \to (2, 8)} x^2 \cdot \lim_{(x, y) \to (2, 8)} y$$

$$+ \sqrt{\lim_{(x, y) \to (2, 8)} x \cdot \lim_{(x, y) \to (2, 8)} y} \qquad \text{Laws 3, 4, 7}$$

$$= 3 \cdot 2^2 \cdot 8 + \sqrt{2 \cdot 8} = 100 \qquad \begin{array}{l}\text{Law 6 and}\\\text{Theorem 15.1}\end{array}$$

Related Exercise 16 ◀

QUICK CHECK 1 Which of the following limits exist?

a. $\lim\limits_{(x,y)\to(1,1)} 3x^{12}y^2$

b. $\lim\limits_{(x,y)\to(0,0)} 3x^{-2}y^2$

c. $\lim\limits_{(x,y)\to(1,2)} \sqrt{x-y^2}$ ◄

Q is a boundary point: Every disk centered at Q contains points in R and points not in R.

P is an interior point: There is a disk centered at P that lies entirely in R.

Figure 15.22

➤ The definitions of *interior point* and *boundary point* apply to regions in $\mathbb{R}^3$ if we replace *disk* by *ball*.

➤ Many sets, such as the annulus $\{(x,y): 2 \le x^2 + y^2 < 5\}$, are neither open nor closed.

QUICK CHECK 2 Give an example of a set that contains none of its boundary points. ◄

➤ Recall that this same method was used with functions of one variable. For example, after the common factor $x-2$ is canceled, the function

$$g(x) = \frac{x^2-4}{x-2}$$

becomes $g(x) = x+2$, provided $x \ne 2$. In this case, 2 plays the role of a boundary point.

In Example 1, the value of the limit equals the value of the function at (a,b); in other words, $\lim\limits_{(x,y)\to(a,b)} f(x,y) = f(a,b)$ and the limit can be evaluated by substitution. This is a property of *continuous* functions, discussed later in this section.

Limits at Boundary Points

This is an appropriate place to make some definitions that are used in the remainder of the text.

DEFINITION Interior and Boundary Points

Let R be a region in $\mathbb{R}^2$. An **interior point** P of R lies entirely within R, which means it is possible to find a disk centered at P that contains only points of R (**Figure 15.22**).

A **boundary point** Q of R lies on the edge of R in the sense that *every* disk centered at Q contains at least one point in R and at least one point not in R.

For example, let R be the points in $\mathbb{R}^2$ satisfying $x^2 + y^2 < 9$. The boundary points of R lie on the circle $x^2 + y^2 = 9$. The interior points lie inside that circle and satisfy $x^2 + y^2 < 9$. Notice that the boundary points of a set need not lie in the set.

DEFINITION Open and Closed Sets

A region is **open** if it consists entirely of interior points. A region is **closed** if it contains all its boundary points.

An example of an open region in $\mathbb{R}^2$ is the open disk $\{(x,y): x^2 + y^2 < 9\}$. An example of a closed region in $\mathbb{R}^2$ is the square $\{(x,y): |x| \le 1, |y| \le 1\}$. Later in the text, we encounter interior and boundary points of three-dimensional sets such as balls, boxes, and pyramids.

Suppose $P_0(a,b)$ is a boundary point of the domain of f. The limit $\lim\limits_{(x,y)\to(a,b)} f(x,y)$ exists, even if P_0 is not in the domain of f, provided $f(x,y)$ approaches the same value as (x,y) approaches (a,b) *along all paths that lie in the domain* (**Figure 15.23**).

Consider the function $f(x,y) = \dfrac{x^2-y^2}{x-y}$ whose domain is $\{(x,y): x \ne y\}$. Provided $x \ne y$, we may cancel the factor $(x-y)$ from the numerator and denominator and write

$$f(x,y) = \frac{x^2-y^2}{x-y} = \frac{(x-y)(x+y)}{x-y} = x+y.$$

The graph of f (**Figure 15.24**) is the plane $z = x+y$, with points corresponding to the line $x=y$ removed.

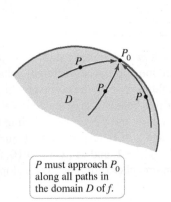

P must approach P_0 along all paths in the domain D of f.

Figure 15.23

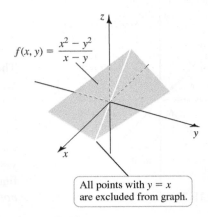

$f(x,y) = \dfrac{x^2-y^2}{x-y}$

All points with $y=x$ are excluded from graph.

Figure 15.24

Now we examine $\displaystyle\lim_{(x,\,y)\to(4,\,4)} \frac{x^2 - y^2}{x - y}$, where $(4, 4)$ is a boundary point of the domain of f but does not lie in the domain. For this limit to exist, $f(x, y)$ must approach the same value along all paths to $(4, 4)$ that lie in the domain of f—that is, all paths approaching $(4, 4)$ that do not intersect $x = y$. To evaluate the limit, we proceed as follows:

$$\lim_{(x,\,y)\to(4,\,4)} \frac{x^2 - y^2}{x - y} = \lim_{(x,\,y)\to(4,\,4)} (x + y) \quad \text{Assume } x \neq y, \text{ cancel } x - y.$$

$$= 4 + 4 = 8. \qquad \text{Same limit along all paths in the domain}$$

To emphasize, we let $(x, y) \to (4, 4)$ along all paths that do not intersect $x = y$, which lies outside the domain of f. Along all admissible paths, the function approaches 8.

QUICK CHECK 3 Can the limit

$$\lim_{(x,\,y)\to(0,\,0)} \frac{x^2 - xy}{x} \text{ be evaluated by}$$

direct substitution? ◄

EXAMPLE 2   Limits at boundary points Evaluate $\displaystyle\lim_{(x,\,y)\to(4,\,1)} \frac{xy - 4y^2}{\sqrt{x} - 2\sqrt{y}}$.

SOLUTION Points in the domain of this function satisfy $x \geq 0$ and $y \geq 0$ (because of the square roots) and $x \neq 4y$ (to ensure the denominator is nonzero). We see that the point $(4, 1)$ lies on the boundary of the domain. Multiplying the numerator and denominator by the algebraic conjugate of the denominator, the limit is computed as follows:

$$\lim_{(x,\,y)\to(4,\,1)} \frac{xy - 4y^2}{\sqrt{x} - 2\sqrt{y}} = \lim_{(x,\,y)\to(4,\,1)} \frac{(xy - 4y^2)(\sqrt{x} + 2\sqrt{y})}{(\sqrt{x} - 2\sqrt{y})(\sqrt{x} + 2\sqrt{y})} \quad \text{Multiply by conjugate.}$$

$$= \lim_{(x,\,y)\to(4,\,1)} \frac{y(x - 4y)(\sqrt{x} + 2\sqrt{y})}{x - 4y} \qquad \text{Simplify.}$$

$$= \lim_{(x,\,y)\to(4,\,1)} y(\sqrt{x} + 2\sqrt{y}) \qquad \begin{array}{l}\text{Cancel } x - 4y,\\ \text{assumed to be nonzero.}\end{array}$$

$$= 4. \qquad\qquad\qquad \text{Evaluate limit.}$$

Because points on the line $x = 4y$ are outside the domain of the function, we assume $x - 4y \neq 0$. Along all other paths to $(4, 1)$, the function values approach 4 (**Figure 15.25**).

Related Exercises 26–27 ◄

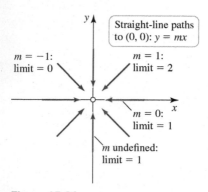

Line $x = 4y$ is not in the domain of f.

$(4, 1)$

$(x, y) \to (4, 1)$ along paths in domain of f.

Figure 15.25

► Notice that if we choose any path of the form $y = mx$, then $y \to 0$ as $x \to 0$. Therefore, $\displaystyle\lim_{(x,\,y)\to(0,\,0)}$ can be replaced by $\displaystyle\lim_{x\to 0}$ along this path. A similar argument applies to paths of the form $y = mx^p$, for $p > 0$.

Straight-line paths to $(0, 0)$: $y = mx$

$m = -1$: limit $= 0$

$m = 1$: limit $= 2$

$m = 0$: limit $= 1$

m undefined: limit $= 1$

Figure 15.26

EXAMPLE 3   Nonexistence of a limit Investigate the limit $\displaystyle\lim_{(x,\,y)\to(0,\,0)} \frac{(x + y)^2}{x^2 + y^2}$.

SOLUTION The domain of the function is $\{(x, y): (x, y) \neq (0, 0)\}$; therefore, the limit is at a boundary point outside the domain. Suppose we let (x, y) approach $(0, 0)$ along the line $y = mx$ for a fixed constant m. Substituting $y = mx$ and noting that $y \to 0$ as $x \to 0$, we have

$$\lim_{\substack{(x,\,y)\to(0,\,0)\\ (\text{along } y = mx)}} \frac{(x + y)^2}{x^2 + y^2} = \lim_{x\to 0} \frac{(x + mx)^2}{x^2 + m^2 x^2} = \lim_{x\to 0} \frac{x^2(1 + m)^2}{x^2(1 + m^2)} = \frac{(1 + m)^2}{1 + m^2}.$$

The constant m determines the direction of approach to $(0, 0)$. Therefore, depending on m, the function approaches different values as (x, y) approaches $(0, 0)$ (**Figure 15.26**). For example, if $m = 0$, the corresponding limit is 1, and if $m = -1$, the limit is 0. The reason for this behavior is revealed if we plot the surface and look at two level curves. The lines $y = x$ and $y = -x$ (excluding the origin) are level curves of the function for $z = 2$ and $z = 0$, respectively. (**Figure 15.27**). Therefore, as $(x, y) \to (0, 0)$ along $y = x$, $f(x, y) \to 2$, and as $(x, y) \to (0, 0)$ along $y = -x$, $f(x, y) \to 0$. Because the function approaches different values along different paths, we conclude that the *limit does not exist.*

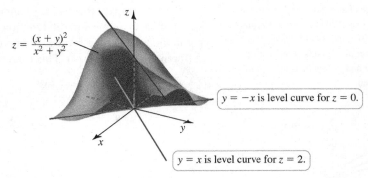

$$z = \frac{(x + y)^2}{x^2 + y^2}$$

$y = -x$ is level curve for $z = 0.$

$y = x$ is level curve for $z = 2.$

Figure 15.27   *Related Exercise 30* ◄

The strategy used in Example 3 is an effective way to prove the nonexistence of a limit.

QUICK CHECK 4 What is the analog of the Two-Path Test for functions of a single variable? ◄

PROCEDURE   Two-Path Test for Nonexistence of Limits

If $f(x, y)$ approaches two different values as (x, y) approaches (a, b) along two different paths in the domain of f, then $\lim\limits_{(x, y)\to(a, b)} f(x, y)$ does not exist.

Continuity of Functions of Two Variables

The following definition of continuity for functions of two variables is analogous to the continuity definition for functions of one variable.

DEFINITION Continuity

The function f is continuous at the point (a, b) provided

1. f is defined at (a, b),

2. $\lim\limits_{(x, y)\to(a, b)} f(x, y)$ exists, and

3. $\lim\limits_{(x, y)\to(a, b)} f(x, y) = f(a, b)$.

A function of two (or more) variables is continuous at a point, provided its limit equals its value at that point (which implies the limit and the value both exist). The definition of continuity applies at boundary points of the domain of f, provided the limits in the definition are taken along all paths that lie in the domain. Because limits of polynomials and rational functions can be evaluated by substitution at points of their domains (that is, $\lim\limits_{(x, y)\to(a, b)} f(x, y) = f(a, b)$), it follows that polynomials and rational functions are continuous at all points of their domains.

EXAMPLE 4   Checking continuity Determine the points at which the following function is continuous.

$$f(x, y) = \begin{cases} \dfrac{3xy^2}{x^2 + y^4} & \text{if } (x, y) \neq (0, 0) \\ 0 & \text{if } (x, y) = (0, 0) \end{cases}$$

SOLUTION The function $\dfrac{3xy^2}{x^2 + y^4}$ is a rational function, so it is continuous at all points of its domain, which consists of all points of $\mathbb{R}^2$ except $(0, 0)$. To determine whether f is continuous at $(0, 0)$, we must show that

$$\lim_{(x, y)\to(0, 0)} \frac{3xy^2}{x^2 + y^4}$$

exists and equals $f(0, 0) = 0$ along all paths that approach $(0, 0)$.

▶ The choice of $x = my^2$ for paths to $(0, 0)$ is not obvious. Notice that if x is replaced with my^2 in f, the result involves the same power of y (in this case, y^4) in the numerator and denominator, which may be canceled.

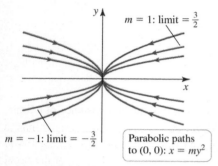

$m = 1$: limit $= \frac{3}{2}$

$m = -1$: limit $= -\frac{3}{2}$

Parabolic paths to $(0, 0)$: $x = my^2$

Figure 15.28

You can verify that as (x, y) approaches $(0, 0)$ along paths of the form $y = mx$, where m is any constant, the function values approach $f(0, 0) = 0$. However, along parabolic paths of the form $x = my^2$ (where m is a nonzero constant), the limit behaves differently (**Figure 15.28**). This time we substitute $x = my^2$ and note that $x \to 0$ as $y \to 0$:

$$\lim_{\substack{(x, y) \to (0, 0) \\ (\text{along } x = my^2)}} \frac{3xy^2}{x^2 + y^4} = \lim_{y \to 0} \frac{3(my^2)y^2}{(my^2)^2 + y^4} \quad \text{Substitute } x = my^2.$$

$$= \lim_{y \to 0} \frac{3my^4}{m^2 y^4 + y^4} \quad \text{Simplify.}$$

$$= \lim_{y \to 0} \frac{3m}{m^2 + 1} \quad \text{Cancel } y^4.$$

$$= \frac{3m}{m^2 + 1}.$$

We see that along parabolic paths, the limit depends on the approach path. For example, with $m = 1$, along the path $x = y^2$ the function values approach $\frac{3}{2}$; with $m = -1$, along the path $x = -y^2$ the function values approach $-\frac{3}{2}$ (**Figure 15.29**). Because $f(x, y)$ approaches two different numbers along two different paths, the limit at $(0, 0)$ does not exist, and f is not continuous at $(0, 0)$.

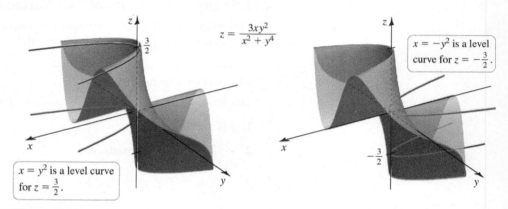

$z = \frac{3xy^2}{x^2 + y^4}$

$x = y^2$ is a level curve for $z = \frac{3}{2}$.

$x = -y^2$ is a level curve for $z = -\frac{3}{2}$.

Figure 15.29

Related Exercises 41–42 ◄

QUICK CHECK 5 Which of the following functions are continuous at $(0, 0)$?

a. $f(x, y) = 2x^2 y^5$

b. $f(x, y) = \dfrac{2x^2 y^5}{x - 1}$

c. $f(x, y) = 2x^{-2} y^5$ ◄

Composite Functions   Recall that for functions of a single variable, compositions of continuous functions are also continuous. The following theorem gives the analogous result for functions of two variables; it is proved in Appendix A.

THEOREM 15.3   Continuity of Composite Functions
If $u = g(x, y)$ is continuous at (a, b) and $z = f(u)$ is continuous at $g(a, b)$, then the composite function $z = f(g(x, y))$ is continuous at (a, b).

With Theorem 15.3, we can easily analyze the continuity of many functions. For example, $\sin x$, $\cos x$, and e^x are continuous functions of a single variable, for all real values of x. Therefore, compositions of these functions with polynomials in x and y (for example, $\sin(x^2 y)$ and $e^{x^4 - y^2}$) are continuous for all real numbers x and y. Similarly, $\sqrt{x}$ is a continuous function of a single variable, for $x \geq 0$. Therefore, $\sqrt{u(x, y)}$ is continuous at (x, y) provided u is continuous at (x, y) and $u(x, y) \geq 0$. As long as we observe restrictions on domains, then compositions of continuous functions are also continuous.

EXAMPLE 5   Continuity of composite functions.   Determine the points at which the following functions are continuous.

a. $h(x, y) = \ln(x^2 + y^2 + 4)$ **b.** $h(x, y) = e^{x/y}$

SOLUTION

a. This function is the composition $f(g(x, y))$, where

$$f(u) = \ln u \quad \text{and} \quad u = g(x, y) = x^2 + y^2 + 4.$$

As a polynomial, g is continuous for all (x, y) in $\mathbb{R}^2$. The function f is continuous for $u > 0$. Because $u = x^2 + y^2 + 4 > 0$ for all (x, y), it follows that h is continuous at all points of $\mathbb{R}^2$.

b. Letting $f(u) = e^u$ and $u = g(x, y) = x/y$, we have $h(x, y) = f(g(x, y))$. Note that f is continuous at all points of $\mathbb{R}$ and g is continuous at all points of $\mathbb{R}^2$ provided $y \neq 0$. Therefore, h is continuous on the set $\{(x, y): y \neq 0\}$.

Related Exercises 48–49 ◄

Functions of Three Variables

The work we have done with limits and continuity of functions of two variables extends to functions of three or more variables. Specifically, the limit laws of Theorem 15.2 apply to functions of the form $w = f(x, y, z)$. Polynomials and rational functions are continuous at all points of their domains, and limits of these functions may be evaluated by direct substitution at all points of their domains. Compositions of continuous functions of the form $f(g(x, y, z))$ are also continuous at points at which $g(x, y, z)$ is in the domain of f.

EXAMPLE 6 Functions of three variables

a. Evaluate $\displaystyle \lim_{(x, y, z) \to (2, \pi/2, 0)} \frac{x^2 \sin y}{z^2 + 4}$.

b. Find the points at which $h(x, y, z) = \sqrt{x^2 + y^2 + z^2 - 1}$ is continuous.

SOLUTION

a. This function consists of products and quotients of functions that are continuous at $(2, \pi/2, 0)$. Therefore, the limit is evaluated by direct substitution:

$$\lim_{(x, y, z) \to (2, \pi/2, 0)} \frac{x^2 \sin y}{z^2 + 4} = \frac{2^2 \sin (\pi/2)}{0^2 + 4} = 1.$$

b. This function is a composition in which the outer function $f(u) = \sqrt{u}$ is continuous for $u \geq 0$. The inner function

$$g(x, y, z) = x^2 + y^2 + z^2 - 1$$

is nonnegative provided $x^2 + y^2 + z^2 \geq 1$. Therefore, h is continuous at all points on or outside the unit sphere in $\mathbb{R}^3$.

Related Exercise 55 ◄

SECTION 15.2 EXERCISES

Getting Started

1. Explain what $\displaystyle \lim_{(x, y) \to (a, b)} f(x, y) = L$ means.

2. Explain why $f(x, y)$ must approach a unique number L as (x, y) approaches (a, b) along *all* paths in the domain in order for $\displaystyle \lim_{(x, y) \to (a, b)} f(x, y)$ to exist.

3. What does it means to say that limits of polynomials may be evaluated by direct substitution?

4. Suppose (a, b) is on the boundary of the domain of f. Explain how you would determine whether $\displaystyle \lim_{(x, y) \to (a, b)} f(x, y)$ exists.

5. Explain how examining limits along multiple paths may prove the nonexistence of a limit.

6. Explain why evaluating a limit along a finite number of paths does not prove the existence of a limit of a function of several variables.

7. What three conditions must be met for a function f to be continuous at the point (a, b)?

8. Let R be the unit disk $\{(x, y): x^2 + y^2 \leq 1\}$ with $(0, 0)$ removed. Is $(0, 0)$ a boundary point of R? Is R open or closed?

9. At what points of $\mathbb{R}^2$ is a rational function of two variables continuous?

10. Evaluate $\displaystyle \lim_{(x, y, z) \to (1, 1, -1)} xy^2z^3$.

11. Evaluate $\displaystyle \lim_{(x, y) \to (5, -5)} \frac{x^2 - y^2}{x + y}$.

12. Let $f(x) = \dfrac{x^2 - 2x - y^2 + 1}{x^2 - 2x + y^2 + 1}$. Use the Two-Path Test to

show that $\displaystyle\lim_{(x,y)\to(1,0)} f(x)$ does not exist. (*Hint:* Examine

$\displaystyle\lim_{\substack{(x,y)\to(1,0)\\ \text{(along } y=0)}} f(x)$ and $\displaystyle\lim_{\substack{(x,y)\to(1,0)\\ \text{(along } x=1)}} f(x)$ first.)

Practice Exercises

13–28. Limits of functions *Evaluate the following limits.*

13. $\displaystyle\lim_{(x,y)\to(2,9)} 101$

14. $\displaystyle\lim_{(x,y)\to(1,-3)} (3x + 4y - 2)$

15. $\displaystyle\lim_{(x,y)\to(-3,3)} (4x^2 - y^2)$

16. $\displaystyle\lim_{(x,y)\to(2,-1)} (xy^8 - 3x^2y^3)$

17. $\displaystyle\lim_{(x,y)\to(0,\pi)} \dfrac{\cos xy + \sin xy}{2y}$

18. $\displaystyle\lim_{(x,y)\to(e^2,4)} \ln\sqrt{xy}$

19. $\displaystyle\lim_{(x,y)\to(2,0)} \dfrac{x^2 - 3xy^2}{x + y}$

20. $\displaystyle\lim_{(u,v)\to(1,-1)} \dfrac{10uv - 2v^2}{u^2 + v^2}$

21. $\displaystyle\lim_{(x,y)\to(6,2)} \dfrac{x^2 - 3xy}{x - 3y}$

22. $\displaystyle\lim_{(x,y)\to(1,-2)} \dfrac{y^2 + 2xy}{y + 2x}$

23. $\displaystyle\lim_{(x,y)\to(3,1)} \dfrac{x^2 - 7xy + 12y^2}{x - 3y}$

24. $\displaystyle\lim_{(x,y)\to(-1,1)} \dfrac{2x^2 - xy - 3y^2}{x + y}$

25. $\displaystyle\lim_{(x,y)\to(2,2)} \dfrac{y^2 - 4}{xy - 2x}$

26. $\displaystyle\lim_{(x,y)\to(4,5)} \dfrac{\sqrt{x + y} - 3}{x + y - 9}$

27. $\displaystyle\lim_{(x,y)\to(1,2)} \dfrac{\sqrt{y} - \sqrt{x+1}}{y - x - 1}$

28. $\displaystyle\lim_{(u,v)\to(8,8)} \dfrac{u^{1/3} - v^{1/3}}{u^{2/3} - v^{2/3}}$

29–34. Nonexistence of limits *Use the Two-Path Test to prove that the following limits do not exist.*

29. $\displaystyle\lim_{(x,y)\to(0,0)} \dfrac{x + 2y}{x - 2y}$

$z = \dfrac{x + 2y}{x - 2y}$

30. $\displaystyle\lim_{(x,y)\to(0,0)} \dfrac{4xy}{3x^2 + y^2}$

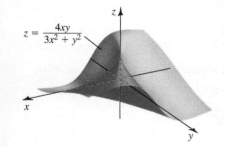

$z = \dfrac{4xy}{3x^2 + y^2}$

31. $\displaystyle\lim_{(x,y)\to(0,0)} \dfrac{y^4 - 2x^2}{y^4 + x^2}$

32. $\displaystyle\lim_{(x,y)\to(0,0)} \dfrac{x^3 - y^2}{x^3 + y^2}$

33. $\displaystyle\lim_{(x,y)\to(0,0)} \dfrac{y^3 + x^3}{xy^2}$

34. $\displaystyle\lim_{(x,y)\to(0,0)} \dfrac{y}{\sqrt{x^2 - y^2}}$

35–54. Continuity *At what points of $\mathbb{R}^2$ are the following functions continuous?*

35. $f(x,y) = x^2 + 2xy - y^3$

36. $f(x,y) = \dfrac{xy}{x^2y^2 + 1}$

37. $p(x,y) = \dfrac{4x^2y^2}{x^4 + y^2}$

38. $S(x,y) = \dfrac{2xy}{x^2 - y^2}$

39. $f(x,y) = \dfrac{2}{x(y^2 + 1)}$

40. $f(x,y) = \dfrac{x^2 + y^2}{x(y^2 - 1)}$

41. $f(x,y) = \begin{cases} \dfrac{xy}{x^2 + y^2} & \text{if } (x,y) \neq (0,0) \\ 0 & \text{if } (x,y) = (0,0) \end{cases}$

42. $f(x,y) = \begin{cases} \dfrac{y^4 - 2x^2}{y^4 + x^2} & \text{if } (x,y) \neq (0,0) \\ 0 & \text{if } (x,y) = (0,0) \end{cases}$

43. $f(x,y) = \sqrt{x^2 + y^2}$

44. $f(x,y) = e^{x^2 + y^2}$

45. $f(x,y) = \sin xy$

46. $g(x,y) = \ln(x - y)$

47. $h(x,y) = \cos(x + y)$

48. $p(x,y) = e^{x-y}$

49. $f(x,y) = \ln(x^2 + y^2)$

50. $f(x,y) = \sqrt{4 - x^2 - y^2}$

51. $g(x,y) = \sqrt[3]{x^2 + y^2 - 9}$

52. $h(x,y) = \dfrac{\sqrt{x - y}}{4}$

53. $f(x,y) = \begin{cases} \dfrac{\sin(x^2 + y^2)}{x^2 + y^2} & \text{if } (x,y) \neq (0,0) \\ 1 & \text{if } (x,y) = (0,0) \end{cases}$

54. $f(x,y) = \begin{cases} \dfrac{1 - \cos(x^2 + y^2)}{x^2 + y^2} & \text{if } (x,y) \neq (0,0) \\ 0 & \text{if } (x,y) = (0,0) \end{cases}$

55–60. Limits of functions of three variables *Evaluate the following limits.*

55. $\displaystyle\lim_{(x,y,z)\to(1,\ln 2,3)} ze^{xy}$

56. $\displaystyle\lim_{(x,y,z)\to(0,1,0)} \ln(1 + y)\,e^{xz}$

57. $\displaystyle\lim_{(x,y,z)\to(1,1,1)} \dfrac{yz - xy - xz - x^2}{yz + xy + xz - y^2}$

58. $\displaystyle\lim_{(x,y,z)\to(1,1,1)} \dfrac{x - \sqrt{xz} - \sqrt{xy} + \sqrt{yz}}{x - \sqrt{xz} + \sqrt{xy} - \sqrt{yz}}$

59. $\displaystyle\lim_{(x,y,z)\to(1,1,1)} \dfrac{x^2 + xy - xz - yz}{x - z}$

60. $\displaystyle\lim_{(x,y,z)\to(1,-1,1)} \dfrac{xz + 5x + yz + 5y}{x + y}$

61. Explain why or why not Determine whether the following statements are true and give an explanation or counterexample.

a. If the limits $\lim\limits_{(x,0)\to(0,0)} f(x, 0)$ and $\lim\limits_{(0,y)\to(0,0)} f(0, y)$ exist and equal L, then $\lim\limits_{(x,y)\to(0,0)} f(x, y) = L$.

b. If $\lim\limits_{(x,y)\to(a,b)} f(x, y)$ equals a finite number L, then f is continuous at (a, b).

c. If f is continuous at (a, b), then $\lim\limits_{(x,y)\to(a,b)} f(x, y)$ exists.

d. If P is a boundary point of the domain of f, then P is in the domain of f.

62–76. Miscellaneous limits *Use the method of your choice to evaluate the following limits.*

62. $\lim\limits_{(x,y)\to(0,0)} \dfrac{y^2}{x^8 + y^2}$

63. $\lim\limits_{(x,y)\to(0,1)} \dfrac{y \sin x}{x(y + 1)}$

64. $\lim\limits_{(x,y)\to(1,1)} \dfrac{x^2 + xy - 2y^2}{2x^2 - xy - y^2}$

65. $\lim\limits_{(x,y)\to(1,0)} \dfrac{y \ln y}{x}$

66. $\lim\limits_{(x,y)\to(0,0)} \dfrac{|xy|}{xy}$

67. $\lim\limits_{(x,y)\to(0,0)} \dfrac{|x - y|}{|x + y|}$

68. $\lim\limits_{(u,v)\to(-1,0)} \dfrac{uve^{-v}}{u^2 + v^2}$

69. $\lim\limits_{(x,y)\to(2,0)} \dfrac{1 - \cos y}{xy^2}$

70. $\lim\limits_{(x,y)\to(4,0)} x^2 y \ln xy$

71. $\lim\limits_{(x,y)\to(1,0)} \dfrac{\sin xy}{xy}$

72. $\lim\limits_{(x,y)\to(0,\pi/2)} \dfrac{1 - \cos xy}{4x^2y^3}$

73. $\lim\limits_{(x,y)\to(0,2)} (2xy)^{xy}$

74. $\lim\limits_{(x,y)\to(3,3)} \dfrac{x^2 + 2xy - 6x + y^2 - 6y}{x + y - 6}$

75. $\lim\limits_{(x,y)\to(1,2)} \dfrac{x^2 + 2xy - x + y^2 - y - 6}{x + y - 3}$

76. $\lim\limits_{(x,y)\to(0,0)} \tan^{-1} \dfrac{(2 + (x + y)^2 + (x - y)^2)}{2e^{x^2+y^2}}$

77. Piecewise function Let

$$f(x, y) = \begin{cases} \dfrac{\sin(x^2 + y^2 - 1)}{x^2 + y^2 - 1} & \text{if } x^2 + y^2 \neq 1 \\ b & \text{if } x^2 + y^2 = 1. \end{cases}$$

Find the value of b for which f is continuous at all points in $\mathbb{R}^2$.

78. Piecewise function Let

$$f(x, y) = \begin{cases} \dfrac{1 + 2xy - \cos xy}{xy} & \text{if } xy \neq 0 \\ a & \text{if } xy = 0. \end{cases}$$

Find the value of a for which f is continuous at all points in $\mathbb{R}^2$.

79–81. Limits using polar coordinates *Limits at $(0, 0)$ may be easier to evaluate by converting to polar coordinates. Remember that the same limit must be obtained as $r \to 0$ along all paths in the domain to $(0, 0)$. Evaluate the following limits or state that they do not exist.*

79. $\lim\limits_{(x,y)\to(0,0)} \dfrac{x^2 y}{x^2 + y^2}$

80. $\lim\limits_{(x,y)\to(0,0)} \dfrac{x - y}{\sqrt{x^2 + y^2}}$

81. $\lim\limits_{(x,y)\to(0,0)} \dfrac{x^2 + y^2 + x^2y^2}{x^2 + y^2}$

Explorations and Challenges

82. Sine limits Verify that $\lim\limits_{(x,y)\to(0,0)} \dfrac{\sin x + \sin y}{x + y} = 1$.

83. Nonexistence of limits Show that $\lim\limits_{(x,y)\to(0,0)} \dfrac{ax^m y^n}{bx^{m+n} + cy^{m+n}}$ does not exist when a, b, and c are nonzero real numbers and m and n are positive integers.

84. Nonexistence of limits Show that $\lim\limits_{(x,y)\to(0,0)} \dfrac{ax^{2(p-n)} y^n}{bx^{2p} + cy^p}$ does not exist when a, b, and c are nonzero real numbers and n and p are positive integers with $p \geq n$.

85. Filling in a function value The domain of $f(x, y) = e^{-1/(x^2+y^2)}$ excludes $(0, 0)$. How should f be defined at $(0, 0)$ to make it continuous there?

86. Limit proof Use the formal definition of a limit to prove that $\lim\limits_{(x,y)\to(a,b)} y = b$. (*Hint:* Take $\delta = \varepsilon$.)

87. Limit proof Use the formal definition of a limit to prove that $\lim\limits_{(x,y)\to(a,b)} (x + y) = a + b$. (*Hint:* Take $\delta = \varepsilon/2$.)

88. Proof of Limit Law 1 Use the formal definition of a limit to prove that

$$\lim\limits_{(x,y)\to(a,b)} (f(x, y) + g(x, y)) = \lim\limits_{(x,y)\to(a,b)} f(x, y) + \lim\limits_{(x,y)\to(a,b)} g(x, y).$$

89. Proof of Limit Law 3 Use the formal definition of a limit to prove that $\lim\limits_{(x,y)\to(a,b)} cf(x, y) = c \lim\limits_{(x,y)\to(a,b)} f(x, y)$.

QUICK CHECK ANSWERS

1. The limit exists only for (a).   **2.** $\{(x, y): x^2 + y^2 < 2\}$
3. If a factor of x is first canceled, then the limit may be evaluated by substitution.   **4.** If the left and right limits at a point are not equal, then the two-sided limit does not exist.
5. (a) and (b) are continuous at $(0, 0)$. ◄

15.3 Partial Derivatives

The derivative of a function of one variable, $y = f(x)$, measures the rate of change of y with respect to x, and it gives slopes of tangent lines. The analogous idea for functions of several variables presents a new twist: Derivatives may be defined with respect to any of the independent variables. For example, we can compute the derivative of $f(x, y)$ with respect to x or y. The resulting derivatives are called *partial derivatives*; they still represent rates of change and they are associated with slopes of tangents. Therefore, much of what you have learned about derivatives applies to functions of several variables. However, much is also different.

Derivatives with Two Variables

Consider a function f defined on a domain D in the xy-plane. Suppose f represents the elevation of the land (above sea level) over D. Imagine that you are on the surface $z = f(x, y)$ at the point $(a, b, f(a, b))$ and you are asked to determine the slope of the surface where you are standing. Your answer should be, *it depends*!

Figure 15.30a shows a function that resembles the landscape in Figure 15.30b. Suppose you are standing at the point $P(0, 0, f(0, 0))$, which lies on the pass or the saddle. The surface behaves differently depending on the direction in which you walk. If you walk east (positive x-direction), the elevation increases and your path takes you upward on the surface. If you walk north (positive y-direction), the elevation decreases and your path takes you downward on the surface. In fact, in every direction you walk from the point P, the function values change at different rates. So how should the slope or the rate of change at a given point be defined?

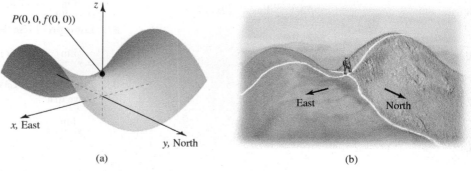

(a)   (b)

Figure 15.30

The answer to this question involves *partial derivatives*, which arise when we hold all but one independent variable fixed and then compute an ordinary derivative with respect to the remaining variable. Suppose we move along the surface $z = f(x, y)$, starting at the point $(a, b, f(a, b))$ in such a way that $y = b$ is fixed and only x varies. The resulting path is a curve (a trace) on the surface that varies in the x-direction (Figure 15.31). This curve is the intersection of the surface with the vertical plane $y = b$; it is described by $z = f(x, b)$, which is a function of the single variable x. We know how to compute the slope of this curve: It is the ordinary derivative of $f(x, b)$ with respect to x. This derivative is called the *partial derivative of f with respect to x*, denoted $\partial f / \partial x$ or f_x. When evaluated at (a, b), its value is defined by the limit

$$f_x(a, b) = \lim_{h \to 0} \frac{f(a + h, b) - f(a, b)}{h},$$

provided this limit exists. Notice that the y-coordinate is fixed at $y = b$ in this limit. If we replace (a, b) with the variable point (x, y), then f_x becomes a function of x and y.

In a similar way, we can move along the surface $z = f(x, y)$ from the point $(a, b, f(a, b))$ in such a way that $x = a$ is fixed and only y varies. Now the result is

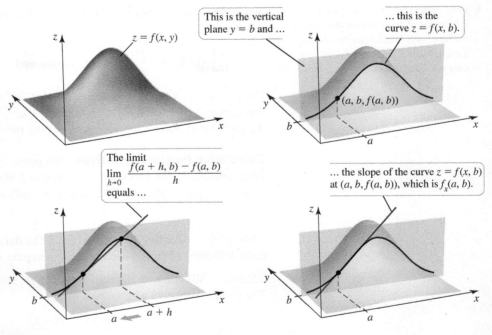

Figure 15.31

a trace described by $z = f(a, y)$, which is the intersection of the surface and the plane $x = a$ (**Figure 15.32**). The slope of this curve at (a, b) is given by the ordinary derivative of $f(a, y)$ with respect to y. This derivative is called the *partial derivative of f with respect to y*, denoted $\partial f / \partial y$ or f_y. When evaluated at (a, b), it is defined by the limit

$$f_y(a, b) = \lim_{h \to 0} \frac{f(a, b + h) - f(a, b)}{h},$$

provided this limit exists. If we replace (a, b) with the variable point (x, y), then f_y becomes a function of x and y.

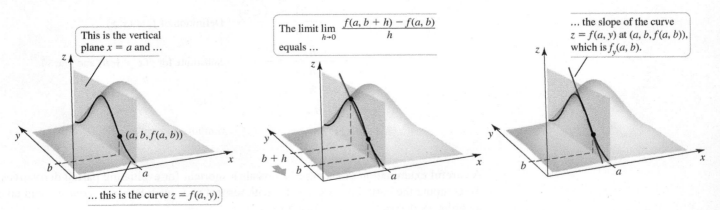

Figure 15.32

DEFINITION Partial Derivatives

The **partial derivative of f with respect to x at the point (a, b)** is

$$f_x(a, b) = \lim_{h \to 0} \frac{f(a + h, b) - f(a, b)}{h}.$$

The **partial derivative of f with respect to y at the point (a, b)** is

$$f_y(a, b) = \lim_{h \to 0} \frac{f(a, b + h) - f(a, b)}{h},$$

provided these limits exist.

▶ Recall that f' is a function, while $f'(a)$ is the value of the derivative at $x = a$. In the same way, f_x and f_y are functions of x and y, while $f_x(a, b)$ and $f_y(a, b)$ are their values at (a, b).

Notation   The partial derivatives evaluated at a point (a, b) are denoted in any of the following ways:

$$\frac{\partial f}{\partial x}(a, b) = \left.\frac{\partial f}{\partial x}\right|_{(a, b)} = f_x(a, b) \quad \text{and} \quad \frac{\partial f}{\partial y}(a, b) = \left.\frac{\partial f}{\partial y}\right|_{(a, b)} = f_y(a, b).$$

Notice that the d in the ordinary derivative df/dx has been replaced with ∂ in the partial derivatives $\partial f/\partial x$ and $\partial f/\partial y$. The notation $\partial/\partial x$ is an instruction or operator: It says, "Take the partial derivative with respect to x of the function that follows."

Calculating Partial Derivatives   We begin by calculating partial derivatives using the limit definition. The procedure in Example 1 should look familiar. It echoes the method used in Chapter 3 when we first introduced ordinary derivatives.

EXAMPLE 1   Partial derivatives from the definition Suppose $f(x, y) = x^2y$. Use the limit definition of partial derivatives to compute $f_x(x, y)$ and $f_y(x, y)$.

SOLUTION We compute the partial derivatives at an arbitrary point (x, y) in the domain. The partial derivative with respect to x is

$$
\begin{aligned}
f_x(x, y) &= \lim_{h \to 0}\frac{f(x + h, y) - f(x, y)}{h} && \text{Definition of } f_x \text{ at } (x, y)\\[2mm]
&= \lim_{h \to 0}\frac{(x + h)^2 y - x^2 y}{h} && \text{Substitute for } f(x + h, y) \text{ and } f(x, y).\\[2mm]
&= \lim_{h \to 0}\frac{(x^2 + 2xh + h^2 - x^2)y}{h} && \text{Factor and expand.}\\[2mm]
&= \lim_{h \to 0}(2x + h)y && \text{Simplify and cancel } h.\\[2mm]
&= 2xy. && \text{Evaluate limit.}
\end{aligned}
$$

In a similar way, the partial derivative with respect to y is

$$
\begin{aligned}
f_y(x, y) &= \lim_{h \to 0}\frac{f(x, y + h) - f(x, y)}{h} && \text{Definition of } f_y \text{ at } (x, y)\\[2mm]
&= \lim_{h \to 0}\frac{x^2(y + h) - x^2 y}{h} && \text{Substitute for } f(x, y + h) \text{ and } f(x, y).\\[2mm]
&= \lim_{h \to 0}\frac{x^2(y + h - y)}{h} && \text{Factor.}\\[2mm]
&= x^2. && \text{Simplify and evaluate limit.}
\end{aligned}
$$

Related Exercise 11 ◀

A careful examination of Example 1 reveals a shortcut for evaluating partial derivatives. To compute the partial derivative of f with respect to x, we treat y as a constant and take an ordinary derivative with respect to x:

$$\frac{\partial}{\partial x}(x^2 y) = y\,\underbrace{\frac{\partial}{\partial x}(x^2)}_{2x} = 2xy. \quad \text{Treat } y \text{ as a constant.}$$

Similarly, we treat x (and therefore x^2) as a constant to evaluate the partial derivative of f with respect to y:

$$\frac{\partial}{\partial y}(x^2 y) = x^2\,\underbrace{\frac{\partial}{\partial y}(y)}_{1} = x^2. \quad \text{Treat } x \text{ as a constant.}$$

The next two examples illustrate the process.

EXAMPLE 2 **Partial derivatives** Let $f(x, y) = x^3 - y^2 + 4$.

a. Compute $\dfrac{\partial f}{\partial x}$ and $\dfrac{\partial f}{\partial y}$.

b. Evaluate each derivative at $(2, -4)$.

SOLUTION

a. We compute the partial derivative with respect to x assuming y is a constant; the Power Rule gives

$$\frac{\partial f}{\partial x} = \frac{\partial}{\partial x}(\underbrace{x^3}_{\text{variable}} - \underbrace{y^2 + 4}_{\substack{\text{constant with} \\ \text{respect to } x}}) = 3x^2 + 0 = 3x^2.$$

The partial derivative with respect to y is computed by treating x as a constant; using the Power Rule gives

$$\frac{\partial f}{\partial y} = \frac{\partial}{\partial y}(\underbrace{x^3}_{\substack{\text{constant} \\ \text{with respect} \\ \text{to } y}} - \underbrace{y^2}_{\text{variable}} + \underbrace{4}_{\text{constant}}) = -2y.$$

b. It follows that $f_x(2, -4) = (3x^2)|_{(2,-4)} = 12$ and $f_y(2, -4) = (-2y)|_{(2,-4)} = 8$.

Related Exercise 16 ◄

QUICK CHECK 1 Compute f_x and f_y for $f(x, y) = 2xy$. ◄

EXAMPLE 3 **Partial derivatives** Compute the partial derivatives of the following functions.

a. $f(x, y) = \sin xy$ **b.** $g(x, y) = x^2 e^{xy}$

SOLUTION

a. Treating y as a constant and differentiating with respect to x, we have

$$\frac{\partial f}{\partial x} = \frac{\partial}{\partial x}(\sin xy) = y \cos xy.$$

Holding x fixed and differentiating with respect to y, we have

$$\frac{\partial f}{\partial y} = \frac{\partial}{\partial y}(\sin xy) = x \cos xy.$$

➤ Recall that

$$\frac{d}{dx}(\sin 2x) = 2 \cos 2x.$$

Replacing 2 with the constant y, we have

$$\frac{\partial}{\partial x}(\sin xy) = y \cos xy.$$

b. To compute the partial derivative with respect to x, we call on the Product Rule. Holding y fixed, we have

$$\frac{\partial g}{\partial x} = \frac{\partial}{\partial x}(x^2 e^{xy})$$

$$= \frac{\partial}{\partial x}(x^2)e^{xy} + x^2 \frac{\partial}{\partial x}(e^{xy}) \quad \text{Product Rule}$$

$$= 2xe^{xy} + x^2 y e^{xy} \quad\quad\quad \text{Evaluate partial derivatives.}$$

$$= xe^{xy}(2 + xy). \quad\quad\quad \text{Simplify.}$$

Treating x as a constant, the partial derivative with respect to y is

$$\frac{\partial g}{\partial y} = \frac{\partial}{\partial y}(x^2 e^{xy}) = x^2 \underbrace{\frac{\partial}{\partial y}(e^{xy})}_{xe^{xy}} = x^3 e^{xy}.$$

➤ Because x and y are *independent* variables,

$$\frac{\partial}{\partial x}(y) = 0 \quad \text{and} \quad \frac{\partial}{\partial y}(x) = 0.$$

Related Exercises 17, 21 ◄

Higher-Order Partial Derivatives

Just as we have higher-order derivatives of functions of one variable, we also have higher-order partial derivatives. For example, given a function f and its partial derivative f_x, we can take the derivative of f_x with respect to x or with respect to y, which accounts for two of the four possible *second-order partial derivatives*. Table 15.3 summarizes the notation for second partial derivatives.

Table 15.3

Notation 1	Notation 2	What we say . . .
$\dfrac{\partial}{\partial x}\left(\dfrac{\partial f}{\partial x}\right) = \dfrac{\partial^2 f}{\partial x^2}$	$(f_x)_x = f_{xx}$	*d squared f dx squared or f-x-x*
$\dfrac{\partial}{\partial y}\left(\dfrac{\partial f}{\partial y}\right) = \dfrac{\partial^2 f}{\partial y^2}$	$(f_y)_y = f_{yy}$	*d squared f dy squared or f-y-y*
$\dfrac{\partial}{\partial x}\left(\dfrac{\partial f}{\partial y}\right) = \dfrac{\partial^2 f}{\partial x \partial y}$	$(f_y)_x = f_{yx}$	*f-y-x*
$\dfrac{\partial}{\partial y}\left(\dfrac{\partial f}{\partial x}\right) = \dfrac{\partial^2 f}{\partial y \partial x}$	$(f_x)_y = f_{xy}$	*f-x-y*

QUICK CHECK 2 Which of the following expressions are equivalent to each other: (a) f_{xy}, (b) f_{yx}, or (c) $\dfrac{\partial^2 f}{\partial y \partial x}$? Write $\dfrac{\partial^2 f}{\partial p \partial q}$ in subscript notation. ◄

The order of differentiation can make a difference in the **mixed partial derivatives** f_{xy} and f_{yx}. So it is important to use the correct notation to reflect the order in which derivatives are taken. For example, the notations $\dfrac{\partial^2 f}{\partial x \partial y}$ and f_{yx} both mean $\dfrac{\partial}{\partial x}\left(\dfrac{\partial f}{\partial y}\right)$; that is, differentiate first with respect to y, then with respect to x.

EXAMPLE 4   Second partial derivatives Find the four second partial derivatives of $f(x, y) = 3x^4 y - 2xy + 5xy^3$.

SOLUTION First, we compute

$$\frac{\partial f}{\partial x} = \frac{\partial}{\partial x}(3x^4 y - 2xy + 5xy^3) = 12x^3 y - 2y + 5y^3$$

and

$$\frac{\partial f}{\partial y} = \frac{\partial}{\partial y}(3x^4 y - 2xy + 5xy^3) = 3x^4 - 2x + 15xy^2.$$

For the second partial derivatives, we have

$$\frac{\partial^2 f}{\partial x^2} = \frac{\partial}{\partial x}\left(\frac{\partial f}{\partial x}\right) = \frac{\partial}{\partial x}(12x^3 y - 2y + 5y^3) = 36x^2 y,$$

$$\frac{\partial^2 f}{\partial y^2} = \frac{\partial}{\partial y}\left(\frac{\partial f}{\partial y}\right) = \frac{\partial}{\partial y}(3x^4 - 2x + 15xy^2) = 30xy,$$

$$\frac{\partial^2 f}{\partial x \partial y} = \frac{\partial}{\partial x}\left(\frac{\partial f}{\partial y}\right) = \frac{\partial}{\partial x}(3x^4 - 2x + 15xy^2) = 12x^3 - 2 + 15y^2, \text{ and}$$

$$\frac{\partial^2 f}{\partial y \partial x} = \frac{\partial}{\partial y}\left(\frac{\partial f}{\partial x}\right) = \frac{\partial}{\partial y}(12x^3 y - 2y + 5y^3) = 12x^3 - 2 + 15y^2.$$

QUICK CHECK 3 Compute f_{xxx} and f_{xxy} for $f(x, y) = x^3 y$. ◄

Related Exercises 39–40 ◄

Equality of Mixed Partial Derivatives Notice that the two mixed partial derivatives in Example 4 are equal; that is, $f_{xy} = f_{yx}$. It turns out that most of the functions we encounter in this text have this property. Sufficient conditions for equality of mixed partial derivatives are given in a theorem attributed to the French mathematician Alexis Clairaut (1713–1765). The proof is found in advanced texts.

> **THEOREM 15.4** (Clairaut) **Equality of Mixed Partial Derivatives**
> Assume f is defined on an open set D of $\mathbb{R}^2$, and that f_{xy} and f_{yx} are continuous throughout D. Then $f_{xy} = f_{yx}$ at all points of D.

Assuming sufficient continuity, Theorem 15.4 can be extended to higher derivatives of f. For example, $f_{xyx} = f_{xxy} = f_{yxx}$.

Functions of Three Variables

Everything we learned about partial derivatives of functions with two variables carries over to functions of three or more variables, as illustrated in Example 5.

EXAMPLE 5 **Partial derivatives with more than two variables** Find f_x, f_y, and f_z when $f(x, y, z) = e^{-xy} \cos z$.

SOLUTION To find f_x, we treat y and z as constants and differentiate with respect to x:

$$\frac{\partial f}{\partial x} = \frac{\partial}{\partial x}(\underbrace{e^{-xy}}_{\substack{y \text{ is} \\ \text{constant}}} \cdot \underbrace{\cos z}_{\text{constant}}) = -ye^{-xy}\cos z.$$

Holding x and z constant and differentiating with respect to y, we have

$$\frac{\partial f}{\partial y} = \frac{\partial}{\partial y}(\underbrace{e^{-xy}}_{\substack{x \text{ is} \\ \text{constant}}} \cdot \underbrace{\cos z}_{\text{constant}}) = -xe^{-xy}\cos z.$$

To find f_z, we hold x and y constant and differentiate with respect to z:

$$\frac{\partial f}{\partial z} = \frac{\partial}{\partial z}(\underbrace{e^{-xy}}_{\text{constant}} \cos z) = -e^{-xy}\sin z.$$

QUICK CHECK 4 Compute f_{xz} and f_{zz} for $f(x, y, z) = xyz - x^2 z + yz^2$. ◄

Related Exercises 55–56 ◄

Applications of Partial Derivatives When functions are used in realistic applications (for example, to describe velocity, pressure, investment fund balance, or population), they often involve more than one independent variable. For this reason, partial derivatives appear frequently in mathematical modeling.

EXAMPLE 6 **Ideal Gas Law** The pressure P, volume V, and temperature T of an ideal gas are related by the equation $PV = kT$, where $k > 0$ is a constant depending on the amount of gas.

a. Determine the rate of change of the pressure with respect to the volume at constant temperature. Interpret the result.

b. Determine the rate of change of the pressure with respect to the temperature at constant volume. Interpret the result.

c. Explain these results using level curves.

▶ Implicit differentiation can also be used with partial derivatives. Instead of solving for P, we could differentiate both sides of $PV = kT$ with respect to V holding T fixed. Using the Product Rule, $P_V V + P = 0$, which implies that $P_V = -P/V$. Substituting $P = kT/V$, we have $P_V = -kT/V^2$.

▶ In the Ideal Gas Law, temperature is a positive variable because it is measured in kelvins.

SOLUTION Expressing the pressure as a function of volume and temperature, we have

$$P = k\frac{T}{V}.$$

a. We find the partial derivative $\partial P/\partial V$ by holding T constant and differentiating P with respect to V:

$$\frac{\partial P}{\partial V} = \frac{\partial}{\partial V}\left(k\frac{T}{V}\right) = kT\frac{\partial}{\partial V}(V^{-1}) = -\frac{kT}{V^2}.$$

Recognizing that P, V, and T are always positive, we see that $\dfrac{\partial P}{\partial V} < 0$, which means that the pressure is a decreasing function of volume at a constant temperature.

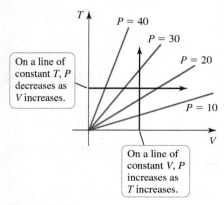

On a line of constant T, P decreases as V increases.

On a line of constant V, P increases as T increases.

Figure 15.33

QUICK CHECK 5 Explain why, in Figure 15.33, the slopes of the level curves increase as the pressure increases. ◄

b. The partial derivative $\partial P/\partial T$ is found by holding V constant and differentiating P with respect to T:

$$\frac{\partial P}{\partial T} = \frac{\partial}{\partial T}\left(k\,\frac{T}{V}\right) = \frac{k}{V}.$$

In this case, $\partial P/\partial T > 0$, which says that the pressure is an increasing function of temperature at constant volume.

c. The level curves (Section 15.1) of the pressure function are curves in the VT-plane that satisfy $k\,\dfrac{T}{V} = P_0$, where P_0 is a constant. Solving for T, the level curves are given by $T = \dfrac{1}{k}\,P_0 V$. Because $\dfrac{P_0}{k}$ is a positive constant, the level curves are lines in the first quadrant (**Figure 15.33**) with slope P_0/k. The fact that $\dfrac{\partial P}{\partial V} < 0$ (from part (a)) means that if we hold $T > 0$ fixed and move in the direction of increasing V on a *horizontal* line, we cross level curves corresponding to decreasing pressures. Similarly, $\dfrac{\partial P}{\partial T} > 0$ (from part (b)) means that if we hold $V > 0$ fixed and move in the direction of increasing T on a *vertical* line, we cross level curves corresponding to increasing pressures.

Related Exercise 69 ◄

Differentiability

We close this section with a technical matter that bears on the remainder of the chapter. Although we know how to compute partial derivatives of a function of several variables, we have not said what it means for such a function to be *differentiable* at a point. It is tempting to conclude that if the partial derivatives f_x and f_y exist at a point, then f is differentiable there. However, it is not so simple.

Recall that a function f of one variable is differentiable at $x = a$ provided the limit

$$f'(a) = \lim_{\Delta x \to 0} \frac{f(a + \Delta x) - f(a)}{\Delta x}$$

exists. If f is differentiable at a, it means that the curve is smooth at the point $(a, f(a))$ (no jumps, corners, or cusps); furthermore, the curve has a unique tangent line at that point with slope $f'(a)$. Differentiability for a function of several variables should carry the same properties: The surface should be smooth at the point in question, and something analogous to a unique tangent line should exist at the point.

Staying with the one-variable case, we define the quantity

$$\varepsilon = \underbrace{\frac{f(a + \Delta x) - f(a)}{\Delta x}}_{\text{slope of secant line}} - \underbrace{f'(a),}_{\substack{\text{slope of}\\ \text{tangent line}}}$$

where ε is viewed as a function of Δx. Notice that ε is the difference between the slopes of secant lines and the slope of the tangent line at the point $(a, f(a))$. If f is differentiable at a, then this difference approaches zero as $\Delta x \to 0$; therefore, $\lim\limits_{\Delta x \to 0} \varepsilon = 0$. Multiplying both sides of the definition of ε by Δx gives

$$\varepsilon\,\Delta x = f(a + \Delta x) - f(a) - f'(a)\,\Delta x.$$

Rearranging, we have the change in the function $y = f(x)$:

$$\Delta y = f(a + \Delta x) - f(a) = f'(a)\,\Delta x + \underbrace{\varepsilon\,\Delta x}_{\varepsilon \to 0 \text{ as } \Delta x \to 0}.$$

► Notice that $f'(a)\,\Delta x$ is the approximate change in the function given by a linear approximation.

This expression says that in the one-variable case, if f is differentiable at a, then the change in f between a and a nearby point $a + \Delta x$ is represented by $f'(a)\,\Delta x$ plus a quantity $\varepsilon\,\Delta x$, where $\lim\limits_{\Delta x \to 0} \varepsilon = 0$.

The analogous requirement with several variables is the definition of differentiability for functions of two (or more) variables.

DEFINITION Differentiability

The function $z = f(x, y)$ is **differentiable at (a, b)** provided $f_x(a, b)$ and $f_y(a, b)$ exist and the change $\Delta z = f(a + \Delta x, b + \Delta y) - f(a, b)$ equals

$$\Delta z = f_x(a, b)\,\Delta x + f_y(a, b)\,\Delta y + \varepsilon_1 \Delta x + \varepsilon_2 \Delta y,$$

where for fixed a and b, ε_1 and ε_2 are functions that depend only on Δx and Δy, with $(\varepsilon_1, \varepsilon_2) \to (0, 0)$ as $(\Delta x, \Delta y) \to (0, 0)$. A function is **differentiable** on an open set R if it is differentiable at every point of R.

Several observations are needed here. First, the definition extends to functions of more than two variables. Second, we show how differentiability is related to linear approximation and the existence of a *tangent plane* in Section 15.6. Finally, the conditions of the definition are generally difficult to verify. The following theorem may be useful in checking differentiability.

THEOREM 15.5 Conditions for Differentiability

Suppose the function f has partial derivatives f_x and f_y defined on an open set containing (a, b), with f_x and f_y continuous at (a, b). Then f is differentiable at (a, b).

As shown in Example 7, the existence of f_x and f_y at (a, b) is not enough to ensure differentiability of f at (a, b). However, by Theorem 15.5, if f_x and f_y are continuous at (a, b) (and defined in an open set containing (a, b)), then we can conclude f is differentiable there. Polynomials and rational functions are differentiable at all points of their domains, as are compositions of exponential, logarithmic, and trigonometric functions with other differentiable functions. The proof of this theorem is given in Appendix A.

We close with the analog of Theorem 3.1, which states that differentiability implies continuity.

THEOREM 15.6 Differentiable Implies Continuous

If a function f is differentiable at (a, b), then it is continuous at (a, b).

Proof: By the definition of differentiability,

$$\Delta z = f_x(a, b)\,\Delta x + f_y(a, b)\,\Delta y + \varepsilon_1 \Delta x + \varepsilon_2 \Delta y,$$

where $(\varepsilon_1, \varepsilon_2) \to (0, 0)$ as $(\Delta x, \Delta y) \to (0, 0)$. Because f is assumed to be differentiable, we see that as Δx and Δy approach 0,

$$\lim_{(\Delta x, \Delta y) \to (0, 0)} \Delta z = 0.$$

Also, because $\Delta z = f(a + \Delta x, b + \Delta y) - f(a, b)$, it follows that

$$\lim_{(\Delta x, \Delta y) \to (0, 0)} f(a + \Delta x, b + \Delta y) = f(a, b),$$

which implies continuity of f at (a, b).   ◀

> Recall that continuity requires that
> $$\lim_{(x, y) \to (a, b)} f(x, y) = f(a, b),$$
> which is equivalent to
> $$\lim_{(\Delta x, \Delta y) \to (0, 0)} f(a + \Delta x, b + \Delta y) = f(a, b).$$

EXAMPLE 7 A nondifferentiable function Discuss the differentiability and continuity of the function

$$f(x, y) = \begin{cases} \dfrac{3xy}{x^2 + y^2} & \text{if } (x, y) \neq (0, 0) \\ 0 & \text{if } (x, y) = (0, 0). \end{cases}$$

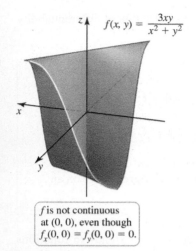

$f(x, y) = \dfrac{3xy}{x^2 + y^2}$

f is not continuous
at $(0, 0)$, even though
$f_x(0, 0) = f_y(0, 0) = 0$.

Figure 15.34

▶ The relationships between the existence and continuity of partial derivatives and whether a function is differentiable are further explored in Exercises 96–97.

SOLUTION As a rational function, f is continuous and differentiable at all points $(x, y) \neq (0, 0)$. The interesting behavior occurs at the origin. Using calculations similar to those in Example 4 in Section 15.2, it can be shown that if the origin is approached along the line $y = mx$, then

$$\lim_{\substack{(x, y) \to (0, 0) \\ (\text{along } y = mx)}} \frac{3xy}{x^2 + y^2} = \frac{3m}{m^2 + 1}.$$

Therefore, the value of the limit depends on the direction of approach, which implies that the limit does not exist, and f is not continuous at $(0, 0)$. By Theorem 15.6, it follows that f is not differentiable at $(0, 0)$. **Figure 15.34** shows the discontinuity of f at the origin. Let's look at the first partial derivatives of f at $(0, 0)$. A short calculation shows that

$$f_x(0, 0) = \lim_{h \to 0} \frac{f(0 + h, 0) - f(0, 0)}{h} = \lim_{h \to 0} \frac{0 - 0}{h} = 0,$$

$$f_y(0, 0) = \lim_{h \to 0} \frac{f(0, 0 + h) - f(0, 0)}{h} = \lim_{h \to 0} \frac{0 - 0}{h} = 0.$$

Despite the fact that its first partial derivatives exist at $(0, 0)$, f is not differentiable at $(0, 0)$. As noted earlier, the existence of first partial derivatives at a point is not enough to ensure differentiability at that point.

Related Exercises 77–78 ◀

SECTION 15.3 EXERCISES

Getting Started

1. Suppose you are standing on the surface $z = f(x, y)$ at the point $(a, b, f(a, b))$. Interpret the meaning of $f_x(a, b)$ and $f_y(a, b)$ in terms of slopes or rates of change.

2. Let $f(x, y) = 3x^2 + y^3$.
   a. Compute f_x and f_y.
   b. Evaluate each derivative at $(1, 3)$.
   c. Find the four second partial derivatives of f.

3. Given the graph of a function $z = f(x, y)$ and its traces in the planes $x = 4$, $y = 1$, and $y = 3$ (see figure), determine whether the following partial derivatives are positive or negative.

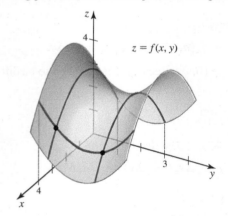

$z = f(x, y)$

a. $f_x(4, 1)$ b. $f_y(4, 1)$ c. $f_x(4, 3)$ d. $f_y(4, 3)$

4. Find f_x and f_y when $f(x, y) = y^8 + 2x^6 + 2xy$.

5. Find f_x and f_y when $f(x, y) = 3x^2y + 2$.

6. Find the four second partial derivatives of $f(x, y) = x^2y^3$.

7. Verify that $f_{xy} = f_{yx}$, for $f(x, y) = 2x^3 + 3y^2 + 1$.

8. Verify that $f_{xy} = f_{yx}$, for $f(x, y) = xe^y$.

9. Find f_x, f_y, and f_z, for $f(x, y, z) = xy + xz + yz$.

10. The volume of a right circular cylinder with radius r and height h is $V = \pi r^2 h$. Is the volume an increasing or decreasing function of the radius at a fixed height (assume $r > 0$ and $h > 0$)?

Practice Exercises

11–14. Evaluating partial derivatives using limits *Use the limit definition of partial derivatives to evaluate $f_x(x, y)$ and $f_y(x, y)$ for the following functions.*

11. $f(x, y) = 5xy$

12. $f(x, y) = x + y^2 + 4$

13. $f(x, y) = \dfrac{x}{y}$

14. $f(x, y) = \sqrt{xy}$

15–37. Partial derivatives *Find the first partial derivatives of the following functions.*

15. $f(x, y) = xe^y$

16. $f(x, y) = 4x^3y^2 + 3x^2y^3 + 10$

17. $f(x, y) = e^{x^2y}$

18. $f(x, y) = (3xy + 4y^2 + 1)^5$

19. $f(w, z) = \dfrac{w}{w^2 + z^2}$

20. $f(s, t) = \dfrac{s - t}{s + t}$

21. $f(x, y) = x \cos xy$

22. $f(x, y) = \tan^{-1} \dfrac{x^2}{y^2}$

23. $s(y, z) = z^2 \tan yz$

24. $g(x, z) = x \ln(z^2 + x^2)$

25. $G(s, t) = \dfrac{\sqrt{st}}{s + t}$

26. $F(p, q) = \sqrt{p^2 + pq + q^2}$

27. $f(x, y) = x^{2y}$

28. $g(x, y) = \cos^5(x^2 y^3)$

29. $h(x, y) = x - \sqrt{x^2 - 4y}$

30. $h(u, v) = \sqrt{\dfrac{uv}{u - v}}$

31. $f(x, y) = \displaystyle\int_x^{y^3} e^{t^2} dt$

32. $g(x, y) = y \sin^{-1} \sqrt{xy}$

33. $f(x, y) = 1 - \tan^{-1}(x^2 + y^2)$

34. $f(x, y) = \ln(1 + e^{-xy})$

35. $h(x, y) = (1 + 2y)^x$

36. $f(x, y) = 1 - \cos(2(x + y)) + \cos^2(x + y)$

37. $f(x, y) = \displaystyle\int_x^y h(s)\, ds$, where h is continuous for all real numbers

38–47. Second partial derivatives *Find the four second partial derivatives of the following functions.*

38. $f(x, y) = x^2 \sin y$

39. $h(x, y) = x^3 + xy^2 + 1$

40. $f(x, y) = 2x^5 y^2 + x^2 y$

41. $f(x, y) = y^3 \sin 4x$

42. $f(x, y) = \sin^2(x^3 y)$

43. $p(u, v) = \ln(u^2 + v^2 + 4)$

44. $Q(r, s) = \dfrac{e^{r^3 s}}{s}$

45. $F(r, s) = re^s$

46. $H(x, y) = \sqrt{4 + x^2 + y^2}$

47. $f(x, y) = \tan^{-1}(x^3 y^2)$

48–53. Equality of mixed partial derivatives *Verify that $f_{xy} = f_{yx}$ for the following functions.*

48. $f(x, y) = 3x^2 y^{-1} - 2x^{-1} y^2$

49. $f(x, y) = e^{x+y}$

50. $f(x, y) = \sqrt{xy}$

51. $f(x, y) = \cos xy$

52. $f(x, y) = e^{\sin xy}$

53. $f(x, y) = (2x - y^3)^4$

54–62. Partial derivatives with more than two variables *Find the first partial derivatives of the following functions.*

54. $G(r, s, t) = \sqrt{rs + rt + st}$

55. $h(x, y, z) = \cos(x + y + z)$

56. $g(x, y, z) = 2x^2 y - 3xz^4 + 10y^2 z^2$

57. $F(u, v, w) = \dfrac{u}{v + w}$

58. $Q(x, y, z) = \tan xyz$

59. $G(r, s, t) = \sqrt{rs^3 t^5}$

60. $g(w, x, y, z) = \cos(w + x) \sin(y - z)$

61. $h(w, x, y, z) = \dfrac{wz}{xy}$

62. $F(w, x, y, z) = w\sqrt{x + 2y + 3z}$

63. Exploiting patterns Let $R(t) = \dfrac{at + b}{ct + d}$ and

$g(x, y, z) = \dfrac{4x - 2y - 2z}{-6x + 3y - 3z}$.

a. Verify that $R'(t) = \dfrac{ad - bc}{(ct + d)^2}$.

b. Use the derivative $R'(t)$ to find the first partial derivatives of g.

64–67. Estimating partial derivatives from a table *The following table shows values of a function $f(x, y)$ for values of x from 2 to 2.5 and values of y from 3 to 3.5. Use this table to estimate the values of the following partial derivatives.*

y \ x	2	2.1	2.2	2.3	2.4	2.5
3	4.243	4.347	4.450	4.550	4.648	4.743
3.1	4.384	4.492	4.598	4.701	4.802	4.902
3.2	4.525	4.637	4.746	4.853	4.957	5.060
3.3	4.667	4.782	4.895	5.005	5.112	5.218
3.4	4.808	4.930	5.043	5.156	5.267	5.376
3.5	4.950	5.072	5.191	5.308	5.422	5.534

64. $f_x(2, 3)$

65. $f_y(2, 3)$

66. $f_x(2.2, 3.4)$

67. $f_y(2.4, 3.3)$

68. Estimating partial derivatives from a graph Use the level curves of f (see figure) to estimate the values of f_x and f_y at $A(0.42, 0.5)$.

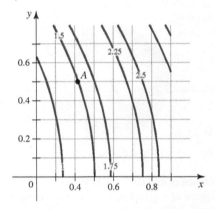

69. Gas law calculations Consider the Ideal Gas Law $PV = kT$, where $k > 0$ is a constant. Solve this equation for V in terms of P and T.

a. Determine the rate of change of the volume with respect to the pressure at constant temperature. Interpret the result.

b. Determine the rate of change of the volume with respect to the temperature at constant pressure. Interpret the result.

c. Assuming $k = 1$, draw several level curves of the volume function, and interpret the results as in Example 6.

70. Body mass index The body mass index (BMI) for an adult human is given by the function $B = \dfrac{w}{h^2}$, where w is the weight measured in kilograms and h is the height measured in meters.

a. Find the rate of change of the BMI with respect to weight at a constant height.

b. For fixed h, is the BMI an increasing or decreasing function of w? Explain.

c. Find the rate of change of the BMI with respect to height at a constant weight.

d. For fixed w, is the BMI an increasing or decreasing function of h? Explain.

71. Resistors in parallel Two resistors in an electrical circuit with resistance R_1 and R_2 wired in parallel with a constant voltage give an effective resistance of R, where $\dfrac{1}{R} = \dfrac{1}{R_1} + \dfrac{1}{R_2}$.

a. Find $\dfrac{\partial R}{\partial R_1}$ and $\dfrac{\partial R}{\partial R_2}$ by solving for R and differentiating.

b. Find $\dfrac{\partial R}{\partial R_1}$ and $\dfrac{\partial R}{\partial R_2}$ by differentiating implicitly.

c. Describe how an increase in R_1 with R_2 constant affects R.

d. Describe how a decrease in R_2 with R_1 constant affects R.

72. Spherical caps The volume of the cap of a sphere of radius r and thickness h is $V = \dfrac{\pi}{3} h^2(3r - h)$, for $0 \le h \le 2r$.

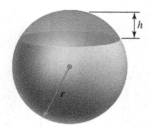

$$V = \tfrac{\pi}{3}h^2(3r - h)$$

a. Compute the partial derivatives V_h and V_r.

b. For a sphere of any radius, is the rate of change of volume with respect to r greater when $h = 0.2r$ or when $h = 0.8r$?

c. For a sphere of any radius, for what value of h is the rate of change of volume with respect to r equal to 1?

d. For a fixed radius r, for what value of h ($0 \le h \le 2r$) is the rate of change of volume with respect to h the greatest?

73–76. Heat equation *The flow of heat along a thin conducting bar is governed by the one-dimensional heat equation (with analogs for thin plates in two dimensions and for solids in three dimensions):*

$$\frac{\partial u}{\partial t} = k \frac{\partial^2 u}{\partial x^2},$$

where u is a measure of the temperature at a location x on the bar at time t and the positive constant k is related to the conductivity of the material. Show that the following functions satisfy the heat equation with $k = 1$.

73. $u(x, t) = 4e^{-4t} \cos 2x$ **74.** $u(x, t) = 10e^{-t} \sin x$

75. $u(x, t) = Ae^{-a^2 t} \cos ax$, for any real numbers a and A

76. $u(x, t) = e^{-t}(2 \sin x + 3 \cos x)$

77–78. Nondifferentiability? *Consider the following functions f.*

a. Is f continuous at $(0, 0)$?

b. Is f differentiable at $(0, 0)$?

c. If possible, evaluate $f_x(0, 0)$ and $f_y(0, 0)$.

d. Determine whether f_x and f_y are continuous at $(0, 0)$.

e. Explain why Theorems 15.5 and 15.6 are consistent with the results in parts (a)–(d).

77. $f(x, y) = \begin{cases} -\dfrac{xy}{x^2 + y^2} & \text{if } (x, y) \ne (0, 0) \\ 0 & \text{if } (x, y) = (0, 0) \end{cases}$

78. $f(x, y) = \begin{cases} \dfrac{2xy^2}{x^2 + y^4} & \text{if } (x, y) \ne (0, 0) \\ 0 & \text{if } (x, y) = (0, 0) \end{cases}$

79. Explain why or why not Determine whether the following statements are true and give an explanation or counterexample.

a. $\dfrac{\partial}{\partial x}(y^{10}) = 10y^9$.

b. $\dfrac{\partial^2}{\partial x \partial y}(\sqrt{xy}) = \dfrac{1}{\sqrt{xy}}$.

c. If f has continuous partial derivatives of all orders, then $f_{xxy} = f_{yxx}$.

80. Mixed partial derivatives

a. Consider the function $w = f(x, y, z)$. List all possible second partial derivatives that could be computed.

b. Let $f(x, y, z) = x^2 y + 2xz^2 - 3y^2 z$ and determine which second partial derivatives are equal.

c. How many second partial derivatives does $p = g(w, x, y, z)$ have?

Explorations and Challenges

81. Partial derivatives and level curves Consider the function $z = x/y^2$.

a. Compute z_x and z_y.

b. Sketch the level curves for $z = 1, 2, 3,$ and 4.

c. Move along the horizontal line $y = 1$ in the xy-plane and describe how the corresponding z-values change. Explain how this observation is consistent with z_x as computed in part (a).

d. Move along the vertical line $x = 1$ in the xy-plane and describe how the corresponding z-values change. Explain how this observation is consistent with z_y as computed in part (a).

82. Volume of a box A box with a square base of length x and height h has a volume $V = x^2 h$.

a. Compute the partial derivatives V_x and V_h.

b. For a box with $h = 1.5$ m, use linear approximation to estimate the change in volume if x increases from $x = 0.5$ m to $x = 0.51$ m.

c. For a box with $x = 0.5$ m, use linear approximation to estimate the change in volume if h decreases from $h = 1.5$ m to $h = 1.49$ m.

d. For a fixed height, does a 10% change in x always produce (approximately) a 10% change in V? Explain.

e. For a fixed base length, does a 10% change in h always produce (approximately) a 10% change in V? Explain.

83. Electric potential function The electric potential in the xy-plane associated with two positive charges, one at $(0, 1)$ with twice the magnitude of the charge at $(0, -1)$, is

$$\varphi(x, y) = \frac{2}{\sqrt{x^2 + (y - 1)^2}} + \frac{1}{\sqrt{x^2 + (y + 1)^2}}.$$

a. Compute φ_x and φ_y.

b. Describe how φ_x and φ_y behave as $x, y \to \pm \infty$.

c. Evaluate $\varphi_x(0, y)$, for all $y \ne \pm 1$. Interpret this result.

d. Evaluate $\varphi_y(x, 0)$, for all x. Interpret this result.

T 84. Cobb-Douglas production function The output Q of an economic system subject to two inputs, such as labor L and capital K, is often modeled by the Cobb-Douglas production function $Q(L, K) = cL^a K^b$. Suppose $a = \frac{1}{3}$, $b = \frac{2}{3}$, and $c = 1$.

a. Evaluate the partial derivatives Q_L and Q_K.
b. Suppose $L = 10$ is fixed and K increases from $K = 20$ to $K = 20.5$. Use linear approximation to estimate the change in Q.
c. Suppose $K = 20$ is fixed and L decreases from $L = 10$ to $L = 9.5$. Use linear approximation to estimate the change in Q.
d. Graph the level curves of the production function in the first quadrant of the LK-plane for $Q = 1, 2,$ and 3.
e. Use the graph of part (d). If you move along the vertical line $L = 2$ in the positive K-direction, how does Q change? Is this consistent with Q_K computed in part (a)?
f. Use the graph of part (d). If you move along the horizontal line $K = 2$ in the positive L-direction, how does Q change? Is this consistent with Q_L computed in part (a)?

85. An identity Show that if $f(x, y) = \dfrac{ax + by}{cx + dy}$, where $a, b, c,$ and d are real numbers with $ad - bc = 0$, then $f_x = f_y = 0$, for all x and y in the domain of f. Give an explanation.

86. Wave on a string Imagine a string that is fixed at both ends (for example, a guitar string). When plucked, the string forms a standing wave. The displacement u of the string varies with position x and with time t. Suppose it is given by $u = f(x, t) = 2 \sin(\pi x) \sin(\pi t/2)$, for $0 \le x \le 1$ and $t \ge 0$ (see figure). At a fixed point in time, the string forms a wave on $[0, 1]$. Alternatively, if you focus on a point on the string (fix a value of x), that point oscillates up and down in time.

a. What is the period of the motion in time?
b. Find the rate of change of the displacement with respect to time at a constant position (which is the vertical velocity of a point on the string).
c. At a fixed time, what point on the string is moving fastest?
d. At a fixed position on the string, when is the string moving fastest?
e. Find the rate of change of the displacement with respect to position at a constant time (which is the slope of the string).
f. At a fixed time, where is the slope of the string greatest?

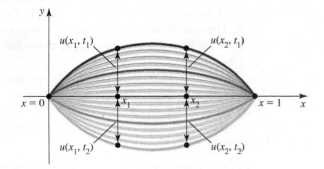

87–89. Wave equation *Traveling waves (for example, water waves or electromagnetic waves) exhibit periodic motion in both time and position. In one dimension, some types of wave motion are governed by the one-dimensional wave equation*

$$\frac{\partial^2 u}{\partial t^2} = c^2 \frac{\partial^2 u}{\partial x^2},$$

where $u(x, t)$ is the height or displacement of the wave surface at position x and time t, and c is the constant speed of the wave. Show that the following functions are solutions of the wave equation.

87. $u(x, t) = \cos(2(x + ct))$
88. $u(x, t) = 5\cos(2(x + ct)) + 3\sin(x - ct)$
89. $u(x, t) = Af(x + ct) + Bg(x - ct)$, where A and B are constants, and f and g are twice differentiable functions of one variable.

90–93. Laplace's equation *A classical equation of mathematics is Laplace's equation, which arises in both theory and applications. It governs ideal fluid flow, electrostatic potentials, and the steady-state distribution of heat in a conducting medium. In two dimensions, Laplace's equation is*

$$\frac{\partial^2 u}{\partial x^2} + \frac{\partial^2 u}{\partial y^2} = 0.$$

*Show that the following functions are **harmonic**; that is, they satisfy Laplace's equation.*

90. $u(x, y) = e^{-x}\sin y$
91. $u(x, y) = x(x^2 - 3y^2)$
92. $u(x, y) = e^{ax}\cos ay$, for any real number a
93. $u(x, y) = \tan^{-1}\left(\dfrac{y}{x - 1}\right) - \tan^{-1}\left(\dfrac{y}{x + 1}\right)$

94–95. Differentiability *Use the definition of differentiability to prove that the following functions are differentiable at $(0, 0)$. You must produce functions ε_1 and ε_2 with the required properties.*

94. $f(x, y) = x + y$ 95. $f(x, y) = xy$

96–97. Nondifferentiability? *Consider the following functions f.*

a. Is f continuous at $(0, 0)$?
b. Is f differentiable at $(0, 0)$?
c. If possible, evaluate $f_x(0, 0)$ and $f_y(0, 0)$.
d. Determine whether f_x and f_y are continuous at $(0, 0)$.
e. Explain why Theorems 15.5 and 15.6 are consistent with the results in parts (a)–(d).

96. $f(x, y) = 1 - |xy|$ 97. $f(x, y) = \sqrt{|xy|}$

98. Cauchy-Riemann equations In the advanced subject of complex variables, a function typically has the form $f(x, y) = u(x, y) + iv(x, y)$, where u and v are real-valued functions and $i = \sqrt{-1}$ is the imaginary unit. A function $f = u + iv$ is said to be *analytic* (analogous to differentiable) if it satisfies the Cauchy-Riemann equations: $u_x = v_y$ and $u_y = -v_x$.

a. Show that $f(x, y) = (x^2 - y^2) + i(2xy)$ is analytic.
b. Show that $f(x, y) = x(x^2 - 3y^2) + iy(3x^2 - y^2)$ is analytic.
c. Show that if $f = u + iv$ is analytic, then $u_{xx} + u_{yy} = 0$ and $v_{xx} + v_{yy} = 0$. Assume u and v satisfy the conditions in Theorem 15.4.

99. Derivatives of an integral Let h be continuous for all real numbers. Find f_x and f_y when $f(x, y) = \int_1^{xy} h(s)\, ds$.

QUICK CHECK ANSWERS

1. $f_x = 2y$; $f_y = 2x$ 2. (a) and (c) are the same; f_{qp}
3. $f_{xxx} = 6y$; $f_{xxy} = 6x$ 4. $f_{xz} = y - 2x$; $f_{zz} = 2y$
5. The equations of the level curves are $T = \dfrac{1}{k}P_0 V$. As the pressure P_0 increases, the slopes of these lines increase. ◄

15.4 The Chain Rule

In this section, we combine ideas based on the Chain Rule (Section 3.7) with what we know about partial derivatives (Section 15.3) to develop new methods for finding derivatives of functions of several variables. To illustrate the importance of these methods, consider the following situation.

Economists modeling manufacturing systems often work with *production functions* that relate the productivity (output) of the system to all the variables on which it depends (input). A simplified production function might take the form $P = F(L, K, R)$, where L, K, and R represent the availability of labor, capital, and natural resources, respectively. However, the variables L, K, and R may be intermediate variables that depend on other variables. For example, it might be that L is a function of the unemployment rate u, K is a function of the prime interest rate i, and R is a function of time t (seasonal availability of resources). Even in this simplified model, we see that productivity, which is the dependent variable, is ultimately related to many other variables (**Figure 15.35**). Of critical interest to an economist is how changes in one variable determine changes in other variables. For instance, if the unemployment rate increases by 0.1% and the interest rate decreases by 0.2%, what is the effect on productivity? In this section, we develop the tools needed to answer such questions.

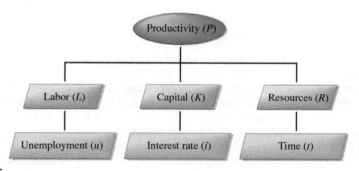

Figure 15.35

The Chain Rule with One Independent Variable

Recall the basic Chain Rule: If y is a function of u and u is a function of t, then $\dfrac{dy}{dt} = \dfrac{dy}{du}\dfrac{du}{dt}$. We first extend the Chain Rule to composite functions of the form $z = f(x, y)$, where x and y are functions of t. What is $\dfrac{dz}{dt}$?

We illustrate the relationships among the variables t, x, y, and z using a *tree diagram* (**Figure 15.36**). To find dz/dt, first notice that z depends on x, which in turn depends on t. The change in z with respect to x is the partial derivative $\partial z/\partial x$, and the change in x with respect to t is the ordinary derivative dx/dt. These derivatives appear on the corresponding branches of the tree diagram. Using the Chain Rule idea, the product of these derivatives gives the change in z with respect to t through x.

Similarly, z also depends on y. The change in z with respect to y is $\partial z/\partial y$, and the change in y with respect to t is dy/dt. The product of these derivatives, which appear on the corresponding branches of the tree, gives the change in z with respect to t through y. Summing the contributions to dz/dt along each branch of the tree leads to the following theorem, the proof of which is found in Appendix A.

THEOREM 15.7   Chain Rule (One Independent Variable)
Let z be a differentiable function of x and y on its domain, where x and y are differentiable functions of t on an interval I. Then

$$\frac{dz}{dt} = \frac{\partial z}{\partial x}\frac{dx}{dt} + \frac{\partial z}{\partial y}\frac{dy}{dt}.$$

Sidebar (left margin)

$$\frac{\partial z}{\partial x} \quad z \quad \frac{\partial z}{\partial y}$$

$$x \qquad y$$

$$\frac{dx}{dt} \qquad \frac{dy}{dt}$$

$$t \qquad t$$

$$\frac{dz}{dt} = \frac{\partial z}{\partial x}\frac{dx}{dt} + \frac{\partial z}{\partial y}\frac{dy}{dt}$$

Figure 15.36

➤ A subtle observation about notation should be made. If $z = f(x, y)$, where x and y are functions of another variable t, it is common to write $z = f(t)$ to show that z ultimately depends on t. However, these two functions denoted f are actually different. We *should* write (or at least remember) that in fact $z = F(t)$, where F is a function other than f. This distinction is often overlooked for the sake of convenience.

QUICK CHECK 1 Explain why Theorem 15.7 reduces to the Chain Rule for a function of one variable in the case that $z = f(x)$ and $x = g(t)$. ◄

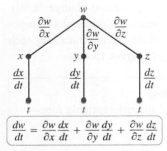

$$\frac{dw}{dt} = \frac{\partial w}{\partial x}\frac{dx}{dt} + \frac{\partial w}{\partial y}\frac{dy}{dt} + \frac{\partial w}{\partial z}\frac{dz}{dt}$$

Figure 15.37

▶ If f, x, and y are simple, as in Example 1, it is possible to substitute $x(t)$ and $y(t)$ into f, producing a function of t only, and then differentiate with respect to t. But this approach quickly becomes impractical with more complicated functions, and the Chain Rule offers a great advantage.

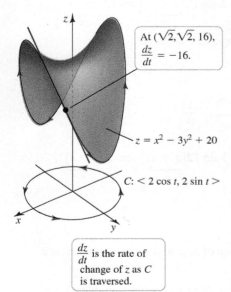

At $(\sqrt{2}, \sqrt{2}, 16)$, $\dfrac{dz}{dt} = -16.$

$z = x^2 - 3y^2 + 20$

$C: < 2\cos t, 2\sin t >$

$\dfrac{dz}{dt}$ is the rate of change of z as C is traversed.

Figure 15.38

Before presenting examples, several comments are in order.

- With $z = f(x(t), y(t))$, the dependent variable is z and the sole independent variable is t. The variables x and y are **intermediate variables**.

- The choice of notation for partial and ordinary derivatives in the Chain Rule is important. We write the ordinary derivatives dx/dt and dy/dt because x and y depend only on t. We write the partial derivatives $\partial z/\partial x$ and $\partial z/\partial y$ because z is a function of both x and y. Finally, we write dz/dt as an ordinary derivative because z ultimately depends only on t.

- Theorem 15.7 generalizes directly to functions of more than two intermediate variables (**Figure 15.37**). For example, if $w = f(x, y, z)$, where x, y, and z are functions of the single independent variable t, then

$$\frac{dw}{dt} = \frac{\partial w}{\partial x}\frac{dx}{dt} + \frac{\partial w}{\partial y}\frac{dy}{dt} + \frac{\partial w}{\partial z}\frac{dz}{dt}.$$

EXAMPLE 1   Chain Rule with one independent variable Let $z = x^2 - 3y^2 + 20$, where $x = 2\cos t$ and $y = 2\sin t$.

a. Find $\dfrac{dz}{dt}$ and evaluate it at $t = \pi/4$.

b. Interpret the result geometrically.

SOLUTION

a. Computing the intermediate derivatives and applying the Chain Rule (Theorem 15.7), we find that

$$\frac{dz}{dt} = \frac{\partial z}{\partial x}\frac{dx}{dt} + \frac{\partial z}{\partial y}\frac{dy}{dt}$$

$$= \underbrace{(2x)}_{\frac{\partial z}{\partial x}} \underbrace{(-2\sin t)}_{\frac{dx}{dt}} + \underbrace{(-6y)}_{\frac{\partial z}{\partial y}} \underbrace{(2\cos t)}_{\frac{dy}{dt}} \quad \text{Evaluate derivatives.}$$

$$= -4x\sin t - 12y\cos t \quad \text{Simplify.}$$

$$= -8\cos t\sin t - 24\sin t\cos t \quad \text{Substitute } x = 2\cos t, y = 2\sin t.$$

$$= -16\sin 2t. \quad \text{Simplify; } \sin 2t = 2\sin t\cos t.$$

Substituting $t = \pi/4$ gives $\dfrac{dz}{dt}\bigg|_{t=\pi/4} = -16.$

b. The parametric equations $x = 2\cos t$, $y = 2\sin t$, for $0 \le t \le 2\pi$, describe a circle C of radius 2 in the xy-plane. Imagine walking on the surface $z = x^2 - 3y^2 + 20$ directly above the circle C consistent with positive (counterclockwise) orientation of C. Your path rises and falls as you walk (**Figure 15.38**); the rate of change of your elevation z with respect to t is given by dz/dt. For example, when $t = \pi/4$, the corresponding point on the surface is $(\sqrt{2}, \sqrt{2}, 16)$. At that point, z decreases at a rate of -16 (by part (a)) as you walk on the surface above C.

Related Exercises 10, 12 ◀

The Chain Rule with Several Independent Variables

The ideas behind the Chain Rule of Theorem 15.7 can be modified to cover a variety of situations in which functions of several variables are composed with one another. For example, suppose z depends on two intermediate variables x and y, each of which depends on the independent variables s and t. Once again, a tree diagram (**Figure 15.39**) helps organize the relationships among variables. The dependent variable z now ultimately depends on the two independent variables s and t, so it makes sense to ask about the rates of change of z with respect to either s or t, which are $\partial z/\partial s$ and $\partial z/\partial t$, respectively.

To compute $\partial z/\partial s$, we note that there are two paths in the tree (in red in Figure 15.39) that connect z to s and contribute to $\partial z/\partial s$. Along one path, z changes with respect to x

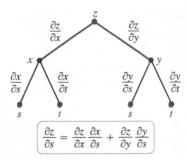

$$\frac{\partial z}{\partial s} = \frac{\partial z}{\partial x}\frac{\partial x}{\partial s} + \frac{\partial z}{\partial y}\frac{\partial y}{\partial s}$$

Figure 15.39

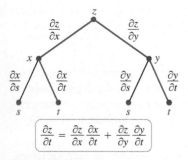

$$\frac{\partial z}{\partial t} = \frac{\partial z}{\partial x}\frac{\partial x}{\partial t} + \frac{\partial z}{\partial y}\frac{\partial y}{\partial t}$$

Figure 15.40

QUICK CHECK 2 Suppose $w = f(x, y, z)$, where $x = g(s, t)$, $y = h(s, t)$, and $z = p(s, t)$. Extend Theorem 15.8 to write a formula for $\partial w/\partial t$. ◄

(with rate of change $\partial z/\partial x$) and x changes with respect to s (with rate of change $\partial x/\partial s$). Along the other path, z changes with respect to y (with rate of change $\partial z/\partial y$) and y changes with respect to s (with rate of change $\partial y/\partial s$). We use a Chain Rule calculation along each path and combine the results. A similar argument leads to $\partial z/\partial t$ (**Figure 15.40**).

THEOREM 15.8   Chain Rule (Two Independent Variables)
Let z be a differentiable function of x and y, where x and y are differentiable functions of s and t. Then

$$\frac{\partial z}{\partial s} = \frac{\partial z}{\partial x}\frac{\partial x}{\partial s} + \frac{\partial z}{\partial y}\frac{\partial y}{\partial s} \quad \text{and} \quad \frac{\partial z}{\partial t} = \frac{\partial z}{\partial x}\frac{\partial x}{\partial t} + \frac{\partial z}{\partial y}\frac{\partial y}{\partial t}.$$

EXAMPLE 2   Chain Rule with two independent variables Let $z = \sin 2x \cos 3y$, where $x = s + t$ and $y = s - t$. Evaluate $\partial z/\partial s$ and $\partial z/\partial t$.

SOLUTION The tree diagram in Figure 15.39 gives the Chain Rule formula for $\partial z/\partial s$: We form products of the derivatives along the red branches connecting z to s and add the results. The partial derivative is

$$\frac{\partial z}{\partial s} = \frac{\partial z}{\partial x}\frac{\partial x}{\partial s} + \frac{\partial z}{\partial y}\frac{\partial y}{\partial s}$$

$$= \underbrace{2 \cos 2x \cos 3y}_{\frac{\partial z}{\partial x}} \cdot \underbrace{1}_{\frac{\partial x}{\partial s}} + \underbrace{(-3 \sin 2x \sin 3y)}_{\frac{\partial z}{\partial y}} \cdot \underbrace{1}_{\frac{\partial y}{\partial s}}$$

$$= 2 \cos \big(2\underbrace{(s + t)}_{x}\big) \cos \big(3\underbrace{(s - t)}_{y}\big) - 3 \sin \big(2\underbrace{(s + t)}_{x}\big) \sin \big(3\underbrace{(s - t)}_{y}\big).$$

Following the branches of Figure 15.40 connecting z to t, we have

$$\frac{\partial z}{\partial t} = \frac{\partial z}{\partial x}\frac{\partial x}{\partial t} + \frac{\partial z}{\partial y}\frac{\partial y}{\partial t}$$

$$= \underbrace{2 \cos 2x \cos 3y}_{\frac{\partial z}{\partial x}} \cdot \underbrace{1}_{\frac{\partial x}{\partial t}} + \underbrace{(-3 \sin 2x \sin 3y)}_{\frac{\partial z}{\partial y}} \cdot \underbrace{-1}_{\frac{\partial y}{\partial t}}$$

$$= 2 \cos \big(2\underbrace{(s + t)}_{x}\big) \cos \big(3\underbrace{(s - t)}_{y}\big) + 3 \sin \big(2\underbrace{(s + t)}_{x}\big) \sin \big(3\underbrace{(s - t)}_{y}\big).$$

Related Exercise 22 ◄

EXAMPLE 3   More variables Let w be a function of x, y, and z, each of which is a function of s and t.

a. Draw a labeled tree diagram showing the relationships among the variables.

b. Write the Chain Rule formula for $\dfrac{\partial w}{\partial s}$.

SOLUTION

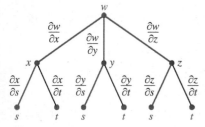

Figure 15.41

QUICK CHECK 3 If Q is a function of w, x, y, and z, each of which is a function of r, s, and t, how many dependent variables, intermediate variables, and independent variables are there? ◄

a. Because w is a function of x, y, and z, the upper branches of the tree (**Figure 15.41**) are labeled with the partial derivatives w_x, w_y, and w_z. Each of x, y, and z is a function of two variables, so the lower branches of the tree also require partial derivative labels.

b. Extending Theorem 15.8, we take the three paths through the tree that connect w to s (red branches in Figure 15.41). Multiplying the derivatives that appear on each path and adding gives the result

$$\frac{\partial w}{\partial s} = \frac{\partial w}{\partial x}\frac{\partial x}{\partial s} + \frac{\partial w}{\partial y}\frac{\partial y}{\partial s} + \frac{\partial w}{\partial z}\frac{\partial z}{\partial s}.$$

Related Exercises 25–26 ◄

It is probably clear by now that we can create a Chain Rule for any set of relationships among variables. The key is to draw an accurate tree diagram and label the branches of the tree with the appropriate derivatives.

EXAMPLE 4 A different kind of tree Let w be a function of z, where z is a function of x and y, and each of x and y is a function of t. Draw a labeled tree diagram and write the Chain Rule formula for dw/dt.

SOLUTION The dependent variable w is related to the independent variable t through two paths in the tree: $w \rightarrow z \rightarrow x \rightarrow t$ and $w \rightarrow z \rightarrow y \rightarrow t$ (**Figure 15.42**). At the top of the tree, w is a function of the single variable z, so the rate of change is the ordinary derivative dw/dz. The tree below z looks like Figure 15.36. Multiplying the derivatives on each of the two branches connecting w to t and adding the results, we have

$$\frac{dw}{dt} = \frac{dw}{dz}\frac{\partial z}{\partial x}\frac{dx}{dt} + \frac{dw}{dz}\frac{\partial z}{\partial y}\frac{dy}{dt} = \frac{dw}{dz}\left(\frac{\partial z}{\partial x}\frac{dx}{dt} + \frac{\partial z}{\partial y}\frac{dy}{dt}\right).$$

Related Exercise 31 ◄

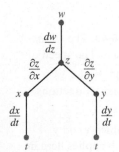

Figure 15.42

Implicit Differentiation

Using the Chain Rule for partial derivatives, the technique of implicit differentiation can be put in a larger perspective. Recall that if x and y are related through an implicit relationship, such as $\sin xy + \pi y^2 = x$, then dy/dx is computed using implicit differentiation (Section 3.8). Another way to compute dy/dx is to define the function $F(x, y) = \sin xy + \pi y^2 - x$. Notice that the original relationship $\sin xy + \pi y^2 = x$ is $F(x, y) = 0$.

To find dy/dx, we treat x as the independent variable and differentiate both sides of $F(x, y(x)) = 0$ with respect to x. The derivative of the right side is 0. On the left side, we use the Chain Rule of Theorem 15.7:

$$\frac{\partial F}{\partial x}\underbrace{\frac{dx}{dx}}_{1} + \frac{\partial F}{\partial y}\frac{dy}{dx} = 0.$$

Noting that $dx/dx = 1$ and solving for dy/dx, we obtain the following theorem.

> ► The question of whether a relationship of the form $F(x, y) = 0$ or $F(x, y, z) = 0$ determines one or more functions is addressed by a theorem of advanced calculus called the Implicit Function Theorem.

THEOREM 15.9 Implicit Differentiation
Let F be differentiable on its domain and suppose $F(x, y) = 0$ defines y as a differentiable function of x. Provided $F_y \neq 0$,

$$\frac{dy}{dx} = -\frac{F_x}{F_y}.$$

EXAMPLE 5 Implicit differentiation Find dy/dx when
$F(x, y) = \sin xy + \pi y^2 - x = 0.$

SOLUTION Computing the partial derivatives of F with respect to x and y, we find that

$$F_x = y \cos xy - 1 \quad \text{and} \quad F_y = x \cos xy + 2\pi y.$$

> ► The method of Theorem 15.9 generalizes to computing $\dfrac{\partial z}{\partial x}$ and $\dfrac{\partial z}{\partial y}$ with functions of the form $F(x, y, z) = 0$ (Exercise 56).

Therefore,

$$\frac{dy}{dx} = -\frac{F_x}{F_y} = -\frac{y \cos xy - 1}{x \cos xy + 2\pi y}.$$

As with many implicit differentiation calculations, the result is left in terms of both x and y. The same result is obtained using the methods of Section 3.8.

Related Exercises 37 ◄

QUICK CHECK 4 Use the method of Example 5 to find dy/dx when $F(x, y) = x^2 + xy - y^3 - 7 = 0$. Compare your solution to Example 3 in Section 3.8. Which method is easier? ◄

Figure 15.43

Figure 15.44

EXAMPLE 6 **Fluid flow** A basin of circulating water is represented by the square region $\{(x, y): 0 \le x \le 1, 0 \le y \le 1\}$, where x is positive in the eastward direction and y is positive in the northward direction. The velocity components of the water are

the east-west velocity $u(x, y) = 2 \sin \pi x \cos \pi y$ and

the north-south velocity $v(x, y) = -2 \cos \pi x \sin \pi y$;

these velocity components produce the flow pattern shown in **Figure 15.43**. The *streamlines* shown in the figure are the paths followed by small parcels of water. The speed of the water at a point (x, y) is given by the function $s(x, y) = \sqrt{u(x, y)^2 + v(x, y)^2}$. Find $\partial s / \partial x$ and $\partial s / \partial y$, the rates of change of the water speed in the x- and y-directions, respectively.

SOLUTION The dependent variable s depends on the independent variables x and y through the intermediate variables u and v (**Figure 15.44**). Theorem 15.8 applies here in the form

$$\frac{\partial s}{\partial x} = \frac{\partial s}{\partial u}\frac{\partial u}{\partial x} + \frac{\partial s}{\partial v}\frac{\partial v}{\partial x} \quad \text{and} \quad \frac{\partial s}{\partial y} = \frac{\partial s}{\partial u}\frac{\partial u}{\partial y} + \frac{\partial s}{\partial v}\frac{\partial v}{\partial y}.$$

The derivatives $\partial s / \partial u$ and $\partial s / \partial v$ are easier to find if we square the speed function to obtain $s^2 = u^2 + v^2$ and then use implicit differentiation. To compute $\partial s / \partial u$, we differentiate both sides of $s^2 = u^2 + v^2$ with respect to u:

$$2s\frac{\partial s}{\partial u} = 2u, \quad \text{which implies that} \quad \frac{\partial s}{\partial u} = \frac{u}{s}.$$

Similarly, differentiating $s^2 = u^2 + v^2$ with respect to v gives

$$2s\frac{\partial s}{\partial v} = 2v, \quad \text{which implies that} \quad \frac{\partial s}{\partial v} = \frac{v}{s}.$$

Now the Chain Rule leads to $\dfrac{\partial s}{\partial x}$:

$$\frac{\partial s}{\partial x} = \frac{\partial s}{\partial u}\frac{\partial u}{\partial x} + \frac{\partial s}{\partial v}\frac{\partial v}{\partial x}$$

$$= \underbrace{\frac{u}{s}}_{\frac{\partial s}{\partial u}}\underbrace{(2\pi \cos \pi x \cos \pi y)}_{\frac{\partial u}{\partial x}} + \underbrace{\frac{v}{s}}_{\frac{\partial s}{\partial v}}\underbrace{(2\pi \sin \pi x \sin \pi y)}_{\frac{\partial v}{\partial x}}$$

$$= \frac{2\pi}{s}(u \cos \pi x \cos \pi y + v \sin \pi x \sin \pi y).$$

A similar calculation shows that

$$\frac{\partial s}{\partial y} = -\frac{2\pi}{s}(u \sin \pi x \sin \pi y + v \cos \pi x \cos \pi y).$$

As a final step, you could replace s, u, and v with their definitions in terms of x and y.

Related Exercises 41–42 ◄

EXAMPLE 7 **Second derivatives** Let $z = f(x, y) = \dfrac{x}{y}$, where $x = s + t^2$ and $y = s^2 - t$. Compute $\dfrac{\partial^2 z}{\partial s^2} = z_{ss}$, $\dfrac{\partial^2 z}{\partial t \partial s} = z_{st}$, and $\dfrac{\partial^2 z}{\partial t^2} = z_{tt}$, and express the results in terms of s and t. We use subscripts for partial derivatives in this example to simplify the notation.

SOLUTION First, we need some ground rules. In this example, it is possible to express f in terms of s and t by substituting, after which the result could be differentiated directly to find the required partial derivatives. Unfortunately, this maneuver is not always possible in practice (see Exercises 72 and 73). Therefore, to make this example as useful as possible, we develop general formulas for the second partial derivatives and make substitutions only in the last step.

Figures 15.39 and 15.40 show the relationships among the variables, and Example 2 demonstrates the calculation of the first partial derivatives. Throughout these calculations, it is important to remember the meaning of differentiation with respect to s and t:

$$(\ )_s = (\ )_x x_s + (\ )_y y_s \text{ and } (\ )_t = (\ )_x x_t + (\ )_y y_t.$$

Let's compute the first partial derivatives:

$$z_s = z_x x_s + z_y y_s \text{ and } z_t = z_x x_t + z_y y_t.$$

Differentiating z_s with respect to s, we have

$$
\begin{aligned}
z_{ss} &= (z_x x_s + z_y y_s)_s \\
&= (z_x)_s x_s + z_x x_{ss} + (z_y)_s y_s + z_y y_{ss} &&\text{Product Rule (twice)} \\
&= \underbrace{(z_{xx} x_s + z_{xy} y_s)}_{(z_x)_s} x_s + z_x x_{ss} &&\begin{array}{l}\text{Differentiate } z_x \text{ and } z_y \text{ with} \\ \text{respect to } s.\end{array} \\
&\quad + \underbrace{(z_{yx} x_s + z_{yy} y_s)}_{(z_y)_s} y_s + z_y y_{ss} \\
&= z_{xx} x_s^2 + 2z_{xy} x_s y_s + z_{yy} y_s^2 + z_x x_{ss} + z_y y_{ss}. &&\text{Simplify with } z_{xy} = z_{yx}.
\end{aligned}
$$

At this point, we substitute

$$z_x = \frac{1}{y}, \ z_y = -\frac{x}{y^2}, \ z_{xx} = 0, \ z_{xy} = -\frac{1}{y^2}, \ z_{yy} = \frac{2x}{y^3}, \ x_s = 1, \ x_{ss} = 0, \ y_s = 2s, \text{ and } y_{ss} = 2$$

and simplify to find that

$$z_{ss} = \frac{2(s^3 + 3st + 3s^2 t^2 + t^3)}{(s^2 - t)^3}.$$

Differentiating z_s with respect to t, a similar procedure produces z_{st}:

$$
\begin{aligned}
z_{st} &= (z_x x_s + z_y y_s)_t \\
&= (z_x)_t x_s + z_x x_{st} + (z_y)_t y_s + z_y y_{st} &&\text{Product Rule (twice)} \\
&= \underbrace{(z_{xx} x_t + z_{xy} y_t)}_{(z_x)_t} x_s + z_x x_{st} &&\begin{array}{l}\text{Differentiate } z_x \text{ and } z_y \\ \text{with respect to } t.\end{array} \\
&\quad + \underbrace{(z_{yx} x_t + z_{yy} y_t)}_{(z_y)_t} y_s + z_y y_{st} \\
&= z_{xx} x_s x_t + z_{xy} x_s y_t + z_{xy} x_t y_s + z_{yy} y_s y_t + z_x x_{st} + z_y y_{st}. &&\text{Simplify with } z_{xy} = z_{yx}.
\end{aligned}
$$

Substituting in terms of s and t with $x_{st} = 0$ and $y_{st} = 0$, we have

$$z_{st} = -\frac{3s^2 + t + 4s^3 t}{(s^2 - t)^3}.$$

An analogous calculation gives

$$z_{tt} = \frac{2s(1 + s^3)}{(s^2 - t)^3}.$$

Related Exercise 45 ◄

SECTION 15.4 EXERCISES

Getting Started

1. Suppose $z = f(x, y)$, where x and y are functions of t. How many dependent, intermediate, and independent variables are there?

2. Let z be a function of x and y, while x and y are functions of t. Explain how to find dz/dt.

3. Suppose w is a function of x, y, and z, which are each functions of t. Explain how to find dw/dt.

4. Let $z = f(x, y)$, $x = g(s, t)$, and $y = h(s, t)$. Explain how to find $\partial z/\partial t$.

5. Given that $w = F(x, y, z)$, and x, y, and z are functions of r and s, sketch a Chain Rule tree diagram with branches labeled with the appropriate derivatives.

6. Suppose $F(x, y) = 0$ and y is a differentiable function of x. Explain how to find dy/dx.

7. Evaluate dz/dt, where $z = x^2 + y^3$, $x = t^2$, and $y = t$, using Theorem 15.7. Check your work by writing z as a function of t and evaluating dz/dt.

8. Evaluate dz/dt, where $z = xy^2$, $x = t^2$, and $y = t$, using Theorem 15.7. Check your work by writing z as a function of t and evaluating dz/dt.

Practice Exercises

9–18. Chain Rule with one independent variable *Use Theorem 15.7 to find the following derivatives.*

9. dz/dt, where $z = x \sin y$, $x = t^2$, and $y = 4t^3$

10. dz/dt, where $z = x^2 y - xy^3$, $x = t^2$, and $y = t^{-2}$

11. dw/dt, where $w = \cos 2x \sin 3y$, $x = t/2$, and $y = t^4$

12. dz/dt, where $z = \sqrt{r^2 + s^2}$, $r = \cos 2t$, and $s = \sin 2t$

13. dz/dt, where $z = (x + 2y)^{10}$, $x = \sin^2 t$, $y = (3t + 4)^5$

14. $\dfrac{dz}{dt}$, where $z = \dfrac{x^{20}}{y^{10}}$, $x = \tan^{-1} t$, $y = \ln(t^2 + 1)$

15. dw/dt, where $w = xy \sin z$, $x = t^2$, $y = 4t^3$, and $z = t + 1$

16. dQ/dt, where $Q = \sqrt{x^2 + y^2 + z^2}$, $x = \sin t$, $y = \cos t$, and $z = \cos t$

17. dV/dt, where $V = xyz$, $x = e^t$, $y = 2t + 3$, and $z = \sin t$

18. $\dfrac{dU}{dt}$, where $U = \dfrac{xy^2}{z^8}$, $x = e^t$, $y = \sin 3t$, and $z = 4t + 1$

19–26. Chain Rule with several independent variables *Find the following derivatives.*

19. z_s and z_t, where $z = x^2 \sin y$, $x = s - t$, and $y = t^2$

20. z_s and z_t, where $z = \sin(2x + y)$, $x = s^2 - t^2$, and $y = s^2 + t^2$

21. z_s and z_t, where $z = xy - x^2 y$, $x = s + t$, and $y = s - t$

22. z_s and z_t, where $z = \sin x \cos 2y$, $x = s + t$, and $y = s - t$

23. z_s and z_t, where $z = e^{x+y}$, $x = st$, and $y = s + t$

24. z_s and z_t, where $z = \sin xy$, $x = s^2 t$, and $y = (s + t)^{10}$

25. w_s and w_t, where $w = \dfrac{x - z}{y + z}$, $x = s + t$, $y = st$, and $z = s - t$

26. w_r, w_s, and w_t, where $w = \sqrt{x^2 + y^2 + z^2}$, $x = st$, $y = rs$, and $z = rt$

27. Changing cylinder The volume of a right circular cylinder with radius r and height h is $V = \pi r^2 h$.

    a. Assume r and h are functions of t. Find $V'(t)$.

    b. Suppose $r = e^t$ and $h = e^{-2t}$, for $t \geq 0$. Use part (a) to find $V'(t)$.

    c. Does the volume of the cylinder in part (b) increase or decrease as t increases?

28. Changing pyramid The volume of a pyramid with a square base x units on a side and a height of h is $V = \dfrac{1}{3} x^2 h$.

    a. Assume x and h are functions of t. Find $V'(t)$.

    b. Suppose $x = \dfrac{t}{t+1}$ and $h = \dfrac{1}{t+1}$, for $t \geq 0$. Use part (a) to find $V'(t)$.

    c. Does the volume of the pyramid in part (b) increase or decrease as t increases?

29–30. Derivative practice two ways *Find the indicated derivative in two ways:*

    a. Replace x and y to write z as a function of t, and differentiate.

    b. Use the Chain Rule.

29. $z'(t)$, where $z = \dfrac{1}{x} + \dfrac{1}{y}$, $x = t^2 + 2t$, and $y = t^3 - 2$

30. $z'(t)$, where $z = \ln(x + y)$, $x = te^t$, and $y = e^t$

31–34. Making trees *Use a tree diagram to write the required Chain Rule formula.*

31. w is a function of z, where z is a function of p, q, and r, each of which is a function of t. Find dw/dt.

32. $w = f(x, y, z)$, where $x = g(t)$, $y = h(s, t)$, and $z = p(r, s, t)$. Find $\partial w/\partial t$.

33. $u = f(v)$, where $v = g(w, x, y)$, $w = h(z)$, $x = p(t, z)$, and $y = q(t, z)$. Find $\partial u/\partial z$.

34. $u = f(v, w, x)$, where $v = g(r, s, t)$, $w = h(r, s, t)$, $x = p(r, s, t)$, and $r = F(z)$. Find $\partial u/\partial z$.

35–40. Implicit differentiation *Use Theorem 15.9 to evaluate dy/dx. Assume each equation implicitly defines y as a differentiable function of x.*

35. $x^2 - 2y^2 - 1 = 0$ **36.** $x^3 + 3xy^2 - y^5 = 0$

37. $2 \sin xy = 1$ **38.** $ye^{xy} - 2 = 0$

39. $\sqrt{x^2 + 2xy + y^4} = 3$ **40.** $y \ln(x^2 + y^2 + 4) = 3$

41–42. Fluid flow *The x- and y-components of a fluid moving in two dimensions are given by the following functions u and v. The speed of the fluid at (x, y) is $s(x, y) = \sqrt{u(x, y)^2 + v(x, y)^2}$. Use the Chain Rule to find $\partial s/\partial x$ and $\partial s/\partial y$.*

41. $u(x, y) = 2y$ and $v(x, y) = -2x$; $x \geq 0$ and $y \geq 0$

42. $u(x, y) = x(1 - x)(1 - 2y)$ and $v(x, y) = y(y - 1)(1 - 2x)$; $0 \leq x \leq 1$, $0 \leq y \leq 1$

43–48. Second derivatives *For the following sets of variables, find all the relevant second derivatives. In all cases, first find general expressions for the second derivatives and then substitute variables at the last step.*

43. $f(x, y) = x^2 y$, where $x = s + t$ and $y = s - t$

44. $f(x, y) = x^2 y - xy^2$, where $x = st$ and $y = s/t$

45. $f(x, y) = y/x$, where $x = s^2 + t^2$ and $y = s^2 - t^2$

46. $f(x, y) = e^{x-y}$, where $x = s^2$ and $y = 3t^2$

47. $f(x, y, z) = xy + xz - yz$, where $x = s^2 - 2s$, $y = 2/s^2$, and $z = 3s^2 - 2$

48. $f(x, y) = xy$, where $x = s + 2t - u$ and $y = s + 2t + u$

49. Explain why or why not Determine whether the following statements are true and give an explanation or counterexample. Assume all partial derivatives exist.

a. If $z = (x + y)\sin xy$, where x and y are functions of s, then $\dfrac{\partial z}{\partial s} = \dfrac{dz}{dx}\dfrac{dx}{ds}$.

b. Given that $w = f(x(s, t), y(s, t), z(s, t))$, the rate of change of w with respect to t is dw/dt.

50–54. Derivative practice *Find the indicated derivative for the following functions.*

50. $\partial z/\partial p$, where $z = x/y$, $x = p + q$, and $y = p - q$

51. dw/dt, where $w = xyz$, $x = 2t^4$, $y = 3t^{-1}$, and $z = 4t^{-3}$

52. $\partial w/\partial x$, where $w = \cos z - \cos x \cos y + \sin x \sin y$, and $z = x + y$

53. $\dfrac{\partial z}{\partial x}$, where $\dfrac{1}{x} + \dfrac{1}{y} + \dfrac{1}{z} = 1$

54. $\partial z/\partial x$, where $xy - z = 1$

55. Change on a line Suppose $w = f(x, y, z)$ and ℓ is the line $\mathbf{r}(t) = \langle at, bt, ct \rangle$, for $-\infty < t < \infty$.

a. Find $w'(t)$ on ℓ (in terms of $a, b, c, w_x, w_y,$ and w_z).
b. Apply part (a) to find $w'(t)$ when $f(x, y, z) = xyz$.
c. Apply part (a) to find $w'(t)$ when $f(x, y, z) = \sqrt{x^2 + y^2 + z^2}$.
d. For a general twice differentiable function $w = f(x, y, z)$, find $w''(t)$.

56. Implicit differentiation rule with three variables Assume $F(x, y, z(x, y)) = 0$ implicitly defines z as a differentiable function of x and y. Extend Theorem 15.9 to show that

$$\frac{\partial z}{\partial x} = -\frac{F_x}{F_z} \quad \text{and} \quad \frac{\partial z}{\partial y} = -\frac{F_y}{F_z}.$$

57–59. Implicit differentiation with three variables *Use the result of Exercise 56 to evaluate $\partial z/\partial x$ and $\partial z/\partial y$ for the following relations.*

57. $xy + xz + yz = 3$
58. $x^2 + 2y^2 - 3z^2 = 1$

59. $xyz + x + y - z = 0$

60. More than one way Let $e^{xyz} = 2$. Find z_x and z_y in three ways (and check for agreement).

a. Use the result of Exercise 56.
b. Take logarithms of both sides and differentiate $xyz = \ln 2$.
c. Solve for z and differentiate $z = \dfrac{\ln 2}{xy}$.

61–64. Walking on a surface *Consider the following surfaces specified in the form $z = f(x, y)$ and the oriented curve C in the xy-plane.*

a. In each case, find $z'(t)$.
b. Imagine that you are walking on the surface directly above the curve C in the direction of positive orientation. Find the values of t for which you are walking uphill (that is, z is increasing).

61. $z = x^2 + 4y^2 + 1$, C: $x = \cos t$, $y = \sin t$; $0 \le t \le 2\pi$

62. $z = 4x^2 - y^2 + 1$, C: $x = \cos t$, $y = \sin t$; $0 \le t \le 2\pi$

63. $z = \sqrt{1 - x^2 - y^2}$, C: $x = e^{-t}$, $y = e^{-t}$; $t \ge \dfrac{1}{2}\ln 2$

64. $z = 2x^2 + y^2 + 1$, C: $x = 1 + \cos t$, $y = \sin t$; $0 \le t \le 2\pi$

65. Conservation of energy A projectile with mass m is launched into the air on a parabolic trajectory. For $t \ge 0$, its horizontal and vertical coordinates are $x(t) = u_0 t$ and $y(t) = -\dfrac{1}{2}gt^2 + v_0 t$, respectively, where u_0 is the initial horizontal velocity, v_0 is the initial vertical velocity, and g is the acceleration due to gravity. Recalling that $u(t) = x'(t)$ and $v(t) = y'(t)$ are the components of the velocity, the energy of the projectile (kinetic plus potential) is

$$E(t) = \frac{1}{2}m(u^2 + v^2) + mgy.$$

Use the Chain Rule to compute $E'(t)$ and show that $E'(t) = 0$, for all $t \ge 0$. Interpret the result.

66. Utility functions in economics Economists use *utility functions* to describe consumers' relative preference for two or more commodities (for example, vanilla vs. chocolate ice cream or leisure time vs. material goods). The Cobb-Douglas family of utility functions has the form $U(x, y) = x^a y^{1-a}$, where x and y are the amounts of two commodities and $0 < a < 1$ is a parameter. Level curves on which the utility function is constant are called *indifference curves*; the preference is the same for all combinations of x and y along an indifference curve (see figure).

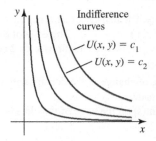

a. The marginal utilities of the commodities x and y are defined to be $\partial U/\partial x$ and $\partial U/\partial y$, respectively. Compute the marginal utilities for the utility function $U(x, y) = x^a y^{1-a}$.
b. The marginal rate of substitution (MRS) is the slope of the indifference curve at the point (x, y). Use the Chain Rule to show that for $U(x, y) = x^a y^{1-a}$, the MRS is $-\dfrac{a}{1 - a}\dfrac{y}{x}$.
c. Find the MRS for the utility function $U(x, y) = x^{0.4} y^{0.6}$ at $(x, y) = (8, 12)$.

67. Constant volume tori The volume of a solid torus is given by $V = \dfrac{\pi^2}{4}(R + r)(R - r)^2$, where r and R are the inner and outer radii and $R > r$ (see figure).

a. If R and r increase at the same rate, does the volume of the torus increase, decrease, or remain constant?
b. If R and r decrease at the same rate, does the volume of the torus increase, decrease, or remain constant?

68. Body surface area One of several empirical formulas that relates the surface area S of a human body to the height h and weight w of the body is the Mosteller formula $S(h, w) = \dfrac{1}{60}\sqrt{hw}$, where h is measured in cm, w is measured in kg, and S is measured in square meters. Suppose h and w are functions of t.

    a. Find $S'(t)$.

    b. Show that the condition under which the surface area remains constant as h and w change is $wh'(t) + hw'(t) = 0$.

    c. Show that part (b) implies that for constant surface area, h and w must be inversely related; that is, $h = C/w$, where C is a constant.

69. The Ideal Gas Law The pressure, temperature, and volume of an ideal gas are related by $PV = kT$, where $k > 0$ is a constant. Any two of the variables may be considered independent, which determines the dependent variable.

    a. Use implicit differentiation to compute the partial derivatives $\dfrac{\partial P}{\partial V}, \dfrac{\partial T}{\partial P},$ and $\dfrac{\partial V}{\partial T}$.

    b. Show that $\dfrac{\partial P}{\partial V}\dfrac{\partial T}{\partial P}\dfrac{\partial V}{\partial T} = -1$. (See Exercise 75 for a generalization.)

70. Variable density The density of a thin circular plate of radius 2 is given by $\rho(x, y) = 4 + xy$. The edge of the plate is described by the parametric equations $x = 2\cos t$, $y = 2\sin t$, for $0 \le t \le 2\pi$.

    a. Find the rate of change of the density with respect to t on the edge of the plate.

    b. At what point(s) on the edge of the plate is the density a maximum?

T 71. Spiral through a domain Suppose you follow the helical path $C: x = \cos t$, $y = \sin t$, $z = t$, for $t \ge 0$, through the domain of the function $w = f(x, y, z) = \dfrac{xyz}{z^2 + 1}$.

    a. Find $w'(t)$ along C.

    b. Estimate the point (x, y, z) on C at which w has its maximum value.

Explorations and Challenges

72. Change of coordinates Recall that Cartesian and polar coordinates are related through the transformation equations

$$\begin{cases} x = r\cos\theta \\ y = r\sin\theta \end{cases} \quad \text{or} \quad \begin{cases} r^2 = x^2 + y^2 \\ \tan\theta = y/x. \end{cases}$$

    a. Evaluate the partial derivatives x_r, y_r, x_θ, and y_θ.

    b. Evaluate the partial derivatives r_x, r_y, θ_x, and θ_y.

    c. For a function $z = f(x, y)$, find z_r and z_θ, where x and y are expressed in terms of r and θ.

    d. For a function $z = g(r, \theta)$, find z_x and z_y, where r and θ are expressed in terms of x and y.

    e. Show that $\left(\dfrac{\partial z}{\partial x}\right)^2 + \left(\dfrac{\partial z}{\partial y}\right)^2 = \left(\dfrac{\partial z}{\partial r}\right)^2 + \dfrac{1}{r^2}\left(\dfrac{\partial z}{\partial\theta}\right)^2$.

73. Change of coordinates continued An important derivative operation in many applications is called the Laplacian; in Cartesian coordinates, for $z = f(x, y)$, the Laplacian is $z_{xx} + z_{yy}$. Determine the Laplacian in polar coordinates using the following steps.

    a. Begin with $z = g(r, \theta)$ and write z_x and z_y in terms of polar coordinates (see Exercise 72).

    b. Use the Chain Rule to find $z_{xx} = \dfrac{\partial}{\partial x}(z_x)$. There should be two major terms, which, when expanded and simplified, result in five terms.

    c. Use the Chain Rule to find $z_{yy} = \dfrac{\partial}{\partial y}(z_y)$. There should be two major terms, which, when expanded and simplified, result in five terms.

    d. Combine parts (b) and (c) to show that

$$z_{xx} + z_{yy} = z_{rr} + \dfrac{1}{r}z_r + \dfrac{1}{r^2}z_{\theta\theta}.$$

74. Geometry of implicit differentiation Suppose x and y are related by the equation $F(x, y) = 0$. Interpret the solution of this equation as the set of points (x, y) that lie on the intersection of the surface $z = F(x, y)$ with the xy-plane ($z = 0$).

    a. Make a sketch of a surface and its intersection with the xy-plane. Give a geometric interpretation of the result that $\dfrac{dy}{dx} = -\dfrac{F_x}{F_y}$.

    b. Explain geometrically what happens at points where $F_y = 0$.

75. General three-variable relationship In the implicit relationship $F(x, y, z) = 0$, any two of the variables may be considered independent, which then determines the dependent variable. To avoid confusion, we may use a subscript to indicate which variable is held fixed in a derivative calculation; for example, $\left(\dfrac{\partial z}{\partial x}\right)_y$ means that y is held fixed in taking the partial derivative of z with respect to x. (In this context, the subscript does *not* mean a derivative.)

    a. Differentiate $F(x, y, z) = 0$ with respect to x, holding y fixed, to show that $\left(\dfrac{\partial z}{\partial x}\right)_y = -\dfrac{F_x}{F_z}$.

    b. As in part (a), find $\left(\dfrac{\partial y}{\partial z}\right)_x$ and $\left(\dfrac{\partial x}{\partial y}\right)_z$.

    c. Show that $\left(\dfrac{\partial z}{\partial x}\right)_y\left(\dfrac{\partial y}{\partial z}\right)_x\left(\dfrac{\partial x}{\partial y}\right)_z = -1$.

    d. Find the relationship analogous to part (c) for the case $F(w, x, y, z) = 0$.

76. Second derivative Let $f(x, y) = 0$ define y as a twice differentiable function of x.

    a. Show that $y''(x) = -\dfrac{f_{xx}f_y^2 - 2f_xf_yf_{xy} + f_{yy}f_x^2}{f_y^3}$.

    b. Verify part (a) using the function $f(x, y) = xy - 1$.

77. Subtleties of the Chain Rule Let $w = f(x, y, z) = 2x + 3y + 4z$, which is defined for all (x, y, z) in $\mathbb{R}^3$. Suppose we are interested in the partial derivative w_x on a subset of $\mathbb{R}^3$, such as the plane P given by $z = 4x - 2y$. The point to be made is that the result is not unique unless we specify which variables are considered independent.

    a. We could proceed as follows. On the plane P, consider x and y as the independent variables, which means z depends on x and y, so we write $w = f(x, y, z(x, y))$. Differentiate with respect to x, holding y fixed, to show that $\left(\dfrac{\partial w}{\partial x}\right)_y = 18$, where the subscript y indicates that y is held fixed.

b. Alternatively, on the plane P, we could consider x and z as the independent variables, which means y depends on x and z, so we write $w = f(x, y(x, z), z)$ and differentiate with respect to x, holding z fixed. Show that $\left(\dfrac{\partial w}{\partial x}\right)_z = 8$, where the subscript z indicates that z is held fixed.

c. Make a sketch of the plane $z = 4x - 2y$ and interpret the results of parts (a) and (b) geometrically.

d. Repeat the arguments of parts (a) and (b) to find $\left(\dfrac{\partial w}{\partial y}\right)_x$, $\left(\dfrac{\partial w}{\partial y}\right)_z$, $\left(\dfrac{\partial w}{\partial z}\right)_x$, and $\left(\dfrac{\partial w}{\partial z}\right)_y$.

QUICK CHECK ANSWERS

1. If $z = f(x(t))$, then $\dfrac{\partial z}{\partial y} = 0$, and the original Chain Rule results. **2.** $\dfrac{\partial w}{\partial t} = \dfrac{\partial w}{\partial x}\dfrac{\partial x}{\partial t} + \dfrac{\partial w}{\partial y}\dfrac{\partial y}{\partial t} + \dfrac{\partial w}{\partial z}\dfrac{\partial z}{\partial t}$

3. One dependent variable, four intermediate variables, and three independent variables **4.** $\dfrac{dy}{dx} = \dfrac{2x + y}{3y^2 - x}$; in this case, using $\dfrac{dy}{dx} = -\dfrac{F_x}{F_y}$ is more efficient. ◄

15.5 Directional Derivatives and the Gradient

Partial derivatives tell us a lot about the rate of change of a function on its domain. However, they do not *directly* answer some important questions. For example, suppose you are standing at a point $(a, b, f(a, b))$ on the surface $z = f(x, y)$. The partial derivatives f_x and f_y tell you the rate of change (or slope) of the surface at that point in the directions parallel to the x-axis and y-axis, respectively. But you could walk in an infinite number of directions from that point and find a different rate of change in every direction. With this observation in mind, we pose several questions.

- Suppose you are standing on a surface and you walk in a direction *other* than a coordinate direction—say, northwest or south-southeast. What is the rate of change of the function in such a direction?

- Suppose you are standing on a surface and you release a ball at your feet and let it roll. In which direction will it roll?

- If you are hiking up a mountain, in what direction should you walk after each step if you want to follow the steepest path?

These questions are answered in this section by introducing the *directional derivative*, followed by one of the central concepts of calculus—the *gradient*.

Directional Derivatives

Let $(a, b, f(a, b))$ be a point on the surface $z = f(x, y)$ and let **u** be a unit vector in the xy-plane (**Figure 15.45**). Our aim is to find the rate of change of f in the direction **u** at $P_0(a, b)$. In general, this rate of change is neither $f_x(a, b)$ nor $f_y(a, b)$ (unless $\mathbf{u} = \langle 1, 0 \rangle$ or $\mathbf{u} = \langle 0, 1 \rangle$), but it turns out to be a combination of $f_x(a, b)$ and $f_y(a, b)$.

Figure 15.46a shows the unit vector $\mathbf{u} = \langle u_1, u_2 \rangle$; its x- and y-components are u_1 and u_2, respectively. The derivative we seek must be computed along the line ℓ in the xy-plane through P_0 in the direction of **u**. A neighboring point P, which is h units from P_0 along ℓ, has coordinates $P(a + hu_1, b + hu_2)$ (**Figure 15.46b**).

Now imagine the plane Q perpendicular to the xy-plane, containing ℓ. This plane cuts the surface $z = f(x, y)$ in a curve C. Consider two points on C corresponding to P_0 and P; they have z-coordinates $f(a, b)$ and $f(a + hu_1, b + hu_2)$ (**Figure 15.47**). The slope of the secant line between these points is

$$\frac{f(a + hu_1, b + hu_2) - f(a, b)}{h}.$$

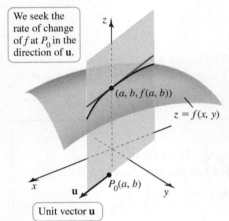

We seek the rate of change of f at P_0 in the direction of **u**.

$(a, b, f(a, b))$

$z = f(x, y)$

$P_0(a, b)$

Unit vector **u**

Figure 15.45

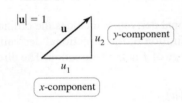

$|\mathbf{u}| = 1$

u

u_2 y-component

u_1

x-component

(a)

ℓ

$P(a + hu_1, b + hu_2)$

u

h

hu_2

$P_0(a, b)$ hu_1

$\mathbf{u} = \langle u_1, u_2 \rangle$

(b)

Figure 15.46

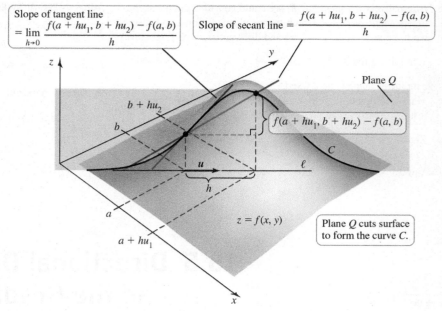

Figure 15.47

The derivative of f in the direction of $\mathbf{u}$ is obtained by letting $h \to 0$; when the limit exists, it is called the *directional derivative of f at (a, b) in the direction of $\mathbf{u}$*. It gives the slope of the line tangent to the curve C in the plane Q.

> The definition of the directional derivative looks like the definition of the ordinary derivative if we write it as
> $$\lim_{P \to P_0} \frac{f(P) - f(P_0)}{|P - P_0|},$$
> where P approaches P_0 along the line ℓ.

DEFINITION Directional Derivative

Let f be differentiable at (a, b) and let $\mathbf{u} = \langle u_1, u_2 \rangle$ be a unit vector in the xy-plane. The **directional derivative of f at (a, b) in the direction of u** is

$$D_{\mathbf{u}} f(a, b) = \lim_{h \to 0} \frac{f(a + hu_1, b + hu_2) - f(a, b)}{h},$$

provided the limit exists.

As motivation, it is instructive to see how the directional derivative includes the ordinary derivative in one variable. Setting $u_2 = 0$ in the definition of the directional derivative and ignoring the second variable gives the rate of change of f in the x-direction. The directional derivative then becomes

$$\lim_{h \to 0} \frac{f(a + hu_1) - f(a)}{h}.$$

Multiplying the numerator and denominator of this quotient by u_1, we have

$$u_1 \underbrace{\lim_{h \to 0} \frac{f(a + hu_1) - f(a)}{hu_1}}_{f'(a)} = u_1 f'(a).$$

Only because $\mathbf{u}$ is a unit vector and $u_1 = 1$ does the directional derivative reduce to the ordinary derivative $f'(a)$ in the x-direction. A similar argument may be used in the y-direction. Choosing $\mathbf{u}$ to be a unit vector gives the simplest formulas for the directional derivative.

As with ordinary derivatives, we would prefer to evaluate directional derivatives without taking limits. Fortunately, there is an easy way to express a directional derivative in terms of partial derivatives.

The key is to define a function that is equal to f along the line ℓ through (a, b) in the direction of the unit vector $\mathbf{u} = \langle u_1, u_2 \rangle$. The points on ℓ satisfy the parametric equations

$$x = a + su_1 \quad \text{and} \quad y = b + su_2,$$

QUICK CHECK 1 Explain why, when $\mathbf{u} = \langle 1, 0 \rangle$ in the definition of the directional derivative, the result is $f_x(a, b)$, and when $\mathbf{u} = \langle 0, 1 \rangle$, the result is $f_y(a, b)$. ◄

> To see that s is an arc length parameter, note that the line ℓ may be written in the form
>
> $$\mathbf{r}(s) = \langle a + su_1, b + su_2 \rangle.$$
>
> Therefore, $\mathbf{r}'(s) = \langle u_1, u_2 \rangle$ and $|\mathbf{r}'(s)| = 1$. It follows by the discussion in Section 14.4 that s is an arc length parameter.

> Note that $g'(s)$ does not correctly measure the slope of f along ℓ unless $\mathbf{u}$ is a unit vector.

where $-\infty < s < \infty$. Because $\mathbf{u}$ is a unit vector, the parameter s corresponds to arc length. As s increases, the points (x, y) move along ℓ in the direction of $\mathbf{u}$ with $s = 0$ corresponding to (a, b). Now we define the function

$$g(s) = f(\underbrace{a + su_1}_{x}, \underbrace{b + su_2}_{y}),$$

which gives the values of f along ℓ. The derivative of f along ℓ is $g'(s)$ (see margin note), and when evaluated at $s = 0$, it is the directional derivative of f at (a, b); that is, $g'(0) = D_{\mathbf{u}} f(a, b)$.

Noting that $\dfrac{dx}{ds} = u_1$ and $\dfrac{dy}{ds} = u_2$, we apply the Chain Rule to find that

$$D_{\mathbf{u}} f(a, b) = g'(0) = \left(\frac{\partial f}{\partial x} \underbrace{\frac{dx}{ds}}_{u_1} + \frac{\partial f}{\partial y} \underbrace{\frac{dy}{ds}}_{u_2} \right)\Bigg|_{s=0} \quad \text{Chain Rule}$$

$$= f_x(a, b)u_1 + f_y(a, b)u_2 \qquad s = 0 \text{ corresponds to } (a, b).$$

$$= \langle f_x(a, b), f_y(a, b) \rangle \cdot \langle u_1, u_2 \rangle. \qquad \text{Identify dot product.}$$

We see that the directional derivative is a weighted average of the partial derivatives $f_x(a, b)$ and $f_y(a, b)$, with the components of $\mathbf{u}$ serving as the weights. In other words, knowing the slope of the surface in the x- and y-directions allows us to find the slope in any direction. Notice that the directional derivative can be written as a dot product, which provides a practical formula for computing directional derivatives.

QUICK CHECK 2 In the parametric description $x = a + su_1$ and $y = b + su_2$, where $\mathbf{u} = \langle u_1, u_2 \rangle$ is a unit vector, show that any positive change Δs in s produces a line segment of length Δs. ◄

THEOREM 15.10 Directional Derivative

Let f be differentiable at (a, b) and let $\mathbf{u} = \langle u_1, u_2 \rangle$ be a unit vector in the xy-plane. The **directional derivative of f at (a, b) in the direction of $\mathbf{u}$** is

$$D_{\mathbf{u}} f(a, b) = \langle f_x(a, b), f_y(a, b) \rangle \cdot \langle u_1, u_2 \rangle.$$

EXAMPLE 1 Computing directional derivatives Consider the paraboloid $z = f(x, y) = \frac{1}{4}(x^2 + 2y^2) + 2$. Let P_0 be the point $(3, 2)$ and consider the unit vectors

$$\mathbf{u} = \left\langle \frac{1}{\sqrt{2}}, \frac{1}{\sqrt{2}} \right\rangle \quad \text{and} \quad \mathbf{v} = \left\langle \frac{1}{2}, -\frac{\sqrt{3}}{2} \right\rangle.$$

a. Find the directional derivative of f at P_0 in the directions of $\mathbf{u}$ and $\mathbf{v}$.

b. Graph the surface and interpret the directional derivatives.

SOLUTION

a. We see that $f_x = x/2$ and $f_y = y$; evaluated at $(3, 2)$, we have $f_x(3, 2) = 3/2$ and $f_y(3, 2) = 2$. The directional derivatives in the directions $\mathbf{u}$ and $\mathbf{v}$ are

$$D_{\mathbf{u}} f(3, 2) = \langle f_x(3, 2), f_y(3, 2) \rangle \cdot \langle u_1, u_2 \rangle$$

$$= \frac{3}{2} \cdot \frac{1}{\sqrt{2}} + 2 \cdot \frac{1}{\sqrt{2}} = \frac{7}{2\sqrt{2}} \approx 2.47 \text{ and}$$

$$D_{\mathbf{v}} f(3, 2) = \langle f_x(3, 2), f_y(3, 2) \rangle \cdot \langle v_1, v_2 \rangle$$

$$= \frac{3}{2} \cdot \frac{1}{2} + 2 \left(-\frac{\sqrt{3}}{2} \right) = \frac{3}{4} - \sqrt{3} \approx -0.98.$$

> It is understood that the line tangent to C in the direction of $\mathbf{u}$ lies in the vertical plane Q containing $\mathbf{u}$.

b. In the direction of $\mathbf{u}$, the directional derivative is approximately 2.47. Because it is positive, the function is increasing at $(3, 2)$ in this direction. Equivalently, if Q is the vertical plane containing $\mathbf{u}$, and C is the curve along which the surface intersects Q, then the slope of the line tangent to C is approximately 2.47 (**Figure 15.48a**). In the direction of $\mathbf{v}$, the directional derivative is approximately -0.98. Because it is negative, the function is decreasing in this direction. In this case, the vertical plane Q contains $\mathbf{v}$ and again C is the curve along which the surface intersects Q; the slope of the line tangent to C is approximately -0.98 (**Figure 15.48b**).

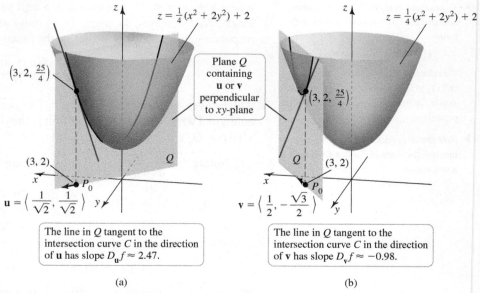

The line in Q tangent to the intersection curve C in the direction of **u** has slope $D_{\mathbf{u}} f \approx 2.47$.

The line in Q tangent to the intersection curve C in the direction of **v** has slope $D_{\mathbf{v}} f \approx -0.98$.

(a)

(b)

Figure 15.48

Related Exercises 11–12 ◀

QUICK CHECK 3 In Example 1, evaluate $D_{-\mathbf{u}} f(3, 2)$ and $D_{-\mathbf{v}} f(3, 2)$. ◀

The Gradient Vector

We have seen that the directional derivative can be written as a dot product: $D_{\mathbf{u}} f(a, b) = \langle f_x(a, b), f_y(a, b) \rangle \cdot \langle u_1, u_2 \rangle$. The vector $\langle f_x(a, b), f_y(a, b) \rangle$ that appears in the dot product is important in its own right and is called the *gradient* of f.

> Recall that the unit coordinate vectors in $\mathbb{R}^2$ are $\mathbf{i} = \langle 1, 0 \rangle$ and $\mathbf{j} = \langle 0, 1 \rangle$. The gradient of f is also written grad f, read *grad f*.

DEFINITION Gradient (Two Dimensions)

Let f be differentiable at the point (x, y). The **gradient** of f at (x, y) is the vector-valued function

$$\nabla f(x, y) = \langle f_x(x, y), f_y(x, y) \rangle = f_x(x, y)\,\mathbf{i} + f_y(x, y)\,\mathbf{j}.$$

With the definition of the gradient, the directional derivative of f at (a, b) in the direction of the unit vector **u** can be written

$$D_{\mathbf{u}} f(a, b) = \nabla f(a, b) \cdot \mathbf{u}.$$

The gradient satisfies sum, product, and quotient rules analogous to those for ordinary derivatives (Exercise 85).

EXAMPLE 2 Computing gradients Find $\nabla f(3, 2)$ for $f(x, y) = x^2 + 2xy - y^3$.

SOLUTION Computing $f_x = 2x + 2y$ and $f_y = 2x - 3y^2$, we have

$$\nabla f(x, y) = \langle 2(x + y), 2x - 3y^2 \rangle = 2(x + y)\,\mathbf{i} + (2x - 3y^2)\,\mathbf{j}.$$

Substituting $x = 3$ and $y = 2$ gives

$$\nabla f(3, 2) = \langle 10, -6 \rangle = 10\mathbf{i} - 6\mathbf{j}.$$

Related Exercises 13–15 ◀

EXAMPLE 3 Computing directional derivatives with gradients Let

$$f(x, y) = 3 - \frac{x^2}{10} + \frac{xy^2}{10}.$$

a. Compute $\nabla f(3, -1)$.

b. Compute $D_{\mathbf{u}} f(3, -1)$, where $\mathbf{u} = \left\langle \dfrac{1}{\sqrt{2}}, -\dfrac{1}{\sqrt{2}} \right\rangle$.

c. Compute the directional derivative of f at $(3, -1)$ in the direction of the vector $\langle 3, 4 \rangle$.

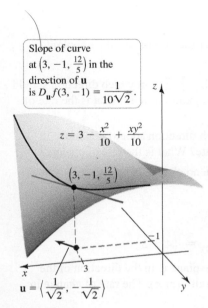

Slope of curve at $\left(3, -1, \frac{12}{5}\right)$ in the direction of $\mathbf{u}$ is $D_{\mathbf{u}}f(3, -1) = \frac{1}{10\sqrt{2}}$.

$z = 3 - \dfrac{x^2}{10} + \dfrac{xy^2}{10}$

$\left(3, -1, \frac{12}{5}\right)$

$\mathbf{u} = \left\langle \dfrac{1}{\sqrt{2}}, -\dfrac{1}{\sqrt{2}} \right\rangle$

Figure 15.49

SOLUTION

a. Note that $f_x = -x/5 + y^2/10$ and $f_y = xy/5$. Therefore,

$$\nabla f(3, -1) = \left\langle -\frac{x}{5} + \frac{y^2}{10}, \frac{xy}{5} \right\rangle \bigg|_{(3, -1)} = \left\langle -\frac{1}{2}, -\frac{3}{5} \right\rangle.$$

b. Before computing the directional derivative, it is important to verify that $\mathbf{u}$ is a unit vector (in this case, it is). The required directional derivative is

$$D_{\mathbf{u}} f(3, -1) = \nabla f(3, -1) \cdot \mathbf{u} = \left\langle -\frac{1}{2}, -\frac{3}{5} \right\rangle \cdot \left\langle \frac{1}{\sqrt{2}}, -\frac{1}{\sqrt{2}} \right\rangle = \frac{1}{10\sqrt{2}}.$$

Figure 15.49 shows the line tangent to the intersection curve in the plane corresponding to $\mathbf{u}$; its slope is $D_{\mathbf{u}} f(3, -1)$.

c. In this case, the direction is given in terms of a nonunit vector. The vector $\langle 3, 4 \rangle$ has length 5, so the unit vector in the direction of $\langle 3, 4 \rangle$ is $\mathbf{u} = \left\langle \frac{3}{5}, \frac{4}{5} \right\rangle$. The directional derivative at $(3, -1)$ in the direction of $\mathbf{u}$ is

$$D_{\mathbf{u}} f(3, -1) = \nabla f(3, -1) \cdot \mathbf{u} = \left\langle -\frac{1}{2}, -\frac{3}{5} \right\rangle \cdot \left\langle \frac{3}{5}, \frac{4}{5} \right\rangle = -\frac{39}{50},$$

which gives the slope of the surface in the direction of $\langle 3, 4 \rangle$ at $(3, -1)$.

Related Exercises 22, 27 ◄

Interpretations of the Gradient

The gradient is important not only in calculating directional derivatives; it plays many other roles in multivariable calculus. Our present goal is to develop some intuition about the meaning of the gradient.

We have seen that the directional derivative of f at (a, b) in the direction of the unit vector $\mathbf{u}$ is $D_{\mathbf{u}} f(a, b) = \nabla f(a, b) \cdot \mathbf{u}$. Using properties of the dot product, we have

$$D_{\mathbf{u}} f(a, b) = \nabla f(a, b) \cdot \mathbf{u}$$
$$= |\nabla f(a, b)| |\mathbf{u}| \cos \theta$$
$$= |\nabla f(a, b)| \cos \theta, \qquad |\mathbf{u}| = 1$$

▶ Recall that $\mathbf{u} \cdot \mathbf{v} = |\mathbf{u}||\mathbf{v}| \cos \theta$, where θ is the angle between $\mathbf{u}$ and $\mathbf{v}$.

where θ is the angle between $\nabla f(a, b)$ and $\mathbf{u}$. It follows that $D_{\mathbf{u}} f(a, b)$ has its maximum value when $\cos \theta = 1$, which corresponds to $\theta = 0$. Therefore, $D_{\mathbf{u}} f(a, b)$ has its maximum value and f has its greatest rate of *increase* when $\nabla f(a, b)$ and $\mathbf{u}$ point in the same direction. Note that when $\cos \theta = 1$, the actual rate of increase is $D_{\mathbf{u}} f(a, b) = |\nabla f(a, b)|$ (Figure 15.50).

▶ It is important to remember (but easy to forget) that $\nabla f(a, b)$ lies in the same plane as the domain of f.

Similarly, when $\theta = \pi$, we have $\cos \theta = -1$, and f has its greatest rate of *decrease* when $\nabla f(a, b)$ and $\mathbf{u}$ point in opposite directions. The actual rate of decrease is $D_{\mathbf{u}} f(a, b) = -|\nabla f(a, b)|$. These observations are summarized as follows: The gradient $\nabla f(a, b)$ points in the *direction of steepest ascent* at (a, b), while $-\nabla f(a, b)$ points in the *direction of steepest descent*.

Notice that $D_{\mathbf{u}} f(a, b) = 0$ when the angle between $\nabla f(a, b)$ and $\mathbf{u}$ is $\pi/2$, which means $\nabla f(a, b)$ and $\mathbf{u}$ are orthogonal (Figure 15.50). These observations justify the following theorem.

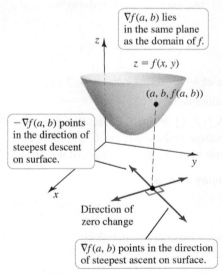

$\nabla f(a, b)$ lies in the same plane as the domain of f.

$z = f(x, y)$

$(a, b, f(a, b))$

$-\nabla f(a, b)$ points in the direction of steepest descent on surface.

Direction of zero change

$\nabla f(a, b)$ points in the direction of steepest ascent on surface.

Figure 15.50

THEOREM 15.11   Directions of Change

Let f be differentiable at (a, b) with $\nabla f(a, b) \neq \mathbf{0}$.

1. f has its maximum rate of increase at (a, b) in the direction of the gradient $\nabla f(a, b)$. The rate of change in this direction is $|\nabla f(a, b)|$.

2. f has its maximum rate of decrease at (a, b) in the direction of $-\nabla f(a, b)$. The rate of change in this direction is $-|\nabla f(a, b)|$.

3. The directional derivative is zero in any direction orthogonal to $\nabla f(a, b)$.

EXAMPLE 4 Steepest ascent and descent Consider the bowl-shaped paraboloid $z = f(x, y) = 4 + x^2 + 3y^2$.

a. If you are located on the paraboloid at the point $\left(2, -\frac{1}{2}, \frac{35}{4}\right)$, in which direction should you move in order to *ascend* on the surface at the maximum rate? What is the rate of change?

b. If you are located at the point $\left(2, -\frac{1}{2}, \frac{35}{4}\right)$, in which direction should you move in order to *descend* on the surface at the maximum rate? What is the rate of change?

c. At the point $(3, 1, 16)$, in what direction(s) is there no change in the function values?

SOLUTION

a. At the point $\left(2, -\frac{1}{2}\right)$, the value of the gradient is

$$\nabla f\left(2, -\tfrac{1}{2}\right) = \langle 2x, 6y \rangle\big|_{(2, -1/2)} = \langle 4, -3 \rangle.$$

Therefore, the direction of steepest ascent in the xy-plane is in the direction of the gradient vector $\langle 4, -3 \rangle$ (or $\mathbf{u} = \frac{1}{5}\langle 4, -3 \rangle$, as a unit vector). The rate of change is $\left|\nabla f\left(2, -\tfrac{1}{2}\right)\right| = |\langle 4, -3 \rangle| = 5$ **(Figure 15.51a)**.

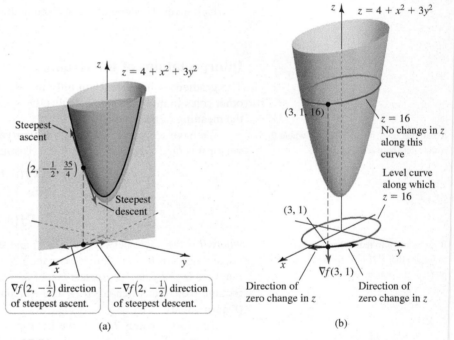

Figure 15.51

b. The direction of steepest *descent* is the direction of $-\nabla f\left(2, -\tfrac{1}{2}\right) = \langle -4, 3 \rangle$ (or $\mathbf{u} = \frac{1}{5}\langle -4, 3 \rangle$, as a unit vector). The rate of change is $-\left|\nabla f\left(2, -\tfrac{1}{2}\right)\right| = -5$.

> ➤ Note that $\langle 6, 6 \rangle$ and $\langle 6, -6 \rangle$ are orthogonal because $\langle 6, 6 \rangle \cdot \langle 6, -6 \rangle = 0$.

c. At the point $(3, 1)$, the value of the gradient is $\nabla f(3, 1) = \langle 6, 6 \rangle$. The function has zero change if we move in either of the two directions orthogonal to $\langle 6, 6 \rangle$; these two directions are parallel to $\langle 6, -6 \rangle$. In terms of unit vectors, the directions of no change are $\mathbf{u} = \dfrac{1}{\sqrt{2}}\langle -1, 1 \rangle$ and $\mathbf{u} = \dfrac{1}{\sqrt{2}}\langle 1, -1 \rangle$ **(Figure 15.51b)**.

Related Exercises 31–32 ◄

EXAMPLE 5 Interpreting directional derivatives Consider the function $f(x, y) = 3x^2 - 2y^2$.

a. Compute $\nabla f(x, y)$ and $\nabla f(2, 3)$.

b. Let $\mathbf{u} = \langle \cos \theta, \sin \theta \rangle$ be a unit vector. At $(2, 3)$, for what values of θ (measured relative to the positive x-axis), with $0 \le \theta < 2\pi$, does the directional derivative have its maximum and minimum values? What are those values?

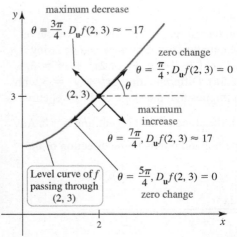

$\theta = \dfrac{3\pi}{4}, D_{\mathbf{u}}f(2,3) \approx -17$

maximum decrease

zero change

$\theta = \dfrac{\pi}{4}, D_{\mathbf{u}}f(2,3) = 0$

θ

(2, 3)

maximum increase

$\theta = \dfrac{7\pi}{4}, D_{\mathbf{u}}f(2,3) \approx 17$

Level curve of f passing through (2, 3)

$\theta = \dfrac{5\pi}{4}, D_{\mathbf{u}}f(2,3) = 0$

zero change

Figure 15.52

SOLUTION

a. The gradient is $\nabla f(x, y) = \langle f_x, f_y \rangle = \langle 6x, -4y \rangle$, and at $(2, 3)$ we have $\nabla f(2, 3) = \langle 12, -12 \rangle$.

b. The gradient $\nabla f(2, 3) = \langle 12, -12 \rangle$ makes an angle of $7\pi/4$ with the positive x-axis. So the maximum rate of change of f occurs in this direction, and that rate of change is $|\nabla f(2, 3)| = |\langle 12, -12 \rangle| = 12\sqrt{2} \approx 17$. The direction of maximum decrease is opposite the direction of the gradient, which corresponds to $\theta = 3\pi/4$. The maximum rate of decrease is the negative of the maximum rate of increase, or $-12\sqrt{2} \approx -17$. The function has zero change in the directions orthogonal to the gradient, which correspond to $\theta = \pi/4$ and $\theta = 5\pi/4$.

Figure 15.52 summarizes these conclusions. Notice that the gradient at $(2, 3)$ appears to be orthogonal to the level curve of f passing through $(2, 3)$. We next see that this is always the case.

Related Exercises 37–38 ◀

The Gradient and Level Curves

Theorem 15.11 states that in any direction orthogonal to the gradient $\nabla f(a, b)$, the function f does not change at (a, b). Recall from Section 15.1 that the curve $f(x, y) = z_0$, where z_0 is a constant, is a *level curve*, on which function values are constant. Combining these two observations, we conclude that the gradient $\nabla f(a, b)$ is orthogonal to the line tangent to the level curve through (a, b).

> **THEOREM 15.12   The Gradient and Level Curves**
> Given a function f differentiable at (a, b), the line tangent to the level curve of f at (a, b) is orthogonal to the gradient $\nabla f(a, b)$, provided $\nabla f(a, b) \neq \mathbf{0}$.

Proof: Consider the function $z = f(x, y)$ and its level curve $f(x, y) = z_0$, where the constant z_0 is chosen so that the curve passes through the point (a, b). Let $\mathbf{r}(t) = \langle x(t), y(t) \rangle$ be a parameterization for the level curve near (a, b) (where it is smooth), and let $\mathbf{r}(t_0)$ correspond to the point (a, b). We now differentiate $f(x, y) = z_0$ with respect to t. The derivative of the right side is 0. Applying the Chain Rule to the left side results in

$$\frac{d}{dt}(f(x, y)) = \frac{\partial f}{\partial x}\frac{dx}{dt} + \frac{\partial f}{\partial y}\frac{dy}{dt}$$

$$= \underbrace{\left\langle \frac{\partial f}{\partial x}, \frac{\partial f}{\partial y} \right\rangle}_{\nabla f(x, y)} \cdot \underbrace{\left\langle \frac{dx}{dt}, \frac{dy}{dt} \right\rangle}_{\mathbf{r}'(t)}$$

$$= \nabla f(x, y) \cdot \mathbf{r}'(t).$$

Level curve: $f(x, y) = z_0$, with parameterization $\mathbf{r}(t) = \langle x(t), y(t) \rangle$

Figure 15.53

Substituting $t = t_0$, we have $\nabla f(a, b) \cdot \mathbf{r}'(t_0) = 0$, which implies that $\mathbf{r}'(t_0)$ (the tangent vector at (a, b)) is orthogonal to $\nabla f(a, b)$. Figure 15.53 illustrates the geometry of the theorem. ◀

QUICK CHECK 4 Draw a circle in the xy-plane centered at the origin and regard it is as a level curve of the surface $z = x^2 + y^2$. At the point (a, a) of the level curve in the xy-plane, the slope of the tangent line is -1. Show that the gradient at (a, a) is orthogonal to the tangent line. ◀

An immediate consequence of Theorem 15.12 is an alternative equation of the tangent line. The curve described by $f(x, y) = z_0$ can be viewed as a level curve in the xy-plane for the surface $z = f(x, y)$. By Theorem 15.12, the line tangent to the curve at (a, b) is orthogonal to $\nabla f(a, b)$. Therefore, if (x, y) is a point on the tangent line, then $\nabla f(a, b) \cdot \langle x - a, y - b \rangle = 0$, which, when simplified, gives an equation of the line tangent to the curve $f(x, y) = z_0$ at (a, b):

$$f_x(a, b)(x - a) + f_y(a, b)(y - b) = 0.$$

EXAMPLE 6   Gradients and level curves Consider the upper sheet $z = f(x, y) = \sqrt{1 + 2x^2 + y^2}$ of a hyperboloid of two sheets.

a. Verify that the gradient at $(1, 1)$ is orthogonal to the corresponding level curve at that point.

b. Find an equation of the line tangent to the level curve at $(1, 1)$.

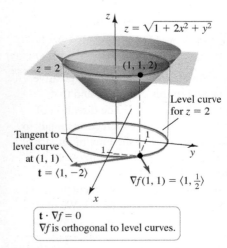

Figure 15.54

▶ The fact that $y' = -2x/y$ may also be obtained using Theorem 15.9: If $F(x, y) = 0$, then $y'(x) = -F_x/F_y$.

SOLUTION

a. You can verify that $(1, 1, 2)$ is on the surface; therefore, $(1, 1)$ is on the level curve corresponding to $z = 2$. Setting $z = 2$ in the equation of the surface and squaring both sides, the equation of the level curve is $4 = 1 + 2x^2 + y^2$, or $2x^2 + y^2 = 3$, which is the equation of an ellipse (**Figure 15.54**). Differentiating $2x^2 + y^2 = 3$ with respect to x gives $4x + 2yy'(x) = 0$, which implies that the slope of the level curve is $y'(x) = -\dfrac{2x}{y}$. Therefore, at the point $(1, 1)$, the slope of the tangent line is -2. Any vector proportional to $\mathbf{t} = \langle 1, -2 \rangle$ has slope -2 and points in the direction of the tangent line.

We now compute the gradient:

$$\nabla f(x, y) = \langle f_x, f_y \rangle = \left\langle \frac{2x}{\sqrt{1 + 2x^2 + y^2}}, \frac{y}{\sqrt{1 + 2x^2 + y^2}} \right\rangle.$$

It follows that $\nabla f(1, 1) = \langle 1, \frac{1}{2} \rangle$ (**Figure 15.54**). The tangent vector $\mathbf{t}$ and the gradient are orthogonal because

$$\mathbf{t} \cdot \nabla f(1, 1) = \langle 1, -2 \rangle \cdot \langle 1, \tfrac{1}{2} \rangle = 0.$$

b. An equation of the line tangent to the level curve at $(1, 1)$ is

$$\underbrace{f_x(1, 1)}_{1}(x - 1) + \underbrace{f_y(1, 1)}_{\frac{1}{2}}(y - 1) = 0,$$

or $y = -2x + 3$.

Related Exercises 49, 52 ◀

EXAMPLE 7 Path of steepest descent The paraboloid $z = f(x, y) = 4 + x^2 + 3y^2$ is shown in **Figure 15.55**. A ball is released at the point $(3, 4, 61)$ on the surface, and it follows the path of steepest descent C to the vertex of the paraboloid.

a. Find an equation of the projection of C in the xy-plane.

b. Find an equation of C on the paraboloid.

SOLUTION

a. The projection of C in the xy-plane points in the direction of $-\nabla f(x, y) = \langle -2x, -6y \rangle$, which means that at the point (x, y), the line tangent to the path has slope $y'(x) = (-6y)/(-2x) = 3y/x$. Therefore, the path in the xy-plane satisfies $y'(x) = 3y/x$ and passes through the initial point $(3, 4)$. You can verify that the solution to this differential equation is $y = 4x^3/27$. Therefore, the projection of the path of steepest descent in the xy-plane is the curve $y = 4x^3/27$. The descent ends at $(0, 0)$, which corresponds to the vertex of the paraboloid (Figure 15.55). At all points of the descent, the projection curve in the xy-plane is orthogonal to the level curves of the paraboloid.

b. To find a parametric description of C, it is easiest to define the parameter $t = x$. Using part (a), we find that

$$y = \frac{4x^3}{27} = \frac{4t^3}{27} \text{ and } z = 4 + x^2 + 3y^2 = 4 + t^2 + \frac{16}{243} t^6.$$

Because $0 \le x \le 3$, the parameter t varies over the interval $0 \le t \le 3$. A parametric description of C is

$$C: \mathbf{r}(t) = \left\langle t, \frac{4t^3}{27}, 4 + t^2 + \frac{16}{243} t^6 \right\rangle, \text{ for } 0 \le t \le 3.$$

With this parameterization, C is traced from $\mathbf{r}(0) = \langle 0, 0, 4 \rangle$ to $\mathbf{r}(3) = \langle 3, 4, 61 \rangle$—in the direction opposite to that of the ball's descent.

Related Exercise 57 ◀

Figure 15.55

QUICK CHECK 5 Verify that $y = 4x^3/27$ satisfies the equation $y'(x) = 3y/x$, with $y(3) = 4$. ◀

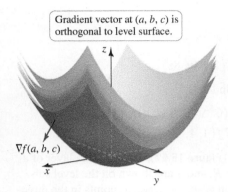

Gradient vector at (a, b, c) is orthogonal to level surface.

$\nabla f(a, b, c)$

Figure 15.56

The Gradient in Three Dimensions

The directional derivative, the gradient, and the idea of a level curve extend immediately to functions of three variables of the form $w = f(x, y, z)$. The main differences are that the gradient is a vector in $\mathbb{R}^3$, and level curves become *level surfaces* (Section 15.1). Here is how the gradient looks when we step up one dimension.

The easiest way to visualize the surface $w = f(x, y, z)$ is to picture its level surfaces—the surfaces in $\mathbb{R}^3$ on which f has a constant value. The level surfaces are given by the equation $f(x, y, z) = C$, where C is a constant (**Figure 15.56**). The level surfaces *can* be graphed, and they may be viewed as layers of the full four-dimensional surface (like layers of an onion). With this image in mind, we now extend the concepts of directional derivative and gradient to three dimensions.

Given the function $w = f(x, y, z)$, we begin just as we did in the two-variable case and define the directional derivative and the gradient.

DEFINITION Directional Derivative and Gradient in Three Dimensions

Let f be differentiable at (a, b, c) and let $\mathbf{u} = \langle u_1, u_2, u_3 \rangle$ be a unit vector. The **directional derivative of f at (a, b, c) in the direction of $\mathbf{u}$** is

$$D_{\mathbf{u}}f(a, b, c) = \lim_{h \to 0} \frac{f(a + hu_1, b + hu_2, c + hu_3) - f(a, b, c)}{h},$$

provided this limit exists.

The **gradient** of f at the point (x, y, z) is the vector-valued function

$$\nabla f(x, y, z) = \langle f_x(x, y, z), f_y(x, y, z), f_z(x, y, z) \rangle$$
$$= f_x(x, y, z)\,\mathbf{i} + f_y(x, y, z)\,\mathbf{j} + f_z(x, y, z)\,\mathbf{k}.$$

An argument similar to that given in two dimensions leads from the definition of the directional derivative to a computational formula. Given a unit vector $\mathbf{u} = \langle u_1, u_2, u_3 \rangle$, the directional derivative of f in the direction of $\mathbf{u}$ at the point (a, b, c) is

$$D_{\mathbf{u}}\,f(a, b, c) = f_x(a, b, c)\,u_1 + f_y(a, b, c)\,u_2 + f_z(a, b, c)\,u_3.$$

As before, we recognize this expression as a dot product of the vector $\mathbf{u}$ and the vector $\nabla f(a, b, c) = \langle f_x(a, b, c), f_y(a, b, c), f_z(a, b, c) \rangle$, which is the gradient evaluated at (a, b, c). These observations lead to Theorem 15.13, which mirrors Theorems 15.10 and 15.11.

> When we introduce the tangent plane in Section 15.6, we can also claim that $\nabla f(a, b, c)$ is orthogonal to the level surface that passes through (a, b, c).

THEOREM 15.13   Directional Derivative and Interpreting the Gradient

Let f be differentiable at (a, b, c) and let $\mathbf{u} = \langle u_1, u_2, u_3 \rangle$ be a unit vector. The directional derivative of f at (a, b, c) in the direction of $\mathbf{u}$ is

$$D_{\mathbf{u}}f(a, b, c) = \nabla f(a, b, c) \cdot \mathbf{u}$$
$$= \langle f_x(a, b, c), f_y(a, b, c), f_z(a, b, c) \rangle \cdot \langle u_1, u_2, u_3 \rangle.$$

Assuming $\nabla f(a, b, c) \neq \mathbf{0}$, the gradient in three dimensions has the following properties.

1. f has its maximum rate of increase at (a, b, c) in the direction of the gradient $\nabla f(a, b, c)$, and the rate of change in this direction is $|\nabla f(a, b, c)|$.

2. f has its maximum rate of decrease at (a, b, c) in the direction of $-\nabla f(a, b, c)$, and the rate of change in this direction is $-|\nabla f(a, b, c)|$.

3. The directional derivative is zero in any direction orthogonal to $\nabla f(a, b, c)$.

QUICK CHECK 6 Compute $\nabla f(-1, 2, 1)$, where $f(x, y, z) = xy/z$. ◄

EXAMPLE 8   Gradients in three dimensions Consider the function $f(x, y, z) = x^2 + 2y^2 + 4z^2 - 1$ and its level surface $f(x, y, z) = 3$.

a. Find and interpret the gradient at the points $P(2, 0, 0)$, $Q(0, \sqrt{2}, 0)$, $R(0, 0, 1)$, and $S\left(1, 1, \frac{1}{2}\right)$ on the level surface.

b. What are the actual rates of change of f in the directions of the gradients in part (a)?

Level surface of $f(x, y, z) = x^2 + 2y^2 + 4z^2 - 1$
$f(x, y, z) = 3$

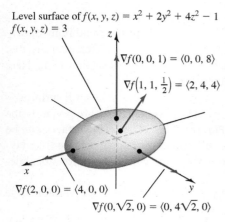

Figure 15.57

SOLUTION

a. The gradient is

$$\nabla f = \langle\, f_x, f_y, f_z \,\rangle = \langle\, 2x, 4y, 8z \,\rangle.$$

Evaluating the gradient at the four points, we find that

$$\nabla f(2, 0, 0) = \langle 4, 0, 0 \rangle, \qquad \nabla f(0, \sqrt{2}, 0) = \langle 0, 4\sqrt{2}, 0 \rangle,$$
$$\nabla f(0, 0, 1) = \langle 0, 0, 8 \rangle, \quad \text{and} \quad \nabla f\!\left(1, 1, \tfrac{1}{2}\right) = \langle 2, 4, 4 \rangle.$$

The level surface $f(x, y, z) = 3$ is an ellipsoid (**Figure 15.57**), which is one layer of a four-dimensional surface. The four points P, Q, R, and S are shown on the level surface with the respective gradient vectors. In each case, the gradient points in the direction that f has its maximum rate of increase. Of particular importance is the fact—to be made clear in the next section—that at each point, the gradient is orthogonal to the level surface.

b. The actual rate of increase of f at (a, b, c) in the direction of the gradient is $|\nabla f(a, b, c)|$. At P, the rate of increase of f in the direction of the gradient is $|\langle 4, 0, 0 \rangle| = 4$; at Q, the rate of increase is $|\langle 0, 4\sqrt{2}, 0 \rangle| = 4\sqrt{2}$; at R, the rate of increase is $|\langle 0, 0, 8 \rangle| = 8$; and at S, the rate of increase is $|\langle 2, 4, 4 \rangle| = 6$.

Related Exercises 59–60 ◄

SECTION 15.5 EXERCISES

Getting Started

1. Explain how a directional derivative is formed from the two partial derivatives f_x and f_y.

2. How do you compute the gradient of the functions $f(x, y)$ and $f(x, y, z)$?

3. Interpret the direction of the gradient vector at a point.

4. Interpret the magnitude of the gradient vector at a point.

5. Given a function f, explain the relationship between the gradient and the level curves of f.

6. The level curves of the surface $z = x^2 + y^2$ are circles in the xy-plane centered at the origin. Without computing the gradient, what is the direction of the gradient at $(1, 1)$ and $(-1, -1)$ (determined up to a scalar multiple)?

7. Suppose f is differentiable at $(3, 4)$, $\nabla f(3, 4) = \langle -\sqrt{3}, 1 \rangle$, and $\mathbf{u} = \left\langle \dfrac{\sqrt{3}}{2}, -\dfrac{1}{2} \right\rangle$. Compute $D_{\mathbf{u}} f(3, 4)$.

8. Suppose f is differentiable at $(9, 9)$, $\nabla f(9, 9) = \langle 3, 1 \rangle$, and $\mathbf{w} = \langle 1, -1 \rangle$. Compute the directional derivative of f at $(9, 9)$ in the direction of the vector $\mathbf{w}$.

9. Suppose f is differentiable at $(3, 4)$. Assume $\mathbf{u}$, $\mathbf{v}$, and $\mathbf{w}$ are unit vectors, $\mathbf{v} = -\mathbf{u}$, $\mathbf{w} \cdot \nabla f(3, 4) = 0$, and $D_{\mathbf{u}} f(3, 4) = 7$. Find $D_{\mathbf{v}} f(3, 4)$ and $D_{\mathbf{w}} f(3, 4)$.

10. Suppose f is differentiable at $(1, 2)$ and $\nabla f(1, 2) = \langle 3, 4 \rangle$. Find the slope of the line tangent to the level curve of f at $(1, 2)$.

Practice Exercises

11. **Directional derivatives** Consider the function
$$f(x, y) = 8 - \frac{x^2}{2} - y^2,$$
whose graph is a paraboloid (see figure).

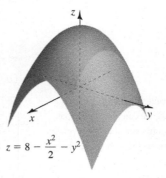

$z = 8 - \dfrac{x^2}{2} - y^2$

	$(a, b) = (2, 0)$	$(a, b) = (0, 2)$	$(a, b) = (1, 1)$
$\mathbf{u} = \left\langle \frac{\sqrt{2}}{2}, \frac{\sqrt{2}}{2} \right\rangle$			
$\mathbf{v} = \left\langle -\frac{\sqrt{2}}{2}, \frac{\sqrt{2}}{2} \right\rangle$			
$\mathbf{w} = \left\langle -\frac{\sqrt{2}}{2}, -\frac{\sqrt{2}}{2} \right\rangle$			

a. Fill in the table with the values of the directional derivative at the points (a, b) in the directions given by the unit vectors $\mathbf{u}$, $\mathbf{v}$, and $\mathbf{w}$.

b. Interpret each of the directional derivatives computed in part (a) at the point $(2, 0)$.

12. **Directional derivatives** Consider the function $f(x, y) = 2x^2 + y^2$, whose graph is a paraboloid (see figure).

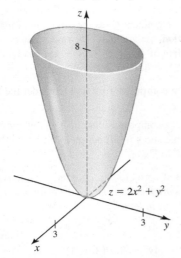

$z = 2x^2 + y^2$

	$(a, b) = (1, 0)$	$(a, b) = (1, 1)$	$(a, b) = (1, 2)$
$\mathbf{u} = \langle 1, 0 \rangle$			
$\mathbf{v} = \langle \frac{\sqrt{2}}{2}, \frac{\sqrt{2}}{2} \rangle$			
$\mathbf{w} = \langle 0, 1 \rangle$			

a. Fill in the table with the values of the directional derivative at the points (a, b) in the directions given by the unit vectors $\mathbf{u}$, $\mathbf{v}$, and $\mathbf{w}$.
b. Interpret each of the directional derivatives computed in part (a) at the point $(1, 0)$.

13–20. Computing gradients *Compute the gradient of the following functions and evaluate it at the given point P.*

13. $f(x, y) = 2 + 3x^2 - 5y^2$; $P(2, -1)$

14. $f(x, y) = 4x^2 - 2xy + y^2$; $P(-1, -5)$

15. $g(x, y) = x^2 - 4x^2y - 8xy^2$; $P(-1, 2)$

16. $p(x, y) = \sqrt{12 - 4x^2 - y^2}$; $P(-1, -1)$

17. $f(x, y) = xe^{2xy}$; $P(1, 0)$

18. $f(x, y) = \sin(3x + 2y)$; $P(\pi, 3\pi/2)$

19. $F(x, y) = e^{-x^2 - 2y^2}$; $P(-1, 2)$

20. $h(x, y) = \ln(1 + x^2 + 2y^2)$; $P(2, -3)$

21–30. Computing directional derivatives with the gradient *Compute the directional derivative of the following functions at the given point P in the direction of the given vector. Be sure to use a unit vector for the direction vector.*

21. $f(x, y) = x^2 - y^2$; $P(-1, -3)$; $\langle \frac{3}{5}, -\frac{4}{5} \rangle$

22. $f(x, y) = 3x^2 + y^3$; $P(3, 2)$; $\langle \frac{5}{13}, \frac{12}{13} \rangle$

23. $f(x, y) = 10 - 3x^2 + \frac{y^4}{4}$; $P(2, -3)$; $\langle \frac{\sqrt{3}}{2}, -\frac{1}{2} \rangle$

24. $g(x, y) = \sin \pi(2x - y)$; $P(-1, -1)$; $\langle \frac{5}{13}, -\frac{12}{13} \rangle$

25. $f(x, y) = \sqrt{4 - x^2 - 2y}$; $P(2, -2)$; $\langle \frac{1}{\sqrt{5}}, \frac{2}{\sqrt{5}} \rangle$

26. $f(x, y) = 13e^{xy}$; $P(1, 0)$; $\langle 5, 12 \rangle$

27. $f(x, y) = 3x^2 + 2y + 5$; $P(1, 2)$; $\langle -3, 4 \rangle$

28. $h(x, y) = e^{-x - y}$; $P(\ln 2, \ln 3)$; $\langle 1, 1 \rangle$

29. $g(x, y) = \ln(4 + x^2 + y^2)$; $g(-1, 2)$; $\langle 2, 1 \rangle$

30. $f(x, y) = \frac{x}{x - y}$; $P(4, 1)$; $\langle -1, 2 \rangle$

31–36. Direction of steepest ascent and descent *Consider the following functions and points P.*

a. *Find the unit vectors that give the direction of steepest ascent and steepest descent at P.*
b. *Find a vector that points in a direction of no change in the function at P.*

31. $f(x, y) = x^2 - 4y^2 - 9$; $P(1, -2)$

32. $f(x, y) = x^2 + 4xy - y^2$; $P(2, 1)$

33. $f(x, y) = x^4 - x^2y + y^2 + 6$; $P(-1, 1)$

34. $p(x, y) = \sqrt{20 + x^2 + 2xy - y^2}$; $P(1, 2)$

35. $F(x, y) = e^{-x^2/2 - y^2/2}$; $P(-1, 1)$

36. $f(x, y) = 2 \sin(2x - 3y)$; $P(0, \pi)$

37–42. Interpreting directional derivatives *A function f and a point P are given. Let θ correspond to the direction of the directional derivative.*

a. *Find the gradient and evaluate it at P.*
b. *Find the angles θ (with respect to the positive x-axis) associated with the directions of maximum increase, maximum decrease, and zero change.*
c. *Write the directional derivative at P as a function of θ; call this function g.*
d. *Find the value of θ that maximizes g(θ) and find the maximum value.*
e. *Verify that the value of θ that maximizes g corresponds to the direction of the gradient. Verify that the maximum value of g equals the magnitude of the gradient.*

37. $f(x, y) = 10 - 2x^2 - 3y^2$; $P(3, 2)$

38. $f(x, y) = 8 + x^2 + 3y^2$; $P(-3, -1)$

39. $f(x, y) = \sqrt{2 + x^2 + y^2}$; $P(\sqrt{3}, 1)$

40. $f(x, y) = \sqrt{12 - x^2 - y^2}$; $P(-1, -1/\sqrt{3})$

41. $f(x, y) = e^{-x^2 - 2y^2}$; $P(-1, 0)$

42. $f(x, y) = \ln(1 + 2x^2 + 3y^2)$; $P(3/4, -\sqrt{3})$

43–46. Directions of change *Consider the following functions f and points P. Sketch the xy-plane showing P and the level curve through P. Indicate (as in Figure 15.52) the directions of maximum increase, maximum decrease, and no change for f.*

43. $f(x, y) = 8 + 4x^2 + 2y^2$; $P(2, -4)$

44. $f(x, y) = -4 + 6x^2 + 3y^2$; $P(-1, -2)$

45. $f(x, y) = x^2 + xy + y^2 + 7$; $P(-3, 3)$

46. $f(x, y) = \tan(2x + 2y)$; $P(\pi/16, \pi/16)$

47–50. Level curves *Consider the paraboloid* $f(x, y) = 16 - \dfrac{x^2}{4} - \dfrac{y^2}{16}$ *and the point P on the given level curve of f. Compute the slope of the line tangent to the level curve at P, and verify that the tangent line is orthogonal to the gradient at that point.*

47. $f(x, y) = 0; P(0, 16)$ **48.** $f(x, y) = 0; P(8, 0)$

49. $f(x, y) = 12; P(4, 0)$ **50.** $f(x, y) = 12; P(2\sqrt{3}, 4)$

51–54. Level curves *Consider the upper half of the ellipsoid*

$$f(x, y) = \sqrt{1 - \dfrac{x^2}{4} - \dfrac{y^2}{16}}$$ *and the point P on the given level curve*

of f. Compute the slope of the line tangent to the level curve at P, and verify that the tangent line is orthogonal to the gradient at that point.

51. $f(x, y) = \dfrac{\sqrt{3}}{2}; P\left(\dfrac{1}{2}, \sqrt{3}\right)$ **52.** $f(x, y) = \dfrac{1}{\sqrt{2}}; P(0, \sqrt{8})$

53. $f(x, y) = \dfrac{1}{\sqrt{2}}; P(\sqrt{2}, 0)$ **54.** $f(x, y) = \dfrac{1}{\sqrt{2}}; P(1, 2)$

55–58. Path of steepest descent *Consider each of the following surfaces and the point P on the surface.*

a. *Find the gradient of f.*
b. *Let C′ be the path of steepest descent on the surface beginning at P, and let C be the projection of C′ on the xy-plane. Find an equation of C in the xy-plane.*
c. *Find parametric equations for the path C′ on the surface.*

55. $f(x, y) = 4 + x$ (a plane); $P(4, 4, 8)$

56. $f(x, y) = y + x$ (a plane); $P(2, 2, 4)$

57. $f(x, y) = 4 - x^2 - 2y^2$ (a paraboloid); $P(1, 1, 1)$

58. $f(x, y) = y + x^{-1}; P(1, 2, 3)$

59–66. Gradients in three dimensions *Consider the following functions f, points P, and unit vectors* **u**.

a. *Compute the gradient of f and evaluate it at P.*
b. *Find the unit vector in the direction of maximum increase of f at P.*
c. *Find the rate of change of the function in the direction of maximum increase at P.*
d. *Find the directional derivative at P in the direction of the given vector.*

59. $f(x, y, z) = x^2 + 2y^2 + 4z^2 + 10; P(1, 0, 4); \left\langle \dfrac{1}{\sqrt{2}}, 0, \dfrac{1}{\sqrt{2}} \right\rangle$

60. $f(x, y, z) = 4 - x^2 + 3y^2 + \dfrac{z^2}{2}; P(0, 2, -1); \left\langle 0, \dfrac{1}{\sqrt{2}}, -\dfrac{1}{\sqrt{2}} \right\rangle$

61. $f(x, y, z) = 1 + 4xyz; P(1, -1, -1); \left\langle \dfrac{1}{\sqrt{3}}, \dfrac{1}{\sqrt{3}}, -\dfrac{1}{\sqrt{3}} \right\rangle$

62. $f(x, y, z) = xy + yz + xz + 4; P(2, -2, 1); \left\langle 0, -\dfrac{1}{\sqrt{2}}, -\dfrac{1}{\sqrt{2}} \right\rangle$

63. $f(x, y, z) = 1 + \sin(x + 2y - z); P\left(\dfrac{\pi}{6}, \dfrac{\pi}{6}, -\dfrac{\pi}{6}\right); \left\langle \dfrac{1}{3}, \dfrac{2}{3}, \dfrac{2}{3} \right\rangle$

64. $f(x, y, z) = e^{xyz-1}; P(0, 1, -1); \left\langle -\dfrac{2}{3}, \dfrac{2}{3}, -\dfrac{1}{3} \right\rangle$

65. $f(x, y, z) = \ln(1 + x^2 + y^2 + z^2); P(1, 1, -1); \left\langle \dfrac{2}{3}, \dfrac{2}{3}, -\dfrac{1}{3} \right\rangle$

66. $f(x, y, z) = \dfrac{x - z}{y - z}; P(3, 2, -1); \left\langle \dfrac{1}{3}, \dfrac{2}{3}, -\dfrac{2}{3} \right\rangle$

67. Explain why or why not Determine whether the following statements are true and give an explanation or counterexample.

a. If $f(x, y) = x^2 + y^2 - 10$, then $\nabla f(x, y) = 2x + 2y$.
b. Because the gradient gives the direction of maximum increase of a function, the gradient is always positive.
c. The gradient of $f(x, y, z) = 1 + xyz$ has four components.
d. If $f(x, y, z) = 4$, then $\nabla f = \mathbf{0}$.

68. Gradient of a composite function Consider the function $F(x, y, z) = e^{xyz}$.

a. Write F as a composite function $f \circ g$, where f is a function of one variable and g is a function of three variables.
b. Relate ∇F to ∇g.

69–72. Directions of zero change *Find the directions in the xy-plane in which the following functions have zero change at the given point. Express the directions in terms of unit vectors.*

69. $f(x, y) = 12 - 4x^2 - y^2; P(1, 2, 4)$

70. $f(x, y) = x^2 - 4y^2 - 8; P(4, 1, 4)$

71. $f(x, y) = \sqrt{3 + 2x^2 + y^2}; P(1, -2, 3)$

72. $f(x, y) = e^{1-xy}; P(1, 0, e)$

73. Steepest ascent on a plane Suppose a long sloping hillside is described by the plane $z = ax + by + c$, where a, b, and c are constants. Find the path in the xy-plane, beginning at (x_0, y_0), that corresponds to the path of steepest ascent on the hillside.

74. Gradient of a distance function Let (a, b) be a given point in $\mathbb{R}^2$, and let $d = f(x, y)$ be the distance between (a, b) and the variable point (x, y).

a. Show that the graph of f is a cone.
b. Show that the gradient of f at any point other than (a, b) is a unit vector.
c. Interpret the direction and magnitude of ∇f.

75–78. Looking ahead—tangent planes *Consider the following surfaces f(x, y, z) = 0, which may be regarded as a level surface of the function w = f(x, y, z). A point P(a, b, c) on the surface is also given.*

a. *Find the (three-dimensional) gradient of f and evaluate it at P.*
b. *The set of all vectors orthogonal to the gradient with their tails at P form a plane. Find an equation of that plane (soon to be called the tangent plane).*

75. $f(x, y, z) = x^2 + y^2 + z^2 - 3 = 0; P(1, 1, 1)$

76. $f(x, y, z) = 8 - xyz = 0; P(2, 2, 2)$

77. $f(x, y, z) = e^{x+y-z} - 1 = 0; P(1, 1, 2)$

78. $f(x, y, z) = xy + xz - yz - 1 = 0; P(1, 1, 1)$

▮ 79. A traveling wave A snapshot (frozen in time) of a set of water waves is described by the function $z = 1 + \sin(x - y)$, where z gives the height of the waves and (x, y) are coordinates in the horizontal plane $z = 0$.

a. Use a graphing utility to graph $z = 1 + \sin(x - y)$.
b. The crests and the troughs of the waves are aligned in the direction in which the height function has zero change. Find the direction in which the crests and troughs are aligned.
c. If you were surfing on one of these waves and wanted the steepest descent from the crest to the trough, in which direction would you point your surfboard (given in terms of a unit vector in the xy-plane)?
d. Check that your answers to parts (b) and (c) are consistent with the graph of part (a).

80. Traveling waves in general Generalize Exercise 79 by considering a set of waves described by the function $z = A + \sin(ax - by)$, where a, b, and A are real numbers.

    a. Find the direction in which the crests and troughs of the waves are aligned. Express your answer as a unit vector in terms of a and b.

    b. Find the surfer's direction—that is, the direction of steepest descent from a crest to a trough. Express your answer as a unit vector in terms of a and b.

Explorations and Challenges

81–83. Potential functions *Potential functions arise frequently in physics and engineering. A potential function has the property that a field of interest (for example, an electric field, a gravitational field, or a velocity field) is the gradient of the potential (or sometimes the negative of the gradient of the potential). (Potential functions are considered in depth in Chapter 17.)*

81. Electric potential due to a point charge The electric field due to a point charge of strength Q at the origin has a potential function $\varphi = kQ/r$, where $r^2 = x^2 + y^2 + z^2$ is the square of the distance between a variable point $P(x, y, z)$ and the charge, and $k > 0$ is a physical constant. The electric field is given by $\mathbf{E} = -\nabla\varphi$, where $\nabla\varphi$ is the gradient in three dimensions.

    a. Show that the three-dimensional electric field due to a point charge is given by

$$\mathbf{E}(x, y, z) = kQ\left\langle \frac{x}{r^3}, \frac{y}{r^3}, \frac{z}{r^3} \right\rangle.$$

    b. Show that the electric field at a point has a magnitude $|\mathbf{E}| = \dfrac{kQ}{r^2}$. Explain why this relationship is called an inverse square law.

82. Gravitational potential The gravitational potential associated with two objects of mass M and m is $\varphi = -GMm/r$, where G is the gravitational constant. If one of the objects is at the origin and the other object is at $P(x, y, z)$, then $r^2 = x^2 + y^2 + z^2$ is the square of the distance between the objects. The gravitational field at P is given by $\mathbf{F} = -\nabla\varphi$, where $\nabla\varphi$ is the gradient in three dimensions. Show that the force has a magnitude $|\mathbf{F}| = GMm/r^2$. Explain why this relationship is called an inverse square law.

83. Velocity potential In two dimensions, the motion of an ideal fluid (an incompressible and irrotational fluid) is governed by a velocity potential φ. The velocity components of the fluid, u in the x-direction and v in the y-direction, are given by $\langle u, v \rangle = \nabla\varphi$. Find the velocity components associated with the velocity potential $\varphi(x, y) = \sin \pi x \sin 2\pi y$.

84. Gradients for planes Prove that for the plane described by $f(x, y) = Ax + By$, where A and B are nonzero constants, the gradient is constant (independent of (x, y)). Interpret this result.

85. Rules for gradients Use the definition of the gradient (in two or three dimensions), assume f and g are differentiable functions on $\mathbb{R}^2$ or $\mathbb{R}^3$, and let c be a constant. Prove the following gradient rules.

    a. Constants Rule: $\nabla(cf) = c\nabla f$

    b. Sum Rule: $\nabla(f + g) = \nabla f + \nabla g$

    c. Product Rule: $\nabla(fg) = (\nabla f)g + f\nabla g$

    d. Quotient Rule: $\nabla\left(\dfrac{f}{g}\right) = \dfrac{g\nabla f - f\nabla g}{g^2}$

    e. Chain Rule: $\nabla(f \circ g) = f'(g)\nabla g$, where f is a function of one variable

86–91. Using gradient rules *Use the gradient rules of Exercise 85 to find the gradient of the following functions.*

86. $f(x, y) = xy \cos(xy)$ **87.** $f(x, y) = \dfrac{x + y}{x^2 + y^2}$

88. $f(x, y) = \ln(1 + x^2 + y^2)$

89. $f(x, y, z) = \sqrt{25 - x^2 - y^2 - z^2}$

90. $f(x, y, z) = (x + y + z)e^{xyz}$

91. $f(x, y, z) = \dfrac{x + yz}{y + xz}$

QUICK CHECK ANSWERS

1. If $\mathbf{u} = \langle u_1, u_2 \rangle = \langle 1, 0 \rangle$ then

$$D_{\mathbf{u}} f(a, b) = \lim_{h \to 0} \frac{f(a + hu_1, b + hu_2) - f(a, b)}{h}$$

$$= \lim_{h \to 0} \frac{f(a + h, b) - f(a, b)}{h} = f_x(a, b).$$

Similarly, when $\mathbf{u} = \langle 0, 1 \rangle$, the partial derivative $f_y(a, b)$ results. **2.** The vector from (a, b) to $(a + \Delta s u_1, b + \Delta s u_2)$ is $\langle \Delta s u_1, \Delta s u_2 \rangle = \Delta s \langle u_1, u_2 \rangle = \Delta s \mathbf{u}$. Its length is $|\Delta s \mathbf{u}| = \Delta s |\mathbf{u}| = \Delta s$. Therefore, s measures arc length. **3.** Reversing (negating) the direction vector negates the directional derivative, so the respective values are approximately -2.47 and 0.98. **4.** The gradient is $\langle 2x, 2y \rangle$, which, evaluated at (a, a), is $\langle 2a, 2a \rangle$. Taking the dot product of the gradient and the vector $\langle -1, 1 \rangle$ (a vector parallel to a line of slope -1), we see that $\langle 2a, 2a \rangle \cdot \langle -1, 1 \rangle = 0$. **6.** $\langle 2, -1, 2 \rangle$ ◄

15.6 Tangent Planes and Linear Approximation

In Section 4.6, we saw that if we zoom in on a point on a smooth curve (one described by a differentiable function), the curve looks more and more like the tangent line at that point. Once we have the tangent line at a point, it can be used to approximate function values and to estimate changes in the dependent variable. In this section, the analogous story is developed in three dimensions. Now we see that differentiability at a point (as discussed in Section 15.3) implies the existence of a tangent *plane* at that point (**Figure 15.58**).

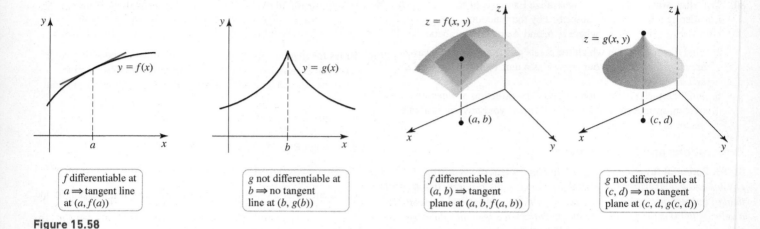

f differentiable at
$a \Rightarrow$ tangent line
at $(a, f(a))$

g not differentiable at
$b \Rightarrow$ no tangent
line at $(b, g(b))$

f differentiable at
$(a, b) \Rightarrow$ tangent
plane at $(a, b, f(a, b))$

g not differentiable at
$(c, d) \Rightarrow$ no tangent
plane at $(c, d, g(c, d))$

Figure 15.58

Consider a smooth surface described by a differentiable function f, and focus on a single point on the surface. As we zoom in on that point (**Figure 15.59**), the surface appears more and more like a plane. The first step is to define this plane carefully; it is called the *tangent plane*. Once we have the tangent plane, we can use it to approximate function values and to estimate changes in the dependent variable.

Tangent Planes

Recall that a surface in $\mathbb{R}^3$ may be defined in at least two different ways:

• **Explicitly** in the form $z = f(x, y)$ or
• **Implicitly** in the form $F(x, y, z) = 0$.

It is easiest to begin by considering a surface defined implicitly by $F(x, y, z) = 0$, where F is differentiable at a particular point. Such a surface may be viewed as a level surface of a function $w = F(x, y, z)$; it is the level surface for $w = 0$.

QUICK CHECK 1 Write the function $z = xy + x - y$ in the form $F(x, y, z) = 0$. ◄

Tangent Planes for $F(x, y, z) = 0$ To find an equation of the tangent plane, consider a smooth curve $C: \mathbf{r}(t) = \langle x(t), y(t), z(t) \rangle$ that lies on the surface $F(x, y, z) = 0$ (**Figure 15.60a**). Because the points of C lie on the surface, we have $F(x(t), y(t), z(t)) = 0$.

As we zoom in on a smooth surface ...

... it appears more like a plane.

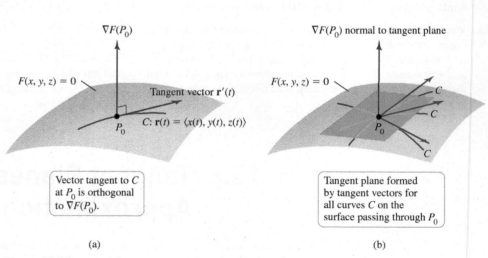

$\nabla F(P_0)$

$F(x, y, z) = 0$

Tangent vector $\mathbf{r}'(t)$

P_0 $C: \mathbf{r}(t) = \langle x(t), y(t), z(t) \rangle$

Vector tangent to C at P_0 is orthogonal to $\nabla F(P_0)$.

(a)

$\nabla F(P_0)$ normal to tangent plane

$F(x, y, z) = 0$

C
C
P_0
C

Tangent plane formed by tangent vectors for all curves C on the surface passing through P_0

(b)

Figure 15.59

Figure 15.60

Differentiating both sides of this equation with respect to t reveals a useful relationship. The derivative of the right side is 0. The Chain Rule applied to the left side yields

$$\frac{d}{dt}\left(F(x(t), y(t), z(t))\right) = \frac{\partial F}{\partial x}\frac{dx}{dt} + \frac{\partial F}{\partial y}\frac{dy}{dt} + \frac{\partial F}{\partial z}\frac{dz}{dt}$$

$$= \underbrace{\left\langle\frac{\partial F}{\partial x}, \frac{\partial F}{\partial y}, \frac{\partial F}{\partial z}\right\rangle}_{\nabla F(x, y, z)} \cdot \underbrace{\left\langle\frac{dx}{dt}, \frac{dy}{dt}, \frac{dz}{dt}\right\rangle}_{\mathbf{r}'(t)}$$

$$= \nabla F(x, y, z) \cdot \mathbf{r}'(t).$$

Therefore, $\nabla F(x, y, z) \cdot \mathbf{r}'(t) = 0$ and at any point on the curve, the tangent vector $\mathbf{r}'(t)$ is orthogonal to the gradient.

Now fix a point $P_0(a, b, c)$ on the surface, assume $\nabla F(a, b, c) \neq \mathbf{0}$, and let C be any smooth curve on the surface passing through P_0. We have shown that any vector tangent to C is orthogonal to $\nabla F(a, b, c)$ at P_0. Because this argument applies to *all* smooth curves on the surface passing through P_0, the tangent vectors for all these curves (with their tails at P_0) are orthogonal to $\nabla F(a, b, c)$; therefore, they all lie in the same plane (**Figure 15.60b**). This plane is called the *tangent plane* at P_0. We can easily find an equation of the tangent plane because we know both a point on the plane $P_0(a, b, c)$ and a normal vector $\nabla F(a, b, c)$; an equation is

$$\nabla F(a, b, c) \cdot \langle x - a, y - b, z - c \rangle = 0.$$

> Recall that an equation of the plane passing though (a, b, c) with a normal vector $\mathbf{n} = \langle n_1, n_2, n_3 \rangle$ is $n_1(x - a) + n_2(y - b) + n_3(z - c) = 0$.

> If $\mathbf{r}$ is a position vector corresponding to an arbitrary point on the tangent plane, and $\mathbf{r}_0$ is a position vector corresponding to a fixed point (a, b, c) on the plane, then an equation of the tangent plane may be written concisely as
>
> $$\nabla F(a, b, c) \cdot (\mathbf{r} - \mathbf{r}_0) = 0.$$
>
> Notice the analogy with tangent lines and level curves (Section 15.5). An equation of the line tangent to $f(x, y) = 0$ at (a, b) is
>
> $$\nabla f(a, b) \cdot \langle x - a, y - b \rangle = 0.$$

DEFINITION Equation of the Tangent Plane for $F(x, y, z) = 0$

Let F be differentiable at the point $P_0(a, b, c)$ with $\nabla F(a, b, c) \neq \mathbf{0}$. The plane tangent to the surface $F(x, y, z) = 0$ at P_0, called the **tangent plane**, is the plane passing through P_0 orthogonal to $\nabla F(a, b, c)$. An equation of the tangent plane is

$$F_x(a, b, c)(x - a) + F_y(a, b, c)(y - b) + F_z(a, b, c)(z - c) = 0.$$

EXAMPLE 1 Equation of a tangent plane Consider the ellipsoid

$$F(x, y, z) = \frac{x^2}{9} + \frac{y^2}{25} + z^2 - 1 = 0.$$

a. Find an equation of the plane tangent to the ellipsoid at $\left(0, 4, \frac{3}{5}\right)$.

b. At what points on the ellipsoid is the tangent plane horizontal?

SOLUTION

a. Notice that we have written the equation of the ellipsoid in the implicit form $F(x, y, z) = 0$. The gradient of F is $\nabla F(x, y, z) = \left\langle\frac{2x}{9}, \frac{2y}{25}, 2z\right\rangle$. Evaluating at $\left(0, 4, \frac{3}{5}\right)$, we have

$$\nabla F\left(0, 4, \frac{3}{5}\right) = \left\langle 0, \frac{8}{25}, \frac{6}{5}\right\rangle.$$

An equation of the tangent plane at this point is

$$0 \cdot (x - 0) + \frac{8}{25}(y - 4) + \frac{6}{5}\left(z - \frac{3}{5}\right) = 0,$$

or $4y + 15z = 25$. The equation does not involve x, so the tangent plane is parallel to (does not intersect) the x-axis (**Figure 15.61**).

b. A horizontal plane has a normal vector of the form $\langle 0, 0, c \rangle$, where $c \neq 0$. A plane tangent to the ellipsoid has a normal vector $\nabla F(x, y, z) = \left\langle\frac{2x}{9}, \frac{2y}{25}, 2z\right\rangle$. Therefore, the ellipsoid has a horizontal tangent plane when $F_x = \frac{2x}{9} = 0$ and $F_y = \frac{2y}{25} = 0$, or

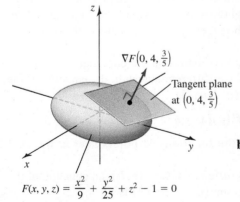

$F(x, y, z) = \dfrac{x^2}{9} + \dfrac{y^2}{25} + z^2 - 1 = 0$

Figure 15.61

when $x = 0$ and $y = 0$. Substituting these values into the original equation for the ellipsoid, we find that horizontal planes occur at $(0, 0, 1)$ and $(0, 0, -1)$.

Related Exercises 14, 16 ◄

▶ This result extends Theorem 15.12, which states that for functions $f(x, y) = 0$, the gradient at a point is orthogonal to the level curve that passes through that point.

The preceding discussion allows us to confirm a claim made in Section 15.5. The surface $F(x, y, z) = 0$ is a level surface of the function $w = F(x, y, z)$ (corresponding to $w = 0$). At any point on that surface, the tangent plane has a normal vector $\nabla F(x, y, z)$. Therefore, the gradient $\nabla F(x, y, z)$ is orthogonal to the level surface $F(x, y, z) = 0$ at all points of the domain at which F is differentiable.

Tangent Planes for $z = f(x, y)$ Surfaces in $\mathbb{R}^3$ are often defined explicitly in the form $z = f(x, y)$. In this situation, the equation of the tangent plane is a special case of the general equation just derived. The equation $z = f(x, y)$ is written as $F(x, y, z) = z - f(x, y) = 0$, and the gradient of F at the point $(a, b, f(a, b))$ is

▶ To be clear, when $F(x, y, z) = z - f(x, y)$, we have $F_x = -f_x$, $F_y = -f_y$, and $F_z = 1$.

$$\nabla F(a, b, f(a, b)) = \langle F_x(a, b, f(a, b)), F_y(a, b, f(a, b)), F_z(a, b, f(a, b)) \rangle$$
$$= \langle -f_x(a, b), -f_y(a, b), 1 \rangle.$$

Using the tangent plane definition, an equation of the plane tangent to the surface $z = f(x, y)$ at the point $(a, b, f(a, b))$ is

$$-f_x(a, b)(x - a) - f_y(a, b)(y - b) + 1(z - f(a, b)) = 0.$$

After some rearranging, we obtain an equation of the tangent plane.

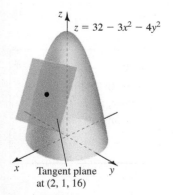

$z = 32 - 3x^2 - 4y^2$

Tangent plane at $(2, 1, 16)$

Figure 15.62

Tangent Plane for $z = f(x, y)$

Let f be differentiable at the point (a, b). An equation of the plane tangent to the surface $z = f(x, y)$ at the point $(a, b, f(a, b))$ is

$$z = f_x(a, b)(x - a) + f_y(a, b)(y - b) + f(a, b).$$

EXAMPLE 2   Tangent plane for $z = f(x, y)$ Find an equation of the plane tangent to the paraboloid $z = f(x, y) = 32 - 3x^2 - 4y^2$ at $(2, 1, 16)$.

SOLUTION   The partial derivatives are $f_x = -6x$ and $f_y = -8y$. Evaluating the partial derivatives at $(2, 1)$, we have $f_x(2, 1) = -12$ and $f_y(2, 1) = -8$. Therefore, an equation of the tangent plane (**Figure 15.62**) is

$$z = f_x(a, b)(x - a) + f_y(a, b)(y - b) + f(a, b)$$
$$= -12(x - 2) - 8(y - 1) + 16$$
$$= -12x - 8y + 48.$$

Related Exercises 17–18 ◄

Linear Approximation

With a function of the form $y = f(x)$, the tangent line at a point often gives good approximations to the function near that point. A straightforward extension of this idea applies to approximating functions of two variables with tangent planes. As before, the method is called *linear approximation*.

Figure 15.63 shows the details of linear approximation in the one- and two-variable cases. In the one-variable case (Section 4.6), if f is differentiable at a, the equation of the line tangent to the curve $y = f(x)$ at the point $(a, f(a))$ is

$$L(x) = f(a) + f'(a)(x - a).$$

The tangent line gives an approximation to the function. At points near a, we have $f(x) \approx L(x)$.

The two-variable case is analogous. If f is differentiable at (a, b), an equation of the plane tangent to the surface $z = f(x, y)$ at the point $(a, b, f(a, b))$ is

$$L(x, y) = f_x(a, b)(x - a) + f_y(a, b)(y - b) + f(a, b).$$

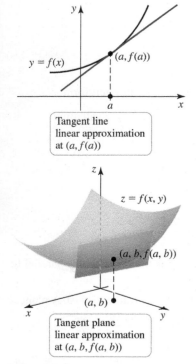

$y = f(x)$

$(a, f(a))$

Tangent line linear approximation at $(a, f(a))$

$z = f(x, y)$

$(a, b, f(a, b))$

(a, b)

Tangent plane linear approximation at $(a, b, f(a, b))$

Figure 15.63

This tangent plane is the linear approximation to f at (a, b). At points near (a, b), we have $f(x, y) \approx L(x, y)$. The pattern established here continues for linear approximations in higher dimensions: For each additional variable, a new term is added to the approximation formula.

> The term *linear approximation* applies in $\mathbb{R}^2$, in $\mathbb{R}^3$, and in higher dimensions. Recall that lines in $\mathbb{R}^2$ and planes in $\mathbb{R}^3$ are described by linear functions of the independent variables. In both cases, we call the linear approximation L.

DEFINITION Linear Approximation

Let f be differentiable at (a, b). The linear approximation to the surface $z = f(x, y)$ at the point $(a, b, f(a, b))$ is the tangent plane at that point, given by the equation

$$L(x, y) = f_x(a, b)(x - a) + f_y(a, b)(y - b) + f(a, b).$$

For a function of three variables, the linear approximation to $w = f(x, y, z)$ at the point $(a, b, c, f(a, b, c))$ is given by

$$L(x, y, z) = f_x(a, b, c)(x - a) + f_y(a, b, c)(y - b)$$
$$+ f_z(a, b, c)(z - c) + f(a, b, c).$$

EXAMPLE 3 Linear approximation Let $f(x, y) = \dfrac{5}{x^2 + y^2}$.

a. Find the linear approximation to the function at the point $(-1, 2, 1)$.

b. Use the linear approximation to estimate the value of $f(-1.05, 2.1)$.

SOLUTION

a. The partial derivatives of f are

$$f_x = -\frac{10x}{(x^2 + y^2)^2} \quad \text{and} \quad f_y = -\frac{10y}{(x^2 + y^2)^2}.$$

Evaluated at $(-1, 2)$, we have $f_x(-1, 2) = \frac{2}{5} = 0.4$ and $f_y(-1, 2) = -\frac{4}{5} = -0.8$. Therefore, the linear approximation to the function at $(-1, 2, 1)$ is

$$L(x, y) = f_x(-1, 2)(x - (-1)) + f_y(-1, 2)(y - 2) + f(-1, 2)$$
$$= 0.4(x + 1) - 0.8(y - 2) + 1$$
$$= 0.4x - 0.8y + 3.$$

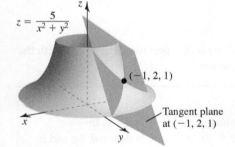

$z = \dfrac{5}{x^2 + y^2}$

$(-1, 2, 1)$

Tangent plane at $(-1, 2, 1)$

Figure 15.64

> Relative error $=$
> $$\dfrac{|\text{approximation} - \text{exact value}|}{|\text{exact value}|}$$

The surface and the tangent plane are shown in **Figure 15.64**.

b. The value of the function at the point $(-1.05, 2.1)$ is approximated by the value of the linear approximation at that point, which is

$$L(-1.05, 2.1) = 0.4(-1.05) - 0.8(2.1) + 3 = 0.90.$$

In this case, we can easily evaluate $f(-1.05, 2.1) \approx 0.907$ and compare the linear approximation with the exact value; the approximation has a relative error of about 0.8%.

Related Exercise 36 ◀

QUICK CHECK 2 Look at the graph of the surface in Example 3 (Figure 15.64) and explain why $f_x(-1, 2) > 0$ and $f_y(-1, 2) < 0$. ◀

Differentials and Change

Recall that for a function of the form $y = f(x)$, if the independent variable changes from x to $x + dx$, the corresponding change Δy in the dependent variable is approximated by the differential $dy = f'(x)\, dx$, which is the change in the linear approximation. Therefore, $\Delta y \approx dy$, with the approximation improving as dx approaches 0.

For functions of the form $z = f(x, y)$, we start with the linear approximation to the surface

$$f(x, y) \approx L(x, y) = f_x(a, b)(x - a) + f_y(a, b)(y - b) + f(a, b).$$

The exact change in the function between the points (a, b) and (x, y) is

$$\Delta z = f(x, y) - f(a, b).$$

Replacing $f(x, y)$ with its linear approximation, the change Δz is approximated by

$$\Delta z \approx \underbrace{L(x, y) - f(a, b)}_{dz} = \underbrace{f_x(a, b)(x - a)}_{dx} + \underbrace{f_y(a, b)(y - b)}_{dy}.$$

▶ Alternative notation for the differential at (a, b) is $dz|_{(a, b)}$ or $df|_{(a, b)}$.

The change in the x-coordinate is $dx = x - a$ and the change in the y-coordinate is $dy = y - b$ (**Figure 15.65**). As before, we let the differential dz denote the change in the linear approximation. Therefore, the approximate change in the z-coordinate is

$$\Delta z \approx dz = \underbrace{f_x(a, b)\, dx}_{\substack{\text{change in } z \text{ due} \\ \text{to change in } x}} + \underbrace{f_y(a, b)\, dy}_{\substack{\text{change in } z \text{ due} \\ \text{to change in } y}}.$$

This expression says that if we move the independent variables from (a, b) to $(x, y) = (a + dx, b + dy)$, the corresponding change in the dependent variable Δz has two contributions—one due to the change in x and one due to the change in y. If dx and dy are small in magnitude, then so is Δz. The approximation $\Delta z \approx dz$ improves as dx and dy approach 0. The relationships among the differentials are illustrated in Figure 15.65.

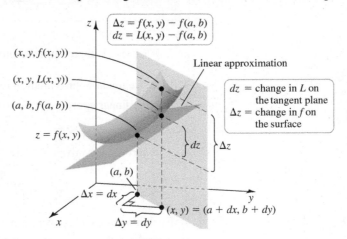

Figure 15.65

QUICK CHECK 3 Explain why, if $dx = 0$ or $dy = 0$ in the change formula for Δz, the result is the change formula for one variable. ◀

More generally, we replace the fixed point (a, b) in the previous discussion with the variable point (x, y) to arrive at the following definition.

DEFINITION The differential dz

Let f be differentiable at the point (x, y). The change in $z = f(x, y)$ as the independent variables change from (x, y) to $(x + dx, y + dy)$ is denoted Δz and is approximated by the differential dz:

$$\Delta z \approx dz = f_x(x, y)\, dx + f_y(x, y)\, dy.$$

EXAMPLE 4 Approximating function change Let $z = f(x, y) = \dfrac{5}{x^2 + y^2}$.

Approximate the change in z when the independent variables change from $(-1, 2)$ to $(-0.93, 1.94)$.

SOLUTION If the independent variables change from $(-1, 2)$ to $(-0.93, 1.94)$, then $dx = 0.07$ (an increase) and $dy = -0.06$ (a decrease). Using the values of the partial derivatives evaluated in Example 3, the corresponding change in z is approximately

$$\begin{aligned}
dz &= f_x(-1, 2)\, dx + f_y(-1, 2)\, dy \\
&= 0.4(0.07) + (-0.8)(-0.06) \\
&= 0.076.
\end{aligned}$$

Again, we can check the accuracy of the approximation. The actual change is $f(-0.93, 1.94) - f(-1, 2) \approx 0.080$, so the approximation has a 5% error.

Related Exercise 40 ◀

EXAMPLE 5   Body mass index   The body mass index (BMI) for an adult human is given by the function $B(w, h) = w/h^2$, where w is weight measured in kilograms and h is height measured in meters.

a. Use differentials to approximate the change in the BMI when weight increases from 55 to 56.5 kg and height increases from 1.65 to 1.66 m.

b. Which produces a greater *percentage* change in the BMI, a 1% change in the weight (at a constant height) or a 1% change in the height (at a constant weight)?

SOLUTION

a. The approximate change in the BMI is $dB = B_w\, dw + B_h\, dh$, where the derivatives are evaluated at $w = 55$ and $h = 1.65$, and the changes in the independent variables are $dw = 1.5$ and $dh = 0.01$. Evaluating the partial derivatives, we find that

$$B_w(w, h) = \frac{1}{h^2}, \qquad\qquad B_w(55, 1.65) \approx 0.37,$$

$$B_h(w, h) = -\frac{2w}{h^3}, \quad \text{and} \quad B_h(55, 1.65) \approx -24.49.$$

Therefore, the approximate change in the BMI is

$$
\begin{aligned}
dB &= B_w(55, 1.65)\, dw + B_h(55, 1.65)\, dh \\
&\approx (0.37)(1.5) + (-24.49)(0.01) \\
&\approx 0.56 - 0.25 \\
&= 0.31.
\end{aligned}
$$

As expected, an increase in weight *increases* the BMI, while an increase in height *decreases* the BMI. In this case, the two contributions combine for a net increase in the BMI.

b. The changes dw, dh, and dB that appear in the differential change formula in part (a) are *absolute changes*. The corresponding *relative*, or *percentage*, changes are $\dfrac{dw}{w}, \dfrac{dh}{h}$, and $\dfrac{dB}{B}$. To introduce relative changes into the change formula, we divide both sides of $dB = B_w\, dw + B_h\, dh$ by $B = w/h^2 = wh^{-2}$. The result is

$$
\begin{aligned}
\frac{dB}{B} &= B_w\,\frac{dw}{wh^{-2}} + B_h\,\frac{dh}{wh^{-2}} \\
&= \frac{1}{h^2}\frac{dw}{wh^{-2}} - \frac{2w}{h^3}\frac{dh}{wh^{-2}} \qquad \text{Substitute for } B_w \text{ and } B_h. \\
&= \underbrace{\frac{dw}{w}}_{\substack{\text{relative} \\ \text{change} \\ \text{in } w}} - 2\underbrace{\frac{dh}{h}}_{\substack{\text{relative} \\ \text{change} \\ \text{in } h}}. \qquad\qquad \text{Simplify.}
\end{aligned}
$$

▶ See Exercises 68–69 for general results about relative or percentage changes in functions.

QUICK CHECK 4 In Example 5, interpret the facts that $B_w > 0$ and $B_h < 0$, for $w, h > 0$. ◄

This expression relates the relative changes in w, h, and B. With h constant ($dh = 0$), a 1% change in w ($dw/w = 0.01$) produces approximately a 1% change of the same sign in B. With w constant ($dw = 0$), a 1% change in h ($dh/h = 0.01$) produces approximately a 2% change in B of the opposite sign. We see that the BMI formula is more sensitive to small changes in h than in w.

Related Exercise 44 ◄

The differential for functions of two variables extends naturally to more variables. For example, if f is differentiable at (x, y, z) with $w = f(x, y, z)$, then

$$dw = f_x(x, y, z)\, dx + f_y(x, y, z)\, dy + f_z(x, y, z)\, dz.$$

The differential dw (or df) gives the approximate change in f at the point (x, y, z) due to changes of dx, dy, and dz in the independent variables.

r = 10 cm

t = 0.1 cm

h = 50 cm

Figure 15.66

EXAMPLE 6 **Manufacturing errors** A company manufactures cylindrical aluminum tubes to rigid specifications. The tubes are designed to have an outside radius of $r = 10$ cm, a height of $h = 50$ cm, and a thickness of $t = 0.1$ cm (**Figure 15.66**). The manufacturing process produces tubes with a maximum error of ± 0.05 cm in the radius and height, and a maximum error of ± 0.0005 cm in the thickness. The volume of the cylindrical tube is $V(r, h, t) = \pi h t (2r - t)$. Use differentials to estimate the maximum error in the volume of a tube.

SOLUTION The approximate change in the volume of a tube due to changes dr, dh, and dt in the radius, height, and thickness, respectively, is

$$dV = V_r \, dr + V_h \, dh + V_t \, dt.$$

The partial derivatives evaluated at $r = 10$, $h = 50$, and $t = 0.1$ are

$$\begin{aligned} V_r(r, h, t) &= 2\pi h t, & V_r(10, 50, 0.1) &= 10\pi, \\ V_h(r, h, t) &= \pi t (2r - t), & V_h(10, 50, 0.1) &= 1.99\pi, \\ V_t(r, h, t) &= 2\pi h (r - t), & V_t(10, 50, 0.1) &= 990\pi. \end{aligned}$$

We let $dr = dh = 0.05$ and $dt = 0.0005$ be the maximum errors in the radius, height, and thickness, respectively. The maximum error in the volume is approximately

$$\begin{aligned} dV &= V_r(10, 50, 0.1) \, dr + V_h(10, 50, 0.1) \, dh + V_t(10, 50, 0.1) \, dt \\ &= 10\pi(0.05) + 1.99\pi(0.05) + 990\pi(0.0005) \\ &\approx 1.57 + 0.31 + 1.56 \\ &= 3.44. \end{aligned}$$

The maximum error in the volume is approximately 3.44 cm³. Notice that the "magnification factor" for the thickness (990π) is roughly 100 and 500 times greater than the magnification factors for the radius and height, respectively. This means that for the same errors in r, h, and t, the volume is far more sensitive to errors in the thickness. The partial derivatives allow us to do a sensitivity analysis to determine which independent (input) variables are most critical in producing change in the dependent (output) variable.

Related Exercise 52 ◄

SECTION 15.6 EXERCISES

Getting Started

1. Suppose **n** is a vector normal to the tangent plane of the surface $F(x, y, z) = 0$ at a point. How is **n** related to the gradient of F at that point?

2. Write the explicit function $z = xy^2 + x^2 y - 10$ in the implicit form $F(x, y, z) = 0$.

3. Write an equation for the plane tangent to the surface $F(x, y, z) = 0$ at the point (a, b, c).

4. Write an equation for the plane tangent to the surface $z = f(x, y)$ at the point $(a, b, f(a, b))$.

5. Explain how to approximate a function f at a point near (a, b), where the values of f, f_x, and f_y are known at (a, b).

6. Explain how to approximate the change in a function f when the independent variables change from (a, b) to $(a + \Delta x, b + \Delta y)$.

7. Write the approximate change formula for a function $z = f(x, y)$ at the point (x, y) in terms of differentials.

8. Write the differential dw for the function $w = f(x, y, z)$.

9–10. *Suppose* $f(1, 2) = 4$, $f_x(1, 2) = 5$, *and* $f_y(1, 2) = -3$.

9. Find an equation of the plane tangent to the surface $z = f(x, y)$ at the point $P_0(1, 2, 4)$.

10. Find the linear approximation to f at $P_0(1, 2, 4)$, and use it to estimate $f(1.01, 1.99)$.

11–12. *Suppose* $F(0, 2, 1) = 0$, $F_x(0, 2, 1) = 3$, $F_y(0, 2, 1) = -1$, *and* $F_z(0, 2, 1) = 6$.

11. Find an equation of the plane tangent to the surface $F(x, y, z) = 0$ at the point $P_0(0, 2, 1)$.

12. Find the linear approximation to the function $w = F(x, y, z)$ at the point $P_0(0, 2, 1)$ and use it to estimate $F(0.1, 2, 0.99)$.

Practice Exercises

13–28. **Tangent planes** *Find an equation of the plane tangent to the following surfaces at the given points (two planes and two equations).*

13. $x^2 + y + z = 3$; $(1, 1, 1)$ and $(2, 0, -1)$

14. $x^2 + y^3 + z^4 = 2$; $(1, 0, 1)$ and $(-1, 0, 1)$

15. $xy + xz + yz - 12 = 0$; $(2, 2, 2)$ and $(2, 0, 6)$

16. $x^2 + y^2 - z^2 = 0$; $(3, 4, 5)$ and $(-4, -3, 5)$

17. $z = 4 - 2x^2 - y^2$; $(2, 2, -8)$ and $(-1, -1, 1)$

18. $z = 2 + 2x^2 + \dfrac{y^2}{2}$; $\left(-\dfrac{1}{2}, 1, 3\right)$ and $(3, -2, 22)$

19. $z = e^{xy}$; $(1, 0, 1)$ and $(0, 1, 1)$

20. $z = \sin xy + 2$; $(1, 0, 2)$ and $(0, 5, 2)$

21. $xy \sin z = 1$; $(1, 2, \pi/6)$ and $(-2, -1, 5\pi/6)$

22. $yze^{xz} - 8 = 0$; $(0, 2, 4)$ and $(0, -8, -1)$

23. $z^2 - x^2/16 - y^2/9 - 1 = 0$; $(4, 3, -\sqrt{3})$ and $(-8, 9, \sqrt{14})$

24. $2x + y^2 - z^2 = 0$; $(0, 1, 1)$ and $(4, 1, -3)$

25. $z = x^2 e^{x-y}$; $(2, 2, 4)$ and $(-1, -1, 1)$

26. $z = \ln(1 + xy)$; $(1, 2, \ln 3)$ and $(-2, -1, \ln 3)$

27. $z = \dfrac{x-y}{x^2 + y^2}$; $\left(1, 2, -\dfrac{1}{5}\right)$ and $\left(2, -1, \dfrac{3}{5}\right)$

28. $z = 2\cos(x - y) + 2$; $\left(\dfrac{\pi}{6}, -\dfrac{\pi}{6}, 3\right)$ and $\left(\dfrac{\pi}{3}, \dfrac{\pi}{3}, 4\right)$

29–32. Tangent planes *Find an equation of the plane tangent to the following surfaces at the given point.*

29. $z = \tan^{-1}(xy)$; $\left(1, 1, \dfrac{\pi}{4}\right)$ **30.** $z = \tan^{-1}(x + y)$; $(0, 0, 0)$

31. $\sin xyz = \dfrac{1}{2}$; $\left(\pi, 1, \dfrac{1}{6}\right)$ **32.** $\dfrac{x + z}{y - z} = 2$; $(4, 2, 0)$

33–38. Linear approximation

a. *Find the linear approximation to the function f at the given point.*
b. *Use part (a) to estimate the given function value.*

33. $f(x, y) = xy + x - y$; $(2, 3)$; estimate $f(2.1, 2.99)$.

34. $f(x, y) = 12 - 4x^2 - 8y^2$; $(-1, 4)$; estimate $f(-1.05, 3.95)$.

35. $f(x, y) = -x^2 + 2y^2$; $(3, -1)$; estimate $f(3.1, -1.04)$.

36. $f(x, y) = \sqrt{x^2 + y^2}$; $(3, -4)$; estimate $f(3.06, -3.92)$.

37. $f(x, y, z) = \ln(1 + x + y + 2z)$; $(0, 0, 0)$; estimate $f(0.1, -0.2, 0.2)$.

38. $f(x, y, z) = \dfrac{x + y}{x - z}$; $(3, 2, 4)$; estimate $f(2.95, 2.05, 4.02)$.

39–42. Approximate function change *Use differentials to approximate the change in z for the given changes in the independent variables.*

39. $z = 2x - 3y - 2xy$ when (x, y) changes from $(1, 4)$ to $(1.1, 3.9)$

40. $z = -x^2 + 3y^2 + 2$ when (x, y) changes from $(-1, 2)$ to $(-1.05, 1.9)$

41. $z = e^{x+y}$ when (x, y) changes from $(0, 0)$ to $(0.1, -0.05)$

42. $z = \ln(1 + x + y)$ when (x, y) changes from $(0, 0)$ to $(-0.1, 0.03)$

43. Changes in torus surface area The surface area of a torus with an inner radius r and an outer radius $R > r$ is $S = 4\pi^2(R^2 - r^2)$.

a. If r increases and R decreases, does S increase or decrease, or is it impossible to say?

b. If r increases and R increases, does S increase or decrease, or is it impossible to say?

c. Estimate the change in the surface area of the torus when r changes from $r = 3.00$ to $r = 3.05$ and R changes from $R = 5.50$ to $R = 5.65$.

d. Estimate the change in the surface area of the torus when r changes from $r = 3.00$ to $r = 2.95$ and R changes from $R = 7.00$ to $R = 7.04$.

e. Find the relationship between the changes in r and R that leaves the surface area (approximately) unchanged.

44. Changes in cone volume The volume of a right circular cone with radius r and height h is $V = \pi r^2 h/3$.

a. Approximate the change in the volume of the cone when the radius changes from $r = 6.5$ to $r = 6.6$ and the height changes from $h = 4.20$ to $h = 4.15$.

b. Approximate the change in the volume of the cone when the radius changes from $r = 5.40$ to $r = 5.37$ and the height changes from $h = 12.0$ to $h = 11.96$.

45. Area of an ellipse The area of an ellipse with axes of length $2a$ and $2b$ is $A = \pi ab$. Approximate the percent change in the area when a increases by 2% and b increases by 1.5%.

46. Volume of a paraboloid The volume of a segment of a circular paraboloid (see figure) with radius r and height h is $V = \pi r^2 h/2$. Approximate the percent change in the volume when the radius decreases by 1.5% and the height increases by 2.2%.

$$V = \frac{\pi}{2} r^2 h$$

47–50. Differentials with more than two variables *Write the differential dw in terms of the differentials of the independent variables.*

47. $w = f(x, y, z) = xy^2 + x^2 z + yz^2$

48. $w = f(x, y, z) = \sin(x + y - z)$

49. $w = f(u, x, y, z) = \dfrac{u + x}{y + z}$

50. $w = f(p, q, r, s) = \dfrac{pq}{rs}$

T 51. Law of Cosines The side lengths of any triangle are related by the Law of Cosines,

$$c^2 = a^2 + b^2 - 2ab \cos \theta.$$

a. Estimate the change in the side length c when a changes from $a = 2$ to $a = 2.03$, b changes from $b = 4.00$ to $b = 3.96$, and θ changes from $\theta = \dfrac{\pi}{3}$ to $\theta = \dfrac{\pi}{3} + \dfrac{\pi}{90}$.

b. If a changes from $a = 2$ to $a = 2.03$ and b changes from $b = 4.00$ to $b = 3.96$, is the resulting change in c greater in magnitude when $\theta = \dfrac{\pi}{20}$ (small angle) or when $\theta = \dfrac{9\pi}{20}$ (close to a right angle)?

52. Travel cost The cost of a trip that is L miles long, driving a car that gets m miles per gallon, with gas costs of $\$p$/gal is $C = Lp/m$ dollars. Suppose you plan a trip of $L = 1500$ mi in a car that gets $m = 32$ mi/gal, with gas costs of $p = \$3.80$/gal.

  a. Explain how the cost function is derived.
  b. Compute the partial derivatives C_L, C_m, and C_p. Explain the meaning of the signs of the derivatives in the context of this problem.
  c. Estimate the change in the total cost of the trip if L changes from $L = 1500$ to $L = 1520$, m changes from $m = 32$ to $m = 31$, and p changes from $p = \$3.80$ to $p = \$3.85$.
  d. Is the total cost of the trip (with $L = 1500$ mi, $m = 32$ mi/gal, and $p = \$3.80$) more sensitive to a 1% change in L, in m, or in p (assuming the other two variables are fixed)? Explain.

53. Explain why or why not Determine whether the following statements are true and give an explanation or counterexample.

  a. The planes tangent to the cylinder $x^2 + y^2 = 1$ in $\mathbb{R}^3$ all have the form $ax + bz + c = 0$.
  b. Suppose $w = xy/z$, for $x > 0$, $y > 0$, and $z > 0$. A decrease in z with x and y fixed results in an increase in w.
  c. The gradient $\nabla F(a, b, c)$ lies in the plane tangent to the surface $F(x, y, z) = 0$ at (a, b, c).

54–57. Horizontal tangent planes *Find the points at which the following surfaces have horizontal tangent planes.*

54. $x^2 + 2y^2 + z^2 - 2x - 2z - 2 = 0$

55. $x^2 + y^2 - z^2 - 2x + 2y + 3 = 0$

56. $z = \sin(x - y)$ in the region $-2\pi \leq x \leq 2\pi$, $-2\pi \leq y \leq 2\pi$

57. $z = \cos 2x \sin y$ in the region $-\pi \leq x \leq \pi$, $-\pi \leq y \leq \pi$

58. Heron's formula The area of a triangle with sides of length a, b, and c is given by a formula from antiquity called Heron's formula:

$$A = \sqrt{s(s - a)(s - b)(s - c)},$$

where $s = \dfrac{1}{2}(a + b + c)$ is the *semiperimeter* of the triangle.

  a. Find the partial derivatives A_a, A_b, and A_c.
  b. A triangle has sides of length $a = 2$, $b = 4$, $c = 5$. Estimate the change in the area when a increases by 0.03, b decreases by 0.08, and c increases by 0.6.
  c. For an equilateral triangle with $a = b = c$, estimate the percent change in the area when all sides increase in length by p%.

59. Surface area of a cone A cone with height h and radius r has a lateral surface area (the curved surface only, excluding the base) of $S = \pi r \sqrt{r^2 + h^2}$.

  a. Estimate the change in the surface area when r increases from $r = 2.50$ to $r = 2.55$ and h decreases from $h = 0.60$ to $h = 0.58$.
  b. When $r = 100$ and $h = 200$, is the surface area more sensitive to a small change in r or a small change in h? Explain.

60. Line tangent to an intersection curve Consider the paraboloid $z = x^2 + 3y^2$ and the plane $z = x + y + 4$, which intersects the paraboloid in a curve C at $(2, 1, 7)$ (see figure). Find the equation of the line tangent to C at the point $(2, 1, 7)$. Proceed as follows.

  a. Find a vector normal to the plane at $(2, 1, 7)$.
  b. Find a vector normal to the plane tangent to the paraboloid at $(2, 1, 7)$.
  c. Argue that the line tangent to C at $(2, 1, 7)$ is orthogonal to both normal vectors found in parts (a) and (b). Use this fact to find a direction vector for the tangent line.

  d. Knowing a point on the tangent line and the direction of the tangent line, write an equation of the tangent line in parametric form.

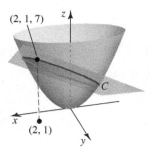

61. Batting averages Batting averages in baseball are defined by $A = x/y$, where $x \geq 0$ is the total number of hits and $y > 0$ is the total number of at-bats. Treat x and y as positive real numbers and note that $0 \leq A \leq 1$.

  a. Use differentials to estimate the change in the batting average if the number of hits increases from 60 to 62 and the number of at-bats increases from 175 to 180.
  b. If a batter currently has a batting average of $A = 0.350$, does the average decrease more if the batter fails to get a hit than it increases if the batter gets a hit?
  c. Does the answer to part (b) depend on the current batting average? Explain.

62. Water-level changes A conical tank with radius 0.50 m and height 2.00 m is filled with water (see figure). Water is released from the tank, and the water level drops by 0.05 m (from 2.00 m to 1.95 m). Approximate the change in the volume of water in the tank. (*Hint:* When the water level drops, both the radius and height of the cone of water change.)

63. Flow in a cylinder Poiseuille's Law is a fundamental law of fluid dynamics that describes the flow velocity of a viscous incompressible fluid in a cylinder (it is used to model blood flow through veins and arteries). It says that in a cylinder of radius R and length L, the velocity of the fluid $r \leq R$ units from the centerline of the cylinder is $V = \dfrac{P}{4\,Lv}(R^2 - r^2)$, where P is the difference in the pressure between the ends of the cylinder, and v is the viscosity of the fluid (see figure). Assuming P and v are constant, the velocity V along the centerline of the cylinder $(r = 0)$ is $V = \dfrac{kR^2}{L}$, where k is a constant that we will take to be $k = 1$.

  a. Estimate the change in the centerline velocity $(r = 0)$ if the radius of the flow cylinder increases from $R = 3$ cm to $R = 3.05$ cm and the length increases from $L = 50$ cm to $L = 50.5$ cm.
  b. Estimate the percent change in the centerline velocity if the radius of the flow cylinder R decreases by 1% and its length L increases by 2%.

c. Complete the following sentence: If the radius of the cylinder increases by $p\%$, then the length of the cylinder must increase by approximately ___% in order for the velocity to remain constant.

Explorations and Challenges

64. Floating-point operations In general, real numbers (with infinite decimal expansions) cannot be represented exactly in a computer by floating-point numbers (with finite decimal expansions). Suppose floating-point numbers on a particular computer carry an error of at most 10^{-16}. Estimate the maximum error that is committed in evaluating the following functions. Express the error in absolute and relative (percent) terms.

a. $f(x, y) = xy$ **b.** $f(x, y) = \dfrac{x}{y}$

c. $F(x, y, z) = xyz$ **d.** $F(x, y, z) = \dfrac{x/y}{z}$

65. Probability of at least one encounter Suppose in a large group of people, a fraction $0 \le r \le 1$ of the people have flu. The probability that in n random encounters you will meet at least one person with flu is $P = f(n, r) = 1 - (1 - r)^n$. Although n is a positive integer, regard it as a positive real number.

a. Compute f_r and f_n.
b. How sensitive is the probability P to the flu rate r? Suppose you meet $n = 20$ people. Approximately how much does the probability P increase if the flu rate increases from $r = 0.1$ to $r = 0.11$ (with n fixed)?
c. Approximately how much does the probability P increase if the flu rate increases from $r = 0.9$ to $r = 0.91$ with $n = 20$?
d. Interpret the results of parts (b) and (c).

66. Two electrical resistors When two electrical resistors with resistance $R_1 > 0$ and $R_2 > 0$ are wired in parallel in a circuit (see figure), the combined resistance R, measured in ohms (Ω), is given by $\dfrac{1}{R} = \dfrac{1}{R_1} + \dfrac{1}{R_2}$.

a. Estimate the change in R if R_1 increases from 2 Ω to 2.05 Ω and R_2 decreases from 3 Ω to 2.95 Ω.
b. Is it true that if $R_1 = R_2$ and R_1 increases by the same small amount as R_2 decreases, then R is approximately unchanged? Explain.
c. Is it true that if R_1 and R_2 increase, then R increases? Explain.
d. Suppose $R_1 > R_2$ and R_1 increases by the same small amount as R_2 decreases. Does R increase or decrease?

67. Three electrical resistors Extending Exercise 66, when three electrical resistors with resistances $R_1 > 0$, $R_2 > 0$, and $R_3 > 0$ are wired in parallel in a circuit (see figure), the combined resistance R, measured in ohms (Ω), is given by $\dfrac{1}{R} = \dfrac{1}{R_1} + \dfrac{1}{R_2} + \dfrac{1}{R_3}$.

Estimate the change in R if R_1 increases from 2 Ω to 2.05 Ω, R_2 decreases from 3 Ω to 2.95 Ω, and R_3 increases from 1.5 Ω to 1.55 Ω.

68. Power functions and percent change Suppose $z = f(x, y) = x^a y^b$, where a and b are real numbers. Let $\dfrac{dx}{x}, \dfrac{dy}{y}$, and $\dfrac{dz}{z}$ be the approximate relative (percent) changes in x, y, and z, respectively. Show that $\dfrac{dz}{z} = \dfrac{a(dx)}{x} + \dfrac{b(dy)}{y}$; that is, the relative changes are additive when weighted by the exponents a and b.

69. Logarithmic differentials Let f be a differentiable function of one or more variables that is positive on its domain.

a. Show that $d(\ln f) = \dfrac{df}{f}$.
b. Use part (a) to explain the statement that the absolute change in $\ln f$ is approximately equal to the relative change in f.
c. Let $f(x, y) = xy$, note that $\ln f = \ln x + \ln y$, and show that relative changes add; that is, $\dfrac{df}{f} = \dfrac{dx}{x} + \dfrac{dy}{y}$.
d. Let $f(x, y) = \dfrac{x}{y}$, note that $\ln f = \ln x - \ln y$, and show that relative changes subtract; that is, $\dfrac{df}{f} = \dfrac{dx}{x} - \dfrac{dy}{y}$.
e. Show that in a product of n numbers, $f = x_1 x_2 \cdots x_n$, the relative change in f is approximately equal to the sum of the relative changes in the variables.

70. Distance from a plane to an ellipsoid (Adapted from 1938 Putnam Exam) Consider the ellipsoid $\dfrac{x^2}{a^2} + \dfrac{y^2}{b^2} + \dfrac{z^2}{c^2} = 1$ and the plane P given by $Ax + By + Cz + 1 = 0$. Let $h = (A^2 + B^2 + C^2)^{-1/2}$ and $m = (a^2 A^2 + b^2 B^2 + c^2 C^2)^{1/2}$.

a. Find the equation of the plane tangent to the ellipsoid at the point (p, q, r).
b. Find the two points on the ellipsoid at which the tangent plane is parallel to P, and find equations of the tangent planes.
c. Show that the distance between the origin and the plane P is h.
d. Show that the distance between the origin and the tangent planes is hm.
e. Find a condition that guarantees the plane P does not intersect the ellipsoid.

QUICK CHECK ANSWERS

1. $F(x, y, z) = z - xy - x + y = 0$ **2.** If you walk in the positive x-direction from $(-1, 2, 1)$, then you walk uphill. If you walk in the positive y-direction from $(-1, 2, 1)$, then you walk downhill.   **3.** If $\Delta x = 0$, then the change formula becomes $\Delta z \approx f_y(a, b) \, \Delta y$, which is the change formula for the single variable y. If $\Delta y = 0$, then the change formula becomes $\Delta z \approx f_x(a, b) \, \Delta x$, which is the change formula for the single variable x.   **4.** The BMI increases with weight w and decreases with height h. ◄

15.7 Maximum/Minimum Problems

In Chapter 4, we showed how to use derivatives to find maximum and minimum values of functions of a single variable. When those techniques are extended to functions of two variables, we discover both similarities and differences. The landscape of a surface is far more complicated than the profile of a curve in the plane, so we see more interesting features when working with several variables. In addition to peaks (maximum values) and hollows (minimum values), we encounter winding ridges, long valleys, and mountain passes. Yet despite these complications, many of the ideas used for single-variable functions reappear in higher dimensions. For example, the Second Derivative Test, suitably adapted for two variables, plays a central role. As with single-variable functions, the techniques developed here are useful for solving practical optimization problems.

Local Maximum/Minimum Values

The concepts of local maximum and minimum values encountered in Chapter 4 extend readily to functions of two variables of the form $z = f(x, y)$. **Figure 15.67** shows a general surface defined on a domain D, which is a subset of $\mathbb{R}^2$. The surface has peaks (local high points) and hollows (local low points) at points in the interior of D. The goal is to locate and classify these extreme points.

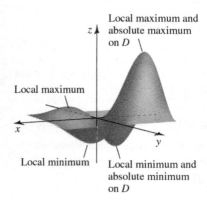

Local maximum and absolute maximum on D

Local maximum

Local minimum

Local minimum and absolute minimum on D

Figure 15.67

> We maintain the convention adopted in Chapter 4 that local maxima or minima occur at interior points of the domain. Recall that an open disk centered at (a, b) is the set of points within a circle centered at (a, b).

DEFINITION Local Maximum/Minimum Values

Suppose (a, b) is a point in a region R on which f is defined. If $f(x, y) \leq f(a, b)$ for all (x, y) in the domain of f and in some open disk centered at (a, b), then $f(a, b)$ is a **local maximum value** of f. If $f(x, y) \geq f(a, b)$ for all (x, y) in the domain of f and in some open disk centered at (a, b), then $f(a, b)$ is a **local minimum value** of f. Local maximum and local minimum values are also called **local extreme values** or **local extrema**.

In familiar terms, a local maximum is a point on a surface from which you cannot walk uphill. A local minimum is a point from which you cannot walk downhill. The following theorem is the analog of Theorem 4.2.

THEOREM 15.14 Derivatives and Local Maximum/Minimum Values

If f has a local maximum or minimum value at (a, b) and the partial derivatives f_x and f_y exist at (a, b), then $f_x(a, b) = f_y(a, b) = 0$.

Proof: Suppose f has a local maximum value at (a, b). The function of one variable $g(x) = f(x, b)$, obtained by holding $y = b$ fixed, also has a local maximum at (a, b). By Theorem 4.2, $g'(a) = 0$. However, $g'(a) = f_x(a, b)$; therefore, $f_x(a, b) = 0$. Similarly, the function $h(y) = f(a, y)$, obtained by holding $x = a$ fixed, has a local maximum at (a, b), which implies that $f_y(a, b) = h'(b) = 0$. An analogous argument is used for the local minimum case. ◄

QUICK CHECK 1 The paraboloid $z = x^2 + y^2 - 4x + 2y + 5$ has a local minimum at $(2, -1)$. Verify the conclusion of Theorem 15.14 for this function. ◄

Suppose f is differentiable at (a, b) (ensuring the existence of a tangent plane) and f has a local extremum at (a, b). Then $f_x(a, b) = f_y(a, b) = 0$, which, when substituted into the equation of the tangent plane, gives the equation $z = f(a, b)$ (a constant). Therefore, if the tangent plane exists at a local extremum, then it is horizontal there.

Recall that for a function of one variable, the condition $f'(a) = 0$ does not guarantee a local extremum at a. A similar precaution must be taken with Theorem 15.14. The conditions $f_x(a, b) = f_y(a, b) = 0$ do not imply that f has a local extremum at (a, b), as we show momentarily. Theorem 15.14 provides *candidates* for local extrema. We call these candidates *critical points*, as we did for functions of one variable. Therefore, the procedure for locating local maximum and minimum values is to find the critical points and then determine whether these candidates correspond to genuine local maximum and minimum values.

DEFINITION Critical Point

An interior point (a, b) in the domain of f is a **critical point** of f if either

1. $f_x(a, b) = f_y(a, b) = 0$, or
2. at least one of the partial derivatives f_x and f_y does not exist at (a, b).

EXAMPLE 1   Finding critical points Find the critical points of
$f(x, y) = xy(x - 2)(y + 3)$.

SOLUTION This function is differentiable at all points of $\mathbb{R}^2$, so the critical points occur only at points where $f_x(x, y) = f_y(x, y) = 0$. Computing and simplifying the partial derivatives, these conditions become

$$f_x(x, y) = 2y(x - 1)(y + 3) = 0$$
$$f_y(x, y) = x(x - 2)(2y + 3) = 0.$$

We must now identify all (x, y) pairs that satisfy both equations. The first equation is satisfied if and only if $y = 0$, $x = 1$, or $y = -3$. We consider each of these cases.

• Substituting $y = 0$, the second equation is $3x(x - 2) = 0$, which has solutions $x = 0$ and $x = 2$. So $(0, 0)$ and $(2, 0)$ are critical points.

• Substituting $x = 1$, the second equation is $-(2y + 3) = 0$, which has the solution $y = -\frac{3}{2}$. So $\left(1, -\frac{3}{2}\right)$ is a critical point.

• Substituting $y = -3$, the second equation is $-3x(x - 2) = 0$, which has roots $x = 0$ and $x = 2$. So $(0, -3)$ and $(2, -3)$ are critical points.

We find that there are five critical points: $(0, 0)$, $(2, 0)$, $\left(1, -\frac{3}{2}\right)$, $(0, -3)$, and $(2, -3)$. Some of these critical points may correspond to local maximum or minimum values. We will return to this example and a complete analysis shortly.

Related Exercises 15, 18 ◄

Second Derivative Test

Critical points are candidates for local extreme values. With functions of one variable, the Second Derivative Test is used to determine whether critical points correspond to local maxima or minima (the test can also be inconclusive). The analogous test for functions of two variables not only detects local maxima and minima, but also identifies another type of point known as a *saddle point*.

DEFINITION Saddle Point

Consider a function f that is differentiable at a critical point (a, b). Then f has a **saddle point** at (a, b) if, in every open disk centered at (a, b), there are points (x, y) for which $f(x, y) > f(a, b)$ and points for which $f(x, y) < f(a, b)$.

If (a, b) is a critical point of f and f has a saddle point at (a, b), then from the point $(a, b, f(a, b))$, it is possible to walk uphill in some directions and downhill in other directions. The function $f(x, y) = x^2 - y^2$ (a hyperbolic paraboloid) is a good example to remember. The surface *rises* from the critical point $(0, 0)$ along the x-axis and *falls* from $(0, 0)$ along the y-axis (**Figure 15.68**). We can easily check that $f_x(0, 0) = f_y(0, 0) = 0$, demonstrating that critical points do not necessarily correspond to local maxima or minima.

QUICK CHECK 2 Consider the plane tangent to a surface at a saddle point. In what direction does the normal to the plane point? ◄

▶ The usual image of a saddle point is that of a mountain pass (or a horse saddle), where you can walk upward in some directions and downward in other directions. The definition of a saddle point given here includes other less common situations. For example, with this definition, the cylinder $z = x^3$ has a line of saddle points along the y-axis.

The hyperbolic paraboloid $z = x^2 - y^2$ has a saddle point at $(0, 0)$.

Figure 15.68

▶ The Second Derivative Test for functions of a single variable states that if a is a critical point with $f'(a) = 0$, then $f''(a) > 0$ implies that f has a local minimum at a and $f''(a) < 0$ implies that f has a local maximum at a; if $f''(a) = 0$, the test is inconclusive. Theorem 15.15 is easier to remember if you notice the parallels between the two second derivative tests.

THEOREM 15.15 Second Derivative Test

Suppose the second partial derivatives of f are continuous throughout an open disk centered at the point (a, b), where $f_x(a, b) = f_y(a, b) = 0$. Let $D(x, y) = f_{xx}(x, y) f_{yy}(x, y) - (f_{xy}(x, y))^2$.

1. If $D(a, b) > 0$ and $f_{xx}(a, b) < 0$, then f has a local maximum value at (a, b).

2. If $D(a, b) > 0$ and $f_{xx}(a, b) > 0$, then f has a local minimum value at (a, b).

3. If $D(a, b) < 0$, then f has a saddle point at (a, b).

4. If $D(a, b) = 0$, then the test is inconclusive.

QUICK CHECK 3 Compute the discriminant $D(x, y)$ of $f(x, y) = x^2 y^2$. ◀

The proof of this theorem is given in Appendix A, but a few comments are in order. The test relies on the quantity $D(x, y) = f_{xx} f_{yy} - (f_{xy})^2$, which is called the **discriminant** of f. It can be remembered as the 2×2 determinant of the **Hessian** matrix $\begin{pmatrix} f_{xx} & f_{xy} \\ f_{yx} & f_{yy} \end{pmatrix}$, where $f_{xy} = f_{yx}$, provided these derivatives are continuous (Theorem 15.4). The condition $D(x, y) > 0$ means that the surface has the same general behavior in all directions near (a, b); either the surface rises in all directions or it falls in all directions. In the case that $D(a, b) = 0$, the test is inconclusive: (a, b) could correspond to a local maximum, a local minimum, or a saddle point.

Finally, another useful characterization of a saddle point can be derived from Theorem 15.15: The tangent plane at a saddle point lies both above and below the surface.

EXAMPLE 2 Analyzing critical points Use the Second Derivative Test to classify the critical points of $f(x, y) = x^2 + 2y^2 - 4x + 4y + 6$.

SOLUTION We begin with the following derivative calculations:

$$f_x = 2x - 4, \qquad f_y = 4y + 4,$$
$$f_{xx} = 2, \qquad f_{xy} = f_{yx} = 0, \quad \text{and} \quad f_{yy} = 4.$$

Setting both f_x and f_y equal to zero yields the single critical point $(2, -1)$. The value of the discriminant at the critical point is $D(2, -1) = f_{xx} f_{yy} - (f_{xy})^2 = 8 > 0$. Furthermore, $f_{xx}(2, -1) = 2 > 0$. By the Second Derivative Test, f has a local minimum at $(2, -1)$; the value of the function at that point is $f(2, -1) = 0$ (**Figure 15.69**).

Related Exercise 24 ◀

$z = x^2 + 2y^2 - 4x + 4y + 6$

$(2, -1, 0)$

Local minimum at $(2, -1)$ where $f_x = f_y = 0$

Figure 15.69

EXAMPLE 3 Analyzing critical points Use the Second Derivative Test to classify the critical points of $f(x, y) = xy(x - 2)(y + 3)$.

SOLUTION In Example 1, we determined that the critical points of f are $(0, 0)$, $(2, 0)$, $\left(1, -\frac{3}{2}\right)$, $(0, -3)$, and $(2, -3)$. The derivatives needed to evaluate the discriminant are

$$f_x = 2y(x - 1)(y + 3), \qquad f_y = x(x - 2)(2y + 3),$$
$$f_{xx} = 2y(y + 3), \qquad f_{xy} = 2(2y + 3)(x - 1), \quad \text{and} \quad f_{yy} = 2x(x - 2).$$

The values of the discriminant at the critical points and the conclusions of the Second Derivative Test are shown in Table 15.4.

Table 15.4

(x, y)	$D(x, y)$	f_{xx}	Conclusion
$(0, 0)$	-36	0	Saddle point
$(2, 0)$	-36	0	Saddle point
$\left(1, -\frac{3}{2}\right)$	9	$-\frac{9}{2}$	Local maximum
$(0, -3)$	-36	0	Saddle point
$(2, -3)$	-36	0	Saddle point

The surface described by f has one local maximum at $\left(1, -\frac{3}{2}\right)$, surrounded by four saddle points (**Figure 15.70a**). The structure of the surface may also be visualized by plotting the level curves of f (**Figure 15.70b**).

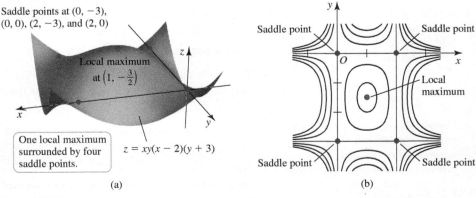

Saddle points at $(0, -3)$, $(0, 0)$, $(2, -3)$, and $(2, 0)$

Local maximum at $\left(1, -\frac{3}{2}\right)$

One local maximum surrounded by four saddle points.

$z = xy(x - 2)(y + 3)$

(a)

Saddle point   Saddle point

Local maximum

Saddle point   Saddle point

(b)

Figure 15.70

Related Exercise 27 ◄

EXAMPLE 4 Inconclusive tests Apply the Second Derivative Test to the following functions and interpret the results.

a. $f(x, y) = 2x^4 + y^4$ **b.** $f(x, y) = 2 - xy^2$

SOLUTION

a. The critical points of f satisfy the conditions

$$f_x = 8x^3 = 0 \quad \text{and} \quad f_y = 4y^3 = 0,$$

so the sole critical point is $(0, 0)$. The second partial derivatives evaluated at $(0, 0)$ are

$$f_{xx}(0, 0) = f_{xy}(0, 0) = f_{yy}(0, 0) = 0.$$

We see that $D(0, 0) = 0$, and the Second Derivative Test is inconclusive. While the bowl-shaped surface (**Figure 15.71**) described by f has a local minimum at $(0, 0)$, the surface also has a broad flat bottom, which makes the local minimum "invisible" to the Second Derivative Test.

b. The critical points of this function satisfy

$$f_x(x, y) = -y^2 = 0 \quad \text{and} \quad f_y(x, y) = -2xy = 0.$$

The solutions of these equations have the form $(a, 0)$, where a is a real number. It is easy to check that the second partial derivatives evaluated at $(a, 0)$ are

$$f_{xx}(a, 0) = f_{xy}(a, 0) = 0 \quad \text{and} \quad f_{yy}(a, 0) = -2a.$$

Therefore, the discriminant is $D(a, 0) = 0$, and the Second Derivative Test is inconclusive. **Figure 15.72** shows that f has a flat ridge above the x-axis that the Second Derivative Test is unable to classify.

Related Exercises 29–30 ◄

Absolute Maximum and Minimum Values

As in the one-variable case, we are often interested in knowing where a function of two or more variables attains its extreme values over its domain (or a subset of its domain).

$z = 2x^4 + y^4$

Local minimum at $(0, 0)$, but the Second Derivative Test is inconclusive.

Figure 15.71

➤ The same "flat" behavior occurs with functions of one variable, such as $f(x) = x^4$. Although f has a local minimum at $x = 0$, the Second Derivative Test is inconclusive.

➤ It is not surprising that the Second Derivative Test is inconclusive in Example 4b. The function has a line of local maxima at $(a, 0)$ for $a > 0$, a line of local minima at $(a, 0)$ for $a < 0$, and a saddle point at $(0, 0)$.

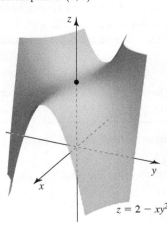

$z = 2 - xy^2$

Second derivative test fails to detect saddle point at $(0, 0)$.

Figure 15.72

DEFINITION Absolute Maximum/Minimum Values

Let f be defined on a set R in $\mathbb{R}^2$ containing the point (a, b). If $f(a, b) \geq f(x, y)$ for every (x, y) in R, then $f(a, b)$ is an **absolute maximum value** of f on R. If $f(a, b) \leq f(x, y)$ for every (x, y) in R, then $f(a, b)$ is an **absolute minimum value** of f on R.

It should be noted that the Extreme Value Theorem of Chapter 4 has an analog in $\mathbb{R}^2$ (or in higher dimensions): A function that is continuous on a closed bounded set in $\mathbb{R}^2$ attains its absolute maximum and absolute minimum values on that set. Absolute maximum and minimum values on a closed bounded set R occur in two ways.

> Recall that a *closed set* in $\mathbb{R}^2$ is a set that includes its boundary. A *bounded set* in $\mathbb{R}^2$ is a set that may be enclosed by a circle of finite radius.

• They may be local maximum or minimum values at interior points of R, where they are associated with critical points.

• They may occur on the boundary of R.

Therefore, the search for absolute maximum and minimum values on a closed bounded set amounts to examining the behavior of the function on the boundary of R and at the interior points of R.

> Example 5 is a *constrained optimization problem*, in which the goal is to maximize the volume subject to an additional condition called a *constraint*. We return to such problems in the next section and present another method of solution.

EXAMPLE 5   Shipping regulations A shipping company handles rectangular boxes provided the sum of the length, width, and height of the box does not exceed 96 in. Find the dimensions of the box that meets this condition and has the largest volume.

SOLUTION Let x, y, and z be the dimensions of the box; its volume is $V = xyz$. The box with the maximum volume must also satisfy the condition $x + y + z = 96$, which is used to eliminate any one of the variables from the volume function. Noting that $z = 96 - x - y$, the volume function becomes

$$V(x, y) = xy(96 - x - y).$$

Notice that because x, y, and $96 - x - y$ are dimensions of the box, they must be nonnegative. The condition $96 - x - y \geq 0$ implies that $x + y \leq 96$. Therefore, among points in the xy-plane, the constraint is met only if (x, y) lies in the triangle bounded by the lines $x = 0$, $y = 0$, and $x + y = 96$ (**Figure 15.73**).

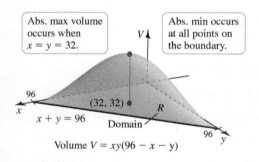

Abs. max volume occurs when $x = y = 32$.

Abs. min occurs at all points on the boundary.

$x + y = 96$ Domain $(32, 32)$ R

Volume $V = xy(96 - x - y)$

Figure 15.73

At this stage, we have reduced the original problem to a related problem: Find the absolute maximum value of $V(x, y) = xy(96 - x - y)$ over the triangular region

$$R = \{(x, y): 0 \leq x \leq 96, 0 \leq y \leq 96 - x\}.$$

The boundaries of R consist of the line segments $x = 0$, $0 \leq y \leq 96$; $y = 0$, $0 \leq x \leq 96$; and $x + y = 96$, $0 \leq x \leq 96$. We find that on these boundary segments, $V = 0$. To determine the behavior of V at interior points of R, we need to find critical points. The critical points of V satisfy

$$V_x = 96y - 2xy - y^2 = y(96 - 2x - y) = 0$$
$$V_y = 96x - 2xy - x^2 = x(96 - 2y - x) = 0.$$

You can check that these two equations have four solutions: $(0, 0)$, $(96, 0)$, $(0, 96)$, and $(32, 32)$. The first three solutions lie on the boundary of the domain, where $V = 0$. At the fourth point, we have $V(32, 32) = 32{,}768 \text{ in}^3$, which is the absolute maximum volume of the box. The dimensions of the box with maximum volume are $x = 32$, $y = 32$, and $z = 96 - x - y = 32$ (it is a cube). We also found that V has an absolute minimum of 0 at every point on the boundary of R.

Related Exercise 43 ◄

We summarize the method of solution given in Example 5 in the following procedure box.

PROCEDURE   Finding Absolute Maximum/Minimum Values on Closed Bounded Sets

Let f be continuous on a closed bounded set R in $\mathbb{R}^2$. To find the absolute maximum and minimum values of f on R:

1. Determine the values of f at all critical points in R.

2. Find the maximum and minimum values of f on the boundary of R.

3. The greatest function value found in Steps 1 and 2 is the absolute maximum value of f on R, and the least function value found in Steps 1 and 2 is the absolute minimum value of f on R.

$f(x, y) = xy - 8x - y^2 + 12y + 160$

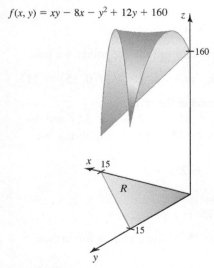

Figure 15.74

The techniques for carrying out Step 1 of this process have been presented. The challenge often lies in locating extreme values on the boundary. Examples 6 and 7 illustrate two approaches to handling the boundary of R. The first expresses the boundary using functions of a single variable, and the second describes the boundary parametrically. In both cases, finding extreme values on the boundary becomes a one-variable problem. In the next section, we discuss an alternative method for finding extreme values on boundaries.

EXAMPLE 6 **Extreme values over a region** Find the absolute maximum and minimum values of $f(x, y) = xy - 8x - y^2 + 12y + 160$ over the triangular region $R = \{(x, y): 0 \le x \le 15, 0 \le y \le 15 - x\}$.

SOLUTION Figure 15.74 shows the graph of f over the region R. The goal is to determine the absolute maximum and minimum values of f over R—including the boundary of R. We begin by finding the critical points of f on the interior of R. The partial derivatives of f are

$$f_x(x, y) = y - 8 \quad \text{and} \quad f_y(x, y) = x - 2y + 12.$$

The conditions $f_x(x, y) = f_y(x, y) = 0$ are satisfied only when $(x, y) = (4, 8)$, which is a point in the interior of R. This critical point is a candidate for the location of an extreme value of f, and the value of the function at this point is $f(4, 8) = 192$.

To search for extrema on the boundary of R, we consider each edge of R separately. Let C_1 be the line segment $\{(x, y): y = 0, \text{ for } 0 \le x \le 15\}$ on the x-axis, and define the single-variable function g_1 to equal f at all points along C_1 (**Figure 15.75**). We substitute $y = 0$ and find that g_1 has the form

$$g_1(x) = f(x, 0) = 160 - 8x.$$

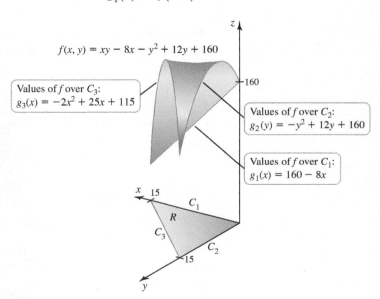

Figure 15.75

Using what we learned in Chapter 4, the candidates for absolute extreme values of g_1 on $0 \le x \le 15$ occur at critical points and endpoints. Specifically, the critical points of g_1 correspond to values where its derivative is zero, but in this case $g_1'(x) = -8$. So there is no critical point, which implies that the extreme values of g_1 occur at the endpoints of the interval $[0, 15]$. At the endpoints, we find that

$$g_1(0) = f(0, 0) = 160 \quad \text{and} \quad g_1(15) = f(15, 0) = 40.$$

Let's set aside this information while we do a similar analysis on the other two edges of the boundary of R.

Let C_2 be the line segment $\{(x, y): x = 0, \text{ for } 0 \le y \le 15\}$ and define g_2 to equal f on C_2 (Figure 15.75). Substituting $x = 0$, we see that

$$g_2(y) = f(0, y) = -y^2 + 12y + 160.$$

The critical points of g_2 satisfy

$$g_2'(y) = -2y + 12 = 0,$$

which has the single root $y = 6$. Evaluating g_2 at this point and the endpoints, we have

$$g_2(6) = f(0, 6) = 196, \quad g_2(0) = f(0, 0) = 160, \quad \text{and} \quad g_2(15) = f(0, 15) = 115.$$

Observe that $g_1(0) = g_2(0)$ because C_1 and C_2 intersect at the origin.

Finally, we let C_3 be the line segment $\{(x, y): y = 15 - x, 0 \le x \le 15\}$ and define g_3 to equal f on C_3 (Figure 15.75). Substituting $y = 15 - x$ and simplifying, we find that

$$g_3(x) = f(x, 15 - x) = -2x^2 + 25x + 115.$$

The critical points of g_3 satisfy

$$g_3'(x) = -4x + 25,$$

whose only root on the interval $0 \le x \le 15$ is $x = 6.25$. Evaluating g_3 at this critical point and the endpoints, we have

$$g_3(6.25) = f(6.25, 8.75) = 193.125, \quad g_3(15) = f(15, 0) = 40, \quad \text{and}$$
$$g_3(0) = f(0, 15) = 115.$$

Observe that $g_3(15) = g_1(15)$ and $g_3(0) = g_2(15)$; the only new candidate for the location of an extreme value is the point $(6.25, 8.75)$.

Collecting and summarizing our work, we have 6 candidates for absolute extreme values:

$$f(4, 8) = 192, \quad f(0, 0) = 160, \quad f(15, 0) = 40, \quad f(0, 6) = 196,$$
$$f(0, 15) = 115, \quad \text{and} \quad f(6.25, 8.75) = 193.125.$$

We see that f has an absolute minimum value of 40 at $(15, 0)$ and an absolute maximum value of 196 at $(0, 6)$. These findings are illustrated in **Figure 15.76**.

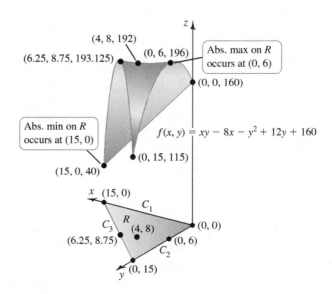

Figure 15.76

Related Exercise 52 ◄

► Finding absolute extrema on a closed set does not require using the Second Derivative Test to classify the critical points. In Example 7, the test *could* be used to show that $\left(\frac{1}{\sqrt{3}}, 0\right)$ and $\left(-\frac{1}{\sqrt{3}}, 0\right)$ correspond to a saddle point and a local maximum, respectively, but that information isn't needed.

EXAMPLE 7   Absolute maximum and minimum values Find the absolute maximum and minimum values of $f(x, y) = \frac{1}{2}(x^3 - x - y^2) + 3$ on the region $R = \{(x, y) : x^2 + y^2 \le 1\}$ (the closed disk centered at $(0, 0)$ with radius 1).

SOLUTION We begin by locating the critical points of f on the interior of R. The critical points satisfy the equations

$$f_x(x, y) = \frac{1}{2}(3x^2 - 1) = 0 \quad \text{and} \quad f_y(x, y) = -y = 0,$$

which have the solutions $x = \pm\dfrac{1}{\sqrt{3}}$ and $y = 0$. The values of the function at these points are $f\left(\dfrac{1}{\sqrt{3}}, 0\right) = 3 - \dfrac{1}{3\sqrt{3}}$ and $f\left(-\dfrac{1}{\sqrt{3}}, 0\right) = 3 + \dfrac{1}{3\sqrt{3}}$.

We now determine the maximum and minimum values of f on the boundary of R, which is a circle of radius 1 described by the parametric equations

$$x = \cos\theta, \quad y = \sin\theta, \quad \text{for} \quad 0 \le \theta \le 2\pi.$$

> ▶ Recall that a parametric description of a circle of radius $|a|$ centered at the origin is $x = a\cos\theta$, $y = a\sin\theta$, for $0 \le \theta \le 2\pi$.

Substituting x and y in terms of θ into the function f, we obtain a new function $g(\theta)$ that gives the values of f on the boundary of R:

$$g(\theta) = \frac{1}{2}(\cos^3\theta - \cos\theta - \sin^2\theta) + 3.$$

Finding the maximum and minimum boundary values is now a one-variable problem. The critical points of g satisfy

$$g'(\theta) = \frac{1}{2}(-3\cos^2\theta\sin\theta + \sin\theta - 2\sin\theta\cos\theta)$$

$$= -\frac{1}{2}\sin\theta(3\cos^2\theta + 2\cos\theta - 1)$$

$$= -\frac{1}{2}\sin\theta(3\cos\theta - 1)(\cos\theta + 1) = 0.$$

> ▶ There are two solutions to the equation $\cos\theta = \frac{1}{3}$ on the interval $(0, 2\pi)$. Recall, however, that using the inverse cosine to solve the equation reveals only the solution $\theta = \cos^{-1}\frac{1}{3}$ because the range of $\cos^{-1}x$ is $[0, \pi]$. The other solution, $\theta = 2\pi - \cos^{-1}\frac{1}{3}$, is found using symmetry.

This condition is satisfied when $\sin\theta = 0$, $\cos\theta = \dfrac{1}{3}$, or $\cos\theta = -1$. The solutions of these equations on the interval $(0, 2\pi)$ are $\theta = \pi$, $\theta = \cos^{-1}\dfrac{1}{3}$, and $\theta = 2\pi - \cos^{-1}\dfrac{1}{3}$, which correspond to the points $(-1, 0)$, $\left(\dfrac{1}{3}, \dfrac{2\sqrt{2}}{3}\right)$, and $\left(\dfrac{1}{3}, -\dfrac{2\sqrt{2}}{3}\right)$ in the xy-plane, respectively. Notice that the endpoints of the interval ($\theta = 0$ and $\theta = 2\pi$) correspond to the same point on the boundary of R, namely $(1, 0)$.

Having completed the first two steps of the procedure, we have six function values to consider:

- $f\left(\dfrac{1}{\sqrt{3}}, 0\right) = 3 - \dfrac{1}{3\sqrt{3}} \approx 2.81$ and $f\left(-\dfrac{1}{\sqrt{3}}, 0\right) = 3 + \dfrac{1}{3\sqrt{3}} \approx 3.19$ (critical points),

- $f(-1, 0) = 3$ (boundary point),

- $f\left(\dfrac{1}{3}, \dfrac{2\sqrt{2}}{3}\right) = f\left(\dfrac{1}{3}, -\dfrac{2\sqrt{2}}{3}\right) = \dfrac{65}{27}$ (boundary points), and

- $f(1, 0) = 3$ (boundary point).

> ▶ Observe that the level curves of f in Figure 15.77b appear to be tangent to the blue curve $x^2 + y^2 = 1$ (the boundary of the region R) at the points corresponding to the maximum and minimum values of f on this boundary. The significance of this observation is explained in Section 15.8.

The greatest value of f on R, $f\left(-\dfrac{1}{\sqrt{3}}, 0\right) = 3 + \dfrac{1}{3\sqrt{3}}$, is the absolute maximum value, and it occurs at an interior point (**Figure 15.77a**). The least value, $f\left(\dfrac{1}{3}, \dfrac{2\sqrt{2}}{3}\right) = f\left(\dfrac{1}{3}, -\dfrac{2\sqrt{2}}{3}\right) = \dfrac{65}{27}$, is the absolute minimum value, and it occurs at two symmetric boundary points. Also revealing is the plot of the level curves of the surface with the boundary of R superimposed (**Figure 15.77b**). As the boundary of R is traversed, the values of f vary, reaching a maximum value of 3 at $(1, 0)$ and $(-1, 0)$, and a minimum value of $\dfrac{65}{27}$ at $\left(\dfrac{1}{3}, \dfrac{2\sqrt{2}}{3}\right)$ and $\left(\dfrac{1}{3}, -\dfrac{2\sqrt{2}}{3}\right)$.

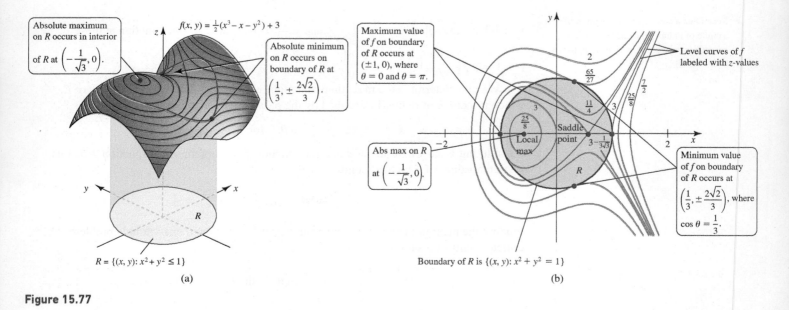

Figure 15.77

Related Exercises 47–48 ◄

Open and/or Unbounded Regions Finding absolute maximum and minimum values of a function on an open region (for example, $R = \{(x, y) = x^2 + y^2 < 9\}$) or an unbounded domain (for example, $R = \{(x, y): x > 0, y > 0\}$) presents additional challenges. Because there is no systematic procedure for dealing with such problems, some ingenuity is generally needed. Notice that absolute extrema may not exist on such regions.

EXAMPLE 8   Absolute extreme values on an open region Find the absolute maximum and minimum values of $f(x, y) = 4 - x^2 - y^2$ on the open disk $R = \{(x, y): x^2 + y^2 < 1\}$ (if they exist).

SOLUTION You should verify that f has a critical point at $(0, 0)$ and it corresponds to a local maximum (on an inverted paraboloid). Moving away from $(0, 0)$ in all directions, the function values decrease, so f also has an absolute maximum value of 4 at $(0, 0)$. The boundary of R is the unit circle $\{(x, y): x^2 + y^2 = 1\}$, which is not contained in R. As (x, y) approaches any point on the unit circle along any path in R, the function values $f(x, y) = 4 - (x^2 + y^2)$ decrease and approach 3 but never reach 3. Therefore, f does not have an absolute minimum on R.

Related Exercise 59 ◄

QUICK CHECK 4 Does the linear function $f(x, y) = 2x + 3y$ have an absolute maximum or minimum value on the open unit square $\{(x, y): 0 < x < 1, 0 < y < 1\}$? ◄

▶ Notice that $\frac{\partial}{\partial x}(d^2) = 2d\frac{\partial d}{\partial x}$ and $\frac{\partial}{\partial y}(d^2) = 2d\frac{\partial d}{\partial y}$. Because $d \geq 0$, d^2 and d have the same critical points.

EXAMPLE 9   Absolute extreme values on an open region Find the point(s) on the plane $x + 2y + z = 2$ closest to the point $P(2, 0, 4)$.

SOLUTION Suppose (x, y, z) is a point on the plane, which means that $z = 2 - x - 2y$. The distance between $P(2, 0, 4)$ and (x, y, z) that we seek to minimize is

$$d(x, y, z) = \sqrt{(x - 2)^2 + y^2 + (z - 4)^2}.$$

It is easier to minimize d^2, which has the same critical points as d. Squaring d and eliminating z using $z = 2 - x - 2y$, we have

$$f(x, y) = (d(x, y, z))^2 = (x - 2)^2 + y^2 + \underbrace{(-x - 2y - 2)^2}_{z - 4}$$

$$= 2x^2 + 5y^2 + 4xy + 8y + 8.$$

The critical points of f satisfy the equations

$$f_x = 4x + 4y = 0 \quad \text{and} \quad f_y = 4x + 10y + 8 = 0,$$

Distance squared:
$f(x, y) = 2x^2 + 5y^2 + 4xy + 8y + 8$

Absolute minimum
$\left(\frac{4}{3}, -\frac{4}{3}, \frac{10}{3}\right)$

Figure 15.78

whose only solution is $x = \frac{4}{3}$, $y = -\frac{4}{3}$. The Second Derivative Test confirms that this point corresponds to a local minimum of f. We now ask: Does $\left(\frac{4}{3}, -\frac{4}{3}\right)$ correspond to the *absolute* minimum value of f over the entire domain of the problem, which is $\mathbb{R}^2$? Because the domain has no boundary, we cannot check values of f on the boundary. Instead, we argue geometrically that there is exactly one point on the plane that is closest to P. We have found a point that is closest to P among nearby points on the plane. As we move away from this point, the values of f increase without bound. Therefore, $\left(\frac{4}{3}, -\frac{4}{3}\right)$ corresponds to the absolute minimum value of f. A graph of f (**Figure 15.78**) confirms this reasoning, and we conclude that the point $\left(\frac{4}{3}, -\frac{4}{3}, \frac{10}{3}\right)$ is the point on the plane nearest P.

Related Exercises 62–63 ◀

SECTION 15.7 EXERCISES

Getting Started

1. Describe the appearance of a smooth surface with a local maximum at a point.

2. Describe the usual appearance of a smooth surface at a saddle point.

3. What are the conditions for a critical point of a function f?

4. If $f_x(a, b) = f_y(a, b) = 0$, does it follow that f has a local maximum or local minimum at (a, b)? Explain.

5. Consider the function $z = f(x, y)$. What is the discriminant of f, and how do you compute it?

6. Explain how the Second Derivative Test is used.

7. What is an absolute minimum value of a function f on a set R in $\mathbb{R}^2$?

8. What is the procedure for locating absolute maximum and minimum values on a closed bounded domain?

9–12. *Assume the second derivatives of f are continuous throughout the xy-plane and $f_x(0, 0) = f_y(0, 0) = 0$. Use the given information and the Second Derivative Test to determine whether f has a local minimum, a local maximum, or a saddle point at $(0, 0)$, or state that the test is inconclusive.*

9. $f_{xx}(0, 0) = 5$, $f_{yy}(0, 0) = 3$, and $f_{xy}(0, 0) = -4$

10. $f_{xx}(0, 0) = -6$, $f_{yy}(0, 0) = -3$, and $f_{xy}(0, 0) = 4$

11. $f_{xx}(0, 0) = 8$, $f_{yy}(0, 0) = 5$, and $f_{xy}(0, 0) = -6$

12. $f_{xx}(0, 0) = -9$, $f_{yy}(0, 0) = -4$, and $f_{xy}(0, 0) = -6$

Practice Exercises

13–22. Critical points *Find all critical points of the following functions.*

13. $f(x, y) = 3x^2 - 4y^2$

14. $f(x, y) = x^2 - 6x + y^2 + 8y$

15. $f(x, y) = 3x^2 + 3y - y^3$

16. $f(x, y) = x^3 - 12x + 6y^2$

17. $f(x, y) = x^4 + y^4 - 16xy$

18. $f(x, y) = \dfrac{x^3}{3} - \dfrac{y^3}{3} + 3xy$

19. $f(x, y) = x^4 - 2x^2 + y^2 - 4y + 5$

20. $f(x, y) = x^3 + 6xy - 6x + y^2 - 2y$

21. $f(x, y) = y^3 + 6xy + x^2 - 18y - 6x$

22. $f(x, y) = e^{8x^2y^2 - 24x^2 - 8xy^4}$

23–40. Analyzing critical points *Find the critical points of the following functions. Use the Second Derivative Test to determine (if possible) whether each critical point corresponds to a local maximum, a local minimum, or a saddle point. If the Second Derivative Test is inconclusive, determine the behavior of the function at the critical points.*

23. $f(x, y) = -4x^2 + 8y^2 - 3$

24. $f(x, y) = x^4 + y^4 - 4x - 32y + 10$

25. $f(x, y) = 4 + 2x^2 + 3y^2$

26. $f(x, y) = xye^{-x-y}$

27. $f(x, y) = x^4 + 2y^2 - 4xy$

28. $f(x, y) = (4x - 1)^2 + (2y + 4)^2 + 1$

29. $f(x, y) = 4 + x^4 + 3y^4$

30. $f(x, y) = x^4y^2$

31. $f(x, y) = \sqrt{x^2 + y^2 - 4x + 5}$

32. $f(x, y) = \tan^{-1}xy$

33. $f(x, y) = 2xye^{-x^2-y^2}$ 34. $f(x, y) = x^2 + xy^2 - 2x + 1$

35. $f(x, y) = \dfrac{x}{1 + x^2 + y^2}$ 36. $f(x, y) = \dfrac{x - 1}{x^2 + y^2}$

37. $f(x, y) = x^4 + 4x^2(y - 2) + 8(y - 1)^2$

38. $f(x, y) = xe^{-x-y}\sin y$, for $|x| \leq 2$, $0 \leq y \leq \pi$

39. $f(x, y) = ye^x - e^y$

40. $f(x, y) = \sin(2\pi x)\cos(\pi y)$, for $|x| \leq \dfrac{1}{2}$ and $|y| \leq \dfrac{1}{2}$

41–42. Inconclusive tests *Show that the Second Derivative Test is inconclusive when applied to the following functions at $(0, 0)$. Describe the behavior of the function at $(0, 0)$.*

41. $f(x, y) = x^2y - 3$ 42. $f(x, y) = \sin(x^2y^2)$

43. **Shipping regulations** A shipping company handles rectangular boxes provided the sum of the height and the girth of the box does not exceed 96 in. (The girth is the perimeter of the smallest side of the box.) Find the dimensions of the box that meets this condition and has the largest volume.

44. **Cardboard boxes** A lidless box is to be made using 2 m² of cardboard. Find the dimensions of the box with the largest possible volume.

45. Cardboard boxes A lidless cardboard box is to be made with a volume of 4 m³. Find the dimensions of the box that requires the least amount of cardboard.

46. Optimal box Find the dimensions of the largest rectangular box in the first octant of the *xyz*-coordinate system that has one vertex at the origin and the opposite vertex on the plane $x + 2y + 3z = 6$.

47–56. Absolute maxima and minima *Find the absolute maximum and minimum values of the following functions on the given region R.*

47. $f(x, y) = x^2 + y^2 - 2y + 1; R = \{(x, y): x^2 + y^2 \le 4\}$

48. $f(x, y) = 2x^2 + y^2; R = \{(x, y): x^2 + y^2 \le 16\}$

49. $f(x, y) = 4 + 2x^2 + y^2;$
$R = \{(x, y): -1 \le x \le 1, -1 \le y \le 1\}$

50. $f(x, y) = 6 - x^2 - 4y^2;$
$R = \{(x, y): -2 \le x \le 2, -1 \le y \le 1\}$

51. $f(x, y) = 2x^2 - 4x + 3y^2 + 2;$
$R = \{(x, y): (x - 1)^2 + y^2 \le 1\}$

52. $f(x, y) = x^2 + y^2 - 2x - 2y; R$ is the closed region bounded by the triangle with vertices $(0, 0)$, $(2, 0)$, and $(0, 2)$.

53. $f(x, y) = -2x^2 + 4x - 3y^2 - 6y - 1;$
$R = \{(x, y): (x - 1)^2 + (y + 1)^2 \le 1\}$

54. $f(x, y) = \sqrt{x^2 + y^2 - 2x + 2};$
$R = \{(x, y): x^2 + y^2 \le 4, y \ge 0\}$

55. $f(x, y) = \dfrac{2y^2 - x^2}{2 + 2x^2y^2}; R$ is the closed region bounded by the lines $y = x, y = 2x$, and $y = 2$.

56. $f(x, y) = \sqrt{x^2 + y^2}; R$ is the closed region bounded by the ellipse $\dfrac{x^2}{4} + y^2 = 1$.

T 57. Pectin Extraction An increase in world production of processed fruit has led to an increase in fruit waste. One way of reducing this waste is to find useful waste byproducts. For example, waste from pineapples is reduced by extracting pectin from pineapple peels (pectin is commonly used as a thickening agent in jam and jellies, and it is also widely used in the pharmaceutical industry). Pectin extraction involves heating and drying the peels, then grinding the peels into a fine powder. The powder is next placed in a solution with a particular pH level H, for $1.5 \le H \le 2.5$, and heated to a temperature T (in degrees Celsius), for $70 \le T \le 90$. The percentage of the powder $F(H, T)$ that becomes extracted pectin is

$$F(H, T) = -0.042T^2 - 0.213TH - 11.219H^2 + 7.327T$$
$$+ 58.729H - 342.684.$$

a. It can be shown that F attains its absolute maximum in the interior of the domain $D = \{(H, T): 1.5 \le H \le 2.5, 70 \le T \le 90\}$. Find the pH level H and temperature T that together maximize the amount of pectin extracted from the powder.

b. What is the maximum percentage of pectin that can be extracted from the powder? Round your answer to the nearest whole number. (*Source: Carpathian Journal of Food Science and Technology*, Dec 2014)

58–61. Absolute extrema on open and/or unbounded regions *If possible, find the absolute maximum and minimum values of the following functions on the region R.*

58. $f(x, y) = x + 3y; R = \{(x, y): |x| < 1, |y| < 2\}$

59. $f(x, y) = x^2 + y^2 - 4; R = \{(x, y): x^2 + y^2 < 4\}$

60. $f(x, y) = x^2 - y^2; R = \{(x, y): |x| < 1, |y| < 1\}$

61. $f(x, y) = 2e^{-x-y}; R = \{(x, y): x \ge 0, y \ge 0\}$

62–66. Absolute extrema on open and/or unbounded regions

62. Find the point on the plane $x + y + z = 4$ nearest the point $P(5, 4, 4)$.

63. Find the point on the plane $x - y + z = 2$ nearest the point $P(1, 1, 1)$.

64. Find the point on the paraboloid $z = x^2 + y^2$ nearest the point $P(3, 3, 1)$.

65. Find the points on the cone $z^2 = x^2 + y^2$ nearest the point $P(6, 8, 0)$.

66. Rectangular boxes with a volume of 10 m³ are made of two materials. The material for the top and bottom of the box costs $10/m² and the material for the sides of the box costs $1/m². What are the dimensions of the box that minimize the cost of the box?

67. Explain why or why not Determine whether the following statements are true and give an explanation or counterexample. Assume f is differentiable at the points in question.

a. The fact that $f_x(2, 2) = f_y(2, 2) = 0$ implies that f has a local maximum, local minimum, or saddle point at $(2, 2)$.

b. The function f could have a local maximum at (a, b) where $f_y(a, b) \neq 0$.

c. The function f could have both an absolute maximum and an absolute minimum at two different points that are not critical points.

d. The tangent plane is horizontal at a point on a smooth surface corresponding to a critical point.

68–69. Extreme points from contour plots *Based on the level curves that are visible in the following graphs, identify the approximate locations of the local maxima, local minima, and saddle points.*

68.

69.

70. Optimal box Find the dimensions of the rectangular box with maximum volume in the first octant with one vertex at the origin and the opposite vertex on the ellipsoid $36x^2 + 4y^2 + 9z^2 = 36$.

Explorations and Challenges

71. Magic triples Let x, y, and z be nonnegative numbers with $x + y + z = 200$.

   a. Find the values of x, y, and z that minimize $x^2 + y^2 + z^2$.

   b. Find the values of x, y, and z that minimize $\sqrt{x^2 + y^2 + z^2}$.

   c. Find the values of x, y, and z that maximize xyz.

   d. Find the values of x, y, and z that maximize $x^2y^2z^2$.

72. Maximum/minimum of linear functions Let R be a closed bounded region in $\mathbb{R}^2$ and let $f(x, y) = ax + by + c$, where a, b, and c are real numbers, with a and b not both zero. Give a geometric argument explaining why the absolute maximum and minimum values of f over R occur on the boundaries of R.

T 73. Optimal locations Suppose n houses are located at the distinct points (x_1, y_1), (x_2, y_2), ..., (x_n, y_n). A power substation must be located at a point such that the *sum of the squares* of the distances between the houses and the substation is minimized.

   a. Find the optimal location of the substation in the case that $n = 3$ and the houses are located at $(0, 0)$, $(2, 0)$, and $(1, 1)$.

   b. Find the optimal location of the substation in the case that $n = 3$ and the houses are located at distinct points (x_1, y_1), (x_2, y_2), and (x_3, y_3).

   c. Find the optimal location of the substation in the general case of n houses located at distinct points (x_1, y_1), (x_2, y_2), ..., (x_n, y_n).

   d. You might argue that the locations found in parts (a), (b) and (c) are not optimal because they result from minimizing the sum of the *squares* of the distances, not the sum of the distances themselves. Use the locations in part (a) and write the function that gives the sum of the distances. Note that minimizing this function is much more difficult than in part (a). Then use a graphing utility to determine whether the optimal location is the same in the two cases. (Also see Exercise 81 about Steiner's problem.)

74–75. Least squares approximation *In its many guises, least squares approximation arises in numerous areas of mathematics and statistics. Suppose you collect data for two variables (for example, height and shoe size) in the form of pairs (x_1, y_1), (x_2, y_2), ..., (x_n, y_n). The data may be plotted as a scatterplot in the xy-plane, as shown in the figure. The technique known as* linear regression *asks the question: What is the equation of the line that "best fits" the data? The least squares*

criterion for best fit requires that the sum of the squares of the vertical distances between the line and the data points be a minimum.

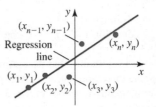

74. Let the equation of the best-fit line be $y = mx + b$, where the slope m and the y-intercept b must be determined using the least squares condition. First assume there are three data points $(1, 2)$, $(3, 5)$, and $(4, 6)$. Show that the function of m and b that gives the sum of the squares of the vertical distances between the line and the three data points is

$$E(m, b) = ((m + b) - 2)^2 + ((3m + b) - 5)^2 + ((4m + b) - 6)^2.$$

Find the critical points of E and find the values of m and b that minimize E. Graph the three data points and the best-fit line.

T 75. Generalize the procedure in Exercise 74 by assuming n data points (x_1, y_1), (x_2, y_2), ..., (x_n, y_n) are given. Write the function $E(m, b)$ (summation notation allows for a more compact calculation). Show that the coefficients of the best-fit line are

$$m = \frac{\left(\sum x_k\right)\left(\sum y_k\right) - n\sum x_k y_k}{\left(\sum x_k\right)^2 - n\sum x_k^2},$$

$$b = \frac{1}{n}\left(\sum y_k - m\sum x_k\right),$$

where all sums run from $k = 1$ to $k = n$.

T 76–77. Least squares practice *Use the results of Exercise 75 to find the best-fit line for the following data sets. Plot the points and the best-fit line.*

76. $(0, 0)$, $(2, 3)$, $(4, 5)$ **77.** $(-1, 0)$, $(0, 6)$, $(3, 8)$

78. Second Derivative Test Suppose the conditions of the Second Derivative Test are satisfied on an open disk containing the point (a, b). Use the test to prove that if (a, b) is a critical point of f at which $f_x(a, b) = f_y(a, b) = 0$ and $f_{xx}(a, b) < 0 < f_{yy}(a, b)$ or $f_{yy}(a, b) < 0 < f_{xx}(a, b)$, then f has a saddle point at (a, b).

79. Maximum area triangle Among all triangles with a perimeter of 9 units, find the dimensions of the triangle with the maximum area. It may be easiest to use Heron's formula, which states that the area of a triangle with side length a, b, and c is $A = \sqrt{s(s - a)(s - b)(s - c)}$, where $2s$ is the perimeter of the triangle.

80. Slicing plane Find an equation of the plane passing through the point $(3, 2, 1)$ that slices off the solid in the first octant with the least volume.

T 81. Steiner's problem for three points Given three distinct noncollinear points A, B, and C in the plane, find the point P in the plane such that the sum of the distances $|AP| + |BP| + |CP|$ is a minimum. Here is how to proceed with three points, assuming the triangle formed by the three points has no angle greater than $2\pi/3$ ($120°$).

   a. Assume the coordinates of the three given points are $A(x_1, y_1)$, $B(x_2, y_2)$, and $C(x_3, y_3)$. Let $d_1(x, y)$ be the distance between $A(x_1, y_1)$ and a variable point $P(x, y)$. Compute the gradient of d_1 and show that it is a unit vector pointing along the line between the two points.

b. Define d_2 and d_3 in a similar way and show that ∇d_2 and ∇d_3 are also unit vectors in the direction of the line between the two points.

c. The goal is to minimize $f(x, y) = d_1 + d_2 + d_3$. Show that the condition $f_x = f_y = 0$ implies that $\nabla d_1 + \nabla d_2 + \nabla d_3 = 0$.

d. Explain why part (c) implies that the optimal point P has the property that the three line segments AP, BP, and CP all intersect symmetrically in angles of $2\pi/3$.

e. What is the optimal solution if one of the angles in the triangle is greater than $2\pi/3$ (just draw a picture)?

f. Estimate the Steiner point for the three points $(0, 0)$, $(0, 1)$, and $(2, 0)$.

T 82. Solitary critical points A function of *one* variable has the property that a local maximum (or minimum) occurring at the only critical point is also the absolute maximum (or minimum) (for example, $f(x) = x^2$). Does the same result hold for a function of *two* variables? Show that the following functions have the property that they have a single local maximum (or minimum), occurring at the only critical point, but the local maximum (or minimum) is not an absolute maximum (or minimum) on $\mathbb{R}^2$.

a. $f(x, y) = 3xe^y - x^3 - e^{3y}$

b. $f(x, y) = (2y^2 - y^4)\left(e^x + \dfrac{1}{1 + x^2}\right) - \dfrac{1}{1 + x^2}$

This property has the following interpretation. Suppose a surface has a single local minimum that is not the absolute minimum. Then water can be poured into the basin around the local minimum and the surface never overflows, even though there are points on the surface below the local minimum.
(*Source: Mathematics Magazine*, May 1985, and *Calculus and Analytical Geometry*, 2nd ed., Philip Gillett, 1984)

T 83. Two mountains without a saddle Show that the following functions have two local maxima but no other extreme points (therefore, no saddle or basin between the mountains).

a. $f(x, y) = -(x^2 - 1)^2 - (x^2 - e^y)^2$

b. $f(x, y) = 4x^2e^y - 2x^4 - e^{4y}$
(*Source: Ira Rosenholtz, Mathematics Magazine*, Feb 1987)

84. Powers and roots Assume $x + y + z = 1$ with $x \geq 0$, $y \geq 0$, and $z \geq 0$.

a. Find the maximum and minimum values of $(1 + x^2)(1 + y^2)(1 + z^2)$.

b. Find the maximum and minimum values of $(1 + \sqrt{x})(1 + \sqrt{y})(1 + \sqrt{z})$.
(*Source: Math Horizons*, Apr 2004)

85. Ellipsoid inside a tetrahedron (1946 Putnam Exam) Let P be a plane tangent to the ellipsoid $\dfrac{x^2}{a^2} + \dfrac{y^2}{b^2} + \dfrac{z^2}{c^2} = 1$ at a point in the first octant. Let T be the tetrahedron in the first octant bounded by P and the coordinate planes $x = 0$, $y = 0$, and $z = 0$. Find the minimum volume of T. (The volume of a tetrahedron is one-third the area of the base times the height.)

QUICK CHECK ANSWERS

1. $f_x(2, -1) = f_y(2, -1) = 0$ **2.** Vertically, in the directions $\langle 0, 0, \pm 1 \rangle$ **3.** $D(x, y) = -12x^2y^2$
4. It has neither an absolute maximum nor an absolute minimum value on this set. ◄

15.8 Lagrange Multipliers

One of many challenges in economics and marketing is predicting the behavior of consumers. Basic models of consumer behavior often involve a *utility function* that expresses consumers' combined preference for several different amenities. For example, a simple utility function might have the form $U = f(\ell, g)$, where ℓ represents the amount of leisure time and g represents the number of consumable goods. The model assumes consumers try to maximize their utility function, but they do so under certain constraints on the variables of the problem. For example, increasing leisure time may increase utility, but leisure time produces no income for consumable goods. Similarly, consumable goods may also increase utility, but they require income, which reduces leisure time. We first develop a general method for solving such constrained optimization problems and then return to economics problems later in the section.

The Basic Idea

We start with a typical constrained optimization problem with two independent variables and give its method of solution; a generalization to more variables then follows. We seek maximum and/or minimum values of a differentiable **objective function** f with the restriction that x and y must lie on a **constraint** curve C in the xy-plane given by $g(x, y) = 0$ (**Figure 15.79**).

Figure 15.80 shows the details of a typical situation in which we assume the (green) level curves of f have increasing z-values moving away from the origin. Now imagine moving along the (red) constraint curve C: $g(x, y) = 0$ toward the point $P(a, b)$. As we approach P (from either side), the values of f evaluated on C increase, and as we move past P along C, the values of f decrease.

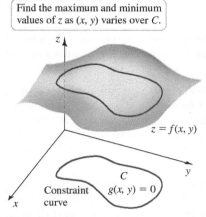

Find the maximum and minimum values of z as (x, y) varies over C.

$z = f(x, y)$

C

Constraint curve $g(x, y) = 0$

Figure 15.79

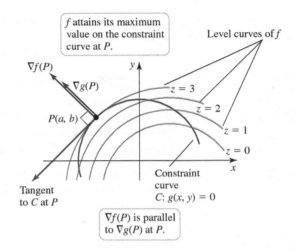

Figure 15.80

What is special about the point P at which f appears to have a local maximum value on C? From Theorem 15.12, we know that at any point $P(a, b)$ on a level curve of f, the line tangent to the level curve at P is orthogonal to $\nabla f(a, b)$. Figure 15.80 also suggests that the line tangent to the level curve of f at P is tangent to the constraint curve C at P. We prove this fact shortly. This observation implies that $\nabla f(a, b)$ is also orthogonal to the line tangent to C at $P(a, b)$.

We need one more observation. The constraint curve C is just one level curve of the function $z = g(x, y)$. Using Theorem 15.12 again, the line tangent to C at $P(a, b)$ is orthogonal to $\nabla g(a, b)$. We have now found two vectors $\nabla f(a, b)$ and $\nabla g(a, b)$ that are both orthogonal to the line tangent to the level curve C at $P(a, b)$. Therefore, these two gradient vectors are parallel. These properties characterize the point P at which f has a local extremum on the constraint curve. They are the basis of the method of *Lagrange multipliers* that we now formalize.

Lagrange Multipliers with Two Independent Variables

The major step in establishing the method of Lagrange multipliers is to prove that Figure 15.80 is drawn correctly; that is, at the point on the constraint curve C where f has a local extreme value, the line tangent to C is orthogonal to $\nabla f(a, b)$ and $\nabla g(a, b)$.

▶ The Greek lowercase ℓ is λ; it is read *lambda*.

> **THEOREM 15.16   Parallel Gradients**
> Let f be a differentiable function in a region of $\mathbb{R}^2$ that contains the smooth curve C given by $g(x, y) = 0$. Assume f has a local extreme value on C at a point $P(a, b)$. Then $\nabla f(a, b)$ is orthogonal to the line tangent to C at P. Assuming $\nabla g(a, b) \neq \mathbf{0}$, it follows that there is a real number λ (called a **Lagrange multiplier**) such that $\nabla f(a, b) = \lambda \nabla g(a, b)$.

Proof: Because C is smooth, it can be expressed parametrically in the form $C: \mathbf{r}(t) = \langle x(t), y(t) \rangle$, where x and y are differentiable functions on an interval in t that contains t_0 with $P(a, b) = (x(t_0), y(t_0))$. As we vary t and follow C, the rate of change of f is given by the Chain Rule:

$$\frac{df}{dt} = \frac{\partial f}{\partial x}\frac{dx}{dt} + \frac{\partial f}{\partial y}\frac{dy}{dt} = \nabla f \cdot \mathbf{r}'(t).$$

At the point $(x(t_0), y(t_0)) = (a, b)$ at which f has a local extreme value, we have $\dfrac{df}{dt}\bigg|_{t=t_0} = 0$, which implies that $\nabla f(a, b) \cdot \mathbf{r}'(t_0) = 0$. Because $\mathbf{r}'(t)$ is tangent to C, the gradient $\nabla f(a, b)$ is orthogonal to the line tangent to C at P.

To prove the second assertion, note that the constraint curve C given by $g(x, y) = 0$ is also a level curve of the surface $z = g(x, y)$. Recall that gradients are orthogonal to level curves. Therefore, at the point $P(a, b)$, $\nabla g(a, b)$ is orthogonal to C at (a, b). Because both $\nabla f(a, b)$ and $\nabla g(a, b)$ are orthogonal to C, the two gradients are parallel, so there is a real number λ such that $\nabla f(a, b) = \lambda \nabla g(a, b)$. ◄

Theorem 15.16 leads directly to the method of Lagrange Multipliers, which produces candidates for *local* maxima and minima of f on the constraint curve. In many problems, however, the goal is to find *absolute* maxima and minima of f on the constraint curve. Much as we did with optimization problems in one variable, we find absolute extrema by examining both local extrema and endpoints. Several different cases arise:

- If the constraint curve is bounded (it lies within a circle of finite radius) and it closes on itself (for example, an ellipse), then we know that the absolute extrema of f exist. In this case, there are no endpoints to consider, and the absolute extrema are found among the local extrema.

- If the constraint curve is bounded and includes its endpoints but does not close on itself (for example, a closed line segment), then the absolute extrema of f exist, and we find them by examining the local extrema and the endpoints.

- In the case that the constraint curve is unbounded (for example, a line or a parabola) or the curve excludes one or both of its endpoints, we have no guarantee that absolute extrema exist. We can find local extrema, but they must be examined carefully to determine whether they are, in fact, absolute extrema (see Example 2 and Exercise 65).

We deal first with the case of finding absolute extrema on closed and bounded constraint curves.

QUICK CHECK 1 It can be shown that the function $f(x, y) = x^2 + y^2$ attains its absolute minimum value on the curve

$$C: g(x, y) = \frac{1}{4}(x - 3)^2 - y = 0$$

at the point $(1, 1)$. Verify that $\nabla f(1, 1)$ and $\nabla g(1, 1)$ are parallel, and that both vectors are orthogonal to the line tangent to C at $(1, 1)$, thereby confirming Theorem 15.16. ◄

PROCEDURE   Lagrange Multipliers: Absolute Extrema on Closed and Bounded Constraint Curves

Let the objective function f and the constraint function g be differentiable on a region of $\mathbb{R}^2$ with $\nabla g(x, y) \neq \mathbf{0}$ on the curve $g(x, y) = 0$. To locate the absolute maximum and minimum values of f subject to the constraint $g(x, y) = 0$, carry out the following steps.

1. Find the values of x, y, and λ (if they exist) that satisfy the equations

$$\nabla f(x, y) = \lambda \nabla g(x, y) \quad \text{and} \quad g(x, y) = 0.$$

2. Evaluate f at the values (x, y) found in Step 1 and at the endpoints of the constraint curve (if they exist). Select the largest and smallest corresponding function values. These values are the absolute maximum and minimum values of f subject to the constraint.

> *In principle*, it is possible to solve a constrained optimization problem by solving the constraint equation for one of the variables and eliminating that variable in the objective function. In practice, this method is often impractical, particularly with three or more variables or two or more constraints.

Notice that $\nabla f = \lambda \nabla g$ is a vector equation: $\langle f_x, f_y \rangle = \lambda \langle g_x, g_y \rangle$. It is satisfied provided $f_x = \lambda g_x$ and $f_y = \lambda g_y$. Therefore, the crux of the method is solving the three equations

$$f_x = \lambda g_x, \qquad f_y = \lambda g_y, \qquad \text{and} \qquad g(x, y) = 0,$$

for the three variables x, y, and λ.

EXAMPLE 1   Lagrange multipliers with two variables Find the absolute maximum and minimum values of the objective function $f(x, y) = x^2 + y^2 + 2$, where x and y lie on the ellipse C given by $g(x, y) = x^2 + xy + y^2 - 4 = 0$.

SOLUTION Because C is closed and bounded, the absolute maximum and minimum values of f exist. **Figure 15.81a** shows the paraboloid $z = f(x, y)$ above the ellipse C in the xy-plane. As the ellipse is traversed, the corresponding function values on the surface vary. The goal is to find the maximum and minimum of these function values. An alternative view is given in **Figure 15.81b**, where we see the level curves of f and the constraint curve C. As the ellipse is traversed, the values of f vary, reaching maximum and minimum values along the way.

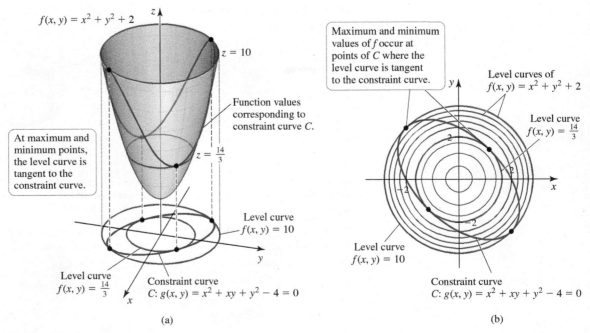

Figure 15.81

Noting that $\nabla f(x, y) = \langle 2x, 2y \rangle$ and $\nabla g(x, y) = \langle 2x + y, x + 2y \rangle$, the equations that result from $\nabla f = \lambda \nabla g$ and the constraint are

$$\underbrace{2x = \lambda(2x + y)}_{f_x = \lambda g_x}, \qquad \underbrace{2y = \lambda(x + 2y)}_{f_y = \lambda g_y}, \qquad \text{and} \qquad \underbrace{x^2 + xy + y^2 - 4 = 0.}_{\text{constraint } g(x, y) = 0}$$

Subtracting the second equation from the first leads to

$$(x - y)(2 - \lambda) = 0,$$

which implies that $y = x$, or $\lambda = 2$. In the case that $y = x$, the constraint equation simplifies to $3x^2 - 4 = 0$, or $x = \pm\frac{2}{\sqrt{3}}$. Therefore, two candidates for locations of extreme values are $\left(\frac{2}{\sqrt{3}}, \frac{2}{\sqrt{3}}\right)$ and $\left(-\frac{2}{\sqrt{3}}, -\frac{2}{\sqrt{3}}\right)$.

Substituting $\lambda = 2$ into the first equation leads to $y = -x$, and then the constraint equation simplifies to $x^2 - 4 = 0$, or $x = \pm 2$. These values give two additional points of interest, $(2, -2)$ and $(-2, 2)$. Evaluating f at each of these points, we find that $f\left(\frac{2}{\sqrt{3}}, \frac{2}{\sqrt{3}}\right) = f\left(-\frac{2}{\sqrt{3}}, -\frac{2}{\sqrt{3}}\right) = \frac{14}{3}$ and $f(2, -2) = f(-2, 2) = 10$. Therefore, the absolute maximum value of f on C is 10, which occurs at $(2, -2)$ and $(-2, 2)$, and the absolute minimum value of f on C is $14/3$, which occurs at $\left(\frac{2}{\sqrt{3}}, \frac{2}{\sqrt{3}}\right)$ and $\left(-\frac{2}{\sqrt{3}}, -\frac{2}{\sqrt{3}}\right)$. Notice that the value of λ is not used in the final result.

Related Exercises 9–10 ◄

QUICK CHECK 2 Choose any point on the constraint curve in Figure 15.81b other than a solution point. Draw ∇f and ∇g at that point and show that they are not parallel. ◄

Lagrange Multipliers with Three Independent Variables

The technique just outlined extends to three or more independent variables. With three variables, suppose an objective function $w = f(x, y, z)$ is given; its level surfaces are surfaces in $\mathbb{R}^3$ (**Figure 15.82a**). The constraint equation takes the form $g(x, y, z) = 0$, which is another surface S in $\mathbb{R}^3$ (**Figure 15.82b**). To find the local maximum and minimum values of f on S (assuming they exist), we must find the points (a, b, c) on S at which $\nabla f(a, b, c)$ is parallel to $\nabla g(a, b, c)$, assuming $\nabla g(a, b, c) \neq \mathbf{0}$ (**Figure 15.82c, d**). In the case where the surface $g(x, y, z) = 0$ is closed and bounded, the procedure for finding the absolute maximum and minimum values of $f(x, y, z)$, where the point (x, y, z) is constrained to lie on S, is similar to the procedure for two variables.

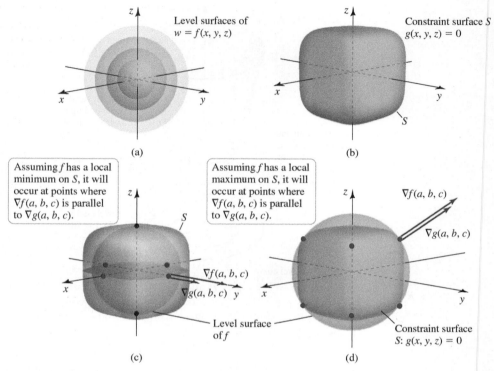

Level surfaces of
$w = f(x, y, z)$

(a)

Constraint surface S
$g(x, y, z) = 0$

(b)

Assuming f has a local minimum on S, it will occur at points where $\nabla f(a, b, c)$ is parallel to $\nabla g(a, b, c)$.

$\nabla f(a, b, c)$
$\nabla g(a, b, c)$

Level surface of f

(c)

Assuming f has a local maximum on S, it will occur at points where $\nabla f(a, b, c)$ is parallel to $\nabla g(a, b, c)$.

$\nabla f(a, b, c)$
$\nabla g(a, b, c)$

Constraint surface
S: $g(x, y, z) = 0$

(d)

Figure 15.82

> If the constraint surface S: $g(x, y, z) = 0$ has a boundary curve C (see figure), then each point on C is a candidate for the location of an absolute maximum or minimum value of f, and these points must be analyzed in Step 2 of the procedure. We avoid this case in the exercise set.

> Problems similar to Example 2 were solved in Section 15.7 using ordinary optimization techniques. These methods may or may not be easier to apply than Lagrange multipliers.

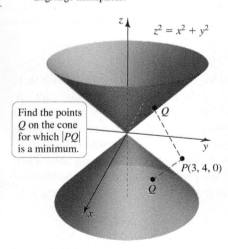

Find the points Q on the cone for which $|PQ|$ is a minimum.

Figure 15.83

PROCEDURE Lagrange Multipliers: Absolute Extrema on Closed and Bounded Constraint Surfaces

Let f and g be differentiable on a region of $\mathbb{R}^3$ with $\nabla g(x, y, z) \neq \mathbf{0}$ on the surface $g(x, y, z) = 0$. To locate the absolute maximum and minimum values of f subject to the constraint $g(x, y, z) = 0$, carry out the following steps.

1. Find the values of x, y, z, and λ that satisfy the equations

$$\nabla f(x, y, z) = \lambda \nabla g(x, y, z) \quad \text{and} \quad g(x, y, z) = 0.$$

2. Among the points (x, y, z) found in Step 1, select the largest and smallest corresponding function values. These values are the absolute maximum and minimum values of f subject to the constraint.

Now there are four equations to be solved for x, y, z, and λ:

$$f_x(x, y, z) = \lambda g_x(x, y, z), \qquad f_y(x, y, z) = \lambda g_y(x, y, z),$$
$$f_z(x, y, z) = \lambda g_z(x, y, z), \quad \text{and} \quad g(x, y, z) = 0.$$

As in the two-variable case, special care must be given to constraint surfaces that are not closed and bounded. We examine one such case in Example 2.

EXAMPLE 2 A geometry problem Find the least distance between the point $P(3, 4, 0)$ and the surface of the cone $z^2 = x^2 + y^2$.

SOLUTION The cone is not bounded, so we begin our calculations recognizing that solutions are only candidates for local extrema. **Figure 15.83** shows both sheets of the cone and the point $P(3, 4, 0)$. Because P is in the xy-plane, we anticipate two solutions, one for each sheet of the cone. The distance between P and any point $Q(x, y, z)$ on the cone is

$$d(x, y, z) = \sqrt{(x - 3)^2 + (y - 4)^2 + z^2}.$$

In many distance problems, it is easier to work with the *square* of the distance to avoid dealing with square roots. This maneuver is allowable because if a point minimizes $(d(x, y, z))^2$, it also minimizes $d(x, y, z)$. Therefore, we define

$$f(x, y, z) = (d(x, y, z))^2 = (x - 3)^2 + (y - 4)^2 + z^2.$$

The constraint is the condition that the point (x, y, z) must lie on the cone, which implies $z^2 = x^2 + y^2$, or $g(x, y, z) = z^2 - x^2 - y^2 = 0$.

Now we proceed with Lagrange multipliers; the conditions are

$$f_x(x, y, z) = \lambda g_x(x, y, z), \quad \text{or} \quad 2(x - 3) = \lambda(-2x), \quad \text{or} \quad x(1 + \lambda) = 3, \qquad (1)$$

$$f_y(x, y, z) = \lambda g_y(x, y, z), \quad \text{or} \quad 2(y - 4) = \lambda(-2y), \quad \text{or} \quad y(1 + \lambda) = 4, \qquad (2)$$

$$f_z(x, y, z) = \lambda g_z(x, y, z), \quad \text{or} \quad 2z = \lambda(2z), \quad \text{or} \quad z = \lambda z, \text{ and} \qquad (3)$$

$$g(x, y, z) = z^2 - x^2 - y^2 = 0. \qquad (4)$$

The solutions of equation (3) (the simplest of the four equations) are either $z = 0$, or $\lambda = 1$ and $z \ne 0$. In the first case, if $z = 0$, then by equation (4), $x = y = 0$; however, $x = 0$ and $y = 0$ do not satisfy (1) and (2). So no solution results from this case.

On the other hand, if $\lambda = 1$ in equation (3), then by (1) and (2), we find that $x = \frac{3}{2}$ and $y = 2$. Using (4), the corresponding values of z are $\pm \frac{5}{2}$. Therefore, the two solutions and the values of f are

$$x = \tfrac{3}{2}, \qquad y = 2, \qquad z = \tfrac{5}{2}, \qquad \text{with } f\left(\tfrac{3}{2}, 2, \tfrac{5}{2}\right) = \tfrac{25}{2}, \text{ and}$$

$$x = \tfrac{3}{2}, \qquad y = 2, \qquad z = -\tfrac{5}{2}, \qquad \text{with } f\left(\tfrac{3}{2}, 2, -\tfrac{5}{2}\right) = \tfrac{25}{2}.$$

You can check that moving away from $\left(\frac{3}{2}, 2, \pm\frac{5}{2}\right)$ in any direction on the cone has the effect of increasing the values of f. Therefore, the points correspond to *local* minima of f. Do these points also correspond to *absolute* minima? The domain of this problem is unbounded; however, one can argue geometrically that f increases without bound moving away from $\left(\frac{3}{2}, 2, \pm\frac{5}{2}\right)$ on the cone with $|x| \to \infty$ and $|y| \to \infty$. Therefore, these points correspond to absolute minimum values and the points on the cone nearest to $(3, 4, 0)$ are

$$\left(\tfrac{3}{2}, 2, \pm\tfrac{5}{2}\right), \text{ at a distance of } \sqrt{\frac{25}{2}} = \frac{5}{\sqrt{2}}. \text{ (Recall that } f = d^2.)$$

Related Exercises 32–34 ◄

Economic Models   In the opening of this section, we briefly described how utility functions are used to model consumer behavior. We now look in more detail at some specific—admittedly simple—utility functions and the constraints that are imposed on them.

As described earlier, a prototype model for consumer behavior uses two independent variables: leisure time ℓ and consumable goods g. A utility function $U = f(\ell, g)$ measures consumer preferences for various combinations of leisure time and consumable goods. The following assumptions about utility functions are commonly made.

1. Utility increases if any variable increases (essentially, *more is better*).

2. Various combinations of leisure time and consumable goods have the same utility; that is, giving up some leisure time for additional consumable goods (or vice versa) results in the same utility.

The level curves of a typical utility function are shown in **Figure 15.84**. Assumption 1 is reflected by the fact that the utility values on the level curves increase as either ℓ or g increases. Consistent with Assumption 2, a single level curve shows the combinations of ℓ and g that have the same utility; for this reason, economists call the level curves *indifference curves*. Notice that if ℓ increases, then g must decrease on a level curve to maintain the same utility, and vice versa.

Economic models assert that consumers maximize utility subject to constraints on leisure time and consumable goods. One assumption that leads to a reasonable constraint is that an increase in leisure time implies a linear decrease in consumable goods. Therefore, the constraint curve is a line with negative slope (**Figure 15.85**). When such a constraint is superimposed on the level curves of the utility function, the optimization problem becomes evident. Among all points on the constraint line, which one maximizes utility? A solution is marked in the figure; at this point, the utility has a maximum value (between 2.5 and 3.0).

▶ With three independent variables, it is possible to impose two constraints. These problems are explored in Exercises 61–64.

QUICK CHECK 3 In Example 2, is there a point that *maximizes* the distance between $(3, 4, 0)$ and the cone? If the point $(3, 4, 0)$ were replaced by $(3, 4, 1)$, how many minimizing solutions would there be? ◄

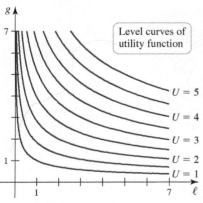

Level curves of utility function

$U = 5$
$U = 4$
$U = 3$
$U = 2$
$U = 1$

Figure 15.84

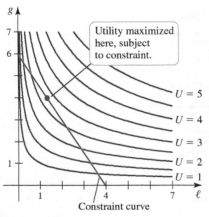

Utility maximized here, subject to constraint.

$U = 5$
$U = 4$
$U = 3$
$U = 2$
$U = 1$

Constraint curve

Figure 15.85

EXAMPLE 3 Constrained optimization of utility Find the absolute maximum value of the utility function $U = f(\ell, g) = \ell^{1/3}g^{2/3}$, subject to the constraint $G(\ell, g) = 3\ell + 2g - 12 = 0$, where $\ell \geq 0$ and $g \geq 0$.

SOLUTION The constraint is closed and bounded, so we expect to find an absolute maximum value of f. The level curves of the utility function and the linear constraint are shown in Figure 15.85. The solution follows the Lagrange multiplier method with two variables. The gradient of the utility function is

$$\nabla f(\ell, g) = \left\langle \frac{\ell^{-2/3}g^{2/3}}{3}, \frac{2\ell^{1/3}g^{-1/3}}{3} \right\rangle = \frac{1}{3}\left\langle \left(\frac{g}{\ell}\right)^{2/3}, 2\left(\frac{\ell}{g}\right)^{1/3} \right\rangle.$$

The gradient of the constraint function is $\nabla G(\ell, g) = \langle 3, 2 \rangle$. Therefore, the equations that must be solved are

$$\frac{1}{3}\left(\frac{g}{\ell}\right)^{2/3} = 3\lambda, \qquad \frac{2}{3}\left(\frac{\ell}{g}\right)^{1/3} = 2\lambda, \qquad \text{and} \qquad G(\ell, g) = 3\ell + 2g - 12 = 0.$$

Eliminating λ from the first two equations leads to the condition $g = 3\ell$, which, when substituted into the constraint equation, gives the solution $\ell = \frac{4}{3}$ and $g = 4$. This point is a candidate for the location of the absolute maximum; the other candidates are the end-points of the constraint curve, $(4, 0)$ and $(0, 6)$. The actual values of the utility function at these points are $U = f\left(\frac{4}{3}, 4\right) = 4/\sqrt[3]{3} \approx 2.8$ and $f(4, 0) = f(0, 6) = 0$. We conclude that the maximum value of f is 2.8; this solution occurs at $\ell = \frac{4}{3}$ and $g = 4$, and it is consistent with Figure 15.85.

Related Exercise 38 ◄

QUICK CHECK 4 In Figure 15.85, explain why, if you move away from the optimal point along the constraint line, the utility decreases. ◄

SECTION 15.8 EXERCISES

Getting Started

1. Explain why, at a point that maximizes or minimizes f subject to a constraint $g(x, y) = 0$, the gradient of f is parallel to the gradient of g. Use a diagram.

2. Describe the steps used to find the absolute maximum value and absolute minimum value of a differentiable function on a circle centered at the origin of the xy-plane.

3. For functions $f(x, y) = x + 4y$ and $g(x, y) = x^2 + y^2 - 1$, write the Lagrange multiplier conditions that must be satisfied by a point that maximizes or minimizes f subject to the constraint $g(x, y) = 0$.

4. For functions $f(x, y, z) = xyz$ and $g(x, y, z) = x^2 + 2y^2 + 3z^2 - 1$, write the Lagrange multiplier conditions that must be satisfied by a point that maximizes or minimizes f subject to the constraint $g(x, y) = 0$.

5–6. The following figures show the level curves of f and the constraint curve $g(x, y) = 0$. Estimate the maximum and minimum values of f subject to the constraint. At each point where an extreme value occurs, indicate the direction of ∇f and a possible direction of ∇g.

5.

6.

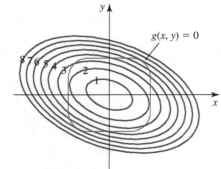

Practice Exercises

7–26. Lagrange multipliers *Each function f has an absolute maximum value and absolute minimum value subject to the given constraint. Use Lagrange multipliers to find these values.*

7. $f(x, y) = x + 2y$ subject to $x^2 + y^2 = 4$

8. $f(x, y) = xy^2$ subject to $x^2 + y^2 = 1$

9. $f(x, y) = x + y$ subject to $x^2 - xy + y^2 = 1$

10. $f(x, y) = x^2 + y^2$ subject to $2x^2 + 3xy + 2y^2 = 7$

11. $f(x, y) = xy$ subject to $x^2 + y^2 - xy = 9$

12. $f(x, y) = x - y$ subject to $x^2 + y^2 - 3xy = 20$

13. $f(x, y) = e^{xy}$ subject to $x^2 + xy + y^2 = 9$

14. $f(x, y) = x^2y$ subject to $x^2 + y^2 = 9$

15. $f(x, y) = 2x^2 + y^2$ subject to $x^2 + 2y + y^2 \doteq 15$

16. $f(x, y) = x^2$ subject to $x^2 + xy + y^2 = 3$

17. $f(x, y, z) = x + 3y - z$ subject to $x^2 + y^2 + z^2 = 4$

18. $f(x, y, z) = xyz$ subject to $x^2 + 2y^2 + 4z^2 = 9$

19. $f(x, y, z) = x$ subject to $x^2 + y^2 + z^2 - z = 1$

20. $f(x, y, z) = x - z$ subject to $x^2 + y^2 + z^2 - y = 2$

21. $f(x, y, z) = x + y + z$ subject to $x^2 + y^2 + z^2 - xy = 5$

22. $f(x, y, z) = x + y + z$ subject to $x^2 + y^2 + z^2 - 2x - 2y = 1$

23. $f(x, y, z) = 2x + z^2$ subject to $x^2 + y^2 + 2z^2 = 25$

24. $f(x, y, z) = xy - z$ subject to $x^2 + y^2 + z^2 - xy = 1$

25. $f(x, y, z) = x^2 + y + z$ subject to $2x^2 + 2y^2 + z^2 = 2$

26. $f(x, y, z) = (xyz)^{1/2}$ subject to $x + y + z = 1$ with $x \geq 0$, $y \geq 0$, $z \geq 0$

27–36. Applications of Lagrange multipliers *Use Lagrange multipliers in the following problems. When the constraint curve is unbounded, explain why you have found an absolute maximum or minimum value.*

27. **Shipping regulations** A shipping company requires that the sum of length plus girth of rectangular boxes not exceed 108 in. Find the dimensions of the box with maximum volume that meets this condition. (The girth is the perimeter of the smallest side of the box.)

28. **Box with minimum surface area** Find the dimensions of the rectangular box with a volume of 16 ft^3 that has minimum surface area.

29. **Extreme distances to an ellipse** Find the minimum and maximum distances between the ellipse $x^2 + xy + 2y^2 = 1$ and the origin.

30. **Maximum area rectangle in an ellipse** Find the dimensions of the rectangle of maximum area with sides parallel to the coordinate axes that can be inscribed in the ellipse $4x^2 + 16y^2 = 16$.

31. **Maximum perimeter rectangle in an ellipse** Find the dimensions of the rectangle of maximum perimeter with sides parallel to the coordinate axes that can be inscribed in the ellipse $2x^2 + 4y^2 = 3$.

32. **Minimum distance to a plane** Find the point on the plane $2x + 3y + 6z - 10 = 0$ closest to the point $(-2, 5, 1)$.

33. **Minimum distance to a surface** Find the point on the surface $4x + y - 1 = 0$ closest to the point $(1, 2, -3)$.

34. **Minimum distance to a cone** Find the points on the cone $z^2 = x^2 + y^2$ closest to the point $(1, 2, 0)$.

35. **Extreme distances to a sphere** Find the minimum and maximum distances between the sphere $x^2 + y^2 + z^2 = 9$ and the point $(2, 3, 4)$.

36. **Maximum volume cylinder in a sphere** Find the dimensions of the right circular cylinder of maximum volume that can be inscribed in a sphere of radius 16.

37–40. Maximizing utility functions *Find the values of ℓ and g with $\ell \geq 0$ and $g \geq 0$ that maximize the following utility functions subject to the given constraints. Give the value of the utility function at the optimal point.*

37. $U = f(\ell, g) = 10\ell^{1/2}g^{1/2}$ subject to $3\ell + 6g = 18$

38. $U = f(\ell, g) = 32\ell^{2/3}g^{1/3}$ subject to $4\ell + 2g = 12$

39. $U = f(\ell, g) = 8\ell^{4/5}g^{1/5}$ subject to $10\ell + 8g = 40$

40. $U = f(\ell, g) = \ell^{1/6}g^{5/6}$ subject to $4\ell + 5g = 20$

41. **Explain why or why not** Determine whether the following statements are true and give an explanation or counterexample.

    a. Suppose you are standing at the center of a sphere looking at a point P on the surface of the sphere. Your line of sight to P is orthogonal to the plane tangent to the sphere at P.

    b. At a point that maximizes f on the curve $g(x, y) = 0$, the dot product $\nabla f \cdot \nabla g$ is zero.

42–47. Alternative method *Solve the following problems from Section 15.7 using Lagrange multipliers.*

42. Exercise 43   43. Exercise 44   44. Exercise 45

45. Exercise 46   46. Exercise 70   47. Exercise 63

48–51. Absolute maximum and minimum values *Find the absolute maximum and minimum values of the following functions over the given regions R. Use Lagrange multipliers to check for extreme points on the boundary.*

48. $f(x, y) = x^2 + 4y^2 + 1; R = \{(x, y): x^2 + 4y^2 \leq 1\}$

49. $f(x, y) = x^2 + y^2 - 2y + 1; R = \{(x, y): x^2 + y^2 \leq 4\}$ (This is Exercise 47, Section 15.7.)

50. $f(x, y) = 2x^2 + y^2; R = \{(x, y): x^2 + y^2 \leq 16\}$ (This is Exercise 48, Section 15.7.)

51. $f(x, y) = 2x^2 - 4x + 3y^2 + 2; R = \{(x, y):(x - 1)^2 + y^2 \leq 1\}$ (This is Exercise 51, Section 15.7.)

52. **Extreme points on flattened spheres** The equation $x^{2n} + y^{2n} + z^{2n} = 1$, where n is a positive integer, describes a flattened sphere. Define the extreme points to be the points on the flattened sphere with a maximum distance from the origin.

    a. Find all the extreme points on the flattened sphere with $n = 2$. What is the distance between the extreme points and the origin?

    b. Find all the extreme points on the flattened sphere for integers $n > 2$. What is the distance between the extreme points and the origin?

    c. Give the location of the extreme points in the limit as $n \to \infty$. What is the limiting distance between the extreme points and the origin as $n \to \infty$?

53–55. Production functions *Economists model the output of manufacturing systems using production functions that have many of the same properties as utility functions. The family of Cobb-Douglas production functions has the form $P = f(K, L) = CK^aL^{1-a}$, where K represents capital, L represents labor, and C and a are positive real numbers with $0 < a < 1$. If the cost of capital is p dollars per unit, the cost of labor is q dollars per unit, and the total available budget is B, then the constraint takes the form $pK + qL = B$. Find the values of K and L that maximize the following production functions subject to the given constraint, assuming $K \geq 0$ and $L \geq 0$.*

53. $P = f(K, L) = K^{1/2}L^{1/2}$ for $20K + 30L = 300$

54. $P = f(K, L) = 10K^{1/3}L^{2/3}$ for $30K + 60L = 360$

55. Given the production function $P = f(K, L) = K^aL^{1-a}$ and the budget constraint $pK + qL = B$, where a, p, q, and B are given, show that P is maximized when $K = \dfrac{aB}{p}$ and $L = \dfrac{(1-a)B}{q}$.

56. Temperature of an elliptical plate The temperature of points on an elliptical plate $x^2 + y^2 + xy \le 1$ is given by $T(x, y) = 25(x^2 + y^2)$. Find the hottest and coldest temperatures on the edge of the plate.

Explorations and Challenges

57–59. Maximizing a sum

57. Find the maximum value of $x_1 + x_2 + x_3 + x_4$ subject to the condition that $x_1^2 + x_2^2 + x_3^2 + x_4^2 = 16$.

58. Generalize Exercise 57 and find the maximum value of $x_1 + x_2 + \cdots + x_n$ subject to the condition that $x_1^2 + x_2^2 + \cdots + x_n^2 = c^2$ for a real number c and a positive integer n.

59. Generalize Exercise 57 and find the maximum value of $a_1 x_1 + a_2 x_2 + \cdots + a_n x_n$ subject to the condition that $x_1^2 + x_2^2 + \cdots + x_n^2 = 1$, for given positive real numbers $a_1, \ldots, a_n$ and a positive integer n.

60. Geometric and arithmetic means Given positive numbers $x_1, \ldots, x_n$, prove that the geometric mean $(x_1 x_2 \cdots x_n)^{1/n}$ is no greater than the arithmetic mean $\dfrac{x_1 + \cdots + x_n}{n}$ in the following cases.

a. Find the maximum value of xyz, subject to $x + y + z = k$, where k is a positive real number and $x > 0$, $y > 0$, and $z > 0$. Use the result to prove that
$$(xyz)^{1/3} \le \frac{x + y + z}{3}.$$

b. Generalize part (a) and show that
$$(x_1 x_2 \cdots x_n)^{1/n} \le \frac{x_1 + \cdots + x_n}{n}.$$

61. Problems with two constraints Given a differentiable function $w = f(x, y, z)$, the goal is to find its absolute maximum and minimum values (assuming they exist) subject to the constraints $g(x, y, z) = 0$ and $h(x, y, z) = 0$, where g and h are also differentiable.

a. Imagine a level surface of the function f and the constraint surfaces $g(x, y, z) = 0$ and $h(x, y, z) = 0$. Note that g and h intersect (in general) in a curve C on which maximum and minimum values of f must be found. Explain why ∇g and ∇h are orthogonal to their respective surfaces.

b. Explain why ∇f lies in the plane formed by ∇g and ∇h at a point of C where f has a maximum or minimum value.

c. Explain why part (b) implies that $\nabla f = \lambda \nabla g + \mu \nabla h$ at a point of C where f has a maximum or minimum value, where λ and μ (the Lagrange multipliers) are real numbers.

d. Conclude from part (c) that the equations that must be solved for maximum or minimum values of f subject to two constraints are $\nabla f = \lambda \nabla g + \mu \nabla h$, $g(x, y, z) = 0$, and $h(x, y, z) = 0$.

62–64. Two-constraint problems *Use the result of Exercise 61 to solve the following problems.*

62. The planes $x + 2z = 12$ and $x + y = 6$ intersect in a line L. Find the point on L nearest the origin.

63. Find the maximum and minimum values of $f(x, y, z) = xyz$ subject to the conditions that $x^2 + y^2 = 4$ and $x + y + z = 1$.

64. Find the maximum and minimum values of $f(x, y, z) = x^2 + y^2 + z^2$ on the curve on which the cone $z^2 = 4x^2 + 4y^2$ and the plane $2x + 4z = 5$ intersect.

65. Check assumptions Consider the function $f(x, y) = xy + x + y + 100$ subject to the constraint $xy = 4$.

a. Use the method of Lagrange multipliers to write a system of three equations with three variables x, y, and λ.

b. Solve the system in part (a) to verify that $(x, y) = (-2, -2)$ and $(x, y) = (2, 2)$ are solutions.

c. Let the curve C_1 be the branch of the constraint curve corresponding to $x > 0$. Calculate $f(2, 2)$ and determine whether this value is an absolute maximum or minimum value of f over C_1. (*Hint:* Let $h_1(x)$, for $x > 0$, equal the values of f over the curve C_1 and determine whether h_1 attains an absolute maximum or minimum value at $x = 2$.)

d. Let the curve C_2 be the branch of the constraint curve corresponding to $x < 0$. Calculate $f(-2, -2)$ and determine whether this value is an absolute maximum or minimum value of f over C_2. (*Hint:* Let $h_2(x)$, for $x < 0$, equal the values of f over the curve C_2 and determine whether h_2 attains an absolute maximum or minimum value at $x = -2$.)

e. Show that the method of Lagrange multipliers fails to find the absolute maximum and minimum values of f over the constraint curve $xy = 4$. Reconcile your explanation with the method of Lagrange multipliers.

QUICK CHECK ANSWERS

1. Note that $\nabla f(1, 1) = \langle 2x, 2y \rangle |_{(1, 1)} = \langle 2, 2 \rangle$ and $\nabla g(1, 1) = \langle \frac{1}{2}(x - 3), -1 \rangle |_{(1, 1)} = \langle -1, -1 \rangle$, which implies the gradients are multiples of one another, and therefore parallel. The equation of the line tangent to C at $(1, 1)$ is $y = -x + 2$; therefore, the vector $\mathbf{v} = \langle 1, -1 \rangle$ is parallel to this tangent line. Because $\nabla f(1, 1) \cdot \mathbf{v} = 0$ and $\nabla g(1, 1) \cdot \mathbf{v} = 0$, both gradients are orthogonal to the tangent line. **3.** The distance between $(3, 4, 0)$ and the cone can be arbitrarily large, so there is no maximizing solution. If the point of interest is not in the xy-plane, there is one minimizing solution. **4.** If you move along the constraint line away from the optimal solution in either direction, you cross level curves of the utility function with decreasing values. ◄

CHAPTER 15 REVIEW EXERCISES

1. **Explain why or why not** Determine whether the following statements are true and give an explanation or counterexample.

   a. The level curves of $g(x, y) = e^{x+y}$ are lines.
   b. The equation $z^2 = 2x^2 - 6y^2$ determines z as a single function of x and y.
   c. If f has continuous partial derivatives of all orders, then $f_{xxy} = f_{yyx}$.
   d. Given the surface $z = f(x, y)$, the gradient $\nabla f(a, b)$ lies in the plane tangent to the surface at $(a, b, f(a, b))$.

2–5. Domains *Find the domain of the following functions. Make a sketch of the domain in the xy-plane.*

2. $f(x, y) = \sin^{-1}\left(\dfrac{x^2 + y^2}{4}\right)$ 3. $f(x, y) = \sqrt{x - y^2}$

4. $f(x, y) = \ln xy$ 5. $f(x, y) = \sqrt{9x^2 + 4y^2 - 36}$

6–7. Graphs *Describe the graph of the following functions, and state the domain and range of the function.*

6. $f(x, y) = -\sqrt{x^2 + y^2}$ 7. $g(x, y) = -\sqrt{x^2 + y^2 - 1}$

8–9. Level curves *Make a sketch of several level curves of the following functions. Label at least two level curves with their z-values.*

8. $f(x, y) = x^2 - y$ 9. $f(x, y) = x^2 + 4y^2$

10. **Matching level curves with surfaces** Match level curve plots a–d with surfaces A–D.

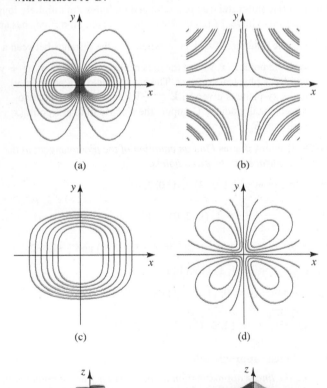

(a)

(b)

(c)

(d)

(A)

(B)

(C)

(D)

11–18. Limits *Evaluate the following limits or determine that they do not exist.*

11. $\displaystyle\lim_{(x, y)\to(4, -2)} (10x - 5y + 6xy)$

12. $\displaystyle\lim_{(x, y)\to(1, 1)} \dfrac{xy}{x + y}$ 13. $\displaystyle\lim_{(x, y)\to(0, 0)} \dfrac{x + y}{xy}$

14. $\displaystyle\lim_{(x, y)\to(0, 0)} \dfrac{\sin xy}{x^2 + y^2}$ 15. $\displaystyle\lim_{(x, y)\to(-1, 1)} \dfrac{x^2 - y^2}{x^2 - xy - 2y^2}$

16. $\displaystyle\lim_{(x, y)\to(0, 0)} \dfrac{25x^6 y^4}{\sin^2(x^3 y^2)}$

17. $\displaystyle\lim_{(x, y, z)\to(2, 2, 3)} \dfrac{x^2 z - 3x^2 - y^2 z + 3y^2}{xz - 3x - yz + 3y}$

18. $\displaystyle\lim_{(x, y, z)\to(3, 4, 7)} \dfrac{\sqrt{x + y} - \sqrt{z}}{x + y - z}$

19–20. Continuity *At what points of $\mathbb{R}^2$ are the following functions continuous?*

19. $f(x, y) = \ln(y - x^2 - 1)$ 20. $g(x, y) = \dfrac{1}{x^2 + y^2}$

21–26. Partial derivatives *Find the first partial derivatives of the following functions.*

21. $f(x, y) = 3x^2 y^5$ 22. $g(x, y, z) = 4xyz^2 - \dfrac{3x}{y}$

23. $f(x, y) = \dfrac{x^2}{x^2 + y^2}$ 24. $g(x, y, z) = \dfrac{xyz}{x + y}$

25. $f(x, y) = xye^{xy}$ 26. $g(u, v) = u\cos v - v\sin u$

27–28. Second partial derivatives *Find the four second partial derivatives of the following functions.*

27. $f(x, y) = e^{2xy}$ 28. $H(p, r) = p^2\sqrt{p + 2r}$

29–30. Laplace's equation *Verify that the following functions satisfy Laplace's equation, $\dfrac{\partial^2 u}{\partial x^2} + \dfrac{\partial^2 u}{\partial y^2} = 0$.*

29. $u(x, y) = y(3x^2 - y^2)$ 30. $u(x, y) = \ln(x^2 + y^2)$

31. **Region between spheres** Two spheres have the same center and radii r and R, where $0 < r < R$. The volume of the region between the spheres is $V(r, R) = \dfrac{4\pi}{3}(R^3 - r^3)$.

   a. First use your intuition. If r is held fixed, how does V change as R increases? What is the sign of V_R? If R is held fixed, how does V change as r increases (up to the value of R)? What is the sign of V_r?
   b. Compute V_r and V_R. Are the results consistent with part (a)?

c. Consider spheres with $R = 3$ and $r = 1$. Does the volume change more if R is increased by $\Delta R = 0.1$ (with r fixed) or if r is decreased by $\Delta r = 0.1$ (with R fixed)?

32–35. Chain Rule *Use the Chain Rule to evaluate the following derivatives.*

32. $w'(t)$, where $w = x \cos yz$, $x = t^2 + 1$, $y = t$, and $z = t^3$

33. $w'(t)$, where $w = z \ln(x^2 + y^2)$, $x = 3e^t$, $y = 4e^t$, and $z = t$

34. w_s, w_t, w_{ss}, w_{tt}, and w_{st}, where $w = xyz$, $x = 2st$, $y = st^2$, and $z = s^2t$

35. w_r, w_s and w_t, where $w = \ln(xy^2)$, $x = rst$, and $y = r + s$

36–37. Implicit differentiation *Find dy/dx for the following implicit relations using Theorem 15.9.*

36. $2x^2 + 3xy - 3y^4 = 2$ **37.** $y \ln(x^2 + y^2) = 4$

38–39. Walking on a surface *Consider the following surfaces and parameterized curves C in the xy-plane.*

a. *In each case, find $z'(t)$ on C.*
b. *Imagine that you are walking on the surface directly above C consistent with the positive orientation of C. Find the values of t for which you are walking uphill.*

38. $z = 4x^2 + y^2 - 2$; $C: x = \cos t$, $y = \sin t$, for $0 \le t \le 2\pi$

39. $z = x^2 - 2y^2 + 4$; $C: x = 2 \cos t$, $y = 2 \sin t$, for $0 \le t \le 2\pi$

40. Constant volume cones Suppose the radius of a right circular cone increases as $r(t) = t^a$ and the height decreases as $h(t) = t^{-b}$, for $t \ge 1$, where a and b are positive constants. What is the relationship between a and b such that the volume of the cone remains constant (that is, $V'(t) = 0$, where $V = (\pi/3)r^2h$)?

41. Directional derivatives Consider the function $f(x, y) = 2x^2 - 4y^2 + 10$, whose graph is shown in the figure.

a. Fill in the table showing the values of the directional derivative at points (a, b) in the directions given by the unit vectors $\mathbf{u}$, $\mathbf{v}$, and $\mathbf{w}$.

	$(a, b) = (0, 0)$	$(a, b) = (2, 0)$	$(a, b) = (1, 1)$
$\mathbf{u} = \left\langle \frac{\sqrt{2}}{2}, \frac{\sqrt{2}}{2} \right\rangle$			
$\mathbf{v} = \left\langle -\frac{\sqrt{2}}{2}, \frac{\sqrt{2}}{2} \right\rangle$			
$\mathbf{w} = \left\langle -\frac{\sqrt{2}}{2}, -\frac{\sqrt{2}}{2} \right\rangle$			

b. Interpret each of the directional derivatives computed in part (a) at the point $(2, 0)$.

42–47. Computing directional derivatives *Compute the gradient of the following functions, evaluate it at the given point P, and evaluate the directional derivative at that point in the direction of the given vector.*

42. $f(x, y) = x^2$; $P(1, 2)$; $\mathbf{u} = \left\langle \frac{1}{\sqrt{2}}, -\frac{1}{\sqrt{2}} \right\rangle$

43. $g(x, y) = x^2y^3$; $P(-1, 1)$; $\mathbf{u} = \left\langle \frac{5}{13}, \frac{12}{13} \right\rangle$

44. $f(x, y) = \frac{x}{y^2}$; $P(0, 3)$; $\mathbf{u} = \left\langle \frac{\sqrt{3}}{2}, \frac{1}{2} \right\rangle$

45. $h(x, y) = \sqrt{2 + x^2 + 2y^2}$; $P(2, 1)$; $\mathbf{u} = \left\langle \frac{3}{5}, \frac{4}{5} \right\rangle$

46. $f(x, y, z) = xe^{1+y^2+z^2}$; $P(0, 1, -2)$; $\mathbf{u} = \left\langle \frac{4}{9}, \frac{1}{9}, \frac{8}{9} \right\rangle$

47. $f(x, y, z) = \sin xy + \cos z$; $P(1, \pi, 0)$; $\mathbf{u} = \left\langle \frac{2}{7}, \frac{3}{7}, -\frac{6}{7} \right\rangle$

48–49. Direction of steepest ascent and descent

a. *Find the unit vectors that give the direction of steepest ascent and steepest descent at P.*
b. *Find a unit vector that points in a direction of no change.*

48. $f(x, y) = \ln(1 + xy)$; $P(2, 3)$

49. $f(x, y) = \sqrt{4 - x^2 - y^2}$; $P(-1, 1)$

50–51. Level curves *Consider the paraboloid $f(x, y) = 8 - 2x^2 - y^2$. For the following level curves $f(x, y) = C$ and points (a, b), compute the slope of the line tangent to the level curve at (a, b) and verify that the tangent line is orthogonal to the gradient at that point.*

50. $f(x, y) = 5$; $(a, b) = (1, 1)$

51. $f(x, y) = 0$; $(a, b) = (2, 0)$

52. Directions of zero change Find the directions in which the function $f(x, y) = 4x^2 - y^2$ has zero change at the point $(1, 1, 3)$. Express the directions in terms of unit vectors.

53. Electric potential due to a charged cylinder An infinitely long charged cylinder of radius R with its axis along the z-axis has an electric potential $V = k \ln \frac{R}{r}$, where r is the distance between a variable point $P(x, y)$ and the axis of the cylinder ($r^2 = x^2 + y^2$) and k is a physical constant. The electric field at a point (x, y) in the xy-plane is given by $\mathbf{E} = -\nabla V$, where ∇V is the two-dimensional gradient. Compute the electric field at a point (x, y) with $r > R$.

54–59. Tangent planes *Find an equation of the plane tangent to the following surfaces at the given points.*

54. $z = 2x^2 + y^2$; $(1, 1, 3)$ and $(0, 2, 4)$

55. $x^2 + \frac{y^2}{4} - \frac{z^2}{9} = 1$; $(0, 2, 0)$ and $\left(1, 1, \frac{3}{2} \right)$

56. $x^2 - 2x + y^2 + 4y + 3z^2 = 2$; $(3, -2, 1)$ and $(-1, -2, 1)$

57. $e^{xy^2z^3-1} = 1$; $(1, 1, 1)$ and $(1, -1, 1)$

58. $z - \tan^{-1} xy = 0$; $(1, 1, \pi/4)$ and $(1, \sqrt{3}, \pi/3)$

59. $\sqrt{\frac{x + y}{z}} = 1$; $(2, 2, 4)$ and $(10, -1, 9)$

T 60–61. Linear approximation

a. *Find the linear approximation to the function f at the point (a, b).*
b. *Use part (a) to estimate the given function value.*

60. $f(x, y) = 4 \cos(2x - y)$; $(a, b) = \left(\frac{\pi}{4}, \frac{\pi}{4} \right)$; estimate $f(0.8, 0.8)$.

61. $f(x, y) = (x + y)e^{xy}$; $(a, b) = (2, 0)$; estimate $f(1.95, 0.05)$.

62. Changes in a function Estimate the change in the function $f(x, y) = -2y^2 + 3x^2 + xy$ when (x, y) changes from $(1, -2)$ to $(1.05, -1.9)$.

63. Volume of a cylinder The volume of a cylinder with radius r and height h is $V = \pi r^2 h$. Find the approximate percentage change in the volume when the radius decreases by 3% and the height increases by 2%.

64. Volume of an ellipsoid The volume of an ellipsoid with axes of length $2a$, $2b$, and $2c$ is $V = \pi abc$. Find the percentage change in the volume when a increases by 2%, b increases by 1.5%, and c decreases by 2.5%.

65. Water level changes A hemispherical tank with a radius of 1.50 m is filled with water to a depth of 1.00 m. Water is released from the tank, and the water level drops by 0.05 m (from 1.00 m to 0.95 m).

  a. Approximate the change in the volume of water in the tank.

  The volume of a spherical cap is $V = \dfrac{1}{3}\pi h^2(3r - h)$, where

  r is the radius of the sphere and h is the thickness of the cap (in this case, the depth of the water).

  b. Approximate the change in the surface area of the water in the tank.

66–69. Analyzing critical points *Identify the critical points of the following functions. Then determine whether each critical point corresponds to a local maximum, local minimum, or saddle point. State when your analysis is inconclusive. Confirm your results using a graphing utility.*

66. $f(x, y) = x^4 + y^4 - 16xy$ **67.** $f(x, y) = \dfrac{x^3}{3} - \dfrac{y^3}{3} + 2xy$

68. $f(x, y) = xy(2 + x)(y - 3)$

69. $f(x, y) = 10 - x^3 - y^3 - 3x^2 + 3y^2$

70–73. Absolute maxima and minima *Find the absolute maximum and minimum values of the following functions on the specified region R.*

70. $f(x, y) = \dfrac{x^3}{3} - \dfrac{y^3}{3} + 2xy$ on the rectangle $R = \{(x, y): 0 \le x \le 3, -1 \le y \le 1\}$

71. $f(x, y) = x^4 + y^4 - 4xy + 1$ on the square $R = \{(x, y): -2 \le x \le 2, -2 \le y \le 2\}$

72. $f(x, y) = x^2 y - y^3$ on the triangle $R = \{(x, y): 0 \le x \le 2, 0 \le y \le 2 - x\}$

73. $f(x, y) = xy$ on the semicircular disk $R = \{(x, y): -1 \le x \le 1, 0 \le y \le \sqrt{1 - x^2}\}$

74. Least distance What point on the plane $x + y + 4z = 8$ is closest to the origin? Give an argument showing you have found an absolute minimum of the distance function.

75–78. Lagrange multipliers *Use Lagrange multipliers to find the absolute maximum and minimum values of f (if they exist) subject to the given constraint.*

75. $f(x, y) = 2x + y + 10$ subject to $2(x - 1)^2 + 4(y - 1)^2 = 1$

76. $f(x, y) = xy$ subject to $3x^2 - 2xy + 3y^2 = 4$

77. $f(x, y, z) = x + 2y - z$ subject to $x^2 + y^2 + z^2 = 1$

78. $f(x, y, z) = x^2 y^2 z$ subject to $2x^2 + y^2 + z^2 = 25$

79. Maximum perimeter rectangle Use Lagrange multipliers to find the dimensions of the rectangle with the maximum perimeter that can be inscribed with sides parallel to the coordinate axes in the ellipse $\dfrac{x^2}{a^2} + \dfrac{y^2}{b^2} = 1$.

80. Minimum surface area cylinder Use Lagrange multipliers to find the dimensions of the right circular cylinder of minimum surface area (including the circular ends) with a volume of 32π in^3. Give an argument showing you have found an absolute minimum.

81. Minimum distance to a cone Find the point(s) on the cone $z^2 - x^2 - y^2 = 0$ that are closest to the point $(1, 3, 1)$. Give an argument showing you have found an absolute minimum of the distance function.

82. Gradient of a distance function Let $P_0(a, b, c)$ be a fixed point in $\mathbb{R}^3$, and let $d(x, y, z)$ be the distance between P_0 and a variable point $P(x, y, z)$.

  a. Compute $\nabla d(x, y, z)$.

  b. Show that $\nabla d(x, y, z)$ points in the direction from P_0 to P and has magnitude 1 for all (x, y, z).

  c. Describe the level surfaces of d and give the direction of $\nabla d(x, y, z)$ relative to the level surfaces of d.

  d. Discuss $\lim\limits_{P \to P_0} \nabla d(x, y, z)$.

83. Minimum distance to a paraboloid Use Lagrange multipliers to find the point on the paraboloid $z = x^2 + y^2$ that lies closest to the point $(5, 10, 3)$. Give an argument showing you have found an absolute minimum of the distance function.

Chapter 15 Guided Projects

Applications of the material in this chapter and related topics can be found in the following Guided Projects. For additional information, see the Preface.

- Traveling waves
- Ecological diversity
- Economic production functions

16

Multiple Integration

Chapter Preview We have now generalized limits and derivatives to functions of several variables. The next step is to carry out a similar process with respect to integration. As you know, single (one-variable) integrals are developed from Riemann sums and are used to compute areas of regions in $\mathbb{R}^2$. In an analogous way, we use Riemann sums to develop double (two-variable) and triple (three-variable) integrals, which are used to compute volumes of solid regions in $\mathbb{R}^3$. These multiple integrals have many applications in statistics, science, and engineering, including calculating the mass, the center of mass, and moments of inertia of solids with a variable density. Another significant development in this chapter is the appearance of cylindrical and spherical coordinates. These alternative coordinate systems often simplify the evaluation of integrals in three-dimensional space. The chapter closes with the two- and three-dimensional versions of the substitution (change of variables) rule. The overall lesson of the chapter is that we can integrate functions over most geometrical objects, from intervals on the x-axis to regions in the plane bounded by curves to complicated three-dimensional solids.

16.1 Double Integrals over Rectangular Regions

In Chapter 15 the concept of differentiation was extended to functions of several variables. In this chapter, we extend integration to multivariable functions. By the close of the chapter, we will have completed Table 16.1, which is a basic road map for calculus.

Table 16.1

	Derivatives	Integrals
Single variable: $f(x)$	$f'(x)$	$\int_a^b f(x)\, dx$
Several variables: $f(x, y)$ **and** $f(x, y, z)$	$\dfrac{\partial f}{\partial x}, \dfrac{\partial f}{\partial y}, \dfrac{\partial f}{\partial z}$	$\iint\limits_R f(x, y)\, dA,\ \iiint\limits_D f(x, y, z)\, dV$

Volumes of Solids

The problem of finding the net area of a region bounded by a curve led to the definite integral in Chapter 5. Recall that we began that discussion by approximating the region with a collection of rectangles and then formed a Riemann sum of the areas of the rectangles. Under appropriate conditions, as the number of rectangles increases, the sum approaches the value of the definite integral, which is the net area of the region.

We now carry out an analogous procedure with surfaces defined by functions of the form $z = f(x, y)$, where, for the moment, we assume $f(x, y) \geq 0$ on a region R in the xy-plane (**Figure 16.1a**). The goal is to determine the volume of the solid bounded by the surface and R. In general terms, the solid is first approximated by *boxes* (**Figure 16.1b**). The sum of the volumes of these boxes, which is a Riemann sum, approximates the volume of the solid. Under appropriate conditions, as the number of boxes increases, the approximations converge to the value of a *double integral*, which is the volume of the solid.

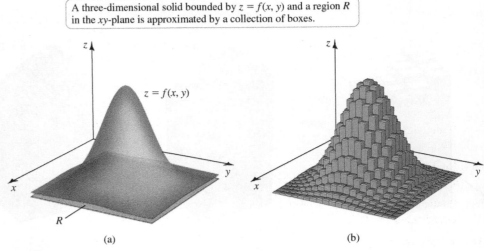

A three-dimensional solid bounded by $z = f(x, y)$ and a region R in the xy-plane is approximated by a collection of boxes.

$z = f(x, y)$

(a)   (b)

Figure 16.1

▶ We adopt the convention that Δx_k and Δy_k are the side lengths of the kth rectangle, for $k = 1, \ldots, n$, even though there are generally fewer than n different values of Δx_k and Δy_k. This convention is used throughout the chapter.

We assume $z = f(x, y)$ is a nonnegative function defined on a *rectangular* region $R = \{(x, y) : a \leq x \leq b, c \leq y \leq d\}$. A **partition** of R is formed by dividing R into n rectangular subregions using lines parallel to the x- and y-axes (not necessarily uniformly spaced). The rectangles may be numbered in any systematic way; for example, left to right and then bottom to top. The side lengths of the kth rectangle are denoted Δx_k and Δy_k, so the area of the kth rectangle is $\Delta A_k = \Delta x_k \Delta y_k$. We also let (x_k^*, y_k^*) be any point in the kth rectangle, for $1 \leq k \leq n$ (**Figure 16.2**).

$z = f(x, y)$

(x_k^*, y_k^*)

Δx_k

Δy_k

(x_k^*, y_k^*) is a point in the kth rectangle, which has area $\Delta A_k = \Delta x_k \Delta y_k$.

Figure 16.2

$z = f(x, y)$

Height of kth box $= f(x_k^*, y_k^*)$

(x_k^*, y_k^*)

Volume of kth box $= f(x_k^*, y_k^*) \Delta A_k$

Figure 16.3

To approximate the volume of the solid bounded by the surface $z = f(x, y)$ and the region R, we construct boxes on each of the n rectangles; each box has a height of $f(x_k^*, y_k^*)$ and a base with area ΔA_k, for $1 \leq k \leq n$ (**Figure 16.3**). Therefore, the volume of the kth box is

$$f(x_k^*, y_k^*) \Delta A_k = f(x_k^*, y_k^*) \Delta x_k \Delta y_k.$$

QUICK CHECK 1 Explain why the sum for the volume is an approximation. How can the approximation be improved? ◄

The sum of the volumes of the n boxes gives an approximation to the volume of the solid:

$$V \approx \sum_{k=1}^{n} f(x_k^*, y_k^*) \, \Delta A_k.$$

We now let Δ be the maximum length of the diagonals of the rectangles in the partition. As $\Delta \to 0$, the areas of *all* the rectangles approach zero ($\Delta A_k \to 0$) and the number of rectangles increases ($n \to \infty$). If the approximations given by these Riemann sums have a limit as $\Delta \to 0$, then we define the volume of the solid to be that limit (**Figure 16.4**).

$n = 4 \times 4 = 16$ $n = 8 \times 8 = 64$ $n = 16 \times 16 = 256$

$n \to \infty$
$\Delta \to 0$

$$\text{Volume} = \lim_{\Delta \to 0} \sum_{k=1}^{n} f(x_k^*, y_k^*) \, \Delta A_k$$

Figure 16.4

▶ The functions that we encounter in this text are integrable. Advanced methods are needed to prove that continuous functions and many functions with finite discontinuities are also integrable.

DEFINITION Double Integrals

A function f defined on a rectangular region R in the xy-plane is **integrable** on R if $\lim_{\Delta \to 0} \sum_{k=1}^{n} f(x_k^*, y_k^*) \Delta A_k$ exists for all partitions of R and for all choices of (x_k^*, y_k^*) within those partitions. The limit is the **double integral of f over R**, which we write

$$\iint\limits_R f(x, y) \, dA = \lim_{\Delta \to 0} \sum_{k=1}^{n} f(x_k^*, y_k^*) \Delta A_k.$$

If f is nonnegative on R, then the double integral equals the volume of the solid bounded by $z = f(x, y)$ and the xy-plane over R. If f is negative on parts of R, the value of the double integral may be zero or negative, and the result is interpreted as a *net volume* (by analogy with *net area* for single variable integrals).

Iterated Integrals

Evaluating double integrals using limits of Riemann sums is tedious and rarely done. Fortunately, there is a practical method for evaluating double integrals that is based on the general slicing method (Section 6.3). An example illustrates the technique.

Suppose we wish to compute the volume of the solid region bounded by the plane $z = f(x, y) = 6 - 2x - y$ over the rectangular region $R = \{(x, y): 0 \le x \le 1, 0 \le y \le 2\}$ (**Figure 16.5**). By definition, the volume is given by the double integral

$$V = \iint\limits_R f(x, y) \, dA = \iint\limits_R (6 - 2x - y) \, dA.$$

If a solid is sliced parallel to the y-axis and perpendicular to the xy-plane, and the cross-sectional area of the slice at the point x is $A(x)$, then the volume of the solid region is

$$V = \int_a^b A(x) \, dx.$$

According to the general slicing method (see margin note and figure), we can compute this volume by taking vertical slices through the solid parallel to the yz-plane (Figure 16.5). The slice at the point x has a cross-sectional area denoted $A(x)$. In general, as x varies, the area $A(x)$ also changes, so we integrate these cross-sectional areas from $x = 0$ to $x = 1$ to obtain the volume

$$V = \int_0^1 A(x)\, dx.$$

The important observation is that for a fixed value of x, $A(x)$ is the area of the plane region under the curve $z = 6 - 2x - y$. This area is computed by integrating f with respect to y from $y = 0$ to $y = 2$, holding x fixed; that is,

$$A(x) = \int_0^2 (6 - 2x - y)\, dy,$$

where $0 \le x \le 1$, and x is treated as a constant in the integration. Substituting for $A(x)$, we have

$$V = \int_0^1 A(x)\, dx = \int_0^1 \left(\underbrace{\int_0^2 (6 - 2x - y)\, dy}_{A(x)} \right) dx.$$

A slice at a fixed value of x has area $A(x)$, where $0 \le x \le 1$.

Figure 16.5

The expression that appears on the right side of this equation is called an **iterated integral** (meaning repeated integral). We first evaluate the inner integral with respect to y holding x fixed; the result is a function of x. Then the outer integral is evaluated with respect to x; the result is a real number, which is the volume of the solid in Figure 16.5. Both of these integrals are ordinary one-variable integrals.

EXAMPLE 1 Evaluating an iterated integral Evaluate $V = \int_0^1 A(x)\, dx$, where $A(x) = \int_0^2 (6 - 2x - y)\, dy$.

SOLUTION Using the Fundamental Theorem of Calculus, holding x constant, we have

$$A(x) = \int_0^2 (6 - 2x - y)\, dy$$

$$= \left(6y - 2xy - \frac{y^2}{2} \right)\Big|_0^2 \quad \text{Evaluate integral with respect to } y;\ x \text{ is constant.}$$

$$= (12 - 4x - 2) - 0 \quad \text{Simplify; limits are in } y.$$

$$= 10 - 4x. \quad \text{Simplify.}$$

Substituting $A(x) = 10 - 4x$ into the volume integral, we have

$$V = \int_0^1 A(x)\, dx$$

$$= \int_0^1 (10 - 4x)\, dx \quad \text{Substitute for } A(x).$$

$$= (10x - 2x^2)\Big|_0^1 \quad \text{Evaluate integral with respect to } x.$$

$$= 8. \quad \text{Simplify.}$$

Related Exercises 10, 25 ◄

A slice at a fixed value of y has area $A(y)$, where $0 \le y \le 2$.

Figure 16.6

EXAMPLE 2 Same double integral, different order Example 1 used slices through the solid parallel to the yz-plane. Compute the volume of the same solid using vertical slices through the solid parallel to the xz-plane, for $0 \le y \le 2$ (**Figure 16.6**).

SOLUTION In this case, $A(y)$ is the area of a slice through the solid for a fixed value of y in the interval $0 \le y \le 2$. This area is computed by integrating $z = 6 - 2x - y$ from $x = 0$ to $x = 1$, holding y fixed; that is,

$$A(y) = \int_0^1 (6 - 2x - y)\, dx,$$

where $0 \le y \le 2$.

Using the general slicing method again, the volume is

$$V = \int_0^2 A(y)\,dy \qquad \text{General slicing method}$$

$$= \int_0^2 \left(\underbrace{\int_0^1 (6 - 2x - y)\,dx}_{A(y)} \right) dy \qquad \text{Substitute for } A(y).$$

$$= \int_0^2 \left((6x - x^2 - yx)\Big|_0^1 \right) dy \qquad \begin{array}{l}\text{Evaluate inner integral with respect}\\ \text{to } x; y \text{ is constant.}\end{array}$$

$$= \int_0^2 (5 - y)\,dy \qquad \text{Simplify; limits are in } x.$$

$$= \left(5y - \frac{y^2}{2} \right)\Big|_0^2 \qquad \text{Evaluate outer integral with respect to } y.$$

$$= 8. \qquad \text{Simplify.}$$

Related Exercise 37 ◄

QUICK CHECK 2 Consider the integral $\int_3^4 \int_1^2 f(x, y)\,dx\,dy$. Give the limits of integration and the variable of integration for the first (inner) integral and the second (outer) integral. Sketch the region of integration. ◄

Several important comments are in order. First, the two iterated integrals give the same value for the double integral. Second, the notation of the iterated integral must be used carefully. When we write $\int_c^d \int_a^b f(x, y)\,dx\,dy$, it means $\int_c^d \left(\int_a^b f(x, y)\,dx \right) dy$. The *inner* integral with respect to x is evaluated first, holding y fixed, and the variable runs from $x = a$ to $x = b$. The result of that integration is a constant or a function of y, which is then integrated in the *outer* integral, with the variable running from $y = c$ to $y = d$. The order of integration is signified by the order of dx and dy.

Similarly, $\int_a^b \int_c^d f(x, y)\,dy\,dx$ means $\int_a^b \left(\int_c^d f(x, y)\,dy \right) dx$. The inner integral with respect to y is evaluated first, holding x fixed. The result is then integrated with respect to x. In both cases, the limits of integration in the iterated integrals determine the boundaries of the rectangular *region of integration*.

Examples 1 and 2 illustrate one version of *Fubini's Theorem*, a deep result that relates double integrals to iterated integrals. The first version of the theorem applies to double integrals over rectangular regions.

► The area of the kth rectangle in the partition is $\Delta A_k = \Delta x_k \Delta y_k$, where Δx_k and Δy_k are the lengths of the sides of that rectangle. Accordingly, the *element of area dA* in the double integral becomes $dx\,dy$ or $dy\,dx$ in the iterated integral.

THEOREM 16.1 (Fubini) Double Integrals over Rectangular Regions
Let f be continuous on the rectangular region $R = \{(x, y): a \le x \le b, c \le y \le d\}$. The double integral of f over R may be evaluated by either of two iterated integrals:

$$\iint\limits_R f(x, y)\,dA = \int_c^d \int_a^b f(x, y)\,dx\,dy = \int_a^b \int_c^d f(x, y)\,dy\,dx.$$

The importance of Fubini's Theorem is twofold: It says that double integrals may be evaluated by iterated integrals. It also says that the order of integration in the iterated integrals does not matter (although in practice, one order of integration is often easier to use than the other).

EXAMPLE 3 A double integral Find the volume of the solid bounded by the surface $f(x, y) = 4 + 9x^2y^2$ over the region $R = \{(x, y): -1 \le x \le 1, 0 \le y \le 2\}$. Use both possible orders of integration.

SOLUTION Because $f(x, y) > 0$ on R, the volume of the region is given by the double integral $\iint_R (4 + 9x^2y^2)\,dA$. By Fubini's Theorem, the double integral is evaluated as an iterated integral. If we first integrate with respect to x, the area of a cross section of the solid for a fixed value of y is given by $A(y)$ (**Figure 16.7a**). The volume of the region is

$$\iint_R (4 + 9x^2y^2)\, dA = \int_0^2 \underbrace{\int_{-1}^{1} (4 + 9x^2y^2)\, dx}_{A(y)}\, dy \qquad \text{Convert to an iterated integral.}$$

$$= \int_0^2 (4x + 3x^3y^2)\Big|_{-1}^{1}\, dy \qquad \begin{array}{l}\text{Evaluate inner integral}\\ \text{with respect to } x.\end{array}$$

$$= \int_0^2 (8 + 6y^2)\, dy \qquad \text{Simplify.}$$

$$= (8y + 2y^3)\Big|_0^2 \qquad \begin{array}{l}\text{Evaluate outer integral}\\ \text{with respect to } y.\end{array}$$

$$= 32. \qquad \text{Simplify.}$$

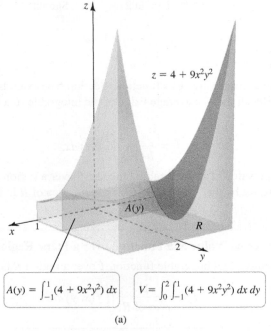

$$A(y) = \int_{-1}^{1}(4 + 9x^2y^2)\, dx \qquad V = \int_0^2\int_{-1}^{1}(4 + 9x^2y^2)\, dx\, dy$$

(a)

$$A(x) = \int_0^2(4 + 9x^2y^2)\, dy \qquad V = \int_{-1}^{1}\int_0^2(4 + 9x^2y^2)\, dy\, dx$$

(b)

Figure 16.7

Alternatively, if we integrate first with respect to y, the area of a cross section of the solid for a fixed value of x is given by $A(x)$ (**Figure 16.7b**). The volume of the region is

$$\iint_R (4 + 9x^2y^2)\, dA = \int_{-1}^{1} \underbrace{\int_0^2 (4 + 9x^2y^2)\, dy}_{A(x)}\, dx \qquad \text{Convert to an iterated integral.}$$

$$= \int_{-1}^{1} (4y + 3x^2y^3)\Big|_0^2\, dx \qquad \begin{array}{l}\text{Evaluate inner integral with}\\ \text{respect to } y.\end{array}$$

$$= \int_{-1}^{1} (8 + 24x^2)\, dx \qquad \text{Simplify.}$$

$$= (8x + 8x^3)\Big|_{-1}^{1} = 32. \qquad \begin{array}{l}\text{Evaluate outer integral with}\\ \text{respect to } x.\end{array}$$

QUICK CHECK 3 Write the iterated integral $\int_{-10}^{10}\int_0^{20}(x^2y + 2xy^3)\, dy\, dx$ with the order of integration reversed. ◄

As guaranteed by Fubini's Theorem, the two iterated integrals are equal, both giving the value of the double integral and the volume of the solid.

Related Exercises 26, 39 ◄

The following example shows that sometimes the order of integration must be chosen carefully either to save work or to make the integration possible.

EXAMPLE 4   Choosing a convenient order of integration Evaluate $\iint_R ye^{xy}\, dA$, where $R = \{(x, y): 0 \le x \le 1, 0 \le y \le \ln 2\}$.

SOLUTION The iterated integral $\int_0^1 \int_0^{\ln 2} ye^{xy}\, dy\, dx$ requires first integrating ye^{xy} with respect to y, which entails integration by parts. An easier approach is to integrate first with respect to x:

$$\int_0^{\ln 2} \int_0^1 ye^{xy}\, dx\, dy = \int_0^{\ln 2} e^{xy}\Big|_0^1\, dy \qquad \text{Evaluate inner integral with respect to } x.$$

$$= \int_0^{\ln 2} (e^y - 1)\, dy \qquad \text{Simplify.}$$

$$= (e^y - y)\Big|_0^{\ln 2} \qquad \text{Evaluate outer integral with respect to } y.$$

$$= 1 - \ln 2. \qquad \text{Simplify.}$$

Related Exercises 41, 43 ◄

Average Value

The concept of the average value of a function (Section 5.4) extends naturally to functions of two variables. Recall that the average value of the integrable function f over the interval $[a, b]$ is

$$\bar{f} = \frac{1}{b - a} \int_a^b f(x)\, dx.$$

To find the average value of an integrable function f over a region R, we integrate f over R and divide the result by the "size" of R, which is the area of R in the two-variable case.

> ▶ The same definition of average value applies to more general regions in the plane.

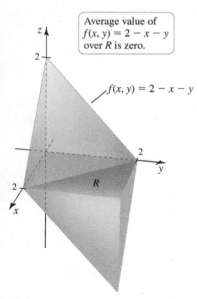

Average value of $f(x, y) = 2 - x - y$ over R is zero.

$f(x, y) = 2 - x - y$

Figure 16.8

> ▶ An average value of 0 means that over the region R, the volume of the solid above the xy-plane and below the surface equals the volume of the solid below the xy-plane and above the surface.

DEFINITION Average Value of a Function over a Plane Region

The **average value** of an integrable function f over a region R is

$$\bar{f} = \frac{1}{\text{area of } R} \iint_R f(x, y)\, dA.$$

EXAMPLE 5   Average value Find the average value of the quantity $2 - x - y$ over the square $R = \{(x, y): 0 \le x \le 2, 0 \le y \le 2\}$ (**Figure 16.8**).

SOLUTION The area of the region R is 4. Letting $f(x, y) = 2 - x - y$, the average value of f is

$$\frac{1}{\text{area of } R} \iint_R f(x, y)\, dA = \frac{1}{4} \iint_R (2 - x - y)\, dA$$

$$= \frac{1}{4} \int_0^2 \int_0^2 (2 - x - y)\, dx\, dy \qquad \text{Convert to an iterated integral.}$$

$$= \frac{1}{4} \int_0^2 \left(2x - \frac{x^2}{2} - xy\right)\Big|_0^2\, dy \qquad \text{Evaluate inner integral with respect to } x.$$

$$= \frac{1}{4} \int_0^2 (2 - 2y)\, dy \qquad \text{Simplify.}$$

$$= 0. \qquad \text{Evaluate outer integral with respect to } y.$$

Related Exercise 46 ◄

SECTION 16.1 EXERCISES

Getting Started

1. Write an iterated integral that gives the volume of the solid bounded by the surface $f(x, y) = xy$ over the square $R = \{(x, y): 0 \le x \le 2, 1 \le y \le 3\}$.

2. Write an iterated integral that gives the volume of a box with height 10 and base $R = \{(x, y): 0 \le x \le 5, -2 \le y \le 4\}$.

3. Write two iterated integrals that equal $\iint_R f(x, y) \, dA$, where $R = \{(x, y): -2 \le x \le 4, 1 \le y \le 5\}$.

4. Consider the integral $\int_1^3 \int_{-1}^1 (2y^2 + xy) \, dy \, dx$. State the variable of integration in the first (inner) integral and the limits of integration. State the variable of integration in the second (outer) integral and the limits of integration.

5. Region $R = \{(x, y): 0 \le x \le 4, 0 \le y \le 6\}$ is partitioned into six equal subregions (see figure, which also shows the level curves of a function f continuous on the region R). Estimate the value of $\iint_R f(x, y) \, dA$ by evaluating the Riemann sum $\sum_{k=1}^6 f(x_k^*, y_k^*)\Delta A_k$, where (x_k^*, y_k^*) is the center of the kth subregion, for $k = 1, \ldots, 6$.

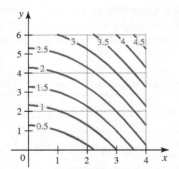

6. Draw a solid whose volume is given by the double integral $\int_0^6 \int_1^2 10 \, dy \, dx$. Then evaluate the integral using geometry.

Practice Exercises

7–24. Iterated integrals *Evaluate the following iterated integrals.*

7. $\int_0^2 \int_0^1 4xy \, dx \, dy$

8. $\int_1^2 \int_0^1 (3x^2 + 4y^3) \, dy \, dx$

9. $\int_1^3 \int_0^2 x^2 y \, dx \, dy$

10. $\int_0^3 \int_{-2}^1 (2x + 3y) \, dx \, dy$

11. $\int_1^3 \int_0^{\pi/2} x \sin y \, dy \, dx$

12. $\int_1^3 \int_1^2 (y^2 + y) \, dx \, dy$

13. $\int_1^4 \int_0^4 \sqrt{uv} \, du \, dv$

14. $\int_0^{\pi/4} \int_0^3 r \sec \theta \, dr \, d\theta$

15. $\int_1^{\ln 5} \int_0^{\ln 3} e^{x+y} \, dx \, dy$

16. $\int_0^{\pi/2} \int_0^1 uv \cos (u^2 v) \, du \, dv$

17. $\int_0^1 \int_0^1 t^2 e^{st} \, ds \, dt$

18. $\int_0^2 \int_0^1 \frac{8xy}{1 + x^4} \, dx \, dy$

19. $\int_1^e \int_0^1 4(p + q) \ln q \, dp \, dq$

20. $\int_0^1 \int_0^\pi y^2 \cos xy \, dx \, dy$

21. $\int_1^2 \int_1^2 \frac{x}{x + y} \, dy \, dx$

22. $\int_0^2 \int_0^1 x^5 y^2 e^{x^3 y^3} \, dy \, dx$

23. $\int_0^1 \int_1^4 \frac{3y}{\sqrt{x + y^2}} \, dx \, dy$

24. $\int_0^1 \int_0^1 x^2 y^2 e^{x^3 y} \, dx \, dy$

25–35. Double integrals *Evaluate each double integral over the region R by converting it to an iterated integral.*

25. $\iint_R (x + 2y) \, dA; R = \{(x, y): 0 \le x \le 3, 1 \le y \le 4\}$

26. $\iint_R (x^2 + xy) \, dA; R = \{(x, y): 1 \le x \le 2, -1 \le y \le 1\}$

27. $\iint_R s^2 t \sin (st^2) \, dA; R = \{(s, t): 0 \le s \le \pi, 0 \le t \le 1\}$

28. $\iint_R \frac{x}{1 + xy} \, dA; R = \{(x, y): 0 \le x \le 1, 0 \le y \le 1\}$

29. $\iint_R \sqrt{\frac{x}{y}} \, dA; R = \{(x, y): 0 \le x \le 1, 1 \le y \le 4\}$

30. $\iint_R xy \sin x^2 \, dA; R = \{(x, y): 0 \le x \le \sqrt{\pi/2}, 0 \le y \le 1\}$

31. $\iint_R e^{x+2y} \, dA; R = \{(x, y): 0 \le x \le \ln 2, 1 \le y \le \ln 3\}$

32. $\iint_R (x^2 - y^2)^2 \, dA; R = \{(x, y): -1 \le x \le 2, 0 \le y \le 1\}$

33. $\iint_R (x^5 - y^5)^2 \, dA; R = \{(x, y): 0 \le x \le 1, -1 \le y \le 1\}$

34. $\iint_R \cos (x\sqrt{y}) \, dA; R = \{(x, y): 0 \le x \le 1, 0 \le y \le \pi^2/4\}$

35. $\iint_R x^3 y \cos (x^2 y^2) \, dA; R = \{(x, y): 0 \le x \le \sqrt{\pi/2}, 0 \le y \le 1\}$

36–39. Volumes of solids *Find the volume of the following solids.*

36. The solid beneath the cylinder $f(x, y) = e^{-x}$ and above the region $R = \{(x, y): 0 \le x \le \ln 4, -2 \le y \le 2\}$

$f(x, y) = e^{-x}$

37. The solid beneath the plane $f(x, y) = 24 - 3x - 4y$ and above the region $R = \{(x, y): -1 \le x \le 3, 0 \le y \le 2\}$

$f(x, y) = 24 - 3x - 4y$

38. The solid in the first octant bounded above by the surface $z = 9xy\sqrt{1 - x^2}\sqrt{4 - y^2}$ and below by the xy-plane

39. The solid in the first octant bounded by the surface $z = xy^2\sqrt{1 - x^2}$ and the planes $z = 0$ and $y = 3$

40–45. Choose a convenient order *When converted to an iterated integral, the following double integrals are easier to evaluate in one order than the other. Find the best order and evaluate the integral.*

40. $\displaystyle\iint_R y \cos xy \, dA; R = \{(x, y): 0 \le x \le 1, 0 \le y \le \pi/3\}$

41. $\displaystyle\iint_R (y + 1)e^{x(y+1)} \, dA; R = \{(x, y): 0 \le x \le 1, -1 \le y \le 1\}$

42. $\displaystyle\iint_R x \sec^2 xy \, dA; R = \{(x, y): 0 \le x \le \pi/3, 0 \le y \le 1\}$

43. $\displaystyle\iint_R 6x^5 e^{x^3 y} \, dA; R = \{(x, y): 0 \le x \le 2, 0 \le y \le 2\}$

44. $\displaystyle\iint_R y^3 \sin(xy^2) \, dA; R = \{(x, y): 0 \le x \le 2, 0 \le y \le \sqrt{\pi/2}\}$

45. $\displaystyle\iint_R \frac{x}{(1 + xy)^2} \, dA; R = \{(x, y): 0 \le x \le 4, 1 \le y \le 2\}$

46–48. Average value *Compute the average value of the following functions over the region R.*

46. $f(x, y) = 4 - x - y; R = \{(x, y): 0 \le x \le 2, 0 \le y \le 2\}$

47. $f(x, y) = e^{-y}; R = \{(x, y): 0 \le x \le 6, 0 \le y \le \ln 2\}$

48. $f(x, y) = \sin x \sin y; R = \{(x, y): 0 \le x \le \pi, 0 \le y \le \pi\}$

49. Average value Find the average squared distance between the points of $R = \{(x, y): -2 \le x \le 2, 0 \le y \le 2\}$ and the origin.

50. Average value Find the average squared distance between the points of $R = \{(x, y): 0 \le x \le 3, 0 \le y \le 3\}$ and the point $(3, 3)$.

51. Explain why or why not Determine whether the following statements are true and give an explanation or counterexample.

a. The region of integration for $\int_4^6 \int_1^3 4 \, dx \, dy$ is a square.
b. If f is continuous on $\mathbb{R}^2$, then
$$\int_4^6 \int_1^3 f(x, y) \, dx \, dy = \int_4^6 \int_1^3 f(x, y) \, dy \, dx.$$
c. If f is continuous on $\mathbb{R}^2$, then
$$\int_4^6 \int_1^3 f(x, y) \, dx \, dy = \int_1^3 \int_4^6 f(x, y) \, dy \, dx.$$

52. Symmetry Evaluate the following integrals using symmetry arguments. Let $R = \{(x, y): -a \le x \le a, -b \le y \le b\}$, where a and b are positive real numbers.

a. $\displaystyle\iint_R xye^{-(x^2+y^2)} \, dA$ **b.** $\displaystyle\iint_R \frac{\sin(x - y)}{x^2 + y^2 + 1} \, dA$

53. Computing populations The population densities in nine districts of a rectangular county are shown in the figure.

a. Use the fact that population = (population density) × (area) to estimate the population of the county.
b. Explain how the calculation of part (a) is related to Riemann sums and double integrals.

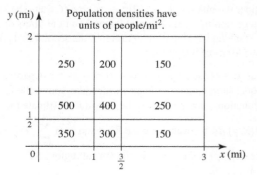

T 54. Approximating water volume The varying depth of an 18 m × 25 m swimming pool is measured in 15 different rectangles of equal area (see figure). Approximate the volume of water in the pool.

y (m)	Depth readings have units of m.				
18					
	0.75	1.25	1.75	2.25	2.75
	1	1.5	2.0	2.5	3.0
	1	1.5	2.0	2.5	3.0
0					25 x (m)

Explorations and Challenges

55. Cylinders Let S be the solid in $\mathbb{R}^3$ between the cylinder $z = f(x)$ and the region $R = \{(x, y): a \le x \le b, c \le y \le d\}$, where $f(x) \ge 0$ on R. Explain why $\int_c^d \int_a^b f(x) \, dx \, dy$ equals the area of the constant cross section of S multiplied by $(d - c)$, which is the volume of S.

56. Product of integrals Suppose $f(x, y) = g(x)h(y)$, where g and h are continuous functions for all real values of x and y.

a. Show that $\int_c^d \int_a^b f(x, y) \, dx \, dy = \left(\int_a^b g(x) \, dx\right)\left(\int_c^d h(y) \, dy\right)$. Interpret this result geometrically.
b. Write $\left(\int_a^b g(x) \, dx\right)^2$ as an iterated integral.
c. Use the result of part (a) to evaluate $\int_0^{2\pi} \int_{10}^{30} e^{-4y^2} \cos x \, dy \, dx$.

57. Solving for a parameter Let $R = \{(x, y): 0 \le x \le \pi, 0 \le y \le a\}$. For what values of a, with $0 \le a \le \pi$, is $\iint_R \sin(x + y) \, dA$ equal to 1?

58–59. Zero average value *Find the value of $a > 0$ such that the average value of the following functions over $R = \{(x, y): 0 \le x \le a, 0 \le y \le a\}$ is zero.*

58. $f(x, y) = x + y - 8$ **59.** $f(x, y) = 4 - x^2 - y^2$

60. Maximum integral Consider the plane $x + 3y + z = 6$ over the rectangle R with vertices at $(0, 0)$, $(a, 0)$, $(0, b)$, and (a, b), where the vertex (a, b) lies on the line where the plane intersects the xy-plane (so $a + 3b = 6$). Find the point (a, b) for which the volume of the solid between the plane and R is a maximum.

61. Density and mass Suppose a thin rectangular plate, represented by a region R in the xy-plane, has a density given by the function $\rho(x, y)$; this function gives the *area density* in units such as grams per square centimeter (g/cm^2). The mass of the plate is $\iint_R \rho(x, y)\, dA$. Assume $R = \{(x, y): 0 \le x \le \pi/2, 0 \le y \le \pi\}$ and find the mass of the plates with the following density functions.

a. $\rho(x, y) = 1 + \sin x$ **b.** $\rho(x, y) = 1 + \sin y$
c. $\rho(x, y) = 1 + \sin x \sin y$

62. Approximating volume Propose a method based on Riemann sums to approximate the volume of the shed shown in the figure (the peak of the roof is directly above the rear corner of the shed). Carry out the method and provide an estimate of the volume.

63. An identity Suppose the second partial derivatives of f are continuous on $R = \{(x, y): 0 \le x \le a, 0 \le y \le b\}$. Simplify

$$\iint_R \frac{\partial^2 f}{\partial x\, \partial y}\, dA.$$

QUICK CHECK ANSWERS

1. The sum gives the volume of a collection of rectangular boxes, and these boxes do not exactly fill the solid region under the surface. The approximation is improved by using more boxes. **2.** Inner integral: x runs from $x = 1$ to $x = 2$; outer integral: y runs from $y = 3$ to $y = 4$. The region is the rectangle $\{(x, y): 1 \le x \le 2, 3 \le y \le 4\}$.
3. $\int_0^{20} \int_{-10}^{10} (x^2 y + 2xy^3)\, dx\, dy$ ◄

16.2 Double Integrals over General Regions

Evaluating double integrals over rectangular regions is a useful place to begin our study of multiple integrals. Problems of practical interest, however, usually involve nonrectangular regions of integration. The goal of this section is to extend the methods presented in Section 16.1 so that they apply to more general regions of integration.

General Regions of Integration

Consider a function f defined over a closed, bounded *nonrectangular* region R in the xy-plane. As with rectangular regions, we use a partition consisting of rectangles, but now, such a partition does not cover R exactly. In this case, only the n rectangles that lie entirely within R are considered to be in the partition (**Figure 16.9**). When f is nonnegative on R, the volume of the solid bounded by the surface $z = f(x, y)$ and the xy-plane over R is approximated by the Riemann sum

$$V \approx \sum_{k=1}^{n} f(x_k^*, y_k^*) \Delta A_k,$$

Figure 16.9

where $\Delta A_k = \Delta x_k \Delta y_k$ is the area of the kth rectangle and (x_k^*, y_k^*) is any point in the kth rectangle, for $1 \le k \le n$. As before, we define Δ to be the maximum length of the diagonals of the rectangles in the partition.

Under the assumptions that f is continuous on R and that the boundary of R consists of a finite number of smooth curves, two things occur as $\Delta \to 0$ and the number of rectangles increases $(n \to \infty)$.

- The rectangles in the partition fill R more and more completely; that is, the union of the rectangles approaches R.
- Over all partitions and all choices of (x_k^*, y_k^*) within a partition, the Riemann sums approach a (unique) limit.

Volume of solid $= \displaystyle\iint_R f(x, y)\, dA$

$= \displaystyle\lim_{\Delta \to 0} \sum_{k=1}^{n} f(x_k^*, y_k^*)\, \Delta A_k$

Figure 16.10

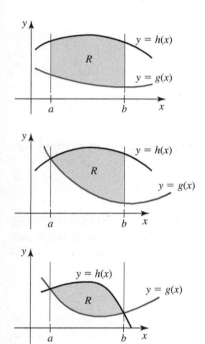

Figure 16.11

The limit approached by the Riemann sums is the **double integral of f over R**; that is,

$$\iint_R f(x, y)\, dA = \lim_{\Delta \to 0} \sum_{k=1}^{n} f(x_k^*, y_k^*)\, \Delta A_k.$$

When this limit exists, f is **integrable** over R. If f is nonnegative on R, then the double integral equals the volume of the solid bounded by the surface $z = f(x, y)$ and the xy-plane over R (**Figure 16.10**).

The double integral $\iint_R f(x, y)\, dA$ has another common interpretation. Suppose R represents a thin plate whose density at the point (x, y) is $f(x, y)$. The units of density are mass per unit area, so the product $f(x_k^*, y_k^*)\, \Delta A_k$ approximates the mass of the kth rectangle in R. Summing the masses of the rectangles gives an approximation to the total mass of R. In the limit as $n \to \infty$ and $\Delta \to 0$, the double integral equals the mass of the plate.

Iterated Integrals

Double integrals over nonrectangular regions are also evaluated using iterated integrals. However, in this more general setting, the order of integration is critical. Most of the double integrals we encounter fall into one of two categories determined by the shape of the region R.

The first type of region has the property that its lower and upper boundaries are the graphs of continuous functions $y = g(x)$ and $y = h(x)$, respectively, for $a \le x \le b$. Such regions have any of the forms shown in **Figure 16.11**.

Once again, we appeal to the general slicing method. Assume for the moment that f is nonnegative on R and consider the solid bounded by the surface $z = f(x, y)$ and R (**Figure 16.12**). Imagine taking vertical slices through the solid parallel to the yz-plane. The cross section through the solid at a fixed value of x extends from the lower curve $y = g(x)$ to the upper curve $y = h(x)$. The area of that cross section is

$$A(x) = \int_{g(x)}^{h(x)} f(x, y)\, dy, \qquad \text{for } a \le x \le b.$$

The volume of the solid is given by a double integral; it is evaluated by integrating the cross-sectional areas $A(x)$ from $x = a$ to $x = b$:

$$\iint_R f(x, y)\, dA = \int_a^b \underbrace{\int_{g(x)}^{h(x)} f(x, y)\, dy}_{A(x)}\, dx.$$

The limits of integration in the iterated integral describe the boundaries of the region of integration R.

Figure 16.12

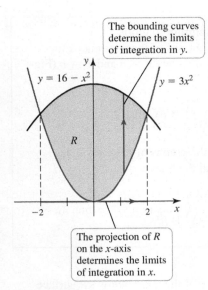

The bounding curves determine the limits of integration in y.

$y = 16 - x^2$

$y = 3x^2$

R

-2 2 x

The projection of R on the x-axis determines the limits of integration in x.

Figure 16.13

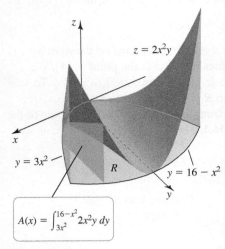

$z = 2x^2y$

$y = 3x^2$

R

$y = 16 - x^2$

y

$A(x) = \int_{3x^2}^{16-x^2} 2x^2y \, dy$

Figure 16.14

QUICK CHECK 1 A region R is bounded by the x- and y-axes and the line $x + y = 2$. Suppose you integrate first with respect to y. Give the limits of the iterated integral over R. ◄

EXAMPLE 1   Evaluating a double integral Express the integral $\iint_R 2x^2y \, dA$ as an iterated integral, where R is the region bounded by the parabolas $y = 3x^2$ and $y = 16 - x^2$. Then evaluate the integral.

SOLUTION The region R is bounded below and above by the graphs of $g(x) = 3x^2$ and $h(x) = 16 - x^2$, respectively. Solving $3x^2 = 16 - x^2$, we find that these curves intersect at $x = -2$ and $x = 2$, which are the limits of integration in the x-direction (**Figure 16.13**).

Figure 16.14 shows the solid bounded by the surface $z = 2x^2y$ and the region R. A typical vertical cross section through the solid parallel to the yz-plane at a fixed value of x has area

$$A(x) = \int_{3x^2}^{16-x^2} 2x^2y \, dy.$$

Integrating these cross-sectional areas between $x = -2$ and $x = 2$, the iterated integral becomes

$$\iint_R 2x^2y \, dA = \int_{-2}^{2} \underbrace{\int_{3x^2}^{16-x^2} 2x^2y \, dy}_{A(x)} \, dx \qquad \text{Convert to an iterated integral.}$$

$$= \int_{-2}^{2} x^2y^2 \Big|_{3x^2}^{16-x^2} \, dx \qquad \begin{array}{l}\text{Evaluate inner integral}\\ \text{with respect to } y.\end{array}$$

$$= \int_{-2}^{2} x^2((16 - x^2)^2 - (3x^2)^2) \, dx \qquad \text{Simplify.}$$

$$= \int_{-2}^{2} (-8x^6 - 32x^4 + 256x^2) \, dx \qquad \text{Simplify.}$$

$$\approx 663.2. \qquad \begin{array}{l}\text{Evaluate outer integral}\\ \text{with respect to } x.\end{array}$$

Because $z = 2x^2y \geq 0$ on R, the value of the integral is the volume of the solid shown in Figure 16.14.

Related Exercises 12, 46 ◄

Change of Perspective   Suppose the region of integration R is bounded on the left and right by the graphs of continuous functions $x = g(y)$ and $x = h(y)$, respectively, on the interval $c \leq y \leq d$. Such regions may take any of the forms shown in **Figure 16.15**.

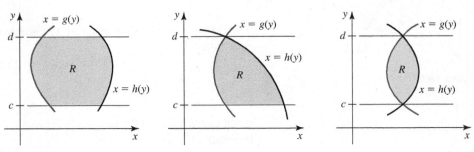

Figure 16.15

To find the volume of the solid bounded by the surface $z = f(x, y)$ and R, we now take vertical slices parallel to the xz-plane. The double integral $\iint_R f(x, y) \, dA$ is then converted to an iterated integral in which the inner integration is with respect to x over the interval $g(y) \leq x \leq h(y)$ and the outer integration is with respect to y over the interval $c \leq y \leq d$. The evaluation of double integrals in these two cases is summarized in the following theorem.

▶ Theorem 16.2 is another version of Fubini's Theorem. With integrals over nonrectangular regions, the order of integration cannot be simply switched; that is,

$$\int_a^b \int_{g(x)}^{h(x)} f(x, y)\, dy\, dx$$

$$\neq \int_{g(x)}^{h(x)} \int_a^b f(x, y)\, dx\, dy.$$

The *element of area dA* corresponds to the area of a small rectangle in the partition. Comparing the double integral to the iterated integral, we see that the element of area is $dA = dy\, dx$ or $dA = dx\, dy$, which is consistent with the area formula for rectangles.

The bounding curves determine the limits of integration in x.

The projection of R on the y-axis determines the limits of integration in y.

Figure 16.16

Figure 16.17

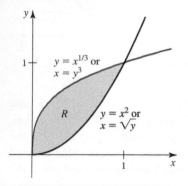

R is bounded above and below, and on the right and left, by curves.

Figure 16.18

> **THEOREM 16.2   Double Integrals over Nonrectangular Regions**
> Let R be a region bounded below and above by the graphs of the continuous functions $y = g(x)$ and $y = h(x)$, respectively, and by the lines $x = a$ and $x = b$ (Figure 16.11). If f is continuous on R, then
>
> $$\iint\limits_R f(x, y)\, dA = \int_a^b \int_{g(x)}^{h(x)} f(x, y)\, dy\, dx.$$
>
> Let R be a region bounded on the left and right by the graphs of the continuous functions $x = g(y)$ and $x = h(y)$, respectively, and the lines $y = c$ and $y = d$ (Figure 16.15). If f is continuous on R, then
>
> $$\iint\limits_R f(x, y)\, dA = \int_c^d \int_{g(y)}^{h(y)} f(x, y)\, dx\, dy.$$

EXAMPLE 2   Computing a volume Find the volume of the solid below the surface $f(x, y) = 2 + \dfrac{1}{y}$ and above the region R in the xy-plane bounded by the lines $y = x$, $y = 8 - x$, and $y = 1$. Notice that $f(x, y) > 0$ on R.

SOLUTION The region R is bounded on the left by $x = y$ and bounded on the right by $y = 8 - x$, or $x = 8 - y$ (**Figure 16.16**). These lines intersect at the point $(4, 4)$. We take vertical slices through the solid parallel to the xz-plane from $y = 1$ to $y = 4$. To visualize these slices, it helps to draw lines through R parallel to the x-axis.

Integrating the cross-sectional areas of slices from $y = 1$ to $y = 4$, the volume of the solid beneath the graph of f and above R (**Figure 16.17**) is given by

$$\iint\limits_R \left(2 + \frac{1}{y}\right) dA = \int_1^4 \int_y^{8-y} \left(2 + \frac{1}{y}\right) dx\, dy \qquad \text{Convert to an iterated integral.}$$

$$= \int_1^4 \left(2 + \frac{1}{y}\right) x \Big|_y^{8-y}\, dy \qquad \text{Evaluate inner integral with respect to } x.$$

$$= \int_1^4 \left(2 + \frac{1}{y}\right)(8 - 2y)\, dy \qquad \text{Simplify.}$$

$$= \int_1^4 \left(14 - 4y + \frac{8}{y}\right) dy \qquad \text{Simplify.}$$

$$= \left(14y - 2y^2 + 8 \ln |y|\right)\Big|_1^4 \qquad \text{Evaluate outer integral with respect to } y.$$

$$= 12 + 8 \ln 4 \approx 23.09. \qquad \text{Simplify.}$$

Related Exercise 74 ◀

QUICK CHECK 2 Could the integral in Example 2 be evaluated by integrating first (inner integral) with respect to y? ◀

Choosing and Changing the Order of Integration

Occasionally, a region of integration is bounded above and below by a pair of curves *and* the region is bounded on the right and left by a pair of curves. For example, the region R in **Figure 16.18** is bounded above by $y = x^{1/3}$ and below by $y = x^2$, and it is bounded on the right by $x = \sqrt{y}$ and on the left by $x = y^3$. In these cases, we can choose either of two orders of integration; however, one order of integration may be preferable. The following examples illustrate the valuable techniques of choosing and changing the order of integration.

Figure 16.19

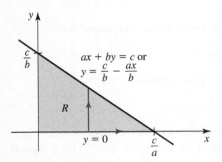

Figure 16.20

▶ In Example 3, it is just as easy to view R as being bounded on the left and the right by the lines $x = 0$ and $x = c/a - by/a$, respectively, and integrating first with respect to x.

EXAMPLE 3 Volume of a tetrahedron Find the volume of the tetrahedron (pyramid with four triangular faces) in the first octant bounded by the plane $z = c - ax - by$ and the coordinate planes ($x = 0$, $y = 0$, and $z = 0$). Assume a, b, and c are positive real numbers (**Figure 16.19**).

SOLUTION Let R be the triangular base of the tetrahedron in the xy-plane; it is bounded by the x- and y-axes and the line $ax + by = c$ (found by setting $z = 0$ in the equation of the plane; **Figure 16.20**). We can view R as being bounded below and above by the lines $y = 0$ and $y = c/b - ax/b$, respectively. The boundaries on the left and right are then $x = 0$ and $x = c/a$, respectively. Therefore, the volume of the solid region between the plane and R is

$$\iint\limits_{R} (c - ax - by)\, dA = \int_0^{c/a} \int_0^{c/b - ax/b} (c - ax - by)\, dy\, dx \qquad \text{Convert to an iterated integral.}$$

$$= \int_0^{c/a} \left(cy - axy - \frac{by^2}{2} \right) \Big|_0^{c/b - ax/b}\, dx \qquad \text{Evaluate inner integral with respect to } y.$$

$$= \int_0^{c/a} \frac{(ax - c)^2}{2b}\, dx \qquad \text{Simplify and factor.}$$

$$= \frac{c^3}{6ab}. \qquad \text{Evaluate outer integral with respect to } x.$$

This result illustrates the volume formula for a tetrahedron. The lengths of the legs of the triangular base are c/a and c/b, which means the area of the base is $c^2/(2ab)$. The height of the tetrahedron is c. The general volume formula is

$$V = \frac{c^3}{6ab} = \underbrace{\frac{1}{3} \frac{c^2}{2ab}}_{\substack{\text{area of} \\ \text{base}}} \cdot \underbrace{c}_{\text{height}} = \frac{1}{3}(\text{area of base})(\text{height}).$$

Related Exercise 73 ◀

EXAMPLE 4 Changing the order of integration Consider the iterated integral $\int_0^{\sqrt{\pi}} \int_y^{\sqrt{\pi}} \sin x^2\, dx\, dy$. Sketch the region of integration determined by the limits of integration and then evaluate the iterated integral.

SOLUTION The region of integration is $R = \{(x, y): y \le x \le \sqrt{\pi}, 0 \le y \le \sqrt{\pi}\}$, which is a triangle (**Figure 16.21a**). Evaluating the iterated integral as given (integrating first with respect to x) requires integrating $\sin x^2$, a function whose antiderivative is not expressible in terms of elementary functions. Therefore, this order of integration is not feasible.

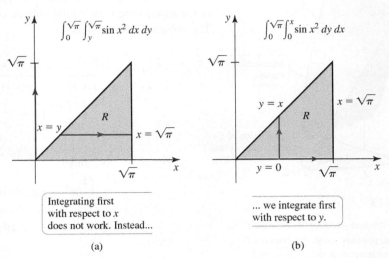

(a)

(b)

Figure 16.21

Instead, we change our perspective (**Figure 16.21b**) and integrate first with respect to y. With this order of integration, y runs from $y = 0$ to $y = x$ in the inner integral, and x runs from $x = 0$ to $x = \sqrt{\pi}$ in the outer integral:

$$\iint\limits_{R} \sin x^2 \, dA = \int_0^{\sqrt{\pi}} \int_0^{x} \sin x^2 \, dy \, dx$$

$$= \int_0^{\sqrt{\pi}} y \sin x^2 \Big|_0^{x} \, dx \qquad \text{Evaluate inner integral with respect to } y; \sin x^2 \text{ is constant.}$$

$$= \int_0^{\sqrt{\pi}} x \sin x^2 \, dx \qquad \text{Simplify.}$$

$$= -\frac{1}{2} \cos x^2 \Big|_0^{\sqrt{\pi}} \qquad \text{Evaluate outer integral with respect to } x.$$

$$= 1. \qquad \text{Simplify.}$$

This example shows that the order of integration can make a practical difference.

Related Exercises 58, 64 ◄

QUICK CHECK 3 Change the order of integration of the integral $\int_0^1 \int_0^y f(x, y) \, dx \, dy$. ◄

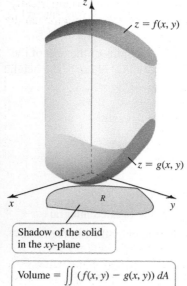

Shadow of the solid in the xy-plane

Volume $= \iint\limits_{R} (f(x, y) - g(x, y)) \, dA$

Figure 16.22

Regions Between Two Surfaces

An extension of the preceding ideas allows us to solve more general volume problems. Let $z = f(x, y)$ and $z = g(x, y)$ be continuous functions with $f(x, y) \geq g(x, y)$ on a region R in the xy-plane. Suppose we wish to compute the volume of the solid between the two surfaces over the region R (**Figure 16.22**). Forming a Riemann sum for the volume, the height of a typical box within the solid is the vertical distance $f(x, y) - g(x, y)$ between the upper and lower surfaces. Therefore, the volume of the solid between the surfaces is

$$V = \iint\limits_{R} (f(x, y) - g(x, y)) \, dA.$$

EXAMPLE 5 Region bounded by two surfaces Find the volume of the solid bounded by the parabolic cylinder $z = 1 + x^2$ and the planes $z = 5 - y$ and $y = 0$ (**Figure 16.23**).

SOLUTION The upper surface bounding the solid is $z = 5 - y$ and the lower surface is $z = 1 + x^2$; these two surfaces intersect along a curve C. Solving $5 - y = 1 + x^2$, we find that $y = 4 - x^2$, which is the projection of C onto the xy-plane. The back wall of the solid is the plane $y = 0$, and its projection onto the xy-plane is the x-axis. This line ($y = 0$) intersects the parabola $y = 4 - x^2$ at $x = \pm 2$. Therefore, the region of integration (**Figure 16.23**) is

$$R = \{(x, y): 0 \leq y \leq 4 - x^2, -2 \leq x \leq 2\}.$$

Notice that both R and the solid are symmetric about the yz-plane. Therefore, the volume of the entire solid is twice the volume of that part of the solid that lies in the first octant. The volume of the solid is

Figure 16.23

▶ To use symmetry to simplify a double integral, you must check that both the region of integration and the integrand have the same symmetry.

$$2 \int_0^2 \int_0^{4-x^2} \underbrace{((5 - y)}_{f(x, y)} - \underbrace{(1 + x^2))}_{g(x, y)} \, dy \, dx$$

$$= 2 \int_0^2 \int_0^{4-x^2} (4 - x^2 - y) \, dy \, dx \qquad \text{Simplify the integrand.}$$

$$= 2 \int_0^2 \left((4 - x^2)y - \frac{y^2}{2} \right) \Big|_0^{4-x^2} \, dx \qquad \text{Evaluate inner integral with respect to } y.$$

$$= \int_0^2 (x^4 - 8x^2 + 16) \, dx \qquad \text{Simplify.}$$

$$= \left(\frac{x^5}{5} - \frac{8x^3}{3} + 16x \right) \Big|_0^2 \qquad \text{Evaluate outer integral with respect to } x.$$

$$= \frac{256}{15}. \qquad \text{Simplify.}$$

Related Exercises 78–79 ◄

Figure 16.24

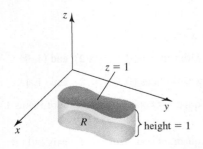

Volume of solid = (area of R) × (height)

= area of R = $\iint\limits_{R} 1 \, dA$

Figure 16.25

▶ We are solving a familiar area problem first encountered in Section 6.2. Suppose R is bounded above by $y = h(x)$ and below by $y = g(x)$, for $a \leq x \leq b$. Using a double integral, the area of R is

$$\iint\limits_{R} dA = \int_{a}^{b} \int_{g(x)}^{h(x)} dy \, dx$$

$$= \int_{a}^{b} (h(x) - g(x)) \, dx,$$

which is a result obtained in Section 6.2.

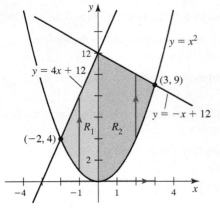

Figure 16.26

Decomposition of Regions

We occasionally encounter regions that are more complicated than those considered so far. A technique called *decomposition* allows us to subdivide a region of integration into two (or more) subregions. If the integrals over the subregions can be evaluated separately, the results are added to obtain the value of the original integral. For example, the region R in **Figure 16.24** is divided into two nonoverlapping subregions R_1 and R_2. By partitioning these regions and using Riemann sums, it can be shown that

$$\iint\limits_{R} f(x, y) \, dA = \iint\limits_{R_1} f(x, y) \, dA + \iint\limits_{R_2} f(x, y) \, dA.$$

This method is illustrated in Example 6. The analog of decomposition with single variable integrals is the property $\int_{a}^{b} f(x) \, dx = \int_{a}^{p} f(x) \, dx + \int_{p}^{b} f(x) \, dx$.

Finding Area by Double Integrals

An interesting application of double integrals arises when the integrand is $f(x, y) = 1$. The integral $\iint_{R} 1 \, dA$ gives the volume of the solid between the horizontal plane $z = 1$ and the region R. Because the height of this solid is 1, its volume equals (numerically) the area of R (**Figure 16.25**). Therefore, we have a way to compute areas of regions in the xy-plane using double integrals.

Areas of Regions by Double Integrals

Let R be a region in the xy-plane. Then

$$\text{area of } R = \iint\limits_{R} dA.$$

EXAMPLE 6   Area of a plane region Find the area of the region R bounded by $y = x^2$, $y = -x + 12$, and $y = 4x + 12$ (**Figure 16.26**).

SOLUTION The region R in its entirety is bounded neither above and below by two curves, nor on the left and right by two curves. However, when decomposed along the y-axis, R may be viewed as two regions R_1 and R_2, each of which is bounded above and below by a pair of curves. Notice that the parabola $y = x^2$ and the line $y = -x + 12$ intersect in the first quadrant at the point $(3, 9)$, while the parabola and the line $y = 4x + 12$ intersect in the second quadrant at the point $(-2, 4)$.

To find the area of R, we integrate the function $f(x, y) = 1$ over R_1 and R_2; the area is

$$\iint\limits_{R_1} 1 \, dA + \iint\limits_{R_2} 1 \, dA \qquad \text{Decompose region.}$$

$$= \int_{-2}^{0} \int_{x^2}^{4x+12} 1 \, dy \, dx + \int_{0}^{3} \int_{x^2}^{-x+12} 1 \, dy \, dx \qquad \text{Convert to iterated integrals.}$$

$$= \int_{-2}^{0} (4x + 12 - x^2) \, dx + \int_{0}^{3} (-x + 12 - x^2) \, dx \qquad \begin{array}{l}\text{Evaluate inner integrals}\\ \text{with respect to } y.\end{array}$$

$$= \left(2x^2 + 12x - \frac{x^3}{3} \right)\Big|_{-2}^{0} + \left(-\frac{x^2}{2} + 12x - \frac{x^3}{3} \right)\Big|_{0}^{3} \qquad \begin{array}{l}\text{Evaluate outer integrals}\\ \text{with respect to } x.\end{array}$$

$$= \frac{40}{3} + \frac{45}{2} = \frac{215}{6}. \qquad \text{Simplify.}$$

Related Exercise 86 ◀

QUICK CHECK 4 Consider the triangle R with vertices $(-1, 0)$, $(1, 0)$, and $(0, 1)$ as a region of integration. If we integrate first with respect to x, does R need to be decomposed? If we integrate first with respect to y, does R need to be decomposed? ◀

SECTION 16.2 EXERCISES

Getting Started

1. Describe and sketch a region that is bounded above and below by two curves.

2. Describe and sketch a region that is bounded on the left and on the right by two curves.

3. Which order of integration is preferable to integrate $f(x, y) = xy$ over $R = \{(x, y): y - 1 \le x \le 1 - y, 0 \le y \le 1\}$?

4. Which order of integration would you use to find the area of the region bounded by the x-axis and the lines $y = 2x + 3$ and $y = 3x - 4$ using a double integral?

5. Change the order of integration in the integral $\int_0^1 \int_{y^2}^{\sqrt{y}} f(x, y)\, dx\, dy$.

6. Sketch the region of integration for $\int_{-2}^{2} \int_{x^2}^{4} e^{xy}\, dy\, dx$.

7. Sketch the region of integration for $\int_0^2 \int_0^{2x} dy\, dx$ and use geometry to evaluate the iterated integral.

8. Describe a solid whose volume equals $\int_{-4}^{4} \int_{-\sqrt{16-x^2}}^{\sqrt{16-x^2}} 10\, dy\, dx$ and evaluate this iterated integral using geometry.

9–10. *Consider the region R shown in the figure and write an iterated integral of a continuous function f over R.*

9.

10.

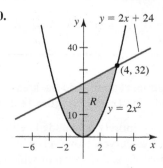

Practice Exercises

11–27. Evaluating integrals *Evaluate the following integrals.*

11. $\int_0^1 \int_x^1 6y\, dy\, dx$

12. $\int_0^1 \int_0^{2x} 15xy^2\, dy\, dx$

13. $\int_0^2 \int_{x^2}^{2x} xy\, dy\, dx$

14. $\int_{-\pi/4}^{\pi/4} \int_{\sin x}^{\cos x} dy\, dx$

15. $\int_{-2}^2 \int_{x^2}^{8-x^2} x\, dy\, dx$

16. $\int_0^{\ln 2} \int_{e^x}^2 dy\, dx$

17. $\int_0^1 \int_0^x 2e^{x^2}\, dy\, dx$

18. $\int_0^{\sqrt[3]{\pi/2}} \int_0^x y\cos x^3\, dy\, dx$

19. $\int_0^{\ln 2} \int_{e^y}^2 \frac{y}{x}\, dx\, dy$

20. $\int_0^4 \int_y^{2y} xy\, dx\, dy$

21. $\int_0^{\pi/2} \int_y^{\pi/2} 6\sin(2x - 3y)\, dx\, dy$

22. $\int_0^{\pi/2} \int_0^{\cos y} e^{\sin y}\, dx\, dy$

23. $\int_0^{\pi/2} \int_0^{y\cos y} dx\, dy$

24. $\int_0^1 \int_{\tan^{-1} x}^{\pi/4} 2x\, dy\, dx$

25. $\int_0^4 \int_{-\sqrt{16-y^2}}^{\sqrt{16-y^2}} 2xy\, dx\, dy$

26. $\int_0^1 \int_0^x 2e^x\, dy\, dx$

27. $\int_{\pi/2}^{\pi} \int_0^{y^2} \cos \frac{x}{y}\, dx\, dy$

28–34. Regions of integration *Sketch each region R and write an iterated integral of a continuous function f over R. Use the order dy dx.*

28. $R = \{(x, y): 0 \le x \le 2, 3x^2 \le y \le -6x + 24\}$

29. $R = \{(x, y): 1 \le x \le 2, x + 1 \le y \le 2x + 4\}$

30. $R = \{(x, y): 0 \le x \le 4, x^2 \le y \le 8\sqrt{x}\}$

31. R is the triangular region with vertices $(0, 0)$, $(0, 2)$, and $(1, 0)$.

32. R is the triangular region with vertices $(0, 0)$, $(0, 2)$, and $(1, 1)$.

33. R is the region in the first quadrant bounded by a circle of radius 1 centered at the origin.

34. R is the region in the first quadrant bounded by the y-axis and the parabolas $y = x^2$ and $y = 1 - x^2$.

35–42. Regions of integration *Write an iterated integral of a continuous function f over the region R. Use the order dy dx. Start by sketching the region of integration if it is not supplied.*

35.

36.

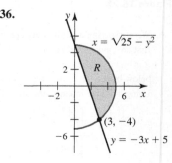

37. R is the region bounded by $y = 4 - x$, $y = 1$, and $x = 0$.

38. $R = \{(x, y): 0 \le x \le y(1 - y)\}$.

39. R is the region bounded by $y = 2x + 3$, $y = 3x - 7$, and $y = 0$.

40. R is the region in quadrants 2 and 3 bounded by the semicircle with radius 3 centered at $(0, 0)$.

41. R is the region bounded by the triangle with vertices $(0, 0)$, $(2, 0)$, and $(1, 1)$.

42. R is the region in the first quadrant bounded by the x-axis, the line $x = 6 - y$, and the curve $y = \sqrt{x}$.

43–56. Evaluating integrals *Evaluate the following integrals. A sketch is helpful.*

43. $\iint_R xy\, dA$; R is bounded by $x = 0$, $y = 2x + 1$, and $y = -2x + 5$.

44. $\iint_R (x + y)\, dA$; R is the region in the first quadrant bounded by $x = 0$, $y = x^2$, and $y = 8 - x^2$.

45. $\iint_R y^2\, dA$; R is bounded by $x = 1$, $y = 2x + 2$, and $y = -x - 1$.

46. $\iint_R x^2 y\, dA$; R is the region in quadrants 1 and 4 bounded by the semicircle of radius 4 centered at $(0, 0)$.

47. $\iint_R 12y\, dA$; R is bounded by $y = 2 - x$, $y = \sqrt{x}$, and $y = 0$.

48. $\iint_R y^2\, dA$; R is bounded by $y = 1$, $y = 1 - x$, and $y = x - 1$.

49. $\iint_R 3xy \, dA$; R is the region in the first quadrant bounded by $y = 2 - x$, $y = 0$, and $x = 4 - y^2$.

50. $\iint_R (x + y) \, dA$; R is bounded by $y = |x|$ and $y = 4$.

51. $\iint_R 3x^2 \, dA$; R is bounded by $y = 0$, $y = 2x + 4$, and $y = x^3$.

52. $\iint_R 8xy \, dA$; $R = \{(x, y): 0 \le y \le \sec x, 0 \le x \le \pi/4\}$

53. $\iint_R (x + y) \, dA$; R is the region bounded by $y = 1/x$ and $y = 5/2 - x$.

54. $\iint_R \dfrac{y}{1 + x + y^2} \, dA$; $R = \{(x, y): 0 \le \sqrt{x} \le y, 0 \le y \le 1\}$

55. $\iint_R x \sec^2 y \, dA$; $R = \{(x, y): 0 \le y \le x^2, 0 \le x \le \sqrt{\pi}/2\}$

56. $\iint_R \dfrac{8xy}{1 + x^2 + y^2} \, dA$; $R = \{(x, y): 0 \le y \le x, 0 \le x \le 2\}$

57–62. Changing order of integration *Reverse the order of integration in the following integrals.*

57. $\displaystyle\int_0^2 \int_{x^2}^{2x} f(x, y) \, dy \, dx$

58. $\displaystyle\int_0^3 \int_0^{6-2x} f(x, y) \, dy \, dx$

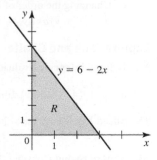

59. $\displaystyle\int_{1/2}^1 \int_0^{-\ln y} f(x, y) \, dx \, dy$

60. $\displaystyle\int_0^1 \int_1^{e^y} f(x, y) \, dx \, dy$

61. $\displaystyle\int_0^1 \int_0^{\cos^{-1} y} f(x, y) \, dx \, dy$

62. $\displaystyle\int_1^e \int_0^{\ln x} f(x, y) \, dy \, dx$

63–68. Changing order of integration *Reverse the order of integration and evaluate the integral.*

63. $\displaystyle\int_0^1 \int_y^1 e^{x^2} \, dx \, dy$

64. $\displaystyle\int_0^\pi \int_x^\pi \sin y^2 \, dy \, dx$

65. $\displaystyle\int_0^{1/2} \int_{y^2}^{1/4} y \cos (16\pi x^2) \, dx \, dy$

66. $\displaystyle\int_0^4 \int_{\sqrt{x}}^2 \dfrac{x}{y^5 + 1} \, dy \, dx$

67. $\displaystyle\int_0^{\sqrt[3]{\pi}} \int_y^{\sqrt[3]{\pi}} x^4 \cos (x^2 y) \, dx \, dy$

68. $\displaystyle\int_0^2 \int_0^{4-x^2} \dfrac{xe^{2y}}{4 - y} \, dy \, dx$

69–70. Two integrals to one *Draw the regions of integration and write the following integrals as a single iterated integral.*

69. $\displaystyle\int_0^1 \int_{e^y}^e f(x, y) \, dx \, dy + \int_{-1}^0 \int_{e^{-y}}^e f(x, y) \, dx \, dy$

70. $\displaystyle\int_{-4}^0 \int_0^{\sqrt{16-x^2}} f(x, y) \, dy \, dx + \int_0^4 \int_0^{4-x} f(x, y) \, dy \, dx$

71–80. Volumes *Find the volume of the following solids.*

71. The solid bounded by the cylinder $z = 2 - y^2$, the xy-plane, the xz-plane, and the planes $y = x$ and $x = 1$

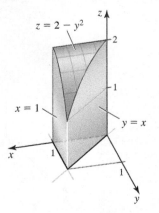

72. The solid bounded between the cylinder $z = 2 \sin^2 x$ and the xy-plane over the region $R = \{(x, y): 0 \le x \le y \le \pi\}$

73. The tetrahedron bounded by the coordinate planes ($x = 0$, $y = 0$, and $z = 0$) and the plane $z = 8 - 2x - 4y$

74. The solid in the first octant bounded by the coordinate planes and the surface $z = 1 - y - x^2$

75. The segment of the cylinder $x^2 + y^2 = 1$ bounded above by the plane $z = 12 + x + y$ and below by $z = 0$

76. The solid S between the surfaces $z = e^{x-y}$ and $z = -e^{x-y}$, where S intersects the xy-plane in the region $R = \{(x, y): 0 \le x \le y, 0 \le y \le 1\}$

77. The solid above the region $R = \{(x, y): 0 \le x \le 1, 0 \le y \le 2 - x\}$ and between the planes $-4x - 4y + z = 0$ and $-2x - y + z = 8$

78. The solid in the first octant bounded by the planes $x = 0$, $y = 0$, $z = 1$, and $z = 2 - y$, and the cylinder $y = 1 - x^2$

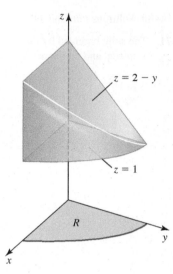

$z = 2 - y$

$z = 1$

R

79. The solid above the region $R = \{(x, y): 0 \le x \le 1, 0 \le y \le 1 - x\}$ bounded by the paraboloids $z = x^2 + y^2$ and $z = 2 - x^2 - y^2$ and the coordinate planes in the first octant

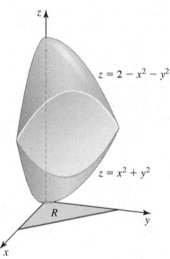

$z = 2 - x^2 - y^2$

$z = x^2 + y^2$

R

80. The solid bounded by the parabolic cylinder $z = x^2 + 1$, and the planes $z = y + 1$ and $y = 1$

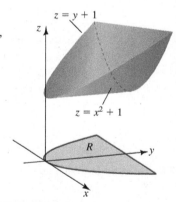

$z = y + 1$

$z = x^2 + 1$

R

⊤ 81–84. Volume using technology *Find the volume of the following solids. Use a computer algebra system to evaluate an appropriate iterated integral.*

81. The column with a square base $R = \{(x, y): |x| \le 1, |y| \le 1\}$ cut by the plane $z = 4 - x - y$

82. The solid between the paraboloid $z = x^2 + y^2$ and the plane $z = 1 - 2y$

83. The wedge sliced from the cylinder $x^2 + y^2 = 1$ by the planes $z = a(2 - x)$ and $z = a(x - 2)$, where $a > 0$

84. The solid bounded by the elliptical cylinder $x^2 + 3y^2 = 12$, the plane $z = 0$, and the paraboloid $z = 3x^2 + y^2 + 1$

85–90. Area of plane regions *Use double integrals to compute the area of the following regions.*

85. The region bounded by the parabola $y = x^2$ and the line $y = 4$

86. The region bounded by the parabola $y = x^2$ and the line $y = x + 2$

87. The region in the first quadrant bounded by $y = e^x$ and $x = \ln 2$

88. The region bounded by $y = 1 + \sin x$ and $y = 1 - \sin x$ on the interval $[0, \pi]$

89. The region in the first quadrant bounded by $y = x^2$, $y = 5x + 6$, and $y = 6 - x$

90. The region bounded by the lines $x = 0$, $x = 4$, $y = x$, and $y = 2x + 1$

91. **Explain why or why not** Determine whether the following statements are true and give an explanation or counterexample.

   a. In the iterated integral $\int_c^d \int_a^b f(x, y)\, dx\, dy$, the limits a and b must be constants or functions of x.
   b. In the iterated integral $\int_c^d \int_a^b f(x, y)\, dx\, dy$, the limits c and d must be functions of y.
   c. Changing the order of integration gives $\int_0^2 \int_1^y f(x, y)\, dx\, dy = \int_1^y \int_0^2 f(x, y)\, dy\, dx$.

Explorations and Challenges

92. **Related integrals** Evaluate each integral.

   a. $\displaystyle\int_0^4 \int_0^4 (4 - x - y)\, dx\, dy$ **b.** $\displaystyle\int_0^4 \int_0^4 |4 - x - y|\, dx\, dy$

93. **Sliced block** Find the volume of the solid bounded by the planes $x = 0$, $x = 5$, $z = y - 1$, $z = -2y - 1$, $z = 0$, and $z = 2$.

94. **Square region** Consider the region $R = \{(x, y): |x| + |y| \le 1\}$ shown in the figure.

   a. Use a double integral to show that the area of R is 2.
   b. Find the volume of the square column whose base is R and whose upper surface is $z = 12 - 3x - 4y$.
   c. Find the volume of the solid above R and beneath the cylinder $x^2 + z^2 = 1$.
   d. Find the volume of the pyramid whose base is R and whose vertex is on the z-axis at $(0, 0, 6)$.

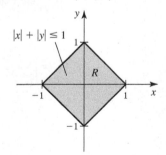

$|x| + |y| \le 1$

R

95–96. Average value *Use the definition for the average value of a function over a region R (Section 16.1), $\overline{f} = \dfrac{1}{\text{area of } R} \iint_R f(x, y)\, dA$.*

95. Find the average value of $a - x - y$ over the region $R = \{(x, y): x + y \le a, x \ge 0, y \ge 0\}$, where $a > 0$.

96. Find the average value of $z = a^2 - x^2 - y^2$ over the region $R = \{(x, y): x^2 + y^2 \le a^2\}$, where $a > 0$.

T 97–98. Area integrals *Consider the following regions R. Use a computer algebra system to evaluate the integrals.*

a. *Sketch the region R.*

b. *Evaluate $\iint_R dA$ to determine the area of the region.*

c. *Evaluate $\iint_R xy \, dA$.*

97. R is the region between both branches of $y = 1/x$ and the lines $y = x + 3/2$ and $y = x - 3/2$.

98. R is the region bounded by the ellipse $x^2/18 + y^2/36 = 1$ with $y \le 4x/3$.

99–102. Improper integrals *Many improper double integrals may be handled using the techniques for improper integrals in one variable (Section 8.9). For example, under suitable conditions on f,*

$$\int_a^\infty \int_{g(x)}^{h(x)} f(x, y) \, dy \, dx = \lim_{b \to \infty} \int_a^b \int_{g(x)}^{h(x)} f(x, y) \, dy \, dx.$$

Use or extend the one-variable methods for improper integrals to evaluate the following integrals.

99. $\displaystyle\int_1^\infty \int_0^{e^{-x}} xy \, dy \, dx$

100. $\displaystyle\int_1^\infty \int_0^{1/x^2} \frac{2y}{x} \, dy \, dx$

101. $\displaystyle\int_0^\infty \int_0^\infty e^{-x-y} \, dy \, dx$

102. $\displaystyle\int_{-\infty}^\infty \int_{-\infty}^\infty \frac{1}{(x^2+1)(y^2+1)} \, dy \, dx$

QUICK CHECK ANSWERS

1. Inner integral: $0 \le y \le 2 - x$; outer integral: $0 \le x \le 2$ **2.** Yes; however, two separate iterated integrals would be required.   **3.** $\int_0^1 \int_x^1 f(x, y) \, dy \, dx$ **4.** No; yes ◄

16.3 Double Integrals in Polar Coordinates

> Recall the conversions between Cartesian and polar coordinates (Section 12.2):
>
> $x = r \cos \theta, \, y = r \sin \theta$, or
> $r^2 = x^2 + y^2, \, \tan \theta = y/x.$

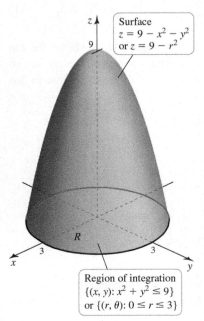

Surface
$z = 9 - x^2 - y^2$
or $z = 9 - r^2$

Region of integration
$\{(x, y): x^2 + y^2 \le 9\}$
or $\{(r, \theta): 0 \le r \le 3\}$

Figure 16.27

In Chapter 12, we explored polar coordinates and saw that in certain situations, they simplify problems considerably. The same is true when it comes to integration over plane regions. In this section, we learn how to formulate double integrals in polar coordinates and how to change double integrals from Cartesian coordinates to polar coordinates.

Moving from Rectangular to Polar Coordinates

Suppose we want to find the volume of the solid bounded by the paraboloid $z = 9 - x^2 - y^2$ and the xy-plane (**Figure 16.27**). The intersection of the paraboloid and the xy-plane $(z = 0)$ is the curve $9 - x^2 - y^2 = 0$, or $x^2 + y^2 = 9$. Therefore, the region of integration R is the disk of radius 3 in the xy-plane, centered at the origin, which, when expressed in Cartesian coordinates, is $R = \{(x, y): -\sqrt{9 - x^2} \le y \le \sqrt{9 - x^2}, -3 \le x \le 3\}$. Using the relationship $r^2 = x^2 + y^2$ for converting Cartesian to polar coordinates, the region of integration expressed in polar coordinates is simply $R = \{(r, \theta): 0 \le r \le 3, 0 \le \theta \le 2\pi\}$. Furthermore, the paraboloid expressed in polar coordinates is $z = 9 - r^2$. This problem (which is solved in Example 1) illustrates how both the integrand and the region of integration in a double integral can be simplified by working in polar coordinates.

The region of integration in this problem is an example of a **polar rectangle**. In polar coordinates, it has the form $R = \{(r, \theta): 0 \le a \le r \le b, \alpha \le \theta \le \beta\}$, where $\beta - \alpha \le 2\pi$ and a, b, α, and β are constants (**Figure 16.28**). Polar rectangles are the analogs of rectangles in Cartesian coordinates. For this reason, the methods used in Section 16.1 for evaluating double integrals over rectangles can be extended to polar rectangles. The goal is to evaluate integrals of the form $\iint_R f(x, y) \, dA$, where f is a continuous function on the polar rectangle R. If f is nonnegative on R, this integral equals the volume of the solid bounded by the surface $z = f(x, y)$ and the region R in the xy-plane.

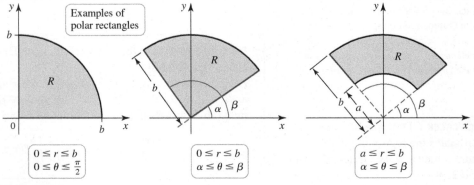

Examples of polar rectangles

$0 \le r \le b$
$0 \le \theta \le \frac{\pi}{2}$

$0 \le r \le b$
$\alpha \le \theta \le \beta$

$a \le r \le b$
$\alpha \le \theta \le \beta$

Figure 16.28

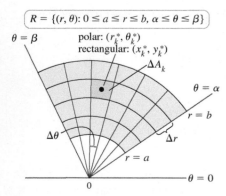

$R = \{(r, \theta): 0 \le a \le r \le b, \alpha \le \theta \le \beta\}$

Figure 16.29

Figure 16.30

➤ Recall that the area of a sector of a circle of radius r subtended by an angle θ is $\frac{1}{2}r^2\theta$.

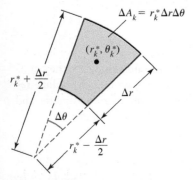

Area $= \frac{1}{2}r^2\theta$

➤ The most common error in evaluating integrals in polar coordinates is to omit the factor r that appears in the integrand. In Cartesian coordinates, the element of area is $dx\, dy$; in polar coordinates, the element of area is $r\, dr\, d\theta$, and without the factor of r, area is not measured correctly.

QUICK CHECK 1 Describe in polar coordinates the region in the first quadrant between the circles of radius 1 and 2. ◄

Our approach is to divide $[a, b]$ into M subintervals of equal length $\Delta r = (b - a)/M$. We similarly divide $[\alpha, \beta]$ into m subintervals of equal length $\Delta\theta = (\beta - \alpha)/m$. Now look at the arcs of the circles centered at the origin with radii

$$r = a, r = a + \Delta r, r = a + 2\Delta r, \ldots, r = b$$

and the rays

$$\theta = \alpha, \theta = \alpha + \Delta\theta, \theta = a + 2\Delta\theta, \ldots, \theta = \beta$$

emanating from the origin (**Figure 16.29**). The arcs and rays divide the region R into $n = Mm$ polar rectangles that we number in a convenient way from $k = 1$ to $k = n$. The area of the kth rectangle is denoted ΔA_k, and we let (r_k^*, θ_k^*) be the polar coordinates of an arbitrary point in that rectangle. Note that this point also has the Cartesian coordinates $(x_k^*, y_k^*) = (r_k^* \cos \theta_k^*, r_k^* \sin \theta_k^*)$. If f is continuous on R, the volume of the solid region beneath the surface $z = f(x, y)$ and above R may be computed with Riemann sums using either ordinary rectangles (as in Sections 16.1 and 16.2) or polar rectangles. Here, we now use polar rectangles.

Consider the "box" whose base is the kth polar rectangle and whose height is $f(x_k^*, y_k^*)$; its volume is $f(x_k^*, y_k^*)\Delta A_k$, for $k = 1, \ldots, n$. Therefore, the volume of the solid region beneath the surface $z = f(x, y)$ with a base R is approximately

$$V = \sum_{k=1}^{n} f(x_k^*, y_k^*)\Delta A_k.$$

This approximation to the volume is a Riemann sum. We let Δ be the maximum value of Δr and $\Delta\theta$. If f is continuous on R, then as $n \to \infty$ and $\Delta \to 0$, the sum approaches a double integral; that is,

$$\iint\limits_{R} f(x, y)\, dA = \lim_{\Delta \to 0} \sum_{k=1}^{n} f(x_k^*, y_k^*)\Delta A_k = \lim_{\Delta \to 0} \sum_{k=1}^{n} f(r_k^* \cos \theta_k^*, r_k^* \sin \theta_k^*)\Delta A_k. \quad (1)$$

The next step is to express ΔA_k in terms of Δr and $\Delta\theta$. **Figure 16.30** shows the kth polar rectangle, with an area of ΔA_k. The point (r_k^*, θ_k^*) (in polar coordinates) is chosen so that the outer arc of the polar rectangle has radius $r_k^* + \Delta r/2$ and the inner arc has radius $r_k^* - \Delta r/2$. The area of the polar rectangle is

$\Delta A_k = $ (area of outer sector) $-$ (area of inner sector)

$$= \frac{1}{2}\left(r_k^* + \frac{\Delta r}{2}\right)^2 \Delta\theta - \frac{1}{2}\left(r_k^* - \frac{\Delta r}{2}\right)^2 \Delta\theta \qquad \text{Area of sector} = \frac{1}{2}r^2\Delta\theta$$

$$= r_k^* \Delta r \Delta\theta. \qquad\qquad \text{Expand and simplify.}$$

Substituting this expression for ΔA_k into equation (1), we have

$$\iint\limits_{R} f(x, y)\, dA = \lim_{\Delta \to 0} \sum_{k=1}^{n} f(x_k^*, y_k^*)\Delta A_k = \lim_{\Delta \to 0} \sum_{k=1}^{n} f(r_k^* \cos \theta_k^*, r_k^* \sin \theta_k^*)r_k^* \Delta r \Delta\theta.$$

This observation leads to a theorem that allows us to write a double integral in x and y as an iterated integral of $f(r \cos \theta, r \sin \theta)r$ in polar coordinates. It is an example of a change of variables, explained more generally in Section 16.7.

THEOREM 16.3 Change of Variables for Double Integrals over Polar Rectangle Regions

Let f be continuous on the region R in the xy-plane expressed in polar coordinates as $R = \{(r, \theta): 0 \le a \le r \le b, \alpha \le \theta \le \beta\}$, where $\beta - \alpha \le 2\pi$. Then f is integrable over R, and the double integral of f over R is

$$\iint\limits_{R} f(x, y)\, dA = \int_{\alpha}^{\beta} \int_{a}^{b} f(r \cos \theta, r \sin \theta)\, r\, dr\, d\theta.$$

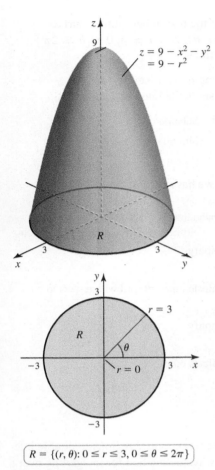

$R = \{(r, \theta): 0 \le r \le 3, 0 \le \theta \le 2\pi\}$

Figure 16.31

QUICK CHECK 2 Express the functions $f(x, y) = (x^2 + y^2)^{5/2}$ and $h(x, y) = x^2 - y^2$ in polar coordinates. ◀

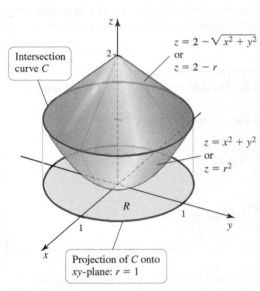

Figure 16.32

QUICK CHECK 3 Give a geometric explanation for the extraneous root $z = 4$ found in Example 2. ◀

EXAMPLE 1 **Volume of a paraboloid cap** Find the volume of the solid bounded by the paraboloid $z = 9 - x^2 - y^2$ and the xy-plane.

SOLUTION Using $x^2 + y^2 = r^2$, the surface is described in polar coordinates by $z = 9 - r^2$. The paraboloid intersects the xy-plane ($z = 0$) when $z = 9 - r^2 = 0$, or $r = 3$. Therefore, the intersection curve is the circle of radius 3 centered at the origin. The resulting region of integration is the disk $R = \{(r, \theta): 0 \le r \le 3, 0 \le \theta \le 2\pi\}$ (**Figure 16.31**). Integrating over R in polar coordinates, the volume is

$$V = \int_0^{2\pi} \int_0^3 \underbrace{(9 - r^2)}_{z} r \, dr \, d\theta \qquad \text{Iterated integral for volume}$$

$$= \int_0^{2\pi} \left(\frac{9r^2}{2} - \frac{r^4}{4} \right) \Big|_0^3 d\theta \qquad \text{Evaluate inner integral with respect to } r.$$

$$= \int_0^{2\pi} \frac{81}{4} \, d\theta = \frac{81\pi}{2}. \qquad \text{Evaluate outer integral with respect to } \theta.$$

Related Exercises 12, 16 ◀

EXAMPLE 2 **Region bounded by two surfaces** Find the volume of the region bounded by the paraboloid $z = x^2 + y^2$ and the cone $z = 2 - \sqrt{x^2 + y^2}$.

SOLUTION As discussed in Section 16.2, the volume of a solid bounded by two surfaces $z = f(x, y)$ and $z = g(x, y)$ over a region R in the xy-plane is given by $\iint_R (f(x, y) - g(x, y)) \, dA$, where $f(x, y) \ge g(x, y)$ over R. Because the paraboloid $z = x^2 + y^2$ lies below the cone $z = 2 - \sqrt{x^2 + y^2}$ (**Figure 16.32**), the volume of the solid bounded by the surfaces is

$$V = \iint_R \left((2 - \sqrt{x^2 + y^2}) - (x^2 + y^2) \right) dA,$$

where the boundary of R is the curve of intersection C of the surfaces projected onto the xy-plane. To find C, we set the equations of the surfaces equal to one another. Writing $x^2 + y^2 = 2 - \sqrt{x^2 + y^2}$ seems like a good start, but it leads to algebraic difficulties. Instead, we write the equation of the cone as $\sqrt{x^2 + y^2} = 2 - z$ and then substitute this equation into the equation for the paraboloid:

$$z = (2 - z)^2 \qquad \begin{array}{l} z = x^2 + y^2 \text{ (paraboloid) and} \\ \sqrt{x^2 + y^2} = 2 - z \text{ (cone)} \end{array}$$

$$z^2 - 5z + 4 = 0 \qquad \text{Simplify.}$$

$$(z - 1)(z - 4) = 0 \qquad \text{Factor.}$$

$$z = 1 \text{ or } z = 4. \qquad \text{Solve for } z.$$

The solution $z = 4$ is an extraneous root (see Quick Check 3). Setting $z = 1$ in the equation of either the paraboloid or the cone leads to $x^2 + y^2 = 1$, which is an equation of the curve C in the plane $z = 1$. Projecting C onto the xy-plane, we conclude that the region of integration (written in polar coordinates) is $R = \{(r, \theta): 0 \le r \le 1, 0 \le \theta \le 2\pi\}$.

Converting the original volume integral into polar coordinates and evaluating it over R, we have

$$V = \iint_R \left((2 - \sqrt{x^2 + y^2}) - (x^2 + y^2) \right) dA \qquad \text{Double integral for volume}$$

$$= \int_0^{2\pi} \int_0^1 (2 - r - r^2) r \, dr \, d\theta \qquad \begin{array}{l} \text{Convert to polar coordinates;} \\ x^2 + y^2 = r^2. \end{array}$$

$$= \int_0^{2\pi} \left(r^2 - \frac{1}{3} r^3 - \frac{1}{4} r^4 \right) \Big|_0^1 d\theta \qquad \text{Evaluate the inner integral.}$$

$$= \int_0^{2\pi} \frac{5}{12} \, d\theta = \frac{5\pi}{6}. \qquad \text{Evaluate the outer integral.}$$

Related Exercises 33, 40 ◀

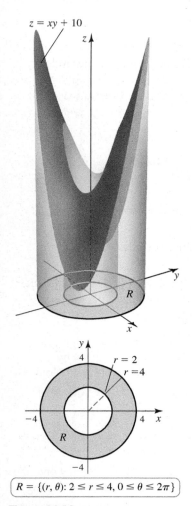

$z = xy + 10$

$R = \{(r, \theta): 2 \le r \le 4, 0 \le \theta \le 2\pi\}$

Figure 16.33

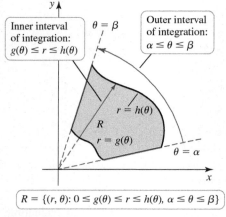

$R = \{(r, \theta): 0 \le g(\theta) \le r \le h(\theta), \alpha \le \theta \le \beta\}$

Figure 16.34

▶ For the type of region described in Theorem 16.4, with the boundaries in the radial direction expressed as functions of θ, the inner integral is always with respect to r.

▶ Recall from Section 12.2 that the polar equation $r = 2a \sin \theta$ describes a circle of radius $|a|$ with center $(0, a)$. The polar equation $r = 2a \cos \theta$ describes a circle of radius $|a|$ with center $(a, 0)$.

EXAMPLE 3   Annular region Find the volume of the region beneath the surface $z = xy + 10$ and above the annular region $R = \{(r, \theta): 2 \le r \le 4, 0 \le \theta \le 2\pi\}$. (An *annulus* is the region between two concentric circles.)

SOLUTION The region of integration suggests working in polar coordinates (**Figure 16.33**). Substituting $x = r \cos \theta$ and $y = r \sin \theta$, the integrand becomes

$$xy + 10 = (r \cos \theta)(r \sin \theta) + 10 \quad \text{Substitute for } x \text{ and } y.$$
$$= r^2 \sin \theta \cos \theta + 10 \quad\quad \text{Simplify.}$$
$$= \tfrac{1}{2} r^2 \sin 2\theta + 10. \quad\quad \sin 2\theta = 2 \sin \theta \cos \theta$$

Substituting the integrand into the volume integral, we have

$$V = \int_0^{2\pi} \int_2^4 \left(\tfrac{1}{2} r^2 \sin 2\theta + 10\right) r \, dr \, d\theta \quad \text{Iterated integral for volume}$$

$$= \int_0^{2\pi} \int_2^4 \left(\tfrac{1}{2} r^3 \sin 2\theta + 10r\right) dr \, d\theta \quad \text{Simplify.}$$

$$= \int_0^{2\pi} \left(\frac{r^4}{8} \sin 2\theta + 5r^2\right) \Big|_2^4 d\theta \quad \text{Evaluate inner integral with respect to } r.$$

$$= \int_0^{2\pi} (30 \sin 2\theta + 60) \, d\theta \quad \text{Simplify.}$$

$$= (15(-\cos 2\theta) + 60\theta) \Big|_0^{2\pi} = 120\pi. \quad \text{Evaluate outer integral with respect to } \theta.$$

Related Exercises 22, 38 ◀

More General Polar Regions

In Section 16.2 we generalized double integrals over rectangular regions to double integrals over nonrectangular regions. In an analogous way, the method for integrating over a polar rectangle may be extended to more general regions. Consider a region (described in polar coordinates) bounded by two rays $\theta = \alpha$ and $\theta = \beta$, where $\beta - \alpha \le 2\pi$, and two curves $r = g(\theta)$ and $r = h(\theta)$ (**Figure 16.34**):

$$R = \{(r, \theta): 0 \le g(\theta) \le r \le h(\theta), \alpha \le \theta \le \beta\}.$$

The double integral $\iint_R f(x, y) \, dA$ is expressed as an iterated integral in which the inner integral has limits $r = g(\theta)$ and $r = h(\theta)$, and the outer integral runs from $\theta = \alpha$ to $\theta = \beta$; the integrand is transformed into polar coordinates as before. If f is nonnegative on R, the double integral gives the volume of the solid bounded by the surface $z = f(x, y)$ and R.

> **THEOREM 16.4   Change of Variables for Double Integrals over More General Polar Regions**
>
> Let f be continuous on the region R in the xy-plane expressed in polar coordinates as
>
> $$R = \{(r, \theta): 0 \le g(\theta) \le r \le h(\theta), \alpha \le \theta \le \beta\},$$
>
> where $0 < \beta - \alpha \le 2\pi$. Then
>
> $$\iint_R f(x, y) \, dA = \int_\alpha^\beta \int_{g(\theta)}^{h(\theta)} f(r \cos \theta, r \sin \theta) \, r \, dr \, d\theta.$$

EXAMPLE 4   Specifying regions Write an iterated integral in polar coordinates for $\iint_R g(r, \theta) \, dA$ for the following regions R in the xy-plane.

a. The region outside the circle $r = 2$ (with radius 2 centered at $(0, 0)$) and inside the circle $r = 4 \cos \theta$ (with radius 2 centered at $(2, 0)$)

b. The region inside both circles of part (a)

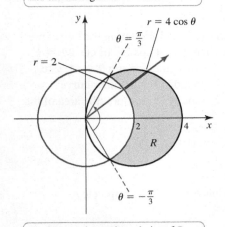

Radial lines enter the region R at $r = 2$ and exit the region at $r = 4 \cos \theta$.

The inner and outer boundaries of R are traversed as θ varies from $-\frac{\pi}{3}$ to $\frac{\pi}{3}$.

Figure 16.35

SOLUTION

a. Equating the two expressions for r, we have $4 \cos \theta = 2$ or $\cos \theta = \frac{1}{2}$, so the circles intersect when $\theta = \pm \pi/3$ (**Figure 16.35**). The inner boundary of R is the circle $r = 2$, and the outer boundary is the circle $r = 4 \cos \theta$. Therefore, the region of integration is $R = \{(r, \theta): 2 \leq r \leq 4 \cos \theta, -\pi/3 \leq \theta \leq \pi/3\}$ and the iterated integral is

$$\iint\limits_{R} g(r, \theta) \, dA = \int_{-\pi/3}^{\pi/3} \int_{2}^{4 \cos \theta} g(r, \theta) \, r \, dr \, d\theta.$$

b. From part (a), we know that the circles intersect when $\theta = \pm \pi/3$. The region R consists of three subregions R_1, R_2, and R_3 (**Figure 16.36a**).

- For $-\pi/2 \leq \theta \leq -\pi/3$, R_1 is bounded by $r = 0$ (inner curve) and $r = 4 \cos \theta$ (outer curve) (**Figure 16.36b**).
- For $-\pi/3 \leq \theta \leq \pi/3$, R_2 is bounded by $r = 0$ (inner curve) and $r = 2$ (outer curve) (**Figure 16.36c**).
- For $\pi/3 \leq \theta \leq \pi/2$, R_3 is bounded by $r = 0$ (inner curve) and $r = 4 \cos \theta$ (outer curve) (**Figure 16.36d**).

Therefore, the double integral is expressed in three parts:

$$\iint\limits_{R} g(r, \theta) \, dA = \int_{-\pi/2}^{-\pi/3} \int_{0}^{4 \cos \theta} g(r, \theta) \, r \, dr \, d\theta + \int_{-\pi/3}^{\pi/3} \int_{0}^{2} g(r, \theta) \, r \, dr \, d\theta$$

$$+ \int_{\pi/3}^{\pi/2} \int_{0}^{4 \cos \theta} g(r, \theta) \, r \, dr \, d\theta.$$

In R_1, radial lines begin at the origin and exit at $r = 4 \cos \theta$.

In R_2, radial lines begin at the origin and exit at $r = 2$.

In R_3, radial lines begin at the origin and exit at $r = 4 \cos \theta$.

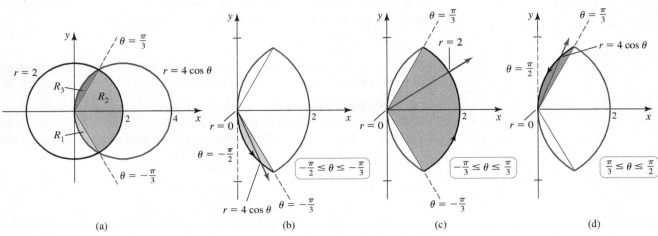

(a)   (b)   (c)   (d)

Figure 16.36

Related Exercise 44 ◄

Areas of Regions

In Cartesian coordinates, the area of a region R in the xy-plane is computed by integrating the function $f(x, y) = 1$ over R; that is, $A = \iint_{R} dA$. This fact extends to polar coordinates.

➤ Do not forget the factor of r in the area integral!

Area of Polar Regions

The area of the polar region $R = \{(r, \theta): 0 \leq g(\theta) \leq r \leq h(\theta), \alpha \leq \theta \leq \beta\}$, where $0 < \beta - \alpha \leq 2\pi$, is

$$A = \iint\limits_{R} dA = \int_{\alpha}^{\beta} \int_{g(\theta)}^{h(\theta)} r \, dr \, d\theta.$$

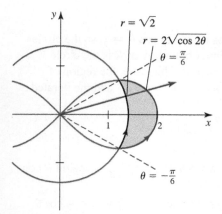

Figure 16.37

QUICK CHECK 4 Express the area of the disk $R = \{(r, \theta): 0 \leq r \leq a,\ 0 \leq \theta \leq 2\pi\}$ in terms of a double integral in polar coordinates. ◄

EXAMPLE 5   Area within a lemniscate Compute the area of the region in the first and fourth quadrants outside the circle $r = \sqrt{2}$ and inside the lemniscate $r^2 = 4\cos 2\theta$ (**Figure 16.37**).

SOLUTION The equation of the circle can be written as $r^2 = 2$. Equating the two expressions for r^2, the circle and the lemniscate intersect when $2 = 4\cos 2\theta$, or $\cos 2\theta = \frac{1}{2}$. The angles in the first and fourth quadrants that satisfy this equation are $\theta = \pm\pi/6$ (**Figure 16.37**). The region between the two curves is bounded by the inner curve $r = g(\theta) = \sqrt{2}$ and the outer curve $r = h(\theta) = 2\sqrt{\cos 2\theta}$. Therefore, the area of the region is

$$
\begin{aligned}
A &= \int_{-\pi/6}^{\pi/6} \int_{\sqrt{2}}^{2\sqrt{\cos 2\theta}} r\, dr\, d\theta \\[2mm]
&= \int_{-\pi/6}^{\pi/6} \left(\frac{r^2}{2}\right)\Bigg|_{\sqrt{2}}^{2\sqrt{\cos 2\theta}} d\theta && \text{Evaluate inner integral with respect to } r. \\[2mm]
&= \int_{-\pi/6}^{\pi/6} (2\cos 2\theta - 1)\, d\theta && \text{Simplify.} \\[2mm]
&= (\sin 2\theta - \theta)\Big|_{-\pi/6}^{\pi/6} && \text{Evaluate outer integral with respect to } \theta. \\[2mm]
&= \sqrt{3} - \frac{\pi}{3}. && \text{Simplify.}
\end{aligned}
$$

Related Exercises 50–51 ◄

Average Value over a Planar Polar Region

We have encountered the average value of a function in several different settings. To find the average value of a function over a region in polar coordinates, we again integrate the function over the region and divide by the area of the region.

EXAMPLE 6   Average y-coordinate Find the average value of the y-coordinates of the points in the semicircular disk of radius a given by $R = \{(r, \theta): 0 \leq r \leq a, 0 \leq \theta \leq \pi\}$.

SOLUTION The double integral that gives the average value we seek is $\bar{y} = \frac{1}{\text{area of } R} \iint_R y\, dA$. We use the facts that the area of R is $\pi a^2/2$ and the y-coordinates of points in the semicircular disk are given by $y = r\sin\theta$. Evaluating the average value integral, we find that

$$
\begin{aligned}
\bar{y} &= \frac{1}{\pi a^2/2} \int_0^{\pi} \int_0^a r\sin\theta\, r\, dr\, d\theta \\[2mm]
&= \frac{2}{\pi a^2} \int_0^{\pi} \sin\theta \left(\frac{r^3}{3}\right)\Bigg|_0^a d\theta && \text{Evaluate inner integral with respect to } r. \\[2mm]
&= \frac{2}{\pi a^2}\frac{a^3}{3} \int_0^{\pi} \sin\theta\, d\theta && \text{Simplify.} \\[2mm]
&= \frac{2a}{3\pi}(-\cos\theta)\Big|_0^{\pi} && \text{Evaluate outer integral with respect to } \theta. \\[2mm]
&= \frac{4a}{3\pi}. && \text{Simplify.}
\end{aligned}
$$

Note that $4/(3\pi) \approx 0.42$, so the average value of the y-coordinates is less than half the radius of the disk.

Related Exercise 53 ◄

SECTION 16.3 EXERCISES

Getting Started

1. Draw the polar region $\{(r, \theta): 1 \le r \le 2, 0 \le \theta \le \pi/2\}$. Why is it called a polar rectangle?

2. Write the double integral $\iint_R f(x, y) \, dA$ as an iterated integral in polar coordinates when $R = \{(r, \theta): a \le r \le b, \alpha \le \theta \le \beta\}$.

3. Sketch in the xy-plane the region of integration for the integral
$$\int_{-\pi/6}^{\pi/6} \int_{1/2}^{\cos 2\theta} g(r, \theta) \, r \, dr \, d\theta.$$

4. Explain why the element of area in Cartesian coordinates $dx \, dy$ becomes $r \, dr \, d\theta$ in polar coordinates.

5. How do you find the area of a polar region $R = \{(r, \theta): g(\theta) \le r \le h(\theta), \alpha \le \theta \le \beta\}$?

6. How do you find the average value of a function over a region that is expressed in polar coordinates?

7–10. Polar rectangles *Sketch the following polar rectangles.*

7. $R = \{(r, \theta): 0 \le r \le 5, 0 \le \theta \le \pi/2\}$

8. $R = \{(r, \theta): 2 \le r \le 3, \pi/4 \le \theta \le 5\pi/4\}$

9. $R = \{(r, \theta): 1 \le r \le 4, -\pi/4 \le \theta \le 2\pi/3\}$

10. $R = \{(r, \theta): 4 \le r \le 5, -\pi/3 \le \theta \le \pi/2\}$

Practice Exercises

11–14. Volume of solids *Find the volume of the solid bounded by the surface $z = f(x, y)$ and the xy-plane.*

11. $f(x, y) = 4 - \sqrt{x^2 + y^2}$

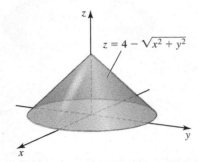

$z = 4 - \sqrt{x^2 + y^2}$

12. $f(x, y) = 16 - 4(x^2 + y^2)$ 13. $f(x, y) = e^{-(x^2+y^2)/8} - e^{-2}$

14. $f(x, y) = \dfrac{20}{1 + x^2 + y^2} - 2$

15–18. Solids bounded by paraboloids *Find the volume of the solid below the paraboloid $z = 4 - x^2 - y^2$ and above the following polar rectangles.*

15. $R = \{(r, \theta): 0 \le r \le 1,$
$0 \le \theta \le 2\pi\}$

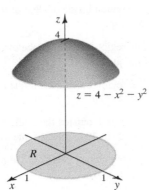

$z = 4 - x^2 - y^2$

R

16. $R = \{(r, \theta): 0 \le r \le 2, 0 \le \theta \le 2\pi\}$

17. $R = \{(r, \theta): 1 \le r \le 2, 0 \le \theta \le 2\pi\}$

18. $R = \{(r, \theta): 1 \le r \le 2, -\pi/2 \le \theta \le \pi/2\}$

19–20. Solids bounded by hyperboloids *Find the volume of the solid below the hyperboloid $z = 5 - \sqrt{1 + x^2 + y^2}$ and above the following polar rectangles.*

19. $R = \{(r, \theta): \sqrt{3} \le r \le 2\sqrt{2}, 0 \le \theta \le 2\pi\}$

20. $R = \{(r, \theta): \sqrt{3} \le r \le \sqrt{15}, -\pi/2 \le \theta \le \pi\}$

21–30. Cartesian to polar coordinates *Evaluate the following integrals using polar coordinates. Assume (r, θ) are polar coordinates. A sketch is helpful.*

21. $\displaystyle\iint_R (x^2 + y^2) \, dA; R = \{(r, \theta): 0 \le r \le 4, 0 \le \theta \le 2\pi\}$

22. $\displaystyle\iint_R 2xy \, dA; R = \{(r, \theta): 1 \le r \le 3, 0 \le \theta \le \pi/2\}$

23. $\displaystyle\iint_R 2xy \, dA; R = \{(x, y): x^2 + y^2 \le 9, y \ge 0\}$

24. $\displaystyle\iint_R \frac{dA}{1 + x^2 + y^2}; R = \{(r, \theta): 1 \le r \le 2, 0 \le \theta \le \pi\}$

25. $\displaystyle\iint_R \frac{dA}{\sqrt{16 - x^2 - y^2}}; R = \{(x, y): x^2 + y^2 \le 4, x \ge 0, y \ge 0\}$

26. $\displaystyle\iint_R e^{-x^2-y^2} \, dA; R = \{(x, y): x^2 + y^2 \le 9\}$

27. $\displaystyle\int_{-1}^{1} \int_{-\sqrt{1-x^2}}^{\sqrt{1-x^2}} (x^2 + y^2)^{3/2} \, dy \, dx$

28. $\displaystyle\int_{0}^{3} \int_{0}^{\sqrt{9-x^2}} \sqrt{x^2 + y^2} \, dy \, dx$

29. $\displaystyle\iint_R \sqrt{x^2 + y^2} \, dA; R = \{(x, y): 1 \le x^2 + y^2 \le 4\}$

30. $\displaystyle\int_{-4}^{4} \int_{0}^{\sqrt{16-y^2}} (16 - x^2 - y^2) \, dx \, dy$

31–40. Volume between surfaces *Find the volume of the following solids.*

31. The solid bounded by the paraboloid $z = x^2 + y^2$ and the plane $z = 9$

$z = 9$

$z = x^2 + y^2$

32. The solid bounded by the paraboloid $z = 2 - x^2 - y^2$ and the plane $z = 1$

33. The solid bounded by the paraboloids $z = x^2 + y^2$ and $z = 2 - x^2 - y^2$

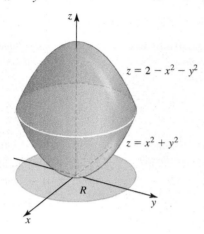

$z = 2 - x^2 - y^2$

$z = x^2 + y^2$

R

34. The solid bounded by the paraboloids $z = 2x^2 + y^2$ and $z = 27 - x^2 - 2y^2$

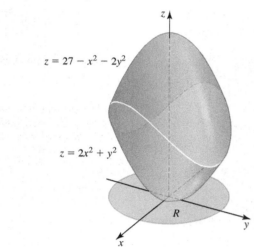

$z = 27 - x^2 - 2y^2$

$z = 2x^2 + y^2$

R

35. The solid bounded below by the paraboloid $z = x^2 + y^2 - x - y$ and above by the plane $x + y + z = 4$

36. The solid bounded by the cylinder $x^2 + y^2 = 4$ and the planes $z = 3 - x$ and $z = x - 3$

37. The solid bounded by the paraboloid $z = 18 - x^2 - 3y^2$ and the hyperbolic paraboloid $z = x^2 - y^2$

38. The solid outside the cylinder $x^2 + y^2 = 1$ that is bounded above by the hyperbolic paraboloid $z = -x^2 + y^2 + 8$ and below by the paraboloid $z = x^2 + 3y^2$

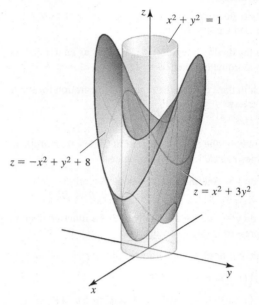

$x^2 + y^2 = 1$

$z = -x^2 + y^2 + 8$

$z = x^2 + 3y^2$

39. The solid outside the cylinder $x^2 + y^2 = 1$ that is bounded above by the sphere $x^2 + y^2 + z^2 = 8$ and below by the cone $z = \sqrt{x^2 + y^2}$

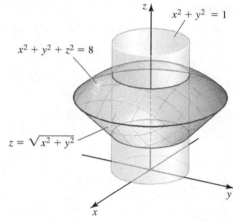

$x^2 + y^2 = 1$

$x^2 + y^2 + z^2 = 8$

$z = \sqrt{x^2 + y^2}$

40. The solid bounded by the cone $z = 2 - \sqrt{x^2 + y^2}$ and the upper half of a hyperboloid of two sheets $z = \sqrt{1 + x^2 + y^2}$

41–46. Describing general regions *Sketch the following regions R. Then express $\iint_R g(r, \theta)\, dA$ as an iterated integral over R in polar coordinates.*

41. The region inside the limaçon $r = 1 + \frac{1}{2}\cos \theta$

42. The region inside the leaf of the rose $r = 2 \sin 2\theta$ in the first quadrant

43. The region inside the lobe of the lemniscate $r^2 = 2 \sin 2\theta$ in the first quadrant

44. The region outside the circle $r = 2$ and inside the circle $r = 4 \sin \theta$

45. The region outside the circle $r = 1$ and inside the rose $r = 2 \sin 3\theta$ in the first quadrant

46. The region outside the circle $r = 1/2$ and inside the cardioid $r = 1 + \cos \theta$

47–52. Computing areas *Use a double integral to find the area of the following regions.*

47. The annular region $\{(r, \theta): 1 \le r \le 2, 0 \le \theta \le \pi\}$

48. The region bounded by the cardioid $r = 2(1 - \sin \theta)$

49. The region bounded by all leaves of the rose $r = 2 \cos 3\theta$

50. The region inside both the cardioid $r = 1 - \cos \theta$ and the circle $r = 1$

51. The region inside both the cardioid $r = 1 + \sin \theta$ and the cardioid $r = 1 + \cos \theta$

52. The region bounded by the spiral $r = 2\theta$, for $0 \le \theta \le \pi$, and the x-axis

53–54. Average values *Find the following average values.*

53. The average distance between points of the disk $\{(r, \theta): 0 \le r \le a\}$ and the origin

54. The average value of $1/r^2$ over the annulus $\{(r, \theta): 2 \le r \le 4\}$

55. **Explain why or why not** Determine whether the following statements are true and give an explanation or counterexample.

   a. Let R be the unit disk centered at $(0, 0)$. Then $\iint_R (x^2 + y^2)\,dA = \int_0^{2\pi}\int_0^1 r^2\,dr\,d\theta$.

   b. The average distance between the points of the hemisphere $z = \sqrt{4 - x^2 - y^2}$ and the origin is 2 (calculus not required).

   c. The integral $\int_0^1 \int_0^{\sqrt{1-y^2}} e^{x^2+y^2}\,dx\,dy$ is easier to evaluate in polar coordinates than in Cartesian coordinates.

56. **Areas of circles** Use integration to show that the circles $r = 2a \cos \theta$ and $r = 2a \sin \theta$ have the same area, which is πa^2.

57. **Filling bowls with water** Which bowl holds the most water when all the bowls are filled to a depth of 4 units?

   • The paraboloid $z = x^2 + y^2$, for $0 \le z \le 4$

   • The cone $z = \sqrt{x^2 + y^2}$, for $0 \le z \le 4$

   • The hyperboloid $z = \sqrt{1 + x^2 + y^2}$, for $1 \le z \le 5$

58. **Equal volumes** To what height (above the bottom of the bowl) must the cone and paraboloid bowls of Exercise 57 be filled to hold the same volume of water as the hyperboloid bowl filled to a depth of 4 units $(1 \le z \le 5)$?

59. **Volume of a hyperbolic paraboloid** Consider the surface $z = x^2 - y^2$.

   a. Find the region in the xy-plane in polar coordinates for which $z \ge 0$.

   b. Let $R = \{(r, \theta): 0 \le r \le a, -\pi/4 \le \theta \le \pi/4\}$, which is a sector of a circle of radius a. Find the volume of the region below the hyperbolic paraboloid and above the region R.

60. **Volume of a sphere** Use double integrals in polar coordinates to verify that the volume of a sphere of radius a is $\frac{4}{3}\pi a^3$.

61. **Volume** Find the volume of the solid bounded by the cylinder $(x - 1)^2 + y^2 = 1$, the plane $z = 0$, and the cone $z = \sqrt{x^2 + y^2}$ (see figure). (*Hint:* Use symmetry.)

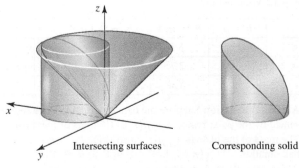

Intersecting surfaces   Corresponding solid

62. **Volume** Find the volume of the solid bounded by the paraboloid $z = 2x^2 + 2y^2$, the plane $z = 0$, and the cylinder $x^2 + (y - 1)^2 = 1$. (*Hint:* Use symmetry.)

Explorations and Challenges

63–64. Miscellaneous integrals *Evaluate the following integrals using the method of your choice. A sketch is helpful.*

63. $\displaystyle\iint_R \frac{dA}{4 + \sqrt{x^2 + y^2}}; R = \left\{(r, \theta): 0 \le r \le 2, \frac{\pi}{2} \le \theta \le \frac{3\pi}{2}\right\}$

64. $\displaystyle\iint_R \frac{x - y}{x^2 + y^2 + 1}\,dA; R$ is the region bounded by the unit circle centered at the origin.

65–68. Improper integrals *Improper integrals arise in polar coordinates when the radial coordinate r becomes arbitrarily large. Under certain conditions, these integrals are treated in the usual way:*

$$\int_\alpha^\beta \int_a^\infty g(r, \theta)\, r\, dr\, d\theta = \lim_{b \to \infty} \int_\alpha^\beta \int_a^b g(r, \theta)\, r\, dr\, d\theta.$$

Use this technique to evaluate the following integrals.

65. $\displaystyle\int_0^{\pi/2} \int_1^\infty \frac{\cos \theta}{r^3}\, r\, dr\, d\theta$

66. $\displaystyle\iint_R \frac{dA}{(x^2 + y^2)^{5/2}}; R = \{(r, \theta): 1 \le r < \infty, 0 \le \theta \le 2\pi\}$

67. $\displaystyle\iint_R e^{-x^2 - y^2}\,dA; R = \left\{(r, \theta): 0 \le r < \infty, 0 \le \theta \le \frac{\pi}{2}\right\}$

68. $\displaystyle\iint_R \frac{dA}{(1 + x^2 + y^2)^2}; R$ is the first quadrant.

T 69. **Slicing a hemispherical cake** A cake is shaped like a hemisphere of radius 4 with its base on the xy-plane. A wedge of the cake is removed by making two slices from the center of the cake outward, perpendicular to the xy-plane and separated by an angle of φ.

   a. Use a double integral to find the volume of the slice for $\varphi = \pi/4$. Use geometry to check your answer.

   b. Now suppose the cake is sliced horizontally at $z = a > 0$ and let D be the piece of cake above the plane $z = a$. For what approximate value of a is the volume of D equal to the volume in part (a)?

T 70. Mass from density data The following table gives the density (in units of g/cm²) at selected points (in polar coordinates) of a thin semicircular plate of radius 3. Estimate the mass of the plate and explain your method.

	$\theta = 0$	$\theta = \pi/4$	$\theta = \pi/2$	$\theta = 3\pi/4$	$\theta = \pi$
$r = 1$	2.0	2.1	2.2	2.3	2.4
$r = 2$	2.5	2.7	2.9	3.1	3.3
$r = 3$	3.2	3.4	3.5	3.6	3.7

71. A mass calculation Suppose the density of a thin plate represented by the polar region R is $\rho(r, \theta)$ (in units of mass per area). The mass of the plate is $\iint_R \rho(r, \theta)\, dA$. Find the mass of the thin half annulus $R = \{(r, \theta): 1 \le r \le 4, 0 \le \theta \le \pi\}$ with a density $\rho(r, \theta) = 4 + r\sin\theta$.

72. Area formula In Section 12.3 it was shown that the area of a region enclosed by the polar curve $r = g(\theta)$ and the rays $\theta = \alpha$ and $\theta = \beta$, where $\beta - \alpha \le 2\pi$, is $A = \frac{1}{2}\int_\alpha^\beta r^2\, d\theta$. Prove this result using the area formula with double integrals.

73. Normal distribution An important integral in statistics associated with the normal distribution is $I = \int_{-\infty}^{\infty} e^{-x^2}\, dx$. It is evaluated in the following steps.

a. In Section 8.9, it is shown that $\int_0^{\infty} e^{-x^2}\, dx$ converges (in the narrative following Example 7). Use this result to explain why $\int_{-\infty}^{\infty} e^{-x^2}\, dx$ converges.

b. Assume

$$I^2 = \left(\int_{-\infty}^{\infty} e^{-x^2}\, dx\right)\left(\int_{-\infty}^{\infty} e^{-y^2}\, dy\right) = \int_{-\infty}^{\infty}\int_{-\infty}^{\infty} e^{-x^2 - y^2}\, dx\, dy,$$

where we have chosen the variables of integration to be x and y and then written the product as an iterated integral. Evaluate this integral in polar coordinates and show that $I = \sqrt{\pi}$. Why is the solution $I = -\sqrt{\pi}$ rejected?

c. Evaluate $\int_0^{\infty} e^{-x^2}\, dx$, $\int_0^{\infty} xe^{-x^2}\, dx$, and $\int_0^{\infty} x^2 e^{-x^2}\, dx$ (using part (a) if needed).

74. Existence of integrals For what values of p does the integral

$$\iint_R \frac{dA}{(x^2 + y^2)^p}$$

exist in the following cases? Assume (r, θ) are polar coordinates.

a. $R = \{(r, \theta): 1 \le r < \infty, 0 \le \theta \le 2\pi\}$
b. $R = \{(r, \theta): 0 \le r \le 1, 0 \le \theta \le 2\pi\}$

75. Integrals in strips Consider the integral

$$I = \iint_R \frac{dA}{(1 + x^2 + y^2)^2}$$

where $R = \{(x, y): 0 \le x \le 1, 0 \le y \le a\}$.

a. Evaluate I for $a = 1$. (*Hint:* Use polar coordinates.)
b. Evaluate I for arbitrary $a > 0$.
c. Let $a \to \infty$ in part (b) to find I over the infinite strip $R = \{(x, y): 0 \le x \le 1, 0 \le y < \infty\}$.

QUICK CHECK ANSWERS

1. $R = \{(r, \theta): 1 \le r \le 2, 0 \le \theta \le \pi/2\}$
2. $r^5, r^2(\cos^2\theta - \sin^2\theta) = r^2\cos 2\theta$
3. $z = 2 - \sqrt{x^2 + y^2}$ is the lower half of the double-napped cone $(2 - z)^2 = x^2 + y^2$. Imagine both halves of this cone in Figure 16.32: It is apparent that the paraboloid $z = x^2 + y^2$ intersects the cone twice, once when $z = 1$ and once when $z = 4$.   4. $\int_0^{2\pi}\int_0^a r\, dr\, d\theta = \pi a^2$ ◄

16.4 Triple Integrals

At this point, you may see a pattern that is developing with respect to integration. In Chapter 5, we introduced integrals of single-variable functions. In the first three sections of this chapter, we moved up one dimension to double integrals of two-variable functions. In this section, we take another step and investigate triple integrals of three-variable functions. There is no end to the progression of multiple integrals. It is possible to define integrals with respect to any number of variables. For example, problems in statistics and statistical mechanics involve integration over regions of many dimensions.

Triple Integrals in Rectangular Coordinates

Consider a function $w = f(x, y, z)$ that is defined on a closed and bounded region D of $\mathbb{R}^3$. The graph of f lies in four-dimensional space and is the set of points $(x, y, z, f(x, y, z))$, where (x, y, z) is in D. Despite the difficulty in representing f in $\mathbb{R}^3$, we may still define the integral of f over D. We first create a partition of D by slicing the region with three sets of planes that run parallel to the xz-, yz-, and xy-planes (**Figure 16.38**). This partition subdivides D into small boxes that are ordered in a convenient way from $k = 1$ to $k = n$. The partition includes all boxes that are wholly contained in D. The kth box has side lengths Δx_k, Δy_k, and Δz_k, and volume $\Delta V_k = \Delta x_k \Delta y_k \Delta z_k$. We let (x_k^*, y_k^*, z_k^*) be an arbitrary point in the kth box, for $k = 1, \ldots, n$.

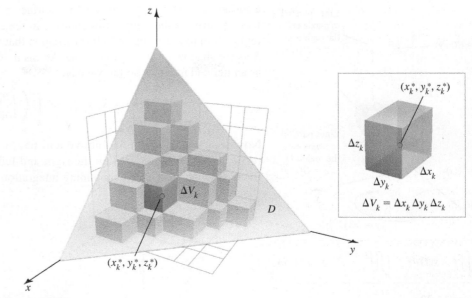

Figure 16.38

A Riemann sum is now formed, in which the kth term is the function value $f(x_k^*, y_k^*, z_k^*)$ multiplied by the volume of the kth box:

$$\sum_{k=1}^{n} f(x_k^*, y_k^*, z_k^*)\Delta V_k.$$

We let Δ denote the maximum length of the diagonals of the boxes. As the number of boxes n increases, while Δ approaches zero, two things happen.

- For commonly encountered regions, the region formed by the collection of boxes approaches the region D.
- If f is continuous, the Riemann sum approaches a limit.

The limit of the Riemann sum is the **triple integral of f over D**, and we write

$$\iiint\limits_{D} f(x, y, z)\, dV = \lim_{\Delta \to 0} \sum_{k=1}^{n} f(x_k^*, y_k^*, z_k^*)\Delta V_k.$$

The kth box in the partition has volume $\Delta V_k = \Delta x_k \Delta y_k \Delta z_k$, where Δx_k, Δy_k, and Δz_k are the side lengths of the box. Accordingly, the *element of volume* in the triple integral, which we denote dV, becomes $dx\, dy\, dz$ (or some rearrangement of dx, dy, and dz) in an iterated integral.

We give two immediate interpretations of a triple integral. First, if $f(x, y, z) = 1$, then the Riemann sum simply adds up the volumes of the boxes in the partition. In the limit as $\Delta \to 0$, the triple integral $\iiint_D dV$ gives the volume of the region D. Second, suppose D is a solid three-dimensional object and its density varies from point to point according to the function $f(x, y, z)$. The units of density are mass per unit volume, so the product $f(x_k^*, y_k^*, z_k^*)\Delta V_k$ approximates the mass of the kth box in D. Summing the masses of the boxes gives an approximation to the total mass of D. In the limit as $\Delta \to 0$, the triple integral gives the mass of the object.

As with double integrals, a version of Fubini's Theorem expresses a triple integral in terms of an iterated integral in x, y, and z. The situation becomes interesting because with three variables, there are *six* possible orders of integration.

▶ Notice the analogy between double and triple integrals:

$$\text{area of } R = \iint\limits_{R} dA \quad \text{and}$$

$$\text{volume of } D = \iiint\limits_{D} dV.$$

The use of triple integrals to compute the mass of an object is discussed in detail in Section 16.6.

QUICK CHECK 1 List the six orders in which the three differentials dx, dy, and dz may be written. ◀

Finding Limits of Integration   We discuss one of the six orders of integration in detail; the others are examined in the examples. Suppose a region D in $\mathbb{R}^3$ is bounded above by

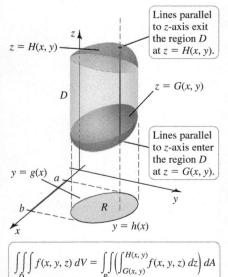

$$\iiint\limits_{D} f(x, y, z)\, dV = \iint\limits_{R}\left(\int_{G(x,y)}^{H(x,y)} f(x, y, z)\, dz\right) dA$$

Figure 16.39

a surface $z = H(x, y)$ and below by a surface $z = G(x, y)$ (**Figure 16.39**). These two surfaces determine the limits of integration in the z-direction. The next step is to project the region D onto the xy-plane to form a region that we call R (**Figure 16.40**). You can think of R as the shadow of D in the xy-plane. At this point, we can begin to write the triple integral as an iterated integral. So far, we have

$$\iiint\limits_{D} f(x, y, z)\, dV = \iint\limits_{R}\left(\int_{G(x,y)}^{H(x,y)} f(x, y, z)\, dz\right) dA.$$

Now assume R is bounded above and below by the curves $y = h(x)$ and $y = g(x)$, respectively, and bounded on the right and left by the lines $x = a$ and $x = b$, respectively (Figure 16.40). The remaining integration over R is carried out as a double integral (Section 16.2).

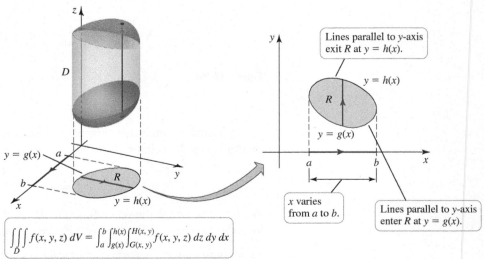

$$\iiint\limits_{D} f(x, y, z)\, dV = \int_{a}^{b}\int_{g(x)}^{h(x)}\int_{G(x,y)}^{H(x,y)} f(x, y, z)\, dz\, dy\, dx$$

Figure 16.40

Table 16.2

Integral	Variable	Interval
Inner	z	$G(x, y) \le z \le H(x, y)$
Middle	y	$g(x) \le y \le h(x)$
Outer	x	$a \le x \le b$

The intervals that describe D are summarized in Table 16.2, which can then be used to formulate the limits of integration. To integrate over all points of D, we carry out the following steps.

1. Integrate with respect to z from $z = G(x, y)$ to $z = H(x, y)$; the result (in general) is a function of x and y.

2. Integrate with respect to y from $y = g(x)$ to $y = h(x)$; the result (in general) is a function of x.

3. Integrate with respect to x from $x = a$ to $x = b$; the result is (always) a real number.

▶ Theorem 16.5 is a version of Fubini's Theorem. Five other versions could be written for the other orders of integration.

THEOREM 16.5   Triple Integrals

Let f be continuous over the region

$$D = \{(x, y, z): a \le x \le b, g(x) \le y \le h(x), G(x, y) \le z \le H(x, y)\},$$

where g, h, G, and H are continuous functions. Then f is integrable over D and the triple integral is evaluated as the iterated integral

$$\iiint\limits_{D} f(x, y, z)\, dV = \int_{a}^{b}\int_{g(x)}^{h(x)}\int_{G(x,y)}^{H(x,y)} f(x, y, z)\, dz\, dy\, dx.$$

We now illustrate this procedure with several examples.

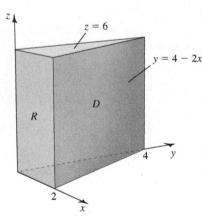

R is the base of the box in the xy-plane.

z varies from 0 to 1.

y varies from 0 to 2.

x varies from 0 to 3.

$$M = \int_0^3 \int_0^2 \int_0^1 (2 - z)\, dz\, dy\, dx$$

Figure 16.41

Table 16.3

Integral	Variable	Interval
Inner	z	$0 \le z \le 1$
Middle	y	$0 \le y \le 2$
Outer	x	$0 \le x \le 3$

QUICK CHECK 2 Write the integral in Example 1 in the orders $dx\, dy\, dz$ and $dx\, dz\, dy$. ◄

EXAMPLE 1 Mass of a box A solid box D is bounded by the planes $x = 0$, $x = 3$, $y = 0$, $y = 2$, $z = 0$, and $z = 1$. The density of the box decreases linearly in the positive z-direction and is given by $f(x, y, z) = 2 - z$. Find the mass of the box.

SOLUTION The mass of the box is found by integrating the density $f(x, y, z) = 2 - z$ over the box. Because the limits of integration for all three variables are constant, the iterated integral may be written in any order. Using the order of integration $dz\, dy\, dx$ (**Figure 16.41**), the limits of integration are shown in Table 16.3.

The mass of the box is

$$
\begin{aligned}
M &= \iiint_D (2 - z)\, dV \\
&= \int_0^3 \int_0^2 \int_0^1 (2 - z)\, dz\, dy\, dx &&\text{Convert to an iterated integral.} \\
&= \int_0^3 \int_0^2 \left(2z - \frac{z^2}{2} \right)\Big|_0^1 dy\, dx &&\text{Evaluate inner integral with respect to } z. \\
&= \int_0^3 \int_0^2 \frac{3}{2}\, dy\, dx &&\text{Simplify.} \\
&= \int_0^3 \left(\frac{3y}{2} \right)\Big|_0^2 dx &&\text{Evaluate middle integral with respect to } y. \\
&= \int_0^3 3\, dx = 9. &&\text{Evaluate outer integral with respect to } x \text{ and simplify.}
\end{aligned}
$$

The result makes sense: The density of the box varies linearly from 1 (at the top of the box) to 2 (at the bottom); if the box had a constant density of 1, its mass would be (volume) × (density) = 6; if the box had a constant density of 2, its mass would be 12. The actual mass is the average of 6 and 12, as you might expect.

Any other order of integration produces the same result. For example, with the order $dy\, dx\, dz$, the iterated integral is

$$M = \iiint_D (2 - z)\, dV = \int_0^1 \int_0^3 \int_0^2 (2 - z)\, dy\, dx\, dz = 9.$$

Related Exercises 8–9 ◄

z = 6

y = 4 − 2x

D

R

4   y

2

x

Figure 16.42

EXAMPLE 2 Volume of a prism Find the volume of the prism D in the first octant bounded by the planes $y = 4 - 2x$ and $z = 6$ (**Figure 16.42**).

SOLUTION The prism may be viewed in several different ways. Letting the base of the prism be in the xz-plane, the upper surface of the prism is the plane $y = 4 - 2x$, and the lower surface is $y = 0$. The projection of the prism onto the xz-plane is the rectangle $R = \{(x, z): 0 \le x \le 2, 0 \le z \le 6\}$. One possible order of integration in this case is $dy\, dx\, dz$.

Inner integral with respect to y: A line through the prism parallel to the y-axis enters the prism through the rectangle R at $y = 0$ and exits the prism at the plane $y = 4 - 2x$. Therefore, we first integrate with respect to y over the interval $0 \le y \le 4 - 2x$ (**Figure 16.43a**).

Middle integral with respect to x: The limits of integration for the middle and outer integrals must cover the region R in the xz-plane. A line parallel to the x-axis enters R at $x = 0$ and exits R at $x = 2$. So we integrate with respect to x over the interval $0 \le x \le 2$ (**Figure 16.43b**).

Outer integral with respect to z: To cover all of R, the line segments from $x = 0$ to $x = 2$ must run from $z = 0$ to $z = 6$. So we integrate with respect to z over the interval $0 \le z \le 6$ (**Figure 16.43b**).

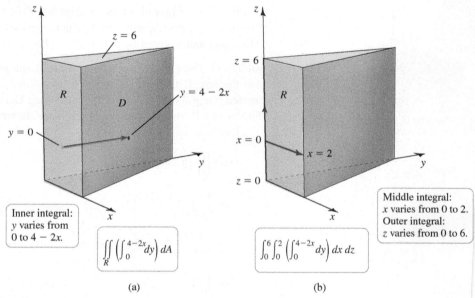

Inner integral:
y varies from
0 to 4 − 2x.

$$\iint_R \left(\int_0^{4-2x} dy \right) dA$$

(a)

Middle integral:
x varies from 0 to 2.
Outer integral:
z varies from 0 to 6.

$$\int_0^6 \int_0^2 \left(\int_0^{4-2x} dy \right) dx\, dz$$

(b)

Figure 16.43

Integrating $f(x, y, z) = 1$, the volume of the prism is

$$V = \iiint_D dV = \int_0^6 \int_0^2 \int_0^{4-2x} dy\, dx\, dz$$

▶ The volume of the prism could also be found using geometry: The area of the triangular base in the xy-plane is 4 and the height of the prism is 6. Therefore, the volume is $6 \cdot 4 = 24$.

$$= \int_0^6 \int_0^2 (4 - 2x)\, dx\, dz \qquad \text{Evaluate inner integral with respect to } y.$$

$$= \int_0^6 (4x - x^2) \Big|_0^2 dz \qquad \text{Evaluate middle integral with respect to } x.$$

QUICK CHECK 3 Write the integral in Example 2 in the orders $dz\, dy\, dx$ and $dx\, dy\, dz$. ◀

$$= \int_0^6 4\, dz \qquad \text{Simplify.}$$

$$= 24. \qquad \text{Evaluate outer integral with respect to } z.$$

Related Exercises 15, 18 ◀

EXAMPLE 3   A volume integral Find the volume of the solid D bounded by the paraboloids $y = x^2 + 3z^2 + 1$ and $y = 5 - 3x^2 - z^2$ (**Figure 16.44a**).

SOLUTION The right boundary of D is the surface $y = 5 - 3x^2 - z^2$ and the left boundary is $y = x^2 + 3z^2 + 1$. These surfaces are functions of x and z, so they determine the limits of integration for the inner integral in the y-direction.

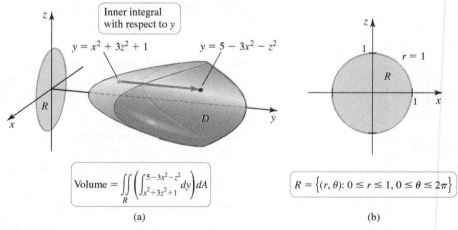

Inner integral with respect to y

$y = x^2 + 3z^2 + 1$

$y = 5 - 3x^2 - z^2$

$$\text{Volume} = \iint_R \left(\int_{x^2+3z^2+1}^{5-3x^2-z^2} dy \right) dA$$

(a)

$r = 1$

$$R = \left\{ (r, \theta) : 0 \le r \le 1, 0 \le \theta \le 2\pi \right\}$$

(b)

Figure 16.44

A key step in the calculation is finding the curve of intersection between the two surfaces and projecting it onto the xz-plane to form the boundary of the region R, where R is the projection of D onto the xz-plane. Equating the y-coordinates of the surfaces, we have $x^2 + 3z^2 + 1 = 5 - 3x^2 - z^2$, which, when simplified, is the equation of a unit circle centered at the origin in the xz-plane:

$$x^2 + z^2 = 1.$$

Observe that a line through the solid parallel to the y-axis enters the solid at $y = x^2 + 3z^2 + 1$ and exits at $y = 5 - 3x^2 - z^2$. Therefore, for fixed values of x and z, we integrate in y over the interval $x^2 + 3z^2 + 1 \le y \le 5 - 3x^2 - z^2$ (Figure 16.44a). After evaluating the inner integral with respect to y, we have

$$V = \iint\limits_R \left(\int_{x^2+3z^2+1}^{5-3x^2-z^2} dy \right) dA = \iint\limits_R \left(y \Big|_{x^2+3z^2+1}^{5-3x^2-z^2} \right) dA = \iint\limits_R 4(1 - x^2 - z^2)\, dA.$$

The region R is bounded by a circle, so it is advantageous to evaluate the remaining double integral in polar coordinates, where θ and r have the same meaning in the xz-plane as they do in the xy-plane. Note that R is expressed in polar coordinates as $R = \{(r, \theta): 0 \le r \le 1, 0 \le \theta \le 2\pi\}$ (**Figure 16.44b**). Using the relationship $x^2 + z^2 = r^2$, we change variables and evaluate the double integral:

$$
\begin{aligned}
V &= \iint\limits_R 4(1 - x^2 - z^2)\, dA \\
&= \int_0^{2\pi} \int_0^1 4(1 - r^2)\, r\, dr\, d\theta &&\text{Convert to polar coordinates: } x^2 + z^2 = r^2 \\
& &&\text{and } dA = r\, dr\, d\theta. \\
&= \int_0^{2\pi} \int_0^1 2u\, du\, d\theta &&\text{Let } u = 1 - r^2 \Rightarrow du = -2r\, dr. \\
&= \int_0^{2\pi} \left(u^2 \Big|_0^1 \right) d\theta &&\text{Evaluate inner integral.} \\
&= \int_0^{2\pi} d\theta = 2\pi. &&\text{Evaluate outer integral.}
\end{aligned}
$$

Related Exercises 23, 25 ◄

Changing the Order of Integration

As with double integrals, choosing an appropriate order of integration may simplify the evaluation of a triple integral. Therefore, it is important to become proficient at changing the order of integration.

EXAMPLE 4 Changing the order of integration Consider the integral

$$\int_0^{\sqrt[4]{\pi}} \int_0^z \int_y^z 12y^2 z^3 \sin x^4 \, dx\, dy\, dz.$$

a. Sketch the region of integration D.

b. Evaluate the integral by changing the order of integration.

SOLUTION

a. We begin by finding the projection of the region of integration D on the appropriate coordinate plane; call the projection R. Because the inner integration is with respect to x, R lies in the yz-plane, and it is determined by the limits on the middle and outer integrals. We see that

$$R = \{(y, z): 0 \le y \le z, 0 \le z \le \sqrt[4]{\pi}\},$$

which is a triangular region in the yz-plane bounded by the z-axis and the lines $y = z$ and $z = \sqrt[4]{\pi}$. Using the limits on the inner integral, for each point in R we let x vary from the plane $x = y$ to the plane $x = z$. In so doing, the points fill an inverted tetrahedron in the first octant with its vertex at the origin, which is D (**Figure 16.45**).

b. It is difficult to evaluate the integral in the given order $(dx\, dy\, dz)$ because the antiderivative of $\sin x^4$ is not expressible in terms of elementary functions. If we integrate first with respect to y, we introduce a factor in the integrand that enables us to use a substitution

▶ How do we know to switch the order of integration so the inner integral is with respect to y? Often we do not know in advance whether a new order of integration will work, and some trial and error is needed. In this case, either y^2 or z^3 is easier to integrate than $\sin x^4$, so either y or z is a likely variable for the inner integral.

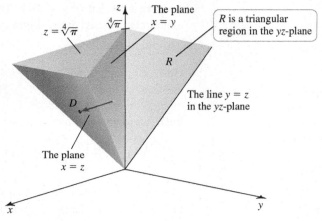

Figure 16.45

to integrate $\sin x^4$. With the order of integration $dy\,dx\,dz$, the bounds of integration for the inner integral extend from the plane $y = 0$ to the plane $y = x$ (**Figure 16.46a**). Furthermore, the projection of D onto the xz-plane is the region R, which must be covered by the middle and outer integrals (**Figure 16.46b**). In this case, we draw a line segment parallel to the x-axis to see that the limits of the middle integral run from $x = 0$ to $x = z$. Then we include all these segments from $z = 0$ to $z = \sqrt[4]{\pi}$ to obtain the outer limits of integration in z. The integration proceeds as follows:

$$\int_0^{\sqrt[4]{\pi}} \int_0^z \int_0^x 12y^2z^3 \sin x^4 \, dy\,dx\,dz = \int_0^{\sqrt[4]{\pi}} \int_0^z (4y^3z^3 \sin x^4)\Big|_0^x dx\,dz \qquad \text{Evaluate inner integral with respect to } y.$$

$$= \int_0^{\sqrt[4]{\pi}} \int_0^z 4x^3z^3 \sin x^4 \, dx\,dz \qquad \text{Simplify.}$$

$$= \int_0^{\sqrt[4]{\pi}} z^3(-\cos x^4)\Big|_0^z dz \qquad \begin{array}{l}\text{Evaluate middle}\\\text{integral with respect to } x; u = x^4.\end{array}$$

$$= \int_0^{\sqrt[4]{\pi}} z^3(1 - \cos z^4)\,dz \qquad \text{Simplify.}$$

$$= \left(\frac{z^4}{4} - \frac{\sin z^4}{4}\right)\Big|_0^{\sqrt[4]{\pi}} \qquad \begin{array}{l}\text{Evaluate outer}\\\text{integral with respect to } z; u = z^4.\end{array}$$

$$= \frac{\pi}{4}. \qquad \text{Simplify.}$$

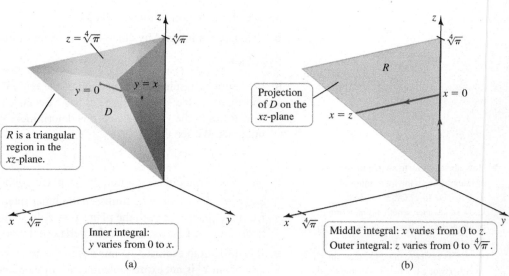

Figure 16.46

Related Exercises 47, 56 ◄

Average Value of a Function of Three Variables

The idea of the average value of a function extends naturally from the one- and two-variable cases. The average value of a function of three variables is found by integrating the function over the region of interest and dividing by the volume of the region.

DEFINITION Average Value of a Function of Three Variables

If f is continuous on a region D of $\mathbb{R}^3$, then the **average value** of f over D is

$$\bar{f} = \frac{1}{\text{volume of } D} \iiint_D f(x, y, z)\, dV.$$

EXAMPLE 5 Average temperature Consider a block of a conducting material occupying the region

$$D = \{(x, y, z): 0 \le x \le 2, 0 \le y \le 2, 0 \le z \le 1\}.$$

Due to heat sources on its boundaries, the temperature in the block is given by $T(x, y, z) = 250xy \sin \pi z$. Find the average temperature of the block.

SOLUTION We must integrate the temperature function over the block and divide by the volume of the block, which is 4. One way to evaluate the temperature integral is as follows:

$$\iiint_D 250xy \sin \pi z\, dV = 250 \int_0^2 \int_0^2 \int_0^1 xy \sin \pi z\, dz\, dy\, dx \qquad \text{Convert to an iterated integral.}$$

$$= 250 \int_0^2 \int_0^2 xy \frac{1}{\pi} (-\cos \pi z) \Big|_0^1 dy\, dx \qquad \begin{array}{l}\text{Evaluate inner integral with}\\ \text{respect to } z.\end{array}$$

$$= \frac{500}{\pi} \int_0^2 \int_0^2 xy\, dy\, dx \qquad \text{Simplify.}$$

$$= \frac{500}{\pi} \int_0^2 x \left(\frac{y^2}{2}\right)\Big|_0^2 dx \qquad \begin{array}{l}\text{Evaluate middle integral with}\\ \text{respect to } y.\end{array}$$

$$= \frac{1000}{\pi} \int_0^2 x\, dx \qquad \text{Simplify.}$$

$$= \frac{1000}{\pi} \left(\frac{x^2}{2}\right)\Big|_0^2 = \frac{2000}{\pi}. \qquad \begin{array}{l}\text{Evaluate outer integral with}\\ \text{respect to } x.\end{array}$$

Dividing by the volume of the region, we find that the average temperature is $(2000/\pi)/4 = 500/\pi \approx 159.2$.

Related Exercises 51–52 ◄

QUICK CHECK 4 Without integrating, what is the average value of $f(x, y, z) = \sin x \sin y \sin z$ on the cube $\{(x, y, z): -1 \le x \le 1, -1 \le y \le 1, -1 \le z \le 1\}$? Use symmetry arguments. ◄

SECTION 16.4 EXERCISES

Getting Started

1. Sketch the region $D = \{(x, y, z): x^2 + y^2 \le 4, 0 \le z \le 4\}$.

2. Write an iterated integral for $\iiint_D f(x, y, z)\, dV$, where D is the box $\{(x, y, z): 0 \le x \le 3, 0 \le y \le 6, 0 \le z \le 4\}$.

3. Write an iterated integral for $\iiint_D f(x, y, z)\, dV$, where D is a sphere of radius 9 centered at $(0, 0, 0)$. Use the order $dz\, dy\, dx$.

4. Sketch the region of integration for the integral

$$\int_0^1 \int_0^{\sqrt{1-z^2}} \int_0^{\sqrt{1-y^2-z^2}} f(x, y, z)\, dx\, dy\, dz.$$

5. Write the integral in Exercise 4 in the order $dy\, dx\, dz$.

6. Write an integral for the average value of $f(x, y, z) = xyz$ over the region bounded by the paraboloid $z = 9 - x^2 - y^2$ and the xy-plane (assuming the volume of the region is known).

Practice Exercises

7–14. Integrals over boxes *Evaluate the following integrals. A sketch of the region of integration may be useful.*

7. $\int_{-2}^2 \int_3^6 \int_0^2 dx\, dy\, dz$

8. $\int_{-1}^1 \int_{-1}^2 \int_0^1 6xyz\, dy\, dx\, dz$

9. $\int_{-2}^2 \int_1^2 \int_1^e \frac{xy^2}{z}\, dz\, dx\, dy$

10. $\int_0^{\ln 4} \int_0^{\ln 3} \int_0^{\ln 2} e^{-x+y+z}\, dx\, dy\, dz$

11. $\int_0^{\pi/2} \int_0^1 \int_0^{\pi/2} \sin \pi x \cos y \sin 2z \, dy \, dx \, dz$

12. $\int_0^2 \int_1^2 \int_0^1 yze^x \, dx \, dz \, dy$

13. $\iiint_D (xy + xz + yz) \, dV;$ $D = \{(x, y, z): -1 \le x \le 1,$

$-2 \le y \le 2, -3 \le z \le 3\}$

14. $\iiint_D xyze^{-x^2-y^2} \, dV;$ $D = \{(x, y, z): 0 \le x \le \sqrt{\ln 2},$

$0 \le y \le \sqrt{\ln 4}, 0 \le z \le 1\}$

15–29. Volumes of solids *Use a triple integral to find the volume of the following solids.*

15. The solid in the first octant bounded by the plane $2x + 3y + 6z = 12$ and the coordinate planes

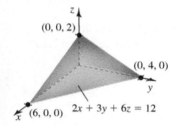

16. The solid in the first octant formed when the cylinder $z = \sin y$, for $0 \le y \le \pi$, is sliced by the planes $y = x$ and $x = 0$

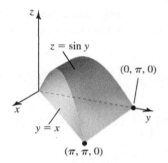

17. The wedge above the xy-plane formed when the cylinder $x^2 + y^2 = 4$ is cut by the planes $z = 0$ and $y = -z$

18. The prism in the first octant bounded by $z = 2 - 4x$ and $y = 8$

19. The solid bounded by the surfaces $z = e^y$ and $z = 1$ over the rectangle $\{(x, y): 0 \le x \le 1, 0 \le y \le \ln 2\}$

20. The wedge bounded by the parabolic cylinder $y = x^2$ and the planes $z = 3 - y$ and $z = 0$

21. The solid between the sphere $x^2 + y^2 + z^2 = 19$ and the hyperboloid $z^2 - x^2 - y^2 = 1$, for $z > 0$

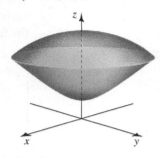

22. The solid bounded below by the cone $z = \sqrt{x^2 + y^2}$ and bounded above by the sphere $x^2 + y^2 + z^2 = 8$

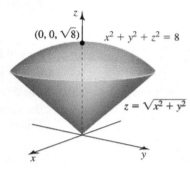

23. The solid bounded by the cylinder $y = 9 - x^2$ and the paraboloid $y = 2x^2 + 3z^2$

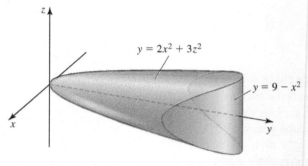

24. The wedge in the first octant bounded by the cylinder $x = z^2$ and the planes $z = 2 - x$, $y = 2$, $y = 0$, and $z = 0$

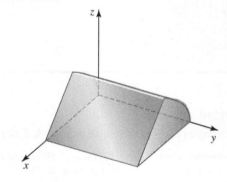

25. The wedge of the cylinder $x^2 + 4z^2 = 4$ created by the planes $y = 3 - x$ and $y = x - 3$

26. The solid bounded by $x = 0$, $x = 2$, $y = 0$, $y = e^{-z}$, $z = 0$, and $z = 1$

27. The solid bounded by $x = 0$, $x = 1 - z^2$, $y = 0$, $z = 0$, and $z = 1 - y$

28. The solid bounded by $x = 0$, $y = z^2$, $z = 0$, and $z = 2 - x - y$

29. The solid bounded by $x = 0$, $x = 2$, $y = z$, $y = z + 1$, $z = 0$, and $z = 4$

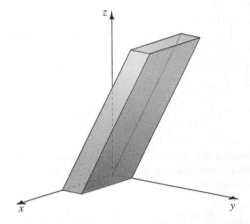

30–35. Six orderings *Let D be the solid in the first octant bounded by the planes $y = 0$, $z = 0$, and $y = x$, and the cylinder $4x^2 + z^2 = 4$. Write the triple integral of $f(x, y, z)$ over D in the given order of integration.*

30. $dz\, dy\, dx$ **31.** $dz\, dx\, dy$ **32.** $dy\, dx\, dz$

33. $dy\, dz\, dx$ **34.** $dx\, dy\, dz$ **35.** $dx\, dz\, dy$

36. All six orders Let D be the solid bounded by $y = x$, $z = 1 - y^2$, $x = 0$, and $z = 0$. Write triple integrals over D in all six possible orders of integration.

37. Changing order of integration Write the integral $\int_0^2 \int_0^1 \int_0^{1-y} dz\, dy\, dx$ in the five other possible orders of integration.

38–46. Triple integrals *Evaluate the following integrals.*

38. $\displaystyle\int_0^\pi \int_0^\pi \int_0^{\sin x} \sin y \, dz\, dx\, dy$

39. $\displaystyle\int_0^2 \int_0^4 \int_{y^2}^4 \sqrt{x}\, dz\, dx\, dy$

40. $\displaystyle\int_0^1 \int_0^{\sqrt{1-x^2}} \int_0^{\sqrt{1-x^2-y^2}} 2xz\, dz\, dy\, dx$

41. $\displaystyle\int_0^1 \int_0^{\sqrt{1-x^2}} \int_0^{\sqrt{1-x^2}} dz\, dy\, dx$

42. $\displaystyle\int_1^6 \int_0^{4-2y/3} \int_0^{12-2y-3z} \frac{1}{y}\, dx\, dz\, dy$

43. $\int_0^3 \int_0^{\sqrt{9-z^2}} \int_0^{\sqrt{1+x^2+z^2}} dy\,dx\,dz$

44. $\int_0^1 \int_y^{2-y} \int_0^{2-x-y} 15xy\,dz\,dx\,dy$

45. $\int_0^{\ln 8} \int_1^{\sqrt{z}} \int_{\ln y}^{\ln 2y} e^{x+y^2-z}\,dx\,dy\,dz$

46. $\int_0^1 \int_0^{\sqrt{1-x^2}} \int_0^{2-x} 4yz\,dz\,dy\,dx$

47–50. Changing the order of integration *Rewrite the following integrals using the indicated order of integration, and then evaluate the resulting integral.*

47. $\int_0^5 \int_{-1}^0 \int_0^{4x+4} dy\,dx\,dz$ in the order $dz\,dx\,dy$

48. $\int_0^1 \int_{-2}^2 \int_0^{\sqrt{4-y^2}} dz\,dy\,dx$ in the order $dy\,dz\,dx$

49. $\int_0^1 \int_0^{\sqrt{1-x^2}} \int_0^{\sqrt{1-x^2}} dy\,dz\,dx$ in the order $dz\,dy\,dx$

50. $\int_0^4 \int_0^{\sqrt{16-x^2}} \int_0^{\sqrt{16-x^2-z^2}} dy\,dz\,dx$ in the order $dx\,dy\,dz$

51–54. Average value *Find the following average values.*

51. The average value of $f(x, y, z) = 8xy \cos z$ over the points inside the box $D = \{(x, y, z): 0 \le x \le 1, 0 \le y \le 2, 0 \le z \le \pi/2\}$

■ 52. The average temperature in the box $D = \{(x, y, z): 0 \le x \le \ln 2, 0 \le y \le \ln 4, 0 \le z \le \ln 8\}$ with a temperature distribution of $T(x, y, z) = 128e^{-x-y-z}$

■ 53. The average of the *squared* distance between the origin and points in the solid cylinder $D = \{(x, y, z): x^2 + y^2 \le 4, 0 \le z \le 2\}$

54. The average z-coordinate of points on and within a hemisphere of radius 4 centered at the origin with its base in the xy-plane

55. Explain why or why not Determine whether the following statements are true and give an explanation or counterexample.

a. An iterated integral of a function over the box $D = \{(x, y, z): 0 \le x \le a, 0 \le y \le b, 0 \le z \le c\}$ can be expressed in eight different ways.

b. One possible iterated integral of f over the prism $D = \{(x, y, z): 0 \le x \le 1, 0 \le y \le 3x - 3, 0 \le z \le 5\}$ is $\int_0^{3x-3} \int_0^1 \int_0^5 f(x, y, z)\,dz\,dx\,dy$.

c. The region $D = \{(x, y, z): 0 \le x \le 1, 0 \le y \le \sqrt{1-x^2}, 0 \le z \le \sqrt{1-x^2}\}$ is a sphere.

Explorations and Challenges

56. Changing the order of integration Use another order of integration to evaluate $\int_1^4 \int_z^{4z} \int_0^{\pi^2} \dfrac{\sin \sqrt{yz}}{x^{3/2}}\,dy\,dx\,dz$.

57–62. Miscellaneous volumes *Use a triple integral to compute the volume of the following regions.*

57. The wedge of the square column $|x| + |y| = 1$ created by the planes $z = 0$ and $x + y + z = 1$

58. The solid common to the cylinders $z = \sin x$ and $z = \sin y$ over the square $R = \{(x, y): 0 \le x \le \pi, 0 \le y \le \pi\}$ (The figure shows the cylinders, but not the common region.)

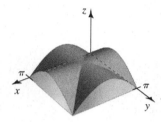

59. The parallelepiped (slanted box) with vertices $(0, 0, 0)$, $(1, 0, 0)$, $(0, 1, 0)$, $(1, 1, 0)$, $(0, 1, 1)$, $(1, 1, 1)$, $(0, 2, 1)$, and $(1, 2, 1)$ (Use integration and find the best order of integration.)

60. The larger of two solids formed when the parallelepiped (slanted box) with vertices $(0, 0, 0)$, $(2, 0, 0)$, $(0, 2, 0)$, $(2, 2, 0)$, $(0, 1, 1)$, $(2, 1, 1)$, $(0, 3, 1)$, and $(2, 3, 1)$ is sliced by the plane $y = 2$

61. The pyramid with vertices $(0, 0, 0)$, $(2, 0, 0)$, $(2, 2, 0)$, $(0, 2, 0)$, and $(0, 0, 4)$

62. The solid in the first octant bounded by the cone $z = 1 - \sqrt{x^2 + y^2}$ and the plane $x + y + z = 1$

63. Two cylinders The x- and y-axes form the axes of two right circular cylinders with radius 1 (see figure). Find the volume of the solid that is common to the two cylinders.

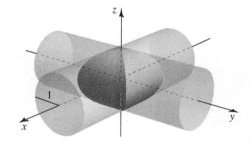

64. Three cylinders The coordinate axes form the axes of three right circular cylinders with radius 1 (see figure). Find the volume of the solid that is common to the three cylinders.

65. Dividing the cheese Suppose a wedge of cheese fills the region in the first octant bounded by the planes $y = z$, $y = 4$, and $x = 4$. You could divide the wedge into two pieces of equal volume by slicing the wedge with the plane $x = 2$. Instead find a with $0 < a < 4$ such that slicing the wedge with the plane $y = a$ divides the wedge into two pieces of equal volume.

66. Partitioning a cube Consider the region
$D_1 = \{(x, y, z): 0 \le x \le y \le z \le 1\}$.

   a. Find the volume of D_1.

   b. Let $D_2, \ldots, D_6$ be the "cousins" of D_1 formed by rearranging x, y, and z in the inequality $0 \le x \le y \le z \le 1$. Show that the volumes of $D_1, \ldots, D_6$ are equal.

   c. Show that the union of $D_1, \ldots, D_6$ is a unit cube.

67–71. General volume formulas *Find equations for the bounding surfaces, set up a volume integral, and evaluate the integral to obtain a volume formula for each region. Assume a, b, c, r, R, and h are positive constants.*

67. Cone Find the volume of a right circular cone with height h and base radius r.

68. Tetrahedron Find the volume of a tetrahedron whose vertices are located at $(0, 0, 0)$, $(a, 0, 0)$, $(0, b, 0)$, and $(0, 0, c)$.

69. Spherical cap Find the volume of the cap of a sphere of radius R with height h.

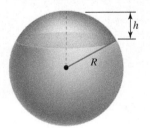

70. Frustum of a cone Find the volume of a truncated cone of height h whose ends have radii r and R.

71. Ellipsoid Find the volume of an ellipsoid with axes of lengths $2a$, $2b$, and $2c$.

72. Exponential distribution The occurrence of random events (such as phone calls or e-mail messages) is often idealized using an exponential distribution. If λ is the average rate of occurrence of such an event, assumed to be constant over time, then the average time between occurrences is λ^{-1} (for example, if phone calls arrive at a rate of $\lambda = 2/\text{min}$, then the mean time between phone calls is $\lambda^{-1} = 1/2$ min). The exponential distribution is given by $f(t) = \lambda e^{-\lambda t}$, for $0 \le t < \infty$.

   a. Suppose you work at a customer service desk and phone calls arrive at an average rate of $\lambda_1 = 0.8/\text{min}$ (meaning the average time between phone calls is $1/0.8 = 1.25$ min). The probability that a phone call arrives during the interval $[0, T]$ is $p(T) = \int_0^T \lambda_1 e^{-\lambda_1 t}\, dt$. Find the probability that a phone call arrives during the first 45 s (0.75 min) that you work at the desk.

   b. Now suppose walk-in customers also arrive at your desk at an average rate of $\lambda_2 = 0.1/\text{min}$. The probability that a phone call *and* a customer arrive during the interval $[0, T]$ is

$$p(T) = \int_0^T \int_0^T \lambda_1 e^{-\lambda_1 t} \lambda_2 e^{-\lambda_2 s}\, dt\, ds.$$

   Find the probability that a phone call and a customer arrive during the first 45 s that you work at the desk.

   c. E-mail messages also arrive at your desk at an average rate of $\lambda_3 = 0.05/\text{min}$. The probability that a phone call *and* a customer *and* an e-mail message arrive during the interval $[0, T]$ is

$$p(T) = \int_0^T \int_0^T \int_0^T \lambda_1 e^{-\lambda_1 t} \lambda_2 e^{-\lambda_2 s} \lambda_3 e^{-\lambda_3 u}\, dt\, ds\, du.$$

   Find the probability that a phone call and a customer and an e-mail message arrive during the first 45 s that you work at the desk.

73. Hypervolume Find the "volume" of the four-dimensional pyramid bounded by $w + x + y + z + 1 = 0$ and the coordinate planes $w = 0$, $x = 0$, $y = 0$, $z = 0$.

74. An identity (Putnam Exam 1941) Let f be a continuous function on $[0, 1]$. Prove that

$$\int_0^1 \int_x^1 \int_x^y f(x)f(y)f(z)\, dz\, dy\, dx = \frac{1}{6}\left(\int_0^1 f(x)\, dx\right)^3.$$

QUICK CHECK ANSWERS

1. $dx\, dy\, dz$, $dx\, dz\, dy$, $dy\, dx\, dz$, $dy\, dz\, dx$, $dz\, dx\, dy$, $dz\, dy\, dx$

2. $\int_0^1 \int_0^2 \int_0^3 (2 - z)\, dx\, dy\, dz$, $\int_0^2 \int_0^1 \int_0^3 (2 - z)\, dx\, dz\, dy$

3. $\int_0^2 \int_0^{4-2x} \int_0^6 dz\, dy\, dx$, $\int_0^6 \int_0^4 \int_0^{2-y/2} dx\, dy\, dz$

4. 0 ($\sin x$, $\sin y$, and $\sin z$ are odd functions.) ◄

16.5 Triple Integrals in Cylindrical and Spherical Coordinates

When evaluating triple integrals, you may have noticed that some regions (such as spheres, cones, and cylinders) have awkward descriptions in Cartesian coordinates. In this section, we examine two other coordinate systems in $\mathbb{R}^3$ that are easier to use when working with certain types of regions. These coordinate systems are helpful not only for integration, but also for general problem solving.

Cylindrical Coordinates

> In cylindrical coordinates, r and θ are the usual polar coordinates, with the additional restriction that $r \geq 0$. Adding the z-coordinate lifts points in the polar plane into $\mathbb{R}^3$.

When we extend polar coordinates from $\mathbb{R}^2$ to $\mathbb{R}^3$, the result is *cylindrical coordinates*. In this coordinate system, a point P in $\mathbb{R}^3$ has coordinates (r, θ, z), where r and θ are polar coordinates for the point P^*, which is the projection of P onto the xy-plane (Figure 16.47). As in Cartesian coordinates, the z-coordinate is the signed vertical distance between P and the xy-plane. Any point in $\mathbb{R}^3$ can be represented by cylindrical coordinates using the intervals $0 \leq r < \infty$, $0 \leq \theta \leq 2\pi$, and $-\infty < z < \infty$.

Many sets of points have simple representations in cylindrical coordinates. For example, the set $\{(r, \theta, z): r = a\}$ is the set of points whose distance from the z-axis is a, which is a right circular cylinder of radius a. The set $\{(r, \theta, z): \theta = \theta_0\}$ is the set of points with a constant θ coordinate; it is a vertical half-plane emanating from the z-axis in the direction $\theta = \theta_0$. Table 16.4 summarizes these and other sets that are ideal for integration in cylindrical coordinates.

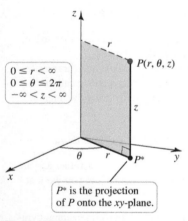

$$0 \leq r < \infty$$
$$0 \leq \theta \leq 2\pi$$
$$-\infty < z < \infty$$

P^* is the projection of P onto the xy-plane.

Figure 16.47

Table 16.4

Name	Description	Example
Cylinder	$\{(r, \theta, z): r = a\},\, a > 0$	
Cylindrical shell	$\{(r, \theta, z): 0 < a \leq r \leq b\}$	
Vertical half-plane	$\{(r, \theta, z): \theta = \theta_0\}$	

Table 16.4 **(Continued)**

Name	Description	Example
Horizontal plane	$\{(r, \theta, z): z = a\}$	
Cone	$\{(r, \theta, z): z = ar\}, a \neq 0$	

(a)

(b)

Figure 16.48

EXAMPLE 1 **Sets in cylindrical coordinates** Identify and sketch the following sets in cylindrical coordinates.

a. $Q = \{(r, \theta, z): 1 \leq r \leq 3, z \geq 0\}$

b. $S = \{(r, \theta, z): z = 1 - r, 0 \leq r \leq 1\}$

SOLUTION

a. The set Q is a cylindrical shell with inner radius 1 and outer radius 3 that extends indefinitely along the positive z-axis (**Figure 16.48a**). Because θ is unspecified, it takes on all values.

b. To identify this surface, it helps to work in steps. The set $S_1 = \{(r, \theta, z): z = r\}$ is a cone that opens *upward* with its vertex at the origin. Similarly, the set $S_2 = \{(r, \theta, z): z = -r\}$ is a cone that opens *downward* with its vertex at the origin. Therefore, S is S_2 shifted vertically upward by 1 unit; it is a cone that opens downward with its vertex at $(0, 0, 1)$. Because $0 \leq r \leq 1$, the base of the cone is on the xy-plane (**Figure 16.48b**).

Related Exercise 11 ◄

Equations for transforming Cartesian coordinates to cylindrical coordinates, and vice versa, are often needed for integration. We simply use the rules for polar coordinates (Section 12.2) with no change in the z-coordinate (**Figure 16.49**).

Figure 16.49

$$\tan \theta = \frac{y}{x}$$
$$r = \sqrt{x^2 + y^2}$$
$$x = r \cos \theta$$
$$y = r \sin \theta$$
$$P(r, \theta, z)$$

QUICK CHECK 1 Find the cylindrical coordinates of the point with rectangular coordinates $(1, -1, 5)$. Find the rectangular coordinates of the point with cylindrical coordinates $(2, \pi/3, 5)$. ◄

Transformations Between Cylindrical and Rectangular Coordinates	
Rectangular → Cylindrical	**Cylindrical → Rectangular**
$r^2 = x^2 + y^2$	$x = r \cos \theta$
$\tan \theta = y/x$	$y = r \sin \theta$
$z = z$	$z = z$

Integration in Cylindrical Coordinates

Among the uses of cylindrical coordinates is the evaluation of triple integrals of the form $\iiint_D f(x, y, z)\, dV$. We begin with a region D in $\mathbb{R}^3$ and partition it into cylindrical wedges formed by changes of Δr, $\Delta \theta$, and Δz in the coordinate directions (**Figure 16.50**). Those wedges that lie entirely within D are labeled from $k = 1$ to $k = n$ in some convenient order. We let $(r_k^*, \theta_k^*, z_k^*)$ be the cylindrical coordinates of an arbitrary point in the kth wedge. This point also has Cartesian coordinates $(x_k^*, y_k^*, z_k^*) = (r_k^* \cos \theta_k^*, r_k^* \sin \theta_k^*, z_k^*)$.

cylindrical: $(r_k^*, \theta_k^*, z_k^*)$
rectangular: (x_k^*, y_k^*, z_k^*)

Δz

$\Delta \theta$

Δr

r_k^*

Base area $\approx r_k^* \Delta r \Delta \theta$

Approximate volume $\Delta V_k = r_k^* \Delta r \Delta \theta \Delta z$

Figure 16.50

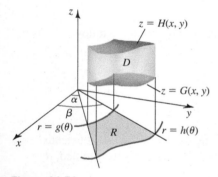

$z = H(x, y)$

D

$z = G(x, y)$

α

β

$r = g(\theta)$

R

$r = h(\theta)$

Figure 16.51

> The order of the differentials specifies the order in which the integrals are evaluated, so we write the volume element dV as $dz\, r\, dr\, d\theta$. Do not lose sight of the factor of r in the integrand. It plays the same role as it does in the area element $dA = r\, dr\, d\theta$ in polar coordinates.

As shown in Figure 16.50, the base of the kth wedge is a polar rectangle with an approximate area of $r_k^* \Delta r \Delta \theta$ (Section 16.3). The height of the wedge is Δz. Multiplying these dimensions together, the approximate volume of the wedge is $\Delta V_k = r_k^* \Delta r \Delta \theta \Delta z$, for $k = 1, \ldots, n$.

We now assume $f(x, y, z)$ is continuous on D and form a Riemann sum over the region by adding function values multiplied by the corresponding approximate volumes:

$$\sum_{k=1}^{n} f(x_k^*, y_k^*, z_k^*) \Delta V_k = \sum_{k=1}^{n} f(r_k^* \cos \theta_k^*, r_k^* \sin \theta_k^*, z_k^*) \Delta V_k.$$

Let Δ be the maximum value of Δr, $\Delta \theta$, and Δz, for $k = 1, \ldots, n$. As $n \to \infty$ and $\Delta \to 0$, the Riemann sums approach a limit called the **triple integral of f over D in cylindrical coordinates**:

$$\iiint_D f(x, y, z)\, dV = \lim_{\Delta \to 0} \sum_{k=1}^{n} f(r_k^* \cos \theta_k^*, r_k^* \sin \theta_k^*, z_k^*) \underbrace{r_k^* \Delta r \Delta \theta \Delta z}_{\Delta V_k}.$$

The rightmost sum tells us how to write a triple integral in x, y, and z as an iterated integral of $f(r \cos \theta, r \sin \theta, z)\, r$ in cylindrical coordinates.

Finding Limits of Integration   We show how to find the limits of integration in one common situation involving cylindrical coordinates. Suppose D is a region in $\mathbb{R}^3$ consisting of points between the surfaces $z = G(x, y)$ and $z = H(x, y)$, where x and y belong to a region R in the xy-plane and $G(x, y) \le H(x, y)$ on R (**Figure 16.51**). Assuming f is continuous on D, the triple integral of f over D may be expressed as the iterated integral

$$\iiint_D f(x, y, z)\, dV = \iint_R \left(\int_{G(x, y)}^{H(x, y)} f(x, y, z)\, dz \right) dA.$$

The inner integral with respect to z runs from the lower surface $z = G(x, y)$ to the upper surface $z = H(x, y)$, leaving an outer double integral over R.

If the region R is described in polar coordinates by

$$\{(r, \theta): g(\theta) \le r \le h(\theta), \alpha \le \theta \le \beta\},$$

then we evaluate the double integral over R in polar coordinates (Section 16.3). The effect is a change of variables from rectangular to cylindrical coordinates. Letting $x = r \cos \theta$ and $y = r \sin \theta$, we have the following result, which is another change of variables formula.

THEOREM 16.6   Change of Variables for Triple Integrals in Cylindrical Coordinates

Let f be continuous over the region D, expressed in cylindrical coordinates as

$$D = \{(r, \theta, z): 0 \le g(\theta) \le r \le h(\theta), \alpha \le \theta \le \beta, G(x, y) \le z \le H(x, y)\}.$$

Then f is integrable over D, and the triple integral of f over D is

$$\iiint_D f(x, y, z)\, dV = \int_\alpha^\beta \int_{g(\theta)}^{h(\theta)} \int_{G(r \cos \theta,\, r \sin \theta)}^{H(r \cos \theta,\, r \sin \theta)} f(r \cos \theta, r \sin \theta, z)\, dz\, r\, dr\, d\theta.$$

Notice that the integrand and the limits of integration are converted from Cartesian to cylindrical coordinates. As with triple integrals in Cartesian coordinates, there are two immediate interpretations of this integral. If $f = 1$, then the triple integral $\iiint_D dV$ equals the volume of the region D. Also, if f describes the density of an object occupying the region D, the triple integral equals the mass of the object.

EXAMPLE 2 Switching coordinate systems Evaluate the integral

$$I = \int_0^{2\sqrt{2}} \int_{-\sqrt{8-x^2}}^{\sqrt{8-x^2}} \int_{-1}^{2} \sqrt{1 + x^2 + y^2} \, dz \, dy \, dx.$$

SOLUTION Evaluating this integral as it is given in Cartesian coordinates requires a tricky trigonometric substitution in the middle integral, followed by an even more difficult integral. Notice that z varies between the planes $z = -1$ and $z = 2$, while x and y vary over half of a disk in the xy-plane. Therefore, D is half of a solid cylinder (**Figure 16.52a**), which suggests a change to cylindrical coordinates.

The limits of integration in cylindrical coordinates are determined as follows:

Inner integral with respect to z A line through the half cylinder parallel to the z-axis enters at $z = -1$ and leaves at $z = 2$, so we integrate over the interval $-1 \le z \le 2$ (**Figure 16.52b**).

Middle integral with respect to r The projection of the half cylinder onto the xy-plane is the half disk R of radius $2\sqrt{2}$ centered at the origin, so r varies over the interval $0 \le r \le 2\sqrt{2}$ (**Figure 16.52c**).

Outer integral with respect to θ The half disk R is swept out by letting θ vary over the interval $-\pi/2 \le \theta \le \pi/2$ (Figure 16.52c).

$$\iint_R \left(\int_{-1}^{2} \sqrt{1 + r^2} \, dz \right) dA$$

In cylindrical coordinates, integrate in z with $-1 \le z \le 2$; ...

$$\int_{-\pi/2}^{\pi/2} \int_0^{2\sqrt{2}} \int_{-1}^{2} \sqrt{1 + r^2} \, dz \, r \, dr \, d\theta$$

... then integrate over R with $0 \le r \le 2\sqrt{2}$, $-\frac{\pi}{2} \le \theta \le \frac{\pi}{2}$.

(a)   (b)   (c)

Figure 16.52

We also convert the integrand to cylindrical coordinates:

$$f(x, y, z) = \sqrt{1 + \underbrace{x^2 + y^2}_{r^2}} = \sqrt{1 + r^2}.$$

The evaluation of the integral in cylindrical coordinates now follows:

$$I = \int_{-\pi/2}^{\pi/2} \int_0^{2\sqrt{2}} \int_{-1}^{2} \sqrt{1 + r^2} \, dz \, r \, dr \, d\theta \qquad \text{Convert to cylindrical coordinates.}$$

$$= 3 \int_{-\pi/2}^{\pi/2} \int_0^{2\sqrt{2}} \sqrt{1 + r^2} \, r \, dr \, d\theta \qquad \text{Evaluate inner integral with respect to } z.$$

$$= \int_{-\pi/2}^{\pi/2} (1 + r^2)^{3/2} \Big|_0^{2\sqrt{2}} \, d\theta \qquad \text{Evaluate middle integral with respect to } r.$$

$$= \int_{-\pi/2}^{\pi/2} 26 \, d\theta = 26\pi. \qquad \text{Evaluate outer integral with respect to } \theta.$$

QUICK CHECK 2 Find the limits of integration for a triple integral in cylindrical coordinates that gives the volume of a cylinder with height 20 and a circular base of radius 10 centered at the origin in the xy-plane. ◄

Related Exercises 17–18 ◄

As illustrated in Example 2, triple integrals given in rectangular coordinates may be more easily evaluated after converting to cylindrical coordinates. Answering the following questions may help you choose the best coordinate system for a particular integral.

• In which coordinate system is the region of integration most easily described?
• In which coordinate system is the integrand most easily expressed?
• In which coordinate system is the triple integral most easily evaluated?

In general, if an integral in one coordinate system is difficult to evaluate, consider using a different coordinate system.

EXAMPLE 3 Mass of a solid paraboloid Find the mass of the solid D bounded by the paraboloid $z = 4 - r^2$ and the plane $z = 0$ (**Figure 16.53a**), where the density of the solid, given in cylindrical coordinates, is $f(r, \theta, z) = 5 - z$ (heavy near the base and light near the vertex).

SOLUTION The z-coordinate runs from the base $z = 0$ to the surface $z = 4 - r^2$ (**Figure 16.53b**). The projection R of the region D onto the xy-plane is found by setting $z = 0$ in the equation of the surface, $z = 4 - r^2$. The positive value of r satisfying the equation $4 - r^2 = 0$ is $r = 2$, so in polar coordinates $R = \{(r, \theta): 0 \le r \le 2, 0 \le \theta \le 2\pi\}$, which is a disk of radius 2 (**Figure 16.53c**).

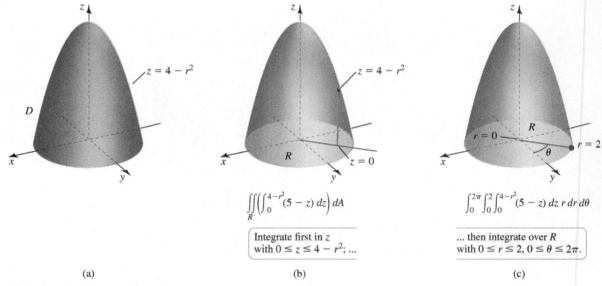

$$\iint_R \left(\int_0^{4-r^2} (5 - z)\, dz \right) dA$$

Integrate first in z
with $0 \le z \le 4 - r^2$; ...

$$\int_0^{2\pi} \int_0^2 \int_0^{4-r^2} (5 - z)\, dz\, r\, dr\, d\theta$$

... then integrate over R
with $0 \le r \le 2, 0 \le \theta \le 2\pi$.

(a)   (b)   (c)

Figure 16.53

▶ In Example 3, the integrand is independent of θ, so the integral with respect to θ could have been done first, producing a factor of 2π.

The mass is computed by integrating the density function over D:

$$\iiint_D f(r, \theta, z)\, dV = \int_0^{2\pi} \int_0^2 \int_0^{4-r^2} (5 - z)\, dz\, r\, dr\, d\theta \qquad \text{Integrate density.}$$

$$= \int_0^{2\pi} \int_0^2 \left(5z - \frac{z^2}{2} \right) \Big|_0^{4-r^2} r\, dr\, d\theta \qquad \begin{array}{l}\text{Evaluate inner integral with} \\ \text{respect to } z.\end{array}$$

$$= \frac{1}{2} \int_0^{2\pi} \int_0^2 (24r - 2r^3 - r^5)\, dr\, d\theta \qquad \text{Simplify.}$$

$$= \int_0^{2\pi} \frac{44}{3}\, d\theta \qquad \begin{array}{l}\text{Evaluate middle integral} \\ \text{with respect to } r.\end{array}$$

$$= \frac{88\pi}{3}. \qquad \begin{array}{l}\text{Evaluate outer integral} \\ \text{with respect to } \theta.\end{array}$$

Related Exercises 25–26 ◀

> Recall that to find the volume of a region D using a triple integral, we set $f = 1$ and evaluate
>
> $$V = \iiint_D dV.$$

EXAMPLE 4   Volume between two surfaces Find the volume of the solid D between the cone $z = \sqrt{x^2 + y^2}$ and the inverted paraboloid $z = 12 - x^2 - y^2$ (**Figure 16.54a**).

SOLUTION Because $x^2 + y^2 = r^2$, the equation of the cone in cylindrical coordinates becomes $z = r$, and the equation of the paraboloid becomes $z = 12 - r^2$. The inner integral in z runs from the cone $z = r$ (the lower surface) to the paraboloid $z = 12 - r^2$ (the upper surface) (**Figure 16.54b**). We project D onto the xy-plane to produce the region R, whose boundary is determined by the intersection of the two surfaces. Equating the z-coordinates in the equations of the two surfaces, we have $12 - r^2 = r$, or $(r - 3)(r + 4) = 0$. Because $r \geq 0$, the relevant root is $r = 3$. Therefore, the projection of D onto the xy-plane is the polar region $R = \{(r, \theta): 0 \leq r \leq 3, 0 \leq \theta \leq 2\pi\}$, which is a disk of radius 3 centered at $(0, 0)$ (**Figure 16.54c**).

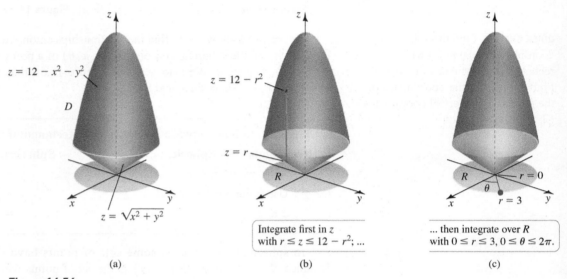

Integrate first in z with $r \leq z \leq 12 - r^2$; ...

... then integrate over R with $0 \leq r \leq 3, 0 \leq \theta \leq 2\pi$.

(a)　　　　(b)　　　　(c)

Figure 16.54

The volume of the region is

$$\iiint_D dV = \int_0^{2\pi} \int_0^3 \int_r^{12-r^2} dz\, r\, dr\, d\theta$$

$$= \int_0^{2\pi} \int_0^3 (12 - r^2 - r)\, r\, dr\, d\theta \qquad \text{Evaluate inner integral with respect to } z.$$

$$= \int_0^{2\pi} \frac{99}{4}\, d\theta \qquad \text{Evaluate middle integral with respect to } r.$$

$$= \frac{99\pi}{2}. \qquad \text{Evaluate outer integral with respect to } \theta.$$

Related Exercises 30–31 ◄

> The coordinate ρ (pronounced "rho") in spherical coordinates should not be confused with r in cylindrical coordinates, which is the distance from P to the z-axis.

Spherical Coordinates

In spherical coordinates, a point P in $\mathbb{R}^3$ is represented by three coordinates (ρ, φ, θ) (**Figure 16.55**).

- ρ is the distance from the origin to P.

- φ is the angle between the positive z-axis and the line OP.

- θ is the same angle as in cylindrical coordinates; it measures rotation about the z-axis relative to the positive x-axis.

> The coordinate φ is called the *colatitude* because it is $\pi/2$ minus the latitude of points in the Northern Hemisphere. Physicists may reverse the roles of θ and φ; that is, θ is the colatitude and φ is the polar angle.

All points in $\mathbb{R}^3$ can be represented by spherical coordinates using the intervals $0 \leq \rho < \infty, 0 \leq \varphi \leq \pi$, and $0 \leq \theta \leq 2\pi$.

Figure 16.55

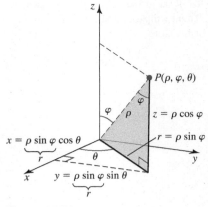

Figure 16.56

QUICK CHECK 3 Find the spherical coordinates of the point with rectangular coordinates $(1, \sqrt{3}, 2)$. Find the rectangular coordinates of the point with spherical coordinates $(2, \pi/4, \pi/4)$. ◄

Figure 16.56 allows us to find the relationships among rectangular and spherical coordinates. Given the spherical coordinates (ρ, φ, θ) of a point P, the distance from P to the z-axis is $r = \rho \sin \varphi$. We also see from Figure 16.56 that $x = r \cos \theta = \rho \sin \varphi \cos \theta$, $y = r \sin \theta = \rho \sin \varphi \sin \theta$, and $z = \rho \cos \varphi$.

Transformations Between Spherical and Rectangular Coordinates	
Rectangular → Spherical	**Spherical → Rectangular**
$\rho^2 = x^2 + y^2 + z^2$	$x = \rho \sin \varphi \cos \theta$
Use trigonometry to find	$y = \rho \sin \varphi \sin \theta$
φ and θ.	$z = \rho \cos \varphi$

In spherical coordinates, some sets of points have simple representations. For instance, the set $\{(\rho, \varphi, \theta): \rho = a\}$ is the set of points whose ρ-coordinate is constant, which is a sphere of radius a centered at the origin. The set $\{(\rho, \varphi, \theta): \varphi = \varphi_0\}$ is the set of points with a constant φ-coordinate; it is a cone with its vertex at the origin and whose sides make an angle φ_0 with the positive z-axis.

EXAMPLE 5 Sets in spherical coordinates Express the following sets in rectangular coordinates and identify the set. Assume a is a positive real number.

a. $\{(\rho, \varphi, \theta): \rho = 2a \cos \varphi, 0 \le \varphi \le \pi/2, 0 \le \theta \le 2\pi\}$

b. $\{(\rho, \varphi, \theta): \rho = 4 \sec \varphi, 0 \le \varphi < \pi/2, 0 \le \theta \le 2\pi\}$

SOLUTION

a. To avoid working with square roots, we multiply both sides of $\rho = 2a \cos \varphi$ by ρ to obtain $\rho^2 = 2a \rho \cos \varphi$. Substituting rectangular coordinates, we have $x^2 + y^2 + z^2 = 2az$. Completing the square results in the equation

$$x^2 + y^2 + (z - a)^2 = a^2.$$

This is the equation of a sphere centered at $(0, 0, a)$ with radius a (**Figure 16.57a**). With the limits $0 \le \varphi \le \pi/2$ and $0 \le \theta \le 2\pi$, the set describes a full sphere.

b. The equation $\rho = 4 \sec \varphi$ is first written $\rho \cos \varphi = 4$. Noting that $z = \rho \cos \varphi$, the set consists of all points with $z = 4$, which is a horizontal plane (**Figure 16.57b**).

Related Exercises 37–38 ◄

(a)

$\rho = 2a \cos \varphi$

(b)

$\rho = 4 \sec \varphi$

Figure 16.57

Table 16.5 summarizes some sets that have simple descriptions in spherical coordinates.

Table 16.5

Name	Description	Example
Sphere, radius a, center $(0, 0, 0)$	$\{(\rho, \varphi, \theta): \rho = a\}, a > 0$	
Cone	$\{(\rho, \varphi, \theta): \varphi = \varphi_0\}, \varphi_0 \neq 0, \pi/2, \pi$	
Vertical half-plane	$\{(\rho, \varphi, \theta): \theta = \theta_0\}$	
Horizontal plane, $z = a$	$a > 0$: $\{(\rho, \varphi, \theta): \rho = a \sec \varphi, 0 \leq \varphi < \pi/2\}$ $a < 0$: $\{(\rho, \varphi, \theta): \rho = a \sec \varphi, \pi/2 < \varphi \leq \pi\}$	
Cylinder, radius $a > 0$	$\{(\rho, \varphi, \theta): \rho = a \csc \varphi, 0 < \varphi < \pi\}$	
Sphere, radius $a > 0$, center $(0, 0, a)$	$\{(\rho, \varphi, \theta): \rho = 2a \cos \varphi, 0 \leq \varphi \leq \pi/2\}$	

▶ Notice that the set of points (ρ, φ, θ) with $\varphi = \pi/2$ is the xy-plane, and if $\pi/2 < \varphi_0 < \pi$, the set of points with $\varphi = \varphi_0$ is a cone that opens downward.

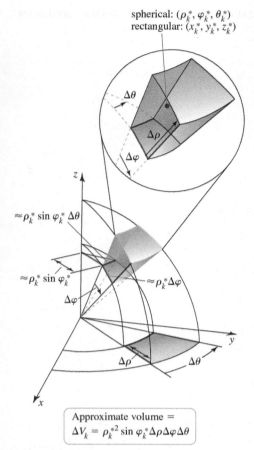

spherical: $(\rho_k^*, \varphi_k^*, \theta_k^*)$
rectangular: (x_k^*, y_k^*, z_k^*)

$\approx \rho_k^* \sin \varphi_k^* \Delta\theta$

$\approx \rho_k^* \sin \varphi_k^*$

$\approx \rho_k^* \Delta\varphi$

$\Delta\varphi$

$\Delta\rho$ $\Delta\theta$

> Approximate volume $=$
> $\Delta V_k = \rho_k^{*2} \sin \varphi_k^* \Delta\rho \Delta\varphi \Delta\theta$

Figure 16.58

▶ Recall that the length s of a circular arc
of radius r subtended by an angle θ is
$s = r\theta$.

Integration in Spherical Coordinates

We now investigate triple integrals in spherical coordinates over a region D in $\mathbb{R}^3$. The region D is partitioned into "spherical boxes" that are formed by changes of $\Delta\rho$, $\Delta\varphi$, and $\Delta\theta$ in the coordinate directions (**Figure 16.58**). Those boxes that lie entirely within D are labeled from $k = 1$ to $k = n$. We let $(\rho_k^*, \varphi_k^*, \theta_k^*)$ be the spherical coordinates for an arbitrary point in the kth box. This point also has Cartesian coordinates

$$(x_k^*, y_k^*, z_k^*) = (\rho_k^* \sin \varphi_k^* \cos \theta_k^*, \rho_k^* \sin \varphi_k^* \sin \theta_k^*, \rho_k^* \cos \varphi_k^*).$$

To approximate the volume of a typical box, note that the length of the box in the ρ-direction is $\Delta\rho$ (Figure 16.58). The approximate length of the kth box in the θ-direction is the length of an arc of a circle of radius $\rho_k^* \sin \varphi_k^*$ subtended by an angle $\Delta\theta$; this length is $\rho_k^* \sin \varphi_k^* \Delta\theta$. The approximate length of the box in the φ-direction is the length of an arc of radius ρ_k^* subtended by an angle $\Delta\varphi$; this length is $\rho_k^* \Delta\varphi$. Multiplying these dimensions together, the approximate volume of the kth spherical box is $\Delta V_k = \rho_k^{*2} \sin \varphi_k^* \Delta\rho \Delta\varphi \Delta\theta$, for $k = 1, \ldots, n$.

We now assume $f(x, y, z)$ is continuous on D and form a Riemann sum over the region by adding function values multiplied by the corresponding approximate volumes:

$$\sum_{k=1}^{n} f(x_k^*, y_k^*, z_k^*) \Delta V_k = \sum_{k=1}^{n} f(\rho_k^* \sin \varphi_k^* \cos \theta_k^*, \rho_k^* \sin \varphi_k^* \sin \theta_k^*, \rho_k^* \cos \varphi_k^*) \Delta V_k.$$

We let Δ denote the maximum value of $\Delta\rho$, $\Delta\varphi$, and $\Delta\theta$. As $n \to \infty$ and $\Delta \to 0$, the Riemann sums approach a limit called the **triple integral of f over D in spherical coordinates**:

$$\iiint_D f(x, y, z)\, dV$$

$$= \lim_{\Delta \to 0} \sum_{k=1}^{n} f(\rho_k^* \sin \varphi_k^* \cos \theta_k^*, \rho_k^* \sin \varphi_k^* \sin \theta_k^*, \rho_k^* \cos \varphi_k^*) \underbrace{\rho_k^{*2} \sin \varphi_k^* \Delta\rho \Delta\varphi \Delta\theta}_{\Delta V_k}.$$

The rightmost sum tells us how to write a triple integral in x, y, and z as an iterated integral of $f(\rho \sin \varphi \cos \theta, \rho \sin \varphi \sin \theta, \rho \cos \varphi)\rho^2 \sin \varphi$ in spherical coordinates.

Finding Limits of Integration   We consider a common situation in which the region of integration D, expressed in spherical coordinates, has the form

$$D = \{(\rho, \varphi, \theta): 0 \le g(\varphi, \theta) \le \rho \le h(\varphi, \theta), a \le \varphi \le b, \alpha \le \theta \le \beta\}.$$

In other words, D is bounded in the ρ-direction by two surfaces given by g and h. In the angular directions, the region lies between two cones ($a \le \varphi \le b$) and two half-planes ($\alpha \le \theta \le \beta$) (**Figure 16.59**).

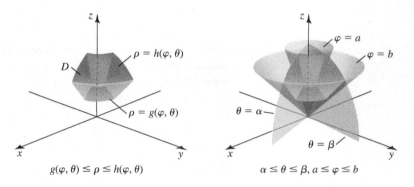

$g(\varphi, \theta) \le \rho \le h(\varphi, \theta)$ $\alpha \le \theta \le \beta, a \le \varphi \le b$

Figure 16.59

For this type of region, the inner integral is with respect to ρ, which varies from $\rho = g(\varphi, \theta)$ to $\rho = h(\varphi, \theta)$. As ρ varies between these limits, imagine letting θ and φ vary over the intervals $a \le \varphi \le b$ and $\alpha \le \theta \le \beta$. The effect is to sweep out all points of D. Notice that the middle and outer integrals, with respect to θ and φ, may be done in either order (**Figure 16.60**).

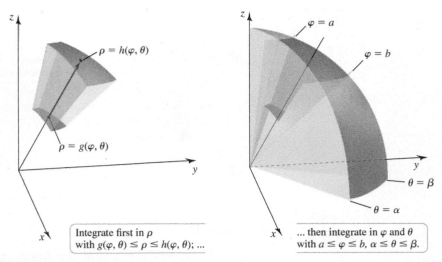

Integrate first in ρ with $g(\varphi, \theta) \le \rho \le h(\varphi, \theta)$; ...

... then integrate in φ and θ with $a \le \varphi \le b$, $\alpha \le \theta \le \beta$.

Figure 16.60

In summary, to integrate over all points of D, we carry out the following steps.

1. Integrate with respect to ρ from $\rho = g(\varphi, \theta)$ to $\rho = h(\varphi, \theta)$; the result (in general) is a function of φ and θ.

2. Integrate with respect to φ from $\varphi = a$ to $\varphi = b$; the result (in general) is a function of θ.

3. Integrate with respect to θ from $\theta = \alpha$ to $\theta = \beta$; the result is (always) a real number.

Another change of variables expresses the triple integral as an iterated integral in spherical coordinates.

> ▶ The element of volume in spherical coordinates is $dV = \rho^2 \sin \varphi \, d\rho \, d\varphi \, d\theta$.

THEOREM 16.7 Change of Variables for Triple Integrals in Spherical Coordinates

Let f be continuous over the region D, expressed in spherical coordinates as

$$D = \{(\rho, \varphi, \theta): 0 \le g(\varphi, \theta) \le \rho \le h(\varphi, \theta), a \le \varphi \le b, \alpha \le \theta \le \beta\}.$$

Then f is integrable over D, and the triple integral of f over D is

$$\iiint_D f(x, y, z) \, dV$$

$$= \int_\alpha^\beta \int_a^b \int_{g(\varphi, \theta)}^{h(\varphi, \theta)} f(\rho \sin \varphi \cos \theta, \rho \sin \varphi \sin \theta, \rho \cos \varphi) \rho^2 \sin \varphi \, d\rho \, d\varphi \, d\theta.$$

If the integrand is given in terms of Cartesian coordinates x, y, and z, it must be expressed in spherical coordinates before integrating. As with other triple integrals, if $f = 1$, then the triple integral equals the volume of D. If f is a density function for an object occupying the region D, then the triple integral equals the mass of the object.

EXAMPLE 6 A triple integral Evaluate $\iiint_D (x^2 + y^2 + z^2)^{-3/2} \, dV$, where D is the region in the first octant between two spheres of radius 1 and 2 centered at the origin.

SOLUTION Both the integrand f and the region D are greatly simplified when expressed in spherical coordinates. The integrand becomes

$$(x^2 + y^2 + z^2)^{-3/2} = (\rho^2)^{-3/2} = \rho^{-3},$$

while the region of integration (**Figure 16.61**) is

$$D = \{(\rho, \varphi, \theta): 1 \le \rho \le 2, 0 \le \varphi \le \pi/2, 0 \le \theta \le \pi/2\}.$$

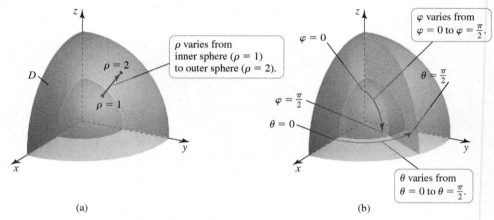

(a)   (b)

Figure 16.61

The integral is evaluated as follows:

$$\iiint_D f(x, y, z)\,dV = \int_0^{\pi/2}\int_0^{\pi/2}\int_1^2 \rho^{-3}\,\rho^2 \sin\varphi\,d\rho\,d\varphi\,d\theta \quad \text{Convert to spherical coordinates.}$$

$$= \int_0^{\pi/2}\int_0^{\pi/2}\int_1^2 \rho^{-1} \sin\varphi\,d\rho\,d\varphi\,d\theta \quad \text{Simplify.}$$

$$= \int_0^{\pi/2}\int_0^{\pi/2} \ln|\rho|\Big|_1^2 \sin\varphi\,d\varphi\,d\theta \quad \text{Evaluate inner integral with respect to } \rho.$$

$$= \ln 2 \int_0^{\pi/2}\int_0^{\pi/2} \sin\varphi\,d\varphi\,d\theta \quad \text{Simplify.}$$

$$= \ln 2 \int_0^{\pi/2} (-\cos\varphi)\Big|_0^{\pi/2}\,d\theta \quad \text{Evaluate middle integral with respect to } \varphi.$$

$$= \ln 2 \int_0^{\pi/2}\,d\theta = \frac{\pi \ln 2}{2}. \quad \text{Evaluate outer integral with respect to } \theta.$$

Related Exercises 41, 43 ◄

EXAMPLE 7 Ice cream cone Find the volume of the solid region D that lies inside the cone $\varphi = \pi/6$ and inside the sphere $\rho = 4$ (**Figure 16.62a**).

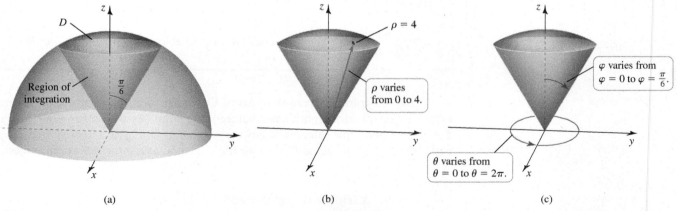

(a)   (b)   (c)

Figure 16.62

SOLUTION To find the volume, we evaluate a triple integral with $f = 1$. In the radial direction, the region extends from the origin $\rho = 0$ to the sphere $\rho = 4$ (**Figure 16.62b**). To sweep out all points of D, φ varies from 0 to $\pi/6$, and θ varies from 0 to 2π (**Figure 16.62c**). The volume of the region is

$$\iiint_D dV = \int_0^{2\pi} \int_0^{\pi/6} \int_0^4 \rho^2 \sin\varphi \, d\rho \, d\varphi \, d\theta \quad \text{Convert to an iterated integral.}$$

$$= \int_0^{2\pi} \int_0^{\pi/6} \frac{\rho^3}{3}\Big|_0^4 \sin\varphi \, d\varphi \, d\theta \quad \text{Evaluate inner integral with respect to } \rho.$$

$$= \frac{64}{3} \int_0^{2\pi} \int_0^{\pi/6} \sin\varphi \, d\varphi \, d\theta \quad \text{Simplify.}$$

$$= \frac{64}{3} \int_0^{2\pi} \underbrace{(-\cos\varphi)\Big|_0^{\pi/6}}_{1 - \sqrt{3}/2} d\theta \quad \text{Evaluate middle integral with respect to } \varphi.$$

$$= \frac{32}{3}(2 - \sqrt{3}) \int_0^{2\pi} d\theta \quad \text{Simplify.}$$

$$= \frac{64\pi}{3}(2 - \sqrt{3}). \quad \text{Evaluate outer integral with respect to } \theta.$$

Related Exercises 49, 51 ◄

SECTION 16.5 EXERCISES

Getting Started

1. Explain how cylindrical coordinates are used to describe a point in $\mathbb{R}^3$.

2. Explain how spherical coordinates are used to describe a point in $\mathbb{R}^3$.

3. Describe the set $\{(r, \theta, z): r = 4z\}$ in cylindrical coordinates.

4. Describe the set $\{(\rho, \varphi, \theta): \varphi = \pi/4\}$ in spherical coordinates.

5. Explain why $dz \, r \, dr \, d\theta$ is the volume of a small "box" in cylindrical coordinates.

6. Explain why $\rho^2 \sin\varphi \, d\rho \, d\varphi \, d\theta$ is the volume of a small "box" in spherical coordinates.

7. Write the integral $\iiint_D w(r, \theta, z) \, dV$ as an iterated integral, where the region D, expressed in cylindrical coordinates, is $D = \{(r, \theta, z): G(r, \theta) \le z \le H(r, \theta), g(\theta) \le r \le h(\theta), \alpha \le \theta \le \beta\}$.

8. Write the integral $\iiint_D w(\rho, \varphi, \theta) \, dV$ as an iterated integral, where the region D, expressed in spherical coordinates, is $D = \{(\rho, \varphi, \theta): g(\varphi, \theta) \le \rho \le h(\varphi, \theta), a \le \varphi \le b, \alpha \le \theta \le \beta\}$.

9. What coordinate system is *suggested* if the integrand of a triple integral involves $x^2 + y^2$?

10. What coordinate system is *suggested* if the integrand of a triple integral involves $x^2 + y^2 + z^2$?

Practice Exercises

11–14. Sets in cylindrical coordinates *Identify and sketch the following sets in cylindrical coordinates.*

11. $\{(r, \theta, z): 0 \le r \le 3, 0 \le \theta \le \pi/3, 1 \le z \le 4\}$

12. $\{(r, \theta, z): 0 \le \theta \le \pi/2, z = 1\}$

13. $\{(r, \theta, z): 2r \le z \le 4\}$

14. $\{(r, \theta, z): 0 \le z \le 8 - 2r\}$

15–22. Integrals in cylindrical coordinates *Evaluate the following integrals in cylindrical coordinates. The figures, if given, illustrate the region of integration.*

15. $\displaystyle\int_0^{2\pi} \int_0^1 \int_{-1}^1 dz \, r \, dr \, d\theta$

16. $\displaystyle\int_0^3 \int_{-\sqrt{9-y^2}}^{\sqrt{9-y^2}} \int_0^{9-3\sqrt{x^2+y^2}} dz \, dx \, dy$

17. $\int_{-1}^{1} \int_{-\sqrt{1-y^2}}^{\sqrt{1-y^2}} \int_{-1}^{1} (x^2 + y^2)^{3/2} \, dz \, dx \, dy$

18. $\int_{-3}^{3} \int_{0}^{\sqrt{9-x^2}} \int_{0}^{2} \dfrac{1}{1 + x^2 + y^2} \, dz \, dy \, dx$

19. $\int_{0}^{4} \int_{0}^{\sqrt{2}/2} \int_{x}^{\sqrt{1-x^2}} e^{-x^2-y^2} \, dy \, dx \, dz$

20. $\int_{-4}^{4} \int_{-\sqrt{16-x^2}}^{\sqrt{16-x^2}} \int_{\sqrt{x^2+y^2}}^{4} dz \, dy \, dx$

21. $\int_{0}^{3} \int_{0}^{\sqrt{9-x^2}} \int_{0}^{\sqrt{x^2+y^2}} (x^2 + y^2)^{-1/2} \, dz \, dy \, dx$

22. $\int_{-1}^{1} \int_{0}^{1/2} \int_{\sqrt{3}y}^{\sqrt{1-y^2}} (x^2 + y^2)^{1/2} \, dx \, dy \, dz$

23–26. Mass from density *Find the mass of the following objects with the given density functions. Assume (r, θ, z) are cylindrical coordinates.*

23. The solid cylinder $D = \{(r, \theta, z): 0 \le r \le 4, 0 \le z \le 10\}$ with density $\rho(r, \theta, z) = 1 + z/2$

24. The solid cylinder $D = \{(r, \theta, z): 0 \le r \le 3, 0 \le z \le 2\}$ with density $\rho(r, \theta, z) = 5e^{-r^2}$

25. The solid cone $D = \{(r, \theta, z): 0 \le z \le 6 - r, 0 \le r \le 6\}$ with density $\rho(r, \theta, z) = 7 - z$

26. The solid paraboloid $D = \{(r, \theta, z): 0 \le z \le 9 - r^2, 0 \le r \le 3\}$ with density $\rho(r, \theta, z) = 1 + z/9$

27. Which weighs more? For $0 \le r \le 1$, the solid bounded by the cone $z = 4 - 4r$ and the solid bounded by the paraboloid $z = 4 - 4r^2$ have the same base in the xy-plane and the same height. Which object has the greater mass if the density of both objects is $\rho(r, \theta, z) = 10 - 2z$?

28. Which weighs more? Which of the objects in Exercise 27 weighs more if the density of both objects is $\rho(r, \theta, z) = \dfrac{8}{\pi} e^{-z}$?

29–34. Volumes in cylindrical coordinates *Use cylindrical coordinates to find the volume of the following solids.*

29. The solid bounded by the plane $z = 0$ and the hyperboloid $z = 3 - \sqrt{1 + x^2 + y^2}$

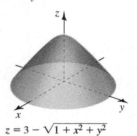

$z = 3 - \sqrt{1 + x^2 + y^2}$

30. The solid bounded by the plane $z = 25$ and the paraboloid $z = x^2 + y^2$

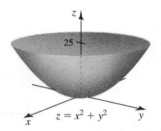

$z = x^2 + y^2$

31. The solid bounded by the plane $z = \sqrt{29}$ and the hyperboloid $z = \sqrt{4 + x^2 + y^2}$

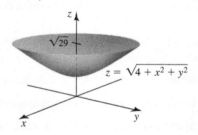

$z = \sqrt{4 + x^2 + y^2}$

32. The solid cylinder whose height is 4 and whose base is the disk $\{(r, \theta): 0 \le r \le 2 \cos \theta\}$

33. The solid in the first octant bounded by the cylinder $r = 1$, and the planes $z = x$ and $z = 0$

34. The solid bounded by the cylinders $r = 1$ and $r = 2$, and the planes $z = 4 - x - y$ and $z = 0$

35–38. Sets in spherical coordinates *Identify and sketch the following sets in spherical coordinates.*

35. $\{(\rho, \varphi, \theta): 1 \le \rho \le 3\}$

36. $\{(\rho, \varphi, \theta): \rho = 2 \csc \varphi, 0 < \varphi < \pi\}$

37. $\{(\rho, \varphi, \theta): \rho = 4 \cos \varphi, 0 \le \varphi \le \pi/2\}$

38. $\{(\rho, \varphi, \theta): \rho = 2 \sec \varphi, 0 \le \varphi < \pi/2\}$

T 39–40. Latitude, longitude, and distances *Assume Earth is a sphere with radius $r = 3960$ miles, oriented in xyz-space so that its center passes through the origin O, the positive z-axis passes through the North Pole, and the xz-plane passes through Greenwich, England (the intersection of Earth's surface and the xz-plane is called the prime meridian). The location of a point on Earth is given by its latitude* *(degrees north or south of the equator) and its longitude (degrees east or west of the prime meridian).*

39. Seattle has a latitude of 47.6° North and a longitude of 122.3° West; Rome, Italy, has a latitude of 41.9° North and a longitude of 12.5° East.

    a. Find the approximate spherical and rectangular coordinates of Seattle. Express the angular coordinates in radians.

    b. Find the approximate spherical and rectangular coordinates of Rome.

    c. Consider the intersection curve of a sphere, and a plane passing through the center of the sphere and two points A and B on the sphere. It can be shown that the arc length of the segment of the intersection curve from A to B is the shortest distance on the sphere from A to B. Find the approximate shortest distance from Seattle to Rome. (*Hint:* Recall that $\cos\theta = \dfrac{\mathbf{u}\cdot\mathbf{v}}{|\mathbf{u}||\mathbf{v}|}$, where $0 \le \theta \le \pi$ is the angle between $\mathbf{u}$ and $\mathbf{v}$; use the arc length formula $s = r\theta$ to find the distance.)

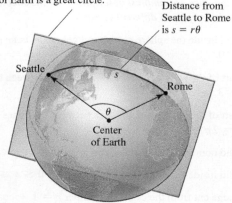

The intersection of Earth with the plane passing through Seattle, Rome, and the center of Earth is a great circle.

Distance from Seattle to Rome is $s = r\theta$

40. Los Angeles has a latitude of 34.05° North and a longitude of 118.24° West, and New York City has a latitude of 40.71° North and a longitude of 74.01° West. Find the approximate shortest distance from Los Angeles to New York City.

41–47. Integrals in spherical coordinates *Evaluate the following integrals in spherical coordinates.*

41. $\displaystyle\iiint_D (x^2 + y^2 + z^2)^{5/2}\,dV;\ D$ is the unit ball.

42. $\displaystyle\iiint_D e^{-(x^2+y^2+z^2)^{3/2}}\,dV;\ D$ is the unit ball.

43. $\displaystyle\iiint_D \frac{dV}{(x^2 + y^2 + z^2)^{3/2}};\ D$ is the solid between the spheres of radius 1 and 2 centered at the origin.

44. $\displaystyle\int_0^{2\pi}\int_0^{\pi/3}\int_0^{4\sec\varphi} \rho^2\sin\varphi\,d\rho\,d\varphi\,d\theta$

45. $\displaystyle\int_0^{\pi}\int_0^{\pi/6}\int_{2\sec\varphi}^{4} \rho^2\sin\varphi\,d\rho\,d\varphi\,d\theta$

46. $\displaystyle\int_0^{2\pi}\int_0^{\pi/4}\int_1^{2\sec\varphi} (\rho^{-3})\,\rho^2\sin\varphi\,d\rho\,d\varphi\,d\theta$

47. $\displaystyle\int_0^{2\pi}\int_{\pi/6}^{\pi/3}\int_0^{2\csc\varphi} \rho^2\sin\varphi\,d\rho\,d\varphi\,d\theta$

48–54. Volumes in spherical coordinates *Use spherical coordinates to find the volume of the following solids.*

48. A ball of radius $a > 0$

49. The solid bounded by the sphere $\rho = 2\cos\varphi$ and the hemisphere $\rho = 1, z \ge 0$

50. The solid cardioid of revolution $D = \{(\rho, \varphi, \theta): 0 \le \rho \le 1 + \cos\varphi, 0 \le \varphi \le \pi, 0 \le \theta \le 2\pi\}$

51. The solid outside the cone $\varphi = \pi/4$ and inside the sphere $\rho = 4\cos\varphi$

52. The solid bounded by the cylinders $r = 1$ and $r = 2$, and the cones $\varphi = \pi/6$ and $\varphi = \pi/3$

53. That part of the ball $\rho \le 4$ that lies between the planes $z = 2$ and $z = 2\sqrt{3}$

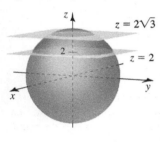

54. The solid lying between the planes $z = 1$ and $z = 2$ that is bounded by the cone $z = (x^2 + y^2)^{1/2}$

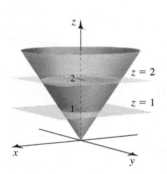

55. Explain why or why not Determine whether the following statements are true and give an explanation or counterexample.

   a. Any point on the z-axis has more than one representation in both cylindrical and spherical coordinates.

   b. The sets $\{(r, \theta, z): r = z\}$ in cylindrical coordinates and the set $\{(\rho, \varphi, \theta): \varphi = \pi/4\}$ in spherical coordinates describe the same set of points.

56. Spherical to rectangular Convert the equation $\rho^2 = \sec 2\varphi$, where $0 \le \varphi < \pi/4$, to rectangular coordinates and identify the surface.

57. Spherical to rectangular Convert the equation $\rho^2 = -\sec 2\varphi$, where $\pi/4 < \varphi \le \pi/2$, to rectangular coordinates and identify the surface.

58–61. Mass from density *Find the mass of the following solids with the given density functions. Note that density is described by the function f to avoid confusion with the radial spherical coordinate ρ.*

58. The ball of radius 4 centered at the origin with a density $f(\rho, \varphi, \theta) = 1 + \rho$

59. The ball of radius 8 centered at the origin with a density $f(\rho, \varphi, \theta) = 2e^{-\rho^3}$

60. The solid cone $\{(r, \theta, z): 0 \le z \le 4, 0 \le r \le \sqrt{3}z, 0 \le \theta \le 2\pi\}$ with a density $f(r, \theta, z) = 5 - z$

61. The solid cylinder $\{(r, \theta, z): 0 \le r \le 2, 0 \le \theta \le 2\pi, -1 \le z \le 1\}$ with a density $f(r, \theta, z) = (2 - |z|)(4 - r)$

62–63. Changing order of integration *If possible, write an iterated integral in cylindrical coordinates of a function $g(r, \theta, z)$ for the following regions in the specified orders. Sketch the region of integration.*

62. The solid outside the cylinder $r = 1$ and inside the sphere $\rho = 5$, for $z \ge 0$, in the orders $dz\,dr\,d\theta$, $dr\,dz\,d\theta$, and $d\theta\,dz\,dr$

63. The solid above the cone $z = r$ and below the sphere $\rho = 2$, for $z \ge 0$, in the orders $dz\,dr\,d\theta$, $dr\,dz\,d\theta$, and $d\theta\,dz\,dr$

64–65. Changing order of integration *If possible, write iterated integrals in spherical coordinates for the following regions in the specified orders. Sketch the region of integration. Assume g is continuous on the region.*

64. $\int_0^{2\pi}\int_0^{\pi/2}\int_0^{4\sec\varphi} g(\rho, \varphi, \theta)\rho^2 \sin\varphi\, d\rho\, d\varphi\, d\theta$ in the orders $d\rho\, d\theta\, d\varphi$ and $d\theta\, d\rho\, d\varphi$

65. $\int_0^{2\pi}\int_{\pi/6}^{\pi/2}\int_{\csc\varphi}^{2} g(\rho, \varphi, \theta)\rho^2 \sin\varphi\, d\rho\, d\varphi\, d\theta$ in the orders $d\rho\, d\theta\, d\varphi$ and $d\theta\, d\rho\, d\varphi$

66–71. Miscellaneous volumes *Choose the best coordinate system and find the volume of the following solids. Surfaces are specified using the coordinates that give the simplest description, but the simplest integration may be with respect to different variables.*

66. The solid inside the sphere $\rho = 1$ and below the cone $\varphi = \pi/4$, for $z \ge 0$

67. That part of the solid cylinder $r \le 2$ that lies between the cones $\varphi = \pi/3$ and $\varphi = 2\pi/3$

68. That part of the ball $\rho \le 2$ that lies between the cones $\varphi = \pi/3$ and $\varphi = 2\pi/3$

69. The solid bounded by the cylinder $r = 1$, for $0 \le z \le x + y$

70. The solid inside the cylinder $r = 2\cos\theta$, for $0 \le z \le 4 - x$

71. The wedge cut from the cardioid cylinder $r = 1 + \cos\theta$ by the planes $z = 2 - x$ and $z = x - 2$

72. Volume of a drilled hemisphere Find the volume of material remaining in a hemisphere of radius 2 after a cylindrical hole of radius 1 is drilled through the center of the hemisphere perpendicular to its base.

73. Density distribution A right circular cylinder with height 8 cm and radius 2 cm is filled with water. A heated filament running along its axis produces a variable density in the water given by $\rho(r) = 1 - 0.05e^{-0.01r^2}$g/cm^3 ($\rho$ stands for density here, not for the radial spherical coordinate). Find the mass of the water in the cylinder. Neglect the volume of the filament.

74. Charge distribution A spherical cloud of electric charge has a known charge density $Q(\rho)$, where ρ is the spherical coordinate. Find the total charge in the cloud in the following cases.

   a. $Q(\rho) = \dfrac{2 \times 10^{-4}}{\rho^4}, 1 \le \rho < \infty$

   b. $Q(\rho) = (2 \times 10^{-4})e^{-0.01\rho^3}, 0 \le \rho < \infty$

Explorations and Challenges

75. Gravitational field due to spherical shell A point mass m is a distance d from the center of a thin spherical shell of mass M and radius R. The magnitude of the gravitational force on the point mass is given by the integral

$$F(d) = \frac{GMm}{4\pi} \int_0^{2\pi} \int_0^{\pi} \frac{(d - R\cos\varphi)\sin\varphi}{(R^2 + d^2 - 2Rd\cos\varphi)^{3/2}} \, d\varphi \, d\theta,$$

where G is the gravitational constant.

  a. Use the change of variable $x = \cos\varphi$ to evaluate the integral and show that if $d > R$, then $F(d) = GMm/d^2$, which means the force is the same as it would be if the mass of the shell were concentrated at its center.

  b. Show that if $d < R$ (the point mass is inside the shell), then $F = 0$.

76. Water in a gas tank Before a gasoline-powered engine is started, water must be drained from the bottom of the fuel tank. Suppose the tank is a right circular cylinder on its side with a length of 2 ft and a radius of 1 ft. If the water level is 6 in above the lowest part of the tank, determine how much water must be drained from the tank.

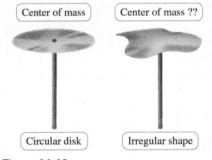

77–80. General volume formulas *Use integration to find the volume of the following solids. In each case, choose a convenient coordinate system, find equations for the bounding surfaces, set up a triple integral, and evaluate the integral. Assume a, b, c, r, R, and h are positive constants.*

77. Cone Find the volume of a solid right circular cone with height h and base radius r.

78. Spherical cap Find the volume of the cap of a sphere of radius R with thickness h.

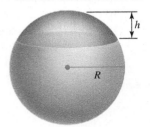

79. Frustum of a cone Find the volume of a truncated solid cone of height h whose ends have radii r and R.

80. Ellipsoid Find the volume of a solid ellipsoid with axes of lengths $2a$, $2b$, and $2c$.

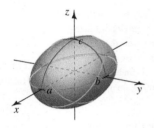

81. Intersecting spheres One sphere is centered at the origin and has a radius of R. Another sphere is centered at $(0, 0, r)$ and has a radius of r, where $r > R/2$. What is the volume of the region common to the two spheres?

QUICK CHECK ANSWERS

1. $\left(\sqrt{2}, 7\pi/4, 5\right), \left(1, \sqrt{3}, 5\right)$

2. $0 \le r \le 10, 0 \le \theta \le 2\pi, 0 \le z \le 20$

3. $\left(2\sqrt{2}, \pi/4, \pi/3\right), \left(1, 1, \sqrt{2}\right)$ ◄

16.6 Integrals for Mass Calculations

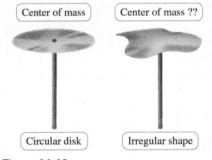

Figure 16.63

Intuition says that a thin circular disk (such as a DVD without a hole) should balance on a pencil placed at the center of the disk (**Figure 16.63**). If, however, you were given a thin plate with an irregular shape, at what point would it balance? This question asks about the *center of mass* of a thin object (thin enough that it can be treated as a two-dimensional region). Similarly, given a solid object with an irregular shape and variable density, where is the point at which all of the mass of the object would be located if it were treated as a point mass? In this section, we use integration to compute the center of mass of one-, two-, and three-dimensional objects.

Sets of Individual Objects

Figure 16.64

Methods for finding the center of mass of an object are ultimately based on a well-known playground principle: If two people with masses m_1 and m_2 sit at distances d_1 and d_2 from the pivot point of a seesaw (with no mass), then the seesaw balances, provided $m_1 d_1 = m_2 d_2$ (**Figure 16.64**).

QUICK CHECK 1 A 90-kg person sits 2 m from the balance point of a seesaw. How far from that point must a 60-kg person sit to balance the seesaw? Assume the seesaw has no mass. ◄

$$\bar{x} = \frac{m_1x_1 + m_2x_2}{m_1 + m_2}$$

Figure 16.65

➤ The center of mass may be viewed as the weighted average of the x-coordinates, with the masses serving as the weights. Notice how the units work out: If x_1 and x_2 have units of meters and m_1 and m_2 have units of kilograms, then $\bar{x}$ has units of meters.

QUICK CHECK 2 Solve the equation $m_1(x_1 - \bar{x}) + m_2(x_2 - \bar{x}) = 0$ for $\bar{x}$ to verify the preceding expression for the center of mass. ◄

Figure 16.66

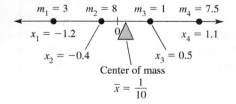

Figure 16.67

To generalize the problem, we introduce a coordinate system with the origin at $x = 0$ (**Figure 16.65**). Suppose the location of the balance point $\bar{x}$ is unknown. The coordinates of the two masses m_1 and m_2 are denoted x_1 and x_2, respectively, with $x_1 > x_2$. The mass m_1 is a distance $x_1 - \bar{x}$ from the balance point (because distance is positive and $x_1 > \bar{x}$). The mass m_2 is a distance $\bar{x} - x_2$ from the balance point (because $\bar{x} > x_2$). The playground principle becomes

$$\underbrace{m_1(x_1 - \bar{x})}_{\substack{\text{distance from} \\ \text{balance point} \\ \text{to } m_1}} = \underbrace{m_2(\bar{x} - x_2)}_{\substack{\text{distance from} \\ \text{balance point} \\ \text{to } m_2}},$$

or $m_1(x_1 - \bar{x}) + m_2(x_2 - \bar{x}) = 0$.

Solving this equation for $\bar{x}$, we find that the balance point or *center of mass* of the two-mass system is located at

$$\bar{x} = \frac{m_1x_1 + m_2x_2}{m_1 + m_2}.$$

The quantities m_1x_1 and m_2x_2 are called *moments about the origin* (or just *moments*). The location of the center of mass is the sum of the moments divided by the sum of the masses.

For example, an 80-kg man standing 2 m to the right of the origin will balance a 160-kg gorilla sitting 4 m to the left of the origin, provided the pivot on their seesaw is placed at

$$\bar{x} = \frac{80(2) + 160(-4)}{80 + 160} = -2,$$

or 2 m to the left of the origin (**Figure 16.66**).

Several Objects on a Line   Generalizing the preceding argument to n objects having masses $m_1, m_2, \ldots,$ and m_n with coordinates $x_1, x_2, \ldots,$ and x_n, respectively, the balance condition becomes

$$m_1(x_1 - \bar{x}) + m_2(x_2 - \bar{x}) + \cdots + m_n(x_n - \bar{x}) = \sum_{k=1}^{n} m_k(x_k - \bar{x}) = 0.$$

Solving this equation for the location of the center of mass, we find that

$$\bar{x} = \frac{m_1x_1 + m_2x_2 + \cdots + m_nx_n}{m_1 + m_2 + \cdots + m_n} = \frac{\displaystyle\sum_{k=1}^{n} m_kx_k}{\displaystyle\sum_{k=1}^{n} m_k}.$$

Again, the location of the center of mass is the sum of the moments $m_1x_1, m_2x_2, \ldots,$ and m_nx_n divided by the sum of the masses.

EXAMPLE 1   Center of mass for four objects Find the point at which the system shown in **Figure 16.67** balances.

SOLUTION The center of mass is

$$\bar{x} = \frac{m_1x_1 + m_2x_2 + m_3x_3 + m_4x_4}{m_1 + m_2 + m_3 + m_4}$$

$$= \frac{3(-1.2) + 8(-0.4) + 1(0.5) + 7.5(1.1)}{3 + 8 + 1 + 7.5}$$

$$= \frac{1}{10}.$$

The balancing point is slightly to the right of the origin.

Related Exercises 7–8 ◄

▶ Density is usually measured in units of *mass per volume*. However, for thin, narrow objects such as rods and wires, linear density with units of *mass per length* is used. For thin, flat objects such as plates and sheets, area density with units of *mass per area* is used.

QUICK CHECK 3 In Figure 16.68, suppose $a = 0, b = 3$, and the density of the rod in g/cm is $\rho(x) = 4 - x$. Where is the rod lightest? Heaviest? ◀

Continuous Objects in One Dimension

Now consider a thin rod or wire with density ρ that varies along the length of the rod (**Figure 16.68**). The density in this case has units of mass per length (for example, g/cm). As before, we want to determine the location $\bar{x}$ at which the rod balances on a pivot.

Using the slice-and-sum strategy, we divide the rod, which corresponds to the interval $a \leq x \leq b$, into n subintervals, each with a width of $\Delta x = \dfrac{b - a}{n}$ (**Figure 16.69**). The corresponding grid points are

$$x_0 = a, x_1 = a + \Delta x, \ldots, x_k = a + k\,\Delta x, \ldots, \text{and } x_n = b.$$

The mass of the kth segment of the rod is approximately the density at x_k multiplied by the length of the interval, or $m_k \approx \rho(x_k)\Delta x$.

Density (mass per unit length) varies with *x*.

Figure 16.68

$$Mass = m_k \approx \rho(x_k)\Delta x$$

Figure 16.69

We now use the center-of-mass formula for several distinct objects to write the approximate center of mass of the rod as

$$\bar{x} = \frac{\sum_{k=1}^{n} m_k x_k}{\sum_{k=1}^{n} m_k} \approx \frac{\sum_{k=1}^{n} (\rho(x_k)\Delta x)\,x_k}{\sum_{k=1}^{n} \rho(x_k)\Delta x}.$$

▶ An object consisting of two different materials that meet at an interface has a discontinuous density function. Physical density functions either are continuous or have a finite number of discontinuities.

▶ We assume the rod has positive mass and the limits in the numerator and denominator exist, so the limit of the quotient is the quotient of the limits.

To model a rod with a continuous density, we let $\Delta x \to 0$ and $n \to \infty$; the center of mass of the rod is

$$\bar{x} = \lim_{\Delta x \to 0} \frac{\sum_{k=1}^{n} (\rho(x_k)\Delta x)\,x_k}{\sum_{k=1}^{n} \rho(x_k)\Delta x} = \frac{\lim_{\Delta x \to 0} \sum_{k=1}^{n} x_k\,\rho(x_k)\Delta x}{\lim_{\Delta x \to 0} \sum_{k=1}^{n} \rho(x_k)\Delta x} = \frac{\int_a^b x\rho(x)\,dx}{\int_a^b \rho(x)\,dx}.$$

As discussed in Section 6.7, the denominator of the last fraction, $\int_a^b \rho(x)\,dx$, is the mass of the rod. The numerator is the "sum" of the moments of the individual pieces of the rod, which is called the *total moment*.

▶ The units of a moment are mass × length. The center of mass is a moment divided by a mass, which has units of length. Notice that if the density is constant, then ρ effectively does not enter the calculation of $\bar{x}$.

DEFINITION Center of Mass in One Dimension

Let ρ be an integrable density function on the interval $[a, b]$ (which represents a thin rod or wire). The **center of mass** is located at the point $\bar{x} = \dfrac{M}{m}$, where the **total moment** M and mass m are

$$M = \int_a^b x\rho(x)\,dx \quad \text{and} \quad m = \int_a^b \rho(x)\,dx.$$

Observe the parallels between the discrete and continuous cases:

$$n \text{ individual objects:} \quad \bar{x} = \frac{\sum_{k=1}^{n} x_k m_k}{\sum_{k=1}^{n} m_k}; \quad \text{continuous object:} \quad \bar{x} = \frac{\int_a^b x\rho(x)\,dx}{\int_a^b \rho(x)\,dx}.$$

EXAMPLE 2 Center of mass of a one-dimensional object Suppose a thin 2-m bar is made of an alloy whose density in kg/m is $\rho(x) = 1 + x^2$, where $0 \le x \le 2$. Find the center of mass of the bar.

SOLUTION The total mass of the bar in kilograms is

$$m = \int_a^b \rho(x)\,dx = \int_0^2 (1 + x^2)\,dx = \left(x + \frac{x^3}{3} \right)\Big|_0^2 = \frac{14}{3}.$$

The total moment of the bar, with units kg-m, is

$$M = \int_a^b x\rho(x)\,dx = \int_0^2 x(1 + x^2)\,dx = \left(\frac{x^2}{2} + \frac{x^4}{4} \right)\Big|_0^2 = 6.$$

> Notice that the density of the bar increases with x. As a consistency check, our calculation must result in a center of mass to the right of the midpoint of the bar.

Therefore, the center of mass is located at $\bar{x} = \dfrac{M}{m} = \dfrac{9}{7} \approx 1.29$ m.

Related Exercises 10–11 ◀

Two-Dimensional Objects

In two dimensions, we start with an integrable density function $\rho(x, y)$ defined over a closed bounded region R in the xy-plane. The density is now an *area density* with units of mass per area (for example, kg/m²). The region represents a thin plate (or *lamina*). The center of mass is the point at which a pivot must be located to balance the plate. If the density is constant, the location of the center of mass depends only on the shape of the plate, in which case the center of mass is called the *centroid*.

For a two- or three-dimensional object, the coordinates for the center of mass are computed independently by applying the one-dimensional argument in each coordinate direction (**Figure 16.70**). The mass of the plate is the integral of the density function over R:

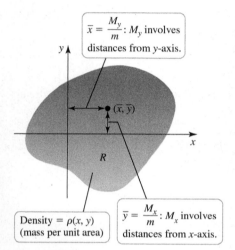

Figure 16.70

$$m = \iint_R \rho(x, y)\,dA.$$

In analogy with the moment calculation in the one-dimensional case, we now define two moments.

> The moment with respect to the y-axis M_y is a weighted average of distances from the y-axis, so it has x in the integrand (the distance between a point and the y-axis). Similarly, the moment with respect to the x-axis M_x is a weighted average of distances from the x-axis, so it has y in the integrand.

QUICK CHECK 4 Explain why the integral for M_y has x in the integrand. Explain why the density drops out of the center-of-mass calculation if it is constant. ◀

DEFINITION Center of Mass in Two Dimensions

Let ρ be an integrable area density function defined over a closed bounded region R in $\mathbb{R}^2$. The coordinates of the center of mass of the object represented by R are

$$\bar{x} = \frac{M_y}{m} = \frac{1}{m}\iint_R x\rho(x, y)\,dA \quad\text{and}\quad \bar{y} = \frac{M_x}{m} = \frac{1}{m}\iint_R y\rho(x, y)\,dA,$$

where $m = \iint_R \rho(x, y)\,dA$ is the mass, and M_y and M_x are the moments with respect to the y-axis and x-axis, respectively. If ρ is constant, the center of mass is called the **centroid** and is independent of the density.

As before, the center-of-mass coordinates are weighted averages of the distances from the coordinate axes. For two- and three-dimensional objects, the center of mass need not lie within the object (Exercises 51, 61, and 62).

EXAMPLE 3 Centroid calculation Find the centroid (center of mass) of the unit-density, dart-shaped region bounded by the y-axis and the curves $y = e^{-x} - \frac{1}{2}$ and $y = \frac{1}{2} - e^{-x}$ (**Figure 16.71**).

SOLUTION Because the region is symmetric about the x-axis and the density is constant, the y-coordinate of the center of mass is $\bar{y} = 0$. This leaves the integrals for m and M_y to evaluate.

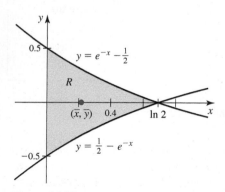

Figure 16.71

The first task is to find the point at which the curves intersect. Solving $e^{-x} - \frac{1}{2} = \frac{1}{2} - e^{-x}$, we find that $x = \ln 2$, from which it follows that $y = 0$. Therefore, the intersection point is $(\ln 2, 0)$. The moment M_y (with $\rho = 1$) is given by

> ▶ The density does not enter the center-of-mass calculation when the density is constant. So it is easiest to set $\rho = 1$.

> ▶ If possible, try to arrange the coordinate system so that at least one of the integrations in the center-of-mass calculation can be avoided by using symmetry. Often the mass (or area) can be found using geometry if the density is constant.

$$
\begin{aligned}
M_y &= \int_0^{\ln 2} \int_{1/2 - e^{-x}}^{e^{-x} - 1/2} x \, dy \, dx && \text{Definition of } M_y \\
&= \int_0^{\ln 2} x \left(\left(e^{-x} - \frac{1}{2} \right) - \left(\frac{1}{2} - e^{-x} \right) \right) dx && \text{Evaluate inner integral.} \\
&= \int_0^{\ln 2} x(2e^{-x} - 1) \, dx. && \text{Simplify.}
\end{aligned}
$$

Using integration by parts for this integral, we find that

$$
\begin{aligned}
M_y &= \int_0^{\ln 2} \underset{u}{x} \underbrace{(2e^{-x} - 1) \, dx}_{dv} \\
&= -x(2e^{-x} + x) \Big|_0^{\ln 2} + \int_0^{\ln 2} (2e^{-x} + x) \, dx && \text{Integration by parts} \\
&= 1 - \ln 2 - \frac{1}{2} \ln^2 2 \approx 0.067. && \text{Evaluate and simplify.}
\end{aligned}
$$

With $\rho = 1$, the mass of the region is given by

$$
\begin{aligned}
m &= \int_0^{\ln 2} \int_{1/2 - e^{-x}}^{e^{-x} - 1/2} dy \, dx && \text{Definition of } m \\
&= \int_0^{\ln 2} (2e^{-x} - 1) \, dx && \text{Evaluate inner integral.} \\
&= (-2e^{-x} - x) \Big|_0^{\ln 2} && \text{Evaluate outer integral.} \\
&= 1 - \ln 2 \approx 0.307. && \text{Simplify.}
\end{aligned}
$$

Therefore, the x-coordinate of the center of mass is $\bar{x} = \dfrac{M_y}{m} \approx 0.217$. The center of mass is located approximately at $(0.217, 0)$.

Related Exercise 18 ◀

EXAMPLE 4   Variable-density plate Find the center of mass of the rectangular plate $R = \{(x, y) \colon -1 \le x \le 1, 0 \le y \le 1\}$ with a density of $\rho(x, y) = 2 - y$ (heavy at the lower edge and light at the top edge; **Figure 16.72**).

SOLUTION Because the plate is symmetric with respect to the y-axis and because the density is independent of x, we have $\bar{x} = 0$. We must still compute m and M_x.

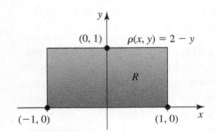

Figure 16.72

> ▶ To verify that $\bar{x} = 0$, notice that to find M_y, we integrate an odd function in x over $-1 \le x \le 1$; the result is zero.

$$
m = \iint_R \rho(x, y) \, dA = \int_{-1}^1 \int_0^1 (2 - y) \, dy \, dx = \frac{3}{2} \int_{-1}^1 dx = 3
$$

$$
M_x = \iint_R y\rho(x, y) \, dA = \int_{-1}^1 \int_0^1 y(2 - y) \, dy \, dx = \frac{2}{3} \int_{-1}^1 dx = \frac{4}{3}
$$

Therefore, the center-of-mass coordinates are

$$
\bar{x} = \frac{M_y}{m} = 0 \quad \text{and} \quad \bar{y} = \frac{M_x}{m} = \frac{4/3}{3} = \frac{4}{9}.
$$

Related Exercise 21 ◀

Three-Dimensional Objects

We now extend the preceding arguments to compute the center of mass of three-dimensional solids. Assume D is a closed bounded region in $\mathbb{R}^3$, on which an integrable density function ρ is defined. The units of the density are mass per volume (for example, g/cm^3). The coordinates of the center of mass depend on the mass of the region, which by

Section 16.4 is the integral of the density function over D. Three moments enter the picture: M_{yz} involves distances from the yz-plane; therefore, it has an x in the integrand. Similarly, M_{xz} involves distances from the xz-plane, so it has a y in the integrand, and M_{xy} involves distances from the xy-plane, so it has a z in the integrand. As before, the coordinates of the center of mass are the total moments divided by the total mass (**Figure 16.73**).

QUICK CHECK 5 Explain why the integral for the moment M_{xy} has z in the integrand. ◄

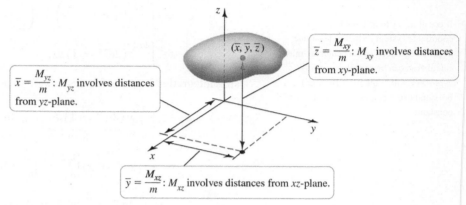

$\bar{x} = \dfrac{M_{yz}}{m}$: M_{yz} involves distances from yz-plane.

$\bar{z} = \dfrac{M_{xy}}{m}$: M_{xy} involves distances from xy-plane.

$\bar{y} = \dfrac{M_{xz}}{m}$: M_{xz} involves distances from xz-plane.

Figure 16.73

DEFINITION Center of Mass in Three Dimensions

Let ρ be an integrable density function on a closed bounded region D in $\mathbb{R}^3$. The coordinates of the center of mass of the region are

$$\bar{x} = \frac{M_{yz}}{m} = \frac{1}{m} \iiint\limits_{D} x\rho(x, y, z)\, dV, \quad \bar{y} = \frac{M_{xz}}{m} = \frac{1}{m} \iiint\limits_{D} y\rho(x, y, z)\, dV, \text{ and}$$

$$\bar{z} = \frac{M_{xy}}{m} = \frac{1}{m} \iiint\limits_{D} z\rho(x, y, z)\, dV,$$

where $m = \iiint_{D} \rho(x, y, z)\, dV$ is the mass, and M_{yz}, M_{xz}, and M_{xy} are the moments with respect to the coordinate planes.

EXAMPLE 5   Center of mass with constant density Find the center of mass of the constant-density solid cone D bounded by the surface $z = 4 - \sqrt{x^2 + y^2}$ and $z = 0$ (**Figure 16.74**).

SOLUTION Because the cone is symmetric about the z-axis and has uniform density, the center of mass lies on the z-axis; that is, $\bar{x} = 0$ and $\bar{y} = 0$. Setting $z = 0$, the base of the cone in the xy-plane is the disk of radius 4 centered at the origin. Therefore, the cone has height 4 and radius 4; by the volume formula, its volume is $\pi r^2 h / 3 = 64\pi / 3$. The cone has a constant density, so we assume $\rho = 1$ and its mass is $m = 64\pi / 3$.

To obtain the value of $\bar{z}$, only M_{xy} needs to be calculated, which is most easily done in cylindrical coordinates. The cone is described by the equation $z = 4 - \sqrt{x^2 + y^2} = 4 - r$. The projection of the cone onto the xy-plane, which is the region of integration in the xy-plane, is the disk $R = \{(r, \theta): 0 \le r \le 4, 0 \le \theta \le 2\pi\}$. The integration for M_{xy} now follows:

$z = 4 - \sqrt{x^2 + y^2}$

$(\bar{x}, \bar{y}, \bar{z}) = (0, 0, 1)$

$R = \{(r, \theta): 0 \le r \le 4, 0 \le \theta \le 2\pi\}$

Figure 16.74

$$\begin{aligned}
M_{xy} &= \iiint\limits_{D} z\, dV && \text{Definition of } M_{xy} \text{ with } \rho = 1 \\[2mm]
&= \int_0^{2\pi} \int_0^4 \int_0^{4-r} z\, dz\, r\, dr\, d\theta && \text{Convert to an iterated integral.} \\[2mm]
&= \int_0^{2\pi} \int_0^4 \frac{z^2}{2}\bigg|_0^{4-r} r\, dr\, d\theta && \text{Evaluate inner integral with respect to } z. \\[2mm]
&= \frac{1}{2} \int_0^{2\pi} \int_0^4 r(4 - r)^2\, dr\, d\theta && \text{Simplify.}
\end{aligned}$$

$$= \frac{1}{2} \int_0^{2\pi} \frac{64}{3} \, d\theta \qquad \text{Evaluate middle integral with respect to } r.$$

$$= \frac{64\pi}{3}. \qquad \text{Evaluate outer integral with respect to } \theta.$$

The z-coordinate of the center of mass is $\bar{z} = \dfrac{M_{xy}}{m} = \dfrac{64\pi/3}{64\pi/3} = 1$, and the center of mass is located at $(0, 0, 1)$. It can be shown (Exercise 55) that the center of mass of a constant-density cone of height h is located $h/4$ units from the base on the axis of the cone, independent of the radius.

Related Exercise 28 ◀

EXAMPLE 6 Center of mass with variable density Find the center of mass of the interior of the hemisphere D of radius a with its base on the xy-plane. The density of the object given in spherical coordinates is $f(\rho, \varphi, \theta) = 2 - \rho/a$ (heavy near the center and light near the outer surface; **Figure 16.75**).

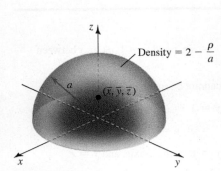

Density $= 2 - \dfrac{\rho}{a}$

Figure 16.75

SOLUTION The center of mass lies on the z-axis because of the symmetry of both the solid and the density function; therefore, $\bar{x} = \bar{y} = 0$. Only the integrals for m and M_{xy} need to be evaluated, and they should be done in spherical coordinates.

The integral for the mass is

$$m = \iiint_D f(\rho, \varphi, \theta) \, dV \qquad \text{Definition of } m$$

$$= \int_0^{2\pi} \int_0^{\pi/2} \int_0^a \left(2 - \frac{\rho}{a}\right) \rho^2 \sin \varphi \, d\rho \, d\varphi \, d\theta \qquad \text{Convert to an iterated integral.}$$

$$= \int_0^{2\pi} \int_0^{\pi/2} \left(\frac{2\rho^3}{3} - \frac{\rho^4}{4a}\right)\Big|_0^a \sin \varphi \, d\varphi \, d\theta \qquad \text{Evaluate inner integral with respect to } \rho.$$

$$= \int_0^{2\pi} \int_0^{\pi/2} \frac{5a^3}{12} \sin \varphi \, d\varphi \, d\theta \qquad \text{Simplify.}$$

$$= \frac{5a^3}{12} \int_0^{2\pi} \underbrace{(-\cos \varphi)\Big|_0^{\pi/2}}_{1} \, d\theta \qquad \text{Evaluate middle integral with respect to } \varphi.$$

$$= \frac{5a^3}{12} \int_0^{2\pi} d\theta \qquad \text{Simplify.}$$

$$= \frac{5\pi a^3}{6}. \qquad \text{Evaluate outer integral with respect to } \theta.$$

In spherical coordinates, $z = \rho \cos \varphi$, so the integral for the moment M_{xy} is

$$M_{xy} = \iiint_D z \, f(\rho, \varphi, \theta) \, dV \qquad \text{Definition of } M_{xy}$$

$$= \int_0^{2\pi} \int_0^{\pi/2} \int_0^a \underbrace{\rho \cos \varphi}_{z} \left(2 - \frac{\rho}{a}\right) \rho^2 \sin \varphi \, d\rho \, d\varphi \, d\theta \qquad \text{Convert to an iterated integral.}$$

$$= \int_0^{2\pi} \int_0^{\pi/2} \left(\frac{\rho^4}{2} - \frac{\rho^5}{5a}\right)\Big|_0^a \sin \varphi \cos \varphi \, d\varphi \, d\theta \qquad \text{Evaluate inner integral with respect to } \rho.$$

$$= \int_0^{2\pi} \int_0^{\pi/2} \frac{3a^4}{10} \underbrace{\sin \varphi \cos \varphi}_{(\sin 2\varphi)/2} \, d\varphi \, d\theta \qquad \text{Simplify.}$$

$$= \frac{3a^4}{10} \int_0^{2\pi} \underbrace{\left(-\frac{\cos 2\varphi}{4}\right)\Big|_0^{\pi/2}}_{1/2} \, d\theta \qquad \text{Evaluate middle integral with respect to } \varphi.$$

$$= \frac{3a^4}{20} \int_0^{2\pi} d\theta$$ Simplify.

$$= \frac{3\pi a^4}{10}.$$ Evaluate outer integral with respect to θ.

The z-coordinate of the center of mass is $\bar{z} = \dfrac{M_{xy}}{m} = \dfrac{3\pi a^4/10}{5\pi a^3/6} = \dfrac{9a}{25} = 0.36a$. It can be shown (Exercise 56) that the center of mass of a uniform-density hemispherical solid of radius a is $3a/8 = 0.375a$ units above the base. In this case, the variable density lowers the center of mass toward the base.

Related Exercise 35 ◀

SECTION 16.6 EXERCISES

Getting Started

1. Explain how to find the balance point for two people on opposite ends of a (massless) plank that rests on a pivot.

2. If a thin 1-m cylindrical rod has a density of $\rho = 1$ g/cm for its left half and a density of $\rho = 2$ g/cm for its right half, what is its mass and where is its center of mass?

3. Explain how to find the center of mass of a thin plate with a variable density.

4. In the integral for the moment M_x of a thin plate, why does y appear in the integrand?

5. Explain how to find the center of mass of a three-dimensional object with a variable density.

6. In the integral for the moment M_{xz} with respect to the xz-plane of a solid, why does y appear in the integrand?

Practice Exercises

7–8. Individual masses on a line *Sketch the following systems on a number line and find the location of the center of mass.*

7. $m_1 = 10$ kg located at $x = 3$ m; $m_2 = 3$ kg located at $x = -1$ m

8. $m_1 = 8$ kg located at $x = 2$ m; $m_2 = 4$ kg located at $x = -4$ m; $m_3 = 1$ kg located at $x = 0$ m

9–14. One-dimensional objects *Find the mass and center of mass of the thin rods with the following density functions.*

9. $\rho(x) = 1 + \sin x$, for $0 \le x \le \pi$

10. $\rho(x) = 1 + x^3$, for $0 \le x \le 1$

11. $\rho(x) = 2 - \dfrac{x^2}{16}$, for $0 \le x \le 4$

12. $\rho(x) = 2 + \cos x$, for $0 \le x \le \pi$

13. $\rho(x) = \begin{cases} 1 & \text{if } 0 \le x \le 2 \\ 1 + x & \text{if } 2 < x \le 4 \end{cases}$

14. $\rho(x) = \begin{cases} x^2 & \text{if } 0 \le x \le 1 \\ x(2 - x) & \text{if } 1 < x \le 2 \end{cases}$

15–20. Centroid calculations *Find the mass and centroid (center of mass) of the following thin plates, assuming constant density. Sketch the region corresponding to the plate and indicate the location of the center of mass. Use symmetry when possible to simplify your work.*

15. The region bounded by $y = \sin x$ and $y = 1 - \sin x$ between $x = \pi/4$ and $x = 3\pi/4$

16. The region in the first quadrant bounded by $x^2 + y^2 = 16$

17. The region bounded by $y = 1 - |x|$ and the x-axis

18. The region bounded by $y = e^x$, $y = e^{-x}$, $x = 0$, and $x = \ln 2$

19. The region bounded by $y = \ln x$, the x-axis, and $x = e$

20. The region bounded by $x^2 + y^2 = 1$ and $x^2 + y^2 = 9$, for $y \ge 0$

21–26. Variable-density plates *Find the center of mass of the following plane regions with variable density. Describe the distribution of mass in the region.*

21. $R = \{(x, y): 0 \le x \le 4, 0 \le y \le 2\}$; $\rho(x, y) = 1 + x/2$

22. $R = \{(x, y): 0 \le x \le 1, 0 \le y \le 5\}$; $\rho(x, y) = 2e^{-y/2}$

23. The triangular plate in the first quadrant bounded by $x + y = 4$ with $\rho(x, y) = 1 + x + y$

24. The upper half ($y \ge 0$) of the disk bounded by the circle $x^2 + y^2 = 4$ with $\rho(x, y) = 1 + y/2$

25. The upper half ($y \ge 0$) of the plate bounded by the ellipse $x^2 + 9y^2 = 9$ with $\rho(x, y) = 1 + y$

26. The quarter disk in the first quadrant bounded by $x^2 + y^2 = 4$ with $\rho(x, y) = 1 + x^2 + y^2$

27–32. Center of mass of constant-density solids *Find the center of mass of the following solids, assuming a constant density of 1. Sketch the region and indicate the location of the centroid. Use symmetry when possible and choose a convenient coordinate system.*

27. The upper half of the ball $x^2 + y^2 + z^2 \le 16$ (for $z \ge 0$)

28. The solid bounded by the paraboloid $z = x^2 + y^2$ and the plane $z = 25$

29. The tetrahedron in the first octant bounded by $z = 1 - x - y$ and the coordinate planes

30. The solid bounded by the cone $z = 16 - r$ and the plane $z = 0$

31. The sliced solid cylinder bounded by $x^2 + y^2 = 1$, $z = 0$, and $y + z = 1$

32. The solid bounded by the upper half ($z \ge 0$) of the ellipsoid $4x^2 + 4y^2 + z^2 = 16$

33–38. Variable-density solids *Find the coordinates of the center of mass of the following solids with variable density.*

33. $R = \{(x, y, z): 0 \le x \le 4, 0 \le y \le 1, 0 \le z \le 1\};$
$\rho(x, y, z) = 1 + x/2$

34. The region bounded by the paraboloid $z = 4 - x^2 - y^2$ and $z = 0$ with $\rho(x, y, z) = 5 - z$

35. The solid bounded by the upper half of the sphere $\rho = 6$ and $z = 0$ with density $f(\rho, \varphi, \theta) = 1 + \rho/4$

36. The interior of the cube in the first octant formed by the planes $x = 1, y = 1,$ and $z = 1$, with $\rho(x, y, z) = 2 + x + y + z$

37. The interior of the prism formed by the planes $z = x, x = 1,$ and $y = 4$, and the coordinate planes, with $\rho(x, y, z) = 2 + y$

38. The solid bounded by the cone $z = 9 - r$ and $z = 0$ with $\rho(r, \theta, z) = 1 + z$

39. **Explain why or why not** Determine whether the following statements are true and give an explanation or counterexample.

  a. A thin plate of constant density that is symmetric about the x-axis has a center of mass with an x-coordinate of zero.
  b. A thin plate of constant density that is symmetric about both the x-axis and the y-axis has its center of mass at the origin.
  c. The center of mass of a thin plate must lie on the plate.
  d. The center of mass of a connected solid region (all in one piece) must lie within the region.

40. **Limiting center of mass** A thin rod of length L has a linear density given by $\rho(x) = 2e^{-x/3}$ on the interval $0 \le x \le L$. Find the mass and center of mass of the rod. How does the center of mass change as $L \to \infty$?

41. **Limiting center of mass** A thin rod of length L has a linear density given by $\rho(x) = \dfrac{10}{1 + x^2}$ on the interval $0 \le x \le L$. Find the mass and center of mass of the rod. How does the center of mass change as $L \to \infty$?

42. **Limiting center of mass** A thin plate is bounded by the graphs of $y = e^{-x}, y = -e^{-x}, x = 0,$ and $x = L$. Find its center of mass. How does the center of mass change as $L \to \infty$?

43–44. Two-dimensional plates *Find the mass and center of mass of the thin constant-density plates shown in the figure.*

43.   44.

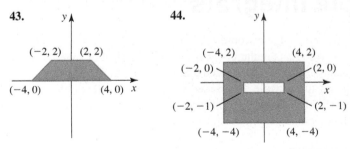

45–50. Centroids *Use polar coordinates to find the centroid of the following constant-density plane regions.*

45. The semicircular disk $R = \{(r, \theta): 0 \le r \le 2, 0 \le \theta \le \pi\}$

46. The quarter-circular disk $R = \{(r, \theta): 0 \le r \le 2, 0 \le \theta \le \pi/2\}$

47. The region bounded by the cardioid $r = 1 + \cos\theta$

48. The region bounded by the cardioid $r = 3 - 3\cos\theta$

49. The region bounded by one leaf of the rose $r = \sin 2\theta$, for $0 \le \theta \le \pi/2$

50. The region bounded by the limaçon $r = 2 + \cos\theta$

51. **Semicircular wire** A thin (one-dimensional) wire of constant density is bent into the shape of a semicircle of radius r. Find the location of its center of mass. (*Hint:* Treat the wire as a thin half-annulus with width Δa, and then let $\Delta a \to 0$.)

52. **Parabolic region** A thin plate of unit density occupies the region between the parabola $y = ax^2$ and the horizontal line $y = b$, where $a > 0$ and $b > 0$. Show that the center of mass is $\left(0, \dfrac{3b}{5}\right)$, independent of a.

53. **Circular crescent** Find the center of mass of the region in the first quadrant bounded by the circle $x^2 + y^2 = a^2$ and the lines $x = a$ and $y = a$, where $a > 0$.

54–57. Centers of mass for general objects *Consider the following two- and three-dimensional regions with variable dimensions. Specify the surfaces and curves that bound the region, choose a convenient coordinate system, and compute the center of mass assuming constant density. All parameters are positive real numbers.*

54. A solid rectangular box has sides of lengths a, b, and c. Where is the center of mass relative to the faces of the box?

55. A solid cone has a base with a radius of a and a height of h. How far from the base is the center of mass?

56. A solid is enclosed by a hemisphere of radius a. How far from the base is the center of mass?

57. A region is enclosed by an isosceles triangle with two sides of length s and a base of length b. How far from the base is the center of mass?

Explorations and Challenges

58. A tetrahedron is bounded by the coordinate planes and the plane $x/a + y/a + z/a = 1$. What are the coordinates of the center of mass?

59. A solid is enclosed by the upper half of an ellipsoid with a circular base of radius r and a height of a. How far from the base is the center of mass?

60. **Geographic vs. population center** Geographers measure the *geographical center* of a country (which is the centroid) and the *population center* of a country (which is the center of mass computed with the population density). A hypothetical country is shown in the figure with the location and population of five towns. Assuming no one lives outside the towns, find the geographical center of the country and the population center of the country.

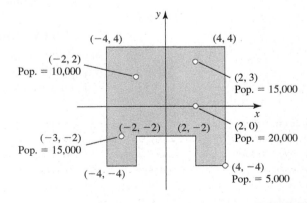

61. Center of mass on the edge Consider the thin constant-density plate $\{(r, \theta): 0 < a \leq r \leq 1, 0 \leq \theta \leq \pi\}$ bounded by two semicircles and the x-axis.

a. Find and graph the y-coordinate of the center of mass of the plate as a function of a.

b. For what value of a is the center of mass on the edge of the plate?

62. Center of mass on the edge Consider the constant-density solid $\{(\rho, \varphi, \theta): 0 < a \leq \rho \leq 1, 0 \leq \varphi \leq \pi/2, 0 \leq \theta \leq 2\pi\}$ bounded by two hemispheres and the xy-plane.

a. Find and graph the z-coordinate of the center of mass of the plate as a function of a.

b. For what value of a is the center of mass on the edge of the solid?

63. Draining a soda can A cylindrical soda can has a radius of 4 cm and a height of 12 cm. When the can is full of soda, the center of mass of the contents of the can is 6 cm above the base on the axis of the can (halfway along the axis of the can). As the can is drained, the center of mass descends for a while. However, when the can is empty (filled only with air), the center of mass is once again 6 cm above the base on the axis of the can. Find the depth of soda in the can for which the center of mass is at its lowest point. Neglect the mass of the can, and assume the density of the soda is 1 g/cm^3 and the density of air is 0.001 g/cm^3.

64. Triangle medians A triangular region has a base that connects the vertices $(0, 0)$ and $(b, 0)$, and a third vertex at (a, h), where $a > 0, b > 0$, and $h > 0$.

a. Show that the centroid of the triangle is $\left(\dfrac{a + b}{3}, \dfrac{h}{3}\right)$.

b. Recall that the three medians of a triangle extend from each vertex to the midpoint of the opposite side. Knowing that the medians of a triangle intersect in a point M and that each median bisects the triangle, conclude that the centroid of the triangle is M.

65. The golden earring A disk of radius r is removed from a larger disk of radius R to form an earring (see figure). Assume the earring is a thin plate of uniform density.

a. Find the center of mass of the earring in terms of r and R. (*Hint:* Place the origin of a coordinate system either at the center of the large disk or at Q; either way, the earring is symmetric about the x-axis.)

b. Show that the ratio $\dfrac{R}{r}$ such that the center of mass lies at the point P (on the edge of the inner disk) is the golden mean $\dfrac{1 + \sqrt{5}}{2} \approx 1.618$.

(*Source:* P. Glaister, *Golden Earrings*, *Mathematical Gazette*, 80, 1996)

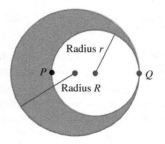

QUICK CHECK ANSWERS

1. 3 m   **3.** It is heaviest at $x = 0$ and lightest at $x = 3$.
4. The distance from the point (x, y) to the y-axis is x. The constant density appears in the integral for the moment, and it appears in the integral for the mass. Therefore, the density cancels when we divide the two integrals.   **5.** The distance from the xy-plane to a point (x, y, z) is z. ◄

16.7 Change of Variables in Multiple Integrals

Converting double integrals from rectangular coordinates to polar coordinates (Section 16.3) and converting triple integrals from rectangular coordinates to cylindrical or spherical coordinates (Section 16.5) are examples of a general procedure known as a *change of variables*. The idea is not new: The Substitution Rule introduced in Chapter 5 with single-variable integrals is also a change of variables. The aim of this section is to show you how to change variables in double and triple integrals.

Recap of Change of Variables

Recall how a change of variables is used to simplify a single-variable integral. For example, to simplify the integral $\int_0^1 2\sqrt{2x + 1} \, dx$, we choose a new variable $u = 2x + 1$, which means that $du = 2 \, dx$. Therefore,

$$\int_0^1 2\sqrt{2x + 1} \, dx = \int_1^3 \sqrt{u} \, du.$$

This equality means that the area under the curve $y = 2\sqrt{2x + 1}$ from $x = 0$ to $x = 1$ equals the area under the curve $y = \sqrt{u}$ from $u = 1$ to $u = 3$ (**Figure 16.76**). The relation $du = 2\, dx$ relates the length of a small interval on the u-axis to the length of the corresponding interval on the x-axis.

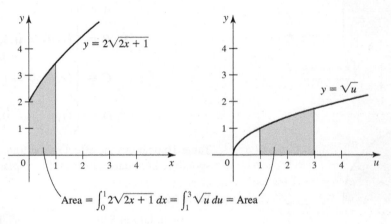

$$\text{Area} = \int_0^1 2\sqrt{2x + 1}\, dx = \int_1^3 \sqrt{u}\, du = \text{Area}$$

Figure 16.76

Similarly, some double and triple integrals can be simplified through a change of variables. For example, the region of integration for

$$\int_0^1 \int_0^{\sqrt{1-x^2}} e^{1-x^2-y^2}\, dy\, dx$$

is the quarter disk $R = \{(x, y): x \geq 0, y \geq 0, x^2 + y^2 \leq 1\}$. Changing variables to polar coordinates with $x = r \cos \theta$, $y = r \sin \theta$, and $dy\, dx = r\, dr\, d\theta$, we have

$$\int_0^1 \int_0^{\sqrt{1-x^2}} e^{1-x^2-y^2}\, dy\, dx \overset{\substack{x = r\cos\theta \\ y = r\sin\theta}}{=} \int_0^{\pi/2} \int_0^1 e^{1-r^2} r\, dr\, d\theta.$$

In this case, the original region of integration R is transformed into a new region $S = \{(r, \theta): 0 \leq r \leq 1, 0 \leq \theta \leq \pi/2\}$, which is a rectangle in the $r\theta$-plane.

Transformations in the Plane

A change of variables in a double integral is a *transformation* that relates two sets of variables, (u, v) and (x, y). It is written compactly as $(x, y) = T(u, v)$. Because it relates pairs of variables, T has two components,

$$T: x = g(u, v) \quad \text{and} \quad y = h(u, v).$$

Geometrically, T takes a region S in the uv-plane and "maps" it point by point to a region R in the xy-plane (**Figure 16.77**). We write the outcome of this process as $R = T(S)$ and call R the **image** of S under T.

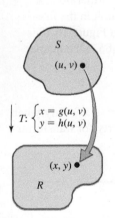

Figure 16.77

EXAMPLE 1   Image of a transformation Consider the transformation from polar to rectangular coordinates given by

$$T: \quad x = g(r, \theta) = r \cos \theta \quad \text{and} \quad y = h(r, \theta) = r \sin \theta.$$

Find the image under this transformation of the rectangle

$$S = \{(r, \theta): 0 \leq r \leq 1, 0 \leq \theta \leq \pi/2\}.$$

▶ In Example 1, we have replaced the coordinates u and v with the familiar polar coordinates r and θ.

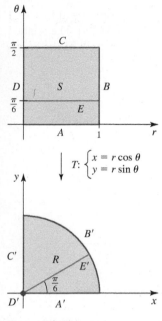

Figure 16.78

SOLUTION If we apply T to every point of S (**Figure 16.78**), what is the resulting set R in the xy-plane? One way to answer this question is to walk around the boundary of S, let's say counterclockwise, and determine the corresponding path in the xy-plane. In the $r\theta$-plane, we let the horizontal axis be the r-axis and the vertical axis be the θ-axis. Starting at the origin, we denote the edges of the rectangle S as follows.

$$A = \{(r, \theta): 0 \le r \le 1, \theta = 0\} \quad \text{Lower boundary}$$

$$B = \left\{(r, \theta): r = 1, 0 \le \theta \le \frac{\pi}{2}\right\} \quad \text{Right boundary}$$

$$C = \left\{(r, \theta): 0 \le r \le 1, \theta = \frac{\pi}{2}\right\} \quad \text{Upper boundary}$$

$$D = \left\{(r, \theta): r = 0, 0 \le \theta \le \frac{\pi}{2}\right\} \quad \text{Left boundary}$$

Table 16.6 shows the effect of the transformation on the four boundaries of S; the corresponding boundaries of R in the xy-plane are denoted A', B', C', and D' (Figure 16.78).

Table 16.6

Boundary of S in $r\theta$-plane	Transformation equations	Boundary of R in xy-plane
A: $0 \le r \le 1, \theta = 0$	$x = r\cos\theta = r,$ $y = r\sin\theta = 0$	A': $0 \le x \le 1, y = 0$
B: $r = 1, 0 \le \theta \le \pi/2$	$x = r\cos\theta = \cos\theta,$ $y = r\sin\theta = \sin\theta$	B': quarter unit circle
C: $0 \le r \le 1, \theta = \pi/2$	$x = r\cos\theta = 0,$ $y = r\sin\theta = r$	C': $x = 0, 0 \le y \le 1$
D: $r = 0, 0 \le \theta \le \pi/2$	$x = r\cos\theta = 0,$ $y = r\sin\theta = 0$	D': single point $(0, 0)$

QUICK CHECK 1 How would the image of S change in Example 1 if $S = \{(r, \theta): 0 \le r \le 1, 0 \le \theta \le \pi\}$? ◄

The image of the rectangular boundary of S is the boundary of R. Furthermore, it can be shown that every point in the interior of R is the image of one point in the interior of S. (For example, the horizontal line segment E in the $r\theta$-plane in Figure 16.78 is mapped to the line segment E' in the xy-plane.) Therefore, the image of S is the quarter disk R in the xy-plane.

Related Exercises 11–12 ◄

Recall that a function f is *one-to-one* on an interval I if $f(x_1) = f(x_2)$ only when $x_1 = x_2$, where x_1 and x_2 are points of I. We need an analogous property for transformations when changing variables.

DEFINITION One-to-One Transformation

A transformation T from a region S to a region R is one-to-one on S if $T(P) = T(Q)$ only when $P = Q$, where P and Q are points in S.

Notice that the polar coordinate transformation in Example 1 is not one-to-one on the rectangle $S = \{(r, \theta): 0 \le r \le 1, 0 \le \theta \le \pi/2\}$ (because all points with $r = 0$ map to the point $(0, 0)$). However, this transformation *is* one-to-one on the interior of S.

We can now anticipate how a transformation (change of variables) is used to simplify a double integral. Suppose we have the integral $\iint_R f(x, y)\, dA$. The goal is to find a transformation to a new set of coordinates (u, v) such that the new equivalent integral $\iint_S f(x(u, v), y(u, v))\, dA$ involves a simple region S (such as a rectangle), a simple integrand, or both. The next theorem allows us to do exactly that, but it first requires a new concept.

▶ The Jacobian is named after the German mathematician Carl Gustav Jacob Jacobi (1804–1851). In some books, the Jacobian is the matrix of partial derivatives. In others, as here, the Jacobian is the determinant of the matrix of partial derivatives. Both $J(u, v)$ and $\dfrac{\partial(x, y)}{\partial(u, v)}$ are used to refer to the Jacobian.

DEFINITION Jacobian Determinant of a Transformation of Two Variables

Given a transformation $T: x = g(u, v), y = h(u, v)$, where g and h are differentiable on a region of the uv-plane, the **Jacobian determinant** (or **Jacobian**) of T is

$$J(u, v) = \frac{\partial(x, y)}{\partial(u, v)} = \begin{vmatrix} \dfrac{\partial x}{\partial u} & \dfrac{\partial x}{\partial v} \\ \dfrac{\partial y}{\partial u} & \dfrac{\partial y}{\partial v} \end{vmatrix} = \frac{\partial x}{\partial u}\frac{\partial y}{\partial v} - \frac{\partial x}{\partial v}\frac{\partial y}{\partial u}.$$

QUICK CHECK 2 Find $J(u, v)$ if $x = u + v, y = 2v$. ◀

The Jacobian is easiest to remember as the determinant of a 2×2 matrix of partial derivatives. With the Jacobian in hand, we can state the change-of-variables rule for double integrals.

THEOREM 16.8   Change of Variables for Double Integrals

Let $T: x = g(u, v), y = h(u, v)$ be a transformation that maps a closed bounded region S in the uv-plane to a region R in the xy-plane. Assume T is one-to-one on the interior of S and g and h have continuous first partial derivatives there. If f is continuous on R, then

$$\iint\limits_{R} f(x, y)\, dA = \iint\limits_{S} f(g(u, v), h(u, v))\, |J(u, v)|\, dA.$$

▶ The condition that g and h have continuous first partial derivatives ensures that the new integrand is integrable.

▶ In the integral over R, dA corresponds to $dx\, dy$. In the integral over S, dA corresponds to $du\, dv$. The relation $dx\, dy = |J|\, du\, dv$ is the analog of $du = g'(x)\, dx$ in a change of variables with one variable.

The proof of this result is technical and is found in advanced texts. The factor $|J(u, v)|$ that appears in the second integral is the absolute value of the Jacobian. Matching the area elements in the two integrals of Theorem 16.8, we see that $dx\, dy = |J(u, v)|\, du\, dv$. This expression shows that the Jacobian is a magnification (or reduction) factor: It relates the area of a small region $dx\, dy$ in the xy-plane to the area of the corresponding region $du\, dv$ in the uv-plane. If the transformation equations are linear, then this relationship is exact in the sense that $\text{area}(T(S)) = |J(u, v)| \cdot \text{area of } S$ (see Exercise 60). The way in which the Jacobian arises is explored in Exercise 61.

EXAMPLE 2   Jacobian of the polar-to-rectangular transformation Compute the Jacobian of the transformation

$$T: \qquad x = g(r, \theta) = r \cos \theta \quad \text{and} \quad y = h(r, \theta) = r \sin \theta.$$

SOLUTION The necessary partial derivatives are

$$\frac{\partial x}{\partial r} = \cos \theta, \qquad \frac{\partial x}{\partial \theta} = -r \sin \theta, \qquad \frac{\partial y}{\partial r} = \sin \theta, \quad \text{and} \quad \frac{\partial y}{\partial \theta} = r \cos \theta.$$

Therefore,

$$J(r, \theta) = \frac{\partial(x, y)}{\partial(r, \theta)} = \begin{vmatrix} \dfrac{\partial x}{\partial r} & \dfrac{\partial x}{\partial \theta} \\ \dfrac{\partial y}{\partial r} & \dfrac{\partial y}{\partial \theta} \end{vmatrix} = \begin{vmatrix} \cos \theta & -r \sin \theta \\ \sin \theta & r \cos \theta \end{vmatrix} = r(\cos^2 \theta + \sin^2 \theta) = r.$$

This determinant calculation confirms the change-of-variables formula for polar coordinates: $dx\, dy$ becomes $r\, dr\, d\theta$.

Related Exercise 20 ◀

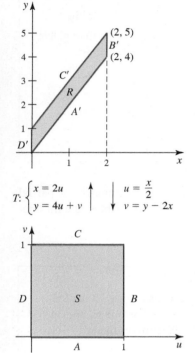

$$T: \begin{cases} x = 2u \\ y = 4u + v \end{cases} \qquad \begin{cases} u = \dfrac{x}{2} \\ v = y - 2x \end{cases}$$

Figure 16.79

▶ The relations that "go the other direction" make up the inverse transformation, usually denoted T^{-1}.

Table 16.7

(x, y)	(u, v)
$(0, 0)$	$(0, 0)$
$(0, 1)$	$(0, 1)$
$(2, 5)$	$(1, 1)$
$(2, 4)$	$(1, 0)$

▶ T is an example of a *shearing transformation*. The greater the u-coordinate of a point, the more that point is displaced in the v-direction. It also involves a uniform stretch in the u-direction.

We are now ready for a change of variables. To transform the integral $\iint_R f(x, y)\, dA$ into $\iint_S f(x(u, v), y(u, v))|J(u, v)|\, dA$, we must find the transformation $x = g(u, v)$ and $y = h(u, v)$, and then use it to find the new region of integration S. The next example illustrates how the region S is found, assuming the transformation is given.

EXAMPLE 3   Double integral with a change of variables given Evaluate the integral $\iint_R \sqrt{2x(y - 2x)}\, dA$, where R is the parallelogram in the xy-plane with vertices $(0, 0)$, $(0, 1)$, $(2, 4)$, and $(2, 5)$ (**Figure 16.79**). Use the transformation

$$T: x = 2u \quad \text{and} \quad y = 4u + v.$$

SOLUTION To what region S in the uv-plane is R mapped? Because T takes points in the uv-plane and assigns them to points in the xy-plane, we must reverse the process by solving $x = 2u$, $y = 4u + v$ for u and v.

$$\text{First equation: } x = 2u \implies u = \frac{x}{2}$$

$$\text{Second equation: } y = 4u + v \implies v = y - 4u = y - 2x$$

Rather than walk around the boundary of R in the xy-plane to determine the resulting region S in the uv-plane, it suffices to find the images of the vertices of R. You should confirm that the vertices map as shown in Table 16.7.

Connecting the points in the uv-plane in order, we see that S is the unit square $\{(u, v): 0 \le u \le 1, 0 \le v \le 1\}$ (Figure 16.79). These inequalities determine the limits of integration in the uv-plane.

Replacing $2x$ with $4u$ and $y - 2x$ with v, the original integrand becomes $\sqrt{2x(y - 2x)} = \sqrt{4uv}$. The Jacobian is

$$J(u, v) = \begin{vmatrix} \dfrac{\partial x}{\partial u} & \dfrac{\partial x}{\partial v} \\ \dfrac{\partial y}{\partial u} & \dfrac{\partial y}{\partial v} \end{vmatrix} = \begin{vmatrix} 2 & 0 \\ 4 & 1 \end{vmatrix} = 2.$$

The integration now follows:

$$\iint_R \sqrt{2x(y - 2x)}\, dA = \iint_S \sqrt{4uv}\ \underbrace{|J(u, v)|}_{2}\, dA \qquad \text{Change variables.}$$

$$= \int_0^1 \int_0^1 \sqrt{4uv}\ 2\, du\, dv \qquad \text{Convert to an iterated integral.}$$

$$= 4 \int_0^1 \frac{2}{3} \sqrt{v}\ (u^{3/2}) \Big|_0^1 dv \qquad \text{Evaluate inner integral.}$$

$$= \frac{8}{3} \cdot \frac{2}{3} (v^{3/2}) \Big|_0^1 = \frac{16}{9}. \qquad \text{Evaluate outer integral.}$$

The effect of the change of variables is illustrated in **Figure 16.80**, where we see the surface $z = \sqrt{2x(y - 2x)}$ over the region R and the surface $w = 2\sqrt{4uv}$ over the region S. The volumes of the solids beneath the two surfaces are equal, but the integral over S is easier to evaluate.

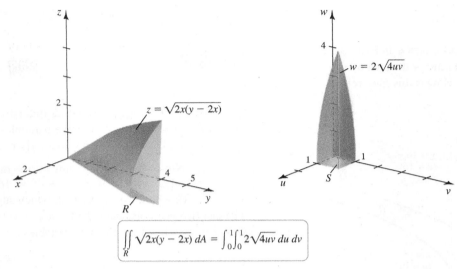

$$\iint\limits_R \sqrt{2x(y - 2x)} \, dA = \int_0^1\int_0^1 2\sqrt{4uv} \, du \, dv$$

Figure 16.80

Related Exercise 29 ◄

QUICK CHECK 3 Solve the equations $u = x + y$, $v = -x + 2y$ for x and y. ◄

Figure 16.81

▶ The transformation in Example 4 is a *rotation*. It rotates the points of R about the origin 45° in the counterclockwise direction (it also increases lengths by a factor of $\sqrt{2}$). In this example, the change of variables $u = x + y$ and $v = x - y$ would work just as well.

▶ An appropriate change of variables for a double integral is not always obvious. Some trial and error is often needed to come up with a transformation that simplifies the integrand and/or the region of integration. Strategies are discussed at the end of this section.

In Example 3, the required transformation was given. More practically, we must deduce an appropriate transformation from the form of either the integrand or the region of integration.

EXAMPLE 4   Change of variables determined by the integrand Evaluate $\displaystyle\iint\limits_R \sqrt{\frac{x - y}{x + y + 1}} \, dA$, where R is the square with vertices $(0, 0)$, $(1, -1)$, $(2, 0)$, and $(1, 1)$ (**Figure 16.81**).

SOLUTION Evaluating the integral as it stands requires splitting the region R into two subregions; furthermore, the integrand presents difficulties. The terms $x + y$ and $x - y$ in the integrand suggest the new variables

$$u = x - y \quad \text{and} \quad v = x + y.$$

To determine the region S in the uv-plane that corresponds to R under this transformation, we find the images of the vertices of R in the uv-plane and connect them in order. The result is the square $S = \{(u, v): 0 \leq u \leq 2, 0 \leq v \leq 2\}$ (Figure 16.81). Before computing the Jacobian, we express x and y in terms of u and v. Adding the two equations and solving for x, we have $x = (u + v)/2$. Subtracting the two equations and solving for y gives $y = (v - u)/2$. The Jacobian now follows:

$$J(u, v) = \begin{vmatrix} \dfrac{\partial x}{\partial u} & \dfrac{\partial x}{\partial v} \\[2mm] \dfrac{\partial y}{\partial u} & \dfrac{\partial y}{\partial v} \end{vmatrix} = \begin{vmatrix} \dfrac{1}{2} & \dfrac{1}{2} \\[2mm] -\dfrac{1}{2} & \dfrac{1}{2} \end{vmatrix} = \dfrac{1}{2}.$$

With the choice of new variables, the original integrand $\sqrt{\dfrac{x - y}{x + y + 1}}$ becomes $\sqrt{\dfrac{u}{v + 1}}$.

The integration in the uv-plane may now be done:

$$\iint\limits_R \sqrt{\frac{x - y}{x + y + 1}} \, dA = \iint\limits_S \sqrt{\frac{u}{v + 1}} \, |J(u, v)| \, dA \qquad \text{Change of variables}$$

$$= \int_0^2\int_0^2 \sqrt{\frac{u}{v + 1}} \frac{1}{2} \, du \, dv \qquad \text{Convert to an iterated integral.}$$

$$= \frac{1}{2}\int_0^2 (v + 1)^{-1/2} \frac{2}{3}\left(u^{3/2}\right)\Big|_0^2 dv \qquad \text{Evaluate inner integral.}$$

$$= \frac{2^{3/2}}{3} 2(v + 1)^{1/2}\Big|_0^2$$ Evaluate outer integral.

$$= \frac{4\sqrt{2}}{3} (\sqrt{3} - 1).$$ Simplify.

Related Exercises 32, 36 ◄

QUICK CHECK 4 In Example 4, what is the ratio of the area of S to the area of R? How is this ratio related to J? ◄

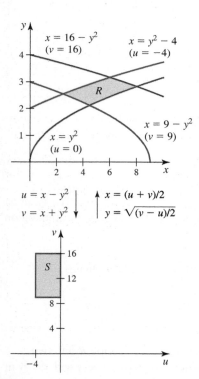

Figure 16.82

EXAMPLE 5 Change of variables determined by the region Let R be the region in the first quadrant bounded by the parabolas $x = y^2$, $x = y^2 - 4$, $x = 9 - y^2$, and $x = 16 - y^2$ (**Figure 16.82**). Evaluate $\iint_R y^2 \, dA$.

SOLUTION Notice that the bounding curves may be written as $x - y^2 = 0$, $x - y^2 = -4$, $x + y^2 = 9$, and $x + y^2 = 16$. The first two parabolas have the form $x - y^2 = C$, where C is a constant, which suggests the new variable $u = x - y^2$. The last two parabolas have the form $x + y^2 = C$, which suggests the new variable $v = x + y^2$. Therefore, the new variables are

$$u = x - y^2 \quad \text{and} \quad v = x + y^2.$$

The boundary curves of S are $u = -4$, $u = 0$, $v = 9$, and $v = 16$. Therefore, the new region is $S = \{(u, v): -4 \le u \le 0, 9 \le v \le 16\}$ (Figure 16.82). To compute the Jacobian, we must find the transformation T by writing x and y in terms of u and v. Solving for x and y, and observing that $y \ge 0$ for all points in R, we find that

$$T: \quad x = \frac{u + v}{2} \quad \text{and} \quad y = \sqrt{\frac{v - u}{2}}.$$

The points of S satisfy $v > u$, so $\sqrt{v - u}$ is defined. Now the Jacobian may be computed:

$$J(u, v) = \begin{vmatrix} \dfrac{\partial x}{\partial u} & \dfrac{\partial x}{\partial v} \\[2ex] \dfrac{\partial y}{\partial u} & \dfrac{\partial y}{\partial v} \end{vmatrix} = \begin{vmatrix} \dfrac{1}{2} & \dfrac{1}{2} \\[2ex] -\dfrac{1}{2\sqrt{2(v - u)}} & \dfrac{1}{2\sqrt{2(v - u)}} \end{vmatrix} = \frac{1}{2\sqrt{2(v - u)}}.$$

The change of variables proceeds as follows:

$$\iint_R y^2 \, dA = \int_9^{16} \int_{-4}^0 \underbrace{\frac{v - u}{2}}_{y^2} \underbrace{\frac{1}{2\sqrt{2(v - u)}}}_{|J(u,v)|} \, du \, dv$$ Convert to an iterated integral.

$$= \frac{1}{4\sqrt{2}} \int_9^{16} \int_{-4}^0 \sqrt{v - u} \, du \, dv$$ Simplify.

$$= \frac{1}{4\sqrt{2}} \frac{2}{3} \int_9^{16} \left(-(v - u)^{3/2}\right)\Big|_{-4}^0 \, dv$$ Evaluate inner integral.

$$= \frac{1}{6\sqrt{2}} \int_9^{16} \left((v + 4)^{3/2} - v^{3/2}\right) dv$$ Simplify.

$$= \frac{1}{6\sqrt{2}} \frac{2}{5} \left((v + 4)^{5/2} - v^{5/2}\right)\Big|_9^{16}$$ Evaluate outer integral.

$$= \frac{\sqrt{2}}{30} \left(32 \cdot 5^{5/2} - 13^{5/2} - 781\right)$$ Simplify.

$$\approx 18.79.$$

Related Exercises 33–34 ◄

Change of Variables in Triple Integrals

With triple integrals, we work with a transformation T of the form

$$T: \quad x = g(u, v, w), \quad y = h(u, v, w), \quad \text{and} \quad z = p(u, v, w).$$

In this case, T maps a region S in uvw-space to a region D in xyz-space. As before, the goal is to transform the integral $\iiint_D f(x, y, z) \, dV$ into a new integral over the region S that is easier to evaluate. First, we need a Jacobian.

➤ Recall that expanding about the first row yields

$$\begin{vmatrix} a_{11} & a_{12} & a_{13} \\ a_{21} & a_{22} & a_{23} \\ a_{31} & a_{32} & a_{33} \end{vmatrix}$$

$$= a_{11}(a_{22}a_{33} - a_{23}a_{32})$$
$$- a_{12}(a_{21}a_{33} - a_{23}a_{31})$$
$$+ a_{13}(a_{21}a_{32} - a_{22}a_{31}).$$

DEFINITION Jacobian Determinant of a Transformation of Three Variables

Given a transformation $T: x = g(u, v, w)$, $y = h(u, v, w)$, and $z = p(u, v, w)$, where g, h, and p are differentiable on a region of uvw-space, the **Jacobian determinant** (or **Jacobian**) of T is

$$J(u, v, w) = \frac{\partial(x, y, z)}{\partial(u, v, w)} = \begin{vmatrix} \dfrac{\partial x}{\partial u} & \dfrac{\partial x}{\partial v} & \dfrac{\partial x}{\partial w} \\[2mm] \dfrac{\partial y}{\partial u} & \dfrac{\partial y}{\partial v} & \dfrac{\partial y}{\partial w} \\[2mm] \dfrac{\partial z}{\partial u} & \dfrac{\partial z}{\partial v} & \dfrac{\partial z}{\partial w} \end{vmatrix}.$$

The Jacobian is evaluated as a 3×3 determinant and is a function of u, v, and w. A change of variables with respect to three variables proceeds in analogy to the two-variable case.

THEOREM 16.9   Change of Variables for Triple Integrals

Let $T: x = g(u, v, w)$, $y = h(u, v, w)$, and $z = p(u, v, w)$ be a transformation that maps a closed bounded region S in uvw-space to a region $D = T(S)$ in xyz-space. Assume T is one-to-one on the interior of S and g, h, and p have continuous first partial derivatives there. If f is continuous on D, then

$$\iiint_D f(x, y, z)\, dV = \iiint_S f(g(u, v, w), h(u, v, w), p(u, v, w))\, |J(u, v, w)|\, dV.$$

➤ If we match the elements of volume in both integrals, then $dx\, dy\, dz = |J(u, v, w)|\, du\, dv\, dw$. As before, the Jacobian is a magnification (or reduction) factor, now relating the volume of a small region in xyz-space to the volume of the corresponding region in uvw-space.

➤ To see that triple integrals in cylindrical and spherical coordinates as derived in Section 16.5 are consistent with this change-of-variables formulation, see Exercises 46 and 47.

EXAMPLE 6   A triple integral Use a change of variables to evaluate $\iiint_D xz\, dV$, where D is a parallelepiped bounded by the planes

$$y = x, \qquad y = x + 2, \qquad z = x, \qquad z = x + 3, \qquad z = 0, \quad \text{and} \quad z = 4$$

(**Figure 16.83a**).

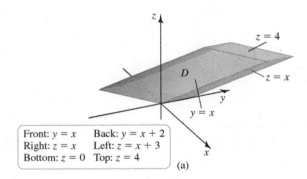

Front: $y = x$ Back: $y = x + 2$
Right: $z = x$ Left: $z = x + 3$
Bottom: $z = 0$ Top: $z = 4$

(a)

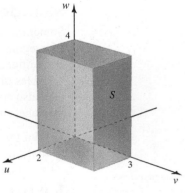

(b)

Figure 16.83

SOLUTION The key is to note that D is bounded by three pairs of parallel planes.

- $y - x = 0$ and $y - x = 2$
- $z - x = 0$ and $z - x = 3$
- $z = 0$ and $z = 4$

These combinations of variables suggest the new variables

$$u = y - x, \qquad v = z - x, \quad \text{and} \quad w = z.$$

With this choice, the new region of integration (**Figure 16.83b**) is the rectangular box

$$S = \{(u, v, w): 0 \le u \le 2, 0 \le v \le 3, 0 \le w \le 4\}.$$

To compute the Jacobian, we must express x, y, and z in terms of u, v, and w. A few steps of algebra lead to the transformation

$$T: \qquad x = w - v, \quad y = u - v + w, \quad \text{and} \quad z = w.$$

The resulting Jacobian is

> ▶ It is easiest to expand the Jacobian determinant in Example 6 about the third row.

$$J(u, v, w) = \begin{vmatrix} \dfrac{\partial x}{\partial u} & \dfrac{\partial x}{\partial v} & \dfrac{\partial x}{\partial w} \\[2mm] \dfrac{\partial y}{\partial u} & \dfrac{\partial y}{\partial v} & \dfrac{\partial y}{\partial w} \\[2mm] \dfrac{\partial z}{\partial u} & \dfrac{\partial z}{\partial v} & \dfrac{\partial z}{\partial w} \end{vmatrix} = \begin{vmatrix} 0 & -1 & 1 \\ 1 & -1 & 1 \\ 0 & 0 & 1 \end{vmatrix} = 1.$$

Noting that the integrand is $xz = (w - v)w = w^2 - vw$, the integral may now be evaluated:

$$\iiint_D xz \, dV = \iiint_S (w^2 - vw)\, |J(u, v, w)| \, dV \qquad \text{Change variables.}$$

$$= \int_0^4 \int_0^3 \int_0^2 (w^2 - vw)\, \underbrace{1}_{|J(u, v, w)|} \, du \, dv \, dw \qquad \text{Convert to an iterated integral.}$$

$$= \int_0^4 \int_0^3 2(w^2 - vw)\, dv \, dw \qquad \text{Evaluate inner integral.}$$

$$= 2 \int_0^4 \left(vw^2 - \frac{v^2 w}{2} \right)\Big|_0^3 \, dw \qquad \text{Evaluate middle integral.}$$

$$= 2 \int_0^4 \left(3w^2 - \frac{9w}{2} \right) dw \qquad \text{Simplify.}$$

QUICK CHECK 5 Interpret a Jacobian with a value of 1 (as in Example 6). ◀

$$= 2\left(w^3 - \frac{9w^2}{4} \right)\Big|_0^4 = 56. \qquad \text{Evaluate outer integral.}$$

Related Exercises 40–41 ◀

Strategies for Choosing New Variables

Sometimes a change of variables simplifies the integrand but leads to an awkward region of integration. Conversely, the new region of integration may be simplified at the expense of additional complications in the integrand. Here are a few suggestions for finding new variables of integration. The observations are made with respect to double integrals, but they also apply to triple integrals. As before, R is the original region of integration in the xy-plane, and S is the new region in the uv-plane.

1. **Aim for simple regions of integration in the uv-plane** The new region of integration in the uv-plane should be as simple as possible. Double integrals are easiest to evaluate over rectangular regions with sides parallel to the coordinate axes.

> ▶ Inverting the transformation means solving for x and y in terms of u and v, or vice versa.

2. **Is $(x, y) \to (u, v)$ or $(u, v) \to (x, y)$ better?** For some problems it is easier to write (x, y) as functions of (u, v); in other cases, the opposite is true. Depending on the

problem, inverting the transformation (finding relations that go in the opposite direction) may be easy, difficult, or impossible.

- If you know (x, y) in terms of (u, v) (that is, $x = g(u, v)$ and $y = h(u, v)$), then computing the Jacobian is straightforward, as is sketching the region R given the region S. However, the transformation must be inverted to determine the shape of S.

- If you know (u, v) in terms of (x, y) (that is, $u = G(x, y)$ and $v = H(x, y)$), then sketching the region S is straightforward. However, the transformation must be inverted to compute the Jacobian.

3. **Let the integrand suggest new variables** New variables are often chosen to simplify the integrand. For example, the integrand $\sqrt{\dfrac{x - y}{x + y}}$ calls for new variables $u = x - y$ and $v = x + y$ (or $u = x + y, v = x - y$). There is, however, no guarantee that this change of variables will simplify the region of integration. In cases in which only one combination of variables appears, let one new variable be that combination and let the other new variable be unchanged. For example, if the integrand is $(x + 4y)^{3/2}$, try letting $u = x + 4y$ and $v = y$.

4. **Let the region suggest new variables** Example 5 illustrates an ideal situation. It occurs when the region R is bounded by two pairs of "parallel" curves in the families $g(x, y) = C_1$ and $h(x, y) = C_2$ (**Figure 16.84**). In this case, the new region of integration is a rectangle $S = \{(u, v): a_1 \leq u \leq a_2, b_1 \leq v \leq b_2\}$, where $u = g(x, y)$ and $v = h(x, y)$.

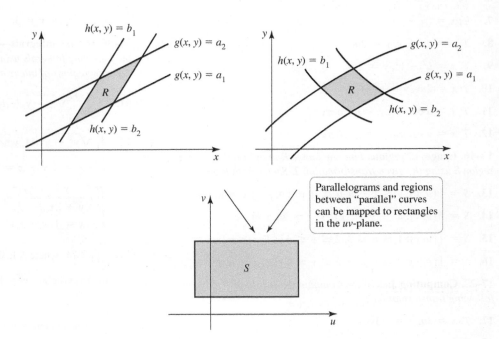

Parallelograms and regions between "parallel" curves can be mapped to rectangles in the uv-plane.

Figure 16.84

As another example, suppose the region is bounded by the lines $y = x$ (or $y/x = 1$) and $y = 2x$ (or $y/x = 2$) and by the hyperbolas $xy = 1$ and $xy = 3$. Then the new variables should be $u = xy$ and $v = y/x$ (or vice versa). The new region of integration is the rectangle $S = \{(u, v): 1 \leq u \leq 3, 1 \leq v \leq 2\}$.

SECTION 16.7 EXERCISES

Getting Started

1. Suppose S is the unit square in the first quadrant of the uv-plane. Describe the image of the transformation $T: x = 2u, y = 2v$.

2. Explain how to compute the Jacobian of the transformation $T: x = g(u, v), y = h(u, v)$.

3. Using the transformation $T: x = u + v, y = u - v$, the image of the unit square $S = \{(u, v): 0 \le u \le 1, 0 \le v \le 1\}$ is a region R in the xy-plane. Explain how to change variables in the integral $\iint_R f(x, y)\, dA$ to find a new integral over S.

4. Suppose S is the unit cube in the first octant of uvw-space with one vertex at the origin. What is the image of the transformation $T: x = u/2, y = v/2, z = w/2$?

Practice Exercises

5–12. Transforming a square *Let $S = \{(u, v): 0 \le u \le 1, 0 \le v \le 1\}$ be a unit square in the uv-plane. Find the image of S in the xy-plane under the following transformations.*

5. $T: x = 2u, y = v/2$

6. $T: x = -u, y = -v$

7. $T: x = (u + v)/2, y = (u - v)/2$

8. $T: x = 2u + v, y = 2u$

9. $T: x = u^2 - v^2, y = 2uv$

10. $T: x = 2uv, y = u^2 - v^2$

11. $T: x = u \cos \pi v, y = u \sin \pi v$

12. $T: x = v \sin \pi u, y = v \cos \pi u$

13–16. Images of regions *Find the image R in the xy-plane of the region S using the given transformation T. Sketch both R and S.*

13. $S = \{(u, v): v \le 1 - u, u \ge 0, v \ge 0\}; T: x = u, y = v^2$

14. $S = \{(u, v): u^2 + v^2 \le 1\}; T: x = 2u, y = 4v$

15. $S = \{(u, v): 1 \le u \le 3, 2 \le v \le 4\}; T: x = u/v, y = v$

16. $S = \{(u, v): 2 \le u \le 3, 3 \le v \le 6\}; T: x = u, y = v/u$

17–22. Computing Jacobians *Compute the Jacobian $J(u, v)$ for the following transformations.*

17. $T: x = 3u, y = -3v$

18. $T: x = 4v, y = -2u$

19. $T: x = 2uv, y = u^2 - v^2$

20. $T: x = u \cos \pi v, y = u \sin \pi v$

21. $T: x = (u + v)/\sqrt{2}, y = (u - v)/\sqrt{2}$

22. $T: x = u/v, y = v$

23–26. Solve and compute Jacobians *Solve the following relations for x and y, and compute the Jacobian $J(u, v)$.*

23. $u = x + y, v = 2x - y$

24. $u = xy, v = x$

25. $u = 2x - 3y, v = y - x$

26. $u = x + 4y, v = 3x + 2y$

27–30. Double integrals—transformation given *To evaluate the following integrals, carry out these steps.*

a. *Sketch the original region of integration R in the xy-plane and the new region S in the uv-plane using the given change of variables.*

b. *Find the limits of integration for the new integral with respect to u and v.*

c. *Compute the Jacobian.*

d. *Change variables and evaluate the new integral.*

27. $\iint_R xy\, dA$, where R is the square with vertices $(0, 0), (1, 1), (2, 0)$, and $(1, -1)$; use $x = u + v, y = u - v$.

28. $\iint_R x^2 y\, dA$, where $R = \{(x, y): 0 \le x \le 2, x \le y \le x + 4\}$; use $x = 2u, y = 4v + 2u$.

29. $\iint_R x^2 \sqrt{x + 2y}\, dA$, where $R = \{(x, y): 0 \le x \le 2, -x/2 \le y \le 1 - x\}$; use $x = 2u, y = v - u$.

30. $\iint_R xy\, dA$, where R is bounded by the ellipse $9x^2 + 4y^2 = 36$; use $x = 2u, y = 3v$.

31–36. Double integrals—your choice of transformation *Evaluate the following integrals using a change of variables. Sketch the original and new regions of integration, R and S.*

31. $\int_0^1 \int_y^{y+2} \sqrt{x - y}\, dx\, dy$

32. $\iint_R \sqrt{y^2 - x^2}\, dA$, where R is the diamond bounded by $y - x = 0$, $y - x = 2, y + x = 0$, and $y + x = 2$

33. $\iint_R \left(\dfrac{y - x}{y + 2x + 1}\right)^4 dA$, where R is the parallelogram bounded by $y - x = 1, y - x = 2, y + 2x = 0$, and $y + 2x = 4$

34. $\iint_R e^{xy}\, dA$, where R is the region in the first quadrant bounded by the hyperbolas $xy = 1$ and $xy = 4$, and the lines $y/x = 1$ and $y/x = 3$

35. $\iint_R xy\, dA$, where R is the region bounded by the hyperbolas $xy = 1$ and $xy = 4$, and the lines $y = 1$ and $y = 3$

36. $\iint_R (x - y)\sqrt{x - 2y}\, dA$, where R is the triangular region bounded by $y = 0, x - 2y = 0$, and $x - y = 1$

37–40. Jacobians in three variables *Evaluate the Jacobians $J(u, v, w)$ for the following transformations.*

37. $x = v + w, y = u + w, z = u + v$

38. $x = u + v - w, y = u - v + w, z = -u + v + w$

39. $x = vw, y = uw, z = u^2 - v^2$

40. $u = x - y, v = x - z, w = y + z$ (*Hint:* Solve for x, y, and z first.)

41–44. Triple integrals *Use a change of variables to evaluate the following integrals.*

41. $\iiint\limits_{D} xy \, dV$; D is bounded by the planes $y - x = 0$, $y - x = 2$, $z - y = 0$, $z - y = 1$, $z = 0$, and $z = 3$.

42. $\iiint\limits_{D} dV$; D is bounded by the planes $y - 2x = 0$, $y - 2x = 1$, $z - 3y = 0$, $z - 3y = 1$, $z - 4x = 0$, and $z - 4x = 3$.

43. $\iiint\limits_{D} z \, dV$; D is bounded by the paraboloid $z = 16 - x^2 - 4y^2$ and the xy-plane. Use $x = 4u \cos v$, $y = 2u \sin v$, $z = w$.

44. $\iiint\limits_{D} dV$; D is bounded by the upper half of the ellipsoid $x^2/9 + y^2/4 + z^2 = 1$ and the xy-plane. Use $x = 3u$, $y = 2v$, $z = w$.

45. Explain why or why not Determine whether the following statements are true and give an explanation or counterexample.

    a. If the transformation $T: x = g(u, v)$, $y = h(u, v)$ is linear in u and v, then the Jacobian is a constant.

    b. The transformation $x = au + bv$, $y = cu + dv$ generally maps triangular regions to triangular regions.

    c. The transformation $x = 2v$, $y = -2u$ maps circles to circles.

46. Cylindrical coordinates Evaluate the Jacobian for the transformation from cylindrical coordinates (r, θ, Z) to rectangular coordinates (x, y, z): $x = r \cos \theta$, $y = r \sin \theta$, $z = Z$. Show that $J(r, \theta, Z) = r$.

47. Spherical coordinates Evaluate the Jacobian for the transformation from spherical to rectangular coordinates: $x = \rho \sin \varphi \cos \theta$, $y = \rho \sin \varphi \sin \theta$, $z = \rho \cos \varphi$. Show that $J(\rho, \varphi, \theta) = \rho^2 \sin \varphi$.

48–52. Ellipse problems *Let R be the region bounded by the ellipse $x^2/a^2 + y^2/b^2 = 1$, where $a > 0$ and $b > 0$ are real numbers. Let T be the transformation $x = au$, $y = bv$.*

48. Find the area of R.

49. Evaluate $\iint\limits_{R} |xy| \, dA$.

50. Find the center of mass of the upper half of R ($y \geq 0$) assuming it has a constant density.

51. Find the average square of the distance between points of R and the origin.

52. Find the average distance between points in the upper half of R and the x-axis.

53–56. Ellipsoid problems *Let D be the solid bounded by the ellipsoid $x^2/a^2 + y^2/b^2 + z^2/c^2 = 1$, where $a > 0$, $b > 0$, and $c > 0$ are real numbers. Let T be the transformation $x = au$, $y = bv$, $z = cw$.*

53. Find the volume of D.

54. Evaluate $\iiint\limits_{D} |xyz| \, dV$.

55. Find the center of mass of the upper half of D ($z \geq 0$) assuming it has a constant density.

56. Find the average square of the distance between points of D and the origin.

Explorations and Challenges

57. Parabolic coordinates Let T be the transformation $x = u^2 - v^2$, $y = 2uv$.

    a. Show that the lines $u = a$ in the uv-plane map to parabolas in the xy-plane that open in the negative x-direction with vertices on the positive x-axis.

    b. Show that the lines $v = b$ in the uv-plane map to parabolas in the xy-plane that open in the positive x-direction with vertices on the negative x-axis.

    c. Evaluate $J(u, v)$.

    d. Use a change of variables to find the area of the region bounded by $x = 4 - y^2/16$ and $x = y^2/4 - 1$.

    e. Use a change of variables to find the area of the curved rectangle above the x-axis bounded by $x = 4 - y^2/16$, $x = 9 - y^2/36$, $x = y^2/4 - 1$, and $x = y^2/64 - 16$.

    f. Describe the effect of the transformation $x = 2uv$, $y = u^2 - v^2$ on horizontal and vertical lines in the uv-plane.

58. Shear transformations in $\mathbb{R}^2$ The transformation T in $\mathbb{R}^2$ given by $x = au + bv$, $y = cv$, where a, b, and c are positive real numbers, is a *shear transformation*. Let S be the unit square $\{(u, v): 0 \leq u \leq 1, 0 \leq v \leq 1\}$. Let $R = T(S)$ be the image of S.

    a. Explain with pictures the effect of T on S.

    b. Compute the Jacobian of T.

    c. Find the area of R and compare it to the area of S (which is 1).

    d. Assuming a constant density, find the center of mass of R (in terms of a, b, and c) and compare it to the center of mass of S, which is $\left(\frac{1}{2}, \frac{1}{2}\right)$.

    e. Find an analogous transformation that gives a shear in the y-direction.

59. Shear transformations in $\mathbb{R}^3$ The transformation T in $\mathbb{R}^3$ given by

$$x = au + bv + cw, \qquad y = dv + ew, \qquad z = w,$$

where a, b, c, d, and e are positive real numbers, is one of many possible shear transformations in $\mathbb{R}^3$. Let S be the unit cube $\{(u, v, w): 0 \leq u \leq 1, 0 \leq v \leq 1, 0 \leq w \leq 1\}$. Let $D = T(S)$ be the image of S.

    a. Explain with pictures and words the effect of T on S.

    b. Compute the Jacobian of T.

    c. Find the volume of D and compare it to the volume of S (which is 1).

    d. Assuming a constant density, find the center of mass of D and compare it to the center of mass of S, which is $\left(\frac{1}{2}, \frac{1}{2}, \frac{1}{2}\right)$.

60. Linear transformations Consider the linear transformation T in $\mathbb{R}^2$ given by $x = au + bv$, $y = cu + dv$, where a, b, c, and d are real numbers, with $ad \neq bc$.

    a. Find the Jacobian of T.

    b. Let S be the square in the uv-plane with vertices $(0, 0)$, $(1, 0)$, $(0, 1)$, and $(1, 1)$, and let $R = T(S)$. Show that $\text{area}(R) = |J(u, v)|$.

    c. Let ℓ be the line segment joining the points P and Q in the uv-plane. Show that $T(\ell)$ (the image of ℓ under T) is the line segment joining $T(P)$ and $T(Q)$ in the xy-plane. (*Hint:* Use vectors.)

    d. Show that if S is a parallelogram in the uv-plane and $R = T(S)$, then $\text{area}(R) = |J(u, v)| \cdot \text{area of } S$. (*Hint:* Without loss of generality, assume the vertices of S are $(0, 0)$, $(A, 0)$, (B, C), and $(A + B, C)$, where A, B, and C are positive, and use vectors.)

61. Meaning of the Jacobian The Jacobian is a magnification (or reduction) factor that relates the area of a small region near the point (u, v) to the area of the image of that region near the point (x, y).

a. Suppose S is a rectangle in the uv-plane with vertices $O(0, 0)$, $P(\Delta u, 0)$, $(\Delta u, \Delta v)$, and $Q(0, \Delta v)$ (see figure). The image of S under the transformation $x = g(u, v)$, $y = h(u, v)$ is a region R in the xy-plane. Let O', P', and Q' be the images of O, P, and Q, respectively, in the xy-plane, where O', P', and Q' do not all lie on the same line. Explain why the coordinates of O', P', and Q' are $(g(0, 0), h(0, 0))$, $(g(\Delta u, 0), h(\Delta u, 0))$, and $(g(0, \Delta v), h(0, \Delta v))$, respectively.

b. Use a Taylor series in both variables to show that

$$g(\Delta u, 0) \approx g(0, 0) + g_u(0, 0)\Delta u,$$
$$g(0, \Delta v) \approx g(0, 0) + g_v(0, 0)\Delta v,$$
$$h(\Delta u, 0) \approx h(0, 0) + h_u(0, 0)\Delta u, \text{ and}$$
$$h(0, \Delta v) \approx h(0, 0) + h_v(0, 0)\Delta v,$$

where $g_u(0, 0)$ is $\dfrac{\partial x}{\partial u}$ evaluated at $(0, 0)$, with similar meanings for g_v, h_u, and h_v.

c. Consider the parallelogram determined by the vectors $\overrightarrow{O'P'}$ and $\overrightarrow{O'Q'}$. Use the cross product to show that the area of the parallelogram is approximately $|J(u, v)|\,\Delta u\,\Delta v$.

d. Explain why the ratio of the area of R to the area of S is approximately $|J(u, v)|$.

62. Open and closed boxes Consider the region R bounded by three pairs of parallel planes: $ax + by = 0$, $ax + by = 1$; $cx + dz = 0$, $cx + dz = 1$; and $ey + fz = 0$, $ey + fz = 1$, where a, b, c, d, e, and f are real numbers. For the purposes of evaluating triple integrals, when do these six planes bound a finite region? Carry out the following steps.

a. Find three vectors $\mathbf{n}_1$, $\mathbf{n}_2$, and $\mathbf{n}_3$ each of which is normal to one of the three pairs of planes.

b. Show that the three normal vectors lie in a plane if their triple scalar product $\mathbf{n}_1 \cdot (\mathbf{n}_2 \times \mathbf{n}_3)$ is zero.

c. Show that the three normal vectors lie in a plane if $ade + bcf = 0$.

d. Assuming $\mathbf{n}_1$, $\mathbf{n}_2$, and $\mathbf{n}_3$ lie in a plane P, find a vector $\mathbf{N}$ that is normal to P. Explain why a line in the direction of $\mathbf{N}$ does not intersect any of the six planes, and therefore the six planes do not form a bounded region.

e. Consider the change of variables $u = ax + by$, $v = cx + dz$, $w = ey + fz$. Show that

$$J(x, y, z) = \frac{\partial(u, v, w)}{\partial(x, y, z)} = -ade - bcf.$$

What is the value of the Jacobian if R is unbounded?

QUICK CHECK ANSWERS

1. The image is a semicircular disk of radius 1.
2. $J(u, v) = 2$ 3. $x = 2u/3 - v/3$, $y = u/3 + v/3$
4. The ratio is 2, which is $1/J(u, v)$.   5. It means that the volume of a small region in xyz-space is unchanged when it is transformed by T into a small region in uvw-space. ◄

CHAPTER 16 REVIEW EXERCISES

1. **Explain why or why not** Determine whether the following statements are true and give an explanation or counterexample.

a. Assuming g is integrable and a, b, c, and d are constants,
$$\int_c^d \int_a^b g(x, y)\,dx\,dy = \left(\int_a^b g(x, y)\,dx\right)\left(\int_c^d g(x, y)\,dy\right).$$

b. The spherical equation $\varphi = \pi/2$, the cylindrical equation $z = 0$, and the rectangular equation $z = 0$ all describe the same set of points.

c. Changing the order of integration in $\displaystyle\iiint_D f(x, y, z)\,dx\,dy\,dz$ from $dx\,dy\,dz$ to $dy\,dz\,dx$ requires also changing the integrand from $f(x, y, z)$ to $f(y, z, x)$.

d. The transformation $T\!: x = v$, $y = -u$ maps a square in the uv-plane to a triangle in the xy-plane.

2–4. Evaluating integrals *Evaluate the following integrals as they are written.*

2. $\displaystyle\int_1^2 \int_1^4 \frac{xy}{(x^2 + y^2)^2}\,dx\,dy$

3. $\displaystyle\int_1^3 \int_1^{e^x} \frac{x}{y}\,dy\,dx$

4. $\displaystyle\int_1^2 \int_0^{\ln x} x^3 e^y\,dy\,dx$

5–7. Changing the order of integration *Assuming f is integrable, change the order of integration in the following integrals.*

5. $\displaystyle\int_{-1}^1 \int_{x^2}^1 f(x, y)\,dy\,dx$

6. $\displaystyle\int_0^2 \int_{y-1}^1 f(x, y)\,dx\,dy$

7. $\displaystyle\int_0^1 \int_0^{\sqrt{1-y^2}} f(x, y)\,dx\,dy$

8–10. Area of plane regions *Use double integrals to compute the area of the following regions. Make a sketch of the region.*

8. The region bounded by the lines $y = -x - 4$, $y = x$, and $y = 2x - 4$

9. The region bounded by $y = |x|$ and $y = 20 - x^2$

10. The region between the curves $y = x^2$ and $y = 1 + x - x^2$

11–16. Miscellaneous double integrals *Choose a convenient method for evaluating the following integrals.*

11. $\displaystyle\iint_R \frac{2y}{\sqrt{x^4 + 1}}\, dA$; R is the region bounded by $x = 1$, $x = 2$, $y = x^{3/2}$, and $y = 0$.

12. $\displaystyle\iint_R x^{-1/2}\, e^y\, dA$; R is the region bounded by $x = 1$, $x = 4$, $y = \sqrt{x}$, and $y = 0$.

13. $\displaystyle\iint_R (x + y)\, dA$; R is the disk bounded by the circle $r = 4 \sin \theta$.

14. $\displaystyle\iint_R (x^2 + y^2)\, dA$; R is the region $\{(x, y): 0 \le x \le 2, 0 \le y \le x\}$.

15. $\displaystyle\int_0^1 \int_{y^{1/3}}^1 x^{10} \cos (\pi x^4 y)\, dx\, dy$

16. $\displaystyle\int_0^2 \int_{y^2}^4 x^8 y \sqrt{1 + x^4 y^2}\, dx\, dy$

17–18. Cartesian to polar coordinates *Evaluate the following integrals over the specified region. Assume (r, θ) are polar coordinates.*

17. $\displaystyle\iint_R 3x^2 y\, dA$; $R = \{(r, \theta): 0 \le r \le 1, 0 \le \theta \le \pi/2\}$

18. $\displaystyle\iint_R \frac{dA}{(1 + x^2 + y^2)^2}$; $R = \{(r, \theta): 1 \le r \le 4, 0 \le \theta \le \pi\}$

19–21. Computing areas *Sketch the following regions and use a double integral to find their areas.*

19. The region bounded by all leaves of the rose $r = 3 \cos 2\theta$

20. The region inside both of the circles $r = 2$ and $r = 4 \cos \theta$

21. The region that lies inside both of the cardioids $r = 2 - 2 \cos \theta$ and $r = 2 + 2 \cos \theta$

22–23. Average values

22. Find the average value of $z = \sqrt{16 - x^2 - y^2}$ over the disk in the xy-plane centered at the origin with radius 4.

23. Find the average distance from the points in the solid cone bounded by $z = 2\sqrt{x^2 + y^2}$ to the z-axis, for $0 \le z \le 8$.

24–26. Changing order of integration *Rewrite the following integrals using the indicated order of integration.*

24. $\displaystyle\int_0^1 \int_0^{\sqrt{1-z^2}} \int_0^{\sqrt{1-x^2}} f(x, y, z)\, dy\, dx\, dz$ in the order $dz\, dy\, dx$

25. $\displaystyle\int_0^1 \int_0^{\sqrt{1-x^2}} \int_{2\sqrt{x^2+y^2}}^2 f(x, y, z)\, dz\, dy\, dx$ in the order $dx\, dz\, dy$

26. $\displaystyle\int_0^2 \int_0^{9-x^2} \int_0^x f(x, y, z)\, dy\, dz\, dx$ in the order $dz\, dx\, dy$

27–31. Triple integrals *Evaluate the following integrals, changing the order of integration if needed.*

27. $\displaystyle\int_0^1 \int_{-z}^z \int_{-\sqrt{1-x^2}}^{\sqrt{1-x^2}} dy\, dx\, dz$

28. $\displaystyle\int_0^\pi \int_0^y \int_0^{\sin x} dz\, dx\, dy$

29. $\displaystyle\int_1^9 \int_0^1 \int_{2y}^2 \frac{4 \sin x^2}{\sqrt{z}}\, dx\, dy\, dz$

30. $\displaystyle\int_0^2 \int_{-\sqrt{2-x^2}/2}^{\sqrt{2-x^2}/2} \int_{x^2+3y^2}^{8-x^2-y^2} dz\, dy\, dx$

31. $\displaystyle\int_0^2 \int_0^{y^{1/3}} \int_0^{y^2} yz^5(1 + x + y^2 + z^6)^2\, dx\, dz\, dy$

32–38. Volumes of solids *Find the volume of the following solids.*

32. The solid beneath the paraboloid $f(x, y) = 12 - x^2 - 2y^2$ and above the region $R = \{(x, y): 1 \le x \le 2, 0 \le y \le 1\}$

$f(x, y) = 12 - x^2 - 2y^2$

33. The solid bounded by the surfaces $x = 0$, $y = 0$, $z = 3 - 2y$, and $z = 2x^2 + 1$

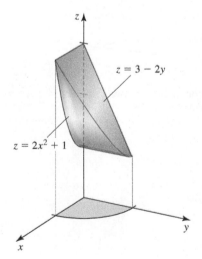

$z = 3 - 2y$

$z = 2x^2 + 1$

34. The prism in the first octant bounded by the planes $y = 3 - 3x$ and $z = 2$

35. One of the wedges formed when the cylinder $x^2 + y^2 = 4$ is cut by the planes $z = 0$ and $y = z$

36. The solid bounded by the parabolic cylinders $z = y^2 + 1$ and $z = 2 - x^2$

37. The solid common to the two cylinders $x^2 + y^2 = 4$ and $x^2 + z^2 = 4$

38. The tetrahedron with vertices $(0, 0, 0)$, $(1, 0, 0)$, $(1, 1, 0)$, and $(1, 1, 1)$

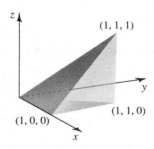

39. Single to double integral Evaluate $\int_0^{1/2}(\sin^{-1} 2x - \sin^{-1} x)\,dx$ by converting it to a double integral.

40. Tetrahedron limits Let D be the tetrahedron with vertices at $(0, 0, 0)$, $(1, 0, 0)$, $(0, 2, 0)$, and $(0, 0, 3)$. Suppose the volume of D is to be found using a triple integral. Give the limits of integration for the six possible orderings of the variables.

41. Polar to Cartesian Evaluate $\int_0^{\pi/4}\int_0^{\sec\theta} r^3\,dr\,d\theta$ using rectangular coordinates, where (r, θ) are polar coordinates.

42–43. Average value

42. Find the average of the *square* of the distance between the origin and the points in the solid paraboloid $D = \{(x, y, z)\colon 0 \le z \le 4 - x^2 - y^2\}$.

43. Find the average x-coordinate of the points in the prism $D = \{(x, y, z)\colon 0 \le x \le 1, 0 \le y \le 3 - 3x, 0 \le z \le 2\}$.

44–45. Integrals in cylindrical coordinates *Evaluate the following integrals in cylindrical coordinates.*

44. $\displaystyle\int_0^3\int_0^{\sqrt{9-x^2}}\int_0^3 (x^2 + y^2)^{3/2}\,dz\,dy\,dx$

45. $\displaystyle\int_{-1}^1\int_{-2}^2\int_0^{\sqrt{1-y^2}} \frac{1}{(1 + x^2 + y^2)^2}\,dx\,dz\,dy$

46–47. Volumes in cylindrical coordinates *Use integration in cylindrical coordinates to find the volume of the following solids.*

46. The solid bounded by the hemisphere $z = \sqrt{9 - x^2 - y^2}$ and the hyperboloid $z = \sqrt{1 + x^2 + y^2}$.

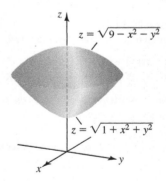

47. The solid cylinder whose height is 4 and whose base is the disk $\{(r, \theta): 0 \le r \le 2\cos\theta\}$

48–49. Integrals in spherical coordinates *Evaluate the following integrals in spherical coordinates.*

48. $\displaystyle\int_0^{2\pi}\int_0^{\pi/2}\int_0^{2\cos\varphi}\rho^2\sin\varphi\,d\rho\,d\varphi\,d\theta$

49. $\displaystyle\int_0^{\pi}\int_0^{\pi/4}\int_{2\sec\varphi}^{4\sec\varphi}\rho^2\sin\varphi\,d\rho\,d\varphi\,d\theta$

50–52. Volumes in spherical coordinates *Use integration in spherical coordinates to find the volume of the following solids.*

50. The solid cardioid of revolution $D = \{(\rho, \varphi, \theta): 0 \le \rho \le (1 - \cos\varphi)/2, 0 \le \varphi \le \pi, 0 \le \theta \le 2\pi\}$

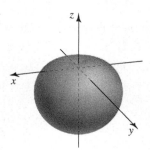

51. The solid rose petal of revolution
$D = \{(\rho, \varphi, \theta): 0 \le \rho \le 4\sin 2\varphi,$
$0 \le \varphi \le \pi/2, 0 \le \theta \le 2\pi\}$

52. The solid above the cone $\varphi = \pi/4$ and inside the sphere $\rho = 4\cos\varphi$

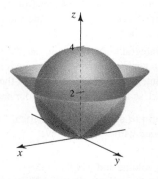

53–56. Center of mass of constant-density plates *Find the center of mass (centroid) of the following thin, constant-density plates. Sketch the region corresponding to the plate and indicate the location of the center of mass. Use symmetry whenever possible to simplify your work.*

53. The region bounded by $y = \sin x$ and $y = 0$ between $x = 0$ and $x = \pi$

54. The region bounded by $y = x^3$ and $y = x^2$ between $x = 0$ and $x = 1$

55. The half-annulus $\{(r, \theta): 2 \le r \le 4, 0 \le \theta \le \pi\}$

56. The region bounded by $y = x^2$ and $y = a^2 - x^2$, where $a > 0$

57–58. Center of mass of constant-density solids *Find the center of mass of the following solids, assuming a constant density. Use symmetry whenever possible and choose a convenient coordinate system.*

57. The paraboloid bowl bounded by $z = x^2 + y^2$ and $z = 36$

58. The tetrahedron bounded by $z = 4 - x - 2y$ and the coordinate planes

59–60. Variable-density solids *Find the coordinates of the center of mass of the following solids with the given density.*

59. The upper half of the ball $\Big\{(\rho, \varphi, \theta): 0 \le \rho \le 16, 0 \le \varphi \le \dfrac{\pi}{2},$
$0 \le \theta \le 2\pi\Big\}$ with density $f(\rho, \varphi, \theta) = 1 + \rho/4$

60. The cube in the first octant bounded by the planes $x = 2$, $y = 2$, and $z = 2$, with $\rho(x, y, z) = 1 + x + y + z$

61–64. Center of mass for general objects *Consider the following two- and three-dimensional regions. Compute the center of mass, assuming constant density. All parameters are positive real numbers.*

61. A solid is bounded by a paraboloid with a circular base of radius R and height h. How far from the base is the center of mass?

62. Let R be the region enclosed by an equilateral triangle with sides of length s. What is the perpendicular distance between the center of mass of R and the edges of R?

63. A sector of a circle in the first quadrant is bounded between the x-axis, the line $y = x$, and the circle $x^2 + y^2 = a^2$. What are the coordinates of the center of mass?

64. An ice cream cone is bounded above by the sphere $x^2 + y^2 + z^2 = a^2$ and below by the upper half of the cone $z^2 = x^2 + y^2$. What are the coordinates of the center of mass?

65. Slicing a conical cake A cake is shaped like a solid cone with radius 4 and height 2, with its base on the xy-plane. A wedge of the cake is removed by making two slices from the axis of the cone outward, perpendicular to the xy-plane and separated by an angle of Q radians, where $0 < Q < 2\pi$.

   a. Use a double integral to find the volume of the slice for $Q = \pi/4$. Use geometry to check your answer.
   b. Use a double integral to find the volume of the slice for any $0 < Q < 2\pi$. Use geometry to check your answer.

66. Volume and weight of a fish tank A spherical fish tank with a radius of 1 ft is filled with water to a level 6 in below the top of the tank.

   a. Determine the volume and weight of the water in the fish tank. (The weight density of water is about 62.5 lb/ft^3.)
   b. How much additional water must be added to completely fill the tank?

67–70. Transforming a square *Let* $S = \{(u, v): 0 \le u \le 1, 0 \le v \le 1\}$ *be a unit square in the uv-plane. Find the image of S in the xy-plane under the following transformations.*

67. $T: x = v, y = u$ **68.** $T: x = -v, y = u$

69. $T: x = 3u + v, y = u + 3v$

70. $T: x = u, y = 2v + 2$

71–74. Computing Jacobians *Compute the Jacobian* $J(u, v)$ *of the following transformations.*

71. $T: x = 4u - v, y = -2u + 3v$

72. $T: x = u + v, y = u - v$

73. $T: x = 3u, y = 2v + 2$ **74.** $T: x = u^2 - v^2, y = 2uv$

75–78. Double integrals—transformation given *To evaluate the following integrals, carry out these steps.*

a. *Sketch the original region of integration R and the new region S using the given change of variables.*

b. *Find the limits of integration for the new integral with respect to u and v.*
c. *Compute the Jacobian.*
d. *Change variables and evaluate the new integral.*

75. $\iint_R xy^2\, dA; R = \{(x, y): y/3 \le x \le (y + 6)/3, 0 \le y \le 3\}$; use $x = u + v/3, y = v$.

76. $\iint_R 3xy^2\, dA; R = \{(x, y): 0 \le x \le 2, x \le y \le x + 4\}$; use $x = 2u, y = 4v + 2u$.

77. $\iint_R (x - y + 1)(x - y)^9\, dy\, dx, R = \{(x, y): 0 \le y \le x \le 1\}$; use $x = u + v, y = v - u$.

78. $\iint_R xy^2\, dA; R$ is the region between the hyperbolas $xy = 1$ and $xy = 4$ and the lines $y = 1$ and $y = 4$; use $x = u/v, y = v$.

79–80. Double integrals *Evaluate the following integrals using a change of variables. Sketch the original and new regions of integration, R and S.*

79. $\iint_R y^4\, dA; R$ is the region bounded by the hyperbolas $xy = 1$ and $xy = 4$ and the lines $y/x = 1$ and $y/x = 3$.

80. $\iint_R (y^2 + xy - 2x^2)\, dA; R$ is the region bounded by the lines $y = x, y = x - 3, y = -2x + 3$, and $y = -2x - 3$.

81–82. Triple integrals *Use a change of variables to evaluate the following integrals.*

81. $\iiint_D yz\, dV; D$ is bounded by the planes $x + 2y = 1, x + 2y = 2, x - z = 0, x - z = 2, 2y - z = 0$, and $2y - z = 3$.

82. $\iiint_D x\, dV; D$ is bounded by the planes $y - 2x = 0, y - 2x = 1, z - 3y = 0, z - 3y = 1, z - 4x = 0$, and $z - 4x = 3$.

Chapter 16 Guided Projects

Applications of the material in this chapter and related topics can be found in the following Guided Projects. For additional information, see the Preface.

- How big are n-balls?
- Electrical field integrals
- The tilted cylinder problem

- The exponential Eiffel Tower
- Moments of inertia
- Gravitational fields

17

Vector Calculus

Chapter Preview This culminating chapter of the text provides a beautiful, unifying conclusion to our study of calculus. Many ideas and themes that have appeared throughout the text come together in these final pages. First, we combine vector-valued functions (Chapter 14) and functions of several variables (Chapter 15) to form *vector fields*. Once vector fields have been introduced and illustrated through their many applications, we explore the calculus of vector fields. Concepts such as limits and continuity carry over directly. The extension of derivatives to vector fields leads to two new operations that underlie this chapter: the *curl* and the *divergence*. When integration is extended to vector fields, we discover new versions of the Fundamental Theorem of Calculus. The chapter ends with a final look at the Fundamental Theorem of Calculus and the several related forms in which it has appeared throughout the text.

17.1 Vector Fields

We live in a world filled with phenomena that can be represented by vector fields. Imagine sitting in a window seat looking out at the wing of an airliner. Although you can't see it, air is rushing over and under the wing. Focus on a point near the wing and visualize the motion of the air at that point at a single instant of time. The motion is described by a velocity vector with three components—for example, east-west, north-south, and up-down. At another point near the wing at the same time, the air is moving at a different speed and direction, and a different velocity vector is associated with that point. In general, at one instant in time, every point around the wing has a velocity vector associated with it (**Figure 17.1**). This collection of velocity vectors—a unique vector for each point in space—is a function called a *vector field*.

Other examples of vector fields include the wind patterns in a hurricane (**Figure 17.2a**) and the circulation of water in a heat exchanger (**Figure 17.2b**). Gravitational, magnetic, and electric force fields are also represented by vector fields (**Figure 17.2c**), as are the stresses and strains in buildings and bridges. Beyond physics and engineering, the transport of a chemical pollutant in a lake and human migration patterns can be modeled by vector fields.

Figure 17.1

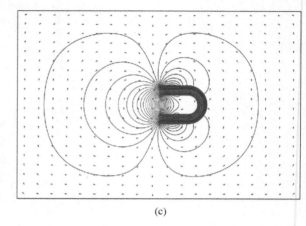

(a)   (b)   (c)

Figure 17.2

Vector Fields in Two Dimensions

To solidify the idea of a vector field, we begin by exploring vector fields in $\mathbb{R}^2$. From there, it is a short step to vector fields in $\mathbb{R}^3$.

▶ Notice that a vector field is both a vector-valued function (Chapter 14) and a function of several (Chapter 15).

> **DEFINITION Vector Fields in Two Dimensions**
>
> Let f and g be defined on a region R of $\mathbb{R}^2$. A **vector field** in $\mathbb{R}^2$ is a function $\mathbf{F}$ that assigns to each point in R a vector $\langle f(x, y), g(x, y) \rangle$. The vector field is written as
>
> $$\mathbf{F}(x, y) = \langle f(x, y), g(x, y) \rangle \quad \text{or}$$
> $$\mathbf{F}(x, y) = f(x, y)\,\mathbf{i} + g(x, y)\,\mathbf{j}.$$
>
> A vector field $\mathbf{F} = \langle f, g \rangle$ is continuous or differentiable on a region R of $\mathbb{R}^2$ if f and g are continuous or differentiable on R, respectively.

A vector field cannot be represented graphically in its entirety. Instead, we plot a representative sample of vectors that illustrates the general appearance of the vector field. Consider the vector field defined by

$$\mathbf{F}(x, y) = \langle 2x, 2y \rangle = 2x\,\mathbf{i} + 2y\,\mathbf{j}.$$

At selected points $P(x, y)$, we plot a vector with its tail at P equal to the value of $\mathbf{F}(x, y)$. For example, $\mathbf{F}(1, 1) = \langle 2, 2 \rangle$, so we draw a vector equal to $\langle 2, 2 \rangle$ with its tail at the point $(1, 1)$. Similarly, $\mathbf{F}(-2, -3) = \langle -4, -6 \rangle$, so at the point $(-2, -3)$, we draw a vector equal to $\langle -4, -6 \rangle$. We can make the following general observations about the vector field $\mathbf{F}(x, y) = \langle 2x, 2y \rangle$.

Lengths of vectors increase with distance from the origin.

Radial vector field
$\mathbf{F} = \langle 2x, 2y \rangle$

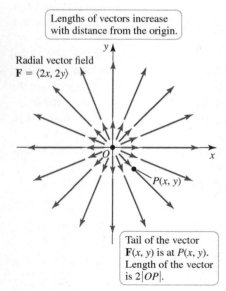

Tail of the vector $\mathbf{F}(x, y)$ is at $P(x, y)$. Length of the vector is $2|OP|$.

Figure 17.3

• For every (x, y) except $(0, 0)$, the vector $\mathbf{F}(x, y)$ points in the direction of $\langle 2x, 2y \rangle$, which is directly outward from the origin.

• The length of $\mathbf{F}(x, y)$ is $|\mathbf{F}| = |\langle 2x, 2y \rangle| = 2\sqrt{x^2 + y^2}$, which increases with distance from the origin.

The vector field $\mathbf{F} = \langle 2x, 2y \rangle$ is an example of a *radial vector field* because its vectors point radially away from the origin (**Figure 17.3**). If $\mathbf{F}$ represents the velocity of a fluid moving in two dimensions, the graph of the vector field gives a vivid image of how a small object, such as a cork, moves through the fluid. In this case, at every point of the vector field, a particle moves in the direction of the arrow at that point with a speed equal to the length of the arrow. For this reason, vector fields are sometimes called *flows*. When sketching vector fields, it is often useful to draw continuous curves that are aligned with the vector field. Such curves are called *flow curves* or *streamlines*; we examine their properties in greater detail later in this section.

Shear vector field
$\mathbf{F} = \langle 0, x \rangle$

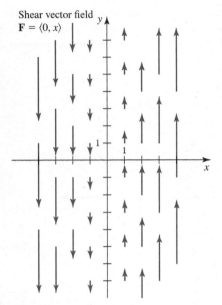

Figure 17.4

➤ Drawing vectors with their actual length often leads to cluttered pictures of vector fields. For this reason, most of the vector fields in this chapter are illustrated with proportional scaling: All vectors are multiplied by a scalar chosen to make the vector field as understandable as possible.

➤ A useful observation for two-dimensional vector fields $\mathbf{F} = \langle f, g \rangle$ is that the slope of the vector at (x, y) is $g(x, y)/f(x, y)$. In Example 1a, the slopes are everywhere undefined; in part (b), the slopes are everywhere 0, and in part (c), the slopes are $-x/y$.

QUICK CHECK 1 If the vector field in Example 1c describes the velocity of a fluid and you place a small cork in the plane at $(2, 0)$, what path will it follow? ◄

EXAMPLE 1 Vector fields Sketch representative vectors of the following vector fields.

a. $\mathbf{F}(x, y) = \langle 0, x \rangle = x\,\mathbf{j}$ (a shear field)

b. $\mathbf{F}(x, y) = \langle 1 - y^2, 0 \rangle = (1 - y^2)\,\mathbf{i}$, for $|y| \le 1$ (channel flow)

c. $\mathbf{F}(x, y) = \langle -y, x \rangle = -y\,\mathbf{i} + x\,\mathbf{j}$ (a rotation field)

SOLUTION

a. This vector field is independent of y. Furthermore, because the x-component of $\mathbf{F}$ is zero, all vectors in the field (for $x \ne 0$) point in the y-direction: upward for $x > 0$ and downward for $x < 0$. The magnitudes of the vectors in the field increase with distance from the y-axis (**Figure 17.4**). The flow curves for this field are vertical lines. If $\mathbf{F}$ represents a velocity field, a particle right of the y-axis moves upward, a particle left of the y-axis moves downward, and a particle on the y-axis is stationary.

b. In this case, the vector field is independent of x and the y-component of $\mathbf{F}$ is zero. Because $1 - y^2 > 0$ for $|y| < 1$, vectors in this region point in the positive x-direction. The x-component of the vector field is zero at the boundaries $y = \pm 1$ and increases to 1 along the center of the strip, $y = 0$. This vector field might model the flow of water in a straight shallow channel (**Figure 17.5**); its flow curves are horizontal lines, indicating motion in the direction of the positive x-axis.

c. It often helps to determine the vector field along the coordinate axes.

• When $y = 0$ (along the x-axis), we have $\mathbf{F}(x, 0) = \langle 0, x \rangle$. With $x > 0$, this vector field consists of vectors pointing upward, increasing in length as x increases. With $x < 0$, the vectors point downward, increasing in length as $|x|$ increases.

• When $x = 0$ (along the y-axis), we have $\mathbf{F}(0, y) = \langle -y, 0 \rangle$. If $y > 0$, the vectors point in the negative x-direction, increasing in length as y increases. If $y < 0$, the vectors point in the positive x-direction, increasing in length as $|y|$ increases.

A few more representative vectors show that this vector field has a counterclockwise rotation about the origin; the magnitudes of the vectors increase with distance from the origin (**Figure 17.6**).

Channel flow
$\mathbf{F} = \langle 1 - y^2, 0 \rangle$

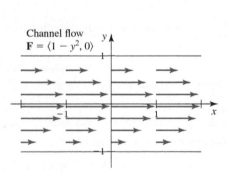

Figure 17.5

Rotation vector field
$\mathbf{F} = \langle -y, x \rangle$

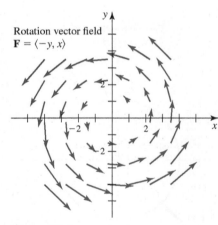

Figure 17.6

Related Exercises 10, 11, 13 ◄

Radial Vector Fields in $\mathbb{R}^2$ Radial vector fields in $\mathbb{R}^2$ have the property that their vectors point directly toward or away from the origin at all points (except the origin), parallel to the position vectors $\mathbf{r} = \langle x, y \rangle$. We will work with radial vector fields of the form

$$\mathbf{F}(x, y) = \frac{\mathbf{r}}{|\mathbf{r}|^p} = \frac{\langle x, y \rangle}{|\mathbf{r}|^p} = \underbrace{\frac{\mathbf{r}}{|\mathbf{r}|}}_{\substack{\text{unit} \\ \text{vector}}} \underbrace{\frac{1}{|\mathbf{r}|^{p-1}}}_{\text{magnitude}},$$

where p is a real number. **Figure 17.7** illustrates radial fields with $p = 1$ and $p = 3$. These vector fields (and their three-dimensional counterparts) play an important role in many applications. For example, central forces, such as gravitational or electrostatic forces between point masses or charges, are described by radial vector fields with $p = 3$. These forces obey an inverse square law in which the magnitude of the force is proportional to $1/|\mathbf{r}|^2$.

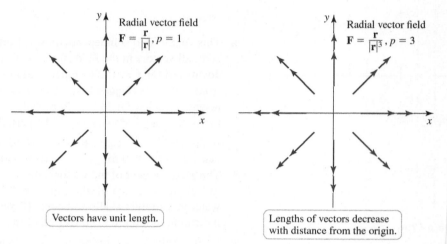

Radial vector field $\mathbf{F} = \dfrac{\mathbf{r}}{|\mathbf{r}|}, p = 1$

Vectors have unit length.

Radial vector field $\mathbf{F} = \dfrac{\mathbf{r}}{|\mathbf{r}|^3}, p = 3$

Lengths of vectors decrease with distance from the origin.

Figure 17.7

DEFINITION Radial Vector Fields in $\mathbb{R}^2$

Let $\mathbf{r} = \langle x, y \rangle$. A vector field of the form $\mathbf{F} = f(x, y)\,\mathbf{r}$, where f is a scalar-valued function, is a **radial vector field**. Of specific interest are the radial vector fields

$$\mathbf{F}(x, y) = \frac{\mathbf{r}}{|\mathbf{r}|^p} = \frac{\langle x, y \rangle}{|\mathbf{r}|^p},$$

where p is a real number. At every point (except the origin), the vectors of this field are directed outward from the origin with a magnitude of $|\mathbf{F}| = \dfrac{1}{|\mathbf{r}|^{p-1}}$.

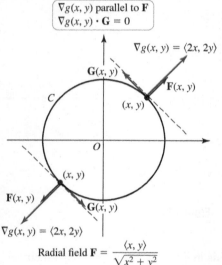

$\nabla g(x, y)$ parallel to $\mathbf{F}$
$\nabla g(x, y) \cdot \mathbf{G} = 0$

$\nabla g(x, y) = \langle 2x, 2y \rangle$

$G(x, y)$

$F(x, y)$

(x, y)

O

(x, y)

$F(x, y)$

$G(x, y)$

$\nabla g(x, y) = \langle 2x, 2y \rangle$

Radial field $\mathbf{F} = \dfrac{\langle x, y \rangle}{\sqrt{x^2 + y^2}}$

Rotation field $\mathbf{G} = \dfrac{\langle -y, x \rangle}{\sqrt{x^2 + y^2}}$

Figure 17.8

QUICK CHECK 2 In Example 2, verify that $\nabla g(x, y) \cdot \mathbf{G}(x, y) = 0$. In parts (a) and (b) of Example 2, verify that $|\mathbf{F}| = 1$ and $|\mathbf{G}| = 1$ at all points excluding the origin. ◀

EXAMPLE 2 Normal and tangent vectors Let C be the circle $x^2 + y^2 = a^2$, where $a > 0$.

a. Show that at each point of C, the radial vector field $\mathbf{F}(x, y) = \dfrac{\mathbf{r}}{|\mathbf{r}|} = \dfrac{\langle x, y \rangle}{\sqrt{x^2 + y^2}}$ is orthogonal to the line tangent to C at that point.

b. Show that at each point of C, the rotation vector field $\mathbf{G}(x, y) = \dfrac{\langle -y, x \rangle}{\sqrt{x^2 + y^2}}$ is parallel to the line tangent to C at that point.

SOLUTION Let $g(x, y) = x^2 + y^2$. The circle C described by the equation $g(x, y) = a^2$ may be viewed as a level curve of the surface $z = x^2 + y^2$. As shown in Theorem 15.12 (Section 15.5), the gradient $\nabla g(x, y) = \langle 2x, 2y \rangle$ is orthogonal to the line tangent to C at (x, y) **(Figure 17.8)**.

a. Notice that $\nabla g(x, y)$ is parallel to $\mathbf{F} = \langle x, y \rangle / |\mathbf{r}|$ at the point (x, y). It follows that $\mathbf{F}$ is also orthogonal the line tangent to C at (x, y).

b. Notice that

$$\nabla g(x, y) \cdot \mathbf{G}(x, y) = \langle 2x, 2y \rangle \cdot \frac{\langle -y, x \rangle}{|\mathbf{r}|} = 0.$$

Therefore, $\nabla g(x, y)$ is orthogonal to the vector field $\mathbf{G}$ at (x, y), which implies that $\mathbf{G}$ is parallel to the tangent line at (x, y).

Related Exercises 27–28 ◀

Vector Fields in Three Dimensions

Vector fields in three dimensions are conceptually the same as vector fields in two dimensions. The vector **F** now has three components, each of which depends on three variables.

DEFINITION Vector Fields and Radial Vector Fields in $\mathbb{R}^3$

Let f, g, and h be defined on a region D of $\mathbb{R}^3$. A **vector field** in $\mathbb{R}^3$ is a function **F** that assigns to each point in D a vector $\langle f(x, y, z), g(x, y, z), h(x, y, z) \rangle$. The vector field is written as

$$\mathbf{F}(x, y, z) = \langle f(x, y, z), g(x, y, z), h(x, y, z) \rangle \quad \text{or}$$
$$\mathbf{F}(x, y, z) = f(x, y, z)\, \mathbf{i} + g(x, y, z)\, \mathbf{j} + h(x, y, z)\, \mathbf{k}.$$

A vector field $\mathbf{F} = \langle f, g, h \rangle$ is continuous or differentiable on a region D of $\mathbb{R}^3$ if f, g, and h are continuous or differentiable on D, respectively. Of particular importance are the **radial vector fields**

$$\mathbf{F}(x, y, z) = \frac{\mathbf{r}}{|\mathbf{r}|^p} = \frac{\langle x, y, z \rangle}{|\mathbf{r}|^p},$$

where p is a real number.

EXAMPLE 3   Vector fields in $\mathbb{R}^3$ Sketch and discuss the following vector fields.

a. $\mathbf{F}(x, y, z) = \langle x, y, e^{-z} \rangle$, for $z \geq 0$

b. $\mathbf{F}(x, y, z) = \langle 0, 0, 1 - x^2 - y^2 \rangle$, for $x^2 + y^2 \leq 1$

SOLUTION

a. First consider the x- and y-components of **F** in the xy-plane ($z = 0$), where $\mathbf{F} = \langle x, y, 1 \rangle$. This vector field looks like a radial field in the first two components, increasing in magnitude with distance from the z-axis. However, each vector also has a constant vertical component of 1. In horizontal planes $z = z_0 > 0$, the radial pattern remains the same, but the vertical component decreases as z increases. As $z \to \infty$, $e^{-z} \to 0$ and the vector field approaches a horizontal radial field (**Figure 17.9**).

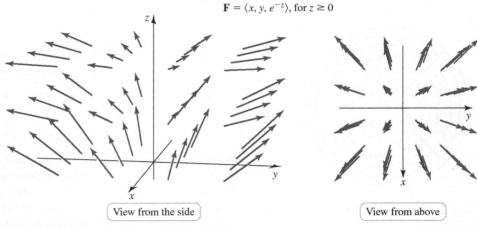

$\mathbf{F} = \langle x, y, e^{-z} \rangle$, for $z \geq 0$

View from the side   View from above

Figure 17.9

$\mathbf{F} = \langle 0, 0, 1 - x^2 - y^2 \rangle$, for $x^2 + y^2 \leq 1$

Cylinder $x^2 + y^2 = 1$

Figure 17.10

b. Regarding **F** as a velocity field for points in and on the cylinder $x^2 + y^2 = 1$, there is no motion in the x- or y-direction. The z-component of the vector field may be written $1 - r^2$, where $r^2 = x^2 + y^2$ is the square of the distance from the z-axis. We see that the z-component increases from 0 on the boundary of the cylinder ($r = 1$) to a maximum value of 1 along the centerline of the cylinder ($r = 0$) (**Figure 17.10**). This vector field models the flow of a fluid inside a tube (such as a blood vessel).

Related Exercises 20, 22 ◄

▶ Physicists often use the convention that a gradient field and its potential function are related by $\mathbf{F} = -\nabla\varphi$ (with a negative sign).

> The vector field $\mathbf{F} = \nabla\varphi$ is orthogonal to the level curves of φ at (x, y).

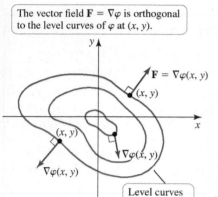

Figure 17.11

Gradient Fields and Potential Functions   One way to generate a vector field is to start with a differentiable scalar-valued function φ, take its gradient, and let $\mathbf{F} = \nabla\varphi$. A vector field defined as the gradient of a scalar-valued function φ is called a *gradient field*, and φ is called a *potential function*.

Suppose φ is a differentiable function on a region R of $\mathbb{R}^2$ and consider the surface $z = \varphi(x, y)$. Recall from Chapter 15 that this function may also be represented by level curves in the xy-plane. At each point (a, b) on a level curve, the gradient $\nabla\varphi(a, b) = \langle \varphi_x(a, b), \varphi_y(a, b) \rangle$ is orthogonal to the level curve at (a, b) (**Figure 17.11**). Therefore, the vectors of $\mathbf{F} = \nabla\varphi$ point in a direction orthogonal to the level curves of φ.

The idea extends to gradients of functions of three variables. If φ is differentiable on a region D of $\mathbb{R}^3$, then $\mathbf{F} = \nabla\varphi = \langle \varphi_x, \varphi_y, \varphi_z \rangle$ is a vector field that points in a direction orthogonal to the level *surfaces* of φ.

Gradient fields are useful because of the physical meaning of the gradient. For example, if φ represents the temperature in a conducting material, then the gradient field $\mathbf{F} = \nabla\varphi$ evaluated at a point indicates the direction in which the temperature increases most rapidly at that point. According to a basic physical law, heat diffuses in the direction of the vector field $-\mathbf{F} = -\nabla\varphi$, the direction in which the temperature *decreases* most rapidly; that is, heat flows "down the gradient" from relatively hot regions to cooler regions. Similarly, water on a smooth surface tends to flow down the elevation gradient.

QUICK CHECK 3 Find the gradient field associated with the function $\varphi(x, y, z) = xyz$. ◀

> A potential function plays the role of an antiderivative of a vector field: Derivatives of the potential function produce the vector field. If φ is a potential function for a gradient field, then $\varphi + C$ is also a potential function for that gradient field, for any constant C.

> Gradient vectors ∇T (not drawn to scale) are orthogonal to the level curves.

Level curves of $T(x, y) = 200 - x^2 - y^2$

Figure 17.12

DEFINITION Gradient Fields and Potential Functions

Let φ be differentiable on a region of $\mathbb{R}^2$ or $\mathbb{R}^3$. The vector field $\mathbf{F} = \nabla\varphi$ is a **gradient field** and the function φ is a **potential function** for $\mathbf{F}$.

EXAMPLE 4   Gradient fields

a. Sketch and interpret the gradient field associated with the temperature function $T = 200 - x^2 - y^2$ on the circular plate $R = \{ (x, y): x^2 + y^2 \leq 25 \}$.

b. Sketch and interpret the gradient field associated with the velocity potential $\varphi = \tan^{-1} xy$.

SOLUTION

a. The gradient field associated with T is

$$\mathbf{F} = \nabla T = \langle -2x, -2y \rangle = -2\langle x, y \rangle.$$

This vector field points inward toward the origin at all points of R except $(0, 0)$. The magnitudes of the vectors,

$$|\mathbf{F}| = \sqrt{(-2x)^2 + (-2y)^2} = 2\sqrt{x^2 + y^2},$$

are greatest on the edge of the disk R, where $x^2 + y^2 = 25$ and $|\mathbf{F}| = 10$. The magnitudes of the vectors in the field decrease toward the center of the plate with $|\mathbf{F}(0, 0)| = 0$. **Figure 17.12** shows the level curves of the temperature function with several gradient vectors, all orthogonal to the level curves. Note that the plate is hottest at the center and coolest on the edge, so heat diffuses *outward*, in the direction opposite that of the gradient.

b. The gradient of a velocity potential gives the velocity components of a two-dimensional flow; that is, $\mathbf{F} = \langle u, v \rangle = \nabla\varphi$, where u and v are the velocities in the x- and y-directions, respectively. Computing the gradient, we find that

$$\mathbf{F} = \langle \varphi_x, \varphi_y \rangle = \left\langle \frac{1}{1 + (xy)^2} \cdot y, \frac{1}{1 + (xy)^2} \cdot x \right\rangle = \left\langle \frac{y}{1 + x^2y^2}, \frac{x}{1 + x^2y^2} \right\rangle.$$

F = ∇φ is orthogonal to level curves.

Level curves of $\varphi(x, y) = \tan^{-1}xy$

Figure 17.13

Notice that the level curves of φ are the hyperbolas $xy = C$ or $y = C/x$. At all points, the vector field is orthogonal to the level curves (**Figure 17.13**).

Related Exercises 38, 47 ◄

Equipotential Curves and Surfaces The preceding example illustrates a beautiful geometric connection between a gradient field and its associated potential function. Let φ be a potential function for the vector field $\mathbf{F}$ in $\mathbb{R}^2$; that is, $\mathbf{F} = \nabla\varphi$. The level curves of a potential function are called **equipotential curves** (curves on which the potential function is constant).

Because the equipotential curves are level curves of φ, the vector field $\mathbf{F} = \nabla\varphi$ is everywhere orthogonal to the equipotential curves (**Figure 17.14**). The vector field may be visualized by drawing continuous **flow curves** or **streamlines** that are everywhere orthogonal to the equipotential curves. These ideas also apply to vector fields in $\mathbb{R}^3$, in which case the vector field is orthogonal to the **equipotential surfaces**.

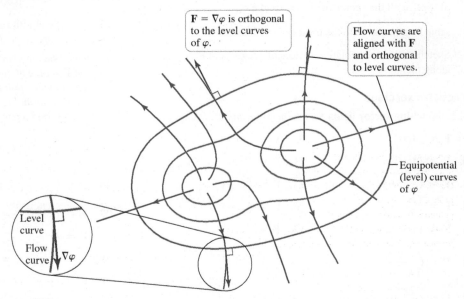

F = ∇φ is orthogonal to the level curves of φ.

Flow curves are aligned with **F** and orthogonal to level curves.

Equipotential (level) curves of φ

Level curve

Flow curve ∇φ

Figure 17.14

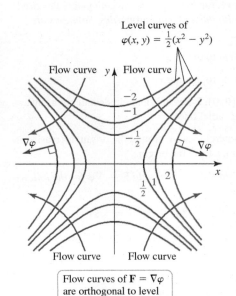

Level curves of $\varphi(x, y) = \frac{1}{2}(x^2 - y^2)$

Flow curve Flow curve

∇φ   ∇φ

Flow curve   Flow curve

Flow curves of **F** = ∇φ are orthogonal to level curves of φ everywhere.

Figure 17.15

► We use the fact that a line with slope a/b points in the direction of the vectors $\langle 1, a/b \rangle$ or $\langle b, a \rangle$.

EXAMPLE 5 Equipotential curves The equipotential curves for the potential function $\varphi(x, y) = (x^2 - y^2)/2$ are shown in green in **Figure 17.15**.

a. Find the gradient field associated with φ and verify that the gradient field is orthogonal to the equipotential curve at $(2, 1)$.

b. Verify that the vector field $\mathbf{F} = \nabla\varphi$ is orthogonal to the equipotential curves at all points (x, y).

SOLUTION

a. The level (or equipotential) curves are the hyperbolas $(x^2 - y^2)/2 = C$, where C is a constant. The slope at any point on a level curve $\varphi(x, y) = C$ (Section 15.4) is

$$\frac{dy}{dx} = -\frac{\varphi_x}{\varphi_y} = \frac{x}{y}.$$

At the point $(2, 1)$, the slope of the level curve is $dy/dx = 2$, so the vector tangent to the curve points in the direction $\langle 1, 2 \rangle$. The gradient field is given by $\mathbf{F} = \nabla\varphi = \langle x, -y \rangle$, so $\mathbf{F}(2, 1) = \nabla\varphi(2, 1) = \langle 2, -1 \rangle$. The dot product of the tangent vector $\langle 1, 2 \rangle$ and the gradient is $\langle 1, 2 \rangle \cdot \langle 2, -1 \rangle = 0$; therefore, the two vectors are orthogonal.

b. In general, the line tangent to the equipotential curve at (x, y) is parallel to the vector $\langle y, x \rangle$, and the vector field at that point is $\mathbf{F} = \langle x, -y \rangle$. The vector field and the tangent vectors are orthogonal because $\langle y, x \rangle \cdot \langle x, -y \rangle = 0$.

Related Exercise 52 ◄

SECTION 17.1 EXERCISES

Getting Started

1. How is a vector field $\mathbf{F} = \langle f, g, h \rangle$ used to describe the motion of air at one instant in time?

2. Sketch the vector field $\mathbf{F} = \langle x, y \rangle$.

3. How do you graph the vector field $\mathbf{F} = \langle f(x, y), g(x, y) \rangle$?

4. Given a differentiable, scalar-valued function φ, why is the gradient of φ a vector field?

5. Interpret the gradient field of the temperature function $T = f(x, y)$.

6. Show that all the vectors in vector field $\mathbf{F} = \dfrac{\sqrt{2}\langle x, y \rangle}{\sqrt{x^2 + y^2}}$ have the same length, and state the length of the vectors.

7. Sketch a few representative vectors of vector field $\mathbf{F} = \langle 0, 1 \rangle$ along the line $y = 2$.

Practice Exercises

8–23. Sketching vector fields *Sketch the following vector fields.*

8. $\mathbf{F} = \langle 1, 0 \rangle$

9. $\mathbf{F} = \langle -1, 1 \rangle$

10. $\mathbf{F} = \langle 1, y \rangle$

11. $\mathbf{F} = \langle x, 0 \rangle$

12. $\mathbf{F} = \langle -x, -y \rangle$

13. $\mathbf{F} = \langle x, -y \rangle$

14. $\mathbf{F} = \langle 2x, 3y \rangle$

15. $\mathbf{F} = \langle y, -x \rangle$

16. $\mathbf{F} = \langle x + y, y \rangle$

17. $\mathbf{F} = \langle x, y - x \rangle$

18. $\mathbf{F} = \left\langle \dfrac{x}{\sqrt{x^2 + y^2}}, \dfrac{y}{\sqrt{x^2 + y^2}} \right\rangle$

19. $\mathbf{F} = \langle e^{-x}, 0 \rangle$

20. $\mathbf{F} = \langle 0, 0, 1 \rangle$

T 21. $\mathbf{F} = \langle 1, 0, z \rangle$

T 22. $\mathbf{F} = \dfrac{\langle x, y, z \rangle}{\sqrt{x^2 + y^2 + z^2}}$

T 23. $\mathbf{F} = \langle y, -x, 0 \rangle$

24. **Matching vector fields with graphs** Match vector fields a–d with graphs A–D.

    a. $\mathbf{F} = \langle 0, x^2 \rangle$ **b.** $\mathbf{F} = \langle x - y, x \rangle$

    c. $\mathbf{F} = \langle 2x, -y \rangle$ **d.** $\mathbf{F} = \langle y, x \rangle$

(A)   (B)

(C)

(D)

25–30. Normal and tangential components *For the vector field* $\mathbf{F}$ *and curve* C, *complete the following:*

a. *Determine the points (if any) along the curve C at which the vector field* $\mathbf{F}$ *is tangent to C.*

b. *Determine the points (if any) along the curve C at which the vector field* $\mathbf{F}$ *is normal to C.*

c. *Sketch C and a few representative vectors of* $\mathbf{F}$ *on C.*

25. $\mathbf{F} = \left\langle \dfrac{1}{2}, 0 \right\rangle$; $C = \{(x, y): y - x^2 = 1\}$

26. $\mathbf{F} = \left\langle \dfrac{y}{2}, -\dfrac{x}{2} \right\rangle$; $C = \{(x, y): y - x^2 = 1\}$

27. $\mathbf{F} = \langle x, y \rangle$; $C = \{(x, y): x^2 + y^2 = 4\}$

28. $\mathbf{F} = \langle y, -x \rangle$; $C = \{(x, y): x^2 + y^2 = 1\}$

29. $\mathbf{F} = \langle x, y \rangle$; $C = \{(x, y): x = 1\}$

30. $\mathbf{F} = \langle y, x \rangle$; $C = \{(x, y): x^2 + y^2 = 1\}$

31–34. Design your own vector field *Specify the component functions of a vector field* $\mathbf{F}$ *in* $\mathbb{R}^2$ *with the following properties. Solutions are not unique.*

31. $\mathbf{F}$ is everywhere normal to the line $y = x$.

32. $\mathbf{F}$ is everywhere normal to the line $x = 2$.

33. At all points except $(0, 0)$, $\mathbf{F}$ has unit magnitude and points away from the origin along radial lines.

34. The flow of $\mathbf{F}$ is counterclockwise around the origin, increasing in magnitude with distance from the origin.

35–42. Gradient fields *Find the gradient field* $\mathbf{F} = \nabla\varphi$ *for the following potential functions* φ.

35. $\varphi(x, y) = x^2 y - y^2 x$

36. $\varphi(x, y) = \sqrt{xy}$

37. $\varphi(x, y) = x/y$

38. $\varphi(x, y) = \tan^{-1}(x/y)$

39. $\varphi(x, y, z) = \dfrac{x^2 + y^2 + z^2}{2}$

40. $\varphi(x, y, z) = \ln(1 + x^2 + y^2 + z^2)$

41. $\varphi(x, y, z) = (x^2 + y^2 + z^2)^{-1/2}$

42. $\varphi(x, y, z) = e^{-z} \sin(x + y)$

43–46. Gradient fields on curves *For the potential function φ and points A, B, C, and D on the level curve $\varphi(x, y) = 0$, complete the following steps.*

a. *Find the gradient field $\mathbf{F} = \nabla\varphi$.*
b. *Evaluate $\mathbf{F}$ at the points A, B, C, and D.*
c. *Plot the level curve $\varphi(x, y) = 0$ and the vectors $\mathbf{F}$ at the points A, B, C, and D.*

43. $\varphi(x, y) = y - 2x$; $A(-1, -2)$, $B(0, 0)$, $C(1, 2)$, and $D(2, 4)$

44. $\varphi(x, y) = \frac{1}{2}x^2 - y$; $A(-2, 2)$, $B(-1, 1/2)$, $C(1, 1/2)$, and $D(2, 2)$

45. $\varphi(x, y) = -y + \sin x$; $A(\pi/2, 1)$, $B(\pi, 0)$, $C(3\pi/2, -1)$, and $D(2\pi, 0)$

T 46. $\varphi(x, y) = \dfrac{32 - x^4 - y^4}{32}$; $A(2, 2)$, $B(-2, 2)$, $C(-2, -2)$, and $D(2, -2)$

T **47–48. Gradient fields** *Find the gradient field $\mathbf{F} = \nabla\varphi$ for the potential function φ. Sketch a few level curves of φ and a few vectors of $\mathbf{F}$.*

47. $\varphi(x, y) = x^2 + y^2$, for $x^2 + y^2 \le 16$

48. $\varphi(x, y) = x + y$, for $|x| \le 2$, $|y| \le 2$

49–52. Equipotential curves *Consider the following potential functions and the graphs of their equipotential curves.*

a. *Find the associated gradient field $\mathbf{F} = \nabla\varphi$.*
b. *Show that the vector field is orthogonal to the equipotential curve at the point $(1, 1)$. Illustrate this result on the figure.*
c. *Show that the vector field is orthogonal to the equipotential curve at all points (x, y).*
d. *Sketch two flow curves representing $\mathbf{F}$ that are everywhere orthogonal to the equipotential curves.*

49. $\varphi(x, y) = 2x + 3y$

50. $\varphi(x, y) = x + y^2$

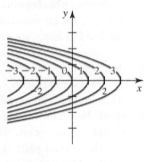

51. $\varphi(x, y) = e^{x-y}$

52. $\varphi(x, y) = x^2 + 2y^2$

53. **Explain why or why not** Determine whether the following statements are true and give an explanation or counterexample.

a. The vector field $\mathbf{F} = \langle 3x^2, 1 \rangle$ is a gradient field for both $\varphi_1(x, y) = x^3 + y$ and $\varphi_2(x, y) = y + x^3 + 100$.

b. The vector field $\mathbf{F} = \dfrac{\langle y, x \rangle}{\sqrt{x^2 + y^2}}$ is constant in direction and magnitude on the unit circle.

c. The vector field $\mathbf{F} = \dfrac{\langle y, x \rangle}{\sqrt{x^2 + y^2}}$ is neither a radial field nor a rotation field.

Explorations and Challenges

54. **Electric field due to a point charge** The electric field in the xy-plane due to a point charge at $(0, 0)$ is a gradient field with a potential function $V(x, y) = \dfrac{k}{\sqrt{x^2 + y^2}}$, where $k > 0$ is a physical constant.

a. Find the components of the electric field in the x- and y-directions, where $\mathbf{E}(x, y) = -\nabla V(x, y)$.
b. Show that the vectors of the electric field point in the radial direction (outward from the origin) and the radial component of $\mathbf{E}$ can be expressed as $E_r = \dfrac{k}{r^2}$, where $r = \sqrt{x^2 + y^2}$.
c. Show that the vector field is orthogonal to the equipotential curves at all points in the domain of V.

55. **Electric field due to a line of charge** The electric field in the xy-plane due to an infinite line of charge along the z-axis is a gradient field with a potential function $V(x, y) = c \ln\left(\dfrac{r_0}{\sqrt{x^2 + y^2}}\right)$, where $c > 0$ is a constant and r_0 is a reference distance at which the potential is assumed to be 0 (see figure).

a. Find the components of the electric field in the x- and y-directions, where $\mathbf{E}(x, y) = -\nabla V(x, y)$.
b. Show that the electric field at a point in the xy-plane is directed outward from the origin and has magnitude $|\mathbf{E}| = c/r$, where $r = \sqrt{x^2 + y^2}$.
c. Show that the vector field is orthogonal to the equipotential curves at all points in the domain of V.

56. **Gravitational force due to a mass** The gravitational force on a point mass m due to a point mass M is a gradient field with potential $U(r) = GMm/r$, where G is the gravitational constant and $r = \sqrt{x^2 + y^2 + z^2}$ is the distance between the masses.

a. Find the components of the gravitational force in the x-, y-, and z-directions, where $\mathbf{F}(x, y, z) = -\nabla U(x, y, z)$.
b. Show that the gravitational force points in the radial direction (outward from point mass M) and the radial component is $F(r) = GMm/r^2$.
c. Show that the vector field is orthogonal to the equipotential surfaces at all points in the domain of U.

57–61. Flow curves in the plane *Let* $\mathbf{F}(x, y) = \langle f(x, y), g(x, y) \rangle$ *be defined on* $\mathbb{R}^2$.

57. Explain why the flow curves or streamlines of $\mathbf{F}$ satisfy
$$y' = \frac{g(x, y)}{f(x, y)}$$
and are everywhere tangent to the vector field.

T 58. Find and graph the flow curves for the vector field $\mathbf{F} = \langle 1, x \rangle$.

T 59. Find and graph the flow curves for the vector field $\mathbf{F} = \langle x, x \rangle$.

T 60. Find and graph the flow curves for the vector field $\mathbf{F} = \langle y, x \rangle$.

Note that $\dfrac{d}{dx}(y^2) = 2yy'(x)$.

T 61. Find and graph the flow curves for the vector field $\mathbf{F} = \langle -y, x \rangle$.

62–63. Unit vectors in polar coordinates

62. Vectors in $\mathbb{R}^2$ may also be expressed in terms of polar coordinates. The standard coordinate unit vectors in polar coordinates are denoted $\mathbf{u}_r$ and $\mathbf{u}_\theta$ (see figure). Unlike the coordinate unit vectors in Cartesian coordinates, $\mathbf{u}_r$ and $\mathbf{u}_\theta$ change their direction depending on the point (r, θ). Use the figure to show that for $r > 0$, the following relationships among the unit vectors in Cartesian and polar coordinates hold:

$$\mathbf{u}_r = \cos\theta \, \mathbf{i} + \sin\theta \, \mathbf{j} \quad \mathbf{i} = \mathbf{u}_r \cos\theta - \mathbf{u}_\theta \sin\theta$$
$$\mathbf{u}_\theta = -\sin\theta \, \mathbf{i} + \cos\theta \, \mathbf{j} \quad \mathbf{j} = \mathbf{u}_r \sin\theta + \mathbf{u}_\theta \cos\theta$$

63. Verify that the relationships in Exercise 62 are consistent when $\theta = 0, \pi/2, \pi$, and $3\pi/2$.

64–66. Vector fields in polar coordinates *A vector field in polar coordinates has the form* $\mathbf{F}(r, \theta) = f(r, \theta)\,\mathbf{u}_r + g(r, \theta)\,\mathbf{u}_\theta$, *where the unit vectors are defined in Exercise 62. Sketch the following vector fields and express them in Cartesian coordinates.*

64. $\mathbf{F} = \mathbf{u}_r$ 65. $\mathbf{F} = \mathbf{u}_\theta$ 66. $\mathbf{F} = r\,\mathbf{u}_\theta$

67. **Cartesian vector field to polar vector field** Write the vector field $\mathbf{F} = \langle -y, x \rangle$ in polar coordinates and sketch the field.

QUICK CHECK ANSWERS

1. The particle follows a circular path around the origin.
3. $\nabla\varphi = \langle yz, xz, xy \rangle$ ◄

17.2 Line Integrals

With integrals of a single variable, we integrate over intervals in $\mathbb{R}$ (the real line). With double and triple integrals, we integrate over regions in $\mathbb{R}^2$ or $\mathbb{R}^3$. *Line integrals* (which really should be called *curve integrals*) are another class of integrals that play an important role in vector calculus. They are used to integrate either scalar-valued functions or vector fields along curves.

Suppose a thin, circular plate has a known temperature distribution and you must compute the average temperature along the edge of the plate. The required calculation involves integrating the temperature function over the *curved* boundary of the plate. Similarly, to calculate the amount of work needed to put a satellite into orbit, we integrate the gravitational force (a vector field) along the curved path of the satellite. Both these calculations require line integrals. As you will see, line integrals take several different forms. It is the goal of this section to distinguish among these various forms and show how and when each form should be used.

Scalar Line Integrals in the Plane

We focus first on line integrals of scalar-valued functions over curves in the xy-plane. Assume C is a smooth curve of finite length given by $\mathbf{r}(t) = \langle x(t), y(t) \rangle$, for $a \le t \le b$. We divide $[a, b]$ into n subintervals using the grid points

$$a = t_0 < t_1 < \cdots < t_{n-1} < t_n = b.$$

This partition of $[a, b]$ divides C into n subarcs (**Figure 17.16**), where the arc length of the kth subarc is denoted Δs_k. Let t_k^* be a point in the kth subinterval $[t_{k-1}, t_k]$, which corresponds to a point $(x(t_k^*), y(t_k^*))$ on the kth subarc of C, for $k = 1, 2, \ldots, n$.

The parameter t resides on the t-axis. As t varies from a to b, the curve C in the xy-plane is generated from $(x(a), y(a))$ to $(x(b), y(b))$.

Figure 17.16

Now consider a scalar-valued function $z = f(x, y)$ defined on a region containing C. Evaluating f at $(x(t_k^*), y(t_k^*))$ and multiplying this value by Δs_k, we form the sum

$$S_n = \sum_{k=1}^{n} f(x(t_k^*), y(t_k^*)) \Delta s_k,$$

which is similar to a Riemann sum. We now let Δ be the maximum value of $\{\Delta s_1, \ldots, \Delta s_n\}$. If the limit of the sum as $n \to \infty$ and $\Delta \to 0$ exists over all partitions, then the limit is called *the line integral of f over C*.

DEFINITION Scalar Line Integral in the Plane

Suppose the scalar-valued function f is defined on a region containing the smooth curve C given by $\mathbf{r}(t) = \langle x(t), y(t) \rangle$, for $a \leq t \leq b$. The **line integral of f over C** is

$$\int_C f(x(t), y(t)) \, ds = \lim_{\Delta \to 0} \sum_{k=1}^{n} f(x(t_k^*), y(t_k^*)) \Delta s_k,$$

provided this limit exists over all partitions of $[a, b]$. When the limit exists, f is said to be **integrable** on C.

The more compact notations $\int_C f(\mathbf{r}(t)) \, ds$, $\int_C f(x, y) \, ds$, and $\int_C f \, ds$ are also used for the line integral of f over C. It can be shown that if f is continuous on a region containing C, then f is integrable over C.

There are several useful interpretations of the line integral of a scalar function. If $f(x, y) = 1$, the line integral $\int_C ds$ gives the length of the curve C, just as the ordinary integral $\int_a^b dx$ gives the length of the interval $[a, b]$, which is $b - a$. If $f(x, y) \geq 0$ on C, then $\int_C f(x, y) \, ds$ can be viewed as the area of one side of the vertical, curtain-like surface that lies between the graphs of f and C (**Figure 17.17**). This interpretation results from regarding the product $f(x(t_k^*), y(t_k^*)) \Delta s_k$ as an approximation to the area of the kth panel of the curtain. Similarly, if f is a density function for a thin wire represented by the curve C, then $\int_C f(x, y) \, ds$ gives the mass of the wire—the product $f(x(t_k^*), y(t_k^*)) \Delta s_k$ is an approximation to the mass of the kth piece of the wire (Exercises 35–36).

Evaluating Line Integrals

The line integral of f over C given in the definition is not an ordinary Riemann integral, because the integrand is expressed as a function of t while the variable of integration is the arc length parameter s. We need a practical way to evaluate such integrals; the key is to use a change of variables to convert a line integral into an ordinary integral. Let C be given by $\mathbf{r}(t) = \langle x(t), y(t) \rangle$, for $a \leq t \leq b$. Recall from Section 14.4 that the length of C over the interval $[a, t]$ is

$$s(t) = \int_a^t |\mathbf{r}'(u)| \, du.$$

Differentiating both sides of this equation and using the Fundamental Theorem of Calculus yields $s'(t) = |\mathbf{r}'(t)|$. We now make a standard change of variables using the relationship

$$ds = s'(t) \, dt = |\mathbf{r}'(t)| \, dt.$$

Relying on a result from advanced calculus, the original line integral with respect to s can be converted into an ordinary integral with respect to t:

$$\int_C f \, ds = \int_a^b f(x(t), y(t)) \underbrace{|\mathbf{r}'(t)| \, dt}_{ds}.$$

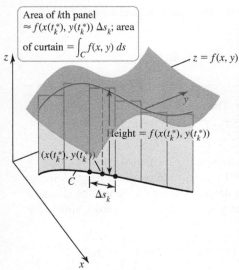

Area of kth panel
$\approx f(x(t_k^*), y(t_k^*)) \Delta s_k$; area
of curtain $= \int_C f(x, y) \, ds$

$z = f(x, y)$

Height $= f(x(t_k^*), y(t_k^*))$

$(x(t_k^*), y(t_k^*))$

C

Δs_k

Figure 17.17

QUICK CHECK 1 Explain mathematically why differentiating the arc length integral leads to $s'(t) = |\mathbf{r}'(t)|$. ◄

▶ If t represents time, then the relationship $ds = |\mathbf{r}'(t)| \, dt$ is a generalization of the familiar formula

distance $=$ speed $\cdot$ time.

▶ The value of a line integral of a scalar-valued function is independent of the parameterization of C and independent of the direction in which C is traversed (Exercises 64–65).

THEOREM 17.1   Evaluating Scalar Line Integrals in $\mathbb{R}^2$

Let f be continuous on a region containing a smooth curve $C: \mathbf{r}(t) = \langle x(t), y(t) \rangle$, for $a \le t \le b$. Then

$$\int_C f \, ds = \int_a^b f(x(t), y(t)) |\mathbf{r}'(t)| \, dt$$

$$= \int_a^b f(x(t), y(t)) \sqrt{x'(t)^2 + y'(t)^2} \, dt.$$

If t represents time and C is the path of a moving object, then $|\mathbf{r}'(t)|$ is the speed of the object. The *speed factor* $|\mathbf{r}'(t)|$ that appears in the integral relates distance traveled along the curve as measured by s to the elapsed time as measured by the parameter t.

Notice that if $f(x, y) = 1$, then the line integral is $\int_a^b \sqrt{x'(t)^2 + y'(t)^2} \, dt$, which is the arc length formula for C. Theorem 17.1 leads to the following procedure for evaluating line integrals.

PROCEDURE   Evaluating the Line Integral $\int_C f \, ds$

1. Find a parametric description of C in the form $\mathbf{r}(t) = \langle x(t), y(t) \rangle$, for $a \le t \le b$.

2. Compute $|\mathbf{r}'(t)| = \sqrt{x'(t)^2 + y'(t)^2}$.

3. Make substitutions for x and y in the integrand and evaluate an ordinary integral:

$$\int_C f \, ds = \int_a^b f(x(t), y(t)) |\mathbf{r}'(t)| \, dt.$$

▶ When we compute the average value by an ordinary integral, we divide by the length of the interval of integration. Analogously, when we compute the average value by a line integral, we divide by the length of the curve L:

$$\bar{f} = \frac{1}{L} \int_C f \, ds.$$

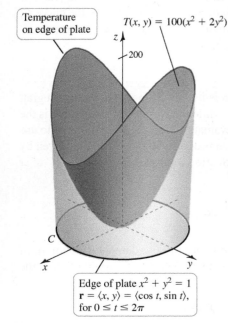

Temperature on edge of plate

$T(x, y) = 100(x^2 + 2y^2)$

Edge of plate $x^2 + y^2 = 1$
$\mathbf{r} = \langle x, y \rangle = \langle \cos t, \sin t \rangle$,
for $0 \le t \le 2\pi$

Figure 17.18

▶ The line integral in Example 1 also gives the area of the vertical cylindrical curtain that hangs between the surface and C in Figure 17.18.

EXAMPLE 1   Average temperature on a circle The temperature of the circular plate $R = \{(x, y): x^2 + y^2 \le 1\}$ is $T(x, y) = 100(x^2 + 2y^2)$. Find the average temperature along the edge of the plate.

SOLUTION Calculating the average value requires integrating the temperature function over the boundary circle $C = \{(x, y): x^2 + y^2 = 1\}$ and dividing by the length (circumference) of C. The first step is to find a parametric description for C. We use the standard parameterization for a unit circle centered at the origin, $\mathbf{r} = \langle x, y \rangle = \langle \cos t, \sin t \rangle$, for $0 \le t \le 2\pi$. Next, we compute the speed factor

$$|\mathbf{r}'(t)| = \sqrt{x'(t)^2 + y'(t)^2} = \sqrt{(-\sin t)^2 + (\cos t)^2} = 1.$$

We substitute $x = \cos t$ and $y = \sin t$ into the temperature function and express the line integral as an ordinary integral with respect to t:

$$\int_C T(x, y) \, ds = \int_0^{2\pi} \underbrace{100(x(t)^2 + 2y(t)^2)}_{T(t)} \underbrace{|\mathbf{r}'(t)|}_{1} \, dt \quad \text{Write the line integral as an ordinary integral with respect to } t; ds = |\mathbf{r}'(t)| \, dt.$$

$$= 100 \int_0^{2\pi} (\cos^2 t + 2 \sin^2 t) \, dt \quad \text{Substitute for } x \text{ and } y.$$

$$= 100 \underbrace{\int_0^{2\pi} (1 + \sin^2 t) \, dt}_{3\pi} \quad \cos^2 t + \sin^2 t = 1$$

$$= 300\pi. \quad \text{Use } \sin^2 t = \frac{1 - \cos 2t}{2} \text{ and integrate.}$$

The geometry of this line integral is shown in **Figure 17.18**. The temperature function on the boundary of C is a function of t. The line integral is an ordinary integral with respect to t over the interval $[0, 2\pi]$. To find the average value, we divide the line integral of the temperature by the length of the curve, which is 2π. Therefore, the average temperature on the boundary of the plate is $300\pi/(2\pi) = 150$.

Related Exercise 17 ◀

QUICK CHECK 2 Suppose $\mathbf{r}(t) = \langle t, 0 \rangle$, for $a \le t \le b$, is a parametric description of C; note that C is the interval $[a, b]$ on the x-axis. Show that $\int_C f(x, y)\, ds = \int_a^b f(t, 0)\, dt$, which is an ordinary, single-variable integral introduced in Chapter 5. ◄

➤ If $f(x, y, z) = 1$, then the line integral gives the length of C.

➤ Recall that a parametric equation of a line is
$$\mathbf{r}(t) = \langle x_0, y_0, z_0 \rangle + t \langle a, b, c \rangle,$$
where $\langle x_0, y_0, z_0 \rangle$ is a position vector associated with a fixed point on the line and $\langle a, b, c \rangle$ is a vector parallel to the line.

Line Integrals in $\mathbb{R}^3$

The argument that leads to line integrals on plane curves extends immediately to three or more dimensions. Here is the corresponding evaluation theorem for line integrals in $\mathbb{R}^3$.

THEOREM 17.2 Evaluating Scalar Line Integrals in $\mathbb{R}^3$

Let f be continuous on a region containing a smooth curve
$C: \mathbf{r}(t) = \langle x(t), y(t), z(t) \rangle$, for $a \le t \le b$. Then

$$\int_C f\, ds = \int_a^b f(x(t), y(t), z(t)) \left| \mathbf{r}'(t) \right|\, dt$$

$$= \int_a^b f(x(t), y(t), z(t)) \sqrt{x'(t)^2 + y'(t)^2 + z'(t)^2}\, dt.$$

EXAMPLE 2 Line integrals in $\mathbb{R}^3$ Evaluate $\int_C (xy + 2z)\, ds$ on the following line segments.

a. The line segment from $P(1, 0, 0)$ to $Q(0, 1, 1)$

b. The line segment from $Q(0, 1, 1)$ to $P(1, 0, 0)$

SOLUTION

a. A parametric description of the line segment from $P(1, 0, 0)$ to $Q(0, 1, 1)$ is

$$\mathbf{r}(t) = \langle 1, 0, 0 \rangle + t \langle -1, 1, 1 \rangle = \langle 1 - t, t, t \rangle, \text{ for } 0 \le t \le 1.$$

The speed factor is

$$\left| \mathbf{r}'(t) \right| = \sqrt{x'(t)^2 + y'(t)^2 + z'(t)^2} = \sqrt{(-1)^2 + 1^2 + 1^2} = \sqrt{3}.$$

Substituting $x = 1 - t$, $y = t$, and $z = t$, the value of the line integral is

$$\int_C (xy + 2z)\, ds = \int_0^1 (\underbrace{(1 - t)}_{x}\, \underbrace{t}_{y} + \underbrace{2t}_{2z}) \sqrt{3}\, dt \quad \text{Substitute for } x, y, \text{ and } z.$$

$$= \sqrt{3} \int_0^1 (3t - t^2)\, dt \qquad \text{Simplify.}$$

$$= \sqrt{3} \left(\frac{3t^2}{2} - \frac{t^3}{3} \right) \Big|_0^1 \qquad \text{Integrate.}$$

$$= \frac{7\sqrt{3}}{6}. \qquad \text{Evaluate.}$$

b. The line segment from $Q(0, 1, 1)$ to $P(1, 0, 0)$ may be described parametrically by

$$\mathbf{r}(t) = \langle 0, 1, 1 \rangle + t \langle 1, -1, -1 \rangle = \langle t, 1 - t, 1 - t \rangle, \text{ for } 0 \le t \le 1.$$

The speed factor is

$$\left| \mathbf{r}'(t) \right| = \sqrt{x'(t)^2 + y'(t)^2 + z'(t)^2} = \sqrt{1^2 + (-1)^2 + (-1)^2} = \sqrt{3}.$$

We substitute $x = t$, $y = 1 - t$, and $z = 1 - t$ and do a calculation similar to that in part (a). The value of the line integral is again $\dfrac{7\sqrt{3}}{6}$, emphasizing the fact that a scalar line integral is independent of the orientation and parameterization of the curve.

Related Exercises 32–33 ◄

EXAMPLE 3 Flight of an eagle An eagle soars on the ascending spiral path

$$C: \mathbf{r}(t) = \langle x(t), y(t), z(t) \rangle = \left\langle 2400 \cos \frac{t}{2}, 2400 \sin \frac{t}{2}, 500t \right\rangle,$$

where x, y, and z are measured in feet and t is measured in minutes. How far does the eagle fly over the time interval $0 \le t \le 10$?

> Because we are finding the length of a curve, the integrand in this line integral is $f(x, y, z) = 1$.

SOLUTION The distance traveled is found by integrating the element of arc length ds along C, that is, $L = \int_C ds$. We now make a change of variables to the parameter t using

$$|\mathbf{r}'(t)| = \sqrt{x'(t)^2 + y'(t)^2 + z'(t)^2}$$

$$= \sqrt{\left(-1200 \sin \frac{t}{2}\right)^2 + \left(1200 \cos \frac{t}{2}\right)^2 + 500^2} \quad \text{Substitute derivatives.}$$

$$= \sqrt{1200^2 + 500^2} = 1300. \qquad \sin^2 \frac{t}{2} + \cos^2 \frac{t}{2} = 1$$

It follows that the distance traveled is

QUICK CHECK 3 What is the speed of the eagle in Example 3? ◄

$$L = \int_C ds = \int_0^{10} |\mathbf{r}'(t)| \, dt = \int_0^{10} 1300 \, dt = 13{,}000 \text{ ft.}$$

Related Exercise 39 ◄

Line Integrals of Vector Fields

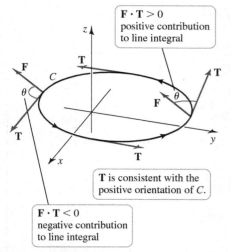

F · T > 0
positive contribution
to line integral

T is consistent with the positive orientation of C.

F · T < 0
negative contribution
to line integral

Figure 17.19

Line integrals along curves in $\mathbb{R}^2$ or $\mathbb{R}^3$ may also have integrands that involve vector fields. Such line integrals are different from scalar line integrals in two respects.

• Recall that an *oriented curve* is a parameterized curve for which a direction is specified. The *positive* orientation is the direction in which the curve is generated as the parameter increases. For example, the positive orientation of the circle $\mathbf{r}(t) = \langle \cos t, \sin t \rangle$, for $0 \le t \le 2\pi$, is counterclockwise. As we will see, vector line integrals must be evaluated on oriented curves, and the value of a line integral depends on the orientation.

• The line integral of a vector field $\mathbf{F}$ along an oriented curve involves a specific component of $\mathbf{F}$ relative to the curve. We begin by defining vector line integrals for the *tangential* component of $\mathbf{F}$, a situation that has many physical applications.

Let C: $\mathbf{r}(s) = \langle x(s), y(s), z(s) \rangle$ be a smooth oriented curve in $\mathbb{R}^3$ parameterized by arc length and let $\mathbf{F}$ be a vector field that is continuous on a region containing C. At each point of C, the unit tangent vector $\mathbf{T}$ points in the positive direction on C (**Figure 17.19**). The component of $\mathbf{F}$ in the direction of $\mathbf{T}$ at a point of C is $|\mathbf{F}| \cos \theta$, where θ is the angle between $\mathbf{F}$ and $\mathbf{T}$. Because $\mathbf{T}$ is a unit vector,

$$|\mathbf{F}| \cos \theta = |\mathbf{F}||\mathbf{T}| \cos \theta = \mathbf{F} \cdot \mathbf{T}.$$

The first line integral of a vector field $\mathbf{F}$ that we introduce is the line integral of the scalar $\mathbf{F} \cdot \mathbf{T}$ along the curve C. When we integrate $\mathbf{F} \cdot \mathbf{T}$ along C, the effect is to add up the components of $\mathbf{F}$ in the direction of C at each point of C.

> The component of $\mathbf{F}$ in the direction of $\mathbf{T}$ is the scalar component of $\mathbf{F}$ in the direction of $\mathbf{T}$, $\text{scal}_\mathbf{T} \mathbf{F}$, as defined in Section 13.3. Note that $|\mathbf{T}| = 1$.

> Some texts let $d\mathbf{s}$ stand for $\mathbf{T} \, ds$. Then the line integral $\int_C \mathbf{F} \cdot \mathbf{T} \, ds$ is written $\int_C \mathbf{F} \cdot d\mathbf{s}$.

DEFINITION Line Integral of a Vector Field

Let $\mathbf{F}$ be a vector field that is continuous on a region containing a smooth oriented curve C parameterized by arc length. Let $\mathbf{T}$ be the unit tangent vector at each point of C consistent with the orientation. The line integral of $\mathbf{F}$ over C is $\int_C \mathbf{F} \cdot \mathbf{T} \, ds$.

Just as we did for line integrals of scalar-valued functions, we need a method for evaluating vector line integrals when the parameter is not the arc length. Suppose C has a parameterization $\mathbf{r}(t) = \langle x(t), y(t), z(t) \rangle$, for $a \le t \le b$. Recall from Section 14.2 that the unit tangent vector at a point on the curve is $\mathbf{T} = \dfrac{\mathbf{r}'(t)}{|\mathbf{r}'(t)|}$. Using the fact that $ds = |\mathbf{r}'(t)| \, dt$, the line integral becomes

$$\int_C \mathbf{F} \cdot \mathbf{T} \, ds = \int_a^b \mathbf{F} \cdot \underbrace{\frac{\mathbf{r}'(t)}{|\mathbf{r}'(t)|}}_{\mathbf{T}} \underbrace{|\mathbf{r}'(t)| \, dt}_{ds} = \int_a^b \mathbf{F} \cdot \mathbf{r}'(t) \, dt.$$

> Keep in mind that $f(t)$ stands for $f(x(t), y(t), z(t))$ with analogous expressions for $g(t)$ and $h(t)$.

This integral may be written in several equivalent forms. If $\mathbf{F} = \langle f, g, h \rangle$, then the line integral is expressed in component form as

$$\int_C \mathbf{F} \cdot \mathbf{T} \, ds = \int_a^b \mathbf{F} \cdot \mathbf{r}'(t) \, dt = \int_a^b (f(t)x'(t) + g(t)y'(t) + h(t)z'(t)) \, dt.$$

Another useful form is obtained by noting that

$$dx = x'(t) \, dt, \qquad dy = y'(t) \, dt, \qquad dz = z'(t) \, dt.$$

Making these replacements in the previous integral results in the form

$$\int_C \mathbf{F} \cdot \mathbf{T} \, ds = \int_C f \, dx + g \, dy + h \, dz.$$

Finally, if we let $d\mathbf{r} = \langle dx, dy, dz \rangle$, then $f \, dx + g \, dy + h \, dz = \mathbf{F} \cdot d\mathbf{r}$, and we have

$$\int_C \mathbf{F} \cdot \mathbf{T} \, ds = \int_C \mathbf{F} \cdot d\mathbf{r}.$$

It is helpful to become familiar with these various forms of the line integral.

Different Forms of Line Integrals of Vector Fields

The line integral $\int_C \mathbf{F} \cdot \mathbf{T} \, ds$ may be expressed in the following forms, where $\mathbf{F} = \langle f, g, h \rangle$ and C has a parameterization $\mathbf{r}(t) = \langle x(t), y(t), z(t) \rangle$, for $a \le t \le b$:

$$\int_a^b \mathbf{F} \cdot \mathbf{r}'(t) \, dt = \int_a^b (f(t)x'(t) + g(t)y'(t) + h(t)z'(t)) \, dt$$

$$= \int_C f \, dx + g \, dy + h \, dz$$

$$= \int_C \mathbf{F} \cdot d\mathbf{r}.$$

For line integrals in the plane, we let $\mathbf{F} = \langle f, g \rangle$ and assume C is parameterized in the form $\mathbf{r}(t) = \langle x(t), y(t) \rangle$, for $a \le t \le b$. Then

$$\int_a^b \mathbf{F} \cdot \mathbf{r}'(t) \, dt = \int_a^b (f(t)x'(t) + g(t)y'(t)) \, dt = \int_C f \, dx + g \, dy = \int_C \mathbf{F} \cdot d\mathbf{r}.$$

> We use the convention that $-C$ is the curve C with the opposite orientation.

EXAMPLE 4   Different paths Evaluate $\int_C \mathbf{F} \cdot \mathbf{T} \, ds$ with $\mathbf{F} = \langle y - x, x \rangle$ on the following oriented paths in $\mathbb{R}^2$ (**Figure 17.20**).

a. The quarter-circle C_1 from $P(0, 1)$ to $Q(1, 0)$

b. The quarter-circle $-C_1$ from $Q(1, 0)$ to $P(0, 1)$

c. The path C_2 from $P(0, 1)$ to $Q(1, 0)$ via two line segments through $O(0, 0)$

SOLUTION

a. Working in $\mathbb{R}^2$, a parametric description of the curve C_1 with the required (clockwise) orientation is $\mathbf{r}(t) = \langle \sin t, \cos t \rangle$, for $0 \le t \le \pi/2$. Along C_1, the vector field is

$$\mathbf{F} = \langle y - x, x \rangle = \langle \cos t - \sin t, \sin t \rangle.$$

The velocity vector is $\mathbf{r}'(t) = \langle \cos t, -\sin t \rangle$, so the integrand of the line integral is

$$\mathbf{F} \cdot \mathbf{r}'(t) = \langle \cos t - \sin t, \sin t \rangle \cdot \langle \cos t, -\sin t \rangle = \underbrace{\cos^2 t - \sin^2 t}_{\cos 2t} - \underbrace{\sin t \cos t}_{\frac{1}{2} \sin 2t}.$$

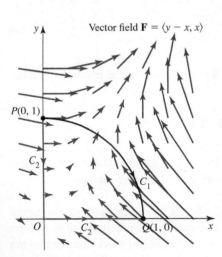

Vector field $\mathbf{F} = \langle y - x, x \rangle$

Figure 17.20

The value of the line integral of $\mathbf{F}$ over C_1 is

$$\int_0^{\pi/2} \mathbf{F} \cdot \mathbf{r}'(t)\, dt = \int_0^{\pi/2} \left(\cos 2t - \frac{1}{2} \sin 2t \right) dt \quad \text{Substitute for } \mathbf{F} \cdot \mathbf{r}'(t).$$

$$= \left(\frac{1}{2} \sin 2t + \frac{1}{4} \cos 2t \right) \Big|_0^{\pi/2} \quad \text{Evaluate integral.}$$

$$= -\frac{1}{2}. \quad \text{Simplify.}$$

b. A parameterization of the curve $-C_1$ from Q to P is $\mathbf{r}(t) = \langle \cos t, \sin t \rangle$, for $0 \le t \le \pi/2$. The vector field along the curve is

$$\mathbf{F} = \langle y - x, x \rangle = \langle \sin t - \cos t, \cos t \rangle,$$

and the velocity vector is $\mathbf{r}'(t) = \langle -\sin t, \cos t \rangle$. A calculation similar to that in part (a) results in

$$\int_{-C_1} \mathbf{F} \cdot \mathbf{T}\, ds = \int_0^{\pi/2} \mathbf{F} \cdot \mathbf{r}'(t)\, dt = \frac{1}{2}.$$

Comparing the results of parts (a) and (b), we see that reversing the orientation of C_1 reverses the sign of the line integral of the vector field.

c. The path C_2 consists of two line segments.

- The segment from P to O is parameterized by $\mathbf{r}(t) = \langle 0, 1 - t \rangle$, for $0 \le t \le 1$. Therefore, $\mathbf{r}'(t) = \langle 0, -1 \rangle$ and $\mathbf{F} = \langle y - x, x \rangle = \langle 1 - t, 0 \rangle$. On this segment, $\mathbf{T} = \langle 0, -1 \rangle$.

- The line segment from O to Q is parameterized by $\mathbf{r}(t) = \langle t, 0 \rangle$, for $0 \le t \le 1$. Therefore, $\mathbf{r}'(t) = \langle 1, 0 \rangle$ and $\mathbf{F} = \langle y - x, x \rangle = \langle -t, t \rangle$. On this segment, $\mathbf{T} = \langle 1, 0 \rangle$.

> Line integrals of vector fields satisfy properties similar to those of ordinary integrals. Suppose C is a smooth curve from A to B, C_1 is the curve from A to P, and C_2 is the curve from P to B, where P is a point on C between A and B. Then
>
> $$\int_C \mathbf{F} \cdot d\mathbf{r} = \int_{C_1} \mathbf{F} \cdot d\mathbf{r} + \int_{C_2} \mathbf{F} \cdot d\mathbf{r}.$$

The line integral is split into two parts and evaluated as follows:

$$\int_{C_2} \mathbf{F} \cdot \mathbf{T}\, ds = \int_{PO} \mathbf{F} \cdot \mathbf{T}\, ds + \int_{OQ} \mathbf{F} \cdot \mathbf{T}\, ds$$

$$= \int_0^1 \langle 1 - t, 0 \rangle \cdot \langle 0, -1 \rangle\, dt + \int_0^1 \langle -t, t \rangle \cdot \langle 1, 0 \rangle\, dt \quad \begin{array}{l}\text{Substitute}\\ \text{for } x, y, \mathbf{r}'.\end{array}$$

$$= \int_0^1 0\, dt + \int_0^1 (-t)\, dt \quad \text{Simplify.}$$

$$= -\frac{1}{2}. \quad \text{Evaluate integrals.}$$

The line integrals in parts (a) and (c) have the same value and run from P to Q, but along different paths. We might ask: For what vector fields are the values of a line integral independent of path? We return to this question in Section 17.3.

Related Exercises 42–43 ◀

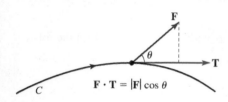

$$\mathbf{F} \cdot \mathbf{T} = |\mathbf{F}| \cos \theta$$

$$\mathbf{F} \cdot \mathbf{T} = |\mathbf{F}| \cos (\pi - \theta)$$
$$= -|\mathbf{F}| \cos \theta$$

Reversing the orientation of C changes the sign of $\mathbf{F} \cdot \mathbf{T}$ at each point on C.

Figure 17.21

The solutions to parts (a) and (b) of Example 4 illustrate a general result that applies to line integrals of vector fields:

$$\int_{-C} \mathbf{F} \cdot \mathbf{T}\, ds = -\int_C \mathbf{F} \cdot \mathbf{T}\, ds.$$

Figure 17.21 provides the justification of this fact: Reversing the orientation of C changes the sign of $\mathbf{F} \cdot \mathbf{T}$ at each point of C, which changes the sign of the line integral.

Work Integrals   A common application of line integrals of vector fields is computing the work done in moving an object in a force field (for example, a gravitational or electric field). First recall (Section 6.7) that if $\mathbf{F}$ is a *constant* force field, the work done in moving an object a distance d along the x-axis is $W = F_x d$, where $F_x = |\mathbf{F}| \cos \theta$ is the component

▶ Remember that the value of $\int_C f\, ds$ (the line integral of a *scalar* function) does not depend on the orientation of C.

QUICK CHECK 4 Suppose a two-dimensional force field is everywhere directed outward from the origin, and C is a circle centered at the origin. What is the angle between the field and the unit vectors tangent to C? ◄

of the force along the x-axis (**Figure 17.22a**). Only the component of **F** in the direction of motion contributes to the work. More generally, if **F** is a *variable* force field, the work done in moving an object from $x = a$ to $x = b$ is $W = \int_a^b F_x\,(x)\, dx$, where again F_x is the component of the force **F** in the direction of motion (parallel to the x-axis, **Figure 17.22b**).

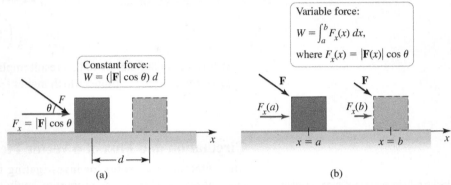

Figure 17.22

We now take this progression one step further. Let **F** be a variable force field defined in a region D of $\mathbb{R}^3$ and suppose C is a smooth oriented curve in D along which an object moves. The direction of motion at each point of C is given by the unit tangent vector **T**. Therefore, the component of **F** in the direction of motion is $\mathbf{F} \cdot \mathbf{T}$, which is the tangential component of **F** along C. Summing the contributions to the work at each point of C, the work done in moving an object along C in the presence of the force is the line integral of $\mathbf{F} \cdot \mathbf{T}$ (**Figure 17.23**).

Figure 17.23

▶ Just to be clear, a work integral is nothing more than a line integral of the tangential component of a force field.

DEFINITION Work Done in a Force Field

Let **F** be a continuous force field in a region D of $\mathbb{R}^3$. Let

$$C\text{: } \mathbf{r}(t) = \langle x(t), y(t), z(t) \rangle, \text{ for } a \le t \le b,$$

be a smooth curve in D with a unit tangent vector **T** consistent with the orientation. The work done in moving an object along C in the positive direction is

$$W = \int_C \mathbf{F} \cdot \mathbf{T}\, ds = \int_a^b \mathbf{F} \cdot \mathbf{r}'(t)\, dt.$$

EXAMPLE 5   An inverse square force Gravitational and electrical forces between point masses and point charges obey inverse square laws: They act along the line joining the centers and they vary as $1/r^2$, where r is the distance between the centers. The force of attraction (or repulsion) of an inverse square force field is given by the vector field $\mathbf{F} = \dfrac{k \langle x, y, z \rangle}{(x^2 + y^2 + z^2)^{3/2}}$, where k is a physical constant. Because $\mathbf{r} = \langle x, y, z \rangle$, this force may also be written $\mathbf{F} = \dfrac{k\mathbf{r}}{|\mathbf{r}|^3}$. Find the work done in moving an object along the following paths.

a. C_1 is the line segment from $(1, 1, 1)$ to (a, a, a), where $a > 1$.

b. C_2 is the extension of C_1 produced by letting $a \to \infty$.

SOLUTION

a. A parametric description of C_1 consistent with the orientation is $\mathbf{r}(t) = \langle t, t, t \rangle$, for $1 \le t \le a$, with $\mathbf{r}'(t) = \langle 1, 1, 1 \rangle$. In terms of the parameter t, the force field is

$$\mathbf{F} = \frac{k \langle x, y, z \rangle}{(x^2 + y^2 + z^2)^{3/2}} = \frac{k \langle t, t, t \rangle}{(3t^2)^{3/2}}.$$

The dot product that appears in the work integral is

$$\mathbf{F} \cdot \mathbf{r}'(t) = \frac{k \langle t, t, t \rangle}{(3t^2)^{3/2}} \cdot \langle 1, 1, 1 \rangle = \frac{3kt}{3\sqrt{3}\, t^3} = \frac{k}{\sqrt{3}\, t^2}.$$

Therefore, the work done is

$$W = \int_1^a \mathbf{F} \cdot \mathbf{r}'(t)\, dt = \frac{k}{\sqrt{3}} \int_1^a t^{-2}\, dt = \frac{k}{\sqrt{3}}\left(1 - \frac{1}{a}\right).$$

b. The path C_2 is obtained by letting $a \to \infty$ in part (a). The required work is

$$W = \lim_{a \to \infty} \frac{k}{\sqrt{3}}\left(1 - \frac{1}{a}\right) = \frac{k}{\sqrt{3}}.$$

If $\mathbf{F}$ is a gravitational field, this result implies that the work required to escape Earth's gravitational field is finite (which makes space flight possible).

Related Exercise 55 ◄

Circulation and Flux of a Vector Field

Line integrals are useful for investigating two important properties of vector fields: *circulation* and *flux*. These properties apply to any vector field, but they are particularly relevant and easy to visualize if you think of $\mathbf{F}$ as the velocity field for a moving fluid.

> ▶ In the definition of circulation, a *closed curve* is a curve whose initial and terminal points are the same, as defined formally in Section 17.3.

> ▶ Although we define circulation integrals for smooth curves, these integrals may be computed on piecewise-smooth curves. We adopt the convention that *piecewise* refers to a curve with finitely many pieces.

Circulation   We assume $\mathbf{F} = \langle f, g, h \rangle$ is a continuous vector field on a region D of $\mathbb{R}^3$, and we take C to be a *closed* smooth oriented curve in D. The *circulation* of $\mathbf{F}$ along C is a measure of how much of the vector field points in the direction of C. More simply, as you travel along C in the positive direction, how much of the vector field is at your back and how much of it is in your face? To determine the circulation, we simply "add up" the components of $\mathbf{F}$ in the direction of the unit tangent vector $\mathbf{T}$ at each point. Therefore, circulation integrals are another example of line integrals of vector fields.

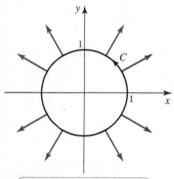

DEFINITION Circulation

Let $\mathbf{F}$ be a continuous vector field on a region D of $\mathbb{R}^3$, and let C be a closed smooth oriented curve in D. The **circulation** of $\mathbf{F}$ on C is $\int_C \mathbf{F} \cdot \mathbf{T}\, ds$, where $\mathbf{T}$ is the unit vector tangent to C consistent with the orientation.

On the unit circle, $\mathbf{F} = \langle x, y \rangle$ is orthogonal to C and has zero circulation on C.

(a)

EXAMPLE 6   Circulation of two-dimensional flows Let C be the unit circle with counterclockwise orientation. Find the circulation on C of the following vector fields.

a. The radial vector field $\mathbf{F} = \langle x, y \rangle$

b. The rotation vector field $\mathbf{F} = \langle -y, x \rangle$

SOLUTION

a. The unit circle with the specified orientation is described parametrically by $\mathbf{r}(t) = \langle \cos t, \sin t \rangle$, for $0 \le t \le 2\pi$. Therefore, $\mathbf{r}'(t) = \langle -\sin t, \cos t \rangle$ and the circulation of the radial field $\mathbf{F} = \langle x, y \rangle$ is

$$\int_C \mathbf{F} \cdot \mathbf{T}\, ds = \int_0^{2\pi} \mathbf{F} \cdot \mathbf{r}'(t)\, dt \qquad \text{Evaluation of a line integral}$$

$$= \int_0^{2\pi} \underbrace{\langle \cos t, \sin t \rangle}_{\mathbf{F} = \langle x, y \rangle} \cdot \underbrace{\langle -\sin t, \cos t \rangle}_{\mathbf{r}'(t)}\, dt \qquad \text{Substitute for } \mathbf{F} \text{ and } \mathbf{r}'.$$

$$= \int_0^{2\pi} 0\, dt = 0. \qquad \text{Simplify.}$$

On the unit circle, $\mathbf{F} = \langle -y, x \rangle$ is tangent to C and has positive circulation on C.

(b)

Figure 17.24

The tangential component of the radial field is zero everywhere on C, so the circulation is zero (**Figure 17.24a**).

b. The circulation for the rotation field $\mathbf{F} = \langle -y, x \rangle$ is

$$\int_C \mathbf{F} \cdot \mathbf{T} \, ds = \int_0^{2\pi} \mathbf{F} \cdot \mathbf{r}'(t) \, dt \qquad \text{Evaluation of a line integral}$$

$$= \int_0^{2\pi} \underbrace{\langle -\sin t, \cos t \rangle}_{\mathbf{F} = \langle -y, x \rangle} \cdot \underbrace{\langle -\sin t, \cos t \rangle}_{\mathbf{r}'(t)} \, dt \qquad \text{Substitute for } \mathbf{F} \text{ and } \mathbf{r}'.$$

$$= \int_0^{2\pi} \underbrace{\left(\sin^2 t + \cos^2 t \right)}_{1} \, dt \qquad \text{Simplify.}$$

$$= 2\pi.$$

In this case, at every point of C, the rotation field is in the direction of the tangent vector; the result is a positive circulation (**Figure 17.24b**).

Related Exercise 57 ◄

EXAMPLE 7 Circulation of a three-dimensional flow Find the circulation of the vector field $\mathbf{F} = \langle z, x, -y \rangle$ on the tilted ellipse $C: \mathbf{r}(t) = \langle \cos t, \sin t, \cos t \rangle$, for $0 \leq t \leq 2\pi$ (**Figure 17.25a**).

SOLUTION We first determine that

$$\mathbf{r}'(t) = \langle x'(t), y'(t), z'(t) \rangle = \langle -\sin t, \cos t, -\sin t \rangle.$$

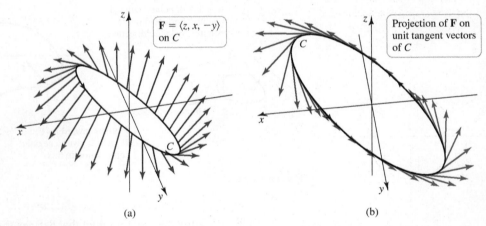

(a)   (b)

Figure 17.25

Substituting $x = \cos t$, $y = \sin t$, and $z = \cos t$ into $\mathbf{F} = \langle z, x, -y \rangle$, the circulation is

$$\int_C \mathbf{F} \cdot \mathbf{T} \, ds = \int_0^{2\pi} \mathbf{F} \cdot \mathbf{r}'(t) \, dt \qquad \begin{array}{l} \text{Evaluation of a} \\ \text{line integral} \end{array}$$

$$= \int_0^{2\pi} \langle \cos t, \cos t, -\sin t \rangle \cdot \langle -\sin t, \cos t, -\sin t \rangle \, dt \qquad \text{Substitute for } \mathbf{F} \text{ and } \mathbf{r}'.$$

$$= \int_0^{2\pi} (-\sin t \cos t + 1) \, dt \qquad \begin{array}{l} \text{Simplify;} \\ \sin^2 t + \cos^2 t = 1. \end{array}$$

$$= 2\pi. \qquad \text{Evaluate integral.}$$

Figure 17.25b shows the projection of the vector field on the unit tangent vectors at various points on C. The circulation is the "sum" of the scalar components associated with these projections, which, in this case, is positive.

Related Exercise 53 ◄

► In the definition of flux, the non-self-intersecting property of C means that C is a *simple* curve, as defined formally in Section 17.3.

Flux of Two-Dimensional Vector Fields Assume $\mathbf{F} = \langle f, g \rangle$ is a continuous vector field on a region R of $\mathbb{R}^2$. We let C be a smooth oriented curve in R that does not intersect itself; C may or may not be closed. To compute the *flux* of the vector field across C, we "add up" the components of $\mathbf{F}$ *orthogonal* or *normal* to C at each point of C. Notice that every point on C has two unit vectors normal to C. Therefore, we let $\mathbf{n}$ denote the unit vector in the xy-plane normal to C in a direction to be defined momentarily. Once the direction of $\mathbf{n}$ is defined, the component of $\mathbf{F}$ normal to C is $\mathbf{F} \cdot \mathbf{n}$, and the flux is the line integral of $\mathbf{F} \cdot \mathbf{n}$ along C, which we denote $\int_C \mathbf{F} \cdot \mathbf{n}\, ds$.

The first step is to define the unit normal vector at a point P of C. Because C lies in the xy-plane, the unit vector $\mathbf{T}$ tangent at P also lies in the xy-plane. Therefore, its z-component is 0, and we let $\mathbf{T} = \langle T_x, T_y, 0 \rangle$. As always, $\mathbf{k} = \langle 0, 0, 1 \rangle$ is the unit vector in the z-direction. Because a unit vector $\mathbf{n}$ in the xy-plane normal to C is orthogonal to both $\mathbf{T}$ and $\mathbf{k}$, we determine the direction of $\mathbf{n}$ by letting $\mathbf{n} = \mathbf{T} \times \mathbf{k}$. This choice has two implications.

- If C is a closed curve oriented counterclockwise (when viewed from above), the unit normal vector points *outward* along the curve (**Figure 17.26a**). When $\mathbf{F}$ also points outward at a point on C, the angle θ between $\mathbf{F}$ and $\mathbf{n}$ satisfies $0 \le \theta < \frac{\pi}{2}$ (**Figure 17.26b**). At all such points, $\mathbf{F} \cdot \mathbf{n} > 0$ and there is a positive contribution to the flux across C. When $\mathbf{F}$ points inward at a point on C, $\frac{\pi}{2} < \theta \le \pi$ and $\mathbf{F} \cdot \mathbf{n} < 0$, which means there is a negative contribution to the flux at that point.

- If C is not a closed curve, the unit normal vector points to the right (when viewed from above) as the curve is traversed in the positive direction.

> ➤ Recall that $\mathbf{a} \times \mathbf{b}$ is orthogonal to both $\mathbf{a}$ and $\mathbf{b}$.

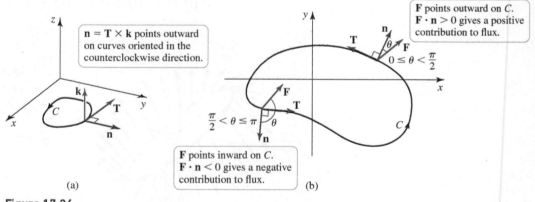

Figure 17.26

QUICK CHECK 5 Sketch a closed curve on a sheet of paper and draw a unit tangent vector $\mathbf{T}$ on the curve pointing in the counterclockwise direction. Explain why $\mathbf{n} = \mathbf{T} \times \mathbf{k}$ is an *outward* unit normal vector. ◄

Calculating the cross product that defines the unit normal vector $\mathbf{n}$, we find that

$$\mathbf{n} = \mathbf{T} \times \mathbf{k} = \begin{vmatrix} \mathbf{i} & \mathbf{j} & \mathbf{k} \\ T_x & T_y & 0 \\ 0 & 0 & 1 \end{vmatrix} = T_y \mathbf{i} - T_x \mathbf{j}.$$

Because $\mathbf{T} = \dfrac{\mathbf{r}'(t)}{|\mathbf{r}'(t)|}$, the components of $\mathbf{T}$ are

$$\mathbf{T} = \langle T_x, T_y, 0 \rangle = \frac{\langle x'(t), y'(t), 0 \rangle}{|\mathbf{r}'(t)|}.$$

We now have an expression for the unit normal vector:

$$\mathbf{n} = T_y \mathbf{i} - T_x \mathbf{j} = \frac{y'(t)}{|\mathbf{r}'(t)|}\mathbf{i} - \frac{x'(t)}{|\mathbf{r}'(t)|}\mathbf{j} = \frac{\langle y'(t), -x'(t) \rangle}{|\mathbf{r}'(t)|}.$$

To evaluate the flux integral $\int_C \mathbf{F} \cdot \mathbf{n}\, ds$, we make a familiar change of variables by letting $ds = |\mathbf{r}'(t)|\, dt$. The flux of $\mathbf{F} = \langle f, g \rangle$ across C is then

$$\int_C \mathbf{F} \cdot \mathbf{n}\, ds = \int_a^b \mathbf{F} \cdot \underbrace{\frac{\langle y'(t), -x'(t) \rangle}{|\mathbf{r}'(t)|}}_{\mathbf{n}} \underbrace{|\mathbf{r}'(t)|\, dt}_{ds} = \int_a^b (f(t)y'(t) - g(t)x'(t))\, dt.$$

This is one useful form of the flux integral. Alternatively, we can note that $dx = x'(t)\,dt$ and $dy = y'(t)\,dt$ and write

$$\int_C \mathbf{F} \cdot \mathbf{n}\, ds = \int_C f\, dy - g\, dx.$$

> Like circulation integrals, flux integrals may be computed on piecewise-smooth curves by finding the flux on each piece and adding the results.

DEFINITION Flux

Let $\mathbf{F} = \langle f, g \rangle$ be a continuous vector field on a region R of $\mathbb{R}^2$. Let $C: \mathbf{r}(t) = \langle x(t), y(t) \rangle$, for $a \le t \le b$, be a smooth oriented curve in R that does not intersect itself. The **flux** of the vector field $\mathbf{F}$ across C is

$$\int_C \mathbf{F} \cdot \mathbf{n}\, ds = \int_a^b (f(t)y'(t) - g(t)x'(t))\, dt,$$

where $\mathbf{n} = \mathbf{T} \times \mathbf{k}$ is the unit normal vector and $\mathbf{T}$ is the unit tangent vector consistent with the orientation. If C is a closed curve with counterclockwise orientation, $\mathbf{n}$ is the outward normal vector, and the flux integral gives the **outward flux** across C.

The concepts of circulation and flux can be visualized in terms of headwinds and crosswinds. Suppose the wind patterns in your neighborhood can be modeled with a vector field $\mathbf{F}$ (that doesn't change with time). Now imagine taking a walk around the block in a counterclockwise direction along a closed path. At different points along your walk, you encounter winds from various directions and with various speeds. The circulation of the wind field $\mathbf{F}$ along your path is the net amount of headwind (negative contribution) and tailwind (positive contribution) that you encounter during your walk. The flux of $\mathbf{F}$ across your path is the net amount of crosswind (positive from your left and negative from your right) encountered on your walk.

EXAMPLE 8 Flux of two-dimensional flows Find the outward flux across the unit circle with counterclockwise orientation for the following vector fields.

a. The radial vector field $\mathbf{F} = \langle x, y \rangle$

b. The rotation vector field $\mathbf{F} = \langle -y, x \rangle$

SOLUTION

a. The unit circle with counterclockwise orientation has a description $\mathbf{r}(t) = \langle x(t), y(t) \rangle = \langle \cos t, \sin t \rangle$, for $0 \le t \le 2\pi$. Therefore, $x'(t) = -\sin t$ and $y'(t) = \cos t$. The components of $\mathbf{F}$ are $f = x(t) = \cos t$ and $g = y(t) = \sin t$. It follows that the outward flux is

$$\int_a^b (f(t)y'(t) - g(t)x'(t))\, dt = \int_0^{2\pi} (\underbrace{\cos t}_{f(t)}\underbrace{\cos t}_{y'(t)} - \underbrace{\sin t}_{g(t)}\underbrace{(-\sin t)}_{x'(t)})\, dt$$

$$= \int_0^{2\pi} 1\, dt = 2\pi. \qquad \cos^2 t + \sin^2 t = 1$$

Because the radial field points outward and is aligned with the unit normal vectors on C, the outward flux is positive (**Figure 17.27a**).

b. For the rotation field, $f = -y(t) = -\sin t$ and $g = x(t) = \cos t$. The outward flux is

$$\int_a^b (f(t)y'(t) - g(t)x'(t))\, dt = \int_0^{2\pi} (\underbrace{-\sin t}_{f(t)}\underbrace{\cos t}_{y'(t)} - \underbrace{\cos t}_{g(t)}\underbrace{(-\sin t)}_{x'(t)})\, dt$$

$$= \int_0^{2\pi} 0\, dt = 0.$$

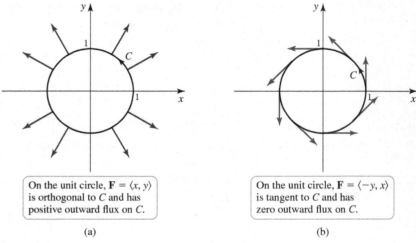

On the unit circle, $\mathbf{F} = \langle x, y \rangle$ is orthogonal to C and has positive outward flux on C.

(a)

On the unit circle, $\mathbf{F} = \langle -y, x \rangle$ is tangent to C and has zero outward flux on C.

(b)

Figure 17.27

Because the rotation field is orthogonal to $\mathbf{n}$ at all points of C, the outward flux across C is zero (**Figure 17.27b**). The results of Examples 6 and 8 are worth remembering: On a unit circle centered at the origin, the *radial* vector field $\langle x, y \rangle$ has outward flux 2π and zero circulation. The *rotation* vector field $\langle -y, x \rangle$ has zero outward flux and circulation 2π.

Related Exercises 59–60 ◄

SECTION 17.2 EXERCISES

Getting Started

1. How does a line integral differ from the single-variable integral $\int_a^b f(x)\,dx$?

2. If a curve C is given by $\mathbf{r}(t) = \langle t, t^2 \rangle$, what is $|\mathbf{r}'(t)|$?

3. Given that C is the curve $\mathbf{r}(t) = \langle \cos t, t \rangle$, for $\pi/2 \le t \le \pi$, convert the line integral $\int_C \dfrac{x}{y}\,ds$ to an ordinary integral. Do not evaluate the integral.

4–7. *Find a parametric description $\mathbf{r}(t)$ for the following curves.*

4. The segment of the curve $x = \sin \pi y$ from $(0, 0)$ to $(0, 3)$

5. The line segment from $(1, 2, 3)$ to $(5, 4, 0)$

6. The quarter-circle from $(1, 0)$ to $(0, 1)$ with its center at the origin

7. The segment of the parabola $x = y^2 + 1$ from $(5, 2)$ to $(17, 4)$

8. Find an expression for the vector field $\mathbf{F} = \langle x - y, y - x \rangle$ (in terms of t) along the unit circle $\mathbf{r}(t) = \langle \cos t, \sin t \rangle$.

9. Suppose C is the curve $\mathbf{r}(t) = \langle t, t^3 \rangle$, for $0 \le t \le 2$, and $\mathbf{F} = \langle x, 2y \rangle$. Evaluate $\int_C \mathbf{F} \cdot \mathbf{T}\,ds$ using the following steps.
   a. Convert the line integral $\int_C \mathbf{F} \cdot \mathbf{T}\,ds$ to an ordinary integral.
   b. Evaluate the integral in part (a).

10. Suppose C is the circle $\mathbf{r}(t) = \langle \cos t, \sin t \rangle$, for $0 \le t \le 2\pi$, and $\mathbf{F} = \langle 1, x \rangle$. Evaluate $\int_C \mathbf{F} \cdot \mathbf{n}\,ds$ using the following steps.
   a. Convert the line integral $\int_C \mathbf{F} \cdot \mathbf{n}\,ds$ to an ordinary integral.
   b. Evaluate the integral in part (a).

11. State two other forms for the line integral $\int_C \mathbf{F} \cdot \mathbf{T}\,ds$ given that $\mathbf{F} = \langle f, g, h \rangle$.

12–13. *Assume f is continuous on a region containing the smooth curve C from point A to point B and suppose $\int_C f\,ds = 10$.*

12. Explain the meaning of the curve $-C$ and state the value of $\int_{-C} f\,ds$.

13. Suppose P is a point on the curve C between A and B, where C_1 is the part of the curve from A to P, and C_2 is the part of the curve from P to B. Assuming $\int_{C_1} f\,ds = 3$, find the value of $\int_{C_2} f\,ds$.

14. Consider the graph of an ellipse C, oriented counterclockwise. Graphs of the vector fields $\mathbf{F}_1$, $\mathbf{F}_2$, $\mathbf{F}_3$, and $\mathbf{F}_4$ along the curve C are given (see figures). Along C, $\mathbf{F}_1$ and $\mathbf{F}_4$ are tangent to C, and $\mathbf{F}_2$ and $\mathbf{F}_3$ are normal to C. Determine whether the following integrals are positive, negative, or equal to 0.

   a. $\displaystyle\int_C \mathbf{F}_1 \cdot \mathbf{T}\,ds$ b. $\displaystyle\int_C \mathbf{F}_2 \cdot \mathbf{T}\,ds$ c. $\displaystyle\int_C \mathbf{F}_3 \cdot \mathbf{T}\,ds$

   d. $\displaystyle\int_C \mathbf{F}_4 \cdot \mathbf{T}\,ds$ e. $\displaystyle\int_C \mathbf{F}_1 \cdot \mathbf{n}\,ds$ f. $\displaystyle\int_C \mathbf{F}_2 \cdot \mathbf{n}\,ds$

   g. $\displaystyle\int_C \mathbf{F}_3 \cdot \mathbf{n}\,ds$ h. $\displaystyle\int_C \mathbf{F}_4 \cdot \mathbf{n}\,ds$

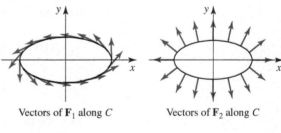

Vectors of $\mathbf{F}_1$ along C Vectors of $\mathbf{F}_2$ along C

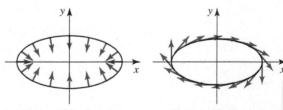

Vectors of $\mathbf{F}_3$ along C Vectors of $\mathbf{F}_4$ along C

15. How is the circulation of a vector field on a closed smooth oriented curve calculated?

16. Given a two-dimensional vector field $\mathbf{F}$ and a smooth oriented curve C, what is the meaning of the flux of $\mathbf{F}$ across C?

Practice Exercises

17–34. Scalar line integrals *Evaluate the following line integrals along the curve C.*

17. $\displaystyle\int_C xy\,ds$; C is the unit circle $\mathbf{r}(t) = \langle \cos t, \sin t \rangle$, for $0 \le t \le 2\pi$.

18. $\displaystyle\int_C (x^2 - 2y^2)\,ds$; C is the line segment $\mathbf{r}(t) = \left\langle \dfrac{t}{\sqrt{2}}, \dfrac{t}{\sqrt{2}} \right\rangle$, for $0 \le t \le 4$.

19. $\displaystyle\int_C (2x + y)\,ds$; C is the line segment $\mathbf{r}(t) = \langle 3t, 4t \rangle$, for $0 \le t \le 2$.

20. $\displaystyle\int_C x\,ds$; C is the curve $\mathbf{r}(t) = \langle t^3, 4t \rangle$, for $0 \le t \le 1$.

21. $\displaystyle\int_C xy^3\,ds$; C is the quarter-circle $\mathbf{r}(t) = \langle 2 \cos t, 2 \sin t \rangle$, for $0 \le t \le \pi/2$.

22. $\displaystyle\int_C 3x \cos y\,ds$; C is the curve $\mathbf{r}(t) = \langle \sin t, t \rangle$, for $0 \le t \le \pi/2$.

23. $\displaystyle\int_C (y - z)\,ds$; C is the helix $\mathbf{r}(t) = \langle 3 \cos t, 3 \sin t, 4t \rangle$, for $0 \le t \le 2\pi$.

24. $\displaystyle\int_C (x - y + 2z)\,ds$; C is the circle $\mathbf{r}(t) = \langle 1, 3 \cos t, 3 \sin t \rangle$, for $0 \le t \le 2\pi$.

25. $\displaystyle\int_C (x^2 + y^2)\,ds$; C is the circle of radius 4 centered at $(0, 0)$.

26. $\displaystyle\int_C (x^2 + y^2)\,ds$; C is the line segment from $(0, 0)$ to $(5, 5)$.

27. $\displaystyle\int_C \dfrac{x}{x^2 + y^2}\,ds$; C is the line segment from $(1, 1)$ to $(10, 10)$.

28. $\displaystyle\int_C (xy)^{1/3}\,ds$; C is the curve $y = x^2$, for $0 \le x \le 1$.

29. $\displaystyle\int_C xy\,ds$; C is a portion of the ellipse $\dfrac{x^2}{4} + \dfrac{y^2}{16} = 1$ in the first quadrant, oriented counterclockwise.

30. $\displaystyle\int_C (2x - 3y)\,ds$; C is the line segment from $(-1, 0)$ to $(0, 1)$ followed by the line segment from $(0, 1)$ to $(1, 0)$.

31. $\displaystyle\int_C (x + y + z)\,ds$; C is the semicircle $\mathbf{r}(t) = \langle 2 \cos t, 0, 2 \sin t \rangle$, for $0 \le t \le \pi$.

32. $\displaystyle\int_C \dfrac{xy}{z}\,ds$; C is the line segment from $(1, 4, 1)$ to $(3, 6, 3)$.

33. $\displaystyle\int_C xz\,ds$; C is the line segment from $(0, 0, 0)$ to $(3, 2, 6)$ followed by the line segment from $(3, 2, 6)$ to $(7, 9, 10)$.

34. $\displaystyle\int_C xe^{yz}\,ds$; C is $\mathbf{r}(t) = \langle t, 2t, -2t \rangle$, for $0 \le t \le 2$.

35–36. Mass and density *A thin wire represented by the smooth curve C with a density ρ (mass per unit length) has a mass $M = \int_C \rho\,ds$. Find the mass of the following wires with the given density.*

T 35. C: $\{(x, y)\colon y = 2x^2, 0 \le x \le 3\}$; $\rho(x, y) = 1 + xy$

36. C: $\mathbf{r}(\theta) = \langle \cos \theta, \sin \theta \rangle$, for $0 \le \theta \le \pi$; $\rho(\theta) = 2\theta/\pi + 1$

37–38. Average values *Find the average value of the following functions on the given curves.*

37. $f(x, y) = x + 2y$ on the line segment from $(1, 1)$ to $(2, 5)$

38. $f(x, y) = xe^y$ on the unit circle centered at the origin

39–40. Length of curves *Use a scalar line integral to find the length of the following curves.*

39. $\mathbf{r}(t) = \left\langle 20 \sin \dfrac{t}{4}, 20 \cos \dfrac{t}{4}, \dfrac{t}{2} \right\rangle$, for $0 \le t \le 2$

40. $\mathbf{r}(t) = \langle 30 \sin t, 40 \sin t, 50 \cos t \rangle$, for $0 \le t \le 2\pi$

41–46. Line integrals of vector fields in the plane *Given the following vector fields and oriented curves C, evaluate $\int_C \mathbf{F} \cdot \mathbf{T}\,ds$.*

41. $\mathbf{F} = \langle x, y \rangle$ on the parabola $\mathbf{r}(t) = \langle 4t, t^2 \rangle$, for $0 \le t \le 1$

42. $\mathbf{F} = \langle -y, x \rangle$ on the semicircle $\mathbf{r}(t) = \langle 4 \cos t, 4 \sin t \rangle$, for $0 \le t \le \pi$

43. $\mathbf{F} = \langle y, x \rangle$ on the line segment from $(1, 1)$ to $(5, 10)$

44. $\mathbf{F} = \langle -y, x \rangle$ on the parabola $y = x^2$ from $(0, 0)$ to $(1, 1)$

45. $\mathbf{F} = \dfrac{\langle x, y \rangle}{(x^2 + y^2)^{3/2}}$ on the curve $\mathbf{r}(t) = \langle t^2, 3t^2 \rangle$, for $1 \le t \le 2$

46. $\mathbf{F} = \dfrac{\langle x, y \rangle}{x^2 + y^2}$ on the line segment $\mathbf{r}(t) = \langle t, 4t \rangle$, for $1 \le t \le 10$

47–48. Line integrals from graphs *Determine whether $\int_C \mathbf{F} \cdot d\mathbf{r}$ along the paths C_1 and C_2 shown in the following vector fields is positive or negative. Explain your reasoning.*

a. $\displaystyle\int_{C_1} \mathbf{F} \cdot d\mathbf{r}$ **b.** $\displaystyle\int_{C_2} \mathbf{F} \cdot d\mathbf{r}$

47.

48.

49–56. Work integrals *Given the force field* **F**, *find the work required to move an object on the given oriented curve.*

49. $\mathbf{F} = \langle y, -x \rangle$ on the line segment from $(1, 2)$ to $(0, 0)$ followed by the line segment from $(0, 0)$ to $(0, 4)$

50. $\mathbf{F} = \langle x, y \rangle$ on the line segment from $(-1, 0)$ to $(0, 8)$ followed by the line segment from $(0, 8)$ to $(2, 8)$

51. $\mathbf{F} = \langle y, x \rangle$ on the parabola $y = 2x^2$ from $(0, 0)$ to $(2, 8)$

52. $\mathbf{F} = \langle y, -x \rangle$ on the line segment $y = 10 - 2x$ from $(1, 8)$ to $(3, 4)$

53. $\mathbf{F} = \langle x, y, z \rangle$ on the tilted ellipse $\mathbf{r}(t) = \langle 4 \cos t, 4 \sin t, 4 \cos t \rangle$, for $0 \leq t \leq 2\pi$

54. $\mathbf{F} = \langle -y, x, z \rangle$ on the helix $\mathbf{r(t)} = \left\langle 2 \cos t, 2 \sin t, \dfrac{t}{2\pi} \right\rangle$, for $0 \leq t \leq 2\pi$

55. $\mathbf{F} = \dfrac{\langle x, y, z \rangle}{(x^2 + y^2 + z^2)^{3/2}}$ on the line segment from $(1, 1, 1)$ to $(10, 10, 10)$

T 56. $\mathbf{F} = \dfrac{\langle x, y, z \rangle}{x^2 + y^2 + z^2}$ on the line segment from $(1, 1, 1)$ to $(8, 4, 2)$

57–58. Circulation *Consider the following vector fields* **F** *and closed oriented curves C in the plane (see figures).*

a. *Based on the picture, make a conjecture about whether the circulation of* **F** *on C is positive, negative, or zero.*

b. *Compute the circulation and interpret the result.*

57. $\mathbf{F} = \langle y - x, x \rangle$; $C: \mathbf{r}(t) = \langle 2 \cos t, 2 \sin t \rangle$, for $0 \leq t \leq 2\pi$

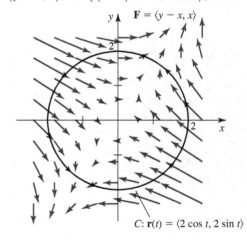

$C: \mathbf{r}(t) = \langle 2 \cos t, 2 \sin t \rangle$

58. $\mathbf{F} = \dfrac{\langle y, -2x \rangle}{\sqrt{4x^2 + y^2}}$; $C: \mathbf{r}(t) = \langle 2 \cos t, 4 \sin t \rangle$, for $0 \leq t \leq 2\pi$

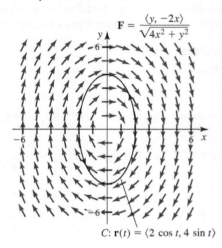

$C: \mathbf{r}(t) = \langle 2 \cos t, 4 \sin t \rangle$

59–60. Flux *Consider the vector fields and curves in Exercises 57–58.*

a. *Based on the picture, make a conjecture about whether the outward flux of* **F** *across C is positive, negative, or zero.*

b. *Compute the flux for the vector fields and curves.*

59. **F** and *C* given in Exercise 57

60. **F** and *C* given in Exercise 58

61. **Explain why or why not** Determine whether the following statements are true and give an explanation or counterexample.

   a. If a curve has a parametric description $\mathbf{r}(t) = \langle x(t), y(t), z(t) \rangle$, where t is the arc length, then $|\mathbf{r}'(t)| = 1$.

   b. The vector field $\mathbf{F} = \langle y, x \rangle$ has both zero circulation along and zero flux across the unit circle centered at the origin.

   c. If at all points of a path a force acts in a direction orthogonal to the path, then no work is done in moving an object along the path.

   d. The flux of a vector field across a curve in $\mathbb{R}^2$ can be computed using a line integral.

62. **Flying into a headwind** An airplane flies in the xz-plane, where x increases in the eastward direction and $z \geq 0$ represents vertical distance above the ground. A wind blows horizontally out of the west, producing a force $\mathbf{F} = \langle 150, 0 \rangle$. On which path between the points $(100, 50)$ and $(-100, 50)$ is more work done overcoming the wind?

   a. The line segment $\mathbf{r}(t) = \langle x(t), z(t) \rangle = \langle -t, 50 \rangle$, for $-100 \leq t \leq 100$

   b. The arc of the circle $\mathbf{r}(t) = \langle 100 \cos t, 50 + 100 \sin t \rangle$, for $0 \leq t \leq \pi$

63. **Flying into a headwind**

   a. How does the result of Exercise 62 change if the force due to the wind is $\mathbf{F} = \langle 141, 50 \rangle$ (approximately the same magnitude, but a different direction)?

   b. How does the result of Exercise 62 change if the force due to the wind is $\mathbf{F} = \langle 141, -50 \rangle$ (approximately the same magnitude, but a different direction)?

64. **Changing orientation** Let $f(x, y) = x + 2y$ and let C be the unit circle.

   a. Find a parameterization of C with a counterclockwise orientation and evaluate $\int_C f \, ds$.

b. Find a parameterization of C with a clockwise orientation and evaluate $\int_C f\,ds$.

c. Compare the results of parts (a) and (b).

65. Changing orientation Let $f(x, y) = x$ and let C be the segment of the parabola $y = x^2$ joining $O(0, 0)$ and $P(1, 1)$.

    a. Find a parameterization of C in the direction from O to P. Evaluate $\int_C f\,ds$.

    b. Find a parameterization of C in the direction from P to O. Evaluate $\int_C f\,ds$.

    c. Compare the results of parts (a) and (b).

66. Work in a rotation field Consider the rotation field $\mathbf{F} = \langle -y, x \rangle$ and the three paths shown in the figure. Compute the work done on each of the three paths. Does it appear that the line integral $\int_C \mathbf{F} \cdot \mathbf{T}\,ds$ is independent of the path, where C is any path from $(1, 0)$ to $(0, 1)$?

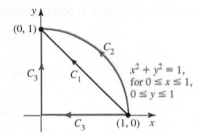

67. Work in a hyperbolic field Consider the hyperbolic force field $\mathbf{F} = \langle y, x \rangle$ (the streamlines are hyperbolas) and the three paths shown in the figure for Exercise 66. Compute the work done in the presence of $\mathbf{F}$ on each of the three paths. Does it appear that the line integral $\int_C \mathbf{F} \cdot \mathbf{T}\,ds$ is independent of the path, where C is any path from $(1, 0)$ to $(0, 1)$?

68–72. Assorted line integrals *Evaluate each line integral using the given curve C.*

68. $\int_C x^2\,dx + dy + y\,dz$; C is the curve $\mathbf{r}(t) = \langle t, 2t, t^2 \rangle$, for $0 \le t \le 3$.

T 69. $\int_C x^3 y\,dx + xz\,dy + (x + y)^2\,dz$; C is the helix $\mathbf{r}(t) = \langle 2t, \sin t, \cos t \rangle$, for $0 \le t \le 4\pi$.

T 70. $\int_C \dfrac{x^2}{y^4}\,ds$; C is the segment of the parabola $x = 3y^2$ from $(3, 1)$ to $(27, 3)$.

71. $\int_C \dfrac{y}{\sqrt{x^2 + y^2}}\,dx - \dfrac{x}{\sqrt{x^2 + y^2}}\,dy$; C is a quarter-circle from $(0, 4)$ to $(4, 0)$.

72. $\int_C (x + y)\,dx + (x - y)\,dy + x\,dz$; C is the line segment from $(1, 2, 4)$ to $(3, 8, 13)$.

73. Flux across curves in a vector field Consider the vector field $\mathbf{F} = \langle y, x \rangle$ shown in the figure.

    a. Compute the outward flux across the quarter-circle $C: \mathbf{r}(t) = \langle 2 \cos t, 2 \sin t \rangle$, for $0 \le t \le \pi/2$.

    b. Compute the outward flux across the quarter-circle $C: \mathbf{r}(t) = \langle 2 \cos t, 2 \sin t \rangle$, for $\pi/2 \le t \le \pi$.

    c. Explain why the flux across the quarter-circle in the third quadrant equals the flux computed in part (a).

d. Explain why the flux across the quarter-circle in the fourth quadrant equals the flux computed in part (b).

e. What is the outward flux across the full circle?

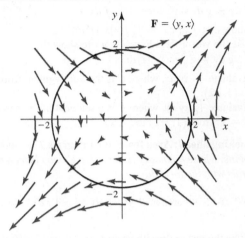

Explorations and Challenges

74–75. Zero circulation fields

74. For what values of b and c does the vector field $\mathbf{F} = \langle by, cx \rangle$ have zero circulation on the unit circle centered at the origin and oriented counterclockwise?

75. Consider the vector field $\mathbf{F} = \langle ax + by, cx + dy \rangle$. Show that $\mathbf{F}$ has zero circulation on any oriented circle centered at the origin, for any a, b, c, and d, provided $b = c$.

76–77. Zero flux fields

76. For what values of a and d does the vector field $\mathbf{F} = \langle ax, dy \rangle$ have zero flux across the unit circle centered at the origin and oriented counterclockwise?

77. Consider the vector field $\mathbf{F} = \langle ax + by, cx + dy \rangle$. Show that $\mathbf{F}$ has zero flux across any oriented circle centered at the origin, for any a, b, c, and d, provided $a = -d$.

78. Heat flux in a plate A square plate $R = \{(x, y): 0 \le x \le 1, 0 \le y \le 1\}$ has a temperature distribution $T(x, y) = 100 - 50x - 25y$.

    a. Sketch two level curves of the temperature in the plate.

    b. Find the gradient of the temperature $\nabla T(x, y)$.

    c. Assume the flow of heat is given by the vector field $\mathbf{F} = -\nabla T(x, y)$. Compute $\mathbf{F}$.

    d. Find the outward heat flux across the boundary $\{(x, y): x = 1, 0 \le y \le 1\}$.

    e. Find the outward heat flux across the boundary $\{(x, y): 0 \le x \le 1, y = 1\}$.

79. Inverse force fields Consider the radial field $\mathbf{F} = \dfrac{\mathbf{r}}{|\mathbf{r}|^p} = \dfrac{\langle x, y, z \rangle}{|\mathbf{r}|^p}$, where $p > 1$ (the inverse square law corresponds to $p = 3$). Let C be the line segment from $(1, 1, 1)$ to (a, a, a), where $a > 1$, given by $\mathbf{r}(t) = \langle t, t, t \rangle$, for $1 \le t \le a$.

    a. Find the work done in moving an object along C with $p = 2$.

    b. If $a \to \infty$ in part (a), is the work finite?

    c. Find the work done in moving an object along C with $p = 4$.

    d. If $a \to \infty$ in part (c), is the work finite?

    e. Find the work done in moving an object along C for any $p > 1$.

    f. If $a \to \infty$ in part (e), for what values of p is the work finite?

80. Line integrals with respect to dx and dy Given a vector field $\mathbf{F} = \langle f, 0 \rangle$ and curve C with parameterization $\mathbf{r}(t) = \langle x(t), y(t) \rangle$, for $a \leq t \leq b$, we see that the line integral $\int_C f\,dx + g\,dy$ simplifies to $\int_C f\,dx$.

   a. Show that $\int_C f\,dx = \int_a^b f(t)x'(t)\,dt$.

   b. Use the vector field $\mathbf{F} = \langle 0, g \rangle$ to show that $\int_C g\,dy = \int_a^b g(t)y'(t)\,dt$.

   c. Evaluate $\int_C xy\,dx$, where C is the line segment from $(0, 0)$ to $(5, 12)$.

   d. Evaluate $\int_C xy\,dy$, where C is a segment of the parabola $x = y^2$ from $(1, -1)$ to $(1, 1)$.

81–82. Looking ahead: Area from line integrals *The area of a region R in the plane, whose boundary is the curve C, may be computed using line integrals with the formula*

$$\text{area of } R = \int_C x\,dy = -\int_C y\,dx.$$

81. Let R be the rectangle with vertices $(0, 0)$, $(a, 0)$, $(0, b)$, and (a, b), and let C be the boundary of R oriented counterclockwise. Use the formula $A = \int_C x\,dy$ to verify that the area of the rectangle is ab.

82. Let $R = \{(r, \theta): 0 \leq r \leq a, 0 \leq \theta \leq 2\pi\}$ be the disk of radius a centered at the origin, and let C be the boundary of R oriented counterclockwise. Use the formula $A = -\int_C y\,dx$ to verify that the area of the disk is πa^2.

QUICK CHECK ANSWERS

1. The Fundamental Theorem of Calculus says that $\frac{d}{dt}\int_a^t f(u)\,du = f(t)$, which applies to differentiating the arc length integral.   **2.** Note that $x = t, y = 0$, and $|\mathbf{r}'(t)| = \sqrt{1^2 + 0^2} = 1$. Therefore,

$$\int_C f(x, y)\,ds = \int_a^b f(t, 0)\,dt.$$ **3.** 1300 ft/min   **4.** $\pi/2$

5. $\mathbf{T}$ and $\mathbf{k}$ are unit vectors, so $\mathbf{n}$ is a unit vector. By the right-hand rule for cross products, $\mathbf{n}$ points outward from the curve. ◀

17.3 Conservative Vector Fields

This is an action-packed section in which several fundamental ideas come together. At the heart of the matter are two questions:

- When can a vector field be expressed as the gradient of a potential function? A vector field with this property will be defined as a *conservative* vector field.

- What special properties do conservative vector fields have?

After some preliminary definitions, we present a test to determine whether a vector field in $\mathbb{R}^2$ or $\mathbb{R}^3$ is conservative. This test is followed by a procedure to find a potential function for a conservative field. We then develop several equivalent properties shared by all conservative vector fields.

Types of Curves and Regions

Many of the results in the remainder of this text rely on special properties of regions and curves. It's best to collect these definitions in one place for easy reference.

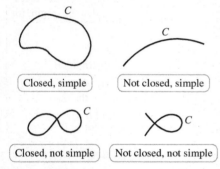

Closed, simple   Not closed, simple

Closed, not simple   Not closed, not simple

Figure 17.28

➤ Recall that all points of an open set are interior points. An open set does not contain its boundary points.

➤ Roughly speaking, connected means that R is all in one piece and simply connected in $\mathbb{R}^2$ means that R has no holes. $\mathbb{R}^2$ and $\mathbb{R}^3$ are themselves connected and simply connected.

DEFINITION Simple and Closed Curves

Suppose a curve C (in $\mathbb{R}^2$ or $\mathbb{R}^3$) is described parametrically by $\mathbf{r}(t)$, where $a \leq t \leq b$. Then C is a **simple curve** if $\mathbf{r}(t_1) \neq \mathbf{r}(t_2)$ for all t_1 and t_2, with $a < t_1 < t_2 < b$; that is, C never intersects itself between its endpoints. The curve C is **closed** if $\mathbf{r}(a) = \mathbf{r}(b)$; that is, the initial and terminal points of C are the same (**Figure 17.28**).

In all that follows, we generally assume that R in $\mathbb{R}^2$ (or D in $\mathbb{R}^3$) is an open region. Open regions are further classified according to whether they are *connected* and whether they are *simply connected*.

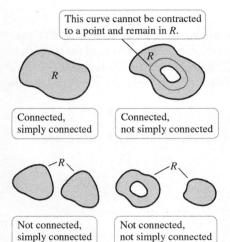

This curve cannot be contracted to a point and remain in R.

Connected, simply connected

Connected, not simply connected

Not connected, simply connected

Not connected, not simply connected

Figure 17.29

QUICK CHECK 1 Is a figure-8 curve simple? Closed? Is a torus connected? Simply connected? ◄

➤ The term *conservative* refers to conservation of energy. See Exercise 66 for an example of conservation of energy in a conservative force field.

➤ Depending on the context and the interpretation of the vector field, the potential function φ may be defined such that $\mathbf{F} = -\nabla\varphi$ (with a negative sign).

DEFINITION Connected and Simply Connected Regions

An open region R in $\mathbb{R}^2$ (or D in $\mathbb{R}^3$) is **connected** if it is possible to connect any two points of R by a continuous curve lying in R. An open region R is **simply connected** if every closed simple curve in R can be deformed and contracted to a point in R (**Figure 17.29**).

Test for Conservative Vector Fields

We begin with the central definition of this section.

DEFINITION Conservative Vector Field

A vector field $\mathbf{F}$ is said to be **conservative** on a region (in $\mathbb{R}^2$ or $\mathbb{R}^3$) if there exists a scalar function φ such that $\mathbf{F} = \nabla\varphi$ on that region.

Suppose the components of $\mathbf{F} = \langle f, g, h \rangle$ have continuous first partial derivatives on a region D in $\mathbb{R}^3$. Also assume $\mathbf{F}$ is conservative, which means by definition that there is a potential function φ such that $\mathbf{F} = \nabla\varphi$. Matching the components of $\mathbf{F}$ and $\nabla\varphi$, we see that $f = \varphi_x$, $g = \varphi_y$, and $h = \varphi_z$. Recall from Theorem 15.4 that if a function has continuous second partial derivatives, the order of differentiation in the second partial derivatives does not matter. Under these conditions on φ, we conclude the following:

- $\varphi_{xy} = \varphi_{yx}$, which implies that $f_y = g_x$,
- $\varphi_{xz} = \varphi_{zx}$, which implies that $f_z = h_x$, and
- $\varphi_{yz} = \varphi_{zy}$, which implies that $g_z = h_y$.

These observations constitute half of the proof of the following theorem. The remainder of the proof is given in Section 17.4.

THEOREM 17.3 Test for Conservative Vector Fields

Let $\mathbf{F} = \langle f, g, h \rangle$ be a vector field defined on a connected and simply connected region D of $\mathbb{R}^3$, where f, g, and h have continuous first partial derivatives on D. Then $\mathbf{F}$ is a conservative vector field on D (there is a potential function φ such that $\mathbf{F} = \nabla\varphi$) if and only if

$$\frac{\partial f}{\partial y} = \frac{\partial g}{\partial x}, \qquad \frac{\partial f}{\partial z} = \frac{\partial h}{\partial x}, \quad \text{and} \quad \frac{\partial g}{\partial z} = \frac{\partial h}{\partial y}.$$

For vector fields in $\mathbb{R}^2$, we have the single condition $\dfrac{\partial f}{\partial y} = \dfrac{\partial g}{\partial x}$.

EXAMPLE 1 Testing for conservative fields Determine whether the following vector fields are conservative on $\mathbb{R}^2$ and $\mathbb{R}^3$, respectively.

a. $\mathbf{F} = \langle e^x \cos y, -e^x \sin y \rangle$ **b.** $\mathbf{F} = \langle 2xy - z^2, x^2 + 2z, 2y - 2xz \rangle$

SOLUTION

a. Letting $f(x, y) = e^x \cos y$ and $g(x, y) = -e^x \sin y$, we see that

$$\frac{\partial f}{\partial y} = -e^x \sin y = \frac{\partial g}{\partial x}.$$

The conditions of Theorem 17.3 are met and $\mathbf{F}$ is conservative.

b. Letting $f(x, y, z) = 2xy - z^2$, $g(x, y, z) = x^2 + 2z$, and $h(x, y, z) = 2y - 2xz$, we have

$$\frac{\partial f}{\partial y} = 2x = \frac{\partial g}{\partial x}, \qquad \frac{\partial f}{\partial z} = -2z = \frac{\partial h}{\partial x}, \qquad \frac{\partial g}{\partial z} = 2 = \frac{\partial h}{\partial y}.$$

By Theorem 17.3, $\mathbf{F}$ is conservative. *Related Exercises 13–14* ◄

Finding Potential Functions

QUICK CHECK 2 Explain why a potential function for a conservative vector field is determined up to an additive constant. ◄

Like antiderivatives, potential functions are determined up to an arbitrary additive constant. Unless an additive constant in a potential function has some physical meaning, it is usually omitted. Given a conservative vector field, there are several methods for finding a potential function. One method is shown in the following example. Another approach is illustrated in Exercise 71.

EXAMPLE 2 **Finding potential functions** Find a potential function for the conservative vector fields in Example 1.

a. $\mathbf{F} = \langle e^x \cos y, -e^x \sin y \rangle$

b. $\mathbf{F} = \langle 2xy - z^2, x^2 + 2z, 2y - 2xz \rangle$

SOLUTION

a. A potential function φ for $\mathbf{F} = \langle f, g \rangle$ has the property that $\mathbf{F} = \nabla\varphi$ and satisfies the conditions

$$\varphi_x = f(x, y) = e^x \cos y \quad \text{and} \quad \varphi_y = g(x, y) = -e^x \sin y.$$

The first equation is integrated with respect to x (holding y fixed) to obtain

$$\int \varphi_x \, dx = \int e^x \cos y \, dx,$$

which implies that

$$\varphi(x, y) = e^x \cos y + c(y).$$

► This procedure may begin with either of the two conditions, $\varphi_x = f$ or $\varphi_y = g$.

In this case, the "constant of integration" $c(y)$ is an arbitrary function of y. You can check the preceding calculation by noting that

$$\frac{\partial \varphi}{\partial x} = \frac{\partial}{\partial x} (e^x \cos y + c(y)) = e^x \cos y = f(x, y).$$

To find the arbitrary function $c(y)$, we differentiate $\varphi(x, y) = e^x \cos y + c(y)$ with respect to y and equate the result to g (recall that $\varphi_y = g$):

$$\varphi_y = -e^x \sin y + c'(y) \quad \text{and} \quad g = -e^x \sin y.$$

We conclude that $c'(y) = 0$, which implies that $c(y)$ is any real number, which we typically take to be zero. So a potential function is $\varphi(x, y) = e^x \cos y$, a result that may be checked by differentiation.

► This procedure may begin with any of the three conditions.

b. The method of part (a) is more elaborate with three variables. A potential function φ must now satisfy these conditions:

$$\varphi_x = f = 2xy - z^2, \quad \varphi_y = g = x^2 + 2z, \quad \text{and} \quad \varphi_z = h = 2y - 2xz.$$

Integrating the first condition with respect to x (holding y and z fixed), we have

$$\varphi = \int (2xy - z^2) \, dx = x^2y - xz^2 + c(y, z).$$

Because the integration is with respect to x, the arbitrary "constant" is a function of y and z. To find $c(y, z)$, we differentiate φ with respect to y, which results in

$$\varphi_y = x^2 + c_y(y, z).$$

Equating φ_y and $g = x^2 + 2z$, we see that $c_y(y, z) = 2z$. To obtain $c(y, z)$, we integrate $c_y(y, z) = 2z$ with respect to y (holding z fixed), which results in $c(y, z) = 2yz + d(z)$. The "constant" of integration is now a function of z, which we call $d(z)$. At this point, a potential function looks like

$$\varphi(x, y, z) = x^2y - xz^2 + 2yz + d(z).$$

To determine $d(z)$, we differentiate φ with respect to z:

$$\varphi_z = -2xz + 2y + d'(z).$$

QUICK CHECK 3 Verify by differentiation that the potential functions found in Example 2 produce the corresponding vector fields. ◀

Equating φ_z and $h = 2y - 2xz$, we see that $d'(z) = 0$, or $d(z)$ is a real number, which we generally take to be zero. Putting it all together, a potential function is

$$\varphi = x^2y - xz^2 + 2yz.$$

Related Exercises 19, 24 ◀

PROCEDURE   **Finding Potential Functions in $\mathbb{R}^3$**

Suppose $\mathbf{F} = \langle f, g, h \rangle$ is a conservative vector field. To find φ such that $\mathbf{F} = \nabla\varphi$, use the following steps:

1. Integrate $\varphi_x = f$ with respect to x to obtain φ, which includes an arbitrary function $c(y, z)$.
2. Compute φ_y and equate it to g to obtain an expression for $c_y(y, z)$.
3. Integrate $c_y(y, z)$ with respect to y to obtain $c(y, z)$, including an arbitrary function $d(z)$.
4. Compute φ_z and equate it to h to get $d(z)$.

A similar procedure beginning with $\varphi_y = g$ or $\varphi_z = h$ may be easier in some cases.

▶ Compare the two versions of the Fundamental Theorem.

$$\int_a^b F'(x)\, dx = F(b) - F(a)$$

$$\int_C \nabla\varphi \cdot d\mathbf{r} = \varphi(B) - \varphi(A)$$

Fundamental Theorem for Line Integrals and Path Independence

Knowing how to find potential functions, we now investigate their properties. The first property is one of several beautiful parallels to the Fundamental Theorem of Calculus.

THEOREM 17.4   **Fundamental Theorem for Line Integrals**

Let R be a region in $\mathbb{R}^2$ or $\mathbb{R}^3$ and let φ be a differentiable potential function defined on R. If $\mathbf{F} = \nabla\varphi$ (which means that $\mathbf{F}$ is conservative), then

$$\int_C \mathbf{F} \cdot \mathbf{T}\, ds = \int_C \mathbf{F} \cdot d\mathbf{r} = \varphi(B) - \varphi(A),$$

for all points A and B in R and all piecewise-smooth oriented curves C in R from A to B.

Proof: Let the curve C in $\mathbb{R}^3$ be given by $\mathbf{r}(t) = \langle x(t), y(t), z(t) \rangle$, for $a \leq t \leq b$, where $\mathbf{r}(a)$ and $\mathbf{r}(b)$ are the position vectors for the points A and B, respectively. By the Chain Rule, the rate of change of φ with respect to t along C is

$$\frac{d\varphi}{dt} = \frac{\partial\varphi}{\partial x}\frac{dx}{dt} + \frac{\partial\varphi}{\partial y}\frac{dy}{dt} + \frac{\partial\varphi}{\partial z}\frac{dz}{dt} \qquad \text{Chain Rule}$$

$$= \left\langle \frac{\partial\varphi}{\partial x}, \frac{\partial\varphi}{\partial y}, \frac{\partial\varphi}{\partial z} \right\rangle \cdot \left\langle \frac{dx}{dt}, \frac{dy}{dt}, \frac{dz}{dt} \right\rangle \qquad \text{Identify the dot product.}$$

$$= \nabla\varphi \cdot \mathbf{r}'(t) \qquad\qquad\qquad \mathbf{r} = \langle x, y, z \rangle$$

$$= \mathbf{F} \cdot \mathbf{r}'(t). \qquad\qquad\qquad \mathbf{F} = \nabla\varphi$$

Evaluating the line integral and using the Fundamental Theorem of Calculus, it follows that

$$\int_C \mathbf{F} \cdot d\mathbf{r} = \int_a^b \mathbf{F} \cdot \mathbf{r}'(t)\, dt$$

$$= \int_a^b \frac{d\varphi}{dt}\, dt \qquad\qquad \mathbf{F} \cdot \mathbf{r}'(t) = \frac{d\varphi}{dt}$$

$$= \varphi(B) - \varphi(A). \qquad\qquad \text{Fundamental Theorem of Calculus; } t = b \text{ corresponds to } B \text{ and } t = a \text{ corresponds to } A.$$
◀

Here is the meaning of Theorem 17.4: If **F** is a conservative vector field, then the value of a line integral of **F** depends only on the endpoints of the path. For this reason, we say the line integral is *independent of path*, which means that to evaluate line integrals of conservative vector fields, we do not need a parameterization of the path.

If we think of φ as an antiderivative of the vector field **F**, then the parallel to the Fundamental Theorem of Calculus is clear. The line integral of **F** is the difference of the values of φ evaluated at the endpoints. Theorem 17.4 motivates the following definition.

DEFINITION Independence of Path

Let **F** be a continuous vector field with domain R. If $\displaystyle\int_{C_1} \mathbf{F} \cdot d\mathbf{r} = \int_{C_2} \mathbf{F} \cdot d\mathbf{r}$ for all

piecewise-smooth curves C_1 and C_2 in R with the same initial and terminal points, then the line integral is **independent of path**.

An important question concerns the converse of Theorem 17.4. With additional conditions on the domain R, the converse turns out to be true.

THEOREM 17.5

Let **F** be a continuous vector field on an open connected region R in $\mathbb{R}^2$. If

$\displaystyle\int_C \mathbf{F} \cdot d\mathbf{r}$ is independent of path, then **F** is conservative; that is, there exists a

potential function φ such that $\mathbf{F} = \nabla\varphi$ on R.

➤ We state and prove Theorem 17.5 in two variables. It is easily extended to three or more variables.

Proof: Let $P(a, b)$ and $Q(x, y)$ be interior points of R and define $\varphi(x, y) = \displaystyle\int_C \mathbf{F} \cdot d\mathbf{r}$,

where C is a piecewise-smooth path from P to Q, and $\mathbf{F} = \langle f, g \rangle$. Because the integral defining φ is independent of path, any piecewise-smooth path in R from P to Q can be used. The goal is to compute the directional derivative $D_{\mathbf{u}}\varphi(x, y)$, where $\mathbf{u} = \langle u_1, u_2 \rangle$ is an arbitrary unit vector, and then show that $\mathbf{F} = \nabla\varphi$. We let $S(x + tu_1, y + tu_2)$ be a point in R near Q and then apply the definition of the directional derivative at Q:

$$D_{\mathbf{u}}\varphi(x, y) = \lim_{t \to 0} \frac{\varphi(x + tu_1, y + tu_2) - \varphi(x, y)}{t}$$

$$= \lim_{t \to 0} \frac{1}{t}\left(\int_P^S \mathbf{F} \cdot d\mathbf{r} - \int_P^Q \mathbf{F} \cdot d\mathbf{r} \right)$$

$$= \lim_{t \to 0} \frac{1}{t} \int_Q^S \mathbf{F} \cdot d\mathbf{r}.$$

Using path independence, we choose the path from Q to S to be a line parameterized by $\mathbf{r}(s) = \langle x + su_1, y + su_2 \rangle$, for $0 \le s \le t$. Noting that $\mathbf{r}'(s) = \mathbf{u}$, the directional derivative is

$$D_{\mathbf{u}}\varphi(x, y) = \lim_{t \to 0} \frac{1}{t} \int_Q^S \mathbf{F} \cdot d\mathbf{r}$$

$$= \lim_{t \to 0} \frac{1}{t} \int_0^t \mathbf{F}(x + su_1, y + su_2) \cdot \mathbf{r}'(s)\, ds \qquad \text{Change line integral to ordinary integral.}$$

$$= \lim_{t \to 0} \frac{\displaystyle\int_0^t \mathbf{F}(x + su_1, y + su_2) \cdot \mathbf{r}'(s)\, ds - \int_0^0 \mathbf{F}(x + su_1, y + su_2) \cdot \mathbf{r}'(s)\, ds}{t}$$

Second integral equals 0.

$$= \frac{d}{dt} \int_0^t \mathbf{F}(x + su_1, y + su_2) \cdot \mathbf{u}\, ds \bigg|_{t=0} \qquad \begin{array}{l}\text{Identify difference quotient;} \\ \mathbf{r}'(s) = \mathbf{u}\end{array}$$

$$= \mathbf{F}(x, y) \cdot \mathbf{u}. \qquad \text{Fundamental Theorem of Calculus}$$

Choosing $\mathbf{u} = \mathbf{i} = \langle 1, 0 \rangle$, we see that $D_\mathbf{i}\varphi(x, y) = \varphi_x(x, y) = \mathbf{F}(x, y) \cdot \mathbf{i} = f(x, y)$. Similarly, choosing $\mathbf{u} = \mathbf{j} = \langle 0, 1 \rangle$, we have $D_\mathbf{j}\varphi(x, y) = \varphi_y(x, y) = \mathbf{F}(x, y) \cdot \mathbf{j} = g(x, y)$. Therefore, $\mathbf{F} = \langle f, g \rangle = \langle \varphi_x, \varphi_y \rangle = \nabla \varphi$, and $\mathbf{F}$ is conservative. ◄

EXAMPLE 3   Verifying path independence Consider the potential function $\varphi(x, y) = (x^2 - y^2)/2$ and its gradient field $\mathbf{F} = \langle x, -y \rangle$.

- Let C_1 be the quarter-circle $\mathbf{r}(t) = \langle \cos t, \sin t \rangle$, for $0 \leq t \leq \pi/2$, from $A(1, 0)$ to $B(0, 1)$.
- Let C_2 be the line $\mathbf{r}(t) = \langle 1 - t, t \rangle$, for $0 \leq t \leq 1$, also from A to B.

Evaluate the line integrals of $\mathbf{F}$ on C_1 and C_2, and show that both are equal to $\varphi(B) - \varphi(A)$.

SOLUTION On C_1, we have $\mathbf{r}'(t) = \langle -\sin t, \cos t \rangle$ and $\mathbf{F} = \langle x, -y \rangle = \langle \cos t, -\sin t \rangle$. The line integral on C_1 is

$$\int_{C_1} \mathbf{F} \cdot d\mathbf{r} = \int_a^b \mathbf{F} \cdot \mathbf{r}'(t)\, dt$$

$$= \int_0^{\pi/2} \underbrace{\langle \cos t, -\sin t \rangle}_{\mathbf{F}} \cdot \underbrace{\langle -\sin t, \cos t \rangle\, dt}_{\mathbf{r}'(t)\, dt} \qquad \text{Substitute for } \mathbf{F} \text{ and } \mathbf{r}'.$$

$$= \int_0^{\pi/2} (-\sin 2t)\, dt \qquad\qquad 2 \sin t \cos t = \sin 2t$$

$$= \left(\frac{1}{2} \cos 2t \right)\Big|_0^{\pi/2} = -1. \qquad\qquad \text{Evaluate the integral.}$$

On C_2, we have $\mathbf{r}'(t) = \langle -1, 1 \rangle$ and $\mathbf{F} = \langle x, -y \rangle = \langle 1 - t, -t \rangle$; therefore,

$$\int_{C_2} \mathbf{F} \cdot d\mathbf{r} = \int_0^1 \underbrace{\langle 1 - t, -t \rangle}_{\mathbf{F}} \cdot \underbrace{\langle -1, 1 \rangle\, dt}_{d\mathbf{r}} \qquad \text{Substitute for } \mathbf{F} \text{ and } d\mathbf{r}.$$

$$= \int_0^1 (-1)\, dt = -1. \qquad\qquad \text{Simplify.}$$

The two line integrals have the same value, which is

$$\varphi(B) - \varphi(A) = \varphi(0, 1) - \varphi(1, 0) = -\frac{1}{2} - \frac{1}{2} = -1.$$

Related Exercises 31–32 ◄

EXAMPLE 4   Line integral of a conservative vector field Evaluate

$$\int_C ((2xy - z^2)\, \mathbf{i} + (x^2 + 2z)\, \mathbf{j} + (2y - 2xz)\, \mathbf{k}) \cdot d\mathbf{r},$$

where C is a simple curve from $A(-3, -2, -1)$ to $B(1, 2, 3)$.

SOLUTION This vector field is conservative and has a potential function $\varphi = x^2 y - xz^2 + 2yz$ (Example 2). By the Fundamental Theorem for line integrals,

$$\int_C ((2xy - z^2)\, \mathbf{i} + (x^2 + 2z)\, \mathbf{j} + (2y - 2xz)\, \mathbf{k}) \cdot d\mathbf{r}$$

$$= \int_C \nabla \underbrace{(x^2 y - xz^2 + 2yz)}_{\varphi} \cdot d\mathbf{r}$$

$$= \varphi(1, 2, 3) - \varphi(-3, -2, -1) = 16.$$

Related Exercise 34 ◄

QUICK CHECK 4 Explain why the vector field $\nabla(xy + xz - yz)$ is conservative. ◄

Line Integrals on Closed Curves

It is a short step to another characterization of conservative vector fields. Suppose C is a simple *closed* piecewise-smooth oriented curve in $\mathbb{R}^2$ or $\mathbb{R}^3$. To distinguish line integrals on closed curves, we adopt the notation $\oint_C \mathbf{F} \cdot d\mathbf{r}$, where the small circle on the integral sign indicates that C is a closed curve. Let A be any point on C and think of A as both the initial point and the final point of C. Assuming $\mathbf{F}$ is a conservative vector field on an open connected region R containing C, it follows by Theorem 17.4 that

$$\oint_C \mathbf{F} \cdot d\mathbf{r} = \varphi(A) - \varphi(A) = 0.$$

Because A is an arbitrary point on C, we see that the line integral of a conservative vector field on a closed curve is zero.

An argument can be made in the opposite direction as well: Suppose $\oint_C \mathbf{F} \cdot d\mathbf{r} = 0$ on all simple closed piecewise-smooth oriented curves in a region R, and let A and B be distinct points in R. Let C_1 denote any curve from A to B, let C_2 be any curve from B to A (distinct from and not intersecting C_1), and let C be the closed curve consisting of C_1 followed by C_2 **(Figure 17.30)**. Then

$$0 = \oint_C \mathbf{F} \cdot d\mathbf{r} = \int_{C_1} \mathbf{F} \cdot d\mathbf{r} + \int_{C_2} \mathbf{F} \cdot d\mathbf{r}.$$

Therefore, $\int_{C_1} \mathbf{F} \cdot d\mathbf{r} = -\int_{C_2} \mathbf{F} \cdot d\mathbf{r} = \int_{-C_2} \mathbf{F} \cdot d\mathbf{r}$, where $-C_2$ is the curve C_2 traversed in the opposite direction (from A to B). We see that the line integral has the same value on two arbitrary paths between A and B. It follows that the value of the line integral is independent of path, and by Theorem 17.5, $\mathbf{F}$ is conservative. This argument is a proof of the following theorem.

> ▶ Notice the analogy to $\int_a^a f(x)\,dx = 0$, which is true of all integrable functions.

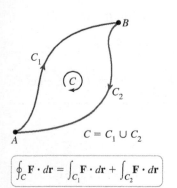

$$\oint_C \mathbf{F} \cdot d\mathbf{r} = \int_{C_1} \mathbf{F} \cdot d\mathbf{r} + \int_{C_2} \mathbf{F} \cdot d\mathbf{r}$$

Figure 17.30

THEOREM 17.6   Line Integrals on Closed Curves
Let R be an open connected region in $\mathbb{R}^2$ or $\mathbb{R}^3$. Then $\mathbf{F}$ is a conservative vector field on R if and only if $\oint_C \mathbf{F} \cdot d\mathbf{r} = 0$ on all simple closed piecewise-smooth oriented curves C in R.

EXAMPLE 5   A closed curve line integral in $\mathbb{R}^3$ Evaluate $\int_C \nabla(-xy + xz + yz) \cdot d\mathbf{r}$ on the curve $C: \mathbf{r}(t) = \langle \sin t, \cos t, \sin t \rangle$, for $0 \le t \le 2\pi$, without using Theorem 17.4 or Theorem 17.6.

SOLUTION The components of the vector field are

$$\mathbf{F} = \nabla(-xy + xz + yz) = \langle -y + z, -x + z, x + y \rangle.$$

Note that $\mathbf{r}'(t) = \langle \cos t, -\sin t, \cos t \rangle$ and $d\mathbf{r} = \mathbf{r}'(t)\,dt$. Substituting values of x, y, and z, the value of the line integral is

$$\oint_C \mathbf{F} \cdot d\mathbf{r} = \oint_C \langle -y + z, -x + z, x + y \rangle \cdot d\mathbf{r} \qquad \text{Substitute for } \mathbf{F}.$$

$$= \int_0^{2\pi} \sin 2t\,dt \qquad \text{Substitute for } x, y, z, d\mathbf{r}.$$

$$= -\frac{1}{2}\cos 2t \Big|_0^{2\pi} = 0. \qquad \text{Evaluate integral.}$$

The line integral of this conservative vector field on the closed curve C is zero. In fact, by Theorem 17.6, the line integral vanishes on any simple closed piecewise-smooth oriented curve.

Related Exercise 50 ◀

Summary of the Properties of Conservative Vector Fields

We have established three equivalent properties of conservative vector fields $\mathbf{F}$ defined on an open connected region R in $\mathbb{R}^2$ (or D in $\mathbb{R}^3$).

- There exists a potential function φ such that $\mathbf{F} = \nabla\varphi$ (definition).
- $\int_C \mathbf{F} \cdot d\mathbf{r} = \varphi(B) - \varphi(A)$ for all points A and B in R and all piecewise-smooth oriented curves C in R from A to B (path independence).
- $\oint_C \mathbf{F} \cdot d\mathbf{r} = 0$ on all simple piecewise-smooth closed oriented curves C in R.

The connections between these properties were established by Theorems 17.4, 17.5, and 17.6 in the following way:

$$\overset{\text{Theorems 17.4 and 17.5}}{\text{Path independence} \iff \mathbf{F} \text{ is conservative } (\nabla\varphi = \mathbf{F})} \overset{\text{Theorem 17.6}}{\iff} \oint_C \mathbf{F} \cdot d\mathbf{r} = 0.$$

SECTION 17.3 EXERCISES

Getting Started

1. Explain with pictures what is meant by a simple curve and a closed curve.

2. Explain with pictures what is meant by a connected region and a simply connected region.

3. How do you determine whether a vector field in $\mathbb{R}^2$ is conservative (has a potential function φ such that $\mathbf{F} = \nabla\varphi$)?

4. How do you determine whether a vector field in $\mathbb{R}^3$ is conservative?

5. Briefly describe how to find a potential function φ for a conservative vector field $\mathbf{F} = \langle f, g \rangle$.

6. If $\mathbf{F}$ is a conservative vector field on a region R, how do you evaluate $\int_C \mathbf{F} \cdot d\mathbf{r}$, where C is a path between two points A and B in R?

7. If $\mathbf{F}$ is a conservative vector field on a region R, what is the value of $\oint_C \mathbf{F} \cdot d\mathbf{r}$, where C is a simple closed piecewise-smooth oriented curve in R?

8. Give three equivalent properties of conservative vector fields.

Practice Exercises

9–16. Testing for conservative vector fields *Determine whether the following vector fields are conservative (in $\mathbb{R}^2$ or $\mathbb{R}^3$).*

9. $\mathbf{F} = \langle 1, 1 \rangle$

10. $\mathbf{F} = \langle x, y \rangle$

11. $\mathbf{F} = \langle -y, x \rangle$

12. $\mathbf{F} = \langle -y, x + y \rangle$

13. $\mathbf{F} = \langle e^{-x}\cos y, e^{-x}\sin y \rangle$

14. $\mathbf{F} = \langle 2x^3 + xy^2, 2y^3 - x^2y \rangle$

15. $\mathbf{F} = \langle yz\cos xz, \sin xz, xy\cos xz \rangle$

16. $\mathbf{F} = \langle ye^{x-z}, e^{x-z}, ye^{x-z} \rangle$

17–30. Finding potential functions *Determine whether the following vector fields are conservative on the specified region. If so, determine a potential function. Let R^* and D^* be open regions of $\mathbb{R}^2$ and $\mathbb{R}^3$, respectively, that do not include the origin.*

17. $\mathbf{F} = \langle x, y \rangle$ on $\mathbb{R}^2$

18. $\mathbf{F} = \langle -y, -x \rangle$ on $\mathbb{R}^2$

19. $\mathbf{F} = \left\langle x^3 - xy, \dfrac{x^2}{2} + y \right\rangle$ on $\mathbb{R}^2$

20. $\mathbf{F} = \dfrac{\langle x, y \rangle}{x^2 + y^2}$ on R^*

21. $\mathbf{F} = \dfrac{\langle x, y \rangle}{\sqrt{x^2 + y^2}}$ on R^*

22. $\mathbf{F} = \langle y, x, x - y \rangle$ on $\mathbb{R}^3$

23. $\mathbf{F} = \langle z, 1, x \rangle$ on $\mathbb{R}^3$

24. $\mathbf{F} = \langle yz, xz, xy \rangle$ on $\mathbb{R}^3$

25. $\mathbf{F} = \langle e^z, e^z, e^z(x - y) \rangle$ on $\mathbb{R}^3$

26. $\mathbf{F} = \langle 1, -z, y \rangle$ on $\mathbb{R}^3$

27. $\mathbf{F} = \langle y + z, x + z, x + y \rangle$ on $\mathbb{R}^3$

28. $\mathbf{F} = \dfrac{\langle x, y, z \rangle}{x^2 + y^2 + z^2}$ on D^*

29. $\mathbf{F} = \dfrac{\langle x, y, z \rangle}{\sqrt{x^2 + y^2 + z^2}}$ on D^*

30. $\mathbf{F} = \langle x^3, 2y, -z^3 \rangle$ on $\mathbb{R}^3$

31–34. Evaluating line integrals *Use the given potential function φ of the gradient field $\mathbf{F}$ and the curve C to evaluate the line integral $\int_C \mathbf{F} \cdot d\mathbf{r}$ in two ways.*

a. *Use a parametric description of C and evaluate the integral directly.*

b. *Use the Fundamental Theorem for line integrals.*

31. $\varphi(x, y) = xy$; $C: \mathbf{r}(t) = \langle \cos t, \sin t \rangle$, for $0 \leq t \leq \pi$

32. $\varphi(x, y) = x + 3y$; $C: \mathbf{r}(t) = \langle 2 - t, t \rangle$, for $0 \leq t \leq 2$

33. $\varphi(x, y, z) = (x^2 + y^2 + z^2)/2$; $C: \mathbf{r}(t) = \langle \cos t, \sin t, t/\pi \rangle$, for $0 \leq t \leq 2\pi$

34. $\varphi(x, y, z) = xy + xz + yz$; $C: \mathbf{r}(t) = \langle t, 2t, 3t \rangle$, for $0 \leq t \leq 4$

35–38. Applying the Fundamental Theorem of Line Integrals *Suppose the vector field $\mathbf{F}$ is continuous on $\mathbb{R}^2$, $\mathbf{F} = \langle f, g \rangle = \nabla\varphi$, $\varphi(1, 2) = 7$, $\varphi(3, 6) = 10$, and $\varphi(6, 4) = 20$. Evaluate the following integrals for the given curve C, if possible.*

35. $\displaystyle\int_C \mathbf{F} \cdot d\mathbf{r}$; $C: \mathbf{r}(t) = \langle 2t - 1, t^2 + t \rangle$, for $1 \leq t \leq 2$

36. $\displaystyle\int_C \mathbf{F} \cdot \mathbf{T}\, ds$; C is a smooth curve from $(1, 2)$ to $(6, 4)$.

37. $\displaystyle\int_C f\, dx + g\, dy$; C is the path consisting of the line segment from $A(6, 4)$ to $B(1, 2)$ followed by the line segment from $B(1, 2)$ to $C(3, 6)$.

38. $\displaystyle\oint_C \mathbf{F} \cdot d\mathbf{r}$; C is a circle, oriented clockwise, starting and ending at the point $A(6, 4)$.

39–44. Using the Fundamental Theorem for line integrals *Verify that the Fundamental Theorem for line integrals can be used to evaluate the given integral, and then evaluate the integral.*

39. $\int_C \langle 2x, 2y \rangle \cdot d\mathbf{r}$, where C is a smooth curve from $(0, 1)$ to $(3, 4)$

40. $\int_C \langle 1, 1, 1 \rangle \cdot d\mathbf{r}$, where C is a smooth curve from $(1, -1, 2)$ to $(3, 0, 7)$

41. $\int_C \nabla(e^{-x} \cos y) \cdot d\mathbf{r}$, where C is the line segment from $(0, 0)$ to $(\ln 2, 2\pi)$

42. $\int_C \nabla(1 + x^2 yz) \cdot d\mathbf{r}$, where C is the helix $\mathbf{r}(t) = \langle \cos 2t, \sin 2t, t \rangle$, for $0 \le t \le 4\pi$

43. $\int_C \cos(2y - z)\, dx - 2x \sin(2y - z)\, dy + x \sin(2y - z)\, dz$, where C is the curve $\mathbf{r}(t) = \langle t^2, t, t \rangle$, for $0 \le t \le \pi$

44. $\int_C e^x y\, dx + e^x\, dy$, where C is the parabola $\mathbf{r}(t) = \langle t + 1, t^2 \rangle$, for $-1 \le t \le 3$

45–50. Line integrals of vector fields on closed curves *Evaluate $\oint_C \mathbf{F} \cdot d\mathbf{r}$ for the following vector fields and closed oriented curves C by parameterizing C. If the integral is not zero, give an explanation.*

45. $\mathbf{F} = \langle x, y \rangle$; C is the circle of radius 4 centered at the origin oriented counterclockwise.

46. $\mathbf{F} = \langle y, x \rangle$; C is the circle of radius 8 centered at the origin oriented counterclockwise.

47. $\mathbf{F} = \langle x, y \rangle$; C is the triangle with vertices $(0, \pm 1)$ and $(1, 0)$ oriented counterclockwise.

48. $\mathbf{F} = \langle y, -x \rangle$; C is the circle of radius 3 centered at the origin oriented counterclockwise.

49. $\mathbf{F} = \langle x, y, z \rangle$; C: $\mathbf{r}(t) = \langle \cos t, \sin t, 2 \rangle$, for $0 \le t \le 2\pi$

50. $\mathbf{F} = \langle y - z, z - x, x - y \rangle$; C: $\mathbf{r}(t) = \langle \cos t, \sin t, \cos t \rangle$, for $0 \le t \le 2\pi$

51–52. Evaluating line integrals using level curves *Suppose the vector field F, whose potential function is φ, is continuous on $\mathbb{R}^2$. Use the curves C_1 and C_2 and level curves of φ (see figure) to evaluate the following line integrals.*

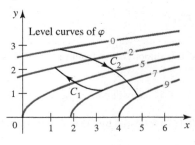

51. $\int_{C_1} \mathbf{F} \cdot d\mathbf{r}$ **52.** $\int_{C_2} \mathbf{F} \cdot d\mathbf{r}$

53–56. Line integrals *Evaluate the following line integrals using a method of your choice.*

53. $\oint_C \mathbf{F} \cdot d\mathbf{r}$, where $\mathbf{F} = \langle 2xy + z^2, x^2, 2xz \rangle$ and C is the circle $\mathbf{r}(t) = \langle 3 \cos t, 4 \cos t, 5 \sin t \rangle$, for $0 \le t \le 2\pi$

54. $\oint_C e^{-x} (\cos y\, dx + \sin y\, dy)$, where C is the square with vertices $(\pm 1, \pm 1)$ oriented counterclockwise

55. $\int_C \nabla(\sin xy) \cdot d\mathbf{r}$, where C is the line segment from $(0, 0)$ to $(2, \pi/4)$

56. $\int_C x^3\, dx + y^3\, dy$, where C is the curve $\mathbf{r}(t) = \langle 1 + \sin t, \cos^2 t \rangle$, for $0 \le t \le \pi/2$

57. **Explain why or why not** Determine whether the following statements are true and give an explanation or counterexample.

   a. If $\mathbf{F} = \langle -y, x \rangle$ and C is the circle of radius 4 centered at $(1, 0)$ oriented counterclockwise, then $\oint_C \mathbf{F} \cdot d\mathbf{r} = 0$.

   b. If $\mathbf{F} = \langle x, -y \rangle$ and C is the circle of radius 4 centered at $(1, 0)$ oriented counterclockwise, then $\oint_C \mathbf{F} \cdot d\mathbf{r} = 0$.

   c. A constant vector field is conservative on $\mathbb{R}^2$.

   d. The vector field $\mathbf{F} = \langle f(x), g(y) \rangle$ is conservative on $\mathbb{R}^2$ (assume f and g are defined for all real numbers).

   e. Gradient fields are conservative.

58. **Closed-curve integrals** Evaluate $\oint_C ds$, $\oint_C dx$, and $\oint_C dy$, where C is the unit circle oriented counterclockwise.

59–62. Work in force fields *Find the work required to move an object in the following force fields along a line segment between the given points. Check to see whether the force is conservative.*

59. $\mathbf{F} = \langle x, 2 \rangle$ from $A(0, 0)$ to $B(2, 4)$

60. $\mathbf{F} = \langle x, y \rangle$ from $A(1, 1)$ to $B(3, -6)$

61. $\mathbf{F} = \langle x, y, z \rangle$ from $A(1, 2, 1)$ to $B(2, 4, 6)$

62. $\mathbf{F} = e^{x+y} \langle 1, 1, z \rangle$ from $A(0, 0, 0)$ to $B(-1, 2, -4)$

63. Suppose C is a circle centered at the origin in a vector field $\mathbf{F}$ (see figure).

   a. If C is oriented counterclockwise, is $\oint_C \mathbf{F} \cdot d\mathbf{r}$ positive, negative, or zero?

   b. If C is oriented clockwise, is $\oint_C \mathbf{F} \cdot d\mathbf{r}$ positive, negative, or zero?

   c. Is $\mathbf{F}$ conservative in $\mathbb{R}^2$? Explain.

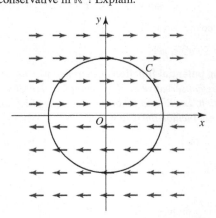

64. A vector field that is continuous in $\mathbb{R}^2$ is given (see figure). Is it conservative?

65. Work by a constant force Evaluate a line integral to show that the work done in moving an object from point A to point B in the presence of a constant force $\mathbf{F} = \langle a, b, c \rangle$ is $\mathbf{F} \cdot \overrightarrow{AB}$.

Explorations and Challenges

66. Conservation of energy Suppose an object with mass m moves in a region R in a conservative force field given by $\mathbf{F} = -\nabla\varphi$, where φ is a potential function in a region R. The motion of the object is governed by Newton's Second Law of Motion, $\mathbf{F} = m\mathbf{a}$, where $\mathbf{a}$ is the acceleration. Suppose the object moves from point A to point B in R.

a. Show that the equation of motion is $m \dfrac{d\mathbf{v}}{dt} = -\nabla\varphi$.

b. Show that $\dfrac{d\mathbf{v}}{dt} \cdot \mathbf{v} = \dfrac{1}{2} \dfrac{d}{dt} (\mathbf{v} \cdot \mathbf{v})$.

c. Take the dot product of both sides of the equation in part (a) with $\mathbf{v}(t) = \mathbf{r}'(t)$ and integrate along a curve between A and B. Use part (b) and the fact that $\mathbf{F}$ is conservative to show that the total energy (kinetic plus potential) $\dfrac{1}{2} m|\mathbf{v}|^2 + \varphi$ is the same at A and B. Conclude that because A and B are arbitrary, energy is conserved in R.

67. Gravitational potential The gravitational force between two point masses M and m is

$$\mathbf{F} = GMm \frac{\mathbf{r}}{|\mathbf{r}|^3} = GMm \frac{\langle x, y, z \rangle}{(x^2 + y^2 + z^2)^{3/2}},$$

where G is the gravitational constant.

a. Verify that this force field is conservative on any region excluding the origin.
b. Find a potential function φ for this force field such that $\mathbf{F} = -\nabla\varphi$.
c. Suppose the object with mass m is moved from a point A to a point B, where A is a distance r_1 from M, and B is a distance r_2 from M. Show that the work done in moving the object is

$$GMm\left(\frac{1}{r_2} - \frac{1}{r_1}\right).$$

d. Does the work depend on the path between A and B? Explain.

68. Radial fields in $\mathbb{R}^3$ are conservative Prove that the radial field $\mathbf{F} = \dfrac{\mathbf{r}}{|\mathbf{r}|^p}$, where $\mathbf{r} = \langle x, y, z \rangle$ and p is a real number, is conservative on any region not containing the origin. For what values of p is $\mathbf{F}$ conservative on a region that contains the origin?

69. Rotation fields are usually not conservative

a. Prove that the rotation field $\mathbf{F} = \dfrac{\langle -y, x \rangle}{|\mathbf{r}|^p}$, where $\mathbf{r} = \langle x, y \rangle$, is not conservative for $p \neq 2$.
b. For $p = 2$, show that $\mathbf{F}$ is conservative on any region not containing the origin.
c. Find a potential function for $\mathbf{F}$ when $p = 2$.

70. Linear and quadratic vector fields

a. For what values of a, b, c, and d is the field $\mathbf{F} = \langle ax + by, cx + dy \rangle$ conservative?
b. For what values of a, b, and c is the field $\mathbf{F} = \langle ax^2 - by^2, cxy \rangle$ conservative?

71. Alternative construction of potential functions in $\mathbb{R}^2$ Assume the vector field $\mathbf{F}$ is conservative on $\mathbb{R}^2$, so that the line integral $\int_C \mathbf{F} \cdot d\mathbf{r}$ is independent of path. Use the following procedure to construct a potential function φ for the vector field $\mathbf{F} = \langle f, g \rangle = \langle 2x - y, -x + 2y \rangle$.

a. Let A be $(0, 0)$ and let B be an arbitrary point (x, y). Define $\varphi(x, y)$ to be the work required to move an object from A to B, where $\varphi(A) = 0$. Let C_1 be the path from A to $(x, 0)$ to B, and let C_2 be the path from A to $(0, y)$ to B. Draw a picture.
b. Evaluate $\int_{C_1} \mathbf{F} \cdot d\mathbf{r} = \int_{C_1} f \, dx + g \, dy$ and conclude that $\varphi(x, y) = x^2 - xy + y^2$.
c. Verify that the same potential function is obtained by evaluating the line integral over C_2.

72–75. Alternative construction of potential functions *Use the procedure in Exercise 71 to construct potential functions for the following fields.*

72. $\mathbf{F} = \langle -y, -x \rangle$

73. $\mathbf{F} = \langle x, y \rangle$

74. $\mathbf{F} = \dfrac{\mathbf{r}}{|\mathbf{r}|}$, where $\mathbf{r} = \langle x, y \rangle$

75. $\mathbf{F} = \langle 2x^3 + xy^2, 2y^3 + x^2y \rangle$

QUICK CHECK ANSWERS

1. A figure-8 is closed but not simple; a torus is connected but not simply connected. **2.** The vector field is obtained by differentiating the potential function. So additive constants in the potential give the same vector field: $\nabla(\varphi + C) = \nabla\varphi$, where C is a constant. **3.** Show that $\nabla(e^x \cos y) = \langle e^x \cos y, -e^x \sin y \rangle$, which is the original vector field. A similar calculation may be done for part (b). **4.** The vector field $\nabla(xy + xz - yz)$ is the gradient of $xy + xz - yz$, so the vector field is conservative. ◄

17.4 Green's Theorem

The preceding section gave a version of the Fundamental Theorem of Calculus that applies to line integrals. In this and the remaining sections of the text, you will see additional extensions of the Fundamental Theorem that apply to regions in $\mathbb{R}^2$ and $\mathbb{R}^3$. All these fundamental theorems share a common feature. Part 2 of the Fundamental Theorem of Calculus (Chapter 5) says

$$\int_a^b \frac{df}{dx}\,dx = f(b) - f(a),$$

which relates the integral of $\dfrac{df}{dx}$ on an interval $[a, b]$ to the values of f on the boundary of $[a, b]$. The Fundamental Theorem for line integrals says

$$\int_C \nabla\varphi \cdot d\mathbf{r} = \varphi(B) - \varphi(A),$$

which relates the integral of $\nabla\varphi$ on a piecewise-smooth oriented curve C to the boundary values of φ. (The boundary consists of the two endpoints A and B.)

The subject of this section is Green's Theorem, which is another step in this progression. It relates the double integral of derivatives of a function over a region in $\mathbb{R}^2$ to function values on the boundary of that region.

Circulation Form of Green's Theorem

Throughout this section, unless otherwise stated, we assume curves in the plane are simple closed piecewise-smooth oriented curves. By a result called the *Jordan Curve Theorem*, such curves have a well-defined interior such that when the curve is traversed in the counterclockwise direction (viewed from above), the interior is on the left. With this orientation, there is a unique outward unit normal vector that points to the right (at points where the curve is smooth). We also assume curves in the plane lie in regions that are both connected and simply connected.

Suppose the vector field $\mathbf{F}$ is defined on a region R whose boundary is the closed curve C. As we have seen, the circulation $\oint_C \mathbf{F} \cdot d\mathbf{r}$ (Section 17.2) measures the net component of $\mathbf{F}$ in the direction tangent to C. It is easiest to visualize the circulation when $\mathbf{F}$ represents the velocity of a fluid moving in two dimensions. For example, let C be the unit circle with a counterclockwise orientation. The vector field $\mathbf{F} = \langle -y, x \rangle$ has a positive circulation of 2π on C (Section 17.2) because the vector field is everywhere tangent to C (**Figure 17.31**). A nonzero circulation on a closed curve says that the vector field must have some property *inside* the curve that produces the circulation. You can think of this property as a *net rotation*.

To visualize the rotation of a vector field, imagine a small paddle wheel, fixed at a point in the vector field, with its axis perpendicular to the xy-plane (Figure 17.31). The strength of the rotation at that point is seen in the speed at which the paddle wheel spins, and the direction of the rotation is the direction in which the paddle wheel spins. At a different point in the vector field, the paddle wheel will, in general, have a different speed and direction of rotation.

The first form of Green's Theorem relates the circulation on C to the double integral, over the region R, of a quantity that measures rotation at each point of R.

Paddle wheel at one point of vector field.

$\mathbf{F} = \langle -y, x \rangle$ has positive (counterclockwise) circulation on C.

Figure 17.31

> The circulation form of Green's Theorem is also called the *tangential*, or *curl*, form.

THEOREM 17.7   Green's Theorem—Circulation Form

Let C be a simple closed piecewise-smooth curve, oriented counterclockwise, that encloses a connected and simply connected region R in the plane. Assume $\mathbf{F} = \langle f, g \rangle$, where f and g have continuous first partial derivatives in R. Then

$$\underbrace{\oint_C \mathbf{F} \cdot d\mathbf{r}}_{\text{circulation}} = \underbrace{\oint_C f\,dx + g\,dy}_{\text{circulation}} = \iint_R \left(\frac{\partial g}{\partial x} - \frac{\partial f}{\partial y}\right) dA.$$

The proof of a special case of the theorem is given at the end of this section. Notice that the two line integrals on the left side of Green's Theorem give the circulation of the vector field on C. The double integral on the right side involves the quantity $\frac{\partial g}{\partial x} - \frac{\partial f}{\partial y}$, which describes the rotation of the vector field *within* C that produces the circulation *on* C. This quantity is called the *two-dimensional curl* of the vector field.

Figure 17.32 illustrates how the curl measures the rotation of a particular vector field at a point P. If the horizontal component of the field decreases in the y-direction at P ($f_y < 0$) and the vertical component increases in the x-direction at P ($g_x > 0$), then $\frac{\partial g}{\partial x} - \frac{\partial f}{\partial y} > 0$, and the field has a counterclockwise rotation at P. The double integral in Green's Theorem computes the net rotation of the field throughout R. The theorem says that the net rotation throughout R equals the circulation on the boundary of R.

QUICK CHECK 1 Compute $\dfrac{\partial g}{\partial x} - \dfrac{\partial f}{\partial y}$ for the radial vector field $\mathbf{F} = \langle x, y \rangle$. What does this tell you about the circulation on a simple closed curve? ◄

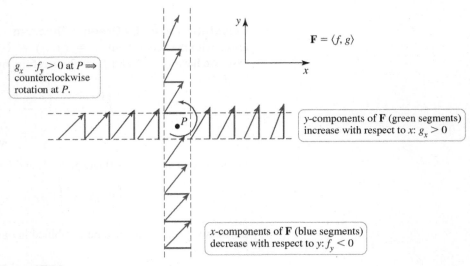

$g_x - f_y > 0$ at $P \Rightarrow$ counterclockwise rotation at P.

$\mathbf{F} = \langle f, g \rangle$

y-components of $\mathbf{F}$ (green segments) increase with respect to x: $g_x > 0$

x-components of $\mathbf{F}$ (blue segments) decrease with respect to y: $f_y < 0$

Figure 17.32

Green's Theorem has an important consequence when applied to a conservative vector field. Recall from Theorem 17.3 that if $\mathbf{F} = \langle f, g \rangle$ is conservative, then its components satisfy the condition $f_y = g_x$. If R is a region of $\mathbb{R}^2$ on which the conditions of Green's Theorem are satisfied, then for a conservative field, we have

$$\oint_C \mathbf{F} \cdot d\mathbf{r} = \iint_R \underbrace{\left(\frac{\partial g}{\partial x} - \frac{\partial f}{\partial y} \right)}_{0} dA = 0.$$

> In some cases, the rotation of a vector field is not obvious. For example, the parallel flow in a channel $\mathbf{F} = \langle 0, 1 - x^2 \rangle$, for $|x| \leq 1$, has a nonzero curl for $x \neq 0$. See Exercise 72.

Green's Theorem confirms the fact (Theorem 17.6) that if $\mathbf{F}$ is a conservative vector field in a region, then the circulation $\oint_C \mathbf{F} \cdot d\mathbf{r}$ is zero on any simple closed curve in the region. A two-dimensional vector field $\mathbf{F} = \langle f, g \rangle$ for which $\frac{\partial g}{\partial x} - \frac{\partial f}{\partial y} = 0$ at all points of a region is said to be *irrotational*, because it produces zero circulation on closed curves in the region. Irrotational vector fields on simply connected regions in $\mathbb{R}^2$ are conservative.

DEFINITION Two-Dimensional Curl

The **two-dimensional curl** of the vector field $\mathbf{F} = \langle f, g \rangle$ is $\frac{\partial g}{\partial x} - \frac{\partial f}{\partial y}$. If the curl is zero throughout a region, the vector field is **irrotational** on that region.

Evaluating circulation integrals of conservative vector fields on closed curves is easy. The integral is always zero. Green's Theorem provides a way to evaluate circulation integrals for nonconservative vector fields.

EXAMPLE 1   Circulation of a rotation field Consider the rotation vector field $\mathbf{F} = \langle -y, x \rangle$ on the unit disk $R = \{(x, y): x^2 + y^2 \leq 1\}$ (Figure 17.31). In Example 6 of Section 17.2, we showed that $\oint_C \mathbf{F} \cdot d\mathbf{r} = 2\pi$, where C is the boundary of R oriented counterclockwise. Confirm this result using Green's Theorem.

SOLUTION Note that $f(x, y) = -y$ and $g(x, y) = x$; therefore, the curl of $\mathbf{F}$ is $\dfrac{\partial g}{\partial x} - \dfrac{\partial f}{\partial y} = 2$. By Green's Theorem,

$$\oint_C \mathbf{F} \cdot d\mathbf{r} = \iint_R \underbrace{\left(\frac{\partial g}{\partial x} - \frac{\partial f}{\partial y} \right)}_{2} dA = \iint_R 2 \, dA = 2 \times \text{area of } R = 2\pi.$$

The curl $\dfrac{\partial g}{\partial x} - \dfrac{\partial f}{\partial y}$ is nonzero on R, which results in a nonzero circulation on the boundary of R.

Related Exercises 18, 20 ◄

Calculating Area by Green's Theorem A useful consequence of Green's Theorem arises with the vector fields $\mathbf{F} = \langle f, g \rangle = \langle 0, x \rangle$ and $\mathbf{F} = \langle f, g \rangle = \langle y, 0 \rangle$. In the first case, we have $g_x = 1$ and $f_y = 0$; therefore, by Green's Theorem,

$$\oint_C \mathbf{F} \cdot d\mathbf{r} = \oint_C x \, dy = \iint_R dA = \text{area of } R.$$

$$\underset{\mathbf{F} \cdot d\mathbf{r}}{} \qquad \underset{\frac{\partial g}{\partial x} - \frac{\partial f}{\partial y} = 1}{}$$

In the second case, $g_x = 0$ and $f_y = 1$, and Green's Theorem says

$$\oint_C \mathbf{F} \cdot d\mathbf{r} = \oint_C y \, dx = -\iint_R dA = -\text{area of } R.$$

These two results may also be combined in one statement to give the following theorem.

THEOREM 17.8   Area of a Plane Region by Line Integrals
Under the conditions of Green's Theorem, the area of a region R enclosed by a curve C is

$$\oint_C x \, dy = -\oint_C y \, dx = \frac{1}{2} \oint_C (x \, dy - y \, dx).$$

A remarkably simple calculation of the area of an ellipse follows from this result.

EXAMPLE 2   Area of an ellipse Find the area of the ellipse $\dfrac{x^2}{a^2} + \dfrac{y^2}{b^2} = 1$.

SOLUTION An ellipse with counterclockwise orientation is described parametrically by $\mathbf{r}(t) = \langle x, y \rangle = \langle a \cos t, b \sin t \rangle$, for $0 \leq t \leq 2\pi$. Noting that $dx = -a \sin t \, dt$ and $dy = b \cos t \, dt$, we have

$$\begin{aligned} x \, dy - y \, dx &= (a \cos t)(b \cos t) \, dt - (b \sin t)(-a \sin t) \, dt \\ &= ab \left(\cos^2 t + \sin^2 t \right) dt \\ &= ab \, dt. \end{aligned}$$

Expressing the line integral as an ordinary integral with respect to t, the area of the ellipse is

$$\frac{1}{2} \oint_C \underbrace{(x \, dy - y \, dx)}_{ab \, dt} = \frac{ab}{2} \int_0^{2\pi} dt = \pi ab.$$

Related Exercises 22–23 ◄

Flux Form of Green's Theorem

Let C be a closed curve enclosing a region R in $\mathbb{R}^2$ and let $\mathbf{F}$ be a vector field defined on R. We assume C and R have the previously stated properties; specifically, C is oriented counterclockwise with an outward normal vector $\mathbf{n}$. Recall that the outward flux of $\mathbf{F}$ across C is $\oint_C \mathbf{F} \cdot \mathbf{n} \, ds$ (Section 17.2). The second form of Green's Theorem relates the flux across C to a property of the vector field within R that produces the flux.

> ▶ The flux form of Green's Theorem is also called the *normal*, or *divergence*, form.

> ▶ The two forms of Green's Theorem are related in the following way: Applying the circulation form of the theorem to $\mathbf{F} = \langle -g, f \rangle$ results in the flux form, and applying the flux form of the theorem to $\mathbf{F} = \langle g, -f \rangle$ results in the circulation form.

THEOREM 17.9　Green's Theorem—Flux Form

Let C be a simple closed piecewise-smooth curve, oriented counterclockwise, that encloses a connected and simply connected region R in the plane. Assume $\mathbf{F} = \langle f, g \rangle$, where f and g have continuous first partial derivatives in R. Then

$$\underbrace{\oint_C \mathbf{F} \cdot \mathbf{n} \, ds}_{\text{outward flux}} = \underbrace{\oint_C f \, dy - g \, dx}_{\text{outward flux}} = \iint_R \left(\frac{\partial f}{\partial x} + \frac{\partial g}{\partial y} \right) dA,$$

where $\mathbf{n}$ is the outward unit normal vector on the curve.

The two line integrals on the left side of Theorem 17.9 give the outward flux of the vector field across C. The double integral on the right side involves the quantity $\dfrac{\partial f}{\partial x} + \dfrac{\partial g}{\partial y}$, which is the property of the vector field that produces the flux across C. This quantity is called the *two-dimensional divergence*.

Figure 17.33 illustrates how the divergence measures the flux of a particular vector field at a point P. If $f_x > 0$ at P, it indicates an expansion of the vector field in the x-direction (if f_x is negative, it indicates a contraction). Similarly, if $g_y > 0$ at P, it indicates an expansion of the vector field in the y-direction. The combined effect of $f_x + g_y > 0$ at a point is a net outward flux across a small circle enclosing P.

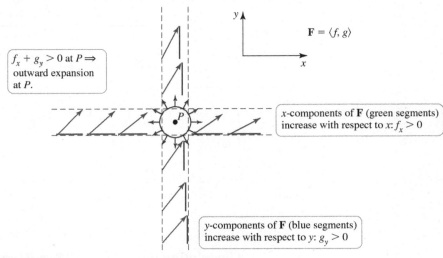

Figure 17.33

If the divergence of $\mathbf{F}$ is zero throughout a region on which $\mathbf{F}$ satisfies the conditions of Theorem 17.9, then the outward flux across the boundary is zero. Vector fields with a zero divergence are said to be *source free*. If the divergence is positive throughout R, the outward flux across C is positive, meaning that the vector field acts as a *source* in R. If the divergence is negative throughout R, the outward flux across C is negative, meaning that the vector field acts as a *sink* in R.

> **QUICK CHECK 2** Compute $\dfrac{\partial f}{\partial x} + \dfrac{\partial g}{\partial y}$ for the rotation field $\mathbf{F} = \langle -y, x \rangle$. What does this tell you about the outward flux of $\mathbf{F}$ across a simple closed curve? ◀

DEFINITION　Two-Dimensional Divergence

The **two-dimensional divergence** of the vector field $\mathbf{F} = \langle f, g \rangle$ is $\dfrac{\partial f}{\partial x} + \dfrac{\partial g}{\partial y}$. If the divergence is zero throughout a region, the vector field is **source free** on that region.

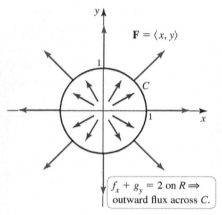

$\mathbf{F} = \langle x, y \rangle$

C

$f_x + g_y = 2$ on $R \Rightarrow$
outward flux across C.

Figure 17.34

EXAMPLE 3   Outward flux of a radial field Use Green's Theorem to compute the outward flux of the radial field $\mathbf{F} = \langle x, y \rangle$ across the unit circle $C = \{(x, y): x^2 + y^2 = 1\}$ (**Figure 17.34**). Interpret the result.

SOLUTION We have already calculated the outward flux of the radial field across C as a line integral and found it to be 2π (Example 8, Section 17.2). Computing the outward flux using Green's Theorem, note that $f(x, y) = x$ and $g(x, y) = y$; therefore, the divergence of $\mathbf{F}$ is $\dfrac{\partial f}{\partial x} + \dfrac{\partial g}{\partial y} = 2$. By Green's Theorem, we have

$$\oint_C \mathbf{F} \cdot \mathbf{n}\, ds = \iint_R \left(\frac{\partial f}{\partial x} + \frac{\partial g}{\partial y} \right) dA = \iint_R \underbrace{2}\, dA = 2 \times \text{area of } R = 2\pi.$$

The positive divergence on R results in an outward flux of the vector field across the boundary of R.

Related Exercise 27 ◀

As with the circulation form, the flux form of Green's Theorem can be used in either direction: to simplify line integrals or to simplify double integrals.

EXAMPLE 4   Line integral as a double integral Evaluate

$$\oint_C (4x^3 + \sin y^2)\, dy - (4y^3 + \cos x^2)\, dx,$$

where C is the boundary of the disk $R = \{(x, y): x^2 + y^2 \le 4\}$ oriented counterclockwise.

SOLUTION Letting $f(x, y) = 4x^3 + \sin y^2$ and $g(x, y) = 4y^3 + \cos x^2$, Green's Theorem takes the form

$$\oint_C \underbrace{(4x^3 + \sin y^2)}_{f}\, dy - \underbrace{(4y^3 + \cos x^2)}_{g}\, dx$$

$$= \iint_R \Big(\underbrace{12x^2}_{f_x} + \underbrace{12y^2}_{g_y}\Big)\, dA \qquad \text{Green's Theorem, flux form}$$

$$= 12 \int_0^{2\pi} \int_0^2 r^2 \underbrace{r\, dr\, d\theta}_{dA} \qquad \text{Polar coordinates; } x^2 + y^2 = r^2$$

$$= 12 \int_0^{2\pi} \frac{r^4}{4}\Big|_0^2\, d\theta \qquad \text{Evaluate inner integral.}$$

$$= 48 \int_0^{2\pi} d\theta = 96\pi. \qquad \text{Evaluate outer integral.}$$

Related Exercises 35–36 ◀

Circulation and Flux on More General Regions

Some ingenuity is required to extend both forms of Green's Theorem to more complicated regions. The next two examples illustrate Green's Theorem on two such regions: a half-annulus and a full annulus.

EXAMPLE 5   Circulation on a half-annulus Consider the vector field $\mathbf{F} = \langle y^2, x^2 \rangle$ on the half-annulus $R = \{(x, y): 1 \le x^2 + y^2 \le 9, y \ge 0\}$, whose boundary is C. Find the circulation on C, assuming it has the orientation shown in **Figure 17.35**.

SOLUTION The circulation on C is

$$\oint_C f\, dx + g\, dy = \oint_C y^2\, dx + x^2\, dy.$$

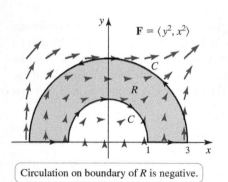

$\mathbf{F} = \langle y^2, x^2 \rangle$

C

R

C

Circulation on boundary of R is negative.

Figure 17.35

With the given orientation, the curve runs counterclockwise on the outer semicircle and clockwise on the inner semicircle. Identifying $f(x, y) = y^2$ and $g(x, y) = x^2$, the circulation form of Green's Theorem converts the line integral into a double integral. The double integral is most easily evaluated in polar coordinates using $x = r \cos \theta$ and $y = r \sin \theta$:

$$\oint_C \underbrace{y^2}_{f} \, dx + \underbrace{x^2}_{g} \, dy = \iint_R (\underbrace{2x}_{g_x} - \underbrace{2y}_{f_y}) \, dA \qquad \text{Green's Theorem, circulation form}$$

$$= 2 \int_0^\pi \int_1^3 (r \cos \theta - r \sin \theta) \underbrace{r \, dr \, d\theta}_{dA} \qquad \text{Convert to polar coordinates.}$$

$$= 2 \int_0^\pi (\cos \theta - \sin \theta) \frac{r^3}{3} \Big|_1^3 \, d\theta \qquad \text{Evaluate inner integral.}$$

$$= \frac{52}{3} \int_0^\pi (\cos \theta - \sin \theta) \, d\theta \qquad \text{Simplify.}$$

$$= -\frac{104}{3}. \qquad \text{Evaluate outer integral.}$$

The vector field (Figure 17.35) suggests why the circulation is negative. The field is roughly *opposed* to the direction of C on the outer semicircle but roughly aligned with the direction of C on the inner semicircle. Because the outer semicircle is longer and the field has greater magnitudes on the outer curve than on the inner curve, the greater contribution to the circulation is negative.

Related Exercise 41 ◄

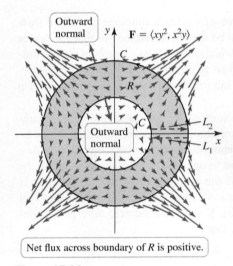

Net flux across boundary of R is positive.

Figure 17.36

➤ Another way to deal with the flux across the annulus is to apply Green's Theorem to the entire disk $|r| \le 2$ and compute the flux across the outer circle. Then apply Green's Theorem to the disk $|r| \le 1$ and compute the flux across the inner circle. Note that the flux *out of* the inner disk is a flux *into* the annulus. Therefore, the difference of the two fluxes gives the net flux for the annulus.

EXAMPLE 6 **Flux across the boundary of an annulus** Find the outward flux of the vector field $\mathbf{F} = \langle xy^2, x^2y \rangle$ across the boundary of the annulus $R = \{(x, y): 1 \le x^2 + y^2 \le 4\}$, which, when expressed in polar coordinates, is the set $\{(r, \theta): 1 \le r \le 2, 0 \le \theta \le 2\pi\}$ (**Figure 17.36**).

SOLUTION Because the annulus R is not simply connected, Green's Theorem does not apply as stated in Theorem 17.9. This difficulty is overcome by defining the curve C shown in Figure 17.36, which is simple, closed, and piecewise smooth. The connecting links L_1 and L_2 below and above the x-axis are traversed in opposite directions. Letting L_1 and L_2 approach the x-axis, the contributions to the line integral cancel on L_1 and L_2. Because of this cancellation, we take C to be the curve that runs counterclockwise on the outer boundary and clockwise on the inner boundary.

Using the flux form of Green's Theorem and converting to polar coordinates, we have

$$\oint_C \mathbf{F} \cdot \mathbf{n} \, ds = \oint_C f \, dy - g \, dx = \oint_C xy^2 \, dy - x^2y \, dx \qquad \text{Substitute for } f \text{ and } g.$$

$$= \iint_R (\underbrace{y^2}_{f_x} + \underbrace{x^2}_{g_y}) \, dA \qquad \text{Green's Theorem, flux form}$$

$$= \int_0^{2\pi} \int_1^2 (r^2) \, r \, dr \, d\theta \qquad \text{Polar coordinates; } x^2 + y^2 = r^2$$

$$= \int_0^{2\pi} \frac{r^4}{4} \Big|_1^2 \, d\theta \qquad \text{Evaluate inner integral.}$$

$$= \frac{15}{4} \int_0^{2\pi} d\theta \qquad \text{Simplify.}$$

$$= \frac{15\pi}{2}. \qquad \text{Evaluate outer integral.}$$

▶ Notice that the divergence of the vector field in Example 6 ($x^2 + y^2$) is positive on R, so we expect an outward flux across C.

▶ Potential function for $\mathbf{F} = \langle f, g \rangle$:
$$\varphi_x = f \quad \text{and} \quad \varphi_y = g$$
Stream function for $\mathbf{F} = \langle f, g \rangle$:
$$\psi_x = -g \quad \text{and} \quad \psi_y = f$$

QUICK CHECK 3 Show that $\psi = \frac{1}{2}(y^2 - x^2)$ is a stream function for the vector field $\mathbf{F} = \langle y, x \rangle$. Show that $\mathbf{F}$ has zero divergence. ◄

▶ In fluid dynamics, velocity fields that are both conservative and source free are called *ideal flows*. They model fluids that are irrotational and incompressible.

▶ Methods for finding solutions of Laplace's equation are discussed in advanced mathematics courses.

Figure 17.36 shows the vector field and explains why the flux across C is positive. Because the field increases in magnitude moving away from the origin, the outward flux across the outer boundary is greater than the inward flux across the inner boundary. Hence, the net outward flux across C is positive.

Related Exercise 42 ◄

Stream Functions

We can now see a wonderful parallel between circulation properties (and conservative vector fields) and flux properties (and source-free fields). We need one more piece to complete the picture; it is the *stream function*, which plays the same role for source-free fields that the potential function plays for conservative fields.

Consider a two-dimensional vector field $\mathbf{F} = \langle f, g \rangle$ that is differentiable on a region R. A **stream function** for the vector field—if it exists—is a function ψ (pronounced *psigh* or *psee*) that satisfies
$$\frac{\partial \psi}{\partial y} = f, \qquad \frac{\partial \psi}{\partial x} = -g.$$

If we compute the divergence of a vector field $\mathbf{F} = \langle f, g \rangle$ that has a stream function and use the fact that $\psi_{xy} = \psi_{yx}$, then
$$\frac{\partial f}{\partial x} + \frac{\partial g}{\partial y} = \underbrace{\frac{\partial}{\partial x}\left(\frac{\partial \psi}{\partial y}\right) + \frac{\partial}{\partial y}\left(-\frac{\partial \psi}{\partial x}\right)}_{\psi_{yx} = \psi_{xy}} = 0.$$

We see that the existence of a stream function guarantees that the vector field has zero divergence or, equivalently, is source free. The converse is also true on simply connected regions of $\mathbb{R}^2$.

As discussed in Section 17.1, the level curves of a stream function are called flow curves or streamlines—and for good reason. It can be shown (Exercise 70) that the vector field $\mathbf{F}$ is everywhere tangent to the streamlines, which means that a graph of the streamlines shows the flow of the vector field. Finally, just as circulation integrals of a conservative vector field are independent of path, flux integrals of a source-free field are also independent of path (Exercise 69).

Vector fields that are both conservative and source free are quite interesting mathematically because they have both a potential function and a stream function. It can be shown that the level curves of the potential and stream functions form orthogonal families; that is, at each point of intersection, the line tangent to one level curve is orthogonal to the line tangent to the other level curve (equivalently, the gradient vector of one function is orthogonal to the gradient vector of the other function). Such vector fields have zero curl ($g_x - f_y = 0$) and zero divergence ($f_x + g_y = 0$). If we write the zero divergence condition in terms of the potential function φ, we find that
$$0 = f_x + g_y = \varphi_{xx} + \varphi_{yy}.$$

Writing the zero curl condition in terms of the stream function ψ, we find that
$$0 = g_x - f_y = -\psi_{xx} - \psi_{yy}.$$

We see that the potential function and the stream function both satisfy an important equation known as **Laplace's equation**:
$$\varphi_{xx} + \varphi_{yy} = 0 \quad \text{and} \quad \psi_{xx} + \psi_{yy} = 0.$$

Any function satisfying Laplace's equation can be used as a potential function or stream function for a conservative, source-free vector field. These vector fields are used in fluid dynamics, electrostatics, and other modeling applications.

Table 17.1 shows the parallel properties of conservative and source-free vector fields in two dimensions. We assume C is a simple piecewise-smooth oriented curve and either is closed or has endpoints A and B.

Table 17.1

Conservative Fields $\mathbf{F} = \langle f, g \rangle$	Source-Free Fields $\mathbf{F} = \langle f, g \rangle$
• curl $= \dfrac{\partial g}{\partial x} - \dfrac{\partial f}{\partial y} = 0$	• divergence $= \dfrac{\partial f}{\partial x} + \dfrac{\partial g}{\partial y} = 0$
• Potential function φ with $\mathbf{F} = \nabla\varphi$ or $f = \dfrac{\partial\varphi}{\partial x}, \quad g = \dfrac{\partial\varphi}{\partial y}$	• Stream function ψ with $f = \dfrac{\partial\psi}{\partial y}, \quad g = -\dfrac{\partial\psi}{\partial x}$
• Circulation $= \oint_C \mathbf{F} \cdot d\mathbf{r} = 0$ on all closed curves C.	• Flux $= \oint_C \mathbf{F} \cdot \mathbf{n}\, ds = 0$ on all closed curves C.
• Evaluation of line integral $\displaystyle\int_C \mathbf{F} \cdot d\mathbf{r} = \varphi(B) - \varphi(A)$	• Evaluation of line integral $\displaystyle\int_C \mathbf{F} \cdot \mathbf{n}\, ds = \psi(B) - \psi(A)$

With Green's Theorem in the picture, we may also give a concise summary of the various cases that arise with line integrals of both the circulation and flux types (Table 17.2).

Table 17.2

Circulation/work integrals: $\displaystyle\int_C \mathbf{F} \cdot \mathbf{T}\, ds = \int_C \mathbf{F} \cdot d\mathbf{r} = \int_C f\, dx + g\, dy$

	C closed	C not closed
F conservative $(\mathbf{F} = \nabla\varphi)$	$\oint_C \mathbf{F} \cdot d\mathbf{r} = 0$	$\displaystyle\int_C \mathbf{F} \cdot d\mathbf{r} = \varphi(B) - \varphi(A)$
F not conservative	Green's Theorem $\oint_C \mathbf{F} \cdot d\mathbf{r} = \displaystyle\iint_R (g_x - f_y)\, dA$	Direct evaluation $\displaystyle\int_C \mathbf{F} \cdot d\mathbf{r} = \int_a^b (fx' + gy')\, dt$

Flux integrals: $\displaystyle\int_C \mathbf{F} \cdot \mathbf{n}\, ds = \int_C f\, dy - g\, dx$

	C closed	C not closed
F source free $(f = \psi_y, g = -\psi_x)$	$\oint_C \mathbf{F} \cdot \mathbf{n}\, ds = 0$	$\displaystyle\int_C \mathbf{F} \cdot \mathbf{n}\, ds = \psi(B) - \psi(A)$
F not source free	Green's Theorem $\oint_C \mathbf{F} \cdot \mathbf{n}\, ds = \displaystyle\iint_R (f_x + g_y)\, dA$	Direct evaluation $\displaystyle\int_C \mathbf{F} \cdot \mathbf{n}\, ds = \int_a^b (fy' - gx')\, dt$

Proof of Green's Theorem on Special Regions

The proof of Green's Theorem is straightforward when restricted to special regions. We consider regions R enclosed by a simple closed smooth curve C oriented in the counterclockwise direction, such that the region can be expressed in two ways (**Figure 17.37**):

> This restriction on R means that lines parallel to the coordinate axes intersect the boundary of R at most twice.

• $R = \{(x, y): a \le x \le b, G_1(x) \le y \le G_2(x)\}$ or

• $R = \{(x, y): H_1(y) \le x \le H_2(y), c \le y \le d\}$.

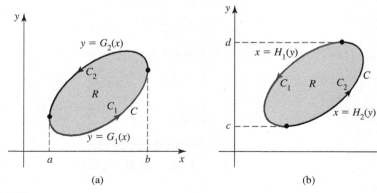

(a)   (b)

Figure 17.37

Under these conditions, we prove the circulation form of Green's Theorem:

$$\oint_C f\,dx + g\,dy = \iint_R \left(\frac{\partial g}{\partial x} - \frac{\partial f}{\partial y}\right) dA.$$

Beginning with the term $\displaystyle\iint_R \frac{\partial f}{\partial y}\,dA$, we write this double integral as an iterated integral, where $G_1(x) \le y \le G_2(x)$ in the inner integral and $a \le x \le b$ in the outer integral (**Figure 17.37a**). The upper curve is labeled C_2 and the lower curve is labeled C_1. Notice that the inner integral of $\dfrac{\partial f}{\partial y}$ with respect to y gives $f(x, y)$. Therefore, the first step of the double integration is

$$\iint_R \frac{\partial f}{\partial y}\,dA = \int_a^b \int_{G_1(x)}^{G_2(x)} \frac{\partial f}{\partial y}\,dy\,dx \qquad \text{Convert to an iterated integral.}$$

$$= \int_a^b \Big(\underbrace{f(x, G_2(x))}_{\text{on } C_2} - \underbrace{f(x, G_1(x))}_{\text{on } C_1}\Big)\,dx.$$

Over the interval $a \le x \le b$, the points $(x, G_2(x))$ trace out the upper part of C (labeled C_2) in the *negative* (clockwise) direction. Similarly, over the interval $a \le x \le b$, the points $(x, G_1(x))$ trace out the lower part of C (labeled C_1) in the *positive* (counterclockwise) direction.

Therefore,

$$\iint_R \frac{\partial f}{\partial y}\,dA = \int_a^b \big(f(x, G_2(x)) - f(x, G_1(x))\big)\,dx$$

$$= \int_{-C_2} f\,dx - \int_{C_1} f\,dx$$

$$= -\int_{C_2} f\,dx - \int_{C_1} f\,dx \qquad\qquad \int_{-C_2} f\,dx = -\int_{C_2} f\,dx$$

$$= -\oint_C f\,dx. \qquad\qquad \int_C f\,dx = \int_{C_1} f\,dx + \int_{C_2} f\,dx$$

A similar argument applies to the double integral of $\dfrac{\partial g}{\partial x}$, except we use the bounding curves $x = H_1(y)$ and $x = H_2(y)$, where C_1 is now the left curve and C_2 is the right curve (**Figure 17.37b**). We have

$$\iint\limits_R \frac{\partial g}{\partial x}\, dA = \int_c^d \int_{H_1(y)}^{H_2(y)} \frac{\partial g}{\partial x}\, dx\, dy \qquad \text{Convert to an iterated integral.}$$

$$= \int_c^d \Big(\underbrace{g(H_2(y), y)}_{C_2} - \underbrace{g(H_1(y), y)}_{-C_1} \Big)\, dy \qquad \int \frac{\partial g}{\partial x}\, dx = g$$

$$= \int_{C_2} g\, dy - \int_{-C_1} g\, dy$$

$$= \int_{C_2} g\, dy + \int_{C_1} g\, dy \qquad\qquad \int_{-C_1} g\, dy = -\int_{C_1} g\, dy$$

$$= \oint_C g\, dy. \qquad\qquad\qquad \int_C g\, dy = \int_{C_1} g\, dy + \int_{C_2} g\, dy$$

Combining these two calculations results in

QUICK CHECK 4 Explain why Green's Theorem proves that if $g_x = f_y$, then the vector field $\mathbf{F} = \langle f, g \rangle$ is conservative. ◀

$$\iint\limits_R \left(\frac{\partial g}{\partial x} - \frac{\partial f}{\partial y} \right) dA = \oint_C f\, dx + g\, dy.$$

As mentioned earlier, with a change of notation (replace g with f and f with $-g$), the flux form of Green's Theorem is obtained. This proof also completes the list of equivalent properties of conservative fields given in Section 17.3: From Green's Theorem, it follows that if $\dfrac{\partial g}{\partial x} = \dfrac{\partial f}{\partial y}$ on a simply connected region R, then the vector field $\mathbf{F} = \langle f, g \rangle$ is conservative on R.

SECTION 17.4 EXERCISES

Getting Started

1. Explain why the two forms of Green's Theorem are analogs of the Fundamental Theorem of Calculus.

2. Referring to both forms of Green's Theorem, match each idea in Column 1 to an idea in Column 2:

Line integral for flux	Double integral of the curl
Line integral for circulation	Double integral of the divergence

3. How do you use a line integral to compute the area of a plane region?

4. Why does a two-dimensional vector field with zero curl on a region have zero circulation on a closed curve that bounds the region?

5. Why does a two-dimensional vector field with zero divergence on a region have zero outward flux across a closed curve that bounds the region?

6. Sketch a two-dimensional vector field that has zero curl everywhere in the plane.

7. Sketch a two-dimensional vector field that has zero divergence everywhere in the plane.

8. Discuss one of the parallels between a conservative vector field and a source-free vector field.

9–14. *Assume C is a circle centered at the origin, oriented counterclockwise, that encloses disk R in the plane. Complete the following steps for each vector field* **F**.

a. *Calculate the two-dimensional curl of* **F**.
b. *Calculate the two-dimensional divergence of* **F**.
c. *Is* **F** *irrotational on R?*
d. *Is* **F** *source free on R?*

9. $\mathbf{F} = \langle x, y \rangle$
10. $\mathbf{F} = \langle y, -x \rangle$
11. $\mathbf{F} = \langle y, -3x \rangle$
12. $\mathbf{F} = \langle x^2 + 2xy, -2xy - y^2 \rangle$
13. $\mathbf{F} = \langle 4x^3 y, xy^2 + x^4 \rangle$
14. $\mathbf{F} = \langle 4x^3 + y, 12xy \rangle$

15. Suppose C is the boundary of region $R = \{(x, y): x^2 \le y \le x \le 1\}$, oriented counterclockwise (see figure); let $\mathbf{F} = \langle 1, x \rangle$.

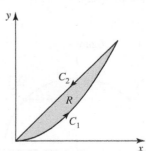

a. Compute the two-dimensional curl of **F** and determine whether **F** is irrotational.

b. Find parameterizations $\mathbf{r}_1(t)$ and $\mathbf{r}_2(t)$ for C_1 and C_2, respectively.

c. Evaluate both the line integral and the double integral in the circulation form of Green's Theorem and check for consistency.

d. Compute the two-dimensional divergence of **F** and use the flux form of Green's Theorem to explain why the outward flux is 0.

16. Suppose C is the boundary of region $R = \{(x, y): 2x^2 - 2x \le y \le 0\}$, oriented counterclockwise (see figure); let $\mathbf{F} = \langle x, 1 \rangle$.

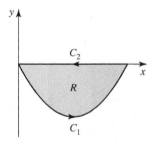

a. Compute the two-dimensional curl of **F** and use the circulation form of Green's Theorem to explain why the circulation is 0.

b. Compute the two-dimensional divergence of **F** and determine whether **F** is source free.

c. Find parameterizations $\mathbf{r}_1(t)$ and $\mathbf{r}_2(t)$ for C_1 and C_2, respectively.

d. Evaluate both the line integral and the double integral in the flux form of Green's Theorem and check for consistency.

Practice Exercises

17–20. Green's Theorem, circulation form *Consider the following regions R and vector fields* **F**.

a. *Compute the two-dimensional curl of the vector field.*

b. *Evaluate both integrals in Green's Theorem and check for consistency.*

17. $\mathbf{F} = \langle 2y, -2x \rangle$; R is the region bounded by $y = \sin x$ and $y = 0$, for $0 \le x \le \pi$.

18. $\mathbf{F} = \langle -3y, 3x \rangle$; R is the triangle with vertices $(0, 0)$, $(1, 0)$, and $(0, 2)$.

19. $\mathbf{F} = \langle -2xy, x^2 \rangle$; R is the region bounded by $y = x(2 - x)$ and $y = 0$.

20. $\mathbf{F} = \langle 0, x^2 + y^2 \rangle$; $R = \{(x, y): x^2 + y^2 \le 1\}$

21–26. Area of regions *Use a line integral on the boundary to find the area of the following regions.*

21. A disk of radius 5

22. A region bounded by an ellipse with major and minor axes of lengths 12 and 8, respectively

23. $\{(x, y): x^2 + y^2 \le 16\}$

24. The region shown in the figure

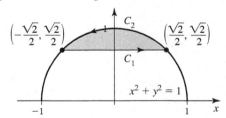

25. The region bounded by the parabolas $\mathbf{r}(t) = \langle t, 2t^2 \rangle$ and $\mathbf{r}(t) = \langle t, 12 - t^2 \rangle$, for $-2 \le t \le 2$

26. The region bounded by the curve $\mathbf{r}(t) = \langle t(1 - t^2), 1 - t^2 \rangle$, for $-1 \le t \le 1$ (*Hint:* Plot the curve.)

27–30. Green's Theorem, flux form *Consider the following regions R and vector fields* **F**.

a. *Compute the two-dimensional divergence of the vector field.*

b. *Evaluate both integrals in Green's Theorem and check for consistency.*

27. $\mathbf{F} = \langle x, y \rangle$; $R = \{(x, y): x^2 + y^2 \le 4\}$

28. $\mathbf{F} = \langle x, -3y \rangle$; R is the triangle with vertices $(0, 0)$, $(1, 2)$, and $(0, 2)$.

29. $\mathbf{F} = \langle 2xy, x^2 \rangle$; $R = \{(x, y): 0 \le y \le x(2 - x)\}$

30. $\mathbf{F} = \langle x^2 + y^2, 0 \rangle$; $R = \{(x, y): x^2 + y^2 \le 1\}$

31–40. Line integrals *Use Green's Theorem to evaluate the following line integrals. Assume all curves are oriented counterclockwise. A sketch is helpful.*

31. $\oint_C \langle 3y + 1, 4x^2 + 3 \rangle \cdot d\mathbf{r}$, where C is the boundary of the rectangle with vertices $(0, 0)$, $(4, 0)$, $(4, 2)$, and $(0, 2)$

32. $\oint_C \langle \sin y, x \rangle \cdot d\mathbf{r}$, where C is the boundary of the triangle with vertices $(0, 0)$, $\left(\frac{\pi}{2}, 0\right)$, and $\left(\frac{\pi}{2}, \frac{\pi}{2}\right)$

33. $\oint_C xe^y \, dx + x \, dy$, where C is the boundary of the region bounded by the curves $y = x^2$, $x = 2$, and the x-axis

34. $\oint_C \frac{1}{1 + y^2} \, dx + y \, dy$, where C is the boundary of the triangle with vertices $(0, 0)$, $(1, 0)$, and $(1, 1)$

35. $\oint_C (2x + e^{y^2}) \, dy - (4y^2 + e^{x^2}) \, dx$, where C is the boundary of the square with vertices $(0, 0)$, $(1, 0)$, $(1, 1)$, and $(0, 1)$

36. $\oint_C (2x - 3y) \, dy - (3x + 4y) \, dx$, where C is the unit circle

37. $\oint_C f \, dy - g \, dx$, where $\langle f, g \rangle = \langle 0, xy \rangle$ and C is the triangle with vertices $(0, 0)$, $(2, 0)$, and $(0, 4)$

38. $\oint_C f \, dy - g \, dx$, where $\langle f, g \rangle = \langle x^2, 2y^2 \rangle$ and C is the upper half of the unit circle and the line segment $-1 \le x \le 1$, oriented clockwise

39. The circulation line integral of $\mathbf{F} = \langle x^2 + y^2, 4x + y^3 \rangle$, where C is the boundary of $\{(x, y): 0 \le y \le \sin x, 0 \le x \le \pi\}$

40. The flux line integral of $\mathbf{F} = \langle e^{x-y}, e^{y-x} \rangle$, where C is the boundary of $\{(x, y): 0 \le y \le x, 0 \le x \le 1\}$

41–48. Circulation and flux *For the following vector fields, compute (a) the circulation on, and (b) the outward flux across, the boundary of the given region. Assume boundary curves are oriented counterclockwise.*

41. $\mathbf{F} = \langle x, y \rangle$; R is the half-annulus $\{(r, \theta); 1 \le r \le 2, 0 \le \theta \le \pi\}$.

42. $\mathbf{F} = \langle -y, x \rangle$; R is the annulus $\{(r, \theta): 1 \le r \le 3, 0 \le \theta \le 2\pi\}$.

43. $\mathbf{F} = \langle 2x + y, x - 4y \rangle$; R is the quarter-annulus $\{(r, \theta): 1 \le r \le 4, 0 \le \theta \le \pi/2\}$.

44. $\mathbf{F} = \langle x - y, -x + 2y \rangle$; R is the parallelogram $\{(x, y): 1 - x \le y \le 3 - x, 0 \le x \le 1\}$.

45. $\mathbf{F} = \nabla\left(\sqrt{x^2 + y^2}\right)$; R is the half-annulus $\{(r, \theta): 1 \le r \le 3, 0 \le \theta \le \pi\}$.

46. $\mathbf{F} = \left\langle \ln(x^2 + y^2), \tan^{-1}\dfrac{y}{x} \right\rangle$; R is the eighth-annulus $\{(r, \theta): 1 \le r \le 2, 0 \le \theta \le \pi/4\}$.

47. $\mathbf{F} = \langle x + y^2, x^2 - y \rangle$; $R = \{(x, y): y^2 \le x \le 2 - y^2\}$.

48. $\mathbf{F} = \langle y \cos x, -\sin x \rangle$; R is the square $\{(x, y): 0 \le x \le \pi/2, 0 \le y \le \pi/2\}$.

49. **Explain why or why not** Determine whether the following statements are true and give an explanation or counterexample.

   a. The work required to move an object around a closed curve C in the presence of a vector force field is the circulation of the force field on the curve.

   b. If a vector field has zero divergence throughout a region (on which the conditions of Green's Theorem are met), then the circulation on the boundary of that region is zero.

   c. If the two-dimensional curl of a vector field is positive throughout a region (on which the conditions of Green's Theorem are met), then the circulation on the boundary of that region is positive (assuming counterclockwise orientation).

50–51. Special line integrals *Prove the following identities, where C is a simple closed smooth oriented curve.*

50. $\oint_C dx = \oint_C dy = 0$

51. $\oint_C f(x)\, dx + g(y)\, dy = 0$, where f and g have continuous derivatives on the region enclosed by C

52. **Double integral to line integral** Use the flux form of Green's Theorem to evaluate $\iint_R (2xy + 4y^3)\, dA$, where R is the triangle with vertices $(0, 0)$, $(1, 0)$, and $(0, 1)$.

53. **Area line integral** Show that the value of

$$\oint_C xy^2\, dx + (x^2 y + 2x)\, dy$$

depends only on the area of the region enclosed by C.

54. **Area line integral** In terms of the parameters a and b, how is the value of $\oint_C ay\, dx + bx\, dy$ related to the area of the region enclosed by C, assuming counterclockwise orientation of C?

55–58. Stream function *Recall that if the vector field $\mathbf{F} = \langle f, g \rangle$ is source free (zero divergence), then a stream function ψ exists such that $f = \psi_y$ and $g = -\psi_x$*

a. *Verify that the given vector field has zero divergence.*

b. *Integrate the relations $f = \psi_y$ and $g = -\psi_x$ to find a stream function for the field.*

55. $\mathbf{F} = \langle 4, 2 \rangle$

56. $\mathbf{F} = \langle y^2, x^2 \rangle$

57. $\mathbf{F} = \langle -e^{-x}\sin y, e^{-x}\cos y \rangle$

58. $\mathbf{F} = \langle x^2, -2xy \rangle$

Explorations and Challenges

59–62. Ideal flow *A two-dimensional vector field describes **ideal flow** if it has both zero curl and zero divergence on a simply connected region.*

a. *Verify that both the curl and the divergence of the given field are zero.*

b. *Find a potential function φ and a stream function ψ for the field.*

c. *Verify that φ and ψ satisfy Laplace's equation*
$$\varphi_{xx} + \varphi_{yy} = \psi_{xx} + \psi_{yy} = 0.$$

59. $\mathbf{F} = \langle e^x\cos y, -e^x\sin y \rangle$

60. $\mathbf{F} = \langle x^3 - 3xy^2, y^3 - 3x^2y \rangle$

61. $\mathbf{F} = \left\langle \tan^{-1}\dfrac{y}{x}, \dfrac{1}{2}\ln(x^2 + y^2) \right\rangle$, for $x > 0$

62. $\mathbf{F} = \dfrac{\langle x, y \rangle}{x^2 + y^2}$, for $x > 0$, $y > 0$

63. **Flow in an ocean basin** An idealized two-dimensional ocean is modeled by the square region $R = \left[-\dfrac{\pi}{2}, \dfrac{\pi}{2}\right] \times \left[-\dfrac{\pi}{2}, \dfrac{\pi}{2}\right]$ with boundary C. Consider the stream function $\psi(x, y) = 4\cos x \cos y$ defined on R (see figure).

   a. The horizontal (east-west) component of the velocity is $u = \psi_y$ and the vertical (north-south) component of the velocity is $v = -\psi_x$. Sketch a few representative velocity vectors and show that the flow is counterclockwise around the region.

   b. Is the velocity field source free? Explain.

   c. Is the velocity field irrotational? Explain.

   d. Let C be the boundary of R. Find the total outward flux across C.

   e. Find the circulation on C assuming counterclockwise orientation.

64. **Green's Theorem as a Fundamental Theorem of Calculus** Show that if the circulation form of Green's Theorem is applied to the vector field $\left\langle 0, \dfrac{f(x)}{c} \right\rangle$, where $c > 0$ and $R = \{(x, y): a \le x \le b, 0 \le y \le c\}$, then the result is the Fundamental Theorem of Calculus,

$$\int_a^b \frac{df}{dx}\, dx = f(b) - f(a).$$

65. Green's Theorem as a Fundamental Theorem of Calculus Show that if the flux form of Green's Theorem is applied to the vector field $\left\langle \dfrac{f(x)}{c}, 0 \right\rangle$, where $c > 0$ and $R = \{(x, y): a \le x \le b, 0 \le y \le c\}$, then the result is the Fundamental Theorem of Calculus,

$$\int_a^b \frac{df}{dx}\, dx = f(b) - f(a).$$

66. What's wrong? Consider the rotation field $\mathbf{F} = \dfrac{\langle -y, x \rangle}{x^2 + y^2}$.

   a. Verify that the two-dimensional curl of $\mathbf{F}$ is zero, which suggests that the double integral in the circulation form of Green's Theorem is zero.

   b. Use a line integral to verify that the circulation on the unit circle of the vector field is 2π.

   c. Explain why the results of parts (a) and (b) do not agree.

67. What's wrong? Consider the radial field $\mathbf{F} = \dfrac{\langle x, y \rangle}{x^2 + y^2}$.

   a. Verify that the divergence of $\mathbf{F}$ is zero, which suggests that the double integral in the flux form of Green's Theorem is zero.

   b. Use a line integral to verify that the outward flux across the unit circle of the vector field is 2π.

   c. Explain why the results of parts (a) and (b) do not agree.

68. Conditions for Green's Theorem Consider the radial field $\mathbf{F} = \langle f, g \rangle = \dfrac{\langle x, y \rangle}{\sqrt{x^2 + y^2}} = \dfrac{\mathbf{r}}{|\mathbf{r}|}$.

   a. Explain why the conditions of Green's Theorem do not apply to $\mathbf{F}$ on a region that includes the origin.

   b. Let R be the unit disk centered at the origin and compute

$$\iint\limits_R \left(\frac{\partial f}{\partial x} + \frac{\partial g}{\partial y} \right) dA.$$

   c. Evaluate the line integral in the flux form of Green's Theorem on the boundary of R.

   d. Do the results of parts (b) and (c) agree? Explain.

69. Flux integrals Assume the vector field $\mathbf{F} = \langle f, g \rangle$ is source free (zero divergence) with stream function ψ. Let C be any smooth simple curve from A to the distinct point B. Show that the flux integral $\int_C \mathbf{F} \cdot \mathbf{n}\, ds$ is independent of path; that is, $\int_C \mathbf{F} \cdot \mathbf{n}\, ds = \psi(B) - \psi(A)$.

70. Streamlines are tangent to the vector field Assume the vector field $\mathbf{F} = \langle f, g \rangle$ is related to the stream function ψ by $\psi_y = f$ and $\psi_x = -g$ on a region R. Prove that at all points of R, the vector field is tangent to the streamlines (the level curves of the stream function).

71. Streamlines and equipotential lines Assume that on $\mathbb{R}^2$, the vector field $\mathbf{F} = \langle f, g \rangle$ has a potential function φ such that $f = \varphi_x$ and $g = \varphi_y$, and it has a stream function ψ such that $f = \psi_y$ and $g = -\psi_x$. Show that the equipotential curves (level curves of φ) and the streamlines (level curves of ψ) are everywhere orthogonal.

72. Channel flow The flow in a long shallow channel is modeled by the velocity field $\mathbf{F} = \langle 0, 1 - x^2 \rangle$, where $R = \{(x, y): |x| \le 1 \text{ and } |y| < 5\}$.

   a. Sketch R and several streamlines of $\mathbf{F}$.

   b. Evaluate the curl of $\mathbf{F}$ on the lines $x = 0$, $x = 1/4$, $x = 1/2$, and $x = 1$.

   c. Compute the circulation on the boundary of the region R.

   d. How do you explain the fact that the curl of $\mathbf{F}$ is nonzero at points of R, but the circulation is zero?

QUICK CHECK ANSWERS

1. $g_x - f_y = 0$, which implies zero circulation on a closed curve.   **2.** $f_x + g_y = 0$, which implies zero flux across a closed curve.   **3.** $\psi_y = y$ is the x-component of $\mathbf{F} = \langle y, x \rangle$, and $-\psi_x = x$ is the y-component of $\mathbf{F}$. Also, the divergence of $\mathbf{F}$ is $y_x + x_y = 0$.   **4.** If the curl is zero on a region, then all closed-path integrals are zero, which is a condition (Section 17.3) for a conservative field. ◀

17.5 Divergence and Curl

Green's Theorem sets the stage for the final act in our exploration of calculus. The last four sections of this chapter have the following goal: to lift both forms of Green's Theorem out of the plane ($\mathbb{R}^2$) and into space ($\mathbb{R}^3$). It is done as follows.

• The circulation form of Green's Theorem relates a line integral over a simple closed oriented curve in the plane to a double integral over the enclosed region. In an analogous manner, we will see that *Stokes' Theorem* (Section 17.7) relates a line integral over a simple closed oriented curve in $\mathbb{R}^3$ to a double integral over a surface whose boundary is that curve.

• The flux form of Green's Theorem relates a line integral over a simple closed oriented curve in the plane to a double integral over the enclosed region. Similarly, the *Divergence Theorem* (Section 17.8) relates an integral over a closed oriented surface in $\mathbb{R}^3$ to a triple integral over the region enclosed by that surface.

In order to make these extensions, we need a few more tools.

• The two-dimensional divergence and two-dimensional curl must be extended to three dimensions (this section).

• The idea of a *surface integral* must be introduced (Section 17.6).

▶ Review: The divergence measures the expansion or contraction of a vector field at each point. The flux form of Green's Theorem implies that if the two-dimensional divergence of a vector field is zero throughout a simply connected plane region, then the outward flux across the boundary of the region is zero. If the divergence is nonzero, Green's Theorem gives the outward flux across the boundary.

▶ In evaluating $\nabla \cdot \mathbf{F}$ as a dot product, each component of ∇ is applied to the corresponding component of $\mathbf{F}$, producing $f_x + g_y + h_z$.

The Divergence

Recall that in two dimensions, the divergence of the vector field $\mathbf{F} = \langle f, g \rangle$ is $\dfrac{\partial f}{\partial x} + \dfrac{\partial g}{\partial y}$. The extension to three dimensions is straightforward. If $\mathbf{F} = \langle f, g, h \rangle$ is a differentiable vector field defined on a region of $\mathbb{R}^3$, the divergence of $\mathbf{F}$ is $\dfrac{\partial f}{\partial x} + \dfrac{\partial g}{\partial y} + \dfrac{\partial h}{\partial z}$. The interpretation of the three-dimensional divergence is much the same as it is in two dimensions. It measures the expansion or contraction of the vector field at each point. If the divergence is zero at all points of a region, the vector field is *source free* on that region.

Recall the *del operator* ∇ that was introduced in Section 15.5 to define the gradient:

$$\nabla = \mathbf{i}\frac{\partial}{\partial x} + \mathbf{j}\frac{\partial}{\partial y} + \mathbf{k}\frac{\partial}{\partial z} = \left\langle \frac{\partial}{\partial x}, \frac{\partial}{\partial y}, \frac{\partial}{\partial z} \right\rangle.$$

This object is not really a vector; it is an operation that is applied to a function or a vector field. Applying it directly to a scalar function f results in the gradient of f:

$$\nabla f = \frac{\partial f}{\partial x}\mathbf{i} + \frac{\partial f}{\partial y}\mathbf{j} + \frac{\partial f}{\partial z}\mathbf{k} = \langle f_x, f_y, f_z \rangle.$$

However, if we form the *dot product* of ∇ and a vector field $\mathbf{F} = \langle f, g, h \rangle$, the result is

$$\nabla \cdot \mathbf{F} = \left\langle \frac{\partial}{\partial x}, \frac{\partial}{\partial y}, \frac{\partial}{\partial z} \right\rangle \cdot \langle f, g, h \rangle = \frac{\partial f}{\partial x} + \frac{\partial g}{\partial y} + \frac{\partial h}{\partial z},$$

which is the divergence of $\mathbf{F}$, also denoted div $\mathbf{F}$. Like all dot products, the divergence is a scalar; in this case, it is a scalar-valued function.

Radial field
$\mathbf{F} = \langle x, y, z \rangle$

$\nabla \cdot \mathbf{F} = 3$ at all points $\Rightarrow$ vector field expands outward at all points.

(a)

> **DEFINITION Divergence of a Vector Field**
>
> The **divergence** of a vector field $\mathbf{F} = \langle f, g, h \rangle$ that is differentiable on a region of $\mathbb{R}^3$ is
>
> $$\text{div } \mathbf{F} = \nabla \cdot \mathbf{F} = \frac{\partial f}{\partial x} + \frac{\partial g}{\partial y} + \frac{\partial h}{\partial z}.$$
>
> If $\nabla \cdot \mathbf{F} = 0$, the vector field is **source free**.

EXAMPLE 1 Computing the divergence Compute the divergence of the following vector fields.

a. $\mathbf{F} = \langle x, y, z \rangle$ (a radial field)

b. $\mathbf{F} = \langle -y, x - z, y \rangle$ (a rotation field)

c. $\mathbf{F} = \langle -y, x, z \rangle$ (a spiral flow)

SOLUTION

a. The divergence is $\nabla \cdot \mathbf{F} = \nabla \cdot \langle x, y, z \rangle = \dfrac{\partial x}{\partial x} + \dfrac{\partial y}{\partial y} + \dfrac{\partial z}{\partial z} = 1 + 1 + 1 = 3.$

Because the divergence is positive, the flow expands outward at all points (**Figure 17.38a**).

b. The divergence is

$$\nabla \cdot \mathbf{F} = \nabla \cdot \langle -y, x - z, y \rangle = \frac{\partial(-y)}{\partial x} + \frac{\partial(x - z)}{\partial y} + \frac{\partial y}{\partial z} = 0 + 0 + 0 = 0,$$

so the field is source free.

c. This field is a combination of the two-dimensional rotation field $\mathbf{F} = \langle -y, x \rangle$ and a vertical flow in the z-direction; the net effect is a field that spirals upward for $z > 0$ and spirals downward for $z < 0$ (**Figure 17.38b**). The divergence is

$$\nabla \cdot \mathbf{F} = \nabla \cdot \langle -y, x, z \rangle = \frac{\partial(-y)}{\partial x} + \frac{\partial x}{\partial y} + \frac{\partial z}{\partial z} = 0 + 0 + 1 = 1.$$

Spiral flow
$\mathbf{F} = \langle -y, x, z \rangle$

(b)

Figure 17.38

QUICK CHECK 1 Show that if a vector field has the form $\mathbf{F} = \langle f(y, z), g(x, z), h(x, y) \rangle$, then div $\mathbf{F} = 0$. ◄

The rotational part of the field in x and y does not contribute to the divergence. However, the z-component of the field produces a nonzero divergence.

Related Exercises 10–11 ◄

Divergence of a Radial Vector Field The vector field considered in Example 1a is just one of many radial fields that have important applications (for example, the inverse square laws of gravitation and electrostatics). The following example leads to a general result for the divergence of radial vector fields.

EXAMPLE 2   Divergence of a radial field Compute the divergence of the radial vector field

$$\mathbf{F} = \frac{\mathbf{r}}{|\mathbf{r}|} = \frac{\langle x, y, z \rangle}{\sqrt{x^2 + y^2 + z^2}}.$$

SOLUTION This radial field has the property that it is directed outward from the origin and all vectors have unit length ($|\mathbf{F}| = 1$). Let's compute one piece of the divergence; the others follow the same pattern. Using the Quotient Rule, the derivative with respect to x of the first component of $\mathbf{F}$ is

$$\frac{\partial}{\partial x} \left(\frac{x}{(x^2 + y^2 + z^2)^{1/2}} \right) = \frac{(x^2 + y^2 + z^2)^{1/2} - x^2 (x^2 + y^2 + z^2)^{-1/2}}{x^2 + y^2 + z^2} \quad \text{Quotient Rule}$$

$$= \frac{|\mathbf{r}| - x^2 |\mathbf{r}|^{-1}}{|\mathbf{r}|^2} \qquad \sqrt{x^2 + y^2 + z^2} = |\mathbf{r}|$$

$$= \frac{|\mathbf{r}|^2 - x^2}{|\mathbf{r}|^3}. \qquad \text{Simplify.}$$

A similar calculation of the y- and z-derivatives yields $\dfrac{|\mathbf{r}|^2 - y^2}{|\mathbf{r}|^3}$ and $\dfrac{|\mathbf{r}|^2 - z^2}{|\mathbf{r}|^3}$, respectively. Adding the three terms, we find that

$$\nabla \cdot \mathbf{F} = \frac{|\mathbf{r}|^2 - x^2}{|\mathbf{r}|^3} + \frac{|\mathbf{r}|^2 - y^2}{|\mathbf{r}|^3} + \frac{|\mathbf{r}|^2 - z^2}{|\mathbf{r}|^3}$$

$$= 3 \frac{|\mathbf{r}|^2}{|\mathbf{r}|^3} - \frac{x^2 + y^2 + z^2}{|\mathbf{r}|^3} \qquad \text{Collect terms.}$$

$$= \frac{2}{|\mathbf{r}|}. \qquad x^2 + y^2 + z^2 = |\mathbf{r}|^2$$

Related Exercise 18 ◄

Examples 1a and 2 give two special cases of the following theorem about the divergence of radial vector fields (Exercise 73).

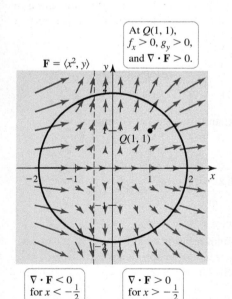

At $Q(1, 1)$, $f_x > 0$, $g_y > 0$, and $\nabla \cdot \mathbf{F} > 0$.

$\mathbf{F} = \langle x^2, y \rangle$

$\nabla \cdot \mathbf{F} < 0$ for $x < -\frac{1}{2}$

$\nabla \cdot \mathbf{F} > 0$ for $x > -\frac{1}{2}$

Figure 17.39

> **THEOREM 17.10   Divergence of Radial Vector Fields**
> For a real number p, the divergence of the radial vector field
> $$\mathbf{F} = \frac{\mathbf{r}}{|\mathbf{r}|^p} = \frac{\langle x, y, z \rangle}{(x^2 + y^2 + z^2)^{p/2}} \quad \text{is} \quad \nabla \cdot \mathbf{F} = \frac{3 - p}{|\mathbf{r}|^p}.$$

EXAMPLE 3   Divergence from a graph To gain some intuition about the divergence, consider the two-dimensional vector field $\mathbf{F} = \langle f, g \rangle = \langle x^2, y \rangle$ and a circle C of radius 2 centered at the origin (**Figure 17.39**).

a. Without computing it, determine whether the two-dimensional divergence is positive or negative at the point $Q(1, 1)$. Why?

b. Confirm your conjecture in part (a) by computing the two-dimensional divergence at Q.

c. Based on part (b), over what regions within the circle is the divergence positive and over what regions within the circle is the divergence negative?

d. By inspection of the figure, on what part of the circle is the flux across the boundary outward? Is the net flux out of the circle positive or negative?

SOLUTION

> To understand the conclusion of Example 3a, note that as you move through the point Q from left to right, the horizontal components of the vectors increase in length ($f_x > 0$). As you move through the point Q in the upward direction, the vertical components of the vectors also increase in length ($g_y > 0$).

a. At $Q(1, 1)$ the x-component and the y-component of the field are increasing ($f_x > 0$ and $g_y > 0$), so the field is expanding at that point and the two-dimensional divergence is positive.

b. Calculating the two-dimensional divergence, we find that

$$\nabla \cdot \mathbf{F} = \frac{\partial}{\partial x}(x^2) + \frac{\partial}{\partial y}(y) = 2x + 1.$$

At $Q(1, 1)$ the divergence is 3, confirming part (a).

c. From part (b), we see that $\nabla \cdot \mathbf{F} = 2x + 1 > 0$, for $x > -\frac{1}{2}$, and $\nabla \cdot \mathbf{F} < 0$, for $x < -\frac{1}{2}$. To the left of the line $x = -\frac{1}{2}$ the field is contracting, and to the right of the line the field is expanding.

QUICK CHECK 2 Verify the claim made in Example 3d by showing that the net outward flux of $\mathbf{F}$ across C is positive. (*Hint:* If you use Green's Theorem to evaluate the integral $\int_C f\, dy - g\, dx$, convert to polar coordinates.) ◄

d. Using Figure 17.39, it appears that the field is tangent to the circle at two points with $x \approx -1$. For points on the circle with $x < -1$, the flow is into the circle; for points on the circle with $x > -1$, the flow is out of the circle. It appears that the net outward flux across C is positive. The points where the field changes from inward to outward may be determined exactly (Exercise 46). *Related Exercises 21–22* ◄

The Curl

> Review: The *two-dimensional curl* $g_x - f_y$ measures the rotation of a vector field at a point. The circulation form of Green's Theorem implies that if the two-dimensional curl of a vector field is zero throughout a simply connected region, then the circulation on the boundary of the region is also zero. If the curl is nonzero, Green's Theorem gives the circulation along the curve.

Just as the divergence $\nabla \cdot \mathbf{F}$ is the dot product of the *del operator* and $\mathbf{F}$, the three-dimensional curl is the cross product $\nabla \times \mathbf{F}$. If we formally use the notation for the cross product in terms of a 3×3 determinant, we obtain the definition of the curl:

$$\nabla \times \mathbf{F} = \begin{vmatrix} \mathbf{i} & \mathbf{j} & \mathbf{k} \\ \dfrac{\partial}{\partial x} & \dfrac{\partial}{\partial y} & \dfrac{\partial}{\partial z} \\ f & g & h \end{vmatrix} \begin{matrix} \leftarrow \text{Unit Vectors} \\ \\ \leftarrow \text{Components of } \nabla \\ \\ \leftarrow \text{Components of } \mathbf{F} \end{matrix}$$

$$= \left(\frac{\partial h}{\partial y} - \frac{\partial g}{\partial z} \right) \mathbf{i} + \left(\frac{\partial f}{\partial z} - \frac{\partial h}{\partial x} \right) \mathbf{j} + \left(\frac{\partial g}{\partial x} - \frac{\partial f}{\partial y} \right) \mathbf{k}.$$

The curl of a vector field, also denoted curl $\mathbf{F}$, is a vector with three components. Notice that the **k**-component of the curl ($g_x - f_y$) is the two-dimensional curl, which gives the rotation in the xy-plane at a point. The **i**- and **j**-components of the curl correspond to the rotation of the vector field in planes parallel to the yz-plane (orthogonal to **i**) and in planes parallel to the xz-plane (orthogonal to **j**) (**Figure 17.40**).

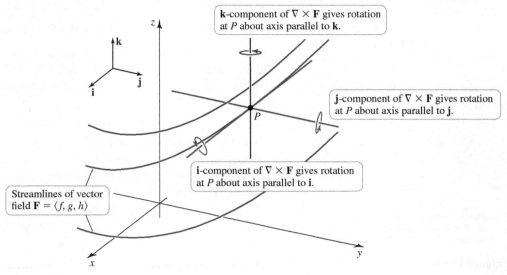

k-component of $\nabla \times \mathbf{F}$ gives rotation at P about axis parallel to **k**.

j-component of $\nabla \times \mathbf{F}$ gives rotation at P about axis parallel to **j**.

i-component of $\nabla \times \mathbf{F}$ gives rotation at P about axis parallel to **i**.

Streamlines of vector field $\mathbf{F} = \langle f, g, h \rangle$

Figure 17.40

DEFINITION Curl of a Vector Field

The **curl** of a vector field $\mathbf{F} = \langle f, g, h \rangle$ that is differentiable on a region of $\mathbb{R}^3$ is

$$\nabla \times \mathbf{F} = \text{curl } \mathbf{F}$$

$$= \left(\frac{\partial h}{\partial y} - \frac{\partial g}{\partial z} \right) \mathbf{i} + \left(\frac{\partial f}{\partial z} - \frac{\partial h}{\partial x} \right) \mathbf{j} + \left(\frac{\partial g}{\partial x} - \frac{\partial f}{\partial y} \right) \mathbf{k}.$$

If $\nabla \times \mathbf{F} = \mathbf{0}$, the vector field is **irrotational**.

Curl of a General Rotation Vector Field   We can clarify the physical meaning of the curl by considering the vector field $\mathbf{F} = \mathbf{a} \times \mathbf{r}$, where $\mathbf{a} = \langle a_1, a_2, a_3 \rangle$ is a nonzero constant vector and $\mathbf{r} = \langle x, y, z \rangle$. Writing out its components, we see that

$$\mathbf{F} = \mathbf{a} \times \mathbf{r} = \begin{vmatrix} \mathbf{i} & \mathbf{j} & \mathbf{k} \\ a_1 & a_2 & a_3 \\ x & y & z \end{vmatrix} = (a_2 z - a_3 y) \mathbf{i} + (a_3 x - a_1 z) \mathbf{j} + (a_1 y - a_2 x) \mathbf{k}.$$

This vector field is a *general rotation field* in three dimensions. With $a_1 = a_2 = 0$ and $a_3 = 1$, we have the familiar two-dimensional rotation field $\langle -y, x \rangle$ with its axis in the **k**-direction. More generally, **F** is the superposition of three rotation fields with axes in the **i**-, **j**-, and **k**-directions. The result is a single rotation field with an axis in the direction of **a** (Figure 17.41).

Three calculations tell us a lot about the general rotation field. The first calculation confirms that $\nabla \cdot \mathbf{F} = 0$ (Exercise 42). Just as with rotation fields in two dimensions, the divergence of a general rotation field is zero.

The second calculation (Exercises 43–44) uses the right-hand rule for cross products to show that the vector field $\mathbf{F} = \mathbf{a} \times \mathbf{r}$ is indeed a rotation field that circles the vector **a** in a counterclockwise direction looking along the length of **a** from head to tail (Figure 17.41).

The third calculation (Exercise 45) says that $\nabla \times \mathbf{F} = 2\mathbf{a}$. Therefore, the curl of the general rotation field is in the direction of the axis of rotation **a** (Figure 17.41). The magnitude of the curl is $|\nabla \times \mathbf{F}| = 2|\mathbf{a}|$. It can be shown (Exercise 52) that if **F** is a velocity field, then $|\mathbf{a}|$ is the constant angular speed of rotation of the field, denoted ω. The angular speed is the rate (radians per unit time) at which a small particle in the vector field rotates about the axis of the field. Therefore, the angular speed is half the magnitude of the curl, or

$$\omega = |\mathbf{a}| = \frac{1}{2} |\nabla \times \mathbf{F}|.$$

The rotation field $\mathbf{F} = \mathbf{a} \times \mathbf{r}$ suggests a related question. Suppose a paddle wheel is placed in the vector field **F** at a point P with the axis of the wheel in the direction of a unit vector **n** (Figure 17.42). How should **n** be chosen so the paddle wheel spins fastest? The scalar component of $\nabla \times \mathbf{F}$ in the direction of **n** is

$$(\nabla \times \mathbf{F}) \cdot \mathbf{n} = |\nabla \times \mathbf{F}| \cos \theta, \quad (|n| = 1)$$

where θ is the angle between $\nabla \times \mathbf{F}$ and **n**. The scalar component is greatest in magnitude and the paddle wheel spins fastest when $\theta = 0$ or $\theta = \pi$; that is, when **n** and $\nabla \times \mathbf{F}$ are parallel. If the axis of the paddle wheel is orthogonal to $\nabla \times \mathbf{F}$ ($\theta = \pm \pi/2$), the wheel doesn't spin.

General Rotation Vector Field

The **general rotation vector field** is $\mathbf{F} = \mathbf{a} \times \mathbf{r}$, where the nonzero constant vector $\mathbf{a} = \langle a_1, a_2, a_3 \rangle$ is the axis of rotation and $\mathbf{r} = \langle x, y, z \rangle$. For all nonzero choices of **a**, $|\nabla \times \mathbf{F}| = 2|\mathbf{a}|$ and $\nabla \cdot \mathbf{F} = 0$. If **F** is a velocity field, then the constant angular speed of the field is

$$\omega = |\mathbf{a}| = \frac{1}{2} |\nabla \times \mathbf{F}|.$$

General rotation field

$$\mathbf{a} = \langle 1, -1, 1 \rangle$$
$$\mathbf{F} = \mathbf{a} \times \mathbf{r}$$
$$\nabla \times \mathbf{F} = 2\mathbf{a}$$

Figure 17.41

> Just as $\nabla f \cdot \mathbf{n}$ is the directional derivative in the direction **n**, $(\nabla \times \mathbf{F}) \cdot \mathbf{n}$ is the directional spin in the direction **n**.

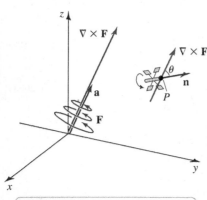

Paddle wheel at P with axis **n** measures rotation about **n**. Rotation is a maximum when $\nabla \times \mathbf{F}$ is parallel to **n**.

Figure 17.42

QUICK CHECK 3 Show that if a vector field has the form $\mathbf{F} = \langle f(x), g(y), h(z) \rangle$, then $\nabla \times \mathbf{F} = \mathbf{0}$. ◄

EXAMPLE 4 Curl of a rotation field Compute the curl of the rotation field $\mathbf{F} = \mathbf{a} \times \mathbf{r}$, where $\mathbf{a} = \langle 2, -1, 1 \rangle$ and $\mathbf{r} = \langle x, y, z \rangle$ (Figure 17.41). What are the direction and the magnitude of the curl?

SOLUTION A quick calculation shows that

$$\mathbf{F} = \mathbf{a} \times \mathbf{r} = (-y - z)\,\mathbf{i} + (x - 2z)\,\mathbf{j} + (x + 2y)\,\mathbf{k}.$$

The curl of the vector field is

$$\nabla \times \mathbf{F} = \begin{vmatrix} \mathbf{i} & \mathbf{j} & \mathbf{k} \\ \dfrac{\partial}{\partial x} & \dfrac{\partial}{\partial y} & \dfrac{\partial}{\partial z} \\ -y - z & x - 2z & x + 2y \end{vmatrix} = 4\mathbf{i} - 2\mathbf{j} + 2\mathbf{k} = 2\mathbf{a}.$$

We have confirmed that $\nabla \times \mathbf{F} = 2\mathbf{a}$ and that the direction of the curl is the direction of $\mathbf{a}$, which is the axis of rotation. The magnitude of $\nabla \times \mathbf{F}$ is $|2\mathbf{a}| = 2\sqrt{6}$, which is twice the angular speed of rotation.

Related Exercises 25–26 ◄

Working with Divergence and Curl

The divergence and curl satisfy some of the same properties that ordinary derivatives satisfy. For example, given a real number c and differentiable vector fields $\mathbf{F}$ and $\mathbf{G}$, we have the following properties.

Divergence Properties	Curl Properties
$\nabla \cdot (\mathbf{F} + \mathbf{G}) = \nabla \cdot \mathbf{F} + \nabla \cdot \mathbf{G}$	$\nabla \times (\mathbf{F} + \mathbf{G}) = (\nabla \times \mathbf{F}) + (\nabla \times \mathbf{G})$
$\nabla \cdot (c\mathbf{F}) = c(\nabla \cdot \mathbf{F})$	$\nabla \times (c\mathbf{F}) = c(\nabla \times \mathbf{F})$

These and other properties are explored in Exercises 65–72.

Additional properties that have importance in theory and applications are presented in the following theorems and examples.

THEOREM 17.11 Curl of a Conservative Vector Field
Suppose $\mathbf{F}$ is a conservative vector field on an open region D of $\mathbb{R}^3$. Let $\mathbf{F} = \nabla\varphi$, where φ is a potential function with continuous second partial derivatives on D. Then $\nabla \times \mathbf{F} = \nabla \times \nabla\varphi = \mathbf{0}$: The curl of the gradient is the zero vector and $\mathbf{F}$ is irrotational.

Proof: We must calculate $\nabla \times \nabla\varphi$:

$$\nabla \times \nabla\varphi = \begin{vmatrix} \mathbf{i} & \mathbf{j} & \mathbf{k} \\ \dfrac{\partial}{\partial x} & \dfrac{\partial}{\partial y} & \dfrac{\partial}{\partial z} \\ \varphi_x & \varphi_y & \varphi_z \end{vmatrix} = \underbrace{(\varphi_{zy} - \varphi_{yz})}_{0}\mathbf{i} + \underbrace{(\varphi_{xz} - \varphi_{zx})}_{0}\mathbf{j} + \underbrace{(\varphi_{yx} - \varphi_{xy})}_{0}\mathbf{k} = \mathbf{0}.$$

The mixed partial derivatives are equal by Clairaut's Theorem (Theorem 15.4).

The converse of this theorem (if $\nabla \times \mathbf{F} = \mathbf{0}$, then $\mathbf{F}$ is a conservative field) is handled in Section 17.7 by means of Stokes' Theorem. ◄

➤ First note that $\nabla \times \mathbf{F}$ is a vector, so it makes sense to take the divergence of the curl.

THEOREM 17.12 Divergence of the Curl
Suppose $\mathbf{F} = \langle f, g, h \rangle$, where f, g, and h have continuous second partial derivatives. Then $\nabla \cdot (\nabla \times \mathbf{F}) = 0$: The divergence of the curl is zero.

Proof: Again, a calculation is needed:

$$\nabla \cdot (\nabla \times \mathbf{F})$$

$$= \frac{\partial}{\partial x}\left(\frac{\partial h}{\partial y} - \frac{\partial g}{\partial z}\right) + \frac{\partial}{\partial y}\left(\frac{\partial f}{\partial z} - \frac{\partial h}{\partial x}\right) + \frac{\partial}{\partial z}\left(\frac{\partial g}{\partial x} - \frac{\partial f}{\partial y}\right)$$

$$= \underbrace{(h_{yx} - h_{xy})}_{0} + \underbrace{(g_{xz} - g_{zx})}_{0} + \underbrace{(f_{zy} - f_{yz})}_{0} = 0.$$

Clairaut's Theorem (Theorem 15.4) ensures that the mixed partial derivatives are equal. ◄

The gradient, the divergence, and the curl may be combined in many ways—some of which are undefined. For example, the gradient of the curl $(\nabla(\nabla \times \mathbf{F}))$ and the curl of the divergence $(\nabla \times (\nabla \cdot \mathbf{F}))$ are undefined. However, a combination that *is* defined and is important is the divergence of the gradient $\nabla \cdot \nabla u$, where u is a scalar-valued function. This combination is denoted $\nabla^2 u$ and is called the **Laplacian** of u; it arises in many physical situations (Exercises 56–58, 62). Carrying out the calculation, we find that

$$\nabla \cdot \nabla u = \frac{\partial}{\partial x}\frac{\partial u}{\partial x} + \frac{\partial}{\partial y}\frac{\partial u}{\partial y} + \frac{\partial}{\partial z}\frac{\partial u}{\partial z} = \frac{\partial^2 u}{\partial x^2} + \frac{\partial^2 u}{\partial y^2} + \frac{\partial^2 u}{\partial z^2}.$$

We close with a result that is useful in its own right but is also intriguing because it parallels the Product Rule from single-variable calculus.

THEOREM 17.13 Product Rule for the Divergence
Let u be a scalar-valued function that is differentiable on a region D and let $\mathbf{F}$ be a vector field that is differentiable on D. Then

$$\nabla \cdot (u\mathbf{F}) = \nabla u \cdot \mathbf{F} + u(\nabla \cdot \mathbf{F}).$$

QUICK CHECK 4 Is $\nabla \cdot (u\mathbf{F})$ a vector function or a scalar function? ◄

The rule says that the "derivative" of the product is the "derivative" of the first function multiplied by the second function plus the first function multiplied by the "derivative" of the second function. However, in each instance, "derivative" must be interpreted correctly for the operations to make sense. The proof of the theorem requires a direct calculation (Exercise 67). Other similar vector calculus identities are presented in Exercises 68–72.

EXAMPLE 5 More properties of radial fields Let $\mathbf{r} = \langle x, y, z \rangle$ and let

$$\varphi = \frac{1}{|\mathbf{r}|} = (x^2 + y^2 + z^2)^{-1/2}$$ be a potential function.

a. Find the associated gradient field $\mathbf{F} = \nabla\left(\dfrac{1}{|\mathbf{r}|}\right)$.

b. Compute $\nabla \cdot \mathbf{F}$.

SOLUTION

a. The gradient has three components. Computing the first component reveals a pattern:

$$\frac{\partial \varphi}{\partial x} = \frac{\partial}{\partial x}(x^2 + y^2 + z^2)^{-1/2} = -\frac{1}{2}(x^2 + y^2 + z^2)^{-3/2}\,2x = -\frac{x}{|\mathbf{r}|^3}.$$

Making a similar calculation for the y- and z-derivatives, the gradient is

$$\mathbf{F} = \nabla\left(\frac{1}{|\mathbf{r}|}\right) = -\frac{\langle x, y, z \rangle}{|\mathbf{r}|^3} = -\frac{\mathbf{r}}{|\mathbf{r}|^3}.$$

This result reveals that $\mathbf{F}$ is an inverse square vector field (for example, a gravitational or electric field), and its potential function is $\varphi = \dfrac{1}{|\mathbf{r}|}$.

b. The divergence $\nabla \cdot \mathbf{F} = \nabla \cdot \left(-\dfrac{\mathbf{r}}{|\mathbf{r}|^3} \right)$ involves a product of the vector function $\mathbf{r} = \langle x, y, z \rangle$ and the scalar function $|\mathbf{r}|^{-3}$. Applying Theorem 17.13, we find that

$$\nabla \cdot \mathbf{F} = \nabla \cdot \left(-\frac{\mathbf{r}}{|\mathbf{r}|^3} \right) = -\nabla \frac{1}{|\mathbf{r}|^3} \cdot \mathbf{r} - \frac{1}{|\mathbf{r}|^3} \nabla \cdot \mathbf{r}.$$

A calculation similar to part (a) shows that $\nabla \dfrac{1}{|\mathbf{r}|^3} = -\dfrac{3\mathbf{r}}{|\mathbf{r}|^5}$ (Exercise 35). Therefore,

$$\nabla \cdot \mathbf{F} = \nabla \cdot \left(-\frac{\mathbf{r}}{|\mathbf{r}|^3} \right) = -\underbrace{\nabla \frac{1}{|\mathbf{r}|^3}}_{-3\mathbf{r}/|\mathbf{r}|^5} \cdot \mathbf{r} - \frac{1}{|\mathbf{r}|^3} \underbrace{\nabla \cdot \mathbf{r}}_{3}$$

$$= \frac{3\mathbf{r}}{|\mathbf{r}|^5} \cdot \mathbf{r} - \frac{3}{|\mathbf{r}|^3} \qquad\qquad \text{Substitute for } \nabla \frac{1}{|\mathbf{r}|^3}.$$

$$= \frac{3|\mathbf{r}|^2}{|\mathbf{r}|^5} - \frac{3}{|\mathbf{r}|^3} \qquad\qquad \mathbf{r} \cdot \mathbf{r} = |\mathbf{r}|^2$$

$$= 0.$$

The result is consistent with Theorem 17.10 (with $p = 3$): The divergence of an inverse square vector field in $\mathbb{R}^3$ is zero. It does not happen for any other radial fields of this form.

Related Exercises 35–36 ◄

Summary of Properties of Conservative Vector Fields

We can now extend the list of equivalent properties of conservative vector fields $\mathbf{F}$ defined on an open connected region. Theorem 17.11 is added to the list given at the end of Section 17.3.

Properties of a Conservative Vector Field

Let $\mathbf{F}$ be a conservative vector field whose components have continuous second partial derivatives on an open connected region D in $\mathbb{R}^3$. Then $\mathbf{F}$ has the following equivalent properties.

1. There exists a potential function φ such that $\mathbf{F} = \nabla\varphi$ (definition).
2. $\int_C \mathbf{F} \cdot d\mathbf{r} = \varphi(B) - \varphi(A)$ for all points A and B in D and all piecewise-smooth oriented curves C in D from A to B.
3. $\oint_C \mathbf{F} \cdot d\mathbf{r} = 0$ on all simple piecewise-smooth closed oriented curves C in D.
4. $\nabla \times \mathbf{F} = \mathbf{0}$ at all points of D.

SECTION 17.5 EXERCISES

Getting Started

1. Explain how to compute the divergence of the vector field $\mathbf{F} = \langle f, g, h \rangle$.

2. Interpret the divergence of a vector field.

3. What does it mean if the divergence of a vector field is zero throughout a region?

4. Explain how to compute the curl of the vector field $\mathbf{F} = \langle f, g, h \rangle$.

5. Interpret the curl of a general rotation vector field.

6. What does it mean if the curl of a vector field is zero throughout a region?

7. What is the value of $\nabla \cdot (\nabla \times \mathbf{F})$?

8. What is the value of $\nabla \times \nabla u$?

Practice Exercises

9–16. Divergence of vectors fields *Find the divergence of the following vector fields.*

9. $\mathbf{F} = \langle 2x, 4y, -3z \rangle$

10. $\mathbf{F} = \langle -2y, 3x, z \rangle$

11. $\mathbf{F} = \langle 12x, -6y, -6z \rangle$ **12.** $\mathbf{F} = \langle x^2yz, -xy^2z, -xyz^2 \rangle$

13. $\mathbf{F} = \langle x^2 - y^2, y^2 - z^2, z^2 - x^2 \rangle$

14. $\mathbf{F} = \langle e^{-x+y}, e^{-y+z}, e^{-z+x} \rangle$

15. $\mathbf{F} = \dfrac{\langle x, y, z \rangle}{1 + x^2 + y^2}$

16. $\mathbf{F} = \langle yz \sin x, xz \cos y, xy \cos z \rangle$

17–20. Divergence of radial fields *Calculate the divergence of the following radial fields. Express the result in terms of the position vector* $\mathbf{r}$ *and its length* $|\mathbf{r}|$. *Check for agreement with Theorem 17.10.*

17. $\mathbf{F} = \dfrac{\langle x, y, z \rangle}{x^2 + y^2 + z^2} = \dfrac{\mathbf{r}}{|\mathbf{r}|^2}$

18. $\mathbf{F} = \dfrac{\langle x, y, z \rangle}{(x^2 + y^2 + z^2)^{3/2}} = \dfrac{\mathbf{r}}{|\mathbf{r}|^3}$

19. $\mathbf{F} = \dfrac{\langle x, y, z \rangle}{(x^2 + y^2 + z^2)^2} = \dfrac{\mathbf{r}}{|\mathbf{r}|^4}$

20. $\mathbf{F} = \langle x, y, z \rangle (x^2 + y^2 + z^2) = \mathbf{r}|\mathbf{r}|^2$

21–22. Divergence and flux from graphs *Consider the following vector fields, the circle C, and two points P and Q.*

a. *Without computing the divergence, does the graph suggest that the divergence is positive or negative at P and Q? Justify your answer.*
b. *Compute the divergence and confirm your conjecture in part (a).*
c. *On what part of C is the flux outward? Inward?*
d. *Is the net outward flux across C positive or negative?*

21. $\mathbf{F} = \langle x, x + y \rangle$

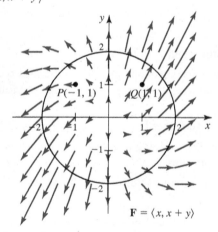

$\mathbf{F} = \langle x, x + y \rangle$

22. $\mathbf{F} = \langle x, y^2 \rangle$

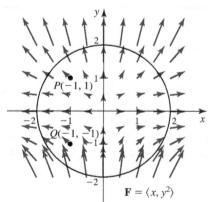

$\mathbf{F} = \langle x, y^2 \rangle$

23–26. Curl of a rotation field *Consider the following vector fields, where* $\mathbf{r} = \langle x, y, z \rangle$.

a. *Compute the curl of the field and verify that it has the same direction as the axis of rotation.*
b. *Compute the magnitude of the curl of the field.*

23. $\mathbf{F} = \langle 1, 0, 0 \rangle \times \mathbf{r}$ **24.** $\mathbf{F} = \langle 1, -1, 0 \rangle \times \mathbf{r}$

25. $\mathbf{F} = \langle 1, -1, 1 \rangle \times \mathbf{r}$ **26.** $\mathbf{F} = \langle 1, -2, -3 \rangle \times \mathbf{r}$

27–34. Curl of a vector field *Compute the curl of the following vector fields.*

27. $\mathbf{F} = \langle x^2 - y^2, xy, z \rangle$ **28.** $\mathbf{F} = \langle 0, z^2 - y^2, -yz \rangle$

29. $\mathbf{F} = \langle x^2 - z^2, 1, 2xz \rangle$ **30.** $\mathbf{F} = \mathbf{r} = \langle x, y, z \rangle$

31. $\mathbf{F} = \dfrac{\langle x, y, z \rangle}{(x^2 + y^2 + z^2)^{3/2}} = \dfrac{\mathbf{r}}{|\mathbf{r}|^3}$

32. $\mathbf{F} = \dfrac{\langle x, y, z \rangle}{(x^2 + y^2 + z^2)^{1/2}} = \dfrac{\mathbf{r}}{|\mathbf{r}|}$

33. $\mathbf{F} = \langle z^2 \sin y, xz^2 \cos y, 2xz \sin y \rangle$

34. $\mathbf{F} = \langle 3xz^3 e^{y^2}, 2xz^3 e^{y^2}, 3xz^2 e^{y^2} \rangle$

35–38. Derivative rules *Prove the following identities. Use Theorem 17.13 (Product Rule) whenever possible.*

35. $\nabla \left(\dfrac{1}{|\mathbf{r}|^3} \right) = -\dfrac{3\mathbf{r}}{|\mathbf{r}|^5}$ (used in Example 5)

36. $\nabla \left(\dfrac{1}{|\mathbf{r}|^2} \right) = -\dfrac{2\mathbf{r}}{|\mathbf{r}|^4}$

37. $\nabla \cdot \nabla \left(\dfrac{1}{|\mathbf{r}|^2} \right) = \dfrac{2}{|\mathbf{r}|^4}$ (*Hint:* Use Exercise 36.)

38. $\nabla (\ln |\mathbf{r}|) = \dfrac{\mathbf{r}}{|\mathbf{r}|^2}$

39. Explain why or why not Determine whether the following statements are true and give an explanation or counterexample.

a. For a function f of a single variable, if $f'(x) = 0$ for all x in the domain, then f is a constant function. If $\nabla \cdot \mathbf{F} = 0$ for all points in the domain, then $\mathbf{F}$ is constant.
b. If $\nabla \times \mathbf{F} = \mathbf{0}$, then $\mathbf{F}$ is constant.
c. A vector field consisting of parallel vectors has zero curl.
d. A vector field consisting of parallel vectors has zero divergence.
e. curl $\mathbf{F}$ is orthogonal to $\mathbf{F}$.

40. Another derivative combination Let $\mathbf{F} = \langle f, g, h \rangle$ and let u be a differentiable scalar-valued function.

a. Take the dot product of $\mathbf{F}$ and the del operator; then apply the result to u to show that

$$(\mathbf{F} \cdot \nabla)u = \left(f \frac{\partial}{\partial x} + g \frac{\partial}{\partial y} + h \frac{\partial}{\partial z} \right) u$$

$$= f \frac{\partial u}{\partial x} + g \frac{\partial u}{\partial y} + h \frac{\partial u}{\partial z}.$$

b. Evaluate $(\mathbf{F} \cdot \nabla)(xy^2 z^3)$ at $(1, 1, 1)$, where $\mathbf{F} = \langle 1, 1, 1 \rangle$.

41. **Does it make sense?** Are the following expressions defined? If so, state whether the result is a scalar or a vector. Assume **F** is a sufficiently differentiable vector field and φ is a sufficiently differentiable scalar-valued function.

    a. $\nabla \cdot \varphi$ b. $\nabla \mathbf{F}$ c. $\nabla \cdot \nabla \varphi$
    d. $\nabla (\nabla \cdot \varphi)$ e. $\nabla (\nabla \times \varphi)$ f. $\nabla \cdot (\nabla \cdot \mathbf{F})$
    g. $\nabla \times \nabla \varphi$ h. $\nabla \times (\nabla \cdot \mathbf{F})$ i. $\nabla \times (\nabla \times \mathbf{F})$

42. **Zero divergence of the rotation field** Show that the general rotation field $\mathbf{F} = \mathbf{a} \times \mathbf{r}$, where **a** is a nonzero constant vector and $\mathbf{r} = \langle x, y, z \rangle$, has zero divergence.

43. **General rotation fields**

    a. Let $\mathbf{a} = \langle 0, 1, 0 \rangle$, let $\mathbf{r} = \langle x, y, z \rangle$, and consider the rotation field $\mathbf{F} = \mathbf{a} \times \mathbf{r}$. Use the right-hand rule for cross products to find the direction of **F** at the points $(0, 1, 1)$, $(1, 1, 0)$, $(0, 1, -1)$, and $(-1, 1, 0)$.
    b. With $\mathbf{a} = \langle 0, 1, 0 \rangle$, explain why the rotation field $\mathbf{F} = \mathbf{a} \times \mathbf{r}$ circles the y-axis in the counterclockwise direction looking along **a** from head to tail (that is, in the negative y-direction).

44. **General rotation fields** Generalize Exercise 43 to show that the rotation field $\mathbf{F} = \mathbf{a} \times \mathbf{r}$ circles the vector **a** in the counterclockwise direction looking along **a** from head to tail.

45. **Curl of the rotation field** For the general rotation field $\mathbf{F} = \mathbf{a} \times \mathbf{r}$, where **a** is a nonzero constant vector and $\mathbf{r} = \langle x, y, z \rangle$, show that curl $\mathbf{F} = 2\mathbf{a}$.

46. **Inward to outward** Find the exact points on the circle $x^2 + y^2 = 2$ at which the field $\mathbf{F} = \langle f, g \rangle = \langle x^2, y \rangle$ switches from pointing inward to pointing outward on the circle, or vice versa.

47. **Maximum divergence** Within the cube $\{(x, y, z): |x| \leq 1, |y| \leq 1, |z| \leq 1\}$, where does div **F** have the greatest magnitude when $\mathbf{F} = \langle x^2 - y^2, xy^2 z, 2xz \rangle$?

48. **Maximum curl** Let $\mathbf{F} = \langle z, 0, -y \rangle$.

    a. Find the scalar component of curl **F** in the direction of the unit vector $\mathbf{n} = \langle 1, 0, 0 \rangle$.
    b. Find the scalar component of curl **F** in the direction of the unit vector $\mathbf{n} = \left\langle \dfrac{1}{\sqrt{3}}, -\dfrac{1}{\sqrt{3}}, \dfrac{1}{\sqrt{3}} \right\rangle$.
    c. Find the unit vector **n** that maximizes $\text{scal}_\mathbf{n} \langle -1, 1, 0 \rangle$ and state the value of $\text{scal}_\mathbf{n} \langle -1, 1, 0 \rangle$ in this direction.

49. **Zero component of the curl** For what vectors **n** is $(\text{curl } \mathbf{F}) \cdot \mathbf{n} = 0$ when $\mathbf{F} = \langle y, -2z, -x \rangle$?

50–51. Find a vector field *Find a vector field* **F** *with the given curl. In each case, is the vector field you found unique?*

50. $\text{curl } \mathbf{F} = \langle 0, 1, 0 \rangle$ 51. $\text{curl } \mathbf{F} = \langle 0, z, -y \rangle$

Explorations and Challenges

52. **Curl and angular speed** Consider the rotational velocity field $\mathbf{v} = \mathbf{a} \times \mathbf{r}$, where **a** is a nonzero constant vector and $\mathbf{r} = \langle x, y, z \rangle$. Use the fact that an object moving in a circular path of radius R with speed $|\mathbf{v}|$ has an angular speed of $\omega = |\mathbf{v}|/R$.

    a. Sketch a position vector **a**, which is the axis of rotation for the vector field, and a position vector **r** of a point P in $\mathbb{R}^3$. Let θ be the angle between the two vectors. Show that the perpendicular distance from P to the axis of rotation is $R = |\mathbf{r}| \sin \theta$.

    b. Show that the speed of a particle in the velocity field is $|\mathbf{a} \times \mathbf{r}|$ and that the angular speed of the object is $|\mathbf{a}|$.

    c. Conclude that $\omega = \dfrac{1}{2} |\nabla \times \mathbf{v}|$.

53. **Paddle wheel in a vector field** Let $\mathbf{F} = \langle z, 0, 0 \rangle$ and let **n** be a unit vector aligned with the axis of a paddle wheel located on the x-axis (see figure).

    a. If the paddle wheel is oriented with $\mathbf{n} = \langle 1, 0, 0 \rangle$, in what direction (if any) does the wheel spin?
    b. If the paddle wheel is oriented with $\mathbf{n} = \langle 0, 1, 0 \rangle$, in what direction (if any) does the wheel spin?
    c. If the paddle wheel is oriented with $\mathbf{n} = \langle 0, 0, 1 \rangle$, in what direction (if any) does the wheel spin?

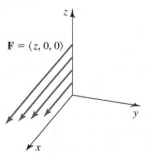

$\mathbf{F} = \langle z, 0, 0 \rangle$

54. **Angular speed** Consider the rotational velocity field $\mathbf{v} = \langle -2y, 2z, 0 \rangle$.

    a. If a paddle wheel is placed in the xy-plane with its axis normal to this plane, what is its angular speed?
    b. If a paddle wheel is placed in the xz-plane with its axis normal to this plane, what is its angular speed?
    c. If a paddle wheel is placed in the yz-plane with its axis normal to this plane, what is its angular speed?

55. **Angular speed** Consider the rotational velocity field $\mathbf{v} = \langle 0, 10z, -10y \rangle$. If a paddle wheel is placed in the plane $x + y + z = 1$ with its axis normal to this plane, how fast does the paddle wheel spin (in revolutions per unit time)?

56–58. Heat flux *Suppose a solid object in $\mathbb{R}^3$ has a temperature distribution given by $T(x, y, z)$. The heat flow vector field in the object is $\mathbf{F} = -k\nabla T$, where the conductivity $k > 0$ is a property of the material. Note that the heat flow vector points in the direction opposite to that of the gradient, which is the direction of greatest temperature decrease. The divergence of the heat flow vector is $\nabla \cdot \mathbf{F} = -k\nabla \cdot \nabla T = -k\nabla^2 T$ (the Laplacian of T). Compute the heat flow vector field and its divergence for the following temperature distributions.*

56. $T(x, y, z) = 100 e^{-\sqrt{x^2 + y^2 + z^2}}$

57. $T(x, y, z) = 100 e^{-x^2 + y^2 + z^2}$

58. $T(x, y, z) = 100 \left(1 + \sqrt{x^2 + y^2 + z^2} \right)$

59. **Gravitational potential** The potential function for the gravitational force field due to a mass M at the origin acting on a mass m is $\varphi = GMm/|\mathbf{r}|$, where $\mathbf{r} = \langle x, y, z \rangle$ is the position vector of the mass m, and G is the gravitational constant.

    a. Compute the gravitational force field $\mathbf{F} = -\nabla \varphi$.
    b. Show that the field is irrotational; that is, show that $\nabla \times \mathbf{F} = \mathbf{0}$.

60. Electric potential The potential function for the force field due to a charge q at the origin is $\varphi = \dfrac{1}{4\pi\varepsilon_0}\dfrac{q}{|\mathbf{r}|}$, where $\mathbf{r} = \langle x, y, z \rangle$ is the position vector of a point in the field, and ε_0 is the permittivity of free space.

a. Compute the force field $\mathbf{F} = -\nabla\varphi$.
b. Show that the field is irrotational; that is, show that $\nabla \times \mathbf{F} = \mathbf{0}$.

61. Navier-Stokes equation The Navier-Stokes equation is the fundamental equation of fluid dynamics that models the flow in everything from bathtubs to oceans. In one of its many forms (incompressible, viscous flow), the equation is

$$\rho\left(\frac{\partial \mathbf{V}}{\partial t} + (\mathbf{V} \cdot \nabla)\mathbf{V}\right) = -\nabla p + \mu(\nabla \cdot \nabla)\mathbf{V}.$$

In this notation, $\mathbf{V} = \langle u, v, w \rangle$ is the three-dimensional velocity field, p is the (scalar) pressure, ρ is the constant density of the fluid, and μ is the constant viscosity. Write out the three component equations of this vector equation. (See Exercise 40 for an interpretation of the operations.)

T 62. Stream function and vorticity The rotation of a three-dimensional velocity field $\mathbf{V} = \langle u, v, w \rangle$ is measured by the **vorticity** $\omega = \nabla \times \mathbf{V}$. If $\omega = \mathbf{0}$ at all points in the domain, the flow is irrotational.

a. Which of the following velocity fields is irrotational: $\mathbf{V} = \langle 2, -3y, 5z \rangle$ or $\mathbf{V} = \langle y, x - z, -y \rangle$?
b. Recall that for a two-dimensional source-free flow $\mathbf{V} = \langle u, v, 0 \rangle$, a stream function $\psi(x, y)$ may be defined such that $u = \psi_y$ and $v = -\psi_x$. For such a two-dimensional flow, let $\zeta = \mathbf{k} \cdot \nabla \times \mathbf{V}$ be the **k**-component of the vorticity. Show that $\nabla^2\psi = \nabla \cdot \nabla\psi = -\zeta$.
c. Consider the stream function $\psi(x, y) = \sin x \sin y$ on the square region $R = \{(x, y): 0 \le x \le \pi, 0 \le y \le \pi\}$. Find the velocity components u and v; then sketch the velocity field.
d. For the stream function in part (c), find the vorticity function ζ as defined in part (b). Plot several level curves of the vorticity function. Where on R is it a maximum? A minimum?

63. Ampère's Law One of Maxwell's equations for electromagnetic waves is $\nabla \times \mathbf{B} = C\dfrac{\partial \mathbf{E}}{\partial t}$, where $\mathbf{E}$ is the electric field, $\mathbf{B}$ is the magnetic field, and C is a constant.

a. Show that the fields
$$\mathbf{E}(z, t) = A \sin(kz - \omega t)\mathbf{i} \text{ and } \mathbf{B}(z, t) = A \sin(kz - \omega t)\mathbf{j}$$
satisfy the equation for constants A, k, and ω, provided $\omega = k/C$.
b. Make a rough sketch showing the directions of $\mathbf{E}$ and $\mathbf{B}$.

64. Splitting a vector field Express the vector field $\mathbf{F} = \langle xy, 0, 0 \rangle$ in the form $\mathbf{V} + \mathbf{W}$, where $\nabla \cdot \mathbf{V} = 0$ and $\nabla \times \mathbf{W} = \mathbf{0}$.

65. Properties of div and curl Prove the following properties of the divergence and curl. Assume $\mathbf{F}$ and $\mathbf{G}$ are differentiable vector fields and c is a real number.

a. $\nabla \cdot (\mathbf{F} + \mathbf{G}) = \nabla \cdot \mathbf{F} + \nabla \cdot \mathbf{G}$
b. $\nabla \times (\mathbf{F} + \mathbf{G}) = (\nabla \times \mathbf{F}) + (\nabla \times \mathbf{G})$
c. $\nabla \cdot (c\mathbf{F}) = c(\nabla \cdot \mathbf{F})$
d. $\nabla \times (c\mathbf{F}) = c(\nabla \times \mathbf{F})$

66. Equal curls If two functions of one variable, f and g, have the property that $f' = g'$, then f and g differ by a constant. Prove or disprove: If $\mathbf{F}$ and $\mathbf{G}$ are nonconstant vector fields in $\mathbb{R}^2$ with curl $\mathbf{F} = $ curl $\mathbf{G}$ and div $\mathbf{F} = $ div $\mathbf{G}$ at all points of $\mathbb{R}^2$, then $\mathbf{F}$ and $\mathbf{G}$ differ by a constant vector.

67–72. Identities *Prove the following identities. Assume φ is a differentiable scalar-valued function and $\mathbf{F}$ and $\mathbf{G}$ are differentiable vector fields, all defined on a region of $\mathbb{R}^3$.*

67. $\nabla \cdot (\varphi\mathbf{F}) = \nabla\varphi \cdot \mathbf{F} + \varphi\nabla \cdot \mathbf{F}$ (Product Rule)

68. $\nabla \times (\varphi\mathbf{F}) = (\nabla\varphi \times \mathbf{F}) + (\varphi\nabla \times \mathbf{F})$ (Product Rule)

69. $\nabla \cdot (\mathbf{F} \times \mathbf{G}) = \mathbf{G} \cdot (\nabla \times \mathbf{F}) - \mathbf{F} \cdot (\nabla \times \mathbf{G})$

70. $\nabla \times (\mathbf{F} \times \mathbf{G}) = (\mathbf{G} \cdot \nabla)\mathbf{F} - \mathbf{G}(\nabla \cdot \mathbf{F}) - (\mathbf{F} \cdot \nabla)\mathbf{G} + \mathbf{F}(\nabla \cdot \mathbf{G})$

71. $\nabla(\mathbf{F} \cdot \mathbf{G}) = (\mathbf{G} \cdot \nabla)\mathbf{F} + (\mathbf{F} \cdot \nabla)\mathbf{G} + \mathbf{G} \times (\nabla \times \mathbf{F}) + \mathbf{F} \times (\nabla \times \mathbf{G})$

72. $\nabla \times (\nabla \times \mathbf{F}) = \nabla(\nabla \cdot \mathbf{F}) - (\nabla \cdot \nabla)\mathbf{F}$

73. Divergence of radial fields Prove that for a real number p, with $\mathbf{r} = \langle x, y, z \rangle$, $\nabla \cdot \dfrac{\langle x, y, z \rangle}{|\mathbf{r}|^p} = \dfrac{3 - p}{|\mathbf{r}|^p}$.

74. Gradients and radial fields Prove that for a real number p, with $\mathbf{r} = \langle x, y, z \rangle$, $\nabla\left(\dfrac{1}{|\mathbf{r}|^p}\right) = \dfrac{-p\mathbf{r}}{|\mathbf{r}|^{p+2}}$.

75. Divergence of gradient fields Prove that for a real number p, with $\mathbf{r} = \langle x, y, z \rangle$, $\nabla \cdot \nabla\left(\dfrac{1}{|\mathbf{r}|^p}\right) = \dfrac{p(p - 1)}{|\mathbf{r}|^{p+2}}$.

QUICK CHECK ANSWERS

1. The x-derivative of the divergence is applied to $f(y, z)$, which gives zero. Similarly, the y- and z-derivatives are zero. **2.** The net outward flux is 4π. **3.** In the curl, the first component of $\mathbf{F}$ is differentiated only with respect to y and z, so the contribution from the first component is zero. Similarly, the second and third components of $\mathbf{F}$ make no contribution to the curl. **4.** The divergence is a scalar-valued function. ◄

17.6 Surface Integrals

We have studied integrals on the real line, on regions in the plane, on solid regions in space, and along curves in space. One situation is still unexplored. Suppose a sphere has a known temperature distribution; perhaps it is cold near the poles and warm near the equator. How do you find the average temperature over the entire sphere? In analogy with other average value calculations, we should expect to "add up" the temperature values over the sphere and divide by the surface area of the sphere. Because the temperature varies continuously over the sphere, adding up means integrating. How do you integrate a function over a surface? This question leads to *surface integrals*.

It helps to keep curves, arc length, and line integrals in mind as we discuss surfaces, surface area, and surface integrals. What we discover about surfaces parallels what we already know about curves—all "lifted" up one dimension.

Parallel Concepts	
Curves	**Surfaces**
Arc length	Surface area
Line integrals	Surface integrals
One-parameter description	Two-parameter description

Parameterized Surfaces

A curve in $\mathbb{R}^2$ is defined parametrically by $\mathbf{r}(t) = \langle x(t), y(t) \rangle$, for $a \le t \le b$; it requires one parameter and two dependent variables. Stepping up one dimension to define a surface in $\mathbb{R}^3$, we need *two* parameters and *three* dependent variables. Letting u and v be parameters, the general parametric description of a surface has the form

$$\mathbf{r}(u, v) = \langle x(u, v), y(u, v), z(u, v) \rangle.$$

We make the assumption that the parameters vary over a rectangle $R = \{(u, v): a \le u \le b, c \le v \le d\}$ (**Figure 17.43**). As the parameters (u, v) vary over R, the vector $\mathbf{r}(u, v) = \langle x(u, v), y(u, v), z(u, v) \rangle$ sweeps out a surface S in $\mathbb{R}^3$.

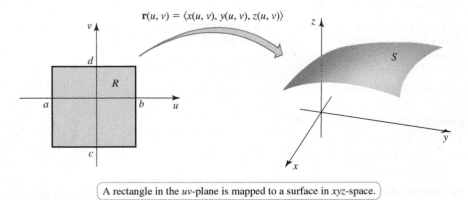

A rectangle in the *uv*-plane is mapped to a surface in *xyz*-space.

Figure 17.43

We work extensively with three surfaces that are easily described in parametric form. As with parameterized curves, a parametric description of a surface is not unique.

Cylinders   In Cartesian coordinates, the set

$$\{(x, y, z): x = a \cos \theta, y = a \sin \theta, 0 \le \theta \le 2\pi, 0 \le z \le h\},$$

where $a > 0$, is a cylindrical surface of radius a and height h with its axis along the z-axis. Using the parameters $u = \theta$ and $v = z$, a parametric description of the cylinder is

$$\mathbf{r}(u, v) = \langle x(u, v), y(u, v), z(u, v) \rangle = \langle a \cos u, a \sin u, v \rangle,$$

where $0 \le u \le 2\pi$ and $0 \le v \le h$ (**Figure 17.44**).

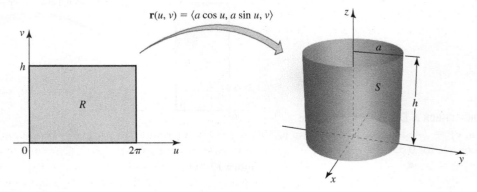

QUICK CHECK 1 Describe the surface $\mathbf{r}(u, v) = \langle 2 \cos u, 2 \sin u, v \rangle$, for $0 \le u \le \pi$ and $0 \le v \le 1$. ◀

Figure 17.44

Cones   The surface of a cone of height h and radius a with its vertex at the origin is described in cylindrical coordinates by

▶ Note that when $r = 0$, $z = 0$ and when $r = a$, $z = h$.

$$\{(r, \theta, z): 0 \le r \le a, 0 \le \theta \le 2\pi, z = rh/a\}.$$

For a fixed value of z, we have $r = az/h$; therefore, on the surface of the cone,

▶ Recall the relationships between polar and rectangular coordinates:

$x = r \cos \theta$, $y = r \sin \theta$, and $x^2 + y^2 = r^2$.

$$x = r \cos \theta = \frac{az}{h} \cos \theta \quad \text{and} \quad y = r \sin \theta = \frac{az}{h} \sin \theta.$$

Using the parameters $u = \theta$ and $v = z$, the parametric description of the conical surface is

$$\mathbf{r}(u, v) = \langle x(u, v), y(u, v), z(u, v) \rangle = \left\langle \frac{av}{h} \cos u, \frac{av}{h} \sin u, v \right\rangle,$$

where $0 \le u \le 2\pi$ and $0 \le v \le h$ (**Figure 17.45**).

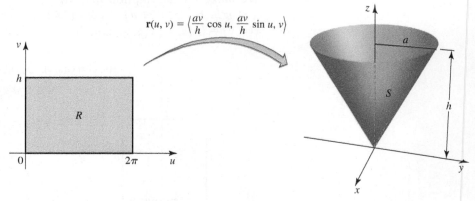

QUICK CHECK 2 Describe the surface $\mathbf{r}(u, v) = \langle v \cos u, v \sin u, v \rangle$, for $0 \le u \le \pi$ and $0 \le v \le 10$. ◀

Figure 17.45

▶ The complete cylinder, cone, and sphere are generated as the angle variable θ varies over the half-open interval $[0, 2\pi)$. As in previous chapters, we will use the closed interval $[0, 2\pi]$.

Spheres   The parametric description of a sphere of radius a centered at the origin comes directly from spherical coordinates:

$$\{(\rho, \varphi, \theta): \rho = a, 0 \le \varphi \le \pi, 0 \le \theta \le 2\pi\}.$$

Recall the following relationships among spherical and rectangular coordinates (Section 16.5):

$$x = a \sin \varphi \cos \theta, \qquad y = a \sin \varphi \sin \theta, \qquad z = a \cos \varphi.$$

When we define the parameters $u = \varphi$ and $v = \theta$, a parametric description of the sphere is

$$\mathbf{r}(u, v) = \langle a \sin u \cos v, a \sin u \sin v, a \cos u \rangle,$$

where $0 \le u \le \pi$ and $0 \le v \le 2\pi$ (**Figure 17.46**).

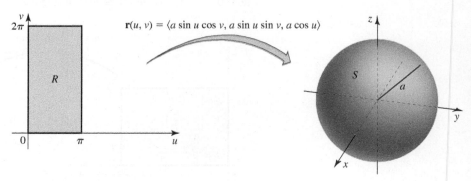

QUICK CHECK 3 Describe the surface $\mathbf{r}(u, v) = \langle 4 \sin u \cos v, 4 \sin u \sin v, 4 \cos u \rangle$, for $0 \le u \le \pi/2$ and $0 \le v \le \pi$. ◀

Figure 17.46

EXAMPLE 1 Parametric surfaces Find parametric descriptions for the following surfaces.

a. The plane $3x - 2y + z = 2$

b. The paraboloid $z = x^2 + y^2$, for $0 \le z \le 9$

SOLUTION

a. Defining the parameters $u = x$ and $v = y$, we find that

$$z = 2 - 3x + 2y = 2 - 3u + 2v.$$

Therefore, a parametric description of the plane is

$$\mathbf{r}(u, v) = \langle u, v, 2 - 3u + 2v \rangle,$$

for $-\infty < u < \infty$ and $-\infty < v < \infty$.

b. Thinking in terms of polar coordinates, we let $u = \theta$ and $v = \sqrt{z}$, which means that $z = v^2$. The equation of the paraboloid is $x^2 + y^2 = z = v^2$, so v plays the role of the polar coordinate r. Therefore, $x = v \cos \theta = v \cos u$ and $y = v \sin \theta = v \sin u$. A parametric description for the paraboloid is

$$\mathbf{r}(u, v) = \langle v \cos u, v \sin u, v^2 \rangle,$$

where $0 \le u \le 2\pi$ and $0 \le v \le 3$.

Alternatively, we could choose $u = \theta$ and $v = z$. The resulting description is

$$\mathbf{r}(u, v) = \langle \sqrt{v} \cos u, \sqrt{v} \sin u, v \rangle,$$

where $0 \le u \le 2\pi$ and $0 \le v \le 9$.   *Related Exercises 9, 12* ◄

Surface Integrals of Scalar-Valued Functions

We now develop the surface integral of a scalar-valued function f defined on a smooth parameterized surface S described by the equation

$$\mathbf{r}(u, v) = \langle x(u, v), y(u, v), z(u, v) \rangle,$$

where the parameters vary over a rectangle $R = \{(u, v): a \le u \le b, c \le v \le d\}$. The functions x, y, and z are assumed to have continuous partial derivatives with respect to u and v. The rectangular region R in the uv-plane is partitioned into rectangles, with sides of length Δu and Δv, that are ordered in some convenient way, for $k = 1, \ldots, n$. The kth rectangle R_k, which has area $\Delta A = \Delta u \Delta v$, corresponds to a curved patch S_k on the surface S (**Figure 17.47**), which has area ΔS_k. We let (u_k, v_k) be the lower-left corner point of R_k. The parameterization then assigns (u_k, v_k) to a point $P(x(u_k, v_k), y(u_k, v_k), z(u_k, v_k))$, or more simply, $P(x_k, y_k, z_k)$, on S_k. To construct the surface integral, we define a Riemann sum, which adds up function values multiplied by areas of the respective patches:

▶ A more general approach allows (u_k, v_k) to be an arbitrary point in the kth rectangle. The outcome of the two approaches is the same.

$$\sum_{k=1}^{n} f(x(u_k, v_k), y(u_k, v_k), z(u_k, v_k)) \Delta S_k.$$

Figure 17.47

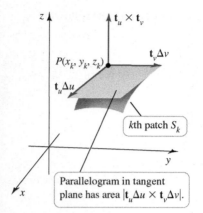

$P(x_k, y_k, z_k)$

Parallelogram in tangent plane has area $|\mathbf{t}_u \Delta u \times \mathbf{t}_v \Delta v|$.

Figure 17.48

➤ In general, the vectors $\mathbf{t}_u$ and $\mathbf{t}_v$ are different for each patch, so they should carry a subscript k. To keep the notation as simple as possible, we have suppressed the subscripts on these vectors with the understanding that they change with k. These tangent vectors are given by partial derivatives because in each case, either u or v is held constant, while the other variable changes.

The crucial step is computing ΔS_k, the area of the kth patch S_k.

Figure 17.48 shows the patch S_k and the point $P(x_k, y_k, z_k)$. Two special vectors are tangent to the surface at P; these vectors lie in the plane tangent to S at P.

- $\mathbf{t}_u$ is a vector tangent to the surface corresponding to a change in u with v constant in the uv-plane.

- $\mathbf{t}_v$ is a vector tangent to the surface corresponding to a change in v with u constant in the uv-plane.

Because the surface S may be written $\mathbf{r}(u, v) = \langle x(u, v), y(u, v), z(u, v) \rangle$, a tangent vector corresponding to a change in u with v fixed is

$$\mathbf{t}_u = \frac{\partial \mathbf{r}}{\partial u} = \left\langle \frac{\partial x}{\partial u}, \frac{\partial y}{\partial u}, \frac{\partial z}{\partial u} \right\rangle.$$

Similarly, a tangent vector corresponding to a change in v with u fixed is

$$\mathbf{t}_v = \frac{\partial \mathbf{r}}{\partial v} = \left\langle \frac{\partial x}{\partial v}, \frac{\partial y}{\partial v}, \frac{\partial z}{\partial v} \right\rangle.$$

Now consider an increment Δu in u with v fixed; the corresponding change in $\mathbf{r}$, which is $\mathbf{r}(u + \Delta u, v) - \mathbf{r}(u, v)$, can be approximated using the definition of the partial derivative of $\mathbf{r}$ with respect to u. Specifically, when Δu is small, we have

$$\frac{\partial \mathbf{r}}{\partial u} \approx \frac{1}{\Delta u}\bigl(\mathbf{r}(u + \Delta u, v) - \mathbf{r}(u, v)\bigr).$$

Multiplying both sides of this equation by Δu and recalling that $\mathbf{t}_u = \dfrac{\partial \mathbf{r}}{\partial u}$, we see that the change in $\mathbf{r}$ corresponding to the increment Δu is approximated by the vector

$$\mathbf{t}_u \Delta u \approx \underbrace{\mathbf{r}(u + \Delta u, v) - \mathbf{r}(u, v)}_{\text{change in } \mathbf{r} \text{ corresponding to } \Delta u}.$$

Using a similar line of reasoning, the change in $\mathbf{r}$ corresponding to the increment Δv (with u fixed) is approximated by the vector

$$\mathbf{t}_v \Delta v \approx \underbrace{\mathbf{r}(u, v + \Delta v) - \mathbf{r}(u, v)}_{\text{change in } \mathbf{r} \text{ corresponding to } \Delta v}.$$

As nonzero scalar multiples of $\mathbf{t}_u$ and $\mathbf{t}_v$, the vectors $\mathbf{t}_u \Delta u$ and $\mathbf{t}_v \Delta v$ are also tangent to the surface. They determine a parallelogram that lies in the plane tangent to S at P (Figure 17.48); the area of this parallelogram approximates the area of the kth patch S_k, which is ΔS_k.

Appealing to the cross product (Section 13.4), the area of the parallelogram is

$$\bigl|\mathbf{t}_u \Delta u \times \mathbf{t}_v \Delta v\bigr| = \bigl|\mathbf{t}_u \times \mathbf{t}_v\bigr| \, \Delta u \, \Delta v \approx \Delta S_k.$$

Note that $\mathbf{t}_u \times \mathbf{t}_v$ is evaluated at (u_k, v_k) and is a vector normal to the surface at P, which we assume to be nonzero at all points of S.

We write the Riemann sum with the observation that the areas of the parallelograms approximate the areas of the patches S_k:

$$\sum_{k=1}^{n} f(x(u_k, v_k), y(u_k, v_k), z(u_k, v_k)) \Delta S_k$$

$$\approx \sum_{k=1}^{n} f(x(u_k, v_k), y(u_k, v_k), z(u_k, v_k)) \underbrace{\bigl|\mathbf{t}_u \times \mathbf{t}_v\bigr| \Delta u \, \Delta v}_{\approx \Delta S_k}.$$

We now assume f is continuous on S. As Δu and Δv approach zero, the areas of the parallelograms approach the areas of the corresponding patches on S. We define the limit

> ▶ The role that the factor $|\mathbf{t}_u \times \mathbf{t}_v|\, dA$ plays in surface integrals is analogous to the role played by $|\mathbf{r}'(t)|\, dt$ in line integrals.

of this Riemann sum to be the surface integral of f over S, which we write $\iint_S f(x, y, z)\, dS$. The surface integral is evaluated as an ordinary double integral over the region R in the uv-plane:

$$\iint_S f(x, y, z)\, dS = \lim_{\Delta u, \Delta v \to 0} \sum_{k=1}^{n} f(x(u_k, v_k), y(u_k, v_k), z(u_k, v_k))|\mathbf{t}_u \times \mathbf{t}_v|\Delta u\, \Delta v$$

$$= \iint_R f(x(u, v), y(u, v), z(u, v))|\mathbf{t}_u \times \mathbf{t}_v|\, dA.$$

If R is a rectangular region, as we have assumed, the double integral becomes an iterated integral with respect to u and v with constant limits. In the special case that $f(x, y, z) = 1$, the integral gives the surface area of S.

> ▶ The condition that $\mathbf{t}_u \times \mathbf{t}_v$ be nonzero means $\mathbf{t}_u$ and $\mathbf{t}_v$ are nonzero and not parallel. If $\mathbf{t}_u \times \mathbf{t}_v \neq \mathbf{0}$ at all points, then the surface is *smooth*. The value of the integral is independent of the parameterization of S.

DEFINITION Surface Integral of Scalar-Valued Functions on Parameterized Surfaces

Let f be a continuous scalar-valued function on a smooth surface S given parametrically by $\mathbf{r}(u, v) = \langle x(u, v), y(u, v), z(u, v) \rangle$, where u and v vary over $R = \{(u, v): a \leq u \leq b, c \leq v \leq d\}$. Assume also that the tangent vectors

$$\mathbf{t}_u = \frac{\partial \mathbf{r}}{\partial u} = \left\langle \frac{\partial x}{\partial u}, \frac{\partial y}{\partial u}, \frac{\partial z}{\partial u} \right\rangle \text{ and } \mathbf{t}_v = \frac{\partial \mathbf{r}}{\partial v} = \left\langle \frac{\partial x}{\partial v}, \frac{\partial y}{\partial v}, \frac{\partial z}{\partial v} \right\rangle$$ are continuous on R and

the normal vector $\mathbf{t}_u \times \mathbf{t}_v$ is nonzero on R. Then the **surface integral of f over S** is

$$\iint_S f(x, y, z)\, dS = \iint_R f(x(u, v), y(u, v), z(u, v))|\mathbf{t}_u \times \mathbf{t}_v|\, dA.$$

If $f(x, y, z) = 1$, this integral equals the surface area of S.

EXAMPLE 2 Surface area of a cylinder and sphere Find the surface area of the following surfaces.

a. A cylinder with radius $a > 0$ and height h (excluding the circular ends)

b. A sphere of radius a

SOLUTION The critical step is evaluating the normal vector $\mathbf{t}_u \times \mathbf{t}_v$. It needs to be done only once for any given surface.

a. As shown before, a parametric description of the cylinder is

$$\mathbf{r}(u, v) = \langle x(u, v), y(u, v), z(u, v) \rangle = \langle a \cos u, a \sin u, v \rangle,$$

where $0 \leq u \leq 2\pi$ and $0 \leq v \leq h$. The required normal vector is

$$\mathbf{t}_u \times \mathbf{t}_v = \begin{vmatrix} \mathbf{i} & \mathbf{j} & \mathbf{k} \\ \dfrac{\partial x}{\partial u} & \dfrac{\partial y}{\partial u} & \dfrac{\partial z}{\partial u} \\ \dfrac{\partial x}{\partial v} & \dfrac{\partial y}{\partial v} & \dfrac{\partial z}{\partial v} \end{vmatrix} \qquad \text{Definition of cross product}$$

$$= \begin{vmatrix} \mathbf{i} & \mathbf{j} & \mathbf{k} \\ -a \sin u & a \cos u & 0 \\ 0 & 0 & 1 \end{vmatrix} \qquad \text{Evaluate derivatives.}$$

$$= \langle a \cos u, a \sin u, 0 \rangle. \qquad \text{Compute cross product.}$$

Notice that this normal vector points outward from the cylinder, away from the z-axis (**Figure 17.49**). It follows that

$$|\mathbf{t}_u \times \mathbf{t}_v| = \sqrt{a^2 \cos^2 u + a^2 \sin^2 u} = a.$$

Normal vector $\langle a \cos u, a \sin u, 0 \rangle$, magnitude $= a$

Cylinder: $\mathbf{r}(u, v) = \langle a \cos u, a \sin u, v \rangle$, $0 \leq u \leq 2\pi$ and $0 \leq v \leq h$

Figure 17.49

Setting $f(x, y, z) = 1$, the surface area of the cylinder is

$$\iint\limits_{S} 1 \, dS = \iint\limits_{R} \underbrace{|\mathbf{t}_u \times \mathbf{t}_v|}_{a} \, dA = \int_0^{2\pi} \int_0^h a \, dv \, du = 2\pi a h,$$

confirming the formula for the surface area of a cylinder (excluding the ends).

b. A parametric description of the sphere is

$$\mathbf{r}(u, v) = \langle a \sin u \cos v, \, a \sin u \sin v, \, a \cos u \rangle,$$

where $0 \le u \le \pi$ and $0 \le v \le 2\pi$. The required normal vector is

▶ Recall that for the sphere, $u = \varphi$ and $v = \theta$, where φ and θ are spherical coordinates. The element of surface area in spherical coordinates is $dS = a^2 \sin \varphi \, d\varphi \, d\theta$.

$$\mathbf{t}_u \times \mathbf{t}_v = \begin{vmatrix} \mathbf{i} & \mathbf{j} & \mathbf{k} \\ a \cos u \cos v & a \cos u \sin v & -a \sin u \\ -a \sin u \sin v & a \sin u \cos v & 0 \end{vmatrix}$$

$$= \langle a^2 \sin^2 u \cos v, \, a^2 \sin^2 u \sin v, \, a^2 \sin u \cos u \rangle.$$

Computing $|\mathbf{t}_u \times \mathbf{t}_v|$ requires several steps (Exercise 70). However, the needed result is quite simple: $|\mathbf{t}_u \times \mathbf{t}_v| = a^2 \sin u$ and the normal vector $\mathbf{t}_u \times \mathbf{t}_v$ points outward from the surface of the sphere (**Figure 17.50**). With $f(x, y, z) = 1$, the surface area of the sphere is

$$\iint\limits_{S} 1 \, dS = \iint\limits_{R} \underbrace{|\mathbf{t}_u \times \mathbf{t}_v|}_{a^2 \sin u} \, dA = \int_0^{2\pi} \int_0^{\pi} a^2 \sin u \, du \, dv = 4\pi a^2,$$

confirming the formula for the surface area of a sphere.

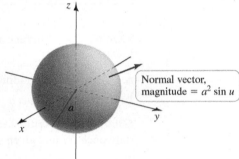

Sphere:
$\mathbf{r}(u, v) = \langle a \sin u \cos v, \, a \sin u \sin v, \, a \cos u \rangle$,
$0 \le u \le \pi$ and $0 \le v \le 2\pi$

Normal vector, magnitude $= a^2 \sin u$

Figure 17.50

Related Exercises 19, 22 ◀

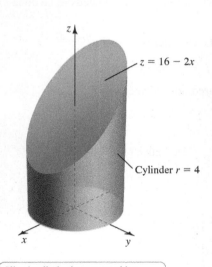

$z = 16 - 2x$

Cylinder $r = 4$

Sliced cylinder is generated by
$\mathbf{r}(u, v) = \langle 4 \cos u, 4 \sin u, v \rangle$, where
$0 \le u \le 2\pi, 0 \le v \le 16 - 8 \cos u$.

Figure 17.51

EXAMPLE 3   Surface area of a partial cylinder   Find the surface area of the cylinder $\{(r, \theta): r = 4, 0 \le \theta \le 2\pi\}$ between the planes $z = 0$ and $z = 16 - 2x$ (excluding the top and bottom surfaces).

SOLUTION   Figure 17.51 shows the cylinder bounded by the two planes. With $u = \theta$ and $v = z$, a parametric description of the cylinder is

$$\mathbf{r}(u, v) = \langle x(u, v), y(u, v), z(u, v) \rangle = \langle 4 \cos u, 4 \sin u, v \rangle.$$

The challenge is finding the limits on v, which is the z-coordinate. The plane $z = 16 - 2x$ intersects the cylinder in an ellipse; along this ellipse, as u varies between 0 and 2π, the parameter v also changes. To find the relationship between u and v along this intersection curve, notice that at any point on the cylinder, we have $x = 4 \cos u$ (remember that $u = \theta$). Making this substitution in the equation of the plane, we have

$$z = 16 - 2x = 16 - 2(4 \cos u) = 16 - 8 \cos u.$$

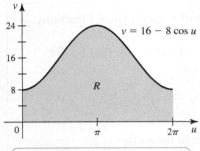

Region of integration in the uv-plane is $R = \{(u, v) : 0 \le u \le 2\pi, 0 \le v \le 16 - 8 \cos u\}$.

Figure 17.52

Substituting $v = z$, the relationship between u and v is $v = 16 - 8 \cos u$ (**Figure 17.52**). Therefore, the region of integration in the uv-plane is

$$R = \{(u, v) : 0 \le u \le 2\pi, 0 \le v \le 16 - 8 \cos u\}.$$

Recall from Example 2a that for the cylinder, $\left| \mathbf{t}_u \times \mathbf{t}_v \right| = a = 4$. Setting $f(x, y, z) = 1$, the surface integral for the area is

$$\iint_S 1 \, dS = \iint_R \underbrace{\left| \mathbf{t}_u \times \mathbf{t}_v \right|}_{4} \, dA$$

$$= \int_0^{2\pi} \int_0^{16 - 8 \cos u} 4 \, dv \, du$$

$$= 4 \int_0^{2\pi} (16 - 8 \cos u) \, du \qquad \text{Evaluate inner integral.}$$

$$= 4(16u - 8 \sin u) \Big|_0^{2\pi} \qquad \text{Evaluate outer integral.}$$

$$= 128\pi. \qquad \text{Simplify.}$$

Related Exercise 24 ◄

EXAMPLE 4   Average temperature on a sphere The temperature on the surface of a sphere of radius a varies with latitude according to the function $T(\varphi, \theta) = 10 + 50 \sin \varphi$, for $0 \le \varphi \le \pi$ and $0 \le \theta \le 2\pi$ (φ and θ are spherical coordinates, so the temperature is $10°$ at the poles, increasing to $60°$ at the equator). Find the average temperature over the sphere.

SOLUTION We use the parametric description of a sphere. With $u = \varphi$ and $v = \theta$, the temperature function becomes $f(u, v) = 10 + 50 \sin u$. Integrating the temperature over the sphere using the fact that $\left| \mathbf{t}_u \times \mathbf{t}_v \right| = a^2 \sin u$ (Example 2b), we have

$$\iint_S (10 + 50 \sin u) \, dS = \iint_R (10 + 50 \sin u) \underbrace{\left| \mathbf{t}_u \times \mathbf{t}_v \right|}_{a^2 \sin u} \, dA$$

$$= \int_0^\pi \int_0^{2\pi} (10 + 50 \sin u) a^2 \sin u \, dv \, du$$

$$= 2\pi a^2 \int_0^\pi (10 + 50 \sin u) \sin u \, du \qquad \text{Evaluate inner integral.}$$

$$= 10\pi a^2 (4 + 5\pi). \qquad \text{Evaluate outer integral.}$$

The average temperature is the integrated temperature $10\pi a^2 (4 + 5\pi)$ divided by the surface area of the sphere $4\pi a^2$, so the average temperature is $(20 + 25\pi)/2 \approx 49.3°$.

Related Exercise 42 ◄

Surface Integrals on Explicitly Defined Surfaces Suppose a smooth surface S is defined not parametrically, but explicitly, in the form $z = g(x, y)$ over a region R in the xy-plane. Such a surface may be treated as a parameterized surface. We simply define the parameters to be $u = x$ and $v = y$. Making these substitutions into the expression for $\mathbf{t}_u$ and $\mathbf{t}_v$, a short calculation (Exercise 71) reveals that $\mathbf{t}_u = \mathbf{t}_x = \langle 1, 0, z_x \rangle$, $\mathbf{t}_v = \mathbf{t}_y = \langle 0, 1, z_y \rangle$, and the required normal vector is

$$\mathbf{t}_x \times \mathbf{t}_y = \langle -z_x, -z_y, 1 \rangle.$$

It follows that

$$\left| \mathbf{t}_x \times \mathbf{t}_y \right| = \left| \langle -z_x, -z_y, 1 \rangle \right| = \sqrt{z_x^2 + z_y^2 + 1}.$$

With these observations, the surface integral over S can be expressed as a double integral over a region R in the xy-plane.

▶ This is a familiar result: A normal to the surface $z = g(x, y)$ at a point is a constant multiple of the gradient of $z - g(x, y)$, which is $\langle -g_x, -g_y, 1 \rangle = \langle -z_x, -z_y, 1 \rangle$. The factor $\sqrt{z_x^2 + z_y^2 + 1}$ is analogous to the factor $\sqrt{f'(x)^2 + 1}$ that appears in arc length integrals.

▶ If the surface S in Theorem 17.14 is generated by revolving a curve in the xy-plane about the x-axis, the theorem gives the standard surface area formula for surfaces of revolution (Exercise 75).

THEOREM 17.14   Evaluation of Surface Integrals of Scalar-Valued Functions on Explicitly Defined Surfaces

Let f be a continuous function on a smooth surface S given by $z = g(x, y)$, for (x, y) in a region R. The surface integral of f over S is

$$\iint_S f(x, y, z)\, dS = \iint_R f(x, y, g(x, y))\sqrt{z_x^2 + z_y^2 + 1}\, dA.$$

If $f(x, y, z) = 1$, the surface integral equals the area of the surface.

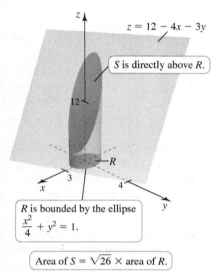

R is bounded by the ellipse $\dfrac{x^2}{4} + y^2 = 1$.

Area of $S = \sqrt{26} \times$ area of R.

Figure 17.53

QUICK CHECK 4 The plane $z = y$ forms a 45° angle with the xy-plane. Suppose the plane is the roof of a room and the xy-plane is the floor of the room. Then 1 ft² on the floor becomes how many square feet when projected on the roof? ◀

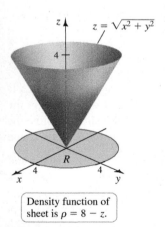

Density function of sheet is $\rho = 8 - z$.

Figure 17.54

EXAMPLE 5   Area of a roof over an ellipse Find the area of the surface S that lies in the plane $z = 12 - 4x - 3y$ directly above the region R bounded by the ellipse $x^2/4 + y^2 = 1$ (**Figure 17.53**).

SOLUTION Because we are computing the area of the surface, we take $f(x, y, z) = 1$. Note that $z_x = -4$ and $z_y = -3$, so the factor $\sqrt{z_x^2 + z_y^2 + 1}$ has the value $\sqrt{(-4)^2 + (-3)^2 + 1} = \sqrt{26}$ (a constant because the surface is a plane). The relevant surface integral is

$$\iint_S 1\, dS = \iint_R \underbrace{\sqrt{z_x^2 + z_y^2 + 1}}_{\sqrt{26}}\, dA = \sqrt{26}\iint_R dA.$$

The double integral that remains is simply the area of the region R bounded by the ellipse. Because the ellipse has semiaxes of length $a = 2$ and $b = 1$, its area is $\pi ab = 2\pi$. Therefore, the area of S is $2\pi\sqrt{26}$.

This result has a useful interpretation. The plane surface S is not horizontal, so it has a greater area than the horizontal region R beneath it. The factor that converts the area of R to the area of S is $\sqrt{26}$. Notice that if the roof *were* horizontal, then the surface would be $z = c$, the area conversion factor would be 1, and the area of the roof would equal the area of the floor beneath it.

Related Exercises 29–30 ◀

EXAMPLE 6   Mass of a conical sheet A thin conical sheet is described by the surface $z = (x^2 + y^2)^{1/2}$, for $0 \le z \le 4$. The density of the sheet in g/cm² is $\rho = f(x, y, z) = (8 - z)$ (decreasing from 8 g/cm² at the vertex to 4 g/cm² at the top of the cone; **Figure 17.54**). What is the mass of the cone?

SOLUTION We find the mass by integrating the density function over the surface of the cone. The projection of the cone on the xy-plane is found by setting $z = 4$ (the top of the cone) in the equation of the cone. We find that $(x^2 + y^2)^{1/2} = 4$; therefore, the region of integration is the disk $R = \{(x, y): x^2 + y^2 \le 16\}$. The next step is to compute z_x and z_y in order to evaluate $\sqrt{z_x^2 + z_y^2 + 1}$. Differentiating $z^2 = x^2 + y^2$ implicitly gives $2zz_x = 2x$, or $z_x = x/z$. Similarly, $z_y = y/z$. Using the fact that $z^2 = x^2 + y^2$, we have

$$\sqrt{z_x^2 + z_y^2 + 1} = \sqrt{(x/z)^2 + (y/z)^2 + 1} = \sqrt{\underbrace{\frac{x^2 + y^2}{z^2}}_{1} + 1} = \sqrt{2}.$$

To integrate the density over the conical surface, we set $f(x, y, z) = 8 - z$. Replacing z in the integrand by $r = (x^2 + y^2)^{1/2}$ and using polar coordinates, the mass in grams is given by

$$\iint_S f(x, y, z)\, dS = \iint_R f(x, y, z)\underbrace{\sqrt{z_x^2 + z_y^2 + 1}}_{\sqrt{2}}\, dA$$

$$= \sqrt{2}\iint_R (8 - z)\, dA \qquad \text{Substitute.}$$

$$= \sqrt{2} \iint\limits_{R} (8 - \sqrt{x^2 + y^2}) \, dA \qquad z = \sqrt{x^2 + y^2}$$

$$= \sqrt{2} \int_{0}^{2\pi} \int_{0}^{4} (8 - r) \, r \, dr \, d\theta \qquad \text{Polar coordinates}$$

$$= \sqrt{2} \int_{0}^{2\pi} \left(4r^2 - \frac{r^3}{3} \right) \Big|_{0}^{4} \, d\theta \qquad \text{Evaluate inner integral.}$$

$$= \frac{128\sqrt{2}}{3} \int_{0}^{2\pi} d\theta \qquad \text{Simplify.}$$

$$= \frac{256\pi\sqrt{2}}{3} \approx 379. \qquad \text{Evaluate outer integral.}$$

As a check, note that the surface area of the cone is $\pi r \sqrt{r^2 + h^2} \approx 71 \text{ cm}^2$. If the entire cone had the maximum density $\rho = 8 \text{ g/cm}^2$, its mass would be approximately 568 g. If the entire cone had the minimum density $\rho = 4 \text{ g/cm}^2$, its mass would be approximately 284 g. The actual mass is between these extremes and closer to the low value because the cone is lighter at the top, where the surface area is greater.

Related Exercise 36 ◄

Table 17.3 summarizes the essential relationships for the explicit and parametric descriptions of cylinders, cones, spheres, and paraboloids. The listed normal vectors are chosen to point away from the z-axis.

Table 17.3

Surface	Explicit Description $z = g(x, y)$		Parametric Description	
	Equation	Normal vector; magnitude $\pm\langle -z_x, -z_y, 1 \rangle; |\langle -z_x, -z_y, 1 \rangle|$	Equation	Normal vector; magnitude $t_u \times t_v; |t_u \times t_v|$
Cylinder	$x^2 + y^2 = a^2$, $0 \le z \le h$	$\langle x, y, 0 \rangle; a$	$\mathbf{r} = \langle a \cos u, a \sin u, v \rangle$, $0 \le u \le 2\pi, 0 \le v \le h$	$\langle a \cos u, a \sin u, 0 \rangle; a$
Cone	$z^2 = x^2 + y^2$, $0 \le z \le h$	$\langle x/z, y/z, -1 \rangle; \sqrt{2}$	$\mathbf{r} = \langle v \cos u, v \sin u, v \rangle$, $0 \le u \le 2\pi, 0 \le v \le h$	$\langle v \cos u, v \sin u, -v \rangle; \sqrt{2}v$
Sphere	$x^2 + y^2 + z^2 = a^2$	$\langle x/z, y/z, 1 \rangle; a/z$	$\mathbf{r} = \langle a \sin u \cos v,$ $a \sin u \sin v, a \cos u \rangle$, $0 \le u \le \pi, 0 \le v \le 2\pi$	$\langle a^2 \sin^2 u \cos v, a^2 \sin^2 u \sin v,$ $a^2 \sin u \cos u \rangle; a^2 \sin u$
Paraboloid	$z = x^2 + y^2$, $0 \le z \le h$	$\langle 2x, 2y, -1 \rangle; \sqrt{1 + 4(x^2 + y^2)}$	$\mathbf{r} = \langle v \cos u, v \sin u, v^2 \rangle$, $0 \le u \le 2\pi, 0 \le v \le \sqrt{h}$	$\langle 2v^2 \cos u, 2v^2 \sin u, -v \rangle; v\sqrt{1 + 4v^2}$

QUICK CHECK 5 Explain why the explicit description for a cylinder $x^2 + y^2 = a^2$ cannot be used for a surface integral over a cylinder, and a parametric description must be used. ◄

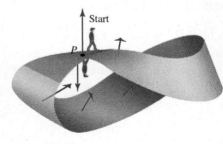

Figure 17.55

Surface Integrals of Vector Fields

Before beginning a discussion of surface integrals of vector fields, we must address two technical issues about surfaces and normal vectors.

The surfaces we consider in this text are called **two-sided**, or **orientable**, surfaces. To be orientable, a surface must have the property that the normal vectors vary continuously over the surface. In other words, when you walk on any closed path on an orientable surface and return to your starting point, your head must point in the same direction it did when you started. A well-known example of a *nonorientable* surface is the Möbius strip (**Figure 17.55**). Suppose you start walking the length of the Möbius strip at a point P with your head pointing upward. When you return to P, your head points in the opposite direction, or downward. Therefore, the Möbius strip is not orientable.

At any point of a parameterized orientable surface, there are two unit normal vectors. Therefore, the second point concerns the orientation of the surface or, equivalently, the direction of the normal vector. Once the direction of the normal vector is determined, the surface becomes **oriented**.

Surfaces that enclose a region are oriented so normal vectors point in the outward direction.

For other surfaces, the orientation of the surface must be specified.

Figure 17.56

We make the common assumption that—unless specified otherwise—a closed orientable surface that fully encloses a region (such as a sphere) is oriented so that the normal vectors point in the *outward direction*. For a surface that does not enclose a region in $\mathbb{R}^3$, the orientation must be specified in some way. For example, we might specify that the normal vectors for a particular surface point in the general direction of the positive z-axis; that is, in an upward direction (**Figure 17.56**).

Now recall that the parameterization of a surface defines a normal vector $\mathbf{t}_u \times \mathbf{t}_v$ at each point. In many cases, the normal vectors are consistent with the specified orientation, in which case no adjustments need to be made. If the direction of $\mathbf{t}_u \times \mathbf{t}_v$ is not consistent with the specified orientation, then the sign of $\mathbf{t}_u \times \mathbf{t}_v$ must be reversed before doing calculations. This process is demonstrated in the following examples.

Flux Integrals   It turns out that the most common surface integral of a vector field is a *flux integral*. Consider a vector field $\mathbf{F} = \langle f, g, h \rangle$, continuous on a region in $\mathbb{R}^3$, that represents the flow of a fluid or the transport of a substance. Given a smooth oriented surface S, we aim to compute the net flux of the vector field across the surface. In a small region containing a point P, the flux across the surface is proportional to the component of $\mathbf{F}$ in the direction of the unit normal vector $\mathbf{n}$ at P. If θ is the angle between $\mathbf{F}$ and $\mathbf{n}$, then this component is $\mathbf{F} \cdot \mathbf{n} = |\mathbf{F}||\mathbf{n}| \cos \theta = |\mathbf{F}| \cos \theta$ (because $|\mathbf{n}| = 1$; **Figure 17.57a**). We have the following special cases.

- If $\mathbf{F}$ and the unit normal vector are aligned at P ($\theta = 0$), then the component of $\mathbf{F}$ in the direction $\mathbf{n}$ is $\mathbf{F} \cdot \mathbf{n} = |\mathbf{F}|$; that is, all of $\mathbf{F}$ flows across the surface in the direction of $\mathbf{n}$ (**Figure 17.57b**).

- If $\mathbf{F}$ and the unit normal vector point in opposite directions at P ($\theta = \pi$), then the component of $\mathbf{F}$ in the direction $\mathbf{n}$ is $\mathbf{F} \cdot \mathbf{n} = -|\mathbf{F}|$; that is, all of $\mathbf{F}$ flows across the surface in the direction opposite that of $\mathbf{n}$ (**Figure 17.57c**).

- If $\mathbf{F}$ and the unit normal vector are orthogonal at P ($\theta = \pi/2$), then the component of $\mathbf{F}$ in the direction $\mathbf{n}$ is $\mathbf{F} \cdot \mathbf{n} = 0$; that is, none of $\mathbf{F}$ flows across the surface at that point (**Figure 17.57d**).

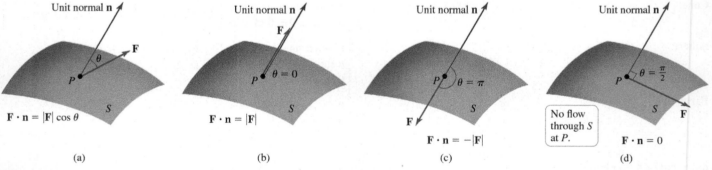

Figure 17.57

The flux integral, denoted $\iint_S \mathbf{F} \cdot \mathbf{n} \, dS$ or $\iint_S \mathbf{F} \cdot d\mathbf{S}$, simply adds up the components of $\mathbf{F}$ normal to the surface at all points of the surface. Notice that $\mathbf{F} \cdot \mathbf{n}$ is a scalar-valued function. Here is how the flux integral is computed.

Suppose the smooth oriented surface S is parameterized in the form

$$\mathbf{r}(u, v) = \langle x(u, v), y(u, v), z(u, v) \rangle,$$

where u and v vary over a region R in the uv-plane. The required vector normal to the surface at a point is $\mathbf{t}_u \times \mathbf{t}_v$, which we assume to be consistent with the orientation of S.

▶ If $\mathbf{t}_u \times \mathbf{t}_v$ is not consistent with the specified orientation, its sign must be reversed.

Therefore, the *unit* normal vector consistent with the orientation is $\mathbf{n} = \dfrac{\mathbf{t}_u \times \mathbf{t}_v}{|\mathbf{t}_u \times \mathbf{t}_v|}$.

Appealing to the definition of the surface integral for parameterized surfaces, the flux integral is

$$\iint_S \mathbf{F} \cdot \mathbf{n}\, dS = \iint_R \mathbf{F} \cdot \mathbf{n} |\mathbf{t}_u \times \mathbf{t}_v|\, dA \qquad \text{Definition of surface integral}$$

$$= \iint_R \mathbf{F} \cdot \underbrace{\frac{\mathbf{t}_u \times \mathbf{t}_v}{|\mathbf{t}_u \times \mathbf{t}_v|}}_{\mathbf{n}} |\mathbf{t}_u \times \mathbf{t}_v|\, dA \qquad \text{Substitute for } \mathbf{n}.$$

$$= \iint_R \mathbf{F} \cdot (\mathbf{t}_u \times \mathbf{t}_v)\, dA. \qquad \text{Convenient cancellation}$$

The remarkable occurrence in the flux integral is the cancellation of the factor $|\mathbf{t}_u \times \mathbf{t}_v|$.

The special case in which the surface S is specified in the form $z = s(x, y)$ follows directly by recalling that the required normal vector is $\mathbf{t}_u \times \mathbf{t}_v = \langle -z_x, -z_y, 1 \rangle$. In this case, with $\mathbf{F} = \langle f, g, h \rangle$, the integrand of the surface integral is $\mathbf{F} \cdot (\mathbf{t}_u \times \mathbf{t}_v) = -fz_x - gz_y + h$.

▶ The value of the surface integral is independent of the parameterization. However, in contrast to a surface integral of a scalar-valued function, the value of a surface integral of a vector field depends on the orientation of the surface. Changing the orientation changes the sign of the result.

DEFINITION Surface Integral of a Vector Field

Suppose $\mathbf{F} = \langle f, g, h \rangle$ is a continuous vector field on a region of $\mathbb{R}^3$ containing a smooth oriented surface S. If S is defined parametrically as $\mathbf{r}(u, v) = \langle x(u, v), y(u, v), z(u, v) \rangle$, for (u, v) in a region R, then

$$\iint_S \mathbf{F} \cdot \mathbf{n}\, dS = \iint_R \mathbf{F} \cdot (\mathbf{t}_u \times \mathbf{t}_v)\, dA,$$

where $\mathbf{t}_u = \dfrac{\partial \mathbf{r}}{\partial u} = \left\langle \dfrac{\partial x}{\partial u}, \dfrac{\partial y}{\partial u}, \dfrac{\partial z}{\partial u} \right\rangle$ and $\mathbf{t}_v = \dfrac{\partial \mathbf{r}}{\partial v} = \left\langle \dfrac{\partial x}{\partial v}, \dfrac{\partial y}{\partial v}, \dfrac{\partial z}{\partial v} \right\rangle$ are continuous on R, the normal vector $\mathbf{t}_u \times \mathbf{t}_v$ is nonzero on R, and the direction of the normal vector is consistent with the orientation of S. If S is defined in the form $z = s(x, y)$, for (x, y) in a region R, then

$$\iint_S \mathbf{F} \cdot \mathbf{n}\, dS = \iint_R (-fz_x - gz_y + h)\, dA.$$

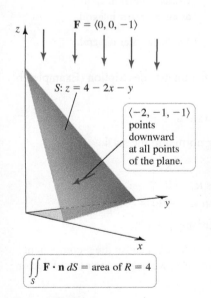

$$\iint_S \mathbf{F} \cdot \mathbf{n}\, dS = \text{area of } R = 4$$

Figure 17.58

EXAMPLE 7 Rain on a roof Consider the vertical vector field $\mathbf{F} = \langle 0, 0, -1 \rangle$, corresponding to a constant downward flow. Find the flux in the downward direction across the surface S, which is the plane $z = 4 - 2x - y$ in the first octant.

SOLUTION In this case, the surface is given explicitly. With $z = 4 - 2x - y$, we have $z_x = -2$ and $z_y = -1$. Therefore, the required normal vector is $\langle -z_x, -z_y, 1 \rangle = \langle 2, 1, 1 \rangle$, which points *upward* (the z-component of the vector is positive). Because we are interested in the *downward* flux of $\mathbf{F}$ across S, the surface must be oriented such that the normal vectors point downward. So we take the normal vector to be $\langle -2, -1, -1 \rangle$ (**Figure 17.58**). Letting R be the region in the xy-plane beneath S and noting that $\mathbf{F} = \langle f, g, h \rangle = \langle 0, 0, -1 \rangle$, the flux integral is

$$\iint_S \mathbf{F} \cdot \mathbf{n}\, dS = \iint_R \langle 0, 0, -1 \rangle \cdot \langle -2, -1, -1 \rangle\, dA = \iint_R dA = \text{area of } R.$$

The base R is a triangle in the xy-plane with vertices $(0, 0)$, $(2, 0)$, and $(0, 4)$, so its area is 4. Therefore, the *downward* flux across S is 4. This flux integral has an interesting interpretation. If the vector field $\mathbf{F}$ represents the rate of rainfall with units of, say, g/m^2

per unit time, then the flux integral gives the mass of rain (in grams) that falls on the surface in a unit of time. This result says that (because the vector field is vertical) the mass of rain that falls on the roof equals the mass that would fall on the floor beneath the roof if the roof were not there. This property is explored further in Exercise 73.

Related Exercises 43–44 ◄

EXAMPLE 8   Flux of the radial field Consider the radial vector field
$\mathbf{F} = \langle f, g, h \rangle = \langle x, y, z \rangle$. Is the upward flux of the field greater across the hemisphere $x^2 + y^2 + z^2 = 1$, for $z \geq 0$, or across the paraboloid $z = 1 - x^2 - y^2$, for $z \geq 0$? Note that the two surfaces have the same base in the xy-plane and the same high point $(0, 0, 1)$. Use the explicit description for the hemisphere and a parametric description for the paraboloid.

SOLUTION The base of both surfaces in the xy-plane is the unit disk $R = \{(x, y): x^2 + y^2 \leq 1\}$, which, when expressed in polar coordinates, is the set $\{(r, \theta): 0 \leq r \leq 1, 0 \leq \theta \leq 2\pi\}$. To use the explicit description for the hemisphere, we must compute z_x and z_y. Differentiating $x^2 + y^2 + z^2 = 1$ implicitly, we find that $z_x = -x/z$ and $z_y = -y/z$. Therefore, the required normal vector is $\langle x/z, y/z, 1 \rangle$, which points upward on the surface. The flux integral is evaluated by substituting for f, g, h, z_x, and z_y; eliminating z from the integrand; and converting the integral in x and y to an integral in polar coordinates:

> ► Recall that the required normal vector for an explicitly defined surface $z = s(x, y)$ is $\langle -z_x, -z_y, 1 \rangle$.

$$\iint_S \mathbf{F} \cdot \mathbf{n} \, dS = \iint_R (-fz_x - gz_y + h) \, dA$$

$$= \iint_R \left(x\frac{x}{z} + y\frac{y}{z} + z \right) dA \qquad \text{Substitute.}$$

$$= \iint_R \left(\frac{x^2 + y^2 + z^2}{z} \right) dA \qquad \text{Simplify.}$$

$$= \iint_R \left(\frac{1}{z} \right) dA \qquad x^2 + y^2 + z^2 = 1$$

$$= \iint_R \left(\frac{1}{\sqrt{1 - x^2 - y^2}} \right) dA \qquad z = \sqrt{1 - x^2 - y^2}$$

$$= \int_0^{2\pi} \int_0^1 \left(\frac{1}{\sqrt{1 - r^2}} \right) r \, dr \, d\theta \quad \text{Polar coordinates}$$

$$= \int_0^{2\pi} \left(-\sqrt{1 - r^2} \right) \Big|_0^1 \, d\theta \qquad \begin{array}{l}\text{Evaluate inner integral}\\\text{as an improper integral.}\end{array}$$

$$= \int_0^{2\pi} d\theta = 2\pi. \qquad \text{Evaluate outer integral.}$$

For the paraboloid $z = 1 - x^2 - y^2$, we use the parametric description (Example 1b or Table 17.3)

$$\mathbf{r}(u, v) = \langle x, y, z \rangle = \langle v \cos u, v \sin u, 1 - v^2 \rangle,$$

for $0 \leq u \leq 2\pi$ and $0 \leq v \leq 1$. The required vector normal to the surface is

$$\mathbf{t}_u \times \mathbf{t}_v = \begin{vmatrix} \mathbf{i} & \mathbf{j} & \mathbf{k} \\ -v \sin u & v \cos u & 0 \\ \cos u & \sin u & -2v \end{vmatrix}$$

$$= \langle -2v^2 \cos u, -2v^2 \sin u, -v \rangle.$$

Notice that the normal vectors point downward on the surface (because the z-component is negative for $0 < v \leq 1$). In order to find the upward flux, we negate the normal vector and use the upward normal vector

$$-(\mathbf{t}_u \times \mathbf{t}_v) = \langle 2v^2 \cos u, 2v^2 \sin u, v \rangle.$$

The flux integral is evaluated by substituting for $\mathbf{F} = \langle x, y, z \rangle$ and $-(\mathbf{t}_u \times \mathbf{t}_v)$ and then evaluating an iterated integral in u and v:

$$\iint_S \mathbf{F} \cdot \mathbf{n} \, dS = \int_0^1 \int_0^{2\pi} \langle v \cos u, v \sin u, 1 - v^2 \rangle \cdot \langle 2v^2 \cos u, 2v^2 \sin u, v \rangle \, du \, dv$$

Substitute for F and $-(\mathbf{t}_u \times \mathbf{t}_v)$.

$$= \int_0^1 \int_0^{2\pi} (v^3 + v) \, du \, dv \qquad \text{Simplify.}$$

$$= 2\pi \left(\frac{v^4}{4} + \frac{v^2}{2} \right) \Big|_0^1 = \frac{3\pi}{2}. \qquad \text{Evaluate integrals.}$$

QUICK CHECK 6 Explain why the upward flux for the radial field in Example 8 is greater for the hemisphere than for the paraboloid. ◄

We see that the upward flux is greater for the hemisphere than for the paraboloid.

Related Exercises 45, 47 ◄

SECTION 17.6 EXERCISES

Getting Started

1. Give a parametric description for a cylinder with radius a and height h, including the intervals for the parameters.

2. Give a parametric description for a cone with radius a and height h, including the intervals for the parameters.

3. Give a parametric description for a sphere with radius a, including the intervals for the parameters.

4. Explain how to compute the surface integral of a scalar-valued function f over a cone using an explicit description of the cone.

5. Explain how to compute the surface integral of a scalar-valued function f over a sphere using a parametric description of the sphere.

6. Explain what it means for a surface to be orientable.

7. Describe the usual orientation of a closed surface such as a sphere.

8. Why is the upward flux of a vertical vector field $\mathbf{F} = \langle 0, 0, 1 \rangle$ across a surface equal to the area of the projection of the surface in the xy-plane?

Practice Exercises

9–14. Parametric descriptions *Give a parametric description of the form* $\mathbf{r}(u, v) = \langle x(u, v), y(u, v), z(u, v) \rangle$ *for the following surfaces. The descriptions are not unique. Specify the required rectangle in the uv-plane.*

9. The plane $2x - 4y + 3z = 16$

10. The cap of the sphere $x^2 + y^2 + z^2 = 16$, for $2\sqrt{2} \le z \le 4$

11. The frustum of the cone $z^2 = x^2 + y^2$, for $2 \le z \le 8$

12. The cone $z^2 = 4(x^2 + y^2)$, for $0 \le z \le 4$

13. The portion of the cylinder $x^2 + y^2 = 9$ in the first octant, for $0 \le z \le 3$

14. The cylinder $y^2 + z^2 = 36$, for $0 \le x \le 9$

15–18. Identify the surface *Describe the surface with the given parametric representation.*

15. $\mathbf{r}(u, v) = \langle u, v, 2u + 3v - 1 \rangle$, for $1 \le u \le 3, 2 \le v \le 4$

16. $\mathbf{r}(u, v) = \langle u, u + v, 2 - u - v \rangle$, for $0 \le u \le 2, 0 \le v \le 2$

17. $\mathbf{r}(u, v) = \langle v \cos u, v \sin u, 4v \rangle$, for $0 \le u \le \pi, 0 \le v \le 3$

18. $\mathbf{r}(u, v) = \langle v, 6 \cos u, 6 \sin u \rangle$, for $0 \le u \le 2\pi, 0 \le v \le 2$

19–24. Surface area using a parametric description *Find the area of the following surfaces using a parametric description of the surface.*

19. The half-cylinder $\{(r, \theta, z): r = 4, 0 \le \theta \le \pi, 0 \le z \le 7\}$

20. The plane $z = 3 - x - 3y$ in the first octant

21. The plane $z = 10 - x - y$ above the square $|x| \le 2, |y| \le 2$

22. The hemisphere $x^2 + y^2 + z^2 = 100$, for $z \ge 0$

23. A cone with base radius r and height h, where r and h are positive constants

24. The cap of the sphere $x^2 + y^2 + z^2 = 4$, for $1 \le z \le 2$

25–28. Surface integrals using a parametric description *Evaluate the surface integral* $\iint_S f \, dS$ *using a parametric description of the surface.*

25. $f(x, y, z) = x^2 + y^2$, where S is the hemisphere $x^2 + y^2 + z^2 = 36$, for $z \ge 0$

26. $f(x, y, z) = y$, where S is the cylinder $x^2 + y^2 = 9, 0 \le z \le 3$

27. $f(x, y, z) = x$, where S is the cylinder $x^2 + z^2 = 1, 0 \le y \le 3$

28. $f(\rho, \varphi, \theta) = \cos \varphi$, where S is the part of the unit sphere in the first octant

29–34. Surface area using an explicit description *Find the area of the following surfaces using an explicit description of the surface.*

29. The part of the plane $z = 2x + 2y + 4$ over the region R bounded by the triangle with vertices $(0, 0)$, $(2, 0)$, and $(2, 4)$

30. The part of the plane $z = x + 3y + 5$ over the region $R = \{(x, y): 1 \le x^2 + y^2 \le 4\}$

31. The cone $z^2 = 4(x^2 + y^2)$, for $0 \le z \le 4$

T 32. The trough $z = \frac{1}{2}x^2$, for $-1 \le x \le 1, 0 \le y \le 4$

33. The paraboloid $z = 2(x^2 + y^2)$, for $0 \le z \le 8$

34. The part of the hyperbolic paraboloid $z = 3 + x^2 - y^2$ above the sector $R = \{(r, \theta): 0 \le r \le \sqrt{2}, 0 \le \theta \le \pi/2\}$

35–38. Surface integrals using an explicit description *Evaluate the surface integral $\iint_S f(x, y, z)\, dS$ using an explicit representation of the surface.*

35. $f(x, y, z) = xy$; S is the plane $z = 2 - x - y$ in the first octant.

36. $f(x, y, z) = x^2 + y^2$; S is the paraboloid $z = x^2 + y^2$, for $0 \le z \le 1$.

37. $f(x, y, z) = 25 - x^2 - y^2$; S is the hemisphere centered at the origin with radius 5, for $z \ge 0$.

38. $f(x, y, z) = e^z$; S is the plane $z = 8 - x - 2y$ in the first octant.

39–42. Average values

39. Find the average temperature on that part of the plane $2x + 2y + z = 4$ over the square $0 \le x \le 1, 0 \le y \le 1$, where the temperature is given by $T(x, y, z) = e^{2x+y+z-3}$.

T 40. Find the average squared distance between the origin and the points on the paraboloid $z = 4 - x^2 - y^2$, for $z \ge 0$.

41. Find the average value of the function $f(x, y, z) = xyz$ on the unit sphere in the first octant.

42. Find the average value of the temperature function $T(x, y, z) = 100 - 25z$ on the cone $z^2 = x^2 + y^2$, for $0 \le z \le 2$.

43–48. Surface integrals of vector fields *Find the flux of the following vector fields across the given surface with the specified orientation. You may use either an explicit or a parametric description of the surface.*

43. $\mathbf{F} = \langle 0, 0, -1 \rangle$ across the slanted face of the tetrahedron $z = 4 - x - y$ in the first octant; normal vectors point upward.

44. $\mathbf{F} = \langle x, y, z \rangle$ across the slanted face of the tetrahedron $z = 10 - 2x - 5y$ in the first octant; normal vectors point upward.

45. $\mathbf{F} = \langle x, y, z \rangle$ across the slanted surface of the cone $z^2 = x^2 + y^2$, for $0 \le z \le 1$; normal vectors point upward.

46. $\mathbf{F} = \langle e^{-y}, 2z, xy \rangle$ across the curved sides of the surface $S = \{(x, y, z): z = \cos y, |y| \le \pi, 0 \le x \le 4\}$; normal vectors point upward.

47. $\mathbf{F} = \mathbf{r}/|\mathbf{r}|^3$ across the sphere of radius a centered at the origin, where $\mathbf{r} = \langle x, y, z \rangle$; normal vectors point outward.

48. $\mathbf{F} = \langle -y, x, 1 \rangle$ across the cylinder $y = x^2$, for $0 \le x \le 1$, $0 \le z \le 4$; normal vectors point in the general direction of the positive y-axis.

49. Explain why or why not Determine whether the following statements are true and give an explanation or counterexample.

   a. If the surface S is given by $\{(x, y, z): 0 \le x \le 1, 0 \le y \le 1, z = 10\}$, then $\iint_S f(x, y, z)\, dS = \int_0^1 \int_0^1 f(x, y, 10)\, dx\, dy$.

   b. If the surface S is given by $\{(x, y, z): 0 \le x \le 1, 0 \le y \le 1, z = x\}$, then $\iint_S f(x, y, z)\, dS = \int_0^1 \int_0^1 f(x, y, x)\, dx\, dy$.

   c. The surface $\mathbf{r} = \langle v \cos u, v \sin u, v^2 \rangle$, for $0 \le u \le \pi$, $0 \le v \le 2$, is the same as the surface $\mathbf{r} = \langle \sqrt{v} \cos 2u, \sqrt{v} \sin 2u, v \rangle$, for $0 \le u \le \pi/2, 0 \le v \le 4$.

   d. Given the standard parameterization of a sphere, the normal vectors $\mathbf{t}_u \times \mathbf{t}_v$ are outward normal vectors.

50–53. Miscellaneous surface integrals *Evaluate the following integrals using the method of your choice. Assume normal vectors point either outward or upward.*

50. $\iint_S \nabla \ln |\mathbf{r}| \cdot \mathbf{n}\, dS$, where S is the hemisphere $x^2 + y^2 + z^2 = a^2$, for $z \ge 0$, and where $\mathbf{r} = \langle x, y, z \rangle$

51. $\iint_S |\mathbf{r}|\, dS$, where S is the cylinder $x^2 + y^2 = 4$, for $0 \le z \le 8$, where $\mathbf{r} = \langle x, y, z \rangle$

52. $\iint_S xyz\, dS$, where S is that part of the plane $z = 6 - y$ that lies in the cylinder $x^2 + y^2 = 4$

53. $\iint_S \dfrac{\langle x, 0, z \rangle}{\sqrt{x^2 + z^2}} \cdot \mathbf{n}\, dS$, where S is the cylinder $x^2 + z^2 = a^2$, $|y| \le 2$

54. Cone and sphere The cone $z^2 = x^2 + y^2$, for $z \ge 0$, cuts the sphere $x^2 + y^2 + z^2 = 16$ along a curve C.

   a. Find the surface area of the sphere below C, for $z \ge 0$.

   b. Find the surface area of the sphere above C.

   c. Find the surface area of the cone below C, for $z \ge 0$.

T 55. Cylinder and sphere Consider the sphere $x^2 + y^2 + z^2 = 4$ and the cylinder $(x - 1)^2 + y^2 = 1$, for $z \ge 0$. Find the surface area of the cylinder inside the sphere.

56. Flux on a tetrahedron Find the upward flux of the field $\mathbf{F} = \langle x, y, z \rangle$ across the plane $\dfrac{x}{a} + \dfrac{y}{b} + \dfrac{z}{c} = 1$ in the first octant, where a, b, and c are positive real numbers. Show that the flux equals c times the area of the base of the region. Interpret the result physically.

57. Flux across a cone Consider the field $\mathbf{F} = \langle x, y, z \rangle$ and the cone $z^2 = \dfrac{x^2 + y^2}{a^2}$, for $0 \le z \le 1$.

   a. Show that when $a = 1$, the outward flux across the cone is zero. Interpret the result.

   b. Find the outward flux (away from the z-axis), for any $a > 0$. Interpret the result.

58. Surface area formula for cones Find the general formula for the surface area of a cone with height h and base radius a (excluding the base).

59. Surface area formula for spherical cap A sphere of radius a is sliced parallel to the equatorial plane at a distance $a - h$ from the equatorial plane (see figure). Find the general formula for the surface area of the resulting spherical cap (excluding the base) with thickness h.

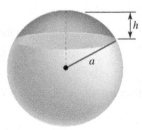

Explorations and Challenges

60. Radial fields and spheres Consider the radial field $\mathbf{F} = \mathbf{r}/|\mathbf{r}|^p$, where $\mathbf{r} = \langle x, y, z \rangle$ and p is a real number. Let S be the sphere of radius a centered at the origin. Show that the outward flux of $\mathbf{F}$ across the sphere is $4\pi/a^{p-3}$. It is instructive to do the calculation using both an explicit and a parametric description of the sphere.

61–63. Heat flux *The heat flow vector field for conducting objects is $\mathbf{F} = -k\nabla T$, where $T(x, y, z)$ is the temperature in the object and $k > 0$ is a constant that depends on the material. Compute the outward flux of $\mathbf{F}$ across the following surfaces S for the given temperature distributions. Assume $k = 1$.*

61. $T(x, y, z) = 100e^{-x-y}$; S consists of the faces of the cube $|x| \leq 1$, $|y| \leq 1$, $|z| \leq 1$.

62. $T(x, y, z) = 100e^{-x^2-y^2-z^2}$; S is the sphere $x^2 + y^2 + z^2 = a^2$.

63. $T(x, y, z) = -\ln(x^2 + y^2 + z^2)$; S is the sphere $x^2 + y^2 + z^2 = a^2$.

64. Flux across a cylinder Let S be the cylinder $x^2 + y^2 = a^2$, for $-L \leq z \leq L$.

  a. Find the outward flux of the field $\mathbf{F} = \langle x, y, 0 \rangle$ across S.

  b. Find the outward flux of the field $\mathbf{F} = \dfrac{\langle x, y, 0 \rangle}{(x^2 + y^2)^{p/2}} = \dfrac{\mathbf{r}}{|\mathbf{r}|^p}$ across S, where $|\mathbf{r}|$ is the distance from the z-axis and p is a real number.

  c. In part (b), for what values of p is the outward flux finite as $a \to \infty$ (with L fixed)?

  d. In part (b), for what values of p is the outward flux finite as $L \to \infty$ (with a fixed)?

65. Flux across concentric spheres Consider the radial fields $\mathbf{F} = \dfrac{\langle x, y, z \rangle}{(x^2 + y^2 + z^2)^{p/2}} = \dfrac{\mathbf{r}}{|\mathbf{r}|^p}$, where p is a real number. Let S consist of the spheres A and B centered at the origin with radii $0 < a < b$, respectively. The total outward flux across S consists of the flux out of S across the outer sphere B minus the flux into S across the inner sphere A.

  a. Find the total flux across S with $p = 0$. Interpret the result.

  b. Show that for $p = 3$ (an inverse square law), the flux across S is independent of a and b.

66–69. Mass and center of mass *Let S be a surface that represents a thin shell with density ρ. The moments about the coordinate planes (see Section 16.6) are $M_{yz} = \iint_S x\rho(x, y, z)\, dS$, $M_{xz} = \iint_S y\rho(x, y, z)\, dS$, and $M_{xy} = \iint_S z\rho(x, y, z)\, dS$. The coordinates of the center of mass of the shell are $\bar{x} = \dfrac{M_{yz}}{m}$, $\bar{y} = \dfrac{M_{xz}}{m}$, and $\bar{z} = \dfrac{M_{xy}}{m}$, where m is the mass of the shell. Find the mass and center of mass of the following shells. Use symmetry whenever possible.*

66. The constant-density hemispherical shell $x^2 + y^2 + z^2 = a^2$, $z \geq 0$

67. The constant-density cone with radius a, height h, and base in the xy-plane

68. The constant-density half-cylinder $x^2 + z^2 = a^2$, $-\dfrac{h}{2} \leq y \leq \dfrac{h}{2}$, $z \geq 0$

69. The cylinder $x^2 + y^2 = a^2$, $0 \leq z \leq 2$, with density $\rho(x, y, z) = 1 + z$

70. Outward normal to a sphere Show that $|\mathbf{t}_u \times \mathbf{t}_v| = a^2 \sin u$ for a sphere of radius a defined parametrically by $\mathbf{r}(u, v) = \langle a \sin u \cos v, a \sin u \sin v, a \cos u \rangle$, where $0 \leq u \leq \pi$ and $0 \leq v \leq 2\pi$.

71. Special case of surface integrals of scalar-valued functions Suppose a surface S is defined as $z = g(x, y)$ on a region R. Show that $\mathbf{t}_x \times \mathbf{t}_y = \langle -z_x, -z_y, 1 \rangle$ and that $\iint_S f(x, y, z)\, dS = \iint_R f(x, y, g(x, y))\sqrt{z_x^2 + z_y^2 + 1}\, dA$.

72. Surfaces of revolution Suppose $y = f(x)$ is a continuous and positive function on $[a, b]$. Let S be the surface generated when the graph of f on $[a, b]$ is revolved about the x-axis.

  a. Show that S is described parametrically by $\mathbf{r}(u, v) = \langle u, f(u) \cos v, f(u) \sin v \rangle$, for $a \leq u \leq b$, $0 \leq v \leq 2\pi$.

  b. Find an integral that gives the surface area of S.

  c. Apply the result of part (b) to the surface generated with $f(x) = x^3$, for $1 \leq x \leq 2$.

  d. Apply the result of part (b) to the surface generated with $f(x) = (25 - x^2)^{1/2}$, for $3 \leq x \leq 4$.

73. Rain on roofs Let $z = s(x, y)$ define a surface over a region R in the xy-plane, where $z \geq 0$ on R. Show that the downward flux of the vertical vector field $\mathbf{F} = \langle 0, 0, -1 \rangle$ across S equals the area of R. Interpret the result physically.

74. Surface area of a torus

  a. Show that a torus with radii $R > r$ (see figure) may be described parametrically by $\mathbf{r}(u, v) = \langle (R + r \cos u) \cos v, (R + r \cos u) \sin v, r \sin u \rangle$, for $0 \leq u \leq 2\pi$, $0 \leq v \leq 2\pi$.

  b. Show that the surface area of the torus is $4\pi^2 Rr$.

75. Surfaces of revolution—single variable Let f be differentiable and positive on the interval $[a, b]$. Let S be the surface generated when the graph of f on $[a, b]$ is revolved about the x-axis. Use Theorem 17.14 to show that the area of S (as given in Section 6.6) is

$$\int_a^b 2\pi f(x)\sqrt{1 + f'(x)^2}\, dx.$$

QUICK CHECK ANSWERS

1. A half-cylinder with height 1 and radius 2 with its axis along the z-axis   **2.** A half-cone with height 10 and radius 10   **3.** A quarter-sphere with radius 4   **4.** $\sqrt{2}$ **5.** The cylinder $x^2 + y^2 = a^2$ does not represent a function, so z_x and z_y cannot be computed.   **6.** The vector field is everywhere orthogonal to the hemisphere, so the hemisphere has maximum flux at every point. ◄

17.7 Stokes' Theorem

▶ Born in Ireland, George Gabriel Stokes (1819–1903) led a long and distinguished life as one of the prominent mathematicians and physicists of his day. He entered Cambridge University as a student and remained there as a professor for most of his life, taking the Lucasian chair of mathematics once held by Sir Isaac Newton. The first statement of Stokes' Theorem was given by William Thomson (Lord Kelvin).

With the divergence, the curl, and surface integrals in hand, we are ready to present two of the crowning results of calculus. Fortunately, all the heavy lifting has been done. In this section, you will see Stokes' Theorem, and in the next section, we present the Divergence Theorem.

Stokes' Theorem

Stokes' Theorem is the three-dimensional version of the circulation form of Green's Theorem. Recall that if C is a closed simple piecewise-smooth oriented curve in the xy-plane enclosing a simply connected region R, and $\mathbf{F} = \langle f, g \rangle$ is a differentiable vector field on R, then Green's Theorem says that

$$\underbrace{\oint_C \mathbf{F} \cdot d\mathbf{r}}_{\text{circulation}} = \iint_R \underbrace{(g_x - f_y)}_{\text{curl or rotation}} dA.$$

The line integral on the left gives the circulation along the boundary of R. The double integral on the right sums the curl of the vector field over all points of R. If $\mathbf{F}$ represents a fluid flow, the theorem says that the cumulative rotation of the flow within R equals the circulation along the boundary.

In Stokes' Theorem, the plane region R in Green's Theorem becomes an oriented surface S in $\mathbb{R}^3$. The circulation integral in Green's Theorem remains a circulation integral, but now over the closed simple piecewise-smooth oriented curve C that forms the boundary of S. The double integral of the curl in Green's Theorem becomes a surface integral of the three-dimensional curl (**Figure 17.59**).

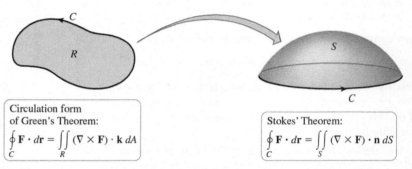

Circulation form
of Green's Theorem:
$$\oint_C \mathbf{F} \cdot d\mathbf{r} = \iint_R (\nabla \times \mathbf{F}) \cdot \mathbf{k} \, dA$$

Stokes' Theorem:
$$\oint_C \mathbf{F} \cdot d\mathbf{r} = \iint_S (\nabla \times \mathbf{F}) \cdot \mathbf{n} \, dS$$

Figure 17.59

Stokes' Theorem involves an oriented curve C and an oriented surface S on which there are two unit normal vectors at every point. These orientations must be consistent and the normal vectors must be chosen correctly. Here is the right-hand rule that relates the orientations of S and C and determines the choice of the normal vectors:

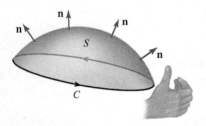

Figure 17.60

> If the fingers of your right hand curl in the positive direction around C, then your right thumb points in the (general) direction of the vectors normal to S (**Figure 17.60**).

A common situation occurs when C has a counterclockwise orientation when viewed from above; then the vectors normal to S point upward.

▶ The right-hand rule tells you which of two normal vectors at a point of S to use. Remember that the direction of normal vectors changes continuously on an oriented surface.

THEOREM 17.15   Stokes' Theorem

Let S be an oriented surface in $\mathbb{R}^3$ with a piecewise-smooth closed boundary C whose orientation is consistent with that of S. Assume $\mathbf{F} = \langle f, g, h \rangle$ is a vector field whose components have continuous first partial derivatives on S. Then

$$\oint_C \mathbf{F} \cdot d\mathbf{r} = \iint_S (\nabla \times \mathbf{F}) \cdot \mathbf{n} \, dS,$$

where $\mathbf{n}$ is the unit vector normal to S determined by the orientation of S.

QUICK CHECK 1 Suppose S is a region in the xy-plane with a boundary oriented counterclockwise. What is the normal to S? Explain why Stokes' Theorem becomes the circulation form of Green's Theorem. ◄

The meaning of Stokes' Theorem is much the same as for the circulation form of Green's Theorem: Under the proper conditions, the accumulated rotation of the vector field over the surface S (as given by the normal component of the curl) equals the net circulation on the boundary of S. An outline of the proof of Stokes' Theorem is given at the end of this section. First, we look at some special cases that give further insight into the theorem.

If $\mathbf{F}$ is a conservative vector field on a domain D, then it has a potential function φ such that $\mathbf{F} = \nabla\varphi$. Because $\nabla \times \nabla\varphi = \mathbf{0}$, it follows that $\nabla \times \mathbf{F} = \mathbf{0}$ (Theorem 17.11); therefore, the circulation integral is zero on all closed curves in D. Recall that the circulation integral is also a work integral for the force field $\mathbf{F}$, which emphasizes the fact that no work is done in moving an object on a closed path in a conservative force field. Among the important conservative vector fields are the radial fields $\mathbf{F} = \mathbf{r}/|\mathbf{r}|^p$, which generally have zero curl and zero circulation on closed curves.

➤ Recall that for a constant nonzero vector $\mathbf{a}$ and the position vector $\mathbf{r} = \langle x, y, z\rangle$, the field $\mathbf{F} = \mathbf{a} \times \mathbf{r}$ is a rotation field. In Example 1,

$$\mathbf{F} = \langle 0, 1, 1\rangle \times \langle x, y, z\rangle.$$

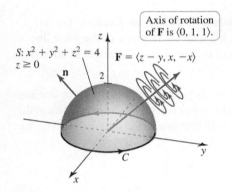

Figure 17.61

EXAMPLE 1 Verifying Stokes' Theorem Confirm that Stokes' Theorem holds for the vector field $\mathbf{F} = \langle z - y, x, -x\rangle$, where S is the hemisphere $x^2 + y^2 + z^2 = 4$, for $z \geq 0$, and C is the circle $x^2 + y^2 = 4$ oriented counterclockwise.

SOLUTION The orientation of C implies that vectors normal to S should point in the outward direction. The vector field is a rotation field $\mathbf{a} \times \mathbf{r}$, where $\mathbf{a} = \langle 0, 1, 1\rangle$ and $\mathbf{r} = \langle x, y, z\rangle$; so the axis of rotation points in the direction of the vector $\langle 0, 1, 1\rangle$ (Figure 17.61). We first compute the circulation integral in Stokes' Theorem. The curve C with the given orientation is parameterized as $\mathbf{r}(t) = \langle 2\cos t, 2\sin t, 0\rangle$, for $0 \leq t \leq 2\pi$; therefore, $\mathbf{r}'(t) = \langle -2\sin t, 2\cos t, 0\rangle$. The circulation integral is

$$\oint_C \mathbf{F} \cdot d\mathbf{r} = \int_0^{2\pi} \mathbf{F} \cdot \mathbf{r}'(t)\, dt \qquad \text{Definition of line integral}$$

$$= \int_0^{2\pi} \langle \underbrace{z - y}_{-2\sin t}, x, -x\rangle \cdot \langle -2\sin t, 2\cos t, 0\rangle\, dt \quad \text{Substitute.}$$

$$= \int_0^{2\pi} 4(\sin^2 t + \cos^2 t)\, dt \qquad \text{Simplify.}$$

$$= 4\int_0^{2\pi} dt \qquad \sin^2 t + \cos^2 t = 1$$

$$= 8\pi. \qquad \text{Evaluate integral.}$$

The surface integral requires computing the curl of the vector field:

$$\nabla \times \mathbf{F} = \nabla \times \langle z - y, x, -x\rangle = \begin{vmatrix} \mathbf{i} & \mathbf{j} & \mathbf{k} \\ \dfrac{\partial}{\partial x} & \dfrac{\partial}{\partial y} & \dfrac{\partial}{\partial z} \\ z - y & x & -x \end{vmatrix} = \langle 0, 2, 2\rangle.$$

Recall from Section 17.6 (Table 17.3) that the required outward normal to the hemisphere is $\langle x/z, y/z, 1\rangle$. The region of integration is the base of the hemisphere in the xy-plane, which is

$$R = \{(x, y): x^2 + y^2 \leq 4\}, \text{ or, in polar coordinates,}$$
$$\{(r, \theta): 0 \leq r \leq 2, 0 \leq \theta \leq 2\pi\}.$$

Combining these results, the surface integral in Stokes' Theorem is

$$\iint_S \underbrace{(\nabla \times \mathbf{F})}_{\langle 0,2,2\rangle} \cdot \mathbf{n}\, dS = \iint_R \langle 0, 2, 2\rangle \cdot \left\langle \frac{x}{z}, \frac{y}{z}, 1\right\rangle dA \quad \begin{array}{l}\text{Substitute and convert to a}\\ \text{double integral over } R.\end{array}$$

$$= \iint_R \left(\frac{2y}{\sqrt{4 - x^2 - y^2}} + 2\right) dA \quad \begin{array}{l}\text{Simplify and use}\\ z = \sqrt{4 - x^2 - y^2}.\end{array}$$

$$= \int_0^{2\pi}\int_0^2 \left(\frac{2r\sin\theta}{\sqrt{4 - r^2}} + 2\right) r\, dr\, d\theta. \quad \text{Convert to polar coordinates.}$$

► In eliminating the first term of this double integral, we note that the improper integral $\int_0^2 \dfrac{r^2}{\sqrt{4 - r^2}}\, dr$ has a finite value.

We integrate first with respect to θ because the integral of $\sin \theta$ from 0 to 2π is zero and the first term in the integral is eliminated. Therefore, the surface integral reduces to

$$\iint_S (\nabla \times \mathbf{F}) \cdot \mathbf{n}\, dS = \int_0^2 \int_0^{2\pi} \left(\frac{2r^2 \sin \theta}{\sqrt{4 - r^2}} + 2r \right) d\theta\, dr$$

$$= \int_0^2 \int_0^{2\pi} 2r\, d\theta\, dr \qquad \qquad \int_0^{2\pi} \sin \theta\, d\theta = 0$$

$$= 4\pi \int_0^2 r\, dr \qquad \qquad \text{Evaluate inner integral.}$$

$$= 8\pi. \qquad \qquad \text{Evaluate outer integral.}$$

Computed either as a line integral or as a surface integral, the vector field has a positive circulation along the boundary of S, which is produced by the net rotation of the field over the surface S.

Related Exercises 5–6 ◄

In Example 1, it was possible to evaluate both the line integral and the surface integral that appear in Stokes' Theorem. Often the theorem provides an easier way to evaluate difficult line integrals.

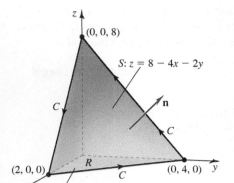

Figure 17.62

► Recall that for an explicitly defined surface S given by $z = s(x, y)$ over a region R with $\mathbf{F} = \langle f, g, h \rangle$,

$$\iint_S \mathbf{F} \cdot \mathbf{n}\, dS = \iint_R (-f z_x - g z_y + h)\, dA.$$

In Example 2, $\mathbf{F}$ is replaced with $\nabla \times \mathbf{F}$.

EXAMPLE 2   Using Stokes' Theorem to evaluate a line integral Evaluate the line integral $\oint_C \mathbf{F} \cdot d\mathbf{r}$, where $\mathbf{F} = z\,\mathbf{i} - z\,\mathbf{j} + (x^2 - y^2)\,\mathbf{k}$ and C consists of the three line segments that bound the plane $z = 8 - 4x - 2y$ in the first octant, oriented as shown in **Figure 17.62**.

SOLUTION Evaluating the line integral directly involves parameterizing the three line segments. Instead, we use Stokes' Theorem to convert the line integral to a surface integral, where S is that portion of the plane $z = 8 - 4x - 2y$ that lies in the first octant. The curl of the vector field is

$$\nabla \times \mathbf{F} = \nabla \times \langle z, -z, x^2 - y^2 \rangle = \begin{vmatrix} \mathbf{i} & \mathbf{j} & \mathbf{k} \\ \dfrac{\partial}{\partial x} & \dfrac{\partial}{\partial y} & \dfrac{\partial}{\partial z} \\ z & -z & x^2 - y^2 \end{vmatrix} = \langle 1 - 2y, 1 - 2x, 0 \rangle.$$

The appropriate vector normal to the plane $z = 8 - 4x - 2y$ is $\langle -z_x, -z_y, 1 \rangle = \langle 4, 2, 1 \rangle$, which points upward, consistent with the orientation of C. The triangular region R in the xy-plane beneath S is found by setting $z = 0$ in the equation of the plane; we find that $R = \{ (x, y) : 0 \le x \le 2, 0 \le y \le 4 - 2x \}$. The surface integral in Stokes' Theorem may now be evaluated:

$$\iint_S (\nabla \times \mathbf{F}) \cdot \mathbf{n}\, dS = \iint_R \underbrace{\langle 1 - 2y, 1 - 2x, 0 \rangle}_{\langle 1 - 2y,\, 1 - 2x,\, 0 \rangle} \cdot \langle 4, 2, 1 \rangle\, dA \qquad \begin{array}{l}\text{Substitute and convert to} \\ \text{a double integral over } R.\end{array}$$

$$= \int_0^2 \int_0^{4-2x} (6 - 4x - 8y)\, dy\, dx \qquad \text{Simplify.}$$

$$= -\frac{88}{3}. \qquad \text{Evaluate integrals.}$$

The circulation around the boundary of R is negative, indicating a net circulation in the clockwise direction on C (looking from above).

Related Exercises 13, 16 ◄

In other situations, Stokes' Theorem may be used to convert a difficult surface integral into a relatively easy line integral, as illustrated in the next example.

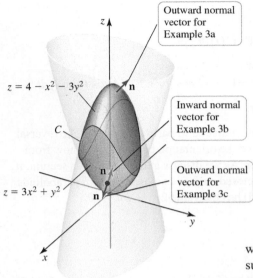

Outward normal vector for Example 3a

$z = 4 - x^2 - 3y^2$

$\mathbf{n}$

C

Inward normal vector for Example 3b

$\mathbf{n}$

$z = 3x^2 + y^2$

$\mathbf{n}$

Outward normal vector for Example 3c

Figure 17.63

EXAMPLE 3   Using Stokes' Theorem to evaluate a surface integral Evaluate $\iint_S (\nabla \times \mathbf{F}) \cdot \mathbf{n} \, dS$, where $\mathbf{F} = -y\,\mathbf{i} + x\,\mathbf{j} + z\,\mathbf{k}$, in the following cases.

a. S is the part of the paraboloid $z = 4 - x^2 - 3y^2$ that lies within the paraboloid $z = 3x^2 + y^2$ (the blue surface in **Figure 17.63**). Assume $\mathbf{n}$ points in the upward direction on S.

b. S is the part of the paraboloid $z = 3x^2 + y^2$ that lies within the paraboloid $z = 4 - x^2 - 3y^2$, with $\mathbf{n}$ pointing in the upward direction on S.

c. S is the surface in part (b), but $\mathbf{n}$ pointing in the downward direction on S.

SOLUTION

a. Finding a parametric description for S is challenging, so we use Stokes' Theorem to convert the surface integral into a line integral along the curve C that bounds S. Note that C is the intersection between the paraboloids $z = 4 - x^2 - 3y^2$ and $z = 3x^2 + y^2$. Eliminating z from these equations, we find that the projection of C onto the xy-plane is the circle $x^2 + y^2 = 1$, which suggests that we choose $x = \cos t$ and $y = \sin t$ for the x- and y-components of the equations for C. To find the z-component, we substitute x and y into the equation of either paraboloid. Choosing $z = 3x^2 + y^2$, we find that a parametric description of C is $\mathbf{r}(t) = \langle \cos t, \sin t, 3\cos^2 t + \sin^2 t \rangle$; note that C is oriented in the counterclockwise direction, consistent with the orientation of S.

To evaluate the line integral in Stokes' Theorem, it is helpful to first compute $\mathbf{F} \cdot \mathbf{r}'(t)$. Along C, the vector field is $\mathbf{F} = \langle -y, x, z \rangle = \langle -\sin t, \cos t, 3\cos^2 t + \sin^2 t \rangle$. Differentiating $\mathbf{r}$ yields $\mathbf{r}'(t) = \langle -\sin t, \cos t, -4\cos t \sin t \rangle$, which leads to

$$\mathbf{F} \cdot \mathbf{r}'(t) = \langle -\sin t, \cos t, 3\cos^2 t + \sin^2 t \rangle \cdot \langle -\sin t, \cos t, -4\cos t \sin t \rangle$$
$$= \underbrace{\sin^2 t + \cos^2 t}_{1} - 12\cos^3 t \sin t - 4\sin^3 t \cos t.$$

Noting that $\sin^2 t + \cos^2 t = 1$, we are ready to evaluate the integral:

$$\iint_S (\nabla \times \mathbf{F}) \cdot \mathbf{n} \, dS = \oint_C \mathbf{F} \cdot d\mathbf{r} \qquad \text{Stokes' Theorem}$$

$$= \int_0^{2\pi} \mathbf{F} \cdot \mathbf{r}'(t) \, dt \qquad \text{Definition of line integral}$$

$$= \int_0^{2\pi} (1 - 12\cos^3 t \sin t - 4\cos t \sin^3 t) \, dt \qquad \text{Substitute.}$$

$$= \int_0^{2\pi} 1 \, dt - \underbrace{\int_0^{2\pi} 12\cos^3 t \sin t \, dt}_{0} - \underbrace{\int_0^{2\pi} 4\cos t \sin^3 t \, dt}_{0}$$

$$\qquad \qquad \qquad \qquad \text{Split integral.}$$

$$= 2\pi. \qquad \text{Evaluate integrals.}$$

A standard substitution in the last two integrals of the final step shows that both integrals equal 0.

b. Because the lower surface ($z = 3x^2 + y^2$) shares the same boundary C with the upper surface ($z = 4 - x^2 - 3y^2$), and because both surfaces have an upward-pointing normal vector, the line integral resulting from an application of Stokes' Theorem is identical to the integral in part (a). For this surface S with its associated normal vector, we conclude that $\iint_S (\nabla \times \mathbf{F}) \cdot \mathbf{n} \, dS = \oint_C \mathbf{F} \cdot d\mathbf{r} = 2\pi$. In fact, the value of this integral is 2π for *any* surface whose boundary is C and whose normal vectors point in the upward direction.

c. In this case, $\mathbf{n}$ points downward. We use the parameterization $\mathbf{r}(t) = \langle \sin t, \cos t, 3\cos^2 t + \sin^2 t \rangle$ for C so that C is oriented in the clockwise direction, consistent with the orientation of S. You should verify that, when duplicating the calculations in part (a) with a new description for C, we have

$$\mathbf{F} \cdot \mathbf{r}'(t) = \underbrace{-\sin^2 t - \cos^2 t}_{-1} - 12\cos^3 t \sin t - 4\sin^3 t \cos t.$$

▶ Recall that $x = \cos t$, $y = \sin t$ is a standard parameterization for the unit circle centered at the origin with counterclockwise orientation. The parameterization $x = \sin t$, $y = \cos t$ reverses the orientation.

Therefore, the required integral is

$$\iint_S (\nabla \times \mathbf{F}) \cdot \mathbf{n} \, dS = \oint_C \mathbf{F} \cdot d\mathbf{r} = \int_0^{2\pi} \mathbf{F} \cdot \mathbf{r}'(t) \, dt$$

$$= \int_0^{2\pi} (-1 - 12 \cos^3 t \sin t - 4 \cos t \sin^3 t) \, dt$$

$$= -2\pi.$$

QUICK CHECK 2 In Example 3a, we used the parameterization $\mathbf{r}(t) = \langle \cos t, \sin t, 3 \cos^2 t + \sin^2 t \rangle$ for C. Confirm that the parameterization C: $\mathbf{r}(t) = \langle \cos t, \sin t, 4 - \cos^2 t - 3 \sin^2 t \rangle$ also results in an answer of 2π. ◄

This result is perhaps not surprising when compared to parts (a) and (b): The reversal of the orientation of S requires a reversal of the orientation of C, and we know from Section 17.2 that $\int_C \mathbf{F} \cdot d\mathbf{r} = -\int_{-C} \mathbf{F} \cdot d\mathbf{r}$. As we discuss at the end of this section, it follows that the surface integral over the closed surface enclosed by *both* paraboloids (with normal vectors everywhere outward) has the value $2\pi - 2\pi = 0$.

Related Exercises 21–22 ◄

Interpreting the Curl

Stokes' Theorem leads to another interpretation of the curl at a point in a vector field. We need the idea of the **average circulation**. If C is the boundary of an oriented surface S, we define the average circulation of $\mathbf{F}$ over S as

$$\frac{1}{\text{area of } S} \oint_C \mathbf{F} \cdot d\mathbf{r} = \frac{1}{\text{area of } S} \iint_S (\nabla \times \mathbf{F}) \cdot \mathbf{n} \, dS,$$

where Stokes' Theorem is used to convert the circulation integral to a surface integral.

First consider a general rotation field $\mathbf{F} = \mathbf{a} \times \mathbf{r}$, where $\mathbf{a} = \langle a_1, a_2, a_3 \rangle$ is a constant nonzero vector and $\mathbf{r} = \langle x, y, z \rangle$. Recall that $\mathbf{F}$ describes the rotation about an axis in the direction of $\mathbf{a}$ with angular speed $\omega = |\mathbf{a}|$. We also showed that $\mathbf{F}$ has a constant curl, $\nabla \times \mathbf{F} = \nabla \times (\mathbf{a} \times \mathbf{r}) = 2\mathbf{a}$. We now take S to be a small circular disk centered at a point P, whose normal vector $\mathbf{n}$ makes an angle θ with the axis $\mathbf{a}$ (**Figure 17.64**). Let C be the boundary of S with a counterclockwise orientation.

The average circulation of this vector field on S is

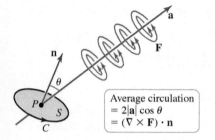

Figure 17.64

> Recall that $\mathbf{n}$ is a unit normal vector with $|\mathbf{n}| = 1$. By definition, the dot product gives $\mathbf{a} \cdot \mathbf{n} = |\mathbf{a}| \cos \theta$.

$$\frac{1}{\text{area of } S} \iint_S \underbrace{(\nabla \times \mathbf{F}) \cdot \mathbf{n}}_{\text{constant}} dS \qquad \text{Definition}$$

$$= \frac{1}{\text{area of } S} (\nabla \times \mathbf{F}) \cdot \mathbf{n} \cdot \text{area of } S \qquad \iint_S dS = \text{area of } S$$

$$= \underbrace{(\nabla \times \mathbf{F})}_{2\mathbf{a}} \cdot \mathbf{n} \qquad \text{Simplify.}$$

$$= 2|\mathbf{a}| \cos \theta. \qquad |\mathbf{n}| = 1, |\nabla \times \mathbf{F}| = 2|\mathbf{a}|$$

If the normal vector $\mathbf{n}$ is aligned with $\nabla \times \mathbf{F}$ (which is parallel to $\mathbf{a}$), then $\theta = 0$ and the average circulation on S has its maximum value of $2|\mathbf{a}|$. However, if the vector normal to the surface S is orthogonal to the axis of rotation ($\theta = \pi/2$), the average circulation is zero.

We see that for a general rotation field $\mathbf{F} = \mathbf{a} \times \mathbf{r}$, the curl of $\mathbf{F}$ has the following interpretations, where S is a small disk centered at a point P with a normal vector $\mathbf{n}$.

• The scalar component of $\nabla \times \mathbf{F}$ at P in the direction of $\mathbf{n}$, which is $(\nabla \times \mathbf{F}) \cdot \mathbf{n} = 2|\mathbf{a}| \cos \theta$, is the average circulation of $\mathbf{F}$ on S.

• The direction of $\nabla \times \mathbf{F}$ at P is the direction that maximizes the average circulation of $\mathbf{F}$ on S. Equivalently, it is the direction in which the axis of a paddle wheel should be oriented to obtain the maximum angular speed.

A similar argument may be applied to a general vector field (with a variable curl) to give an interpretation of the curl at a point (Exercise 48).

EXAMPLE 4 Horizontal channel flow Consider the velocity field
$\mathbf{v} = \langle 0, 1 - x^2, 0 \rangle$, for $|x| \le 1$ and $|z| \le 1$, which represents a horizontal flow in the
y-direction (**Figure 17.65a**).

a. Suppose you place a paddle wheel at the point $P\left(\frac{1}{2}, 0, 0\right)$. Using physical arguments,
in which of the coordinate directions should the axis of the wheel point in order for the
wheel to spin? In which direction does it spin? What happens if you place the wheel at
$Q\left(-\frac{1}{2}, 0, 0\right)$?

b. Compute and graph the curl of $\mathbf{v}$ and provide an interpretation.

(a)   (b)

Figure 17.65

SOLUTION

a. If the axis of the wheel is aligned with the x-axis at P, the flow strikes the upper and
lower halves of the wheel symmetrically and the wheel does not spin. If the axis of the
wheel is aligned with the y-axis, the flow is parallel to the axis of the wheel and the
wheel does not spin. If the axis of the wheel is aligned with the z-axis at P, the flow in
the y-direction is greater for $x < \frac{1}{2}$ than it is for $x > \frac{1}{2}$. Therefore, a wheel located at
$P\left(\frac{1}{2}, 0, 0\right)$ spins in the clockwise direction, looking from above (Figure 17.65a). Using a similar argument, we conclude that a vertically oriented paddle wheel placed at
$Q\left(-\frac{1}{2}, 0, 0\right)$ spins in the counterclockwise direction (when viewed from above).

b. A short calculation shows that

$$\nabla \times \mathbf{v} = \begin{vmatrix} \mathbf{i} & \mathbf{j} & \mathbf{k} \\ \dfrac{\partial}{\partial x} & \dfrac{\partial}{\partial y} & \dfrac{\partial}{\partial z} \\ 0 & 1 - x^2 & 0 \end{vmatrix} = -2x\,\mathbf{k}.$$

QUICK CHECK 3 In Example 4, explain
why a paddle wheel with its axis
aligned with the z-axis does not spin
when placed on the y-axis. ◄

As shown in **Figure 17.65b**, the curl points in the z-direction, which is the direction of
the paddle wheel axis that gives the maximum angular speed of the wheel. Consider
the z-component of the curl, which is $(\nabla \times \mathbf{v}) \cdot \mathbf{k} = -2x$. At $x = 0$, this component
is zero, meaning the wheel does not spin at any point along the y-axis when its axis is
aligned with the z-axis. For $x > 0$, we see that $(\nabla \times \mathbf{v}) \cdot \mathbf{k} < 0$, which corresponds
to clockwise rotation of the vector field. For $x < 0$, we have $(\nabla \times \mathbf{v}) \cdot \mathbf{k} > 0$, corresponding to counterclockwise rotation.

Related Exercise 26 ◄

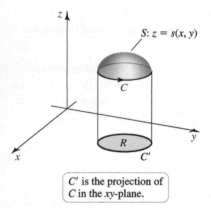

C' is the projection of
C in the xy-plane.

Figure 17.66

Proof of Stokes' Theorem

The proof of the most general case of Stokes' Theorem is intricate. However, a proof of a special case is instructive and relies on several previous results.

Consider the case in which the surface S is the graph of the function $z = s(x, y)$, defined on a region in the xy-plane. Let C be the curve that bounds S with a counterclockwise orientation, let R be the projection of S in the xy-plane, and let C' be the projection of C in the xy-plane (**Figure 17.66**).

Letting $\mathbf{F} = \langle f, g, h \rangle$, the line integral in Stokes' Theorem is

$$\oint_C \mathbf{F} \cdot d\mathbf{r} = \oint_C f\,dx + g\,dy + h\,dz.$$

The key observation for this integral is that along C (which is the boundary of S), $dz = z_x\,dx + z_y\,dy$. Making this substitution, we convert the line integral on C to a line integral on C' in the xy-plane:

$$\oint_C \mathbf{F} \cdot d\mathbf{r} = \oint_{C'} f\,dx + g\,dy + \underbrace{h(z_x\,dx + z_y\,dy)}_{dz}$$

$$= \oint_{C'} \underbrace{(f + hz_x)}_{M(x,\,y)}\,dx + \underbrace{(g + hz_y)}_{N(x,\,y)}\,dy.$$

We now apply the circulation form of Green's Theorem to this line integral with $M(x, y) = f + hz_x$ and $N(x, y) = g + hz_y$; the result is

$$\oint_{C'} M\,dx + N\,dy = \iint_R (N_x - M_y)\,dA.$$

A careful application of the Chain Rule (remembering that z is a function of x and y, Exercise 49) reveals that

$$M_y = f_y + f_z z_y + hz_{xy} + z_x(h_y + h_z z_y) \quad \text{and}$$
$$N_x = g_x + g_z z_x + hz_{yx} + z_y(h_x + h_z z_x).$$

Making these substitutions in the line integral and simplifying (note that $z_{xy} = z_{yx}$ is needed), we have

$$\oint_C \mathbf{F} \cdot d\mathbf{r} = \iint_R (z_x(g_z - h_y) + z_y(h_x - f_z) + (g_x - f_y))\,dA. \tag{1}$$

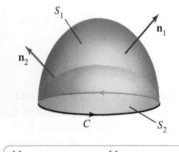

$$\iint_{S_1} (\nabla \times \mathbf{F}) \cdot \mathbf{n}_1\,dS = \iint_{S_2} (\nabla \times \mathbf{F}) \cdot \mathbf{n}_2\,dS$$

(a)

Now let's look at the surface integral in Stokes' Theorem. The upward vector normal to the surface is $\langle -z_x, -z_y, 1 \rangle$. Substituting the components of $\nabla \times \mathbf{F}$, the surface integral takes the form

$$\iint_S (\nabla \times \mathbf{F}) \cdot \mathbf{n}\,dS = \iint_R ((h_y - g_z)(-z_x) + (f_z - h_x)(-z_y) + (g_x - f_y))\,dA,$$

which upon rearrangement becomes the integral in (1).

Two Final Notes on Stokes' Theorem

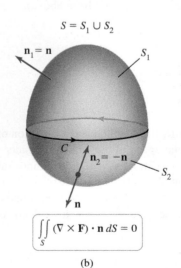

$S = S_1 \cup S_2$

$$\iint_S (\nabla \times \mathbf{F}) \cdot \mathbf{n}\,dS = 0$$

(b)

Figure 17.67

1. Stokes' Theorem allows a surface integral $\iint_S (\nabla \times \mathbf{F}) \cdot \mathbf{n}\,dS$ to be evaluated using only the values of the vector field on the boundary C. This means that if a closed curve C is the boundary of two different smooth oriented surfaces S_1 and S_2, which both have an orientation consistent with that of C, then the integrals of $(\nabla \times \mathbf{F}) \cdot \mathbf{n}$ on the two surfaces are equal; that is,

$$\iint_{S_1} (\nabla \times \mathbf{F}) \cdot \mathbf{n}_1\,dS = \iint_{S_2} (\nabla \times \mathbf{F}) \cdot \mathbf{n}_2\,dS,$$

where $\mathbf{n}_1$ and $\mathbf{n}_2$ are the respective unit normal vectors consistent with the orientation of the surfaces (**Figure 17.67a**; see Example 3).

Now let's take a different perspective. Suppose S is a *closed* surface consisting of S_1 and S_2 with a common boundary curve C (**Figure 17.67b**). Let $\mathbf{n}$ represent the outward unit normal vector for the entire surface S. It follows that $\mathbf{n}$ points in the same direction as $\mathbf{n}_1$ and in the direction opposite to that of $\mathbf{n}_2$ (Figure 17.67b). Therefore, $\iint_{S_1} (\nabla \times \mathbf{F}) \cdot \mathbf{n} \, dS$ and $\iint_{S_2} (\nabla \times \mathbf{F}) \cdot \mathbf{n} \, dS$ are equal in magnitude and of opposite sign, from which we conclude that

$$\iint_S (\nabla \times \mathbf{F}) \cdot \mathbf{n} \, dS = \iint_{S_1} (\nabla \times \mathbf{F}) \cdot \mathbf{n} \, dS + \iint_{S_2} (\nabla \times \mathbf{F}) \cdot \mathbf{n} \, dS = 0.$$

This argument can be adapted to show that $\iint_S (\nabla \times \mathbf{F}) \cdot \mathbf{n} \, dS = 0$ over any closed oriented surface S (Exercise 50).

2. We can now resolve an assertion made in Section 17.5. There we proved (Theorem 17.11) that if $\mathbf{F}$ is a conservative vector field, then $\nabla \times \mathbf{F} = \mathbf{0}$; we claimed, but did not prove, that the converse is true. The converse follows directly from Stokes' Theorem.

THEOREM 17.16   Curl F $= 0$ Implies F Is Conservative
Suppose $\nabla \times \mathbf{F} = \mathbf{0}$ throughout an open simply connected region D of $\mathbb{R}^3$. Then $\oint_C \mathbf{F} \cdot d\mathbf{r} = 0$ on all closed simple smooth curves C in D, and $\mathbf{F}$ is a conservative vector field on D.

Proof: Given a closed simple smooth curve C, an advanced result states that C is the boundary of at least one smooth oriented surface S in D. By Stokes' Theorem,

$$\oint_C \mathbf{F} \cdot d\mathbf{r} = \iint_S \underbrace{(\nabla \times \mathbf{F})}_{0} \cdot \mathbf{n} \, dS = 0.$$

Because the line integral equals zero over all such curves in D, the vector field is conservative on D by Theorem 17.6. ◄

SECTION 17.7 EXERCISES

Getting Started

1. Explain the meaning of the integral $\oint_C \mathbf{F} \cdot d\mathbf{r}$ in Stokes' Theorem.

2. Explain the meaning of the integral $\iint_S (\nabla \times \mathbf{F}) \cdot \mathbf{n} \, dS$ in Stokes' Theorem.

3. Explain the meaning of Stokes' Theorem.

4. Why does a conservative vector field produce zero circulation around a closed curve?

Practice Exercises

5–10. Verifying Stokes' Theorem *Verify that the line integral and the surface integral of Stokes' Theorem are equal for the following vector fields, surfaces S, and closed curves C. Assume C has counterclockwise orientation and S has a consistent orientation.*

5. $\mathbf{F} = \langle y, -x, 10 \rangle$; S is the upper half of the sphere $x^2 + y^2 + z^2 = 1$ and C is the circle $x^2 + y^2 = 1$ in the xy-plane.

6. $\mathbf{F} = \langle 0, -x, y \rangle$; S is the upper half of the sphere $x^2 + y^2 + z^2 = 4$ and C is the circle $x^2 + y^2 = 4$ in the xy-plane.

7. $\mathbf{F} = \langle x, y, z \rangle$; S is the paraboloid $z = 8 - x^2 - y^2$, for $0 \le z \le 8$, and C is the circle $x^2 + y^2 = 8$ in the xy-plane.

8. $\mathbf{F} = \langle 2z, -4x, 3y \rangle$; S is the cap of the sphere $x^2 + y^2 + z^2 = 169$ above the plane $z = 12$ and C is the boundary of S.

9. $\mathbf{F} = \langle y - z, z - x, x - y \rangle$; S is the cap of the sphere $x^2 + y^2 + z^2 = 16$ above the plane $z = \sqrt{7}$ and C is the boundary of S.

10. $\mathbf{F} = \langle -y, -x - z, y - x \rangle$; S is the part of the plane $z = 6 - y$ that lies in the cylinder $x^2 + y^2 = 16$ and C is the boundary of S.

11–16. Stokes' Theorem for evaluating line integrals *Evaluate the line integral $\oint_C \mathbf{F} \cdot d\mathbf{r}$ by evaluating the surface integral in Stokes' Theorem with an appropriate choice of S. Assume C has a counterclockwise orientation.*

11. $\mathbf{F} = \langle 2y, -z, x \rangle$; C is the circle $x^2 + y^2 = 12$ in the plane $z = 0$.

12. $\mathbf{F} = \langle y, xz, -y \rangle$; C is the ellipse $x^2 + y^2/4 = 1$ in the plane $z = 1$.

13. $\mathbf{F} = \langle x^2 - z^2, y, 2xz \rangle$; C is the boundary of the plane $z = 4 - x - y$ in the first octant.

14. $\mathbf{F} = \langle x^2 - y^2, z^2 - x^2, y^2 - z^2 \rangle$; C is the boundary of the square $|x| \le 1$, $|y| \le 1$ in the plane $z = 0$.

15. $F = \langle y^2, -z^2, x \rangle$; C is the circle $r(t) = \langle 3 \cos t, 4 \cos t, 5 \sin t \rangle$, for $0 \le t \le 2\pi$.

16. $F = \langle 2xy \sin z, x^2 \sin z, x^2 y \cos z \rangle$; C is the boundary of the plane $z = 8 - 2x - 4y$ in the first octant.

17–24. Stokes' Theorem for evaluating surface integrals *Evaluate the line integral in Stokes' Theorem to determine the value of the surface integral $\iint_S (\nabla \times F) \cdot n \, dS$. Assume n points in an upward direction.*

17. $F = \langle x, y, z \rangle$; S is the upper half of the ellipsoid $x^2/4 + y^2/9 + z^2 = 1$.

18. $F = r/|r|$; S is the paraboloid $x = 9 - y^2 - z^2$, for $0 \le x \le 9$ (excluding its base), and $r = \langle x, y, z \rangle$.

19. $F = \langle 2y, -z, x - y - z \rangle$; S is the cap of the sphere $x^2 + y^2 + z^2 = 25$, for $3 \le x \le 5$ (excluding its base).

20. $F = \langle x + y, y + z, z + x \rangle$; S is the tilted disk enclosed by $r(t) = \langle \cos t, 2 \sin t, \sqrt{3} \cos t \rangle$.

21. $F = \langle y, z - x, -y \rangle$; S is the part of the paraboloid $z = 2 - x^2 - 2y^2$ that lies within the cylinder $x^2 + y^2 = 1$.

22. $F = \langle 4x, -8z, 4y \rangle$; S is the part of the paraboloid $z = 1 - 2x^2 - 3y^2$ that lies within the paraboloid $z = 2x^2 + y^2$.

23. $F = \langle y, 1, z \rangle$; S is the part of the surface $z = 2\sqrt{x}$ that lies within the cone $z = \sqrt{x^2 + y^2}$.

24. $F = \langle e^x, 1/z, y \rangle$; S is the part of the surface $z = 4 - 3y^2$ that lies within the paraboloid $z = x^2 + y^2$.

25–28. Interpreting and graphing the curl *For the following velocity fields, compute the curl, make a sketch of the curl, and interpret the curl.*

25. $v = \langle 0, 0, y \rangle$ **26.** $v = \langle 1 - z^2, 0, 0 \rangle$

27. $v = \langle -2z, 0, 1 \rangle$ **28.** $v = \langle 0, -z, y \rangle$

29. Explain why or why not Determine whether the following statements are true and give an explanation or counterexample.

   a. A paddle wheel with its axis in the direction $\langle 0, 1, -1 \rangle$ would not spin when put in the vector field $F = \langle 1, 1, 2 \rangle \times \langle x, y, z \rangle$.

   b. Stokes' Theorem relates the flux of a vector field F across a surface to values of F on the boundary of the surface.

   c. A vector field of the form $F = \langle a + f(x), b + g(y), c + h(z) \rangle$, where a, b, and c are constants, has zero circulation on a closed curve.

   d. If a vector field has zero circulation on all simple closed smooth curves C in a region D, then F is conservative on D.

30–33. Conservative fields *Use Stokes' Theorem to find the circulation of the following vector fields around any simple closed smooth curve C.*

30. $F = \langle 2x, -2y, 2z \rangle$ **31.** $F = \nabla(x \sin ye^z)$

32. $F = \langle 3x^2y, x^3 + 2yz^2, 2y^2z \rangle$

33. $F = \langle y^2z^3, 2xyz^3, 3xy^2z^2 \rangle$

34–38. Tilted disks *Let S be the disk enclosed by the curve C: $r(t) = \langle \cos \varphi \cos t, \sin t, \sin \varphi \cos t \rangle$, for $0 \le t \le 2\pi$, where $0 \le \varphi \le \pi/2$ is a fixed angle.*

34. What is the area of S? Find a vector normal to S.

35. What is the length of C?

36. Use Stokes' Theorem and a surface integral to find the circulation on C of the vector field $F = \langle -y, x, 0 \rangle$ as a function of φ. For what value of φ is the circulation a maximum?

37. What is the circulation on C of the vector field $F = \langle -y, -z, x \rangle$ as a function of φ? For what value of φ is the circulation a maximum?

38. Consider the vector field $F = a \times r$, where $a = \langle a_1, a_2, a_3 \rangle$ is a constant nonzero vector and $r = \langle x, y, z \rangle$. Show that the circulation is a maximum when a points in the direction of the normal to S.

39. Circulation in a plane A circle C in the plane $x + y + z = 8$ has a radius of 4 and center $(2, 3, 3)$. Evaluate $\oint_C F \cdot dr$ for $F = \langle 0, -z, 2y \rangle$, where C has a counterclockwise orientation when viewed from above. Does the circulation depend on the radius of the circle? Does it depend on the location of the center of the circle?

40. No integrals Let $F = \langle 2z, z, 2y + x \rangle$, and let S be the hemisphere of radius a with its base in the xy-plane and center at the origin.

   a. Evaluate $\iint_S (\nabla \times F) \cdot n \, dS$ by computing $\nabla \times F$ and appealing to symmetry.

   b. Evaluate the line integral using Stokes' Theorem to check part (a).

41. Compound surface and boundary Begin with the paraboloid $z = x^2 + y^2$, for $0 \le z \le 4$, and slice it with the plane $y = 0$. Let S be the surface that remains for $y \ge 0$ (including the planar surface in the xz-plane) (see figure). Let C be the semicircle and line segment that bound the cap of S in the plane $z = 4$ with counterclockwise orientation. Let $F = \langle 2z + y, 2x + z, 2y + x \rangle$.

   a. Describe the direction of the vectors normal to the surface that are consistent with the orientation of C.

   b. Evaluate $\iint_S (\nabla \times F) \cdot n \, dS$.

   c. Evaluate $\oint_C F \cdot dr$ and check for agreement with part (b).

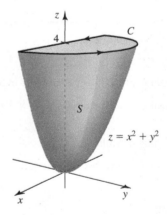

42. Ampère's Law The French physicist André-Marie Ampère (1775–1836) discovered that an electrical current I in a wire produces a magnetic field B. A special case of Ampère's Law relates the current to the magnetic field through the equation $\oint_C B \cdot dr = \mu I$, where C is any closed curve through which the wire passes and μ is a physical constant. Assume the current I is given in terms of the current density J as $I = \iint_S J \cdot n \, dS$, where S is an oriented surface with C as a boundary. Use Stokes' Theorem to show that an equivalent form of Ampère's Law is $\nabla \times B = \mu J$.

43. Maximum surface integral Let S be the paraboloid $z = a(1 - x^2 - y^2)$, for $z \ge 0$, where $a > 0$ is a real number. Let $F = \langle x - y, y + z, z - x \rangle$. For what value(s) of a (if any) does $\iint_S (\nabla \times F) \cdot n \, dS$ have its maximum value?

Explorations and Challenges

44. Area of a region in a plane Let R be a region in a plane that has a unit normal vector $\mathbf{n} = \langle a, b, c \rangle$ and boundary C. Let $\mathbf{F} = \langle bz, cx, ay \rangle$.

a. Show that $\nabla \times \mathbf{F} = \mathbf{n}$.

b. Use Stokes' Theorem to show that

$$\text{area of } R = \oint_C \mathbf{F} \cdot d\mathbf{r}.$$

c. Consider the curve C given by $\mathbf{r} = \langle 5 \sin t, 13 \cos t, 12 \sin t \rangle$, for $0 \leq t \leq 2\pi$. Prove that C lies in a plane by showing that $\mathbf{r} \times \mathbf{r}'$ is constant for all t.

d. Use part (b) to find the area of the region enclosed by C in part (c). (*Hint:* Find the unit normal vector that is consistent with the orientation of C.)

45. Choosing a more convenient surface The goal is to evaluate $A = \iint_S (\nabla \times \mathbf{F}) \cdot \mathbf{n} \, dS$, where $\mathbf{F} = \langle yz, -xz, xy \rangle$ and S is the surface of the upper half of the ellipsoid $x^2 + y^2 + 8z^2 = 1$ ($z \geq 0$).

a. Evaluate a surface integral over a more convenient surface to find the value of A.

b. Evaluate A using a line integral.

46. Radial fields and zero circulation Consider the radial vector fields $\mathbf{F} = \mathbf{r}/|\mathbf{r}|^p$, where p is a real number and $\mathbf{r} = \langle x, y, z \rangle$. Let C be any circle in the xy-plane centered at the origin.

a. Evaluate a line integral to show that the field has zero circulation on C.

b. For what values of p does Stokes' Theorem apply? For those values of p, use the surface integral in Stokes' Theorem to show that the field has zero circulation on C.

47. Zero curl Consider the vector field

$$\mathbf{F} = -\frac{y}{x^2 + y^2}\mathbf{i} + \frac{x}{x^2 + y^2}\mathbf{j} + z\,\mathbf{k}.$$

a. Show that $\nabla \times \mathbf{F} = \mathbf{0}$.

b. Show that $\oint_C \mathbf{F} \cdot d\mathbf{r}$ is not zero on a circle C in the xy-plane enclosing the origin.

c. Explain why Stokes' Theorem does not apply in this case.

48. Average circulation Let S be a small circular disk of radius R centered at the point P with a unit normal vector $\mathbf{n}$. Let C be the boundary of S.

a. Express the average circulation of the vector field $\mathbf{F}$ on S as a surface integral of $\nabla \times \mathbf{F}$.

b. Argue that for small R, the average circulation approaches $(\nabla \times \mathbf{F})|_P \cdot \mathbf{n}$ (the component of $\nabla \times \mathbf{F}$ in the direction of $\mathbf{n}$ evaluated at P) with the approximation improving as $R \to 0$.

49. Proof of Stokes' Theorem Confirm the following step in the proof of Stokes' Theorem. If $z = s(x, y)$ and f, g, and h are functions of x, y, and z, with $M = f + hz_x$ and $N = g + hz_y$, then

$$M_y = f_y + f_z z_y + hz_{xy} + z_x(h_y + h_z z_y) \quad \text{and}$$
$$N_x = g_x + g_z z_x + hz_{yx} + z_y(h_x + h_z z_x).$$

50. Stokes' Theorem on closed surfaces Prove that if $\mathbf{F}$ satisfies the conditions of Stokes' Theorem, then $\iint_S (\nabla \times \mathbf{F}) \cdot \mathbf{n} \, dS = 0$, where S is a smooth surface that encloses a region.

51. Rotated Green's Theorem Use Stokes' Theorem to write the circulation form of Green's Theorem in the yz-plane.

QUICK CHECK ANSWERS

1. If S is a region in the xy-plane, $\mathbf{n} = \mathbf{k}$ and $(\nabla \times \mathbf{F}) \cdot \mathbf{n}$ becomes $g_x - f_y$. **3.** The vector field is symmetric about the y-axis. ◄

17.8 Divergence Theorem

Vector fields can represent electric or magnetic fields, air velocities in hurricanes, or blood flow in an artery. These and other vector phenomena suggest movement of a "substance." A frequent question concerns the amount of a substance that flows across a surface—for example, the amount of water that passes across the membrane of a cell per unit time. Such flux calculations may be done using flux integrals as in Section 17.6. The Divergence Theorem offers an alternative method. In effect, it says that instead of integrating the flow into and out of a region across its boundary, you may also add up all the sources (or sinks) of the flow throughout the region.

Divergence Theorem

► Circulation form of Green's Theorem → Stokes' Theorem

Flux form of Green's Theorem → Divergence Theorem

The Divergence Theorem is the three-dimensional version of the flux form of Green's Theorem. Recall that if R is a region in the xy-plane, C is the simple closed piecewise-smooth oriented boundary of R, and $\mathbf{F} = \langle f, g \rangle$ is a vector field, then Green's Theorem says that

$$\underbrace{\oint_C \mathbf{F} \cdot \mathbf{n} \, ds}_{\text{flux across } C} = \iint_R \underbrace{(f_x + g_y)}_{\text{divergence}} \, dA.$$

The line integral on the left gives the flux across the boundary of R. The double integral on the right measures the net expansion or contraction of the vector field within R. If $\mathbf{F}$ represents a fluid flow or the transport of a material, the theorem says that the cumulative effect of the sources (or sinks) of the flow within R equals the net flow across its boundary.

The Divergence Theorem is a direct extension of Green's Theorem. The plane region in Green's Theorem becomes a solid region D in $\mathbb{R}^3$, and the closed curve in Green's Theorem becomes the oriented surface S that encloses D. The flux integral in Green's Theorem becomes a surface integral over S, and the double integral in Green's Theorem becomes a triple integral over D of the three-dimensional divergence (**Figure 17.68**).

Flux form of
Green's Theorem:

$$\oint_C \mathbf{F} \cdot \mathbf{n}\, ds = \iint_R \operatorname{div} \mathbf{F}\, dA$$

Divergence Theorem:

$$\iint_S \mathbf{F} \cdot \mathbf{n}\, dS = \iiint_D \operatorname{div} \mathbf{F}\, dV$$

Figure 17.68

THEOREM 17.17   Divergence Theorem
Let $\mathbf{F}$ be a vector field whose components have continuous first partial derivatives in a connected and simply connected region D in $\mathbb{R}^3$ enclosed by an oriented surface S. Then

$$\iint_S \mathbf{F} \cdot \mathbf{n}\, dS = \iiint_D \nabla \cdot \mathbf{F}\, dV,$$

where $\mathbf{n}$ is the outward unit normal vector on S.

QUICK CHECK 1 Interpret the Divergence Theorem in the case that $\mathbf{F} = \langle a, b, c \rangle$ is a constant vector field and D is a ball. ◄

The surface integral on the left gives the flux of the vector field across the boundary; a positive flux integral means there is a net flow of the field out of the region. The triple integral on the right is the cumulative expansion or contraction of the field over the region D. The proof of a special case of the theorem is given later in this section.

EXAMPLE 1   Verifying the Divergence Theorem Consider the radial field $\mathbf{F} = \langle x, y, z \rangle$ and let S be the sphere $x^2 + y^2 + z^2 = a^2$ that encloses the region D. Assume $\mathbf{n}$ is the outward unit normal vector on the sphere. Evaluate both integrals of the Divergence Theorem.

SOLUTION The divergence of $\mathbf{F}$ is

$$\nabla \cdot \mathbf{F} = \frac{\partial}{\partial x}(x) + \frac{\partial}{\partial y}(y) + \frac{\partial}{\partial z}(z) = 3.$$

Integrating over D, we have

$$\iiint_D \nabla \cdot \mathbf{F}\, dV = \iiint_D 3\, dV = 3 \times \text{volume of } D = 3 \cdot \frac{4}{3}\pi a^3 = 4\pi a^3.$$

To evaluate the surface integral, we parameterize the sphere (Section 17.6, Table 17.3) in the form

$$\mathbf{r} = \langle x, y, z \rangle = \langle a \sin u \cos v, a \sin u \sin v, a \cos u \rangle,$$

where $R = \{(u, v): 0 \leq u \leq \pi, 0 \leq v \leq 2\pi\}$ (u and v are the spherical coordinates φ and θ, respectively). The surface integral is

$$\iint_S \mathbf{F} \cdot \mathbf{n}\, dS = \iint_R \mathbf{F} \cdot (\mathbf{t}_u \times \mathbf{t}_v)\, dA,$$

where the required vector normal to the surface is

$$\mathbf{t}_u \times \mathbf{t}_v = \langle a^2 \sin^2 u \cos v, a^2 \sin^2 u \sin v, a^2 \sin u \cos u \rangle.$$

Substituting for $\mathbf{F} = \langle x, y, z \rangle$ and $\mathbf{t}_u \times \mathbf{t}_v$, we find after simplifying that $\mathbf{F} \cdot (\mathbf{t}_u \times \mathbf{t}_v) = a^3 \sin u$. Therefore, the surface integral becomes

▶ See Exercise 32 for an alternative evaluation of the surface integral.

$$\iint\limits_{S} \mathbf{F} \cdot \mathbf{n} \, dS = \iint\limits_{R} \underbrace{\mathbf{F} \cdot (\mathbf{t}_u \times \mathbf{t}_v)}_{a^3 \sin u} \, dA$$

$$= \int_{0}^{2\pi} \int_{0}^{\pi} a^3 \sin u \, du \, dv \quad \text{Substitute for } \mathbf{F} \text{ and } \mathbf{t}_u \times \mathbf{t}_v.$$

$$= 4\pi a^3. \qquad\qquad \text{Evaluate integrals.}$$

The two integrals of the Divergence Theorem are equal.

Related Exercise 9 ◀

EXAMPLE 2 **Divergence Theorem with a rotation field** Consider the rotation field

$$\mathbf{F} = \mathbf{a} \times \mathbf{r} = \langle 1, 0, 1 \rangle \times \langle x, y, z \rangle = \langle -y, x - z, y \rangle.$$

Let S be the hemisphere $x^2 + y^2 + z^2 = a^2$, for $z \ge 0$, together with its base in the xy-plane. Find the net outward flux across S.

SOLUTION To find the flux using surface integrals, two surfaces must be considered (the hemisphere and its base). The Divergence Theorem gives a simpler solution. Note that

$$\nabla \cdot \mathbf{F} = \frac{\partial}{\partial x}(-y) + \frac{\partial}{\partial y}(x - z) + \frac{\partial}{\partial z}(y) = 0.$$

We see that the flux across the hemisphere is zero.

Related Exercise 13 ◀

With Stokes' Theorem, rotation fields are noteworthy because they have a nonzero curl. With the Divergence Theorem, the situation is reversed. As suggested by Example 2, pure rotation fields of the form $\mathbf{F} = \mathbf{a} \times \mathbf{r}$ have zero divergence (Exercise 16). However, with the Divergence Theorem, radial fields are interesting and have many physical applications.

EXAMPLE 3 **Computing flux with the Divergence Theorem** Find the net outward flux of the field $\mathbf{F} = xyz \langle 1, 1, 1 \rangle$ across the boundaries of the cube $D = \{(x, y, z): 0 \le x \le 1, 0 \le y \le 1, 0 \le z \le 1\}$.

SOLUTION Computing a surface integral involves the six faces of the cube. The Divergence Theorem gives the outward flux with a single integral over D. The divergence of the field is

$$\nabla \cdot \mathbf{F} = \frac{\partial}{\partial x}(xyz) + \frac{\partial}{\partial y}(xyz) + \frac{\partial}{\partial z}(xyz) = yz + xz + xy.$$

The integral over D is a standard triple integral:

$$\iiint\limits_{D} \nabla \cdot \mathbf{F} \, dV = \iiint\limits_{D} (yz + xz + xy) \, dV$$

$$= \int_{0}^{1} \int_{0}^{1} \int_{0}^{1} (yz + xz + xy) \, dx \, dy \, dz \quad \text{Convert to a triple integral.}$$

$$= \frac{3}{4}. \qquad\qquad \text{Evaluate integrals.}$$

On three faces of the cube (those that lie in the coordinate planes), we see that $\mathbf{F}(0, y, z) = \mathbf{F}(x, 0, z) = \mathbf{F}(x, y, 0) = \mathbf{0}$, so there is no contribution to the flux on these faces (**Figure 17.69**). On the other three faces, the vector field has components out of the cube. Therefore, the net outward flux is positive, as calculated.

Related Exercises 18–19 ◀

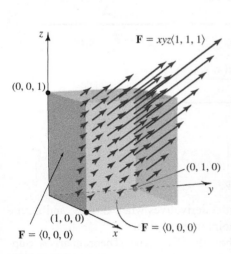

$\mathbf{F} = xyz\langle 1, 1, 1 \rangle$

$(0, 0, 1)$

$(0, 1, 0)$

$(1, 0, 0)$

$\mathbf{F} = \langle 0, 0, 0 \rangle$

$\mathbf{F} = \langle 0, 0, 0 \rangle$

Figure 17.69

▶ The mass transport is also called the *flux density*; when multiplied by an area, it gives the flux. We use the convention that flux has units of mass per unit time.

▶ Check the units: If **F** has units of mass/(area · time), then the flux has units of mass/time (**n** has no units).

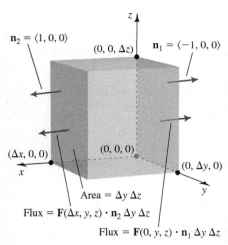

Flux = $\mathbf{F}(\Delta x, y, z) \cdot \mathbf{n}_2 \, \Delta y \, \Delta z$

Flux = $\mathbf{F}(0, y, z) \cdot \mathbf{n}_1 \, \Delta y \, \Delta z$

Figure 17.70

Interpretation of the Divergence Theorem Using Mass Transport Suppose **v** is the velocity field of a material, such as water or molasses, and ρ is its constant density. The vector field $\mathbf{F} = \rho\mathbf{v} = \langle f, g, h \rangle$ describes the **mass transport** of the material, with units of (mass/vol) $\times$ (length/time) = mass/(area · time); typical units of mass transport are $g/m^2/s$. This means that **F** gives the mass of material flowing past a point (in each of the three coordinate directions) per unit of surface area per unit of time. When **F** is multiplied by an area, the result is the **flux**, with units of mass/unit time.

Now consider a small cube located in the vector field with its faces parallel to the coordinate planes. One vertex is located at $(0, 0, 0)$, the opposite vertex is at $(\Delta x, \Delta y, \Delta z)$, and (x, y, z) is an arbitrary point in the cube (**Figure 17.70**). The goal is to compute the approximate flux of material across the faces of the cube. We begin with the flux across the two parallel faces $x = 0$ and $x = \Delta x$.

The outward unit vectors normal to the faces $x = 0$ and $x = \Delta x$ are $\mathbf{n}_1 = \langle -1, 0, 0 \rangle$ and $\mathbf{n}_2 = \langle 1, 0, 0 \rangle$, respectively. Each face has area $\Delta y \, \Delta z$, so the approximate net flux across these faces is

$$\underbrace{\mathbf{F}(\Delta x, y, z)}_{x = \Delta x \text{ face}} \cdot \underbrace{\mathbf{n}_2}_{\langle 1, 0, 0 \rangle} \Delta y \, \Delta z + \underbrace{\mathbf{F}(0, y, z)}_{x = 0 \text{ face}} \cdot \underbrace{\mathbf{n}_1}_{\langle -1, 0, 0 \rangle} \Delta y \, \Delta z$$

$$= (f(\Delta x, y, z) - f(0, y, z)) \, \Delta y \, \Delta z.$$

Note that if $f(\Delta x, y, z) > f(0, y, z)$, the net flux across these two faces of the cube is positive, which means the net flow is *out* of the cube. Letting $\Delta V = \Delta x \, \Delta y \, \Delta z$ be the volume of the cube, we rewrite the net flux as

$$(f(\Delta x, y, z) - f(0, y, z)) \, \Delta y \, \Delta z$$

$$= \frac{f(\Delta x, y, z) - f(0, y, z)}{\Delta x} \, \Delta x \, \Delta y \, \Delta z \quad \text{Multiply by } \frac{\Delta x}{\Delta x}.$$

$$= \frac{f(\Delta x, y, z) - f(0, y, z)}{\Delta x} \, \Delta V. \qquad \Delta V = \Delta x \, \Delta y \, \Delta z$$

A similar argument can be applied to the other two pairs of faces. The approximate net flux across the faces $y = 0$ and $y = \Delta y$ is

$$\frac{g(x, \Delta y, z) - g(x, 0, z)}{\Delta y} \, \Delta V,$$

and the approximate net flux across the faces $z = 0$ and $z = \Delta z$ is

$$\frac{h(x, y, \Delta z) - h(x, y, 0)}{\Delta z} \, \Delta V.$$

Adding these three individual fluxes gives the approximate net flux out of the cube:

$$\text{net flux out of cube} \approx \left(\underbrace{\frac{f(\Delta x, y, z) - f(0, y, z)}{\Delta x}}_{\approx \frac{\partial f}{\partial x}(0, 0, 0)} + \underbrace{\frac{g(x, \Delta y, z) - g(x, 0, z)}{\Delta y}}_{\approx \frac{\partial g}{\partial y}(0, 0, 0)} \right.$$

$$\left. + \underbrace{\frac{h(x, y, \Delta z) - h(x, y, 0)}{\Delta z}}_{\approx \frac{\partial h}{\partial z}(0, 0, 0)} \right) \Delta V$$

$$\approx \left. \left(\frac{\partial f}{\partial x} + \frac{\partial g}{\partial y} + \frac{\partial h}{\partial z} \right) \right|_{(0, 0, 0)} \Delta V$$

$$= (\nabla \cdot \mathbf{F})(0, 0, 0) \, \Delta V.$$

Notice how the three quotients approximate partial derivatives when Δx, Δy, and Δz are small. A similar argument may be made at any point in the region.

Taking one more step, we show informally how the Divergence Theorem arises. Suppose the small cube we just analyzed is one of many small cubes of volume ΔV that fill a region D. We label the cubes $k = 1, \ldots, n$ and apply the preceding argument to each cube, letting $(\nabla \cdot \mathbf{F})_k$ be the divergence evaluated at a point in the kth cube. Adding the individual contributions to the net flux from each cube, we obtain the approximate net flux across the boundary of D:

$$\text{net flux out of } D \approx \sum_{k=1}^{n} (\nabla \cdot \mathbf{F})_k \, \Delta V.$$

▶ In making this argument, notice that for two adjacent cubes, the flux into one cube equals the flux out of the other cube across the common face. Therefore, there is a cancellation of fluxes throughout the interior of D.

Letting the volume of the cubes ΔV approach 0, and letting the number of cubes n increase, we obtain an integral over D:

$$\text{net flux out of } D = \lim_{n \to \infty} \sum_{k=1}^{n} (\nabla \cdot \mathbf{F})_k \, \Delta V = \iiint_D \nabla \cdot \mathbf{F} \, dV.$$

QUICK CHECK 3 Draw the unit cube $D = \{(x, y, z) \colon 0 \le x \le 1,\ 0 \le y \le 1, 0 \le z \le 1\}$ and sketch the vector field $\mathbf{F} = \langle x, -y, 2z \rangle$ on the six faces of the cube. Compute and interpret div $\mathbf{F}$. ◀

The net flux across the boundary of D is also given by $\iint_S \mathbf{F} \cdot \mathbf{n} \, dS$. Equating the surface integral and the volume integral gives the Divergence Theorem. Now we look at a formal proof.

Proof of the Divergence Theorem

We prove the Divergence Theorem under special conditions on the region D. Let R be the projection of D in the xy-plane (**Figure 17.71**); that is,

$$R = \{(x, y) \colon (x, y, z) \text{ is in } D\}.$$

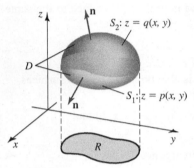

Figure 17.71

Assume the boundary of D is S and let $\mathbf{n}$ be the unit vector normal to S that points outward.

Letting $\mathbf{F} = \langle f, g, h \rangle = f\,\mathbf{i} + g\,\mathbf{j} + h\,\mathbf{k}$, the surface integral in the Divergence Theorem is

$$\iint_S \mathbf{F} \cdot \mathbf{n} \, dS = \iint_S (f\,\mathbf{i} + g\,\mathbf{j} + h\,\mathbf{k}) \cdot \mathbf{n} \, dS$$

$$= \iint_S f\,\mathbf{i} \cdot \mathbf{n} \, dS + \iint_S g\,\mathbf{j} \cdot \mathbf{n} \, dS + \iint_S h\,\mathbf{k} \cdot \mathbf{n} \, dS.$$

The volume integral in the Divergence Theorem is

$$\iiint_D \nabla \cdot \mathbf{F} \, dV = \iiint_D \left(\frac{\partial f}{\partial x} + \frac{\partial g}{\partial y} + \frac{\partial h}{\partial z} \right) dV.$$

Matching terms of the surface and volume integrals, the theorem is proved by showing that

$$\iint_S f\,\mathbf{i} \cdot \mathbf{n} \, dS = \iiint_D \frac{\partial f}{\partial x} \, dV, \tag{1}$$

$$\iint_S g\,\mathbf{j} \cdot \mathbf{n} \, dS = \iiint_D \frac{\partial g}{\partial y} \, dV, \text{ and} \tag{2}$$

$$\iint_S h\,\mathbf{k} \cdot \mathbf{n} \, dS = \iiint_D \frac{\partial h}{\partial z} \, dV. \tag{3}$$

We work on equation (3) assuming special properties for D. Suppose D is bounded by two surfaces $S_1 \colon z = p(x, y)$ and $S_2 \colon z = q(x, y)$, where $p(x, y) \le q(x, y)$ on R (Figure 17.71). The Fundamental Theorem of Calculus is used in the triple integral to show that

$$\iiint_D \frac{\partial h}{\partial z} \, dV = \iint_R \int_{p(x, y)}^{q(x, y)} \frac{\partial h}{\partial z} \, dz \, dx \, dy$$

$$= \iint_R \big(h(x, y, q(x, y)) - h(x, y, p(x, y)) \big) \, dx \, dy. \quad \text{Evaluate inner integral.}$$

Now let's turn to the surface integral in equation (3), $\iint_S h\,\mathbf{k} \cdot \mathbf{n} \, dS$, and note that S consists of three pieces: the lower surface S_1, the upper surface S_2, and the vertical sides S_3 of the surface (if they exist). The normal to S_3 is everywhere orthogonal to $\mathbf{k}$, so $\mathbf{k} \cdot \mathbf{n} = 0$

and the S_3 integral makes no contribution. What remains is to compute the surface integrals over S_1 and S_2.

The required outward normal to S_2 (which is the graph of $z = q(x, y)$) is $\langle -q_x, -q_y, 1 \rangle$. The outward normal to S_1 (which is the graph of $z = p(x, y)$) points *downward*, so it is given by $\langle p_x, p_y, -1 \rangle$. The surface integral of (3) becomes

$$\iint\limits_S h\,\mathbf{k} \cdot \mathbf{n}\, dS = \iint\limits_{S_2} h(x, y, z)\,\mathbf{k} \cdot \mathbf{n}\, dS + \iint\limits_{S_1} h(x, y, z)\,\mathbf{k} \cdot \mathbf{n}\, dS$$

$$= \iint\limits_R h(x, y, q(x, y))\,\underbrace{\mathbf{k} \cdot \langle -q_x, -q_y, 1 \rangle}_{1}\, dx\, dy$$

$$+ \iint\limits_R h(x, y, p(x, y))\,\underbrace{\mathbf{k} \cdot \langle p_x, p_y, -1 \rangle}_{-1}\, dx\, dy$$

$$= \iint\limits_R h(x, y, q(x, y))\, dx\, dy - \iint\limits_R h(x, y, p(x, y))\, dx\, dy. \quad \text{Simplify.}$$

Observe that both the volume integral and the surface integral of (3) reduce to the same integral over R. Therefore, $\iint_S h\,\mathbf{k} \cdot \mathbf{n}\, dS = \iiint_D \dfrac{\partial h}{\partial z}\, dV$.

Equations (1) and (2) are handled in a similar way.

- To prove (1), we make the special assumption that D is also bounded by two surfaces, $S_1: x = s(y, z)$ and $S_2: x = t(y, z)$, where $s(y, z) \le t(y, z)$.
- To prove (2), we assume D is bounded by two surfaces, $S_1: y = u(x, z)$ and $S_2: y = v(x, z)$, where $u(x, z) \le v(x, z)$.

When combined, equations (1), (2), and (3) yield the Divergence Theorem. ◄

Divergence Theorem for Hollow Regions

The Divergence Theorem may be extended to more general solid regions. Here we consider the important case of hollow regions. Suppose D is a region consisting of all points inside a closed oriented surface S_2 and outside a closed oriented surface S_1, where S_1 lies within S_2 (**Figure 17.72**). Therefore, the boundary of D consists of S_1 and S_2, which we denote S. (Note that D is simply connected.)

We let $\mathbf{n}_1$ and $\mathbf{n}_2$ be the outward unit normal vectors for S_1 and S_2, respectively. Note that $\mathbf{n}_1$ points into D, so the outward normal to S on S_1 is $-\mathbf{n}_1$. With this observation, the Divergence Theorem takes the following form.

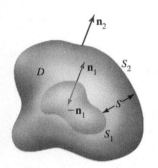

$\mathbf{n}_1$ is the outward unit normal to S_1 and points into D. The outward unit normal to S on S_1 is $-\mathbf{n}_1$.

Figure 17.72

➤ It's important to point out again that $\mathbf{n}_1$ is the unit normal that we would use for S_1 alone, independent of S. It is the outward unit normal to S_1, but it points into D.

> **THEOREM 17.18   Divergence Theorem for Hollow Regions**
> Suppose the vector field $\mathbf{F}$ satisfies the conditions of the Divergence Theorem on a region D bounded by two oriented surfaces S_1 and S_2, where S_1 lies within S_2. Let S be the entire boundary of D ($S = S_1 \cup S_2$) and let $\mathbf{n}_1$ and $\mathbf{n}_2$ be the outward unit normal vectors for S_1 and S_2, respectively. Then
>
> $$\iiint\limits_D \nabla \cdot \mathbf{F}\, dV = \iint\limits_S \mathbf{F} \cdot \mathbf{n}\, dS = \iint\limits_{S_2} \mathbf{F} \cdot \mathbf{n}_2\, dS - \iint\limits_{S_1} \mathbf{F} \cdot \mathbf{n}_1\, dS.$$

This form of the Divergence Theorem is applicable to vector fields that are not differentiable at the origin, as is the case with some important radial vector fields.

EXAMPLE 4   Flux for an inverse square field Consider the inverse square vector field

$$\mathbf{F} = \frac{\mathbf{r}}{|\mathbf{r}|^3} = \frac{\langle x, y, z \rangle}{(x^2 + y^2 + z^2)^{3/2}}.$$

▶ Recall that an inverse square force is proportional to $1/|\mathbf{r}|^2$ multiplied by a unit vector in the radial direction, which is $\mathbf{r}/|\mathbf{r}|$. Combining these two factors gives $\mathbf{F} = \mathbf{r}/|\mathbf{r}|^3$.

a. Find the net outward flux of $\mathbf{F}$ across the surface of the region
$D = \{(x, y, z): a^2 \le x^2 + y^2 + z^2 \le b^2\}$ that lies between concentric spheres with radii a and b.

b. Find the outward flux of $\mathbf{F}$ across any sphere that encloses the origin.

SOLUTION

a. Although the vector field is undefined at the origin, it is defined and differentiable in D, which excludes the origin. In Section 17.5 (Exercise 73) it was shown that the divergence of the radial field $\mathbf{F} = \dfrac{\mathbf{r}}{|\mathbf{r}|^p}$ with $p = 3$ is 0. We let S be the union of S_2, the larger sphere of radius b, and S_1, the smaller sphere of radius a. Because $\iiint_D \nabla \cdot \mathbf{F} \, dV = 0$, the Divergence Theorem implies that

$$\iint_S \mathbf{F} \cdot \mathbf{n} \, dS = \iint_{S_2} \mathbf{F} \cdot \mathbf{n}_2 \, dS - \iint_{S_1} \mathbf{F} \cdot \mathbf{n}_1 \, dS = 0.$$

Therefore, the next flux across S is zero.

b. Part (a) implies that

$$\underbrace{\iint_{S_2} \mathbf{F} \cdot \mathbf{n}_2 \, dS}_{\text{out of } D} = \underbrace{\iint_{S_1} \mathbf{F} \cdot \mathbf{n}_1 \, dS}_{\text{into } D}.$$

We see that the flux out of D across S_2 equals the flux into D across S_1. To find that flux, we evaluate the surface integral over S_1 on which $|\mathbf{r}| = a$. (Because the fluxes are equal, S_2 could also be used.)

The easiest way to evaluate the surface integral is to note that on the sphere S_1, the unit outward normal vector is $\mathbf{n}_1 = \mathbf{r}/|\mathbf{r}|$. Therefore, the surface integral is

$$\iint_{S_1} \mathbf{F} \cdot \mathbf{n}_1 \, dS = \iint_{S_1} \frac{\mathbf{r}}{|\mathbf{r}|^3} \cdot \frac{\mathbf{r}}{|\mathbf{r}|} \, dS \qquad \text{Substitute for } \mathbf{F} \text{ and } \mathbf{n}_1.$$

$$= \iint_{S_1} \frac{|\mathbf{r}|^2}{|\mathbf{r}|^4} \, dS \qquad \mathbf{r} \cdot \mathbf{r} = |\mathbf{r}|^2$$

$$= \iint_{S_1} \frac{1}{a^2} \, dS \qquad |\mathbf{r}| = a$$

$$= \frac{4\pi a^2}{a^2} \qquad \text{Surface area} = 4\pi a^2$$

$$= 4\pi.$$

The same result is obtained using S_2 or any smooth surface enclosing the origin. The flux of the inverse square field across *any* surface enclosing the origin is 4π. As shown in Exercise 46, among radial fields, this property holds only for the inverse square field $(p = 3)$.

Related Exercises 26–27 ◀

Gauss' Law

Applying the Divergence Theorem to electric fields leads to one of the fundamental laws of physics. The electric field due to a point charge Q located at the origin is given by the inverse square law,

$$\mathbf{E}(x, y, z) = \frac{Q}{4\pi\varepsilon_0} \frac{\mathbf{r}}{|\mathbf{r}|^3},$$

where $\mathbf{r} = \langle x, y, z \rangle$ and ε_0 is a physical constant called the *permittivity of free space*.

According to the calculation of Example 4, the flux of the field $\dfrac{\mathbf{r}}{|\mathbf{r}|^3}$ across any surface that encloses the origin is 4π. Therefore, the flux of the electric field across any surface enclosing the origin is $\dfrac{Q}{4\pi\varepsilon_0} \cdot 4\pi = \dfrac{Q}{\varepsilon_0}$ (**Figure 17.73a**). This is one statement of Gauss' Law: If S is a surface that encloses a point charge Q, then the flux of the electric field across S is

$$\iint\limits_{S} \mathbf{E} \cdot \mathbf{n}\, dS = \frac{Q}{\varepsilon_0}.$$

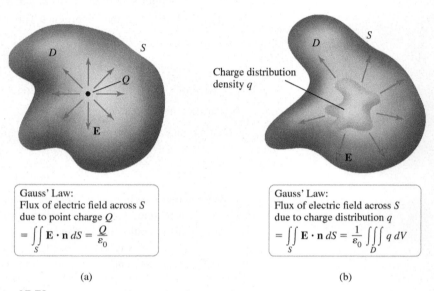

Gauss' Law:
Flux of electric field across S
due to point charge Q

$= \iint\limits_{S} \mathbf{E} \cdot \mathbf{n}\, dS = \dfrac{Q}{\varepsilon_0}$

Charge distribution
density q

Gauss' Law:
Flux of electric field across S
due to charge distribution q

$= \iint\limits_{S} \mathbf{E} \cdot \mathbf{n}\, dS = \dfrac{1}{\varepsilon_0} \iiint\limits_{D} q\, dV$

(a)   (b)

Figure 17.73

In fact, Gauss' Law applies to more general charge distributions (Exercise 39). If $q(x, y, z)$ is a charge density (charge per unit volume) defined on a region D enclosed by S, then the total charge within D is $Q = \iiint_{D} q(x, y, z)\, dV$ (**Figure 17.73b**). Replacing Q with this triple integral, Gauss' Law takes the form

$$\iint\limits_{S} \mathbf{E} \cdot \mathbf{n}\, dS = \frac{1}{\varepsilon_0} \underbrace{\iiint\limits_{D} q(x, y, z)\, dV}_{Q}.$$

Gauss' Law applies to other inverse square fields. In a slightly different form, it also governs heat transfer. If T is the temperature distribution in a solid body D, then the heat flow vector field is $\mathbf{F} = -k\nabla T$. (Heat flows down the temperature gradient.) If $q(x, y, z)$ represents the sources of heat within D, Gauss' Law says

$$\iint\limits_{S} \mathbf{F} \cdot \mathbf{n}\, dS = -k \iint\limits_{S} \nabla T \cdot \mathbf{n}\, dS = \iiint\limits_{D} q(x, y, z)\, dV.$$

We see that, in general, the flux of material (fluid, heat, electric field lines) across the boundary of a region is the cumulative effect of the sources within the region.

A Final Perspective

Table 17.4 offers a look at the progression of fundamental theorems of calculus that have appeared throughout this text. Each theorem builds on its predecessors, extending the same basic idea to a different situation or to higher dimensions.

In all cases, the statement is effectively the same: The cumulative (integrated) effect of the *derivatives* of a function throughout a region is determined by the values of the function on the boundary of that region. This principle underlies much of our understanding of the world around us.

Table 17.4

Fundamental Theorem of Calculus	$\int_a^b f'(x)\,dx = f(b) - f(a)$	
Fundamental Theorem for Line Integrals	$\int_C \nabla f \cdot d\mathbf{r} = f(B) - f(A)$	
Green's Theorem (Circulation form)	$\iint_R (g_x - f_y)\,dA = \oint_C f\,dx + g\,dy$	
Stokes' Theorem	$\iint_S (\nabla \times \mathbf{F}) \cdot \mathbf{n}\,dS = \oint_C \mathbf{F} \cdot d\mathbf{r}$	
Divergence Theorem	$\iiint_D \nabla \cdot \mathbf{F}\,dV = \iint_S \mathbf{F} \cdot \mathbf{n}\,dS$	

SECTION 17.8 EXERCISES

Getting Started

1. Explain the meaning of the surface integral in the Divergence Theorem.

2. Interpret the volume integral in the Divergence Theorem.

3. Explain the meaning of the Divergence Theorem.

4. What is the net outward flux of the rotation field $\mathbf{F} = \langle 2z + y, -x, -2x \rangle$ across the surface that encloses any region?

5. What is the net outward flux of the radial field $\mathbf{F} = \langle x, y, z \rangle$ across the sphere of radius 2 centered at the origin?

6. What is the divergence of an inverse square vector field?

7. Suppose div $\mathbf{F} = 0$ in a region enclosed by two concentric spheres. What is the relationship between the outward fluxes across the two spheres?

8. If div $\mathbf{F} > 0$ in a region enclosed by a small cube, is the net flux of the field into or out of the cube?

Practice Exercises

9–12. Verifying the Divergence Theorem *Evaluate both integrals of the Divergence Theorem for the following vector fields and regions. Check for agreement.*

9. $\mathbf{F} = \langle 2x, 3y, 4z \rangle$; $D = \{(x, y, z): x^2 + y^2 + z^2 \le 4\}$

10. $\mathbf{F} = \langle -x, -y, -z \rangle$; $D = \{(x, y, z): |x| \le 1, |y| \le 1, |z| \le 1\}$

11. $\mathbf{F} = \langle z - y, x, -x \rangle$; $D = \{(x, y, z): x^2/4 + y^2/8 + z^2/12 \le 1\}$

12. $\mathbf{F} = \langle x^2, y^2, z^2 \rangle$; $D = \{(x, y, z): |x| \le 1, |y| \le 2, |z| \le 3\}$

13–16. Rotation fields

13. Find the net outward flux of the field $\mathbf{F} = \langle 2z - y, x, -2x \rangle$ across the sphere of radius 1 centered at the origin.

14. Find the net outward flux of the field $\mathbf{F} = \langle z - y, x - z, y - x \rangle$ across the boundary of the cube $\{(x, y, z): |x| \le 1, |y| \le 1, |z| \le 1\}$.

15. Find the net outward flux of the field $\mathbf{F} = \langle bz - cy, cx - az, ay - bx \rangle$ across any smooth closed surface in $\mathbb{R}^3$, where a, b, and c are constants.

16. Find the net outward flux of $\mathbf{F} = \mathbf{a} \times \mathbf{r}$ across any smooth closed surface in $\mathbb{R}^3$, where $\mathbf{a}$ is a constant nonzero vector and $\mathbf{r} = \langle x, y, z \rangle$.

17–24. Computing flux *Use the Divergence Theorem to compute the net outward flux of the following fields across the given surface S.*

17. $\mathbf{F} = \langle x, -2y, 3z \rangle$; S is the sphere $\{(x, y, z): x^2 + y^2 + z^2 = 6\}$.

18. $\mathbf{F} = \langle x^2, 2xz, y^2 \rangle$; S is the surface of the cube cut from the first octant by the planes $x = 1$, $y = 1$, and $z = 1$.

19. $\mathbf{F} = \langle x, 2y, z \rangle$; S is the boundary of the tetrahedron in the first octant formed by the plane $x + y + z = 1$.

20. $\mathbf{F} = \langle x^2, y^2, z^2 \rangle$; S is the sphere $\{(x, y, z): x^2 + y^2 + z^2 = 25\}$.

21. $\mathbf{F} = \langle y - 2x, x^3 - y, y^2 - z \rangle$; S is the sphere $\{(x, y, z): x^2 + y^2 + z^2 = 4\}$.

22. $\mathbf{F} = \langle y + z, x + z, x + y \rangle$; S consists of the faces of the cube $\{(x, y, z): |x| \le 1, |y| \le 1, |z| \le 1\}$.

23. $\mathbf{F} = \langle x, y, z \rangle$; S is the surface of the paraboloid $z = 4 - x^2 - y^2$, for $z \ge 0$, plus its base in the xy-plane.

24. $\mathbf{F} = \langle x, y, z \rangle$; S is the surface of the cone $z^2 = x^2 + y^2$, for $0 \le z \le 4$, plus its top surface in the plane $z = 4$.

25–30. Divergence Theorem for more general regions *Use the Divergence Theorem to compute the net outward flux of the following vector fields across the boundary of the given regions D.*

25. $\mathbf{F} = \langle z - x, x - y, 2y - z \rangle$; D is the region between the spheres of radius 2 and 4 centered at the origin.

26. $\mathbf{F} = \mathbf{r}|\mathbf{r}| = \langle x, y, z \rangle \sqrt{x^2 + y^2 + z^2}$; D is the region between the spheres of radius 1 and 2 centered at the origin.

27. $\mathbf{F} = \dfrac{\mathbf{r}}{|\mathbf{r}|} = \dfrac{\langle x, y, z \rangle}{\sqrt{x^2 + y^2 + z^2}}$; D is the region between the spheres of radius 1 and 2 centered at the origin.

28. $\mathbf{F} = \langle z - y, x - z, 2y - x \rangle$; D is the region between two cubes: $\{(x, y, z): 1 \le |x| \le 3, 1 \le |y| \le 3, 1 \le |z| \le 3\}$.

29. $\mathbf{F} = \langle x^2, -y^2, z^2 \rangle$; D is the region in the first octant between the planes $z = 4 - x - y$ and $z = 2 - x - y$.

30. $\mathbf{F} = \langle x, 2y, 3z \rangle$; D is the region between the cylinders $x^2 + y^2 = 1$ and $x^2 + y^2 = 4$, for $0 \le z \le 8$.

31. **Explain why or why not** Determine whether the following statements are true and give an explanation or counterexample.

   a. If $\nabla \cdot \mathbf{F} = 0$ at all points of a region D, then $\mathbf{F} \cdot \mathbf{n} = 0$ at all points of the boundary of D.

   b. If $\iint_S \mathbf{F} \cdot \mathbf{n}\, dS = 0$ on all closed surfaces in $\mathbb{R}^3$, then $\mathbf{F}$ is constant.

   c. If $|\mathbf{F}| < 1$, then $\left| \iiint_D \nabla \cdot \mathbf{F}\, dV \right|$ is less than the area of the surface of D.

32. **Flux across a sphere** Consider the radial field $\mathbf{F} = \langle x, y, z \rangle$ and let S be the sphere of radius a centered at the origin. Compute the outward flux of $\mathbf{F}$ across S using the representation $z = \pm\sqrt{a^2 - x^2 - y^2}$ for the sphere (either symmetry or two surfaces must be used).

33–35. Flux integrals *Compute the outward flux of the following vector fields across the given surfaces S. You should decide which integral of the Divergence Theorem to use.*

33. $\mathbf{F} = \langle x^2 e^y \cos z, -4x e^y \cos z, 2x e^y \sin z \rangle$; S is the boundary of the ellipsoid $x^2/4 + y^2 + z^2 = 1$.

34. $\mathbf{F} = \langle -yz, xz, 1 \rangle$; S is the boundary of the ellipsoid $x^2/4 + y^2/4 + z^2 = 1$.

35. $\mathbf{F} = \langle x \sin y, -\cos y, z \sin y \rangle$; S is the boundary of the region bounded by the planes $x = 1$, $y = 0$, $y = \pi/2$, $z = 0$, and $z = x$.

36. **Radial fields** Consider the radial vector field
$$\mathbf{F} = \frac{\mathbf{r}}{|\mathbf{r}|^p} = \frac{\langle x, y, z \rangle}{(x^2 + y^2 + z^2)^{p/2}}.$$ Let S be the sphere of radius a centered at the origin.

   a. Use a surface integral to show that the outward flux of $\mathbf{F}$ across S is $4\pi a^{3-p}$. Recall that the unit normal to the sphere is $\mathbf{r}/|\mathbf{r}|$.

   b. For what values of p does $\mathbf{F}$ satisfy the conditions of the Divergence Theorem? For these values of p, use the fact (Theorem 17.10) that $\nabla \cdot \mathbf{F} = \dfrac{3 - p}{|\mathbf{r}|^p}$ to compute the flux across S using the Divergence Theorem.

37. **Singular radial field** Consider the radial field
$$\mathbf{F} = \frac{\mathbf{r}}{|\mathbf{r}|} = \frac{\langle x, y, z \rangle}{(x^2 + y^2 + z^2)^{1/2}}.$$

   a. Evaluate a surface integral to show that $\iint_S \mathbf{F} \cdot \mathbf{n}\, dS = 4\pi a^2$, where S is the surface of a sphere of radius a centered at the origin.

   b. Note that the first partial derivatives of the components of $\mathbf{F}$ are undefined at the origin, so the Divergence Theorem does not apply directly. Nevertheless, the flux across the sphere as computed in part (a) is finite. Evaluate the triple integral of the Divergence Theorem as an improper integral as follows. Integrate div $\mathbf{F}$ over the region between two spheres of radius a and $0 < \varepsilon < a$. Then let $\varepsilon \to 0^+$ to obtain the flux computed in part (a).

38. **Logarithmic potential** Consider the potential function
$$\varphi(x, y, z) = \frac{1}{2} \ln (x^2 + y^2 + z^2) = \ln |\mathbf{r}|, \text{ where } \mathbf{r} = \langle x, y, z \rangle.$$

   a. Show that the gradient field associated with φ is
$$\mathbf{F} = \frac{\mathbf{r}}{|\mathbf{r}|^2} = \frac{\langle x, y, z \rangle}{x^2 + y^2 + z^2}.$$

   b. Show that $\iint_S \mathbf{F} \cdot \mathbf{n}\, dS = 4\pi a$, where S is the surface of a sphere of radius a centered at the origin.

   c. Compute div $\mathbf{F}$.

   d. Note that $\mathbf{F}$ is undefined at the origin, so the Divergence Theorem does not apply directly. Evaluate the volume integral as described in Exercise 37.

39. **Gauss' Law for electric fields** The electric field due to a point charge Q is $\mathbf{E} = \dfrac{Q}{4\pi\varepsilon_0} \dfrac{\mathbf{r}}{|\mathbf{r}|^3}$, where $\mathbf{r} = \langle x, y, z \rangle$, and ε_0 is a constant.

   a. Show that the flux of the field across a sphere of radius a centered at the origin is $\iint_S \mathbf{E} \cdot \mathbf{n}\, dS = \dfrac{Q}{\varepsilon_0}$.

   b. Let S be the boundary of the region between two spheres centered at the origin of radius a and b, respectively, with $a < b$. Use the Divergence Theorem to show that the net outward flux across S is zero.

   c. Suppose there is a distribution of charge within a region D. Let $q(x, y, z)$ be the charge density (charge per unit volume). Interpret the statement that
$$\iint_S \mathbf{E} \cdot \mathbf{n}\, dS = \frac{1}{\varepsilon_0} \iiint_D q(x, y, z)\, dV.$$

   d. Assuming $\mathbf{E}$ satisfies the conditions of the Divergence Theorem on D, conclude from part (c) that $\nabla \cdot \mathbf{E} = \dfrac{q}{\varepsilon_0}$.

   e. Because the electric force is conservative, it has a potential function φ. From part (d), conclude that $\nabla^2 \varphi = \nabla \cdot \nabla \varphi = \dfrac{q}{\varepsilon_0}$.

40. Gauss' Law for gravitation The gravitational force due to a point mass M at the origin is proportional to $\mathbf{F} = GM\mathbf{r}/|\mathbf{r}|^3$, where $\mathbf{r} = \langle x, y, z \rangle$ and G is the gravitational constant.

   a. Show that the flux of the force field across a sphere of radius a centered at the origin is $\iint_S \mathbf{F} \cdot \mathbf{n} \, dS = 4\pi GM$.

   b. Let S be the boundary of the region between two spheres centered at the origin of radius a and b, respectively, with $a < b$. Use the Divergence Theorem to show that the net outward flux across S is zero.

   c. Suppose there is a distribution of mass within a region D. Let $\rho(x, y, z)$ be the mass density (mass per unit volume). Interpret the statement that

$$\iint_S \mathbf{F} \cdot \mathbf{n} \, dS = 4\pi G \iiint_D \rho(x, y, z) \, dV.$$

   d. Assuming $\mathbf{F}$ satisfies the conditions of the Divergence Theorem on D, conclude from part (c) that $\nabla \cdot \mathbf{F} = 4\pi G\rho$.

   e. Because the gravitational force is conservative, it has a potential function φ. From part (d), conclude that $\nabla^2 \varphi = 4\pi G\rho$.

41–45. Heat transfer *Fourier's Law of heat transfer (or heat conduction) states that the heat flow vector $\mathbf{F}$ at a point is proportional to the negative gradient of the temperature; that is, $\mathbf{F} = -k\nabla T$, which means that heat energy flows from hot regions to cold regions. The constant $k > 0$ is called the conductivity, which has metric units of $J/(m\text{-}s\text{-}K)$. A temperature function for a region D is given. Find the net outward heat flux $\iint_S \mathbf{F} \cdot \mathbf{n} \, dS = -k \iint_S \nabla T \cdot \mathbf{n} \, dS$ across the boundary S of D. In some cases, it may be easier to use the Divergence Theorem and evaluate a triple integral. Assume $k = 1$.*

41. $T(x, y, z) = 100 + x + 2y + z$; $D = \{(x, y, z): 0 \le x \le 1, 0 \le y \le 1, 0 \le z \le 1\}$

42. $T(x, y, z) = 100 + x^2 + y^2 + z^2$; $D = \{(x, y, z): 0 \le x \le 1, 0 \le y \le 1, 0 \le z \le 1\}$

43. $T(x, y, z) = 100 + e^{-z}$; $D = \{(x, y, z): 0 \le x \le 1, 0 \le y \le 1, 0 \le z \le 1\}$

44. $T(x, y, z) = 100 + x^2 + y^2 + z^2$; D is the unit sphere centered at the origin.

T 45. $T(x, y, z) = 100e^{-x^2-y^2-z^2}$; D is the sphere of radius a centered at the origin.

Explorations and Challenges

46. Inverse square fields are special Let $\mathbf{F}$ be a radial field $\mathbf{F} = \mathbf{r}/|\mathbf{r}|^p$, where p is a real number and $\mathbf{r} = \langle x, y, z \rangle$. With $p = 3$, $\mathbf{F}$ is an inverse square field.

   a. Show that the net flux across a sphere centered at the origin is independent of the radius of the sphere only for $p = 3$.

   b. Explain the observation in part (a) by finding the flux of $\mathbf{F} = \mathbf{r}/|\mathbf{r}|^p$ across the boundaries of a spherical box $\{(\rho, \varphi, \theta): a \le \rho \le b, \varphi_1 \le \varphi \le \varphi_2, \theta_1 \le \theta \le \theta_2\}$ for various values of p.

47. A beautiful flux integral Consider the potential function $\varphi(x, y, z) = G(\rho)$, where G is any twice differentiable function and $\rho = \sqrt{x^2 + y^2 + z^2}$; therefore, G depends only on the distance from the origin.

   a. Show that the gradient vector field associated with φ is

$$\mathbf{F} = \nabla \varphi = G'(\rho)\frac{\mathbf{r}}{\rho}, \text{ where } \mathbf{r} = \langle x, y, z \rangle \text{ and } \rho = |\mathbf{r}|.$$

   b. Let S be the sphere of radius a centered at the origin and let D be the region enclosed by S. Show that the flux of $\mathbf{F}$ across S is $\iint_S \mathbf{F} \cdot \mathbf{n} \, dS = 4\pi a^2 G'(a)$.

   c. Show that $\nabla \cdot \mathbf{F} = \nabla \cdot \nabla \varphi = \dfrac{2G'(\rho)}{\rho} + G''(\rho)$.

   d. Use part (c) to show that the flux across S (as given in part (b)) is also obtained by the volume integral $\iiint_D \nabla \cdot \mathbf{F} \, dV$. (*Hint:* Use spherical coordinates and integrate by parts.)

48. Integration by parts (Gauss' Formula) Recall the Product Rule of Theorem 17.13: $\nabla \cdot (u\mathbf{F}) = \nabla u \cdot \mathbf{F} + u(\nabla \cdot \mathbf{F})$.

   a. Integrate both sides of this identity over a solid region D with a closed boundary S, and use the Divergence Theorem to prove an integration by parts rule:

$$\iiint_D u(\nabla \cdot \mathbf{F}) \, dV = \iint_S u\mathbf{F} \cdot \mathbf{n} \, dS - \iiint_D \nabla u \cdot \mathbf{F} \, dV.$$

   b. Explain the correspondence between this rule and the integration by parts rule for single-variable functions.

   c. Use integration by parts to evaluate $\iiint_D (x^2y + y^2z + z^2x) \, dV$, where D is the cube in the first octant cut by the planes $x = 1$, $y = 1$, and $z = 1$.

49. Green's Formula Write Gauss' Formula of Exercise 48 in two dimensions—that is, where $\mathbf{F} = \langle f, g \rangle$, D is a plane region R and C is the boundary of R. Show that the result is Green's Formula:

$$\iint_R u(f_x + g_y) \, dA = \oint_C u(\mathbf{F} \cdot \mathbf{n}) \, ds - \iint_R (fu_x + gu_y) \, dA.$$

Show that with $u = 1$, one form of Green's Theorem appears. Which form of Green's Theorem is it?

50. Green's First Identity Prove Green's First Identity for twice differentiable scalar-valued functions u and v defined on a region D:

$$\iiint_D (u\nabla^2 v + \nabla u \cdot \nabla v) \, dV = \iint_S u\nabla v \cdot \mathbf{n} \, dS,$$

where $\nabla^2 v = \nabla \cdot \nabla v$. You may apply Gauss' Formula in Exercise 48 to $\mathbf{F} = \nabla v$ or apply the Divergence Theorem to $\mathbf{F} = u\nabla v$.

51. Green's Second Identity Prove Green's Second Identity for scalar-valued functions u and v defined on a region D:

$$\iiint_D (u\nabla^2 v - v\nabla^2 u) \, dV = \iint_S (u\nabla v - v\nabla u) \cdot \mathbf{n} \, dS.$$

(*Hint:* Reverse the roles of u and v in Green's First Identity.)

52–54. Harmonic functions *A scalar-valued function φ is harmonic on a region D if $\nabla^2 \varphi = \nabla \cdot \nabla \varphi = 0$ at all points of D.*

52. Show that the potential function $\varphi(x, y, z) = |\mathbf{r}|^{-p}$ is harmonic provided $p = 0$ or $p = 1$, where $\mathbf{r} = \langle x, y, z \rangle$. To what vector fields do these potentials correspond?

53. Show that if φ is harmonic on a region D enclosed by a surface S, then $\iint_S \nabla \varphi \cdot \mathbf{n} \, dS = 0$.

54. Show that if u is harmonic on a region D enclosed by a surface S, then $\iint_S u\nabla u \cdot \mathbf{n} \, dS = \iiint_D |\nabla u|^2 \, dV$.

55. Miscellaneous integral identities Prove the following identities.

   a. $\iiint_D \nabla \times \mathbf{F} \, dV = \iint_S (\mathbf{n} \times \mathbf{F}) \, dS$ (*Hint:* Apply the Divergence Theorem to each component of the identity.)

   b. $\iint_S (\mathbf{n} \times \nabla \varphi) \, dS = \oint_C \varphi \, d\mathbf{r}$ (*Hint:* Apply Stokes' Theorem to each component of the identity.)

QUICK CHECK ANSWERS

1. If **F** is constant, then div **F** $= 0$, so $\iiint_D \nabla \cdot \mathbf{F} \, dV = \iint_S \mathbf{F} \cdot \mathbf{n} \, dS = 0$. This means that all the "material" that flows into one side of D flows out of the other side of D.
2. The vector field and the divergence are positive throughout D.   **3.** The vector field has no flow into or out of the

cube on the faces $x = 0$, $y = 0$, and $z = 0$ because the vectors of **F** on these faces are parallel to the faces. The vector field points out of the cube on the $x = 1$ and $z = 1$ faces and into the cube on the $y = 1$ face. div **F** $= 2$, so there is a net flow out of the cube. ◄

CHAPTER 17   REVIEW EXERCISES

1. **Explain why or why not** Determine whether the following statements are true and give an explanation or counterexample.

   a. The rotation field $\mathbf{F} = \langle -y, x \rangle$ has zero curl and zero divergence.
   b. $\nabla \times \nabla \varphi = \mathbf{0}$
   c. Two vector fields with the same curl differ by a constant vector field.
   d. Two vector fields with the same divergence differ by a constant vector field.
   e. If $\mathbf{F} = \langle x, y, z \rangle$ and S encloses a region D, then $\iint_S \mathbf{F} \cdot \mathbf{n} \, dS$ is three times the volume of D.

2. **Matching vector fields** Match vector fields a–f with the graphs A–F. Let $\mathbf{r} = \langle x, y \rangle$.

   a. $\mathbf{F} = \langle x, y \rangle$ **b.** $\mathbf{F} = \langle -2y, 2x \rangle$
   c. $\mathbf{F} = \mathbf{r}/|\mathbf{r}|$ **d.** $\mathbf{F} = \langle y - x, x \rangle$
   e. $\mathbf{F} = \langle e^{-y}, e^{-x} \rangle$ **f.** $\mathbf{F} = \langle \sin \pi x, \sin \pi y \rangle$

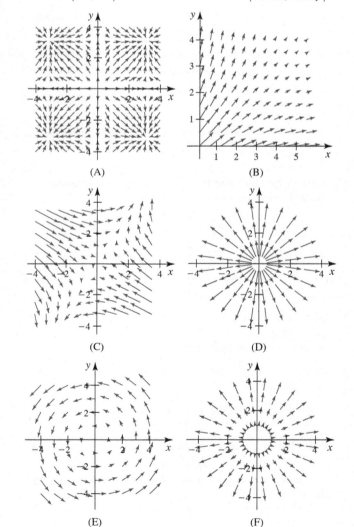

(A)

(B)

(C)

(D)

(E)

(F)

3–4. Gradient fields in $\mathbb{R}^2$ *Find the vector field* $\mathbf{F} = \nabla \varphi$ *for the following potential functions. Sketch a few level curves of φ and a few vectors of* **F** *along the level curves.*

3. $\varphi(x, y) = x^2 + 4y^2$, for $|x| \le 5$, $|y| \le 5$

4. $\varphi(x, y) = (x^2 - y^2)/2$, for $|x| \le 2$, $|y| \le 2$

5–6. Gradient fields in $\mathbb{R}^3$ *Find the vector field* $\mathbf{F} = \nabla \varphi$ *for the following potential functions.*

5. $\varphi(x, y, z) = 1/|\mathbf{r}|$, where $\mathbf{r} = \langle x, y, z \rangle$

6. $\varphi(x, y, z) = \dfrac{1}{2} e^{-x^2 - y^2 - z^2}$

7. **Normal component** Let C be the circle of radius 2 centered at the origin with counterclockwise orientation. Give the unit outward normal vector at any point (x, y) on C.

8–10. Line integrals *Evaluate the following line integrals.*

8. $\displaystyle\int_C (x^2 - 2xy + y^2) \, ds$; C is the upper half of a circle $\mathbf{r}(t) = \langle 5 \cos t, 5 \sin t \rangle$, for $0 \le t \le \pi$.

9. $\displaystyle\int_C y e^{-xz} \, ds$; C is the path $\mathbf{r}(t) = \langle 2t, 3t, -6t \rangle$, for $0 \le t \le 2$.

10. $\displaystyle\int_C (xz - y^2) \, ds$; C is the line segment from $(0, 1, 2)$ to $(-3, 7, -1)$.

11. **Two parameterizations** Verify that $\oint_C (x - 2y + 3z) \, ds$ has the same value when C is given by $\mathbf{r}(t) = \langle 2 \cos t, 2 \sin t, 0 \rangle$, for $0 \le t \le 2\pi$, and by $\mathbf{r}(t) = \langle 2 \cos t^2, 2 \sin t^2, 0 \rangle$, for $0 \le t \le \sqrt{2\pi}$.

12. **Work integral** Find the work done in moving an object from $P(1, 0, 0)$ to $Q(0, 1, 0)$ in the presence of the force $\mathbf{F} = \langle 1, 2y, -4z \rangle$ along the following paths.

   a. The line segment from P to Q
   b. The line segment from P to $O(0, 0, 0)$ followed by the line segment from O to Q
   c. The arc of the quarter circle from P to Q
   d. Is the work independent of the path?

13–14. Work integrals in $\mathbb{R}^3$ *Given the force field* **F**, *find the work required to move an object on the given curve.*

13. $\mathbf{F} = \langle -y, z, x \rangle$ on the path consisting of the line segment from $(0, 0, 0)$ to $(0, 1, 0)$ followed by the line segment from $(0, 1, 0)$ to $(0, 1, 4)$

14. $\mathbf{F} = \dfrac{\langle x, y, z \rangle}{(x^2 + y^2 + z^2)^{3/2}}$ on the path $\mathbf{r}(t) = \langle t^2, 3t^2, -t^2 \rangle$, for $1 \le t \le 2$

15–18. Circulation and flux *Find the circulation and the outward flux of the following vector fields for the curve* $\mathbf{r}(t) = \langle 2\cos t, 2\sin t \rangle$, *for* $0 \le t \le 2\pi$.

15. $\mathbf{F} = \langle y - x, y \rangle$

16. $\mathbf{F} = \langle x, y \rangle$

17. $\mathbf{F} = \mathbf{r}/|\mathbf{r}|^2$, where $\mathbf{r} = \langle x, y \rangle$

18. $\mathbf{F} = \langle x - y, x \rangle$

19. Flux in channel flow Consider the flow of water in a channel whose boundaries are the planes $y = \pm L$ and $z = \pm 1/2$. The velocity field in the channel is $\mathbf{v} = \langle v_0(L^2 - y^2), 0, 0 \rangle$. Find the flux across the cross section of the channel at $x = 0$ in terms of v_0 and L.

20–23. Conservative vector fields and potentials *Determine whether the following vector fields are conservative on their domains. If so, find a potential function.*

20. $\mathbf{F} = \langle y^2, 2xy \rangle$

21. $\mathbf{F} = \langle y, x + z^2, 2yz \rangle$

22. $\mathbf{F} = \langle e^x \cos y, -e^x \sin y \rangle$

23. $\mathbf{F} = e^z \langle y, x, xy \rangle$

24–27. Evaluating line integrals *Evaluate the line integral* $\int_C \mathbf{F} \cdot d\mathbf{r}$ *for the following vector fields* $\mathbf{F}$ *and curves* C *in two ways.*

a. *By parameterizing* C

b. *By using the Fundamental Theorem for line integrals, if possible*

24. $\mathbf{F} = \nabla(x^2 y)$; C: $\mathbf{r}(t) = \langle 9 - t^2, t \rangle$, for $0 \le t \le 3$

25. $\mathbf{F} = \nabla(xyz)$; C: $\mathbf{r}(t) = \langle \cos t, \sin t, t/\pi \rangle$, for $0 \le t \le \pi$

26. $\mathbf{F} = \langle x, -y \rangle$; C is the square with vertices $(\pm 1, \pm 1)$ with counterclockwise orientation.

27. $\mathbf{F} = \langle y, z, -x \rangle$; C: $\mathbf{r}(t) = \langle \cos t, \sin t, 4 \rangle$, for $0 \le t \le 2\pi$

28. Radial fields in $\mathbb{R}^2$ **are conservative** Prove that the radial field $\mathbf{F} = \mathbf{r}/|\mathbf{r}|^p$, where $\mathbf{r} = \langle x, y \rangle$ and p is a real number, is conservative on $\mathbb{R}^2$ with the origin removed. For what value of p is $\mathbf{F}$ conservative on $\mathbb{R}^2$ (including the origin)?

29–32. Green's Theorem for line integrals *Use either form of Green's Theorem to evaluate the following line integrals.*

29. $\oint_C xy^2 \, dx + x^2 y \, dy$; C is the triangle with vertices $(0, 0)$, $(2, 0)$, and $(0, 2)$ with counterclockwise orientation.

30. $\oint_C (-3y + x^{3/2}) \, dx + (x - y^{2/3}) \, dy$; C is the boundary of the half-disk $\{(x, y): x^2 + y^2 \le 2, y \ge 0\}$ with counterclockwise orientation.

31. $\oint_C (x^3 + xy) \, dy + (2y^2 - 2x^2 y) \, dx$; C is the square with vertices $(\pm 1, \pm 1)$ with counterclockwise orientation.

32. $\oint_C 3x^3 \, dy - 3y^3 \, dx$; C is the circle of radius 4 centered at the origin with *clockwise* orientation.

33–34. Areas of plane regions *Find the area of the following regions using a line integral.*

33. The region enclosed by the ellipse $x^2 + 4y^2 = 16$

34. The region bounded by the hypocycloid $\mathbf{r}(t) = \langle \cos^3 t, \sin^3 t \rangle$, for $0 \le t \le 2\pi$

35–36. Circulation and flux *Consider the following vector fields.*

a. *Compute the circulation on the boundary of the region* R *(with counterclockwise orientation).*

b. *Compute the outward flux across the boundary of* R.

35. $\mathbf{F} = \mathbf{r}/|\mathbf{r}|$, where $\mathbf{r} = \langle x, y \rangle$ and R is the half-annulus $\{(r, \theta): 1 \le r \le 3, 0 \le \theta \le \pi\}$

36. $\mathbf{F} = \langle -\sin y, x \cos y \rangle$, where R is the square $\{(x, y): 0 \le x \le \pi/2, 0 \le y \le \pi/2\}$

37. Parameters Let $\mathbf{F} = \langle ax + by, cx + dy \rangle$, where a, b, c, and d are constants.

a. For what values of a, b, c, and d is $\mathbf{F}$ conservative?

b. For what values of a, b, c, and d is $\mathbf{F}$ source free?

c. For what values of a, b, c, and d is $\mathbf{F}$ conservative and source free?

38–41. Divergence and curl *Compute the divergence and curl of the following vector fields. State whether the field is source free or irrotational.*

38. $\mathbf{F} = \langle yz, xz, xy \rangle$

39. $\mathbf{F} = \mathbf{r}|\mathbf{r}| = \langle x, y, z \rangle \sqrt{x^2 + y^2 + z^2}$

40. $\mathbf{F} = \langle \sin xy, \cos yz, \sin xz \rangle$

41. $\mathbf{F} = \langle 2xy + z^4, x^2, 4xz^3 \rangle$

42. Identities Prove that $\nabla\left(\dfrac{1}{|\mathbf{r}|^4}\right) = -\dfrac{4\mathbf{r}}{|\mathbf{r}|^6}$ and use the result to prove that $\nabla \cdot \nabla\left(\dfrac{1}{|\mathbf{r}|^4}\right) = \dfrac{12}{|\mathbf{r}|^6}$.

43. Maximum curl Let $\mathbf{F} = \langle z, x, -y \rangle$.

a. What is the scalar component of curl $\mathbf{F}$ in the direction of $\mathbf{n} = \langle 1, 0, 0 \rangle$?

b. What is the scalar component of curl $\mathbf{F}$ in the direction of $\mathbf{n} = \langle 0, -1/\sqrt{2}, 1/\sqrt{2} \rangle$?

c. In the direction of what unit vector $\mathbf{n}$ is the scalar component of curl $\mathbf{F}$ a maximum?

44. Paddle wheel in a vector field Let $\mathbf{F} = \langle 0, 2x, 0 \rangle$ and let $\mathbf{n}$ be a unit vector aligned with the axis of a paddle wheel located on the y-axis.

a. If the axis of the paddle wheel is aligned with $\mathbf{n} = \langle 1, 0, 0 \rangle$, how fast does it spin?

b. If the axis of the paddle wheel is aligned with $\mathbf{n} = \langle 0, 0, 1 \rangle$, how fast does it spin?

c. For what direction $\mathbf{n}$ does the paddle wheel spin fastest?

45–48. Surface areas *Use a surface integral to find the area of the following surfaces.*

45. The hemisphere $x^2 + y^2 + z^2 = 9$, for $z \ge 0$

46. The frustum of the cone $z^2 = x^2 + y^2$, for $2 \le z \le 4$ (excluding the bases)

47. The plane $z = 6 - x - y$ above the square $|x| \le 1$, $|y| \le 1$

48. The surface $f(x, y) = \sqrt{2}\, xy$ above the polar region $\{(r, \theta): 0 \le r \le 2, 0 \le \theta \le 2\pi\}$

49–51. Surface integrals *Evaluate the following surface integrals.*

49. $\iint_S (1 + yz) \, dS$; S is the plane $x + y + z = 2$ in the first octant.

50. $\iint_S \langle 0, y, z \rangle \cdot \mathbf{n} \, dS$; S is the curved surface of the cylinder $y^2 + z^2 = a^2$, for $|x| \le 8$, with outward normal vectors.

51. $\iint_S (x - y + z) \, dS$; S is the entire surface, including the base, of the hemisphere $x^2 + y^2 + z^2 = 4$, for $z \ge 0$.

52–53. Flux integrals *Find the flux of the following vector fields across the given surface. Assume the vectors normal to the surface point outward.*

52. $\mathbf{F} = \langle x, y, z \rangle$ across the curved surface of the cylinder $x^2 + y^2 = 1$, for $|z| \le 8$

53. $\mathbf{F} = \mathbf{r}/|\mathbf{r}|$ across the sphere of radius a centered at the origin, where $\mathbf{r} = \langle x, y, z \rangle$

54. Three methods Find the surface area of the paraboloid $z = x^2 + y^2$, for $0 \le z \le 4$, in three ways.

    a. Use an explicit description of the surface.

    b. Use the parametric description $\mathbf{r} = \langle v \cos u, v \sin u, v^2 \rangle$.

    c. Use the parametric description $\mathbf{r} = \langle \sqrt{v} \cos u, \sqrt{v} \sin u, v \rangle$.

55. Flux across hemispheres and paraboloids Let S be the hemisphere $x^2 + y^2 + z^2 = a^2$, for $z \ge 0$, and let T be the paraboloid $z = a - (x^2 + y^2)/a$, for $z \ge 0$, where $a > 0$. Assume the surfaces have outward normal vectors.

    a. Verify that S and T have the same base $(x^2 + y^2 \le a^2)$ and the same high point $(0, 0, a)$.

    b. Which surface has the greater area?

    c. Show that the flux of the radial field $\mathbf{F} = \langle x, y, z \rangle$ across S is $2\pi a^3$.

    d. Show that the flux of the radial field $\mathbf{F} = \langle x, y, z \rangle$ across T is $3\pi a^3/2$.

56. Surface area of an ellipsoid Consider the ellipsoid $x^2/a^2 + y^2/b^2 + z^2/c^2 = 1$, where a, b, and c are positive real numbers.

    a. Show that the surface is described by the parametric equations

$$\mathbf{r}(u, v) = \langle a \cos u \sin v, b \sin u \sin v, c \cos v \rangle$$

    for $0 \le u \le 2\pi, 0 \le v \le \pi$.

    b. Write an integral for the surface area of the ellipsoid.

57–58. Stokes' Theorem for line integrals *Evaluate the line integral $\oint_C \mathbf{F} \cdot d\mathbf{r}$ using Stokes' Theorem. Assume C has counterclockwise orientation.*

57. $\mathbf{F} = \langle xz, yz, xy \rangle$; C is the circle $x^2 + y^2 = 4$ in the xy-plane.

58. $\mathbf{F} = \langle x^2 - y^2, x, 2yz \rangle$; C is the boundary of the plane $z = 6 - 2x - y$ in the first octant.

59–60. Stokes' Theorem for surface integrals *Use Stokes' Theorem to evaluate the surface integral $\iint_S (\nabla \times \mathbf{F}) \cdot \mathbf{n} \, dS$. Assume $\mathbf{n}$ is the outward normal.*

59. $\mathbf{F} = \langle -z, x, y \rangle$, where S is the hyperboloid $z = 10 - \sqrt{1 + x^2 + y^2}$, for $z \ge 0$

60. $\mathbf{F} = \langle x^2 - z^2, y^2, xz \rangle$, where S is the hemisphere $x^2 + y^2 + z^2 = 4$, for $y \ge 0$

61. Conservative fields Use Stokes' Theorem to find the circulation of the vector field $\mathbf{F} = \nabla(10 - x^2 + y^2 + z^2)$ around any smooth closed curve C with counterclockwise orientation.

62–64. Computing fluxes *Use the Divergence Theorem to compute the outward flux of the following vector fields across the given surfaces S.*

62. $\mathbf{F} = \langle -x, x - y, x - z \rangle$; S is the surface of the cube cut from the first octant by the planes $x = 1$, $y = 1$, and $z = 1$.

63. $\mathbf{F} = \langle x^3, y^3, z^3 \rangle /3$; S is the sphere $\{(x, y, z) : x^2 + y^2 + z^2 = 9\}$.

64. $\mathbf{F} = \langle x^2, y^2, z^2 \rangle$; S is the cylinder $\{(x, y, z) : x^2 + y^2 = 4, 0 \le z \le 8\}$.

65–66. General regions *Use the Divergence Theorem to compute the outward flux of the following vector fields across the boundary of the given regions D.*

65. $\mathbf{F} = \langle x^3, y^3, 10 \rangle$; D is the region between the hemispheres of radius 1 and 2 centered at the origin with bases in the xy-plane.

66. $\mathbf{F} = \dfrac{\mathbf{r}}{|\mathbf{r}|^3} = \dfrac{\langle x, y, z \rangle}{(x^2 + y^2 + z^2)^{3/2}}$; D is the region between two spheres with radii 1 and 2 centered at $(5, 5, 5)$.

67. Flux integrals Compute the outward flux of the field $\mathbf{F} = \langle x^2 + x \sin y, y^2 + 2 \cos y, z^2 + z \sin y \rangle$ across the surface S that is the boundary of the prism bounded by the planes $y = 1 - x$, $x = 0$, $y = 0$, $z = 0$, and $z = 4$.

68. Stokes' Theorem on a compound surface Consider the surface S consisting of the quarter-sphere $x^2 + y^2 + z^2 = a^2$, for $z \ge 0$ and $x \ge 0$, and the half-disk in the yz-plane $y^2 + z^2 \le a^2$, for $z \ge 0$. The boundary of S in the xy-plane is C, which consists of the semicircle $x^2 + y^2 = a^2$, for $x \ge 0$, and the line segment $[-a, a]$ on the y-axis, with a counterclockwise orientation. Let $\mathbf{F} = \langle 2z - y, x - z, y - 2x \rangle$.

    a. Describe the direction in which the normal vectors point on S.

    b. Evaluate $\oint_C \mathbf{F} \cdot d\mathbf{r}$

    c. Evaluate $\iint_S (\nabla \times \mathbf{F}) \cdot \mathbf{n} \, dS$ and check for agreement with part (b).

Chapter 17 Guided Projects

Applications of the material in this chapter and related topics can be found in the following Guided Projects. For additional information, see the Preface.

- Ideal fluid flow
- Maxwell's equations
- Planimeters and vector fields
- Vector calculus in other coordinate systems

A

Appendix

Proofs of Selected Theorems

> **THEOREM 2.3   Limit Laws**
> Assume $\lim\limits_{x \to a} f(x)$ and $\lim\limits_{x \to a} g(x)$ exist. The following properties hold, where c is a real number, and $n > 0$ is an integer.
>
> **1. Sum** $\lim\limits_{x \to a} (f(x) + g(x)) = \lim\limits_{x \to a} f(x) + \lim\limits_{x \to a} g(x)$
>
> **2. Difference** $\lim\limits_{x \to a} (f(x) - g(x)) = \lim\limits_{x \to a} f(x) - \lim\limits_{x \to a} g(x)$
>
> **3. Constant multiple** $\lim\limits_{x \to a} (cf(x)) = c\lim\limits_{x \to a} f(x)$
>
> **4. Product** $\lim\limits_{x \to a} (f(x)g(x)) = (\lim\limits_{x \to a} f(x))(\lim\limits_{x \to a} g(x))$
>
> **5. Quotient** $\lim\limits_{x \to a} \dfrac{f(x)}{g(x)} = \dfrac{\lim\limits_{x \to a} f(x)}{\lim\limits_{x \to a} g(x)}$, provided $\lim\limits_{x \to a} g(x) \neq 0$
>
> **6. Power** $\lim\limits_{x \to a} (f(x))^n = (\lim\limits_{x \to a} f(x))^n$
>
> **7. Root** $\lim\limits_{x \to a} (f(x))^{1/n} = (\lim\limits_{x \to a} f(x))^{1/n}$, provided $f(x) > 0$, for x near a, if n is even

Proof: The proof of Law 1 is given in Example 6 of Section 2.7. The proof of Law 2 is analogous to that of Law 1; the triangle inequality in the form $|x - y| \leq |x| + |y|$ is used (Exercise 43, Section 2.7). The proof of Law 3 is outlined in Exercise 44 of Section 2.7. The proofs of Laws 4 and 5 are given below. The proof of Law 6 involves the repeated use of Law 4. Law 7 is a special case of Theorem 2.12, whose proof is given in this appendix (p. AP-2). ◄

Proof of Product Law: Let $L = \lim\limits_{x \to a} f(x)$ and $M = \lim\limits_{x \to a} g(x)$. Using the definition of a limit, the goal is to show that given any $\varepsilon > 0$, it is possible to specify a $\delta > 0$ such that $|f(x)g(x) - LM| < \varepsilon$ whenever $0 < |x - a| < \delta$. Notice that

$$
\begin{aligned}
|f(x)g(x) - LM| &= |f(x)g(x) - Lg(x) + Lg(x) - LM| && \text{Add and subtract } Lg(x). \\
&= |(f(x) - L)g(x) + (g(x) - M)L| && \text{Group terms.} \\
&\leq |(f(x) - L)g(x)| + |(g(x) - M)L| && \text{Triangle inequality} \\
&= |f(x) - L||g(x)| + |g(x) - M||L|. && |xy| = |x||y|
\end{aligned}
$$

▶ Real numbers x and y obey the triangle inequality $|x + y| \le |x| + |y|$.

We now use the definition of the limits of f and g, and we note that L and M are fixed real numbers. Given $\varepsilon > 0$, there exist $\delta_1 > 0$ and $\delta_2 > 0$ such that

$$|f(x) - L| < \frac{\varepsilon}{2(|M| + 1)} \quad \text{and} \quad |g(x) - M| < \frac{\varepsilon}{2(|L| + 1)}$$

▶ $|g(x) - M| < 1$ implies that $g(x)$ is less than 1 unit from M. Therefore, whether $g(x)$ and M are positive or negative, $|g(x)| < |M| + 1$.

whenever $0 < |x - a| < \delta_1$ and $0 < |x - a| < \delta_2$, respectively. Furthermore, by the definition of the limit of g, there exits a $\delta_3 > 0$ such that $|g(x) - M| < 1$ whenever $0 < |x - a| < \delta_3$. It follows that $|g(x)| < |M| + 1$ whenever $0 < |x - a| < \delta_3$. Now take δ to be the minimum of δ_1, δ_2, and δ_3. Then for $0 < |x - a| < \delta$, we have

$$|f(x)g(x) - LM| \le \underbrace{|f(x) - L|}_{<\frac{\varepsilon}{2(|M| + 1)}} \underbrace{|g(x)|}_{<(|M| + 1)} + \underbrace{|g(x) - M|}_{<\frac{\varepsilon}{2(|L| + 1)}} |L|$$

$$< \frac{\varepsilon}{2} + \frac{\varepsilon}{2} \underbrace{\frac{|L|}{|L| + 1}}_{<1} < \frac{\varepsilon}{2} + \frac{\varepsilon}{2} = \varepsilon.$$

It follows that $\lim_{x \to a} (f(x)g(x)) = LM$.   ◀

Proof of Quotient Law: We first prove that if $\lim_{x \to a} g(x) = M$ exists, where $M \ne 0$, then $\lim_{x \to a} \frac{1}{g(x)} = \frac{1}{M}$. The Quotient Law then follows when we replace g with $1/g$ in the Product Law. Therefore, the goal is to show that given any $\varepsilon > 0$, it is possible to specify a $\delta > 0$ such that $\left| \frac{1}{g(x)} - \frac{1}{M} \right| < \varepsilon$ whenever $0 < |x - a| < \delta$. First note that $M \ne 0$ and $g(x)$ can be made arbitrarily close to M. For this reason, there exists a $\delta_1 > 0$ such that $|g(x)| > |M|/2$, or equivalently, $1/|g(x)| < 2/|M|$, whenever $0 < |x - a| < \delta_1$. Furthermore, using the definition of the limit of g, given any $\varepsilon > 0$, there exists a $\delta_2 > 0$ such that $|g(x) - M| < \frac{\varepsilon|M|^2}{2}$ whenever $0 < |x - a| < \delta_2$. Now take δ to be the minimum of δ_1 and δ_2. Then for $0 < |x - a| < \delta$, we have

▶ Note that if $|g(x)| > |M|/2$, then $1/|g(x)| < 2/|M|$.

$$\left| \frac{1}{g(x)} - \frac{1}{M} \right| = \left| \frac{M - g(x)}{Mg(x)} \right| \qquad \text{Common denominator}$$

$$= \frac{1}{|M|} \underbrace{\frac{1}{|g(x)|}}_{<\frac{2}{|M|}} \underbrace{|g(x) - M|}_{<\frac{\varepsilon|M|^2}{2}} \qquad \text{Rewrite.}$$

$$< \frac{1}{|M|} \frac{2}{|M|} \frac{\varepsilon|M|^2}{2} = \varepsilon. \qquad \text{Simplify.}$$

By the definition of a limit, we have $\lim_{x \to a} \frac{1}{g(x)} = \frac{1}{M}$. The proof can be completed by applying the Product Law with g replaced with $1/g$.

THEOREM 2.12   Limits of Composite Functions

1. If g is continuous at a and f is continuous at $g(a)$, then

$$\lim_{x \to a} f(g(x)) = f(\lim_{x \to a} g(x)).$$

2. If $\lim_{x \to a} g(x) = L$ and f is continuous at L, then

$$\lim_{x \to a} f(g(x)) = f(\lim_{x \to a} g(x)).$$

Proof: Part 1 was proved in Section 2.6 (it is also a direct consequence of Part 2). Let $\varepsilon > 0$ be given. We need to show that there exists $\delta > 0$ such that $\left|f(g(x)) - f(L)\right| < \varepsilon$ whenever $0 < \left|x - a\right| < \delta$. Because f is continuous at L, there exists $\delta_1 > 0$ such that

$$\left|f(y) - f(L)\right| < \varepsilon \text{ whenever } \left|y - L\right| < \delta_1.$$

We also know that $\lim\limits_{x \to a} g(x) = L$, so there exists $\delta > 0$ such that

$$\left|g(x) - L\right| < \delta_1 \text{ whenever } 0 < \left|x - a\right| < \delta.$$

Combining these results, if $0 < \left|x - a\right| < \delta$, then $\left|g(x) - L\right| < \delta_1$, and the inequality $\left|g(x) - L\right| < \delta_1$ implies that $\left|f(g(x)) - f(L)\right| < \varepsilon$. Therefore, $\left|f(g(x)) - f(L)\right| < \varepsilon$ whenever $0 < \left|x - a\right| < \delta$. ◀

THEOREM 10.20   Ratio Test

Let $\sum a_k$ be an infinite series, and let $r = \lim\limits_{k \to \infty} \left|\dfrac{a_{k+1}}{a_k}\right|$.

1. If $r < 1$, the series converges absolutely, and therefore it converges (Theorem 10.19).
2. If $r > 1$ (including $r = \infty$), the series diverges.
3. If $r = 1$, the test is inconclusive.

Proof: We consider three cases: $r < 1$, $r > 1$, and $r = 1$.

1. Assume $r < 1$ and choose a number R such that $r < R < 1$. Because the sequence $\left\{\left|\dfrac{a_{k+1}}{a_k}\right|\right\}$ converges to a number r less than R, eventually all the terms in the tail of the sequence $\left\{\left|\dfrac{a_{k+1}}{a_k}\right|\right\}$ are less than R. That is, there is a positive integer N such that $\left|\dfrac{a_{k+1}}{a_k}\right| < R$, for all $k > N$. Multiplying both sides of this inequality by $\dfrac{|a_k|}{R^{k+1}}$, we have $\dfrac{|a_{k+1}|}{R^{k+1}} < \dfrac{|a_k|}{R^k}$, for all $k > N$. So the sequence $\left\{\dfrac{|a_k|}{R^k}\right\}_{k=N+1}^{\infty}$ is decreasing and it follows that $\dfrac{|a_k|}{R^k} < \dfrac{|a_{N+1}|}{R^{N+1}}$, for all $k \geq N + 1$. By letting $c = \dfrac{|a_{N+1}|}{R^{N+1}}$, we have

$$0 < |a_k| \leq cR^k, \text{ for all } k \geq N + 1.$$ Let S_n represent the nth partial sum of $\sum\limits_{k=N+1}^{\infty} |a_k|$;

note that the partial sums of this series are bounded by a convergent geometric series:

$$
\begin{aligned}
S_n &= |a_{N+1}| + |a_{N+2}| + \cdots + |a_{N+n}| \\
&\leq cR^{N+1} + cR^{N+2} + \cdots + cR^{N+n} \\
&< cR^{N+1} + cR^{N+2} + \cdots + cR^{N+n} + \cdots \\
&= \frac{cR^{N+1}}{1 - R}.
\end{aligned}
$$

Because the sequence $\{S_n\}$ is increasing (each partial sum in the sequence consists of positive terms) and is bounded above by $\dfrac{cR^{N+1}}{1 - R}$, it converges by the Bounded Monotonic Sequences Theorem (Theorem 10.5). Therefore, $\sum\limits_{k=N+1}^{\infty} |a_k|$ converges and we conclude that $\sum\limits_{k=1}^{\infty} |a_k|$ converges (Theorem 10.8), which in turn implies that $\sum\limits_{k=1}^{\infty} a_k$ converges (Theorem 10.19).

2. If $r > 1$, there is a positive integer N for which $\left|\dfrac{a_{k+1}}{a_k}\right| > 1$, or equivalently $|a_{k+1}| > |a_k|$, for all $k > N$. So every term in the sequence $\{|a_k|\}_{k=N+1}^{\infty}$ is greater than or equal to the positive number $|a_{N+1}|$, which implies that $\lim\limits_{n\to\infty} a_n \neq 0$. Therefore, the series $\sum\limits_{k=N+1}^{\infty} a_k$ diverges by the Divergence Test, and we conclude that $\sum\limits_{k=1}^{\infty} a_k$ diverges (Theorem 10.8).

3. In the case that $r = 1$, the series $\sum\limits_{k=1}^{\infty} a_k$ may or may not converge. For example, both the divergent harmonic series $\sum\limits_{k=1}^{\infty} \dfrac{1}{k}$ and the convergent p-series $\sum\limits_{k=1}^{\infty} \dfrac{1}{k^2}$ produce a value of $r = 1$. ◄

THEOREM 10.21   Root Test

Let $\sum a_k$ be an infinite series, and let $\rho = \lim\limits_{k\to\infty} \sqrt[k]{|a_k|}$.

1. If $\rho < 1$, the series converges absolutely, and therefore it converges (Theorem 10.19).

2. If $\rho > 1$ (including $\rho = \infty$), the series diverges.

3. If $\rho = 1$, the test is inconclusive.

Proof: We consider three cases: $\rho < 1$, $\rho > 1$, and $\rho = 1$.

1. Assume $\rho < 1$ and choose a number R such that $\rho < R < 1$. Because the sequence $\left\{\sqrt[k]{|a_k|}\right\}$ converges to a number less than R, there is a positive integer N such that $\sqrt[k]{|a_k|} < R$, or equivalently $|a_k| < R^k$, for all $k > N$. Let S_n represent the nth partial sum of $\sum\limits_{k=N+1}^{\infty} |a_k|$; note that the partial sums of this series are bounded by a convergent geometric series:

$$\begin{aligned}
S_n &= |a_{N+1}| + |a_{N+2}| + \cdots + |a_{N+n}| \\
&\leq R^{N+1} + R^{N+2} + \cdots + R^{N+n} \\
&< R^{N+1} + R^{N+2} + \cdots + R^{N+n} + \cdots \\
&= \frac{R^{N+1}}{1-R}.
\end{aligned}$$

Because the sequence $\{S_n\}$ is nondecreasing (each partial sum in the sequence consists of nonnegative terms) and is bounded above by $\dfrac{R^{N+1}}{1-R}$, it converges by the Bounded Monotonic Sequences Theorem (Theorem 10.5). Therefore, $\sum\limits_{k=N+1}^{\infty} |a_k|$ converges, and we conclude that $\sum\limits_{k=1}^{\infty} |a_k|$ converges (Theorem 10.8), which in turn implies that $\sum\limits_{k=1}^{\infty} a_k$ converges (Theorem 10.19).

2. If $\rho > 1$, there is an integer N for which $\sqrt[k]{|a_k|} > 1$, or equivalently $|a_k| > 1$, for all $k > N$. So every term in the sequence $\{|a_k|\}_{k=N+1}^{\infty}$ is greater than or equal to 1, which implies that $\lim\limits_{n\to\infty} a_n \neq 0$. Therefore, the series $\sum\limits_{k=N+1}^{\infty} a_k$ diverges by the Divergence Test, and we conclude that $\sum\limits_{k=1}^{\infty} a_k$ diverges (Theorem 10.8).

3. If $\rho = 1$, the series $\sum\limits_{k=1}^{\infty} a_k$ may or may not converge. For example, both the divergent harmonic series $\sum\limits_{k=1}^{\infty} \dfrac{1}{k}$ and the convergent p-series $\sum\limits_{k=1}^{\infty} \dfrac{1}{k^2}$ produce a value of $\rho = 1$. ◄

> **THEOREM 11.3 Convergence of Power Series**
>
> A power series $\sum_{k=0}^{\infty} c_k(x-a)^k$ centered at a converges in one of three ways:
>
> 1. The series converges absolutely for all x. It follows, by Theorem 10.19, that the series converges, in which case the interval of convergence is $(-\infty, \infty)$ and the radius of convergence is $R = \infty$.
>
> 2. There is a real number $R > 0$ such that the series converges absolutely (and therefore converges) for $|x - a| < R$ and diverges for $|x - a| > R$, in which case the radius of convergence is R.
>
> 3. The series converges only at a, in which case the radius of convergence is $R = 0$.

Proof: Without loss of generality, we take $a = 0$. (If $a \neq 0$, the following argument may be shifted so it is centered at $x = a$.) The proof hinges on a preliminary result:

If $\sum_{k=0}^{\infty} c_k x^k$ converges for $x = b \neq 0$, then it converges absolutely, for

$|x| < |b|$. If $\sum_{k=0}^{\infty} c_k x^k$ diverges for $x = d$, then it diverges, for $|x| > |d|$.

To prove these facts, assume that $\sum_{k=0}^{\infty} c_k b^k$ converges, which implies that $\lim_{k \to \infty} c_k b^k = 0$. Then there exists a real number $M > 0$ such that $|c_k b^k| < M$, for $k = 0, 1, 2, 3, \ldots$. It follows that

$$\sum_{k=0}^{\infty} |c_k x^k| = \sum_{k=0}^{\infty} \underbrace{|c_k b^k|}_{<M} \left|\frac{x}{b}\right|^k < M \sum_{k=0}^{\infty} \left|\frac{x}{b}\right|^k.$$

If $|x| < |b|$, then $|x/b| < 1$ and $\sum_{k=0}^{\infty} \left|\frac{x}{b}\right|^k$ is a convergent geometric series. Therefore, $\sum_{k=0}^{\infty} |c_k x^k|$ converges by the comparison test, which implies that $\sum_{k=0}^{\infty} c_k x^k$ converges absolutely for $|x| < |b|$. The second half of the preliminary result is proved by supposing the series diverges at $x = d$. The series cannot converge at a point x_0 with $|x_0| > |d|$ because by the preceding argument, it would converge for $|x| < |x_0|$, which includes $x = d$. Therefore, the series diverges for $|x| > |d|$.

Now we may deal with the three cases in the theorem. Let S be the set of real numbers for which the series converges, which always includes 0. If $S = \{0\}$, then we have Case 3. If S consists of all real numbers, then we have Case 1. For Case 2, assume that $d \neq 0$ is a point at which the series diverges. By the preliminary result, the series diverges for $|x| > |d|$. Therefore, if x is in S, then $|x| < |d|$, which implies that S is bounded. By the Least Upper Bound Property for real numbers, S has a least upper bound R, such that $x \leq R$, for all x in S. If $|x| > R$, then x is not in S and the series diverges. If $|x| < R$, then x is not the least upper bound of S and there exists a number b in S with $|x| < b \leq R$.

Because the series converges at $x = b$, by the preliminary result, $\sum_{k=0}^{\infty} |c_k x^k|$ converges for $|x| < |b|$. Therefore, the series $\sum_{k=0}^{\infty} c_k x^k$ converges absolutely (which, by Theorem 10.19, implies the series converges) for $|x| < R$ and diverges for $|x| > R$. ◄

> The Least Upper Bound Property for real numbers states that if a nonempty set S is bounded (that is, there exists a number M, called an *upper bound*, such that $x \leq M$ for all x in S), then S has a *least upper bound L*, which is the smallest of the upper bounds.

THEOREM 12.3   Eccentricity-Directrix Theorem
Suppose ℓ is a line, F is a point not on ℓ, and e is a positive real number. Let C be the set of points P in a plane with the property that $\dfrac{|PF|}{|PL|} = e$, where $|PL|$ is the perpendicular distance from P to ℓ.

1. If $e = 1$, C is a **parabola**.
2. If $0 < e < 1$, C is an **ellipse**.
3. If $e > 1$, C is a **hyperbola**.

$$|PF| = e|PL|$$

Figure A.1

Proof: If $e = 1$, then the defining property becomes $|PF| = |PL|$, which is the standard definition of a parabola (Section 12.4). We prove the result for ellipses $(0 < e < 1)$; a small modification handles the case of hyperbolas $(e > 1)$.

Let E be the curve whose points satisfy $|PF| = e\,|PL|$ and suppose $0 < e < 1$; the goal is to show that E is an ellipse. There are two points on the x-axis (the *vertices*), call them V and V', that satisfy $|VF| = e|VL|$ and $|V'F| = e|V'L|$. (Note that $|VL|$ is the perpendicular distance from V to the line ℓ, not the distance from the points labeled V and L in **Figure A.1**. $|V'L|$ should also be interpreted in this way.) We choose the origin such that V and V' have coordinates $(a, 0)$ and $(-a, 0)$, respectively (Figure A.1). We locate the point F (a *focus*) at $(c, 0)$ and let ℓ (a *directrix*) be the line $x = d$, where $c > 0$ and $d > 0$. These choices place the center of E at the origin. Notice that we have four parameters (a, c, d, and e) that must be related.

Because the vertex $V(a, 0)$ is on E, it satisfies the defining property $|PF| = e\,|PL|$, with $P = V$. This condition implies that $a - c = e(d - a)$. Because the vertex $V'(-a, 0)$ is on the curve E, it also satisfies the defining property $|PF| = e\,|PL|$, with $P = V'$. This condition implies that $a + c = e(d + a)$. Solving these two equations for c and d, we find that $c = ae$ and $d = a/e$. To summarize, the parameters a, c, d, and e are related by the equations

$$c = ae \quad \text{and} \quad a = de.$$

Because $e < 1$, it follows that $c < a < d$.

We now use the property $|PF| = e\,|PL|$ with an arbitrary point $P(x, y)$ on the curve E. Figure A.1 shows the geometry with the focus $(c, 0) = (ae, 0)$ and the directrix $x = d = a/e$. The condition $|PF| = e\,|PL|$ becomes

$$\sqrt{(x - ae)^2 + y^2} = e\left(\frac{a}{e} - x\right).$$

The goal is to find the simplest possible relationship between x and y. Squaring both sides and collecting terms, we have

$$(1 - e^2)x^2 + y^2 = a^2(1 - e^2).$$

Dividing through by $a^2(1 - e^2)$ gives the equation of the standard ellipse:

$$\frac{x^2}{a^2} + \frac{y^2}{a^2(1 - e^2)} = \frac{x^2}{a^2} + \frac{y^2}{b^2} = 1, \quad \text{where} \quad b^2 = a^2(1 - e^2).$$

This is the equation of an ellipse centered at the origin with vertices and foci on the x-axis.

The preceding proof is now applied with $e > 1$. The argument for ellipses with $0 < e < 1$ led to the equation

$$\frac{x^2}{a^2} + \frac{y^2}{a^2(1 - e^2)} = 1.$$

With $e > 1$, we have $1 - e^2 < 0$, so we write $(1 - e^2) = -(e^2 - 1)$. The resulting equation describes a hyperbola centered at the origin with the foci on the x-axis:

$$\frac{x^2}{a^2} - \frac{y^2}{b^2} = 1, \quad \text{where} \quad b^2 = a^2(e^2 - 1). \qquad \blacktriangleleft$$

THEOREM 15.3   Continuity of Composite Functions

If $u = g(x, y)$ is continuous at (a, b) and $z = f(u)$ is continuous at $g(a, b)$, then the composite function $z = f(g(x, y))$ is continuous at (a, b).

Proof: Let P and P_0 represent the points (x, y) and (a, b), respectively. Let $u = g(P)$ and $u_0 = g(P_0)$. The continuity of f at u_0 means that $\lim_{u \to u_0} f(u) = f(u_0)$. This limit implies that given any $\varepsilon > 0$, there exists a $\delta^* > 0$ such that

$$|f(u) - f(u_0)| < \varepsilon \quad \text{whenever} \quad 0 < |u - u_0| < \delta^*.$$

The continuity of g at P_0 means that $\lim_{P \to P_0} g(P) = g(P_0)$. Letting $|P - P_0|$ denote the distance between P and P_0, this limit implies that given any $\delta^* > 0$, there exists a $\delta > 0$ such that

$$|g(P) - g(P_0)| = |u - u_0| < \delta^* \quad \text{whenever} \quad 0 < |P - P_0| < \delta.$$

We now combine these two statements. Given any $\varepsilon > 0$, there exists a $\delta > 0$ such that

$$|f(g(P)) - f(g(P_0))| = |f(u) - f(u_0)| < \varepsilon \quad \text{whenever} \quad 0 < |P - P_0| < \delta.$$

Therefore, $\lim_{(x,y) \to (a,b)} f(g(x, y)) = f(g(a, b))$ and $z = f(g(x, y))$ is continuous at (a, b). ◄

THEOREM 15.5   Conditions for Differentiability

Suppose the function f has partial derivatives f_x and f_y defined on an open set containing (a, b), with f_x and f_y continuous at (a, b). Then f is differentiable at (a, b).

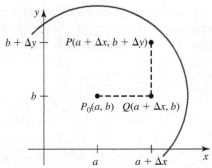

Figure A.2

Proof: Figure A.2 shows a region on which the conditions of the theorem are satisfied containing the points $P_0(a, b)$, $Q(a + \Delta x, b)$, and $P(a + \Delta x, b + \Delta y)$. By the definition of differentiability of f at P_0, we must show that

$$\Delta z = f(P) - f(P_0) = f_x(a, b)\Delta x + f_y(a, b)\Delta y + \varepsilon_1 \Delta x + \varepsilon_2 \Delta y,$$

where ε_1 and ε_2 depend only on $a, b, \Delta x$, and Δy, with $(\varepsilon_1, \varepsilon_2) \to (0, 0)$ as $(\Delta x, \Delta y) \to (0, 0)$. We can view the change Δz taking place in two stages:

- $\Delta z_1 = f(a + \Delta x, b) - f(a, b)$ is the change in z as (x, y) moves from P_0 to Q.
- $\Delta z_2 = f(a + \Delta x, b + \Delta y) - f(a + \Delta x, b)$ is the change in z as (x, y) moves from Q to P.

Applying the Mean Value Theorem to the first variable and noting that f is differentiable with respect to x, we have

$$\Delta z_1 = f(a + \Delta x, b) - f(a, b) = f_x(c, b)\Delta x,$$

where c lies in the interval $(a, a + \Delta x)$. Similarly, applying the Mean Value Theorem to the second variable and noting that f is differentiable with respect to y, we have

$$\Delta z_2 = f(a + \Delta x, b + \Delta y) - f(a + \Delta x, b) = f_y(a + \Delta x, d)\Delta y,$$

where d lies in the interval $(b, b + \Delta y)$. We now express Δz as the sum of Δz_1 and Δz_2:

$$\begin{aligned}
\Delta z &= \Delta z_1 + \Delta z_2 \\
&= f_x(c, b)\Delta x + f_y(a + \Delta x, d)\Delta y \\
&= \underbrace{(f_x(c, b) - f_x(a, b)}_{\varepsilon_1} + f_x(a, b))\Delta x \qquad \text{Add and subtract } f_x(a, b). \\
&\quad + \underbrace{(f_y(a + \Delta x, d) - f_y(a, b)}_{\varepsilon_2} + f_y(a, b))\Delta y \quad \text{Add and subtract } f_y(a, b). \\
&= (f_x(a, b) + \varepsilon_1)\Delta x + (f_y(a, b) + \varepsilon_2)\Delta y.
\end{aligned}$$

Note that as $\Delta x \to 0$ and $\Delta y \to 0$, we have $c \to a$ and $d \to b$. Because f_x and f_y are continuous at (a, b), it follows that as $\Delta x \to 0$ and $\Delta y \to 0$,

$$\varepsilon_1 = f_x(c, b) - f_x(a, b) \to 0 \quad \text{and} \quad \varepsilon_2 = f_y(a + \Delta x, d) - f_y(a, b) \to 0.$$

Therefore, the condition for differentiability of f at (a, b) has been proved.   ◄

THEOREM 15.7   **Chain Rule (One Independent Variable)**

Let $z = f(x, y)$ be a differentiable function of x and y on its domain, where x and y are differentiable functions of t on an interval I. Then

$$\frac{dz}{dt} = \frac{\partial f}{\partial x}\frac{dx}{dt} + \frac{\partial f}{\partial y}\frac{dy}{dt}.$$

Proof: Assume $(a, b) = (x(t), y(t))$ is in the domain of f, where t is in I. Let $\Delta x = x(t + \Delta t) - x(t)$ and $\Delta y = y(t + \Delta t) - y(t)$. Because f is differentiable at (a, b), we know (Section 15.3) that

$$\Delta z = \frac{\partial f}{\partial x}(a, b)\Delta x + \frac{\partial f}{\partial y}(a, b)\Delta y + \varepsilon_1 \Delta x + \varepsilon_2 \Delta y,$$

where $(\varepsilon_1, \varepsilon_2) \to (0, 0)$ as $(\Delta x, \Delta y) \to (0, 0)$. Dividing this equation by Δt gives

$$\frac{\Delta z}{\Delta t} = \frac{\partial f}{\partial x}\frac{\Delta x}{\Delta t} + \frac{\partial f}{\partial y}\frac{\Delta y}{\Delta t} + \varepsilon_1\frac{\Delta x}{\Delta t} + \varepsilon_2\frac{\Delta y}{\Delta t}.$$

As $\Delta t \to 0$, several things occur. First, because $x = g(t)$ and $y = h(t)$ are differentiable on I, $\dfrac{\Delta x}{\Delta t}$ and $\dfrac{\Delta y}{\Delta t}$ approach $\dfrac{dx}{dt}$ and $\dfrac{dy}{dt}$, respectively. Similarly, $\dfrac{\Delta z}{\Delta t}$ approaches $\dfrac{dz}{dt}$ as $\Delta t \to 0$. The fact that x and y are continuous on I (because they are differentiable there) means that $\Delta x \to 0$ and $\Delta y \to 0$ as $\Delta t \to 0$. Therefore, because $(\varepsilon_1, \varepsilon_2) \to (0, 0)$ as $(\Delta x, \Delta y) \to (0, 0)$, it follows that $(\varepsilon_1, \varepsilon_2) \to (0, 0)$ as $\Delta t \to 0$. Letting $\Delta t \to 0$, we have

$$\underbrace{\lim_{\Delta t \to 0} \frac{\Delta z}{\Delta t}}_{\frac{dz}{dt}} = \frac{\partial f}{\partial x}\underbrace{\lim_{\Delta t \to 0} \frac{\Delta x}{\Delta t}}_{\frac{dx}{dt}} + \frac{\partial f}{\partial y}\underbrace{\lim_{\Delta t \to 0} \frac{\Delta y}{\Delta t}}_{\frac{dy}{dt}} + \underbrace{\lim_{\Delta t \to 0} \varepsilon_1}_{\to 0}\underbrace{\frac{\Delta x}{\Delta t}}_{\to \frac{dx}{dt}} + \underbrace{\lim_{\Delta t \to 0} \varepsilon_2}_{\to 0}\underbrace{\frac{\Delta y}{\Delta t}}_{\to \frac{dy}{dt}}$$

or

$$\frac{dz}{dt} = \frac{\partial f}{\partial x}\frac{dx}{dt} + \frac{\partial f}{\partial y}\frac{dy}{dt}.$$

  ◄

THEOREM 15.15   **Second Derivative Test**

Suppose the second partial derivatives of f are continuous throughout an open disk centered at the point (a, b), where $f_x(a, b) = f_y(a, b) = 0$. Let $D(x, y) = f_{xx}(x, y) f_{yy}(x, y) - (f_{xy}(x, y))^2$.

1. If $D(a, b) > 0$ and $f_{xx}(a, b) < 0$, then f has a local maximum value at (a, b).
2. If $D(a, b) > 0$ and $f_{xx}(a, b) > 0$, then f has a local minimum value at (a, b).
3. If $D(a, b) < 0$, then f has a saddle point at (a, b).
4. If $D(a, b) = 0$, then the test is inconclusive.

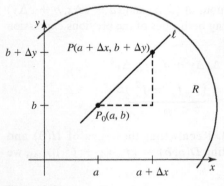

Figure A.3

Proof: The proof relies on a two-variable version of Taylor's Theorem, which we prove first. **Figure A.3** shows the open disk R on which the conditions of Theorem 15.15 are satisfied; it contains the points $P_0(a, b)$ and $P(a + \Delta x, b + \Delta y)$. The line ℓ through $P_0 P$ has a parametric description

$$\langle x(t), y(t) \rangle = \langle a + t\Delta x, b + t\Delta y \rangle,$$

where $t = 0$ corresponds to P_0 and $t = 1$ corresponds to P.

We now let $F(t) = f(a + t\Delta x, b + t\Delta y)$ be the value of f along that part of ℓ that lies in R. By the Chain Rule, we have

$$F'(t) = f_x \underbrace{x'(t)}_{\Delta x} + f_y \underbrace{y'(t)}_{\Delta y} = f_x \Delta x + f_y \Delta y.$$

Differentiating again with respect to t (f_x and f_y are differentiable), we use $f_{xy} = f_{yx}$ to obtain

$$F''(t) = \frac{\partial F'}{\partial x} \underbrace{x'(t)}_{\Delta x} + \frac{\partial F'}{\partial y} \underbrace{y'(t)}_{\Delta y}$$

$$= \frac{\partial}{\partial x}(f_x \Delta x + f_y \Delta y)\Delta x + \frac{\partial}{\partial y}(f_x \Delta x + f_y \Delta y)\Delta y$$

$$= f_{xx} \Delta x^2 + 2f_{xy} \Delta x\Delta y + f_{yy} \Delta y^2.$$

Noting that F meets the conditions of Taylor's Theorem for one variable with $n = 1$, we write

$$F(t) = F(0) + F'(0)(t - 0) + \frac{1}{2}F''(c)(t - 0)^2,$$

where c is between 0 and t. Setting $t = 1$, it follows that

$$F(1) = F(0) + F'(0) + \frac{1}{2}F''(c),$$

where $0 < c < 1$. Recalling that $F(t) = f(a + t\Delta x, b + t\Delta y)$ and invoking the condition $f_x(a, b) = f_y(a, b) = 0$, we have

$$F(1) = f(a + \Delta x, b + \Delta y)$$

$$= f(a, b) + \underbrace{f_x(a, b)\Delta x + f_y(a, b)\Delta y}_{F'(0) = 0}$$

$$+ \frac{1}{2}(f_{xx} \Delta x^2 + 2f_{xy} \Delta x\Delta y + f_{yy} \Delta y^2)\Big|_{(a+c\Delta x,\, b+c\Delta y)}$$

$$= f(a, b) + \frac{1}{2}\underbrace{(f_{xx} \Delta x^2 + 2f_{xy} \Delta x\Delta y + f_{yy} \Delta y^2)\Big|_{(a+c\Delta x,\, b+c\Delta y)}}_{H(c)}$$

$$= f(a, b) + \frac{1}{2}H(c).$$

The existence and type of extreme point at (a, b) are determined by the sign of $f(a + \Delta x, b + \Delta y) - f(a, b)$ (for example, if $f(a + \Delta x, b + \Delta y) - f(a, b) \geq 0$ for all Δx and Δy near 0, then f has a local minimum at (a, b)). Note that $f(a + \Delta x, b + \Delta y) - f(a, b)$ has the same sign as the quantity we have denoted $H(c)$. Assuming $H(0) \neq 0$, for Δx and Δy sufficiently small and nonzero, the sign of $H(c)$ is the same as the sign of

$$H(0) = f_{xx}(a, b)\Delta x^2 + 2f_{xy}(a, b)\Delta x\Delta y + f_{yy}(a, b)\Delta y^2$$

(because the second partial derivatives are continuous at (a, b) and $(a + c\Delta x, b + c\Delta y)$ can be made arbitrarily close to (a, b)). Multiplying both sides of the previous expression by f_{xx} and rearranging terms leads to

$$f_{xx} H(0) = f_{xx}^2 \, \Delta x^2 + 2f_{xy} \, f_{xx} \, \Delta x \Delta y + f_{yy} \, f_{xx} \, \Delta y^2$$

$$= \underbrace{(f_{xx}\Delta x + f_{xy}\Delta y)^2}_{\geq 0} + \underbrace{(f_{xx} f_{yy} - f_{xy}^2)}_{D}\Delta y^2,$$

where all derivatives are evaluated at (a, b). Recall that the signs of $H(0)$ and $f(a + \Delta x, b + \Delta y) - f(a, b)$ are the same. Letting $D(a, b) = (f_{xx} f_{yy} - f_{xy}^2)|_{(a, b)}$, we reach the following conclusions:

- If $D(a, b) > 0$ and $f_{xx}(a, b) < 0$, then $H(0) < 0$ (for Δx and Δy sufficiently close to 0) and $f(a + \Delta x, b + \Delta y) - f(a, b) < 0$. Therefore, f has a local maximum value at (a, b).

- If $D(a, b) > 0$ and $f_{xx}(a, b) > 0$, then $H(0) > 0$ (for Δx and Δy sufficiently close to 0) and $f(a + \Delta x, b + \Delta y) - f(a, b) > 0$. Therefore, f has a local minimum value at (a, b).

- If $D(a, b) < 0$, then $H(0) > 0$ for some small nonzero values of Δx and Δy (implying $f(a + \Delta x, b + \Delta y) > f(a, b)$), *and* $H(0) < 0$ for other small nonzero values of Δx and Δy (implying $f(a + \Delta x, b + \Delta y) < f(a, b)$). (The relative sizes of $(f_{xx} \, \Delta x + f_{xy} \, \Delta y)^2$ and $(f_{xx} f_{yy} - f_{xy}^2) \, \Delta y^2$ can be adjusted by varying Δx and Δy.) Therefore, f has a saddle point at (a, b).

- If $D(a, b) = 0$, then $H(0)$ may be zero, in which case the sign of $H(c)$ cannot be determined. Therefore, the test is inconclusive. ◄

Answers

Section 10.1 Exercises, pp. 647–649

1. A sequence is an ordered list of numbers. Example: $1, \frac{1}{3}, \frac{1}{9}, \frac{1}{27}, \ldots$
3. $1, 1, 2, 6, 24$ **5.** $a_n = (-1)^{n+1} n$, for $n = 1, 2, 3, \ldots$;
$a_n = (-1)^n (n + 1)$, for $n = 0, 1, 2, \ldots$ (Answers may vary.)

7. e **9.** $1, 5, 14, 30$ **11.** $\displaystyle\sum_{k=1}^{\infty} 10$ (Answer is not unique.)

13. $\frac{1}{10}, \frac{1}{100}, \frac{1}{1000}, \frac{1}{10,000}$ **15.** $-\frac{1}{2}, \frac{1}{4}, -\frac{1}{8}, \frac{1}{16}$ **17.** $\frac{4}{3}, \frac{8}{5}, \frac{16}{9}, \frac{32}{17}$

19. $2, 1, 0, 1$ **21.** $2, 4, 8, 16$ **23.** $10, 18, 42, 114$ **25.** $1, \dfrac{1}{2}, \dfrac{2}{3}, \dfrac{3}{5}$

27. a. $\dfrac{1}{32}, \dfrac{1}{64}$ **b.** $a_1 = 1, a_{n+1} = \dfrac{1}{2} a_n$, for $n \geq 1$ **c.** $a_n = \dfrac{1}{2^{n-1}}$,
for $n \geq 1$ **29. a.** $32, 64$ **b.** $a_1 = 1, a_{n+1} = 2a_n$, for $n \geq 1$
c. $a_n = 2^{n-1}$, for $n \geq 1$ **31. a.** $243, 729$ **b.** $a_1 = 1, a_{n+1} = 3a_n$,
for $n \geq 1$ **c.** $a_n = 3^{n-1}$, for $n \geq 1$ **33. a.** $-5, 5$ **b.** $a_1 = -5$,
$a_{n+1} = -a_n$, for $n \geq 1$ **c.** $a_n = (-1)^n \cdot 5$, for $n \geq 1$
35. $9, 99, 999, 9999$; diverges **37.** $\frac{1}{10}, \frac{1}{100}, \frac{1}{1000}, \frac{1}{10,000}$;
converges to 0 **39.** $2, 4, 2, 4$; diverges **41.** $2, 2, 2, 2$; converges to 2
43. $54.545, 54.959, 54.996, 55.000$; converges to 55

45.

n	a_n
1	0.83333333
2	0.96153846
3	0.99206349
4	0.99840256
5	0.99968010
6	0.99993600
7	0.99998720
8	0.99999744
9	0.99999949
10	0.99999990

The limit appears to be 1.

47.

n	a_n
1	2
2	6
3	12
4	20
5	30
6	42
7	56
8	72
9	90
10	110

The sequence appears to diverge.

49. a. $\frac{5}{2}, \frac{9}{4}, \frac{17}{8}, \frac{33}{16}$ **b.** 2

51.

n	a_n
1	3.00000000
2	3.50000000
3	3.75000000
4	3.87500000
5	3.93750000
6	3.96875000
7	3.98437500
8	3.99218750
9	3.99609375
10	3.99804688

The limit appears to be 4.

53.

n	a_n
0	1
1	5
2	21
3	85
4	341
5	1365
6	5461
7	21,845
8	87,381
9	349,525

The sequence appears to diverge.

55.

n	a_n
1	8.00000000
2	4.41421356
3	4.05050150
4	4.00629289
5	4.00078630
6	4.00009828
7	4.00001229
8	4.00000154
9	4.00000019
10	4.00000002

The limit appears to be 4.

57. a. $20, 10, 5, \frac{5}{2}$
b. $h_n = 20\left(\frac{1}{2}\right)^n$, for $n \ge 0$
59. a. $30, \frac{15}{2}, \frac{15}{8}, \frac{15}{32}$
b. $h_n = 30\left(\frac{1}{4}\right)^n$, for $n \ge 0$
61. $S_1 = 0.3, S_2 = 0.33, S_3 = 0.333,$
$S_4 = 0.3333; \frac{1}{3}$
63. $S_1 = 4, S_2 = 4.9,$
$S_3 = 4.99, S_4 = 4.999; 5$
65. $S_1 = 0.6, S_2 = 0.66, S_3 = 0.666,$
$S_4 = 0.6666; \frac{2}{3}$ **67. a.** $\frac{2}{3}, \frac{4}{5}, \frac{6}{7}, \frac{8}{9}$
b. $S_n = \frac{2n}{2n+1}; \frac{10}{11}, \frac{12}{13}, \frac{14}{15}, \frac{16}{17}$ **c.** 1

69. a. $9, 9.9, 9.99, 9.999$ **b.** $S_n = 10 - (0.1)^{n-1}; 9.9999, 9.99999,$
$9.999999, 9.9999999$ **c.** 10 **71. a.** True **b.** False **c.** True
73. a. $20, 10, 5, \frac{5}{2}, \frac{5}{4}$ **b.** $M_n = 20\left(\frac{1}{2}\right)^n$, for $n \ge 0$ **c.** $M_0 = 20,$
$M_{n+1} = \frac{1}{2}M_n$, for $n \ge 0$ **d.** $\lim_{n \to \infty} M_n = 0$ **75. a.** $200, 190, 180.5,$
$171.475, 162.90125$ **b.** $d_n = 200(0.95)^n$, for $n \ge 0$
c. $d_0 = 200, d_{n+1} = (0.95)d_n$, for $n \ge 0$ **d.** $\lim_{n \to \infty} d_n = 0$
77. a. $40, 70, 92.5, 109.375$ **b.** 160 **79.** 0.739

Section 10.2 Exercises, pp. 659–662

1. $a_n = \frac{1}{n}, n \ge 1$ **3.** $a_n = \frac{n}{n+1}, n \ge 1$ **5.** Converges for
$-1 < r \le 1$, diverges otherwise **7.** Diverges monotonically
9. Converges, oscillates; 0 **11.** $\{e^{n/100}\}$ grows faster then $\{n^{100}\}$.

13. 0 **15.** $\frac{3}{2}$ **17.** $\frac{\pi}{4}$ **19.** 2 **21.** 0 **23.** $\frac{1}{4}$ **25.** 2 **27.** 0
29. 0 **31.** 3 **33.** Diverges **35.** $\frac{\pi}{2}$ **37.** 0 **39.** e^2 **41.** e^3
43. $e^{1/4}$ **45.** 0 **47.** 1 **49.** 0 **51.** 6
53.
55. 0 **57.** Diverges
59. Diverges **61.** 0 **63.** 0
65. 0 **67.** 0 **69.** 0
71. a. $d_{n+1} = \frac{1}{2}d_n + 80$, for $n \ge 1$
b. 160 mg

Diverges

73. a. $\$0, \$100, \$200.75, \$302.26, \$404.53$
b. $B_{n+1} = 1.0075B_n + 100$, for $n \ge 0$ **c.** Approx. 43 months
75. 0 **77.** Diverges **79.** 0 **81.** 1 **83. a.** True **b.** False
c. True **d.** True **e.** False **f.** True **85. a.** Nondecreasing
b. $\frac{1}{2}$ **87. a.** Nonincreasing **b.** 2 **89. a.** $d_{n+1} = 0.4d_n + 75$;
$d_1 = 75$ **c.** 125; in the long run there will be approximately 125 mg
of medication in the blood. **91.** 0.607 **93. b.** 9
95. a. $\sqrt{2}, \sqrt{2 + \sqrt{2}}, \sqrt{2 + \sqrt{2 + \sqrt{2}}},$
$\sqrt{2 + \sqrt{2 + \sqrt{2 + \sqrt{2}}}}$, or $1.41421, 1.84776, 1.96157, 1.99037$
c. 2 **97. a.** $1, 1, 2, 3, 5, 8, 13, 21, 34, 55$ **b.** No **99. b.** $1, 2,$
$1.5, 1.6667, 1.6$ **c.** Approx. 1.618 **e.** $\frac{a + \sqrt{a^2 + 4b}}{2}$
101. Given a tolerance $\varepsilon > 0$, look beyond a_N, where $N > 1/\varepsilon$.
103. Given a tolerance $\varepsilon > 0$, look beyond a_N, where $N > \frac{1}{4}\sqrt{3/\varepsilon}$,
provided $\varepsilon < \frac{3}{4}$. **105.** Given a tolerance $\varepsilon > 0$, look beyond a_N,
where $N > c/(\varepsilon b^2)$. **107.** $a < 1$ **109.** $\{n^2 + 2n - 17\}_{n=3}^{\infty}$
111. $n = 4, n = 6, n = 25$

Section 10.3 Exercises, pp. 668–671

1. The constant r in the series $\sum_{k=0}^{\infty} ar^k$ **3.** No **5. a.** $a = \frac{2}{3}; r = \frac{1}{5}$
b. $a = \frac{1}{27}; r = -\frac{1}{3}$ **7.** $S_n = \frac{1}{4} - \frac{1}{n+4}; S_{36} = \frac{9}{40}$ **9.** 9841
11. Approx. 1.1905 **13.** Approx. 0.5392 **15.** $\frac{1093}{2916}$
17. $\$15,920.22$ **19. a.** $\frac{7}{9}$ **21.** $\frac{4}{3}$ **23.** $\frac{10}{19}$ **25.** 10 **27.** Diverges
29. $\frac{1}{e^2 - 1}$ **31.** $\frac{1}{7}$ **33.** $\frac{1}{500}$ **35.** $\frac{3\pi}{\pi + 1}$ **37.** $\frac{\pi}{\pi - e}$ **39.** $\frac{9}{460}$
41. $\frac{4}{11}$ **43.** $A_5 = 266.406; A_{10} = 266.666; A_{30} = 266.667;$
$\lim_{n \to \infty} A_n = 266\frac{2}{3}$ mg, which is the steady-state level. **45.** 400 mg
47. $0.\overline{3} = \sum_{k=1}^{\infty} 3(0.1)^k = \frac{1}{3}$ **49.** $0.\overline{037} = \sum_{k=1}^{\infty} 37(0.001)^k = \frac{1}{27}$
51. $0.\overline{456} = \sum_{k=0}^{\infty} 0.456(0.001)^k = \frac{152}{333}$
53. $0.00\overline{952} = \sum_{k=0}^{\infty} 0.00952(0.001)^k = \frac{238}{24,975}$
55. $S_n = \frac{n}{2n + 4}; \frac{1}{2}$ **57.** $S_n = \frac{1}{7} - \frac{1}{n + 7}; \frac{1}{7}$
59. $S_n = \frac{1}{9} - \frac{1}{4n + 9}; \frac{1}{9}$ **61.** $S_n = \ln(n + 1)$; diverges

63. $S_n = \dfrac{1}{p+1} - \dfrac{1}{n+p+1}; \dfrac{1}{p+1}$

65. $S_n = \left(\dfrac{1}{\sqrt{2}} + \dfrac{1}{\sqrt{3}}\right) - \left(\dfrac{1}{\sqrt{n+2}} + \dfrac{1}{\sqrt{n+3}}\right); \dfrac{1}{\sqrt{2}} + \dfrac{1}{\sqrt{3}}$

67. $S_n = \dfrac{13}{12} - \dfrac{1}{n+2} - \dfrac{3}{n+3} - \dfrac{1}{n+4}; \dfrac{13}{12}$

69. $S_n = \tan^{-1}(n+1) - \tan^{-1}1; \dfrac{\pi}{4}$ **71. a, b.** $\dfrac{4}{3}$ **73.** $-\dfrac{1}{4}$

75. $\dfrac{2500}{19}$ **77.** $-\dfrac{2}{15}$ **79.** $\dfrac{1}{\ln 2}$ **81.** -2 **83.** $\dfrac{113}{30}$ **85.** $\dfrac{17}{10}$

87. **a.** True **b.** True **c.** False **d.** False **e.** True **f.** False
g. True **89. a.** $\dfrac{1}{5}$ **b.** Approx. 0.19999695 **91.** Approx. 0.96

95. 462 months **99.** $\displaystyle\sum_{k=0}^{\infty}\left(\dfrac{1}{4}\right)^k A_1 = \dfrac{A_1}{1-1/4} = \dfrac{4}{3}A_1$

101. **a.** $L_n = 3\left(\dfrac{4}{3}\right)^n$, so $\lim_{n\to\infty} L_n = \infty$ **b.** $\lim_{n\to\infty} A_n = \dfrac{2\sqrt{3}}{5}$

103. $R_n = S - S_n = \dfrac{1}{1-r} - \left(\dfrac{1-r^n}{1-r}\right) = \dfrac{r^n}{1-r}$ **105. a.** 60

b. 9 **107. a.** 13 **b.** 15 **109. a.** $1, \dfrac{5}{6}, \dfrac{2}{3}$, undefined, undefined

b. $(-1, 1)$ **111.** Converges for x in $(-\infty, -2)$ or $(0, \infty)$; $x = \dfrac{1}{2}$

Section 10.4 Exercises, pp. 680–683

1. The series diverges. **3.** $\lim_{k\to\infty} a_k = 0$ **5.** Converges for $p > 1$ and
diverges for $p \le 1$ **7.** $R_n = S - S_n$ **9.** Diverges
11. Inconclusive **13.** Diverges **15.** Diverges **17.** Diverges
19. Diverges **21.** Converges **23.** Diverges **25.** Converges
27. Diverges **29.** Converges **31.** Converges **33.** Converges
35. Diverges **37.** Diverges **39. a.** $S \approx S_2 = 1.0078125$
b. $R_2 < 0.0026042$ **c.** $L_2 = 1.0080411; U_2 = 1.0104167$

41. **a.** $\dfrac{1}{5n^5}$ **b.** 3 **c.** $L_n = S_n + \dfrac{1}{5(n+1)^5}; U_n = S_n + \dfrac{1}{5n^5}$

d. $(1.017342754, 1.017343512)$ **43. a.** $\dfrac{3^{-n}}{\ln 3}$ **b.** 7

c. $L_n = S_n + \dfrac{3^{-n-1}}{\ln 3}; U_n = S_n + \dfrac{3^{-n}}{\ln 3}$

d. $(0.499996671, 0.500006947)$ **45.** 1.0083 **47. a.** False
b. True **c.** False **d.** True **e.** False **f.** False **49.** Converges
51. Converges **53.** Diverges **55.** Diverges **57.** Diverges
59. Converges **61.** Converges **63.** Converges **65. a.** $p > 1$

b. $\displaystyle\sum_{k=2}^{\infty} \dfrac{1}{k(\ln k)^2}$ converges faster. **67.** $\zeta(3) \approx 1.202, \zeta(5) \approx 1.037$

69. $\dfrac{\pi^2}{8}$ **73. a.** $\dfrac{1}{2}, \dfrac{7}{12}, \dfrac{37}{60}$

Section 10.5 Exercises, pp. 687–688

1. Find an appropriate comparison series. Then take the limit of the
ratio of the terms of the given series and the comparison series as
$n \to \infty$. The value of the limit determines whether the series converges.

3. $\displaystyle\sum_{k=1}^{\infty} \dfrac{1}{k^2}$ **5.** $\displaystyle\sum_{k=1}^{\infty}\left(\dfrac{2}{3}\right)^k$ **7.** $\displaystyle\sum_{k=1}^{\infty} \dfrac{1}{k}$ **9.** Converges
11. Diverges **13.** Converges **15.** Converges **17.** Converges
19. Diverges **21.** Diverges **23.** Converges **25.** Diverges
27. Converges **29.** Diverges **31.** Diverges **33.** Diverges
35. Converges **37. a.** False **b.** True **c.** True **d.** True

39. Converges **41.** Diverges **43.** Diverges **45.** Diverges
47. Converges **49.** Diverges **51.** Converges **53.** Converges
55. Diverges **57.** Converges **59.** Diverges **61.** Converges

Section 10.6 Exercises, pp. 694–696

1. Because $S_{n+1} - S_n = (-1)^n a_{n+1}$ alternates sign
3. Because the remainder $R_n = S - S_n$ alternates sign
5. $|R_n| = |S - S_n| \le |S_{n+1} - S_n| = a_{n+1}$ **7.** No; if a series
of positive terms converges, it does so absolutely and not conditionally.

9. Yes, $\displaystyle\sum_{k=1}^{\infty} \dfrac{(-1)^k}{k^2}$ has this property. **11.** Converges **13.** Diverges
15. Converges **17.** Converges **19.** Diverges **21.** Diverges
23. Diverges **25.** Diverges **27.** Converges **29.** $S_4 = -0.945939$;
$|S - S_4| \le 0.0016$ **31.** $S_5 = 0.70696; |S - S_5| \le 0.001536$
33. 10,000 **35.** 5000 **37.** 10 **39.** -0.973 **41.** -0.269
43. -0.783 **45.** Converges conditionally **47.** Converges absolutely
49. Converges absolutely **51.** Converges absolutely **53.** Diverges
55. Converges conditionally **57.** Diverges **59.** Converges absolutely **61.** Converges conditionally **63.** Converges absolutely
65. **a.** False **b.** True **c.** True **d.** True **e.** False **f.** True
g. True **69.** x and y are divergent series.

71. **b.** $S_{2n} = \displaystyle\sum_{k=1}^{n}\left(\dfrac{1}{k^2} - \dfrac{1}{k}\right)$

Section 10.7 Exercises, pp. 699–700

1. Take the limit of the magnitude of the ratio of consecutive terms
of the series as $k \to \infty$. The value of the limit determines whether
the series converges absolutely or diverges. **3.** 999,000

5. $\dfrac{1}{(k+2)(k+1)}$ **7.** Ratio Test **9.** Converges absolutely
11. Diverges **13.** Converges absolutely **15.** Converges absolutely
17. Diverges **19.** Diverges **21.** Converges absolutely
23. Converges absolutely **25.** Diverges **27.** Converges absolutely
29. Diverges **31. a.** False **b.** True **c.** True **d.** True
33. Converges absolutely **35.** Diverges **37.** Converges absolutely
39. Converges conditionally **41.** Converges absolutely
43. Converges absolutely **45.** Converges conditionally
47. Converges absolutely **49.** Converges conditionally
51. $p > 1$ **53.** $p > 1$ **55.** $p < 1$ **57.** Diverges for all p
59. $-1 < x < 1$ **61.** $-1 \le x \le 1$ **63.** $-2 < x < 2$

Section 10.8 Exercises, pp. 703–704

1. Root Test **3.** Divergence Test **5.** p-series Test or Limit Comparison Test **7.** Comparison Test or Limit Comparison Test
9. Alternating Series Test **11.** Diverges **13.** Diverges
15. Converges **17.** Diverges **19.** Converges **21.** Converges
23. Converges **25.** Converges **27.** Converges **29.** Diverges
31. Converges **33.** Diverges **35.** Converges **37.** Diverges
39. Diverges **41.** Converges **43.** Diverges **45.** Converges
47. Diverges **49.** Converges **51.** Diverges **53.** Converges
55. Diverges **57.** Converges **59.** Converges **61.** Diverges
63. Diverges **65.** Converges **67.** Converges **69.** Diverges
71. Converges **73.** Converges **75.** Converges **77.** Diverges
79. Diverges **81.** Converges **83.** Converges **85.** Converges
87. **a.** False **b.** True **c.** True **d.** False **89.** Diverges
91. Diverges **93.** Diverges

Chapter 10 Review Exercises, pp. 704–707

1. a. False **b.** False **c.** True **d.** False **e.** True **f.** False
g. False **h.** True **3.** Approx. 1.25; approx. 0.05 **5.** $\lim_{k\to\infty} a_k = 0$,
$\lim_{n\to\infty} S_n = 8$ **7.** $a_k = \dfrac{1}{k}$ **9. a.** 0 **b.** $\dfrac{5}{9}$ **11. a.** Yes; $\lim_{k\to\infty} a_k = 1$
b. No; $\lim_{k\to\infty} a_k \neq 0$ **13.** Diverges **15.** 5 **17.** 0 **19.** 0 **21.** $1/e$

23. Diverges **25. a.** 80, 48, 32, 24, 20 **b.** 16 **27.** Diverges

29. Diverges **31.** Diverges **33.** $\dfrac{3\pi}{4}$ **35.** 3 **37.** 2/9

39. $\dfrac{311}{990}$ **41.** 200 mg **43.** Diverges **45.** Diverges **47.** Converges
49. Converges **51.** Converges **53.** Converges **55.** Converges
57. Diverges **59.** Converges **61.** Converges **63.** Converges
65. Converges **67.** Converges **69.** Converges **71.** Converges
73. Diverges **75.** Diverges **77.** Converges conditionally
79. Converges absolutely **81.** Diverges **83.** Converges absolutely
85. Converges absolutely **87.** Diverges **89. a.** Approx. 1.03666
b. 0.0004 **c.** $L_5 = 1.03685$; $U_5 = 1.03706$ **91.** 0.0067

93. 100 **95. a.** 803 m, 1283 m, $2000(1 - 0.95^N)$ m **b.** 2000 m

97. a. $\dfrac{\pi}{2^{n-1}}$ **b.** 2π **99. a.** $T_1 = \dfrac{\sqrt{3}}{16}$, $T_2 = \dfrac{7\sqrt{3}}{64}$

b. $T_n = \dfrac{\sqrt{3}}{4}\left(1 - \left(\dfrac{3}{4}\right)^n\right)$ **c.** $\lim_{n\to\infty} T_n = \dfrac{\sqrt{3}}{4}$ **d.** 0

101. $\sqrt{\dfrac{20}{g}\left(\dfrac{1 + \sqrt{p}}{1 - \sqrt{p}}\right)}$ s

CHAPTER 11

Section 11.1 Exercises, pp. 718–721

1. $f(0) = p_2(0)$, $f'(0) = p_2{}'(0)$, and $f''(0) = p_2{}''(0)$
3. 1, 1.05, 1.04875 **5.** $p_3(x) = 1 + x^2 + x^3$; 1.048
7. $p_3(x) = 1 + (x - 2) + 2(x - 2)^3$; 0.898
9. a. $p_1(x) = 8 + 12(x - 1)$
b. $p_2(x) = 8 + 12(x - 1) + 3(x - 1)^2$ **c.** 9.2; 9.23
11. a. $p_1(x) = 1 - 2x$ **b.** $p_2(x) = 1 - 2x + 2x^2$ **c.** 0.8, 0.82
13. a. $p_1(x) = 1 - x$ **b.** $p_2(x) = 1 - x + x^2$ **c.** 0.95, 0.9525
15. a. $p_1(x) = 2 + \frac{1}{12}(x - 8)$
b. $p_2(x) = 2 + \frac{1}{12}(x - 8) - \frac{1}{288}(x - 8)^2$ **c.** $1.958\overline{3}$, 1.95747
17. $p_1(x) = 1, p_2(x) = p_3(x) = 1 - 18x^2, p_4(x) = 1 - 18x^2 + 54x^4$
19. $p_3(x) = 1 - 3x + 6x^2 - 10x^3$,
$p_4(x) = 1 - 3x + 6x^2 - 10x^3 + 15x^4$
21. $p_1(x) = 1 + 3(x - 1), p_2(x) = 1 + 3(x - 1) + 3(x - 1)^2$,
$p_3(x) = 1 + 3(x - 1) + 3(x - 1)^2 + (x - 1)^3$
23. $p_3(x) = 1 + \dfrac{1}{e}(x - e) - \dfrac{1}{2e^2}(x - e)^2 + \dfrac{1}{3e^3}(x - e)^3$

25. a. $p_1(x) = -x, p_2(x) = -x - \dfrac{x^2}{2}$

b.

27. a. $p_1(x) = \dfrac{\sqrt{2}}{2} + \dfrac{\sqrt{2}}{2}\left(x - \dfrac{\pi}{4}\right)$,
$p_2(x) = \dfrac{\sqrt{2}}{2} + \dfrac{\sqrt{2}}{2}\left(x - \dfrac{\pi}{4}\right) - \dfrac{\sqrt{2}}{4}\left(x - \dfrac{\pi}{4}\right)^2$

b.

29. a. 1.0247 **b.** 7.6×10^{-6} **31. a.** 0.8613 **b.** 5.4×10^{-4}
33. a. 1.1274988 **b.** Approx. 8.85×10^{-6} (Answers may vary if
intermediate calculations are rounded.) **35. a.** Approx. -0.10033333
b. Approx. 1.34×10^{-6} (Answers may vary if intermediate calculations are rounded.) **37. a.** 1.0295635 **b.** Approx. 4.86×10^{-7}
(Answers may vary if intermediate calculations are rounded.)
39. a. Approx. 0.52083333 **b.** Approx. 2.62×10^{-4} (Answers
may vary if intermediate calculations are rounded.)

41. $R_n(x) = \dfrac{\sin^{(n+1)}(c)}{(n + 1)!} x^{n+1}$, for c between x and 0

43. $R_n(x) = \dfrac{(-1)^{n+1}e^{-c}}{(n + 1)!} x^{n+1}$, for c between x and 0

45. $R_n(x) = \dfrac{\sin^{(n+1)}(c)}{(n + 1)!}\left(x - \dfrac{\pi}{2}\right)^{n+1}$, for c between x and $\dfrac{\pi}{2}$

47. 2.0×10^{-5} **49.** 1.6×10^{-5} $(e^{0.25} < 2)$ **51.** 2.6×10^{-4}
53. With $n = 4$, $|\text{error}| \leq 2.5 \times 10^{-3}$
55. With $n = 2$, $|\text{error}| \leq 4.2 \times 10^{-2}$ $(e^{0.5} < 2)$
57. With $n = 2$, $|\text{error}| \leq 5.4 \times 10^{-3}$ **59.** 4 **61.** 3 **63.** 1
65. a. False **b.** True **c.** True **d.** True **67. a.** C **b.** E
c. A **d.** D **e.** B **f.** F **69. a.** 0.1; 1.7×10^{-4} **b.** 0.2;
1.3×10^{-3} **71. a.** 0.995; 4.2×10^{-6} **b.** 0.98; 6.7×10^{-5}
73. a. 1.05; 1.3×10^{-3} **b.** 1.1; 5×10^{-3} **75. a.** 1.1; 10^{-2}
b. 1.2; 4×10^{-2}

77. a.

x	$|\sec x - p_2(x)|$	$|\sec x - p_4(x)|$
-0.2	3.39×10^{-4}	5.51×10^{-6}
-0.1	2.09×10^{-5}	8.51×10^{-8}
0.0	0	0
0.1	2.09×10^{-5}	8.51×10^{-8}
0.2	3.39×10^{-4}	5.51×10^{-6}

b. The errors decrease as $|x|$ decreases.

79. a.

x	$|e^{-x} - p_1(x)|$	$|e^{-x} - p_2(x)|$
-0.2	2.14×10^{-2}	1.40×10^{-3}
-0.1	5.17×10^{-3}	1.71×10^{-4}
0.0	0	0
0.1	4.84×10^{-3}	1.63×10^{-4}
0.2	1.87×10^{-2}	1.27×10^{-3}

b. The errors decrease as $|x|$ decreases.

81. Centered at $x = 0$, for all n

83. a. $y = f(a) + f'(a)(x - a)$ **85. a.** $p_5(x) = x - \dfrac{x^3}{6} + \dfrac{x^5}{120}$;

$q_5(x) = -(x - \pi) + \dfrac{1}{6}(x - \pi)^3 - \dfrac{1}{120}(x - \pi)^5$

b.

p_5 is a better approximation on $[-\pi, \pi/2]$; q_5 is a better approximation on $(\pi/2, 2\pi]$.

c.

x	$|\sin x - p_5(x)|$	$|\sin x - q_5(x)|$
$\pi/4$	3.6×10^{-5}	7.4×10^{-2}
$\pi/2$	4.5×10^{-3}	4.5×10^{-3}
$3\pi/4$	7.4×10^{-2}	3.6×10^{-5}
$5\pi/4$	2.3	3.6×10^{-5}
$7\pi/4$	20	7.4×10^{-2}

d. p_5 is a better approximation at $x = \pi/4$; at $x = \pi/2$ the errors are equal.

87. a. $p_1(x) = 6 + \dfrac{1}{12}(x - 36)$; $q_1(x) = 7 + \dfrac{1}{14}(x - 49)$

b.

x	$|\sqrt{x} - p_1(x)|$	$|\sqrt{x} - q_1(x)|$
37	5.7×10^{-4}	6.0×10^{-2}
39	5.0×10^{-3}	4.1×10^{-2}
41	1.4×10^{-2}	2.5×10^{-2}
43	2.6×10^{-2}	1.4×10^{-2}
45	4.2×10^{-2}	6.1×10^{-3}
47	6.1×10^{-2}	1.5×10^{-3}

c. p_1 is a better approximation at $x = 37, 39$, and 41.

Section 11.2 Exercises, pp. 729–730

1. $c_0 + c_1 x + c_2 x^2 + c_3 x^3$ **3.** Ratio and Root Tests   **5.** The radius of convergence does not change. The interval of convergence may change.   **7.** $R = 10$; $[-8, 12]$ **9.** $R = \frac{1}{2}$; $\left(-\frac{1}{2}, \frac{1}{2}\right)$
11. $R = 0$; $\{x : x = 0\}$ **13.** $R = \infty$; $(-\infty, \infty)$ **15.** $R = 3$; $(-3, 3)$
17. $R = \infty$; $(-\infty, \infty)$ **19.** $R = 2$; $(-2, 2)$ **21.** $R = \infty$; $(-\infty, \infty)$
23. $R = 1$; $(0, 2]$ **25.** $R = \frac{1}{4}$; $\left[0, \frac{1}{2}\right]$ **27.** $R = 5$; $(-3, 7)$
29. $R = \infty$; $(-\infty, \infty)$ **31.** $R = \sqrt{3}$; $(-\sqrt{3}, \sqrt{3})$ **33.** $R = 1$; $(0, 2)$
35. $R = \infty$; $(-\infty, \infty)$ **37.** $R = e$ **39.** $R = e^4$

41. $\displaystyle\sum_{k=0}^{\infty} (3x)^k$; $\left(-\frac{1}{3}, \frac{1}{3}\right)$ **43.** $2\displaystyle\sum_{k=0}^{\infty} x^{k+3}$; $(-1, 1)$

45. $4\displaystyle\sum_{k=0}^{\infty} x^{k+12}$; $(-1, 1)$ **47.** $-\displaystyle\sum_{k=1}^{\infty} \dfrac{(3x)^k}{k}$; $\left[-\frac{1}{3}, \frac{1}{3}\right)$

49. $-2\displaystyle\sum_{k=1}^{\infty} \dfrac{x^{k+6}}{k}$; $[-1, 1)$ **51.** $g(x) = 2\displaystyle\sum_{k=1}^{\infty} k(2x)^{k-1}$; $\left(-\frac{1}{2}, \frac{1}{2}\right)$

53. $g(x) = \displaystyle\sum_{k=1}^{\infty} (-1)^k k x^{k-1}$; $(-1, 1)$

55. $g(x) = -\displaystyle\sum_{k=1}^{\infty} \dfrac{3^k x^k}{k}$; $\left[-\frac{1}{3}, \frac{1}{3}\right)$

57. $\displaystyle\sum_{k=1}^{\infty} (-1)^{k+1} 2k x^{2k-1}$; $(-1, 1)$ **59.** $\displaystyle\sum_{k=0}^{\infty} \left(-\dfrac{x}{3}\right)^k$; $(-3, 3)$

61. $\ln 2 - \dfrac{1}{2}\displaystyle\sum_{k=1}^{\infty} \dfrac{x^{2k}}{k \, 4^k}$; $(-2, 2)$ **63. a.** True   **b.** True   **c.** True

d. True   **65.** $|x - a| < R$ **67.** $f(x) = \dfrac{1}{3 - \sqrt{x}}$; $1 < x < 9$

69. $f(x) = \dfrac{e^x}{e^x - 1}$; $0 < x < \infty$ **71.** $f(x) = \dfrac{3}{4 - x^2}$; $-2 < x < 2$

73. $\displaystyle\sum_{k=0}^{\infty} \dfrac{(-3x)^k}{k!}$; $-\infty < x < \infty$

75. $\displaystyle\lim_{k \to \infty} \left| \dfrac{c_{k+1} x^{k+1}}{c_k x^k} \right| = \lim_{k \to \infty} \left| \dfrac{c_{k+1} x^{k+m+1}}{c_k x^{k+m}} \right|$, so by the Ratio Test, the two series have the same radius of convergence.
77. a. $f(x) g(x) = c_0 d_0 + (c_0 d_1 + c_1 d_0)x +$
$(c_0 d_2 + c_1 d_1 + c_2 d_0)x^2$ **b.** $\displaystyle\sum_{k=0}^{n} c_k d_{n-k}$

Section 11.3 Exercises, pp. 740–742

1. The nth Taylor polynomial is the nth partial sum of the corresponding Taylor series.   **3.** $\displaystyle\sum_{k=0}^{\infty} \dfrac{(x - 2)^k}{k!}$ **5.** Replace x with x^2 in the Taylor series for $f(x)$; $|x| < 1$.   **7.** The Taylor series for a function f converges to f on an interval if, for all x in the interval, $\displaystyle\lim_{n \to \infty} R_n(x) = 0$, where $R_n(x)$ is the remainder at x.
9. a. $1 - 2(x - 1) + 3(x - 1)^2 - 4(x - 1)^2$
b. $\displaystyle\sum_{k=0}^{\infty} (-1)^k (k + 1)(x - 1)^k$ **c.** $(0, 2)$ **11. a.** $1 - x + \dfrac{x^2}{2!} - \dfrac{x^3}{3!}$
b. $\displaystyle\sum_{k=0}^{\infty} \dfrac{(-1)^k x^k}{k!}$ **c.** $(-\infty, \infty)$ **13. a.** $2 + 6x + 12x^2 + 20x^3$
b. $\displaystyle\sum_{k=0}^{\infty} (k + 1)(k + 2)x^k$ **c.** $(-1, 1)$ **15. a.** $1 - x^2 + x^4 - x^6$
b. $\displaystyle\sum_{k=0}^{n} (-1)^k x^{2k}$ **c.** $(-1, 1)$ **17. a.** $1 + 2x + \dfrac{(2x)^2}{2!} + \dfrac{(2x)^3}{3!}$
b. $\displaystyle\sum_{k=0}^{\infty} \dfrac{(2x)^k}{k!}$ **c.** $(-\infty, \infty)$ **19. a.** $\dfrac{x}{2} - \dfrac{x^3}{3 \cdot 2^3} + \dfrac{x^5}{5 \cdot 2^5} - \dfrac{x^7}{7 \cdot 2^7}$
b. $\displaystyle\sum_{k=0}^{\infty} \dfrac{(-1)^k x^{2k+1}}{(2k + 1)2^{2k+1}}$ **c.** $[-2, 2]$
21. a. $1 + (\ln 3)x + \dfrac{\ln^2 3}{2}x^2 + \dfrac{\ln^3 3}{6}x^3$ **b.** $\displaystyle\sum_{k=0}^{\infty} \dfrac{\ln^k 3}{k!}x^k$ **c.** $(-\infty, \infty)$
23. a. $1 + \dfrac{(3x)^2}{2} + \dfrac{(3x)^4}{24} + \dfrac{(3x)^6}{720}$ **b.** $\displaystyle\sum_{k=0}^{\infty} \dfrac{(3x)^{2k}}{(2k)!}$ **c.** $(-\infty, \infty)$
25. a. $(x - 3) - \dfrac{1}{2}(x - 3)^2 + \dfrac{1}{3}(x - 3)^3 - \dfrac{1}{4}(x - 3)^4$
b. $\displaystyle\sum_{k=1}^{\infty} \dfrac{(-1)^{k+1}(x - 3)^k}{k}$ **c.** $(2, 4]$
27. a. $1 - \dfrac{(x - \pi/2)^2}{2!} + \dfrac{(x - \pi/2)^4}{4!} - \dfrac{(x - \pi/2)^6}{6!}$
b. $\displaystyle\sum_{k=0}^{\infty} \dfrac{(-1)^k}{(2k)!}(x - \pi/2)^{2k}$
29. a. $1 - (x - 1) + (x - 1)^2 - (x - 1)^3$ **b.** $\displaystyle\sum_{k=0}^{\infty} (-1)^k (x - 1)^k$
31. a. $\ln 3 + \dfrac{(x - 3)}{3} - \dfrac{(x - 3)^2}{3^2 \cdot 2} + \dfrac{(x - 3)^3}{3^3 \cdot 3}$
b. $\ln 3 + \displaystyle\sum_{k=1}^{\infty} \dfrac{(-1)^{k+1}(x - 3)^k}{k 3^k}$

33. a. $2 + 2(\ln 2)(x - 1) + (\ln^2 2)(x - 1)^2 + \dfrac{\ln^3 2}{3}(x - 1)^3$

b. $\displaystyle\sum_{k=0}^{\infty} \dfrac{2(x-1)^k \ln^k 2}{k!}$ **35.** $x^2 - \dfrac{x^4}{2} + \dfrac{x^6}{3} - \dfrac{x^8}{4}$

37. $1 + 2x + 4x^2 + 8x^3$ **39.** $1 + \dfrac{x}{2} + \dfrac{x^2}{6} + \dfrac{x^3}{24}$

41. $1 - x^4 + x^8 - x^{12}$ **43.** $x^2 + \dfrac{x^6}{6} + \dfrac{x^{10}}{120} + \dfrac{x^{14}}{5040}$

45. a. $1 - 2x + 3x^2 - 4x^3$ **b.** 0.826

47. a. $1 + \dfrac{1}{4}x - \dfrac{3}{32}x^2 + \dfrac{7}{128}x^3$ **b.** 1.029

49. a. $1 - \dfrac{2}{3}x + \dfrac{5}{9}x^2 - \dfrac{40}{81}x^3$ **b.** 0.895 **51.** $1 + \dfrac{x^2}{2} - \dfrac{x^4}{8} + \dfrac{x^6}{16}$;

$[-1, 1]$ **53.** $3 - \dfrac{3x}{2} - \dfrac{3x^2}{8} - \dfrac{3x^3}{16}$; $[-1, 1]$

55. $a + \dfrac{x^2}{2a} - \dfrac{x^4}{8a^3} + \dfrac{x^6}{16a^5}$; $|x| \le a$

57. $1 - 8x + 48x^2 - 256x^3$ **59.** $\dfrac{1}{16} - \dfrac{x^2}{32} + \dfrac{3x^4}{256} - \dfrac{x^6}{256}$

61. $\dfrac{1}{9} - \dfrac{2}{9}\left(\dfrac{4x}{3}\right) + \dfrac{3}{9}\left(\dfrac{4x}{3}\right)^2 - \dfrac{4}{9}\left(\dfrac{4x}{3}\right)^3$

63. $R_n(x) = \dfrac{f^{(n+1)}(c)}{(n+1)!} x^{n+1}$, where c is between 0 and x and

$f^{(n+1)}(c) = \pm \sin c$ or $\pm \cos c$. Therefore, $|R_n(x)| \le \dfrac{|x|^{n+1}}{(n+1)!} \to 0$

as $n \to \infty$, for $-\infty < x < \infty$. **65.** $R_n(x) = \dfrac{f^{(n+1)}(c)}{(n+1)!} x^{n+1}$,

where c is between 0 and x and $f^{(n+1)}(c) = (-1)^n e^{-c}$. Therefore,

$|R_n(x)| \le \dfrac{|x|^{n+1}}{e^c(n+1)!} \to 0$ as $n \to \infty$, for $-\infty < x < \infty$.

67. a. False **b.** True **c.** False **d.** False **e.** True

69. a. $1 + \dfrac{x^2}{2!} + \dfrac{x^4}{4!} + \dfrac{x^6}{6!}$ **b.** $R = \infty$

71. a. $1 - \dfrac{2}{3}x^2 + \dfrac{5}{9}x^4 - \dfrac{40}{81}x^6$ **b.** $R = 1$

73. a. $1 - \dfrac{1}{2}x^2 - \dfrac{1}{8}x^4 - \dfrac{1}{16}x^6$ **b.** $R = 1$

75. a. $1 - 2x^2 + 3x^4 - 4x^6$ **b.** $R = 1$ **77.** Approx. 3.9149

79. Approx. 1.8989 **85.** $\displaystyle\sum_{k=0}^{\infty} \left(\dfrac{x-4}{2}\right)^k$ **87.** Use three terms of the

Taylor series for $\cos x$ centered at $a = \pi/4$; 0.766 **89. a.** Use

three terms of the Taylor series for $\sqrt[3]{125 + x}$ centered at $a = 0$;

5.03968 **b.** Use three terms of the Taylor series for $\sqrt[3]{x}$ centered at

$a = 125$; 5.03968 **c.** Yes

Section 11.4 Exercises, pp. 748–750

1. Replace f and g with their Taylor series centered at a and evaluate
the limit. **3.** Substitute $x = -0.6$ into the Taylor series for e^x
centered at 0. Because the resulting series is an alternating series, the
error can be estimated easily. **7.** 1 **9.** $\dfrac{1}{2}$ **11.** 2 **13.** $\dfrac{2}{3}$ **15.** $\dfrac{1}{3}$

17. $\dfrac{3}{5}$ **19.** $-\dfrac{8}{5}$ **21.** 1 **23.** $\dfrac{3}{4}$ **25. a.** $1 + x + \dfrac{x^2}{2!} + \cdots + \dfrac{x^n}{n!} + \cdots$

b. e^x **c.** $-\infty < x < \infty$

27. a. $1 - x + x^2 - \cdots (-1)^{n-1}x^{n-1} + \cdots$ **b.** $\dfrac{1}{1+x}$ **c.** $|x| < 1$

29. a. $-2 + 4x - 8 \cdot \dfrac{x^2}{2!} + \cdots + (-2)^n \dfrac{x^{n-1}}{(n-1)!} + \cdots$

b. $-2e^{-2x}$ **c.** $-\infty < x < \infty$ **31. a.** $1 - x^2 + x^4 - \cdots$

b. $\dfrac{1}{1+x^2}$ **c.** $-1 < x < 1$

33. a. $2 + 2t + \dfrac{2t^2}{2!} + \cdots + \dfrac{2t^n}{n!} + \cdots$ **b.** $y(t) = 2e^t$

35. a. $2 + 16t + 24t^2 + 24t^3 + \cdots + \dfrac{3^{n-1} \cdot 16}{n!} t^n + \cdots$

b. $y(t) = \dfrac{16}{3}e^{3t} - \dfrac{10}{3}$ **37.** 0.2448 **39.** 0.6958

41. 0.0600 **43.** 0.4994 **45.** $1 + 2 + \dfrac{2^2}{2!} + \dfrac{2^3}{3!}$

47. $1 - 2 + \dfrac{2}{3} - \dfrac{4}{45}$ **49.** $\dfrac{1}{2} - \dfrac{1}{8} + \dfrac{1}{24} - \dfrac{1}{64}$ **51.** $e - 1$

53. $\displaystyle\sum_{k=1}^{\infty} \dfrac{(-1)^{k+1}x^k}{k}$, for $-1 < x \le 1$; $\ln 2$ **55.** $\dfrac{2}{2 - x}$ **57.** $\dfrac{4}{4 + x^2}$

59. $-\ln(1 - x)$ **61.** $-\dfrac{3x^2}{(3+x)^2}$ **63.** $\dfrac{6x^2}{(3-x)^3}$

65. a. False **b.** False **c.** True **67.** $\dfrac{a}{b}$ **69.** $e^{-1/6}$

71. $f^{(3)}(0) = 0$; $f^{(4)}(0) = 4e$ **73.** $f^{(3)}(0) = 2$; $f^{(4)}(0) = 0$

75. 2 **77.** 1.575 using four terms **79. a.** $S'(x) = \sin x^2$;

$C'(x) = \cos x^2$ **b.** $\dfrac{x^3}{3} - \dfrac{x^7}{7 \cdot 3!} + \dfrac{x^{11}}{11 \cdot 5!} - \dfrac{x^{15}}{15 \cdot 7!}$;

$x - \dfrac{x^5}{5 \cdot 2!} + \dfrac{x^9}{9 \cdot 4!} - \dfrac{x^{13}}{13 \cdot 6!}$ **c.** $S(0.05) \approx 0.00004166664807$;

$C(-0.25) \approx -0.2499023614$ **d.** 1 **e.** 2

81. a. $1 - \dfrac{x^2}{4} + \dfrac{x^4}{64} - \dfrac{x^6}{2304}$ **b.** $R = \infty$, $-\infty < x < \infty$

Chapter 11 Review Exercises, pp. 750–752

1. a. True **b.** False **c.** True **d.** True **e.** True

3. $p_2(x) = 1 - \dfrac{3}{2}x^2$ **5.** $p_2(x) = 1 - (x - 1) + \dfrac{3}{2}(x - 1)^2$

7. $p_2(x) = 1 - \dfrac{1}{2}(x - 1)^2$

9. $p_3(x) = \dfrac{5}{4} + \dfrac{3}{4}(x - \ln 2) + \dfrac{5}{8}(x - \ln 2)^2 + \dfrac{1}{8}(x - \ln 2)^3$

11. a. $p_1(x) = 1 + x$; $p_2(x) = 1 + x + \dfrac{x^2}{2}$

b.

n	$p_n(x)$	Error
1	0.92	3.1×10^{-3}
2	0.9232	8.4×10^{-5}

13. a. $p_1(x) = \dfrac{\sqrt{2}}{2} + \dfrac{\sqrt{2}}{2}\left(x - \dfrac{\pi}{4}\right)$;

$p_2(x) = \dfrac{\sqrt{2}}{2} + \dfrac{\sqrt{2}}{2}\left(x - \dfrac{\pi}{4}\right) - \dfrac{\sqrt{2}}{4}\left(x - \dfrac{\pi}{4}\right)^2$

b.

n	$p_n(x)$	Error
1	0.5960	8.2×10^{-3}
2	0.5873	4.7×10^{-4}

15. $|R_3| < \dfrac{\pi^4}{4!}$ **17.** $R = \infty$, $(-\infty, \infty)$ **19.** $R = \infty$, $(-\infty, \infty)$

21. $R = 9$, $(-9, 9)$ **23.** $R = 2$, $[-4, 0)$ **25.** $R = \dfrac{3}{2}$, $[-2, 1]$

27. $R = \dfrac{1}{27}$ **29.** $\displaystyle\sum_{k=0}^{\infty} x^{2k}$; $(-1, 1)$ **31.** $\displaystyle\sum_{k=0}^{\infty}(-5x)^k$; $\left(-\dfrac{1}{5}, \dfrac{1}{5}\right)$

33. $\sum_{k=1}^{\infty} k(10x)^{k-1}$; $\left(-\frac{1}{10}, \frac{1}{10}\right)$ **35.** $1 + 3x + \frac{9x^2}{2!}$; $\sum_{k=0}^{\infty} \frac{(3x)^k}{k!}$

37. $-(x - \pi/2) + \frac{(x - \pi/2)^3}{3!} - \frac{(x - \pi/2)^5}{5!}$;

$\sum_{k=0}^{\infty} (-1)^{k+1} \frac{(x - \pi/2)^{2k+1}}{(2k + 1)!}$

39. $4x - \frac{(4x)^3}{3} + \frac{(4x)^5}{5}$; $\sum_{k=0}^{\infty} \frac{(-1)^k(4x)^{2k+1}}{2k + 1}$

41. $1 + 2(x - 1)^2 + \frac{2}{3}(x - 1)^4$; $\sum_{k=0}^{\infty} \frac{4^k(x - 1)^{2k}}{(2k)!}$

43. $1 + \frac{x}{3} - \frac{x^2}{9} + \cdots$ **45.** $1 - \frac{3}{2}x + \frac{3}{2}x^2 - \cdots$

47. $R_n(x) = \frac{(\sinh c + \cosh c) x^{n+1}}{(n + 1)!}$, where c is between 0 and x;

$\lim_{n \to \infty} |R_n(x)| = |\sinh c + \cosh c| \lim_{n \to \infty} \frac{|x|^{n+1}}{(n + 1)!} = 0$ because

$|x|^{n+1} \ll (n + 1)!$ for any fixed value of x.

49. $\frac{1}{24}$ **51.** $\frac{1}{8}$ **53.** $\frac{1}{6}$ **55.** Approx. 0.4615   **57.** Approx. 0.3819

59. $11 - \frac{1}{11} - \frac{1}{2 \cdot 11^3} - \frac{1}{2 \cdot 11^5}$ **61.** $-\frac{1}{3} + \frac{1}{3 \cdot 3^3} - \frac{1}{5 \cdot 3^5} + \frac{1}{7 \cdot 3^7}$

63. $y = 4 + 4x + \frac{4^2}{2!}x^2 + \frac{4^3}{3!}x^3 + \cdots + \frac{4^n}{n!}x^n + \cdots = 3 + e^{4x}$

65. a. $\sum_{k=1}^{\infty} \frac{(-1)^{k+1}}{k}$ **b.** $\sum_{k=1}^{\infty} \frac{1}{k2^k}$ **c.** $2\sum_{k=0}^{\infty} \frac{x^{2k+1}}{2k + 1}$

d. $x = \frac{1}{3}$; $2\sum_{k=0}^{\infty} \frac{1}{3^{2k+1}(2k + 1)}$ **e.** Series in part (d)

CHAPTER 12

Section 12.1 Exercises, pp. 763–767

1. Plotting $\{(f(t), g(t)): a \le t \le b\}$ generates a curve in the xy-plane.
3. $x = R \cos(\pi t/5)$, $y = -R \sin(\pi t/5)$ **5.** $x = t^2, y = t$,

$-\infty < t < \infty$ **7.** $-\frac{1}{2}$ **9.** $x = t, y = t, 0 \le t \le 6$; $x = 2t, y = 2t$,

$0 \le t \le 3$; $x = 3t, y = 3t, 0 \le t \le 2$ (answers will vary)

11. a.

t	-10	-4	0	4	10
x	-20	-8	0	8	20
y	-34	-16	-4	8	26

b.

c. $y = \frac{3}{2}x - 4$
d. A line segment rising to the right as t increases

13. a.

t	-5	-2	0	2	5
x	11	8	6	4	1
y	-18	-9	-3	3	12

b.

c. $y = -3x + 15$
d. A line segment rising to the left as t increases

15. a. $y = -x + 4$ **b.** A line segment starting at $(3, 1)$ and ending at $(4, 0)$ **17. a.** $y = 3x - 12$ **b.** A line segment starting at $(4, 0)$ and ending at $(8, 12)$ **19. a.** $x^2 + y^2 = 9$ **b.** Lower half of a circle of radius 3 centered at $(0, 0)$; starts at $(-3, 0)$ and ends at $(3, 0)$
21. a. $y = 1 - x^2, -1 \le x \le 1$ **b.** A parabola opening downward with a vertex at $(0, 1)$ starting at $(1, 0)$ and ending at $(-1, 0)$
23. a. $x^2 + (y - 1)^2 = 1$ **b.** A circle of radius 1 centered at $(0, 1)$; generated counterclockwise, starting and ending at $(1, 1)$
25. a. $y = (x + 1)^3$ **b.** A cubic curve rising to the right as r increases   **27. a.** $x^2 + y^2 = 49$ **b.** A circle of radius 7 centered at $(0, 0)$; generated counterclockwise, starting and ending at $(-7, 0)$
29. a. $y = 1, -\infty < x < \infty$ **b.** A horizontal line with y-intercept 1, generated from left to right   **31.** $x^2 + y^2 = 4$ **33.** $y = \sqrt{4 - x^2}$
35. $y = x^2$ **37.** $x = 4 \cos t, y = 4 \sin t, 0 \le t \le 2\pi$
39. $x = \cos t + 2, y = \sin t + 3, 0 \le t \le 2\pi$
41. $x = 2t, y = 8t; 0 \le t \le 1$
43. $x = t, y = 2t^2 - 4; -1 \le t \le 5$ **45.** $x = 2, y = t; 3 \le t \le 9$
47. $x = 4t - 2, y = -6t + 3; 0 \le t \le 1$ and
$x = t + 1, y = 8t - 11; 1 \le t \le 2$
49. $x = 1 + 2t, y = 1 + 4t; -\infty < t < \infty$
51. $x = t^2, y = t; t \ge 0$

53. $x = 400 \cos\left(\frac{4\pi t}{3}\right), y = 400 \sin\left(\frac{4\pi t}{3}\right); 0 \le t \le 1.5$

55. $x = 50 \cos\left(\frac{\pi t}{12}\right), y(t) = 50 \sin\left(\frac{\pi t}{12}\right); 0 \le t \le 24$

57.

59.

61.

63. Plot $x = 1 + \cos^2 t - \sin^2 t$, $y = t$.
65. Approx. 2857 m

67. a. $\dfrac{dy}{dx} = -2; -2$

b.

(10, −12)

69. a. $\dfrac{dy}{dx} = -8 \cot t; 0$

b.

(0, 8) $y = 8$

71. a. $\dfrac{dy}{dx} = \dfrac{t^2 + 1}{t^2 - 1}, t \neq 0;$ undefined

b.

$x = 2$ (2, 0)

73. $y = \dfrac{13}{4}x + \dfrac{1}{4}$ **75.** $y = x - \dfrac{\pi\sqrt{2}}{4}$ **77.** $\left(-\dfrac{4}{\sqrt{5}}, \dfrac{8}{\sqrt{5}}\right)$ and $\left(\dfrac{4}{\sqrt{5}}, -\dfrac{8}{\sqrt{5}}\right)$ **79.** There is no such point.   **81.** 10   **83.** $\pi\sqrt{2}$

85. $\dfrac{1}{3}(5\sqrt{5} - 8)$ **87.** $\dfrac{3}{2}$ **89. a.** False   **b.** True   **c.** False   **d.** True   **e.** True

91. $0 \leq t \leq 2\pi$

93. $x = 3\cos t, y = \dfrac{3}{2}\sin t,$

$0 \leq t \leq 2\pi; \left(\dfrac{x}{3}\right)^2 + \left(\dfrac{2y}{3}\right)^2 = 1;$

in the counterclockwise direction

95. a. Lines intersect at $(1, 0)$.   **b.** Lines are parallel.
c. Lines intersect at $(4, 6)$.

97.

99.

101.

a. $(0, 2)$ and $(0, -2)$
b. $(1, \sqrt{2}), (1, -\sqrt{2}),$
$(-1, \sqrt{2}), (-1, -\sqrt{2})$

103. 27π **105.** $\dfrac{3\pi}{8}$ **107. a.** A circle of radius 3 centered at $(0, 4)$

b. A torus (doughnut); $48\pi^2$ **109.** $\dfrac{64\pi}{3}$

111. $\displaystyle\int_0^1 2\pi(e^{3t} + 1)\sqrt{4e^{4t} + 9e^{6t}}\, dt \approx 1445.9$

113. a.

b.

Section 12.2 Exercises, pp. 775–779

1.

$(-3, -\frac{\pi}{2})$ $(2, \frac{\pi}{6})$ $\frac{\pi}{6}$

$(-2, -5\pi/6), (2, 13\pi/6);$
$(3, \pi/2), (3, 5\pi/2)$

3. $r^2 = x^2 + y^2, \tan\theta = \dfrac{y}{x}$
5. $r\cos\theta = 5$ or $r = 5\sec\theta$
7. x-axis symmetry occurs if (r, θ) on the graph implies $(r, -\theta)$ is on the graph. y-axis symmetry occurs if (r, θ) on the graph implies $(r, \pi - \theta) = (-r, -\theta)$ is on the graph. Symmetry about the origin occurs if (r, θ) on the graph implies $(-r, \theta) = (r, \theta + \pi)$ is on the graph.

9.

$(2, \frac{\pi}{4})$ $\frac{\pi}{4}$

$(-2, -3\pi/4), (2, 9\pi/4)$

11.

$(-1, -\frac{\pi}{3})$ $-\frac{\pi}{3}$

$(1, 2\pi/3), (1, 8\pi/3)$

13.

$(-4, \frac{3\pi}{2})$ $\frac{3\pi}{2}$

$(4, \pi/2), (4, 5\pi/2)$

15.

17.

19.

69. $[0, 8\pi]$

71. $[0, 2\pi]$

21.

23. $\left(100, -\dfrac{\pi}{4}\right)$

25. $(3\sqrt{2}/2, 3\sqrt{2}/2)$

27. $(1/2, -\sqrt{3}/2)$

29. $(2\sqrt{2}, -2\sqrt{2})$

31. $(2\sqrt{2}, \pi/4), (-2\sqrt{2}, 5\pi/4)$

33. $(2, \pi/3), (-2, 4\pi/3)$

35. $(8, 2\pi/3), (-8, -\pi/3)$

73. $[0, 5\pi]$

75. $[0, 2\pi]$

37. $x = -4$; vertical line passing through $(-4, 0)$

39. $x^2 + y^2 = 4$; circle of radius 2 centered at $(0, 0)$

41. $(x - 1)^2 + (y - 1)^2 = 2$; circle of radius $\sqrt{2}$ centered at $(1, 1)$

43. $(x - 3)^2 + (y - 4)^2 = 25$; circle of radius 5 centered at $(3, 4)$

45. $x^2 + (y - 1)^2 = 1$; circle of radius 1 centered at $(0, 1)$ and $x = 0$ **47.** $x^2 + (y - 4)^2 = 16$; circle of radius 4 centered at $(0, 4)$

49. $r = \tan\theta \sec\theta$ **51.** $r^2 = \sec\theta \csc\theta$ or $r^2 = 2\csc 2\theta$

77. a. True   **b.** True   **c.** False   **d.** True   **e.** True

81. A circle of radius 4 and center $(2, \pi/3)$ (polar coordinates)

83. A circle of radius 4 centered at $(2, 3)$ (Cartesian coordinates)

53.

55.

85. a. 132.3 miles   **b.** 264.6 mi/hr

57.

59.

87.

$y = -\dfrac{x}{\sqrt{3}} + 2\sqrt{3}$

89. $y = 4x + 3$

91. a. A   **b.** C   **c.** B   **d.** D   **e.** E   **f.** F

61.

63.

93.

95.

65.

$r = 1 - 2\sin 3\theta$

Origin:
B, D, F, H, J, L

67.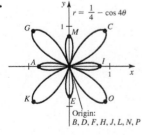

$r = \dfrac{1}{4} - \cos 4\theta$

Origin:
B, D, F, H, J, L, N, P

97.

99.

103.

For $a = -1$, the spiral winds inward toward the origin.

105. a.

107. a.

109. Symmetry about the x-axis   **111.** $(x^2 + y^2)^2 = a^2(x^2 - y^2)$

Section 12.3 Exercises, pp. 786–788

1. $x = f(\theta)\cos\theta$, $y = f(\theta)\sin\theta$ **3.** The slope of the tangent line is the rate of change of the vertical coordinate with respect to the horizontal coordinate.   **5.** $\sqrt{3}$ **7.** $\dfrac{\pi^2}{4}$ **9.** Both curves pass through the origin, but for different values of θ.   **11.** 0   **13.** $-\sqrt{3}$
15. Undefined, undefined   **17.** 0 at $(-4, \pi/2)$ and $(-4, 3\pi/2)$, undefined at $(4, 0)$ and $(4, \pi)$ **19.** ± 1 **21.** $\theta = \dfrac{3\pi}{4}$; $m = -1$
23. a. $[0, \pi]$

     b. $\theta = \dfrac{\pi}{4}$, $m = 1$; $\theta = \dfrac{7\pi}{12}$, $m \approx -3.73$;
     $\theta = \dfrac{11\pi}{12}$, $m \approx -0.27$

25. Horizontal: $(2\sqrt{2}, \pi/4)$, $(-2\sqrt{2}, 3\pi/4)$; vertical: $(0, \pi/2)$, $(4, 0)$
27. Horizontal: $(0, 0)$ $(0.943, 0.955)$, $(-0.943, 2.186)$, $(0.943, 4.097)$, $(-0.943, 5.328)$; vertical: $(0, 0)$, $(0.943, 0.615)$, $(-0.943, 2.526)$, $(0.943, 3.757)$, $(-0.943, 5.668)$ **29.** $(2, 0)$ and $(0, 0)$
31. $\left(1, \dfrac{\pi}{12}\right)$, $\left(1, \dfrac{5\pi}{12}\right)$, $\left(1, \dfrac{7\pi}{12}\right)$, $\left(1, \dfrac{11\pi}{12}\right)$,

$\left(1, \dfrac{13\pi}{12}\right)$, $\left(1, \dfrac{17\pi}{12}\right)$, $\left(1, \dfrac{19\pi}{12}\right)$, and $\left(1, \dfrac{23\pi}{12}\right)$

33. 1

35. 16π

37. $9\pi/2$

39. $\dfrac{\pi}{12}$

41. a. $(0, 0)$, $\left(\dfrac{3}{\sqrt{2}}, \dfrac{\pi}{4}\right)$ **b.** $\dfrac{9}{8}(\pi - 2)$

43. a. $\left(1 + \dfrac{1}{\sqrt{2}}, \dfrac{\pi}{4}\right)$, $\left(1 - \dfrac{1}{\sqrt{2}}, \dfrac{5\pi}{4}\right)$, $(0, 0)$ **b.** $\dfrac{3\pi}{2} - 2\sqrt{2}$

45. $\dfrac{1}{24}(3\sqrt{3} + 2\pi)$ **47.** $\dfrac{1}{4}(2 - \sqrt{3}) + \dfrac{\pi}{12}$ **49.** $\pi/20$
51. $4(4\pi/3 - \sqrt{3})$ **53.** $2\pi/3 - \sqrt{3}/2$ **55.** $9\pi + 27\sqrt{3}$
57. 6   **59.** 18π **61.** Intersection points: $\left(3, \pm\dfrac{\pi}{3}\right)$; area of region A $= 6\sqrt{3} - 2\pi$; area of region B $= 5\pi - 6\sqrt{3}$; area of region C $= 4\pi + 6\sqrt{3}$ **63.** πa **65.** $\dfrac{8}{3}\big((1 + \pi^2)^{3/2} - 1\big)$
67. 32   **69.** $63\sqrt{5}$ **71.** $\dfrac{2\pi - 3\sqrt{3}}{8}$ **73.** 26.73
75. a. False   **b.** False   **c.** True
77. Horizontal: $(0, 0)$, $(4.05, 2.03)$, $(9.83, 4.91)$; vertical: $(1.72, 0.86)$, $(6.85, 3.43)$, $(12.87, 6.44)$ **79.** $\dfrac{\sqrt{1 + a^2}}{a}$
81. a. $A_n = \dfrac{1}{4e^{(4n+2)\pi}} - \dfrac{1}{4e^{4n\pi}} - \dfrac{1}{4e^{(4n-2)\pi}} + \dfrac{1}{4e^{(4n-4)\pi}}$ **b.** 0
c. $e^{-4\pi}$ **85.** $(a^2 - 2)\theta^* + \pi - \sin 2\theta^*$, where $\theta^* = \cos^{-1}(a/2)$
87. $a^2(\pi/2 + a/3)$

Section 12.4 Exercises, pp. 797–800

1. A parabola is the set of all points in a plane equidistant from a fixed point and a fixed line.   **3.** A hyperbola is the set of all points in a plane whose distances from two fixed points have a constant difference.

5. Parabola:

Hyperbola:

Ellipse:

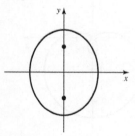

7. $\left(\dfrac{x}{a}\right)^2 + \dfrac{y^2}{a^2 - c^2} = 1$ **9.** $(\pm ae, 0)$ **11.** $y = \pm\dfrac{b}{a}x$

13. Parabola

15. Ellipse

Vertices: $(\pm 2, 0)$; foci: $(\pm\sqrt{3}, 0)$; major axis has length 4; minor axis has length 2.

17. Parabola

19. Hyperbola

Vertices: $(\pm 2, 0)$; foci: $(\pm\sqrt{5}, 0)$; asymptotes: $y = \pm\dfrac{1}{2}x$

21. Hyperbola

23. Parabola

25. Ellipse

Vertices: $(0, \pm 4)$; foci: $(0, \pm 2\sqrt{3})$; major axis has length 8; minor axis has length 4.

27. Ellipse

Vertices: $(0, \pm\sqrt{7})$; foci: $(0, \pm\sqrt{2})$; major axis has length $2\sqrt{7}$; minor axis has length $2\sqrt{5}$.

29. Hyperbola

31. $y^2 = 16x$ **33.** $y^2 = 12x$

35. $x^2 = -\dfrac{2}{3}y$

37. $y^2 = 4(x + 1)$

39. $\dfrac{x^2}{16} + \dfrac{y^2}{9} = 1$

41. $\dfrac{x^2}{16} - \dfrac{y^2}{20} = 1$

43. $\dfrac{x^2}{25} + y^2 = 1$ **45.** $\dfrac{x^2}{4} - \dfrac{y^2}{9} = 1$ **47.** $\dfrac{x^2}{4} + \dfrac{y^2}{9} = 1$

49. $\dfrac{x^2}{16} - \dfrac{y^2}{9} = 1$ **51. a.** True **b.** True **c.** True **d.** True

53. $\dfrac{x^2}{81} + \dfrac{y^2}{72} = 1$

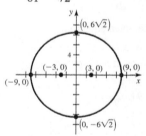

Directrices: $x = \pm 27$

55. $x^2 - \dfrac{y^2}{8} = 1$

57.

Vertex: $(2, 0)$; focus: $(0, 0)$; directrix: $x = 4$

59.

Vertices: $(1, 0)$, $\left(-\dfrac{1}{3}, 0\right)$; center: $\left(\dfrac{1}{3}, 0\right)$; foci: $(0, 0)$, $\left(\dfrac{2}{3}, 0\right)$; directrices: $x = -1$, $x = \dfrac{5}{3}$

61.

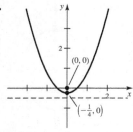

Vertex: $\left(0, -\frac{1}{4}\right)$; focus: $(0, 0)$;
directrix: $y = -\frac{1}{2}$

63.

The parabola starts at $(1, 0)$ and goes through quadrants I, II, and III for θ in $[0, 3\pi/2]$; then it approaches $(1, 0)$ by traveling through quadrant IV on $(3\pi/2, 2\pi)$.

65.

The parabola begins in the first quadrant and passes through the points $(0, 3)$, $\left(-\frac{3}{2}, 0\right)$, and $(0, -3)$ as θ ranges from 0 to 2π.

67. The parabolas open to the left due to the presence of a positive $\cos\theta$ term in the denominator. As d increases, the directrix $x = d$ moves to the right, resulting in wider parabolas.

69. $y = 2x + 6$ **71.** $y = -\frac{3}{40}x - \frac{4}{5}$ **73.** $\frac{dy}{dx} = -\frac{b^2 x}{a^2 y}$, so

$\frac{y - y_0}{x - x_0} = -\frac{b^2 x_0}{a^2 y_0}$, which is equivalent to the given equation.

75. $r = \frac{4}{1 - 2\sin\theta}$ **79.** $\frac{4\pi b^2 a}{3}$; $\frac{4\pi a^2 b}{3}$; yes, if $a \neq b$

81. a. $\frac{\pi b^2}{3a^2}(a - c)^2(2a + c)$ **b.** $\frac{4\pi b^4}{3a}$ **91.** $2p$

97. a. $u(m) = \frac{2m^2 - \sqrt{3m^2 + 1}}{m^2 - 1}$; $v(m) = \frac{2m^2 + \sqrt{3m^2 + 1}}{m^2 - 1}$;

2 intersection points for $|m| > 1$ **b.** $\frac{5}{4}$, ∞ **c.** 2, 2
d. $2\sqrt{3} - \ln(\sqrt{3} + 2)$

Chapter 12 Review Exercises, pp. 800–803

1. a. False   **b.** False   **c.** True   **d.** False   **e.** True   **f.** True

3. $\frac{x^2}{16} + \frac{y^2}{9} = 1$; ellipse generated counterclockwise

5. Segment of the parabola $y = \sqrt{x}$ starting at $(4, 2)$ and ending at $(9, 3)$

7. a. $(x - 1)^2 + (y - 2)^2 = 64$ **9.** $x = 5(t - 1)(t - 2)\sin t$,
b. $-\frac{1}{\sqrt{3}}$ $y = t$

c.

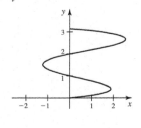

11. a. $x^2 + (y + 1)^2 = 9$ **b.** Lower half of a circle of radius 3 centered at $(0, -1)$, starting at $(3, -1)$ and ending at $(-3, -1)$ **c.** 0

13. At $t = \pi/6$: $y = (2 + \sqrt{3})x + \left(2 - \frac{\pi}{3} - \frac{\pi\sqrt{3}}{6}\right)$; at

$t = \frac{2\pi}{3}$: $y = \frac{x}{\sqrt{3}} + 2 - \frac{2\pi}{3\sqrt{3}}$ **15.** $x = -1 + 2t$, $y = t$,

for $0 \leq t \leq 1$; $x = 1 - 2t$, $y = 1 - t$, for $0 \leq t \leq 1$
17. $x = 3\sin t$, $y = 3\cos t$, for $0 \leq t \leq 2\pi$

19. $\frac{4}{15}$ **21.** 9.1   **23.** $4 - 2\sqrt{2}$

25.

27.

29.

31.

33. Liz should choose $r = 1 - \sin\theta$.

35. $(x - 3)^2 + (y + 1)^2 = 10$; a circle of radius $\sqrt{10}$ centered at $(3, -1)$
37. $r = 1 + \cos\theta$; a cardioid
39. $r = 8\cos\theta$, $0 \leq \theta \leq \pi$

41. a. Horizontal: $\left(\frac{1}{2}, \frac{\pi}{6}\right)$, $\left(\frac{1}{2}, \frac{5\pi}{6}\right)$, $\left(2, \frac{3\pi}{2}\right)$;

vertical: $\left(\frac{3}{2}, \frac{7\pi}{6}\right)$, $\left(\frac{3}{2}, \frac{11\pi}{6}\right)$, $\left(0, \frac{\pi}{2}\right)$ **b.** Undefined

c.

43. $\left(\frac{\pi}{12}, \frac{1}{2^{1/4}}\right)$, $\left(\frac{3\pi}{4}, \frac{1}{2^{1/4}}\right)$, $\left(\frac{17\pi}{12}, \frac{1}{2^{1/4}}\right)$, $(0, 0)$

45. $\pi - \frac{3\sqrt{3}}{2}$ **47.** $2\sqrt{3} - \frac{2\pi}{3}$ **49.** 4   **51.** 40.09

53. a. Hyperbola
b. Foci $(\pm\sqrt{3},0)$, vertices
$(\pm 1,0)$, directrices $x=\pm\dfrac{1}{\sqrt{3}}$
c. $e=\sqrt{3}$

d.

$y=-\sqrt{2}x$ $y=\sqrt{2}x$

55. a. Hyperbola
b. Foci $(0,\pm 2\sqrt{5})$,
vertices $(0,\pm 4)$, directrices
$y=\pm\dfrac{8}{\sqrt{5}}$ **c.** $e=\dfrac{\sqrt{5}}{2}$

d.

$y=-2x$ $y=2x$

57. a. Ellipse
b. Foci $(\pm\sqrt{2},0)$,
vertices $(\pm 2,0)$, directrices
$x=\pm 2\sqrt{2}$ **c.** $e=\dfrac{\sqrt{2}}{2}$

d.

59. $y=\frac{3}{2}x-2$

61. $e=1$

$y=2$ (0,1) (0,0)

63. $e=\frac{1}{2}$

$x=-\frac{20}{3}$ $x=4$ $\left(-\frac{8}{3},0\right)$ (0,0) $(-4,0)$ $\left(\frac{4}{3},0\right)$

65. $\dfrac{y^2}{4}-\dfrac{x^2}{12}=1$

(0,2) $y=\frac{x}{2}$ $y=1$ $y=-1$ (0,-2) $y=-\frac{x}{2}$

67. $\dfrac{y^2}{16}+\dfrac{25x^2}{336}=1$; foci: $\left(0,\pm\dfrac{8}{5}\right)$

$y=10$ $\left(-\frac{4\sqrt{21}}{5},0\right)$ (0,4) $\left(\frac{4\sqrt{21}}{5},0\right)$ (0,-4) $y=-10$

69. $e=2/3, y=\pm 9, (\pm 2\sqrt{5},0)$ **71.** $m=\dfrac{b}{a}$

75. a. $x=\pm a\cos^{2/n}t,\ y=\pm b\sin^{2/n}t$
c. The curve becomes more rectangular as n increases.

CHAPTER 13

Section 13.1 Exercises, pp. 813–816

3. There are infinitely many vectors with the same direction and length as **v**.   **5.** $\mathbf{u}+\mathbf{v}=\langle u_1+v_1, u_2+v_2\rangle$ **7.** No
9. $|\langle v_1, v_2\rangle|=\sqrt{v_1^2+v_2^2}$ **11.** If P has coordinates (x_1,y_1) and Q has coordinates (x_2,y_2), then the magnitude of $\overrightarrow{PQ}$ is given by $\sqrt{(x_2-x_1)^2+(y_2-y_1)^2}$.   **13.** a, c, e   **15. a.** 3**v** **b.** 2**u**
c. $-3\mathbf{u}$ **d.** $-2\mathbf{u}$ **e.** **v**   **17. a.** $3\mathbf{u}+3\mathbf{v}$ **b.** $\mathbf{u}+2\mathbf{v}$ **c.** $2\mathbf{u}+5\mathbf{v}$
d. $-2\mathbf{u}+3\mathbf{v}$ **e.** $3\mathbf{u}+2\mathbf{v}$ **f.** $-3\mathbf{u}-2\mathbf{v}$ **g.** $-2\mathbf{u}-4\mathbf{v}$
h. $\mathbf{u}-4\mathbf{v}$ **i.** $-\mathbf{u}-6\mathbf{v}$

19. a.

$\overrightarrow{OP}=\langle 3,2\rangle=3\mathbf{i}+2\mathbf{j}$
$|\overrightarrow{OP}|=\sqrt{13}$

b.

$\overrightarrow{QP}=\langle -1,0\rangle=-\mathbf{i}$
$|\overrightarrow{QP}|=1$

c.

$\overrightarrow{RQ}=\langle 10,3\rangle=10\mathbf{i}+3\mathbf{j}$
$|\overrightarrow{RQ}|=\sqrt{109}$

21. $\overrightarrow{QU}=\langle 7,2\rangle, \overrightarrow{PT}=\langle 7,3\rangle, \overrightarrow{RS}=\langle 2,3\rangle$

23. $\overrightarrow{QT}$ **25.** $\langle -4, 10 \rangle$ **27.** $\langle 52, -30 \rangle$ **29.** $2\sqrt{2}$

31. $\mathbf{w} - \mathbf{u}$ **33.** $13\left\langle -\dfrac{5}{13}, \dfrac{12}{13} \right\rangle$ **35.** $\langle 3, 3\sqrt{3} \rangle$

37. $\left\langle \dfrac{15}{13}, -\dfrac{36}{13} \right\rangle$ **39.** $\left\langle \dfrac{30}{\sqrt{13}}, -\dfrac{20}{\sqrt{13}} \right\rangle$ **41.** $-\mathbf{i} + 10\mathbf{j}$

43. $\pm\dfrac{1}{\sqrt{61}} \langle 6, 5 \rangle$ **45.** $\left\langle -\dfrac{28}{\sqrt{74}}, \dfrac{20}{\sqrt{74}} \right\rangle, \left\langle \dfrac{28}{\sqrt{74}}, -\dfrac{20}{\sqrt{74}} \right\rangle$

47. a. $\left\langle \dfrac{3}{5}, -\dfrac{4}{5} \right\rangle, \left\langle -\dfrac{3}{5}, \dfrac{4}{5} \right\rangle$ **b.** $b = \pm\dfrac{2\sqrt{2}}{3}$ **c.** $a = \pm\dfrac{3}{\sqrt{10}}$

49. $\langle -4\sqrt{3}, 4 \rangle$ **51.** $\langle 15\sqrt{3}, -15 \rangle$ **53. a.** $\mathbf{v}_a = \langle -320, 0 \rangle$;
$\mathbf{w} = \langle -20\sqrt{2}, -20\sqrt{2} \rangle$; $\mathbf{v}_g = \langle -320 - 20\sqrt{2}, -20\sqrt{2} \rangle$
b. Approx. 349.4 mi/hr; approx. 4.6° south of west
55. Approx. 490.3 mi/hr with a heading of about 1.2° west of north
57. $5\sqrt{65}$ km/hr $\approx$ 40.3 km/hr   **59.** 1 m/s in the direction 30° east
of north   **61. a.** $\langle 20, 20\sqrt{3} \rangle$ **b.** Yes   **c.** No   **63.** $250\sqrt{2}$ lb
65. a. True   **b.** True   **c.** False   **d.** False   **e.** False   **f.** False
g. False   **h.** True   **67.** $\mathbf{x} = \left\langle \dfrac{1}{5}, -\dfrac{3}{10} \right\rangle$ **69.** $\mathbf{x} = \left\langle \dfrac{4}{3}, -\dfrac{11}{3} \right\rangle$

71. $4\mathbf{i} - 8\mathbf{j}$ **73.** $\langle a, b \rangle = \left(\dfrac{a+b}{2} \right)\mathbf{u} + \left(\dfrac{b-a}{2} \right)\mathbf{v}$

75. a. $\mathbf{0}$ **b.** The 6:00 vector   **c.** Sum any six consecutive vectors.
d. A vector pointing from 12:00 to 6:00 with a length 12 times the
radius of the clock

77. $\mathbf{u} + \mathbf{v} = \langle u_1, u_2 \rangle + \langle v_1, v_2 \rangle = \langle u_1 + v_1, u_2 + v_2 \rangle$
$\qquad = \langle v_1 + u_1, v_2 + u_2 \rangle = \langle v_1, v_2 \rangle + \langle u_1, u_2 \rangle$
$\qquad = \mathbf{v} + \mathbf{u}$

79. $a(c\mathbf{v}) = a(c\langle v_1, v_2 \rangle) = a\langle cv_1, cv_2 \rangle$
$\qquad = \langle acv_1, acv_2 \rangle = \langle (ac)v_1, (ac)v_2 \rangle$
$\qquad = ac\langle v_1, v_2 \rangle = (ac)\mathbf{v}$

81. $(a + c)\mathbf{v} = (a + c)\langle v_1, v_2 \rangle$
$\qquad = \langle (a+c)v_1, (a+c)v_2 \rangle$
$\qquad = \langle av_1 + cv_1, av_2 + cv_2 \rangle$
$\qquad = \langle av_1, av_2 \rangle + \langle cv_1, cv_2 \rangle$
$\qquad = a\langle v_1, v_2 \rangle + c\langle v_1, v_2 \rangle$
$\qquad = a\mathbf{v} + c\mathbf{v}$

85. a. $\{\mathbf{u}, \mathbf{v}\}$ are linearly dependent. $\{\mathbf{u}, \mathbf{w}\}$ and $\{\mathbf{v}, \mathbf{w}\}$ are linearly
independent.   **b.** Two linearly dependent vectors are parallel. Two
linearly independent vectors are not parallel.   **87. a.** $\frac{5}{3}$ **b.** -15

Section 13.2 Exercises, pp. 823–827

1. Move 3 units from the origin in the direction of the positive x-axis,
then 2 units in the direction of the negative y-axis, and then 1 unit in
the direction of the positive z-axis.   **3.** It is parallel to the yz-plane
and contains the point $(4, 0, 0)$.   **5.** $\mathbf{u} + \mathbf{v} = \langle 9, 0, -6 \rangle$;
$3\mathbf{u} - \mathbf{v} = \langle 3, 20, -22 \rangle$ **7.** $(0, 0, -4)$ **9.** $A(3, 0, 5), B(3, 4, 0)$,
$C(0, 4, 5)$ **11.** $A(3, -4, 5), B(0, -4, 0), C(0, -4, 5)$
13. a.

b.

c.

15.

17.

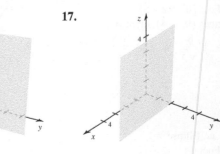

19.   **21.**

23. $(x - 1)^2 + (y - 2)^2 + (z - 3)^2 = 16$
25. $(x + 2)^2 + y^2 + (z - 4)^2 \le 1$
27. $\left(x - \frac{3}{2} \right)^2 + \left(y - \frac{3}{2} \right)^2 + (z - 7)^2 = \frac{13}{2}$ **29.** A sphere centered
at $(1, 0, 0)$ with radius 3   **31.** A sphere centered at $(0, 1, 2)$ with
radius 3   **33.** All points on or outside the sphere with center $(0, 7, 0)$
and radius 6   **35.** The ball centered at $(4, 7, 9)$ with radius 15
37. The single point $(1, -3, 0)$ **39. a.** $\langle 12, -7, 2 \rangle$
b. $\langle 16, -13, -1 \rangle$ **c.** 5   **41. a.** $\langle -4, 5, -4 \rangle$ **b.** $\langle -9, 3, -9 \rangle$
c. $3\sqrt{2}$ **43. a.** $\langle -15, 23, 22 \rangle$ **b.** $\langle -31, 49, 33 \rangle$ **c.** $3\sqrt{5}$
45. a. $\overrightarrow{PQ} = \langle 2, 6, 2 \rangle = 2\mathbf{i} + 6\mathbf{j} + 2\mathbf{k}$ **b.** $|\overrightarrow{PQ}| = 2\sqrt{11}$
c. $\left\langle \dfrac{1}{\sqrt{11}}, \dfrac{3}{\sqrt{11}}, \dfrac{1}{\sqrt{11}} \right\rangle$ and $\left\langle -\dfrac{1}{\sqrt{11}}, -\dfrac{3}{\sqrt{11}}, -\dfrac{1}{\sqrt{11}} \right\rangle$
47. a. $\overrightarrow{PQ} = \langle 0, -5, 1 \rangle = -5\mathbf{j} + \mathbf{k}$ **b.** $|\overrightarrow{PQ}| = \sqrt{26}$
c. $\left\langle 0, -\dfrac{5}{\sqrt{26}}, \dfrac{1}{\sqrt{26}} \right\rangle$ and $\left\langle 0, \dfrac{5}{\sqrt{26}}, -\dfrac{1}{\sqrt{26}} \right\rangle$
49. a. $\overrightarrow{PQ} = \langle -2, 4, -2 \rangle = -2\mathbf{i} + 4\mathbf{j} - 2\mathbf{k}$ **b.** $|\overrightarrow{PQ}| = 2\sqrt{6}$
c. $\left\langle -\dfrac{1}{\sqrt{6}}, \dfrac{2}{\sqrt{6}}, -\dfrac{1}{\sqrt{6}} \right\rangle$ and $\left\langle \dfrac{1}{\sqrt{6}}, -\dfrac{2}{\sqrt{6}}, \dfrac{1}{\sqrt{6}} \right\rangle$
51. a. $20\mathbf{i} + 20\mathbf{j} - 10\mathbf{k}$;   **b.** 30 mi/hr
53. The speed of the plane is approximately 220 mi/hr; the direction
is slightly south of east and upward.

55. $5\sqrt{6}$ knots to the east, $5\sqrt{6}$ knots to the north, 10 knots upward
57. a. False **b.** False **c.** False **d.** True **59.** All points in $\mathbb{R}^3$ except those on the coordinate axes **61.** A circle of radius 1 centered at $(0, 0, 0)$ in the xy-plane

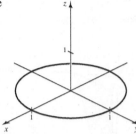

63. A circle of radius 2 centered at $(0, 0, 1)$ in the horizontal plane $z = 1$ **65.** $(x - 2)^2 + (z - 1)^2 = 9, y = 4$ **67.** $6\left\langle \frac{1}{3}, -\frac{2}{3}, \frac{2}{3} \right\rangle$

69. $\left\langle -\frac{15}{4}, \frac{5}{2}, -\frac{5\sqrt{3}}{4} \right\rangle$ **71.** $\langle 12, -16, 0 \rangle, \langle -12, 16, 0 \rangle$

73. $\langle -\sqrt{3}, -\sqrt{3}, \sqrt{3} \rangle, \langle \sqrt{3}, \sqrt{3}, -\sqrt{3} \rangle$ **75. a.** Collinear; Q is between P and R. **b.** Collinear; P is between Q and R.

c. Noncollinear **d.** Noncollinear **77.** $\left\langle \frac{500\sqrt{3}}{9}, 0, -\frac{500}{3} \right\rangle$,

$\left\langle -\frac{250\sqrt{3}}{9}, -\frac{250}{3}, -\frac{500}{3} \right\rangle, \left\langle -\frac{250\sqrt{3}}{9}, \frac{250}{3}, -\frac{500}{3} \right\rangle$
79. $(3, 8, 9), (-1, 0, 3), (1, 0, -3)$

Section 13.3 Exercises, pp. 833–837

1. $\mathbf{u} \cdot \mathbf{v} = |\mathbf{u}||\mathbf{v}| \cos \theta$ **3.** -40 **5.** $\cos \theta = \dfrac{\mathbf{u} \cdot \mathbf{v}}{|\mathbf{u}||\mathbf{v}|}$, so

$\theta = \cos^{-1}\left(\dfrac{\mathbf{u} \cdot \mathbf{v}}{|\mathbf{u}||\mathbf{v}|} \right)$ **7.** $\left\langle -\frac{4}{3}, \frac{2}{3}, \frac{4}{3} \right\rangle$ **9.** -1 **11.** 2

13. $\dfrac{\pi}{2}; 0$ **15.** $100; \dfrac{\pi}{4}$ **17.** $\dfrac{1}{2}$ **19.** $0; \dfrac{\pi}{2}$ **21.** $1; \pi/3$

23. $-2; 93.2°$ **25.** $2; 87.2°$ **27.** $-4; 104°$ **29.** $\angle P = 78.8°$, $\angle Q = 47.2°, \angle R = 54.0°$ **31.** $\langle 3, 0 \rangle; 3$ **33.** $\langle 0, 3 \rangle; 3$

35. $\dfrac{6}{5}\langle -2, 1 \rangle; \dfrac{6}{\sqrt{5}}$ **37.** $\dfrac{14}{19}\langle -1, -3, 3 \rangle; -\dfrac{14}{\sqrt{19}}$

39. $-\mathbf{i} + \mathbf{j} - 2\mathbf{k}; \sqrt{6}$ **41.** $750\sqrt{3}$ ft-lb **43.** $25\sqrt{2}$ J

45. 400 J **47.** $\dfrac{1}{2}\langle 5\sqrt{3}, -15 \rangle, \dfrac{1}{2}\langle -5\sqrt{3}, -5 \rangle$ **49.** $\langle 490, -490 \rangle$, $\langle -490, -490 \rangle$ **51. a.** False **b.** True **c.** True **d.** False

e. False **f.** True **53.** $c = \dfrac{4}{9}$ **55.** $\langle 1, a, 4a - 2 \rangle, a$ real

57. a. $\text{proj}_{\mathbf{k}}\mathbf{u} = |\mathbf{u}| \cos 60° \left(\dfrac{\mathbf{k}}{|\mathbf{k}|} \right) = \dfrac{1}{2}\mathbf{k}$, for all such $\mathbf{u}$ **b.** Yes

59. The heads of the vectors lie on the line $y = 3 - x$.
61. The heads of the vectors lie on the plane $z = 3$.

63. $\mathbf{u} = \left\langle -\dfrac{4}{5}, -\dfrac{2}{5} \right\rangle + \left\langle -\dfrac{6}{5}, \dfrac{12}{5} \right\rangle$

65. $\mathbf{u} = \left\langle 1, \dfrac{1}{2}, \dfrac{1}{2} \right\rangle + \left\langle -2, \dfrac{3}{2}, \dfrac{5}{2} \right\rangle$ **67.** $3x - 7y = -36$

69. $-\dfrac{5}{3}$ **71.** $\mathbf{I} = \dfrac{1}{\sqrt{2}}\mathbf{i} + \dfrac{1}{\sqrt{2}}\mathbf{j}, \mathbf{J} = -\dfrac{1}{\sqrt{2}}\mathbf{i} + \dfrac{1}{\sqrt{2}}\mathbf{j};$

$\mathbf{i} = \dfrac{1}{\sqrt{2}}(\mathbf{I} - \mathbf{J}), \mathbf{j} = \dfrac{1}{\sqrt{2}}(\mathbf{I} + \mathbf{J})$ **73. a.** $|\mathbf{I}| = |\mathbf{J}| = |\mathbf{K}| = 1$

b. $\mathbf{I} \cdot \mathbf{J} = 0, \mathbf{I} \cdot \mathbf{K} = 0, \mathbf{J} \cdot \mathbf{K} = 0$ **c.** $\langle 1, 0, 0 \rangle = \dfrac{1}{2}\mathbf{I} - \dfrac{1}{\sqrt{2}}\mathbf{J} + \dfrac{1}{2}\mathbf{K}$

75. a. The faces on $y = 0$ and $z = 0$ **b.** The faces on $y = 1$ and $z = 1$ **c.** The faces on $x = 0$ and $x = 1$ **d.** 0 **e.** 1 **f.** 2

77. a. $\left(\dfrac{2}{\sqrt{3}}, 0, \dfrac{2\sqrt{2}}{\sqrt{3}} \right)$ **b.** $\mathbf{r}_{OP} = \langle \sqrt{3}, -1, 0 \rangle, \mathbf{r}_{OQ} = \langle \sqrt{3}, 1, 0 \rangle$,

$\mathbf{r}_{PQ} = \langle 0, 2, 0 \rangle, \mathbf{r}_{OR} = \left\langle \dfrac{2}{\sqrt{3}}, 0, \dfrac{2\sqrt{2}}{\sqrt{3}} \right\rangle, \mathbf{r}_{PR} = \left\langle -\dfrac{\sqrt{3}}{3}, 1, \dfrac{2\sqrt{2}}{\sqrt{3}} \right\rangle$

83. a. $\cos^2 \alpha + \cos^2 \beta + \cos^2 \gamma$

$= \left(\dfrac{\mathbf{v} \cdot \mathbf{i}}{|\mathbf{v}||\mathbf{i}|} \right)^2 + \left(\dfrac{\mathbf{v} \cdot \mathbf{j}}{|\mathbf{v}||\mathbf{j}|} \right)^2 + \left(\dfrac{\mathbf{v} \cdot \mathbf{k}}{|\mathbf{v}||\mathbf{k}|} \right)^2$

$= \dfrac{a^2}{a^2 + b^2 + c^2} + \dfrac{b^2}{a^2 + b^2 + c^2} + \dfrac{c^2}{a^2 + b^2 + c^2} = 1$

b. $\langle 1, 1, 0 \rangle, 90°$ **c.** $\left\langle \dfrac{1}{\sqrt{2}}, \dfrac{1}{\sqrt{2}}, 1 \right\rangle, 45°$ **d.** No. If so,

$\left(\dfrac{\sqrt{3}}{2} \right)^2 + \left(\dfrac{\sqrt{3}}{2} \right)^2 + \cos^2 \gamma = 1$, which has no solution. **e.** $54.7°$

85. $|\mathbf{u} \cdot \mathbf{v}| = 33 = \sqrt{33} \cdot \sqrt{33} < \sqrt{70} \cdot \sqrt{74} = |\mathbf{u}||\mathbf{v}|$

Section 13.4 Exercises, pp. 842–844

1. 0 **3. a.** $\mathbf{u}$ is orthogonal to $\mathbf{v}$. **b.** $\mathbf{u}$ is parallel to $\mathbf{v}$. **5.** $\sqrt{2}/2$

7. $-3\mathbf{i} - 2\mathbf{j} + 7\mathbf{k}$ **9.** $\mathbf{u} \times \mathbf{v} = \begin{vmatrix} \mathbf{i} & \mathbf{j} & \mathbf{k} \\ u_1 & u_2 & u_3 \\ v_1 & v_2 & v_3 \end{vmatrix}$ **11.** $15\mathbf{k}$

13. 0

15. 18

17. $\mathbf{i}$

19. $-\mathbf{i}$

21. $6\mathbf{j}$

23. $\mathbf{u} \times \mathbf{v} = \langle -30, 18, 9 \rangle, \mathbf{v} \times \mathbf{u} = \langle 30, -18, -9 \rangle$
25. $\mathbf{u} \times \mathbf{v} = \langle 6, 11, 5 \rangle, \mathbf{v} \times \mathbf{u} = \langle -6, -11, -5 \rangle$
27. $\mathbf{u} \times \mathbf{v} = \langle 8, 4, 10 \rangle, \mathbf{v} \times \mathbf{u} = \langle -8, -4, -10 \rangle$ **29.** 11
31. $3\sqrt{10}$ **33.** $\sqrt{11}/2$ **35.** $4\sqrt{2}$ **37.** $9\sqrt{2}$ **41.** Not collinear
43. $\langle 3, -4, 2 \rangle$ **45.** $\langle 0, 20, -20 \rangle$ **47.** The force $\mathbf{F} = 5\mathbf{i} - 5\mathbf{k}$
produces the greater torque. **49.** $5/\sqrt{2}$ N-m **51.** $|\tau| = 13.2$ N-m; direction: into the page

53. The magnitude is $20\sqrt{2}$ at a $135°$ angle with the positive x-axis in the xy-plane.

55. 4.53×10^{-14} kg-m/s^2 **57. a.** False   **b.** False   **c.** False
d. True   **e.** False   **59.** $\langle u_1, u_1 + 2, u_1 + 1 \rangle$, u_1 real

61. $\dfrac{\sqrt{(ab)^2 + (ac)^2 + (bc)^2}}{2}$

63. $|\mathbf{u} \cdot (\mathbf{v} \times \mathbf{w})| = |\mathbf{u}||\mathbf{v} \times \mathbf{w}||\cos\theta|$, where $|\mathbf{v} \times \mathbf{w}|$ is the area of the base of the parallelepiped and $|\mathbf{u}||\cos\theta|$ is its height.
67. 1.76×10^7 m/s

Section 13.5 Exercises, pp. 852–855

1. $\langle 4, -8, 9 \rangle$ **3.** $\langle x_1 - x_0, y_1 - y_0, z_1 - z_0 \rangle$ **5.** Perpendicular
7. A point and a normal vector   **9.** $(-6, 0, 0), (0, -4, 0), (0, 0, 3)$
11. $x = 4t, y = 7t, z = 1; \mathbf{r} = \langle 0, 0, 1 \rangle + t\langle 4, 7, 0 \rangle$
13. $x = 0, y = t, z = 1; \mathbf{r} = \langle 0, 0, 1 \rangle + t\langle 0, 1, 0 \rangle$
15. $x = t, y = 2t, z = 3t; \mathbf{r} = t\langle 1, 2, 3 \rangle$
17. $x = -2t, y = 8t, z = -4t; \mathbf{r} = t\langle -2, 8, -4 \rangle$
19. $x = -2t, y = -t, z = t; \mathbf{r} = t\langle -2, -1, 1 \rangle$
21. $x = -2, y = 5 - 2t, z = 3 - t; \mathbf{r} = \langle -2, 5, 3 \rangle +$
$t\langle 0, -2, -1 \rangle$ **23.** $x = 1 - 4t, y = 2 + 6t, z = 3 + 14t;$
$\mathbf{r} = \langle 1, 2, 3 \rangle + t\langle -4, 6, 14 \rangle$ **25.** $x = 4, y = 3 - 9t, z = 3 + 6t;$
$\mathbf{r} = \langle 4, 3, 3 \rangle + t\langle 0, -9, 6 \rangle$ **27.** $x = t, y = 2t, z = 3t, 0 \le t \le 1$
29. $x = 2 + 5t, y = 4 + t, z = 8 - 5t, 0 \le t \le 1$ **31.** Intersect
at $(1, 3, 2)$ **33.** Skew   **35.** Same line   **37.** Parallel, distinct lines
39. 13   **41. a.** Yes   **b.** No   **c.** $13.16° < \theta < 18.12°$
43. $x + y - z = 4$ **45.** $2x + y - 2z = -2$
47. $x + 4y + 7z = 0$ **49.** $7x + 2y + z = 10$
51. $-x + 2y - 4z = -17$ **53.** $3y - 2z = 0$
55. $8x - 7y + 2z = 0$ **57.** $x + 3y - z = -3$
59. Yes; $2x - y = -1$

61. Intercepts
$x = 2, y = -3, z = 6;$
$3x - 2y = 6, z = 0;$
$-2y + z = 6, x = 0;$ and
$3x + z = 6, y = 0$

63. Intercepts
$x = 30, y = 10, z = -6;$
$x + 3y = 30, z = 0;$
$x - 5z = 30, y = 0;$ and
$3y - 5z = 30, x = 0$

65. Orthogonal   **67.** Neither   **69.** Q and T are identical; $Q, R,$ and T are parallel; S is orthogonal to $Q, R,$ and T.
71. $\mathbf{r} = \langle 2 + 2t, 1 - 4t, 3 + t \rangle$
73. $x = t, y = 1 + 2t, z = -1 - 3t$

75. $x = \frac{7}{5} + 2t, y = \frac{9}{5} + t, z = -t$ **77.** $(3, 3, 3)$ **79.** $(1, 1, 2)$
81. a. True   **b.** False   **c.** False   **d.** True   **e.** False   **f.** False
g. True   **83.** 6   **85.** $\dfrac{x - 1}{4} = \dfrac{y - 2}{7} = \dfrac{z}{2}$ **87.** Approx. $43°$
89. $6x - 4y + z = d$ **91.** The planes intersect in the point $(3, 6, 0)$.
93. a.

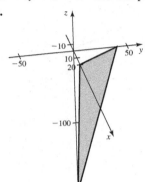

b. Positive
c. $2x + y = 40$, line in the xy-plane

Section 13.6 Exercises, pp. 863–865

1. z-axis; x-axis; y-axis   **3.** Intersection of the surface with a plane parallel to one of the coordinate planes   **5.** Ellipsoid
7. a. x-axis
b.

9. a. y-axis
b.

11. a. z-axis
b.

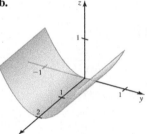

13. a. x-axis
b.

15. Ellipsoid; xy-trace: $x^2 + y^2 = 1$ (circle); xz-trace:
$x^2 + \dfrac{z^2}{25} = 1$ (ellipse); yz-trace: $y^2 + \dfrac{z^2}{25} = 1$ (ellipse)
17. Paraboloid; xy-trace: $(0, 0, 0)$ (a single point); xz-trace:
$z = 25x^2$ (parabola); yz-trace: $z = 25y^2$ (parabola)   **19.** Hyperboloid
of two sheets; xz-trace: $z^2 - 25x^2 = 25$ (hyperbola);
yz-trace: $z^2 - 25y^2 = 25$ (hyperbola)   **21.** Hyperbolic paraboloid
23. Elliptic paraboloid   **25.** Hyperbolic cylinder
27. Elliptic paraboloid

29. a. $x = \pm 1, y = \pm 2,$ $z = \pm 3$
b. $x^2 + \dfrac{y^2}{4} = 1, x^2 + \dfrac{z^2}{9} = 1,$ $\dfrac{y^2}{4} + \dfrac{z^2}{9} = 1$
c. Ellipsoid

31. a. $x = y = z = 0$
b. $x = y^2, x = z^2,$ origin
c. Elliptic paraboloid

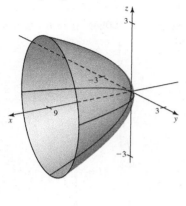

37. a. $x = y = z = 0$
b. Origin, $\dfrac{y^2}{4} = z^2, x^2 = z^2$
c. Elliptic cone

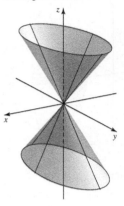

39. a. $x = \pm 3, y = \pm 1, z = \pm 6$
b. $\dfrac{x^2}{3} + 3y^2 = 3, \dfrac{x^2}{3} + \dfrac{z^2}{12} = 3,$ $3y^2 + \dfrac{z^2}{12} = 3$
c. Ellipsoid

33. a. $x = \pm 5, y = \pm 3$, no z-intercept
b. $\dfrac{x^2}{25} + \dfrac{y^2}{9} = 1, \dfrac{x^2}{25} - z^2 = 1, \dfrac{y^2}{9} - z^2 = 1$
c. Hyperboloid of one sheet

41. a. $x = y = z = 0$
b. Origin, $x - 9y^2 = 0, 9x - \dfrac{z^2}{4} = 0$
c. Elliptic paraboloid

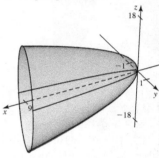

43. a. No x-intercept, $y = \pm 12, z = \pm\dfrac{1}{2}$
b. $-\dfrac{x^2}{4} + \dfrac{y^2}{16} = 9,$ $-\dfrac{x^2}{4} + 36z^2 = 9, \dfrac{y^2}{16} + 36z^2 = 9$
c. Hyperboloid of one sheet

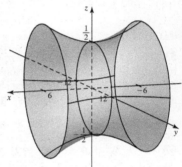

35. a. $x = y = z = 0$ **b.** $\dfrac{x^2}{9} - y^2 = 0, z = \dfrac{x^2}{9}, z = -y^2$
c. Hyperbolic paraboloid

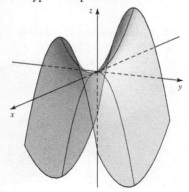

45. a. $x = y = z = 0$
b. $5x - \dfrac{y^2}{5} = 0, 5x + \dfrac{z^2}{20} = 0,$ $-\dfrac{y^2}{5} + \dfrac{z^2}{20} = 0$
c. Hyperbolic paraboloid

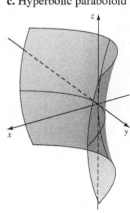

47. a. $x = y = z = 0$
b. $\dfrac{y^2}{18} = 2x^2, \dfrac{z^2}{32} = 2x^2,$ origin
c. Elliptic cone

49. a. No x-intercept, $y = \pm 2$, no z-intercept

b. $-x^2 + \dfrac{y^2}{4} = 1$, no xz-trace, $\dfrac{y^2}{4} - \dfrac{z^2}{9} = 1$

c. Hyperboloid of two sheets

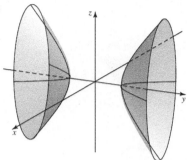

51. a. No x-intercept, $y = \pm\dfrac{\sqrt{3}}{3}$, no z-intercept

b. $-\dfrac{x^2}{3} + 3y^2 = 1$, no xz-trace, $3y^2 - \dfrac{z^2}{12} = 1$

c. Hyperboloid of two sheets

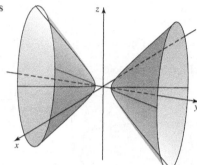

53. The graph of the ellipsoid $x^2 + 4y^2 + 9z^2 + 54z = 19$ is obtained by shifting the graph of the ellipsoid $x^2 + 4y^2 + 9z^2 = 100$ down 3 units.   **55.** Hyperboloid of one sheet   **57.** Hyperboloid of two sheets

59. a. True   **b.** True   **c.** True   **d.** False   **e.** False   **61.** All except the hyperbolic paraboloid   **63.** 8   **65. b.** $\dfrac{x^2 + z^2}{(10.55/\pi)^2} + \dfrac{y^2}{(5.55)^2} = 1$

67. $4x^2 + 8y^2 + 4(z - 3)^2 = 9$, $3 \le z \le 4.5$

Chapter 13 Review Exercises, pp. 865–867

1. a. True   **b.** False   **c.** True   **d.** False   **e.** True   **f.** True

3. $\langle 3, -6 \rangle$ **5.** $\langle -5, 8 \rangle$ **7.** $\sqrt{221}$ **9.** $12\left\langle \dfrac{1}{3}, -\dfrac{2}{3}, \dfrac{2}{3} \right\rangle$

11. $\left\langle \dfrac{10}{3}, -\dfrac{20}{3}, \dfrac{20}{3} \right\rangle$ **13.** $\langle 58, 26, 44 \rangle$ **15.** $a = -3$

17. a. $\mathbf{v} = -275\sqrt{2}\mathbf{i} + 275\sqrt{2}\mathbf{j}$ **b.** $-275\sqrt{2}\mathbf{i} + (275\sqrt{2} + 40)\mathbf{j}$

19. $\{(x, y, z): (x - 1)^2 + y^2 + (z + 1)^2 = 16\}$

21. $\{(x, y, z): x^2 + (y - 1)^2 + z^2 > 4\}$ **23.** A ball centered at $\left(\frac{1}{2}, -2, 3\right)$ of radius $\frac{3}{2}$ **25.** All points outside a sphere of radius 10 centered at $(3, 0, 10)$ **27.** 50.15 m/s; 85.4° below the horizontal in the northerly horizontal direction   **29.** 50 lb; 36.9° north of east

31. A circle of radius 1 centered at $(0, 2, 0)$ in the vertical plane $y = 2$

33. a. 0.68 radian   **b.** $\dfrac{7}{9}\langle 1, 2, 2 \rangle$; $\dfrac{7}{3}$ **c.** $\dfrac{7}{3}\langle -1, 2, 2 \rangle$; 7

35. $250\sqrt{2}$ ft-lb   **37.** $90\sqrt{3}$ lb; 90 lb   **39.** 11

41. $\pm\left\langle \dfrac{12}{\sqrt{197}}, \dfrac{7}{\sqrt{197}}, \dfrac{2}{\sqrt{197}} \right\rangle$ **43.** $\langle -10, 10, 10 \rangle$

45. $|\tau|(\theta) = 39.2 \sin\theta$ has a maximum value of 39.2 N-m (when $\theta = \pi/2$) and a minimum value of 0 N-m (when $\theta = 0$). Direction does *not* change.   **47.** $\mathbf{r} = \langle 0, -3, 9 \rangle + t\langle 2, -5, -8 \rangle$, $0 \le t \le 1$

49. $\mathbf{r} = \langle t, 1 + 6t, 1 + 2t \rangle$

51. a. $18x - 9y + 2z = 6$ **b.** $x = \frac{1}{3}, y = -\frac{2}{3}, z = 3$

c.

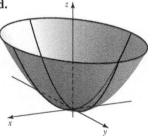

53. $x = t, y = 12 - 9t, z = -6 + 6t$ **55.** $4x + 2y + 13z = 39$

57. $3x + y + 7z = 4$ **59.** 3

61. a. Hyperbolic paraboloid

b. $y^2 = 4x^2, z = \dfrac{x^2}{36}, z = -\dfrac{y^2}{144}$

c. $x = y = z = 0$

d.

63. a. Elliptic cone

b. $y^2 = 4x^2$, origin, $y^2 = \dfrac{z^2}{25}$

c. Origin

d.

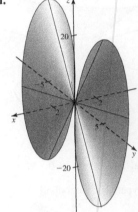

65. a. Elliptic paraboloid

b. Origin, $z = \dfrac{x^2}{16}, z = \dfrac{y^2}{36}$

c. Origin

d.

67. a. Hyperboloid of one sheet

b. $y^2 - 2x^2 = 1, 4z^2 - 2x^2 = 1$, $y^2 + 4z^2 = 1$ **c.** No x-intercept, $y = \pm 1, z = \pm\frac{1}{2}$

d.

69. a. Hyperboloid of one sheet

b. $\dfrac{x^2}{4} + \dfrac{y^2}{16} = 4, \dfrac{x^2}{4} - z^2 = 4, \dfrac{y^2}{16} - z^2 = 4$

c. $x = \pm 4, y = \pm 8$, no z-intercept

d.

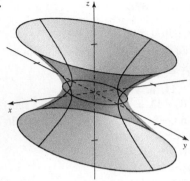

71. a. Ellipsoid **b.** $\dfrac{x^2}{4} + \dfrac{y^2}{16} = 4, \dfrac{x^2}{4} + z^2 = 4, \dfrac{y^2}{16} + z^2 = 4$

c. $x = \pm 4, y = \pm 8, z = \pm 2$
d.

73. a. Elliptic cone **b.** Origin, $\dfrac{x^2}{9} = \dfrac{z^2}{64}, \dfrac{y^2}{49} = \dfrac{z^2}{64}$ **c.** Origin
d.

75. a. A **b.** D **c.** C **d.** B

CHAPTER 14

Section 14.1 Exercises, pp. 873–875

1. One **3.** Its output is a vector.
5. $\lim\limits_{t \to a} \mathbf{r}(t) = \left\langle \lim\limits_{t \to a} f(t), \lim\limits_{t \to a} g(t), \lim\limits_{t \to a} h(t) \right\rangle$
7. $\mathbf{r}(t) = t\,\mathbf{i} + 2t\,\mathbf{j} + 3t\,\mathbf{k}$
9. $\mathbf{r}(t) = \langle 2 + 2t, 3 + 3t, 7 - 4t \rangle$
11. $\mathbf{r}(t) = \langle 3 + 2t, 4, 5 - t \rangle$
13. $\mathbf{r}(t) = \langle 1 - t, 2, 1 + 2t \rangle$, for $0 \le t \le 1$
15.

17.

19.

21.

23.

25.

27.

29. When viewed from above,
the curve is a portion of the
parabola $y = x^2$.

31. $-\mathbf{i} - 4\mathbf{j} + \mathbf{k}$ **33.** $-2\mathbf{j} + \dfrac{\pi}{2}\mathbf{k}$ **35.** $\mathbf{i}$ **37. a.** True **b.** False
c. True **d.** True **39.** $\{t : |t| \le 2\}$ **41.** $\{t : 0 \le t \le 2\}$
43. $(4, 8, 16)$ **45. a.** E **b.** D **c.** F **d.** C **e.** A **f.** B
47. $\mathbf{r}(t) = \langle 2 \cos t, 2 \sin t, 4 \rangle$
49. $\mathbf{r}(t) = \langle 5 \cos t, 5 \sin t, 10 \cos t + 10 \sin t \rangle$
51. a. Ball has a parabolic trajectory in the yz-plane; 1200 ft
b. Approx. 1199.7 ft **c.** 1196 ft **53.** Hyperboloid of one sheet
55. Ellipsoid **57.** $(4, 2, 2); \sqrt{179}$
59.

The curve lies on the sphere
$x^2 + y^2 + z^2 = 1$.

61. $\dfrac{2\pi}{(m, n)}$, where $(m, n) = $ greatest common factor of m and n

63. a.

b. Curve is a tilted circle of radius 1 centered at the origin.

65. $\langle cf - ed, be - af, ad - bc \rangle$ or any scalar multiple

Section 14.2 Exercises, pp. 881–883

1. $\mathbf{r}'(t) = \langle f'(t), g'(t), h'(t) \rangle$ **3.** $\mathbf{T}(t) = \dfrac{\mathbf{r}'(t)}{|\mathbf{r}'(t)|}$

5. $\int \mathbf{r}(t)\, dt = \left(\int f(t)\, dt \right) \mathbf{i} + \left(\int g(t)\, dt \right) \mathbf{j} + \left(\int h(t)\, dt \right) \mathbf{k}$

7. $\mathbf{C} = \langle -1, -3, -10 \rangle$ **9.** $\langle -\sin t, 2t, \cos t \rangle$

11. $\left\langle 6t^2, \dfrac{3}{\sqrt{t}}, -\dfrac{3}{t^2} \right\rangle$ **13.** $e^t \mathbf{i} - 2e^{-t} \mathbf{j} - 8e^{2t} \mathbf{k}$

15. $\langle e^{-t}(1 - t), 1 + \ln t, \cos t - t \sin t \rangle$

17. $\langle 1, 6, 3 \rangle$ **19.** $\langle 1, 0, 0 \rangle$ **21.** $8\mathbf{i} + 9\mathbf{j} - 10\mathbf{k}$

23. $\langle 2/3, 2/3, 1/3 \rangle$

25. $\dfrac{\langle 0, -\sin 2t, 2\cos 2t \rangle}{\sqrt{1 + 3\cos^2 2t}}$ **27.** $\dfrac{t^2}{\sqrt{t^4 + 4}} \left\langle 1, 0, -\dfrac{2}{t^2} \right\rangle$

29. $\langle 0, 0, -1 \rangle$ **31.** $\left\langle \dfrac{2}{\sqrt{5}}, 0, -\dfrac{1}{\sqrt{5}} \right\rangle$

33. $\langle 30t^{14} + 24t^3, 14t^{13} - 12t^{11} + 9t^2 - 3, -96t^{11} - 24 \rangle$

35. $4t(2t^3 - 1)(t^3 - 2) \langle 3t(t^3 - 2), 1, 0 \rangle$

37. $e^t(2t^3 + 6t^2) - 2e^{-t}(t^2 - 2t - 1) - 16e^{-2t}$

39. 11 **41.** $\langle 0, 7, 1 \rangle$ **43.** $\langle 2e^{2t}, -2e^t, 0 \rangle$ **45.** $\left\langle 4, -\dfrac{2}{\sqrt{t}}, 0 \right\rangle$

47. $\langle 1 + 6t^2, 4t^3, -2 - 3t^2 \rangle$ **49.** $5te^t(t + 2) - 6t^2 e^{-t}(t - 3)$

51. $-3t^2 \sin t + 6t \cos t + 2\sqrt{t} \cos 2t + \dfrac{1}{2\sqrt{t}} \sin 2t$

53. $\langle 2, 0, 0 \rangle, \langle 0, 0, 0 \rangle$ **55.** $\langle -9 \cos 3t, -16 \sin 4t, -36 \cos 6t \rangle$, $\langle 27 \sin 3t, -64 \cos 4t, 216 \sin 6t \rangle$

57. $\left\langle -\dfrac{1}{4}(t + 4)^{-3/2}, -2(t + 1)^{-3}, 2e^{-t^2}(1 - 2t^2) \right\rangle$, $\left\langle \dfrac{3}{8}(t + 4)^{-5/2}, 6(t + 1)^{-4}, -4te^{-t^2}(3 - 2t^2) \right\rangle$

59. $\left\langle \dfrac{t^5}{5} - \dfrac{3t^2}{2}, t^2 - t, 10t \right\rangle + \mathbf{C}$

61. $\left\langle 2 \sin t, -\dfrac{2}{3} \cos 3t, \dfrac{1}{2} \sin 8t \right\rangle + \mathbf{C}$

63. $\frac{1}{3} e^{3t} \mathbf{i} + \tan^{-1} t\, \mathbf{j} - \sqrt{2t}\, \mathbf{k} + \mathbf{C}$

65. $\mathbf{r}(t) = \langle e^t + 1, 3 - \cos t, \tan t + 2 \rangle$

67. $\mathbf{r}(t) = \langle t + 3, t^2 + 2, t^3 - 6 \rangle$

69. $\mathbf{r}(t) = \langle \frac{1}{2} e^{2t} + \frac{1}{2}, 2e^{-t} + t - 1, t - 2e^t + 3 \rangle$

71. $\langle 2, 0, 2 \rangle$ **73.** $\mathbf{i}$ **75.** $\langle 0, 0, 0 \rangle$

77. $(e^2 + 1)\langle 1, 2, -1 \rangle$ **79. a.** False   **b.** True   **c.** True

81. $\langle 2 - t, 3 - 2t, \pi/2 + t \rangle$ **83.** $\langle 2 + 3t, 9 + 7t, 1 + 2t \rangle$

85. $(1, 0)$ **87.** $(1, 0, 0)$ **89.** $\mathbf{r}(t) = \langle a_1 t, a_2 t, a_3 t \rangle$ or $\mathbf{r}(t) = \langle a_1 e^{kt}, a_2 e^{kt}, a_3 e^{kt} \rangle$, where a_i and k are real numbers

Section 14.3 Exercises, pp. 892–896

1. $\mathbf{v}(t) = \mathbf{r}'(t)$, speed $= |\mathbf{r}'(t)|$, $\mathbf{a}(t) = \mathbf{r}''(t)$ **3.** $m\mathbf{a}(t) = \mathbf{F}$

5. $\mathbf{v}(t) = \int \mathbf{a}(t)\, dt = \langle v_1(t), v_2(t) \rangle + \mathbf{C}$. Use initial conditions to find $\mathbf{C}$.   **7. a.** $t = 3$ s   **b.** $\mathbf{r}(t) = \langle 60t, -16t^2 + 96t + 3 \rangle$

9. a. $\langle 6t, 8t \rangle$, $10t$ **b.** $\langle 6, 8 \rangle$ **11. a.** $\mathbf{v}(t) = \langle 2, -4 \rangle$, $|\mathbf{v}(t)| = 2\sqrt{5}$ **b.** $\mathbf{a}(t) = \langle 0, 0 \rangle$ **13. a.** $\mathbf{v}(t) = \langle 8 \cos t, -8 \sin t \rangle$, $|\mathbf{v}(t)| = 8$ **b.** $\mathbf{a}(t) = \langle -8 \sin t, -8 \cos t \rangle$ **15. a.** $\langle 2t, 2t, t \rangle$, $3t$ **b.** $\langle 2, 2, 1 \rangle$ **17. a.** $\mathbf{v}(t) = \langle 1, -4, 6 \rangle$, $|\mathbf{v}(t)| = \sqrt{53}$ **b.** $\mathbf{a}(t) = \langle 0, 0, 0 \rangle$ **19. a.** $\mathbf{v}(t) = \langle 0, 2t, -e^{-t} \rangle$, $|\mathbf{v}(t)| = \sqrt{4t^2 + e^{-2t}}$ **b.** $\mathbf{a}(t) = \langle 0, 2, e^{-t} \rangle$

21. a. $[c, d] = [0, 1]$ **b.** $\langle 1, 2t \rangle, \langle 2, 8t \rangle$

c.

23. a. $\left[0, \frac{2\pi}{3} \right]$ **b.** $\mathbf{V}_r(t) = \langle -\sin t, 4 \cos t \rangle$, $\mathbf{V}_R(t) = \langle -3 \sin 3t, 12 \cos 3t \rangle$

c.

25. a. $[1, e^{36}]$ **b.** $\mathbf{V}_r(t) = \langle 2t, -8t^3, 18t^5 \rangle$, $\mathbf{V}_R(t) = \left\langle \dfrac{1}{t}, -\dfrac{4}{t} \ln t, \dfrac{9}{t} \ln^2 t \right\rangle$

c.

27. a.

b. $\langle -20 \sin t - 50 \sin 5t, 20 \cos t + 50 \cos 5t \rangle$

c.   **d.** 70 ft/s; 30 ft/s

29. $\mathbf{r}(t)$ lies on a circle of radius 8;
$\langle -16 \sin 2t, 16 \cos 2t \rangle \cdot \langle 8 \cos 2t, 8 \sin 2t \rangle = 0.$ **31.** $\mathbf{r}(t)$ lies
on a sphere of radius 2; $\langle \cos t - \sqrt{3} \sin t, \sqrt{3} \cos t + \sin t \rangle \cdot$
$\langle \sin t + \sqrt{3} \cos t, \sqrt{3} \sin t - \cos t \rangle = 0.$ **33.** 5

35. $\mathbf{v}(t) = \langle 2, t + 3 \rangle, \mathbf{r}(t) = \left\langle 2t, \dfrac{t^2}{2} + 3t \right\rangle$

37. $\mathbf{v}(t) = \langle 0, 10t + 5 \rangle, \mathbf{r}(t) = \langle 1, 5t^2 + 5t - 1 \rangle$
39. $\mathbf{v}(t) = \langle \sin t, -2 \cos t + 3 \rangle,$
$\mathbf{r}(t) = \langle -\cos t + 2, -2 \sin t + 3t \rangle$
41. a. $\mathbf{v}(t) = \langle 30, -9.8t + 6 \rangle, \mathbf{r}(t) = \langle 30t, -4.9t^2 + 6t \rangle$
b.   **c.** $T \approx 1.22$ s, range ≈ 36.7 m
    d. 1.84 m

43. a. $\mathbf{v}(t) = \langle 80, 10 - 32t \rangle, \mathbf{r}(t) = \langle 80t, -16t^2 + 10t + 6 \rangle$
b.   **c.** 1 s, 80 ft
    d. Max height ≈ 7.56 ft

45. a. $\mathbf{v}(t) = \langle 125, -32t + 125\sqrt{3} \rangle,$
$\mathbf{r}(t) = \langle 125t, -16t^2 + 125\sqrt{3}t + 20 \rangle$
b.   **c.** 13.6 s, 1702.5 ft   **d.** 752.4 ft

47. $\mathbf{v}(t) = \langle 1, 5, 10t \rangle, \mathbf{r}(t) = \langle t, 5t + 5, 5t^2 \rangle$
49. $\mathbf{v}(t) = \langle -\cos t + 1, \sin t + 2, t \rangle,$

$\mathbf{r}(t) = \left\langle -\sin t + t, -\cos t + 2t + 1, \dfrac{t^2}{2} \right\rangle$

51. a. $\mathbf{v}(t) = \langle 200, 200, -9.8t \rangle, \mathbf{r}(t) = \langle 200t, 200t, -4.9t^2 + 1 \rangle$
b.   **c.** 0.452 s, 127.8 m   **d.** 1 m

53. a. $\mathbf{v}(t) = \langle 60 + 10t, 80, 80 - 32t \rangle,$
$\mathbf{r}(t) = \langle 60t + 5t^2, 80t, 80t - 16t^2 + 3 \rangle$
b.   **c.** 5.04 s, 589 ft
    d. 103 ft

55. a. $\mathbf{v}(t) = \langle 300, 2.5t + 400, -9.8t + 500 \rangle,$
$\mathbf{r}(t) = \langle 300t, 1.25t^2 + 400t, -4.9t^2 + 500t + 10 \rangle$
b.   **c.** 102.1 s, 61,941.5 m
    d. 12,765.1 m

57. a. False   **b.** True   **c.** False   **d.** True   **e.** False   **f.** True
g. True   **59.** 15.3 s, 1988.3 m, 287.0 m   **61.** 21.7 s, 4330.1 ft, 1875 ft
63. Approx. 27.4° and 62.6°
65. a. $\mathbf{v}(t) = \langle -a \sin t, b \cos t \rangle; |\mathbf{v}(t)| = \sqrt{a^2 \sin^2 t + b^2 \cos^2 t}$
b.

c. Yes   **d.** Max $\left\{ \dfrac{a}{b}, \dfrac{b}{a} \right\}$ **67.** Approx. 23.5° or 59.6°
69. 113.4 ft/s   **71. a.** 1.2 ft, 0.46 s   **b.** 0.88 ft/s   **c.** 0.85 ft
d. More curve in the second half   **e.** $c = 28.17$ ft/s^2

73. $T = \dfrac{|\mathbf{v}_0| \sin \alpha + \sqrt{|\mathbf{v}_0|^2 \sin^2 \alpha + 2gy_0}}{g},$

range $= |\mathbf{v}_0| (\cos \alpha) T,$ max height $= y_0 + \dfrac{|\mathbf{v}_0|^2 \sin^2 \alpha}{2g}$

75. a. $\left[0, \dfrac{2\pi}{\omega} \right]$ **b.** $\mathbf{v}(t) = \langle -A\omega \sin \omega t, A\omega \cos \omega t \rangle$ is not constant;
$|\mathbf{v}(t)| = |A\omega|$ is constant.   **c.** $\mathbf{a}(t) = \langle -A\omega^2 \cos \omega t, -A\omega^2 \sin \omega t \rangle$

d. $\mathbf{r}$ and $\mathbf{v}$ are orthogonal; $\mathbf{r}$ and $\mathbf{a}$ are in opposite directions.

e.

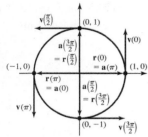

77. a. $r(t) = \langle 5 \sin(\pi t/6), 5 \cos(\pi t/6) \rangle$
b. $r(t) = \langle 5 \sin\left(\frac{1 - e^{-t}}{5}\right), 5 \cos\left(\frac{1 - e^{-t}}{5}\right) \rangle$

79. $\{(\cos t, \sin t, c \sin t): t \in \mathbb{R}\}$ satisfies the equations $x^2 + y^2 = 1$ and $z - cy = 0$ so that $\langle \cos t, \sin t, c \sin t \rangle$ lies on the intersection of a right circular cylinder and a plane, which is an ellipse.
83. a. The direction of r does not change.   **b.** Constant in direction, not in magnitude

Section 14.4 Exercises, pp. 900–902

1. $\sqrt{5}(b - a)$ **3.** $\int_a^b |v(t)|\, dt$ **5.** 20π **7.** If the parameter t used to describe a trajectory also measures the arc length s of the curve that is generated, we say the curve has been parameterized by its arc length.

9. 5   **11.** 3π **13.** $\frac{\pi^2}{8}$ **15.** $5\sqrt{34}$ **17.** $4\pi\sqrt{65}$ **19.** 9   **21.** $\frac{3}{2}$

23. $3t^2\sqrt{30}$; $64\sqrt{30}$ **25.** 26; 26π
27. Approx. 66,626 mi/hr   **29.** 19.38

31. 32.50   **33.** Yes   **35.** No; $r(s) = \langle \frac{s}{\sqrt{5}}, \frac{2s}{\sqrt{5}} \rangle, 0 \le s \le 3\sqrt{5}$

37. No; $r(s) = \langle 2 \cos \frac{s}{2}, 2 \sin \frac{s}{2} \rangle, 0 \le s \le 4\pi$

39. No; $r(s) = \langle \cos s, \sin s \rangle, 0 \le s \le \pi$

41. No; $r(s) = \langle \frac{s}{\sqrt{3}} + 1, \frac{s}{\sqrt{3}} + 1, \frac{s}{\sqrt{3}} + 1 \rangle, s \ge 0$

43. a. True   **b.** True   **c.** True   **d.** False   **45. a.** If $a^2 = b^2 + c^2$, then $|r(t)|^2 = (a \cos t)^2 + (b \sin t)^2 + (c \sin t)^2 = a^2$ so that $r(t)$ is a circle centered at the origin of radius $|a|$.   **b.** $2\pi a$
c. If $a^2 + c^2 + e^2 = b^2 + d^2 + f^2$ and $ab + cd + ef = 0$, then $r(t)$ is a circle of radius $\sqrt{a^2 + c^2 + e^2}$ and its arc length is $2\pi\sqrt{a^2 + c^2 + e^2}$.   **47. a.** $\int_a^b \sqrt{(Ah'(t))^2 + (Bh'(t))^2}\, dt$
$= \int_a^b \sqrt{(A^2 + B^2)(h'(t))^2}\, dt = \sqrt{A^2 + B^2}\int_a^b |h'(t)|\, dt$
b. $64\sqrt{29}$ **c.** $\frac{7\sqrt{29}}{4}$ **49. a.** 5.102 s
b. $\int_0^{5.102} \sqrt{400 + (25 - 9.8t)^2}\, dt$ **c.** 124.43 m   **d.** 102.04 m
51. $|v(t)| = \sqrt{a^2 + b^2 + c^2} = 1$, if $a^2 + b^2 + c^2 = 1$
53. $\int_a^b |r'(t)|\, dt = \int_a^b \sqrt{(cf'(t))^2 + (cg'(t))^2}\, dt$
$= |c| \int_a^b \sqrt{(f'(t))^2 + (g'(t))^2}\, dt = |c|L$

Section 14.5 Exercises, pp. 913–915

1. 0   **3.** $\kappa = \frac{1}{|v|}\left|\frac{dT}{dt}\right|$ or $\kappa = \frac{|v \times a|}{|v|^3}$ **5.** $N = \frac{dT/dt}{|dT/dt|}$

7. These three unit vectors are mutually orthogonal at all points of the curve.   **9.** The torsion measures the rate at which the curve rises or twists out of the **TN**-plane at a point.   **11.** $T = \frac{\langle 1, 2, 3 \rangle}{\sqrt{14}}, \kappa = 0$

13. $T = \frac{\langle 1, 2 \cos t, -2 \sin t \rangle}{\sqrt{5}}, \kappa = \frac{1}{5}$

15. $T = \frac{\langle \sqrt{3} \cos t, \cos t, -2 \sin t \rangle}{2}, \kappa = \frac{1}{2}$

17. $T = \frac{\langle 1, 4t \rangle}{\sqrt{1 + 16t^2}}, \kappa = \frac{4}{(1 + 16t^2)^{3/2}}$

19. $T = \langle \cos\left(\frac{\pi t^2}{2}\right), \sin\left(\frac{\pi t^2}{2}\right) \rangle, \kappa = \pi t$

21. $\frac{1}{3}$ **23.** $\frac{2}{(4t^2 + 1)^{3/2}}$ **25.** $\frac{2\sqrt{5}}{(20 \sin^2 t + \cos^2 t)^{3/2}}$

27. $T = \langle \cos t, -\sin t \rangle, N = \langle -\sin t, -\cos t \rangle$

29. $T = \frac{\langle t, -3, 0 \rangle}{\sqrt{t^2 + 9}}, N = \frac{\langle 3, t, 0 \rangle}{\sqrt{t^2 + 9}}$

31. $T = \langle -\sin t^2, \cos t^2 \rangle, N = \langle -\cos t^2, -\sin t^2 \rangle$

33. $T = \frac{\langle 2t, 1 \rangle}{\sqrt{4t^2 + 1}}, N = \frac{\langle 1, -2t \rangle}{\sqrt{4t^2 + 1}}$ **35.** $a_N = a_T = 0$

37. $a_T = \sqrt{3}e^t; a_N = \sqrt{2}e^t$ **39.** $a = \frac{6t}{\sqrt{9t^2 + 4}}N + \frac{18t^2 + 4}{\sqrt{9t^2 + 4}}T$

41. $B(t) = \langle 0, 0, -1 \rangle, \tau = 0$ **43.** $B(t) = \langle 0, 0, 1 \rangle, \tau = 0$

45. $B(t) = \frac{\langle -\sin t, \cos t, 2 \rangle}{\sqrt{5}}, \tau = -\frac{1}{5}$

47. $B(t) = \frac{\langle 5, 12 \sin t, -12 \cos t \rangle}{13}, \tau = \frac{12}{169}$ **49. a.** False
b. False   **c.** False   **d.** True   **e.** False   **f.** False   **g.** False

51. $\kappa = \frac{2}{(1 + 4x^2)^{3/2}}$ **53.** $\kappa = \frac{x}{(x^2 + 1)^{3/2}}$

57. $\kappa = \frac{|ab|}{(a^2 \cos^2 t + b^2 \sin^2 t)^{3/2}}$ **59.** $\kappa = \frac{2|a|}{(1 + 4a^2 t^2)^{3/2}}$

61. b. $v_A(t) = \langle 1, 2, 3 \rangle, a_A(t) = \langle 0, 0, 0 \rangle$ and $v_B(t) = \langle 2t, 4t, 6t \rangle$, $a_B(t) = \langle 2, 4, 6 \rangle$; A has constant velocity and zero acceleration, while B has increasing speed and constant acceleration.
c. $a_A(t) = 0N + 0T, a_B(t) = 0N + 2\sqrt{14}T$; both normal components are zero since the path is a straight line ($\kappa = 0$).
63. b. $v_A(t) = \langle -\sin t, \cos t \rangle, a_A(t) = \langle -\cos t, -\sin t \rangle$
$v_B(t) = \langle -2t \sin t^2, 2t \cos t^2 \rangle$
$a_B(t) = \langle -4t^2 \cos t^2 - 2 \sin t^2, -4t^2 \sin t^2 + 2 \cos t^2 \rangle$
c. $a_A(t) = N + 0T, a_B(t) = 4t^2 N + 2T$; for A, the acceleration is always normal to the curve, but this is not true for B.

65. b. $\kappa = \frac{1}{2\sqrt{2}(1 - \cos t)}$ **c.**

d. Minimum curvature at $t = \pi$

67. b. $\kappa = \frac{1}{t(1 + t^2)^{3/2}}$ **c.**

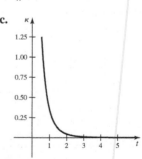

d. No maximum or minimum curvature

69. $\kappa = \dfrac{e^x}{(1 + e^{2x})^{3/2}}, \left(-\dfrac{\ln 2}{2}, \dfrac{1}{\sqrt{2}}\right), \dfrac{2\sqrt{3}}{9}$

71. $\dfrac{1}{\kappa} = \dfrac{1}{2}; x^2 + \left(y - \dfrac{1}{2}\right)^2 = \dfrac{1}{4}$

73. $\dfrac{1}{\kappa} = 4; (x - \pi)^2 + (y + 2)^2 = 16$

75. $\kappa\left(\dfrac{\pi}{2n}\right) = n^2; \kappa$ increases as n increases.

77. a. Speed $= \sqrt{V_0^2 - 2V_0 gt \sin\alpha + g^2 t^2}$

b. $\kappa(t) = \dfrac{gV_0 \cos\alpha}{(V_0^2 - 2V_0 gt \sin\alpha + g^2 t^2)^{3/2}}$

c. Speed has a minimum at $t = \dfrac{V_0 \sin\alpha}{g}$ and $\kappa(t)$ has a maximum at

$t = \dfrac{V_0 \sin\alpha}{g}$. **79.** $\kappa = \dfrac{1}{|\mathbf{v}|} \cdot \left|\dfrac{d\mathbf{T}}{dt}\right|$, where $\mathbf{T} = \dfrac{\langle b, d, f \rangle}{\sqrt{b^2 + d^2 + f^2}}$

and $b, d,$ and f are constant. Therefore, $\dfrac{d\mathbf{T}}{dt} = \mathbf{0}$ so $\kappa = 0$.

81. a. $\kappa_1(x) = \dfrac{2}{(1 + 4x^2)^{3/2}}$

$\kappa_2(x) = \dfrac{12x^2}{(1 + 16x^6)^{3/2}}$

$\kappa_3(x) = \dfrac{30x^4}{(1 + 36x^{10})^{3/2}}$

b.

c. κ_1 has its maximum at $x = 0$, κ_2 has its maxima at $x = \pm\sqrt[6]{\tfrac{1}{56}}$, and κ_3 has its maxima at $x = \pm\sqrt[10]{\tfrac{1}{99}}$. **d.** $\lim\limits_{n\to\infty} z_n = 1$; the graphs of $y = f_n(x)$ show that as $n \to \infty$, the point corresponding to maximum curvature gets arbitrarily close to the point $(1, 0)$.

Chapter 14 Review Exercises, pp. 916–918

1. a. False **b.** True **c.** True **d.** True **e.** False **f.** False

3.

5.

7. $x^2 + y^2 + z^2 = 2; y = z;$ a tilted circle of radius $\sqrt{2}$ centered at $(0, 0, 0)$ **9.** $\mathbf{r}(t) = \langle 4 + 15t, -2 - t, 3 - 5t \rangle$

11. $\mathbf{r}(t) = \langle 2, 3\cos t, 4\sin t \rangle$, for $0 \le t \le 2\pi$

13. $\mathbf{r}(t) = \langle \cos t, \sin t, \sin t \rangle$, for $0 \le t \le 2\pi$

15. $\mathbf{r}(t) = \langle 3\cos t, \sin t, \sin t \rangle$, for $0 \le t \le 2\pi$

17. a. $\langle 1, 0 \rangle; \langle 0, 1 \rangle$ **b.** $\left\langle -\dfrac{2}{(2t + 1)^2}, \dfrac{1}{(t + 1)^2} \right\rangle; \langle -2, 1 \rangle$

c. $\left\langle \dfrac{8}{(2t + 1)^3}, -\dfrac{2}{(t + 1)^3} \right\rangle$

d. $\left\langle \dfrac{1}{2}\ln|2t + 1|, t - \ln|t + 1| \right\rangle + \mathbf{C}$

19. a. $\langle 0, 3, 0 \rangle$; does not exist

b. $\langle 2\cos 2t, -12\sin 4t, 1 \rangle; \langle 2, 0, 1 \rangle$ **c.** $\langle -4\sin 2t, -48\cos 4t, 0 \rangle$

d. $\left\langle -\dfrac{1}{2}\cos 2t, \dfrac{3}{4}\sin 4t, \dfrac{1}{2}t^2 \right\rangle + \mathbf{C}$ **21.** $2\mathbf{j} + \pi\mathbf{k}$

23. $23\mathbf{i} - 41\mathbf{k}$ **25.** $\mathbf{r}(t) = \left\langle t + 2, -\dfrac{1}{2}\cos 2t + \dfrac{5}{2}, \tan t + 2 \right\rangle$

27. $\mathbf{r}(t) = \langle 4\tan^{-1} t - \pi, t^2 + t - 2, t^3 - 1 \rangle$

29. $\mathbf{T}(t) = \left\langle \dfrac{2e^t}{2e^{2t} + 1}, \dfrac{2e^{2t}}{2e^{2t} + 1}, \dfrac{1}{2e^{2t} + 1} \right\rangle; \left\langle \dfrac{2}{3}, \dfrac{2}{3}, \dfrac{1}{3} \right\rangle$

31. a. $\langle 4e^{4t}, 4e^{4t}, 2e^{4t} \rangle; 6e^{4t}$ **b.** $\langle 16e^{4t}, 16e^{4t}, 8e^{4t} \rangle$

33. $\mathbf{v}(t) = \langle 2 + \sin t, 3 - 2\cos t \rangle;$

$\mathbf{r}(t) = \langle 2t + 2 - \cos t, 3t + 2 - 2\sin t \rangle$

35. a. $\mathbf{v}(t) = \langle 40, -32t + 40\sqrt{3} \rangle;$

$\mathbf{r}(t) = \langle 40t, -16t^2 + 40\sqrt{3}t + 3 \rangle$

b. Approx. 4.37 s; approx. 174.9 ft **c.** 78 ft

37. a. $\mathbf{v}(t) = \langle 4t + 40, 20, 40 - 32t \rangle;$

$\mathbf{r}(t) = \langle 2t^2 + 40t, 20t, -16t^2 + 40t + 2 \rangle$ **b.** 2.549 s

c. 126 ft **39. a.** $(116, 30)$ **b.** 39.1 ft **c.** 2.315 s

d. $\displaystyle\int_0^{2.315} \sqrt{50^2 + (-32t + 50)^2}\, dt$ **e.** 129 ft **f.** 41.4° to 79.4°

41. $(1.47, 3.15, 4.4)$ **43.** 12 **45.** Approx. 6.42

47. a. $\mathbf{v}(t) = \mathbf{i} + t\sqrt{2}\,\mathbf{j} + t^2\mathbf{k}$ **b.** 12

49. $\mathbf{r}(s) = \left\langle (\sqrt{1 + s} - 1)^2, \dfrac{4\sqrt{2}}{3}(\sqrt{1 + s} - 1)^{3/2}, \right.$

$\left. 2(\sqrt{1 + s} - 1) \right\rangle$, for $s \ge 0$ **51. a.** $\mathbf{v} = \langle -6\sin t, 3\cos t \rangle,$

$\mathbf{T} = \dfrac{\langle -2\sin t, \cos t \rangle}{\sqrt{1 + 3\sin^2 t}}$ **b.** $\kappa(t) = \dfrac{2}{3(1 + 3\sin^2 t)^{3/2}}$

c. $\mathbf{N} = \left\langle -\dfrac{\cos t}{\sqrt{1 + 3\sin^2 t}}, -\dfrac{2\sin t}{\sqrt{1 + 3\sin^2 t}} \right\rangle$

d. $|\mathbf{N}| = \sqrt{\dfrac{\cos^2 t + 4\sin^2 t}{1 + 3\sin^2 t}} = \sqrt{\dfrac{(\cos^2 t + \sin^2 t) + 3\sin^2 t}{1 + 3\sin^2 t}} = 1;$

$\mathbf{T} \cdot \mathbf{N} = \dfrac{2\sin t\cos t - 2\sin t\cos t}{1 + 3\sin^2 t} = 0$

e.

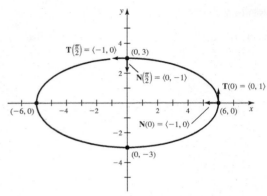

53. a. $\mathbf{v}(t) = \langle -\sin t, -2 \sin t, \sqrt{5} \cos t \rangle$,

$\mathbf{T}(t) = \left\langle -\dfrac{1}{\sqrt{5}} \sin t, -\dfrac{2}{\sqrt{5}} \sin t, \cos t \right\rangle$ **b.** $\kappa(t) = \dfrac{1}{\sqrt{5}}$

c. $\mathbf{N}(t) = \left\langle -\dfrac{1}{\sqrt{5}} \cos t, -\dfrac{2}{\sqrt{5}} \cos t, -\sin t \right\rangle$

d. $|\mathbf{N}(t)| = \sqrt{\dfrac{1}{5} \cos^2 t + \dfrac{4}{5} \cos^2 t + \sin^2 t} = 1$;

$\mathbf{T} \cdot \mathbf{N} = \left(\dfrac{1}{5} \cos t \sin t + \dfrac{4}{5} \cos t \sin t \right) - \sin t \cos t = 0$

e.

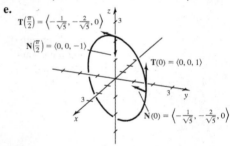

55. a. $\mathbf{a}(t) = 2\mathbf{N} + 0\mathbf{T} = 2\langle -\cos t, -\sin t \rangle$

b.

57. a. $a_T = \dfrac{2t}{\sqrt{t^2 + 1}}$ and $a_N = \dfrac{2}{\sqrt{t^2 + 1}}$

b.

59. $\mathbf{B}(1) = \dfrac{\langle 3, -3, 1 \rangle}{\sqrt{19}}$; $\tau = \dfrac{3}{19}$

61. a. $\mathbf{T}(t) = \dfrac{1}{5}\langle 3 \cos t, -3 \sin t, 4 \rangle$

b. $\mathbf{N}(t) = \langle -\sin t, -\cos t, 0 \rangle$; $\kappa = \dfrac{3}{25}$

c.

d. Yes

e. $\mathbf{B}(t) = \dfrac{1}{5}\langle 4 \cos t, -4 \sin t, -3 \rangle$

f. See graph in part (c).

g. Check that $\mathbf{T}$, $\mathbf{N}$, and $\mathbf{B}$ have unit length and are mutually orthogonal.

h. $\tau = -\dfrac{4}{25}$

63. a. Consider first the case where $a_3 = b_3 = c_3 = 0$, and show that for all $s \neq t$ in I, $\mathbf{r}(t) \times \mathbf{r}(s)$ is a multiple of the constant vector $\langle b_1 c_2 - b_2 c_1, a_2 c_1 - a_1 c_2, a_1 b_2 - a_2 b_1 \rangle$, which implies $\mathbf{r}(t) \times \mathbf{r}(s)$ is always orthogonal to the same vector, and therefore the vectors $\mathbf{r}(t)$ must all lie in the same plane. When a_3, b_3, and c_3 are not necessarily 0, the curve still lies in a plane because these constants represent a simple translation of the curve to a different location in $\mathbb{R}^3$.

b. Because the curve lies in a plane, $\mathbf{B}$ is always normal to the plane and has length 1. Therefore, $\dfrac{d\mathbf{B}}{ds} = \mathbf{0}$ and $\tau = 0$.

CHAPTER 15

Section 15.1 Exercises, pp. 927–930

1. Independent: x and y; dependent: z
3. $D = \{(x, y) : x \neq 0 \text{ and } y \neq 0\}$ **5.** Three   **7.** 3; 4
9. a. 1300 ft   **b.** Katie; Katie is 100 ft higher than Zeke.   **11.** Circles
13. $n = 6$ **15.** $D = \mathbb{R}^2$ **17.** $D = \{(x, y) : x^2 + y^2 \leq 25\}$
19. $D = \{(x, y) : y \neq 0\}$ **21.** $D = \{(x, y) : y < x^2\}$
23. $D = \{(x, y) : xy \geq 0, (x, y) \neq (0, 0)\}$
25. Plane; $D = \mathbb{R}^2$, $R = \mathbb{R}$

27. Hyperbolic paraboloid; $D = \mathbb{R}^2$, $R = \mathbb{R}$

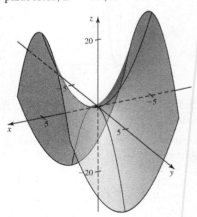

29. Lower part of a hyperboloid of two sheets;
$D = \mathbb{R}^2, R = (-\infty, -1]$

31. Upper half of a hyperboloid of one sheet;
$D = \{(x, y): x^2 + y^2 \geq 1\}, R = [0, \infty)$

33. Upper half of an elliptical cylinder;
$D = \{(x, y): -2 \leq x \leq 2, -\infty < y < \infty\}, R = [0, 4]$

35. a. A **b.** D **c.** B **d.** C

37.

39.

41.

43.

45. a.

b. $\mathbb{R}^2$ excluding the points $(0, 1)$ and $(0, -1)$

c. $\varphi(2, 3)$ is greater. **d.**

47. a.

b. $R(10, 10) = 5$
c. $R(x, y) = R(y, x)$

49. a. $P = \dfrac{20{,}000r}{(1+r)^{240} - 1}$ **b.** $P = \dfrac{Br}{(1+r)^{240} - 1}$, with

$B = 5000, 10{,}000, 15{,}000, 25{,}000$

51. $D = \{(x, y, z): x \neq z\}$; all points not on the plane $x = z$
53. $D = \{(x, y, z): y \geq z\}$; all points on or below the plane $y = z$
55. $D = \{(x, y, z): x^2 \leq y\}$; all points on the side of the vertical cylinder $y = x^2$ that contains the positive y-axis   **57. a.** False
b. False   **c.** True

59. a.

b.

c.

61. a.

b. $(0, 0), (-5, 3), (4, -1)$
c. $f(0, 0) = 10.17, f(-5, 3) = 5.00, f(4, -1) = 4.00$
63. a. $D = \mathbb{R}^2, R = [0, \infty)$
b.

65. a. $D = \{(x, y): x \neq y\}, R = \mathbb{R}$
b.

67. a. $D = \{(x, y): y \neq x + \pi/2 + n\pi \text{ for any integer } n\}$,
$R = [0, \infty)$ **b.**

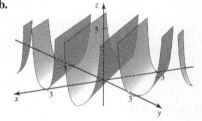

69. Peak at the origin   **71.** Depression at $(1, 0)$ **73.** The level curves are $ax + by = d - cz_0$, where z_0 is a constant, which are lines with slope $-a/b$ if $b \neq 0$ or vertical lines if $b = 0$.
75. $z = x^2 + y^2 - C$; paraboloids with vertices at $(0, 0, -C)$
77. $x^2 + 2z^2 = C$; elliptic cylinders parallel to the y-axis
79. $D = \{(x, y): x - 1 \leq y \leq x + 1\}$
81. $D = \{(x, y, z): (x \leq z \text{ and } y \geq -z) \text{ or } (x \geq z \text{ and } y \leq -z)\}$

Section 15.2 Exercises, pp. 937–939

1. The values of $f(x, y)$ are arbitrarily close to L for all (x, y) sufficiently close to (a, b).   **3.** Because polynomials of n variables are continuous on all of $\mathbb{R}^n$, limits of polynomials can be evaluated with direct substitution.   **5.** If the function approaches different values along different paths, the limit does not exist.   **7.** f must be defined, the limit must exist, and the limit must equal the function value.
9. At any point where the denominator is nonzero   **11.** 10   **13.** 101
15. 27   **17.** $1/(2\pi)$ **19.** 2   **21.** 6   **23.** -1 **25.** 2
27. $1/(2\sqrt{2}) = \sqrt{2}/4$ **29.** $L = 1$ along $y = 0$, and $L = -1$ along $x = 0$ **31.** $L = 1$ along $x = 0$, and $L = -2$ along $y = 0$
33. $L = 2$ along $y = x$, and $L = 0$ along $y = -x$ **35.** $\mathbb{R}^2$
37. All points except $(0, 0)$ **39.** $\{(x, y): x \neq 0\}$ **41.** All points except $(0, 0)$ **43.** $\mathbb{R}^2$ **45.** $\mathbb{R}^2$ **47.** $\mathbb{R}^2$ **49.** All points except $(0, 0)$ **51.** $\mathbb{R}^2$ **53.** $\mathbb{R}^2$ **55.** 6   **57.** -1 **59.** 2   **61. a.** False

b. False   **c.** True   **d.** False   **63.** $\frac{1}{2}$ **65.** 0   **67.** Does not exist
69. $\frac{1}{4}$ **71.** 1   **73.** 1   **75.** 5   **77.** $b = 1$ **79.** 0   **81.** 1   **85.** 0

Section 15.3 Exercises, pp. 948–951

1. $f_x(a, b)$ is the slope of the surface in the direction parallel to the positive x-axis, $f_y(a, b)$ is the slope of the surface in the direction parallel to the positive y-axis, both taken at (a, b).   **3. a.** Negative
b. Negative   **c.** Negative   **d.** Positive   **5.** $f_x(x, y) = 6xy$;
$f_y(x, y) = 3x^2$ **7.** $f_{xy} = 0 = f_{yx}$ **9.** $f_x(x, y, z) = y + z$;
$f_y(x, y, z) = x + z$; $f_z(x, y, z) = x + y$

11. $f_x(x, y) = 5y$; $f_y(x, y) = 5x$ **13.** $f_x(x, y) = \dfrac{1}{y}$; $f_y(x, y) = -\dfrac{x}{y^2}$

15. $f_x(x, y) = e^y$; $f_y(x, y) = xe^y$

17. $f_x(x, y) = 2xye^{x^2y}$; $f_y(x, y) = x^2 e^{x^2y}$

19. $f_w(w, z) = \dfrac{z^2 - w^2}{(w^2 + z^2)^2}$; $f_z(w, z) = -\dfrac{2wz}{(w^2 + z^2)^2}$

21. $f_x(x, y) = \cos xy - xy \sin xy$; $f_y(x, y) = -x^2 \sin xy$

23. $s_y(y, z) = z^3 \sec^2 yz$; $s_z(y, z) = 2z \tan yz + yz^2 \sec^2 yz$

25. $G_s(s, t) = \dfrac{\sqrt{st}(t - s)}{2s(s + t)^2}$; $G_t(s, t) = \dfrac{\sqrt{st}(s - t)}{2t(s + t)^2}$

27. $f_x(x, y) = 2yx^{2y-1}$; $f_y(x, y) = 2x^{2y} \ln x$

29. $h_x(x, y) = \dfrac{\sqrt{x^2 - 4y} - x}{\sqrt{x^2 - 4y}}$; $h_y(x, y) = \dfrac{2}{\sqrt{x^2 - 4y}}$

31. $f_x(x, y) = -e^{x^2}$; $f_y(x, y) = 3y^2 e^{y^6}$

33. $f_x(x, y) = -\dfrac{2x}{1 + (x^2 + y^2)^2}$; $f_y(x, y) = -\dfrac{2y}{1 + (x^2 + y^2)^2}$

35. $h_x(x, y) = (1 + 2y)^x \ln (1 + 2y)$; $h_y(x, y) = 2x(1 + 2y)^{x-1}$
37. $f_x(x, y) = -h(x)$; $f_y(x, y) = h(y)$
39. $h_{xx}(x, y) = 6x$; $h_{xy}(x, y) = 2y = h_{yx}(x, y)$; $h_{yy}(x, y) = 2x$
41. $f_{xx}(x, y) = -16y^3 \sin 4x$; $f_{xy}(x, y) = 12y^2 \cos 4x = f_{yx}(x, y)$;
$f_{yy}(x, y) = 6y \sin 4x$

43. $p_{uu}(u, v) = \dfrac{-2u^2 + 2v^2 + 8}{(u^2 + v^2 + 4)^2}$;

$p_{uv}(u, v) = -\dfrac{4uv}{(u^2 + v^2 + 4)^2} = p_{vu}(u, v)$;

$p_{vv}(u, v) = \dfrac{2u^2 - 2v^2 + 8}{(u^2 + v^2 + 4)^2}$

45. $F_{rr}(r, s) = 0$; $F_{rs}(r, s) = e^s = F_{sr}(r, s)$; $F_{ss}(r, s) = re^s$

47. $f_{xx}(x, y) = \dfrac{6xy^2(1 - 2x^6y^4)}{(1 + x^6y^4)^2}$;

$f_{xy}(x, y) = \dfrac{6x^2y(1 - x^6y^4)}{(1 + x^6y^4)^2} = f_{yx}(x, y)$;

$f_{yy}(x, y) = \dfrac{2x^3(1 - 3x^6y^4)}{(1 + x^6y^4)^2}$

49. $f_{xy} = e^{x+y} = f_{yx}$ **51.** $f_{xy} = -(xy \cos xy + \sin xy) = f_{yx}$

53. $f_{xy}(x, y) = -72y^2(2x - y^3)^2 = f_{yx}(x, y)$
55. $h_x(x, y, z) = h_y(x, y, z) = h_z(x, y, z) = -\sin (x + y + z)$

57. $F_u(u, v, w) = \dfrac{1}{v + w}$; $F_v(u, v, w) = F_w(u, v, w) = -\dfrac{u}{(v + w)^2}$

59. $G_r(r, s, t) = \dfrac{s^3 t^5}{2\sqrt{rs^3 t^5}}$;

$G_s(r, s, t) = \dfrac{3rs^2 t^5}{2\sqrt{rs^3 t^5}}$;

$G_t(r, s, t) = \dfrac{5rs^3 t^4}{2\sqrt{rs^3 t^5}}$

61. $h_w(w, x, y, z) = \dfrac{z}{xy}$; $h_x(w, x, y, z) = -\dfrac{wz}{x^2 y}$;

$h_y(w, x, y, z) = -\dfrac{wz}{xy^2}$; $h_z(w, x, y, z) = \dfrac{w}{xy}$

63. b. $g_x(x, y, z) = -\dfrac{8z}{3(2x - y + z)^2}$;

$g_y(x, y, z) = \dfrac{4z}{3(2x - y + z)^2}$;

$g_z(x, y, z) = \dfrac{4(2x - y)}{3(2x - y + z)^2}$

65. 1.41   **67.** 1.55 (answer will vary)   **69. a.** $\dfrac{\partial V}{\partial P} = -\dfrac{kT}{P^2}$; volume
decreases with pressure at fixed temperature.   **b.** $\dfrac{\partial V}{\partial T} = \dfrac{k}{P}$; volume
increases with temperature at fixed pressure.

c.

71. a. $\dfrac{\partial R}{\partial R_1} = \dfrac{R_2^2}{(R_1 + R_2)^2}$; $\dfrac{\partial R}{\partial R_2} = \dfrac{R_1^2}{(R_1 + R_2)^2}$

b. $\dfrac{\partial R}{\partial R_1} = \dfrac{R^2}{R_1^2}$; $\dfrac{\partial R}{\partial R_2} = \dfrac{R^2}{R_2^2}$ **c.** Increase   **d.** Decrease

73. $u_t = -16e^{-4t} \cos 2x = u_{xx}$ **75.** $u_t = -a^2 Ae^{-a^2 t} \cos ax = u_{xx}$
77. a. No   **b.** No   **c.** $f_x(0, 0) = f_y(0, 0) = 0$ **d.** f_x and f_y are
not continuous at $(0, 0)$.   **79. a.** False   **b.** False   **c.** True

81. a. $z_x(x, y) = \dfrac{1}{y^2}$; $z_y(x, y) = -\dfrac{2x}{y^3}$

b.

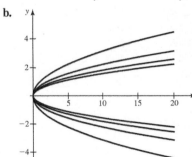

c. z increases as x increases.
d. z increases as y increases when $y < 0$, z is undefined for $y = 0$, and z decreases as y increases for $y > 0$.

83. a. $\varphi_x(x, y) = -\dfrac{2x}{(x^2 + (y - 1)^2)^{3/2}} - \dfrac{x}{(x^2 + (y + 1)^2)^{3/2}}$;

$\varphi_y(x, y) = -\dfrac{2(y - 1)}{(x^2 + (y - 1)^2)^{3/2}} - \dfrac{y + 1}{(x^2 + (y + 1)^2)^{3/2}}$

b. They both approach zero.   **c.** $\varphi_x(0, y) = 0$

d. $\varphi_y(x, 0) = \dfrac{1}{(x^2 + 1)^{3/2}}$

87. $\dfrac{\partial^2 u}{\partial t^2} = -4c^2 \cos(2(x + ct)) = c^2 \dfrac{\partial^2 u}{\partial x^2}$

89. $\dfrac{\partial^2 u}{\partial t^2} = c^2 Af''(x + ct) + c^2 Bg''(x - ct) = c^2 \dfrac{\partial^2 u}{\partial x^2}$

91. $u_{xx} = 6x;\ u_{yy} = -6x$

93. $u_{xx} = \dfrac{2(x - 1)y}{((x - 1)^2 + y^2)^2} - \dfrac{2(x + 1)y}{((x + 1)^2 + y^2)^2};$

$u_{yy} = -\dfrac{2(x - 1)y}{((x - 1)^2 + y^2)^2} + \dfrac{2(x + 1)y}{((x + 1)^2 + y^2)^2}$

95. $\varepsilon_1 = \Delta y, \varepsilon_2 = 0$ or $\varepsilon_1 = 0, \varepsilon_2 = \Delta x$ **97. a.** f is continuous at $(0, 0)$.   **b.** f is not differentiable at $(0, 0)$.
c. $f_x(0, 0) = f_y(0, 0) = 0$ **d.** f_x and f_y are not continuous at $(0, 0)$.
e. Theorem 15.5 does not apply because f_x and f_y are not continuous at $(0, 0)$; Theorem 15.6 does not apply because f is not differentiable at $(0, 0)$.   **99.** $f_x(x, y) = yh(xy);\ f_y(x, y) = xh(xy)$

Section 15.4 Exercises, pp. 957–961

1. One dependent, two intermediate, and one independent variable
3. Multiply each of the partial derivatives of w by the t-derivative of the corresponding function and add all these expressions.
5.

7. $4t^3 + 3t^2$
9. $z'(t) = 2t \sin 4t^3 + 12t^4 \cos 4t^3$

11. $w'(t) = -\sin t \sin 3t^4 + 12t^3 \cos t \cos 3t^4$
13. $z'(t) = 20(\sin^2 t + 2(3t + 4)^5)^9 (\sin t \cos t + 15(3t + 4)^4)$
15. $w'(t) = 20t^4 \sin(t + 1) + 4t^5 \cos(t + 1)$
17. $V'(t) = e^t((2t + 5)\sin t + (2t + 3)\cos t)$
19. $z_s = 2(s - t)\sin t^2;\ z_t = 2(s - t)(t(s - t)\cos t^2 - \sin t^2)$
21. $z_s = 2s - 3s^2 - 2st + t^2;\ z_t = -s^2 - 2t + 2st + 3t^2$
23. $z_s = (t + 1)e^{st+s+t};\ z_t = (s + 1)e^{st+s+t}$
25. $w_s = -\dfrac{2t(t + 1)}{(st + s - t)^2};\ w_t = \dfrac{2s}{(st + s - t)^2}$
27. a. $V'(t) = 2\pi r(t)h(t)r'(t) + \pi r(t)^2 h'(t)$ **b.** $V'(t) = 0$
c. The volume remains constant.
29. $z'(t) = -\dfrac{2t + 2}{(t + 2t)} - \dfrac{3t^2}{(t^3 - 2)}$

31.

$\dfrac{dw}{dt} = \dfrac{dw}{dz}\dfrac{\partial z}{\partial p}\dfrac{dp}{dt} + \dfrac{dw}{dz}\dfrac{\partial z}{\partial q}\dfrac{dq}{dt} + \dfrac{dw}{dz}\dfrac{\partial z}{\partial r}\dfrac{dr}{dt}$

33.

$\dfrac{\partial u}{\partial z} = \dfrac{du}{dv}\left(\dfrac{\partial v}{\partial w}\dfrac{dw}{dz} + \dfrac{\partial v}{\partial x}\dfrac{\partial x}{\partial z} + \dfrac{\partial v}{\partial y}\dfrac{\partial y}{\partial z}\right)$

35. $\dfrac{dy}{dx} = \dfrac{x}{2y}$ **37.** $\dfrac{dy}{dx} = -\dfrac{y}{x}$ **39.** $\dfrac{dy}{dx} = -\dfrac{x + y}{2y^3 + x}$

41. $\dfrac{\partial s}{\partial x} = \dfrac{2x}{\sqrt{x^2 + y^2}};\ \dfrac{\partial s}{\partial y} = \dfrac{2y}{\sqrt{x^2 + y^2}}$

43. $f_{ss} = 2(3s + t);\ f_{st} = 2(s - t);\ f_{tt} = -2(s + 3t)$

45. $f_{ss} = \dfrac{4t^2(-3s^2 + t^2)}{(s^2 + t^2)^3};\ f_{st} = \dfrac{8st(s^2 - t^2)}{(s^2 + t^2)^3};\ f_{tt} = -\dfrac{4(s^4 - 3s^2t^2)}{(s^2 + t^2)^3}$

47. $f''(s) = 4\left(\dfrac{6}{s^4} - \dfrac{2}{s^3} - 1 - 9s + 9s^2\right)$ **49. a.** False   **b.** False

51. $w'(t) = 0$ **53.** $\dfrac{\partial z}{\partial x} = -\dfrac{z^2}{x^2}$ **55. a.** $w'(t) = af_x + bf_y + cf_z$

b. $w'(t) = ayz + bxz + cxy = 3abct^2$

c. $w'(t) = \sqrt{a^2 + b^2 + c^2}\,\dfrac{t}{|t|}$

d. $w''(t) = a^2 f_{xx} + b^2 f_{yy} + c^2 f_{zz} + 2ab f_{xy} + 2ac f_{xz} + 2bc f_{yz}$

57. $\dfrac{\partial z}{\partial x} = -\dfrac{y + z}{x + y};\ \dfrac{\partial z}{\partial y} = -\dfrac{x + z}{x + y}$ **59.** $\dfrac{\partial z}{\partial x} = -\dfrac{yz + 1}{xy - 1};$

$\dfrac{\partial z}{\partial y} = -\dfrac{xz + 1}{xy - 1}$ **61. a.** $z'(t) = -2x \sin t + 8y \cos t = 3 \sin 2t$

b. $0 < t < \pi/2$ and $\pi < t < 3\pi/2$

63. a. $z'(t) = \dfrac{(x + y)e^{-t}}{\sqrt{1 - x^2 - y^2}} = \dfrac{2e^{-2t}}{\sqrt{1 - 2e^{-2t}}}$ **b.** All $t \geq \frac{1}{2}\ln 2$

65. $E'(t) = mx'x'' + my'y'' + mgy' = 0$
67. a. The volume increases.   **b.** The volume decreases.

69. a. $\dfrac{\partial P}{\partial V} = -\dfrac{P}{V};\ \dfrac{\partial T}{\partial P} = \dfrac{V}{k};\ \dfrac{\partial V}{\partial T} = \dfrac{k}{P}$ **b.** Follows directly from part (a)

71. a. $w'(t) = \dfrac{2t(t^2 + 1)\cos 2t - (t^2 - 1)\sin 2t}{2(t^2 + 1)^2}$

b. Max value of $t \approx 0.838$, $(x, y, z) \approx (0.669, 0.743, 0.838)$

73. a. $z_x = \dfrac{x}{r}z_r - \dfrac{y}{r^2}z_\theta;\ z_y = \dfrac{y}{r}z_r + \dfrac{x}{r^2}z_\theta$

b. $z_{xx} = \dfrac{x^2}{r^2}z_{rr} + \dfrac{y^2}{r^4}z_{\theta\theta} - \dfrac{2xy}{r^3}z_{r\theta} + \dfrac{y^2}{r^3}z_r + \dfrac{2xy}{r^4}z_\theta$

c. $z_{yy} = \dfrac{y^2}{r^2}z_{rr} + \dfrac{x^2}{r^4}z_{\theta\theta} + \dfrac{2xy}{r^3}z_{r\theta} + \dfrac{x^2}{r^3}z_r - \dfrac{2xy}{r^4}z_\theta$

75. a. $\left(\dfrac{\partial z}{\partial x}\right)_y = -\dfrac{F_x}{F_z}$ **b.** $\left(\dfrac{\partial y}{\partial z}\right)_x = -\dfrac{F_z}{F_y};\ \left(\dfrac{\partial x}{\partial y}\right)_z = -\dfrac{F_y}{F_x}$

d. $\left(\dfrac{\partial w}{\partial x}\right)_{y,z}\left(\dfrac{\partial z}{\partial w}\right)_{x,y}\left(\dfrac{\partial y}{\partial z}\right)_{x,w}\left(\dfrac{\partial x}{\partial y}\right)_{z,w} = 1$

77. a. $\left(\dfrac{\partial w}{\partial x}\right)_y = f_x + f_z\dfrac{dz}{dx} = 18$ **b.** $\left(\dfrac{\partial w}{\partial x}\right)_z = f_x + f_y\dfrac{dy}{dx} = 8$

d. $\left(\dfrac{\partial w}{\partial y}\right)_x = -5;\ \left(\dfrac{\partial w}{\partial y}\right)_z = 4;\ \left(\dfrac{\partial w}{\partial z}\right)_x = \dfrac{5}{2};\ \left(\dfrac{\partial w}{\partial z}\right)_y = \dfrac{9}{2}$

Section 15.5 Exercises, pp. 970–973

1. Form the dot product between the unit direction vector **u** and the gradient of the function.   **3.** Direction of steepest ascent
5. The gradient is orthogonal to the level curves of f.
7. -2 **9.** $-7; 0$
11. a.

	$(a,b) = (2,0)$	$(a,b) = (0,2)$	$(a,b) = (1,1)$
$\mathbf{u} = \left\langle \frac{\sqrt{2}}{2}, \frac{\sqrt{2}}{2} \right\rangle$	$-\sqrt{2}$	$-2\sqrt{2}$	$-3\sqrt{2}/2$
$\mathbf{v} = \left\langle -\frac{\sqrt{2}}{2}, \frac{\sqrt{2}}{2} \right\rangle$	$\sqrt{2}$	$-2\sqrt{2}$	$-\sqrt{2}/2$
$\mathbf{w} = \left\langle -\frac{\sqrt{2}}{2}, -\frac{\sqrt{2}}{2} \right\rangle$	$\sqrt{2}$	$2\sqrt{2}$	$3\sqrt{2}/2$

b. The function is decreasing at $(2, 0)$ in the direction of **u** and increasing at $(2, 0)$ in the directions of **v** and **w**.
13. $\nabla f(x, y) = \langle 6x, -10y \rangle$, $\nabla f(2, -1) = \langle 12, 10 \rangle$
15. $\nabla g(x, y) = \langle 2(x - 4xy - 4y^2), -4x(x + 4y) \rangle$, $\nabla g(-1, 2) = \langle -18, 28 \rangle$ **17.** $\nabla f(x, y) = e^{2xy} \langle 1 + 2xy, 2x^2 \rangle$, $\nabla f(1, 0) = \langle 1, 2 \rangle$ **19.** $\nabla F(x, y) = -2e^{-x^2 - 2y^2} \langle x, 2y \rangle$, $\nabla F(-1, 2) = 2e^{-9} \langle 1, -4 \rangle$ **21.** -6 **23.** $\frac{27}{2} - 6\sqrt{3}$
25. $-\dfrac{2}{\sqrt{5}}$ **27.** -2 **29.** 0   **31. a.** Direction of steepest ascent:
$\dfrac{1}{\sqrt{65}} \langle 1, 8 \rangle$; direction of steepest descent: $-\dfrac{1}{\sqrt{65}} \langle 1, 8 \rangle$

b. $\langle -8, 1 \rangle$ **33. a.** Direction of steepest ascent: $\dfrac{1}{\sqrt{5}} \langle -2, 1 \rangle$; direction of steepest descent: $\dfrac{1}{\sqrt{5}} \langle 2, -1 \rangle$ **b.** $\langle 1, 2 \rangle$
35. a. Direction of steepest ascent: $\dfrac{1}{\sqrt{2}} \langle 1, -1 \rangle$; direction of steepest descent: $\dfrac{1}{\sqrt{2}} \langle -1, 1 \rangle$ **b.** $\langle 1, 1 \rangle$
37. a. $\nabla f(3, 2) = -12\mathbf{i} - 12\mathbf{j}$
b. Direction of max increase: $\theta = \dfrac{5\pi}{4}$; direction of max decrease: $\theta = \dfrac{\pi}{4}$; directions of no change: $\theta = \dfrac{3\pi}{4}, \dfrac{7\pi}{4}$
c. $g(\theta) = -12\cos\theta - 12\sin\theta$ **d.** $\theta = \frac{5\pi}{4}, g\left(\frac{5\pi}{4}\right) = 12\sqrt{2}$
e. $\nabla f(3, 2) = 12\sqrt{2} \left\langle \cos\frac{5\pi}{4}, \sin\frac{5\pi}{4} \right\rangle$, $|\nabla f(3, 2)| = 12\sqrt{2}$
39. a. $\nabla f(\sqrt{3}, 1) = \frac{\sqrt{6}}{6} \langle \sqrt{3}, 1 \rangle$ **b.** Direction of max increase: $\theta = \frac{\pi}{6}$; direction of max decrease: $\theta = \frac{7\pi}{6}$; directions of no change: $\theta = \frac{2\pi}{3}, \frac{5\pi}{3}$ **c.** $g(\theta) = \frac{\sqrt{2}}{2}\cos\theta + \frac{\sqrt{6}}{6}\sin\theta$ **d.** $\theta = \frac{\pi}{6}, g\left(\frac{\pi}{6}\right) = \frac{\sqrt{6}}{3}$
e. $\nabla f(\sqrt{3}, 1) = \frac{\sqrt{6}}{3} \langle \cos\frac{\pi}{6}, \sin\frac{\pi}{6} \rangle$, $|\nabla f(\sqrt{3}, 1)| = \frac{\sqrt{6}}{3}$
41. a. $\nabla F(-1, 0) = \frac{2}{e}\mathbf{i}$ **b.** Direction of max increase: $\theta = 0$; direction of max decrease: $\theta = \pi$; directions of no change: $\theta = \pm\frac{\pi}{2}$
c. $g(\theta) = \frac{2}{e}\cos\theta$ **d.** $\theta = 0, g(0) = \frac{2}{e}$
e. $\nabla F(-1, 0) = \frac{2}{e} \langle \cos 0, \sin 0 \rangle$, $|\nabla F(-1, 0)| = \frac{2}{e}$
43.

45.

47. $y' = 0$
49. Vertical tangent
51. $y' = -2/\sqrt{3}$
53. Vertical tangent

55. a. $\nabla f = \langle 1, 0 \rangle$ **b.** $x = 4 - t, y = 4, t \geq 0$
c. $\mathbf{r}(t) = \langle 4 - t, 4, 8 - t \rangle$, for $t \geq 0$
57. a. $\nabla f = \langle -2x, -4y \rangle$ **b.** $y = x^2, x \geq 1$
c. $\mathbf{r}(t) = \langle t, t^2, 4 - t^2 - 2t^4 \rangle$, for $t \geq 1$
59. a. $\nabla f(x, y, z) = 2x\mathbf{i} + 4y\mathbf{j} + 8z\mathbf{k}$, $\nabla f(1, 0, 4) = 2\mathbf{i} + 32\mathbf{k}$
b. $\dfrac{1}{\sqrt{257}}(\mathbf{i} + 16\mathbf{k})$ **c.** $2\sqrt{257}$ **d.** $17\sqrt{2}$
61. a. $\nabla f(x, y, z) = 4yz\mathbf{i} + 4xz\mathbf{j} + 4xy\mathbf{k}$,
$\nabla f(1, -1, -1) = 4\mathbf{i} - 4\mathbf{j} - 4\mathbf{k}$ **b.** $\dfrac{1}{\sqrt{3}}(\mathbf{i} - \mathbf{j} - \mathbf{k})$ **c.** $4\sqrt{3}$
d. $\dfrac{4}{\sqrt{3}}$ **63. a.** $\nabla f(x, y, z) = \cos(x + 2y - z)(\mathbf{i} + 2\mathbf{j} - \mathbf{k})$,
$\nabla f\left(\frac{\pi}{6}, \frac{\pi}{6}, -\frac{\pi}{6}\right) = -\frac{1}{2}\mathbf{i} - \mathbf{j} + \frac{1}{2}\mathbf{k}$ **b.** $\dfrac{1}{\sqrt{6}}(-\mathbf{i} - 2\mathbf{j} + \mathbf{k})$
c. $\sqrt{6}/2$ **d.** $-\dfrac{1}{2}$
65. a. $\nabla f(x, y, z) = \dfrac{2}{1 + x^2 + y^2 + z^2}(x\mathbf{i} + y\mathbf{j} + z\mathbf{k})$,
$\nabla f(1, 1, -1) = \frac{1}{2}\mathbf{i} + \frac{1}{2}\mathbf{j} - \frac{1}{2}\mathbf{k}$ **b.** $\dfrac{1}{\sqrt{3}}(\mathbf{i} + \mathbf{j} - \mathbf{k})$
c. $\dfrac{\sqrt{3}}{2}$ **d.** $\dfrac{5}{6}$ **67. a.** False   **b.** False   **c.** False   **d.** True
69. $\pm\dfrac{1}{\sqrt{5}}(\mathbf{i} - 2\mathbf{j})$ **71.** $\pm\dfrac{1}{\sqrt{2}}(\mathbf{i} + \mathbf{j})$
73. $x = x_0 + at, y = y_0 + bt$ **75. a.** $\nabla f(x, y, z) = \langle 2x, 2y, 2z \rangle$, $\nabla f(1, 1, 1) = \langle 2, 2, 2 \rangle$ **b.** $x + y + z = 3$
77. a. $\nabla f(x, y, z) = e^{x+y-z}\langle 1, 1, -1 \rangle$, $\nabla f(1, 1, 2) = \langle 1, 1, -1 \rangle$
b. $x + y - z = 0$
79. a.

b. $\mathbf{v} = \pm\langle 1, 1 \rangle$
c. $\mathbf{v} = \pm\langle 1, -1 \rangle$

83. $\langle u, v \rangle = \langle \pi\cos\pi x\sin 2\pi y, 2\pi\sin\pi x\cos 2\pi y \rangle$
87. $\nabla f(x, y) = \dfrac{1}{(x^2 + y^2)^2} \langle y^2 - x^2 - 2xy, x^2 - y^2 - 2xy \rangle$
89. $\nabla f(x, y, z) = -\dfrac{1}{\sqrt{25 - x^2 - y^2 - z^2}} \langle x, y, z \rangle$
91. $\nabla f(x, y, z) = \dfrac{(y + xz)\langle 1, z, y \rangle - (x + yz)\langle z, 1, x \rangle}{(y + xz)^2}$
$= \dfrac{1}{(y + xz)^2} \langle y(1 - z^2), x(z^2 - 1), y^2 - x^2 \rangle$

Section 15.6 Exercises, pp. 980–983

1. The gradient of f is a multiple of **n**.
3. $F_x(a, b, c)(x - a) + F_y(a, b, c)(y - b) + F_z(a, b, c)(z - c) = 0$
5. Multiply the change in x by $f_x(a, b)$ and the change in y by $f_y(a, b)$, and add both terms to f.   **7.** $dz = f_x(x, y)\,dx + f_y(x, y)\,dy$
9. $z = 5x - 3y + 5$ **11.** $3x - y + 6z = 4$ **13.** $2x + y + z = 4$; $4x + y + z = 7$ **15.** $x + y + z = 6$; $3x + 4y + z = 12$
17. $z = -8x - 4y + 16$ and $z = 4x + 2y + 7$ **19.** $z = y + 1$
and $z = x + 1$ **21.** $x + \dfrac{1}{2}y + \sqrt{3}z = 2 + \dfrac{\sqrt{3}\pi}{6}$ and
$\dfrac{1}{2}x + y + \sqrt{3}z = \dfrac{5\sqrt{3}\pi}{6} - 2$ **23.** $\dfrac{1}{2}x + \dfrac{2}{3}y + 2\sqrt{3}z = -2$ and
$x - 2y + 2\sqrt{14}z = 2$ **25.** $z = 8x - 4y - 4$ and $z = -x - y - 1$
27. $z = \dfrac{7}{25}x - \dfrac{1}{25}y - \dfrac{2}{5}$ and $z = -\dfrac{7}{25}x + \dfrac{1}{25}y + \dfrac{6}{5}$
29. $z = \dfrac{1}{2}x + \dfrac{1}{2}y + \dfrac{\pi}{4} - 1$
31. $\dfrac{1}{6}(x - \pi) + \dfrac{\pi}{6}(y - 1) + \pi\left(z - \dfrac{1}{6}\right) = 0$
33. a. $L(x, y) = 4x + y - 6$ **b.** $L(2.1, 2.99) = 5.39$
35. a. $L(x, y) = -6x - 4y + 7$ **b.** $L(3.1, -1.04) = -7.44$
37. a. $L(x, y, z) = x + y + 2z$ **b.** $L(0.1, -0.2, 0.2) = 0.3$
39. $dz = -6dx - 5dy = -0.1$ **41.** $dz = dx + dy = 0.05$
43. a. The surface area decreases.   **b.** Impossible to say
c. $\Delta S \approx 53.3$ **d.** $\Delta S \approx 33.95$ **e.** $R\,dR = r\,dr$ **45.** $\dfrac{\Delta A}{A} \approx 3.5\%$
47. $dw = (y^2 + 2xz)\,dx + (2xy + z^2)\,dy + (x^2 + 2yz)\,dz$
49. $dw = \dfrac{dx}{y + z} - \dfrac{u + x}{(y + z)^2}\,dy - \dfrac{u + x}{(y + z)^2}\,dz + \dfrac{du}{y + z}$
51. a. $\Delta c \approx 0.035$ **b.** When $\theta = \dfrac{\pi}{20}$ **53. a.** True   **b.** True
c. False   **55.** $(1, -1, 1)$ and $(1, -1, -1)$
57. Points with $x = 0, \pm\dfrac{\pi}{2}, \pm\pi$ and $y = \pm\dfrac{\pi}{2}$, or points with
$x = \pm\dfrac{\pi}{4}, \pm\dfrac{3\pi}{4}$ and $y = 0, \pm\pi$
59. a. $\Delta S \approx 0.749$ **b.** More sensitive to changes in r
61. a. $\Delta A \approx \dfrac{2}{1225} = 0.00163$ **b.** No. The batting average increases
more if the batter gets a hit than it decreases if he fails to get a hit.
c. Yes. The answer depends on whether A is less than 0.500 or greater
than 0.500.   **63. a.** $\Delta V \approx \dfrac{21}{5000} = 0.0042$ **b.** $\dfrac{\Delta V}{V} \approx -4\%$ **c.** $2p$
65. a. $f_r = n(1 - r)^{n-1}, f_n = -(1 - r)^n \ln(1 - r)$
b. $\Delta P \approx 0.027$ **c.** $\Delta P \approx 2 \times 10^{-20}$ **67.** $\Delta R \approx 7/540 \approx 0.013\Omega$
69. a. Apply the Chain Rule.   **b.** Follows directly from (a)
c. $d(\ln(xy)) = \dfrac{dx}{x} + \dfrac{dy}{y}$ **d.** $d(\ln(x/y)) = \dfrac{dx}{x} - \dfrac{dy}{y}$
e. $\dfrac{df}{f} = \dfrac{dx_1}{x_1} + \dfrac{dx_2}{x_2} + \cdots + \dfrac{dx_n}{x_n}$

Section 15.7 Exercises, pp. 993–996

1. The local maximum occurs at the highest point on the surface; you
cannot get to a higher point in any direction.   **3.** The partial derivatives are both zero or do not exist.   **5.** The discriminant is a determinant; it is defined as $D(a, b) = f_{xx}(a, b) f_{yy}(a, b) - f_{xy}^2(a, b)$.
7. f has an absolute minimum value on R at (a, b) if $f(a, b) \leq f(x, y)$
for all (x, y) in R.   **9.** Saddle point   **11.** Local min   **13.** $(0, 0)$
15. $(0, 1), (0, -1)$ **17.** $(0, 0), (2, 2)$, and $(-2, -2)$
19. $(0, 2), (\pm 1, 2)$ **21.** $(3, 0), (-15, 6)$ **23.** Saddle point at $(0, 0)$

25. Local min at $(0, 0)$ **27.** Saddle point at $(0, 0)$; local min at $(1, 1)$
and $(-1, -1)$ **29.** $(0, 0)$; Second Derivative Test is inconclusive;
absolute min of 4 at $(0, 0)$ **31.** Local min at $(2, 0)$
33. Saddle point at $(0, 0)$; local max at $\left(\dfrac{1}{\sqrt{2}}, \dfrac{1}{\sqrt{2}}\right)$ and
$\left(-\dfrac{1}{\sqrt{2}}, -\dfrac{1}{\sqrt{2}}\right)$; local min at $\left(\dfrac{1}{\sqrt{2}}, -\dfrac{1}{\sqrt{2}}\right)$ and $\left(-\dfrac{1}{\sqrt{2}}, \dfrac{1}{\sqrt{2}}\right)$
35. Local min at $(-1, 0)$; local max at $(1, 0)$ **37.** Saddle point at
$(0, 1)$; local min at $(\pm 2, 0)$ **39.** Saddle point at $(0, 0)$ **41.** Saddle
point   **43.** Height $= 32$ in, base is 16 in $\times$ 16 in; volume is 8192 in^3
45. 2 m $\times$ 2 m $\times$ 1 m   **47.** Absolute min: $0 = f(0, 1)$; absolute
max: $9 = f(0, -2)$ **49.** Absolute min: $4 = f(0, 0)$; absolute max:
$7 = f(\pm 1, \pm 1)$ **51.** Absolute min: $0 = f(1, 0)$; absolute max:
$3 = f(1, 1) = f(1, -1)$ **53.** Absolute min:
$1 = f(1, -2) = f(1, 0)$; absolute max: $4 = f(1, -1)$
55. Absolute min: $0 = f(0, 0)$; absolute max: $\dfrac{7}{8} = f\left(\dfrac{1}{\sqrt{2}}, \sqrt{2}\right)$
57. a. 1.83; 82.6°   **b.** 14%   **59.** Absolute min: $-4 = f(0, 0)$; no
absolute max on R **61.** Absolute max: $2 = f(0, 0)$; no absolute min
on R **63.** $P\left(\dfrac{4}{3}, \dfrac{2}{3}, \dfrac{4}{3}\right)$ **65.** $(3, 4, 5), (3, 4, -5)$
67. a. True   **b.** False   **c.** True   **d.** True
69. Local min at $(0.3, -0.3)$; saddle point at $(0, 0)$
71. a.–d. $x = y = z = \dfrac{200}{3}$
73. a. $P\left(1, \dfrac{1}{3}\right)$ **b.** $P\left(\dfrac{1}{3}(x_1 + x_2 + x_3), \dfrac{1}{3}(y_1 + y_2 + y_3)\right)$
c. $P(\bar{x}, \bar{y})$, where $\bar{x} = \dfrac{1}{n}\displaystyle\sum_{k=1}^{n} x_k$ and $\bar{y} = \dfrac{1}{n}\displaystyle\sum_{k=1}^{n} y_k$
d. $d(x, y) = \sqrt{x^2 + y^2} + \sqrt{(x - 2)^2 + y^2} + $
$\sqrt{(x - 1)^2 + (y - 1)^2}$. Absolute max: $1 + \sqrt{3} = f\left(1, \dfrac{1}{\sqrt{3}}\right)$
77. $y = \dfrac{22}{13}x + \dfrac{46}{13}$ **79.** $a = b = c = 3$
81. a. $\nabla d_1(x, y) = \dfrac{x - x_1}{d_1(x, y)}\mathbf{i} + \dfrac{y - y_1}{d_1(x, y)}\mathbf{j}$
b. $\nabla d_2(x, y) = \dfrac{x - x_2}{d_2(x, y)}\mathbf{i} + \dfrac{y - y_2}{d_2(x, y)}\mathbf{j}$;
$\nabla d_3(x, y) = \dfrac{x - x_3}{d_3(x, y)}\mathbf{i} + \dfrac{y - y_3}{d_3(x, y)}\mathbf{j}$
c. Follows from $\nabla f = \nabla d_1 + \nabla d_2 + \nabla d_3$ **d.** Three unit vectors
add to zero.   **e.** P is the vertex at the large angle.
f. $P(0.255457, 0.304504)$ **83. a.** Local max at $(1, 0), (-1, 0)$
b. Local max at $(1, 0)$ and $(-1, 0)$ **85.** $\dfrac{abc\sqrt{3}}{2}$

Section 15.8 Exercises, pp. 1002–1004

1. The level curve of f is tangent to the curve $g = 0$ at the optimal
point; therefore, the gradients are parallel.
3. $1 = 2\lambda x, 4 = 2\lambda y, x^2 + y^2 - 1 = 0$
5. Abs. min: 1; abs. max: 8   **7.** Abs. min: $-2\sqrt{5}$ at
$\left(-\dfrac{2}{\sqrt{5}}, -\dfrac{4}{\sqrt{5}}\right)$; abs. max: $2\sqrt{5}$ at $\left(\dfrac{2}{\sqrt{5}}, \dfrac{4}{\sqrt{5}}\right)$
9. Abs. min: -2 at $(-1, -1)$; abs. max: 2 at $(1, 1)$
11. Abs. min: -3 at $(-\sqrt{3}, \sqrt{3})$ and $(\sqrt{3}, -\sqrt{3})$;
abs. max: 9 at $(3, 3)$ and $(-3, -3)$
13. Abs. min: e^{-9} at $(-3, 3)$ and $(3, -3)$; abs. max: e^3 at $(\sqrt{3}, \sqrt{3})$
and $(-\sqrt{3}, -\sqrt{3})$ **15.** Abs. min: 9 at $(0, 3)$; abs. max: 34 at
$(-\sqrt{15}, -2)$ and $(\sqrt{15}, -2)$ **17.** Abs. min: $-2\sqrt{11}$ at
$\left(-\dfrac{2}{\sqrt{11}}, -\dfrac{6}{\sqrt{11}}, \dfrac{2}{\sqrt{11}}\right)$; abs. max: $2\sqrt{11}$ at $\left(\dfrac{2}{\sqrt{11}}, \dfrac{6}{\sqrt{11}}, -\dfrac{2}{\sqrt{11}}\right)$

19. Abs. min: $-\dfrac{\sqrt{5}}{2}$ at $\left(-\dfrac{\sqrt{5}}{2}, 0, \dfrac{1}{2}\right)$; abs. max: $\dfrac{\sqrt{5}}{2}$ at $\left(\dfrac{\sqrt{5}}{2}, 0, \dfrac{1}{2}\right)$

21. Abs. min: -5 at $(-2, -2, -1)$; abs. max: 5 at $(2, 2, 1)$

23. Abs. min: -10 at $(-5, 0, 0)$; abs. max: $\dfrac{29}{2}$ at $\left(2, 0, \pm\sqrt{\dfrac{21}{2}}\right)$

25. Abs. min: $-\sqrt{3}$ at $\left(0, -\dfrac{1}{\sqrt{3}}, -\dfrac{2}{\sqrt{3}}\right)$; abs. max: $\dfrac{7}{4}$ at $\left(\dfrac{1}{2}, \dfrac{1}{2}, 1\right)$

and $\left(-\dfrac{1}{2}, \dfrac{1}{2}, 1\right)$ **27.** 18 in $\times$ 18 in $\times$ 36 in   **29.** Abs. min: 0.6731;

abs. max: 1.1230   **31.** 2×1 **33.** $\left(-\dfrac{3}{17}, \dfrac{29}{17}, -3\right)$

35. Abs. min: $\sqrt{38 - 6\sqrt{29}}$ (or $\sqrt{29} - 3$); abs. max:
$\sqrt{38 + 6\sqrt{29}}$ (or $\sqrt{29} + 3$)   **37.** $\ell = 3$ and $g = \dfrac{3}{2}$; $U = 15\sqrt{2}$
39. $\ell = \dfrac{16}{5}$ and $g = 1$; $U = 20.287$ **41. a.** True   **b.** False

43. $\dfrac{\sqrt{6}}{3}$ m $\times \dfrac{\sqrt{6}}{3}$ m $\times \dfrac{\sqrt{6}}{6}$ m   **45.** $2 \times 1 \times \dfrac{2}{3}$ **47.** $P\left(\dfrac{4}{3}, \dfrac{2}{3}, \dfrac{4}{3}\right)$

49. Abs. min: 1; abs. max: 9   **51.** Abs. min: 0; abs. max: 3
53. $K = 7.5$ and $L = 5$ **57.** Abs. max: 8   **59.** Abs. max:
$\sqrt{a_1^2 + a_2^2 + a_3^2 + \cdots + a_n^2}$ **61. a.** Gradients are perpendicular
to level surfaces.   **b.** If the gradient were not in the plane spanned
by ∇g and ∇h, f could be increased (decreased) by moving the point
slightly.   **c.** ∇f is a linear combination of ∇g and ∇h, since it belongs
to the plane spanned by these two vectors.   **d.** The gradient condition
from part (c), as well as the constraints, must be satisfied.
63. Abs. min: $2 - 4\sqrt{2}$; abs. max: $2 + 4\sqrt{2}$
65. a. $y + 1 = \lambda y, x + 1 = \lambda x, xy = 4$ **c.** Abs. min of 108
over the curve C_1 **d.** Abs. max of 100 over the curve C_2
e. The constraint curve is unbounded, so there is no guarantee that an
abs. min or max occurs over the curve $xy = 4$.

Chapter 15 Review Exercises, pp. 1005–1007

1. a. True   **b.** False   **c.** False   **d.** False
3. $D = \{(x, y): x \geq y^2\}$

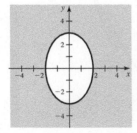

5. $D = \left\{(x, y): \dfrac{x^2}{4} + \dfrac{y^2}{9} \geq 1\right\}$

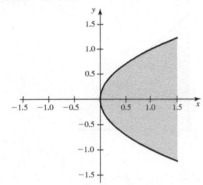

7. $D = \{(x, y): x^2 + y^2 \geq 1\}$; $R = (-\infty, 0]$; lower half of the
hyperboloid of one sheet $x^2 + y^2 - z^2 = 1$

9.

11. 2   **13.** Does not exist
15. $\dfrac{2}{3}$ **17.** 4
19. $\{(x, y): y > x^2 + 1\}$
21. $f_x = 6xy^5$; $f_y = 15x^2y^4$

23. $f_x = \dfrac{2xy^2}{(x^2 + y^2)^2}$; $f_y = -\dfrac{2x^2y}{(x^2 + y^2)^2}$

25. $f_x = y(1 + xy)e^{xy}$; $f_y = x(1 + xy)e^{xy}$ **27.** $f_{xx} = 4y^2e^{2xy}$;
$f_{xy} = 2e^{2xy}(2xy + 1) = f_{yx}$; $f_{yy} = 4x^2e^{2xy}$

29. $\dfrac{\partial^2 u}{\partial x^2} = 6y = -\dfrac{\partial^2 u}{\partial y^2}$ **31. a.** V increases with R if r is fixed,

$V_R > 0$; V decreases if r increases and R is fixed, $V_r < 0$.
b. $V_r = -4\pi r^2$; $V_R = 4\pi R^2$ **c.** The volume increases more if R is
increased.   **33.** $4t + 2 \ln 5$

35. $w_r = \dfrac{3r + s}{r(r + s)}$; $w_s = \dfrac{r + 3s}{s(r + s)}$; $w_t = \dfrac{1}{t}$

37. $\dfrac{dy}{dx} = -\dfrac{2xy}{2y^2 + (x^2 + y^2) \ln (x^2 + y^2)}$

39. a. $z'(t) = -24 \sin t \cos t = -12 \sin 2t$

b. $z'(t) > 0$ for $\dfrac{\pi}{2} < t < \pi$ and $\dfrac{3\pi}{2} < t < 2\pi$

41. a.

	$(a, b) = (0, 0)$	$(a, b) = (2, 0)$	$(a, b) = (1, 1)$
$\mathbf{u} = \left\langle \dfrac{\sqrt{2}}{2}, \dfrac{\sqrt{2}}{2} \right\rangle$	0	$4\sqrt{2}$	$-2\sqrt{2}$
$\mathbf{v} = \left\langle -\dfrac{\sqrt{2}}{2}, \dfrac{\sqrt{2}}{2} \right\rangle$	0	$-4\sqrt{2}$	$-6\sqrt{2}$
$\mathbf{w} = \left\langle -\dfrac{\sqrt{2}}{2}, -\dfrac{\sqrt{2}}{2} \right\rangle$	0	$-4\sqrt{2}$	$2\sqrt{2}$

b. The function is increasing at $(2, 0)$ in the direction of $\mathbf{u}$ and
decreasing at $(2, 0)$ in the directions of $\mathbf{v}$ and $\mathbf{w}$.
43. $\nabla g = \langle 2xy^3, 3x^2y^2 \rangle$; $\nabla g(-1, 1) = \langle -2, 3 \rangle$; $D_{\mathbf{u}} g(-1, 1) = 2$

45. $\nabla h = \left\langle \dfrac{x}{\sqrt{2 + x^2 + 2y^2}}, \dfrac{2y}{\sqrt{2 + x^2 + 2y^2}} \right\rangle$;

$\nabla h(2, 1) = \left\langle \dfrac{\sqrt{2}}{2}, \dfrac{\sqrt{2}}{2} \right\rangle$; $D_{\mathbf{u}} h(2, 1) = \dfrac{7\sqrt{2}}{10}$

47. $\nabla f = \langle y \cos xy, x \cos xy, -\sin z \rangle$; $\nabla f(1, \pi, 0) = \langle -\pi, -1, 0 \rangle$;

$D_{\mathbf{u}} f(1, \pi, 0) = -\dfrac{1}{7}(3 + 2\pi)$

49. a. Direction of steepest ascent: $\mathbf{u} = \dfrac{\sqrt{2}}{2}\mathbf{i} - \dfrac{\sqrt{2}}{2}\mathbf{j}$; direction of

steepest descent: $\mathbf{u} = -\dfrac{\sqrt{2}}{2}\mathbf{i} + \dfrac{\sqrt{2}}{2}\mathbf{j}$

b. No change: $\mathbf{u} = \pm\left(\dfrac{\sqrt{2}}{2}\mathbf{i} + \dfrac{\sqrt{2}}{2}\mathbf{j}\right)$

51. Tangent line is vertical; $\nabla f(2, 0) = -8\mathbf{i}$

53. $E = \dfrac{kx}{x^2 + y^2}\mathbf{i} + \dfrac{ky}{x^2 + y^2}\mathbf{j}$

55. $y = 2$ and $12x + 3y - 2z = 12$
57. $x + 2y + 3z = 6$ and $x - 2y + 3z = 6$
59. $x + y - z = 0$ and $x + y - z = 0$
61. a. $L(x, y) = x + 5y$ **b.** $L(1.95, 0.05) = 2.2$ **63.** Approx. -4%
65. a. $\Delta V \approx -0.1\pi$ m^3 **b.** $\Delta S \approx -0.05\pi$ m^2
67. Saddle point at $(0, 0)$; local min at $(2, -2)$

69. Saddle points at $(0, 0)$ and $(-2, 2)$; local max at $(0, 2)$; local min at $(-2, 0)$ **71.** Abs. min: $-1 = f(1, 1) = f(-1, -1)$; abs. max: $49 = f(2, -2) = f(-2, 2)$

73. Abs. min: $-\dfrac{1}{2} = f\left(-\dfrac{1}{\sqrt{2}}, \dfrac{1}{\sqrt{2}}\right)$; abs. max: $\dfrac{1}{2} = f\left(\dfrac{1}{\sqrt{2}}, \dfrac{1}{\sqrt{2}}\right)$

75. Abs. min: $\dfrac{23}{2} = f\left(\dfrac{1}{3}, \dfrac{5}{6}\right)$ abs. max: $\dfrac{29}{2} = f\left(\dfrac{5}{3}, \dfrac{7}{6}\right)$;

77. Abs. min: $-\sqrt{6} = f\left(-\dfrac{\sqrt{6}}{6}, -\dfrac{\sqrt{6}}{3}, \dfrac{\sqrt{6}}{6}\right)$;

abs. max: $\sqrt{6} = f\left(\dfrac{\sqrt{6}}{6}, \dfrac{\sqrt{6}}{3}, -\dfrac{\sqrt{6}}{6}\right)$

79. $\dfrac{2a^2}{\sqrt{a^2 + b^2}}$ by $\dfrac{2b^2}{\sqrt{a^2 + b^2}}$

81. $x = \dfrac{1}{2} + \dfrac{\sqrt{10}}{20}, y = \dfrac{3}{2} + \dfrac{3\sqrt{10}}{20} = 3x, z = \dfrac{1}{2} + \dfrac{\sqrt{10}}{2} = \sqrt{10}x$

83. $(1, 2, 5)$

CHAPTER 16

Section 16.1 Exercises, pp. 1015–1017

1. $\displaystyle\int_0^2\int_1^3 xy \, dy \, dx$ or $\displaystyle\int_1^3\int_0^2 xy \, dx \, dy$ **3.** $\displaystyle\int_{-2}^4\int_1^5 f(x, y) \, dy \, dx$ or

$\displaystyle\int_1^5\int_{-2}^4 f(x, y) \, dx \, dy$ **5.** 48   **7.** 4   **9.** $\dfrac{32}{3}$ **11.** 4   **13.** $\dfrac{224}{9}$

15. $10 - 2e$ **17.** $\dfrac{1}{2}$ **19.** $e^2 + 3$ **21.** $\dfrac{1}{2}$ **23.** $10\sqrt{5} - 4\sqrt{2} - 14$

25. $\dfrac{117}{2}$ **27.** $\dfrac{\pi^2}{4} + 1$ **29.** $\dfrac{4}{3}$ **31.** $\dfrac{9 - e^2}{2}$ **33.** $\dfrac{4}{11}$ **35.** $\dfrac{1}{4}$

37. 136   **39.** 3   **41.** $e^2 - 3$ **43.** $e^{16} - 17$ **45.** $\ln\dfrac{5}{3}$ **47.** $\dfrac{1}{2\ln 2}$

49. $\dfrac{8}{3}$ **51. a.** True   **b.** False   **c.** True   **53. a.** 1475   **b.** The sum of products of population densities and areas is a Riemann sum.

55. $\int_c^d\int_a^b f(x) \, dy \, dx = (c - d)\int_a^b f(x) \, dx$. The integral is the area of the cross section of S.   **57.** $a = \pi/6, 5\pi/6$ **59.** $a = \sqrt{6}$

61. a. $\frac{1}{2}\pi^2 + \pi$ **b.** $\frac{1}{2}\pi^2 + \pi$ **c.** $\frac{1}{2}\pi^2 + 2$

63. $f(a, b) - f(a, 0) - f(0, b) + f(0, 0)$

Section 16.2 Exercises, pp. 1024–1027

1.

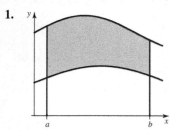

3. $dx \, dy$ **5.** $\displaystyle\int_0^1\int_{x^2}^{\sqrt{x}} f(x, y) \, dy \, dx$ **7.** 4   **9.** $\displaystyle\int_0^2\int_{x^3}^{4x} f(x, y) \, dy \, dx$

11. 2   **13.** $\dfrac{8}{3}$ **15.** 0   **17.** $e - 1$ **19.** $\dfrac{\ln^3 2}{6}$

21. 2   **23.** $\dfrac{\pi}{2} - 1$ **25.** 0   **27.** $\pi - 1$

29.

$\displaystyle\int_1^2\int_{x+1}^{2x+4} f(x, y) \, dy \, dx$

31.

$\displaystyle\int_0^1\int_0^{-2x+2} f(x, y) \, dy \, dx$

33.

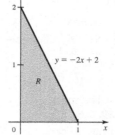

$\displaystyle\int_0^1\int_0^{\sqrt{1-x^2}} f(x, y) \, dy \, dx$

35. $\displaystyle\int_0^{18}\int_{y/2}^{(y+9)/3} f(x, y) \, dx \, dy$

37.

$\displaystyle\int_1^4\int_0^{4-y} f(x, y) \, dx \, dy$

39.

$\displaystyle\int_0^{23}\int_{(y-3)/2}^{(y+7)/3} f(x, y) \, dx \, dy$

41.

$\displaystyle\int_0^1\int_y^{2-y} f(x, y) \, dx \, dy$

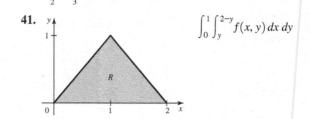

43. 2 **45.** 12 **47.** 5 **49.** 14 **51.** 32 **53.** $\frac{9}{8}$ **55.** $\frac{1}{4}\ln 2$

57. $\int_0^4 \int_{y/2}^{\sqrt{y}} f(x,y)\, dx\, dy$ **59.** $\int_0^{\ln 2}\int_{1/2}^{e^{-x}} f(x,y)\, dy\, dx$

61. $\int_0^{\pi/2}\int_0^{\cos x} f(x,y)\, dy\, dx$ **63.** $\frac{1}{2}(e-1)$ **65.** 0 **67.** $\frac{2}{3}$

69.

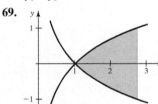

$\int_1^e \int_{-\ln x}^{\ln x} f(x,y)\, dy\, dx$

71. $\frac{11}{12}$ **73.** $\frac{32}{3}$ **75.** 12π **77.** $\frac{43}{6}$ **79.** $\frac{2}{3}$ **81.** 16 **83.** $4a\pi$

85. $\frac{32}{3}$ **87.** 1 **89.** $\frac{140}{3}$ **91. a.** False **b.** False **c.** False

93. 30 **95.** $\frac{a}{3}$ **97. a.**

b. $\frac{15}{4}+4\ln 2$ **c.** $2\ln 2 - \frac{5}{64}$ **99.** $\frac{3}{8e^2}$ **101.** 1

Section 16.3 Exercises, pp. 1033–1036

1. It is called a polar rectangle because r and θ vary between two constants.

3.

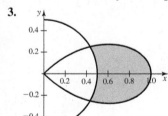

5. Evaluate the integral $\int_\alpha^\beta \int_{g(\theta)}^{h(\theta)} r\, dr\, d\theta$.

7.

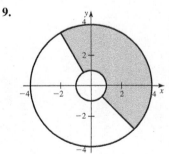

9.

11. $\frac{64\pi}{3}$ **13.** $(8-24e^{-2})\pi$ **15.** $\frac{7\pi}{2}$ **17.** $\frac{9\pi}{2}$ **19.** $\frac{37\pi}{3}$

21. 128π **23.** 0 **25.** $(2-\sqrt{3})\pi$ **27.** $2\pi/5$ **29.** $\frac{14\pi}{3}$

31. $\frac{81\pi}{2}$ **33.** π **35.** 8π **37.** 81π **39.** $\frac{2\pi}{3}(7\sqrt{7}-15)$

41.

$\int_0^{2\pi}\int_0^{1+\frac{1}{2}\cos\theta} g(r,\theta)r\, dr\, d\theta$

43.

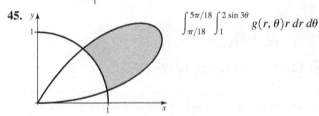

$\int_0^{\pi/2}\int_0^{\sqrt{2\sin 2\theta}} g(r,\theta)r\, dr\, d\theta$

45.

$\int_{\pi/18}^{5\pi/18}\int_1^{2\sin 3\theta} g(r,\theta)r\, dr\, d\theta$

47. $3\pi/2$ **49.** π **51.** $\frac{3\pi}{2}-2\sqrt{2}$ **53.** $2a/3$

55. a. False **b.** True **c.** True **57.** The hyperboloid $\left(V=\frac{112\pi}{3}\right)$

59. a. $R=\{(r,\theta): -\pi/4 \le \theta \le \pi/4 \text{ or } 3\pi/4 \le \theta \le 5\pi/4\}$

b. $\frac{a^4}{4}$ **61.** $\frac{32}{9}$ **63.** $2\pi\left(1-2\ln\frac{3}{2}\right)$ **65.** 1

67. $\pi/4$ **69. a.** $\frac{16\pi}{3}$ **b.** 2.78 **71.** $30\pi + 42$

73. c. $\sqrt{\pi}/2, 1/2,$ and $\sqrt{\pi}/4$ **75. a.** $I=\frac{\sqrt{2}}{2}\tan^{-1}\frac{\sqrt{2}}{2}$

b. $I=\frac{\sqrt{2}}{4}\tan^{-1}\frac{\sqrt{2}a}{2}+\frac{a}{2\sqrt{a^2+1}}\tan^{-1}\frac{1}{\sqrt{a^2+1}}$ **c.** $\frac{\sqrt{2}\pi}{8}$

Section 16.4 Exercises, pp. 1043–1047

1.

3. $\int_{-9}^9 \int_{-\sqrt{81-x^2}}^{\sqrt{81-x^2}} \int_{-\sqrt{81-x^2-y^2}}^{\sqrt{81-x^2-y^2}} f(x,y,z)\, dz\, dy\, dx$

5. $\int_0^1 \int_0^{\sqrt{1-z^2}} \int_0^{\sqrt{1-z^2-x^2}} f(x, y, z)\, dy\, dx\, dz$ **7.** 24   **9.** 8   **11.** $\dfrac{2}{\pi}$

13. 0   **15.** 8   **17.** $\dfrac{16}{3}$ **19.** $1 - \ln 2$

21. $\dfrac{2\pi(1 + 19\sqrt{19} - 20\sqrt{10})}{3}$ **23.** $\dfrac{27\pi}{2}$ **25.** 12π

27. $\dfrac{5}{12}$ **29.** 8   **31.** $\int_0^1 \int_y^1 \int_0^{2\sqrt{1-x^2}} f(x, y, z)\, dz\, dx\, dy$

33. $\int_0^1 \int_0^{2\sqrt{1-x^2}} \int_0^x f(x, y, z)\, dy\, dz\, dx$

35. $\int_0^1 \int_0^{2\sqrt{1-y^2}} \int_{y^2}^{\frac{1}{2}\sqrt{4-z^2}} f(x, y, z)\, dx\, dz\, dy$

37. $\int_0^1 \int_0^2 \int_0^{1-y} dz\, dx\, dy,\ \int_0^2 \int_0^1 \int_0^{1-z} dy\, dz\, dx,\ \int_0^1 \int_0^2 \int_0^{1-z} dy\, dx\, dz,$

$\int_0^1 \int_0^{1-y} \int_0^2 dx\, dz\, dy,\ \int_0^1 \int_0^{1-z} \int_0^2 dx\, dy\, dz$ **39.** $\dfrac{256}{9}$ **41.** $\dfrac{2}{3}$

43. $(10\sqrt{10} - 1)\dfrac{\pi}{6}$ **45.** $\dfrac{3\ln 2}{2} + \dfrac{e}{16} - 1$

47. $\int_0^4 \int_{y/4-1}^0 \int_0^5 dz\, dx\, dy = 10$ **49.** $\int_0^1 \int_0^{\sqrt{1-x^2}} \int_0^{\sqrt{1-x^2}} dz\, dy\, dx = \dfrac{2}{3}$

51. $\dfrac{8}{\pi}$ **53.** $\dfrac{10}{3}$ **55. a.** False   **b.** False   **c.** False   **57.** 2

59. 1   **61.** $\dfrac{16}{3}$ **63.** $\dfrac{16}{3}$ **65.** $a = 2\sqrt{2}$ **67.** $V = \dfrac{\pi r^2 h}{3}$

69. $V = \dfrac{\pi h^2}{3}(3R - h)$ **71.** $V = \dfrac{4\pi abc}{3}$ **73.** $\dfrac{1}{24}$

Section 16.5 Exercises, pp. 1059–1063

1. r measures the distance from the point to the z axis, θ is the angle that the segment from the point to the z-axis makes with the positive xz-plane, and z is the directed distance from the point to the xy-plane.
3. A cone   **5.** It approximates the volume of the cylindrical wedge formed by the changes Δr, $\Delta \theta$, and Δz.

7. $\int_\alpha^\beta \int_{g(\theta)}^{h(\theta)} \int_{G(r,\theta)}^{H(r,\theta)} w(r, \theta, z)\, r\, dz\, dr\, d\theta$ **9.** Cylindrical coordinates

11. Wedge

13. Solid bounded by cone and plane

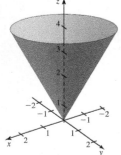

15. 2π **17.** $4\pi/5$ **19.** $\pi(1 - e^{-1})/2$ **21.** $9\pi/4$

23. 560π **25.** 396π **27.** The paraboloid $(V = 44\pi/3)$

29. $\dfrac{20\pi}{3}$ **31.** $\dfrac{(16 + 17\sqrt{29})\pi}{3}$ **33.** $\dfrac{1}{3}$

35. Hollow ball

37. Sphere of radius $r = 2$, centered at $(0, 0, 2)$

39. a. $(3960, 0.74, -2.13),\ (-1426.85, -2257.05, 2924.28)$
b. $(3960, 0.84, 0.22),\ (2877.61, 637.95, 2644.62)$ **c.** 5666 mi

41. $\dfrac{\pi}{2}$ **43.** $4\pi \ln 2$ **45.** $\pi\left(\dfrac{188}{9} - \dfrac{32\sqrt{3}}{3}\right)$ **47.** $\dfrac{32\pi\sqrt{3}}{9}$

49. $\dfrac{5\pi}{12}$ **51.** $\dfrac{8\pi}{3}$ **53.** $\dfrac{8\pi}{3}(9\sqrt{3} - 11)$ **55. a.** True   **b.** True

57. $z = \sqrt{x^2 + y^2 - 1}$; upper half of a hyperboloid of one sheet

59. $\dfrac{8\pi}{3}(1 - e^{-512}) \approx \dfrac{8\pi}{3}$ **61.** 32π

63.

$\int_0^{2\pi} \int_0^{\sqrt{2}} \int_r^{\sqrt{4-r^2}} g(r, \theta, z)\, r\, dz\, dr\, d\theta,$

$\int_0^{2\pi} \int_0^{\sqrt{2}} \int_0^z g(r, \theta, z)\, r\, dr\, dz\, d\theta$

$+ \int_0^{2\pi} \int_{\sqrt{2}}^2 \int_0^{\sqrt{4-z^2}} g(r, \theta, z)\, r\, dr\, dz\, d\theta,$

$\int_0^{\sqrt{2}} \int_r^{\sqrt{4-r^2}} \int_0^{2\pi} g(r, \theta, z)\, r\, d\theta\, dz\, dr$

65.

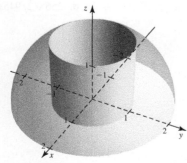

$$\int_{\pi/6}^{\pi/2}\int_0^{2\pi}\int_{\csc\varphi}^2 g(\rho,\varphi,\theta)\rho^2\sin\varphi\,d\rho\,d\theta\,d\varphi,$$

$$\int_{\pi/6}^{\pi/2}\int_{\csc\varphi}^2\int_0^{2\pi} g(\rho,\varphi,\theta)\rho^2\sin\varphi\,d\theta\,d\rho\,d\varphi$$

67. $32\sqrt{3}\pi/9$ **69.** $2\sqrt{2}/3$ **71.** $7\pi/2$ **73.** 95.6036

77. $V=\dfrac{\pi r^2 h}{3}$ **79.** $V=\dfrac{\pi}{3}(R^2+rR+r^2)h$

81. $V=\dfrac{\pi R^3(8r-3R)}{12r}$

Section 16.6 Exercises, pp. 1070–1072

1. The pivot should be located at the center of mass of the system.
3. Use a double integral. Integrate the density function over the region occupied by the plate.   **5.** Use a triple integral to find the mass of the object and the three moments.

7.

$\dfrac{27}{13}$ **9.** Mass is $2+\pi$; $\bar{x}=\dfrac{\pi}{2}$

11. Mass is $\dfrac{20}{3}$; $\bar{x}=\dfrac{9}{5}$ **13.** Mass is 10; $\bar{x}=\dfrac{8}{3}$

15.

$\left(\dfrac{\pi}{2},\dfrac{1}{2}\right)$

17.

$\left(0,\dfrac{1}{3}\right)$

19.

$\left(\dfrac{1}{4}(e^2+1),\dfrac{e}{2}-1\right)\approx(2.10,0.36)$

21. $\left(\dfrac{7}{3},1\right)$; density increases to the right.   **23.** $\left(\dfrac{16}{11},\dfrac{16}{11}\right)$; density increases toward the hypotenuse of the triangle.

25. $\left(0,\dfrac{16+3\pi}{16+12\pi}\right)\approx(0,0.4735)$; density increases away from the x-axis.

27.

$\left(0,0,\dfrac{3}{2}\right)$

29.

$\left(\dfrac{1}{4},\dfrac{1}{4},\dfrac{1}{4}\right)$

31.

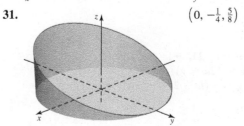

$\left(0,-\dfrac{1}{4},\dfrac{5}{8}\right)$

33. $\left(\dfrac{7}{3},\dfrac{1}{2},\dfrac{1}{2}\right)$ **35.** $\left(0,0,\dfrac{198}{85}\right)$ **37.** $\left(\dfrac{2}{3},\dfrac{7}{3},\dfrac{1}{3}\right)$ **39. a.** False
b. True   **c.** False   **d.** False   **41.** $\bar{x}=\dfrac{\ln(1+L^2)}{2\tan^{-1}L}$, $\displaystyle\lim_{L\to\infty}\bar{x}=\infty$

43. $\left(0,\dfrac{8}{9}\right)$ **45.** $\left(0,\dfrac{8}{3\pi}\right)$ **47.** $\left(\dfrac{5}{6},0\right)$ **49.** $\left(\dfrac{128}{105\pi},\dfrac{128}{105\pi}\right)$

51. On the line of symmetry, $2a/\pi$ units above the diameter
53. $\left(\dfrac{2a}{3(4-\pi)},\dfrac{2a}{3(4-\pi)}\right)$ **55.** $h/4$ units
57. $h/3$ units, where h is the height of the triangle   **59.** $3a/8$ units

61. a. $\left(0,\dfrac{4(1+a+a^2)}{3(1+a)\pi}\right)$

b. $a=\dfrac{1}{2}\left(-1+\sqrt{1+\dfrac{16}{3\pi-4}}\right)\approx0.4937$

63. Depth $=\dfrac{40\sqrt{10}-4}{333}$ cm ≈0.3678 cm

65. a. $(\bar{x},\bar{y})=\left(\dfrac{-r^2}{R+r},0\right)$ (origin at center of large circle);

$(\bar{x},\bar{y})=\left(\dfrac{R^2+Rr+r^2}{R+r},0\right)$ (origin at common point of the circles)
b. *Hint:* Solve $\bar{x}=R-2r$.

Section 16.7 Exercises, pp. 1082–1084

1. The image of S is the 2×2 square with vertices at $(0,0)$, $(2,0)$, $(2,2)$, and $(0,2)$.   **3.** $\displaystyle\int_0^1\int_0^1 f(u+v,u-v)\,2\,du\,dv$
5. The rectangle with vertices at $(0,0)$, $(2,0)$, $\left(2,\dfrac{1}{2}\right)$, and $\left(0,\dfrac{1}{2}\right)$
7. The square with vertices at $(0,0)$, $\left(\dfrac{1}{2},\dfrac{1}{2}\right)$, $(1,0)$, and $\left(\dfrac{1}{2},-\dfrac{1}{2}\right)$
9. The region above the x-axis and bounded by the curves
$y^2=4\pm4x$ **11.** The upper half of the unit circle

13.

15.

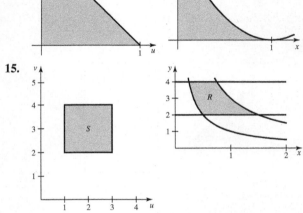

17. -9 **19.** $-4(u^2 + v^2)$ **21.** -1

23. $x = (u + v)/3, y = (2u - v)/3; -1/3$

25. $x = -(u + 3v), y = -(u + 2v); -1$

27. a.

b. $0 \le u \le 1, 0 \le v \le 1$ **c.** $J(u, v) = -2$ **d.** 0

29. a.

b. $0 \le u \le 1, 0 \le v \le 1 - u$ **c.** $J(u, v) = 2$ **d.** $256\sqrt{2}/945$

31. $4\sqrt{2}/3$

33. $3844/5625$

35. $\dfrac{15 \ln 3}{2}$

37. 2   **39.** $2w(u^2 - v^2)$ **41.** 5   **43.** $1024\pi/3$

45. a. True   **b.** True   **c.** True

47. *Hint:* $J(\rho, \varphi, \theta) = \begin{vmatrix} \sin\varphi\cos\theta & \rho\cos\varphi\cos\theta & -\rho\sin\varphi\sin\theta \\ \sin\varphi\sin\theta & \rho\cos\varphi\sin\theta & \rho\sin\varphi\cos\theta \\ \cos\varphi & -\rho\sin\varphi & 0 \end{vmatrix}$

49. $a^2 b^2/2$ **51.** $(a^2 + b^2)/4$ **53.** $4\pi abc/3$

55. $(\bar{x}, \bar{y}, \bar{z}) = \left(0, 0, \dfrac{3c}{8}\right)$ **57. a.** $x = a^2 - \dfrac{y^2}{4a^2}$

b. $x = \dfrac{y^2}{4b^2} - b^2$ **c.** $J(u, v) = 4(u^2 + v^2)$ **d.** $\dfrac{80}{3}$ **e.** 160

f. Vertical lines become parabolas opening downward with vertices on the positive y-axis, and horizontal lines become parabolas opening upward with vertices on the negative y-axis.   **59. a.** S is stretched in the positive u- and v-directions but not in the w-direction. The amount of stretching increases with u and v.   **b.** $J(u, v, w) = ad$

c. Volume $= ad$ **d.** $\left(\dfrac{a + b + c}{2}, \dfrac{d + e}{2}, \dfrac{1}{2} \right)$

Chapter 16 Review Exercises, pp. 1084–1088

1. a. False **b.** True **c.** False **d.** False **3.** $\frac{26}{3}$

5. $\displaystyle\int_0^1 \int_{-\sqrt{y}}^{\sqrt{y}} f(x, y)\, dx\, dy$ **7.** $\displaystyle\int_0^1 \int_0^{\sqrt{1-x^2}} f(x, y)\, dy\, dx$

9. $\dfrac{304}{3}$

11. $\dfrac{\sqrt{17} - \sqrt{2}}{2}$

13. 8π **15.** $\dfrac{2}{7\pi^2}$ **17.** $\dfrac{1}{5}$

19. $\dfrac{9\pi}{2}$

21. $6\pi - 16$

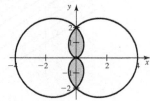

23. 2   **25.** $\displaystyle\int_0^1 \int_{2y}^2 \int_0^{\sqrt{z^2 - 4y^2}/2} f(x, y, z)\, dx\, dz\, dy$ **27.** $\pi - \dfrac{4}{3}$

29. $8 \sin^2 2 = 4(1 - \cos 4)$ **31.** $\dfrac{848}{9}$ **33.** $\dfrac{8}{15}$ **35.** $\dfrac{16}{3}$

37. $\dfrac{128}{3}$ **39.** $\dfrac{\pi}{6} - \dfrac{\sqrt{3}}{2} + \dfrac{1}{2}$ **41.** $\dfrac{1}{3}$ **43.** $\dfrac{1}{3}$ **45.** π

47. 4π **49.** $\dfrac{28\pi}{3}$ **51.** $\dfrac{2048\pi}{105}$

53.

$(\bar{x}, \bar{y}) = \left(\dfrac{\pi}{2}, \dfrac{\pi}{8} \right)$

55.

$(\bar{x}, \bar{y}) = \left(0, \dfrac{56}{9\pi} \right)$

57. $(\bar{x}, \bar{y}, \bar{z}) = (0, 0, 24)$ **59.** $(\bar{x}, \bar{y}, \bar{z}) = \left(0, 0, \dfrac{63}{10} \right)$

61. $\dfrac{h}{3}$ **63.** $\left(\dfrac{4\sqrt{2}a}{3\pi}, \dfrac{4(2 - \sqrt{2})a}{3\pi} \right)$ **65. a.** $\dfrac{4\pi}{3}$ **b.** $\dfrac{16Q}{3}$

67. $R = \{ (x, y) : 0 \le x \le 1, 0 \le y \le 1 \}$

69. The parallelogram with vertices $(0, 0)$, $(3, 1)$, $(4, 4)$, and $(1, 3)$

71. 10   **73.** 6

75. a.

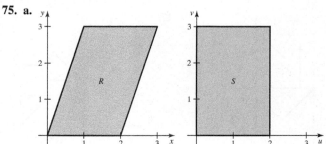

b. $0 \le u \le 2, 0 \le v \le 3$ **c.** $J(u, v) = 1$ **d.** $\frac{63}{2}$

77. a.

b. $u \le v \le 1 - u, 0 \le u \le \dfrac{1}{2}$ **c.** $J(u, v) = 2$ **d.** $\dfrac{1}{60}$

79. 42

81. $-\dfrac{7}{16}$

CHAPTER 17

Section 17.1 Exercises, pp. 1096–1098

1. $\mathbf{F} = \langle f, g, h \rangle$ evaluated at (x, y, z) is the velocity vector of an air particle at (x, y, z) at a fixed point in time.   **3.** At selected points (a, b), plot the vector $\langle f(a, b), g(a, b) \rangle$.   **5.** It shows the direction in which the temperature increases the fastest and the amount of increase.

7.

9.

11.

13.

15.

17.

19.

21.

23.

25. a. $(0, 1)$ **b.** None
c.

27. a. None
b. At all points on C
c.

29. a. None
b. $(1, 0)$
c.

31. $\mathbf{F} = \langle -y, x \rangle$ or $\mathbf{F} = \langle -1, 1 \rangle$

33. $\mathbf{F}(x, y) = \dfrac{\langle x, y \rangle}{\sqrt{x^2 + y^2}} = \dfrac{\mathbf{r}}{|\mathbf{r}|}, \mathbf{F}(0, 0) = \mathbf{0}$

35. $\nabla\varphi(x, y) = \langle 2xy - y^2, x^2 - 2xy \rangle$

37. $\nabla\varphi(x, y) = \langle 1/y, -x/y^2 \rangle$ **39.** $\nabla\varphi(x, y, z) = \langle x, y, z \rangle = \mathbf{r}$

41. $\nabla\varphi(x, y, z) = -(x^2 + y^2 + z^2)^{-3/2} \langle x, y, z \rangle = -\dfrac{\mathbf{r}}{|\mathbf{r}|^3}$

43. a. $\mathbf{F} = \langle -2, 1 \rangle$
b. $\mathbf{F}(-1, -2) = \mathbf{F}(0, 0) = \mathbf{F}(1, 2) = \mathbf{F}(2, 4) = \langle -2, 1 \rangle$
c.

45. a. $\mathbf{F} = \langle \cos x, -1 \rangle$
b. $\mathbf{F}\left(\dfrac{\pi}{2}, 1\right) = \langle 0, -1 \rangle; \mathbf{F}(\pi, 0) = \langle -1, -1 \rangle;$
$\mathbf{F}\left(\dfrac{3\pi}{2}, -1\right) = \langle 0, -1 \rangle; \mathbf{F}(2\pi, 0) = \langle 1, -1 \rangle$
c.

47. $\nabla\varphi(x, y) = 2\langle x, y \rangle$

49. a. $\nabla\varphi(x, y) = \langle 2, 3 \rangle$
b. $y' = -2/3, \langle 1, -\frac{2}{3} \rangle \cdot \nabla\varphi(1, 1) = 0$
c. $y' = -2/3, \langle 1, -\frac{2}{3} \rangle \cdot \nabla\varphi(x, y) = 0$
d.

51. a. $\nabla\varphi(x, y) = \langle e^{x-y}, -e^{x-y} \rangle = e^{x-y}\langle 1, -1 \rangle$
b. $y' = 1, \langle 1, 1 \rangle \cdot \nabla\varphi(1, 1) = 0$
c. $y' = 1, \langle 1, 1 \rangle \cdot \nabla\varphi(x, y) = 0$
d.

53. a. True   **b.** False   **c.** True   **55. a.** $\mathbf{E} = \dfrac{c}{x^2 + y^2}\langle x, y \rangle$

b. $|\mathbf{E}| = \left|\dfrac{c}{|\mathbf{r}|^2}\mathbf{r}\right| = \dfrac{c}{r}$ **c.** *Hint:* The equipotential curves are circles centered at the origin.   **57.** The slope of the streamline at (x, y) is $y'(x)$, which equals the slope of the vector $\mathbf{F}(x, y)$, which is g/f. Therefore, $y'(x) = g/f$.

59.

$$y = x + C$$

61.

$$x^2 + y^2 = C$$

63. For $\theta = 0$: $\mathbf{u}_r = \mathbf{i}$ and $\mathbf{u}_\theta = \mathbf{j}$
for $\theta = \frac{\pi}{2}$: $\mathbf{u}_r = \mathbf{j}$ and $\mathbf{u}_\theta = -\mathbf{i}$
for $\theta = \pi$: $\mathbf{u}_r = -\mathbf{i}$ and $\mathbf{u}_\theta = -\mathbf{j}$
for $\theta = \frac{3\pi}{2}$: $\mathbf{u}_r = -\mathbf{j}$ and $\mathbf{u}_\theta = \mathbf{i}$

65.

$$\mathbf{F} = \frac{1}{\sqrt{x^2 + y^2}} \langle -y, x \rangle$$

67.

$$\mathbf{F} = r\,\mathbf{u}_\theta$$

Section 17.2 Exercises, pp. 1110–1114

1. A line integral is taken along a curve; an ordinary single-variable integral is taken along an interval. **3.** $\int_{\pi/2}^{\pi} \frac{1}{t} \cos t \sqrt{\sin^2 t + 1}\, dt$
5. $\mathbf{r}(t) = \langle 1 + 4t, 2 + 2t, 3 - 3t \rangle$, for $0 \le t \le 1$
7. $\mathbf{r}(t) = \langle t^2 + 1, t \rangle$, for $2 \le t \le 4$
9. a. $\int_0^2 (t + 6t^5)\, dt$ **b.** 66 **11.** $\int_C \mathbf{F} \cdot d\mathbf{r}$ and $\int_C f\, dx + g\, dy + h\, dz$
13. 7 **15.** Take the line integral of $\mathbf{F} \cdot \mathbf{T}$ along the curve with arc length as the parameter. **17.** 0 **19.** 100 **21.** 8 **23.** $-40\pi^2$
25. 128π **27.** $\frac{\sqrt{2}}{2} \ln 10$ **29.** $\frac{112}{9}$ **31.** 8 **33.** 414 **35.** 409.5
37. $\frac{15}{2}$ **39.** $\sqrt{101}$ **41.** $\frac{17}{2}$ **43.** 49 **45.** $\frac{3}{4\sqrt{10}}$ **47. a.** Negative
b. Positive **49.** 0 **51.** 16 **53.** 0 **55.** $\frac{3\sqrt{3}}{10}$ **57. b.** 0
59. a. Negative **b.** -4π **61. a.** True **b.** True **c.** True **d.** True
63. a. Both paths require the same work: $W = 28,200$.
b. Both paths require the same work: $W = 28,200$.
65. a. $\frac{5\sqrt{5} - 1}{12}$ **b.** $\frac{5\sqrt{5} - 1}{12}$ **c.** The results are identical.
67. The work equals zero for all three paths.
69. $8\pi(48 + 7\pi - 128\pi^2) \approx -29{,}991.4$ **71.** 2π **73. a.** 4 **b.** -4
e. 0 **75.** *Hint:* Show that $\int_C \mathbf{F} \cdot \mathbf{T}\, ds = \pi r^2(c - b)$.
77. *Hint:* Show that $\int_C \mathbf{F} \cdot \mathbf{n}\, ds = \pi r^2(a + d)$. **79. a.** $\ln a$ **b.** No
c. $\frac{1}{6}\left(1 - \frac{1}{a^2}\right)$ **d.** Yes **e.** $W = \frac{3^{1-p/2}}{2 - p}(a^{2-p} - 1)$, for $p \ne 2$;
otherwise, $W = \ln a$. **f.** $p > 2$ **81.** ab

Section 17.3 Exercises, pp. 1121–1123

1. A simple curve has no self-intersections; the initial and terminal points of a closed curve are identical. **3.** Test for equality of partial derivatives as given in Theorem 17.3. **5.** Integrate f with respect to x and make the constant of integration a function of y to obtain $\varphi = \int f\, dx + h(y)$; finally, set $\frac{\partial \varphi}{\partial y} = g$ to determine h. **7.** 0
9. Conservative **11.** Not conservative **13.** Conservative
15. Conservative **17.** $\varphi(x, y) = \frac{1}{2}(x^2 + y^2)$ **19.** Not conservative
21. $\varphi(x, y) = \sqrt{x^2 + y^2}$ **23.** $\varphi(x, y, z) = xz + y$
25. Not conservative **27.** $\varphi(x, y, z) = xy + yz + zx$
29. $\varphi(x, y) = \sqrt{x^2 + y^2 + z^2}$ **31. a, b.** 0 **33. a, b.** 2 **35.** 3
37. -10 **39.** 24 **41.** $-\frac{1}{2}$ **43.** $-\pi^2$ **45.** 0 **47.** 0 **49.** 0
51. -5 **53.** 0 **55.** 1 **57. a.** False **b.** True **c.** True **d.** True
e. True **59.** 10 **61.** 25 **63. a.** Negative **b.** Positive **c.** No
67. a. Compare partial derivatives.
b. $\varphi(x, y, z) = \frac{GMm}{\sqrt{x^2 + y^2 + z^2}} = \frac{GMm}{|\mathbf{r}|}$
c. $\varphi(B) - \varphi(A) = GMm\left(\frac{1}{r_2} - \frac{1}{r_1}\right)$ **d.** No
69. a. $\dfrac{\partial}{\partial y}\left(\dfrac{-y}{(x^2 + y^2)^{p/2}}\right) = \dfrac{-x^2 + (p-1)y^2}{(x^2 + y^2)^{1+p/2}}$ and
$\dfrac{\partial}{\partial x}\left(\dfrac{x}{(x^2 + y^2)^{p/2}}\right) = \dfrac{(1 - p)x^2 + y^2}{(x^2 + y^2)^{1+p/2}}$
b. The two partial derivatives in (a) are equal if $p = 2$.
c. $\varphi(x, y) = \tan^{-1}(y/x)$ **73.** $\varphi(x, y) = \frac{1}{2}(x^2 + y^2)$
75. $\varphi(x, y) = \frac{1}{2}(x^4 + x^2y^2 + y^4)$

Section 17.4 Exercises, pp. 1133–1136

1. In both forms, the integral of a *derivative* is computed from boundary data. **3.** Area $= \frac{1}{2}\oint_C x\, dy - y\, dx$, where C encloses the region **5.** The integral in the flux form of Green's Theorem vanishes.
7. $\mathbf{F} = \langle y, x \rangle$

9. a. 0 **b.** 2 **c.** Yes **d.** No **11. a.** -4 **b.** 0 **c.** No **d.** Yes
13. a. y^2 **b.** $12x^2y + 2xy$ **c.** No **d.** No **15. a.** 1; no
b. $\mathbf{r}_1(t) = \langle t, t^2 \rangle$, for $0 \le t \le 1$, and $\mathbf{r}_2(t) = \langle 1 - t, 1 - t \rangle$, for
$0 \le t \le 1$ (answers may vary) **c.** Both integrals equal $\frac{1}{6}$. **d.** 0
17. a. -4 **b.** Both integrals equal -8. **19. a.** $4x$ **b.** $\frac{16}{3}$
21. 25π **23.** 16π **25.** 32 **27. a.** 2 **b.** Both integrals equal 8π.
29. a. $2y$ **b.** $\frac{16}{15}$ **31.** 104 **33.** $\frac{31 - 3e^4}{6}$ **35.** 6 **37.** $\frac{8}{3}$
39. $8 - \frac{\pi}{2}$ **41. a.** 0 **b.** 3π **43. a.** 0 **b.** $-\frac{15\pi}{2}$ **45. a.** 0
b. 2π **47. a.** $\frac{16}{3}$ **b.** 0 **49. a.** True **b.** False **c.** True

51. Note: $\dfrac{\partial f}{\partial y} = 0 = \dfrac{\partial g}{\partial x}$ **53.** The integral becomes $\iint_R 2\,dA$.

55. a. $f_x = g_y = 0$ **b.** $\psi(x, y) = -2x + 4y$

57. a. $f_x = e^{-x} \sin y = -g_y$ **b.** $\psi(x, y) = e^{-x} \cos y$

59. a. *Hint:* $f_x = e^x \cos y,\ f_y = -e^x \sin y,$

$g_x = -e^x \sin y,\ g_y = -e^x \cos y$

b. $\varphi(x, y) = e^x \cos y,\ \psi(x, y) = e^x \sin y$

61. a. *Hint:* $f_x = -\dfrac{y}{x^2 + y^2},\ f_y = \dfrac{x}{x^2 + y^2},$

$g_x = \dfrac{x}{x^2 + y^2},\ g_y = \dfrac{y}{x^2 + y^2}$

b. $\varphi(x, y) = x \tan^{-1} \dfrac{y}{x} + \dfrac{y}{2} \ln(x^2 + y^2) - y,$

$\psi(x, y) = y \tan^{-1} \dfrac{y}{x} - \dfrac{x}{2} \ln(x^2 + y^2) + x$

63. a.

$\mathbf{F} = \langle -4 \cos x \sin y,\ 4 \sin x \cos y \rangle$ **b.** Yes, the divergence equals zero.
c. No, the two-dimensional curl equals $8 \cos x \cos y$.   **d.** 0   **e.** 32
67. c. The vector field is undefined at the origin.

69.

Basic ideas: Let C_1 and C_2 be two smooth simple curves from A to B.

$\displaystyle \int_{C_1} \mathbf{F} \cdot \mathbf{n}\,ds - \int_{C_2} \mathbf{F} \cdot \mathbf{n}\,ds = \oint_C \mathbf{F} \cdot \mathbf{n}\,ds = \iint_R (f_x + g_y)\,dA = 0$

and $\displaystyle \int_{C_1} \mathbf{F} \cdot \mathbf{n}\,ds = \int_{C_1} \psi_x\,dx + \psi_y\,dy = \int_{C_1} d\psi = \psi(B) - \psi(A)$

71. Use $\nabla\varphi \cdot \nabla\psi = \langle f, g \rangle \cdot \langle -g, f \rangle = 0$.

Section 17.5 Exercises, pp. 1143–1146

1. Compute $f_x + g_y + h_z$.   **3.** There is no source or sink.
5. It indicates the axis and the angular speed of the circulation at a
point.   **7.** 0   **9.** 3   **11.** 0   **13.** $2(x + y + z)$

15. $\dfrac{x^2 + y^2 + 3}{(1 + x^2 + y^2)^2}$ **17.** $\dfrac{1}{|\mathbf{r}|^2}$ **19.** $-\dfrac{1}{|\mathbf{r}|^4}$ **21. a.** Positive for

both points   **b.** div $\mathbf{F} = 2$ **c.** Outward everywhere   **d.** Positive
23. a. curl $\mathbf{F} = 2\mathbf{i}$ **b.** $|\text{curl } \mathbf{F}| = 2$
25. a. curl $\mathbf{F} = 2\mathbf{i} - 2\mathbf{j} + 2\mathbf{k}$ **b.** $|\text{curl } \mathbf{F}| = 2\sqrt{3}$ **27.** $3y\,\mathbf{k}$
29. $-4z\,\mathbf{j}$ **31.** 0   **33.** 0   **35.** Follows from partial differentiation of
$\dfrac{1}{(x^2 + y^2 + z^2)^{3/2}}$ **37.** Combine Exercise 36 with Theorem 17.10.

39. a. False   **b.** False   **c.** False   **d.** False   **e.** False   **41. a.** No
b. No   **c.** Yes, scalar function   **d.** No   **e.** No   **f.** No   **g.** Yes,
vector field   **h.** No   **i.** Yes, vector field   **43. a.** At $(0, 1, 1)$,
$\mathbf{F}$ points in the positive x-direction; at $(1, 1, 0)$, $\mathbf{F}$ points in the negative z-direction; at $(0, 1, -1)$, $\mathbf{F}$ points in the negative x-direction;
and at $(-1, 1, 0)$, $\mathbf{F}$ points in the positive z-direction. These vectors

circle the y-axis in the counterclockwise direction looking along $\mathbf{a}$
from head to tail.   **b.** The argument in part (a) can be repeated in
any plane perpendicular to the y-axis to show that the vectors of $\mathbf{F}$
circle the y-axis in the counterclockwise direction looking along $\mathbf{a}$
from head to tail. Alternatively, computing the cross product, we
find that $\mathbf{F} = \mathbf{a} \times \mathbf{r} = \langle z, 0, -x \rangle$, which is a rotation field in any
plane perpendicular to $\mathbf{a}$.   **45.** Compute an explicit expression for
$\mathbf{a} \times \mathbf{r}$ and then take the required partial derivatives.   **47.** div $\mathbf{F}$ has
a maximum value of 6 at $(1, 1, 1)$, $(1, -1, -1)$, $(-1, 1, 1)$, and
$(-1, -1, -1)$.   **49.** $\mathbf{n} = \langle a, b, 2a + b \rangle$, where a and b are real
numbers   **51.** $\mathbf{F} = \frac{1}{2}(y^2 + z^2)\,\mathbf{i}$; no   **53. a.** The wheel does not spin.
b. Clockwise, looking in the positive y-direction   **c.** The wheel does
not spin.   **55.** $\omega = \dfrac{10}{\sqrt{3}}$, or $\dfrac{5}{\sqrt{3}\pi} \approx 0.9189$ revolution per unit time
57. $\mathbf{F} = -200ke^{-x^2 + y^2 + z^2}(-x\,\mathbf{i} + y\,\mathbf{j} + z\,\mathbf{k})$
$\nabla \cdot \mathbf{F} = -200k(1 + 2(x^2 + y^2 + z^2))e^{-x^2 + y^2 + z^2}$

59. a. $\mathbf{F} = -\dfrac{GMm\mathbf{r}}{|\mathbf{r}|^3}$ **b.** See Theorem 17.11.

61. $\rho\left(\dfrac{\partial u}{\partial t} + u\dfrac{\partial u}{\partial x} + v\dfrac{\partial u}{\partial y} + w\dfrac{\partial u}{\partial z}\right) = -\dfrac{\partial p}{\partial x} + \mu\left(\dfrac{\partial^2 u}{\partial x^2} + \dfrac{\partial^2 u}{\partial y^2} + \dfrac{\partial^2 u}{\partial z^2}\right)$

$\rho\left(\dfrac{\partial v}{\partial t} + u\dfrac{\partial v}{\partial x} + v\dfrac{\partial v}{\partial y} + w\dfrac{\partial v}{\partial z}\right) = -\dfrac{\partial p}{\partial y} + \mu\left(\dfrac{\partial^2 v}{\partial x^2} + \dfrac{\partial^2 v}{\partial y^2} + \dfrac{\partial^2 v}{\partial z^2}\right)$

$\rho\left(\dfrac{\partial w}{\partial t} + u\dfrac{\partial w}{\partial x} + v\dfrac{\partial w}{\partial y} + w\dfrac{\partial w}{\partial z}\right) = -\dfrac{\partial p}{\partial z} + \mu\left(\dfrac{\partial^2 w}{\partial x^2} + \dfrac{\partial^2 w}{\partial y^2} + \dfrac{\partial^2 w}{\partial z^2}\right)$

63. a. Use $\nabla \times \mathbf{B} = -Ak \cos(kz - \omega t)\,\mathbf{i}$ and
$\dfrac{\partial \mathbf{E}}{\partial t} = -A\omega \cos(kz - \omega t)\,\mathbf{i}$.   **b.**

Section 17.6 Exercises, pp. 1159–1161

1. $\mathbf{r}(u, v) = \langle a \cos u, a \sin u, v \rangle, 0 \le u \le 2\pi, 0 \le v \le h$
3. $\mathbf{r}(u, v) = \langle a \sin u \cos v, a \sin u \sin v, a \cos u \rangle, 0 \le u \le \pi,$
$0 \le v \le 2\pi$ **5.** Use the parameterization from Exercise 3 and

compute $\displaystyle \int_0^\pi \int_0^{2\pi} f(a \sin u \cos v, a \sin u \sin v, a \cos u)\, a^2 \sin u \, dv\, du$.
7. The normal vectors point outward.   **9.** $\langle u, v, \frac{1}{3}(16 - 2u + 4v)\rangle$,
$|u| < \infty, |v| < \infty$ **11.** $\langle v \cos u, v \sin u, v \rangle, 0 \le u \le 2\pi,$
$2 \le v \le 8$ **13.** $\langle 3 \cos u, 3 \sin u, v \rangle, 0 \le u \le \dfrac{\pi}{2}, 0 \le v \le 3$

15. The plane $z = 2x + 3y - 1$ **17.** Part of the upper half of the
cone $z^2 = 16x^2 + 16y^2$ of height 12 and radius 3 (with $y \ge 0$)
19. 28π **21.** $16\sqrt{3}$ **23.** $\pi r\sqrt{r^2 + h^2}$ **25.** 1728π **27.** 0

29. 12   **31.** $4\pi\sqrt{5}$ **33.** $\dfrac{(65\sqrt{65} - 1)\pi}{24}$ **35.** $\dfrac{2\sqrt{3}}{3}$ **37.** $\dfrac{1250\pi}{3}$

39. $e - 1$ **41.** $\dfrac{1}{4\pi}$ **43.** -8 **45.** 0   **47.** 4π **49. a.** True

b. False   **c.** True   **d.** True   **51.** $8\pi(4\sqrt{17} + \ln(\sqrt{17} + 4))$
53. $8\pi a$ **55.** 8   **57. a.** 0   **b.** 0; the flow is tangent to the surface
(radial flow).   **59.** $2\pi ah$ **61.** $-400\left(e - \dfrac{1}{e}\right)^2$ **63.** $8\pi a$

65. a. $4\pi(b^3 - a^3)$ **b.** The net flux is zero.   **67.** $\left(0, 0, \frac{2}{3}h\right)$

69. $\left(0, 0, \frac{7}{6}\right)$ **73.** Flux $= \iint\limits_{S} \mathbf{F} \cdot \mathbf{n}\, dS = \iint\limits_{R} dA$

Section 17.7 Exercises, pp. 1169–1171

1. The integral measures the circulation along the closed curve C.
3. Under certain conditions, the accumulated rotation of the vector field over the surface S equals the net circulation on the boundary of S.
5. Both integrals equal -2π.   **7.** Both integrals equal zero.   **9.** Both integrals equal -18π.   **11.** -24π **13.** $-\frac{128}{3}$ **15.** 15π **17.** 0   **19.** 0   **21.** -2π **23.** -4π **25.** $\nabla \times \mathbf{v} = \langle 1, 0, 0 \rangle$; a paddle wheel with its axis aligned with the x-axis will spin with maximum angular speed counterclockwise (looking in the negative x-direction) at all points.   **27.** $\nabla \times \mathbf{v} = \langle 0, -2, 0 \rangle$; a paddle wheel with its axis aligned with the y-axis will spin with maximum angular speed clockwise (looking in the negative y-direction) at all points.   **29. a.** False   **b.** False   **c.** True   **d.** True   **31.** 0   **33.** 0   **35.** 2π **37.** $\pi(\cos\varphi - \sin\varphi)$; maximum for $\varphi = 0$ **39.** The circulation is 48π; it depends on the radius of the circle but not on the center.   **41. a.** The normal vectors point toward the z-axis on the curved surface of S and in the direction of $\langle 0, 1, 0 \rangle$ on the flat surface of S.   **b.** 2π **c.** 2π **43.** The integral is π for all a.   **45. a.** 0   **b.** 0   **47. b.** 2π for any circle of radius r centered at the origin   **c.** $\mathbf{F}$ is not differentiable along the z-axis.

49. Apply the Chain Rule.   **51.** $\displaystyle\int\limits_{C} \mathbf{F} \cdot d\mathbf{r} = \iint\limits_{R} \left(\frac{\partial h}{\partial y} - \frac{\partial g}{\partial z} \right) dA$

Section 17.8 Exercises, pp. 1179–1182

1. The surface integral measures the flow across the boundary.
3. The flux across the boundary equals the cumulative expansion or contraction of the vector field inside the region.   **5.** 32π
7. The outward fluxes are equal.   **9.** Both integrals equal 96π.
11. Both integrals equal zero.   **13.** 0   **15.** 0   **17.** $16\sqrt{6}\pi$ **19.** $\frac{2}{3}$
21. $-\frac{128}{3}\pi$ **23.** 24π **25.** -224π **27.** 12π **29.** 20
31. a. False   **b.** False   **c.** True   **33.** 0   **35.** $\frac{3}{2}$ **37. b.** The net flux between the two spheres is $4\pi(a^2 - \varepsilon^2)$.   **39. b.** Use $\nabla \cdot \mathbf{E} = 0$.

c. The flux across S is the sum of the contributions from the individual charges.   **d.** For an arbitrary volume, we find

$$\frac{1}{\varepsilon_0} \iiint\limits_{D} q(x, y, z)\, dV = \iint\limits_{S} \mathbf{E} \cdot \mathbf{n}\, dS = \iiint\limits_{D} \nabla \cdot \mathbf{E}\, dV.$$

e. Use $\nabla^2 \varphi = \nabla \cdot \nabla \varphi$.   **41.** 0   **43.** $e^{-1} - 1$ **45.** $800\pi a^3 e^{-a^2}$

Chapter 17 Review Exercises, pp. 1182–1184

1. a. False   **b.** True   **c.** False   **d.** False   **e.** True
3. $\nabla \varphi = \langle 2x, 8y \rangle$

5. $-\dfrac{\mathbf{r}}{|\mathbf{r}|^3}$ **7.** $\mathbf{n} = \frac{1}{2}\langle x, y \rangle$ **9.** $\dfrac{7}{8}(e^{48} - 1)$ **11.** Both integrals equal zero.   **13.** 0   **15.** The circulation is -4π; the outward flux is zero.   **17.** The circulation is zero; the outward flux is 2π.

19. $\dfrac{4v_0 L^3}{3}$ **21.** $\varphi(x, y, z) = xy + yz^2$ **23.** $\varphi(x, y, z) = xye^z$

25. 0 for both methods   **27. a.** $-\pi$ **b.** $\mathbf{F}$ is not conservative.
29. 0   **31.** $\frac{20}{3}$ **33.** 8π **35.** The circulation is zero; the outward flux equals 2π.   **37. a.** $b = c$ **b.** $a = -d$ **c.** $a = -d$ and $b = c$
39. $\nabla \cdot \mathbf{F} = 4\sqrt{x^2 + y^2 + z^2} = 4|\mathbf{r}|$, $\nabla \times \mathbf{F} = \mathbf{0}$, $\nabla \cdot \mathbf{F} \neq 0$; irrotational but not source free   **41.** $\nabla \cdot \mathbf{F} = 2y + 12xz^2$, $\nabla \times \mathbf{F} = \mathbf{0}$, $\nabla \cdot \mathbf{F} \neq 0$; irrotational but not source free

43. a. -1 **b.** 0   **c.** $\mathbf{n} = \dfrac{1}{\sqrt{3}}\langle -1, 1, 1 \rangle$ **45.** 18π **47.** $4\sqrt{3}$

49. $\dfrac{8\sqrt{3}}{3}$ **51.** 8π **53.** $4\pi a^2$ **55. a.** Use $x = y = 0$ to confirm the highest point; use $z = 0$ to confirm the base.   **b.** The hemisphere S has the greater surface area—$2\pi a^2$ for S versus $\dfrac{5\sqrt{5} - 1}{6}\pi a^2$ for T.

57. 0   **59.** 99π **61.** 0   **63.** $\dfrac{972}{5}\pi$ **65.** $\dfrac{124}{5}\pi$ **67.** $\dfrac{32}{3}$

Index

Note:
- Italics indicate figures or margin notes.
- "t" indicates a table.
- "e" indicates an exercise.
- GP indicates a Guided Project (located online in MyLab Math).
- AP-, B-, and C- indicate Appendices A, B, and C respectively (Appendices B and C are online at goo.gl/6DCbbM and goo.gl/1bW164).

A

Absolute convergence, 692–694, 695e, 701t
Absolute error, 568
Absolute extreme value, 241
  on any interval, 262
  on closed intervals, 246
Absolute growth rate, 493, 499e
Absolute maxima, 241–243
  on closed bounded sets, 988
  on constraint curves, 998
  on constraint surfaces, 1000
  defining, 987
  finding, 990–991
  locating, 245–247
Absolute minima, 241–243
  on closed bounded sets, 988
  on constraint curves, 998
  on constraint surfaces, 1000
  defining, 987
  finding, 990–991
  locating, 245–247
Absolute values, B-3, B-4, 335e
  functions and, 114e, 249, 352e
  integrals of, 360
Acceleration, 179, 255e, 501e, 883
  for circular motion, 884–885
  on circular path, 908
  components of, 907–909, 913e
  defining, 180, 884
  formulas for, 912
  gravity and, *331*
  position from, 333e, 412e
  velocity and, *180*, 408, 412e
Accumulation
  over finite intervals, 647t
  over infinite intervals, 647t
Acid, noise, and earthquakes, GP8
Addition
  associative property of, 811
  commutative property of, 811
  of vectors, 806–807
Addition Rule, 344
Additive identity, 811
Additive inverse, 811
Algebra, B-1, B-8
Algebraic functions, 14
  end behavior of, 98

Alternating harmonic series, 688
  convergent, 690
Alternating series, remainder in, 691–692
Alternating Series Test, 689–690, 694, 701t
Alternating sums, errors in, 706e
Ampère's Law, 1146, 1170e
Amplitude, GP6, 43, 50e
Analytical methods, 15, 520
  defining, 313
  technology and, 279e
Angles
  half-angle formula, *533*
  half-angle identities, 42
  between planes, 855
  projections and, 866e
  sag, 515e
  small-angle approximation, 294
  of triangle, 833e
Angular coordinates, 767
Angular frequencies, 755
Angular speed, 1145e
Annual growth rate, 496
Annual percentage yield (APY), 495
  determining, 499e
Annuities, perpetual, 591
Annular regions, 1030
Annulus, flux across, 1129
Antiderivatives, 321–331, 406
  Constant Multiple Rule and, 324
  defining, 321
  family of, 322
  finding, 323, 332e
  motion problems and, 329–331
  particular, 332e
  Sum Rule and, 324
  of vector-valued functions, 880
  of velocity, *329*
Antidifferentiation, 321
Antiquity, problems of, 387e
Anvil of hyperbolas, 800e
Approximation, GP11, GP17, GP22, GP32, GP33, GP46, GP49, GP51, GP52, GP63
APY. *See* Annual percentage yield
Arbelos, 291e
Arccosine, 44
Arches, height of, 386e
Archimedes, GP26, GP46
Arc length, 517e, 519e, 595e, 765e, 897–898, GP30, GP54
  approximation of, 901e
  astroid, 762
  calculation of, 456, 901e
  of ellipse, 581
  of exponential curve, 453
  formula, 452, 454, 762, 785
  functions from, 456e, 899–900
  integrals, 901e

  of parabolas, 541, 567e
  as parameter, 898–900
  parametric equations and, 760–763
  for polar curves, 785, 788e
  speed and, 901e
  spiral, 787e
  for vector functions, 897
  for $x = g(y)$, 454–455
  for $y = f(x)$, 451–453
Arcsine, 44
Area. *See also* Surface area
  approximation of, from graphs, 347e, 348e
  of circles, 538–539, 544e, 1035e
  computation of, 784, 1035e
  between curves, 417, 418, 422e
  by double integrals, 1023
  of ellipses, 544e, 981e, 1126
  equal, 397e
  of parabolas, 424e, 482e
  formula, 417, *418*, 1036e
  by geometry, 399e
  Green's Theorem for, 1126
  of hyperbola sector, 799e
  intersection points and, 514e
  within lemniscates, 1032
  net, 353, 364e
    defining, 354
    definite integrals and, 373–374
    maximum, 380e
    minimum, 380e
    zero, 381e
  of parallelograms, 838, 843e, 866e
  of plane region, 1026e, 1126
  of polar regions, 1031–1032
  of rectangles, 285e
  of regions, 379e, 478e, 518e, 555e, 786e, 802e, 1134e
    bounded by parametric curves, 766e
    bounded by polar curves, 781–784
    intersection points and, 783
  Riemann sums for, 340–344
  signed, *354*
  of trapezoids, *358*, 369
  of triangle, 839–840, 843e, 844e, 995e
  under curves, 338–346
  under velocity curve, 338–340
  volume and, 479e, 545e
Area function, 19, 25e, 480e
  comparing, 368–369
  for constant functions, 378e
  cross-sectional, *427*
  defining, 368
  Fundamental Theorem of Calculus and, 367–370
  by geometry, 400e
  from graphs, 380e
  for linear functions, 378e

TABLE OF INTEGRALS

Substitution Rule	Integration by Parts
$\int f(g(x))g'(x)\,dx = \int f(u)\,du \quad (u = g(x))$	$\int u\,dv = uv - \int v\,du$
$\int_a^b f(g(x))g'(x)\,dx = \int_{g(a)}^{g(b)} f(u)\,du$	$\int_a^b uv'\,dx = uv\Big|_a^b - \int_a^b vu'\,dx$

Basic Integrals

1. $\int x^n\,dx = \dfrac{1}{n+1}x^{n+1} + C;\ n \neq -1$

2. $\int \dfrac{dx}{x} = \ln|x| + C$

3. $\int \cos ax\,dx = \dfrac{1}{a}\sin ax + C$

4. $\int \sin ax\,dx = -\dfrac{1}{a}\cos ax + C$

5. $\int \tan x\,dx = \ln|\sec x| + C$

6. $\int \cot x\,dx = \ln|\sin x| + C$

7. $\int \sec x\,dx = \ln|\sec x + \tan x| + C$

8. $\int \csc x\,dx = -\ln|\csc x + \cot x| + C$

9. $\int e^{ax}\,dx = \dfrac{1}{a}e^{ax} + C$

10. $\int b^{ax}\,dx = \dfrac{1}{a\ln b}b^{ax} + C;\ b > 0, b \neq 1$

11. $\int \ln x\,dx = x\ln x - x + C$

12. $\int \log_b x\,dx = \dfrac{1}{\ln b}(x\ln x - x) + C$

13. $\int \dfrac{dx}{x^2 + a^2} = \dfrac{1}{a}\tan^{-1}\dfrac{x}{a} + C$

14. $\int \dfrac{dx}{\sqrt{a^2 - x^2}} = \sin^{-1}\dfrac{x}{a} + C, a > 0$

15. $\int \dfrac{dx}{x\sqrt{x^2 - a^2}} = \dfrac{1}{a}\sec^{-1}\left|\dfrac{x}{a}\right| + C, a > 0$

16. $\int \sin^{-1} x\,dx = x\sin^{-1} x + \sqrt{1 - x^2} + C$

17. $\int \cos^{-1} x\,dx = x\cos^{-1} x - \sqrt{1 - x^2} + C$

18. $\int \tan^{-1} x\,dx = x\tan^{-1} x - \dfrac{1}{2}\ln(1 + x^2) + C$

19. $\int \sec^{-1} x\,dx = x\sec^{-1} x - \ln(x + \sqrt{x^2 - 1}) + C$

20. $\int \sinh x\,dx = \cosh x + C$

21. $\int \cosh x\,dx = \sinh x + C$

22. $\int \operatorname{sech}^2 x\,dx = \tanh x + C$

23. $\int \operatorname{csch}^2 x\,dx = -\coth x + C$

24. $\int \operatorname{sech} x \tanh x\,dx = -\operatorname{sech} x + C$

25. $\int \operatorname{csch} x \coth x\,dx = -\operatorname{csch} x + C$

26. $\int \tanh x\,dx = \ln\cosh x + C$

27. $\int \coth x\,dx = \ln|\sinh x| + C$

28. $\int \operatorname{sech} x\,dx = \tan^{-1}\sinh x + C = \sin^{-1}\tanh x + C$

29. $\int \operatorname{csch} x\,dx = \ln|\tanh(x/2)| + C$

Trigonometric Integrals

30. $\int \cos^2 x\,dx = \dfrac{x}{2} + \dfrac{\sin 2x}{4} + C$

31. $\int \sin^2 x\,dx = \dfrac{x}{2} - \dfrac{\sin 2x}{4} + C$

32. $\int \sec^2 ax\,dx = \dfrac{1}{a}\tan ax + C$

33. $\int \csc^2 ax\,dx = -\dfrac{1}{a}\cot ax + C$

34. $\int \tan^2 x\,dx = \tan x - x + C$

35. $\int \cot^2 x\,dx = -\cot x - x + C$

36. $\int \cos^3 x\,dx = -\dfrac{1}{3}\sin^3 x + \sin x + C$

37. $\int \sin^3 x\,dx = \dfrac{1}{3}\cos^3 x - \cos x + C$

38. $\int \sec^3 x\, dx = \dfrac{1}{2}\sec x \tan x + \dfrac{1}{2}\ln\left|\sec x + \tan x\right| + C$

39. $\int \csc^3 x\, dx = -\dfrac{1}{2}\csc x \cot x - \dfrac{1}{2}\ln\left|\csc x + \cot x\right| + C$

40. $\int \tan^3 x\, dx = \dfrac{1}{2}\tan^2 x - \ln\left|\sec x\right| + C$

41. $\int \cot^3 x\, dx = -\dfrac{1}{2}\cot^2 x - \ln\left|\sin x\right| + C$

42. $\int \sec^n ax \tan ax\, dx = \dfrac{1}{na}\sec^n ax + C;\ n \neq 0$

43. $\int \csc^n ax \cot ax\, dx = -\dfrac{1}{na}\csc^n ax + C;\ n \neq 0$

44. $\int \dfrac{dx}{1 + \sin ax} = -\dfrac{1}{a}\tan\left(\dfrac{\pi}{4} - \dfrac{ax}{2}\right) + C$

45. $\int \dfrac{dx}{1 - \sin ax} = \dfrac{1}{a}\tan\left(\dfrac{\pi}{4} + \dfrac{ax}{2}\right) + C$

46. $\int \dfrac{dx}{1 + \cos ax} = \dfrac{1}{a}\tan\dfrac{ax}{2} + C$

47. $\int \dfrac{dx}{1 - \cos ax} = -\dfrac{1}{a}\cot\dfrac{ax}{2} + C$

48. $\int \sin mx \cos nx\, dx = -\dfrac{\cos(m+n)x}{2(m+n)} - \dfrac{\cos(m-n)x}{2(m-n)} + C;\ m^2 \neq n^2$

49. $\int \sin mx \sin nx\, dx = \dfrac{\sin(m-n)x}{2(m-n)} - \dfrac{\sin(m+n)x}{2(m+n)} + C;\ m^2 \neq n^2$

50. $\int \cos mx \cos nx\, dx = \dfrac{\sin(m-n)x}{2(m-n)} + \dfrac{\sin(m+n)x}{2(m+n)} + C;\ m^2 \neq n^2$

Reduction Formulas for Trigonometric Functions

51. $\int \cos^n x\, dx = \dfrac{1}{n}\cos^{n-1} x \sin x + \dfrac{n-1}{n}\int \cos^{n-2} x\, dx$

52. $\int \sin^n x\, dx = -\dfrac{1}{n}\sin^{n-1} x \cos x + \dfrac{n-1}{n}\int \sin^{n-2} x\, dx$

53. $\int \tan^n x\, dx = \dfrac{\tan^{n-1} x}{n-1} - \int \tan^{n-2} x\, dx;\ n \neq 1$

54. $\int \cot^n x\, dx = -\dfrac{\cot^{n-1} x}{n-1} - \int \cot^{n-2} x\, dx;\ n \neq 1$

55. $\int \sec^n x\, dx = \dfrac{\sec^{n-2} x \tan x}{n-1} + \dfrac{n-2}{n-1}\int \sec^{n-2} x\, dx;\ n \neq 1$

56. $\int \csc^n x\, dx = -\dfrac{\csc^{n-2} x \cot x}{n-1} + \dfrac{n-2}{n-1}\int \csc^{n-2} x\, dx;\ n \neq 1$

57. $\int \sin^m x \cos^n x\, dx = -\dfrac{\sin^{m-1} x \cos^{n+1} x}{m+n} + \dfrac{m-1}{m+n}\int \sin^{m-2} x \cos^n x\, dx;\ m \neq -n$

58. $\int \sin^m x \cos^n x\, dx = \dfrac{\sin^{m+1} x \cos^{n-1} x}{m+n} + \dfrac{n-1}{m+n}\int \sin^m x \cos^{n-2} x\, dx;\ m \neq -n$

59. $\int x^n \sin ax\, dx = -\dfrac{x^n \cos ax}{a} + \dfrac{n}{a}\int x^{n-1} \cos ax\, dx;\ a \neq 0$

60. $\int x^n \cos ax\, dx = \dfrac{x^n \sin ax}{a} - \dfrac{n}{a}\int x^{n-1} \sin ax\, dx;\ a \neq 0$

Integrals Involving $a^2 - x^2;\ a > 0$

61. $\int \sqrt{a^2 - x^2}\, dx = \dfrac{x}{2}\sqrt{a^2 - x^2} + \dfrac{a^2}{2}\sin^{-1}\dfrac{x}{a} + C$

62. $\int \dfrac{dx}{x\sqrt{a^2 - x^2}} = -\dfrac{1}{a}\ln\left|\dfrac{a + \sqrt{a^2 - x^2}}{x}\right| + C$

63. $\int \dfrac{dx}{x^2\sqrt{a^2 - x^2}} = -\dfrac{\sqrt{a^2 - x^2}}{a^2 x} + C$

64. $\int x^2\sqrt{a^2 - x^2}\, dx = \dfrac{x}{8}(2x^2 - a^2)\sqrt{a^2 - x^2} + \dfrac{a^4}{8}\sin^{-1}\dfrac{x}{a} + C$

65. $\int \dfrac{\sqrt{a^2 - x^2}}{x^2}\, dx = -\dfrac{1}{x}\sqrt{a^2 - x^2} - \sin^{-1}\dfrac{x}{a} + C$

66. $\int \dfrac{x^2}{\sqrt{a^2 - x^2}}\, dx = -\dfrac{x}{2}\sqrt{a^2 - x^2} + \dfrac{a^2}{2}\sin^{-1}\dfrac{x}{a} + C$

67. $\int \dfrac{dx}{a^2 - x^2} = \dfrac{1}{2a}\ln\left|\dfrac{x + a}{x - a}\right| + C$

Integrals Involving $x^2 - a^2;\ a > 0$

68. $\int \sqrt{x^2 - a^2}\, dx = \dfrac{x}{2}\sqrt{x^2 - a^2} - \dfrac{a^2}{2}\ln\left|x + \sqrt{x^2 - a^2}\right| + C$

69. $\int \dfrac{dx}{\sqrt{x^2 - a^2}} = \ln\left|x + \sqrt{x^2 - a^2}\right| + C$

70. $\int \dfrac{dx}{x^2\sqrt{x^2 - a^2}} = \dfrac{\sqrt{x^2 - a^2}}{a^2 x} + C$

71. $\int x^2\sqrt{x^2 - a^2}\, dx = \dfrac{x}{8}(2x^2 - a^2)\sqrt{x^2 - a^2} - \dfrac{a^4}{8}\ln\left|x + \sqrt{x^2 - a^2}\right| + C$

72. $\int \dfrac{\sqrt{x^2 - a^2}}{x^2}\, dx = \ln\left|x + \sqrt{x^2 - a^2}\right| - \dfrac{\sqrt{x^2 - a^2}}{x} + C$

73. $\int \dfrac{x^2}{\sqrt{x^2 - a^2}}\, dx = \dfrac{a^2}{2}\ln\left|x + \sqrt{x^2 - a^2}\right| + \dfrac{x}{2}\sqrt{x^2 - a^2} + C$

74. $\int \dfrac{dx}{x^2 - a^2} = \dfrac{1}{2a}\ln\left|\dfrac{x - a}{x + a}\right| + C$

75. $\int \dfrac{dx}{x(x^2 - a^2)} = \dfrac{1}{2a^2}\ln\left|\dfrac{x^2 - a^2}{x^2}\right| + C$

Integrals Involving $a^2 + x^2$; $a > 0$

76. $\displaystyle \int \sqrt{a^2 + x^2}\, dx = \frac{x}{2}\sqrt{a^2 + x^2} + \frac{a^2}{2}\ln\left(x + \sqrt{a^2 + x^2}\right) + C$

77. $\displaystyle \int \frac{dx}{\sqrt{a^2 + x^2}} = \ln\left(x + \sqrt{a^2 + x^2}\right) + C$

78. $\displaystyle \int \frac{dx}{x\sqrt{a^2 + x^2}} = \frac{1}{a}\ln\left|\frac{a - \sqrt{a^2 + x^2}}{x}\right| + C$

79. $\displaystyle \int \frac{dx}{x^2\sqrt{a^2 + x^2}} = -\frac{\sqrt{a^2 + x^2}}{a^2 x} + C$

80. $\displaystyle \int x^2\sqrt{a^2 + x^2}\, dx = \frac{x}{8}\left(a^2 + 2x^2\right)\sqrt{a^2 + x^2} - \frac{a^4}{8}\ln\left(x + \sqrt{a^2 + x^2}\right) + C$

81. $\displaystyle \int \frac{\sqrt{a^2 + x^2}}{x^2}\, dx = \ln\left|x + \sqrt{a^2 + x^2}\right| - \frac{\sqrt{a^2 + x^2}}{x} + C$

82. $\displaystyle \int \frac{x^2}{\sqrt{a^2 + x^2}}\, dx = -\frac{a^2}{2}\ln\left(x + \sqrt{a^2 + x^2}\right) + \frac{x\sqrt{a^2 + x^2}}{2} + C$

83. $\displaystyle \int \frac{\sqrt{a^2 + x^2}}{x}\, dx = \sqrt{a^2 + x^2} - a\ln\left|\frac{a + \sqrt{a^2 + x^2}}{x}\right| + C$

84. $\displaystyle \int \frac{dx}{(a^2 + x^2)^{3/2}} = \frac{x}{a^2\sqrt{a^2 + x^2}} + C$

85. $\displaystyle \int \frac{dx}{x(a^2 + x^2)} = \frac{1}{2a^2}\ln\left(\frac{x^2}{a^2 + x^2}\right) + C$

Integrals Involving $ax \pm b$; $a \neq 0, b > 0$

86. $\displaystyle \int (ax + b)^n\, dx = \frac{(ax + b)^{n+1}}{a(n + 1)} + C;\ n \neq -1$

87. $\displaystyle \int (\sqrt{ax + b})^n\, dx = \frac{2}{a}\frac{(\sqrt{ax + b})^{n+2}}{n + 2} + C;\ n \neq -2$

88. $\displaystyle \int \frac{dx}{x\sqrt{ax - b}} = \frac{2}{\sqrt{b}}\tan^{-1}\sqrt{\frac{ax - b}{b}} + C$

89. $\displaystyle \int \frac{dx}{x\sqrt{ax + b}} = \frac{1}{\sqrt{b}}\ln\left|\frac{\sqrt{ax + b} - \sqrt{b}}{\sqrt{ax + b} + \sqrt{b}}\right| + C$

90. $\displaystyle \int \frac{x}{ax + b}\, dx = \frac{x}{a} - \frac{b}{a^2}\ln|ax + b| + C$

91. $\displaystyle \int \frac{x^2}{ax + b}\, dx = \frac{1}{2a^3}\left((ax + b)^2 - 4b(ax + b) + 2b^2\ln|ax + b|\right) + C$

92. $\displaystyle \int \frac{dx}{x^2(ax + b)} = -\frac{1}{bx} + \frac{a}{b^2}\ln\left|\frac{ax + b}{x}\right| + C$

93. $\displaystyle \int x\sqrt{ax + b}\, dx = \frac{2}{15a^2}(3ax - 2b)(ax + b)^{3/2} + C$

94. $\displaystyle \int \frac{x}{\sqrt{ax + b}}\, dx = \frac{2}{3a^2}(ax - 2b)\sqrt{ax + b} + C$

95. $\displaystyle \int x(ax + b)^n\, dx = \frac{(ax + b)^{n+1}}{a^2}\left(\frac{ax + b}{n + 2} - \frac{b}{n + 1}\right) + C;\ n \neq -1, -2$

96. $\displaystyle \int \frac{dx}{x(ax + b)} = \frac{1}{b}\ln\left|\frac{x}{ax + b}\right| + C$

Integrals with Exponential and Trigonometric Functions

97. $\displaystyle \int e^{ax}\sin bx\, dx = \frac{e^{ax}(a\sin bx - b\cos bx)}{a^2 + b^2} + C$

98. $\displaystyle \int e^{ax}\cos bx\, dx = \frac{e^{ax}(a\cos bx + b\sin bx)}{a^2 + b^2} + C$

Integrals with Exponential and Logarithmic Functions

99. $\displaystyle \int \frac{dx}{x\ln x} = \ln|\ln x| + C$

100. $\displaystyle \int x^n\ln x\, dx = \frac{x^{n+1}}{n + 1}\left(\ln x - \frac{1}{n + 1}\right) + C;\ n \neq -1$

101. $\displaystyle \int xe^x\, dx = xe^x - e^x + C$

102. $\displaystyle \int x^n e^{ax}\, dx = \frac{1}{a}x^n e^{ax} - \frac{n}{a}\int x^{n-1}e^{ax}\, dx;\ a \neq 0$

103. $\displaystyle \int \ln^n x\, dx = x\ln^n x - n\int \ln^{n-1} x\, dx$

Miscellaneous Formulas

104. $\displaystyle \int x^n\cos^{-1} x\, dx = \frac{1}{n + 1}\left(x^{n+1}\cos^{-1}x + \int \frac{x^{n+1}dx}{\sqrt{1 - x^2}}\right);\ n \neq -1$

105. $\displaystyle \int x^n\sin^{-1} x\, dx = \frac{1}{n + 1}\left(x^{n+1}\sin^{-1} x - \int \frac{x^{n+1}\, dx}{\sqrt{1 - x^2}}\right);\ n \neq -1$

106. $\displaystyle \int x^n\tan^{-1} x\, dx = \frac{1}{n + 1}\left(x^{n+1}\tan^{-1} x - \int \frac{x^{n+1}\, dx}{x^2 + 1}\right);\ n \neq -1$

107. $\displaystyle \int \sqrt{2ax - x^2}\, dx = \frac{x - a}{2}\sqrt{2ax - x^2} + \frac{a^2}{2}\sin^{-1}\left(\frac{x - a}{a}\right) + C;\ a > 0$

108. $\displaystyle \int \frac{dx}{\sqrt{2ax - x^2}} = \sin^{-1}\left(\frac{x - a}{a}\right) + C;\ a > 0$

ALGEBRA

Exponents and Radicals

$$x^a x^b = x^{a+b} \qquad \frac{x^a}{x^b} = x^{a-b} \qquad x^{-a} = \frac{1}{x^a} \qquad (x^a)^b = x^{ab} \qquad \left(\frac{x}{y}\right)^a = \frac{x^a}{y^a}$$

$$x^{1/n} = \sqrt[n]{x} \qquad x^{m/n} = \sqrt[n]{x^m} = (\sqrt[n]{x})^m \qquad \sqrt[n]{xy} = \sqrt[n]{x}\sqrt[n]{y} \qquad \sqrt[n]{x/y} = \sqrt[n]{x}/\sqrt[n]{y}$$

Factoring Formulas

$$a^2 - b^2 = (a - b)(a + b) \qquad\qquad a^2 + b^2 \text{ does not factor over real numbers.}$$
$$a^3 - b^3 = (a - b)(a^2 + ab + b^2) \qquad a^3 + b^3 = (a + b)(a^2 - ab + b^2)$$
$$a^n - b^n = (a - b)(a^{n-1} + a^{n-2}b + a^{n-3}b^2 + \cdots + ab^{n-2} + b^{n-1})$$

Binomials

$$(a \pm b)^2 = a^2 \pm 2ab + b^2$$
$$(a \pm b)^3 = a^3 \pm 3a^2b + 3ab^2 \pm b^3$$

Binomial Theorem

$$(a + b)^n = a^n + \binom{n}{1}a^{n-1}b + \binom{n}{2}a^{n-2}b^2 + \cdots + \binom{n}{n-1}ab^{n-1} + b^n,$$

$$\text{where } \binom{n}{k} = \frac{n(n-1)(n-2)\cdots(n-k+1)}{k(k-1)(k-2)\cdots 3 \cdot 2 \cdot 1} = \frac{n!}{k!(n-k)!}$$

Quadratic Formula

The solutions of $ax^2 + bx + c = 0$ are

$$x = \frac{-b \pm \sqrt{b^2 - 4ac}}{2a}.$$

GEOMETRY

Parallelogram

$$A = bh$$

Triangle

$$A = \frac{1}{2}bh$$

Trapezoid

$$A = \frac{1}{2}(a + b)h$$

Circle

$$A = \pi r^2$$
$$C = 2\pi r$$

Sector

$$A = \frac{1}{2}r^2\theta$$
$$s = r\theta \ (\theta \text{ in radians})$$

Cylinder

$$V = \pi r^2 h$$
$$S = 2\pi rh$$
(lateral surface area)

Cone

$$V = \frac{1}{3}\pi r^2 h$$
$$S = \pi r\ell$$
(lateral surface area)

Sphere

$$V = \frac{4}{3}\pi r^3$$
$$S = 4\pi r^2$$

Equations of Lines and Circles

$$m = \frac{y_2 - y_1}{x_2 - x_1} \qquad \text{slope of line through } (x_1, y_1) \text{ and } (x_2, y_2)$$

$$y - y_1 = m(x - x_1) \qquad \text{point–slope form of line through } (x_1, y_1) \text{ with slope } m$$

$$y = mx + b \qquad \text{slope–intercept form of line with slope } m \text{ and } y\text{-intercept } (0, b)$$

$$(x - h)^2 + (y - k)^2 = r^2 \qquad \text{circle of radius } r \text{ with center } (h, k)$$

$$(x - h)^2 + (y - k)^2 = r^2$$

TRIGONOMETRY

$$\cos\theta = \frac{\text{adj}}{\text{hyp}} \qquad \sin\theta = \frac{\text{opp}}{\text{hyp}} \qquad \tan\theta = \frac{\text{opp}}{\text{adj}}$$

$$\sec\theta = \frac{\text{hyp}}{\text{adj}} \qquad \csc\theta = \frac{\text{hyp}}{\text{opp}} \qquad \cot\theta = \frac{\text{adj}}{\text{opp}}$$

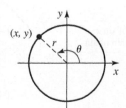

$$\cos\theta = \frac{x}{r} \qquad \sec\theta = \frac{r}{x}$$

$$\sin\theta = \frac{y}{r} \qquad \csc\theta = \frac{r}{y}$$

$$\tan\theta = \frac{y}{x} \qquad \cot\theta = \frac{x}{y}$$

(Continued)

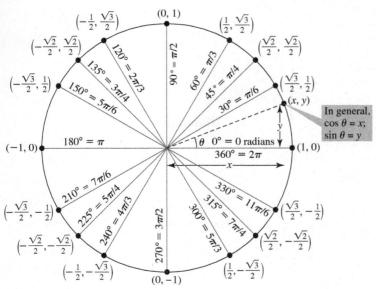

Reciprocal Identities

$$\tan\theta = \frac{\sin\theta}{\cos\theta} \quad \cot\theta = \frac{\cos\theta}{\sin\theta} \quad \sec\theta = \frac{1}{\cos\theta} \quad \csc\theta = \frac{1}{\sin\theta}$$

Pythagorean Identities

$$\sin^2\theta + \cos^2\theta = 1 \quad \tan^2\theta + 1 = \sec^2\theta \quad 1 + \cot^2\theta = \csc^2\theta$$

Sign Identities

$$\sin(-\theta) = -\sin\theta \quad \cos(-\theta) = \cos\theta \quad \tan(-\theta) = -\tan\theta$$
$$\csc(-\theta) = -\csc\theta \quad \sec(-\theta) = \sec\theta \quad \cot(-\theta) = -\cot\theta$$

Double-Angle Identities

$$\sin 2\theta = 2\sin\theta\cos\theta \qquad \cos 2\theta = \cos^2\theta - \sin^2\theta$$
$$= 2\cos^2\theta - 1$$
$$\tan 2\theta = \frac{2\tan\theta}{1 - \tan^2\theta} \qquad = 1 - 2\sin^2\theta$$

Half-Angle Identities

$$\cos^2\theta = \frac{1 + \cos 2\theta}{2} \qquad \sin^2\theta = \frac{1 - \cos 2\theta}{2}$$

Addition Formulas

$$\sin(\alpha + \beta) = \sin\alpha\cos\beta + \cos\alpha\sin\beta \qquad \sin(\alpha - \beta) = \sin\alpha\cos\beta - \cos\alpha\sin\beta$$
$$\cos(\alpha + \beta) = \cos\alpha\cos\beta - \sin\alpha\sin\beta \qquad \cos(\alpha - \beta) = \cos\alpha\cos\beta + \sin\alpha\sin\beta$$
$$\tan(\alpha + \beta) = \frac{\tan\alpha + \tan\beta}{1 - \tan\alpha\tan\beta} \qquad \tan(\alpha - \beta) = \frac{\tan\alpha - \tan\beta}{1 + \tan\alpha\tan\beta}$$

Law of Sines

$$\frac{\sin\alpha}{a} = \frac{\sin\beta}{b} = \frac{\sin\gamma}{c}$$

Law of Cosines

$$a^2 = b^2 + c^2 - 2bc\cos\alpha$$

Graphs of Trigonometric Functions and Their Inverses

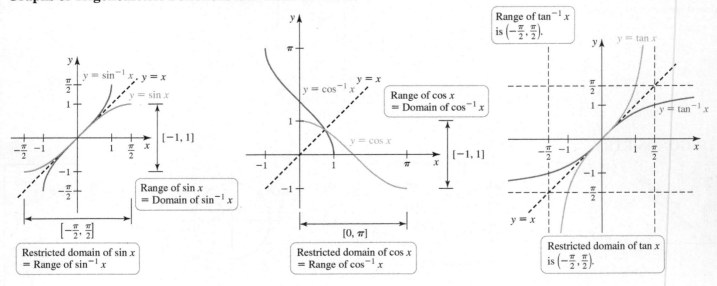

GRAPHS OF ELEMENTARY FUNCTIONS

Linear functions

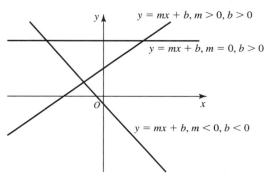

$y = mx + b, m > 0, b > 0$

$y = mx + b, m = 0, b > 0$

$y = mx + b, m < 0, b < 0$

Quadratic functions

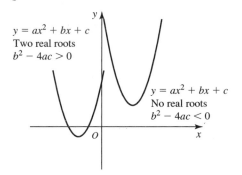

$y = ax^2 + bx + c$
Two real roots
$b^2 - 4ac > 0$

$y = ax^2 + bx + c$
No real roots
$b^2 - 4ac < 0$

Positive even powers

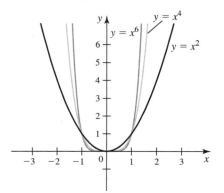

$y = x^4$
$y = x^6$
$y = x^2$

Positive odd powers

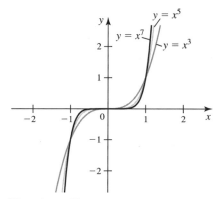

$y = x^5$
$y = x^7$
$y = x^3$

Negative even powers

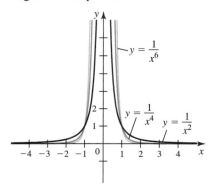

$y = \dfrac{1}{x^6}$
$y = \dfrac{1}{x^4}$
$y = \dfrac{1}{x^2}$

Negative odd powers

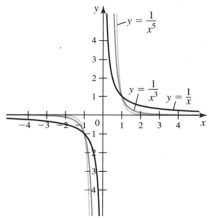

$y = \dfrac{1}{x^5}$
$y = \dfrac{1}{x^3}$
$y = \dfrac{1}{x}$

Exponential functions

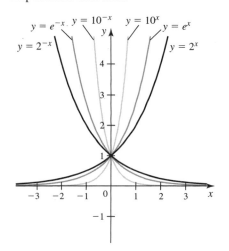

$y = e^{-x}$ $y = 10^{-x}$ $y = 10^x$ $y = e^x$
$y = 2^{-x}$ $y = 2^x$

Natural logarithmic and exponential functions

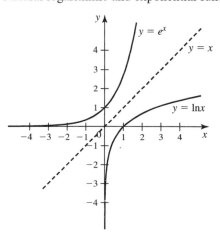

$y = e^x$
$y = x$
$y = \ln x$

DERIVATIVES

General Formulas

$$\frac{d}{dx}(c) = 0$$

$$\frac{d}{dx}(cf(x)) = cf'(x)$$

$$\frac{d}{dx}(f(x) + g(x)) = f'(x) + g'(x)$$

$$\frac{d}{dx}(f(x) - g(x)) = f'(x) - g'(x)$$

$$\frac{d}{dx}(f(x)g(x)) = f'(x)g(x) + f(x)g'(x)$$

$$\frac{d}{dx}\left(\frac{f(x)}{g(x)}\right) = \frac{g(x)f'(x) - f(x)g'(x)}{(g(x))^2}$$

$$\frac{d}{dx}(x^n) = nx^{n-1}, \text{ for real numbers } n$$

$$\frac{d}{dx}(f(g(x))) = f'(g(x)) \cdot g'(x)$$

Trigonometric Functions

$$\frac{d}{dx}(\sin x) = \cos x$$

$$\frac{d}{dx}(\cos x) = -\sin x$$

$$\frac{d}{dx}(\tan x) = \sec^2 x$$

$$\frac{d}{dx}(\cot x) = -\csc^2 x$$

$$\frac{d}{dx}(\sec x) = \sec x \tan x$$

$$\frac{d}{dx}(\csc x) = -\csc x \cot x$$

Inverse Trigonometric Functions

$$\frac{d}{dx}(\sin^{-1} x) = \frac{1}{\sqrt{1 - x^2}}$$

$$\frac{d}{dx}(\cos^{-1} x) = -\frac{1}{\sqrt{1 - x^2}}$$

$$\frac{d}{dx}(\tan^{-1} x) = \frac{1}{1 + x^2}$$

$$\frac{d}{dx}(\cot^{-1} x) = -\frac{1}{1 + x^2}$$

$$\frac{d}{dx}(\sec^{-1} x) = \frac{1}{|x|\sqrt{x^2 - 1}}$$

$$\frac{d}{dx}(\csc^{-1} x) = -\frac{1}{|x|\sqrt{x^2 - 1}}$$

Exponential and Logarithmic Functions

$$\frac{d}{dx}(e^x) = e^x$$

$$\frac{d}{dx}(b^x) = b^x \ln b$$

$$\frac{d}{dx}(\ln |x|) = \frac{1}{x}$$

$$\frac{d}{dx}(\log_b |x|) = \frac{1}{x \ln b}$$

Hyperbolic Functions

$$\frac{d}{dx}(\sinh x) = \cosh x$$

$$\frac{d}{dx}(\cosh x) = \sinh x$$

$$\frac{d}{dx}(\tanh x) = \text{sech}^2 x$$

$$\frac{d}{dx}(\coth x) = -\text{csch}^2 x$$

$$\frac{d}{dx}(\text{sech } x) = -\text{sech } x \tanh x$$

$$\frac{d}{dx}(\text{csch } x) = -\text{csch } x \coth x$$

Inverse Hyperbolic Functions

$$\frac{d}{dx}(\sinh^{-1} x) = \frac{1}{\sqrt{x^2 + 1}}$$

$$\frac{d}{dx}(\cosh^{-1} x) = \frac{1}{\sqrt{x^2 - 1}} \quad (x > 1)$$

$$\frac{d}{dx}(\tanh^{-1} x) = \frac{1}{1 - x^2} \quad (|x| < 1)$$

$$\frac{d}{dx}(\coth^{-1} x) = \frac{1}{1 - x^2} \quad (|x| > 1)$$

$$\frac{d}{dx}(\text{sech}^{-1} x) = -\frac{1}{x\sqrt{1 - x^2}} \quad (0 < x < 1)$$

$$\frac{d}{dx}(\text{csch}^{-1} x) = -\frac{1}{|x|\sqrt{1 + x^2}} \quad (x \neq 0)$$

FORMS OF THE FUNDAMENTAL THEOREM OF CALCULUS

Fundamental Theorem of Calculus	$\int_a^b f'(x)\,dx = f(b) - f(a)$
Fundamental Theorem for Line Integrals	$\int_C \nabla f \cdot d\mathbf{r} = f(B) - f(A)$ (A and B are the initial and final points of C.)
Green's Theorem	$\iint_R (g_x - f_y)\,dA = \oint_C f\,dx + g\,dy$ $\iint_R (f_x + g_y)\,dA = \oint_C f\,dy - g\,dx$
Stokes' Theorem	$\iint_S (\nabla \times \mathbf{F}) \cdot \mathbf{n}\,dS = \oint_C \mathbf{F} \cdot d\mathbf{r}$
Divergence Theorem	$\iiint_D \nabla \cdot \mathbf{F}\,dV = \iint_S \mathbf{F} \cdot \mathbf{n}\,dS$

FORMULAS FROM VECTOR CALCULUS

Assume $\mathbf{F}(x, y, z) = f(x, y, z)\mathbf{i} + g(x, y, z)\mathbf{j} + h(x, y, z)\mathbf{k}$, where f, g, and h are differentiable on a region D of $\mathbb{R}^3$.

Gradient: $\quad \nabla f(x, y, z) = \dfrac{\partial f}{\partial x}\mathbf{i} + \dfrac{\partial f}{\partial y}\mathbf{j} + \dfrac{\partial f}{\partial z}\mathbf{k}$

Divergence: $\quad \nabla \cdot \mathbf{F}(x, y, z) = \dfrac{\partial f}{\partial x} + \dfrac{\partial g}{\partial y} + \dfrac{\partial h}{\partial z}$

Curl: $\quad \nabla \times \mathbf{F}(x, y, z) = \begin{vmatrix} \mathbf{i} & \mathbf{j} & \mathbf{k} \\ \dfrac{\partial}{\partial x} & \dfrac{\partial}{\partial y} & \dfrac{\partial}{\partial z} \\ f & g & h \end{vmatrix}$

$\nabla \times (\nabla f) = \mathbf{0} \qquad \nabla \cdot (\nabla \times \mathbf{F}) = 0$

$\mathbf{F}$ conservative on $D \Leftrightarrow \mathbf{F} = \nabla \varphi \quad$ for some potential function φ

$\qquad \Leftrightarrow \oint_C \mathbf{F} \cdot d\mathbf{r} = 0$ over closed paths C in D

$\qquad \Leftrightarrow \int_C \mathbf{F} \cdot d\mathbf{r}$ is independent of path for C in D

$\qquad \Leftrightarrow \nabla \times \mathbf{F} = \mathbf{0}$ on D